ENCYCLOPEDIA OF

ETHICAL, LEGAL,
AND
POLICY ISSUES IN BIOTECHNOLOGY

VOLUME 2

WILEY BIOTECHNOLOGY ENCYCLOPEDIAS

Encyclopedia of Bioprocess Technology: Fermentation, Biocatalysis, and Bioseparation
Edited by Michael C. Flickinger and Stephen W. Drew

Encyclopedia of Molecular Biology
Edited by Thomas E. Creighton

Encyclopedia of Cell Technology
Edited by Raymond E. Spier

Encyclopedia of Ethical, Legal, and Policy Issues in Biotechnology
Edited by Thomas H. Murray and Maxwell J. Mehlman

ENCYCLOPEDIA OF ETHICAL, LEGAL, AND POLICY ISSUES IN BIOTECHNOLOGY
EDITORIAL BOARD

ENCYCLOPEDIA OF

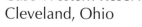

ETHICAL, LEGAL,
—AND—
POLICY ISSUES IN
BIOTECHNOLOGY

VOLUME 2

Thomas H. Murray
The Hastings Center
Garrison, New York

Maxwell J. Mehlman
Case Western Reserve University
Cleveland, Ohio

A Wiley-Interscience Publication

John Wiley & Sons, Inc.

New York / Chichester / Weinheim / Brisbane / Singapore / Toronto

This book is printed on acid-free paper. ∞

Copyright © 2000 by John Wiley & Sons, Inc. All rights reserved.

Published simultaneously in Canada.

For ordering and customer service, call 1-800-CALL-WILEY.

Library of Congress Cataloging in Publication Data:
Murray, Thomas H. (Thomas Henry), 1946–
    Encyclopedia of ethical, legal, and policy issues in biotechnology / Thomas Murray, Maxwell J. Mehlman.
        p.   cm.
    Includes index.
    ISBN 0-471-17612-5 (cloth : alk. paper)
    1. Biotechnology – Moral and ethical aspects – Encyclopedias. 2. Biotechnology industries – Law and legislation – Encyclopedias. 3. Biotechnology – government policy – United States – Encyclopedias. I. Mehlman, Maxwell J. II. Title.

TP248.16.M87   2000
174'.96606 – dc21
                                                                            00-021383

Printed in the United States of America.

10 9 8 7 6 5 4 3 2 1

ENCYCLOPEDIA OF

ETHICAL, LEGAL,
AND
POLICY ISSUES IN BIOTECHNOLOGY

VOLUME 2

HUMAN ENHANCEMENT USES OF BIOTECHNOLOGY: OVERVIEW

ROBERT WACHBROIT
University of Maryland
College Park, Maryland

OUTLINE

INTRODUCTION

There is little doubt that the most controversial issue regarding biotechnology is the prospect of employing it for the purpose of human enhancement. The following discussion is intended to serve as a roadmap to the various questions and topics raised by the prospect of enhancement, with special emphasis on the conceptual issues. A detailed examination of the ethical issues is the topic of another article.

We will begin by examining the so-called demarcation problem: What is enhancement and what is it being contrasted with? We will then survey some of the types of modifications that lead to enhancement. Although the primary modification people have in mind is genetic, it is worth looking at nongenetic modifications — biotechnological and nonbiotechnological — in order to place the concerns with genetic modification in a broader landscape. Finally, we will examine some general approaches for assessing genetic enhancement.

THE DEMARCATION PROBLEM

Enhancement modifications are typically defined along-side therapeutic modifications. A therapeutic modification is one that brings a trait that was below a recognizable, specieswide norm up to that norm. (The term "traits" is meant in its broadest sense, including physical attributes, mental or physical abilities, dispositions, and capabilities.) As a first approximation, we can characterize an enhancement modification in contrast as one that is a nontherapeutic improvement. The norm referred to here is the one that separates conditions of health from those of disease. The distinction between enhancement and therapy is therefore linked to the distinction between health and disease.

Two important points should be raised about this linkage. First, while the various proposed theories of health and disease yield corresponding accounts of what enhancement and therapeutic modification means, controversies and obscurities in the former will translate into

the latter. Problems with particular theories of health will have counterparts in problems with the corresponding account of enhancement modification. Indeed, skepticism about there being an objective contrast between health and disease will translate into a corresponding skepticism about the distinction between enhancement and therapy. Consequently the health/disease distinction is of limited use in *explaining* the enhancement/therapy distinction. None of this, however, undermines the link. The first distinction will be as clear and useful as the second. Thus, while it is true to say that therapeutic modifications attempt to treat disease whereas enhancement modifications attempt to improve a trait that is not diseased, there can be considerable debate over whether a particular modification therefore constitutes an enhancement and why.

Perhaps the most debated issue regarding theories of health and disease is whether or not the distinction — what constitutes the norm — is value-free: Is the judgment that someone is diseased — that someone's condition falls below a norm — an objective discernment of a biological state or a value judgment? Is a particular condition a disease independent of whether we think it is bad or undesirable? Can a condition be a disease in one culture and not in another? Although this dispute is not particularly salient in discussions over enhancement, these discussions have typically proceeded with the idea of the norm being fixed and not relative to individuals or cultures. If what constitutes enhancement varies with individual or culture — if enhancement is in the eye of the beholder — then it is not clear that we can sensibly articulate an (ethical) issue about enhancement as such. Nonrelativistic conceptions of normality tend to favor objective theories of health and disease, though that still leaves considerable latitude over how to conceive of normality, from statistical conceptions (1) to biological conceptions (2). Nonetheless, objective conceptions are not the only kind of nonrelativistic conception of normality. Norms that are recognized to be arbitrary and conventional can still frame the issue, as discussions over the problems of enhancements in sports demonstrates. Indeed, even a normative conception of norms could be invoked, as long as the relevant values are themselves understood to be nonrelativistic.

The second point to note is that acknowledging the link between health/disease and enhancement/therapy can suggest that the latter is a medical matter. Ethical issues regarding enhancement modification should then be seen in terms of the ethics of medicine and the professional duties and responsibilities of health professionals. As plausible as this suggestion may be, we need to distinguish at least theoretically between questions regarding the ethics of enhancement modifications and questions regarding the ethics of physicians performing enhancements modifications — for example, whether a particular enhancement modification is ethically objectionable from whether it is unethical for a physicians to perform such a procedure. It may well be that the answer to the second determines the answer to the first—for example, that the ethical questions regarding enhancements comes down to questions about the role morality and professional ethics of physicians, but a claim like that requires more of an argument than pointing to the connection between the enhancement/therapy distinction and the health/disease distinction (3, Introduction).

An immediate challenge to any account of the enhancement/therapy distinction is the presence of apparent borderline cases or exceptions to the classification scheme. These fall into two classes.

One class of cases is modifications that, strictly speaking are enhancements, but whose purpose is to respond to (the threat of) a disease. For example, a modification that improves people's resistance to particular diseases beyond the normal capacity would count as an enhancement but its purpose would be disease prevention and so arguably therapeutic.

A different class of borderline cases or exceptions arises from an ambiguity in the idea of "normal traits." It can mean a trait whose appearance and function is normal, but it could also mean a trait whose appearance, function, *and* development is normal. Moreover normality itself often refers to a range within a trait rather than to a sharp line. Thus, imagine two people, both of whose height is five feet. While the first person has short parents, the second has tall parents but suffers from a growth disorder. Both people have a height that falls within the normal range, but the second person's height is the result of a disease. A modification that brought the second person's height from five to six feet would be a modification within the normal range *and* a response to a disease (4).

Both types of cases indicates that a classification scheme generated by outcomes *and* by purposes is a scheme driven by one too many criteria. Since there is no direct line between outcomes and purposes, we should not be surprised if there are cases that fit each criterion differently. Which criterion we should use will depend on what we are trying to classify. If we are trying to classify modifications, then outcomes would be the better choice; if we are instead trying to classify practices or aims, then purposes might be a better criterion. Using both criteria is inevitably confusing in that not only can the same procedure and outcome be associated with different purposes but the same event often is shaped by multiple purposes. Ambiguities in classifying particular cases will inevitably arise. (See Juengst's article in Ref. 3 for a discussion of various other alternatives and their problems.)

The point about normality not being a sharp line raises an important distinction within the category of enhancement modifications: There could be modifications that raise a trait above the norm and there could be modifications that raise a trait from one point within the normal range of that trait to a higher point in that range. This suggests that the classification of modifications should be tripartite: therapeutic, intranormal, and (proper) enhancement. Nevertheless, we should note that many commentators understand enhancement to mean any improvement of a normal trait, thereby collapsing the second and third categories. Cosmetic surgeries, which can often be regarded as intranormal modifications, are thus placed in the same category as genetic modifications

to create superpeople. Whether fewer distinctions or categories is better will depend on how the issues are analyze and whether one classification clarifies matters more than the other. As we will suggest below, it is better to keep intranormal modifications, which are differences of degree, distinct from enhancements proper, which are differences of kind.

TYPES OF MODIFICATIONS

Biotechnology covers a range of technologies and procedures, many of which could conceivably be employed for enhancement. Drugs could be designed to interact with the body's chemistry in such a way as to alter behavior, biological functioning, structure, or affect. Even without introducing drugs, special procedures—such as transfusing a person with their own blood or "blood doping"—can affect traits or behavior. But the most discussed enhancement technology is one in which a person's genome is altered.

It is an empirical question which traits can be enhanced by modifying an individual's genes. And it may turn out that enhancing certain traits requires not only genetic modifications but also certain alterations in the individual's environment. That is to say, a particular genetic modification might not by itself bring about an enhanced trait; it might give the person a capacity to developed the enhanced trait whose realization demands a special exercise regime, diet, or other efforts. The idea of genetic enhancement technologies therefore does not rest on an assumption of genetic determinism—that a genetic alteration alone is sufficient to bring about a particular trait. While a popular image of genetic enhancements is that of some magic-wand transformation in which the person is a passive recipient, the matter can be more complex. Realizing a genetic enhancement might involve hard work. This point will become important later when we consider assessments of enhancements.

Genetic modifications are often separated into two kinds—somatic and germ line. The difference is whether the particular genetic modification affects the individual's gametes so that the modification can be passed on to the individual's offspring. The object of a somatic modification is a modified individual, but the object of a germ-line modification is a modification that becomes part of the individual's legacy or inheritance. In saying that there are these two kinds of genetic modifications, we are *not* claiming that of any particular genetic enhancement there is a somatic version and a germ-line version. That is entirely an empirical matter. It may well be that certain kinds of enhancements can only be done as somatic while others can only be done as germ line. For example, a modification may only be somatic because it interferes with the individual's ability to reproduce. A modification may only be germ line because the only feasible way of delivering the modification to all the relevant cells requires inserting the modification in the few cells of the embryo stage, which would then likely affect the individual's germ cells. Although the distinction between somatic and germ-line modification is conceptually clear, it may not be applicable everywhere.

Nevertheless, many commentators find the distinction useful. It is reasonable to assume that somatic enhancements are simpler as far as ethics and public policy is concerned. Germ-line enhancements appear to raise all the issues of somatic enhancement and then some. And so it would seem that we should first examine the acceptability of somatic enhancements and only after settling that should we proceed to an examination of germline enhancements. This strategy however is difficult to sustain if we allow for the possibility that enhancements might not stand or fall as a group. If we are open to the possibility that some enhancements might be acceptable and others not and we acknowledge that some enhancements may have only a somatic or a germ-line version, the strategy of considering first the somatic case and then the germ-line case may not always be applicable. Some enhancements might as a matter of technology not have a somatic version.

(Of course there is one way of ensuring a genetic enhancement is not in effect germ line—combining the modification with one that also renders the individual infertile. But this possibility is probably not worth dwelling on: It is difficult to conceive of a case where, as a matter of ethics or public policy, an enhancement would be acceptable but only if the individual agrees to sterilization.)

ASSESSING ENHANCEMENTS

There are two broad approaches to determining the acceptability of an enhancement: the assessment can be based on what the enhancement is—the product—or on how the enhancement is achieved—the process. We will briefly provide an overview of the questions each of these raise in turn, leaving a more detailed discussion of the ethics for a separate article. For this entire discussion, we will be assuming that the modification is safe and effective so as to target our inquiry on the acceptability of the enhancement modification itself rather than on side issues regarding the acceptability of risky biotechnological procedures.

Product Assessment

Recall the earlier contrast between enhancements, properly called, and intranormal modifications. An assessment that focuses only on products will pass on intranormal modifications, since there is presumably nothing wrong with the result of an intranormal modification, at least at the level of the individual. For example, there is nothing wrong in itself with being six feet tall, so an intranormal modification that renders a person six feet tall cannot be unacceptable because of the result.

Confining ourselves therefore to only proper enhancements, the first kind of product assessment is directed at trade-offs the modification allegedly imposes. For example, suppose that a modification enhances a people's memory capacity but with the result that the speed in accessing memory is considerably slower. Or suppose that an enhanced memory capacity results in greater irritability. How should these trade-offs be assessed? Should it be a matter of individual choice or public policy?

A second kind of product assessment is directed at the "humanity" of the modification. According to this approach there is something wrong in itself and not because of alleged trade-offs in having a particular trait enhanced beyond what is (normally) human. Indeed, because of the enhanced trait, the individual might not be regarded as human. The problems arising from the various racial and ethnic divisions of humanity might well carry over to this new kind of division. In addition some people might regard enhanced individuals as an insult to the integrity of the species or, seen religiously, an insult to God's creation. How should these concerns be addressed in a pluralistic society?

A third kind of product assessment is directed at the widespread use of enhancements. Even if there is nothing wrong with any particular enhancement use, problems arise when many people or certain sectors of the population primarily make use of this enhancement. For example, suppose genetic enhancement of memory were possible and it resulted in memory-enhanced individuals being significantly more successful in several aspects of life. If only the wealthy had access to this technology, genetic enhancements would create or exacerbate troublesome inequalities. But even if the technology were made available to everyone, problems could arise. The desirability of some traits arguably rests on their not being common or widespread; if everyone is a blonde, then blondes will not have more fun. Furthermore people who do not want to be enhanced might nevertheless feel under some considerable pressure to avail themselves of it because many other people are doing so.

Process Assessments

When the issue becomes the process, then intranormal modifications are as much a subject for examination as enhancements, properly called. Indeed, it might be argued that we only need to examine intranormal modifications: If an intranormal modification is unacceptable from the standpoint of process, it would seem that extending that modification to the point of enhancement would also be unacceptable from the standpoint of process. The obverse would also seem to be true, though we should emphasize that acceptability from the standpoint of process does not entail acceptability from the standpoint of product.

The first kind of process assessment is directed at the suggestion that using biotechnology to effect an improvement is wrong because it is artificial. This concern need not be one that crudely equates natural with good and nonnatural with bad, raising concerns even about ordinary medical interventions. The worry here is a "commodification" of certain traits, and the people who have them, because of their being made to order, so to speak. This type of assessment is often linked with the concern about the humanity of the modification mentioned above.

The second kind of process assessment arises from a concern that using biotechnology in order to effect an improvement undermines the value of the improvement. The value we place on certain achievements may depend on the struggle and effort required to achieve them. If they could be made effortless—at least on the part of the individual—and common, we might well cease to value them. As we noted earlier, some (genetic) enhancements may only result in enhanced capacities; realizing them may still require effort, discipline, and luck on the part of the individual. Is the kind of effort relevant to the value we place on certain achievements?

The third kind of process assessment is directed at the suggestion that using biotechnology to enhance people is not the sort of thing physicians should do. The values or aims of the medical profession are held to be incompatible with performing enhancements. This can be a parochial concern in that a judgment that physicians should not perform enhancements leaves the question of the ethics of enhancement untouched. One can consistently be a supporter of capital punishment and yet hold that physicians should not be involved in either administering lethal injections or making the official pronouncement of death. In order to make this type of assessment have broader significance, one must argue that any profession that provides enhancements has suspect aims or values.

BIBLIOGRAPHY

1. C. Boorse, *Philos. Sci.* **44**, 542–573 (1977).
2. R. Wachbroit, *Philos. Sci.* **61**, 579–591 (1994).
3. E. Parens, *Enhancing Human Traits: Ethical and Social Implications*, Georgetown University Press, Washington, DC, 1998.
4. D. Allen and N. Fost, *J. Pediatrics* **117**, 16–21 (1990).

See other entries BEHAVIORAL GENETICS, HUMAN; GENE THERAPY, ETHICS, GERM CELL GENE TRANSFER; GENETIC DETERMINISM, GENETIC REDUCTIONISM, AND GENETIC ESSENTIALISM; see also HUMAN ENHANCEMENT USES OF BIOTECHNOLOGY entries.

HUMAN GENOME DIVERSITY PROJECT

HENRY T. GREELY
Stanford University Law School
Stanford, California

OUTLINE

INTRODUCTION

The Human Genome Diversity Project (HGDP), first proposed in 1991, has thus far been responsible for more controversy than research. It has raised many of the same concerns as the well-established Human Genome Project (HGP) but at the level of human groups rather than that of individuals. As of the date of this article, the future of HGDP remains uncertain, but it is certain that the study of human population genetics, with its implications for human groups, will continue. The history of HGDP and the discussions of ethical, legal, and political issues it has stimulated are a source of useful lessons in either case, lessons about the complexity of social consequences of genetic research on human groups.

HGDP

HGP plans to publish the nucleotide sequence of the human genome in the next few years. The underlying point to HGDP is that there is no one human genome; instead there are about six billion existing human genomes, one for every living member of *Homo sapiens*, each of whose genome is separate and at least slightly distinct from all others. (Even monozygotic, "identical," twins, who make up under 1 percent of our species will show some genetic differences caused by mutations during development.) HGDP seeks to create a resource for studying this diversity. Its goals, which are set out most clearly in the report of its founding meeting in September 1993 in Alghero, Sardinia, include collecting and preserving genetic samples from approximately 500 different human populations around the world, performing some genetic analysis of the samples, and making both the results of those analyses and portions of the samples themselves available to interested researchers (1). HGDP, and the scientists behind the project have pursued these goals with limited success since 1991.

Scientific Background

The human genome is made up of approximately three billion base pairs of DNA, spread over 46 chromosomes. Individual human genomes do not vary by much. Genomes from two people, from anywhere in the world, vary on average at about one base in a thousand along any given stretch of their DNA. Within the regions of the genome that code for protein, the variation is closer to one in ten thousand. Most human genetic variation falls in regions of the genome that have no known function. Variations in these regions may be completely without consequence. Even variations that fall within the coding region of genes may be unimportant, either because they do not change the protein product, as when a single nucleotide substitution does not change the amino acid coded for, or because they change in the protein product in ways that make no apparent difference—that do not change the individual's phenotype. Other variations will affect the person's phenotype: in ways that may be negative, such as a genetic disease; in positive ways, such as increased resistance to a disease; or in ways that, as far as can be seen, are neutral, such as eye color.

The consequences of these variations will often depend on the individual's environment; having one copy of the gene for sickle cell anemia may cause mild health problems but provides some protection against malaria. In areas where malaria is rare, it may be a disadvantage; in areas where malaria is common, an advantage. The study of variations in individual genomes that have phenotypic consequences is the traditional study of human genetics.

If humans did not vary genetically, classical genetics could not see anything. Hair color, eye color, blood type, strongly genetic diseases — all these genetic traits are linked to the existence in individuals of genetic variations.

Particular genetic variations, or genetic "markers," are not distributed randomly among the world's peoples. The percentage of any human group with a particular genetic variant, or marker, may differ from the percentage in another group. The genetic contribution to skin coloration, for example, clearly varies among humans in ways that correlate, albeit imperfectly, with culturally defined ethnic groups. The same is true of other genetic variation, some of which has observable consequences, such as blood types, and most of which does not. The study of the patterns of genetic variation among human groups is called human population genetics. After the rediscovery of Mendel's work at the beginning of the twentieth century it became possible to study human variation at the genetic level even though human genes, at that point, remained abstractions. The key was to find some observable physical characteristics that were inherited in the manner described by Mendel and hence, presumably, determined by genes. By observing the variations in those characteristics, one was observing variations in the underlying genes.

This research began around 1915 with studies of the prevalence of ABO blood groups in different populations (2,3). And different patterns were discovered (4). Type A blood was most common in Northern Europe (although not found in a majority of the population even there). The peoples of Southern and Eastern Europe had a higher percentage of type B blood than those of Northern Europe, although, again, only a minority of them had type B blood. Native Americans overwhelmingly carried type O blood. The classification of humans into different *biological* groups based on their observable physical variations had been undertaken in Europe at least since Linneaus began biological classification of animals in the eighteenth century. The usual result was "scientific proof" that Europeans were "biologically superior." Human population genetics, which appeared to offer a way of classifying human groups based on characteristics that were not environmental, could be enlisted into such an effort. Thus some Nazi propaganda talked of the importance of blood type A Nordic peoples holding back the flood of blood type B blood from the "inferior" peoples of south and east.

In fact, over time studies of human genetic variation using classical markers have revealed that humans are not genetically very distinct and that most of the variation that exists is found within human populations. An estimated 85 percent of all human genetic variation exists within populations; only 15 percent reflects statistical differences between populations (5,6). Where differences among groups exist, they are usually *not* in the presence or absence of particular variations, but in their frequency. These differences are the products of statistical analysis and meaningful only for the groups, not for the individuals within them. For example, all human populations seem to include some members with each of the ABO blood types: O, A, B, and AB. Around the world, the A blood type is found in about 30 percent of humans. In some populations, particularly in the Americas, it is found in only a few percent. Among Armenians, it is found in just under 50 percent of the population. Any one person with an A blood type is highly unlikely to be Armenian; any one Armenian is more likely than not to carry some other blood type. If, however, 50 percent of a town of 3000 people have the A blood type, the inference that most of the town is Armenian — or made up of people related to the Armenians — may be worth investigating.

For most of the twentieth century, a shortage of classical genetic markers — phenotypical variation that is inherited in a Mendelian manner — limited direct empirical research into human genetic diversity. The theory of population genetics, however, both human and nonhuman, fruitfully expanded through the middle of the twentieth century. Then, the finding by Avery that genes were made of DNA and the discovery by Watson and Crick of the structure of DNA led ultimately to an era where much more genetic variation was much more directly observable through analysis of variations in the DNA itself. This not only allowed examination of more genetic traits but permitted for the first time the examination of genetic variation in regions of the genome that had no effects on phenotype. As these regions contained the vast bulk of human genetic variation, the ability to study such variation was greatly enhanced.

Such studies could have many uses. Perhaps the most interesting would be to provide additional evidence concerning human history and evolution. It will rarely be the case that one bit of genetic variation will be very informative, but by analyzing the patterns of variations in many different locations in the genome, population geneticists can, in some cases, estimate the closeness of the relationships between human groups. Eventually they may be able to generate a "phylogenetic tree," showing the relationships between human populations as putative descendants of common ancestors. This kind of information could be used as evidence of human migrations. The evidence can be used for very general questions, such as testing the "Out of Africa" hypothesis of human history, or for very narrow ones, such as exploring the history of the Japanese. The migrations studied might be recent, as in the Native American movements to the Great Plains after the acquisition of the horse made bison a more easily exploited resource. They can be more distant, such as, perhaps, the spread of Polynesians through the Pacific, Bantu-speakers through sub-Saharan Africa, or Indo-European speakers through Europe. In some of these cases, the evidence may be negative, showing that these changes in culturally defined groupings involved changes in cultures but did not result from migrations of genetically related people, which would itself be an interesting finding. Or the migrations can be still more distant, such as the evolution and spread of humans across the globe from their African origins.

This general approach of using variations to trace history is neither novel nor foolproof. The same kind of approach has been used to trace the history and changes of different texts, the development of languages, and, with molecular and nonmolecular evidence, the evolution of many different living things. This phylogenetic approach may prove particularly difficult with humans, who did not separate into reproductively isolated populations. Genetic variations have flowed between human groups prodigiously in the last few centuries; it is certain that some level of gene flow between many groups has long existed. Today's culturally defined "populations" may not have substantial genealogical, and hence genetic, connections (7,8). The absence of such connections, though interesting on its own account, would undercut the historical value of population genetics (and, as noted later, cause some ethical problems). Genetic evidence of history and evolution is one line of evidence, to be considered with historical, ethnographic, linguistic, archaeological, and other kinds of evidence. It is, in the abstract, no more or less powerful or accurate than any of the others. But it does offer a different, and independent, line of evidence, which is of great interest.

The study of human genetic diversity has other uses. It may be used to answer some questions in cultural anthropology, such as patterns of marriage in the caste system in India. It might be used to test more general propositions in population genetics — after all, we know far more about the history and mating habits of Homo sapiens than we know about Drosophila in the wild. Or it could be of some help in biomedical research. If, for example, a variation in a "candidate gene" is identified as linked to a particular disease, the ability to check the prevalence of that variation in a population where the incidence of the disease is known might provide useful hints for further research.

A final, and more symbolic, value concerns the definition of the "normal" human genome. Genetic variation within the human species is small — especially variation that clusters along population lines — but it does exist. If "the" human genome were defined by the genetic variations most common in the countries where HGP is taking place, it would greatly overrepresent people of European ancestry, who make up less than 20 percent of the world's population. Variations not found in those of European ancestry, even if common around the world, might be viewed as "abnormal." For example, the ability to digest lactose, the sugar found in milk, is very rare among adult mammals. Among humans, most North Americans, and most Northern Europeans, can digest milk well as adults, but that ability is not common in the rest of the world's humans (9). Adult lactose intolerance is considered an abnormal condition in the United States, even though it is found in most humans. A specieswide resource of human genetic variation could help counteract similar parochial misunderstandings of the human genome.

Human genetic variation can serve these goals only if it is known. Current efforts to study it run into the so-called empty matrix problem (5,10). Genetic variation has been studied in many populations around the world for most of the twentieth century, either as classical genetic markers (like blood types) or, more recently, as DNA variations. But genetic variations that have been examined in one population will not have been analyzed in others, so those populations cannot be compared. Additionally, for many populations the samples examined were quite small. Human population geneticists ideally would like to have

a large number of samples from a large number of populations, all analyzed for the same large set of genetic variations. It was this desire that led to the birth of HGDP.

HGDP Forms, 1991 to 1993

HGDP's parents were two population geneticists: Luca Cavalli-Sforza at Stanford and Allan Wilson at University of California, Berkeley. Their discussions of the "empty matrix" problem took on new life at in the early 1990s when the Human Genome Project began operating. They conceived the idea of HGDP as an important, and relatively inexpensive, supplement to HGP. With Charles Cantor, Robert Cook-Deegan, and Mary-Claire King, they wrote the first call for the project, published in Genomics in 1991 (10).

Wilson's illness and subsequent untimely death from leukemia in July 1991 prevented him from playing a large role in the project, which had an important consequence for its shape. Wilson and Cavalli-Sforza had taken very different views over HGDP's sampling strategy. Wilson had wanted to sample on a grid basis, taking a certain number of randomly chosen people from each of a series of squares laid down over a world map. This sampling method was used by population geneticists studying wild populations of drosophila or other nonhuman species. Cavalli-Sforza was more interested in sampling based on existing populations, both for logistical reasons and for the information such sampling could provide about population history. Cavalli-Sforza might well have won this argument in any case, but Wilson's death ensured that result (1).

The nascent project began to attract other supporters, from anthropology as well as genetics, and from overseas as well as from the United States (11–13). Sir Walter Bodmer, then President of the Human Genome Organisation (HUGO), appointed a committee to study the idea of a Human Genome Diversity Project in 1992. Cavalli-Sforza chaired the committee (1).

In early 1992 the American members of this committee — Cavalli-Sforza; Marcus W. Feldman, a population biologist from Stanford; Kenneth K. Kidd, a geneticist from Yale; Mary-Claire King, a geneticist then at the University of California at Berkeley (and now at the University of Washington); and Kenneth M. Weiss, a physical anthropologist at Pennsylvania State University — received funding from the U.S. federal government for planning HGDP (1). The funds, amounting to about $55,000, came from the National Science Foundation (NSF), the Department of Energy (DOE), and from two parts of the National Institutes of Health (NIH): the National Institute for General Medical Science and what was then the National Center for Human Genome Research (now the National Human Genome Research Institute). The committee used the planning funds to sponsor three workshops: one at Stanford in July 1992, one at Penn State in October 1992, and one at NIH in February 1993.

The first planning workshop focused on collection methods. In considered what kinds of samples would be needed and how many samples should be obtained from each population. A crucial issue at this stage was whether the project should collect DNA samples or lymphoblast cell-lines. DNA, purified from blood or from samples scraped from the interior of the cheek, was easy to process and inexpensive, but would provide only a small amount of DNA from each participant. The latter, white blood cells transformed by viral infection to have an indefinite life span, were more expensive and technically more difficult but held the promise of an inexhaustible supply of DNA from each sample. The workshop recommended a combined strategy of acquiring a small number of cell-lines from each population along with a much larger set of DNA samples (1,14).

The second planning meeting focused on anthropology and sampling strategy. Anthropologists met at Penn State to discuss what anthropological questions could be addressed by a resource of human genetic diversity and what *kinds* of populations would be most useful to sample. The participants talked about wanting to get samples from populations that represented major linguistic groups in a region, that had interesting cultural or linguistic aspects, or that could be used to answer specific anthropological questions. They also noted the value of collecting samples from isolated populations that were rapidly disappearing as distinctive cultures, largely through assimilation. These populations they referred to as "isolates of historical interest." Using these guidelines, the workshop divided into groups based on geographical specialization, with orders to produce a list of 500 populations as examples of the types of populations such a project might want to sample. The workshop felt that the list would be useful in demonstrating to funding agencies that the project thought through these sampling issues. [In the event, the anthropologists did not get their list below 700 populations (1,15).]

The third workshop, held for two and a half days in February 1993 at NIH, had three parts. The first day was returned to issues of sample types and the debate over cell-lines. The second day was devoted to a discussion of ethical and human rights issues raised by the project (16,17). On the third morning, the project organizers met with representatives from possible federal funding sources, including NSF, DOE, and several institutes of NIH. The morning started under a cloud, as the previous evening had featured President Clinton's first State of the Union address, in which he stressed the importance of reducing government expenditures. It did not get better for the project organizers, as the federal funding agencies listened to their plans for HGDP with interest but made no commitments.

In September 1993, HGDP held what became its founding meeting, in Alghero, Sardinia. This four-day meeting was supported by the remaining funding for planning workshops, with additional support from the Porto Conte Research and Training Laboratories Foundation, the European Commission, the Soros Foundation, and HUGO Europe. The researchers gathered at this meeting agreed on an organization and a substantive outline for HGDP (1).

The organization was to work at two levels. An international executive committee, affiliated with HUGO, would exercise oversight over the entire project. Thirteen members were appointed to this committee, from four continents. The international executive committee could

have subcommittees and would, the meeting decided, establish at least two: one on informatics and one on ethics. Both the fund-raising and the actual operations of the project were to take place at a regional level, run by committees made up of scientists living and working in those regions. The regions were envisioned as having continental or near-continental scale.

The substantive work for these committees was to fall into three main categories: collection, preservation, and analysis (with subsequent data base entry) of DNA samples. The project adopted as an interim goal the collection of samples from 500 different human populations. Using the number of distinct languages as a rough proxy for the number of populations, this would be a sample of around 5 to 10 percent of human populations. The samples from each participating population were to include about 15 cell-lines and many more samples of purified DNA. Basic ethnographic information, derived from a standard questionnaire, would also be obtained from each individual providing samples.

These samples would then be preserved at repositories. The group at Alghero concluded that there should be more than one repository, in order to provide backup storage for samples, and that regional repositories should be considered. Samples of DNA from the repositories would be provided at cost to qualified researchers on request. What qualifications were necessary was not entirely settled, but the intent was to prevent the wide spread of these samples to cranks.

The samples were also to be the subjects of analysis, at the repositories and elsewhere. The repositories were expected to analyze samples for a standard set of markers. Researchers who accepted samples from the repositories were to do so subject to a condition that they return the results of their analyses of the samples to the repositories. The analysis of the samples was to be placed into a database by the project, which was also to be open to all qualified researchers.

The Alghero meeting endorsed an estimate that, worldwide, the project would take about five to seven years and cost about $5 million to $7 million per year. By the end of 1993, regional committees had formed in North America, South America, Europe, and Africa, while organizing efforts had begun in Oceania. The structure for HGDP proposed at the Alghero meeting was adopted by HUGO in January 1994. But attention had already been drawn to HGDP and opposition had started to build. For the rest of the decade, HGDP would make very little progress.

HGDP Stalls, 1994 to 1999

HGDP began to attract press attention in the 1992 planning workshops. The October 1992 workshop at Penn State in particularly was featured in *Science* magazine, where one of the sidebar stories had the inflammatory headline about "Endangered Populations" (18). Press attention brought less welcome attention, notably from a nongovernmental organization headquartered in Canada called the Rural Advancement Foundation International (RAFI) (19). RAFI had begun in the 1970s as a nonprofit group focusing on developing world agriculture, largely in opposition to the so-called Green Revolution. Over the years RAFI had become concerned about what it termed "bio-piracy," collections of plant material from developing regions by Western seed and pharmaceutical companies. These collections, RAFI and others charged, were often used to produce new products with no financial or other sharing with the peoples from whom the plants had been taken. Starting at a Pan American Health Organization meeting in spring 1993, RAFI began to spread the word among nongovernmental associations concerned with indigenous peoples that HGDP was another, deeper attempt at bio-piracy, taking, this time, not indigenous people's plants but their own human genes (20). RAFI took the sample list of populations, compiled at the Penn State workshop in October 1992, and publicized it on the Internet as HGDP's "hit list" (21). This caused understandable concern to groups that suddenly read of themselves as the "targets" of an international genetics project from whom they had never heard.

RAFI also tried to tie HGDP to examples of what it claimed was bio-piracy by the U.S. government. In September 1993, RAFI protested a patent application filed by the United States on a white blood cell (lymphoblast) cell-line derived from a woman from the Panamanian Guaymí people. The cell-line was infected with a human retrovirus called HTLV-2, which made it interesting as a source for virus and viral antibodies. In October 1995, RAFI protested against a patent that had been granted to the United States for a cell-line derived from the white blood cells of a man from the Hagahai population in Papua New Guinea. This cell-line contained a related human retrovirus, HTLV-1. In both cases RAFI won. The United States abandoned the Guaymí cell-line patent application and abandoned the Hagahai cell-line patent application. In both cases RAFI tried to connect HGDP to these applications, claiming, in the Hagahai case, that "The thin veneer of the HGDP as an academic, non-commercial exercise has been shattered by the US government patenting an indigenous person from Papua New Guinea" (22,23).

RAFI's actions sparked a surge of condemnations of HGDP by various nongovernmental organizations, including some representing, or purporting to represent, indigenous peoples (One writer has counted 13 such resolutions (24); another, earlier article had a list of 11 (25).) Opposition to HGDP came from organizations concerned about indigenous peoples, about patent rights and the developing world, and about genetic technologies more generally. Most of the condemnations were taken after hearing only from opponents of the HGDP and they often demanded that the Project, which had not yet begun any collections, stop or reverse its work. Thus, for example, in 1994 the Foundation for Economic Trends, founded and controlled by Jeremy Rifkin, a well-known opponent of biotechnology, formally petitioned NIH to cease all of its (nonexistent) funding for the HGDP's collecting activities (26).

At least in part as a reaction to this activist opposition, the HGDP's North American Committee decided to spell out its ethical positions more fully. This was not the first time HGDP had discussed ethical issues. It had sponsored a full-day workshop on these issues at NIH in February 1993 (16). The organization's founding meeting in Alghero, Sardinia in September 1993 included a discussion of ethics and required that the International Executive Committee create an ethics subcommittee (1). The meeting's report devoted several pages to ethical concerns. It classified as major concerns collection issues, intellectual property; racism, xenophobia, and hypernationalism; and public understanding. The report included 10 proposed ethical guidelines that the meeting had accepted:

1. The HGD project and its participating researchers must always respect the human of the sampled individual and the cultural integrity of the sampled population....

2. Informed consent is both an ethical imperative and a legal requirement. The HGD project must satisfy both conditions....

3. Researchers should actively seek ways in which participation in the HGD project can bring benefits to the sampled individual and their communities. Examples of such benefit include health screening, medical treatment or educational resources.

4. One way to avoid bringing harm to the sampled individuals or their communities is by protecting the confidentiality of those sampled and, in some cases, of their entire community.

5. Although very unlikely, it is neverless [sic] possible that the results of the HGD project may lead to the production of commercially beneficial pharmaceuticals or other products. Should a patent be granted on any specific product, the project must work to ensure that the sampled population benefit from the financial return from sales.

6. Human history — and the human present — is full of racism, xenophobia, hypernationalism, and other tragedies stemming from beliefs about human populations. In the past, some of those tragedies have been perpetrated by, or aided by, the misuse of scientific information. All those involved in the HGD project must accept a responsible to strive, in every way possible, to avoid misuse of the project data.

7. Many people in the world have, at best, a limited understanding of human genetics. Some fear the consequences of human genetic research, in part because of the limits of their understanding. To scientists involved in the HGDP project, such fears may not seem justified or even, in some cases, fully rational by the concerns are very real of the people involved and they must be addressed. It is essential that a worldwide "public awareness" program is included within the project to educate people about its aims, methods and results.

8. Inevitably, the ethical issues faced by the project will evolve over time. The issues must therefore be kept under continual review. The widest possible consideration of the issues should be encouraged.

9. The transfer of technology to developing regions of the world, which is an integral part of the

proposed project, should contribute positively to the development of self-sufficiency in these regions. The help given should not be superficial and of only short-term usefulness.

10. There should be a feed-back of information to populations that participate in the HGD project, most especially about any aspect of the project in which a particular interest had been expressed.

The report from the Alghero meeting was not published until early 1995, and even then, it received little attention. It also became clear to the North American Committee that the project, at least in North America, needed a more concrete position on a number of ethical, legal, and social issues. As a result, in 1995, using funds provided by a grant from the John D. and Catherine T. MacArthur Foundation, the North American Committee developed its own draft "Model Ethical Protocol for the Collection of DNA Samples" (27). The Model Ethical Protocol was largely completed by the fall of 1995 (17). It was posted on the project's Web site in 1996 and published in a law review in 1997. As discussed in more detail below, the 15,000-word Model Ethical Protocol offers detailed guidance on many aspects of DNA collection for HGDP. It introduced two particularly significant innovations into this kind of research: the concept of "group consent" and the use of contracts to give participating populations some control over subsequent uses of their materials and derived data. The Model Ethical Protocol has been adopted by the project's North American Committee to guide collections in North America; it has not adopted by the project overall.

The ethical, legal, and social issues raised by HGDP began to be discussed outside HGDP and its opponents. UNESCO's International Bioethics Committee heard discussion of HGDP in the fall of 1994 and appointed a committee to study the issue. That committee, chaired by Dr. Darryl Macer, gave a mixed report in November 1995 (27–29). HUGO, which had appointed the original HGDP committee in 1992 and had adopted the Alghero meeting's recommendations in 1994, asked its Ethical, Legal, and Social Implications Committee to study the issues raised by the project. The Committee, chaired by Professor Bartha Knoppers, considered the project at an October 1995 meeting and produced a set of ethical principles, subsequently adopted by HUGO, on both HGP and HGDP (30). The ethical issues of HGDP were discussed at a workshop at Mt. Kisco, New York, in November 1993 (sponsored by the Wenner-Gren Foundation), at conferences at Stanford in November 1995, and in Montreal in September 1996 (31), at the meeting of the International Association for Bioethics in San Francisco in November 1996, at the annual meeting of the American Association for the Advancement of Science in February 1998 (32), and at a conference at the University of Wisconsin-Milwaukee in February 1999 (33), among other venues. Gradually a literature began to build about the ethical, legal, and social issues raised by HGDP specifically and by human population genetics more generally.

While the ethical (and political) conversation about the project was moving forward, the project itself was largely stalled for lack of funding. The project's organizers, having rejected the idea of any commercial funding, looked primarily to the U.S. government for funds. (HUGO itself never had sufficient funds to consider supporting HGDP and, in fact, has struggled to sustain itself.) The early interest shown by that government, including a largely favorable congressional subcommittee hearing in April 1993 (34), did not translate into substantive funding. DOE's part of HGP made clear quite early its lack of interest in HGDP. NIH was less immediately dismissive. Its portion of HGP, the National Human Genome Research Center (since 1997, the National Human Genome Research Institute) acknowledged the eventual importance of studying genetic variation but did not choose to invest any funds in HGDP. The National Institute for General Medical Science, which seemed more open to HGDP, also failed to fund the project. Only NSF, and in particular, its program in physical anthropology, was encouraging about funding HGDP. The entire annual budget for the physical anthropology program at NSF in the mid-1990s, however, was only about $2 million. It could only fund HGDP if it received a major infusion of funds.

As they had done a decade earlier with HGP, the federal agencies with the most interest in HGDP decided to ask the U.S. National Research Council to report on the idea of the project. The National Research Council, the report-writing arm of the U.S. National Academy of Sciences, National Academy of Engineering, and Institute of Medicine, requires funding for its reports. The NSF, NIGMS, and the office of the Director of the NIH agreed in late 1994 to contribute about $400,000 to fund an NRC committee report on HGDP in the expectation, shared by HGDP advocates, that the Committee's report would either launch or bury the project. The Committee's report, which was finally released in November 1997, did neither.

In the event, the NRC committee was not appointed until early 1996 (35). Chaired by prominent geneticist Dr. Jack Schull, the 17-member committee included three ethicists, Professor George Annas, Dr. Eric Juengst, and Dr. Katherine Mosely. The committee held public meetings in April, July, and September 1996 at which both organizers and opponents of HGDP spoke. In November 1996 it began its deliberations, which took 12 months to yield a report. The report, when published, satisfied neither the proponents nor the opponents of the project—and may have confused many of its readers. The world's two leading scientific magazines, *Science* and *Nature*, reported on the committee's work with diametrically opposite headlines: "NRC OKs long-delayed survey of human genome diversity," claimed *Science* (36), while *Nature* wrote "Diversity project 'does not merit federal funding'" (37), (a headline *Nature* later retracted) (38).

The committee's chair, Dr. Schull, ended up writing correction letters to *both* journals (39,40), but, in fact, the committee's report seemed to provide, in different sections, support for both journals, particularly as diversely interpreted by some committee members. The committee stated that the plans for HGDP were too vague to be the subjects of a specific evaluation, and as a result it would evaluate the overall concept. It concluded that although there were serious reasons for ethical concern, the federal government might fund such research but only through U.S. researchers who would be subject to U.S. rules on protection of human subjects. The overall message of the long-awaited NRC report was so equivocal, however, that it failed to achieve its sponsors' overall goal — to provide a clear verdict, up or down, on HGDP.

As of early 2000, HGDP — proposed in 1991 and organized in 1993 — remains largely unfunded. Its committee structure has in large part withered. The annual meetings of its International Executive Committee proposed at Alghero in 1993 became, as a result of lack of funds, one meeting in London in September 1994 and one meeting as part of a Cold Spring Harbor conference in October 1997. The North American Committee has remained active as a committee, thanks in part to the funds provided by the 1994 MacArthur grant. HGDP Regional Committees in Southwest Asia and in China, with local funding, have collected some DNA samples from their regions. The South American Committee disbanded and the others are inactive.

The NSF provided some funding for HGDP-related activities in 1997, when it awarded about $600,000 in grant funding for "pilot projects" on various issues related to the HGDP. These funds were explicitly not to be used for collecting DNA. Some continued interest in HGDP and its goals has been sustained by two conferences at Cold Spring Harbor Laboratories on Human Evolution, in October 1997 and April 1999. And, in 1999, HGDP has agreed with the Centre pour l'Étude du Polymorphisme Humain (CEPH) in Paris that CEPH will store 1000 cell-lines from existing collections and make DNA samples available to qualified researchers, thus advancing some of the goals of HGDP.

Outside HGDP the subject of human genetic diversity has become of greater interest. In 1998 the NHGRI, with the end of the sequencing phase of the HGP in sight, created an initiative to use samples from diverse populations to search for "single nucleotide polymorphisms," or SNPs. These SNPs are expected to be useful for locating genes of medical interest, and NIH sought to create a resource of SNPs in the public domain before private firms patented too many of them. (A consortium of pharmaceutical companies and foundations has embarked on a similar effort, again with the goal of putting SNPs in the public domain (41).) The NIH samples, selected from Americans of European, African, Asian, and Native American ancestry, are available from a public repository but with no identifying ethnic information, even at the continental level (42). In fact, investigators taking samples from the repository are required to promise that they will not attempt to identify the ethnic background of the person who gave the sample (43).

Meanwhile, human population genetics continues to be done, for anthropological and other purposes. Recently published research has used the analysis of DNA samples to provide evidence about the origins of Native American (44), Japanese (45), and Chinese populations, among others (46,47). But the empty matrix problem remains — there still exists no broadly derived set of human DNA samples that have been analyzed for a standard set of markers. And no special ethical oversight exists for the ongoing efforts to collect samples for the study of human population genetics.

ETHICAL, LEGAL, AND SOCIAL ISSUES RAISED BY HGDP

As a large-scale genetics research project, aimed at collecting DNA samples from thousands of people, HGDP has raised most of the issues that are raised in similar research. These concerns include, among other things, ensuring the confidentiality of individual information, avoiding undue inducements for research subject participation, the use of previously collected samples, and dealing with the possible return of medically significant information to research participants. The wide variety of cultural backgrounds of anticipated participants in HGDP makes resolving those issues unusually complex, but the basic problems are the same ones faced by any research related to human genetics. Answers to those problems are the same ones faced by any research related to human genetics. Answers to those problems in any context may be contentious, but the HGDP's answers, as proposed in the Alghero report and especially in the Model Ethical Protocol, are unusually detailed but not extraordinary (27). One difficult area not covered at length in the Model Ethical Protocol is the use of previously collected samples, an issue that remains as vexing and complicated for HGDP (48) as it does for medical research more generally (49,50).

The issues unique to the HGDP arise from the nature of the "research subjects" of HGDP: not primarily individual humans, but human groups. The group nature of this research that makes the ethical, legal, and social implications of HGDP so fascinating and so difficult. It also makes the proposed solutions to those issues, particularly the North American Committee's requirement of "group consent," complex and controversial, in some respects perhaps more controversial than the research itself.

Group Concerns

A host of concerns have been raised on behalf of the populations that might participate in the project. Although many different issues are involved, they all fall into two main categories: fears that genetic information will harm the groups that take part and concerns that the groups will be financially exploited with respect to their genetic resources. These concerns have, not surprisingly, been raised largely on behalf of indigenous groups that have suffered from oppression and continue to exist under European domination. Thus the concerns have been raised most actively by groups speaking for Native Americans throughout the Western hemisphere, Australian aborigines, and the Maori, the Polynesian minority in New Zealand. Although the concern about exploitation seems to exist in some other settings, the fears of harm are strongest among the populations with the least power.

It is worth noting that HGDP does not intend to sample solely populations that are indigenous, small, or powerless.

Its goal is to collect a roughly proportionate sample of the world's human populations. Some of them will be small; on the other hand, the Han (ethnic Chinese) are a population of great interest to population geneticists, and they make up about 20 percent of all humans. Thus much of the intended work of HGDP could be achieved without sampling the populations that might be most concerned about the project.

Direct Harms. Critics of HGDP have argued that the project could bring a wide range of harms to participating populations. These range from harms to the population's cultural, to harms in its members relationships to the broader society, to political costs. Each is discussed below.

Two kinds of possible cultural harms have been suggested. One is that participation in this kind of project may violate either a particular group's culture or, more broadly, general indigenous norms (51–54). It has been argued that the assertion of domination and control over nature implicit in modern science is, in itself, antithetical to indigenous cultures. More specifically it is also claimed that some aspects of the DNA collection process will violate cultural norms, such as those that may relate to the treatment of blood or hair. The great degree of variation in human cultures makes it prudent to be skeptical of claims of "universal" cultural norms for indigenous peoples, but certainly some groups will hold such views.

A second type of cultural harm looks instead to the results of the project. Some indigenous activists have argued that scientific evidence from HGDP, by contradicting local cultural histories and origin stories, may undermine the authority and power of the culture. Thus many Native American populations have oral histories that place their origin in American locations. Genetic evidence that the ancestors of Native Americans migrated to the Western Hemisphere from Siberia could shake community members' faith both in that origin story and in the entire culture (51). Supporters of the project counter that these claims need to be examined carefully. Anthropologists have long argued, based on many lines of evidence, that Native Americans migrated to the Western Hemisphere from Asia. HGDP might add some evidence to that conclusion, but its supplemental effect seems unlikely to be great. In addition, it is urged, many people have shown a great ability to disregard scientific evidence that conflicts with their origin myths. In the United States, which provides support for extensive research into the evolution of modern humans, over 40 percent of the population continues to believe that humanity was created in the Garden of Eden about 6000 years ago. Finally, one may question whether cultures necessarily want to preserve their myths unchanged. Nonetheless, in some circumstances this may be a realistic threat to a population's culture.

Those concerned about HGDP also point to broader social harms revolving around concerns about discrimination. As is the case with individuals, groups might be considered genetically susceptible to particular diseases or conditions. This could lead to discrimination against group members in employment, insurance, or other social activities. It might stigmatize the group or support a racist belief in the group's inferiority.

HGDP's supporters use several arguments to try to undercut this criticism. First, HGDP is not looking for disease-related genes; it will focus on random markers that will not reveal this kind of information. Second, the relevance of these concerns they depend crucially on a population's situation. Commentators have discussed in detail the possible and actual existence and significance of discrimination in insurance and in employment in the United States. (And, in the American context, the motivation for employment discrimination often stems from the employer's payment of health insurance costs (55).) In societies where every individual is guaranteed health coverage — which comprise all wealthy countries other than the United States as well as some middle- and lower-income countries — this discrimination becomes irrelevant. In some traditional societies where Western medicine itself may be unavailable, both health insurance and employment discrimination may be irrelevant.

More fundamentally, though, supporters of the project contend that discrimination fears exist in an individual context because genetic information might reveal something about risk that cannot otherwise be known. At the level of human populations, genetic analysis would not often provide that kind of information. Such an analysis might indicate that from genetic causes, the Irish have several times the average levels of the genetic variation that, when an individual inherits two copies of the variant, causes the genetic disease phenylketonuria. It would not indicate which Irish people were at risk and which were not. And, more important, the levels of these risks will already be known directly from public health statistics and epidemiological research. One does not need to know the distribution of genetic variations associated with disease to know disease rates in populations; those rates can be examined directly. It is conceivable that for some disorders, cases linked to genetic variations will have differences from nongenetic cases that would be important to insurers or employers. One could hypothesize that for example, breast cancer cases related to mutations in BRCA1 were less easily and cheaply treatable than other breast cancers. In that case, knowledge that a population had an unusually high rate of those mutations might be relevant. But, for the most part, the prevalence in a population of genetic variations linked to disease will add nothing to the risk information already available from observation of the disease incidence.

The issue of stigmatization is more complicated. Project supporters point out that it is hard to think of examples any populations that are stigmatized because of higher rates of disease. Europeans have higher than average rates of cystic fibrosis, many Africans and Southeast Asians have higher rates of hemoglobinopathies, Ashkenazic Jews have unusually high rates of Tay-Sachs disease. Alcoholism may be one of the few examples of a disease or condition that has stigmatized some populations, such as Irish, Russians, and Native Americans. Again, one might think that if a population is going to suffer stigmatization because it carries disease-associated genetic variations at an unusual level, then it would already been stigmatized for having a high rate of the disease itself. The fact that the cause is genetic, though, might lead to a higher social impact. The concept of "genetic essentialism"—the idea that one's genes are one's essence (56,57)—could lead people to believe that a population with a high rate of genes associated with a disease is somehow inherently flawed, or more flawed, than a population with a low rate of that genetic condition. Of course, every population is likely to have somewhat higher and lower genetic risks for different diseases, but one might focus on a particularly stigmatizing disorder, such as schizophrenia. On the other hand, a genetic explanation might have opposite effects. If a higher rate of disease in a population is believed to be genetic, the individuals suffering from the disease may be viewed more favorably on the theory that the disease is not the result of their actions. (This relies on the general popular assumption, often false, that diseases with genetic "causes" do not also have, in the same individual, environmental "causes.") This issue has been interestingly discussed in the context of a possible genetic association with sexual preference (58); whether a conclusion that a population's higher rate of a condition has genetic roots would increase, decrease, or leave the same any stigma attached to the group because of the condition is unclear.

The asserted political harms vary. One set of concerns revolves around land claims. If rights to land are affected, legally or politically, by the history of a population's occupation of the land, genetic evidence that the population migrated from elsewhere, or direct evidence from ancient DNA that a different population occupied the land in the past, might have some political relevance (8). Project supporters argue, on the other hand, that few if any peoples are thought to be truly indigenous to any location—under current thinking about human evolution, only some African groups could even possibly make that claim. In many regions of the world, migrations and changes of homeland have happened within historical memories—in too many places, within the living memories. For legal and political significance, the relevant time frame is important; it is hard to imagine many situations where DNA evidence would have current relevance. And, if it were, it might be thought as likely, in a given situation, to favor a population's claims as to harm them. But if one takes the view—which might be quite reasonable for some subordinated populations—that the dominant culture will twist any new "scientific" evidence to its political benefit and your detriment, this concern becomes more understandable. Even if genetic evidence is not directly relevant, it might be used to weaken the political standing for some land claims. This fear of this kind of general political effect may be one of the sources of the controversy over the so-called Kennewick man, an ancient skeleton found in the northwestern United States that some allege has "European" features.

Also related to land claims, concrete or broad, is a possible concern about membership. Some worried that an outside government might impose a genetic test for membership on the population, adding and subtracting members without the group's consent.

Critics of HGDP cite the possibility that genetic variations might be used for biological warfare against sampled populations as yet another political concern.

Project supporters claim that such ethnically targeted biological warfare seems scientifically implausible for two reasons. First, the enormous overlap in genetic variations between populations would mean that such a weapon would not affect many in the targeted group and would affect many outside the targeted group. The twentieth century has seen all too many more discriminating and "efficient" methods for genocide. Second, scientists do not know how to kill a cell based on variations in its genetic material. If they did, all infectious disease organisms and all tumors could be handily defeated. Nonetheless, some recent publications have fed this fear (59–61).

Exploitation. Issues of exploitation of participating populations are somewhat different from those of direct harms to those populations. These concerns focus on the possibility that participating populations might be robbed of something of value as a result of the project—"their genes" (20). This view is fed by the great interest, popular and financial, in biotechnology. It is also exacerbated by stories, which can reach the level of myths, about past "victims" of predatory commercial biotechnology, such as the tale of John Moore's spleen (62) or the patent applications for cell-lines derived from indigenous Panamanians and Papuans. There are also parallels, close and distant. The term "bio-piracy" was coined, apparently by RAFI, to describe the use of plants and plant genes from the developing world by commercial firms from the developed world for pharmaceutical and agricultural profit. HGDP may have appeared just an extension of that process to human genes. More generally, though, many indigenous cultures have suffered from exploitation by outsiders in recent history. The slogan that "You stole our land, you stole our resources, and now you want to steal our very genes" has power. Finally, this fear of exploitation flourishes as a result of limited understanding of human genetics. People are ready to believe that their population has its own "genes," let alone its own special variants of human genes. From the identification of a gene in a small number of people, to an understanding of its function, to its application as a profitable pharmaceutical may appear an easy series of events to the outsider. The idea that one's group has "special genes" of great value and power is also a flattering one. For all these reasons people concerned with indigenous groups or, more broadly, the developing world might conclude that HGDP was an effort to exploit the commercial value of human genes from outside the developed world—after all, if those genes did not have value, why would anyone look for them?

The reality is less financially promising and more complicated. Populations do not have unique "genes," although they may have an unusually high (or low) percentage of certain variants of human genes. Identifying a genetic variation as associated with a disease (or with protection from a disease) is a long and complicated process, requiring extensive medical as well as genetic work with individuals and their families, both affected and unaffected with the disease. The jump from the discovery of a disease-related variation to a commercial product is enormous and has rarely been made. None of these activities has any relationship to HGDP. The project has disclaimed any commercial interests, backing, or connections. But, as critics of the project point out, scientists with commercial connections would, under HGDP's initial plan, have access to samples and data collected by HGDP.

In fact, neither a renegade HGDP nor scientific third parties would be likely to be able to derive information with strong medical, and hence commercial, value from the HGDP samples. HGDP will not obtain the kind of medical data about participants that would be necessary to make the samples it collects useful for this kind of medical research. Without knowing which participating individuals had diabetes and which did not, for example, research into connections between genetic variations and diabetes is not possible. HGDP's organizers have stated that some information of value to medical researchers might be created by the project. For example, if a researcher had identified a "candidate gene" for a particular disease, HGDP's resources might be able to tell her whether the high risk variant was more or less common in populations with known epidemiological risk. But this kind of information seems likely to have little direct commercial value — it will neither prove nor disprove a genetic connection but might narrow the possibilities. Thus HGDP has consistently evaluated the likely commercial value of the samples and data it collects as quite low. It has not, however, been able to rule out the faint possibility that some samples or data could end up having commercial value for some third party user of the information. It is this indirect and unlikely contingency that could give rise to concern about the financial exploitation of participating populations.

HGDP's Responses to Group Concerns

HGDP has been aware of some of the group concerns raised by its research since at least early 1993. It tried to address many of them in the Alghero meeting and report. Some of its solutions in that report include respect for the cultural integrity of participating populations; the resolution to protect individual and, in some cases, group confidentiality; and the renunciation of commercial connections; and, when appropriate, a sharing of financial benefits. The Alghero report expressly recognized that "the ethical issues faced by the Project will evolve over time" and called for their continual review. The most detailed such review from within the project has come from the North American Committee and its Model Ethical Protocol. This document accepted, for North American HGDP work, two broad responses to group concerns: group consent and contractual restrictions on subsequent uses of samples and data. Its idea of group consent has spawned nearly as much controversy in ethical circles as HGDP itself — and more academic literature.

Group Consent. The Model Ethical Protocol requires that where feasible, those collecting DNA samples in North America for HGDP obtain not only the informed consent of individual participants but also of their population (27,63). This collective consent is to be sought from the group's "culturally appropriate authorities," as defined by the group itself. Where permission is denied, the project would not accept any samples from members of the population. The protocol recognizes that such consent will not be feasible for groups that do not have an authority structure, like Irish-Americans or Ashkenazic Jews. In those cases it requires full discussion within the direct community in which the collection is to be done — the local town or religious or cultural association — as well as dissemination of the facts about the planned study broadly to those who identify with the group to enable them to try to influence their colleagues to participate or not in the project.

The North American Committee put forward several arguments in favor of group consent. At the most basic level, its claim is that human population genetics research, by its very nature, has the population as its subject as well as the participating individuals from that population. Every member of that population may be affected, positively or negatively, by the results of the research and thus the population, collectively, should have a chance to decide on its participation. Just as individual informed consent lets people decide whether a particular medical procedure or research project meets *their* balance between costs and benefits, under group consent the group as a whole could hear about the possible harms discussed above and decide whether to take those risks. If a group decides that HGDP involves too great a risk of cultural disruption, negative land claim evidence, or any other harms, it will simply not participate. Individual informed consent is also justified as respecting the autonomy and personhood of the patient or research subject. Group consent acknowledges the cultural reality of ethnic groups. This may have particular force with populations that have been dominated by others. By seeking and respecting their decision, one respects their cultural autonomy. This kind of respect may be welcome by almost any group, but is particularly appropriate with Native American tribes in the United States. Legally, federally recognized tribes are sovereign governments existing in a complex political relationship with the federal government. Culturally non-Indian peoples often do not recognize that legal reality. By expressly giving them a chance to say no to the research with their populations, whether or not conducted on the reservation and hence within their political jurisdiction, the Model Ethical Protocol would uphold and extend that autonomy. And, as HGDP seeks to sample only about 5 to 10 percent of the world's human populations, many groups can decide not to participate without jeopardizing the project's goals.

This idea of group consent is not new. Ethnographers and epidemiologists have long known that the kinds of intensive studies they conduct in the field often require, as a practical matter, a great deal of community consensus building and the approval of those with authority, formal or informal, within the group. In the United States many federally recognized Native American tribes, which are sovereign governments within their reservation boundaries, have in recent years established their own institutional review boards for assessing research proposed on the reservation (64). The idea of collective approval is new to some areas of research, but not to all. In its August 1999 report on the use of human biological materials, the United States National Bioethics

Advisory Commission recognized the significance of group interests and the possible value of group consent, but noted that current law regulating human subjects research does not require that these collective issues be addressed (50).

Group consent has precedents, but it also has problems, both practical and philosophical. Perhaps the largest problem, and one that fits both categories, revolves around the definition of the "group" whose consent is required. Eric Juengst has argued that this is a fatal flaw (8,65). Genetic information about one Navajo village, for example, may have implications all Navajos, on the reservation or off. It may also say something about other speakers of languages in the Na Dene language family, which is spread widely in western North America. It could demonstrate something distinctive about all Native Americans. Or it may have no implications for any groups larger than the individuals involved. Until the research is completed—along with complementary research on other populations—the scope of its effects cannot be known. If the justification of group consent is to have the approval of those who may be affected by the research, it cannot succeed. The relevant group, which would be population defined by a relative closeness of genetic relationship, cannot be defined in advance other than to say, based on the existing knowledge of population genetics, that it does not coincide perfectly with any culturally defined ethnic group.

Juengst's criticism is clearly correct, though its impact is not as clear. Defenders of the group consent can argue that although the definition of the relevant group will never be perfect, it will be a good start. The Model Ethical Protocol says that researchers should determine the relevant group in consultation with the community with whom they are working. If they view a larger grouping as necessarily involved in the research, then it should be so involved. Using the relevant community's own definition upholds, at least, the autonomy justification for group consent even if it does not fully encompass the consent of those affected. Juengst suggests replacing group consent with consideration of group interests during individual consent. Thus, rather than ask the group for consent, individuals asked to participate in the research would be told that the research may have implications for specific groups they belong to and that they should consider those implications when deciding whether to participate. This kind of directed consideration could play a useful role, although it too would be imperfect. The particular individuals approached for the research might not reflect well the concerns of the larger community. And the process of consent—the discussion and debate—may well be much broader (as well as longer and more arduous) in group meetings than in individual informed consent sessions.

Group consent has been subject to a number of other concerns, as well as arguments about the significance of the problem it attempts to solve (66–68). The determination of the "culturally relevant authorities," whose approval will be required may prove difficult (69). Federally recognized Indian tribes, in the United States and in Canada, have governmental structures that will provide a starting place for such authorities. Other communities will not usually have formal governments and finding any "culturally appropriate authority" may be difficult. Even recognized Indian tribes may have a variety of nongovernmental structures claiming authority. If there is a dispute within the population about which people or organizations have authority, how should researchers resolve it? This may be particularly difficult if participation in the research becomes, itself, a divisive issue within the population. One can even imagine one faction using the researcher's acceptance of them as authoritative as a weapon in a power struggle within the group. In some cases, where the structure of authority is sufficiently unclear, researchers may have to walk away from the research entirely. Other criticisms are less pragmatic. Juengst has suggested that group consent is suspect because it asks cultures to become complicit in research that may harm them, which seems a strained way to view giving choices to groups (67). Presumably the interests of the groups would not be better served by a process that imposed research on them whether they liked it or not. Some researchers, including some on the North American Committee, have been troubled that group consent would deprive individuals of their "right" to take part in research on the basis of the group's fears, even if the individuals had removed themselves from the group. Finally, it has been pointed out that the logic of group consent may extend uncomfortably far. Traditional research on human genetics uses "groups" made up both of people suffering from disease and of families. Should disease organizations or extended families also be accorded the power of group consent (63)? The issues raised by a group consent requirement are indeed broad, so broad that it has not been formally adopted yet by any HGDP regions other than the North American Committee. And a Canadian effort to revise their codes of research ethics, though initially proposing a strong "group consent" requirement, removed the discussion from their final version, leaving only a short section on special issues in dealing with the native peoples of Canada (70).

Some have argued that the process of group consultation and discussion should be required even if strict group "consent" is not sought. In a series of articles, Morris Foster and coauthors have argued for a form of "community review," which involves the community where the research is proposed without formally requiring community approval (71–73). Foster has combined this approach with a model agreement for regulating the rights of communities that agree to be the subjects of research (74).

One other response to the risks of group harms deserves mention. Foster has at least occasionally suggested that the identities of populations participating in genetic research should be hidden (33). If no one knows what group provided the samples, the information cannot be used against them by outsiders: in employment, insurance, or politics. (The internally derived harms, such as the undermining of the population's culture, could still occur.) HGDP proposed a limited form of this response, through "fuzzing" the precise identity of a participating group. For example, rather than identify a specific village, perhaps even by use of the global positioning system, HGDP could say that the samples came from a village within a broader region. Of course, withholding information about the group's identity may have its own bad consequences, both for science, which may not proceed accurately without sufficient specificity, and for the group, which could lose some benefits of the research it not identified. Besides, if the level of group anonymity were too great, it would render impossible some of the historical research that is a key purpose of HGDP. Nonetheless, some amount of group anonymity is a protective tactic that should be considered in this kind of research—and perhaps should itself be a subject of discussion in the consent process, group or individual.

Contractual Defenses against Exploitation. The founding meeting of the HGDP endorsed the idea that participating populations should share in the financial gains, if any, resulting from the HGDP. Its report states: "Although very unlikely, it is neverless [sic] possible that the results of the HGD Project may lead to the production of commercially beneficial pharmaceuticals or other products. Should a patent be granted on any specific product, the Project must work to ensure that the sampled population benefit from the financial return from sales" (1). The Alghero report, however, provided no suggestions for how the Project might ensure such a result.

Again, the Model Ethical Protocol sought to implement that principle. It did so through the medium of contracts (27). HGDP plans to operate both DNA sample and cell-line repositories and a database containing the analyses of HGDP samples. Samples and information in both facilities would be available to qualified researchers. The Model Ethical Protocol states that those facilities should only allow access to researchers who agree, by contract, to restricted commercial uses of the information they obtain from the repository or database. Contracts governing this kind of access, called "materials transfer agreements" in the case of samples and "database access agreements" for the computerized information, have long existed, initially to protect the repositories and databases from liability. The Model Ethical Protocol would expand those agreements to provide that all of the HGDP's samples and information could only be used in ways consistent with terms and conditions limiting their use. These terms and conditions, the protocol suggests, would usually be set by the participating population as part of the process of group consent. Thus a population might choose to forbid any commercial use or any patenting of their samples, the information, or products derived from the samples or information. Alternatively, it might authorize such uses on the payment of a specified royalty to the group or on the completion of a subsequent express written agreement. Or it might allow any uses. Anyone who wanted to use samples or information provided by that group would have to agree to abide by its terms.

Like group consent, contractual limitations on commercial use also have problems, both in their inception and in their implementation, but, unlike group consent, both the idea and its problems have been little discussed. At least four topics need to be raised. First, the topic of possible commercial gains would need to be discussed very carefully in the group (or individual) informed consent. Otherwise, the (very faint) hope of financial rewards might unfairly influence a group to participate in research. Undue inducements to participate in human subjects research are not

allowed; despite the researchers' disclaimers, a group might wrongly believe that it has a strong chance of becoming wealthy from its "gene royalties," which, as the researchers would recognize, would be very unlikely ever to exist. Second, some default standards would have to apply to populations for which group consent was not obtained because it was infeasible. These standards would not only have to determine what would be a "fair" return but would also have to consider the problem of who should receive or spend any such sums, as, by definition, no "culturally appropriate authority" would exist for a group from which no consent was sought. Third, attention would have to be paid to who could sue to enforce these contractual clauses. The population might have the best incentive to sue, but it would have poor access to information about the use of its materials and often poor access to the legal system. HGDP would have better access to both but less incentive to protect the population's rights. Finally, would these contractual clauses actually be enforced? Although the status of DNA as "property" remains unclear, there seems no reason to think that courts would not enforce this kind of contract. The main problem is whether anyone would recognize that a pharmaceutical company had started marketing a product based, at least in part, on research done many years earlier using some HGDP samples? This kind of enforcement problem could be enormous, but it might find a solution in the very structure of the pharmaceutical market. Bringing a drug to market is a very long and extremely expensive proposition, costing in the United States an average of several hundred million dollars. The existence of even a possible claim that such a product breached a contract with HGDP and participating populations might throw such a "cloud on the title" of the pharmaceutical company as to ensure that it would not proceed without negotiations.

Racism

In the nineteenth and early twentieth century, the biological sciences, including the newly born field of genetics, helped legitimize and reinforce racism by providing "scientific proof" of the inferiority of disfavored human groups (75–77). This history, shameful to contemporary geneticists, combines with the continued persistence of racist stereotypes to make genetic research into "racial" or ethnic characteristics still politically and ethically charged (78–80). HGDP risked being caught in these controversies around the genetics of race. Although its organizers denounce the idea that human races have *any* biological or genetic meaning, let alone that one "race" is genetically "superior" or "inferior" (6), the project's very name emphasizes that it is looking for genetic differences—and its research agenda makes it clear that those are differences between human populations. The jump from talking about "populations" to being seen as talking about "races" is short. This issue affected HGDP in at least three different ways.

First, some critics accused HGDP of being itself in the thrall, consciously or not, of nineteenth-century visions of genetically defined races. Its plan to compare the patterns of genetic variation frequencies of different culturally defined human groupings was seen as reflecting a belief by the project organizers that such "pure" groups, or, in a term used by the project itself initially, population "isolates" really existed (81–84). Critics viewed this as both scientifically naive and socially harmful. Even structuring the project around an examination of these culturally defined populations would, in their view, necessarily bias the results. From this perspective, Allan Wilson's "grid" plan of sampling had great benefits. Second, other commentators pointed out that even if the HGDP organizers held a scientifically correct vision of the limited connection between cultural groups and genetic populations, the public did not (8). HGDP, by even looking at genetic differences between group, would reinforce for the public the ideas that such differences existed and that they were important. Finally, HGDP organizers themselves recognized that, however meritless scientists considered "scientific racism," the project's results could be misused for racist ends. In an event used at the time of the greatest discussion of this point, Bosnian Serbs might seize on some slight genetic variation between themselves, as a group, and Bosnian Muslims, as a group, and claim "scientific" support for ethnic cleansing.

The project's supporters disclaimed any belief in genetically "pure" populations, let alone in genetically defined human races. They countered that, although culturally defined groups were not the same as genetically defined populations, they were often genealogically more closely related to each other than to outsiders. Thus the statistical analysis of patterns of genetic variation between different groups had led to useful results in prior research. Further they called the Wilson grid plan impracticable—after all, which one person would represent the genetic variations found in New York or in San Francisco? And, they urged, because the project's data would all be public, the data could be analyzed by those skeptical about the sampling strategy to detect any bias that might exist. The project's organizers admitted the second and third risks: the unconscious reinforcement of a false concept of "genetic race" in the public mind and the possibility of racist or nationalist misuse of the project's data or findings. To counter those problems, the organizers called both for general education and for "ready response teams" of scientists prepared to refute false claims (1,27). The project's supporters went farther and argued that HGDP's results would make "scientific racism" even more untenable by demonstrating the great genetic similarities among humans. This response drew its own reply, accusing HGDP of exaggerating the effects its findings would have in countering racism (7).

One other claim, at least related to racism, has been advanced against HGDP. Its opponents argue that it did not take indigenous groups seriously and did not include them in its planning. This, it is urged, reflects anything from racism and colonialism (51,52), to a scientific arrogance toward research subjects (7). HGDP's organizers respond that indigenous peoples were not excluded from the project planning and point, among other things, to numerous meetings with indigenous groups and to two Native Americans on the project's North American Committee. It is true, though, that at least in North America, the project decided not to emphasize contacts with indigenous activist groups. Instead, it decided to focus its contacts on people living in native communities, after the project was funded to begin collection and so concrete measures could be discussed. It would, in that way, build connections with the very people whose participation in the project it would seek. And it would avoid encounters with political activist groups, often distant from the Native American communities, whose opposition to the project was viewed as highly likely. Whether this strategy was wise remains unclear, particularly when the funding that would have allowed concrete contacts with local communities—and, the project had hoped, would have provided examples of successful collaborations—did not arrive.

FUTURE OF HGDP

HGDP's future remains unclear. As of early 2000, no funding for substantive work had yet been received from the U.S. federal government. No DNA samples have been collected under the auspices of HGDP except in China and in Southwest Asia, with local funding to those regional committees. Discussions are under way for the creation of general repository for some genetic samples, but no database for the results of analysis of HGDP samples had been created. The project continues to inspire controversy, among some ethicists, some nongovernmental organizations, and some indigenous groups. It also continues to inspire its organizers' hopes. Only time will tell whether it thrives or shrivels.

Whatever happens to HGDP, the future of studies of human genetic variation seems quite clear. The kinds of isolated, individual studies that have been going on for 80 years are continuing and expanding. Perhaps more important, the importance of genetic variation for understanding genetic links to disease is increasingly recognized. Once HGP completes sequencing "the" human genome, two next steps seem obvious: determining the function of the identified genes and understanding the consequences of the variations found in the six billion human genomes. Already NIH has created a resource of 450 DNA samples from ethnically diverse residents of the United States to be used to find SNPs (single nucleotide polymorphisms). This kind of variation is mainly important to provide landmarks in the genome; other examples of human genetic variation will almost certainly be of medical and scientific interest. In the long run, as the studies of genetic variation for nonanthropological reasons get broader, the samples and data that HGDP seeks to gather may appear from other sources.

This kind of future, however, would be a continuation of the past. Unless DNA samples were available for analysis or were analyzed against a standard set of markers, the empty matrix problem that led to the organizers to propose HGDP would continue. And supporters of the project argue that the ethical, legal, and social concerns would be heightened, not diminished, by the end of HGDP. Rather than having a project that insisted on compliance with certain ethical standards, DNA collection and research would continue to be done in hundreds of different laboratories, using many different sets of ethical rules. Opponents, on the other hand, can hope that communities can more effectively resist many small laboratories than one worldwide "project." And they may believe that an end to HGDP will lead to protections better than those HGDP claimed, unconvincingly to them, to plan to impose. One way or another, the issues will remain as relevant. Whether they would be as visible and as fully debated in the absence of a formal HGDP is, at best, unclear. But, whether it ever is completed or not, HGDP and the controversy around it has at least helped illuminate the complex issues — ethical, legal, social, and political (85,86) — raised when human genetics moves from the individual or family to the population.

ACKNOWLEDGMENT

Professor Greely has been associated with the HGDP since 1993 and chairs the Ethics Subcommittee of the Project's North American Committe. In that role he was the principal author of that Committee's Model Ethical Protocol.

BIBLIOGRAPHY

1. Human Genome Diversity Committee of HUGO, *Human Genome Diversity (HGD) Project, Summary Document*, HUGO Europe, London, 1995, Available at *http://www.stanford.edu/group/morrinst/HGDP.hmt*

2. L. Hirschfeld and H. Hirschfeld, *Lancet* **2**, 675–695 (1919).

3. W.H. Schneider, *History and Philosophy of the Life Sciences* **18**, 7–33 (1996).

4. A.E. Mourant, *The ABO Blood Groups: Comprehensive Tables and Maps*, Blackwell Scientific, Oxford, 1958.

5. L.L. Cavalli-Sforza, P. Menozzi, and A. Piazza, *History and Geography of Human Genes*, Princeton University Press, Princeton, NJ, 1994.

6. L.L. Cavalli-Sforza and F. Cavalli-Sforza, *The Great Human Diasporas: The History of Variation and Evolution*, Addison-Wesley, Reading, MA, 1995.

7. M. Lock, in B.M. Knoppers, ed., *Human DNA Sampling: Law and Policy — International and Comparative Perspectives*, Kluwer Law, The Hague, 1997.

8. E.T. Juengst, *Kennedy Inst. Ethics J.* **8**, 183–200 (1998).

9. W.H. Durham, *Coevolution: Genes, Culture, and Human Diversity*, Stanford University Press, Stanford, CA, 1991, pp. 226–285.

10. L.L. Cavalli-Sforza et al., *Genomics* **11**, 490–491 (1991).

11. K.M. Weiss, K.K. Kidd, and J.R. Kidd, *Evolu. Anthropol.* **1**, 78–80 (1992).

12. J.R. Kidd, K.K. Kidd, and K.M. Weiss, *Hum. Biol.* **65**, 1–6 (1993).

13. L. Roberts, *Science* **252**, 1614–1617 (1991).

14. L. Roberts, *Science* **257**, 1204–1205 (1992).

15. L. Roberts, *Science* **258**, 1300–1301 (1992).

16. H.T. Greely, *Human Genome Diversity Project — Summary of Planning Workshop 3(B): Ethical and Human Rights Implications*, 1993.

17. H.T. Greely, in B.M. Knoppers, ed., *Human DNA Sampling: Law and Policy — International and Comparative Perspectives*, Kluwer Law, The Hague, 1997.

18. L. Roberts, *Science* **258**, 1300–1301 (1992).

19. Rural Advancement Foundation International, *Home Page*, Available at: *http://rafi.ca*

20. Rural Advancement Foundation International, *RAFI Communiqué*, May 1993.

21. Rural Advancement Foundation International, *HGDP Hit List*, Available at: *http://rafi.ca*

22. Rural Advancement Foundation International, *Indigenous Person from Papua New Guinea Claimed in U.S. Government Patent* (1995), Available at: *http://www.rafi.ca*

23. H.T. Greely, *Cultural Survival Q.* **20**(Summer), 54–58 (1996).

24. L.A. Whitt, *Oklahoma City Univ. L. Rev.* **23**, 211–259 (1998).

25. D.R.J. Macer, in B.M. Knoppers, ed., *Human DNA Sampling: Law and Policy — International and Comparative Perspectives*, Kluwer Law, The Hague, 1997.

26. Foundation on Economic Trends, the Rural Advancement Foundation International, the World Council of Indigenous Peoples, the Third World Network, and the Cordillera Peoples' Alliance, *Petition to Dr. Harold Varmus*, June 30, 1994.

27. North American Regional Committee of the Human Genome Diversity Project, *Houston Law Rev.* **33**, 1431–1473 (1997).

28. UNESCO, International Bioethics Committee, Subcommittee on Bioethics and Population Genetics, *Bioethics and Human Population Genetics Research* (November 15, 1995), Available at: *http://www.biol.tsukuba.ac.jp/~macer/PG.html*

29. D.R.J. Macer et al., *Nature* **379**, 11 (1996).

30. B.M. Knoppers, M. Hirtle, and S. Lormeau, *Genomics* **50**(3), 385 (1998).

31. B.M. Knoppers, ed., *Human DNA Sampling: Law and Policy — International and Comparative Perspectives*, Kluwer Law, The Hague, 1997.

32. M.S. Frankel and A. Herzog, *Human Genome Diversity Project: The Model Ethical Protocol as a Guide to Researchers*, American Association for the Advancement of Science, Washington, DC, 1998.

33. Workshop on Anthropology, Genetic Diversity, and Ethics Workshop, *Meeting Summaries* (1999), Available at: *http://www.uwm.edu/Dept/20th/projects/GeneticDiversity/index.html*

34. Senate Committee on Governmental Affairs, *Human Genome Diversity Project: Hearing before the Committee on Governmental Affairs*, United States Senate, 103rd Congr. 1st Sess., April 26, 1993.

35. National Research Council, *Evaluating Human Genetic Diversity*, National Academy Press, Washington, DC, 1997.

36. E. Pennisi, *Science* **278**, 568 (1997).

37. C. MacIlwain, *Nature* **389**, 774 (1997).

38. *Nature* **390**, 221 (1997).

39. W.J. Schull, *Science* **279**, 10 (1998).

40. W.J. Schull, *Nature* **390**, 221 (1997).

41. E. Marshall, *Science* **284**, 406 (1999).

42. F.S. Collins, L.D. Brooks, and A. Chakravarti, *Genome Res.* **8**, 1229 (1998).

43. Available at: *http://locus.umdnj.edu/nigms/comm/order/assurance.html*

44. D. Hurtado and R. Braginski, *Science* **283**, 1439 (1999).

45. D. Normile, *Science* **283**, 1426 (1999).

46. L.L. Cavalli-Sforza, *Proc. Nat. Acad. Sci. USA* **95**, 11501 (1998).

47. A. Spaeth, *Time*, January 17, 45 (2000).

48. H.T. Greely, in R. Weir, ed., *Stored Tissue Samples: Ethical, Legal, and Policy Implications*, University of Iowa Press, Iowa City, IA, 1998.

49. E.W. Clayton et al., *J. Amer. Med. Assoc.* **274**, 1786–1792 (1995).

50. U.S. National Bioethics Advisory Commission, *The Use of Human Biological Materials in Research*, National Bioethics Advisory Commission, Bethesda, MD, 1999.

51. D. Harry, *Abya Yala News* **8**, 8 (1994).

52. J. Armstrong, *Nat. Cath. Rep.*, January 11–12, 27 (1995).

53. A.T.P. Mead, *Cultural Survival Q.* **20**(Summer), 46–51 (1996).

54. R. Liloqula, *Cultural Survival Q.* **20**(Summer), 42–45 (1996).

55. H.T. Greely, in D.J. Kevles and L. Hood, eds., *The Code of Codes*, Harvard University Press, Cambridge, MA, 1992.

56. D. Nelkin and M.S. Lindee, *The DNA Mystique: The Gene as Cultural Icon*, W.H. Freeman, New York, 1995.

57. S. Wolf, *J. Law Med. Ethics* **23**, 345 (1995).

58. J. Halley, *Stanford Law Rev.* **46**, 503 (1994).

59. M. Dando, *Biotechnology, Weapons, and Humanity*, British Medical Association, London, 1999, pp. 53–67.

60. U. Mahnaimi and M. Colvin, *Sunday Times (London)*, November 1 (1998).

61. W.J. Broad and J. Miller, *N.Y. Times*, Sunday, December 27, Sec. 4, p. 1 (1998).

62. *Moore v. Regents of the University of California*, 51 Cal. 3d 120; 793 P.2d 479; 271 Cal.Rptr. 146 (1990).

63. H.T. Greely, *Houston Law Rev.* **33**, 1397–1430 (1997).

64. W.L. Freeman, in R.F. Weir, ed., *Stored Tissue Samples: Ethical, Legal, and Policy Implications*, University of Iowa Press, Iowa City, IA, 1998.

65. E. Juengst, *Am. J. Hum. Genet.* **63**, 673–677 (1998).

66. C. Weijer, G. Goldsand, and E.J. Emanuel, *Nature Genet.* **23**, 275 (1999).

67. P.R. Reilly, *Am. J. Hum. Genet* **63**, 682–685 (1998).

68. P.R. Reilly and D.C. Page, *Nature Genet.* **20**, 15–17 (1998).

69. R.A. Grounds, *Cultural Survival Q.* **20**(Summer), 64–68 (1996).

70. Canada Tri-Council Working Group on Ethics, *Code of Conduct for Research Involving Humans*, Minister of Supply and Services, Ottawa, 1997.

71. M.W. Foster et al., *Am. J. Hum. Genet.* **64**, 1719–1727 (1999).

72. M.W. Foster, A.J. Eisenbraum, and T.H. Carter, *Genet. Test.* **1**(4), 269–274 (1997).

73. M.W. Foster and R.R. Sharp, *J. Law Med. Ethics* **28**, 41–51 (2000).

74. M.W. Foster, D. Bernsten, and T.H. Carter, *Am. J. Hum. Genet.* **63**, 696–702 (1998).

75. S.J. Gould, *The Mismeasure of Man*, Norton, New York, 1981.

76. D.J. Kevles, *In the Name of Eugenics*, Knopf, New York, 1985.

77. D.B. Paul, *Controlling Human Heredity: 1865 to the Present*, Humanities Press International, Atlantic Highlands, NJ, 1995.

78. A.L. Caplan, in T.F. Murphy and M.A. Lappe, *Justice and the Human Genome Project*, University of California Press, Berkeley, CA, 1994.

79. P.A. King, in J. Kahn, A. Mastroianni, and J. Sugarman, *Beyond Consent: Seeking Justice in Research*, Oxford University Press, New York, 1998.

80. P.A. Baird, *Perspect. Biol. Med.* **38**(2), 159–166 (1995).

81. J. Marks, *Anthrop. News* **72**, April (1995).

82. J. Marks, *Human Biodiversity: Genes, Race, and History*, Aldine De Gruyter, New York, 1995.

83. M. Lock, *Soc. Sci. Med.* **39**, 603–606 (1994).

84. M. Lock, *Chicago-Kent Law Rev.* **75**, 83 (1999).

85. D.B. Resnik, *Politics Life Sci.* **18**, 15 (1999).

86. H.T. Greely, *Politics Life Sci.* **18**, (2000).

See other entries OWNERSHIP OF HUMAN BIOLOGICAL MATERIAL; see also PATENTS AND LICENSING entries.

HUMAN SUBJECTS RESEARCH, ETHICS, AND INTERNATIONAL CODES ON GENETIC RESEARCH

BARTHA MARIA KNOPPERS
University of Montreal
Montreal, Canada

DOMINIQUE SPRUMONT
University of Neuchâtel
Neuchâtel, Switzerland

OUTLINE

INTRODUCTION

At least theoretically, medicine has benefited and continues to benefit from the participation of sick or healthy volunteers, in human research. Medical practice and research are inseparable elements of modern medicine. Yet, despite the unprecedented success of medicine and biotechnology in the second half of the twentieth century, human subjects research remains associated with the atrocities committed by Nazi doctors during World War II (1), the dreadful biological warfare experiments conducted by the Unit 731 of the Imperial Japanese Army (2), not to mention the infamous radiation experiments conducted and/or financed by the U.S. government and by governments of other countries (i.e., UK, Switzerland) at the height of the cold war (3). As Jean Bernard, the first chairman of the French National Ethics Committee, has said "Human experimentation is morally necessary and necessarily immoral." The high hopes and expectations in the constant progress of medicine, as expressed by the growing efficacy and quality of health care, have not overridden the fears of society that human beings could be or are being grossly abused for the sake of science. As noted by Jay Katz in the introduction to his comprehensive case book on human experimentation: "When science takes man as its subjects, tensions arise between two values basic to Western society: freedom of scientific inquiry and protection of individual inviolability" (4, p. 1). The primary goal of the regulation of human experiments is indeed to ensure protection of the rights and welfare of human subjects.

As in many other fields of medical practice today, there is a clear shift toward legal regulation (5). By way of national and international legislation, a growing number of detailed guidelines must be followed by all participants in research activities; this means investigators, sponsors, monitors, ethics review boards, research institutions (hospitals, universities, etc.), local and national authorities. The complex regulatory framework is aimed both at protecting the interests of the persons participating in research and at ensuring the quality of the research's results will meet with general public approval, indeed, that the research is "necessarily moral." Before considering in more depth the basic rules that apply to human subjects research as they are presented in international codes, we will review the nature and scope of current research. It is also important to understand how these rules evolved historically as well as the underlying ethical, professional, and legal issues.

It is against this background that the basic principles of research in human genetics should be examined. Of particular interest is the emergence of international ethical norms governing human genetics in the area of human subjects research in biotechnology (6). Positions and proposals have emanated from international and regional bodies such as the United Nations Education Science and Culture Organization (UNESCO), the World Health Organization (WHO), the Human Genome Organization (HUGO), the Council of Europe (CE), the European Commission (EEC), and the Latin American Human Genome Program (PLAGH). We will limit our discussion to the last decade and demonstrate that in the gradual evolution of ethical norms governing human subjects, the area of human genetics research is witnessing a movement from general principles to more refined and complex approaches, sometimes in contradiction with one another. Moreover, because of the personal, familial, and social nature of genetic information, some controversial areas of application are "gen-ethics," which test the founding ethical principles governing human subjects research. The immediate conclusion to be drawn is that the increasing multiplicity, complexity, and specificity of ethics in human subjects research may well lead to losing sight of the fundamental ethical principles, if not undermine them.

NOTIONS AND TERMINOLOGY

Ties Between Medical Practice and Research

In the now classical essay "Training for Uncertainty," Renée Fox analyzed the process by which medical students learn to cope with their uncertainty which, by and large, remains the main (if not only) certainty in medical knowledge. Indeed, the proportion of medical care which relies on solid scientific or empirical evidence varies from only 10 to 50 percent depending on the authors (7). Three basic types of uncertainty can be identified:

> The first results from incomplete or imperfect mastery of available knowledge.... The second depends upon limitations in current medical knowledge.... A third source of uncertainty derives from the first two. This consists of difficulty in distinguishing between personal ignorance or ineptitude and the limitations of present medical knowledge (8, p. 20).

The first and third types of uncertainty can be limited by the proper selection, training, and continuing education of physicians, while the second type calls for research and the collection of empirical data. Research appears in the latter sense as a means to limit uncertainty and to acquire new, generalizable knowledge. Paradoxically, newly acquired knowledge is per se a source of uncertainty as it often raises more questions than it answers. The constant quest for new knowledge and the expected progress that should result from it is indeed an intrinsic element of scientific thinking (9).

Without elaborating on this point further, it is important to note that research is the basis of medical knowledge, and it is an activity that should be better promoted to give everyone the chance to benefit from it. Even if the recent history of medical research has been full of scandals and abuses, the positive dimension of research is beginning to be accepted. There has been a dramatic change in the general perception of research especially since the 1980s. The major change came from the AIDS epidemic which caused the gay community to put pressure on the scientific community and the government authorities to funnel research into that area. Feminist groups have also successfully promoted the idea that women should no longer be systematically excluded from research protocols, which formerly biased results. Finally, developing countries have asked to receive more support in research activities that are specific to their needs.

Distinctions Between Medical Practice and Research

In 1974 the U.S. Congress created the National Commission for the Protection of Human Subjects of Biomedical and Behavioral Research (the Commission). Its primary task was to establish "the boundaries between biomedical and behavioral research involving human subjects and the accepted and routine practice of medicine" (10). After extensive consideration of the topic, the Commission concluded:

> For the most part, the term "practice" refers to interventions that are designed solely to enhance the well-being of an individual patient or client and that have a reasonable expectation of success. The purpose of medical and behavioral practice is to provide diagnosis, preventive treatment or therapy to particular individuals. By contrast, the term "research" designates an activity designed to test a hypothesis, permit conclusions to be drawn and thereby to develop or contribute to generalizable knowledge (expressed, for example, in theories, principles and statements of relationships). Research is usually described in a formal protocol that sets forth an objective and a set of procedures designed to reach that objective.
>
> When a clinician departs in a significant way from standard or accepted practice, the innovation does not, in and of itself, constitute research. The fact that a procedure is "experimental," in the sense of new, untested or different, does not automatically place it in the category of research. Radically new procedures of this description should, however, be made the object of formal research at an early stage in order to determine whether they are safe and effective. Thus, it is the responsibility of medical practice committees, for example, to insist that a major innovation be incorporated into a formal research project.
>
> Research and practice may be carried on together when research is designed to evaluate the safety and efficacy of therapy. This need not cause any confusion regarding whether or not the activity requires review; the general rule is that if there is any element of research in an activity, that activity should undergo review for the protection of human subjects (11, pp. 2–3).

This characterization of medical practice and research makes three basic distinctions. First, the primary goal of practice is to enhance the health and/or the well-being of an individual patient. By contrast, the investigator's goals include those of the research itself. He or she is somehow a double agent whose two masters are research and practice. Even where one does not behave contrary to the interests of one's subjects, the investigator does not act exclusively in their interests. Levine speaks in this regard, of "practice for the benefits of others" (12) which includes not only research with human subjects but also organ or tissue donation, vaccine programs, and participation in the training of health professionals. Second, the doctor–patient relationship is highly personal in the sense that all the activities of the practitioner should be based exclusively on the specific needs and interests of the patient. By contrast, an investigator must strictly follow the procedures fixed in the research protocol. Third, research should be based, in principle, on a written protocol defining its purpose, goals, and means. This research protocol is essential in assessing scientific validity at every step of the research process. It is necessary not only to guarantee the quality and reliability of the research results but also to protect the human subjects against unnecessary and unpredicted risks and burdens.

Questionable Terminology: Therapeutic Research

In 1984 Taylor and his colleagues published an enlightening study analyzing the reasons physicians tend to avoid including all eligible patients in relevant research (13). Of particular import was their discovery that only 27 percent of participating physicians maximized inclusion. The authors cited the following as the most common reasons for this finding:

> (1) concern that the doctor–patient relationship would be affected by a randomized clinical trial (73 percent), (2) difficulty with informed consent (38 percent), (3) dislike of open discussions involving uncertainty (22 percent), (4) perceived conflict between the role of scientist and clinician (18 percent), (5) practical difficulties in following procedures (9 percent), and (6) feelings of personal responsibility if the treatment were found to be unequal (8 percent) (13, p. 21).

This research points out the uneasiness of doctors acting as investigators to the detriment of their role as healers. If the term "therapeutic research" helps physicians to better accept their ambivalent role, it is a source of confusion for the patients–subjects. Research and therapy are fundamentally different. It is contradictory to speak of "therapeutic research," and this term should be avoided (14). The term is ambiguous as it implies some therapeutic benefits for the research subjects, regardless of the fact that benefits are, by definition, hypothetical. It creates confusion about the exact role of physicians involved in research activities who do not clearly disclose the fact that in the research setting they are not acting as healers but as investigators. By the term "therapeutic research," the physician is inviting the patient to participate in research, and the emphasis is on his or her latter role rather than the former. The patient may not be aware of this fact, that he or she is involved in research and that a valid consent is needed.

Categories of Research with Human Beings

The activities covered by the expression "human subjects research" are in fact very broad. A first distinction can be drawn from the term "human subjects." Does it refer solely to research with persons, meaning a person between the moment of birth and the instant of death or should it be extended to research done on fetuses and embryos as well as on human cadavers? Research with fetuses and embryos is, in principle, treated separately from research with human subjects. In general, vulnerable groups are provided special attention and protection in the regulation of research. This is the case for children, psychiatric patients, and prisoners. Minority members are also subject to specific regulation, not only to protect them from research risks and abuses but also to ensure that they have access to research protocols specific to their needs. In summary, the regulation of human subjects research varies according to the vulnerability of the group of subjects involved as well as the nature and degree of expected risks and discomfort.

Concerning the nature of research risks, a distinction should be made between (1) clinical research, namely research done "at the bed side," implying direct contact between the research subjects and the investigator, (2) epidemiological research which is based on medical data usually collected for other purposes than the research itself, and (3) research on biological material of human origin. Human subject research usually refers to clinical research and the regulation of research is generally designed to address the specific problems raised by this type of research. Yet there is increasing attention being directed toward the two other types of research. Epidemiological research raises particular problems for the respect of privacy and confidentially. Its regulation is linked to the problem of data protection. It also provides directives on the way to collect and assess the informed consent of the subjects. Research on biological material lies somewhere in the middle, depending on whether the material being used has been collected for research purposes. We will see that the development of gene therapy and the mapping of the human genome feed a growing concern about how to regulate this type of research.

ETHICS, PROFESSIONAL RULES, AND LAW

After World War II, 23 doctors, physicians, and high ranking officials of the Ministry of Health in the Nazi government were brought before an international court founded under the principles of public international law. During this trial, which was held in Nuremberg, ten basic principles to be followed in the conduct of research were enumerated. This became known as the Nuremberg Code. Yet there has remained uncertainty as to the Code's specific nature whether its principles are ethical, legal, or both. The Nuremberg Code is usually presented as the first international code of ethics in the field of research. It has had tremendous influence on the codes and rules that followed especially in the drafting of the Declaration of Helsinki adopted by the World Medical Association in 1964. One question remains unanswered, however, at least for the medical profession. Is it strictly a codification of ethical principles or is it legally binding? Interestingly most lawyers would agree on the legal nature of the Nuremberg Code (15), while physicians would rather consider it an ethical code.

Historically two sets of rules dealing with human research had been promulgated earlier in Germany, and they were, to some extent, more detailed than the Nuremberg Code. They are the 1900 Directives to the Directors of Clinics, Out-Patient Clinics and Other Medical Facilities formulated by the Prussian Ministry of Religious, Educational, and Medical Affairs and the 1931 Guidelines on Innovative Therapy and Scientific Experimentation of the Reich Minister of the Interior. These rules were either simply not applied by doctors or were applied in such a confidential way that they could hardly be considered as the expression of a common practice by the medical profession. The Nuremberg Code was also largely inspired by guidelines published only a few months earlier by the American Medical Association. But while the judges who elaborated the Nuremberg Code felt that they were only codifying or summarizing the common rules of the medical profession, physicians have long refused to admit that this intrusion in the regulation

of their research activities had any legal consequences. By coining the Nuremberg Code as an ethical code, they were emphasizing that it had no binding force, and therefore that its application was open to interpretation. Its only binding force was derived from the fact that it was to be considered as the expression of the state of the art by the professionals. From the beginning of modern regulation of human research subjects, rules were confined by the medical profession to the restrictive scope of professional ethics. Such interpretation was necessary, if not indispensable, to maintaining the independence of the researchers from any unwanted intervention by lawyers and the legislature.

We agree with Dr. Leo Alexander when he states: "The Nuremberg Code covers all these contingencies in a much more specific manner than subsequent formulations such as the Helsinki Resolution (sic)" (16, p. 396). In fact the first goal of the Helsinki Declaration was not to protect the freedom and rights of the human subjects but rather to allow the continuation of human experimentation: "Because it is essential that the results of laboratory experiments be applied to human beings to further scientific knowledge and to help suffering humanity, the WMA has prepared the following recommendations...." This difference in apprehending the regulation of human subjects research whether it is viewed in a legal or in a medical perspective should be better acknowledged. To a great extent there is a misunderstanding about the fact that when there is no specific legal norm, one should apply the general principles of the law. In short, the absence of statutory laws does not mean the absence of law. This has created the illusion that "ethical codes" were filling a lacunae in the law, while they were only making the regulatory framework for research more complete and precise.

INTERNATIONAL CODES IN THE REGULATION OF HUMAN SUBJECTS RESEARCH

Introduction

The regulation of human subjects research is characterized by myriad national and international rules that can be qualified as ethical, professional, or legal. Two of the most preeminent of these norms have already been mentioned, namely the Nuremberg Code and the Declaration of Helsinki. Concerning the latter, it should be noted that it was revised in 1975 (Tokyo), 1983 (Venice), 1989 (Hong Kong), and 1996 (Somerset West). Other documents of reference at the international level should also be cited: Article 7 of the United Nations Covenant on Civil and Political Rights (1966), the International Guidelines for Biomedical Research Involving Human Subjects first adopted in 1982 and revised in 1993 by the Council for International Organizations of Medical Sciences (CIOMS) in collaboration with WHO. The so-called Belmont Report issued in 1978 by the U.S. National Commission for the Protection of Human Subjects of Biomedical and Behavioral Research should also be mentioned. It defends a conception of bioethics that has greatly influenced modern ethical reasoning and is based on the balancing of four basic ethical principles: the principles of justice, respect for persons, beneficence, and nonmaleficence. This is only a short list of the existing guidelines (18). There are guidelines that concern specific types of clinical trials, for instance, the Declaration of Madrid adopted in 1996 by the World Psychiatric Association or the various guidelines for good clinical research practice (19) primarily designed for drug trials such as the WHO Guidelines for Good Clinical Practice (GCP) for trials on pharmaceutical products (1995) (17) and the "International Conference on Harmonisation" (ICH) Guideline for Good Clinical Practice (ICH–GCP). Other guidelines have only a regional scope, for instance, the Nordic Guidelines for Good Clinical Trial Practice (1989), or the present draft European Union Directive on Drug Trials (20) as well as the CE Convention on Human Rights and Biomedicine (21), not to mention the numerous national laws and codes of ethics. To complete this broad picture of the regulation of biomedical research, two observations still need to be made: First, these codes and guidelines, such as the Declaration of Helsinki and the CIOMS guidelines, are being regularly revised; second, if some of these codes and guidelines should be considered as mainly ethical and/or professional norms, others, such as most of the GCP guidelines, are more or less statutory rules.

Even if guidelines and laws are very heterogeneous, a consensus emerges from them all on some basic requirements regarding the protection of human research subjects during clinical trials. Among theses, the most commonly identified are the following:

- Review of the research project by a competent ethics review board (ERB)
- Fair selection of the human research subjects (special protection of vulnerable population)
- Free, informed consent of the human research subjects
- Respect for the privacy of the human research subjects and for the confidentiality of clinical data
- Favorable balance of harms and benefits
- Compensation for research-induced injury
- Sound, scientific design of the research based on sufficient data from previous nonclinical and clinical studies
- Qualification and experience of the investigator
- Adequacy of the resources available (time, staff, facilities, finances)

These norms are the core elements of all regulation concerning the protection of human research subjects. To better understand the nature and scope of the aforementioned codes, we propose to analyze them in light of the rules of informed consent and ethical review mechanisms. This will also serve as an occasion to clarify the specific links as well as discrepancies between these codes.

Informed Consent

There are few texts like the 'Nuremberg Code' that express with such strength and precision the principle of informed consent freely given by human subjects. It is worth quoting *in extenso*:

> The voluntary consent of the human subject is absolutely essential. This means that the person involved should have legal capacity to give consent; should be so situated as to be able to exercise free power of choice, without the intervention of any element of force, fraud, deceit, duress, overreaching, or other ulterior form of constraint or coercion; and should have sufficient knowledge and comprehension of the elements of the subject matter involved as to enable him to make an understanding and enlightened decision. This latter element requires that before the acceptance of an affirmative decision by the experimental subject there should be made known to him the nature, duration, and purpose of the experiment; the method and means by which it is to be conducted; all inconveniences and hazards reasonably to be expected; and the effects upon his health or person which may possibly come from his participation in the experiment.
>
> The duty and responsibility for ascertaining the quality of the consent rests upon each individual who initiates, directs, or engages in the experiment. It is a personal duty and responsibility which may not be delegated to another with impunity.

It is interesting to note that the first version of the Declaration of Helsinki, adopted in 1964 by the World Medical Association and claiming a relationship to the 'Nuremberg Code', does not mention the rule of informed consent among its basic principles of research ethics. Furthermore, though it provides that the informed consent of human subjects should be obtained before beginning a study with no direct therapeutic benefit, the Helsinki Declaration only requires that "if at all possible, consistent with patient psychology, the doctor should obtain the patient's freely given consent after the patient has been given a full explanation" (23, p. 473).

This wording seems to admit the possibility of applying the therapeutic privilege in the field of clinical trials. By therapeutic privilege, we mean the physician's ability to withhold some information from a patient based on the rationale that it could be detrimental for the patient. Such an interpretation contradicts the first principle of the 'Nuremberg Code'. It has been partly abandoned in the later versions of the Helsinki Declaration, the 1996 version stating clearly that "in any research on human beings, each potential subject must be adequately informed.... The physician should then obtain the subjects' freely-given informed consent, preferably in writing." Yet a 1999 proposed revision raises some serious concerns as it would allow a waiver of written informed consent if the ethics review board determined that the research risks are slight or if the procedures to be used in the research are customarily used in medical practice without informed consent (24). If adopted, this would represent a step backward for the respect of the principle of free and informed consent of the human research subjects (25).

The respect of the human subject's autonomy, and thus of the principle of informed consent, is also a cornerstone of the CIOMS guidelines. Of particular interest is the fact that the CIOMS guidelines recognize that the Declaration of Helsinki is "the fundamental document in the field of ethics in biomedical research," and a copy of the 1989 version of this text can be found in their annex. Nevertheless, they differ on several points, in particular, on the requirement of the subject's informed consent. Guideline 8 of the CIOMS guidelines which deals with the question of research involving subjects in developing communities provides that: "every effort will be made to secure the ethical imperative that the consent of individual subjects be informed...." Yet, in the commentary following guideline 8, it states: "For example, when because of communication difficulties investigator cannot make prospective subjects sufficiently aware of the implications of participation to give adequately informed consent, the decision of each prospective subject should be elicited through a reliable intermediary such as trusted community leader. In some cases other mechanisms, approved by an ethical review committee, may be more suitable." Thus, after stressing the need to respect individual informed in accordance with the basic principle I, 9 of the Declaration of Helsinki as amended in 1996, the CIOMS guidelines open the door for a broad exception in case of research conducted in developing countries (26). Which norm should apply in a given situation? Should it be the first principle of the Nuremberg Code, the basic principle I, 9 of the Declaration of Helsinki, and also guideline 1 of the CIOMS guidelines, or on the contrary, guideline 8 of the CIOMS guidelines interpreted in light of its commentary? The situation becomes more complicated by the fact that the CIOMS guidelines do not make explicitly reference to the Declaration of Helsinki. To what extent then should the new amendments of the Declaration of Helsinki be taken into consideration by someone referring to the CIOMS guidelines? Such questions arise for many other codes and guidelines that refer to the Declaration of Helsinki or to the CIOMS guidelines, if not to both (27).

In short, the great variety of rules of conduct concerning the protection of human research subjects at the international level does not always contribute to a better understanding of what ethical rule should finally prevail in a given situation. The need for a more effective coordination of these numerous codes is a growing concern, especially at a time several of them are being revised. The very existence of so many documents of reference raises some questions on their role and objectives. As stated by Jay Katz concerning ethical codes of conduct more than 30 years ago:

> The proliferation of such codes testifies to the difficulty of promulgating a set of rules which do not immediately raise more questions than they answer. By necessity these codes have to be succinctly worded and, being devoid of commentary, their meaning is subject to a variety of interpretations. Moreover, since they generally aspire to ideal practices, they invite judicious and injudicious neglect. Consequently, as long as they remain unelaborated tablets of exhortation, codes will at best have limited usefulness in guiding the daily behavior of investigators (28).

If these codes are still relevant for the diminishing number of countries and regions of the world where there is presently no statutory regulation of clinical trials, there is concern about their effectiveness, especially in view of their built-in contradictions. In the near future, it is foreseeable that the issue will become more a matter of the proper training of all participants in clinical trials, the investigators, members of ethical review boards, and the human subjects. The existing codes could then be regarded more as a source of inspiration than of regulation. The need to frequently revise the codes to meet the demand of the research would be less important than the need to assess that everyone involved in research with human subjects is working according to high ethical standards. The question then would be not so much revising the codes but ensuring that the fundamental ethical principles that underlie them are effectively implemented.

GENETHICS

The continuing drive to greater levels of specificity has led to the proclamation of "gen-ethics," which is quickly moving toward more detailed "gen-policies." Some examples of the latter are the legal status of DNA, DNA banking, patentability, genetic research involving children, and, confidentiality. Before turning to specific "gen-policies," we will review the common interpretation of the ethical principles discussed in the first part of this article within the context of human genetics. Broadly speaking they are autonomy, privacy, justice, equity, and quality, all relating to respect for human dignity (29).

Autonomy

The principle of autonomy has found its most recent expression in UNESCO's 1997 Universal Declaration on the Human Genome and Human Rights. The simple yes–no of participation has moved to include not only the notion of free and informed consent to participation [Article 5(b)] but also the choice to be informed or not of the results [Article 5(c)]. The probabilistic nature of most predictive genetic information (to say nothing of its personal, familial, and social nature) requires that the expression of individual autonomy include this choice. The problem with this new approach, however, is that little is known of how much information must be communicated for this right to be exercised. Would too much information unduly restrict or undermine the "right" not to know, and too little that of a truly informed consent not to know? Another addition to the second requirement of consent is that of requiring a specific consent for the banking of a DNA sample, often collected and conserved in the past without such consent (30). Only newborn screening for treatable disorders constitutes a valid exception to individual informed consent according to WHO (31). Finally, another "new" consent appears in the preamble to the European Directive on the Legal Protection of Biotechnological Inventions (32). The preamble states that a patent application on an invention using human biological material of human origin must be from a person who has had the "opportunity of expressing [a] free and informed consent thereto, in accordance with national law" (para. 26). While laudatory in its transparency, the weakness of this formulation lies in the latter part that refers to national law. This is because only few national laws have addressed the issue of payment to DNA "donors," although this may be covered by human tissue gift legislation that does not specifically exclude DNA from its mandate. No law specifically requires an explicit consent to eventual commercialization (33). Furthermore it remains to be seen if such consent will be necessary for each possible commercial application (impossible?) or whether a simple notification will suffice and, if so, constitute a preliminary condition of participation.

Privacy

Two genetic-specific interpretations of the principle of privacy merit mention here. The first is the increased emphasis on the confidentiality of genetic information especially as concerns third parties (owners and employers). Article 9 of the UNESCO Declaration maintains that exceptions to the confidentiality of genetic information can only be prescribed by law, thus excluding individual consent. Indeed, this could be because consent to access by insurers and employers is not totally free. Thus WHO has proposed: "Genetic information should not be used as the basis for refusing employment or insurance. Exceptions would have to be legally defined (34). The formulation found in the European Convention on Biomedicine (21) specifying that genetic tests should only be performed for "health purposes" (Article 12), that is, medical indications, is particularly enlightening. Yet, if this information is in the medical record due to participation in research, an individual applying for insurance or employment will certainly have "consented" to access to such confidential information. Hence the importance of keeping participation in genetic research out of the medical record.

Finally, WHO's 1997 Proposed International Guidelines (31) and HUGO's Statement on DNA Sampling: Control and Access (30) make a specific exception for professional disclosure to at-risk family members for serious, treatable, or preventable conditions where the patient or research participant refuses to do so. This issue merits further discussion.

Justice

The principle of justice includes the notion of distributive justice, and in the context of genetics this involves not only living persons (e.g., vulnerable populations) but also future generations. Both the inclusion and protection of vulnerable populations in genetic research is reinforced through the requirement of a legal authorisation by a third party and a direct benefit to the incompetent person or, in the absence of the latter, only research of minimal risk Article 17 (2)(ii) (22). Ironically, this protection, while necessary, may inadvertently create indirect discrimination on economic grounds. This is because only those with the legal foresight and financial wherewithal to provide in advance for such a legal mandate specific to research can be included. Thus a competent person would have to undertake the legal procedures to name a future legal representative with a mandate specifically for participation in research. It goes without saying that in the absence of this advance

mandate, it is very difficult (emotionally and financially) for family members to deliberately have an incompetent family member legally declared incompetent so that person can benefit from participation in research!

As for the interests of future generations, Article 24 of the UNESCO Declaration mentions that germ-line therapy, which could irrevocably affect future generations "could be contrary to human dignity" (35). This addition was only added in July 1997 when the government representatives met to approve the Declaration prepared by the IBC. The Declaration of Manzanillo would not grant legal representatives the power to authorise interventions in the human genome but would entrust such decisions to a neutral body (Principle 6(e)).

In contrast, neither the 1997 Proposed International Guidelines of the WHO nor its 1999 Draft Guidelines on Bioethics refers to germ-line therapy. As mentioned, the Declaration of Manzanillo (1996) would establish a "neutral" multidisciplinary authorizing body to consider such requests (Principle 6 (d)) (37). This is a subject that should be at the forefront of public debate, or as was the case with cloning, legislation will be inadequate (due to hasty preparation), too comprehensive, or unduly restrictive.

Equity of Access

Both the Declaration of Manzanillo and the WHO Proposed International Guidelines stress the need to respect equity of access to services, that is, access according to need and not economic means. Closely related to the principle of justice, this principle could also serve as the ethical underpinning for the concept of benefit-sharing first found in HUGO's statement on the Principled Conduct of Genetic Research (33). Likewise the 1997 Proposed International Guidelines by WHO maintain that if genetic knowledge leads to the development of "a diagnostic test or new therapies, equity requires that the donors, or the community generally, should receive some benefit" (31). Finally, while not specifically mentioning the principle of equity or access to genetic research or testing, UNESCO's Universal Declaration does provide in Article 12 that the "benefits" of advances in biology, genetics and medicine should be available to all (35).

Quality Control

Building on the principle of "basic science" of the Helsinki Declaration (25), the emergence of the notion of competence and quality control as a prerequisite for permitting research in a given area is a welcome development. Both HUGO in its 1996 Statement on the Principled Conduct of Genetic Research and WHO's 1997 Proposed International Guidelines see the provision of quality control through accreditation and surveillance of laboratory services. Requiring a medical indication for access would also serve to limit the number of tests being offered or advertised by those not trained in the all-important communications or counseling areas. With the advent of predictive testing, counseling takes on extraordinary importance. Still missing, however (and this is not unique to genetics research) are the quality control aspects (e.g., monitoring/surveillance), once a protocol has been approved.

In short, the translation of general ethical principles governing human research to human genetics has led not only to their further refinement but, at least as a leitmotiv, to their amplification. Indeed, in the area of consent, the emergence of the right not to know, of the need to obtain consent for DNA banking, and perhaps to patenting, if not at least to eventual commercialization, are welcome developments. The principle of the privacy of one's genetic makeup while reinforced as concerns employers and insurers is nevertheless severely tested by the legitimate needs of at-risk family members. In contrast, overprotection of incompetent research subjects in the requirement of a legally recognized representative for research purposes may discriminate against certain families and those with late-onset genetic conditions. Finally, the last two principles, of equity and quality, merit further definition in the context of genetic research. It would be salutary, to say the least, if it was the fear of abuse of testing and misuse of genetic information that has led to the further development of guidelines in this area, since quality control and ongoing surveillance and monitoring are problems endemic to all research.

GEN-POLICIES

The specific translation of ethical principles into gen-ethics further contextualizes into policies concerning issues, such as, the DNA banking, patenting, research involving children, and confidentality. At this more specific domestic level, the interpretation of gen-ethics is less harmonious. Diversity is a cultural and political fact, and it raises problems for multicentered, international trials.

Status

The issue of the status of human genetic material has both ethical and legal significance. Ethics are involved because status distinguishes the human "source" from other elements of the body or products used in research. This is especially evident in the "common" or "collective" characterization at the level of the whole human genome and that of "familial" even at the level of the individual. The law is involved because of the classical division between persons and things (37).

The characterization as "common heritage" at the collective level can influence the treatment of human genetic material in international trade and patenting disputes. The characterization as "familial" can serve to sensitize researchers and research participants to the familial implications of genetic testing. Such distinctions form the underpinning for the construction of an ethical framework that is not solely individualistic in nature, although the choice to participate in research or to be tested must necessarily remain so.

DNA Banking

Nowhere has respect for individual choice based on personal values blossomed more than in the options offered in the context of DNA banking. From the blanket yes or no of a decade ago, participants are now not only offered the choice to be part of disease-specific or more extensive research collaboration, to have their DNA anonymized or destroyed, or to set limits on length of storage, but are also warned of socioeconomic risks (e.g., employment/insurance), possible stigmatization, and commercialization. It is interesting that regardless of the status of the human genetic material (property or person), the same choices are offered.

Theoretically any person with a property right, as proposed in the model Genetic Privacy Act of the United States, must either renounce future financial interests or claim a benefit. Absent an extremely rare genotype, the latter approach would not be practical. The time it takes for a discovery, an invention, and possible profits usually spans a decade and involves thousands of participants in different countries. Only recently has there been any move to simplify and standardize DNA banking choices (38). More interesting is the suggestion (albeit with some dissent) by the National Bioethics Advisory Commission of the United States to allow participants to agree to other future research without further specification (39). This recommendation does not require anonymization as a condition of such an open-ended choice. Last, the international Ethics Committee of HUGO has proposed notification for the anonymous use of genetic material removed as part of routine care (30). This evolution is significant in that the "reification" and "sacralization" extremes in DNA banking policies seem now to have reached a compromise that respects individual choice but does not accord a higher status to DNA than to the person.

Patenting

UNESCO's Universal Declaration on the Human Genome and Human Rights epitomizes both the ethical and legal positions concerning the trend toward the commercialization of genetic research around the world. Article 4 maintains that "the human genome in its natural state shall not give rise to financial gains" (35). While genetic material in its natural state cannot be bought or sold, this should not be confused with the issue of patentability. As the 1998 European Directive on the Legal Protection of Biotechnological Inventions stated (Article 5):

> 1. The human body, at the various stages of its formation and development, and the simple discovery of one of its elements, including the sequence or partial sequence of a gene, cannot constitute patentable inventions (32).

Nevertheless,

> 2. An element isolated from the human body or otherwise produced by means of a technical process, including the sequence or partial sequence of a gene, may constitute a patentable invention, even if the structure of that element is identical to that of a natural element (33).

This ruling is buttressed by two international instruments that bear mentioning. Both the General Agreement on Tariffs and Trade (GATT) agreement and the North American Free Trade Agreement (NAFTA) include the ethical filter of excluding inventions whose commercial exploitation would be contrary to public policy, a criterion also found in the European Patent Convention of 1973 (40).

Nevertheless, despite both the potential of this ethical filter and the clarification provided by the European Directive, it was not until 1999 that the U.S. Patent and Trademark Office stated that applicants should "explicitly identify a specific and substantial utility." Therefore, for genes and gene products, there is a need "to specify an immediate and identifiable benefit to the public" (41). Such clarification is necessary because patent applications on sequences and on genes have been seriously hampering research and possible therapeutic applications (33).

Genetic Research Involving Children

The earlier discussion of the principles governing the inclusion of vulnerable populations, such as, incompetent adults, raised the issue of potential discrimination by automatic exclusion in the absence of a "legally" appointed representative. When guidelines do not differentiate between incompetent adults and children, the same issue of possible exclusion occurs with children. In the majority of legislation or policy statements, the parents are legal guardians by law. Thus, while the potential inclusion of children is possible (and may even lead to overparticipation), provided their best interests are served by parental authorization, the door is not completely open. Indeed, not only the best interest criterion but also the notions of benefit and risk come into play.

As concerns benefit, the requirement is that in principle, some therapeutic benefit must be expected as well as a favorable risks–benefit ratio. Even in the absence of direct benefit, when the results are "capable of conferring benefit to other persons in the same age category or afflicted with the same disease or disorder or having the same condition ... [and] entails only minimal risk and minimal burden" [Article 17(2)] (21), the research may proceed. This is important in that many genetic conditions are specific to childhood (43) and the automatic exclusion from research would seriously harm both affected children and those potentially at risk. Indeed, overprotection (43), while an understandable reaction to the abuses of the past or even as a precautionary measure when faced with "genetic" unknowns, is becoming a major ethical issue. That being said, there are specific limitations on the genetic testing of children in the absence of available treatment or prevention. These limitations include no testing for the benefit of other family members or for carrier status, nor for late-onset, susceptibility, or predictive purposes.

These restrictions, however, imply the availability of a test, whereas most research aims to find a gene that is either a cause or a determinant among other factors. Thus, in theory, the restrictions above would apply in the situation where DNA is taken with a view toward developing tests that situate the issue within the wider realm of the legitimacy of genetic testing for carrier status or prediction, in general, and where there is no therapeutic potential (prevention, reproductive information, or treatment). It is here that anonymity (removal of all identifiers with some clinical

and demographic information remaining) of DNA samples taken in research involving children might best serve their interests and still respect the above-mentioned principles. Samples that cannot be traced can still serve the pediatric community writ large. When the child reaches maturity/majority and treatment becomes available or even reproductive genetic information, the choice to be tested will be an autonomous and informed one.

Confidentiality

Equally sacrosanct to the principle of consent is that of confidentiality. The principle of confidentiality has acquired greater importance in human genetics where the risk of socioeconomic discrimination and stigmatization could take hold. Yet genetic information is at once personal, familial, and social. This raises two further thorny issues: the protection of genetic information in the research setting and disclosure to at-risk family members. The first remains to be settled, and the second is slowly emerging into a consensus.

Currently the protection of genetic information in the research setting is no different from other research findings. When relevant to treatment or prevention, genetic information is placed in the medical record. But in the absence of such medical significance, patient consent is sought for inclusion in the medical record. Indeed, because of the sensitivity of genetic information, most research records are kept under secure conditions. Increased legislative protection of medical records is needed before this practice will change. Any eventual "normalization" of genetic information as medical information depends on such legislative protection. As genetics is integrated more into mainstream medicine, it may well force this issue of medical confidentiality (45).

On disclosure to family members in the situation where an at-risk family member refuses to warn another member, there has been some international agreement about the ethical obligations, although the law is not clear. Drawing on the notions of genetic material or information as familial, on the obligation to rescue, on professional ethics, or on a legal basis for disclosure, the consensus can be summarized as follows: (1) if the patient has repeatedly refused to warn an at-risk identifiable family member, (2) if that member is at high probability of a serious disorder, and (3) if the disorder in question can be prevented or treated, the physician could ethically breach confidentiality but is not legally obliged. At a minimum, however, such disclosure could be considered a legal privilege or a defense to an accusation of breach of confidentiality (45). It goes without saying that the normalization of genetic conditions and information may well solve this serious dilemma of breaching medical confidentiality and allow families to be less stigmatized and more open.

CONCLUSION

The 1997 Universal Declaration on the Human Genome and Human Rights maintains in Article 12(b) that "Freedom of research, which is necessary for the progress of knowledge, is part of freedom of thought. ..." (35). There is no doubt that freedom is fundamental to progress and it should be protected and promoted even in a controversial area like human genetics. Still the therapeutic and intellectual benefits of research are challenged by the ethical principles put in place to ensure that voluntary participation is kept in balance with risk–benefit ratios. The major issue of today might well be not that of adding more "rules" but rather constructing the processes and structures that allow us to remember and to realize the basic principles in the first place. The ethical principles should hold true (whatever their translation into more specific policies) in all scientific domains. Like basic legal principles, the ethical foundations of research have survived the test of time. The procedures in place for the implementation of ethical principles must be developed and observed in this time of rapid scientific upheaval.

BIBLIOGRAPHY

1. R.J. Lifton, *The Nazi Doctors, Medical Killing and Psychology of Genocide*, Basic Books, New York, 1986.

2. S.H. Harris, *Factories of Death*, Routledge, London, 1994.

3. *Final Report of the Advisory Committee on Human Radiation Experiments*, U.S. Government Printing Office, Washington, DC, 1995, Available at: *http://www.seas.gwu.edu/narchive/radiation*

4. J. Katz, *Experimentation with Human Beings*, Russel Sage Foundation, New York, 1972.

5. *Bioethics: From Ethics to Law, From Law to Ethics*, Publications of the Swiss Institute of Comparative Law No. 30, Schulthess, Zurich, 1997.

6. S. LeBris, B.M. Knoppers, and L. Luther, *Houston Law Rev.* **33**, 1363–1395 (1997).

7. J. Ellis et al., *Lancet* **346**, 407–410 (1995).

8. R.C. Fox, in *Essays in Medical Sociology, Journeys into the Field*, 2nd ed., Transaction Books, New Brunschwig and Oxford, UK, 1988, p. 20 (first published in *The Student-Physician*, Robert K. Merton et al., eds., Harvard University Press, Cambridge, MA, 1957, pp. 207–241).

9. J.F. Lyotard, *La condition postmoderne*. Editions de Minuit, Paris, 1979.

10. Public Law 93–348.

11. National Commission for the Protection of Human Subjects of Biomedical and Behavioral Research, *The Belmont Report: Ethical Principles and Guidelines for the Protection of Human Subjects of Research*, Publication No. (OS) 78-0012, 2–3, DHEW, Washington, DC, 1978.

12. R.J. Levine, *Ethics and Regulation of Clinical Research*, 2nd ed., Urban & Schwarzenberg, Baltimore, MD, 1986, p. 7.

13. K.M. Taylor, R.G. Margolese, and C.L. Soskolne, *Engl. J. Med.* **310**(21), 1363–1367 (1984).

14. K. Lebacqz, *Villanova Law Rev.* **22**, 357–366 (1976–1977).

   F. Rolleston and J.R. Miller, *IRB: Rev. Hum. Subjects Res.* **3**(7), 1–3 (1981).

   A.M. Capron and R.J. Levine, *IRB: Rev. Hum. Subjects Res.* **4**(1), 10–11 (1982).

   R.J. Levine, *Ethics and Regulation of Clinical Research*, 2nd ed., Urban & Schwarzenberg, Baltimore, MD, 1986, pp. 8–10.

   D. Sprumont, *La protection des sujets de recherche*, Staempfli, Berne, Switzerland, 1993, pp. 33–37.

15. C. Bassiouni, T.G. Baffes, and J.T. Evrard, *Revue Int. Droit Pénal* (also exists in english) **51**, 285 (1980).

   M. Belanger, *Droit international de la santé*, Paris, 1983, p. 43.

   J. Katz, *J. Am. Med. Assoc.* **276**, 1662 (1996).

   G.A. Grodin, *Health Hum. Rights* **2**(1), 7 ff (1996).

   P. Arnold and D. Sprumont, in U. Tröhler and S. Reiter-Theil, eds., *Ethics Codes in Medicine: Foundations and Achievements of Codification since 1947*, Ashgate, Aldershot, UK, 1998, pp. 84–96.

16. L. Alexander, in *Ann. N.Y. Acad. Sci.* **169**(344), 346 (1970).

17. *WHO Technical Report Series*, No. 850, 1995, pp. 97–131.

18. *Informal listing of selected international codes, declarations, guidelines, etc., on medical ethics/bioethics/health care ethics/human rights aspects of health*, Periodically updated by Sev S. Fluss, c/o CIOMS, WHO, 1211 Geneva 27.

19. E.F. Hvidberg, *Drugs* **45**(2), 172 (1993).

20. United Nations Commission for the Protection of Human Subjects of Biomedical and Behavioral Research, *Good Clinical Practice*, (O.J. 97/C 306/10).

21. Council of Europe, *Int. Dig. Hlth. Leg.* **48**(1), 99 (1997).

22. H.D.C. Roscam Abbing, *European J. Health Law* **1**(2), 148 (1994).

23. World Medical Association, *N. Eng. J. Med.* **271**(9), 473 (1964).

24. World Medical Association, *Proposed Revision of the World Medical Association Declaration of Helsinki*, Santiago, Chile, 1999 (WMA document: 17.C/Rev 1/98).

25. See T.A. Brennan, *N. Engl. J. Med.* **341**(7), 527 (1999).

26. J. Legemaate, *Eur. J. Health Law* **1**(2), 161–165 (1994).

27. WHO GCP, provision 1.2, *Ethical principles*.

28. J. Katz, *Daedalus* **98**, 482–483 (1969), also published in: *Experimentation with Human Subject*, P.A. Freund, ed., George Braziller, New York, 1970, pp. 293–314.

29. B.M. Knoppers and R. Chadwick, *Science* **265**, 2035–2036 (1994).

30. HUGO Ethics Committee, *Genome Digest* **6**(1), 8 (1999).

31. World Health Organization, *Proposed International Guidelines on Ethical Issues in Medical Genetics and Genetic Services*, Geneva, Switzerland, 1997.

32. European Union, *Official J. Eur. Communities* **13**, L 213 (July 6, 1998).

33. B.M. Knoppers, *Nat. Gen.* **22**, 23 (1999).

34. World Health Organization, *Draft Guidelines on Bioethics*, Geneva, Switzerland, 1999.

35. United Nations Educational, Scientific and Cultural Organization, International Bioethics Committee, *Universal Declaration on the Human Genome and Human Rights*, Paris, November 11, 1997.

36. Latin American Human Genome Program, *Int. Dig. Hlth. Leg.* **48**(3–4), 424 (1997).

37. B.M. Knoppers et al., *Genomics* **50**(3), 385–401 (1998).

38. B.M. Knoppers, M. Hirtle, and K. Cranley Glass, *Science* **286**(5448), 2277–2278 (1999).

39. National Bioethics Advisory Commission, *Research Involving Human Biological Materials: Ethical Issues and Policy Guidance*, Rockville, MD, August 1999.

40. European Patent Organization (Administrative Council), *European Patent Convention*, 1973.

41. Paul Smaglik, *Nature* **403**, 6765 (2000).

42. World Health Organization, *Human Genetics and Noncommunicable Diseases*, Fact Sheet No. 209, 1999, *http://www.who.int/inf-fs/en/fact209.html*

43. M. Hirtle, *Health Law J.* **6**, 43–81 (1998).

44. Mark A. Rothstein, ed., *Genetic Secrets: Protecting Privacy and Confidentiality in the Genetic Era*, Yale University Press, New Haven, CT, 1997.

45. American Society of Human Genetics, Social Issues Subcommittee, *Am. J. Hum. Genet.* **62**(2), 474–483 (1998).

See other HUMAN SUBJECTS RESEARCH entries.

HUMAN SUBJECTS RESEARCH, ETHICS, AND RESEARCH ON CHILDREN

LORETTA M. KOPELMAN
Brody School of Medicine at East Carolina University
Greenville, North Carolina

OUTLINE

INTRODUCTION

Research with children as subjects can help them individually and collectively. Those who participate can benefit through access to programs or treatments otherwise unavailable to them. In addition research can be socially useful to gain better information about children's diagnoses, therapies, and prognoses. Without research with children as subjects, science cannot address children's unique needs, conditions, or reactions. Doctors may be reluctant to prescribe therapies tested only on adults for children, and insurance companies may refuse to pay for their use because safety and efficacy has not been shown for children. Insufficient information can result in undertreatment of children's conditions or unexpected adverse drug reactions.

What should be done, however, when it is socially useful to conduct a study that will provide extremely worthwhile information, yet some of the children will not be helped and might even be harmed? Should we say that a child's well-being always takes precedent, even if the harm to the child represents only a slight inconvenience, and the information to be gained is urgently needed? On the other hand, what risks are too great to be tolerated, and how should we assess and balance potential harms and benefits? For example, some genetic research involves screening people for certain conditions by taking a small blood sample. While this carries almost no physical risk, a variety of psychosocial harms could befall the participants. Testing may reveal many things people would prefer to remain unknown. They may find that the person's biological father is not the person everyone supposes or detect diseases that are likely to occur later in life. People may find that they are likely to get a late-onset disease like Huntington's disease, schizophrenia, breast cancer, Alzheimer's disease, and so on, and their lives will be changed forever by the test results. Even though they are entirely healthy and can do nothing to prevent the disease, a positive test for one of these late-onset diseases may expose them to prejudice and discrimination. They may be stigmatized, finding people unwilling to marry them, give them jobs, or insure them. Who should decide when children should participate in these or other studies and what standards should be used to base these decisions?

CHANGING ATTITUDES ABOUT RESEARCH PARTICIPATION

Until recently, research was regarded as a dangerous activity and certain vulnerable groups, especially children, were often excluded from studies for their protection. Because children were kept out of studies, however, they often could not obtain untested or experimental therapies, or investigational new drugs (INDs). Such restrictions were criticized as unfair for unjustly excluding some groups from receiving needed care and treatment. For example, regulatory obstacles including federal and state rules restricted the use of untested drugs such as AZT on children with AIDS until some of the rules were changed. However well meaning, these protective measures kept many children with AIDS from getting the only drug that could help them (1). In some cases people's best chance for good medical care is through participation in trials and access to these INDs.

Another issue of justice arose from the practice of automatically excluding certain groups, such as women and children, in order to make the subject population homogeneous. Investigators sometimes restricted those whom they allowed in studies to make it easier to analyze results. People of different ages, races, ethnic groups, or genders sometimes react to interventions differently because of their unique cultural, biological, or behavioral variations. The more homogeneous the subjects, the easier it is for investigators to analyze findings. It also helps hold down costs because studies with similar people require fewer subjects. This practice, however, made it difficult to generalize the findings of studies to other groups. For example, test results of drugs used to treat depression on white, middle-class men might not apply to other groups, especially children, because the causes of depression vary.

While children and adults often have the same diseases, if studies are not done with children, it is uncertain if the results apply to them. Yet they are often treated with the same drugs. The American Academy of Pediatrics concluded that only a small number of drugs and interventions used on both adults and children have had clinical trials performed on pediatric populations. Moreover a majority of drugs on the market have not even been labeled for pediatric populations (2). The National Institutes of Health (NIH) found that 10 to 20 percent of research inappropriately excluded pediatric populations (3). Insufficient information about products can result in undertreatment of children's conditions and adverse drug reactions, because clinicians must choose between prescribing drugs that do not have good information about safety and efficacy, or using therapies that are potentially less effective. Careful studies show that drugs commonly used among adults sometimes harm children, and this could not be discovered until testing was done on this population. For example, Chloramphenicol, a common antibiotic, caused many deaths in neonates because their immature livers were unable to tolerate it (4). Including children in research, then, may provide opportunities for the individual child, as well as benefits for children as a group.

In recent years, federal agencies tried to address problems of inappropriate "protection" and inadequate testing of pediatric drugs and interventions. The NIH adopted a policy that "children (i.e., individuals under age 21) must be included in all human-subjects research, conducted or supported by the NIH, unless there are scientific and ethical reasons not to include them. This policy applies to all NIH conducted or supported research involving human subjects..." (3, p. 2). Justification for excluding children from the research would be that the topic is irrelevant to children's health and well-being, that legal or regulatory restrictions exist prohibiting the inclusion of children, that information about them is already available, or that a separate study for children would be inappropriate due to the rarity of the condition in this population, the few number of the children who would be affected, or because there is insufficient data to include them.

In addition the Food and Drug Administration (FDA) (5) now requires drug manufacturers to develop information to help provide better care and treatment of children, especially data regarding the most frequently used pediatric drugs. Like the NIH, the FDA requires manufacturers to test interventions on pediatric populations unless they show that the research would be unlikely to be useful to the treatment of children in any great numbers, or that including children in studies would be impractical, unsafe, unlikely to be unsuccessful, or otherwise unreasonable.

MORAL ASSUMPTIONS

In discussing what sort of research should be permitted using children as subjects, several assumptions are generally made (5). First, competent adults have responsibilities to ensure that children are not abused, neglected, exploited, or denied access to basic care. Second, different policy solutions regarding children should be evaluated in terms of their adherence to primary moral values. Among the most important of these are beneficence (what will benefit children), social utility (what will fulfill some social good), and justice (what is fair). Policies are judged superior when they are likely to benefit children and ensure that they are not exploited. Good standards fairly promote children's interests, well-being, and opportunities to flourish, as well as help children to develop their potential to become empowered and self-fulfilled. Promoting children's well-being and helping them develop their potential as self-determined individuals demonstrates a concern for their welfare and a willingness to address inequalities of the "natural lottery" (the inequalities caused by nature such as the inherited intelligence or diseases) and "social lottery" (the inequalities caused by social factors such as environment or illness that also affect intelligence and abilities).

To introduce the complexity of the moral issues and disputes about children's research, consider two examples. In each, important values must be ranked in order to assess what ought to be done. These values include the importance of pursuing knowledge, respecting people's rights, protecting minors' well-being, and conducting socially needed research.

Case 1

Dr. D, an endocrinologist, first met Martin two years ago as a happy-go-lucky 6-year-old when his concerned parents brought him to Dr. D's office. Martin was extremely short for his age, in the lower 2 percent for height, with a projected adult height of 5 feet 3 inches. The doctor concluded this is a normal genetic consequence of the boy's parents' short stature. They wanted their son to escape the discrimination they had suffered and sought to get growth hormone therapy for him. They were disappointed to learn they could not afford the $20,000 a year or more for up to 10 years that it would cost for this intervention. They had no insurance to cover such matters. Dr. D told them about a double-blind placebo controlled study at the NIH. In this study all the children receive injections three times a week until they reach their full height, perhaps 10 years later. Half the children get the growth hormone, but half get only a placebo (salt-water) injection, resulting in approximately 1000 injections over the years. The children are brought to NIH each year with expenses paid for, and have physical and psychological examinations, nude photographs, and X rays. Martin's parents enrolled him in the study because he has a 50 percent chance of getting the growth hormone. Two years later they return sad and frustrated because Martin, who hates the injections, is still in the lower 2 percent for height and has undergone no growth spurt. The study is underway to test the theory that the children getting the growth hormone will be taller as adults. While clinicians did not know if final growth is affected, they do know if someone has not had a growth spurt, it is unlikely he is getting the growth hormone. Dr. D suspects if he tells the parents this, they will remove Martin from the study. What should Dr. D say when the parents ask if Martin's lack of a growth spurt means he is probably not getting growth hormone but is in the control group getting injections of salt solution?

This case raises issues about the sometimes divided loyalties of clinicians toward the patients in studies, and about how they balance gaining information by supporting the study and acting in the best interest of the subject. The child, although short, was perfectly happy initially but is now sad perhaps because he thinks he needs treatments for his shortness. Having the opportunity to be in the study has arguably harmed him by making him feel there is something wrong with him that needs to be "fixed" by medicine. The general problem is whether doctors' obligations to their patients change when they are enrolled as subjects in research studies. If the physician's first duty is to act in the patient's best interest, and to be truthful, then perhaps Dr. D. should tell the parents that he does not believe that their son is getting the growth hormone because he has had no growth spurt. He knows that if many subjects withdraw from the study this could cause problems for the investigators and even undermine the study.

Case 2

Dr. M is a pediatrician who's taking care of Barbara, a healthy 12-year-old. Her paternal grandfather died of Huntington's disease after a long and debilitating illness where he became increasingly demented and unable to control himself. Barbara, who was horrified by her grandfather's decline, tells her parents she wants to enroll in a research project in which they test people who may get this inherited condition. She longs to know if she is free of this genetic condition and "like everyone else" or has to plan her life to do things she wants before she reaches middle age. Her father does not want to be tested, although he understands that if his daughter has a genetic predisposition for Huntington's disease, then he does as well. He agrees to her being tested because it is important to her but prefers not to know the test results. While she and her parents agree that she should be tested, her pediatrician, Dr. M, and the genetic counselors strongly object to testing her or recommending her as a research subject for this study. They cite policies barring minors from testing for late-onset genetic conditions (6) and have effectively opposed the investigators who would like to include minors in their study. The clinicians oppose late-onset testing and research with minors because they argue that if children test positively for diseases such as Huntington's, they may feel differently about themselves, suppose they are already sick when they are not, find it harder to get jobs or health insurance, and face other psychosocial risks of harm. Should restrictions be placed on parental authority to assess what is in their child's best interest and to give consent for their child to be tested or participate in studies?

The case raises concerns about parental consent, minor's assent, and genetic testing and screening. Especially difficult to assess are the harms and benefits of testing for diseases that will appear many years later. Who should decide whether testing should be done for these "late-onset" conditions? Are some studies too risky to permit minors to participate, even if the parents consent? Many adults handle information of late-onset disease badly, and this could cause children even more harm. Who should assess the potential benefits and psychosocial harms such as loss of self-esteem, bias, prejudice, and

discrimination in employment or insurance? While many policies bar minors from testing for late-onset disease, some parents object to such interference with their authority to decide what is best for the children. Should late-onset disease testing be prohibited on minors, or left to the judgment of parents about whether it is in her best interest to get this testing? In addition, what role should older children have in making decisions? In case 2, the parents and the 12-year-old wish to have the child participate in research, but the clinicians strongly object. If the minor is 17 or on the threshold of full civil rights, her views might carry more weight. As children get older their views gain increasing moral importance.

The transition from dependence to independence is gradual in childhood. Infants should be fully protected, while well-adjusted minors on the verge of majority have a justified claim to understand and express themselves in actions affecting their lives. As children become older and can comprehend more, they should participate in important decisions about their lives, including whether they will serve as research subjects. The problem of whether to include children in decision making becomes acute when children are neither clearly incompetent nor clearly as competent as most adults. For this reason the children's assent is sought as they mature.

As with competent people, the ethical basis for research policy with persons lacking capacity to give informed consent concerns promoting their self-determination, fair treatment, and well-being. The difference is that children cannot be expected to promote their own self-interests. Consequently we have to consider when others, usually parents, will be permitted to consent to have their children participate in research.

CONSENSUS ON INFORMED CONSENT

The right to consent for treatment or research is a civil right, and people must achieve a certain degree of maturity before they gain it. As a matter of administrative convenience an arbitrary age such as 18 is selected as the time when a person is assumed to be sufficiently mature to be granted full civil rights. Until minors reach the age of majority, adults generally have legal authority to make decisions for them. Guardians have this authority because they are usually best suited to protect, identify, and act in children's best interest, and foster development to the point at which children can take responsibility for themselves. Yet guardians do not have an unqualified right to control their children's destiny, and their approval of research projects is insufficient to include minors in research.

A moral, legal, and medical consensus now exists that the competent, free and informed choices of adults must generally be followed in devising treatment plans or enrolling them or their children as research subjects (7,8). Ideally, people who are patients getting standard or experimental care share decision making with physicians as part of an ongoing process, where doctors clarify options and make recommendations. When the subject or patient is a minor, their parental consent is sought. Yet parental views do not carry the same weight when the consent is given for dependents and the professionals believe that the choice for a minor are abusive or neglectful.

There is a moral, legal, clinical, and regulatory consensus about how to understand the meaning of informed consent. To give informed consent, persons must (*1*) have had disclosed to them all information material to the decision, (*2*) understand or comprehend the information disclosed, (*3*) act or agree voluntarily, (*4*) be competent to act or make the decision, and (*5*) authorize or consent to the procedure, act, or intervention (9). Overcoming this presumptive duty to gain informed consent may be accomplished by demonstrating that the study or treatment constitutes a well-established exception to the duty to gain consent. Incompetent patients cannot give informed consent, so this is a well-established exception. Other morally and legally valid exceptions include medical emergencies, public health emergencies, and patient's waiver of consent (8).

Unlike competent adults who can generally decide what constitutes their own best interest, children need others to protect them. For this reason children are not authorized to decide if they will participate in research. Children are vulnerable subjects. Potential research subjects are called "vulnerable" if they lack capacity to give informed consent or are likely to be coerced or manipulated to participate. Children and those severely impaired by mental illness or retardation are typically regarded as vulnerable because they lack capacity to give informed consent. (Others such as institutionalized subjects, prisoners, the poor, those desperately ill, members of the military, students, hospital staff, and laboratory assistants may be called vulnerable because, while able to give informed consent, they are vulnerable to coercion or manipulation.) The informed participation of vulnerable subjects is problematic, and enrolling them in research protocols often requires special justification or safeguards.

FOUR DIFFERENT POLICY INITIATIVES

There are several important research policy options offering different approaches to balancing what is fair, most protective of incompetent people's well-being, and most respectful of whatever self-determination they have or may develop (5,10). These four policies represent different regulative ideals because they balance these primary values differently and because they offer different authority principles (stating who decides) and guidance principles (substantive directions about how decisions should be made). The remaining discussion will be focused on these options.

Let Surrogates Decide Using Their Own Values

One policy allows guardians to give consent for children under their care as they do for themselves, without any additional restrictions. If they can consent for themselves to take part in a research program testing for late-onset genetic diseases (see case 2 in the discussion above), then, according to this policy, they should be able to authorize this for their children. Guardians are typically knowledgeable and concerned, defenders argue,

and society should intrude as little as possible into the privacy of family life. Since guardians have the authority to make such life-shaping choices as to their child's religion and schooling, then, according to this view, guardians should also determine whether or not their child should participate in research.

Early in the twentieth century, it was assumed that guardians had almost unqualified rights to control their children's futures until adulthood unless they were emancipated through military service, marriage, or independence (11). One reason for this policy was that minors were judged incompetent to make rational decisions, so complete adult direction was regarded as necessary to help them develop their potential. Parents, usually the father, had this right. No consideration in the courts was given to minors' emergent rationality, hopes, or plans. Their views were not material to a decision, unless they were emancipated. It was considered unnecessary and irrelevant to find out, for example, how the child wanted to be treated (12).

In addition, minors were thought to virtually belong to their guardians until they reached the age of 21 (11). When there were disagreements about the treatment or placement of a child and the dispute went to the courts, the courts' role would be to decide who "owned" the child, since that person had authority to make decisions. It was a great tragedy if the guardian was abusive or neglectful, and as a result a child died or was maimed for life, but the courts took the view that the decision maker, usually the parents, had a right to decide (12). The attitude was that competent persons had a right to make even an unfortunate decision for themselves and their wards, and there was little to be done beyond trying to persuade them to act otherwise.

Critics began to challenge this position in the late nineteenth century, when a major legal and social shift occurred and as children's interests were increasingly considered when choices were made about them. They were granted entitlements to be protected from abusive or neglect as well as liberties independent of their parents. More recently this included setting research policy restricting when children could participate in studies, even if parents wanted them to be in the research. In addition the child's assent was sought as a precondition of research participation.

Guardians now have authority insofar as they promote the well-being of those under their care, and prevent, remove or minimize harms to them. Volunteering to put oneself in harm's way to gain knowledge may be morally admirable. Volunteering to put another in harm's way is not admirable, and may violate the guardian's protective role. Consequently this first policy option—to let parents decide as they would for themselves—is rarely defended today. Its echoes, however, may be heard in case 2, discussed above, where parents resent state restrictions on research with minors.

Nuremberg: Require Consent for All Research

Another policy requires consent from competent persons to enroll them as research subjects. This view is found in the first written international research code, the Nuremberg code (13). Composed at the end of World War II, the

Nuremberg code stands as a international response to the horrible, involuntary medical studies done by the Nazis in which many people, including children, were killed or permanently maimed. The Nuremberg code states: "The voluntary consent of the human subject is absolutely essential." It goes on to define consent in a way that has become fairly standard, as requiring legal capacity, free choice, and understanding of "the nature, duration, and purpose of the experiment; the methods and means by which it is conducted; all inconveniences and hazards reasonably to be expected; any effects upon his health or person which may possibly come from participation in the research" (13, pp. 181–182).

The Nuremberg code was a response to invasive, harmful and sometimes-deadly studies done on unwilling subjects. If taken as a general code for research (and it may not have been intended as such), this policy excludes subjects who lack capacity to give informed consent. The reason sometimes given for adopting this policy is that it violates people's rights of self-determination to use them in research if they cannot give consent for themselves to participate.

One difficulty with this policy is that research for children and other people who cannot give consent will stop and, consequently, so will advances in their care. Children have unique medical problems, and the results from studies with adult subjects may be inapplicable to children. Adults cannot be used to test the safety and efficacy of drugs for childhood schizophrenia, or for premature infants with respiratory distress and infections. To test the safety and efficacy of many standard, innovative, or investigational treatments for distinctive groups, some members of the groups have to be subjects.

The initial justification for excluding persons who lack capacity to give informed consent for research is to honor their rights and protect their welfare. This restriction, however, may prohibit studies that benefit children. Some are denied access to projects or investigational therapies that could help them as well. All children are denied good information about their conditions because there are no results applicable to pediatric populations. Thus it may not promote their welfare, individually or collectively, to forbid their participation. In addition participation in controlled research does not always violate the rights of persons who lack capacity to give informed consent. By excluding persons who are unable to give consent from relatively safe studies that advance knowledge or provide therapeutic benefit, children's needs and opportunities are not given full consideration. This policy would exclude all children from being enrolled in any research, however low the risk of harm. Naturally it would disallow studies outlined in each of the two cases mentioned earlier. More troubling is that it would not permit even safe therapeutic studies. A dying child could not, for example, obtain an experimental drug that might save her life.

Helsinki: Only Therapeutic Research Without Subject's Consent

A third policy option holds that persons who lack the capacity to give informed consent may be enrolled only in therapeutic studies. This view is represented in the next major international code for research to follow the Nuremberg code, the World Medical Association's Declaration of Helsinki. It states: "In the case of legal incompetence, informed consent should be obtained from the legal guardian in accordance with national legislation" (14, p. I, 11). The incompetent person must agree as well, when able to do so. It allows research with incompetent people, but "…only to the extent that medical research is justified by its potential diagnostic or therapeutic value for the patient" (14, p. II, 6). If, however, the medical research is nontherapeutic, then: "The subjects should be volunteers — either healthy persons or patients for whom the experimental design is not related to the patient's illness" (14, p. III, 2). People who cannot volunteer therefore cannot be subjects in nontherapeutic studies.

This policy option distinguishes clinical or therapeutic research (studies seeking generalizable knowledge but intending to provide medically acceptable therapy for the individual) from nontherapeutic biomedical research (studies seeking generalizable knowledge but not intended as therapy to benefit the individual directly). Therapy is designed to benefit the person, so drawing the line at therapeutic research for people who lack the capacity to give informed consent might seem a good solution.

This approach has difficulties. First, classifying research as either therapeutic or nontherapeutic may be misleading and arbitrary. Studies often have features that are not routine therapy such as extra procedures, tests, or visits to the doctor. Second, important medical research may hold out benefits other than therapy to subjects. Small children who are not patients may enjoy participating in studies where they are asked to do such things as stack similar blocks or identify animals from sounds that they made. Older children might, for example, enjoy an outing to a research facility or find the study interesting in itself.

The Helsinki policy, however, forbids nontherapeutic studies, even though some are important and safe. For example, it is very important to learn if children are developmentally delayed. But to determine this, investigators need to establish a normal range against which to compare results. This standard is obtained by collecting data from many children about their height, weight, vision, hearing, and so on. These children are not sick, so the studies cannot be regarded as therapeutic, and therefore on this policy cannot be permitted even if they are extremely important and safe.

Third, the distinction between therapeutic and nontherapeutic studies focuses on direct benefits to the individual. Yet nontherapeutic studies may indirectly benefit persons unable to give consent. When these studies obtain valuable information about them as a group, and involve little or no risk, it seems problematic to prohibit them. As we indicated, standards of typical growth and development help all children yet require testing on large numbers of normal children to generate data that help distinguish developmental delays or impairments in children from normal growth and development. If children ride tricycles, it is not research; if investigators observe when and how they do it to make generalizations about children, it is research that may not be burdensome to the child.

The Helsinki policy is controversial then because it excludes many important, low-risk studies. It also excludes, of course, the research studies suggested in two cases discussed above. In case 1 the child is normal, not sick, so this is not a therapeutic study; in case 2 no therapy is planned.

The initial justification for excluding persons who lack the capacity to give informed consent from nontherapeutic research was to honor their rights and protect their welfare. Safe, nontherapeutic research, however, seems neither unfair, nor a violation of the rights or welfare of people who lack the capacity to give consent. Failing to do safe, but important, studies might be unfair and violate their rights and welfare, since it fails to consider all their needs. Thus, when nontherapeutic studies are not potentially harmful or inconvenient and when the subjects want to participate, it is not clear that their rights and welfare are always violated or that they are treated unfairly.

U.S. Rules, CIOMS and Other Guidelines: Assess the Ratio of Likely Harms to Benefits

A fourth approach allows research with incompetent persons if it holds out benefit or does not place them at unwarranted risk of harm, discomfort, or inconvenience. To try to balance the social utility of research with respect and protection of incompetent people, this option stipulates that the greater the risk, the more rigorous and elaborate are the procedural protection and consent requirements. The U.S. federal government (15, §46.404–7) reflects this policy option in its codes for research, involving children as well as adults. The Council of International Organizations of Medical Science (CIOMS) (16) and health policy groups in other countries have also adopted this general approach (10).

There are advantages to focusing directly on the benefits and harms of procedures or interventions in distinguishing the permissibility of research. It avoids the three problems discussed earlier concerning other options. It offers special protections for vulnerable subjects but permits some research with children as subjects. It also allows that there may be benefits other than therapy to consider in assessing potential harms and benefits to children participating in studies. It does not make the sometimes troubling distinction between therapeutic and nontherapeutic research in order to determine which studies are acceptable for children as recommended in the Helsinki approach. As we saw above, classifying studies as therapeutic or beneficial can be misleading or arbitrary. It can be arbitrary because studies have potential harms unrelated to this classification such as extra tests or visits. It can also be misleading if people assume therapeutic studies are always safe or beneficial. Calling something "therapeutic" may hide potential harms, disadvantages, or other nonbeneficial features, creating an inappropriate bias for participation that would be revealed by careful risk assessment.

In contrast, the fourth policy option focuses on risk assessment to balance the social utility of encouraging studies with the protection of people's rights and well-being. It stipulates that, whenever possible, the incompetent persons should give their assent or affirmative agreement to participate.

In using the likely harms to benefit calculation, the U.S. regulations specify four categories of research with children (15). As risks increase, the regulations require increasingly more rigorous documentation of appropriate parental consent, children's assent, direct benefits to each child, and benefits to other children with similar conditions. Local Institutional Review Boards (IRBs) can approve studies only in the first three categories.

The *first* category permits research with no greater than a minimal risk provided that the study makes adequate provisions for consent from at least one parent and the child's assent. The *second* category of research permits approval of studies with greater than a minimal risk if the risk is justified by the anticipated benefit to each subject, the risks in relation to these benefits are at least as favorable to each subject as available alternatives, and provisions are made for consent from one parent and the child's assent. The *third* category permits research with a minor increase over minimal risk that holds out no prospect of direct benefit to the individual subject where the study is like the child's actual or expected medical, dental, psychological, or educational situation, is likely to result in very important information about the child's disorder or condition, and provisions are made for parental consent, typically consent from both parents, and the child's assent. Investigators using this category might be permitted to conduct, for example, additional lumbar punctures on children with leukemia to help study their disease. Consent from both parents is required if practicable.

Research that cannot be approved under the first three categories might be approved under a *fourth* category if it presents a reasonable opportunity to understand, prevent, or alleviate a serious problem affecting the health or welfare of children, and the study is approved by the secretary of the Department of Health and Human Services (HHS) after consulting with a panel of experts about the study's value and ethics and determining that adequate provisions have been made for the parental consent, typically consent from both parents, and the child's assent. Using this category, investigators might, for example, gain approval to conduct studies on normal, healthy children intended to prevent expression of late-onset genetic diseases. In the United States, IRBs cannot approve studies that have more than a minor increase over a minimal risk which do not hold out benefit for the children. As Table 1 shows, IRBs must seek approval from the federal government to conduct them.

Unfortunately, this fourth policy leaves key terms poorly defined and consequently allows broad interpretations about what risks of harm are warranted. The pivotal concepts of "minimal risk" and "a minor increase over a minimal risk" are problematic. The regulations state: " 'Minimal risks' means that the risks anticipated in the proposed research are not greater, considering probability and magnitude, than those ordinarily encountered in daily life or during the performance of routine physical or psychological examinations or tests." (US 45 CFR46 102i) (15).

Table 1. U.S. Regulations Requirements for Research with Children

Research Category	Risk-of-Harm Category	Requirement for Children's Participation
I (46.404)	Minimal risk	1. IRB approval 2. Child's assent, and 3. Informed consent from at least one parent or guardian
II (46.405)	More than minimal risk, with prospects of direct benefit for each subject	1. IRB approval 2. Child's assent 3. Informed consent from at least one parent or guardian 4. Risk justified by anticipated benefit to each subject, and 5. Anticipated benefits to each subject at least as favorable as that presented by available alternate approaches
III (46.406)	More than minimal risk, without prospect of direct benefits to each subject	1. IRB approval 2. Child's assent 3. Informed consent from both parents or guardians 4. Risk represents a minor increase over minimal risk 5. Likely to yield generalizable knowledge about child's disorder or condition that is important for the understanding or ameliorating of disorder or condition, and 6. Intervention or procedure presents experiences to child that are reasonably commensurate with those in child's actual or expected medical, dental, psychological, social, or educational situations
IV (46.407)	Research not otherwise approvable	1. IRB approval 2. Child's assent 3. Informed consent from both parents or guardians 4. IRB finds that research presents a reasonable opportunity to further the understanding, prevention, or alleviation of a serious problem affecting the health or welfare of children, and 5. Study gains the approval of the Secretary of HHS after consultation with a panel of experts in pertinent fields and following opportunity for public review and comment

Source: CRF **45**, 46, 404–7 (1991).

The first part of the definition is vague because daily risks include dangers from riding in cars, flying in airplanes, and living in a world filled with nuclear and conventional weapons. How well do we know the nature, probability, and magnitude of these "everyday" risks and why should they serve as a baseline to estimate minimal research risk? Depending on where one lives, daily risks can be life-threatening. It seems considerably easier to determine that having a 4-year-old stack blocks is a minimal risk study than to determine the nature and magnitude of whatever risks people normally encounter (10).

The second part of the definition seems to set a standard for physical interventions that have a minimal risk. The test is whether the activity is like that of a routine examination. Accordingly, IRB or Research Ethics Committee (REC) members may not approve as minimal risk research with such procedures as X radiography, bronchoscopy, spinal taps, or cardiac catheterization because they are not part of routine examinations. IRBs and RECs, however, can approve studies which have a minor increase over minimal risk, and some of these procedures have been approved as having only a minor increase over a minimal risk. Without standards for risk assessment, how effective are these guidelines? Not surprisingly there are considerable differences of opinion about whether procedures such as lumbar punctures, multiple placebo injections,

arterial punctures, and gastric and intestinal intubations are regarded as risky (10).

Moreover this definition of "minimal risk" and subsequent judgments about what constitutes a minor increase over minimal risk offers no guidance about how to assess psychosocial risks. These include invasion of privacy, breach of confidentiality, labeling, and stigmatization. In "routine" visits doctors and nurses "ordinarily encounter" discussions of family abuse, sexual preference, and diagnoses that could affecting people's abilities to get jobs or insurance. As we saw in our two examples, these can be grave risks.

Freedman, Fuks, and Weijer (17) respond that identifying everyday risks we all encounter is not difficult, although it may be hard to quantify them. Yet they acknowledge the potential problem of using different standards for subject inclusion because of intercultural variation regarding "everyday risks." Their solution is to promote "intercultural ethics" where the norms of *all* the cultures participating would have to be honored to conduct any crosscultural studies. Clearly, however, this proposed solution has not been adopted. This is obvious from the multinational AIDS research conducted in developing countries but funded by nations that would not be able to conduct them in their own countries because they violate research policy. Nations from North America and Europe, especially the United States, supported this research and tried to justify it in terms of local standards and conditions. There is an ongoing debate over permitting greater consideration of local standards or conditions (18,19).

Second, their assumption that a consensus exists regarding what paradigms to use in assessing the crucial upper levels of justifiable risks of harm also seems unwarranted. There are considerable differences among pediatric experts, in both treatment and research settings, about how to assess the risk of such procedures as lumbar punctures, multiple placebo injections, venipunctures, arterial punctures, and gastric and intestinal intubations (20). Investigators and others have concluded that better standards of risk assessment in children's research need to be formulated (21,22,10). More recently this lack of consensus and regulatory guidance within the research community was extensively discussed in testimony before National Bioethics Advisory Commission (NBAC) recorded in "Regulatory Understanding of Minimal Risk" where members of NBAC agreed this was a problem (22).

The fourth approach represented by the U.S. research rules, CIOMS regulations, and those of other countries seems to many to balance the need to protect the rights and welfare of people who lack the capacity to give informed consent with the need to encourage research. It may, however, be popular because it is so vague that it permits different understandings of what constitutes acceptable risks of harm to them. Each of the other three policy options, although flawed, offer clear guidance about what is permissible. This is not the case with the fourth standard, and debates arise about how to use this policy in certain cases.

DEBATES ABOUT BALANCING HARMS AND BENEFITS

The general acceptance of the fourth policy alternative, with its attendant ambiguities about what constitutes a warranted risk of harm, has engendered disputes about how to understand and balance harms and benefits in permitting research involving children. Recall that in case 1, the parents focus on what is best for their child, and they balance their views about what helps him against the harms that will fall to him if he is in the study but does not get the growth hormone. It is not unreasonable for parents to want to improve their child's life, using the values they think most important. Martin's parents are willing to take risks to have him avoid the discrimination they faced because they are short, but unwilling to have him suffer without a chance of some advantages. If they decide there are insufficient benefits to him from being in the study, then the harms from the hundreds of placebo injections, inconvenience, medicalizing of his shortness, and so on, tip the balance and, using their values, justify taking him out of the study.

The investigators, however, might consider a different set of values, harms, and benefits. They might want to consider not just the harms and benefits to a few people in the study but to all people in the future who might take this drug. Investigators might want to emphasize the virtues of keeping one's bargains of staying in the study until the end. Investigators sometimes stress the long-term benefits of their research in possibly developing new and better therapies in the future. On the other hand, if we speculate on possible long-term benefits from a not yet developed therapy, then we probably should also consider and balance them against the possible long-term harms from it as well. This calculation becomes further complicated because it is hard to factor in special interests, biases, prejudices, and wishful thinking. In addition, if open information makes those in one arm of a study want to withdraw, one must question if there is a genuine balance of harms and benefits in the other arms of the study.

The research rules give little guidance about what goals to use to determine harms and benefits and how to balance them. For example, in case 1 it is crucial to determine the benefits of a good study against the harms to the child, including getting so many placebo injections, and the long- and short-term harms to the doctor–patient relationship. How do we establish the different goals and assessments about harms and benefits? One might conduct an opinion poll about this. If so, what groups should be surveyed? Should they be among parents, doctors, short parents, investigators, or the general public? Clearly we can influence the results of the opinion poll by the group we pick to survey.

Different goals and assessments of harms and benefits are also apparent in case 2, where the parents mean well and want to help their child find out if she has a late-onset disease, hoping, of course, that she will find out that she does not. Unlike case 1, where parents can withdraw their child from a research study, in case 2 they want to get their child into a study if that is the only way that they can get her tested for a late-onset disease. There are different assessments of harms and benefits, with the investigators, parents, and the potential subject minimizing the importance of the predicted harms from testing for Huntington's disease and the pediatrician and genetic counselor refusing to accept their assessment.

These debates arise because reasonable and informed people of good will have different goals and views about what constitutes a harm and a benefit, and about how to rank or balance them. In addition they have different expectations of the future. In what follows, two areas of debate about goals, benefits and harms are discussed: (1) distinguishing treatment, prevention, and enhancement, and (2) research to develop germ-line gene therapies.

Distinguishing Treatment, Prevention, and Enhancement

New genetic technologies create problems about how to use them. One proposal is to distinguish treatment, prevention, and enhancement and permit use of these new techniques only to prevent or treat medical conditions like sickle cell disease; but not to enhance normal people's height (as in case 1) or their beauty, strength, and intelligence, and the like. On this view we should simply prohibit development of new technologies for the purposes of enhancement. Yet this viewpoint is questionable. In some sense to prevent or treat a disease already enhances one's life, making it difficult to say how these distinctions should be determined. For example, putting fluoride in the water enhances people's lives by reducing dental pathology. Treating infectious diseases also enhances lives by making people better and preventing others from getting sick. Since such enhancements are not controversial, it appears there are no principled objection to enhancement understood as the prevention or treatment of disease. On the other hand, making normal children taller, as in case 1, is more controversial because shortness is not a disease like dental pathologies or infections. Martin's parents understand the projections are that he will be only 5 feet 3 inches, and while possibly agreeing that shortness is not a disease, they see this is a disadvantage for a man in our society.

Some enhancements are even more controversial than those aiming to bring someone into normal range of height so he will escape discrimination, as the parents in case 1 desired. For example, some enhancements seek to give people special advantages. For example, parents might wish to improve their son's opportunities to become a professional basketball player and are disappointed to learn their child's projected height is only 6 feet 3 inches. They want him to get growth hormone so he can have a better chance of becoming the next Michael Jordan. Many people might object to this use of technologies on several grounds. First, not everyone has access to this technology, so it is unfair for some to get this advantage. Second, if everyone had access to the technology, no one would have the sought-for advantage. It is not possible to enhance the height of all short, normal children because no matter what resources are used, there would still be children in the lowest 2 percent for height. Third, it exposes the child to some risks for controversial or questionable motives. Finally, it is a poor use of an expensive technology when there is so much need in the world. The debate continues over what limits should be placed on parental authority, investigators, and the use of state funds to develop, test, and use new technologies for enhancement, especially where they involve possible harms to the child.

In trying to resolve the debate by prohibiting these technologies for enhancement but permitting them for prevention and treatment, consider the following example. Two boys have a projected height of 5 feet 3 inches. One has short parents, while the other has a tumor. Should one be refused growth hormone therapy because it would constitute enhancement, while the other gets it because it is treatment? Some would argue they both should get the therapy so they both can reach normal height, gain a normal opportunity range, and escape the social disadvantages from prejudice (7). Some might object, however, to either of them getting it, not on the ground that this may be enhancement rather than treatment but because both are in the normal opportunity range for height; they might agree if either child's projected height were far less.

Trying to settle the debate in terms of someone's normal opportunity range raises difficult problems about what constitutes a normal range, as well as the limits of social obligations to find the money to get children to some norm (23). Some might object to giving children a very costly drug using state money when they have a projected heights of 5 feet 3 inches, since being 5 feet 3 inches is not, in their view, abnormal. That is, most options are still open

to people with a projected height of 5 feet 3 inches. Many women get along perfectly well at this height, finding no great hardship in being 5 feet 3 inches. Moreover, if the problem is social prejudice, we should fix that, not the child, especially given the great needs to use state resources for basic health care for children. Little girls with Turner syndrome have extremely short stature if they do not get growth hormone, so public funds are routinely used in treating their short stature with growth hormone. Arguably, if any child were projected to have such short statue as these girls, they too should have the growth hormone. But some uses of these new technologies seem like cheating. The use of anabolic steroids, for example, to enhance athletic performance is prohibited as seeking unfair advantage over competitors.

Ultimately, resolution of these debates may involve factors other than finding precise delineations of enhancement, prevention, and treatment. Not only is it very difficult to draw clear lines between them but, in the end, some believe it may not be very important (7). What may be more important is whether it permits someone to gain a normal opportunity range, without undue risk of harm, at a cost society can afford (7,23). The issue may turn on how we rank values and on the balancing of potential harms, benefits, and costs.

Research to Develop Germ-Line Gene Therapies

Human germ-line therapies are controversial because they might modify traits that offspring would inherit from parents. They can be beneficial for all, for example, by curing a condition such as sickle cell disease for patients and their descendants. Or, they might harm the person and subsequent generations. How do we assess the goals and potential harms and benefits of research that could harm not only the individual but his or her offspring? The debate also raises issues about the proper moral restraints on science. Eric Juengst (24) summarizes arguments regarding the goals and potential benefits and harms in developing germ-line gene therapies. The first is its potential to be medically useful. Research into germ-line gene therapy will develop ever more therapies and techniques to cure conditions, rather than merely treat the symptoms. Second, developing these techniques may be the only effective way to treat some of these conditions when palliative or symptomatic therapeutic interventions are unavailable. Third, such measures may be efficient in that by developing germ-line gene therapies, one would prevent the transmission of diseases between generations and therefore avoid treatments for many others in the future. Finally, many resist limiting these therapeutic techniques on the ground that it is an unwarranted restriction of scientific freedom. The federal regulations delineate the limits of research involving human subjects, and if research falls within these regulations, it is unreasonable to limit free inquiry.

On the other hand, there are arguments opposing development of germ-line gene therapy techniques, especially with children as subjects. The first is the unknown, unexplored, and unpredictable risks to the persons and their offspring. Given the uncertainty of these risks of harm, there is no way to determine exactly whether the harm–benefit ratio would be justified or not. A second issue concerns the problem of whether we are developing techniques for therapy or for enhancement. As we discussed earlier, it is difficult to distinguish between medical and nonmedical uses of these techniques. Third, some object that since these techniques could harm future generations as they try to perfect the development of these techniques, they would be putting future individuals, who gave no consent, at risk. Fourth, there is the issue of the allocation of resources. It would be costly to develop these new germ-line gene therapy techniques, and some argue that we have higher social priorities. Finally, some are concerned that we would change the germ line for all time by these techniques, altering inheritance intentionally and perhaps capriciously.

To conclude, research involving children can foster different and important values including benefit for the children who are subjects, medical advances, prudent use of public resources, health care planning, and public health. With more information, better diagnoses, treatments, and prognoses, as well as informed consent, are possible. In some cases, however, these values can conflict, posing difficult choices on how to acknowledge people's rights, protect their interests, gain good information, and do what is useful socially. Different research policies offer different solutions on how to balance these values, and these policies have their own different strengths and weaknesses. Solutions were judged superior when they fairly promote children's well-being and opportunities to develop and flourish.

BIBLIOGRAPHY

1. L.M. Kopelman, in J.F. Monagle and D.C. Thomasma, eds., *Health Care Ethics: Critical Issues*, Aspen Publishing, Gaithersburg, MD, 1994, pp. 199–209.

2. American Academy of Pediatrics, Committee on Drugs, *Pediatrics* **95**(2:2), 286–294 (1995).

3. U.S. National Institutes of Health, *Policy and Guidelines on the Inclusion of Children as Participants in Research Involving Human Subjects*, March 6, 1998. Available at: *http://grants.nih.gov,grants/guide.notice-files-98-024.html*.

4. U.S. Department of Health and Human Services Food and Drug Administration, *Regulations Requiring Manufacturers to Assess the Safety and Effectiveness of New Drugs and Biological Products in Pediatric Patients*, [FR Doc. 98-31902] OC 98142, Docket No. 97N-0165, pp. 66631–66672. Available at: *http://www.verity.fda.gov/search97cgi/s97_cgi.exe?action= View&VdkVgwKey*.

5. A.E. Buchanan and D.W. Brock, *Deciding for Others: The Ethics of Surrogate Decision Making*, Cambridge University Press, Cambridge, UK, 1989.

6. E.W. Clayton et al., *J. Am. Med. Assoc.* **274**(22), 1786–1792 (1995).

7. D.W. Brock, *Bioethics* **9**(3, 4), 269–275 (1995).

8. R.R. Faden and T.L. Beauchamp, *History and Theory of Informed Consent*, Oxford University Press, New York, 1986.

9. T.L. Beauchamp et al., *J. Clin. Epidemiol.* **44S**(1), 151S–169S (1991).

10. L.M. Kopelman, *J. Med. Philos.* (2000), forthcoming.

11. L.S. McGough, in *Encyclopedia of Bioethics*, rev. ed., vol. 1, Simon & Schuster, New York, 1995, pp. 371–378.

12. A. Holder, *Legal Issues in Pediatrics and Adolescent Medicine*, 2nd ed., Yale University Press, New Haven, CT, 1985.

13. The Nuremberg Code, Nuremberg Military Tribunals, 1949, pp. 181–182.

14. World Medical Association, Declaration of Helsinki [1964] 1989, *Recommendations Guiding Medical Doctors and Biomedical Research Involving Human Subjects*, Adopted by the 18th World Medical Assembly, Helsinki, Finland, and amended in 1975, 1983, and 1989.

15. U.S. Federal Regulations, 45 CFR §46 (1991).

16. Council for International Organizations of Medical Science (CIOMS), *International Ethical Guidelines for Biomedical Research Involving Human Subjects*, Geneva, Switzerland, 1993.

17. B. Freedman, A. Fuks, and C. Weijer, *Hastings Center Rep.* **23**(2), 13–19 (1993).

18. R.J. Levine, *N. Engl. J. Med.* **341**(7), 531–534 (1999).

19. T.A. Brennan, *N. Engl. J. Med.* **341**(7), 527–531 (1999).

20. J. Janofsky and B. Starfield, *J. Pediatrics* **98**, 842–846 (1981).

21. A.D. Lascari, *J. Pediatrics* **98**, 759, 760 (1981).

22. National Bioethics Advisory Board, Testimony before the Human Subject Subcommittee of the National Bioethics Advisory Commission Meetings, January 8, 1998, in Arlington, VA. Availability at: *http://bioethics.gov/transcripts/index# jan98*.

23. N. Daniels, *Just Health Care*, Cambridge University Press, Cambridge, UK, 1985.

24. E.T. Juengst, *J. Med. Philos.* **16**(6), 587–592 (1991).

See other HUMAN SUBJECTS RESEARCH entries.

HUMAN SUBJECTS RESEARCH, ETHICS, COMPENSATION OF SUBJECTS FOR INJURY

TERRENCE F. ACKERMAN
University of Tennessee
Memphis, Tennessee

OUTLINE

Introduction

Some Initial Distinctions

History of the Compensation Debate

Moral Basis of Compensation

Extent of Research Injuries

Elements of a Compensation Program

Summary

Bibliography

INTRODUCTION

Participation in clinical research sometimes results in pain, physical disability, or even death for human subjects. For several decades scholarly commentators, medical researchers, and government advisory committees have debated the advisability of establishing a national program to provide compensation for injured research subjects. The substance of this debate has focused on several critical questions. What is the basis and extent of society's moral obligation to compensate subjects for research injuries? Does the incidence and magnitude of research injuries justify the establishment of a national compensation program? What are the essential components of a morally acceptable and practicably workable compensation plan? Disagreements about our moral obligations, the paucity of empirical data about research injuries, difficulties in devising a workable system, as well as the press of other public priorities, have thwarted any clear social consensus on the merits of a compensation program. Nevertheless, the unavoidable fact that some persons are injured in medical research for the benefit of society has sustained the policy controversy.

SOME INITIAL DISTINCTIONS

In exploring the issues regarding compensation, several distinctions are crucial to the analysis. First, it is important to distinguish research injuries for which subjects possess legal means for securing redress and those for which legal recourse is not available (1,2). Injuries may occur as a result of researcher negligence when subjects have not been adequately informed of the risks involved or when due care has not been exercised in protecting them from risks. In these instances injured persons may bring civil suits for damages against investigators. In other limited cases, such as workers' compensation for federal employees, persons injured in research may use existing administrative mechanisms for securing compensation. By contrast, most research injuries occur without any fault on the part of researchers. There are several reasons. Most interventions employed in clinical research, such as chemotherapy or tissue biopsy, possess a definite risk of harm even when competently performed. Oftentimes the use of new drugs or medical devices results in harm that could not be anticipated based on prior knowledge. Moreover, many research subjects have diseases or disabilities that result in a heightened susceptibility to the harmful effects of medical interventions. Yet very few research institutions have formal compensation programs. As a result the debate about compensation has focused on research injuries that occur without researcher negligence and for which no redress is legally available.

Important to the contours of the debate is a second distinction between therapeutic and nontherapeutic research procedures (3). Therapeutic research procedures are performed in order to benefit the individual subject, as well as to produce generalizable knowledge. For example, administration of a new drug in a randomized clinical trial comparing it to standard treatment is a therapeutic research procedure. By contrast, when a medical intervention is performed solely to produce generalizable knowledge, without being intended to benefit the subject, it is a nontherapeutic research procedure. For example, serial venipunctures performed to profile the pharmacokinetics of a new drug in healthy volunteers constitute nontherapeutic research procedures. This distinction is crucial to disputes about the limits of the moral obligation to compensate injuries,

the extent of the need for a program, and the practicality of any scheme for identifying compensable injuries.

A third distinction concerns the reasons for which subjects participate in clinical research. Some may participate primarily to promote their own interests. Other subjects may participate in research studies primarily to contribute to the welfare of society. This distinction cuts across the prior one between therapeutic and nontherapeutic research procedures. Most subjects enter trials of therapeutic procedures to secure personal benefits, especially the opportunity to receive a new treatment that may be more efficacious or safe than existing therapies. But participation in a clinical trial may offer no special advantage, as when investigators are comparing the efficacy and safety of two standard treatments. On the other hand, subjects who participate in studies involving only nontherapeutic research procedures anticipate no medical benefit. Yet in many cases they may receive an attractive payment for their participation. An important point of contention is whether personal sacrifice for the common good is a necessary condition for a valid claim against society for compensation of injuries.

A final distinction concerns the difference between medical research that is undertaken at the behest of society and research not similarly sanctioned. In the former category there is research conducted by employees of the federal government, such as physician-investigators employed by the National Institutes of Health (NIH). In addition much medical research in the United States is supported through grants from the federal government to investigators working in public and private academic institutions. There is also considerable medical research conducted by private medical companies required by regulations of the Food and Drug Administration (FDA) for the development of new drugs and medical devices. Whether medical research is federally conducted, supported, or regulated, it is commissioned through the official agencies of society. By contrast, considerable medical research is funded through private sources, particularly charitable foundations. While such research contributes to the common good, it is not officially sanctioned. An important issue is whether an obligation to compensate injuries should be limited to subjects in research sanctioned by the official agencies of society or should extend to all research that contributes to the common welfare.

HISTORY OF THE COMPENSATION DEBATE

Between 1945 and 1965 annual expenditures of the NIH increased exponentially from $701,800 to $436,000,000. This increase reflected an exploding confidence of the American public that medical research could yield enormous advances in the prevention and treatment of disease. By 1960 governmental support for clinical research sustained a vast infrastructure of research programs at medical colleges in the United States.

Despite the proliferation of medical research employing human subjects, there was only sporadic discussion of issues related to the rights and welfare of human subjects. Although the ethical guidelines for the conduct of human research formulated at Nuremberg were considered relevant to the American enterprise, it was generally believed that the moral conscientiousness of individual investigators was sufficient to assure that subjects were adequately protected. The absence of public reports of research injuries reinforced the informal approach.

This public and professional confidence began to unravel in the 1960s with a series of revelations about studies in which persons were uninformed that they were subjects and/or exposed to unjustifiable risks of harm (4). Public concern was awakened after severely deformed infants were born to mothers who took the sedative thalidomide during pregnancy without being aware of its investigational status. This tragedy helped solidify support for the 1962 Harris-Kefauver Amendments to the Food, Drug and Cosmetic Act, which included the requirement that informed consent be secured from prospective subjects for the testing of investigational drugs. In 1964 it was revealed that two New York physicians, while studying cancer immunology, had injected live cancer cells under the skin of elderly, debilitated patients without their knowledge. This was followed in 1966 by the publication, in the *New England Journal of Medicine*, of Henry Beecher's paper, "Ethics in Clinical Research," which described 22 studies published in leading medical journals that involved the exposure of subjects to unjustifiable risks of harm (5). Finally, in 1972, front-page newspaper articles detailed the infamous Tuskegee Syphilis Study in which 400 poor, black males in Alabama suffering from syphilis had been left untreated for 40 years to study the natural history of the disease (6). Medical research was now acknowledged to involve palpable human costs.

These developments fostered a growing recognition that adequate protection for the rights and welfare of human subjects required the creation of formal social controls over clinical research. Beginning in 1966, the Public Health Service (PHS) established protective mechanisms involving two main components (7). First, human research activities conducted or supported by PHS were required to undergo prior review by committees established at each institution engaging in clinical research. Second, these committees were to approve research only if adequate provision was made for securing the informed consent of subjects or their legally authorized representatives and only if the risks were justified by the anticipated benefits to subjects and others. These essential elements of social control were refined over the next decade and a half, assuming much of their present form by 1981 with the implementation of the recommendations of the National Commission for the Protection of Human Subjects of Biomedical and Behavioral Research (hereafter, the National Commission) (8).

At the same time, a third mechanism of social control was being explored in the legal and medical literature: no-fault compensation for injured research subjects. Some early commentators viewed compensation as a means of fulfilling a societal obligation to those injured in contributing to the common good (9). Initially, however, compensation was often conceptualized as a mechanism for controlling risks (10,11). Compensation

programs funded by research institutions would encourage more careful scrutiny of the risk/benefit profile of proposed research studies and would necessitate closer surveillance of the safety of subjects during their participation. Moreover, if injured subjects were fully compensated for medical expenses, lost wages, and related expenses, society would be forced to acknowledge the full costs of medical research.

Beginning in the 1970s a series of government advisory committees formally examined the compensation issue. With the strengthening of other mechanisms for controlling research risks, these groups began to focus on the moral obligation of society to compensate research injuries. For example, the Tuskegee Syphilis Study Ad Hoc Advisory Panel in 1973 endorsed the concept of compensation as follows:

> No policy for the compensation of research subjects ... has been formulated, despite the fact that no matter how careful investigators may be, unavoidable injury to a few is the price society must pay for the privilege of engaging in research which ultimately benefits the many. Remitting injured subjects to the uncertainties of the law court is not a solution (12).

Based on this rationale, the committee recommended establishment of a no-fault compensation program for injured research subjects. Similarly NIH accepted this rationale in three proposals submitted to the Secretary of Health, Education, and Welfare (HEW) in the early 1970s for implementation of a no-fault compensation program administered by the government for injuries sustained in research supported by federal funds. However, acceptance of these proposals was considered premature. Their fiscal implications had not been assessed, alternatives to a federally administered program had not been explored, and the need had not been clearly established (13).

These shortcomings in prior proposals prompted the Secretary of HEW in 1974 to create a task force to examine in greater detail "whether and how to compensate subjects injured in the course of research." The HEW Secretary's Task Force on the Compensation of Injured Research Subjects (hereafter, the HEW Secretary's Task Force) conducted an 18-month study in which it elicited papers on the ethical and legal aspects of compensation, commissioned an empirical study on the extent of research injuries, and consulted with officials from the insurance industry regarding the feasibility of a compensation program underwritten by private companies. In its final report, the Task Force strongly endorsed the proposition that society has a moral obligation to compensate injured research subjects (13). Moreover it maintained that this obligation extended to subjects of both therapeutic and nontherapeutic research procedures. While the Task Force report recommended that subjects in research conducted by PHS be included under the provisions of the Federal Employees Compensation Act, it suggested that compensation for subjects in PHS-supported research be provided by the research institutions receiving financial support. A similar recommendation was made regarding compensation for subjects of research on new drugs and medical devices regulated by FDA.

Although the recommendations of the Task Force were endorsed in June 1977 by the National Commission, they were not readily embraced by the research community (14). This hesitancy did not focus on the ethical argument. Rather, widespread concern was voiced regarding the ability of research institutions to secure private insurance which met their needs for protection, while also not imposing costs or limitations on coverage that would thwart valuable research. These concerns were reflected in the report of the Task Force itself, which found private insurers extremely reluctant to consider compensation coverage, given the diversity of research institutions, studies and subjects, as well as the absence of actuarial data on which to formulate premiums.

While the controversy continued regarding the practical contours of a compensation program, the National Commission addressed the issue from another standpoint. In its report on institutional review boards, the Commission outlined those items of information that a reasonable person in the prospective subject's position would need to know in deciding about participation in research. The Commission proposed that a reasonable person would need to know whether medical treatment and compensation is available in the event of injury (15). In January 1979, this recommendation was adopted by the Department of Health, Education, and Welfare (DHEW), which now required that, for research involving more than minimal risk, institutions receiving federal funds for research must inform prospective subjects regarding the provision of medical treatment for injuries, the availability of monetary compensation, and the identity of the appropriate individual to contact in the event of injury (16). This requirement of informed consent disclosure remains in effect today.

With the substantive issue of a compulsory compensation program still unresolved, the President's Commission for the Study of Ethical Problems in Medicine and Biomedical and Behavioral Research (hereafter, the President's Commission) decided in 1980 to undertake a fresh investigation of the issue. In developing its recommendations, the Commission reviewed papers on the ethical and legal issues, reports on the few existing compensation programs, an analysis of the risks of common research procedures, and assessments of alternative program designs (14). The conclusions it reached were more narrowly drawn and tentative than those of the HEW Secretary's Task Force. The Commission affirmed that society has an obligation of justice to compensate injured research subjects. However, it considered this a prima facie obligation that might be outweighed by other duties of social justice if the need for a program was negligible or the costs of administering it were disproportionate to the need. Moreover the Commission restricted the scope of the obligation to injuries incurred as a result of nontherapeutic research procedures. With respect to therapeutic research procedures, the Commission found little evidence that subjects who seek therapy in the research setting expose themselves to risk exceeding that presented in the nonresearch setting. In addition the Commission asserted that distinguishing excess injury in subjects who are seriously ill patients posed administrative burdens that could not be justified in

light of the relatively weaker moral claim to compensation for injuries resulting from therapeutic research procedures. The Commission also identified numerous practical aspects of administering a program whose appropriate form could not be settled without some limited experience in implementing such a program. Lastly, the Commission believed that insufficient data on research injuries existed to establish a compelling need for a program. Based on these factors, the Commission recommended that the (now) Department of Health and Human Services (DHHS) conduct "a small, controlled experiment to determine whether a formal program is needed and, if so, the most fair and efficient means of providing compensation." The report of the President's Commission was conveyed to President Reagan in June 1982. In an era marked by shrinking federal programs, no action was taken on its recommendations.

In the intervening years the issue of compensation for research injuries has been revisited only once at the federal level. In 1995 the Advisory Committee on Human Radiation Experiments issued its report on radiation research on human subjects sponsored by the federal government and conducted during the period 1944 through 1974 (17). The Advisory Committee recommended that the federal government financially compensate subjects or next of kin in cases where the government deliberately withheld information from individuals or families regarding the nature of the research. It also recommended compensation for subjects who were physically injured in studies involving either no prospect of direct benefit to the subjects, or the use of controversial interventions that were misrepresented as standard practice. However, the Advisory Committee did not examine general issues regarding implementation of a no-fault, federal compensation program for injured research subjects.

MORAL BASIS OF COMPENSATION

Moral arguments favoring compensation of injured research subjects follow two strategies. Some proponents have argued that establishing a compensation program will have consequences that improve the overall cost–benefit ratio of clinical research. Other analysts have contended that compensation is owed to injured research subjects as a moral right, regardless of its beneficial consequences. This moral right reflects the obligation of society to justly distribute the benefits and burdens of cooperative social activities like clinical research.

The view that a compensation program is justified by its cost/benefit consequences was articulated by early theorists who viewed it as a mechanism for the social control of research (10,11). Several potential benefits of no-fault compensation were envisioned. One is that compensation would encourage investigators to engage in less dangerous research, because the costs of compensating injured research subjects would be drawn from the total funds available for research. Also reduction of risk in planning research would be facilitated by accumulating data regarding the compensation cost profile of common research interventions used in specific subject groups. In addition compulsory compensation would encourage the allocation of more resources for the early identification and treatment of injuries incurred by subjects. Another goal achieved would be increased public support for the research enterprise, deriving from awareness that appropriate resources are devoted to the needs of injured subjects. Moreover, availability of compensation would encourage participation in research by persons who might demur in the absence of protection against personal loss. Finally, a compensation program might facilitate the conduct of some clinical research that is unusually risky but promises significant advances in medical knowledge. Thus, on this view, society ought to provide compensation for research injuries because it will lead to safer research, encourage societal support and participation, and facilitate the conduct of some risky but highly promising studies.

There are serious shortcomings in this approach to grounding a societal obligation to provide compensation. One problem is that the argument depends on unproven factual assumptions about the incentives for researchers and subjects created by a compensation program. For example, the argument assumes that the availability of compensation would increase recruitment of subjects, even though there is no evidence that persons now decline participation due to the absence of this guarantee. Similarly the claim that research will be practiced more safely if compensation costs are included in the total funding for studies presupposes that current review mechanisms and the conscientiousness of investigators are inefficient in minimizing risks. A second problem is that the incentives created by a compensation program may depend on its exact administrative configuration, and particular schemes may inconsistently both promote and undermine specific goals cited. For example, if a federal compensation program is financed through funds separate from research budgets, it may have little deterrent effect on investigators contemplating the use of risky research procedures. At the same time, greater public awareness about research injuries might discourage research participation, irrespective of the availability of compensation. A third problem is the most decisive. While the incentives of a compensation program may decrease the total harm caused to human subjects and encourage public support for and participation in clinical research, the moral claim of injured subjects against society is valid even if these beneficial consequences are not realized. That is, it is the injury itself that seems to trigger the obligation to provide compensation. Therefore the moral basis of the obligation to compensate injured research subjects must reside in considerations other than the valuable consequences of a compensation program.

The second major approach to delineating a moral basis for compensation proceeds from this important insight. This view asserts that compensation is owed to injured research subjects as a matter of social justice. According to one common interpretation, justice requires that the benefits and burdens of certain cooperative social endeavors be distributed in ways that give all persons an equal opportunity for a good life. When persons are injured in these activities, compensation may help to restore the equality of their opportunity vis-à-vis other members of society.

A crucial question is what cooperative social endeavors create this obligation of justice. In a position that strongly influenced both the HEW Secretary's Task Force and the President's Commission, Childress argues that three features of an activity create a societal obligation of compensatory justice (18). First, the injured party has accepted or has been compelled to accept a position of risk. Objective risks that the party would not otherwise have encountered are created by this position. Second, the activity is for the benefit of society, although any particular person's motives may not be to benefit society. Third, society, through its government or agencies conducts, sponsors or mandates the practices in question. For example, these conditions obtain with respect to persons who serve in the military forces and create an obligation to compensate service-connected injuries. They also apply to the injuries incurred by some research subjects.

Several crucial points should be noted about the scope of the obligation of compensatory justice as specified by Childress. The obligation to compensate research subjects is triggered by the injury without regard to negligence by investigators. Compensation should be provided on a no-fault basis. In addition the obligation to compensate pertains to injuries derived from assuming a position of objective risk that benefits society. The subject's motives to secure a promising new treatment or to receive an attractive cash payment are irrelevant to the right of compensation for injury. Moreover, the obligation to compensate applies to injuries associated with both therapeutic and nontherapeutic research procedures, because both involve a position of risk for the benefit of society. Lastly, the societal obligation to injured subjects applies only to cooperative endeavors officially conducted, supported or mandated by society. Thus research conducted or supported federal agencies, or regulated by FDA are covered by the obligation of compensatory justice. Research privately supported or conducted is not officially endorsed by society and does not create a societal obligation to redress the injuries of subjects.

Despite its intuitive appeal, there are two serious difficulties with this argument. The first relates to the role of consent by prospective subjects to participation in research. When persons knowingly and freely participate in cooperative ventures possessing a risk of harm, other parties are not normally held responsible for injuries that materialize without negligent behavior. This notion is captured by the maxim, *volenti non fit injuria* — there is no injury to one who consents. Moreover, as required by current federal regulations, prospective subjects must be advised as to the availability of compensation. If subjects accept participation in research knowing that it is not available, then their consent is secured with disclosure of the information that a reasonable person would need to know. The choice to participate without this guarantee would seem to relieve any prima facie obligation to redress resulting injuries.

Some commentators respond that the actual process of informed consent is too imperfect to absolve society's obligation to compensate research injuries. The report of the President's Commission delineates some of these imperfections (14). One is that in some research the risk profile of interventions is not yet established. Prospective subjects must decide about participation without being aware of the nature of unknown or unanticipated risks. In addition experimental evidence suggests that persons tend to discount the importance of substantial harms with a very low probability of occurrence. Finally, it is difficult to convey to prospective subjects a full appreciation for the ways in which specific injuries might be harmful to their interests. In light of these imperfections in knowledgeable decision making, it might be argued that subjects do not really intend to waive their security against the personal losses imposed by injuries.

The argument from imperfections in the consent process is badly weakened by its paradoxical implications. As Engelhardt has pointed out, we are asked to imagine a consent process in which the decision making of prospective subjects is adequate enough to validate their decision to participate in research but not adequate enough to validate their willingness to forgo compensation for injuries (19). Yet the decision to expose oneself to the injury in the first place does not seem to require less ability to comprehend facts and to deliberate about their consequences than the decision to accept research participation without the promise of compensation. Thus, if prospective subjects have sufficient capacity to competently decide about participation in research, they are also competent to accept participation without compensation for injuries.

Another response to the argument that consent relieves the obligation to compensate focuses on the morality of the request rather than the validity of consent (20). We may grant that knowing and willing agreement by subjects relieves society of its obligation. Nevertheless, it may be morally wrong of society to even make such a request. If justice requires that subjects be compensated for injuries incurred in officially sanctioned medical research, then society should not be free to ask persons to forgo compensation. The request for a waiver of this right should be limited to situations in which weightier obligations constrain the use of societal funds for compensation. While persons should be free to decline compensation after they have been injured, society must make the offer as a matter of compensatory justice.

A more serious problem with Childress' argument for the obligation of compensatory justice relates to its scope. For Childress the obligation to compensate research injuries is triggered when persons assume a position of risk in activities that benefit society and are sanctioned through its official agencies. Injuries resulting from both therapeutic and nontherapeutic research procedures are compensable. Likewise injuries are compensable whether the primary motivation of subjects is to contribute to the interests of others or to promote their own interests. The essential difficulty is explaining why the assumption of risk triggers an obligation to compensate injuries when subjects' participation is premised on the judgment that the risks are outweighed by anticipated personal benefits. If subjects are not making a sacrifice for society, then compensation for their losses is not owed them as a matter of justice. The scope of the obligation must be restricted

to circumstances in which the positional risk is assumed without the anticipation of offsetting personal benefits.

An example illustrates the problem (21). The federal government sponsors a national lottery to fund educational programs. These programs contribute to the general welfare. Participants in the lottery assume a position of risk. As a cooperative social enterprise, the lottery satisfies the three conditions for compensation of injury cited by Childress. If a losing participant has purchased so many tickets that he faces financial ruin, it follows that he ought to be compensated for his losses. If this implication is unacceptable, then it appears that the scope of the obligation to compensate injuries as sketched by Childress is too broad.

The same difficulty arises with regard to clinical research involving the evaluation of therapeutic interventions (22). Childress assumes that subjects are placed at special risk by participation. However, it is generally agreed that it is not morally permissible to undertake a trial of a new therapy unless prior evidence suggests that it is likely to be at least as efficacious and safe as alternative treatments acceptable to the subject. Moreover, because shortcomings in the efficacy and safety of standard treatments provide the stimulus for evaluating new therapies, it is usually the case that preliminary evidence suggests that the new treatment may be superior. In addition treatment in the research context often includes more intensive monitoring and nursing care for subjects. In many research centers there are also more numerous and specialized ancillary personnel available, including allied health professionals, social workers, and psychologists. Lastly, new treatments being investigated are often provided without charge to the subjects. Thus subjects assume a position of risk to secure overriding personal advantages associated with participation. If there is no sacrifice involved in the subject's participation, then an obligation of compensatory justice does not apply in the event of injury.

Most commentators have assumed that the same factors do not apply to injuries resulting from nontherapeutic research procedures. Subjects injured by these interventions cannot anticipate commensurate medical benefits. If justice requires that the impact of these burdens be ameliorated, then it is appropriate to redress these injuries. But even here, the motivation of subjects complicates the moral equation. In some cases subjects participate in research involving nontherapeutic procedures to secure an attractive payment. Rather than being an act of sacrifice, participation reflects a calculation of overriding personal benefit. Nevertheless, in many cases, research involving nontherapeutic procedures does not involve payment for participation, and subsequent injuries to subjects create the societal obligation to provide compensation.

If many subjects do not incur special risks, then the obligation to compensate injuries is more narrowly circumscribed than envisioned by Childress and the HEW Secretary's Task Force. However, there are ways to blunt the force of this argument. First, the risk–benefit profile of a new therapy undergoing evaluation is, *ex hypothesi*, not yet clearly established. Although preliminary evidence may suggest that it will be at least as favorable as alternative treatments, the new therapy may prove inferior in a controlled clinical trial. Second, there is always the danger that unknown or unanticipated risks of new therapies will emerge during or after the conduct of clinical trials. Participating subjects expose themselves to these uncertainties that are not present in the nonresearch setting. Third, for subjects who incur injuries in excess of what might be anticipated in the nonresearch setting, it can be maintained they have clearly made a sacrifice that contributes to the general welfare of society. If these considerations are decisive, then the obligation to compensate research injuries should cover therapeutic as well as nontherapeutic interventions.

Even if the argument for an obligation of compensatory justice is compelling, it establishes only a prima facie societal obligation to compensate research injuries. There may be other societal needs, or other obligations of social justice, which have moral priority in the use of limited societal resources. Thus the case for compensation may depend on the extent of the problem of research injuries, as well as the importance of this problem relative to other pressing social needs.

EXTENT OF RESEARCH INJURIES

In the period right after World War II, medical research was frequently depicted as a perilous enterprise exposing subjects to grave risks of harm that would not otherwise be encountered. However, this assumption was not based on empirical data about the frequency and magnitude of injuries. Even today there is only one large descriptive study of injuries in clinical research and the accumulated experience of a few small compensation programs. Available information raises serious questions about whether the frequency and magnitude of research injuries establishes a compelling need for a national compensation program.

A large empirical study was performed by Cardon and colleagues for the HEW Secretary's Task Force (23). Data were compiled by telephone from 331 investigators conducting research on 133,000 human subjects. Overall, injuries were reported in 3.7 percent of all subjects. Eighty percent of these injuries were classified as trivial by the investigator, while nearly all the remaining were temporarily disabling. Permanently disabling and fatal injuries together accounted for about 1 percent of the total injuries.

Separate analyses were performed for therapeutic and nontherapeutic research procedures. Of 39,216 subjects undergoing therapeutic research procedures, 10.8 percent were reported injured. Trivial injuries were incurred by 8.3 percent of subjects and temporarily disabling injuries by 2.4 percent, with less than 0.1 percent suffering either permanently disabling injuries or death. In actual numbers, 13 subjects were permanently disabled and 43 died. Of 93,399 subjects undergoing nontherapeutic research procedures, 0.8 percent were injured. Trivial injuries accounted for 0.7 percent of all subjects, temporarily disabling injuries 0.1 percent, and permanently disabling injuries less than 0.1 percent. Only one subject was permanently disabled, and there were no fatalities.

The investigators also determined that injuries resulting from therapeutic procedures were clustered in clinical trials of cancer therapies. These trials accounted for 37 of 43 fatalities, 9 of 13 permanently disabling injuries, and 648 of 937 temporarily disabling injuries. Except for these cancer trials, there were few serious and/or irreversible injuries reported. Moreover, investigators were asked to enumerate actual injuries, without regard to whether the same results might have been expected from treatment in the nonresearch setting. This point is significant with respect to research on cancer treatments. Standard treatments for cancer carry substantial risks of disabling injuries or death. The study design did not permit investigators to determine whether the frequency and magnitude of injuries in cancer trials was greater or less than expected in the nonresearch setting.

Finally, investigators broached the question of how the risks of harm to subjects undergoing nontherapeutic research procedures compare to the risks of everyday life. Annual rates of accidental injuries that are temporarily disabling, permanently disabling and fatal, per 100,000 Americans, were found to be about 50, 2, and 0.6, respectively. Rates for the same types of injuries in approximately 93,00 research subjects were 37, 1, and 0. If the average duration of participation in research employing nontherapeutic procedures is 3 days, or 1/100 of a year, then these rates do not seem significantly greater than the risks of everyday life.

The President's Commission also reviewed data on research injuries from existing compensation programs in the United States. McCann and Pettit analyzed data from the University of Washington for 1972 to 1981 during which an estimated 356,000 subjects were covered by the university's compensation program (24). In the year prior to initiation of the program, 10 out of 14,942 subjects (0.07 percent) experienced adverse effects. None were partially or permanently disabled, and no deaths were reported. For the first eight years of the compensation program, 144 out of 356,000 subjects (0.04 percent) experienced temporary disability, none were permanently disabled, and two died. In addition a report was received from the Quincy Research Center in Kansas City, Missouri, whose subjects were covered by a workers' compensation program (25). The data covered 2596 normal volunteers and 2478 patients. Clinically adverse events resulted in the hospitalization of 1.43 percent of patients and 0.2 percent of normal volunteers, while fatalities occurred in six patients and no normal volunteers. Analysis suggested that no deaths in the patient group, and only 7 of 36 hospitalizations could be related to the research interventions. Finally, the Commission reviewed the results of the drug testing program using normal volunteers in Michigan prisons for the years 1964 to 1976 (26). The authors reviewed 805 protocols involving 29,162 research participants. They reported that 64 subjects experienced adverse reactions (0.2 percent), but there was only one permanently disabling injury and one fatality (in a patient receiving a placebo). Thus data from these research programs corroborate the findings of Cardon and associates that serious research injuries are very infrequent.

An important policy issue is whether the extent of research injuries is sufficient to transform a prima facie moral obligation to offer compensation into a legitimate social priority. As outlined earlier, the moral argument for compensating research injuries is most compelling with regard to subjects harmed by nontherapeutic procedures. However, the evidence generated by Cardon et al. suggests that the frequency and magnitude of injuries resulting from nontherapeutic procedures is no greater than the risks of accidental injury in daily life. The moral argument for compensating injuries caused by therapeutic procedures is less compelling, because subjects often accept the risks of harm to secure the offsetting benefits of therapy in the research setting. Here the Cardon study found a greater frequency and magnitude of injuries to subjects. Nevertheless, these injuries were clustered in cancer treatment trials, and no attempt was made to determine whether the extent of injuries exceeded that anticipated in the nonresearch setting. Thus available empirical data does not provide convincing evidence that the prima facie obligation to compensate injured subjects should be a social priority.

The President's Commission concluded that existing evidence was insufficient to establish a compelling need for a national compensation program (14). In recommending experimental, pilot programs for compensating injured subjects, the Commission indicated that gathering more comprehensive data on the extent of research injuries should be a crucial component of this endeavor. In this way the data necessary to an informed social policy decision on the relative need for a compensation program could be generated.

ELEMENTS OF A COMPENSATION PROGRAM

Even if the moral argument for compensation is persuasive and the prevalence of research injuries significant, numerous questions remain regarding the design of a morally acceptable program. A leading question involves how compensable injuries are to be defined. The challenge is to isolate those harms to subjects that exceed what would have occurred if they had not participated in research. With regard to nontherapeutic research procedures, the resolution is relatively straightforward. Any injuries that are reasonably related to the interventions should be compensable. Because the procedures are performed for reasons unrelated to the welfare of subjects, any injuries suffered would not have occurred but for their participation in the research.

The problem is considerably more complex in dealing with therapeutic research procedures. Subjects undergoing therapeutic procedures are patients. As patients, they may incur harms from the progression of their disease. In addition, even if they do not participate in a research study, they receive treatments that may cause harm. Therefore it becomes difficult to specify which injuries would not have occurred but for participation in research.

In its recommendations, the HEW Secretary's Task Force proposed the following criterion for compensable injuries:

Human subjects who suffer physical, psychological, or social injury ... should be compensated if (1) the injury is proximately caused by such research, and (2) the injury on balance exceeds that reasonably associated with such illness from which the subject may be suffering, as well as with treatment usually associated with such illness at the time the subject began participation in the research (13).

Although intuitively plausible, application of the "on balance" test in the clinical setting presents at least four serious problems. These problems are well-illustrated by clinical trials of cancer treatments, where physical injuries often occur (22).

First, it is frequently not possible to determine whether the disease or the treatment has caused an identified harm. For example, in treating acute leukemia, both the leukemic process and chemotherapy may suppress bone marrow function in a way that disposes to serious infection or bleeding. Second, the "on balance" test relies on a distinction between investigational and standard treatment. However, in some forms of disseminated adult cancer, no standard treatment of choice has emerged. It may not be clear with what treatment an investigational therapy should be compared to determine excess injury. Third, the "on balance" test requires comparison of the overall frequency of various harms and benefits associated with research and conventional therapies. But it is uncertain how to compare incommensurable harms and benefits. For example, does compensable injury occur when an investigational chemotherapy for bone cancer increases the median length of survival by four months but involves a higher incidence of acute hearing loss and severe nausea and vomiting than standard therapy? A final problem is that all chemotherapies for disseminated cancer involve serious risks and fail to achieve remission in a definite percentage of cases. Thus, although the harm–benefit ratio of an investigational therapy in a series of subjects may be less favorable than with standard treatment, it is not possible to determine whether the outcome for a particular subject has, "on balance," resulted in more harm than standard treatment. Thus the President's Commission found that this test for identifying compensable injuries posed "enormous burdens of administration" that could not be justified in light of the less compelling moral claim to compensation of subjects undergoing therapeutic research procedures (14). Of course, if compensable injuries are restricted to nontherapeutic research procedures, these problems are avoided.

A second important issue involves the types of injuries for which compensation will be provided. The HEW Secretary's Task Force recommended coverage for physical, psychological, and social injuries. However, DHEW's interim rule restricted informed consent disclosure regarding the availability of compensation to physical injuries. The President's Commission argued that compensation should focus solely on physical and psychological injuries. The Commission defined "social" injuries as harms to subjects' reputation, personal relationships, or legal status resulting from unauthorized disclosure of personal information gathered in research. It suggested that problems related to confidentiality could be handled effectively in the prospective review of protocols (14).

The moral analysis of the obligation of compensatory justice provides some guidance in selecting relevant categories of injuries. The principle of justice requires that all persons have equal opportunity to pursue their life plans. Categories of injuries are pertinent if failure to compensate them might seriously compromise the opportunity of injured subjects vis-à-vis other persons. Serious and irreversible physical injuries obviously satisfy this criterion. Likewise serious psychological injuries, such as increased frequency of clinical depression resulting from an experimental drug for bipolar affective disorder, may substantially impair the opportunity of subjects to pursue their life plans. Moreover, even if we restrict "social" injuries to unauthorized disclosure of confidential information, there is no reason in principle to make these injuries ineligible for compensation. Injuries to reputation, personal relationships, or legal status may gravely impair the ability of persons to pursue their goals.

Nevertheless, compensatory justice constitutes only a prima facie societal obligation. Other requirements of social justice, or other compelling social needs may present weightier moral demands on limited resources. These limitations may necessitate that the relative burdens imposed by different categories of research injuries be ranked for funding priority. If the frequency and magnitude of "social" injuries is far less significant in abridging the opportunity of subjects to pursue their life plans, then compensation for these harms may not achieve funding priority.

A third crucial component in the design of a compensation program concerns the nature and extent of benefits for compensable injuries. Benefits might include medical care for injuries, lost wages, monetary awards for pain and suffering, and death benefits for survivors. At present, some academic institutions and pharmaceutical companies provide short-term medical care for injuries incurred by research subjects. With few exceptions, other benefits are not provided.

Our understanding of the obligation of compensatory justice also provides guidance in determining relevant types of benefits. The function of compensation benefits is to restore any deficit in the ability of injured subjects to pursue their life plans compared to citizens who are not injured in research (20). This suggests that benefits should be provided to redress any negative consequences that flow from a compensable injury and that impair the ability of injured subjects to pursue their life plans. These negative consequences obviously include the costs of short-term and long-term medical care, earnings lost due to time in treatment and residual disabilities, as well as loss of income support for families in the event of the premature death of subjects.

However, monetary awards for pain and suffering are more controversial. On one hand, it is clear that pain and suffering may result from compensable injuries and may seriously erode the capacity of persons to pursue their goals. On the other hand, there are practical difficulties in devising a benefits schedule for pain and suffering. One problem is that degrees of pain and suffering can differ substantially for different persons with similar compensable injuries. Another is measuring the pain and suffering experienced by specific individuals. A third problem is that, when pain and suffering is genuinely disabling, much of its harmful impact will be ameliorated by benefits for items such as the cost of medical care and lost wages. Isolating the residual impact of pain and suffering is exceedingly difficult. These problems have led some commentators to suggest that providing compensation for pain and suffering poses insuperable administrative burdens. Nevertheless, one simple solution would be to provide a fixed percentage of the other total benefits as a payment for pain and suffering. This would avoid the administrative complexities of individual determinations, while acknowledging the obligation to rectify its harmful consequences (27).

The status of compensatory justice as a prima facie societal obligation also affects selection of the types and amount of benefits. Other obligations of social justice may limit total government funds that are properly allocated to the compensation of injured subjects. As a result there may be overriding moral reasons for limiting the types or amounts of benefits for injured subjects. Once again, such limitations would require establishing priorities among the types of benefits to be provided.

Another relevant variable in the design of a compensation program concerns the scope of covered research activities. According to Childress, the societal obligation of compensatory justice arises only for research done at the behest of society, as sanctioned through official government agencies (18). This condition reflects the notion that society should redress sacrifices that it compels or encourages for the common good. In accepting this proposition, the HEW Secretary's Task Force recommended that a compensation program cover injuries to subjects in research conducted or supported by the federal government. It also recommended that FDA consider legislation to require compensation in research conducted under regulations for testing the efficacy and safety of new drugs and medical devices (13). The President's Commission agreed that the societal obligation to offer compensation is strongest when the nexus between the government and the research activity is closest. However, the Commission recommended that the pilot compensation program include only research conducted or supported by the federal government, while requesting voluntary participation by private research institutions conducting FDA-regulated research (14). Thus research funded by private philanthropic organizations is not covered by the societal obligation of compensatory justice in the view of either advisory body.

This conclusion can be challenged by reconsidering the factors that determine whether research is conducted with the official sanction of society (1). Privately funded, non-regulated medical research may produce generalizable medical knowledge. The results are typically published in peer-reviewed, academic medical journals. The information is available to other investigators in designing their own research studies. Insofar as the use of these results is permitted in designing research conducted, supported or regulated by the federal government, it might be concluded that these privately funded research studies are officially sanctioned by society. If this argument is compelling, then equity requires that subjects injured in privately funded medical research be covered by society's obligation of compensatory justice. Legitimate worries about the prospective social control of such research could be addressed by extending the use of institutional review boards to all human subjects research occurring within the United States.

A final critical design variable concerns the source of funding for the compensation program. There are two basic alternatives: governmental or nongovernmental funding. A federally funded program would likely be administered by a government agency, although an alternative might involve provision of funds in research grants for institutions to secure private insurance. Nongovernmental funding would involve statutory or regulatory standards for compensation programs, with research institutions being required to provide benefits through self-insurance, private insurance, or cooperative insurance pools. Some combination of funding might also be possible, with research institutions securing basic coverage for compensation benefits and the government providing backup insurance against catastrophic losses.

There are complex issues regarding what funding mechanism would permit the most efficient and effective administration for a compensation program (28). The moral implications of these alternatives are more limited. If the moral justification for a compensation program lies in its beneficial consequences, then alternative funding mechanisms may create incentives that differentially impact on achievement of these goals. For example, if a compensation program is justified as a mechanism for controlling exposure of subjects to risks and for encouraging surveillance of their safety, then a program in which research institutions purchase private insurance is more likely to provide incentives for safe practice. The costs of coverage for research institutions will depend partly on their success in limiting the frequency and magnitude of compensable injuries to subjects. An alternative moral justification for redressing research injuries is based on a societal obligation of compensatory justice. In this case the primary goal is to ensure that benefits are provided to injured subjects. Funding by the government rather than research institutions is more likely to encourage investigators to assist subjects in identifying compensable injuries and securing adequate benefits. If investigators must worry about institutional expenditures for compensation provided through private insurance, then incentives to discourage legitimate claims are created. The modest level of claims in the few compensation programs funded by research institutions suggests that cost consciousness inhibits advocacy for subjects with compensable injuries.

SUMMARY

Although initial discussions of compensation for research injuries were stimulated by research whose conduct violated the rights or welfare of subjects, subsequent focus has been on injuries occurring without negligence.

Numerous government advisory panels have accepted the moral claim that society has an obligation to compensate human subjects for injuries incurred when they assume a position of risk in officially sanctioned research activities contributing to the common good. Despite the intuitive plausibility of this proposition, implementation of a national compensation program has been stymied by a series of vexing issues. The applicability of the obligation to subjects injured by therapeutic research procedures has been widely challenged. A compelling need for a federal program has never been clearly established. Complex issues arise in defining compensable injuries, determining appropriate benefits, clarifying the scope of covered research, and designing a suitable funding mechanism. Implementation of federal regulations for the prospective review of research protocols has substantially reduced incidences in which the plight of injured subjects has received public attention. In an era of shrinking federal programs and without public perception of a palpable need, the prospects for a national compensation program are not promising.

BIBLIOGRAPHY

1. B.R. Adams and M. Shea-Stonum, *Case Western Reserve Law Rev.* **25**, 604–648 (1975).

2. G.R. Smith, in HEW Secretary's Task Force on the Compensation of Injured Research Subjects, *Report*, Appendix A, DHEW Publ. No. (OS) 77-004, U.S. Department of Health, Education, and Welfare, Washington, DC, 1977, pp. 27–42.

3. R.J. Levine, *Ethics and Regulation of Clinical Research*, 2nd ed., Urban & Schwarzenberg, Baltimore and Munich, 1986.

4. D.J. Rothman, *Strangers at the Bedside*, Basic Books, New York, 1991.

5. H.K. Beecher, *N. Engl. J. Med.* **74**, 1354–1360 (1966).

6. J. Jones, *Bad Blood: The Tuskegee Syphilis Experiment*, 2nd ed., Free Press, New York, 1993.

7. W.J. Curran, in P.A. Freund, ed., *Experimentation With Human Subjects*, George Braziller, New York, 1969, pp. 402–454.

8. U.S. Department of Health and Human Services, *Fed. Regist.* **46**(16), 8366–8392 (1981).

9. I. Ladimer, *J. Chronic Dis.* **16**, 1229–1235 (1963).

10. C.C. Havighurst, *Science* **169**, 153–157 (1970).

11. G. Calabresi, in P. Freund, ed., *Experimentation with Human Subjects*, George Braziller, New York, 1969, pp. 178–196.

12. Tuskegee Syphilis Study Ad Hoc Advisory Panel, *Final Report*, U.S. Department of Health, Education and Welfare, Washington, DC, 1973.

13. HEW Secretary's Task Force on the Compensation of Injured Research Subjects, *Report*, DHEW Publ. No. (OS) 77-003, U.S. Department of Health, Education, and Welfare, Washington, DC, 1977.

14. President's Commission for the Study of Ethical Problems in Medicine and Biomedical and Behavioral Research, *Compensating for Research Injuries*, vol. 1: Report, Stock No. 040-000-00455-6, U.S. Government Printing Office, Washington, DC, 1982.

15. National Commission for the Protection of Human Subjects of Biomedical and Behavioral Research, *Report and Recommendations: Institutional Review Boards*, DHEW Publ. No.

(OS) 78-0008, U.S. Government Printing Office, Washington, DC, 1978.

16. U.S. Department of Health, Education and Welfare, *Fed. Regist.* **43**(214), 41449 (1978).

17. Advisory Committee on Human Radiation Experiments, *Final Report*, Stock No. 061-000-00-848-9, U.S. Government Printing Office, Washington, DC, 1995.

18. J.F. Childress, *Hastings Cent. Rep.* **6**(6), 21–27 (1976).

19. H.T. Engelhardt, Jr., in HEW Secretary's Task Force on the Compensation of Injured Research Subjects, *Report*, Appendix A, DHEW Publ. No. (OS) 77-004, U.S. Department of Health, Education and Welfare, Washington, DC, 1977, pp. 45–63.

20. B.R. Boxill, in President's Commission for the Study of Ethical Problems in Medicine and Biomedical and Behavioral Research, *Compensating for Research Injuries*, vol. 2: Appendices, U.S. Government Printing Office, Washington, DC, 1982, pp. 41–55.

21. H.M. Smith, in President's Commission for the Study of Ethical Problems in Medicine and Biomedical and Behavioral Research, *Compensating for Research Injuries*, vol. 2: Appendices, U.S. Government Printing Office, Washington, DC, 1982, pp. 19–39.

22. T.F. Ackerman and A.M. Mauer, *N. Engl. J. Med.* **305**, 760–763 (1981).

23. P.V. Cardon, F.W. Dommel, and R.R. Trumble, *N. Engl. J. Med.* **295**, 650–654 (1976).

24. D. McCann and J.R. Pettit, in President's Commission for the Study of Ethical Problems in Medicine and Biomedical and Behavioral Research, *Compensating for Research Injuries*, vol. 2: Appendices, U.S. Government Printing Office, Washington, DC, 1982, pp. 241–274.

25. J.D. Arnold, in President's Commission for the Study of Ethical Problems in Medicine and Biomedical and Behavioral Research, *Compensating for Research Injuries*, vol. 2: Appendices, U.S. Government Printing Office, Washington, DC, 1982, pp. 275–302.

26. C.J. Zarafonetis et al., *Clini. Pharmacol. Ther.* **24**, 127–132 (1978).

27. C.C. Havighurst, in HEW Secretary's Task Force on the Compensation of Injured Research Subjects, *Report*, Appendix A, DHEW Pub. No. (OS) 77-004, U.S. Department of Health, Education, and Welfare, Washington, DC, 1977, pp. 81–132.

28. R. Zeckhauser, in HEW Secretary's Task Force on the Compensation of Injured Research Subjects, *Report*, Appendix A, DHEW Publ. No. (OS) 77-004, U.S. Department of Health, Education, and Welfare, Washington, DC, 1977, pp. 155–166.

See other HUMAN SUBJECTS RESEARCH entries.

HUMAN SUBJECTS RESEARCH, ETHICS, FAMILY, AND PEDIGREE STUDIES

GEORGIA L. WIESNER
SUSAN LEWIS
JENNIFER SCOTT
Case Western Reserve University
University Hospitals of Cleveland
Cleveland, Ohio

INTRODUCTION

Our current understanding of human genetic disease is a direct result of family studies research and the explosion of genetic technology over the last half-century. The benefits of this research have been enormous, and range from the discovery of specific genes causing common diseases, such as cancer or heart disease, to a deeper understanding of the mechanics of how environmental factors interact with the human genome or set of genes. However, these advances in genetic technology often alarm the lay public. Indeed, one harsh view is that scientists "have given no more thought to the potential social applications of genome mapping and sequencing than Victor Frankenstein had given to the consequences of creating his monster..." (1, p. 660). Nevertheless, a revolution in biomedical technology and genetics has taken place in recent years, so that genetic studies based on family research are now the mainstays of biomedical research.

The genetic revolution now enjoys governmental support through the Human Genome Project (HGP) (2), a large international effort to sequence and map each of the 24 human chromosomes as well as to understand the underlying genetic variation and function of the genome (3,4). Indeed, HGP is a monumental undertaking that seeks to unravel the human "genetic code" discovered only 47 years ago by Watson and Crick (5). Francis Collins, the current head of the National Human Genome Research Institute (NHGRI) in the National Institutes of Health (NIH) describes HGP as the "single most important project in biology and the biomedical sciences — one that will permanently change biology and medicine" (3).

Both supporters and detractors of HGP predicted that this undertaking will have unprecedented effects on the social fabric of the human community (1). In response to these concerns, the analysis of the ethical, legal, and social implications of this new genetic knowledge has become a vital component of HGP. According to Eric Juengst, the first chief of HGP's Ethical, Legal and Social Implications (ELSI) branch (italics added):

> ...the Human Genome Project represents geneticists' growing ability to explore human heredity. It thus generates a wide range of charged questions about how our society's genetic explorations should proceed and how their results should be used. These questions include a set of *unresolved issues regarding the conduct of research involving human subjects.* While these issues predate the Human Genome Project and would continue to exist without it, they are becoming increasingly important as genomic tools and *genetic strategies become pervasive* in biomedical research (6, p. 401).

Juengst and others have thus identified two essential themes for human genetics research (6,7). First, researchers who conduct genetic studies will face new ethical challenges that will require the implementation of specific strategies in the study protocol addressing these issues. Second, because more common diseases and traits will be studied in the future, genetic research will increase in scope and complexity (6,8).

Many human genetic studies such as gene-searching projects sponsored by HGP are based on the analysis of the family group. These are collectively known as pedigree or family studies and are an integral component of modern genetic research. Family pedigree studies were previously considered to have little, if any, inherent ethical conflicts by researchers in the field because the studies were often based on observations of a rare condition in a small number of families. In joining a proposed research study, these families were eager to discover the genetic link to their condition, and often worked closely with the research team. The potential impact of this research on such rare conditions was limited to a few families with these unusual disorders. In contrast, present day family pedigree studies are often focused on the familial nature of common diseases and are conducted on a national and international scale. Thus these studies can potentially impact many more people in communities across the country. Contemporary family pedigree projects are a hybrid between standard epidemiological research and molecular analysis of gene function, in which large numbers of individuals and their family members are enrolled to identify genes that cause human disease.

A crucial difference between family pedigree studies and other biomedical research lies in the collection and analysis of information on family groups rather than unrelated individual volunteers (9). Epidemiologic protocols analyze clinical information on a large number of cases, and thus the risks and benefits of these studies stem from the potential impact on an individual volunteer. In contrast, in family pedigree research, the information gained from one family member can lead to consequences for other family members (6,9). Holtzman and Andrews further argue that genetic research "is different because it often involves testing, and thus *creates* genetic information about individuals and groups that did not exist before" (10). Thus the ethical issues arising from family pedigree research may be inherently more difficult for researchers in the biomedical community. This difficulty may stem from the lay public's view of genes and heredity. Henry T. Greeley, professor of law at Stanford University writes:

> Rightly or wrongly, many people are convinced that genes are special, that they contain and reveal a person's, or a people's, essence, which has enormous value, spiritual and commercial. This exaggerated emphasis on the importance of individual genetic variation makes human genomic research particularly sensitive (11, p. 625).

Participating in genetic research can have significant psychosocial consequences for research volunteers (10). By simply enrolling in a pedigree research study, participants may experience unforeseen psychosocial anxiety and depression. These feelings stem from concern over the propensity to develop a disease, potential genetic discrimination, and the consequences of medical decisions based on perceived risk for disease (6,7,10). Human geneticists who study rare monogenic disorders have recognized these psychosocial issues for many years (12). Current and future researchers aided by the enormous growth in genetic research will focus on common disease processes. In this way a greater proportion of the general population could potentially learn they are at risk for health problems because a relative participated in a family pedigree study. Thus the potential for unforeseen harm to participants and their family members will increase, as genetic family-based studies become more prevalent.

The explosion in the number of family pedigree projects has generated a lively debate in the biomedical community as to the proper ethical conduct of genetic research. Most ethicists agree that the Belmont Report defines the moral guidelines for researchers who enroll human subjects in biomedical research. The three overarching principles contained in this document are respect for persons, beneficence, and justice for human subjects (7,13). However, the authors of the Belmont Report did not specifically address issues relating to use of genetic material and the potential impact of family pedigree research. Several organizations, such as the American Society of Human Genetics (ASHG), the National Bioethics Advisory Commission (NBAC), and the Institute of Medicine have recognized the need for a standardized policy for the conduct of genetic research (14–18). The Institute of Medicine's Committee on Assessing Genetic Risks summarized the relevant ethical principles for genetic research subjects and recommended "vigorous protection be given to autonomy, privacy, confidentiality and equity" (15).

Using unconfirmed research findings in medical decision-making is also a major ethical concern for genetic researchers and has prompted several policy statements by interested groups (17,19). This issue gained national attention in the mid-1990s when BRCA1, the first susceptibility gene for early-onset breast cancer was identified and cloned (20). Most researchers and lay persons viewed the application of genetic markers to forecast the risk of future disease as a positive advance in biomedical sciences (21). However, some groups felt that health decisions based on unsubstantiated genetic research without confirmatory clinical trails could be harmful. Recognizing the uncertainty arising from the use of the new genetic BRCA1 research results, ASHG strongly advised (italics added) "it was *premature* to offer population screening" for BRCA1 gene testing (17). This issue has been raised with other gene discoveries, such as the ApoE link with Alzheimer's disease, since the health implications of genetic research are not limited to breast cancer research (22). Most authors now recommend that genetic results should ideally be provided within the context of clinical care after the health implications are studied, although the pressure to hasten the transition of new genetic findings into the medical arena is pervasive (17,23,24). The blending of scientific discoveries with the potential health benefits of the new research represents a potential conflict for human genetic researchers, who are then cast in the dual role of healer and scientist.

The time is long past when scientists were able to conduct genetic research involving individuals and their families isolated from the ethical and social impact of the research process and eventual findings of the study. In discussing the transition of research findings to patient care, Ray White from the University of Utah states:

> Human geneticists have a problem. Finally, after years of effort, we are beginning to resolve and identify the genetic components of a number of genetically transmitted disorders and predispositions. On the eve of this scientific triumph, however, at a time when we should be delivering this new knowledge to affected individuals, we have instead discovered that this delivery is compromised by social, economic, and ethical issues (25, p. 173).

It is thus imperative to incorporate ethical strategies into all aspects of study design. These strategies should consider the ascertainment of subjects and family members, information supplied to study participants, control and databasing of study data, disclosure of results, publication of data, and communication among researchers.

This article addresses the many ethical dilemmas faced by genetic researchers who perform family pedigree research. There is a growing realization that conflicts arising from family studies will be encountered in other types of genomic research projects. These genomic studies explore a variety of genetic topics such as the translation of genetic discoveries into clinical practice or analyses of genetic variation between specific human ethnic groups and subpopulations (26). Thus equal attention to the medical and psychosocial impact of the genetic research must continue in parallel with the exciting biomedical and genomic research of the future.

GENETIC RESEARCH ON HUMAN POPULATIONS

A major goal of genetic research is to understand the hereditary factors that cause human disease. Studies based on the family unit have been essential to this research over the last half-century and will continue to form the foundation for future investigations (27). Several older methods, such as twin and family pedigree studies, were developed before the introduction of molecular technology, emphasizing the notion that important genetic

information can be gleaned from an individual's family history alone. Geneticists then developed sophisticated study methods, such as linkage analysis and marker association studies, to exploit newly discovered molecular markers for the analysis of family information. While the types of analyses and research goals differ, a common feature of these study methods is the collection and statistical analysis of a trait, or phenotype, in multiple family members. Although most biomedical genetic research focus on specific disease process, the amount of clinical information gathered from each family member and the extent of the family history is variable for each study design (Table 1). For example, twin studies gather complete information about both twins but may not collect information about other relatives. In contrast, family pedigree or linkage studies collect clinical information about many members in the extended family.

Twin studies have been used extensively by researchers to support or refute the genetic nature of a specific disease process or physical attribute, such as height. This method compares the prevalence of a disease in identical or monozygotic (MZ) twins and in fraternal or dizygotic (DZ) twins. Since MZ twins share 100 percent of their genetic material, it is logical to expect that both members of an MZ twin-pair would develop the condition if an underlying gene causes the disease. However, both members of a DZ twin-pair would be less likely to be affected as they only share about 50 percent of their genes. This approach is the basis of a 1991 twin study demonstrating a strong genetic component for the development of asthma (28). MZ twins participating in the study both suffered from asthma or allergies 80 percent of the time compared to 0 percent of the DZ twin volunteers. This asthma study illustrates that genetic information can be gained from the analysis of phenotypic and family information alone, since the researchers did not analyze genetic material from the twin participants.

The risk of a family member developing a specific disease can also be estimated from family-based studies without the analysis of DNA markers. One example of the clinical usefulness of family information is from the

Table 1. Types of Human Studies Used by Researchers to Determine the Genetic Component of Human Disease

Study Design	Human Subjects	Use of Family Information	Use of Genetic Material
Twin	Monozygotic and dizygotic twins	Minimal	No
Segregation analysis	Multigenerational families	Extensive	No
Linkage analysis	Multigenerational families	Extensive	Yes
	or		
	Relative pairs (i.e., sibling pairs)	Immediate family members	Yes
Molecular epidemiologic	Individual cases and nonrelative controls	Minimal	Yes

National Polyp Study, a multicentered clinical trial that examined family history as a risk factor for colon cancer (29,30). The study showed that a family history of colon cancer and colon polyps increased the risk of developing colon cancer two- to threefold over cases with no family history of the disease (30). This study compared the trait in question in affected persons to the occurrence of the trait in biological family members. Importantly, the results from this study were then used to develop risk profiles for colon cancer, illustrating the far-reaching effect of genetic research in clinical practice.

Researchers use the twin and family history study methods to investigate whether hereditary factors play a role in a particular condition, but these studies cannot be used to establish a pattern of inheritance based on Mendel's laws. Family pedigree studies are required in order to determine whether a trait is segregating in a specific pattern of inheritance in the pedigree. Most hereditary conditions typically follow an autosomal dominant, autosomal recessive, X-linked dominant or X-linked recessive pattern, although nontraditional patterns have been described. In order to determine the inheritance pattern or segregation of a gene or trait in a family group, researchers must rely on studies of the extended family (27). An essential ingredient of such studies is the construction of a family tree from the proband or other informants that includes detailed information about affected and unaffected family members over several generations. The research group then uses this multigenerational information to examine whether a gene is responsible for the trait segregating in the affected kindreds. This type of study, called segregation analysis, is a genetic epidemiological research method designed to measure the likelihood, or chance, that a hereditary factor causes the trait or disease in question (27,31). Because entire family groups are analyzed, segregation analysis can determine the inheritance pattern of a disease. This type of study is often the first piece of evidence tying hereditary factors to a disease process. For example, segregation analyses of several hundred families with breast cancer in the female relatives first suggested the existence of a gene or genes responsible for familial breast cancer (32). On the basis of these studies, researchers were then able to estimate that the gene would be carried by 1 in 500 individuals and would be transmitted as an autosomal dominant trait to other family members (33,34). Likewise segregation analyses showed that a major gene segregating in high-risk families most likely caused Hirschsprung's disease, a form of congenital megacolon (35).

Neither twin studies or segregation analyses are used to identify the exact gene causing a particular disease. However, the chromosomal location for the disease-causing gene can be found when family pedigree information is jointly analyzed with genetic markers, or molecular signposts, from multiple members of a family (36). This powerful genetic method is called linkage analysis because statistical tests are used to "link" the disease exhibited by affected family members to known molecular markers found at regular intervals along each human chromosome. These markers are now easily analyzed in a small sample of genetic material, or DNA, from the person being tested. Since the researchers know the chromosomal location of each marker, the disease-causing gene can be "mapped" to a specific genetic region. Linkage studies, otherwise known as mapping studies, rely on the ability to distinguish between a chromosomal region inherited from one's father and the same chromosomal region inherited from one's mother. The strength of this type of study stems from the ability to track these genetic regions between the parent and child within the family. One of the important advances in molecular technology has been the development of highly informative molecular markers for genetic mapping studies that allow the researcher to distinguish the paternal and the maternal copies of a particular genetic region.

Up to 400 separate genetic markers can be used to blanket the entire genome for mapping studies searching for disease causing genes. The markers can be used for candidate gene analysis or whole scale genome scan. Candidate gene analysis is used when the researcher suspects a known gene causes the disorder. The researchers must also know the chromosomal location of the suspected gene, and will test markers that are physically close to the suspected gene. In contrast, a genome scan is used when the location of the gene or genes is not known. These scans test numerous polymorphic markers diffusely located across the entire genome, resulting in the generation of enormous numbers of genotypes derived from each individual and family group.

Several computational methods are employed for linkage analysis each differing in the amount of family information required for the study. Thus the family groups used for linkage analysis range from large multigenerational kindreds to a smaller number of relatives, such as the affected sibling pairs method (31,36). Contemporary family pedigree linkage studies are thus very complex and require experts from a wide variety of disciplines in order to be successful (9). Model-dependent linkage analysis has been very successful in identifying the genes causing many diseases, such as cystic fibrosis, breast cancer susceptibility, and Huntington's disease. In this study design, clinical and genetic information on multigenerational families are studied under assumptions of monogenic autosomal recessive or autosomal dominant inheritance. The results are provided in a log of odds (LOD) score that signifies whether the genetic region is linked or unlinked to the disorder under study. Model-independent methods have also been developed that analyze the genotypes of pairs of family relatives, such as sibling pairs, or parent–child pairs. LOD expresses the chances that a marker is associated, or linked, to the phenotype under study. LOD scores of greater than three signifies that a particular marker has 1000 to 1 odds of occurring by chance alone and is generally used as evidence for linkage. A LOD score of negative 2 or less is generally accepted as evidence against linkage, and LOD scores between 3 and negative 2 is considered unclear for linkage (36).

The components of a linkage study are diagramed in Figure 1 and consist of family recruitment, molecular laboratory, and statistical analysis groups. The recruitment

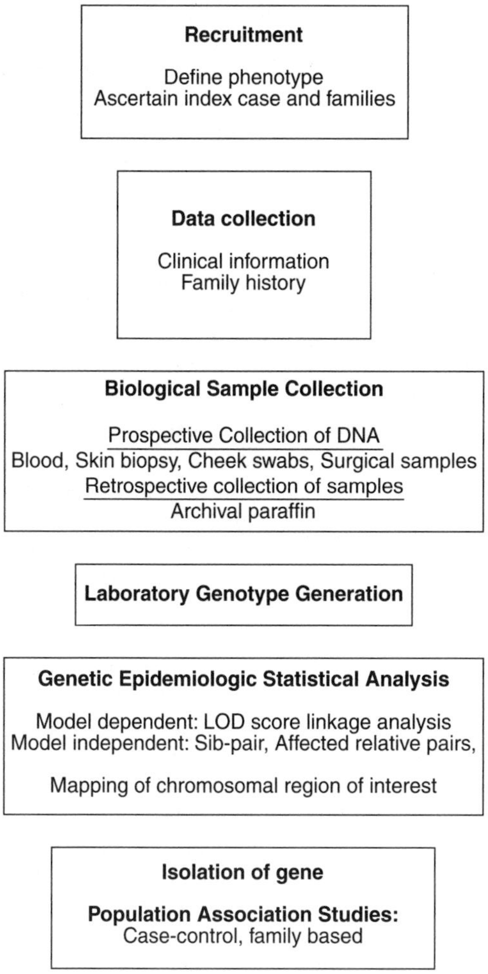

Recruitment

Define phenotype
Ascertain index case and families

Data collection

Clinical information
Family history

Biological Sample Collection

Prospective Collection of DNA
Blood, Skin biopsy, Cheek swabs, Surgical samples
Retrospective collection of samples
Archival paraffin

Laboratory Genotype Generation

Genetic Epidemiologic Statistical Analysis

Model dependent: LOD score linkage analysis
Model independent: Sib-pair, Affected relative pairs,

Mapping of chromosomal region of interest

Isolation of gene

Population Association Studies:
Case-control, family based

Figure 1. Components of a genetic pedigree or linkage study.

component includes experts in family contact, medical information retrieval, and sample collection. The laboratory component includes technical experts in sample processing and molecular analysis. Experts in genetic epidemiologic methods then carry out the linkage analysis in the statistical analysis component. All these components must be coordinated so that the medical, family, and genetic information can be used to identify disease-causing genes. The common feature of family pedigree linkage studies is the analysis of clinical information and blood samples from several family members in order to assess whether a genetic region is linked to the disease under study.

After a linkage study identifies a possible chromosomal location of a disease-causing gene, the gene is then isolated and identified with more refined genetic techniques and analysis, including positional cloning and sequencing (37). A full explanation of these techniques is beyond the scope of this chapter, except to note that genes are analyzed by a number of technologies. One method, called *DNA sequencing*, determines the exact order of the chemical building blocks, or nucleotides, of the gene. DNA sequencing also allows the researchers to identify genetic changes in an individual compared to the usual sequence found in the general population. Some of these

changes are true deleterious mutations in the genetic code, in that the normal function of the gene is altered in the person carrying the mutation (37). However, other gene changes, called *polymorphisms*, may not have real functional significance. Polymorphisms in DNA sequences are very common and are thought to have little impact on human diseases.

Determining whether a gene change is a deleterious mutation or a polymorphism is a common problem for genetic researchers. If a potential polymorphism or mutation is found in a putative disease-causing gene, researchers use several techniques to determine the true effect on gene function. One of these techniques is to study multigenerational families with the disease to test whether the gene change correlates with the phenotype of affected individuals. A second method for researchers to study the phenotypic effect of a potential mutation is to perform molecular association studies examining the occurrence of the genetic marker in larger populations. The goal of these studies is to correlate the potential mutation in groups of unrelated affected and unaffected individuals (38). While most molecular epidemiologic studies do not use information about family relatives, these studies collect and analyze biologic samples for DNA analysis. Molecular epidemiologic studies illustrate the expanding role of genetic techniques in biomedical research, which can be used to provide information about disease-related risks in clinical practice. As knowledge about the human genome increases, future studies will focus on the interaction of several genes as well as the interaction of specific genes and environmental influences as necessary steps for disease development. Thus the methodology and technology of genetic research will become more complex in the future.

Accurate family data and information is the first necessary step in understanding human hereditary. As HGP nears its goal of determining the exact DNA sequence of the human genome, gene identification will be streamlined so that the causal genes will be identified at a faster rate in the future. Indeed, many genes will be identified without knowledge of the biological function or role so that researchers will continue to rely on family participation to understand the gene function in the population. Thus, with the advent of DNA markers for linkage analysis and genomic research, family studies have become the workhorse of modern genetic discovery. While future genetic research will continue to study rare monogenic disorders, the research focus will shift to include the study of genes that contribute to more common diseases, such as cancer, diabetes, cardiovascular disease, and aging. In this way family pedigree studies will then become more pervasive in the biomedical community (6).

POTENTIAL HARMS FOR SUBJECTS ENROLLED IN FAMILY PEDIGREE STUDIES

The wealth of information derived from genetic research is unprecedented and has given the biomedical community many new tools to understand and treat human disease. Genetic studies are no different from other forms of biomedical research, in which the benefits should outweigh the potential harms to each volunteer. However, as Phillip Reilly notes, "that gene-discovery studies posed the threat of genetic discrimination—that there is a risk of *informational harm* associated with participating in studies that elicit genetic information from which one might infer health status" (39, p. 683). This "informational harm" is different from the usual physical harms that may result from participation in other types of biomedical research, such as clinical trials of new treatments or drugs. Informational harm from genetic research may cause emotional or social difficulties for research volunteers, including the impact of new genetic knowledge on health risks, family dynamics, and possible social stigmatization. Participants in genetic family pedigree studies are also at risk for unintended consequences of the research process. These unintended consequences can include misidentified parentage, learning private information about a family relative, having a research sample entered in additional studies without their knowledge or consent, or finding risk for additional diseases not part of the initial study (9,10,40).

Genetic Discrimination

Genetic discrimination is viewed by the general public as a major threat for individuals who enter genetic research studies (39). In particular, there is concern that genetic information obtained through the analysis of human tissue samples could be used to discriminate against individuals by insurers or employers (41). While the fear of stigmatization has generated a national debate on genetic privacy, it should be noted that few studies have scientifically examined this issue (14,42,43). In one of the first studies to document the occurrence of employment stigmatization and insurance abuse, Billings and colleagues reviewed 41 separate instances of discrimination submitted by genetic professional and patient advocacy groups (44). These authors defined genetic discrimination as "discrimination directed against an individual or family based solely on an apparent or perceived genetic variation from the 'normal' human genotype" and concluded that genetic discrimination is found in many social institutions (44). This report was controversial, as representatives of the health insurance industry argued that the number of discrimination reports were relatively small compared to the thousands of policies issued every year on a national basis (45). Other groups felt the reported instances of discrimination were primarily anecdotal and unsubstantiated by the authors, or that the cited examples did not conform to the authors proposed definition of genetic discrimination (46).

Nevertheless, the level of concern about potential discrimination is very high as shown by a 1997 study of over 1000 geneticist and primary care physicians, who reported over 550 instances of employment or life insurance refusal (43). This report echoes the sentiments of 332 members of the genetic support groups affiliated with the Alliance of Genetic Support Groups (47). This survey documented that up to 43 percent of respondents felt they experienced some form of discrimination by health insurers, life insurers, and employers, including refusal of life or health insurance or employment denial.

There is also concern over genetic discrimination on an international level, as evidenced by a similar survey of genetic support groups in the United Kingdom. This survey found that one-third of the study respondents had difficulty when applying for life insurance compared with 5 percent of the control participants (48). The respondents who perceived themselves as suffering from discrimination reported that they experienced rate increases or outright refusals for insurance.

It must be remembered that the public's fear of genetic discrimination is not unfounded, as past abuses and stigmatization based on eugenics and physical disabilities are a matter of public record (1). In the United States these abuses ranged from the forced sterilization programs for persons with physical disabilities to the ill-advised sickle cell anemia screening program for African-Americans (15,49). Holtzman and Rothstein point out that the sentiments of the eugenic movement in the early part of this century still resonate in today's social institutions (50). In fact a 1998 U.S. government report estimated that 15 percent of employers plan to inquire into the genetic status of employment applicants (14). Thus the fear of marginalizing individuals on the basis of their unique genotype prompted policy review and recommendations on a federal level. Several governmental working groups and special commissions have been established to formulate specific policy agenda items relating to the social impact of genetic information (14,51). One of the first working groups was the Task Force on Genetic Information and Insurance sponsored by NIH and Department of Energy (42,51,52). NIH charged this group with studying the social implications of genetic discrimination by health insurance companies. Their 1993 report warned "people will be asked to provide information about their genetic risks to insurers" (51). The Task Force also noted that the risk of losing health insurance coverage for "preexisting" conditions may prevent people from obtaining predictive genetic information that could be used to improve the health and welfare of the person and family.

The Task Force concerns were based on the health insurance risks for people with a genetic condition or a family history of the disease, and most instances of documented discrimination have followed genetic diagnoses made in the clinical setting. However, some authors have suggested that study volunteers might also be required to disclose results from genetic studies in the research setting (53,54). Thus genetic information gained through voluntary participation in a family pedigree study might place the subject at risk for economic harm. This is an issue of *distributive justice* according to Thomas H. Murray, who writes that:

> Human genetics is, from this perspective, a science of human inequality. The principal ethical problem created by such scientific pronouncements of human difference is the task of reconciling such differences with our central moral, political and legal commitments to treating people as equals. That is, we must reconcile the ever-increasing evidences of human inequality with our vital commitment to moral equality (55, p. 80D).

For the research community it is essential to recognize the potential harms of genetic discrimination due to the scientific process. Researchers should develop appropriate study protocols to alert research participants to the possibility of informational harm. However, the magnitude of this risk is currently unknown, and may be part of an "urban myth" (56). A recent abstract presented to the plenary session of the 1999 ASHG national meeting found little evidence for genetic discrimination in a review of applications to 143 health insurance agents (57,58). In addition the study found no indication that insurers were using genetic information for health prediction and risk stratification. However, there have been no systematic surveys of the underwriting practices of life or disability insurance providers, hiring practices of employers, or services provided by social agencies such as housing or adoption. Thus the extent of insurance and employment discrimination based on genetic grounds is currently unknown. The research community must remain vigilant as to the potential economic and social harms to their study participants from inadvertent or premature disclosure of results from genetic studies.

Impact of Susceptibility Gene Identification

Pedigree linkage studies have been very successful in identifying genes responsible for human disease. Physicians are now able to use newly developed DNA tests to diagnose a suspected genetic condition, illustrating one of the benefits of genomic research. One important example in which a gene-based test has supplanted older biochemical testing is RET gene testing for multiple endocrine neoplasia type 2 (MEN2) (59). Gene carriers of this autosomal dominant condition have a 90 percent chance of developing medullary thyroid cancer, a potentially lethal cancer that can strike in childhood and early adulthood. Ninety-five percent of all cases of MEN2 are caused by deleterious mutations in the RET gene. A highly accurate DNA test for MEN2 is commercially available and has replaced the older biochemical calcitonin test used to diagnose this disease. DNA testing can be now offered to healthy at risk family members to determine if they also have a RET gene mutation. RET gene testing also illustrates the benefits of family pedigree research, in that genetic scientists used information from family pedigrees to identify a gene causing lethal thyroid cancer, paving the way for a simple medical test for the entire community. It should be remembered, however, that in order for such studies to be successful, many human volunteers donated their clinical information and blood samples to scientists studying MEN2. In addition volunteers helped the researchers to develop the clinical standards of care before the test could be provided to the general public.

While RET gene discovery is a clear example of the benefits of gene discovery, it is important to recognize that genetic tests are not absolute predictors of health or disease (40). This notion is in stark contrast to the commonly held view of genetic determinism, which suggests a person's genetic makeup is an unalterable blueprint for the future. Genetic determinism ignores the interaction of multiple genes, or environmental factors in developing disease (6,40,60). In addition proponents of this view neglect the underlying uncertainty about the health risks associated with susceptibility genes, in that the risks are more probabilistic in nature rather than an unequivocal link to the development of disease (6). The probabilistic health risks are true even for highly accurate gene tests, such as RET testing for MEN2 where most, but not all, gene carriers will develop thyroid cancer.

An additional reason that a person's genetic code can not be equated with a predetermined outcome comes from an understanding of the mechanism of action for deleterious genetic mutations. A deleterious mutation alters the expression of a single protein product produced by the cell. However, the cell may have several pathways that provide the same function, so that the effect of the deleterious mutation can be masked by the normal proteins encoded by other genes. Thus a one-to-one correspondence between gene mutation and phenotype, which is termed the *genotype–phenotype correlation*, is rarely seen for common conditions due to the complexity of the biological cellular pathways encoded by the genome. As such, disease-causing mutations will usually produce a recognizable phenotype. However, other gene changes, termed *susceptibility genes*, act to increase the propensity for the development of disease. Environmental factors are thought to interact with susceptibility genes to cause a specific disease. Thus the phenotypic effects of most genetic alterations falls somewhere on a continuum of risk for disease rather than an absolute cause of disease.

While categorizing mutations as disease-causing or susceptibility-causing is helpful in describing the potential effect of gene mutations, these labels are overly simplistic and thus are poor predictors for clinical disease. For example, individuals with neurofibromatosis type 1 (NF1) harbor a deleterious "disease-causing" copy of the NF1 gene on chromosome 17 (61). However, when the clinical phenotype of individuals with NF1 are compared, some exhibit the typical skin neurofibromas as teenagers, while others never develop this type of skin manifestation (62). This variation of clinical phenotype is known as *variable expression* of the gene and is thought to be a result of the interaction of the NF1 gene with unknown genetic or environmental factors. Nevertheless, the NF1 gene is fully penetrant. In other words, 100 percent of individuals who carry a deleterious NF1 gene mutation will exhibit symptoms of this disorder to some degree. However, because of variable expression, persons carrying a deleterious NF1 gene may exhibit very mild to severe health problems.

In contrast, other susceptibility gene mutations, such as mutations in BRCA1 and 2 breast cancer genes, will increase the propensity to develop breast, ovarian, or other cancers. Researchers estimate that between 50 and 85 percent of female BRCA1/2 gene carriers will develop breast cancer at some time in their lives, compared to general population risk of approximately 10 percent (63). In other words, since BRCA1/2 gene mutations are not fully penetrant, up to half of female gene carriers will remain cancer free and will be unaffected by their constitutional genotype. Other changes in genes, such as metabolic polymorphisms, are associated with minor functional effects of the protein product and can have mild to moderate effect on the risk for disease. One well-known example is the CYP gene family, which is associated with differing ability to metabolize drugs and medications and other ingested compounds. Several studies have shown metabolic polymorphisms are linked with an increased rate of cancers due to the differing metabolic rates encoded by the different CYP genes (38,64). Most experts caution against the use of such polymorphisms for risk prediction, but the pharmaceutical industry has recognized that these polymorphisms will be important in drug development and have invested heavily in this area of pharmocogenetics (65).

Identifying disease-causing genes will continue to be enormously beneficial in the clinical setting. However, in the research setting, there is a growing recognition that gene identification can have harmful effects on participants and their families. Researchers may find themselves in a situation in which the very success of the project can have a negative impact on the participating individuals and family members. The major liability of pedigree studies stems from the inference of health states from the untested genetic discoveries. Volunteers may receive preliminary and unproven health information about a specific genetic marker, since the necessary follow-up studies on the function and clinical impact of the newly discovered gene have yet to be conducted. Thus there is a potential for research volunteers to receive preliminary, incomplete, and potentially harmful information from participating in pedigree research projects. In addition participants may never learn about important future developments in clinical trials or may learn about important health risks months to years after entering a study. It should be noted, however, that there is little evidence documenting harm to volunteers from genetic research at this time.

Volunteers for genetic pedigree studies are thus participating in the first step of scientific discovery in which there is a potential for finding disease-causing genes. However, since genetic mutations cause a wide range of phenotypic effects, researchers must use discretion when deciding whether to disclose preliminary study results to volunteers. In some cases the participant may wish to be informed about their genetic result prior to the completion of clinical trials. The researchers may also wish to share the experimental results with certain participants believing that more harm would come from withholding research results (6,66). Disclosing research results is a difficult dilemma, placing the researcher's duty to nonmaleficence in conflict with the subject's autonomy. The potential uses of genetic testing for disease prediction was the overriding concern of the researchers who released premature BRCA1 genetic results to a young woman intending to undergo a prophylactic mastectomy based on her family history of breast cancer (66). After being informed of her genetic research results, she found that she did not carry the family's high-risk gene mutation and was able to avoid prophylactic surgery. Before providing her with these results, the researchers weighed the potential psychosocial harms of releasing untested preliminary information with the harm of undergoing an unneeded medical procedure. The researchers were also concerned

about alleviating the anxiety associated with genetic risks for disease. Anxiety and concern about health risks are well-recognized among family members at risk for Huntington's disease, Alzheimer's disease, and familial cancer syndromes (67).

Other researchers have cautioned that experimental findings from genetic studies should not be disclosed to the research subjects. The Children Cancer Group declined to disclose individual genetic research results to the physicians and parents of children enrolled in a genetic study of the p53 gene and childhood osteosarcoma (68). The p53 gene had been identified as the cause of cancer in families with the Li-Fraumeni syndrome in which multiple family members with osteosarcoma, leukemia, adrenal carcinomas, and other tumors (69). After careful consideration, this group chose to publish the genetic test results only in aggregate, as originally planned. They cited the lack of proven clinical utility of constitutional p53 gene mutations in nonfamilial cases, the potential for stigmatization and discrimination, and the subject's age as minors as factors in their decision. This group and other authors are concerned that predictive knowledge can be emotionally burdensome and can stem from knowledge of the family history as well as known carrier status from genetic testing (6,70). Thus the potential harm to research subjects from genetic knowledge cannot be neglected, and must be weighed when researchers are confronted with requests to divulge genetic research information.

Unintended Consequences of the Research Process

The problem of unexpected detection of new or secret information is also a concern for researchers who conduct family pedigree studies. The discovery of such knowledge by the researcher is an "unintended consequence" of the scientific process since this information is not related to the research goals of the study. This unexpected information can surface as a result of molecular analysis of the family blood samples and generally involves the inadvertent discovery of information that may not be known by all members in the family. For example, a serious, yet unintended, finding in genetic research is the discovery of mistaken parentage for a study volunteer, such as in nonpaternity or secret adoption. Since linkage analysis is dependent on distinguishing the maternal and paternal copies of each genetic marker, researchers can easily identify inconsistencies in the family when tracking the marker from parent to child. In this way the transmission of genetic markers is followed through subsequent generations and a family secret could potentially be discovered. While discrepancies in genetic markers can result from laboratory sample mix-up or other technical mistakes, a significant proportion is due to nonpaternity.

The rate of nonpaternity is estimated to be between 2.8 and 28 percent depending on the population group under study (71–73). While accurate figures are not known, it is not uncommon for researchers to confront this situation in family pedigree studies. Juengst outlines the dilemma of such a nonpaternity discovery for a research team studying a rare skin disorder (70,74). Subsequent publication of the pedigree revealed inconsistencies in the parental genetic markers showing nonpaternity in two children in the kindred. Reilly also described a situation in which nonpaternity was discovered in a gene–mapping family study (75). Following discussion with outside consultants, the researchers decided that they would not disclose this sensitive information to the family. In addition this group chose to omit the details of parental inconsistencies in a subsequent publication (75).

Another unintended research finding is the discovery of a genetic mutation associated with a risk or susceptibility for a disorder that is distinct from the original focus of the study. In this way a person may be surprised to find that they are at risk for unforeseen health problems. In fact this type of incidental finding has occurred in genetic studies of common health problems, such as cardiovascular disease, as well as studies involving rare disorders, such as Hirschsprung's disease. Greely summarized this complicated aspect of genetic research by stating:

> One gene may be associated with multiple diseases. Therefore, a person who takes a genetic test to learn something about one disease may end up with information, possibly unwanted or harmful, about another disease (22).

Cardiovascular researchers have known for years that carriers of some forms of ApoE, a lipid transporting protein in the circulating blood stream, moderately increases the risk for heart disease. Many people underwent ApoE screening on a research and clinical basis to determine if they were at a greater risk for cardiovascular problems. In 1993, a surprising association was made between the APOE gene and the risk for Alzheimer's disease, where APOE4 carriers were two to three times as likely to develop dementia later in life. Individuals who accepted ApoE testing in the context of their cardiovascular health then discovered, perhaps unwillingly, information regarding their risk for Alzheimer's disease later in life (reviewed in Ref. 22).

A similar dilemma occurred when researchers attempted to isolate the genes involved in Hirschsprung's disease, or congenital megacolon. As early as 1982 families with Hirschsprung's disease and a rare form of medullary thyroid cancer were identified (76). Medullary thyroid cancer is one of the cardinal features of MEN2 and can develop in young children or in early adulthood (59). As previously discussed, the gene responsible for MEN2 is the RET gene located on human chromosome 10. In 1994 linkage studies based on families with multiple cases Hirschsprung's disease found that one of the genes causing this disease was localized to chromosome 10 at the same location as the RET gene (77,78). Further studies confirmed that mutations in the RET gene were also responsible for some familial cases of Hirschsprung's disease (79,80). Volunteers in Hirschsprung family studies were identified to be at risk for a lethal form of thyroid cancer in addition to the childhood form of megacolon. Thus the APOE and RET gene discoveries were complicated by the unexpected detection of an increased risk for more than one disease process, each with different health implications.

Entire subpopulations can also be identified to have unforeseen health risks as an unintended consequence of the research process. One example of this is the identification of three founder mutations in the BRCA1 and BRCA2 genes in the Ashkenazi Jewish families with breast and ovarian cancer. Population studies subsequently found that three specific mutations, 185delAG and 5382insC in BRCA1 and 6174delT in BRCA2, are carried by about 2 percent of all Ashkenazi Jews regardless of a history of cancer in the family (81). Female carriers of any one of these mutations have a 50 percent chance to develop breast cancer as well as a 20 percent chance to develop ovarian cancer over their lifetimes. The effect of this gene discovery on people with Ashkenazi Jewish ancestry has been immense, since an entire ethnic subpopulation learned that certain members are at a higher risk for cancer (42,82). Because of the high prevalence of these founder mutations, some feel that strong consideration should be given for genetic testing for breast cancer risk based on Ashkenazi Jewish background. Others feel stigmatized by this discovery. Thus, because of family pedigree research and gene discovery, the autonomy and decision making for individual members of this subgroup was subverted. The unexpected detection of disease-causing genes will become a prominent issue for genetic research, as gene discovery in identifiable populations will become more prevalent in the coming years (26).

Breeches of Confidentiality

Researchers using clinical and genetic material from human subjects have a duty to respect autonomy and privacy of each participant (14). As previously discussed, this becomes problematic when the context of the study is the family unit in that "family secrets" such nonpaternity or private difficulties can be inadvertently disclosed to other family members, co-investigators, personal physicians, or employers (6,83). Since genetic linkage studies are generally multifaceted (Fig. 1) and composed of several discrete areas of expertise, the possibility of inadvertent disclosure through the process of the study is a distinct possibility. For example, recruiters for genetic studies may ask, with the participant's permission, to gather medically related materials from the subject's personal physician and medical file. Researchers conducting a linkage study for colon cancer susceptibility genes learned that a subject's personal physician noted in the subject's medical chart that the patient was "in a genetic study for colon cancer" (84). This illustrates the dissemination of information about study volunteers due to the interaction with healthcare workers rather than from the results or knowledge directly gained from the study. Likewise the research team may include identifying information about the participants in reports and data analysis generated by the study, leading to the suggestion that only some investigators in the research project team should have access to sensitive materials.

Once several family members enter a family pedigree study, considerable care must be taken to ensure that medical or genetic information of one family member is not accidentally revealed to other relatives. Researchers must keep in mind that individual family members expect their information to remain confidential, since there may be " family secrets" shared with only a few members in the kindred (9,83). Psychiatric research is one area of genetic research in which inadvertent disclosure is a concern as the clinical data collected on family members may include potentially stigmatizing information. Clinical information is not always freely shared among family members and can include the severity of psychiatric symptoms, alcohol or drug abuse, or criminal activity. Indeed, researchers may have to share background information about the proband when enrolling other family members in a family pedigree study. Thus some information may not remain private. Juengst notes, for example, that it is particularly difficult to approach a distantly related family member about enrolling in a psychiatric linkage study without sharing the fact that someone in the family is affected with a psychiatric disorder (6). In these cases Shore and colleagues feel that "Subjects will need to be informed, when agreeing to participate in genetic research, that their relatives may also be asked to participate as subjects. It should be made clear to psychiatric patients whose relatives will be contacted exactly what information about a subject will be provided to those relatives" (83).

Family registries and databases of genetic material are rapidly increasing in number. NBAC estimates that over 282 million specimens of human biological materials are currently stored in the United States, illustrating the magnitude of potential genetic information available to researchers (14). These databases contain sensitive genetic information that could be accessed by nonresearch team individuals, such as computer hackers, or social and governmental agencies requesting information. Simply removing names or social security numbers as identifiers may not guarantee that the data will remain anonymous or unidentifiable, since electronic databases are proliferating at a fast pace (85). Schulte and Sweeney point out: "Although the records of government-sponsored or funded studies will be maintained according to the Privacy Act of 1974 (P.L. 93-579), this does not ensure that records will never be disclosed" (86). For investigators accepting federal funds for specific research projects, the Privacy Act permits the release of identifiable research information in some circumstances. Particularly relevant for family pedigree studies, researchers may be obliged to respond to a court order seeking information that would be used to protect the health and safety of other persons. Thus the law can require that the investigator disclose confidential information (19,53,86).

In light of this potential ethical conflict between the duty to ensure the privacy and confidentiality of volunteers and a federally mandated court order to divulge sensitive information, Earley and Strong suggest that genetic researchers use a little known Certificate of Confidentiality (53,87). NIH established the certificate as a means to provide protection for federal research projects investigating the extent of illegal drug use in the 1970s. The 1974 and 1988 amendments expanded the certificate protection to cover other research including mental health research and genetics (87). It is important to note that the Certificate does not provide protection to individual participants in biomedical research. Rather, the Certificate protects the investigators from being compelled to disclose results to outside interests. The Certificate is issued to the principal investigator and provides protection for the life of the study. In most cases the Certificate can be extended beyond the funding cycle for the project. To date, there are no published studies addressing the efficacy of the Certificate of Confidentiality for protecting the privacy of participants in genetic pedigree, linkage, or biomedical studies. Thus it is unclear whether these Certificates will provide adequate protection for investigators or their research subjects.

Breeches of confidentiality and privacy can also occur through the publication of family pedigrees when the researchers report their results to peer-reviewed journals or to other investigators. A unique aspect of family pedigree studies is that each subject's clinical and genetic information is analyzed within the context of his or her family. The pedigree diagram is a valuable tool allowing an investigator to convey phenotypic and genotypic information while maintaining the biological relationship between each research subject in the family. The pedigree also includes such personal information such as age, gender, and birth order of the family members. Importantly, the pedigree diagram is a visual aid enabling the reader to quickly assess the mode of transmission of a disease gene, DNA marker, or trait (88). Recently concern for maintaining the anonymity of families in publications containing pedigree diagrams has been raised because of the comprehensive nature of the information (89,90). This concern has increased within the last few years partially due to the expanding interest and education of the lay public in the field of genetics. Increasing access to research articles via the Internet may also be a contributing factor (91).

In the past, journal editors have treated the publication of pedigrees similarly to the publication of traditional case studies, by simply withholding the names of the subjects depicted in the pedigree diagram (88). However, concerns have been raised that a pedigree diagram depicting the family structure with a description of the disease could pose a risk to the privacy and confidentiality of the participating family (88,90). Publication of such information may adversely impact the members of the kindred in several ways. Pedigree diagrams often contain medical and social information that is highly personal and may be not have been shared with other family members. As discussed earlier, instances of "paternal genotype inconsistencies" included in published pedigrees have caused repercussions within the family under study (6). Additionally pedigree diagrams can indicate medical illnesses, reproductive history, and adoption status of the family members (90). Publication can thus result in the disclosure of private information to the proband, other family members, or outside acquaintances. Furthermore, if genotypes are included in the pedigree diagram, information regarding disease status, carrier status, or disease susceptibility could be unintentionally communicated to the study participant and family members. In addition Byers and Ashkenas concluded that there is a remote possibility of discrimination through the inadvertent disclosure to third parties such as insurers or employers (89). Such disclosure directly violates the study volunteer's right to privacy, regardless of the level of harm. However, it is unclear whether such disclosure to a third party is a substantial risk at this time, since no published studies have directly examined this question.

Future Use of Research Samples

By collecting medical and family information along with a blood sample, most family pedigree and linkage studies also serve as a repository of genetic material, in which most of the individuals are related in family units. The genetic material used in family pedigree research is usually DNA taken from the volunteer's white blood cells, but it can be DNA from immortalized cell lines, buccal swabs, or paraffin embedded tissue from surgical samples. In most cases the DNA samples are from living individuals who provided their consent to the investigators to use their sample when enrolled in the study. Researchers may also use genetic material from tissue from deceased family members, since adequate amounts of DNA can be removed from surgical samples stored in hospital pathology departments. Since DNA is very stable when properly stored, it is possible for scientist to use the DNA samples for ancillary studies, or for unrelated studies long after the primary study is completed.

Using research samples from previous genetic studies is a common practice, since family pedigree projects will often have unused genetic material at the completion of the initial study. Some investigators stress that such repositories are extremely valuable for future scientific discovery, in which the cost of replicating the collection would be prohibitively expensive (92). In addition others have vigorously protested limitations on studying DNA from archival and family DNA banks, fearing that scientific progress will be impeded if previously collected samples cannot be used in future projects (93–95). However, ethicists and consumer groups suggest that participants may not realize that a sample of their genetic material could be used in future research projects that study other disease processes. Participants may not understand that their sample will remain part of a larger collection of genetic material, since research samples are not routinely destroyed at the completion of the project. Thus the future use of biologic materials has become a controversial and contentious topic among researchers, clinicians, ethicists, and patient advocacy groups (14,16,94,96–98).

The fact that tissue collected for genetic research may be used in future studies complicates the researcher's ability to protect the participant's right to autonomy and privacy. For example, study participants in one survey raised concerns that the previously collected sample could later be included in a research project that the participant would not have supported, such as fetal research or cloning experiments (99). In addition the subsequent research projects may yield clinically relevant information from material from DNA banks or other tissue sources. In this case the participant would be unprepared for information about risk for other severe but preventable disease, since he or she would not have been aware of this possibility when enrolling in the initial study (6,83). Concerns have also been raised about the potential for misuse of genetic information and the possibility of lost health insurance or employment opportunities resulting from unanticipated genetic research on previously collected DNA (38,50,98).

In light of these concerns, it is important to note that two surveys have shown a high proportion of participants enrolled in genetic research studies are willing to have their sample used in future research projects. Lewis and colleagues reported a survey of 416 subjects enrolled in a colon cancer linkage study demonstrating a very small percentage (<3 percent) of respondents refused to allow their sample to be used for future unrelated genetic studies (99). These individuals were primarily concerned about privacy and confidentiality of their personal information if their sample was used in this fashion. Fifty-one percent of the remaining participants allowed their sample to be used if their confidentiality was maintained, while 46 percent indicated that they wanted the researchers to contact them to learn more about the subsequent research before permitting future use. A second survey of 263 subjects enrolled in three separate genetic protocols sponsored by NHGRI found similar results (100). Four percent refused further use of their DNA sample, while the overwhelming majority allowed their sample to be used either after being recontacted (73 percent) or after stripping personal identifiers from the sample (26 percent) (101). The results of these studies indicate that the vast majority of research participants are willing to allow their DNA sample to be used in future research, although potential discrimination is a concern for most participants.

The pace and complexity of human genomic research will continue to grow in the near future and benefit the entire community. However, numerous volunteers in family pedigree studies will be potentially exposed to genetic discrimination from the results or conduct of the study. These harms can simultaneously affect the individual volunteer and their family during the life of the study as well as in the future. The potential harms include confidentiality, genetic discrimination, clinical impact and unintended consequences of gene discovery, and the future use of DNA samples. Thus researchers need to address these issues as they conduct family pedigree studies.

PROFESSIONAL CHALLENGES FOR THE CONDUCT OF RESEARCH

While the exciting advances in molecular technology promise a better understanding of many common diseases plaguing humankind, a consensus among researchers, bioethicists, and legal experts on the ethical dilemmas posed by pedigree studies is noticeably absent (6,8,10,22). Several prominent genome scientists have openly called for the development of standardized policies for pedigree and genome research (3). Some of the proposed policies have been controversial, and may place additional burdens on the research team in the time and resources required to conduct the research project (101). Others have responded that the duty of the researcher is clear and that "ethical research is good quality research" (102). Since one of the inherent ethical conflicts in family pedigree research stems from the blending of basic bench science with the potential use of unvalidated tests for clinical health decisions, each research group should establish protocols

Table 2. Policy Areas for the Conduct of Genetic Family Studies

1. Recruitment and ascertainment
2. Privacy and confidentiality of medical, family, and genetic information
3. Disclosure of experimental results
4. Future use of DNA
5. Informed consent

that are specific for the genetic condition under study. There are five basic areas that the research community should address in developing such policies for genome research (Table 2) including recruitment, confidentiality, disclosure of results, future use of DNA samples, and informed consent.

Recruitment and Ascertainment

Identifying eligible participants who have the correct family structure for a specific study is one of the most important aspects of family pedigree research. Many recruitment issues are no different than other human subjects research, in that subjects should be informed about the goals of the study, what participation entails, and the risks and benefits of entering the study. The NIH Office for Protection from Research Risks (OPRR) published an Institutional Review Board Guidebook in 1993 listing the guidelines for protecting human research subjects (103). While the Common Rule as outlined in the Federal statutes clearly state that human subjects research must be approved by each institution's Institutional Review Board (IRB), it has only been recently that additional guidelines have been put in place for genetic pedigree studies (103, p. A58; 104). Recognizing the unique position of recruiters for genetic studies, the OPPR guidebook states that "The familial nature of the research cohorts involved in pedigree studies can pose challenges for ensuring that recruitment procedures are free of elements that unduly influence decisions to participate" (103).

Recruitment protocols for family-based studies are more complex that those used for standard epidemiologic research projects, since each family member, although biologically related, must be separately enrolled in the study. Thus it is important to recognize that the recruitment process can exert undue pressure on family members to enroll. Pressure can come from the research team, since the study actively searches for kindreds with multiple members affected with the disease under study as these families provide more genetic information to the project (6,9). Coercion can occur within the family unit, especially if a volunteer expects that his or her family will directly benefit from the results of a study. "Thus the pressure of compelling familial relationships may simply replace the researcher's influence in recruiting potential subjects" (6, p. 407). Some researchers suggest approaching families in large groups or through organizations like support groups and allowing interested members to contact the researcher if they wish to join the study. This may serve to increase the autonomy of some people's decision. However, as Juengst points out

"...family members may actually feel less free to demur in large group settings, and lay-led support groups vary in expertise, understanding, and objectivity" (6, p. 407).

Consensus has not been reached on how IRBs should require researchers to protect the privacy interests of family members. OPPR suggests that researchers might collect only publicly available facts about family members, such as names and addresses, from the proband (103). Although Juengst acknowledges that obtaining detailed medical and family pedigree information from probands "...is a practice so traditional as to be ethically invisible within the community," he suggests that researchers should follow OPRR's suggestion to collect only publicly available information about family members from probands, and then "...convert this tree into a genetic pedigree by soliciting relevant health data from each relative directly" (6, p. 405). However, this recruitment practice will place additional burdens on the research team to gather this information within the research budget allocated to the project.

Pedigree studies will use many sources for referral, such as support groups, health care providers, clinics hospital databases and family members previously enrolled by the study group. Cohen and Wolpert describe several methods of family enrollment and caution that people may pressure their kin to enroll in the study (9). They also suggest that one family member must first inform other members of the family about the study and provide permission for the recruiters to separately reach each member. The researcher then contacts the family member to further discuss the project and offer participation. Other recruitment strategies in the future will be to use data from family registries that have been developed for research. One such Quebec registry for familial Alzheimer's disease developed a recruitment strategy that relied upon local health care professionals as well as leaflet advertisements in hospitals and clinics (105). This group also developed ethical guidelines for the conduct of the entire research project, incorporating protections for family members and incapacitated adults. Researchers must also determine whether minor children or incapacitated adults should be enrolled in a family study (6,9). Parents enrolling their child must use substituted judgment for the child and not be influenced by other family concerns. Thus specific protections should be in place to include the minor child's assent to join the study.

Ensuring and Maintaining Confidentiality

As previously discussed, researchers have a duty to uphold the privacy and confidentiality of participants in family pedigree research, including medical information and DNA genetic results. Pedigree studies pose additional challenges for the researcher to ensure that the data remains confidential, to prevent private information about some family members from being inadvertently disseminated within the extended pedigree, and to limit potential breeches in confidentiality through the process of publication (7,9,70). Standard approaches have been developed for coding and tracking health and genotype information for study enrollees to aid in managing confidential information (9). While there are no federal laws that guarantee protection of genetic data, the Certificate of Confidentiality has been proposed to protect researchers from being compelled to submit genetic data to outside agencies (16,53). As previously noted, this protection is afforded to the researcher and not to the research participant. Again, there are no specific guidelines on this issue from OPPR or another regulatory agency, although patient advocacy groups have proposed that similar protections be developed for research subjects (19,96). The National Action Plan on Breast Cancer (NAPBC) focused on developing specific strategies to ensure privacy for participants in genetic studies (19). Stating that "privacy protections for experimental research data in which health care is not delivered should exceed the protections established for medical records" and recommended that identifiable genetic research data should not be included in a person's medical file.

Researchers should guard against providing identifiable data to the public at large through the publication of pedigrees. Representing the family medical history in a pedigree format is an essential part of data for publication but could possibly disclose the familial condition in an identifiable format. Thus a practice of altering pedigree information has been developed to provide anonymity for the family members, although it is debatable whether it affords true protection and may undermine the scientific validity of the study (70,89). OPRR recommends that written consent be obtained from participants as to the release of personal information (103). However, there may be no reason to assume that all family members depicted in the diagram had enrolled in the study (89). The International Committee of Medical Journal Editors (ICMJE) issued guidelines in 1995 for protecting the privacy of research subjects in scientific publications, and recommended that "Identifying details should be omitted if they are not essential, but patient data should never be altered or falsified in an attempt to attain anonymity" (88,106).

Disclosure

The genetic revolution has enabled researchers to locate disease-causing genes which as paved the way to the development of new genetic tests for clinical care. While these advances benefit the entire community, the individuals who donated their clinical and genetic material to family pedigree researchers may wish to know their personal study results. As previously discussed, these research results are experimental, and may not be clinically valid for health concerns. Indeed, the research genetic test result may be technically inaccurate, as these tests are performed in the research laboratory rather than in the clinical laboratory (15). NAPBC also points out that research laboratories have a higher tolerance level for inaccurate experimental results than do clinical laboratories (19). Thus falsely negative or falsely positive test results could be provided to a research volunteer.

Federal regulations in the form of the Clinical Laboratory Improvements Amendments (CLIA) of 1988 were enacted to codify the specific requirements for clinical laboratories providing test results that will be used in clinical management of patients (107). CLIA standards help ensure sample integrity and clinical validity of the test results. While the CLIA statute was developed for all type of laboratory tests, there are limitations in the requirements for testing and monitoring genetic testing (108). Nevertheless, these standards apply to all laboratories that test samples for clinical decision making including research laboratories that supply genetic tests for rare disorders at no cost (15). Thus researchers may be in violation of the CLIA statute when disclosing test results to their participants or to clinicians caring for the subject.

The Institute of Medicine report on genetic testing noted that research laboratories may offer the only available genetic tests for rare disorders, since it is impractical for general clinical laboratories to develop tests that would be infrequently used. The authors of this report recommended the establishment of a central repository and genetic CLIA approved laboratory to offer these tests to patients and family members. Other research groups, such as the newly formed International Gastric Cancer Linkage Consortium, have instituted specific protocols for offering genetic testing to research participants (109). This group recommended that clinical genetic counseling be offered to family members in which a mutation in the E-cadherin gene is found in research subjects. E-cadherin is cellular adhesion molecule and persons with constitutional deleterious mutations are at risk for an aggressive form of gastric cancer. In addition the consortium arranged to have the experimental research findings validated by a CLIA approved molecular laboratory. Thus individuals and their family members will be able to learn their research results while the risk of inaccurate test results are minimized. Several groups have developed guidelines as to the proper avenue of disclosure of research information to volunteers. For example, the ASHG strongly recommends that research results only be communicated "by persons able to provide genetic counseling" (13).

NBAC recognized that disclosing results to research subjects is controversial (14). In a 1999 report the commission recommended that disclosure should occur only where the findings are scientifically valid, have significant health implications for the subject and a treatment is available for the disorder in question (14). The authors of this report also assumed that disclosure would be a rare circumstance for the researcher, although this may not be true for researchers who identify highly prevalent genetic changes for a common disorder. Patient advocacy groups, such as NAPBC, feel that research participants should have access to experimental findings, except when the results have unproven clinical validity. They also recommended withholding research data when the results could harm the subject, interfere with the study, or cause harm to another individual (19). Mac Kay suggests that results should not be disclosed to research volunteers as "a more equitable way of dealing with the possibly conflicting views of family members as well as avoiding the problems of information whose reliability is not yet established" (7, p. 489). However, it is important to recognize that many subjects will be interested in their personal genetic results. Thus, researchers must specifically address whether experimental information will be provided to their study volunteers and, if so, how disclosure will take place.

Future Use of Samples

One of the most difficult issues for genetic researchers is the development of an ethical framework for the future use of the DNA samples collected for the research study. Members of the research community have hotly debated specific guidelines for research using previously collected or archival tissues, since the future use of genetic material was not considered when many of these repositories were established (14). Most authors recommend contacting and obtaining the subject's consent for research on projects that require the use of specific personal identifiers (67). An alternative approach would be to use samples from retrospective DNA repositories as long as the sample is anonymized and stripped of all identifiers (102).

Several organizations and consumer groups have made recommendations for the future use of genetic samples. In a 1996 policy statement ASHG classified biological samples into one of four groups (16). First, " anonymous" biological samples are defined as samples that were originally collected without any specific identifiers from the person who donated the sample. Thus, linking an anonymous sample to the original source is impossible. Second, "anonymized" samples are defined as samples which were initially collected with specific identifiers, were subsequently stripped of all these identifiers. Anonymized samples are thus irreversibly removed from any link to their source, except that the samples can remain linked with clinical, pathological, and demographic information as long as the amount and type of this linked information does not breech anonymity. Third, "identifiable" samples are linked to sources by a confidential code developed by the original investigator. While a member of the original research team can decode these samples, the person's identity can not be revealed to persons outside of the study. Fourth, "identified" samples are those tissue samples associated with the participant's name, hospital number, or pedigree location and are available to the researchers. Hospital pathology departments are an example of repositories of identifiable tissue samples. A similar classification for research samples has been proposed by NBAC, where samples are categorized into unidentified, unlinked, coded and identified samples (14).

The ASHG report on informed consent for genetic research made several recommendations regarding future use of biological samples. Regarding permission to use the sample in other unspecified studies, this group stated "...It is inappropriate to ask a subject to grant blanket consent for all future unspecified genetic research projects on any disease or in any area if the samples are identifiable in those subsequent studies." This group also recommends that researchers ask the volunteer to "indicate if unused portions of the samples may be shared with other researchers." The report also recommended that the subject should indicate whether subsequent researchers should "receive their samples as anonymous or identifiable specimens" (16).

Several other groups have made recommendations for future use of research samples. In general, most groups agree with the OPRR guidelines calling for researchers to "obtaining consent from the participants for any use of the data (and samples) that is not strictly within the original uses to which the participants agree" (103). Clayton and colleagues made further recommendations, by suggesting that researchers inform the subjects "about the scope and potential consequences of the projects" (98). These authors also suggest that subjects should be asked if they are wish to have their sample anonymized, as well as if they would allow their sample to be used by investigators outside the institution or outside the original research project. The American College of Medical Genetics (ACMG) guidelines propose that researchers request permission for future use of the sample from the volunteer at the time the sample is collected (97). This group also recommended that researchers develop a specific policy about whether subjects will be recontacted if permission to use the sample in the future was not obtained at the time the sample was collected.

Informed Consent

The ethical complexities of genetic research studies prompted one ethicist to write that the "steps to obtaining consent from members of a pedigree can be tortuous" (7). However, the evolving duty of the genetic researchers has been recently described:

> In sum, the people whose genetic and clinical data will be essential for the next phase of human genomics research need to be treated not merely as "subjects" but more as (somewhat limited) partners. Researchers must realize that these people have interests beyond safety; ethicists must recognize that, when well informed, they have the right to participate even in broadly defined research. The goal of this approach is not to prevent research but to prevent research subjects from feeling cheated, powerless, misled, or betrayed (22).

Most researchers recognize that a trusting relationship with the study volunteers is essential for the success of the entire project. Indeed, it is through the informed consent process that such a relationship is first developed. In this way the informed consent process is more significant than the signed document detailing the proposed research and can be a blueprint for the ethical conduct of the research to be performed. Hence the study protocol should include detailed plan for informing the volunteer about the study and obtaining consent prior to enrollment.

Since family pedigree research involves the collection of clinical information and biological samples on multiple family numbers, these studies fall under the Common Rule, which requires that human subjects research supported by federal agencies be reviewed by an IRB (104). NBAC extended this recommendation for IRB oversight for all human subjects research, regardless of federal support (14). The ASHG statement on informed consent for genetic research encouraged researchers to develop procedures to obtain informed consent for both prospective and retrospective studies (16). In addition this report suggested that specific protocols be developed for maintaining confidentiality, disclosure of expected

Table 3. Elements of Informed Consent Document for Family Pedigree Studies

1. Purpose of study
2. Participation is voluntary
3. Costs and/or reimbursement
4. Benefits of participation
5. Disclosure of experimental results
7. Risks or informational harm from participating
8. Ensuring confidentiality: "Certificates of Confidentiality"
9. Future use of DNA sample

and unexpected experimental data, and deposition of samples.

The elements to be included in the investigator's informed consent document should follow standard formats with special additions relevant to genetic studies (Table 3). As with all human subject research, informed consent requires that the consent document be written in language understandable to average readers. It must also include a description of the project and the purpose of the research. The potential participant must be given the option of withdrawing from the research at anytime without penalty. The researcher must also identify any costs related to participating in the project and include an estimate of the amount of time. In addition to these points, informed consent documents must also contain information on benefits, disclosure of results, risks, confidentiality, and future use of samples. Researchers who are composing consent documents for genetic family pedigree studies must pay particular attention to these components to insure that participants are informed of the potential harms specific to genetic studies (103).

Researchers must clearly state that there may be no direct benefits for the participant or family members. However, as discussed previously, researchers must also indicate whether experimental results will be communicated to study volunteers. If the research team is planning to withhold results, the informed consent document should indicate that no research results will be given to the participant. On the other hand, if the research team is planning to share experimental results, the disclosure protocol should be explained to the participant in advance. In this case additional costs that may be incurred by the participant, such as CLIA laboratory confirmation or genetic counseling, should be included in the consent document. In addition to the physical risks that may be incurred in the genetic study the risks associated with "informational harms" must be disclosed (16,39). As previously discussed, these include potential psychological harms from learning preliminary health information directly from the study, or from new and unintended information which may not be related to the initial focus of the study. IRB guidelines state: "Prospective subjects should be informed during the consent process that the discovery of such information is possible" (103). Identification of nonpaternity or undisclosed adoption in a family is another unintended consequence of genetic studies that may cause psychological harm to participants. ASHG recommends that researchers consider including a statement in the

informed consent document that mistaken parentage will not be disclosed (16).

A description of steps that will be taken by the researcher to protect the privacy of the study participants should be included in the informed consent document. Most volunteers will want to know that personal information or study results will not be disclosed to third parties, including employers, insurers, and family members, unless there is written consent from the participant. If the researchers follow the ASHG recommendations and obtain a Certificate of Confidentiality, a brief description of the protections it affords should be included in the consent form (16). The researcher should describe how confidentiality will be maintained if the research results or the family pedigree will be published. In regards to publication of family pedigrees, the ICMJE guidelines state: "Identifying information should not be published in written descriptions, photographs, or pedigrees unless the information is essential for scientific purposes and the patient (or parent or guardian) gives written informed consent for publication" (106). This requires the consent of all family members depicted in the pedigree, which can be a daunting task for researchers. Finally, researchers must clarify use of the samples in future research. Many groups have made recommendations about the use of genetic samples for future genetic studies, and most groups agree with the recommendation of the 1995 ASHG report that it is inappropriate to ask a participant to provide unrestricted consent to the future use of a sample when the risks of the future project are unknown (16).

Successful family pedigree studies are dependent upon the generous donation of clinical information, family information and biologic samples by volunteer participants. Recruitment of potential subjects for genetic studies requires a trusting relationship between the volunteer and researcher. Thus the protocol that a researcher develops for the informed consent process is a major component of the study; it will require policies addressing the potential benefits and risks specific to genetic studies. In order to continue the enormous success of the genetic revolution, researchers must develop guidelines that will ensure the ethical code of conduct for their genetic studies.

BIBLIOGRAPHY

1. G.J. Annas, *Emory Law J.* **39**, 630–664 (1990).
2. National Human Genome Research Institute in NIH, 1998, Available at: *http://www.nhgri.nih.gov/HGP/*
3. F.S. Collins et al., *Science* **282**(23), 682–689 (1998).
4. J. Watson, *Science* **248**(4951), 44–49 (1990).
5. J.D. Watson and F.H.C. Crick, *Nature* **171**(4356), 737–738 (1953).
6. E.T. Juengst, in H.Y. Vanderpool, ed., *The Ethics of Research Involving Human Subjects: Facing the Twenty-first Century*, University Publishing Group, Frederick, MD, 1996.
7. C.R. Mac Kay, *Hum. Genet. Theory* **4**(4), 477–495 (1993).
8. E.T. Juengst, *Am. J. Hum. Genet.* **54**(1), 121–128 (1994).
9. P. Cohen and C. Wolpert, in J.L. Ed Haines and M.A. Pericak-Vance, eds., *Approaches to Gene Mapping in*

Complex Human Disease, Wiley-Liss, New York, 1998, pp. 131–159.

10. N.A. Holtzman and L.B. Andrews, *Epidemiol. Rev.* **19**(1), 163–174 (1997).

11. H.T. Greeley, *Science* **282**(5389), 625 (1998).

12. A. Baum, A.L. Friedman, and S.G. Zakowski, *Health Psychol.* **16**(1), 8–19 (1997).

13. National Commission for the Protection of Human Subjects of Biomedical and Behavioral Research, *The Belmont Report: Ethical Principles and Guidelines in the Conduct of Research Involving Human Subjects*, Department of Heath, Education and Welfare, U.S. Government Printing Office, Washington, DC, 1979.

14. National Bioethics Advisory Commission, *Research Involving Human Biological Materials: Ethical Issues and Policy Guidance*, vol. 1, U.S. Government Printing Office, Rockville, MD, 1999.

15. L.B. Andrews et al., eds., *Assessing Genetic Risks: Implications for Health and Social Policy*, National Academy Press, Washington, DC, 1994.

16. American Society of Human Genetics, *Am. J. Hum. Genet.* **59**(2), 471–474 (1996).

17. American Society of Human Genetics Ad hoc Committee on Breast and Ovarian Cancer Screening, *Am. J. Hum. Genet.* **55**(5), i–iv (1994).

18. American Society of Human Genetics social Issues Subcommittee on Familial Disclosure, *Am. J. Hum. Genet.* **62**(2), 474–483 (1998).

19. B.P. Fuller et al., *Science* **285**(5432), 1359–1361 (1999).

20. Y. Miki et al., *Science* **266**(5182), 66–71 (1994).

21. R. Nowak, *Science* **266**(5190), 1470 (1994).

22. H.T. Greeley, *Genet. Test.* **3**(1), 115–119 (1999).

23. E. Kodish et al., *J. Am. Med. Assoc.* **279**(3), 179–181 (1998).

24. American Society of Clinical Oncology, *J. Clin. Oncol.* **14**(5), 1730–1736 (1996).

25. R. White and C.T. Caskey, in G.J. Annas and S. Elias, eds., *Gene Mapping: Using Law and Ethics and Guides*, Oxford University Press, New York, 1992, pp. 173–183.

26. M.W. Foster and W.L. Freeman, *Genome Res.* **8**(8), 755–757 (1998).

27. J.S. Risch, in D.L. Rimoin, M.J. Connor, and R.E. Pyeritz, eds., *Emery and Rimoin's Principles and Practice of Medical Genetics*, Churchill Livingstone, New York, 1997, pp. 371–382.

28. B. Hanson et al., *Am. J. Hum. Genet.* **48**(5), 873–879 (1991).

29. S.J. Winawer et al., *Cancer* **70**(5), 1236–1245 (1992).

30. S.J. Winawer et al., *N. Engl. J. Med.* **334**(2), 82–87 (1996).

31. R.C. Elston and J. Stewart, *Hum. Hered.* **21**, 523–542 (1971).

32. R.C. Go et al., *J. Nat. Cancer Inst.* **71**(3), 455–461 (1983).

33. M.C. King et al., *J. Nat. Cancer Inst.* **71**(3), 463–467 (1983).

34. B. Newman et al., *Proc. Nat. Acad. Sci.* **85**(9), 3044–3048 (1988).

35. J. Badner et al., *Am. J. Hum. Genet.* **46**(3), 568–580 (1990).

36. E.S. Lander and N.J. Schork, *Science* **265**(5181), 2037–2048 (1994).

37. A.L. Beaudet et al., in C.R. Scriver et al., eds., *The Metabolic and Molecular Bases of Inherited Disease*, McGraw-Hill, New York, 1995, pp. 53–228.

38. P.A. Schulte, G.P. Lomax, and E.M. Ward, *J. Occup. Environ. Med.* **41**(8), 639–646 (1999).

39. P.R. Reilly, *Am. J. Hum. Genet.* **63**(3), 682–685 (1998).

40. N.A. Holtzman and D. Shapiro, *Br. Med. J.* **316**(7134), 852–856 (1998).

41. Ad hoc Committee on Genetic Testing/Insurance Issues, *Am. J. Hum. Genet.* **55**(1), 327–331 (1995).

42. K.H. Rothenberg, *Health Matrix* **7**(1), 97–124 (1997).

43. J.C. Fletcher and D.C. Wertz, *Am. J. Hum. Genet.* **61**(4), A56 (1997).

44. P.R. Billings et al., *Am. J. Hum. Genet.* **50**(3), 476–482 (1992).

45. J.A. Lowden, *Am. J. Hum. Genet.* **51**(4), 901–903 (1992).

46. E.B. Hook, *Am. J. Hum. Genet.* **51**(4), 899–901 (1992).

47. E.V. Lapham, C. Koxma, and J.O. Weiss, *Science* **274**(5287), 621–624 (1996).

48. L. Low, S. King, and T. Wilkie, *Br. Med. J.* **317**(7173), 1632–1635 (1998).

49. P.S. Harper, *Am. J. Hum. Genet.* **50**(3), 460–464 (1992).

50. N.A. Holtzman and M.A. Rothstein, *Am. J. Hum. Genet.* **50**(3), 457–459 (1992).

51. NIH/DOE Task Force on Genetic Information and Insurance, *Genetic Information and Health Insurance. Report of the Task Force on Genetic Information and Insurance*, U.S. Government Printing Office, Washington, DC, 1993.

52. M.R. Natowicz, J.K. Alper, and J.S. Alper, *Am. J. Hum. Genet.* **50**(3), 465–475 (1992).

53. C.L. Earley and L.C. Strong, *Am. J. Hum. Genet.* **57**(3), 727–731 (1995).

54. T.H. Murray, *Hastings Center Rep.* **22**(6), 12–17 (1992).

55. T.H. Murray, *Am. J. Cardiol.* **72**(10), 80D–84D (1993).

56. J. Stephenson, *J. Am. Med. Assoc.* **282**(23), 2197–2198 (1999).

57. M.A. Hall and S.S. Rich, *Am. J. Hum. Genet.* **65A**(4), A3 (1999).

58. M.A. Hall and S.S. Rich, *Am. J. Hum. Genet.* **66**(1), 293–307 (2000).

59. G.L. Wiesner, *The Genetics of Multiple Endocrine Neoplasia, Type II in GeneClinics: Medical Genetics Knowledge Base*, University of Washington School of Medicine and National Institutes of Health, Washington, DC, 1999.

60. G.J. Annas, *N. Engl. J. Med.* **331**(5), 1027–1030 (1994).

61. R.M. Cawthon et al., *Cell* **62**(1), 193–201 (1990).

62. J.M. Friedman and P.H. Birch, *Am. J. Med. Genet.* **70**(2), 138–143 (1997).

63. D.F. Easton et al., *Am. J. Hum. Genet.* **56**, 356–371 (1995).

64. H. Raunio et al., *Gene* **159**(1), 113–121 (1995).

65. S. Lehrman, *Nature* **389**(6647), 107 (1997).

66. B.B. Biesecker et al., *J. Am. Med. Assoc.* **269**(15), 1970–1974 (1993).

67. P.S. Harper, *Br. Med. J.* **306**(6889), 1391–1394 (1993).

68. E. Kodish, T.H. Murray, and S. Shurin, *Clin. Res.* **42**(3), 396–402 (1994).

69. F.P. Li et al., *Cancer Res.* **48**(18), 5358–5362 (1988).

70. E.T. Juengst, *Genome Sci. Technol.* **1**(1), 21–36 (1995).

71. M.-G.L. Roux et al., *Lancet* **340**, 607 (1992).

72. S. Macintyre and A. Sooman, *Lancet* **338**, 869–871 (1991).

73. Parentage Testing Standards Program Unit, *Annual Report Summary for 1998*, vol. 4 (1998).

74. J.G. Compton et al., *Nat. Genet.* **1**(4), 301–305 (1992).

75. P.R. Reilly, M.F. Boshar, and S.H. Holtzman, *Nat. Genet.* **15**(1), 16–20 (1997).

76. M. Verdy et al., *J. Pediatr. Gastroenterol. Nutr.* **1**(4), 603–607 (1982).

77. S. Lyonnet et al., *Nat. Genet.* **4**(4), 346–350 (1993).

78. M. Angrist et al., *Nat. Genet.* **4**(4), 351–356 (1993).

79. P. Edery et al., *Nature* **367**(6461), 378–380 (1994).

80. G. Romeo et al., *Nature* **367**(6461), 377–378 (1994).

81. B.B. Roa et al., *Nat. Genet.* **14**(2), 185–187 (1996).

82. G.L. Wiesner, *Health Matrix* **7**(1), 3–30 (1997).

83. D. Shore et al., *Am. J. Med. Genet.* **48**, 17–21 (1993).

84. G.L. Wiesner, Personal communication, 1999.

85. J.J. McQueen, *Clin. Chem. Lab Med.* **36**(8), 545–549 (1998).

86. P.A. Schulte and M.H. Sweeney, *Environ. Health Perspect.* **103**(3), 69–74 (1995).

87. National Institutes of Health, *Privacy Protection for Research Subjects, Certificates of Confidentiality*, 1999. Available at: *http://grants.nih.gov/grants/oprr/human subjects/guidance/certconpriv.htm*

88. P.H. Byers and J. Ashkenas, *Am. J. Hum. Genet.* **63**(3), 678–681 (1998).

89. M. Powers, *Inst. Rev. Board* **15**(9), 7–11 (1993).

90. J.R. Botkin et al., *J. Am. Med. Assoc.* **279**(22), 1808–1812 (1998).

91. Editorial, *Nat. Genet.* **19**(3), 207–208 (1998).

92. E. Marshall, *Science* **282**, 2165 (1998).

93. W.W. Grody, *Diag. Mol. Pathol.* **4**(3), 155–157 (1995).

94. W.W. Grody and M.E. Sobel, *Diag. Mol. Pathol.* **5**(2), 79–80 (1996).

95. E. Marshall, *Science* **271**, 440 (1996).

96. Alliance of Genetic Support Groups, *Informed Consent: Participation in Genetic Research Studies*, Chevy Chase, MD, 1993.

97. American College of Medical Genetics Storage of Genetics Materials Committee, *Am. J. Hum. Genet.* **57**(6), 1499–1500 (1995).

98. E.W. Clayton et al., *J. Am. Med. Assoc.* **274**(22), 1786–1792 (1995).

99. S. Lewis et al., *Am. J. Hum. Genet.* **63**(4), 28 (1998).

100. L.A. Middelton et al., *Am. J. Hum. Genet.* **61**(4), A58 (1997).

101. M. Bondy and C. Mastromarino, *Ann. Epidemiol.* **7**(5), 363–366 (1997).

102. P. Hainaut and K. Vahakangas, in W. Ryder, ed., *Metabolic Polymorphisms and Susceptibility to Cancer*, IARC Science Publication, Lyon, France, 1999, pp. 395–402.

103. (Human Genetic Research) OPRR, *Protecting Human Research Subjects: Institute Review Board Guide Book*, U.S. Government Printing Office, Bethesda, MD, 1993, pp. 5–49.

104. 56 Federal Register, *Code of Federal Regulations (CFR) 45* (1999).

105. B. Godard et al., *Alzheimer Dis. Assoc. Disorders* **8**(2), 79–93 (1994).

106. International Committee of Medical Journal Editors, *Br. Med. J.* **311**, 1272 (1995).

107. M.K. Schwartz, *Clin. Chem.* **45**(5), 739–745 (1999).

108. N.A. Holtzman and M.S. Watson, eds., *Promoting Safe and Effective Genetic Testing in the United States. Final Report of the Task Force on Genetic Testing*, U.S. Government Printing Office, Washington, DC, 1997.

109. C. Caldas et al., *J. Med. Genet.* **36**, 873–880 (1999).

See other HUMAN SUBJECTS RESEARCH entries.

HUMAN SUBJECTS RESEARCH, ETHICS, INFORMED CONSENT IN RESEARCH

NANCY E. KASS
Johns Hopkins University
Baltimore, Maryland

OUTLINE

INTRODUCTION

It is only in the latter half of this century that significant intellectual and regulatory attention has been devoted to human subjects protection. In response to isolated, but horrific, examples of unethical research studies, codes of ethics for human subjects research were developed, regulations were passed, and standards of informed consent were established. And yet as often is the case, while this branch of ethics has made tremendous progress in just a few decades, new challenges continue to emerge. It is the purpose of this entry to provide a brief history of human subjects protections, to describe specific elements of informed consent as they apply to research, and to discuss specific examples in which upholding standards of informed consent remains particularly challenging.

HISTORY OF HUMAN SUBJECTS PROTECTIONS

The Nuremberg Code of 1948 usually is considered the first code of research ethics (1). This Code grew out of the Nazi war crime tribunals, during which descriptions had been revealed of experiments conducted on concentration camp prisoners. These experiments, conducted through force and coercion, studied such questions as how long humans can be immersed in ice water before dying of hypothermia, the effects of ingesting poisons, and the effects of being injected with viruses. The Nuremberg Code, intended to guide all future research with humans, was developed as part of the judgment in *United States v. Karl Brandt* et al. (1a). Given the context out of which the Nuremberg Code emerged, it is not surprising that its first stipulation is that "the voluntary consent of the human subject is absolutely essential" (1, p. 181). It elaborated that the subject should have the legal capacity to consent, should be able to exercise free choice without any coercion or deceit, and should have sufficient knowledge and understanding of the experiment to enable an "enlightened decision." Ensuring that the consent is voluntary and informed is the "personal responsibility" of the investigator. The Code further states that the experiment should be expected to yield important results for society that cannot be obtained through other methods, should be based on previous animal research and a knowledge of the problem, should avoid all unnecessary physical and mental suffering and injury, should not be conducted when there is a priori reason to believe death or disabling injury will result, should not involve a level of risk that exceeds the importance of the problem, should be conducted only by qualified persons, should guarantee that the subject has the right to stop participating at any point, and should be terminated early should there be reason to believe that the experiment is unduly risky.

Shortly afterward, in 1953, the clinical center of the National Institutes of Health (NIH) opened. The NIH clinical center is a research hospital, funded by the federal government, where all "care" provided to patients is part of a research protocol. When the clinical center opened, the NIH decided to require informed consent of healthy volunteers who entered studies, but not of patient-subjects, who were presumed to have reason to want to participate in clinical research (2). Nonetheless, this was one of the first times that an entity of the U.S. government required informed consent for any type of human subjects research.

The first code regarding the ethical conduct of research put forth by a professional medical body was from the World Medical Assembly in 1964 (3). The Declaration of Helsinki, as it was called, echoed many of the tenets of the Nuremberg Code but included additional elements of relevance to doctors who conduct research. It reminds doctors that despite the importance of conducting research, "the health of my patient will be my first consideration." It adds that the responsibility for the welfare of the subject always rests with the investigator and not with the subject, despite the subject having given voluntary consent. Further, "concern for the interests of the subject must always prevail over the interest of science and society." The Declaration further states that experiments not conducted in accordance with the proposed ethical requirements should not be considered publishable regardless of their scientific findings, and that the doctor should be particularly "cautious" if the subject is in a dependent relationship with him or her, in which case a different member of the research staff should obtain consent from the subject.

In the 1960s and 1970s several events in this country brought attention to human subjects research and the abuses that potentially can be associated with it. The Willowbrook hepatitis study was conducted from 1956 to 1970. The Willowbrook School, where this study occurred, is an institution for mentally retarded children. There were poor sanitary conditions at the school, and most children contracted hepatitis A at some point after being sent to live there. Researchers wanted to study the natural history of hepatitis A and the possibility of creating a vaccine for the disease. They decided to inject children who were newly admitted to Willowbrook with the strain of hepatitis that was rampant there. The "study" was justified by saying that the children probably would have become infected anyway and that, scientifically, more can be learned about the natural history of the disease if it is known precisely when the child became infected.

In 1963 the public became aware of experiments conducted at the Jewish Chronic Disease Hospital in New York. The Jewish Chronic Disease Hospital was an institution for elderly, chronically ill adults. In the experiments, residents of the hospital were injected with live cancer cells without their knowledge or consent. The experiments were justified by arguing that these patients would have died soon anyway.

In 1966 Henry Beecher, a well-respected Harvard physician, published in *The New England Journal of Medicine* an article that has become one of the classic pieces in research ethics (4). Beecher, who earlier had called for "a long, straight look at our current practices" (5), now conducted a review of articles published in top medical research journals. In the article he described 22 articles gleaned from medical literature of the time in which ethically questionable practices had been involved. Among his examples were placebos being substituted for an established treatment without patient-subjects' knowledge, studies of vulnerable subjects, and/or studies with a high degree or risk relative to benefit. Dr. Beecher's article received considerable attention, in great part, because his examples were drawn from *published* and therefore well-sanctioned research, and also because he implied that the research described was not necessarily unrepresentative nor unusual.

In response to this series of events, the United States Public Health Service established guidelines for research in 1966 (6). These guidelines required that each institution conducting human subjects research funded by the U.S. Public Health Service establish an Institutional Review Board (IRB) that would review projects in advance. The IRB would determine whether (1) the rights and welfare of study subjects are protected, (2) the methods to obtain informed consent are appropriate, and (3) the risks and potential benefits of the investigation are clear, and the potential benefits outweigh the risks (6). Therefore research review was to be prospective decentralized, based at the researchers' institution, and required to include informed consent.

In 1971 the public's attention turned to yet another horrific example in the history of research ethics. The Tuskegee syphilis experiment had been conducted by the American government from 1932 to 1972 (7). In this example, the Public Health Service was studying the natural history of syphilis. They chose as their subjects poor, black men from the rural south, 400 of whom had syphilis and 200 of whom served as controls. None of the men were told that a study was being conducted. Rather, they were led to believe that the government doctors were providing medical care for them, something that poor, rural men were eager to find. When certain diagnostic procedures (e.g., spinal taps) were conducted for research purposes, the men were told that they were receiving treatment, and when efficacious antibiotic treatment became available in the 1940s, these men were denied therapy. The justification given for the study was that researchers simply were observing the disease that men already had, and since most of these men had no access to care, not providing care was no worse than what they already would have experienced. This study violated all ethical requirements of research, in that the men were not told they were participating in research and consequently could not provide meaningful consent. The research itself was unreasonably risky, particularly after penicillin became available; and the research singled out certain segments of the populations who were the poorest, were from a racial minority, and experienced none of the study's benefits. After press reports in the early 1970s exposed the horrors of the Tuskegee study, the Department of Health, Educations, and Welfare (DHEW) appointed the Tuskegee Study Ad hoc Panel to review the study, as well as to review the Department's policies for the conduct of human subjects research (8). The panel noted that, although DHEW guidelines had been in place since 1966, it was a journalist, rather than a review committee, that brought the conduct of this study to light. (for further discussion, see Ref. 9). The panel recommended that the Tuskegee study be stopped immediately, and also that a permanent body to regulate human subjects research be established by Congress. While such a proposal was introduced before Congress (9). it was not successful. However, two other responses to this decade of research exposes were successful: New regulations were promulgated by DHEW designed to build upon and strengthen the 1966 guidelines, and the National Commission for the Protection of Human Subjects of Biomedical and Behavioral Research (National Commission) was established (10).

The National Commission, in existence from 1974 to 1978, was charged by Congress to investigate the ethics of human subjects research, particularly research with vulnerable populations, such as prisoners, children, and mentally disabled adults. In addition to producing 17 separate reports on research with each of these populations, the National Commission created The Belmont Report (11). *The Belmont Report* laid out three principles of bioethics to help examine the ethics of any research endeavor that involves human subjects, principles that remain extremely influential in contemporary bioethics: beneficence, respect for persons, and justice (for further discussion of these principles, see Ref. 12). Briefly, beneficence requires us to look out for the welfare of others. In the context of research, this means that harms to potential subjects must be minimized, and balancing of harms and benefits must occur. Respect for persons requires us to treat individuals as autonomous agents and, when persons have diminished capacity, requires us to protect them from harm. Therefore this principle requires us to respect the decisions and judgments of others, even if we may disagree with them. It is out of the principle of respect for persons that we

are required to engage in a process of informed consent with research subjects. Justice requires us to be fair in the distribution of research burdens and benefits. Because of justice, we cannot disproportionately target vulnerable populations for enrollment in research to bear its risks, nor can we allow only those who are well-to-do and sophisticated to reap the benefits of research participation.

The 1974 DHEW regulations in many ways formalized the 1966 PHS guidelines that had required IRB review at local research institutions. The regulations went further, however, by delineating the elements that must be included in the informed consent process conducted with research subjects. These elements will be described below. All institutions that receive federal funding remain subject to these regulations. In 1975 the original regulations (Subpart A) were supplemented by another set of regulations (Subpart B) pertaining to research with pregnant women and fetuses. In 1978 Subpart C was added, providing regulations for the conduct of research with prisoners, and in 1983 Subpart D was added to provide oversight for research with children. In 1991 fifteen other federal agencies adopted Subpart A to apply also to their own research, calling it "The Common Rule." *The Common Rule* is in existence today, providing a common set of regulations for almost all federal agencies that either sponsor or conduct human subjects research in the United States.

Clearly, attention to the adequacy of human subjects protections usually has occurred in the context of a specific example that raised concern. The most recent such instance prompted the creation of the President's Advisory Committee on Human Radiation Experiments (ACHRE). ACHRE was formed in 1994 in response to allegations that radiation-related research was conducted on Americans between the 1940s and 1970s (during the cold war) without the participants' knowledge or consent. The Advisory Committee investigated these allegations and their extensive surrounding history. Moreover ACHRE did work examining the ethics of contemporary human subjects research, understanding that accusations of past abuses would raise questions in the minds of Americans about how much trust ought be put in current research practices and to establish a basis in fact on which to make recommendations for change in the future. Among ACHRE's recommendations were that research ethics training should be required of all research students and trainees, and that competency in research ethics should be required of all individual and institutional federal research grant recipients (9). Further ACHRE recommended that IRBs develop mechanisms to allocate their time more appropriately to riskier and more complex research, that information provided to patient-subjects clearly distinguish research from treatment, and not overestimate potential benefits. ACHRE also recommended that oversight of research be improved and that sanctions be created for those who do not comply with federal regulations (9, pp. 524–526).

Mention also should be given to the code that addresses specifically the conduct of human subjects research in the international setting. Put forth by the Council for International Organizations in the Medical Sciences

(CIOMS), the CIOMS guidelines were established in 1982. Among its provisions were that community consent, while often appropriate and necessary to obtain, cannot serve as a substitute for individual consent, and that research must be responsive to the health needs of the community in which the research occurs.

ELEMENTS OF INFORMED CONSENT

This section will describe both the specific elements of informed consent as delineated in the Common Rule and also will provide broader discussion of concepts inherent to the theory of informed consent. Informed consent has a history in medical practice that precedes but clearly influences its history in research. The clinical history starts with a series of cases at the beginning of the twentieth Century brought by patients who had not given their *consent* to certain procedures. Later in the century are cases brought by patients for not having been adequately *informed* about the procedures to which they were providing consent. In 1906, in *Pratt v. Davis*, a doctor performed a hysterectomy on a woman without her consent (13). The defense had been that when a patient enlists a doctor or surgeon's services, the doctor is given "implied license to do whatever in the exercise of his judgment may be necessary." The defense was rejected, and the case was decided in the patient's favor. Perhaps the most famous case on consent was *Schloendorff v. Society of New York Hospitals* in 1914 (14). In this case the patient had given consent for exploratory abdominal surgery but had specifically requested no further surgery. The surgeon, upon finding a fibroid tumor during the surgery, had gone ahead and removed it. One of the justices ruling on the case, Judge Benjamin Cordozo, wrote in his historic opinion, "Every human being of adult years and sound mind has a right to determine what shall be done with his own body; and a surgeon who performs an operation without his patient's consent commits an assault, for which he is liable in damages." The next fifty years brought a series of cases that led to a new requirement of patients also being informed. A landmark case, perhaps because it coined the phrase "informed consent," was in 1957, *Salgo v. Leland Stanford Jr. University Board of Trustees* (15). In this case Martin Salgo had undergone translumbar aortography which resulted in permanent paralysis, a known potential risk of the surgery. Mr. Salgo sued physicians for failing to warn him that a potential risk of the procedure was paralysis. The court found, in the patient's favor, that physicians had a duty to disclose "any facts which are necessary to form the basis of an intelligent consent by the patient to proposed treatment." This evolved into a requirement that all issues that would be pertinent to a patient when making a decision — such as the nature, consequences, risks, benefits, and alternatives to a proposed treatment — be disclosed before a patient makes a decision.

Extrapolating to the research context, informed consent requires both informing the research participant about the research and obtaining the participant's consent. *Informing* a participant requires disclosing pertinent information and ensuring at least some threshold

level of understanding. Of course, fulfilling the former is considerably easier than fulfilling the latter, and consequently significantly more attention in the literature and the regulations exist concerning disclosure. For *consent* to occur, the participant must be competent and must make the decision voluntarily. The concepts of disclosure, understanding, competence, and voluntariness will be discussed below.

Disclosure

While understanding is what ultimately is required for valid informed consent, it is typically through disclosure that a research subject learns enough about the research project to understand it. If, indeed, a research subject were to understand the research other than through disclosure (e.g., through a video tape or from prior familiarity with the research), then substantially less disclosure would be necessary. Most discussions of informed consent, however, rightly assume that subjects know little about the research before they enter into a research relationship. Consequently the regulations governing human subjects research (the Common Rule) lay out in detail the elements of informed consent that must be disclosed (16):

1. A statement that the study involves research, an explanation of the purposes of the research and the expected duration of the subject's participation, a description of the procedures to be followed, and identification of any procedures that are experimental.

2. A description of any reasonably foreseeable risks or discomforts to the subject.

3. A description of any benefits to the subject or to others which may reasonably be expected from the research.

4. A disclosure of appropriate alternative procedures or courses of treatment, if any, that might be advantageous to the subject.

5. A statement describing the extent, if any, to which confidentiality of records identifying the subject will be maintained.

6. For research involving more than minimal risk, an explanation as to whether any compensation and an explanation as to whether any medical treatments are available if injury occurs, and if so, what they consist of, or where further information may be obtained.

7. An explanation of whom to contact for answers to pertinent questions about the research and research subjects' rights, and whom to contact in the event of a research-related injury to the subject.

8. A statement that participation is voluntary, refusal to participate will involve no penalty or loss of benefits to which the subject is otherwise entitled, and the subject may discontinue participation at any time without penalty or loss of benefits to which the subject is otherwise entitled.

Ruth Faden and Tom Beauchamp, in their book on informed consent (17), discuss that potential subjects want to know information that is *material* to them. The federal regulations are based on assumptions about what would be material to most persons when considering participation. Faden and Beauchamp remind us, however, that certain additional details about a project may be important for a given individual, whereas they may be irrelevant to others. For this reason it is important that in addition to providing the basic elements of disclosure, there also be the opportunity for an informed consent discussion in which potential subjects can raise additional questions. Faden and Beauchamp note that information that is material is not necessarily information that is required for the *decision*, but rather it may be important for the potential subject to feel s/he has a good understanding of the research. For example, it may be important to a potential subject to know whether study hours can be arranged at night, even if the potential subject knows that s/he will participate regardless of the answer.

Understanding

It has been written that "assent to [research] given by a [subject] who actually has not understood disclosed information is not valid authorization" (18, p. 59). This reminds us that while adequate disclosure in most instances is necessary for there to be understanding, it in no way guarantees it. Many factors can contribute to inadequate understanding of research. Investigators may use jargon or language that is difficult for potential subjects to understand (19–21). Subjects who also are patients may be anxious about their medical condition and unable to focus on the specific information provided. Information also might be provided quickly and in large amounts, with little or no time for discussion, such that subjects simply cannot remember or integrate all they were told. Sometimes the words used in research are qualitative and vague. Nakao and Axelrod (22) as well as Fetting et al. (23) found imprecision in many terms used to describe research risks and benefits (e.g., "rare," "infrequent"), and recommend that numeric estimates be used when possible. In their review Silva and Sorrell report many factors that influence comprehension of research information (24). Several studies report that often too much information is given, and that subjects better understand the research when smaller amounts of information are provided. Who delivers information can also be relevant. Muss et al. found that more information about chemotherapy side effects was retained when risks and benefits were described by personnel other than the doctor (25). Not surprising, higher education is associated with greater comprehension of informed consent information (25). Understanding also is unavoidably compromised by lack of experience with a situation. That is, it can be difficult for subjects to imagine how they would react to a certain side effect until they have experienced it. While not a guarantee of understanding by any means, most commentators recommend using some measure to assess subjects' understanding of what they have been told before embarking on the research. This includes not simply seeing whether they can recall what are the study procedures but, at least as relevant, if they can recall what the *purpose* of the research is.

One area where understanding about research may be particularly challenging is clinical research. Here patient-subjects often enroll because of a hope that the research will be of medical benefit to them personally. This can result in research subjects losing sight of the fact that the study is investigational, the intervention has not been shown to be a valid medical treatment, and the purpose of the activity is not primarily to treat their own medical condition. Given that many clinical research investigators also are physicians, it can be easy for patient-subjects confuse clinical research participation with medical care. Studies have demonstrated that some proportion of patients enrolled in research do not understand, or at least, do not remember, that they are enrolled in research. Riecken and Ravich found that 28 percent of patients enrolled in research studies through Veterans Administration hospitals were not aware of their participation in research, despite having signed consent documents, remembering that the intervention had been explained to them and believing that they had been given sufficient information (26). The rate of unawareness dropped the longer patients were enrolled in studies, and patients were more likely to be unaware of their participation if a staff member other than the investigator had explained the study. Penman et al. similarly found that nearly one-fourth of patients receiving investigational chemotherapy did not recall that it was investigational (27).

Yet more subtle clouding of the boundary between treatment and research is demonstrated by many patients who are perfectly aware that they are participating in research but nonetheless view the research as a treatment intervention to improve their underlying disease state. This clouding is critical to issues of informed consent, since a key tenet of informed consent is understanding by the patient (17,18), and a key element of understanding in clinical research is appreciating what is meant by investigational.

Some years ago Appelbaum et al. described this threat to understanding as the "therapeutic misconception" (28). Appelbaum et al. report a psychiatrist approaching a patient to consider participating in a research study. The patient responds, "Yes, I'm willing to do anything that might help me." The patient further says that he understands what is meant by the trial's randomized placebo-controlled design but then goes on to say that he believes *he* will receive the study medication most likely to help him.

Competence

In order for informed consent to be valid, the person providing consent must be considered competent to consent. When a person is *not* competent to consent, a surrogate must consent on the subject's behalf, and IRBs usually engage in a higher level of oversight. That is, there is even greater scrutiny of the risk–benefit ratio, with an assumption that individuals who cannot understand the research in which they will be participating cannot be subject to as much risk as could a person who fully understands. A key principle in discussion of competence here is that individuals are not necessarily uniformly competent or uniformly incompetent. Rather, the relevant question is whether the individual is competent to understand what is required of him or her by participating, and whether the individual is competent to exercise autonomous decision-making capacity in consenting to enroll. Individuals, for example, may be able to understand that they are being asked to be in research and what would be required of them, while having no recollection of what day of the week it is or who is President. Conversely, some individuals may function well in day-to-day activities but have no understanding of who doctors or researchers are, and cannot comprehend an informed consent discussion. Discussion concerning three specific populations whose full consent often is impaired — children, persons in emergency situations, and persons of limited decision-making capacity — will be provided below.

Voluntariness

Faden and Beauchamp write that "a fundamental condition of personal autonomy is that actions ... are free of ... controls on the person" (17, p. 256). Persons may be *influenced* in their thinking (e.g., by family members, information, or doctors), which ethically is quite consistent with valid informed consent, in contrast to being *controlled* by others (e.g., they are being forced or coerced), whereby they are not acting autonomously. In research, potential subjects may be most likely to feel "controlled" by others if they believe that other opportunities are dependent on their participation. For example, if a patient believes that her doctor will not treat her as well if she refuses to participate, then her decision cannot be considered voluntary. Similarly, if parole comes more quickly to prisoners who agree to participate in research, their decisions about participation may not be fully voluntary. All research consent discussions and forms must therefore emphasize that access to other opportunities will not be affected by potential subjects' decisions about participation. IRBs may decide that certain recruitment strategies are unacceptable because they would raise questions of compromised voluntariness. For example, professors may be told that they cannot solicit research participation from their own graduate students, or physicians from their own patients; in the latter case another physician can seek consent from patients, but the patient's own doctor may be perceived as exerting too strong an influence. Generally, the law "has long recognized a consent or refusal coerced by threats or manipulated by misrepresentation [to be] invalid" (12, p. 163).

CHALLENGES TO INFORMED CONSENT: SPECIAL POPULATIONS

Children

Tension exists when conducting research with children. On the one hand, children are assumed to be unable to fully appreciate the consequences of their actions, and therefore are not considered to be fully autonomous for the purposes of consenting to research. On the other hand, there are conditions that uniquely affect children, or affect children in ways that are different from how they affect adults, and therefore the medical or psychological care of children cannot be improved without research. Concerns based in justice dictate both that children not be used in research when fully competent adults could provide the same answers, yet also that children as a class not be denied the benefits of research knowledge. That is, children as a class ultimately are harmed if they are treated with drugs that never have been tested properly for safety, efficacy, or dosing in children. Consequently children can be used in research ethically only when the research question relates uniquely to them or if it is not greater than minimal risk research. Further, given that children cannot fully consent, we require a higher standard of beneficence. It is assumed that more protection for their welfare should be provided, with IRBs being more paternalistic than they are with competent adults. When children are approached for participation, it is required both that their parent or guardian sign written permission and that the child "assent" at an age-appropriate level. That is, researchers are required to explain certain relevant pieces of the research to the child in language the child will understand to see if the child is willing to participate. This assent undoubtedly will not include all information about the research. It may be as simple as saying to the child, "to learn more about your health, we would like to take some blood from your arm and ask you some questions. We're asking all the kids who come into the clinic today to do this. It's OK if you decide you don't want us to do this. Is it OK with you for us to do this?" In certain instances where the research in expected to be beneficial for the child, parents may overrule a child's lack of assent.

The Departments of Health and Human Services (DHHS) regulations regarding human subjects research were amended in 1983 to include Subpart D, "Additional Protections for Children Involved as Subjects in Research." The regulations define assent as a child's "affirmative agreement to participate in research. Mere failure to object should not, absent affirmative agreement, be construed as an assent" (29). The regulations say that minimal risk research with children is ethically acceptable, assuming that permission from parents and assent of child have been obtained. Research involving greater than minimal risk is allowed only if the risk is justified by anticipated benefit, if the benefit anticipated is at least as great as that offered by alternatives to participation, and if assent and permission are obtained. Research offering more than minimal risk and *not* offering direct benefit to the individual subjects is allowable only if the risk anticipated is minimally more than the risk the child would have experienced through his or her illness or through ordinary medical treatment, and if the expected generalizable knowledge is clear. Research involving more than minimal risk over what the child otherwise would experience, with no anticipated individual benefit, is unlikely ever to be approved. To be approved, the Secretary of DHHS would need to determine, after consultation with "a panel of experts in pertinent disciplines, for example, science, medicine, education, ethics, law, and following opportunity for public review and comment" (30) that the research is expected to yield great understanding for the treatment of prevention of a serious problem affecting the health or welfare of children, that sound ethical principles are otherwise followed, and that permission and assent are obtained. Two other changes have occurred recently with regard to inclusion of children in research. The NIH issued "Policy and Guidelines on the Inclusion of Children as Participants in Research Involving Human Subjects" in 1998 (31). The policy mandates that research conducted or funded by the NIH include children in all studies unless there are scientific or ethical reasons to exclude children (for further discussion, see Ref. 32). The Food and Drug Administration (FDA) passed similar regulations in December 1998 requiring manufacturers of new and marketed drugs to evaluate the safety of those products in pediatric patients if the product is likely to be used in children (33).

Emergency Consent. In October 1996 the Federal Register published for the first time a waiver to informed consent requirements when conducting research in the emergency setting and when certain conditions apply (34). The regulations apply to Subparts A and D of the federal human subjects regulations (i.e., research with adults and with children) but not to Subparts B and C (research with pregnant women/fetuses and research with prisoners). The rationale for the new regulations was similar to rationales for wanting to be more inclusive of other populations of persons with limited decisional capacity: While it is imperative that the welfare and interests of individual vulnerable persons always be protected, it also is important that interventions that potentially might help such persons be tested and identified. Consequently, where research questions uniquely affect persons in emergency situations, it is sometimes appropriate, in balancing risks and benefits, that research be conducted. The conditions laid out in the new requirements include that eligible patients must be in a life-threatening condition, the available (nonexperimental) treatments are unsatisfactory, there is a need to collect scientific evidence to test new interventions, and an authorized representative for the patient cannot be found in the short time frame required. The investigators are required under the new waiver to document their efforts to find the patient's representative before enrolling the patient without consent. Further there must be procedures in place to inform the subject or his/her legally authorized representative, at the "earliest feasible opportunity," of the subject's inclusion in research and relevant details. If the trial remains ongoing at the time when the subject becomes aware, or the family is contacted, they have the right to terminate participation immediately. It also is required that the relevant IRB approve both the research activity and the waiver of consent. The waiver states explicitly that when appropriate, placebo-controlled trials are allowable under the new policy.

While in general, there has been considerable support for the emergency waiver, some have taken issue with the specific language included that could, again, lead to a misunderstanding about the distinctions between research and treatment. Jay Katz, for example, objected

to the insinuation in the new regulations that emergency research has therapeutic intent for the individual patient: "In its emphasis on therapeutic benefits, the FDA obscures the fact that some of the permissible research activities either hold out no promise for therapeutic benefit or are so vaguely defined that potential therapeutic benefit can be inferred when research is the predominant intent.... Research must be stripped of the therapeutic illusion which misleads patient-subjects into believing that they are receiving the most advanced and beneficial treatments available, when instead they are being asked to serve the interests of science.... Research is not treatment" (35).

Other Persons With Limited Decision-Making Capacity

It is in many ways surprising that except for research with children and with persons in emergency situations, there are no federal regulations governing research with persons with questionable capacity to consent. Instead, it is left to IRBs to determine whether the research ought to go forward and whether consent and other safeguards are sufficient. Again, the challenge when evaluating research of this nature is wanting to enable more knowledge to accrue in this area, which clearly will benefit persons with mental disabilities, yet not wanting to compromise the interests of the individuals who participate in the studies. In order to determine if the participant can provide consent for him/herself, investigators should devise a method for evaluating whether participants have the ability to "understand, appreciate, and reason about the experimental situation" (36). Where they cannot, someone else must give permission on their behalf. Moreover, again, assent to whatever degree is possible, must be sought. If the potential subject will not assent, generally speaking, s/he should not be included in the research. Exceptions sometimes are made when there is clear likelihood that individual benefit would come to the participant as a result of enrolling, but given that research is by definition testing the efficacy of interventions, this may be difficult to prove. When deciding on behalf of someone else, two different standards can be used. If the person with limited decisional capacity previously had capacity, and had made relevant preferences known, then a substituted judgment standard can be used, meaning that the surrogate is simply voicing what s/he believes are (or previously have been) the wishes of potential subject. In contrast, if the subject has never had relevant capacity, then a best interest standard must be used. Essentially a caring person charged with the responsibility of guardianship for the subject decides what is best for the subject, based on how much anxiety it could provoke, its safety and invasiveness, and expected benefits and burdens (36,37). The obligation remains throughout the study to monitor ongoing effects in order to determine whether the subject's participation should be stopped at any time.

Persons With Serious Illnesses

As described earlier, a challenge for researchers is conducting research with persons with serious illnesses. These persons often are vulnerable by virtue of their illness. They may be so eager to participate in anything that they, rightly or wrongly, believe may help them, that their judgment may be clouded. Moreover physician-investigators who sincerely care about the well-being of their patients, may encourage their patients to participate in research of unknown or little value because they too do not want to admit that few other options remain. While participation of such persons also may occur for altruistic reasons and while research in certain circumstances requires the participation of such persons, IRBs and investigators must be sensitive to these vulnerabilities, vulnerabilities that are not recognized the way being a prisoner, a child, or a person with limited decisional capacity routinely are.

Studies have been done that highlight how persons with serious illnesses often overestimate the benefit they could get from the research and/or forget or ignore altogether, as described earlier, that the research intervention is investigational. For example, Penman et al. found that the primary reasons 144 cancer patients accepted investigational chemotherapy were trust in their physician and belief that the treatment would help (27), and that one-quarter of the patient-subjects interviewed did not recall that the chemotherapy was investigational. Cassileth et al. documented that respondents describe research in different terms depending on whether they are speaking about research generally or their own participation. Patients and members of the general public reported that people generally should participate in research, in order to benefit others and increase scientific knowledge, but that they themselves would participate primarily in order to help "get the best medical care" (38). In a study conducted by the Advisory Committee on Human Radiation Experiments, cardiology and oncology outpatients who had been research participants said in closed-ended interviews that they viewed research as a way to help others (76 percent) and joined to get better treatment (67 percent) and because the research gave them hope (61 percent) (39). When interviewed in greater depth using more open-ended questions, however, these patients with serious illnesses reported that they joined research studies either because their doctors had recommended it, or because they believed they would gain additional medical benefit. For example, one patient-subject said, "When you reach that stage . . . and somebody offered that something that could probably save you, you sort of make a grab of it, and that's what I did" (40).

Concern about patient-subjects' vulnerabilities ought be most acute in the context of Phase I trials where the chance of personal medical benefit is minimal at best (41–43). Studies with patient-subjects enrolled in Phase I research, however, echo other findings. Rodenhuis et al. interviewed 44 patients who had agreed to participate in a Phase I cancer trial. They report that for all patients who participated in the study, "the hope for stabilization, improvement, or even cure of their diseases was the major motivation" (44). Further they report that doing *something* seemed to be of psychological benefit to patients in and of itself: "By continuing to receive medical attention and some form of treatment, they were able to cope with their incurable

diseases and deny or postpone more easily the realization of impending death." Daugherty et al. asked patient-subjects enrolled in Phase I cancer research specifically whether they *expected* therapeutic benefit as a result of participation (45). Twenty-two percent of patients said they believed they would receive therapeutic benefit from their participation.

Genetics Research Informed Consent

Increasingly, researchers are conducting genetic research that raises new issues related to genetics research informed consent. Among these are issues surrounding use of stored tissue samples, the challenge of learning uninterpretable information, the potential for learning information about others who did not consent to the research, the potential for learning potentially harmful or damaging information, and the potential for risk to communities as well as to individuals.

Stored Tissue Samples. Tissue samples that are obtained from individuals, either as part of a research study or as part of clinical care, can be stored indefinitely. As such, previously collected samples of blood, particularly those that can be linked to certain demographic and clinical characteristics of the source individuals, are of great interest to other researchers. The ethics question becomes when and under what circumstances those samples may be used by future researchers for purposes quite unrelated to those for which they originally were collected and for purposes never disclosed to source individuals. The concern is that material risks to individuals can occur when genetic information about them is shared (see the discussion below); moreover, gathering information about someone that does *not* result in material harm still can wrong them if it is done without their knowledge and consent. In deciding individual cases, it is important to examine for what purposes consent originally was obtained from the source individuals and whether future researchers want the samples to remain identifiable. Many persons believe that it is inappropriate for individuals ever to be asked to provide blanket consent for all unknown future purposes if the samples remain identifiable (46). Rather, individuals may be asked willingness to provide consent for focused future purposes of their identifiable samples (e.g., future studies also related to Alzheimer's research). Bartha Knoppers has developed a core list of elements to include in a consent discussion when DNA samples will be stored (47). She suggests that individuals be required to agree or disagree with specific uses of data, including, for example, whether to undergo diagnostic tests, whether to permit consultation of their medical records, and whether to be contacted if genetic disorders are identified. Alternatively, samples may be stripped of identifiers. When samples are made anonymous, most ethics concerns disappear, since most risks and harms to individuals can never occur. The use of anonymous samples requires that the samples already existed when the new research plan was proposed and that it is impossible to go back and link samples to identifiable persons (48). Generally, when genetic information is anonymous, the potential benefits of scientific research are thought to outweigh the risks to individual integrity even without permission from the source individuals.

Uninterpretable Information Learned through Genetic Research. Genetic testing, particularly in its early stages, often is probabilistic, rather than predictive, in nature. Through genetic research, markers may be identified that are associated with certain conditions, but their presence does not guarantee that the individual will become clinically affected, nor will all clinically affected persons possess the marker. Rather, additional markers and/or environmental factors will need to be identified that increase the predictive value of the genetic tests. This fact poses the challenge of, first, explaining complex probabilities to research participants who are more accustomed to extremely sensitive and specific diagnostic tests, and, in addition, determining when the level of uncertainty is so great that it is inappropriate to provide genetic testing results conducted in research to research participants. Indeed, some research at the initial stages of identifying a genetic marker may be considered far too premature to provide information to subjects. Researchers should consider the implications of ambiguous information before the study is initiated and determine *in advance* whether or not research test results will be made available to study subjects. If researchers determine that the results will not be made available, this must be communicated very clearly to subjects during the informed consent process. If results are to be made available, it is best to have a genetic counselor provide and interpret the information to subjects (46). Occasionally researchers may find themselves in the position of having informed subjects in advance that no information gained through the research would be disclosed, but then coming to believe that the information has greater clinical relevance than anticipated. Deciding whether to change disclosure procedures from that which was originally described is a difficult decision and should be made on a case-by-case basis with careful consideration and consultation from others, such as the IRB (49).

In addition to information being of questionable predictive value, there may be little to do as a result of learning it. There are many genetic conditions for which tests now are available yet for which there are no good treatments. It must be highlighted to research participants in such circumstances that the individual benefit to gaining the information (should it be disclosed) is psychological, rather than clinical. Further, when information will not be disclosed, it must be made clear that the purpose of the study is *not* to provide clinical benefit to the individuals who enroll. That is, the study may be conducted in order to compare the prevalence of a marker among different populations, or to determine the sensitivity and specificity of a recently developed test. Ellen Wright Clayton et al. (48) discuss possible legal liability from knowing information about a person that is not disclosed; she suggests that the risk of this is small if truly clinical decisions would not have been made differently had access to the information been available. It must be remembered that in some instances, however, the information *is* of clinical value to the individual or

his/her relatives. In such instances, if researchers have a threshold level of confidence in the validity of their data, research participants should be given the choice of whether to learn the information discovered about them through their participation. Further, given that subjects often expect to be told information learned about them, or expect that hearing nothing means that no abnormalities were found, researchers must be very clear in circumstances in which research findings are not disclosed that this does not in any way reveal whether or not a marker was identified for that person.

Potential for Learning Information About Others without their Consent. Inherent to genetic information is that it usually is hereditary, and usually reveals at least some amount of information about other members of one's family. The type of study often conducted to identify a genetic marker is called a pedigree study. In pedigree studies researchers do genetic analyses of blood samples of many members of an extended family known either to be affected or unaffected by a hereditary condition in order to identify a genetic marker that exists uniquely among the affected individuals. Although researchers seek the consent of as many family members as possible, some cannot be found, and others refuse to participate. The awkwardness arises that through the testing of those who *do* provide consent, researchers may de facto learn genetic information about individuals who were not involved in the study. For example, an adult child may refuse to be in a study in which both of her parents agreed to participate, and the child's makeup is inevitably knowable; or one parent may refuse to participate while the other parent and children consent, again inevitably revealing part of the genotype of the parent who refused participation. It is up to IRBs and researchers to determine whether this ability to identify individuals, which ultimately is an invasion of their privacy, is acceptable, taking into consideration the sensitivity of the information revealed. In rare instances, pedigree studies may only be allowable when all members of an extended family agree to participate. Again, it is the role of genetic counselors, acting on behalf of investigators, to make this knowledge clear to participants in the informed consent process before a pedigree or family study is initiated.

Issues of family identification and confidentiality also become relevant in the context of research presentation and publication. While it always is true that research findings must be presented in ways that do not reveal the identities of the individuals who participated, researchers may forget that simply deleting names of individual research subjects does not guarantee anonymity. Pedigree studies may reveal the identities of individuals if, for example, the disease is rare and the family has other unusual characteristics (e.g., it includes triplets), and/or if the town in which many of the family live is identified in the report. It also is possible that individuals reading a published pedigree in a medical journal will recognize the family described as their own and will learn genetic information about other members of their extended family. There is great debate about whether it is ethically and scientifically appropriate for researchers to modify the pedigree slightly in published reports in order to make the family no longer identifiable, for example, by randomly adding a few family members to the tree who never existed.

Gathering Potentially Harmful or Damaging Information, and Issues of Confidentiality. The information gathered through genetic testing may put individuals at risk of psychological or material harm. Individuals differ greatly in how they react to information that they often can do nothing about it. While some individuals find the knowledge beneficial, others may be made more anxious and upset about something over which they have no control. To the degree possible, counselors should try to walk individuals through different scenarios and try to determine how they think they would react to different types of information before they decide to be in studies in which disclosure could occur.

Moreover, impaired access to certain opportunities, such as health or life insurance and employment, may result as a consequence of participation in genetic research (50). While this rarely is due to researchers' negligence in maintaining confidentiality—and indeed some researchers have Certificates of Confidentiality from the federal government to provide additional confidentiality protections (51,52)—information can become available to insurers in other ways. For example, an individual may mention to her personal physician that she is enrolled in the pedigree study. If the physician makes a note of this in the patient's medical record, the information—even if the test results are not documented—could cause a red flag for insurers who might for other reasons be examining the patient's medical record. In such an instance, the insurer might insist on obtaining a copy of the research records before determining whether the individual was eligible for health, life, or disability insurance. Also persons applying for individual health, life, or disability insurance policies might be asked on a general health question about any medical testing or adverse findings of which they are aware. There is not clear consensus concerning whether information that individuals did not obtain through the clinical setting and was not obtained out of clinical concern must be disclosed to insurance companies in this context.

In some studies researchers bill health insurance companies for genetic testing conducted as part of the research. Obviously this alerts companies to the fact that the testing was done and, usually, that the individual is from a family of higher than average risk of disease. It may be appropriate for researchers, out of respect for the welfare of their participants, not to bill insurance companies in this manner.

A further potential harm arises from learning unanticipated information. While uncommon, it sometimes happens when conducting genetic testing that information that was not sought or anticipated becomes apparent. For example, researchers may become aware of mistaken paternity or of anomalies in the sex chromosomes (e.g., XXY genotypes). This often is troubling to researchers who are caught completely off guard and wonder if they have a responsibility to share the information with participants. In general, it is best for researchers to make it clear in advance that such incidental information will

not be shared with participants. The exception may be information that researchers, *in consultation with others*, believe holds clinical relevance to the individual would lead the individual to act differently in some relevant way were he aware of it.

Information of Relevance to Communities, in Addition to Individuals. Often specific communities are targeted for genetic research. This may be a community of family members, or it may be an ethnic or religious group (e.g., Ashkenazi Jews), or it may be a group defined by where it lives, such as research conducted as part of the Human Genome Diversity Project, which seeks "information on human genetic diversity, the origins and migration of human populations, and genetic factors related either to resistance or susceptibility to disease" (53, p. 7). In such contexts, risks and potential benefits to communities must be considered, and a person or persons able to provide consent on behalf of the community must be identified. It has been suggested that a fourth principle of bioethics is needed to supplement those described earlier, a principle of "respect for communities" (53). Risks to communities from genetic research are real, since even research that strips samples of individual identifiers often still identifies the sample by "kindred, locality, or ethnicity" (54). In the past, material harms came to persons with sickle cell trait (rather than disease) as a result of being identified (54), and harmful stereotyping or self-perception can result from sweeping conclusions that sometimes are made in research. Foster et al. suggest a process of "communal discourse" to supplement and inform individual informed consent (54). This would occur through public meetings with representatives of the community in which opinions and suggestions are sought and communicated to investigators. Moreover, consulting an advisory board through the duration of the study can be helpful in ensuring that two-way communication remains ongoing.

In terms of recommendations, researchers should engage in an extensive consent discussion and dialogue with potential participants before and, often, during the research. All practical and procedural issues must be covered (amount of time required, purpose of results, etc.). Moreover, a genetic counselor should try to "walk" potential participants through different scenarios to try to discern how they would react to hearing ambiguous or probabilistic information. As a practical matter, given the risks to confidentiality that exist, investigators should apply for a certificate of confidentiality in genetic studies and also should minimize the likelihood that information they learn through studies can become part of the medical record. This means maintaining separate research and clinical records, not filing claims with health insurance companies for tests conducted solely for research purposes, and counseling participants about circumstances that might lead to information about them becoming part of their medical record.

CONCLUSION

Research as an enterprise has grown exponentially in the last several decades, and the field of research ethics has grown with it. It is unfortunate that it takes examples of unethical practice to prompt us to develop appropriate guidelines and regulations, but the standards we currently have certainly are far more likely to protect the interests of potential subjects than was true decades before. As more research is conducted in the areas of genetics and incapacity, surely the field of research ethics similarly will move forward. In the meantime there is no better protection for subjects than a conscientious and humble researcher, who is aware that he or she may be easily misunderstood, and who is aware that the welfare of the subject always is of greater importance than any particular research question.

BIBLIOGRAPHY

1. *Trials of War Criminals before the Nuremberg Military Tribunals under Control Council Law No. 10*, vol. 2, U.S. Government Printing Office, Washington, DC, 1949, pp. 181–182.

1a. *United States v. Karl Brandt* et al., *The Medical Case, Trials of War Criminals before the Nuremberg Military Tribunals under Control Council Law No. 10*, U.S. Government Printing Office, Washington, D.C., 1949.

2. National Institutes of Health, Policy statement of November 17, 1953: *Group Consideration of Clinical Research Procedures Deviating from Accepted Medical Practice or Involving Unusual Hazard*, NIH, Washington, DC, 1953.

3. World Medical Association, *Declaration of Helsinki: Recommendations Guiding Medical Doctors in Biomedical Research Involving Human Subjects*, Adopted by the 18th World Medical Assembly, Helsinki, Finland, 1964 and as revised by the World Medical Assembly to Tokyo, Japan in 1975, in Venice, Italy in 1983, and in Hong Kong in 1989.

4. H.K. Beecher, *N. Engl. J. Med.* **274**, 1354–1360 (1966).

5. H.K. Beecher, *JAMA, J. Am. Med. Assoc.* **169**, 461–478 (1959).

6. Surgeon General, Public Health Service, to the Heads of the Institutions Conducting Research with Public Health Service Grants, February 8, 1966.

7. J.H. Jones, *Bad Blood*, Free Press, New York, 1993.

8. U.S. Department of Health, Education, and Welfare, *Final Report of the Tuskegee Syphilis Study Ad hoc Panel*, U.S. Government Printing Office, Washington, DC, 1973.

9. Advisory Committee on Human Radiation Experiments, *Final Report*, Oxford University Press, New York, 1996, p. 103.

10. National Research Act of 1974. P.L. 348, 93rd Cong., 2d Sess. (July 12, 1974).

11. National Commission for the Protection of Human Subjects of Biomedical and Behavioral Research, *The Belmont Report: Ethical Principles and Guidelines for the Protection of Human Subjects of Research*, April 18, 1979, U.S. Government Printing Office, Washington, DC.

12. T.L. Beauchamp and J.F. Childress, *Principles of Biomedical Ethics*, 4th ed., Oxford University Press, New York, 1994.

13. *Pratt v. Davis*, 118 Ill. App. 161 (1905) *aff'd* 79 N.E. 562 (Ill. 1906).

14. *Schloendorff v. Society of New York Hospitals*, 105 N.E. 92 (N.Y. 1914).

15. *Salgo v. Leland Stanford Jr. University Board of Trustees*, 317 P.2d 170 (Cal. Ct. App. 1957).

16. Title 45, Code of Federal Regulations, Part 46, §46.116, General requirements for informed consent.

17. R.R. Faden and T.L. Beauchamp, *A History and Theory of Informed Consent*, Oxford University Press, New York, 1986.

18. P.S. Appelbaum, C.W. Lidz, and A. Meisel, *Informed Consent: Legal Theory and Clinical Practice*, Oxford University Press, New York, 1987.

19. K.D. Hopper, T.R. TenHave, and J. Harzel, *Am. J. Radiol.* **164**, 493–496 (1995).

20. C.D. Meade and D.M. Howser, *Oncol. Nurs. Forum* **19**, 1523–1528 (1992).

21. M. Hochhauser, *IRB: Rev. Hum. Subj. Res.* September–October, 5–9 (1997).

22. M.A. Nakao and S. Axelrod, *Am. J. Med.* **74**, 1061–1065 (1983).

23. J.H. Fetting et al., *J. Clin. Oncol.* **8**, 1476–1482 (1990).

24. M.C. Silva and J.M. Sorrell, *Int. J. Nurs. Stud.* **21**, 233–240 (1984).

25. H.B. Muss et al., *Cancer (Philadalphia)* **43**, 1549–1556 (1979).

26. H.W. Riecken and R. Ravich, *JAMA, J. Am. Med. Assoc.* **248**, 344–348 (1982).

27. D.T. Penman et al., *J. Clin. Oncol.* **7**, 849–855 (1984).

28. P.S. Appelbaum et al., *Hastings Cent. Rep.* **17**, 20–24 (1987).

29. Title 45, Code of Federal Regulations, Part 46, §46.402.

30. Title 45, Code of Federal Regulations, Part 46, §46.407.

31. National Institute of Health (NIH), *Guide for Grants and Contracts*, vol. 27, NIH, Washington, DC, 1998.

32. J. Sugarman, A.C. Mastroianni, and J.P. Kahn, *Ethics of Research with Human Subjects*, University Publishing Group, Frederick, MD, 1998, pp. 169–175.

33. U.S. Department of Health and Human Services, Food and Drug Administration, *Fed. Regis.* **63**(231), 66631–66672 (1998).

34. U.S. Department of Health and Human Services, *Fed. Regis.* **61**, 51531–51533 (1996).

35. J. Katz, *Hastings Cent. Rep.* **27**, 9–11 (1997).

36. R. Dresser, *JAMA, J. Am. Med. Assoc.* **276**, 67–72 (1996).

37. E. DeRenzo, *Cambridge Q. Healthcare Ethics* **3**, 539–548 (1994).

38. B.R. Cassileth, E.J. Lusk, D.S. Miller, and S. Hurwitz, *JAMA, J. Am. Med. Assoc.* **248**, 968–970 (1982).

39. Advisory Committee on Human Radiation Experiments, *Final Report*, Oxford University press, New York, 1995.

40. N. Kass, J. Sugarman, R. Faden, and M. Schoch-Spana, *Hastings Cent. Rep.* **26**, 25–29 (1996).

41. D.D. von Hoff and J. Turner, *Invest. New Drugs* **9**, 114–122 (1991).

42. E. Estey et al., *Proc. Am. Soc. Clin. Oncol.* **3**, 35 (1984).

43. E. Estey et al., *Cancer Treat. Rep.* **70**, 1105–1115 (1986).

44. S. Rodenhuis et al., *Eur. J. Cancer Clin. Oncol.* **20**, 457–462 (1984).

45. C. Daugherty et al., *J. Clin. Oncol.* **13**, 1062–1072 (1995).

46. American Society of Human Genetics, *Am. J. Hum. Genet.* **59**, 471–474 (1996).

47. B.M. Knoppers and C.M. Laberge, *JAMA, J. Am. Med. Assoc.* **274**, 1806–1807 (1995).

48. E. Wright Clayton et al., *JAMA, J. Am. Med. Assoc.* **274**, 1786–1792 (1995).

49. E. Kodish, T.H. Murray, and S. Shurin, *Clin. Res.* **42**(3), 396–402 (1994).

50. N.E. Kass, *IRB: Rev. Hum. Subj. Res.* **15**, 7–10 (1993).

51. C.L. Early and L.C. Strong, *Am. J. Hum. Genet.* **57**, 727–731 (1994).

52. Public Health Service Act, S.301(d), 42 U.S.C.

53. K. Cranley Glass et al., *IRB: Rev. Hum. Subj. Res.* **19**, 1–13 (1997).

54. M.W. Foster, A.J. Eisenbraun, and T.H. Carter, *Nat. Genet.* **17**, 277–279 (1997).

See other HUMAN SUBJECTS RESEARCH entries.

HUMAN SUBJECTS RESEARCH, ETHICS, PRINCIPLES GOVERNING RESEARCH WITH HUMAN SUBJECTS

ROBERT J. LEVINE
Yale University School of Medicine
New Haven, Connecticut

OUTLINE

INTRODUCTION

In 1974 the United States Congress passed the National Research Act (Public Law 93-348) which established the National Commission for the Protection of Human Subjects of Biomedical and Behavioral Research (hereafter, the National Commission). This Commission was charged by Congress to "identify the basic ethical principles which should underlie the conduct of ... research involving human subjects (and) to develop guidelines which should be followed in such research to assure that it is conducted in accordance with such principles...." The principles identified by the National Commission were published in 1978 in its *Belmont Report* (1). This document proved to be highly influential; its principles provide the basis for virtually all commentary on the ethics of research

involving human subjects. Moreover these principles have been adopted in major international documents that contain guidance for the ethical conduct of research involving human subjects such as the Council of International Organizations of Medical Sciences' (CIOMS) *International Ethical Guidelines for Biomedical Research Involving Human Subjects* (2).

The guidelines developed by the National Commission were published in a series of reports (3). They have been adopted by the U.S. federal government as the regulations for the protection of human subjects and have also influenced the development of policy in many other nations.

The National Commission defined the basic ethical principle as "a general judgment that serves as a basic justification for the many particular prescriptions for and evaluations of human actions" (1, p. 4). Such a principle is taken as an ultimate foundation for any second-order principles, rules, and norms; it is not derived from any other statement of ethical values. The National Commission identified three basic ethical principles as particularly relevant to the ethics of research involving human subjects: respect for persons, beneficence, and justice. The norms and procedures presented in regulations and ethical codes are derived from and are intended to uphold these fundamental principles. As the National Commission observed in an early draft of the *Belmont Report*:

> Reliance on these three fundamental underlying principles is consonant with the major traditions of western ethical, political and theological thought represented in the pluralistic society of the United States, as well as being compatible with the results of an experimentally based scientific analysis of human behavior... (3, p. 18).

Thus, in the National Commission's view, these principles pertain to human behavior in general; it is through the development of norms that they are made peculiarly relevant to specific classes of activities such as research and the practice of medicine.

Some of the language used by the National Commission may imply an endorsement of one or another foundational ethical theory. For example, the term "respect for persons" suggests a reference to Kantian theory. It is clear, however, that the National Commission did not embrace any such theory (3). As observed by Abram and Wolf:

> Aware of Kantian (deontological), utilitarian, and Aristotelian traditions, for instance, the commission nonetheless refrained from relying on any one of them for the legitimacy of its conclusions. Agreement on a fundamental moral system was not sought or needed (4).

Although these authors discussed the President's Commission, they made it clear that it was patterned after the National Commission, which similarly refrained from relying exclusively on any particular moral system.

ETHICAL PRINCIPLES

Respect for Persons

The principle of respect for persons was stated formally by Immanuel Kant: "So act as to treat humanity, whether in thine own person or in that of any other, in every case as an end withal, never as a means only." However, what it means to treat a person as an end and not merely as a means to an end may be variously interpreted. The National Commission concluded that:

> Respect for persons incorporates at least two basic ethical convictions: First, that individuals should be treated as autonomous agents, and second, that persons with diminished autonomy and thus in need of protection are entitled to such protections (1).

An autonomous person is "...an individual capable of deliberation about personal goals and of acting under the direction of such deliberation" (1). To show respect for autonomous persons requires that we leave them alone, even to the point of allowing them to choose activities that might be harmful (e.g., mountain climbing) unless they agree or consent that we may do otherwise. We are not to touch them or to encroach upon their private spaces unless such touching or encroachment accords with their wishes. Our actions should be designed to affirm their authority and enhance their capacity to be self-determining; we are not to obstruct their actions unless they are clearly detrimental to others. We show disrespect for autonomous persons when we either repudiate their considered judgments or deny them the freedom to act on those judgments in the absence of compelling reasons to do so.

Clearly, not every human being is capable of self-determination. The capacity for self-determination matures during a person's life; some lose this capacity partially or completely owing to illness or mental disability or in situations that severely restrict liberty, such as in prisons. Respect for the immature or the incapacitated may require one to offer protection to them as they mature or while they are incapacitated.

Beneficence

The principle of beneficence is firmly embedded in the ethical tradition of medicine. It is commonly said that the first principle of medical ethics is "Do no harm." This principle is often stated in Latin, *primum non nocere*, which translated literally means "first (or above all) do no harm." Moreover this statement of principle is commonly and incorrectly attributed to Hippocrates.

Parenthetically, the closest approximation of this statement that can be found in the Hippocratic writings is in the book entitled *Epidemics*: "As to diseases, make a habit of two things—to help, or at least to do no harm" (5). If the first principle of medicine were truly "above all, do no harm," this would rule out virtually all medical therapy; almost all therapies present to the patient a risk of injury. The statement from *Epidemics* is much more compatible with the modern emphasis on trying to achieve a favorable balance of harms and benefits.

In the Hippocratic Oath, the principle of beneficence is expressed in several statements such as:

> I will apply dietetic measures for the benefit of the sick according to my ability and judgment. ...I will neither give a deadly drug to anybody if asked for it, nor will I make a suggestion to this effect.

In biomedical research the leading ethical codes such as the World Medical Association's Declaration of Helsinki enjoin the physician-investigator not only to secure the well-being of individuals (research subjects and patients) but also to develop information that will form the basis of being better able to do so in the future. And, according to the Nuremberg Code, the risks of research must be justified by "the humanitarian importance of the problem to be solved by the experiment."

The National Commission defined 'beneficence' as follows:

> The term, *beneficence*, is often understood to cover acts of kindness or charity that go beyond strict obligation. In this document, beneficence is understood in a stronger sense, as an obligation. Two general rules have been formulated as complementary expressions of beneficent actions in this sense: (*1*) Do no harm and (*2*) maximize possible benefits and minimize possible harms (1).

The first of these two "general rules" proscribes the deliberate infliction of serious injury on an identified individual for research purposes. That is, one may not impose a 100 percent probability of disabling injury on a human subject with no justification other than to solve a research problem (6). "Do no harm" as envisioned by the National Commission does not mean "above all, do no harm." Rather, it permits exposing research subjects to a statistical probability of harm if such exposure is justified in terms of the anticipated benefits, among other considerations (3).

The second general rule of beneficence calls upon investigators to design all of their work so as to maximize the probability and magnitude of benefit to individual research subjects as well as to society. It further requires investigators to minimize the probability and magnitude of injury to subjects and of harm to the interests of human collectives (3). Among the ethical norms that are grounded primarily in the second general rule are the requirements for good research design and competent investigators (3) and the requirement for a favorable relation of risks to anticipated benefits (3, pp. 37ff).

Some authors argue that separation of these two general rules into two fundamental ethical principles, beneficence (do good) and nonmaleficence (do no harm), would tend to decrease confusion (7). For the present purposes, however, it is more convenient to treat beneficence as a single principle; it is generally necessary to consider harms and benefits in relation to each other.

Frankena identifies four obligations that derive from the principle of beneficence (8); listed in decreasing order of ethical force, these are:

A. One ought not to inflict evil or harm.

B. One ought to prevent evil or harm.

C. One ought to remove evil.

D. One ought to do or promote good.

Statement A is a straightforward articulation of the National Commission's first general rule, do no harm (also known as principle of nonmaleficence). Few, if any, would argue that the injunction against inflicting harm or evil is not at least a very strong prima facie duty. Statement D, by contrast, is not regarded generally as a duty or obligation but rather, as the National Commission pointed out, an exhortation to act kindly or charitably. Statement D becomes a duty in the strict sense most typically when one consents or contracts to be bound by it. Physicians, for example, pledge themselves to act for the benefit of patients. Similarly, researchers who accept public support for their work assume a contractual obligation to promote good by contributing to the development of new knowledge. Moreover, without regard to funding, when one invites human beings to participate in activities that expose them to risk of injury, it is generally necessary to offer them in return something they find valuable; in the context of research involving human subjects, the most generally acceptable item of value that can be offered is a promise to pursue benefits either to the individual subjects or to others with whom they feel a bond of kinship (3). Such promises are made explicit in the process of informed consent. In the light of these considerations the National Commission determined that investigators who performed research involving human subjects incurred a strict duty or obligation to do or promote good.

Justice

Justice requires that we treat persons fairly and we give each person what he or she is due or owed (3). Justice is either comparative or noncomparative. Comparative justice is concerned with determining what a person is due by weighing his or her claims against the competing claims of others. Noncomparative justice is concerned with identifying what persons are due without regard to the claims of others (e.g., never punish an innocent person). The concerns addressed by the National Commission under the rubric of justice are exclusively concerns of distributive justice, a type of comparative justice. This article follows the National Commission in using the term "justice" to mean "distributive justice."

Distributive justice is concerned with the distribution of scarce benefits where there is competition for these benefits. If there is no scarcity, there is no need to consider just systems of distribution. Distributive justice is also concerned with the distribution of burdens, specifically when it is necessary to impose burdens on fewer than all members of a seemingly similar class of persons.

Justice requires a fair sharing of burdens and benefits; however, just what constitutes a fair sharing is a matter of considerable controversy. To determine who deserves to receive which benefits and which burdens, we must identify morally relevant criteria for distinguishing unequals. Various criteria have been proposed (9). Is it fair for persons to be treated differently on the basis of their needs? Their accomplishments? Their purchasing power? Their social worth? Their past records or future potential?

There are those who argue that the fairest distribution of burdens and benefits is precisely that which creates the most benefits for society at large. This is the classical utilitarian argument, which harmonizes the principles of justice and beneficence by stipulating that there

is no conflict. To create goods is to do justice; just institutions act so as to produce the greatest good for the greatest number. The National Commission rejected this formulation because it does not accord either with Western concepts of the fundamental equality of persons (e.g., before the law) or with the very strong tradition that interprets fairness to require extra protection for those who are weaker, more vulnerable, or less advantaged than others. This latter interpretation is reflected in such disparate sources as the injunction in the Judeo-Christian tradition to protect widows and orphans, the Marxist dicta "from each according to ability; to each according to need," and more recently, Rawls's contractual derivation of principles of justice (10).

The National Commission's interpretation of the requirements of justice is embodied in one of its statements on the relevance of this principle to the problem of selection of subjects:

> [T]he selection of research subjects needs to be scrutinized in order to determine whether some classes (e.g., welfare patients, particular racial and ethnic minorities, or persons confined to institutions) are being systematically selected simply because of their easy availability, their compromised position, or their manipulability, rather than for reasons directly related to the problem being studied. Finally, whenever research supported by public funds leads to the development of therapeutic devices and procedures, justice demands both that these not provide advantages only to those who can afford them and that such research should not unduly involve persons from groups unlikely to be among the beneficiaries of subsequent applications of the research (1, pp. 9–10).

Each of the three basic ethical principles identified by the National Commission is intended to have equal moral force. In this way the ethical system differs from those that have a single overarching superprinciple such as "justice" or *agape*. It further differs from those that assign priority to principles — to rank them in order of moral forcefulness — to resolve disputes engendered by conflicting requirements of two or more principles (10). The three principles give rise to norms that often create conflicting requirements. For example, the principle of justice, as articulated by the National Commission, creates requirements that are incompatible with some of those created by its principle of respect for persons (3). Similarly norms derived from the principle of beneficence inevitably engender conflicts with those arising from the principle of respect for persons. Implicitly, then, the National Commission also endorsed the notion of prima facie rules; these are rules that are binding unless they are in conflict with other stronger rules or unless in specific situations there is ethical justification for overriding the rule's requirements (3).

ETHICAL NORMS

An ethical norm is a statement that actions of a certain type ought (or ought not) to be done. If reasons are supplied for these behavioral prescriptions (or proscriptions), they are that these acts are morally right (or wrong).

Statements of ethical norms commonly include the words "should" or "ought," but in some cases there are stronger terms such as "must" or "forbidden." A typical statement of an ethical norm is: Research should be conducted only by scientifically qualified persons. The behavior-prescribing statements contained in the various codes and regulations on research involving human subjects may be regarded as variants of five general ethical norms (11). There should be (1) good research design, (2) competent investigators, (3) a favorable balance of harm and benefit, (4) informed consent, and (5) equitable selection of subjects. In addition a sixth general ethical norm appears in some international guidelines: (6) there should be compensation for research-induced injury. The purpose of these ethical norms is to indicate how the requirements of the three fundamental ethical principles may be met in the conduct of research involving human subjects.

Because statements of the ethical norms in codes and regulations tend to be rather vague, they permit a variety of interpretations; it is sometimes difficult to know exactly how to apply them to particular cases. When faced with such uncertainty, it is generally helpful to look behind the norm to examine the fundamental ethical principle or principles it is intended to uphold or embody. Accordingly, the discussion of each ethical norm will call attention to the fundamental ethical principle or principles it is designed to serve.

Good Research Design

The experiment should be so designed and based on the results of animal experimentation and a knowledge of the natural history of the disease or other problem under study that the anticipated results will justify the performance of the experiment (Nuremberg 3).

Biomedical research involving human subjects must conform to generally accepted scientific principles and should be based on adequately performed laboratory and animal experimentation and on a thorough knowledge of the scientific literature (Helsinki I. 1).

[S]cientifically unsound research on human subjects is *ipso facto* unethical in that it may expose subjects to risk or inconvenience to no purpose (CIOMS 14, commentary).

These are typical expressions of the ethical requirement that research must be sufficiently well-designed to achieve its purposes; otherwise, it is not justified. The primary purpose of this norm is to uphold the principle of beneficence. If the research is not well designed, there will be no benefits; investigators who conduct badly designed research are not responsive to the obligation to do good or to develop generalizable knowledge that is sufficiently important to justify the expenditure of public funds, to impose upon human subjects risks of physical or psychological harm, and so on.

This norm is also responsive to the principle of respect for persons. Persons who agree to participate in research as subjects are entitled to assume that something of value will come of their participation. Poorly designed research wastes the time of the subjects and frustrates their desire to participate in a meaningful activity.

Competence of the Investigators

The experiment should be conducted only by scientifically qualified persons. The highest degree of skill and care should be required through all stages of the experiment of those who conduct or engage in the experiment (Nuremberg 8).

Biomedical research involving human subjects should be conducted only by scientifically qualified persons and under the supervision of a clinically competent medical person. The responsibility for the human subject must rest with a medically qualified person . . . (Helsinki I.3).

This norm requires that the investigators be competent in at least two respects. They should have adequate scientific training and skill to accomplish the purposes of the research. The purpose of this component of the norm is precisely the same as that requiring good research design; it is responsive primarily to the obligations to produce benefits to society through the development of important knowledge. It is also responsive to the obligation to show respect for research subjects by not wasting their time or frustrating their wishes to participate in meaningful activities. In addition investigators are expected to be sufficiently competent to care for the subject. The Declaration of Helsinki, as an instrument of the World Medical Association, is addressed only to medical research. Therefore, it places responsibility with ". . .a medically qualified person." The Nuremberg Code, on the other hand, is addressed more generally to research; consequently, it does not call for medical qualification.

Competence to care for the subjects of most clinical research requires that at least one member of the research team be responsible for observing the subject with a view toward early detection of adverse effects of his or her participation or other evidence that the subject should be removed from the study. The investigator should have the competence to assess the subjects' symptoms, signs, and laboratory results. There should further be the competence to intervene as necessary in the interests of minimizing any harm, such as by prompt administration of an antidote to a toxic substance.

Balance of Harms and Benefits

There are normative statements in the ethical codes and regulations that require a favorable balance between harm and benefit. Without such a favorable balance there is no justification for beginning or continuing the research.

The degree of risk to be taken should never exceed that determined by the humanitarian importance of the problem to be solved by the experiment (Nuremberg 6).

Biomedical research involving human subjects cannot legitimately be carried out unless the importance of the objective is in proportion to the inherent risk to the subject (Helsinki I.4).

Risks to the subjects [must be] reasonable in relation to anticipated benefits, if any, to subjects, and the importance of the knowledge that may reasonably be expected to result (CFR. 46. 111a(2)).

There are additional norms in codes and regulations that call for vigilance on the part of those conducting or supervising the research. At any point along the way, the balance of harms and benefits may become unfavorable; under these circumstances the research should be terminated.

> During the course of the experiment the scientist in charge must be prepared to terminate the experiment at any stage, if he has probable cause to believe ... that a continuation ... is likely to result in injury, disability or death to the experimental subject (Nuremberg 10).
>
> The investigator ... should discontinue the research if in his/her ... judgment it may, if continued, be harmful to the individual (Helsinki III. 3).
>
> Where appropriate, the research plan makes adequate provision for monitoring the data collected to insure the safety of the subjects (CFR. 111a (6)).
>
> An IRB shall have authority to suspend or terminate approval of research ... that has been associated with unexpected serious harm to subjects (CFR. 113).

The requirement that research be justified on the basis of a favorable balance of harms and benefits is derived primarily from the ethical principle of beneficence. In addition a thorough and accurate compilation of the risks and hoped-for benefits of a research proposal also facilitates responsiveness to the requirements of the principles of respect for persons and justice. A clear and accurate presentation of risks and benefits is necessary in the negotiations with the subject for informed consent. Similarly such a compilation of burdens and benefits facilitates discussions of how they might be distributed equitably.

Ethical codes and regulations require not only that risks be justified by being in a favorable relationship to hoped-for benefits but also that they be minimized.

> The experiment should be so conducted as to avoid all unnecessary physical and mental suffering and injury (Nuremberg 4).
>
> The IRB shall determine that ... risks to subjects are minimized: 1) By using procedures which are consistent with sound research design and which do not unnecessarily expose subjects to risk, and (ii) whenever appropriate, by using procedures already being performed on the subjects for diagnostic or treatment purposes (CFR. 111a (1)).

Informed Consent

Principle I of the Nuremberg Code provides the definition of consent from which the definitions contained in all subsequent codes and regulations are derivative:

> The *voluntary* consent of the human subject is absolutely essential.
>
> This means that the person involved should have *legal capacity* to give consent; should be so situated as to be able to exercise *free power of choice*, without the intervention of any element of force, fraud, deceit, duress, over-reaching or other ulterior form of constraint or coercion; and should have sufficient *knowledge* and *comprehension* of the elements of the subject matter involved as to enable him to make an understanding and enlightened decision. This latter element requires that before the acceptance of an affirmative decision by the experimental subject there should be made known to him the nature, duration, and purpose of the experiment;

the method and means by which it is to be conducted; all inconveniences and hazards reasonably to be expected; and the effects upon his health or person which may possibly come from his participation in the experiment [*emphasis supplied*].

Thus the consent of the subject in order to be recognized as valid must have four essential attributes. It must be competent (legally), voluntary, informed, and comprehending (or understanding).

It is through informed consent that the investigator and the subject enter into a relationship, defining mutual expectations and their limits. This relationship differs from ordinary commercial transactions in which each party is responsible for informing himself or herself of the terms and implications of any of their agreements. Professionals who intervene in the lives of others are held to higher standards. They are obligated to inform the lay person of the consequences of their mutual agreements.

According to the President's Commission, "Although the informed consent doctrine has substantial foundations in law, it is essentially an ethical imperative" (12, p. 2). The President's Commission refers repeatedly to "ethically valid consent" and in this way reflects a perspective differing with that of federal regulations, which refer to "legally effective informed consent."

The requirement for informed consent is designed to uphold the ethical principle of respect for persons (3, pp. 95ff). It is through informed consent that we make operational our duty to respect the rights of others to be self-determining, namely to be left alone or to make free choices. We are not to touch others or to enter their private spaces without permission. As stated by Justice Cardozo, "Every human being of adult years and sound mind has a right to determine what will be done with his own body..." (13, p. 526).

Privacy and Confidentiality

Closely related to the norms calling for informed consent and minimization of risk are the requirements found in ethical codes and regulations that protect privacy and confidentiality. Privacy is "the freedom of the individual to pick and choose for himself the time and circumstances under which, and most importantly, the extent to which, his attitudes, beliefs, behavior and opinions are to be shared with or withheld from others" (14). In general, investigators are not permitted to intrude into individuals' privacy without their informed consent. When an informed person allows an investigator into his or her private space, there is no invasion.

"Confidentiality," a term that is often and incorrectly used interchangeably with "privacy," refers to a mode of management of private information; if a subject shares private information with (confides in) an investigator, the investigator is expected to refrain from sharing this information with others without the subject's authorization or some other justification.

The ethical grounding for the requirement to respect the privacy of persons may be found in the principle of respect for persons. The ethical requirement for the maintenance of confidentiality of private information is grounded in the norm calling for the minimization of harms. Breaches of confidentiality may result in such social injuries as loss of personal autonomy, valued relationships, or eligibility for insurance or employment. Moreover, maintenance of confidentiality is essential to the successful practice of various professions; for example, sick people would not consult physicians unless they were confident that their private information would be kept secret (15).

Equitable Selection of Subjects

> Individuals ... to be invited to be subjects of research should be selected in such a way that the burdens and benefits of the research will be equitably distributed. Special justification is required for inviting vulnerable individuals ... (CIOMS 10).
>
> [T]he IRB shall determine that.... Selection of subjects is equitable ... (CFR. 111a (3)).

This requirement is derived from the principle of justice, which requires equitable distribution of both the burdens and the benefits of research. Until the 1970s codes of ethics and regulations were relatively silent on this matter; however, the preamble to the Nuremberg Code reflected an concern with issues of social justice. It pointed out that the "crimes against humanity" were particularly egregious in that they were perpetrated on "non-German nationals, both prisoners of war and civilians, including Jews and 'asocial' persons" Implicit in this statement is the perception that because these subjects were not considered persons in the full sense of the word, they were not accorded the respect due to fully enfranchised persons. As a consequence principle I of the Nuremberg Code established the high standards for consent discussed earlier. When thoroughly honest offers are made to fully autonomous persons, they are presumed capable of defending their own interests and of selecting themselves as research subjects. Because the Nuremberg Code does not entertain the possibility of involving less than fully autonomous subjects, no requirements for their selection are provided.

Ethical codes and regulations in the field of research involving human subjects project an attitude of protectionism. Their dominant concerns are the protection of individuals from injury and from exploitation. There are important historical reasons for this protectionistic attitude. These documents were written with the aim of ensuring that there would never be a repetition of atrocities like those committed by the Nazi physician-researchers, calamities like the thalidomide experience, or ethical violations like those of the Tuskegee syphilis study. In recent years society's perception of biomedical research has shifted dramatically. Now, largely as a consequence of the efforts of the AIDS activists, biomedical research is widely perceived as benign and beneficial (16).

In the era of protectionism, which lasted from the mid-1940s through the mid-1980s, those who wrote policy for the protection of research subjects interpreted the principle of justice to require protection of vulnerable or disadvantaged persons from bearing an unfair share of the burdens of serving as research subjects. Since the mid-1980s, as a reflection of the shift in society's attitude toward research, policies and practices are being revised; the same principle of justice is being interpreted to require

assurance that the vulnerable and disadvantaged will enjoy equitable access to the benefits of participation in research.

Populations who were excluded previously because they were considered vulnerable include children, women who have the biological capacity to conceive, and members of racial and ethnic minorities; in recent years federal policies have been revised to require adequate representation of each of these populations in research (17).

Compensation for Research-Induced Injury

> Research subjects who suffer physical injury as a result of their participation are entitled to such financial or other assistance as would compensate equitably for any temporary or permanent impairment or disability. In case of death, their dependants are entitled to material compensation. The right to compensation may not be waived (CIOMS 13).

During the 1970s commentators on the ethics of research reached a consensus that subjects who are injured as a consequence of their participation in research are entitled to compensation. The ethical arguments to support this entitlement are grounded in considerations of compensatory justice (18). Compensatory justice consists in giving injured persons their due by taking account of their previous conditions and attempting to restore them. Sometimes it is possible to literally restore injured persons to their previous conditions, such as through medical therapy for the research-induced injury or illness. On other occasions, when literal restoration is not feasible, a monetary substitute is about the best we can do. Most discussions of compensation for research-induced injury focus on the provision of monetary substitutes in cases in which there is temporary or permanent disability or death.

In the commentary under its guideline number 13, CIOMS notes:

> In some societies the right to compensation for accidental injury is not acknowledged. Therefore, when giving their informed consent to participate, research subjects should be told whether there is a provision for compensation in case of physical injury, and the circumstances in which they or their dependants would receive it.

The United States is one society in which the right to such compensation is not acknowledged. Federal regulations require that prospective research subjects receive:

> an explanation as to whether any compensation and an explanation as to whether any medical treatments are available if injury occurs and, if so, what they consist of, or where further information may be obtained ... (46.116a).

PROCEDURAL NORMS

Ethical codes and regulations contain, in addition to the substantive ethical norms, descriptions of procedures (procedural norms) that are to be followed to assure that investigators comply with the requirements of the substantive norms. The most important general procedural requirement that is relevant to all research involving human subjects conducted in the United States is the review by an Institutional Review Board (IRB).

Federal regulations require that research involving human subjects be reviewed and approved by an IRB before it may be initiated. The criteria for approval are to ensure that plans are adequate for compliance with each of the substantive norms other than those calling for good research design and competent investigators; responsibility for these determinations is assigned to other agents or agencies (19). In other countries the same assignment is issued to research ethics committees or research ethics boards (20). Article I.2 of the Declaration of Helsinki requires only that the "experimental protocol ... be ... transmitted for consideration, comment and guidance to a committee independent of the investigator and sponsor...." CIOMS guideline 14, by contrast, requires "review and approval by one or more independent ethical and scientific review committees."

Another general procedural requirement (general in that it is designed to ensure compliance with all of the substantive norms) is concerned with the publication of the results of research that appears to have been done unethically. US federal regulations are silent on this issue. There are differing positions in the international documents:

> Reports of experimentation not in accordance with the principles laid down in this Declaration should not be accepted for publication (Helsinki I.8).

CIOMS offers the following commentary under Guideline 15:

> Refusal to publish the results of research conducted unethically ... may be considered, as may refusal to accept unethically obtained data submitted in support of an application for drug registration. However, these sanctions deprive of benefit not only the errant investigator or sponsor but also that segment of society intended to benefit from the research; such possible consequences merit careful consideration.
>
> publication of reports of the results of research ... should include, when appropriate, a statement that the research was conducted in accordance with these guidelines. Departures, if any, should be explained and justified....

There are also specific procedural norms such as the requirement for documentation of informed consent which is designed to ensure compliance with the substantive norm that calls for informed consent.

VULNERABLE PERSONS AS RESEARCH SUBJECTS

When the federal government published its first proposals to develop regulations providing additional protections for especially vulnerable populations of research subjects, it designated them as persons having "limited capacities to consent" (3). The choice of this label highlights the nature of the fundamental problem in justifying their use as research subjects. Because the Nuremberg Code identifies voluntary consent as "absolutely essential," it is clearly problematic to involve subjects who lack free power of

choice (e.g., prisoners), the legal capacity to consent (e.g., children), or the ability to comprehend (e.g., the mentally infirm). (Another term commonly used for "those having limited capacities to consent" is "the special populations.")

The National Commission concluded that persons having limited capacity to consent are vulnerable or disadvantaged in ways that are morally relevant to their involvement as subjects of research (3). Therefore the principle of justice is interpreted as requiring that we facilitate activities that are designed to yield direct benefit to the subjects and that we encourage research designed to develop knowledge that will be of benefit to the class of persons of which the subject is a representative. However, we should generally refrain from involving the special populations in research that is irrelevant to their conditions as individuals or at least as a class of persons. Respect for persons is interpreted as requiring that we show respect for a potential subject's capacity for self-determination to the extent that it exists. Some who cannot consent can register knowledgeable agreements (assents) or deliberate objections. In most instances the assent of an individual who cannot consent must be supplemented by the permission of that individual's parent or guardian (in the language of federal regulations, the "legally authorized representative").

To the extent that the capacity for self-determination is limited, respect is shown by protection from harm. Thus the Commission recommends that the authority accorded to members of the special populations or their legally authorized representatives to accept risk be strictly limited; any proposal to exceed the threshold of "minimal risk" requires special justification.

The reports of the National Commission on each of the "special populations" were followed by the promulgation of federal regulations providing "additional protections" for the fetus and pregnant women and for human in vitro fertilization (IVF); for prisoners, and for children (3). Regulations also were proposed for "those institutionalized as mentally disabled" but these have not been promulgated as final regulations (3).

Federal regulations also require:

> Where some or all of the subjects are likely to be vulnerable to coercion or undue influence, such as persons with acute or severe physical or mental illness, or persons who are economically or educationally disadvantaged, appropriate additional safeguards (should be) included in the study to protect the rights and welfare of these subjects (46.111b).

CIOMS, in the commentary under guideline 10, provides an extensive list of the types of persons who may in certain circumstances be considered vulnerable.

ARE THE BASIC ETHICAL PRINCIPLES UNIVERSAL?

The question of whether the basic ethical principles are universal will elicit radically different responses from adherents of ethical universalism than it will from the cultural pluralists. The tension between these two positions has existed since classical times and it is unlikely to be resolved in the foreseeable future (21).

Ethical universalists believe there is a universal set of ethical principles that are applicable to all human beings regardless of their situations in particular cultures. The task of the moral philosopher, then, is to *discover* those universal principles that apply in all times and in all places. Variations across cultures indicate that some societies are ahead of others in the degree of "moral progress" they have accomplished (22).

Ethical pluralists, by contrast, recognize that all ethical principles are developed in the course of discussions held within particular cultures and that these discussions necessarily reflect the unique histories and other circumstances of particular cultures. On this view ethical principles are *invented* rather than discovered. Pluralists further acknowledge the inevitability and recognize the legitimacy of variation across cultures of ethical norms and principles.

This debate has practical import in the increasingly common circumstances in which research protocols are designed by investigators and sponsors in technologically developed countries and then carried out in developing (or underdeveloped) countries or communities. What if there are differences in ethical values in the two cultures? Whose ethics should apply?

It has been argued that there should be a compromise between the two extremes (21). Some ethical principles seem to be universally valid. For example, there is a universal proscription against inflicting injury on a person without justification. But what counts as justification in various societies differs substantially. The principle of respect for persons, when stated at a sufficient level of abstraction, enjoining people to treat persons as ends and not merely as means, is universally applicable. However, when this principle is elaborated to require that all persons are to be treated as self-determining, it loses its relevance to some cultures in which individual self-determination is less highly valued than it is in the United States.

American universalists would argue that persons in such cultures must be educated; they must be taught to value self-determination as much as we do. Some might add that they must learn to value and protect the right to be self-determining or else they will remain vulnerable to exploitation by those who have decision-making authority. Pluralists counter this argument by pointing out that the society seems functional as it is; if we impose on it our ethical standards, it may have a destructive effect on the culture. Furthermore we should show respect for a society by allowing it to be self-determining.

The CIOMS *International Ethical Guidelines* reflect what may be a satisfactory compromise position, holding that some ethical standards are universal while recognizing the legitimacy of some degree of ethical pluralism. These guidelines set forth procedures to be followed when research is initiated and financed in one country (the external sponsoring country) and carried out by investigators from the external sponsoring country in another country (the host country) involving as subjects residents of the host country:

> In short, ethical review in the external sponsoring country may be limited to ensuring compliance with broadly stated ethical standards, on the understanding that ethical review

committees in the host country will have greater competence in reviewing the detailed plans for compliance in view of their better understanding of the cultural and moral values of the population in which the research is to be conducted (Commentary under guideline 15).

ACKNOWLEDGMENT

Passages of this article have been adapted or excerpted from previous writings of the author (3,16,21). This work was funded in part by grant number PO1 MH/DA 56 826–01A1 from the National Institute of Mental Health and the National Institute on Drug Abuse. References to U.S. federal regulations are to the "Common Rule," which is applicable to all federal agencies in the executive branch that conduct or support research. It is customary to cite these in the form "_____CFR_____.101"; in this article the dashes are omitted resulting in the following form: "CFR.101."

BIBLIOGRAPHY

1. National Commission for the Protection of Human Subjects of Biomedical and Behavioral Research, *The Belmont Report: Ethical Principles and Guidelines for the Protection of Human Subjects of Research*, DHEW Publ. No. (OS) 78-0012, U.S. Government Printing Office, Washington, DC, 1978.

2. Council for International Organizations of Medical Sciences (CIOMS) in collaboration with the World Health Organization (WHO), *International Ethical Guidelines for Biomedical Research Involving Human Subjects*, CIOMS, Geneva, 1993.

3. R.J. Levine, *Ethics and Regulation of Clinical Research*, 2nd ed., Urban & Schwarzenberg, Baltimore and Munich, 1986.

4. M.B. Abram and S.M. Wolf, *N. Engl. J. Med.* **310**, 627–632 (1984).

5. A.R. Jonsen, *Ann. Intern. Med.* **99**, 261–264 (1983).

6. R.J. Levine, *Hastings Cent. Rep.* **16**(4), 32 (1986).

7. T.L. Beauchamp and J.F. Childress, *Principles of Biomedical Ethics*, 4th ed., Oxford University Press, New York, 1996.

8. W.K. Frankena, *Ethics*, 2nd ed., Prentice-Hall, Englewood Cliffs, NJ, 1973.

9. G. Outka, *Perspect. Biol. Med.* **18**, 185–203 (1975).

10. J. Rawls, *A Theory of Justice*, Harvard University Press, Cambridge, MA, 1971.

11. R.J. Levine and K. Lebacqz, *Clin. Pharmacol. Ther.* **25**, 728–741 (1970).

12. President's Commission for the Study of Ethical Problems in Medicine and Biomedical and Behavioral Research, *Making Health Care Decisions: The Ethical and Legal Implications of Informed Consent in the Patient-Practitioner Relationship*, Stock No. 040-000-00459-9, U.S. Government Printing Office, Washington, DC, 1982.

13. J. Katz, *Experimentation with Human Beings*, Russell Sage Foundation, New York, 1972.

14. H.C. Kelman, *J. Soc. Issues* **33**, 169–195 (1977).

15. S. Bok, *Secrets: On the Ethics of Concealment and Revelation*, Pantheon, New York, 1982.

16. R.J. Levine, *Kennedy Inst. Ethics J.* **4**(2), 93–98 (1994).

17. U.S. Department of Health and Human Services, *Fed. Regist.* **59**, 11146–11151 (1994).

18. J.F. Childress, *Hastings Cent. Rep.* **6**(6), 21–27 (1976).

19. R.J. Levine, *Int. J. Pharm. Med.* (in press).

20. R.J. Levine, in S.S. Coughlin and T.L. Beauchamp, eds., *Ethics in Epidemiology*, Oxford University Press, New York, 1996, pp. 257–273.

21. R.J. Levine, in H.Y. Vanderpool, ed., *The Ethics of Research Involving Human Subjects: Facing the 21st Century*, University Publishing Group, Frederick, MD, 1996, pp. 235–259.

22. R. Macklin, in G.J. Annas and M.A. Grodin, eds., *The Nazi Doctors and the Nuremberg Code: Human Rights in Human Experimentation*, Oxford University Press, New York and Oxford, 1992, pp. 240–257.

See other HUMAN SUBJECTS RESEARCH entries.

HUMAN SUBJECTS RESEARCH, ETHICS, RESEARCH ON HUMAN EMBRYOS

RONALD GREEN
Dartmouth College
Hanover, New Hampshire

OUTLINE

INTRODUCTION

Background

The development of the technology of in vitro fertilization (IVF) during the 1970s by Edwards, Steptoe, and colleagues (1,2) made possible the first systematic study of the live, developing human embryo from fertilization onward. At the same time the relatively low success rates of IVF increased demand for more systematic research on fertilization and embryo development. Together these factors have made the issue of embryo research increasingly important in law, ethics, and public policy. Embryo research is also one of the most controversial issues in biomedical ethics and law today. Wherever it has been discussed, there have been significant disagreements about the moral status of the embryo and what constitute legitimate reasons for putting it at risk.

DEFINITIONS

Embryo

The term "human embryo" is used in many ways, some opposed to one another. In ordinary speech it is used to refer to the developing human being following conception. As defined in medical literature, the term is usually applied more strictly to the product of conception from the end of the second week after fertilization to the end of the seventh or eighth week when the fetus is said to exist (3). Before two weeks, the terms zygote (the one-cell conceptus), morula (= mulberry), and blastocyst are used for the developing entity in its various stages.

In the context of ethical and legal discussions of the human embryo, the term usually refers to the product of conception (zygote, morula, and blastocyst respectively) during the first two to three weeks of development outside the womb. Normally this means an embryo that has been created by IVF, although it can also refer to a fertilized ovum that has been flushed from a uterus shortly after conception and kept alive in vitro for purposes of study. The defining features of this entity are its existence ex utero and its early stage of development, usually before or just up to the first processes of cellular differentiation, tissue formation, and the appearance of rudimentary bodily form. In both respects, embryo research differs in U.S. law from fetal research, which typically involves an embryo or later stage fetus in utero or following abortion (4). In the mid-1980s the term "pre-embryo" was introduced for the human embryo at the earliest stage of development (5), although some have objected to it as possibly finessing the discussion of complex moral questions (6,7).

The National Institutes of Health's (NIH) Human Embryo Research Panel used the term "preimplantation embryo." This is accurate since the embryo involved in research has not yet been transferred back to a womb for implantation and, in current research, is also at the earliest or "preimplantation" stages of development. However, since implantation normally begins at 6 to 7 days in vivo, it is possible that, as our ability to sustain embryos in vitro progresses, some human embryo research, as defined here, will involve "preimplantation" embryos in the first sense (embryos that have never implanted in a womb) but not in the second sense (embryos that are less than 6 to 7 days old).

Conception/Fertilization

Further complicating the definition of the embryo is the question of what is meant by the term "conception" or "fertilization." The embryo is not regarded as coming in existence until after sex cells have joined at conception/fertilization, but conception/fertilization occurs over a period of time (8). It spans a period of at least 22 hours from the sperm's initial contact with the outer membrane (*zona pellucida*) of the egg to syngamy, the alignment on the mitotic spindle of the chromosomes derived from the male and female pronuclei (9). The first appearance of a new diploid nucleus within its own nuclear membrane occurs only after the first embryonic cell division (cleavage) three or more hours later, and the first activation of paternal genes occurs only after the second cleavage division, 12 or more hours later. A specific research program beginning with conception/fertilization therefore might involve gametes or an embryo, depending on which definition of conception/fertilization is accepted and where the cutoff point for the study is established following the admixture of sperm and egg.

In legal jurisdictions where embryo research is banned or stringently controlled, this definitional matter can be of great importance. Although the issue is often ignored in most legislation, two jurisdictions exhibit contrasting approaches. Australia's state of Victoria does not define fertilization, but its law banning human embryo research permits an exception for research "from the point of sperm penetration prior to but not including the point of syngamy" (10). Britain's Human Fertilisation and Embryology Act is quite specific, defining the embryo as present only when fertilization is complete at "the appearance of a two cell zygote" (11). These two items of legislation evidence how it is possible to select different moments in the process of fertilization. The Australian legislation also shows how important this definition can be in determining whether some types of contraceptive research aimed at blocking fertilization are permitted (12). For the balance of this discussion, fertilization will be taken to mean sperm penetration of the egg, since this is the reference point used in most chronological accounts of embryological development.

Human

The term "human" in the definition of the embryo also raises questions. Should a parthenote, an ovum artificially stimulated to begin the earliest stages of cell division, be considered a human embryo in this sense? Is an animal-human hybrid resulting from the fertilization of a human by a nonhuman gamete a *human* embryo? Are chimeric or transgenic embryos, involving the admixture of human and nonhuman embryonic cells or human and nonhuman genetic material, human embryos? Is a cloned embryo, one produced by the insertion of a nucleus from one cell into an enucleated egg cell, a human embryo (13)? Many persons who object to embryo research will predictably find research involving these other forms of embryos to be morally offensive. Current U.S. law, for example, prohibits federal funding for research on parthenotes as well as human embryos created by cloning. For this reason, research on embryos resulting from parthenogenesis, cloning, or transgenic manipulations involving nuclear human DNA should probably be considered under the heading of human embryo research.

Description

As defined, the human embryo involved in research, ranges in size from a one-cell fertilized ovum to an embryo of several thousand cells. In the words of Jones and Tefler, "The recurring motifs of embryonic development are: gradually decreasing potentiality, increasing determination and differentiation, and increasing complexity and interaction. This development occurs smoothly rather than in quantum leaps and its object is the transformation of a single fertilized ovum into a complex organism" (14, p. 45). During the earliest cleavage stages, the cells (or blastomeres) exist in a small, loosely packed mass. Each cell is undifferentiated and totipotent: taken from the cellular mass, it has the capacity to develop on its own into a full human being. Twinning can occur spontaneously at this stage, and the removal of one or more blastomeres does not interfere with the embryo's normal development. As the number of cells continues to increase, some specialization occurs. The outer layer of cells forms the trophectoderm, which at implantation give rise to the tissues that begin formation of the placenta (15). Inside the outer layer of trophectoderm cells, fluid accumulates to form a cavity, the blastocoele, and the resulting entity at 4 to 5 days following fertilization is called a blastocyst. By 6 or 7 days, when implantation normally has begun in vivo, the embryo consists of approximately 100 cells and is roughly 130 μm in diameter. Outer or trophoblast cells surround an inner group of about 20 to 30 undifferentiated cells.

One week later, by 14 days development, two fluid-filled cavities have formed, the amniotic sac and yoke sac, with a two-layered embryonic disk about 0.5 mm in diameter between them. The cells of this disk, now about 2,000 in number, remain undifferentiated. At this time the process of gastrulation begins with the appearance of the primitive streak and the establishment of left-right, head-tail orientation (16). From this point onward, the embryonic disk is committed to forming a single individual with twinning no longer possible unless two primitive streaks have formed (17). Over the next 3 days, cells migrate to form three layers within the embryo and become pluripotent rather than totipotent: They are able to form broad categories of specialized tissues and organs but no longer an entire individual. At 17 days the primitive nervous system begins to develop with formation of the neural plate that soon develops into the neural tube. By the beginning of the fourth week of development, the neural tube closes and begins to differentiate into distinct regions of the nervous system.

In the two weeks from fertilization to the appearance of the primitive streak, therefore, the human embryo as discussed in this context is marked by very small size, the absence of bodily form, and (with the exception of placental material), the lack of differentiated tissues, organs, or a nervous system. In all these respects it differs from the postgastrulation embryo and the fetus. Currently it is not possible to sustain a human embryo in vitro until the point of gastrulation, so embryo research, as discussed here, usually refers to an undifferentiated entity of this sort. This may change in the future. Some kinds of embryo research that have been proposed, such as research to find genetic or biochemical markers for gastrulation, may extend human embryo research into the third week of development.

WHY CONDUCT HUMAN EMBRYO RESEARCH?

Limits of Animal Models

Many features of embryological development have been conserved in the course of evolution. It is now known that among vertebrates the basic principles of embryonic development, and even the genes regulating it, are similar. This means that much valuable research can be done on animals without the need to use human embryos. Nevertheless, there are significant limits to animal models. Early rubella vaccination research in monkeys indicated that the vaccination did not cross the placenta, yet subsequent research showed otherwise, rendering the vaccine unsafe for human use (18,19). Subtle matters like these can have dramatic effects on the safety of drugs or procedures when findings based on animal models are applied to human beings. From a scientific perspective some human embryo research is therefore necessary.

Assisted Reproduction

There are many areas where human embryo research can be of value. One, already suggested, is research aimed at enhancing the efficiency and safety of assisted reproductive technologies (ARTs). Despite nearly 20 years of clinical utilization and some strides forward in improving success rates, the expense and emotional toll of these procedures are enormous (20–22). Multiple safety concerns have also been raised. When used on healthy women, the drugs that stimulate development of multiple follicles only rarely have serious immediate side effects, but there is an unresolved controversy about the long-term impact of multiple cycles of superovulation on a woman's health, particularly cancer risks (23–25). Little is known about the effects of these drugs on the ova or embryos

subjected to them, although it has been suggested that they may be implicated in the high rates of chromosomal anomalies (aneuploidies) found in human occytes used in IVF (26).

The need to transfer multiple embryos to enhance the chances of a pregnancy carries considerable risks in its own right. In Europe and the United States, IVF has resulted in an epidemic of higher order multiple births with its associated toll of miscarriage and prematurity (27). A better understanding of normal embryological development and improved ability to identify "implantation competent" embryos can help reduce the need for multiple embryo transfers. Because of a lack of resources for coordinated multi-center studies of assisted reproductive techniques, some newer procedures like Intracytoplasmic Sperm Injection (ICSI) have been introduced with little or no previous research into their safety for the resulting offspring. One recent study indicates that there is a slight but significant increase in the rate of spontaneous sex-chromosome anomalies among children born as a result of ICSI as compared with the general neonatal population (28). Here, as elsewhere, studies utilizing human embryos can help answer safety questions and improve the success rates of existing procedures.

Embryo research may also lead to entirely new methods of assisted reproduction. One promising set of technologies involves the cryopreservation of immature oocytes (or of ovarian tissue) followed by in vitro maturation and fertilization (29–32). Currently, multiple follicles must be matured in vivo, exposing a woman to the potent drugs used for this. Such stimulation is hazardous to women with polycystic ovarian disease, estrogen sensitive cancers of the breast, or other estrogen-sensitive disorders (31). If oocytes could be matured in vitro, these risks could be eliminated. A host of new donor sources of ova might also be made available, including women who would not wish to be exposed to stimulatory medication. Research on cryopreservation of ova is currently underway (32). Its development, along with in vitro maturation, would provide new reproductive options for women facing cancer treatments or other impediments to the future utilization of their eggs. All these promising techniques require human embryo research involving fertilization and verification in vitro of normal development.

Implantation Research

Many pregnancies fail because fertilized ova never implant in the womb. Improved understanding of the complex process of implantation can help improve pregnancy outcomes for many infertile women. Embryo research in this area can also assist our understanding of related biological processes. Implantation involves the penetration of host tissues by a rapidly growing and highly invasive foreign body. Close analysis based of the growth factors and gene expression associated with implantation can advance our understanding of similar processes that occur in cancer and tumor metastasis. Improved insight into how both fertilization and implantation occur can also lead to new techniques for preventing them. This promises development of improved contraceptive methods.

Normal/Abnormal Development

Embryo research can advance our ability to recognize key events or processes associated with normal or abnormal development. Many serious birth defects and genetic disorders that express themselves months or years following birth have their beginning in the earliest phases of embryogenesis, when cells are rapidly dividing and the basic plan of the body is being established by the operation of genes that act for only short periods of time. It is well known, for example, that viral diseases like rubella can have a devastating effect on early development. The recent discovery of the important role of folic acid deficiency early in pregnancy in contributing to neural tube defects is another example. Embryo research aimed at deepening our understanding of the patterns of normal and abnormal development may thus lead to new preventative measures or therapies.

Genetic Diagnosis

Human embryo research can help improve the newly developed technology of preimplantation genetic diagnosis (PGD). This involves coupling an IVF procedure with genetic biopsy of a single cell (blastomere) from each of the resulting embryos. It permits parents whose offspring are at risk for a genetic disease to avoid transferring embryos that exhibit the mutation or abnormality (33). Although currently much more expensive than other prenatal testing techniques like chorionic villus sampling (CVS) or amniocentesis, PGD offers parents an alternative to later term genetic abortion which is needed to avoid a birth when CVS or amniocentesis are employed.

Cell Differentiation

Human embryo research can contribute to our understanding of the complex processes of cell differentiation and can lead to new techniques for tissue transplantation and repair. Mouse research has already led to the development of techniques for producing in that species immortalized, pluripotent embryonic stem cell lines (34). In this research the inner cell mass of fertilized embryos is dissociated into single cells and dispersed into another dish with a rich culture medium. The embryonic cells continue to grow rapidly and indefinitely. Because of their low immune competence and pluripotentiality, they might be inserted back into other embryos or more mature individuals to replace defective or missing nerve, blood, skin, bone, and germ cells. Recent research by James Thomson and others suggests that this technology can be applied to the development of immortalized, canonical human stem cell lines (35,36). Somewhat differently, nuclear transfer or cloning technologies at the embryonic level have been shown to be an efficient means of introducing altered genes into every cell of the resulting organism. Employed by Wilmut and others to develop the transgenic sheep Polly, this technology also holds out the prospect of new gametic or embryonic gene therapies that might eliminate the need to discard embryos or genetic abortion (37). Human embryo research remains the essential precondition for the development and clinical deployment of all these technologies.

LAWS AND REGULATIONS

Other Countries

A review of existing national laws and regulations reveals significant disagreements about the acceptability of research involving embryos (38,39). Norway prohibits all research on fertilized ova (40); Australia's states of Victoria (41) and Western Australia (42), Austria (43), Germany (44), and Switzerland (45) forbid all embryo research other than that aimed at enhancing the survival of the embryo being studied or used in an IVF procedure. These restrictions rule out research on embryos remaining from infertility procedures that are otherwise destined to be discarded. French law specifies that research "may not be harmful to the embryo" (46), while Denmark permits embryo research only when its purpose is "to improve in vitro fertilization in order to bring about pregnancy" (47). Spain permits research employing nonviable embryos remaining from infertility procedures (48), while Sweden permits embryo research in very general terms (49). Italy and Greece have virtually no legislation in this area (39).

Britain's legislation, the Human Fertilisation and Embryology Act 1990 (50), is the most extensive and most permissive, including a provision allowing the deliberate fertilization of ova for research purposes. Similar legislation was proposed in Canada but much more stringent rules were subsequently enacted (51,52). European disagreements about embryo research are reflected in article 18 of the Convention on Human Rights and Biomedicine of the Council of Europe. This seeks to accommodate very diverse positions on embryo research by calling vaguely for "adequate protection of the embryo" when it is used in research. The convention does, however, explicity prohibits the creation of embryos for research purposes only (53). Because of its different legislation on this matter, Great Britain has entered a reservation against this article, rendering it inapplicable in that country.

Areas of Agreement

Despite the differences there are some broad areas of agreement in this legislation. These include strict requirements of informed consent for gamete or embryo donors; discouragements or prohibitions on the commercialization of gametes or embryos; and widespread prohibition of cloning, the creation of animal-human hybrids or chimeras, or the transfer of human embryos to animal wombs. Where embryo research is permitted, as in Britain and Sweden, a 14-day time limit on such research has been imposed. Several countries have also established special regulatory bodies that must approve or license human embryo research protocols. Britain's Human Fertilisation and Embryo Authority is an example. Arising out of the extensive debate occasioned by the Warnock Committee Report (54), it is charged with the extensive oversight of infertility clinics and the licensing of specific embryo research programs.

U.S. State Law

In the United States only 10 states have legislation on embryo research, nine of them very restrictive (55).

Louisiana's law defines the embryo as a "juridical person" and states that it is to be used in research "solely for the support and contribution of the complete development of human in utero implantation" (56). Federal courts have raised questions about whether this kind of restrictive legislation violates the constitutionally protected right to privacy, and there are also questions as to whether it opposes the right to freedom of expression (as this pertains to scientific research) (57), but the matter has not yet reached the Supreme Court.

Early Federal Initiatives

At the national level there have been significant regulatory and legal initiatives. In the late 1970s, shortly after the development and introduction of IVF, a special body, the Ethics Advisory Board (EAB), was formed by an act of Congress to review and provide guidance for federally funded research on the human embryo. In late 1979 the EAB issued its report containing a broad permission for such research under federal auspices subject to guidelines and limitations indicated in the report and to be implemented by the Board itself (58). However, before the EAB's recommendations could be put into effect, there was a change in adminstrations. During the Reagan and Bush years, funding for the EAB and nominations to its membership, were halted. The legal requirement of EAB approval of all such research and the absence of an EAB created a de facto moratorium on funding for embryo research in the United States.

Human Embryo Research Panel

In June 1993 Congress passed legislation nullifying the earlier requirement of EAB approval (59). To provide ethical guidance for this area, NIH established a special body, the Human Embryo Research panel. Beginning work in January 1994, the Panel held five monthly meetings in Washington open to the public and issued its report in September of that year (60). In broad terms, the report's recommendations paralleled existing legislation in Great Britain and proposals under consideration in Canada (51). Responding to its charge, the Panel divided research into three areas: (1) acceptable for federal funding, (2) unacceptable for federal support, and (3) warranting additional review.

In the first category the Panel placed research aimed at enhancing the safety and efficiency of infertility procedures, at preventing disease conditions arising during prenatal development, and at improving our understanding of the human embryo. Permitted research was governed by a series of stringent guidelines. They included the requirements of demonstrated scientific validity and merit, informed consent on the part of gamete or embryo donors, a prohibition on the purchase or sale of gametes or embryos, and a 14-day age limit on the use of embryos in research.

The category of research that was unacceptable for federal funding included research on embryo cloning (whether by splitting existing embryos to multiply the number of genetically identical embryos, or by somatic cell nuclear transfer to replicate an existing individual) where transfer to a womb was intended; research involving the fertilization of fetal oocytes with intent to transfer; parthenogenesis research where transfer was intended, and the creation of animal-human chimeras. Among those items placed in the category of warranting additional review were cloning by embryo splitting and the use of fetal oocytes where no transfer to a womb was intended.

Subsequent Developments

Among the most controversial of the Panel's proposals, but paralleling existing regulations in Great Britain, was a permission to fertilize oocytes for research purposes without the intention to transfer. Late on December 2, 1994, following a morning meeting in which the Advisory Council to the NIH director unanimously accepted the Panel's report, President Clinton issued a directive overruling this recommendation. Eventually this limited dissent from the Panel's recommendations was overtaken by congressional initiatives barring federal funding of any embryo research that threatened the embryo's survival (61). In May 1999, in the wake of several new research reports showing the possibility of establishing canonical embryonic stem (ES) cell lines from human embryos, the National Bioethics Advisory Commission, once again recommended establishing federal support for human embryo research (62).

Thus the United States during this period has exhibited roughly the same polarization of opinion and legislation on this issue as has occurred in Europe. In one important respect, however, the United States differs from Europe, including Great Britain, where regulations govern all research on the embryo. In the United States, federal legislation and guidelines affect only federally funded research. To date, there has been no federal legislation prohibiting or limiting research with private funds.

ETHICAL ISSUES

Range of Views

The legal turmoil over human embryo research reflects deep ethical disagreements. Foremost among these is disagreement about the moral status of the human embryo in the earliest stages of development. At one end of the spectrum of opinion are those who regard the embryo as fully a human being meriting all the protections due any other human research subject. This position has been publicly defended by the Roman Catholic Church (63,64). At the other end of the spectrum are those who see the embryo as meriting little or no protection and who believe that the primary consideration in human embryo research is whether it benefits or harms adult human beings or children. Someone holding this view, for example, might argue that if embryo research could improve contraceptive alternatives for women or reduce the risks of current methods, then it is not only morally permitted but morally required. Between these polar positions lie those who, for different reasons, accord the embryo some measure of moral respect and believe that any kind of harmful research on the embryo requires stringent moral justification. This issue of the moral status of the preimplantation embryo is at the center of these debates. The question becomes on what grounds a being or entity achieves moral status.

Genetic Individuality and Moral Status

Some who defend the full moral status of the embryo answer this question by focusing on the embryo's possession of what they regard as the essential biological criterion of humanness in a moral and spiritual sense: a diploid human genome with the inherent and active potential to develop on its own, under normal conditions, into a child or adult human being (65,66). Those who hold this view believe that biological humanness in this sense represents a "bright-line" distinction providing maximum protection for all who are human from possible abuses that may come about if humanness is defined in less determinative ways (67).

Biological Problems

The full moral status view, although seemingly clear, is challenged by biological knowledge on several fronts. One problem has to do with the so-called moment of fertilization, which marks the cherished bright line separating embryos, as protectable, from gametes that are regarded as having no greater moral status than other expendable bodily tissue. Strictly speaking, there is no single "moment" of fertilization but a series of events—a process. This raises the question of whether references to biological phenomena really do provide the bright lines that defenders of this position seek.

A second problem has to do with the realities of twinning and recombination. In a small number of cases, single embryos spontaneously split into one or more embryos, resulting in the birth of monozygotic (identical) twins if the pregnancy proceeds to term (68). Different embryos also sometimes combine to form chimaeras containing one or more fused genomes (69). For some ethicists and theologians, these rare but naturally occurring events challenge the notion that the appearance of a single, discrete genome following fertilization is the marker for moral and spiritual individuality (8,70). The possibility of twinning and the fact that most cells of the early embryo develop into nonembryonic supporting tissue also lead some to question whether we can speak of the early embryo as a single "being" with the "potential" to develop into an adult person (71). Others question whether the high degree of human intervention necessary for IVF procedures does not efface any distinction between embryos and gametes in terms of their "inherent" potential to develop to maturity (72). The fact that cloning technology renders any cell in the body able to become an embryo further complicates the position of those who stress the presence of a diploid genome and raises further questions of what "potentiality for development" means in this context.

Finally, the clarity of this biological view is troubled by the very large proportion of embryos—estimates are as high as two thirds to three quarters of all fertilized ova (73,74)—that do not successfully implant in the womb. If the likelihood of development to term, an important consideration on some accounts of

potentiality (67), takes a significant upward turn as gametes becomes an embryo, still another increment in likelihood of birth takes place at implantation. This raises the question of whether this point, or perhaps the later establishment of individuality at gastrulation (after which twinning cannot occur), is not the better "bright" line on which to base protected humanness.

Personhood Views

In addition to these biological challenges, critics of the position that locates full moral protectability at fertilization also offer many objections of a philosophical nature, some of which also appear in the context of the abortion debate. The issue, some of these critics contend, is not biological humanness but moral "personhood." By this they mean the status of being a moral subject worthy of all the rights and protections normally accorded human beings (75). They point out that some individuals who are biologically human (e.g., adults who are brainstem dead or anencephalic infants) may not be persons in a moral sense and may ethically be treated with less than equal respect. It is also possible that in the future we will encounter intelligent life forms that are not biologically human but that nevertheless are persons in a moral sense.

Those holding this view tend to espouse a variety of distinct, but sometimes overlapping, positions regarding the qualities that are needed for personhood. Some believe that highly developed qualities, such as consciousness and a sense of self are needed (76–78). Others stress the beginnings of brain activity or brain function, arguing that if the cessation of brain activity at the end of life is an appropriate marker of death, it should also be a marker for the commencement of moral personhood (79–82). Still others point to sentience, the ability to experience pain (83,84). Although those adhering to the brain activity or sentience views may come to a less permissive position on early term abortion than those holding the consciousness view, all those holding these views tend to agree about the status of the early embryo, whose lack of differentiated nervous system tissue makes it unreasonable to regard it as capable of feeling or thinking.

Symbolic Concerns

Complicating this typology of positions, are the views of those who believe that the significant moral issue here is not the status or rights of the embryo so much as the implications of its treatment for society as a whole. Some who hold this view fear that funding embryo research creates a "slippery slope" that could lead to undermining current protections for human subjects in research, erode respect for persons with disabilities, or encourage eugenic practices (85). Others emphasize the symbolic importance of the embryo, rather than its intrinsic moral claims, and ask how its treatment (or mistreatment) might affect respect for life generally (86,87).

PUBLIC POLICY

The status of the embryo and the question of what degree of respect it deserves are difficult to settle partly because they involve intensely personal matters of moral conviction or religious belief. Where law or public policy is concerned, however, debate must focus primarily on those considerations that are appropriate for setting policy in a religiously and morally pluralistic democracy. As one writer puts it, "law is not really concerned with the enforcement of morality but rather with providing a framework of peace and order within which people may exercise their personal liberty ... and make their own personal moral choices and engage in what John Stuart Mill calls their own 'experiments in living' " (88, p. 149).

Public Reasoning

This context imposes limits on what can be brought to public debate about issues significantly affecting the lives, welfare, and basic liberties of citizens. The philosopher John Rawls argues that basic public policy in such contexts must employ "public reasoning" that is understandable in terms that are not dependent on particular religious, theological, or philosophical perspectives. It should appeal "only to presently accepted general beliefs and forms of reasoning found in common sense, and the methods and conclusions of science when these are not controversial" (89, p. 224). From this perspective, religious teachings that are unable to sustain themselves without resort to special metaphysical premises belong to the realm of private decision and should not direct public policy where basic matters of citizens' welfare and liberties are concerned.

Two considerations qualify this conclusion. One is that some religiously based views can be articulated in terms of widely shared moral values. When this is done, and the arguments are persuasive, there is no reason to set aside a view merely because it originates in a religious context or is held by a community of religious believers. Second, even when a religious position cannot commend itself to a wider society, it is good public policy to respect this view to the extent that doing so is compatible with the protection of public healthy and safety.

Policy Analysis

Because there is no clear consensus on the moral status of the embryo itself, the most relevant arguments surrounding embryo research from a public policy perspective have to do with the indirect and symbolic impacts on society and people's respect for the sanctity of human life. Will using embryos in research start us down a slippery slope and contribute to the neglect or injury of other protectable human subjects? Will use of embryos lead to a "cheapening" of the value of human life? Will embryo research damage the way we regard human parenting or diminish the protection we accord our young? These questions are all enormously difficult to answer. If we keep in mind that similar questions have been raised with the advent each new reproductive technology, we see how marked this whole area is by deep, if largely speculative, concerns.

On one side of the balance, therefore, are a set of hard-to-assess symbolic concerns to which we might add the very determined opposition of a large number of citizens,

some influenced by personal religious beliefs. On the other side is the great promise human embryo research holds out for reducing illness and improving human health.

Policy Conclusions

Taking all these considerations into account, and trying to balance the avoidance of immediate harms against symbolic and other concerns, expert panels like the Warnock Committee, the NIH Human Embryo Research Panel, and Canada's Royal Commission on New Reproductive Technologies have all recommended permitting embryo research under stringent control and limitations, including a 14-day time limit. Apart from Great Britain, however, these expert public assessments have had little impact. Political opposition to embryo research, some of it influenced by religious views or traumatic national experiences with biomedical research and eugenics, as in the case of Germany, has had more effect than the kind of moral analysis and public reasoning employed by expert panels (90).

SPECIAL QUESTIONS

Those who accept the possibility of embryo research must resolve a series of additional ethical questions. Substantial consensus exists about the answers to some of these questions, while others remain very controversial.

Time Limits

As mentioned above, one area in which there has been widespread agreement is the need to set a limit, usually 14 days, on the time during which research can be conducted after fertilization. Since embryo development in vitro usually proceeds more slowly than in vivo, this ensures that embryos used in research have not entered the phase of gastrulation and have not yet developed the primitive streak. It is important to note that the 14-day limit should not be misconstrued, as it has sometimes been (91), as a statement about the definitive commencement of moral personhood at this time. Rather, those who accept this limit usually hold that the embryo cannot definitively be held to be a person before this point, although they may disagree about the time when personhood or enhanced protectability is subsequently established. It is also taken as a reasonable compromise between the moral claims surrounding the embryo and the needs of researchers.

Understanding the 14-day point as a reasonable but not necessarily absolute limit, the NIH Human Embryo Research panel urged that some research beyond this point (in vitro) be permitted in exceptional circumstances when the goal was to assist in the identification in the laboratory of the appearance of the primitive streak. Some who regard the beginning of gastrulation as a morally definitive step would not agree with this exception. Others see exceptions of this sort as propelling us down a slippery slope leading to the abuse of other vulnerable research subjects (85).

Intent to Transfer

Many believe that the question of whether or not researchers intend to transfer an embryo to a uterus is of great moral consequence in evaluating research proposals. Where transfer is in prospect, another class of human subjects becomes involved: children born as a result of these procedures. Since born children are recognized in law and ethics as having a right to protection from injuries inflicted before their birth, this means that researchers must demonstrate that manipulations of the embryo impose no greater risks than would be encountered in a normal pregnancy or, where the purpose is helping a couple have a child, of accepted assisted reproductive procedures. If also means that new assisted reproductive techniques should not initially be applied to human embryos that are to be transferred with the aim of establishing a pregnancy. Before transfer is considered, intermediate studies should be conducted on human embryos that are not intended for transfer. This approach was adopted by Steptoe and Edwards in their initial development of the clinical protocols used today in all IVF programs throughout the world (92,93). Risks to children born as a result of cloning procedures also played a large role in the U.S. National Bioethics Advisory Commission's June 1997 recommendation of a five-year legal prohibition of all attempts at cloning a human being (although the Commission did not at that time bring under review the existing ban on federal funding for the embryo research that would be needed to reduce these risks) (94).

Philosophical Complexities

There are interesting philosophical complexities associated with this high standard of safety for research involving embryos that are to be transferred. Some have argued that a child who would not otherwise exist cannot be morally wronged by procedures needed to bring it into being, even if these same procedures seriously impair its health or well-being (95,96). Without these harmful procedures, it is contended, the child would not have been and, at least up to the point where the harm is not so great as to make it reasonable to wish to not be alive, the child cannot morally complain about its treatment. Those who accept this argument might permit research on embryos for transfer that jeopardizes the well-being of a resulting child if this is the only way that the child can be brought into being.

Not everyone accepts this line of reasoning. Some doubt that being brought into existence should be reckoned as a benefit, and maintain that it is always the obligation of parents (and those who assist them in having children) to strive to ensure that the child has a healthy start in life (97,98). Fortunately these issues do not have to be resolved to establish public policy in this area. Governmental agencies supporting research are chartered to promote public health and to protect the welfare of children and other vulnerable subjects involved in research. Such agencies therefore have good reason to insist on the most stringent guarantees of safety where research anticipates transfer of embryos to a womb.

These protections are less important where transfer is not intended and the embryo is destined to be discarded in any case. It is useful to note here a difference between embryo research and fetal research. In U.S. law, fetuses intended for abortion are accorded the same level of moral protection in research (a requirement of no harm or minimal harm) as fetuses that are intended to be brought to term (99). However, it is not inconsistent to treat embryo research differently because fetal research poses a special set of problems. If researchers were permitted to perform harmful studies on the fetus of a woman intending abortion and she subsequently changes her mind and insists on continuing with the pregnancy, this would create a situation of unavoidable harm for the resulting child because it is ethically unacceptable to force her to terminate the pregnancy. A similar conflict situation cannot arise where embryo research is involved.

Donor Rights and Donor Sources

Donors of the gametes or embryos used in research are an important moral constituency whose rights and welfare must be respected. Three considerations shape thinking in this area: the requirement of informed consent, concern for the impact on family and the succession of generations in the use of certain donor sources, and the implications for society when financial incentives create a market in gametes or embryos used in research.

The requirement that researchers obtain the full, free, and informed consent of those who donate their reproductive material for research purposes means, among other things, that consent must be specific to the type of research proposed. Because donors can reasonably find one kind of embryo research morally acceptable and another repugnant, general consent to the use of gametes or embryos is not appropriate. Behind this requirement of specific consent lies the understanding that gametes and embryos are different from other bodily tissues. They contain the possibility of and association with people's dreams of offspring. Some individuals might be willing to donate reproductive material where transfer is intended, hoping to directly assist others to start a family. Other donors might be unwilling to see children of theirs raised outside the context of their family but would be willing to donate gametes or embryos for research not involving transfer. Despite their different feelings in this regard, all donors have the right to freely understand and consent to how their reproductive material is used in research.

Because oocyte donation for purposes of embryo research now requires follicle maturation in vivo, these procedures are also not without risk to the donor. Although these risks are routinely accepted by women who wish to donate oocytes to an infertile couple, some believe they are disproportionate to any benefits in research contexts where no transfer to a womb is anticipated. The NIH Human Embryo Research Panel came to this conclusion, although it permitted such donation by women about to undergo pelvic surgery who were properly informed about the additional risks of drug stimulation and superovulation. If and when embryo research goes forward in the future, Institutional Review Boards and other groups assigned to assess research risks will have to balance the values involved in permitting donation (including respect for donors' wishes) against the specific risks to donors and their vulnerabilities to coercion or pressure from researchers or physicians. The clinical IVF context merits special attention in this regard. Although women undergoing treatment for infertility are a constituency that has a significant stake in the benefits arising from embryo research, this context also creates special pressures that run counter to the requirement of full, free, and informed consent. Among these are the hope (or promise) of price reductions or additional support from one's treating physician in return for gamete or embryo donation.

Concern for the impact on family and the succession of generation suggests limits on donor sources. For example, although the use of cadavers as sources of oocytes might be appropriate for research where no transfer is intended (assuming that the women or her appropriate surrogates consent), this source raises many questions where transfer is intended. Among other things we can ask whether it is ethical to bring a child into the world whose genetic mother is deceased. Apart from the psychological risks of this practice, the child conceived in this way is possibly cut off from access to the deceased parent's medical history. Similar questions arise in connection with the idea of using aborted female fetuses as a source of oocytes for in vitro fertilization and embryo research. No only does this undermine the idea of donor consent (a problem not entirely eased by the consent of the abortus's mother to the use of her fetus in this way), it also implicates embryo research in the abortion and fetal tissue debate. Where transfer is intended, it raises the disturbing prospect, in terms of our ideas of generational succession, of the birth of a child whose genetic mother was aborted. For all these reasons, at least where transfer is intended, this donor source is best avoided.

Commercialization

It is also wise for government agencies involved in funding embryo research to avoid creating a market in gametes or embryos used for this purpose. In the United States substantial commercialization already exists in the area of egg or embryo donation, but payment for gametes or embryos used in embryo research poses special problems. Introducing government funding for the purchase of eggs or embryos would create a large source of revenue that might be particularly attractive to poor women, a constituency largely untouched by present commercial practices. In view of the special risks associated with egg donation and the deep symbolic importance of reproductive material, the purchase and sale of eggs and embryos is a practice that could tarnish the whole field of embryo research. This does not preclude the compensation of donors for their expenses in participating in a research program.

"SPARE" VERSUS "RESEARCH EMBRYOS"

Moral Issues

Probably the most vexing ethical issue in embryo research is the question of whether it is ever appropriate deliberately to fertilize oocytes for research purposes when there is no intention of transferring the resulting embryos (100). The alternative to such "research embryos" is to utilize only "spare" or "surplus" embryos remaining from infertility procedures. Although many who regard the embryo as a fully protectable being would object to the use of even surplus embryos in research, some who place substantial moral weight on the early embryo believe that spare embryos can ethically be used, since most are destined to perish anyway. This position has been subjected to multiple criticisms (101). For example, it is not clear why the fact that surplus embryos might otherwise continue in storage or be destroyed warrants their use in research. Those who believe the embryo is fully a moral person may therefore be unable to sustain a distinction between spare and research embryos and may be required by their position to oppose all embryo research.

Some who oppose the creation and use of "research" embryos base their views less on harm done the embryo than on symbolic concerns, believing that it is objectionable to use potential persons in such an instrumental way. These arguments, too, raise many questions. Among them are the question of why it should be regarded as acceptable to create more embryos than can be transferred in an IVF procedure but impermissible deliberately to create embryos for research purposes. Whatever the reasons behind it, opposition to the creation and use of "research embryos" is widespread and cuts across familiar political lines, drawing criticism even from those who accept embryo research generally (102).

Scientific Issues

In practical terms, there are many reasons for wanting to develop embryos for research purposes only. In many types of research existing surplus embryos cannot be used. Limiting research to spare embryos, for example, precludes most research involving study of the process of fertilization, since spare embryos have already been fertilized. Similarly research on in vitro oocyte maturation and oocyte cryopreservation would require fertilization and surveillance of the resulting embryos to establish the efficacy and safety of these procedures. It is true that if fertilization is defined as occurring at some point *after* the penetration the egg by the sperm, some of this research could go forward without being regarded as "embryo research" in a technical sense. This points up the importance of these definitional matters. Nevertheless, in many cases demonstrating the safety of new techniques will require the observation of the resulting embryos for some period of time.

Embryos developed in vitro during infertility procedures, often derived from older eggs, also evidence high rates of chromosomal abnormalities that may explain their inability to implant (103,104). This means that surplus embryos are a less than ideal population for broad classes of research on normal embryo development, and their use may actually produce misleading findings in many studies. To the extent that research requires a "normal" embryo population, whether for direct study or as controls, some research embryos may be needed. For all these reasons, the stakes here are high. Those who believe that embryos merit significant moral protection will not be persuaded that these benefits justify the deliberate creation of embryos for research purposes only. However, those who oppose this idea of "research embryos" on symbolic grounds, or merely because it strikes them as offensive, will have to reexamine their opposition in the light of its longer-term and possibly important negative impact on medical progress.

CONCLUSION

Human embryo research provides a vivid illustration of the way emerging scientific and medical capabilities are continually challenging the application of established ethical ideas and norms. Despite the elaboration of a significant body of ethical and legal rules governing the use of human subjects in research, new reproductive technologies have brought to the fore an entity, the early human embryo, whose existence raises basic questions about who is a human subject and what should be the limits of the research enterprise. The political turmoil surrounding embryo research is a sign of the difficulty and troubling nature of these questions. Few areas of scientific inquiry evoke so many emotions and differences of opinion. Although prohibited in many European nations and deprived of federal funding in the United States, research involving the human embryo continues. New scientific developments that arise from embryo research or that require it will predictably sustain public interest in this scientific and ethical frontier and stimulate continued debate.

BIBLIOGRAPHY

1. R.G. Edwards, P.C. Steptoe, and J.M. Purdy, *Nature (London)* **227**, 1307–1309 (1970).
2. P.C. Steptoe, R.G. Edwards, and J.M. Purdy, *Nature (London)* **229**, 132–133 (1971).
3. *Dorland's Illustrated Medical Dictionary*, 28th ed., Saunders, Philadelphia, PA, 1994.
4. 45 Code of Federal Regulations 46.203(c).
5. A. McClaren, in G. Bock and M. O'Connor, eds., *Human Embryo Research: Yes or No?* Tavistock, London, 1986, pp. 5–23.
6. F.E. Baylis, *Bioethics* **4**, 311–329 (1990).
7. D. Heyd, *Bioethics* **10**(4), 292–309 (1996).
8. N.M. Ford, *When Did I Begin?* Cambridge University Press, Cambridge, UK, 1988, pp. 102–118.
9. K. Moore, *The Developing Human: Clinically Oriented Embryology*, 3rd ed., Saunders, Philadelphia, PA, 1982.
10. Australia (Victoria), The Infertility (Medical Procedures) (Amendment) Act 1987, 9a(1).
11. Britain's Human Fertilisation and Embryology Act 1990, Section 1(1)(a)(b).
12. S. Buckle, K. Dawson, and P. Singer, in P. Singer et al., eds., *Embryo Experimentation: Ethical Legal and Social Issues*, Cambridge University Press, Cambridge, UK, 1990, pp. 213–225.
13. *Nature (London)* **386**, 98 (1997).
14. D.G. Jones and B. Tefler, *Bioethics* **9**(1), 32–49 (1995).

15. C.R. Austin, *Human Embryos*, Oxford University Press, Oxford, UK, 1989.
16. M.J.T. FitzGerald and M. FitzGerald, *Human Embryology*, Baillière Tindall, London, 1994.
17. Ethics Committee of the American Fertility Society, *Fertil. Steril.* **53**(6, Suppl. 2), 32SS (1990).
18. National Commission for the Protection of Human Subjects of Biomedical and Behavioral Research, *Report and Recommendations: Research on the Fetus*, U.S. Government Printing Office, Washington, DC, 1975, reprinted in *Fed. Regist.* **40**, 33, 530 (1976).
19. A. Vaheri et al., *N. Engl. J. Med.* **286**(20), 1071–1074 (1972).
20. P.J. Neumann, S.D. Gharib, and M.C. Weinstein, *N. Engl. J. Med.* **331**(4), 239–243 (1994).
21. H.F. Hodder, *Harv. Mag.*, November/December, pp. 54–99 (1997).
22. S.G. Stolberg, *N.Y. Times*, Sunday, December 14, pp. A1, A36 (1997).
23. B.R. Spirtas and S.J. Alexander, *Fertil. Steril.* **59**, 291–293 (1993).
24. A.S. Whittemore et al., *Am. J. Epidemiol.* **136**, 1175–1220 (1993).
25. M.A. Rossing et al., *N. Engl. J. Med.* **331**(12), 771–776 (1994).
26. L. Gras, J. McBain, A.O. Trounson, and I. Kola, *Hum. Reprod.* **7**, 1396–1401 (1992).
27. P. Belluck, *N.Y. Times*, January 3, pp. A1, A9 (1998).
28. A.C. Van Steirteghem, *N. Engl. J. Med.* **338**, 194–195 (1998).
29. K.Y. Cha et al., *Fertil. Steril.* **55**, 109–113 (1991).
30. T.L. Toth et al., *Fertil. Steril.* **61**, 1077–1082 (1994).
31. C.E. Wood, J.M. Shaw, and A.O. Trounson, *Med. J. Aus.* **166**, 366–369 (1997).
32. G. Kolata, *N.Y. Times*, October 17, pp. A1, A16 (1997).
33. A.H. Handyside et al., *N. Engl. J. Med.* **327**, 905–909 (1992).
34. T.C. Doetschman et al., *J. Embryol. Exp. Morphol.* **87**, 27–45 (1987).
35. A.J. Thomson et al., *Science* **282**, 1145–1147 (1998).
36. Shamblott et al., *Proc. Nat. Acad. USA* **95**, 13726–13731 (1998).
37. G. Kolata, *Clone: The Road to Dolly and the Path Ahead*, Wm., Morrow, New York, 1998.
38. L.B. Andrews and N. Elster, *Papers Commissioned for the Human Embryo Research Panel*, NIH Publ. No. 95-3916, National Institutes of Health, Bethesda, MD, 1994, pp. 251–264.
39. L. Nielsen, in D. Evans, ed., *Conceiving the Embryo: Ethics, Law and Practice in Human Embryology*, Martinus Nijhoff Publishers, The Hague, 1996, pp. 325–338.
40. Law No. 56 of 5 August, 1994 on the Medical Use of Biotechnology, *Nor. Lovtidend*, Section 3-1; *Int. Dig. Health Legis.* **46**(1), 51–54 (1995).
41. *The Infertility (Medical Procedures) Act, 1984* and *The Infertility (Medical Procedures) (Amendment) Act 1987*.
42. *The Human Reproductive Technology Act 1991*, Preamble, paragraphs B and C; *Int. Dig. Health Legis.* **44**(4), 630–631 (1993).
43. Federal Law of 1992 (Serial No. 275), Art. I, Section 9; *Int. Dig. Health Legis.* **44**(2), 248–249 (1993).
44. Law of 13 December for the Protection of Embryos (the Embryo Protection Law), *Bundesgesetzblatt*, Part I, pp. 2746–2748 (1990); *Int. Dig. Health Legis.* **42**(1), 60–69 (1991).

45. *Genetics and Assisted Procreation*, Amendment of Federal Constitution, dated 13 August 1992, *Rec. Off. Lois Féd.* **32**, 1579, 1580 (1992); *Int. Dig. Health Legis.* **44**(4), 745–746 (1993).

46. Law No. 94-654 of 29 July 1994 on the Donation and Use of Parts and Elements and Products of the Human Body, Medically Assisted Procreation, and Prenatal Diagnostics, *J. Off. Répub. Fr. Lois Décrets* **175**, 11060–11068 (Article I, 152, 153) (1994); *Int. Dig. Health Legis.* **45**, 473–482 (1994).

47. Law No. 503 of 24 June, 1992 on the Scientific Ethics Committee System and the Examination of Biomedical research Projects, Chapter 4, Section 14, *Lovtidende* **84** (Part A), 2017–2020 (1992); *Int. Dig. Health Legis.* **43**(4), 758–760 (1993).

48. Law No. 35/1988 of 22 November 1988 on Assisted Reproduction Procedures, *Bol. Of. Estado* **282**, 33373–33378 (1988); Law No. 42/1988 of 28 December 1988 on the Donation and Use of Human Embryos and Fetuses or Their Cells and Tissues, or Organs, *Bol. of. Estado* **314**, 36766–36767 (1988).

49. Law No. 115 of 14 March 1989 Concerning Measures for the Purpose of Research or Treatment in Connection with Fertilized Human Oocytes, *Sven. Författningssamling*, March 26, pp. 1–2 (1991); *Int. Dig. Health Legis.* **44**(1), 58–59 (1993).

50. The Human Fertilization and Embryology Act 1990, *Int. Dig. Health Legis.* **42**(1), 69–85 (1990).

51. *Proceed with Care*, Final Report of the Royal Commission on New Reproductive Technologies, vol. I, Minister of Government Services, Ottawa, 1993, pp. 607–659.

52. W. Kondro, *Lancet* **347**, 1758 (1996).

53. Council of Europe, *Convention for the Protection of Human Rights and Dignity of the Human Being with Regard to the Application of Biology and Medicine, Convention on Human Rights and Biomedicine*, Council of Europe, Strasbourg, 1996.

54. M. Warnock, *A Question of Life: The Warnock Report on Human Fertilisation and Embryology*, Blackwell, Oxford, 1985.

55. L.B. Andrews, in *Papers Commissioned for the Human Embryo Research Panel*, NIH Publ. No. 95-3916, National Institutes of Health, Bethesda, MD, 1994, pp. 297–330.

56. La. Rev. Stat. Ann. §9 : 123, 122, *et seq.* (West 1991).

57. L.B. Andrews, *Papers Commissioned for the Human Embryo Research Panel*. NIH Publ. No. 95-3916. National Institutes of Health, Bethesda, MD, 1994, pp. 303–305.

58. *HEW Support and Research Involving Human In vitro Fertilization and Embryo Transfer*, Report and Conclusions, U.S. Department of Health Education and Welfare, Washington, DC, 1979; Ethics Advisory Board, Summary and Conclusions, *Fed. Regis.* **44**(18), 35057 (1979).

59. National Institutes of Health (NIH) Revitalization Act of 1993 (P.L. 103-43, Section 121 (c)).

60. *Report of the Human Embryo Research Panel*, National Institutes of Health, Bethesda, MD, 1994.

61. HR 3019-PL 104-34, *Congressional Quarterly*, 1874, (June 19, 1996).

62. N. Wade, *N.Y. Times*, June 29, p. A13 (1999).

63. Pope John Paul II, *On the Value and Inviolability of Human Life (Evangelium Vitae), Encyclical Letter Addressed by the Supreme Pontiff*, 1995, Section 63.

64. The United States Catholic Conference of Bishops, *Documentation on Abortion and the Right to Life*, U.S. Catholic Conference, Washington, DC, 1976, p. 39.

65. R.J. White, Testimony before the National Institutes of Health Human Embryo Research Panel, June 21, 1994; reprinted as *America*, September 14, pp. 4–5 (1996).

66. H. Watt, *J. Med. Ethics* **22**, 222–226 (1996).

67. J.T. Noonan, Jr., in J.T. Noonan, Jr., ed., *The Morality of Abortion: Legal and Historical Perspectives*, Harvard University Press, Cambridge, MA, 1970, pp. 51–59.

68. Ethics Committee of the American Fertility Society, *Fertil. Steril.* **46**(3, Suppl. 1), 26S–28S (1986).

69. L. Strain et al., *N. Engl. J. Med.* **338**, 166–169 (1998).

70. R.A. McCormick, *Kennedy Inst. Ethics J.* **1**, 1–15 (1991).

71. S. Buckle, in P. Singer et al., eds., *Embryo Experimentation: Ethical Legal and Social Issues*, Cambridge University Press, Cambridge, UK, 1990, pp. 90–108.

72. P. Singer and K. Dawson, in P. Singer and K. Dawson, eds., *Embryo Experimentation: Ethical, Legal and Social Issues*, Cambridge University Press, Cambridge, UK, 1990, pp. 76–89.

73. British Royal College of Obstetricians and Gynaecologists (RCOG), *Report of the RCOG Ethics Committee on In vitro Fertilization and Embryo Replacement or Transfer*, RCOG, London, 1983.

74. C.J. Roberts and C.R. Lowe, *Lancet* **1**, 498–499 (1975).

75. H.T. Englehardt, Jr., *The Foundations of Bioethics*, Oxford University Press, New York, 1996.

76. M. Tooley, *Abortion and Infanticide*, Oxford University Press, New York, 1983.

77. M.A. Warren, *The Monist* **57**(1), 43–61 (1973).

78. B. Steinbock, *Life Before Birth: The Moral and Legal Status of Embryos and Fetuses*, Oxford University Press, New York, 1992.

79. B. Brody, *Abortion and the Sanctity of Human Life*, MIT Press, Cambridge, MA, 1975, p. 114.

80. J.M. Goldenring, *J. Med. Ethics* **11**, 200 (1985).

81. H.M. Sass, *J. Med. Philos.* **14**, 45–59 (1989).

82. R.M. Veatch, in M.W. Shaw and A.E. Doudera, eds., *Defining Human Life: Medical, Legal, and Ethical Implications*, ALPHA Press, Ann Arbor, MI, 1983, pp. 99–113.

83. P. Singer and D. Wells, *Making Babies*, Scribner's, New York, 1985.

84. P. Singer, *Animal Liberation*, Avon Books, New York, 1990.

85. J.S. Freeman, *J. Med. Philos.* **21**, 61–81 (1986).

86. J. Robertson, *Hastings Cent. Rep.* **25**(1), 37–38 (1995).

87. R. Dworkin, *Life's Dominion*, Knopf, New York, 1993.

88. M. Charlesworth, in P. Singer et al., eds., *Embryo Experimentation: Ethical Legal and Social Issues*, Cambridge University Press, Cambridge, UK, 1990, pp. 147–152.

89. J. Rawls, *Political Liberalism*, Columbia University Press, New York, 1993.

90. I.H. Carmen, *Hum. Gene Ther.* **7**, 97–108 (1996).

91. D. Callahan, *Hastings Cent. Rep.* **25**(1), 39–40 (1995).

92. P.R. Braude, V.N. Bolton, and M.H. Johnson, in G. Bock and M. O'Connor, eds., *Human Embryo Research: Yes or No?* Tavistock, London, 1986, pp. 63–82.

93. R. Edwards, in A. Dyson and J. Harris, eds., *Experiments on Embryos*, London, Routldge, 1990, pp. 42–54.

94. Report and Recommendations of the National Bioethics Advisory Commission, *Cloning Human Beings*, Rockville, MD, 1997.

95. R. Macklin, Testimony before the National Bioethics Advisory Commission, March 14, 1997, quoted in G. Kolata (37, p. 20).

96. D. Brock, *Bioethics* **9**(3/4), 269–275 (1995).

97. R.M. Green, *J. Law, Med. Ethics* **25**(1), 5–15 (1997).

98. C.B. Cohen, in N. Fotion and J.C. Heller, eds., *Contingent Future Persons*, Kluwer Academic Dordrecht, The Netherlands, 1997, pp. 27–40.

99. The Public Health Service Act as Amended by the Health Research Extension Act of 1985, Public Law 99-158, November 20, 1985, Section 498(a)(3).

100. D.S. Davis, *Kennedy Inst. Ethics J.* **5**(4), 343–354 (1995).

101. C.A. Tauer, in N. Fotion and J.C. Heller, eds., *Contingent Future Persons*, Kluwer Academic Dordrecht, The Netherlands, 1996, pp. 171–189.

102. Editorial, *Washington Post*, Sunday, October 2, p. C8 (1994).

103. Z. Rosenwaks, O.K. Davis, and M.A. Damario, *Hum. Reprod.* **10**(Suppl. 1), 165–173 (1995).

104. C.M. Strom, *J. Assist. Reprod. Genet.* **17**(7), 592–593 (1996).

See other HUMAN SUBJECTS RESEARCH entries.

HUMAN SUBJECTS RESEARCH, ETHICS, RESEARCH ON VULNERABLE POPULATIONS

PATRICIA BACKLAR
Portland State University
Oregon Health Sciences University
Portland, Oregon

OUTLINE

INTRODUCTION

Adult individuals characterized as vulnerable usually are persons who, for a variety of reasons, are incapable of protecting their own interests. Biomedical research subjects may be considered vulnerable because of illness, mental disorder, or particular circumstances (1). In general, individuals are recognized as vulnerable when they are cognitively impaired, or when their circumstances subject them to intimidation and exploitation, thereby limiting their freedom to exercise autonomous choice. Some persons with mental disorders are prototypical of individuals who are recognized as especially vulnerable. Not only is their decision making compromised by illness, but often they also are socially disadvantaged and devalued for reasons of poverty, institutionalization, or stigmatization. Historically, vulnerable and devalued populations, often unwittingly, were forced to serve as research subjects in studies not relevant to their own conditions, but which benefited the health of more privileged members of society. Yet measures designed to protect these populations have come to be seen as overly exclusionary and unjust. Indeed, populations judged vulnerable are now considered to be at risk of being relegated to a class of persons for whom little or no therapeutic benefit may be available. Exclusion may reinforce their vulnerability. However, powerless and impoverished persons may be especially vulnerable to the popular therapeutic misconception that research protocols provide beneficial treatment. Inability to adequately engage in the process of voluntary and informed consent — for whatever reasons — raises the research subject's degree of risk and vulnerability. Increased regulatory protections that provide for research anticipatory planning, surrogacy, independent health care supervision, and other safeguards may allow vulnerable research subjects to be enrolled in studies that address their conditions while shielding them from exposure to harms and wrongs.

VULNERABLE POPULATIONS IN BIOMEDICAL RESEARCH

Among persons with mental illness there are those who share at least four characteristics that are considered prototypical of other devalued and vulnerable populations. First, their disorders put them at risk for loss of decision-making capacity. Second, they are likely to be poor. Even though persons may be afflicted with mental disorders regardless of socioeconomic level, there is a significant association between poverty and serious mental illness due to the incapacitating effects of the disease itself (2). Third, they are little understood, often demeaned, and unjustifiably feared. Stigma is a significant and widespread feature of mental disorders (3,4). Fourth, like other undervalued populations who are regarded as less important than more powerful and privileged members of society, they are at increased risk of being exploited in research. Abuse of human research subjects has long been associated with members of socially devalued populations. During World War II, a euthanasia program implemented by Nazi physicians killed thousands of patients in German mental hospitals, and set the stage in the concentration camps for the infamous human subject research on little valued groups like jews, gypsies, and homosexuals (5). Such eugenics programs also were proposed in the United States. A 1942 article in the *American Journal of Psychiatry* suggested killing retarded children (6).

Members of socially devalued groups are singularly vulnerable to exploitation in human subject research. Persons in such groups are more likely to be poor, to be welfare patients, to be in need of health care services, to be high users of institutional facilities like hospitals, group homes, jails, and prisons, to lack social support networks, and, in the United States, to have limited access to health care services. Such populations are apt to be both oppressed and exploited because society discounts them. Their well-being, their rights, and their welfare are disregarded by the society in which they live (7).

Some of the most egregious research studies have targeted specific groups of persons who were considered, in some way, to be inferior. The notorious Public Health Service Tuskegee Syphilis Study, which began in the 1930s and continued for 42 years, used poor and uneducated African-American men as study subjects. The researchers compared the health and longevity of an untreated syphilitic population with a nonsyphilitic but similar population. These men did not even know that they were in a study. They were never told that they had syphilis, nor were they given appropriate medication for their disease when it became available. In 1956 researchers took young residents at the Willowbrook State School who were severely developmentally disabled children, and deliberately infected them with isolated strains of the infectious hepatitis virus. And, another study in 1963, at the Brooklyn Jewish Chronic Disease Hospital dying, impoverished, senile, elderly were — unbeknown to them — deliberately injected with live cancer cells (8).

Genetic Research with Vulnerable Populations

In general, genetic research is minimally physically invasive and does not involve hazardous procedures. Vulnerable populations and the general public may encounter similar kinds of issues in genetic research. Nonetheless, the discrete ethical, social, and legal concerns inherent in genetic research may have greater impact on vulnerable populations. Disorders identified by genetic research are distinctive because they affect not only individuals but also groups of related persons, and groups of unrelated persons. Genetic information about an individual may reveal particular or probabilistic information about his or her living relatives, dead relatives, and future unborn offspring (3). When disclosure of genetic information occurs, individuals, their families, and their ascriptive groups may be affected by a loss of privacy, breaches of confidentiality, familial conflict, and by psychosocial harms.

Studies that make specific comparisons between racial and ethnic groups or involve behavioral genetics may be used to stigmatize and discriminate against members of such populations (9). Genetic information may affect the ways such individuals, their relatives, and their cohorts are viewed by others. On the positive side, knowledge of genetic information relevant to the biological basis of a disorder like Alzheimer's disease appears to have helped dispel myths and reduce stigma associated with the condition. However, some vulnerable and devalued populations, like persons with mental disorders, who already may suffer stigma identified with their disorders and experience unfair discrimination in housing, employment, or insurance, may endure even more intolerance and prejudice. Commentators suggest that strong protections of privacy and confidentiality should be developed in order not to add to the burdens of already disadvantaged groups (3).

Subject Selection Predicament — Exclusion or Inclusion?

A protectionist stance that shields vulnerable groups from research participation may deny them the benefits garnered from scientific research that are available to other disease populations. Paradoxically, such exclusion may reinforce their vulnerability. According to commentators, when diseases that affect women, minorities, and other undervalued populations are not addressed by research, knowledge that could rectify prevailing ineffectual or harmful routine medical care is never produced (10,11). Even though women (and particularly pregnant women), minorities, and children have been systematically excluded from biomedical research, U.S. federal policies have encouraged the research participation of all societal groups. There has been a continuing pressure to increase fair access to, and opportunity to participate in clinical trials (12). Indeed, there has been a paradigm shift in the way enrollment in research trials is viewed since acquired immunodeficiency syndrome (AIDS) was first identified in 1981 (13). Participation in research trials previously had been considered as unavoidably risky and burdensome.

Historically devalued populations, like prisoners and persons with mental disorders, served as subjects and bore the brunt of biomedical research with no forthcoming advantage to themselves or to the populations that they represented, while the more powerful members of society reaped the health benefits gleaned from the research (9,14). Thus vulnerable research subjects were seen as inherently needing protection and an exclusionary model of protection came to be employed. But with the emergence of AIDS this traditional model of protection for selection of human subjects for research trials was transformed. The energetic advocacy of AIDS activists for inclusion in research protocols, and the subsequent clamor of advanced cancer patients to gain admittance to cancer trials encouraged a tight connection between research and treatment. Research came to be seen as a pathway to better medical care for individuals and their particular disease populations. Currently a more expansive and inclusionary model is advocated and greater weight is given to sharing the benefits of research.

Perils of the Therapeutic Misconception: Making the Distinction Between Clinical Treatment and Biomedical Research

Another kind of hazard for vulnerable populations may lurk in a standard that argues for wider inclusion in research trials. Persons who suffer from disabling and damaging diseases, for which no cure exists, understandably are desirous of relief. However, individuals whose incapacitating diseases may be inadequately treated are likely to be susceptible to the popular therapeutic misconception that research protocols provide beneficial treatment, whether or not there is any prospect of benefit. Indeed, such persons may be especially vulnerable if they are powerless and impoverished members of devalued populations (15).

The distribution of healthcare services in the United States is unequal. Participation in a research protocol for some people may be their only way to gain access to an essential therapy that addresses their particular condition. For example, many persons with mental disorders have no or only inadequate access to any mental health care service due to discriminatory and inequitable health insurance policies that anachronistically differentiate between "physical" health and "mental" health and deny parity. Persons in such circumstances may be willing to bear the risks of research when it appears to offer a prospect for beneficial treatment (12). But biomedical research is not clinical treatment. In clinical treatment the patient's welfare and "personal care" is the physician's first consideration (16). In contrast, in biomedical research the investigator's primary purpose is to develop or contribute to generalizable knowledge, not to benefit any particular research subject (17). The distinction between clinical practice and biomedical research often is misunderstood by research subjects (18). When the demarcation lines are blurred between patient and subject, between physician and investigator, and between treatment and research, patient-subjects are likely to believe, as is the case with treatment, that the research is designed to directly benefit them.

According to Jay Katz, the misconception has grown out of the recent practice of conflating the role of the physician with the role of the research investigator. Physicians who also are investigators compound the potential subject's confusion. It has become common practice for physicians to enroll their patients in their own research studies. Many patients enrolled in such studies believe that they will be the beneficiaries of effective treatments, even though there may be minimal or no likelihood of benefit. Patients misconstrue their physician's invitation to take part in research as a treatment recommendation (19). Patients trust their physicians and permit them to hold enormous power; in exchange, patients expect their physicians only to serve their therapeutic needs (20).

Because the language of many consent forms fosters the belief that a therapeutic benefit will be forthcoming, some commentators propose that the research community take measures to help potential subjects appreciate that they are unlikely to benefit from research participation (21). Other writers express concern that if the therapeutic misconception is not adequately dealt with during the informed consent process, subjects are likely to feel that they have been deceived and public trust of researchers, and the health care system in general, may be eroded. They recommend that a trained, neutral educator provide information about the study and advise subjects when it might not be in their best interests to take part in

the study. Additionally these writers suggest that the therapeutic misconception may be lessened if investigators emphasize the substance of the disclosure, explain the scientific methodology, and maintain an ongoing consent process (22).

IMPACT AND INFLUENCE OF INTERNATIONAL CODES, U.S. REGULATORY POLICIES, AND BIOETHICS COMMISSIONS ON RESEARCH WITH VULNERABLE SUBJECTS

International Codes

The central tenet of the Hippocratic Oath—the primary obligation of physicians is to benefit their patients—has prevailed for more than 2400 years in Western clinical medicine. This fundamental ethic recognizes the patient's potential for vulnerability and thus requires that physicians act solely on behalf of their patients' best interests. However, because in biomedical research the scientific method requires that the researcher engender generalizable knowledge, the needs and interests of the vulnerable subject may be compromised. Thus the clinician's obligations to the patient stand in direct conflict with the researcher's duties. These incommensurable goals were rendered potentially compatible by the first set of international research principles, the Nuremberg Code. By making mandatory the voluntary informed consent of all research subjects, the Code strove to shelter the research endeavor (23). The authors of the Nuremberg Code—by employing the moral principal of respect for autonomy—established two kinds of freedom for competent research subjects; the freedom not to be interfered with by others, and the freedom to make their own choices (24).

The Code's dominant principle, "The voluntary consent of the human subject is absolutely necessary. This means that the person involved shall have the legal capacity to consent," was delineated in response to the cruel research conducted by Nazi doctors on prisoners who, because of their circumstance, were not in a position to give their consent, whether or not they were competent (25). Special protections for mentally ill persons, which allowed for proxy consent, were specified in a final memorandum by the chief medical advisor to the Nuremberg judges but were not included in the final document (26). Thus the Code's affirmation of the primacy of consent effectively excludes many persons with disorders affecting their decision-making capacity from participation in drug trials and other sorts of research.

The international codes and regulations that were developed after the Nuremberg Code have attempted to reconcile society's twin responsibilities to adequately protect vulnerable research subjects and to ensure that vulnerable populations receive the benefits of research. The various drafts of the World Medical Association's Declaration of Helsinki, first issued in 1964, endeavor to move along and ease the consent requirements via the mechanism of a legal guardian for persons who lack capacity to consent to research. The Declaration classifies research as "therapeutic" and "nontherapeutic." But the Declaration seems to forbid the enrollment of incapable subjects in research protocols that do not offer subjects the probability of direct benefit. Research that advances generalized knowledge solely for the benefit of others only may be carried out with volunteers (27).

Another document, the *International Ethical Guidelines for Biomedical Research Involving Human Subjects*, which considers, among other concerns, research involving persons who are incapable of giving adequately informed consent, was issued in 1993 by the Council for International Organizations of Medical Sciences (CIOMS) in collaboration with the World Health Organization (WHO) (28). In the case of incompetent subjects, the CIOMS/WHO guidelines allow for proxy consent: Informed consent may be obtained from a legal guardian or other duly authorized person. The investigator, however, is directed to obtain the consent of each subject to the extent of the subject's capabilities, and also to always respect the prospective subject's refusal to participate in "nonclinical" research.

Evolution of U.S. Policies for Protection of Vulnerable Human Research Subjects

Existing federal regulations provide special protections for research subjects regarded as vulnerable. These regulations apply to research involving children, fetuses, pregnant women, human in vitro fertilization, and prisoners (1). When the federal regulations were first promulgated, special protections for persons, then described as mentally infirm and institutionalized, were proposed but never enacted into law. Although the regulations mention that additional safeguards must be included in the study to protect the rights and welfare of mentally disabled persons, these directions do not (1) elucidate criteria, methods, procedures, or limitations to the research design; (2) specify the particularities of informed consent processes, or of appointments, powers, and education of surrogate decision makers; (3) describe appropriate pathways between the research study and the subject's ongoing clinical care; or (4) make provisions for aftercare health arrangements that may be required when the research is concluded (29–31). How did this state of affairs come about?

During and following the Second World War, this country's medical and behavioral research expanded and flourished unchecked, for the most part, by any supervision. Reports of numerous and varied research scandals were exposed (32). Initially the federal government was slow to react despite the efforts of a few members of Congress and some government scientists (33). The government's own Public Health Service's Tuskegee Syphilis Study continued unrestrained until exposed in a 1972 newspaper article. Only then did a flurry of congressional activity occur regarding human research subject protection. In May 1974, regulations protecting human subjects became effective. These regulations raised to statutory status the National Institutes of Health (NIH) Policies for the Protection of Human Subjects, which were first issued in 1966. Forty-three days later Congress created the National Commission for the Protection of Human Subjects of Biomedical and Behavioral Research. Because of concerns about the use psychosurgical experimental

procedures like pre-frontal lobotomy over which no peer group or federal regulatory body was in a position to regulate, decisionally impaired persons were among the special populations the National Commission was expected to consider (34).

The National Commission in its classic document, the Belmont Report, advanced the principle of respect for persons as autonomous agents, recognized that persons with diminished autonomy are entitled to protection, and applied this basic ethical precept to the process of voluntary and informed consent, as the first principle appropriate to the conduct of research (17). Thus the National Commission echoed the primary tenet of the Nuremberg Code. However, the National Commission allowed a less restrictive approach to involving incapable subjects in research. In its 1978 Report and Recommendations, *Research Involving Those Institutionalized as Mentally Infirm*, the National Commission recognized that indirect harms are likely to occur should research involving persons with mental disorders be forbidden:

> [S]ince some research involving the mentally infirm cannot be undertaken with any other group, and since this research may yield significant knowledge about the causes and treatment of mental disabilities, it is necessary to consider the consequences of prohibiting such research. Some argue that prohibiting such research might harm the class of mentally infirm persons as a whole by depriving them of the benefits they could have received if the research had proceeded (35, p. 58).

This opinion signaled a critical development in the conceptual framework for addressing the moral issues relevant to human subject research in general and to vulnerable human subjects in particular. The National Commission moved away from the position that procedures that hold no prospect of direct benefit to the noncompetent subject are not morally permissible (36). Instead, the National Commission was persuaded by the view that when risks are not unreasonable everyone, with or without decision-making capacity, has an obligation to benefit society, and that within the scope of that obligation research may be acceptable (37). This position allowed for the weighing or balancing of moral interests. The overlapping goals of benefiting the class of incapable persons, while at the same time safeguarding incapable individual subjects from unacceptable harm could be satisfied by balancing the risk of harm with the likelihood of benefit (38).

The National Commission recommended that researchers not involve vulnerable populations in research that was unconnected to their conditions as individuals or as a class of persons. They proposed a ranking of research classifications that instituted to a more precise degree substantive and procedural standards for research protocols involving more than minimal risk to incapable subjects. The National Commission appreciated that not all persons with mental disorders are incapable of giving voluntary and informed consent. They also recommended a process that allowed incapable subjects to assent or object with a simple yes or no when asked about their choice about being enrolled in a study. However, the National Commission was concerned about the vulnerability of this population and advised that when research protocols involved greater than minimal risk, that institutional review boards (IRBs) be allowed to use their discretionary judgment to appoint a consent auditor to oversee and guarantee the adequacy of the research protocol's consent process. Only in greater than minimal risk research protocols where there was no prospect of direct benefit for the subjects should the presence of consent auditors be compulsory. Because incapable adults lack the legal guardian that most children have, the National Commission suggested it might be necessary for a court-appointed guardian to approve research participation.

The U.S. Code of Federal Regulations reflects many of the National Commission's other recommendations involving populations regarded as vulnerable such as children, pregnant women, and prisoners. Yet the recommendations in its report on *Research Involving Those Institutionalized as Mentally Infirm*, were never transposed into final regulations. In 1978, after the work of National Commission had been completed and the Commission was dissolved, the proposed regulations were published in the Federal Register. The Department of Health, Education and Welfare (now known as the Department of Health and Human Services) regulation writers significantly altered the original recommendations and inserted proposals that were more demanding. They suggested that a consent auditor monitor all research including that which involved no more than minimal risk (39). There was a strong negative response from those who perceived that such regulations would significantly limit valuable research. Human rights advocates had another perspective and voiced concern that clinicians, researchers, and IRBs might not adequately respect the interests of persons with mental disorders.

The next bioethics commission, the President's Commission for the Study of Ethical Problems in Medicine and Biomedical Research, which was established in 1980 by Congress, insisted in 1981 and again in 1983 that the proposed regulations be made official. The Secretary of the Department of Health and Human Services (DHHS) countered that the proposed rules produced a "lack of consensus," and furthermore that the "basic regulations on human subjects research adequately respond to the recommendations made by the National Commission to protect persons institutionalized as mentally disabled" (40). Despite the Secretary's judgment, as of this writing, it is generally agreed that the Code of Federal Regulations does not provide adequate guidance for research with adult persons who have disorders involving some degree of cognitive impairment (30,41–43).

Recent Efforts to Establish Appropriate Safeguards for Vulnerable Subjects

The Advisory Committee on Human Radiation Experiments (ACHRE) was created in 1994 by the President in response to concerns about possible past unethical research conduct by the U.S. government and institutions funded by the government. ACHRE was charged not only to "tell the full story about [human radiation research] to the American public," but also to "examine the present, to determine how the conduct of human radiation research

today compares with the past and to assess whether, in light of this inquiry, changes need to be made in the policies of the federal government" for the protection of human subjects in research (44, p. 1).

Among the contemporary research protocols examined by ACHRE were four studies involving diagnostic imaging with cognitively impaired adult subjects. ACHRE's final report noted that in these studies, where subjects' movements were severely restricted, there had been no discussion in the documents or consent forms with the subjects about the implications of these potentially anxiety-provoking conditions. The ACHRE report also mentioned that there was no discussion of the subjects' capacity to consent or evidence that appropriate surrogate decisionmakers had given permission for the subjects' participation. According to the report:

> The question of whether or under what conditions adults with questionable decision-making capacity can be used as subjects of research that offers no prospect of benefit to them is unresolved in both research ethics and regulation. When such research puts potentially incompetent people at greater than minimal risk of harm, it is even more ethically problematic (44, p. 707).

Moreover ACHRE voiced concern in regard to the current system of research oversight. "Without guidance from the federal government, and perhaps regulatory relief, IRBs may not have the flexibility necessary to concentrate their efforts where subjects are in greatest need of protection—on the proposals that pose the greatest risks to subjects" (44, p. 819).

The DHHS Office of the Inspector General (OIG), which conducted a broad inquiry into current IRB practice, reached similar conclusions. According to the 1998 OIG report, IRBs have vulnerabilities that threaten their effectiveness in protecting human subjects (45). The OIG report drew attention to the fact that IRBs rarely conduct ongoing review of active research. Such lapses in oversight may have particularly serious implications when research protocols involve vulnerable subjects who have limited capacity for decisionmaking and thus a limited ability to protect their own interests.

> ... [This] is a serious national issue because it compromises... [the IRBs'] protection of human subjects. It inhibits their capacity to identify and address situations where unacceptable risks emerge, or research results prove to be too favorable to continue, or protocol stray beyond approved limits. It also inhibits their capacity to ensure that the subjects have sufficient understanding of the risks they may incur in the research process (45, p. iii).

In October 1995 the National Bioethics Advisory Commission (NBAC) was established by executive order. The first meeting took place in October 1996. The immediate charge to NBAC was to respond to the ACHRE recommendations to focus on the protection of the rights and welfare of human research subjects. Primary among other tasks, NBAC was to complete the National Commission's unfinished business to consider how ethically acceptable research may be conducted with human subjects who suffer from mental disorders that may affect their decision-making capacity. The NBAC report and recommendations, *Research Involving Persons with Mental Disorders that May Affect Decisionmaking Capacity*, was published in December 1998 (46). Specifically, the NBAC report "focused its attention on those who may be primarily considered for research because it is their particular mental disorder that is being studied" (46, p. 5). The report identified four types of limitations that may be experienced by persons with mental disorders:

> First, some individuals might have fluctuating capacity, what is often called waxing and waning ability to make decisions, as in schizophrenia, bipolar disorders, depressive disorders, and some dementias. Second, decisionmaking deficits can be predicted in some individuals due to the course of their disease or the nature of the treatment. Although these individuals may be decisionally capable in the early stages of the disease progression, such as in Alzheimer's disease, they have prospective incapacity. Third, most persons with limited capacity are in some way still able to object or assent to research, as in the case of more advanced Alzheimer's disease. Fourth, persons who have permanently lost the ability to make nearly any decision that involves any significant degree of reflection are decisionally incapable, as in the later stages of Alzheimer's disease and profound dementia (46, p. 10).

NBAC commissioners were concerned with the lack of specificity in the federal regulations in regard to safeguards that should be included in research protocols to protect the rights and welfare of adult subjects with illnesses involving some degree of cognitive impairment. Problems surrounding the informed consent process with vulnerable or potentially vulnerable subjects, and challenging moral concerns in regard to research design, distinct to this population, were addressed. Because decisional impairment, some forms of attention deficit, and incapacity may occur more often among "some people with certain mental disorders than in the general population" (46, p. 58), the commission recommended that IRBs require, in studies that pose more than minimal risk, that a qualified professional, independent of the research, assess the person's capacity to consent, even when the potential subject appears to have capacity for decision making. However, the NBAC report allowed for less formal capacity assessment approaches to be used if the researcher can establish good reasons for doing so.

The report also elaborates upon the concept of anticipatory research planning, and elucidates and expands the role of the legally authorized representative. The concept of obtaining a subject's consent to research participation in advance had been examined in 1989 by the American College of Physicians (47). And in 1996 new regulations adopted by the Food and Drug Administration (FDA) and National Institutes of Health (NIH) clarified rules in regard to research involving incapable subjects in emergent and life-threatening situations in emergency settings, and fostered the concept of research anticipatory planning (48). The new rule yields a narrow exception to federal informed consent requirements and permits research to proceed when it is not feasible to get informed consent from a potential subject or the subject's legally authorized representative. The rules allow researchers to obtain a waiver when they cannot reasonably obtain

consent in advance or at the time of the subject's enrollment. The regulations also expand the definition of a family member to encompass "any individual related by blood or affinity whose close association with the subject is the equivalent of a family relationship" (49). By supporting the concept of consent in advance and opening up the meaning of "family relationship," these regulations are relevant to and provide guidance for research in general involving cognitively impaired persons (50). It should be noted that NBAC commissioners did "not endorse the idea of authorizing third parties to enroll incapable subjects in research involving greater than minimal risk without the prospect of direct medical benefit" (46, p. 63). For research protocols falling into this category, where subjects have not given specific consent in advance, NBAC proposed that the Secretary of the DHHS create a national panel to review individual protocols that cannot otherwise be approved.

During the 18 months of deliberation that preceded the NBAC report, the commissioners heard testimony from former research subjects, their families, IRB members, and researchers who described their experience with protocols involving subjects with mental disorders. From these public discussions the commissioners learned that certain types of protocols may be expected to escalate subjects' symptoms, relapse, and suffering. The NBAC commissioners reviewed a small selection of such research protocols and consent forms that recently had been conducted in the United States. According to the findings from this brief review, NBAC recommended that IRBs pay special attention and "heightened scrutiny" (46, p. 56) to protocols that incorporate an ethically controversial design. The report identified such studies as those "that are designed to provoke symptoms, to withdraw subjects rapidly from therapies, and to use placebo controls" (46, p. 64). NBAC proposed that in any such studies judged to be clearly critical to the development of scientific knowledge, IRBs, who according to current regulations have the authority to continue observation of approved studies, should exercise that prerogative in studies where subjects may be at risk of relapse.

Ethically Controversial Research Design

Use of placebo in medication trials involving persons with schizophrenia has drawn considerable attention to the ethics of research design. Persons with mental disorders involving decisional impairments, such as schizophrenia and major depression, are likely to suffer painful symptoms that can be life-threatening to themselves or others. Nowadays many of these symptoms can be well controlled by medication. Thus the ethical question arises as to whether it is ever appropriate to withdraw medication from individuals who rely on a specific medical therapy for their continued good health. Even though placebo controlled randomized clinical trials (RCT) generally are judged to be the gold standard for evaluating therapeutic efficacy, the use of placebo controls in RCTs have in certain circumstances come to be viewed as morally problematic. In particular, trials that enroll persons who have cognitive impairments that may affect their decision-making capacity raise serious ethical concerns (51). Such persons may not adequately understand the concept of a RCT and may be especially confused about the distinction between research (which is to advance and generate generalizable research findings) and clinical treatment (which is to advance the welfare of particular patients). Indeed, as the recent ACHRE studies indicate, persons who were considered to be competent also suffered from the therapeutic misconception and had difficulties distinguishing medical research from clinical treatment (44, p. 761).

There are a variety of opinions regarding placebo use in research protocols. Some supporters of placebo use, including the U.S. Food and Drug Administration (FDA), consider that in the majority of trials with investigational drugs placebos may be necessary in order for the study's conclusions to be reliable (52). Another commentator agrees that the use of placebo controls in RCTs can be highly valuable in certain circumstances, but also points out that such RCTs may be used more often than is necessary (53). In another paper this commentator acknowledges that even when potential subjects are adequately informed, rational individuals are unlikely to agree to participate in RCTs (54). Some writers posit that despite the methodological difficulties, standard therapy should be used as the control in new drug investigations where subjects are at risk for relapse (51). The NBAC propose the employment of three criteria for excluding prospective subjects in placebo arms of studies: (1) when "an individualized assessment reveals that certain patients would be at high risk for relapse if a current or prospective therapeutic regimen were discontinued"; (2) when "a washout period would not be contemplated for these patients if they were not enrolling in the study"; or (3) when "standard therapy has previously proven to be effective" (46, p. 56).

Nevertheless, because unanticipated circumstances can occur, danger may remain for some persons who do not fit the exclusionary criteria and are capable of consenting to participate in drug-free research, challenge studies, or long-term protocols. Subjects who have fluctuating or prospective decision-making impairments whose symptoms are apt to increase are likely to be particularly vulnerable to the exigencies of high-risk protocols. Thus, when symptoms worsen, such individuals may no longer have the capacity for decision making that they were capable of when they initially were enrolled in the study. Such subjects are at risk of becoming vulnerable at precisely the point in a study when they most need to understand, to be aware of, and to make judgments about the use of safeguards—such as their right to withdraw from the study—that have been put place for their protection. Some commentators note that a person's autonomous and voluntary choice to enroll in a research protocol, secured by an informed consent document, by itself may not provide a sufficient safeguard against risks of harm (12,55).

For vulnerable subjects at risk for loss of decision-making capacity, there are other means by which they may provide protections for themselves. According to a number of commentators, anticipatory planning in the form of research advance directives provides not only a mechanism by which potential subjects may choose and appoint surrogate decision makers to act on their behalf should they lose their ability to make decisions for themselves, but also an important method of respecting individual choices. The prior authorization of the surrogate decision maker, and the precise delineation of appropriate subject protections, may make the employment of research advance directives desirable for subjects who may be competent to consent when a study begins, but who may lose their decision-making capacity while participating in the protocol (21,29,46,47,56,57).

RESEARCH ANTICIPATORY PLANNING

Genealogy of Research Advance Directives

In the United States there is some familiarity with anticipatory planning for end-of-life health care. The concept of anticipatory planning was embraced in order to encourage competent individuals to make autonomous choices, in the present, about the medical treatment they would or would not want should they, in the future, lose their capacity for health care decision making. It was hoped that such planning would prompt and enhance a dialogue between doctors and their patients, aid patients to be better informed, and foster the appointment of surrogate decision makers. In truth, this kind of advance planning was seen as a way to forbid heroic measures in the event of a terminal illness (58). However, the employment of such advance directives has been far less successful than numerous bioethicists had at one time anticipated (59,60). Notwithstanding this lack of success, there appears to be a burgeoning interest in psychiatric advance directives. Psychiatric advance directives and advance directives for end-of-life health care are shaped by the same concept—anticipatory planning for a time when the principal may no longer have the capacity to make treatment decisions. But the two kinds of documents differ in substance. The document attentive to end-of-life health care mainly addresses circumstances immediately preceding a singular event—the principal's death. In contrast, the document created for psychiatric treatment endeavors to secure, for a specific population of individuals, a good life (61). Psychiatric advance directives are intended for persons who already have experienced the sort of crisis that they anticipate may recur. Thus they are able to use their past experience to better plan for their needs in similar situations in the future.

The research advance directive is a direct descendant of advance directives for end-of-life care, and has been strongly influenced by the psychiatric advance directive (29). Commentators suggest that substantive and procedural research advance directives, which allow for specific instructions and the appointment of legally authorized representatives (surrogate decision makers), may afford a method to provide protection for some vulnerable research subjects. Research advance directives may be particularly suitable for potential subjects whose decision-making capacity may change during the course of the research. Such persons already may have experienced fluctuating periods of decision-making incapacity or are in the early stages of progressive diseases like Alzheimer's, dementia, and Huntington's disease. Although research anticipatory planning has been little employed, the concept has been discussed and considered since the 1980s (62,63). In 1996 the FDA and the Office for Protection from Research Risk (OPRR) endorsed the concept of advance informed consent for emergency research (49).

There are critical distinctions between the three types of directives. The end-of-life advance directive was designed primarily to refuse treatment—in the United States the right to refuse treatment appears to be fundamental. In contrast, the purpose of the psychiatric advance directive is both to reject *and to elect* treatment—yet the right to demand treatment is not protected. The research advance directive introduces another kind of anomaly. Such documents, unless carefully regulated, could be used to authorize interventions that may not benefit the research subject. The American College of Physicians in their 1989 position paper, anticipate such a circumstance and propose that these directives may be abrogated if the research would unreasonably endanger the subject's welfare (47). The NBAC commissioners strongly urge that such documents should not be prepared as a blank check for future protocols without regard to risk and benefit:

> Prospective authorization cannot be a "blank check" for research participation.... NBAC limits valid Prospective Authorization to a "particular class of research" and then only if the potential subject, while capable, understood the "risks, potential direct and indirect benefits, and other pertinent conditions" of this particular class of research (46, p. 61).

Healthy elderly persons who participated in a Canadian study were concerned about the use of research advance directives. They believed that procedures and treatment not envisioned at the time the directive was prepared should be prohibited (57). Because the author of the document could inadvertently direct the surrogate decision maker to be an agent in harming them, other commentators also advocate that limits be set on what the principal may request. These commentators recommend that the surrogate decision maker may never—under any circumstances—overrule the principal's objections to participate in the research or any section of the research. Furthermore they suggest that should the subject lose capacity for decision making and be at risk of harm due to some aspect of the study itself, the surrogate is obligated to overturn the subject's instructions and to withdraw the subject from the protocol (64). Yet, other writers argue that because the surrogate's obligation is to implement the principle's wishes, the subject's withdrawal only may occur should the study itself has changed substantially (65).

Unanticipated Circumstances and the Appointment of a Surrogate Decision Maker

Research by its very nature is an activity designed to test a hypothesis and is characterized by uncertainty. Always there is the possibility that unanticipated circumstances will occur. The real prospect of future unknown situations prompts some commentators to argue that research advance directives should only be valid if subjects personally chose, and record, their selection of a surrogate decision maker (21). Some writers believe that if advance directives are to adequately safeguard the principal, the appointment of appropriate and reliable surrogates may prove more important for protection than the principal's ability to give detailed instructions (66,67). Findings from a study on ethical aspects of dementia research reveal that more than half of the study's cognitively impaired patients had the capacity to designate a surrogate decision maker, even though they did not have the capacity to understand a detailed protocol. According to the study's researchers, these patients were capable of identifying surrogates whom they trusted (68).

In 1987 the NIH Clinical Center developed a durable power of attorney mechanism for surrogate decision making in research with persons who were at risk for becoming decisionally impaired. Because surrogate decision makers only assume surrogacy responsibilities during the period when subjects lose their decision making capacities, some writers propose that surrogates participate with prospective subjects in the informed consent process and also co-sign the consent form. Thus surrogate decision makers would be educated about the protocol along with the potential subjects and consequently be privy to the subjects' concerns and wishes in regard to their research participation. According to these writers, surrogate decision-maker participation in the informed consent process may put to rest apprehensions on the part of researchers in regard to sharing the subject's medical information and also bypass the need to pinpoint when services of the surrogate might be required (69).

Nevertheless, there may be obstacles that impede the appointment of an appropriate surrogate decision maker. A designated surrogate, even though well trusted by the principal, may not be capable of doing the job adequately. Some surrogates may not properly pursue a subject's best interests. Who should decide if an appointed surrogate is "appropriate"? One writer suggests that this should not be a judgment made by the researchers alone. This commentator proposes that researchers and IRB committee members educate proxies on the correct ethical standards to be used when making decisions about a decisionally incapable person's research participation (70).

Capacity to Engage in Anticipatory Planning

Making judgments about another person's decision-making capacity may be problematic (70). Society appears to be unable to agree on the degree of impairment it is willing to countenance before it deems that a person lacks adequate decision-making capacity (71). Nonetheless, all potential human research subjects are presumed to be capable of making decisions for themselves, unless there are specific reasons and conditions that lead to the belief that a capacity assessment is required. A careful appraisal of each prospective subject's clinical condition, particular circumstances, and the design of the research protocol are necessary factors in capacity determination. Decision-making capacity customarily is considered to be task specific. A person may lack decision-making capacity in one area but have capacity to make decisions in other areas.

Evaluating a potential subject's capacity typically has consisted of subjective judgments. Now, however, there are beginning to be some tested approaches for assessing capacity to consent to research more objectively (72,73). Many persons with mental disorders whose capacity may fluctuate will have intermittent periods when they have decision-making capacity. It is morally correct, and also usually possible, to approach such potential subjects about participation in relevant research at a time when they are competent.

When consent to participate in research is obtained in conjunction with a research advance directive for persons with limited or fluctuating periods of capacity, or with prospective incapacity, the immediate "task at hand" for such potential research subjects is to understand the concepts involved in anticipatory planning. Commentators consider that capacity to prepare an advance directive is distinct from the capacity to consent to treatment, to research, or even to complete a testamentary will (74). These writers propose that potential subjects should be able to grasp and understand that their consent to participate in a specific research protocol, made in the present, constitutes their agreement to take part in a study that will occur over a specified and perhaps extended period of time. In other words, subjects should be able to discern that some of their choices made in the present may be acted upon in the future. Subjects also should be aware that some of their decisions, when relevant, may involve their agreement to medical procedures. Furthermore potential subjects should clearly appreciate that their appointed surrogate will make decisions for them, should they at a future time while participating in the research protocol become incapable of making decisions for themselves (74). Other commentators maintain that subjects should comprehend that whatever they may have recorded in their research advance directive, that with — or without — decision-making capacity, they may object and withdraw from the study (64).

Research Risk Assessment Ambiguities That Reinforce Need for Anticipatory Planning

Specifying criteria and developing policy that assists people to make accurate judgments about risks of harm in research protocols continues to be difficult. The National Commission's 1978 *Belmont Report* acknowledged this problem but did not resolve it: "It is commonly said that benefits and risks must be 'balanced' and shown to be 'in a favorable ratio.' The metaphorical character of these terms draws attention to the difficulty of making precise judgments" (17, p. 7). Assessments of risk of harm in research protocols must attempt to measure the harm's duration, consequences, potential damage, and how the harm might be considered from a subjective point of view. Even though certain types of risks may be precisely and objectively quantified, many risks of harm only may be qualified because they may be of a more subjective kind (75). It is this tension between objective and subjective considerations that makes it so difficult to fashion procedures and policy that do "justice to the equal importance of all persons, without making unacceptable demands on individuals" (76, p. 5).

The phrase "minimal risk" was advanced by the National Commission as an attempt to establish a baseline measure, and it is the standard used in current U.S. federal regulations. "Minimal risk" is defined as meaning "that the probability and magnitude of harm and discomfort anticipated in research is not greater than those ordinarily encountered in daily life or during the performance of routine or psychological examination" (1, §46.102i). Many commentators recognize that this definition is ambiguous. Writers question whether the "harm and discomfort" is that which may be encountered by healthy people in their everyday lives, or whether ordinary "harm and discomfort" is meant to describe that which may be endured in the daily lives of any population of research subjects who have a particular condition or disease. Some writers propose that if the definition of minimal risk is bound to a subjects's disease or condition, it may be easier to more accurately evaluate the level of risk (77). Others argue that in order to appreciate the meaning of minimal risk in the research context, it must be examined in its specific employment (78). The National Commission in its 1978 report on *Research Involving Those Institutionalized as Mentally Infirm* suggests that the "IRB may determine that prospective subjects who are institutionalized as mentally infirm are likely to react more severely than normal persons to certain routine procedures; in such instances, the procedures present more than minimal risk to the subjects" (35, pp. 8–9).

Making judgments about the risk of harm in research protocols is imprecise. A commentator recommends that when research involves vulnerable or potentially vulnerable subjects, investigators and research institutions must be held to a high standard. Not only must the scientific and ethical justifications be especially sound, but researchers should specify what extra safeguards will be put in place to protect subjects' rights and safeguard their welfare (43). In protocols involving greater than minimal risk, research advance directives may provide a practical way to specify safeguards and guarantee protections for vulnerable subjects. Commentators posit that when a legally authorized representative co-signs the consent form along with the potential subject who is competent, not only is the potential subject's autonomy respected but all parties — the subject, the researchers, and the surrogate decision makers — acknowledge the stipulated protections that must be complied with during (and in some cases after) the study period (62,69).

CONCLUSION

Biomedical research involving human subjects is a distinct kind of undertaking and is essentially different to the routine practice of medicine. In order for research with vulnerable subjects to be ethically permissible, this difference should not only be clarified, but special procedural protections should also be employed. However, the concept of protection for research subjects should no longer mean that vulnerable populations must be excluded from research participation. Rather, protection should signify that a constellation of safeguards are provided that will guarantee the rights and welfare of all subjects enrolled in studies. The assurance that such protections will be put in place may allow vulnerable populations to volunteer and participate in research protocols designed to study their particular conditions without fear that they will be subjected to research abuse.

BIBLIOGRAPHY

1. U.S. Department of Health and Human Services, Title 45 CFR Part 46, U.S. Government Printing Office, Washington, DC, 1991.

2. H.P. Lefley, *Family Caregiving in Mental Illness*, Sage Publ., Thousand Oaks, CA, 1996.

3. Nuffield Council on Bioethics, *Mental Disorders and Genetics*, Nuffield Council on Bioethics, London, 1998.

4. G. Albrecht, V. Walker, and L. Levy, *Soc. Sci. Med.* **16**, 1319–1327 (1982).

5. R.N. Proctor, in G.J. Annas and M.A. Grodin, eds., *The Nazi Doctors and the Nuremberg Code: Human Rights in Human Experimentation*, Oxford University Press, New York, 1992, pp. 17–31.

6. F. Kennedy, *Am. J. Psychiatry* **99**(13–14), 141–143 (1942).

7. F. Baylis, J. Downie, and S. Sherwin, in S. Sherwin, coordinator, *The Politics of Women's Health: Exploring Agency and Autonomy*, The Feminists Health Care Ethics Research Network, Temple University Press, Philadelphia, PA, 1998, pp. 234–259.

8. R.R. Faden and T.L. Beauchamp, in collaboration with N.M.P. King, *A History and Theory of Informed Consent*, Oxford University Press, New York, 1986.

9. National Bioethics Advisory Commission, *The Use of Human Biological Materials in Research: Ethical Issues and Policy Guidance*, National Bioethics Advisory Commission, Rockville, MD, 1999.

10. R. Dresser, *Hastings Cent. Rep.* **22**(1), 24–29 (1992).

11. A. Charo, *S. Louis Univ. Law J.* **38**, 135–167 (1993).

12. B. Freedman, in H.Y. Vanderpool, ed., *The Ethics of Research Involving Human Subjects: Facing the 21st Century*, University Publishing Group, Frederick, MD, 1996, pp. 319–338.

13. C. Levine, in H.Y. Vanderpool, ed., *The Ethics of Research Involving Human Subjects: Facing the 21st Century*, University Publishing Group, Frederick, MD, 1996, pp. 105–126.

14. A.M. Hornblum, *Acres of Skin: Human Experiments at Holmesburg Prison*, Routledge, New York, 1998.

15. Statement of Commissioner Alexander M. Capron, in Report and Recommendations of the National Bioethics Advisory Commission, *Research Involving Persons with Mental Disorders that May Affect Decisionmaking Capacity*, Appendix VI, National Bioethics Advisory Commission, Rockville, MD, 1998, pp. 86–88.

16. C. Fried, *Medical Experimentation: Personal Integrity and Social Policy*, American Elsevier, New York, 1974.

17. National Commission for the Protection of Human Subjects of Biomedical and Behavioral Research, *The Belmont Report: Ethical Principles and Guidelines for the Protection of Human Subjects of Research*, U.S. Government Printing Office, Washington, DC, 1979.

18. P.S. Appelbaum, L.H. Roth, and C.W. Lidz, *Int. J. Law Psychiatry* **5**, 319–329 (1982).

19. J. Katz, *S. Louis Univ. Law J.* **38**, 7–54 (1993).

20. M.A. Rodwin, *Medicine, Money, and Morals: Physicians' Conflicts of Interests*, Oxford University Press, New York, 1993.

21. J. Moreno, A.L. Caplan, P.R. Wolpe, and the Members of the Project on Informed Consent, Human Research Ethics Group, *JAMA, J. Am. Med. Assoc.* **280**(22), 1951–1958 (1998).

22. P.S. Appelbaum et al., *Hastings Cent. Rep.* **70**, April, pp. 20–24 (1987).

23. R.M. Veatch, in H.Y. Vanderpool, ed., *The Ethics of Research Involving Human Subjects: Facing the 21st Century*, University Publishing Group, Frederick, MD, 1996, pp. 45–58.

24. I. Berlin, in H. Hardy and R. Hausheer, eds., *The Proper Study of Mankind: An Anthology of Essays*, Farrar, Strauss & Giroux, New York, 1997, pp. 191–242.

25. *Tribunals of War Criminals before the Nuremberg Military Tribunal under Control Council Law No. 10*, Vols. I and II, U.S. Government Printing Office, Washington, DC, 1949, vol. II, p. 181.

26. M.A. Grodin, in G.J. Annas and M. Grodin, eds., *The Nazi Doctors of the Nuremberg Code*, Oxford University Press, New York, 1992, pp. 129–131.

27. World Medical Association, in A.R. Jonsen, R.M. Veatch, and L. Walters, eds., *Source Book in Bioethics: A Documentary History*, Georgetown University Press, Washington, DC, 1998, pp. 13–15.

28. Council for International Organizations of Medical Sciences (CIOMS) in Collaboration with the World Health Organization (WHO), *International Ethical Guidelines for Biomedical Research Involving Human Subjects*, in H.Y. Vanderpool, ed., *The Ethics of Research Involving Human Subjects: Facing the 21st Century*, University Publishing Group, Frederick, MD, 1996, pp. 501–510.

29. P. Backlar, *Commun. Ment. Health J.* **34**(3), 229–240 (1998).

30. R.J. Bonnie, *Arch. Gen. Psychiatry* **54**, 105–111 (1997).

31. E.G. DeRenzo, *IRB: Rev. Hum. Subj. Res.* **16**(6), 7–11 (1994).

32. H.K. Beecher, *N. Engl. J. Med.* **74**, 1354–1360 (1996).

33. J.C. Fletcher and F.G. Miller, in H.Y. Vanderpool, ed., *The Ethics of Research Involving Human Subjects*, University Publishing Group, Frederick, MD, 1996, pp. 155–184.

34. D. Rothman, *Strangers at the Bedside: A History of How Law and Bioethics Transformed Medical Decision Making*, Basic Books, New York, 1991.

35. National Commission for the Protection of Human Subjects of Biomedical and Behavioral Research, Report and Recommendations, *Research Involving those Institutionalized as Mentally Infirm*, DHEW Publ. No. (OS)78-0006, U.S. Government Printing Office, Washington, DC, 1978.

36. P. Ramsey, *The Patient as Person*, Yale University Press, New Haven, CT, 1970.

37. R. McCormick, *Perspect. Biol. Med.* **18**, 2–20 (1974).

38. R. Dresser, *Research Involving Persons with Mental Disorders that may Affect Decisionmaking Capacity*, vol. II, National Bioethics Advisory Committee, Rockville, MD, 1998, pp. 5–28.

39. U.S. Department of Health, Education and Welfare, *Fed. Regist.* **43**(223), 53950–53956 (1978).

40. President's Commission for the Study of Ethical Problems in Medicine and Biomedical Research, *Implementing Human Research Regulations: The Adequacy and Uniformity of Federal Rules and Their Implementation*, Stock No. 040-000-00471-8, U.S. Government Printing Office, Washington, DC, 1983, p. 26.

41. J.W. Berg, *J. Law, Med. Ethics* **24**, 18–35 (1996).

42. R.J. Levine, *IRB: Rev. Hum. Subj. Res.* **18**(5), 1–5 (1996).

43. A. Wichman, *J. Health Care Law Policy* **1**, 88–104 (1998).

44. White House Advisory Committee on Human Radiation Experiments, The Final Report, 061-000-00-848-9, U.S. Government Printing Office, Washington, DC, 1995.

45. U.S. Department of Health and Human Services, Office of Inspector General, *Institutional Review Boards: Their Role in Reviewing Approved Research*, OEI-01-97-00190, U.S. Government Printing Office, Washington, DC, 1998.

46. National Bioethics Advisory Commission, *Research Involving Persons with Mental Disorders that may Affect Decision-making Capacity*, National Bioethics Advisory Commission, Rockville, MD, 1998.

47. American College of Physicians, *Ann. Inter. Med.* **111**(10), 843–848 (1989).

48. U.S. Department of Health and Human Services, Food and Drug Administration, *Fed. Regist.* **61**, 51498 (1996).

49. Food and Drug Administration, Rules and Regulations 21 Code of Federal Regulations 50.3(n), FDA, Washington, DC, 1996.

50. J.H.T. Karlawish and G.A. Sachs, *Am. Geriat. Soc.* **45**, 474–481 (1997).

51. K.J. Rothman and K.B. Michels, *N. Engl. J. Med.* **331**, 394–398 (1994).

52. R. Temple, *Accountability Res.* **4**, 276–275 (1996).

53. R.J. Levine, *Ethics and Regulation of Clinical Research*, 2nd ed., Yale University Press, New Haven, CT, 1988.

54. R.J. Levine, *Law, Med. Health Care* **16**(3–4), 174–182 (1998).

55. J.F. Childress, *Hastings Cent. Rep.* **20**(1), 12–17 (1990).

56. P.S. Appelbaum, *Psychiatr. Serv.* **48**(7), 873–874, 882 (1997).

57. E.W. Keyserlingk, K. Glass, S. Kogan, and S. Gauthier, *Perspect. Biol. Med.* **38**(2), 319–362 (1995).

58. D. Callahan, *Hastings Cent. Rep.* **25**(6), 25–31 (1995).

59. R. Dresser, *Hastings Cent. Rep.* **24**(6, Spec. Suppl.), S2–S5 (1994).

60. SUPPORT Principal Investigators, *JAMA, J. Am. Med. Assoc.* **274**(20), 1591–1598 (1995).

61. P. Backlar, *Commun. Ment. Health J.* **33**(4), 261–268 (1997).

62. B.I. Miller, in V.L. Melnick and N.N. Dubler, eds., *Alzheimer's Dementia: Dilemmas in Clinical Research*, Humana Press, Totowa, NJ, 1985, pp. 239–263.

63. J. Fletcher, F.W. Dommel, and D.D. Cowell, *IRB: Rev. Hum. Subj. Res.* **7**, 1–6 (1985).

64. A. Moorhouse and D.N. Weisstub, *Int. J. Law Psychiatry* **19**(2), 107–141 (1996).

65. J.W. Berg, *J. Law, Med. Ethics* **24**, 18–35 (1996).

66. D.M. High, *Gerontologist* **33**, 342–349 (1993).

67. J. Lynn, *Law, Med. Health* **19**, 101–104 (1991).

68. G.A. Sachs et al., *Clin. Res.* **42**, 403–412 (1994).

69. R. Dukoff and T. Sunderland, *Am. J. Psychiatry* **154**, 1070–1075 (1997).

70. R. Dresser, *JAMA, J. Am. Med. Assoc.* **267**, 67–72 (1996).

71. P. Backlar, *Commun. Ment. Health J.* **32**(4), 321–325 (1996).

72. T. Grisso and P.S. Appelbaum, *Law Hum. Behav.* **19**, 149–174 (1995).

73. D.A. Wirshing et al., *Am. J. Psychiatry* **155**(11), 1508–1511 (1998).

74. M. Silberfeld, C. Nash, and P. Singer, *J. Am. Geriatr. Soc.* **41**, 1141–1143 (1993).

75. E. Meslin, *IRB: Rev. Hum. Subj. Res.* **12**(1), 7–10 (1990).

76. T. Nagel, *Equality and Partiality*, Oxford University Press, New York, 1991.

77. E.G. DeRenzo, *J. Law, Med. Ethics* **25**(2&3), 139–149 (1997).

78. B. Freedman, A. Fuks, and C. Weijer, *Hastings Cent. Rep.* **23**(2), 13–19 (1993).

See other HUMAN SUBJECTS RESEARCH entries.

HUMAN SUBJECTS RESEARCH, ETHICS, STOPPING RULES FOR RANDOMIZED CLINICAL TRIALS

ROBERT J. LEVINE
Yale University School of Medical
New Haven, Connecticut

SUSAN S. ELLENBERG
U.S. Food and Drug Administration
Washington, District of Columbia

OUTLINE

Introduction

Statistical Approaches to Data Monitoring

    Sequential Designs

    Monitoring for Lack of Effect

    Monitoring for Safety

Issues in the Implementation of Stopping Rules

    Need for Judgment

    Concerns About Designs That Permit Early Termination of Clinical Trials

Acknowledgment

Bibliography

INTRODUCTION

"Stopping rule" is a term that refers to a statistical criterion for termination of a randomized clinical trial (RCT). Typically stopping rules are based on p values, namely the probability that an observed difference between two arms of a RCT (presumed under the null hypothesis to be equal in efficacy) with regard to an outcome measure would have occurred by chance. When the difference in outcomes appears quite large at some interim point in the trial, such that this probability is extremely small, one might conclude that the question of whether the two arms have equivalent effects can be answered definitely in the negative, and the trial can be terminated at that point. Although they are called stopping rules, they are for good reasons usually treated as "stopping guidelines," criteria for giving serious thought to ending a RCT before its planned termination date.

Stopping rules are generally developed only for primary outcome measures, which, in turn, are almost always expressions of efficacy. A typical stopping rule might be represented in the following way for a particular interim analysis: We will consider recommending that the trial be terminated if at the second interim analysis the difference between arm A and arm B with regard to the primary outcome measure is significant at a level of $p < 0.0005$.

Responsibility for monitoring the data developed in the course of conducting a RCT is often assigned to a data and safety monitoring board (DSMB), especially when the trial is evaluating treatment effects on mortality or major morbidity. DSMB members have access to data regarding safety and, when appropriate, to efficacy; such data are kept highly confidential and neither the sponsors nor the investigators are permitted access to them. The DSMB generally conducts periodic assessments of the data regarding the outcome measures (*interim analyses*); one of its major responsibilities is to decide, based on the previously specified stopping rule, whether to recommend to the steering committee (or other group having the responsibility and authority to take such actions) that the RCT be terminated or modified.

The ethical functions served by stopping rules are related to the fundamental ethical principle of beneficence, a formal statement of which was provided by the National Commission for the Protection of Human Subjects of Biomedical and Behavioral Research: (*1*) Do no harm and (*2*) maximize possible benefits and minimize possible harms (1). Skillful design and implementation of stopping rules maximizes benefits by increasing the efficiency of clinical trials. When efficacy is demonstrated before the planned termination of a RCT, the therapy that was found effective may be provided to its intended beneficiaries that much earlier. Efficiency in a different sense is also served by conserving resources that would have been wasted by continuing expensive and unnecessary clinical trials. Effective use of stopping rules also serves to minimize harms by reducing the time of exposure of research subjects in the control arm to the ineffective or less effective therapy.

STATISTICAL APPROACHES TO DATA MONITORING

Sequential Designs

When data are reviewed on multiple occasions over the course of a study, the chances of observing a statistically significant result ($p < 0.05$) on at least one occasion can be substantially greater than 5 percent. This results from having multiple opportunities to observe the event of interest, thereby increasing the overall chance of ever observing the event. To take a more everyday example, the chance of drawing the ace of spades from a complete deck of cards is 1 in 52, but the chance of drawing the ace of spades at least once if one draws one card each day for 10 days is substantially greater than 1 in 52.

McPherson, writing in the *New England Journal of Medicine* nearly 30 years ago, showed that, in a case in which there was no difference in outcomes between the treatment and control groups, the probability of ever observing a difference significant at the 0.05 level was actually about 14 percent if the data were reviewed a total of 5 times, and about 19 percent if there were 10 interim reviews (2). Thus, if the risk of a false positive finding (or type I error) is to be kept under 5 percent for the experiment as a whole, statistical designs that preserve the

5 percent level of error (or 1 percent, or whatever other level has been predetermined as appropriate) must be employed.

Study designs that provide for interim analyses of accumulating data while maintaining the overall type I error at the desired level are called *sequential designs*. The simplest approach to preserving type I error in a sequential design is to determine the number of times one wishes to examine the accumulating data during the course of the study, and to then determine the threshold significance level that, if applied at each interim analysis, would lead to a type I error of 5 percent for the experiment as a whole. This problem was studied by Pocock, who showed how to calculate these values (3). For example, he showed that if there were to be a total of five analyses, a *p*-value of 0.0158 would have to be used at each analysis in order to ensure that the false positive rate for the entire study did not exceed 5 percent. A problem with this approach, however, was the difficulty in interpreting the final analysis when the significance level fell between the threshold of 0.0158 and the nominal value of 0.05. However mathematically correct the threshold value was, it was disconcerting to declare a study with a final *p*-value of 0.03 as a nondefinitive result because of the number of times the data had been reviewed prior to the final analysis.

A few years later O'Brien and Fleming at the Mayo Clinic developed an alternative design that provided for varying threshold levels as the study proceeded (4). In this design the first interim analysis is performed using an exceedingly small threshold value. At each successive interim analysis the threshold value is increased by a modest increment. This design allows the final threshold value to be close to the nominal value of 0.05. Using the same example as above, with a total of 5 interim analyses, O'Brien and Fleming showed that the following sequence of threshold values would produce an overall type I error rate of 5 percent: 0.0000005, 0.0013, 0.0085, 0.0228, and 0.0417. This design was appealing because it decreased the chance of stopping the study very early, when most investigators would want to be especially conservative, and also decreased the chance that the final "correct" analysis (based on a *p*-value of 0.0417) would lead to a different conclusion than a final analysis based on the nominal 0.05 level of error.

More recently a number of variations on the O'Brien-Fleming design have appeared. Perhaps the most important is that developed by Lan and DeMets (5), who showed that type I error could be preserved even if the number of interim analyses to be performed were not specified at the beginning of the study. These statisticians recognized that ensuring patient safety sometimes required that additional interim analyses beyond those that were originally planned, and it was important to be able to maintain the validity of study conclusions about efficacy in those circumstances. They introduced the concept of an *alpha-spending function* that specifies how rapidly the type I (alpha) error is to be used up (or "spent") but does not require the number of analyses to be determined in advance. Thus the incorporation of spending functions into the sequential analysis of clinical trials has provided added flexibility without jeopardizing the control of the type I error level. O'Brien-Fleming designs that incorporate Lan-DeMets alpha-spending functions are probably the most commonly used type of sequential statistical designs in today's RCTs. An excellent discussion of the use of alpha-spending functions can be found in DeMets and Lan (6).

Monitoring for Lack of Effect

In addition to monitoring for definitive establishment of treatment effect, it is sometimes important to monitor for lack of effect, and to have the opportunity to terminate a trial early when the accumulating data are highly inconsistent with the existence of a clinically important treatment effect. Because such monitoring may increase the number of times that a treatment could be declared ineffective, but not the number of times it could be declared effective, the concern with these designs is not increase of the false positive rate but rather an increase in the rate of false negatives (type II errors). Monitoring procedures of this type (often referred to as *stochastically curtailed testing*) have been shown to reduce substantially the number of patients treated on clinical trials of ineffective treatments, while having minimal impact on type II error (7). These procedures are based on a re-calculation of the statistical power of the trial (the complement of the type II error, i.e., the probability that a true benefit will be statistically detected), given the data that have already been observed. Because the analyses take into account the already-observed data, they are often called *conditional power analyses*. Stochastically curtailed testing can be implemented in conjunction with a standard sequential design as described above.

Monitoring for Safety

Monitoring for safety is usually very different from efficacy monitoring. With the latter, the variable identified as the primary study endpoint — such as mortality, disease recurrence, occurrence of another undesirable clinical event — is prespecified, as is the analytical method that will be used to assess it. While it is often possible to prespecify certain safety outcomes of particular concern, interim safety monitoring must cover all types of adverse events, whether or not they are anticipated. Further, in many cases safety concerns arising in RCTs, even without the definitive probabilistic framework that would be demanded for an efficacy endpoint, will lead to modification or even early termination of the trial. A strong suggestion of harm, even without definitive proof, would likely be sufficient to warrant such actions when the intended benefit of the treatment is relatively modest.

ISSUES IN THE IMPLEMENTATION OF STOPPING RULES

Need for Judgment

Sequential designs have proved very useful in the conduct of RCTs fostering careful monitoring of interim data to ensure that patients are being treated appropriately and safely without sacrificing the validity of the statistical conclusions that will ultimately be drawn. This point notwithstanding, it is essential to recognize the impossibility of developing a design that will account for all

contingencies that might occur in the trial. For example, suppose that the interim data at, say, the third look, show a positive effect of treatment that exceeds the threshold for early termination, but at the same time, unexpected safety concerns have emerged. A DSMB in that situation probably should not recommend stopping at that point on the basis that efficacy had been demonstrated; rather, it probably should recommend collection of additional data to help clarify the risk–benefit considerations. In some cases information about findings from other trials or related studies might affect the DSMB's perspective on the data from the trial being monitored. DSMBs must be prepared to consider all the available information pertaining to the safety and efficacy of the treatment being studied prior to making its recommendations, and cannot rely entirely on the statistical stopping thresholds as the basis for decision making. As stated by DeMets et al.:

> Although sophisticated statistical methods have been developed to assess the quantitative strength of trial results, it is important to note that statistical methods alone are not adequate to guide early termination decisions. The collective experience and judgment of the [data monitoring committee] is necessary. Making decisions about early termination requires consideration of many additional factors such as results on secondary outcomes, safety data, degree of compliance to the protocol, possible sources of bias in outcome evaluation, completeness and currency of data, and internal and external consistency of the data as well as emerging data from other trials (8).

Concerns About Designs That Permit Early Termination of Clinical Trials

While statisticians have paid much attention to the development of methods for interim analysis of clinical trials that preserve type I error at the desired level, it must be remembered that the only occasion for concern about type I error is when one might terminate a study and draw a final conclusion based on its interim results. Early termination options are most often important for trials of therapies that may, as compared with other available treatments, improve the likelihood of survival or reduce the probability or magnitude of disability. In many clinical trials, however, there may be good reasons to continue the trial to the planned conclusion even if an interim analysis demonstrates definitive benefit with respect to the primary efficacy variable. When the new treatment is aimed at relief of symptoms, or any endpoint that is neither serious nor irreversible, the need for a fuller safety database becomes more compelling; we must have adequate data on the possible risks before being able to conclude that the observed benefit outweighs these risks. (Adverse outcomes are caused by many factors, not just investigational treatments, so safety data collected in a randomized, controlled, blinded trial are much more reliable indicators of the potential adverse consequences of treatment than safety data collected using less rigorous methods such as uncontrolled case series or retrospective data base analyses.) Thus, in trials in which the outcome measures are neither death nor disability, there may be no need for sequential designs;

monitoring for safety, however, would of course remain important.

Some investigators have argued that even in circumstances when mortality or serious morbidity is the trial endpoint, we ought not to be terminating trials early except in the rarest of situations. Their rationale is that the results of a trial with truncated enrollment may be less convincing to the medical community than results based on a larger patient population, and that a positive trial that does not lead physicians to change their practices has not accomplished any worthwhile purpose and might as well not have been performed. Concerns about the adequacy of estimates of treatment effect from trials stopped early and the consequent concern about the ability to derive reliable cost-benefit considerations for the treatment studied, may threaten the acceptance of results from such trials (9).

The counterargument is that it is unethical to continue enrolling patients in RCT when one of the treatments being compared has already been demonstrated to be superior. To do so entails a deliberate withholding of the therapy known to be superior from those subjects who are assigned to the other (inferior therapy) arm of the trial. It is hard to imagine that fully informed subjects would agree to cooperate with such a plan. Deliberate withholding of known effective therapy without the consent of the subjects is unethical. It at least appears to violate the standard set forth in Article 1.5 of the Declaration of Helsinki: "Concern for the interests of the subject must always prevail over the interests of science and society" (10).

The fundamental controversy here is perhaps more accurately focused on the level of evidence that should be required for a trial to be considered positive, rather than whether interim analysis, possibly leading to early termination, should be performed. If the threshold for statistical significance were to be set at $p < 0.01$ (or lower), the required sample size would be substantially larger (30 to 60 percent larger, for power in the typically acceptable range) than for a trial using the more conventional $p < 0.05$ threshold, and the O'Brien-Fleming stopping boundaries that would result would require much more extreme evidence of benefit for early stopping — possibly even enough to be persuasive to a highly skeptical medical community.

ACKNOWLEDGMENT

R.J. Levine's work was funded, in part, by grant number PO1 MH/DA 56 826-01A1 from the National Institute of Mental Health and the National Institute on Drug Abuse.

BIBLIOGRAPHY

1. National Commission for the Protection of Human Subjects of Biomedical and Behavioral Research, *The Belmont Report: Ethical Principles and Guidelines for the Protection of Human Subjects of Research*, DHEW Publ. No. (OS) 78–0012, U.S. Government Printing Office, Washington, DC, 1978.

2. K. McPherson, *N. Engl. J. Med.* **290**, 501–502 (1974).

3. S.J. Pocock, *Biometrika* **64**, 191–199 (1977).

4. P.C. O'Brien and T.R. Fleming, *Biometrics* **35**, 549–556 (1979).

5. K.K.G. Lan and D.L. DeMets, *Biometrika* **70**, 659–663 (1983).

6. D.L. DeMets and K.K.G. Lan, *Stat. Med.* **13**, 1341–1352 (1994).

7. K.K.G. Lan, R. Simon, and M. Halperin, *Commun. Stat. Sequent Anal.* **1**, 207–219 (1982).

8. D.L. DeMets et al., *Controlled Clin. Trials* **16**, 408–421 (1995).

9. R.L. Souhami, *Stat. Med.* **13**, 1293–1295 (1994).

10. World Medical Association (WMA), *Declaration of Helsinki: Recommendations Guiding Physicians in Biomedical Research Involving Human Subjects*, WMA, as amended by the World Medical Assembly in South Africa, 1996.

See other HUMAN SUBJECTS RESEARCH entries.

HUMAN SUBJECTS RESEARCH, LAW, COMMON LAW OF HUMAN EXPERIMENTATION

WENDY K. MARINER
Boston University
Boston, Massachusetts

OUTLINE

Introduction
    The Common Law
    Classification of Common Law Claims
    Injuries to Research Subjects
Common Law Causes of Action Against Investigators and Research Organizations
    Using Human Subjects Without Their Knowledge or Consent
    Liability for the Design and Implementation of Research
    Defenses Against Liability
Conclusion
Bibliography

INTRODUCTION

The Common Law

There is little common law on research with human subjects in the United States. Common law is distinct from statutes enacted by legislatures and from regulations issued by government agencies. Unlike statutes and regulations, common law rules have no single official text. They consist of principles defining general rights and obligations that are summarized from the reasoning of court decisions in lawsuits. This gives the common law flexibility to adapt itself to new circumstances without the need for rewriting a law in its entirety. It also means that the principles and their application are subject to interpretation, and often dispute, which permits a degree of uncertainty that can be discomfiting to those who seek absolute predictability in the law.

Common law in the United States developed from English common law applied in the colonies before Independence (1). It remains within the jurisdiction of the states, each of which is free to develop its own principles and rules, thereby eroding some of the commonality of the common law. Nonetheless, at the level of generality with which we are concerned here, there is enough consistency in principle and doctrine to permit useful generalizations. For detailed application of common law rules in any one state, it is essential to consult that state's specific laws.

For decades, the common law remained in the background of research policy. The fact that common law imposed duties of care on researchers and granted rights to human subjects of research did not prevent scandalous abuses of research subjects. Perhaps the most comprehensive statement of researcher's duties in what may be considered international common law is the Nuremberg Code, the 10 principles set forth in the 1947 judgment against Nazi physicians convicted of crimes against humanity (murder and torture) under the guise of medical experimentation (2). The judgment, by American judges in a military tribunal established by the United States Military Government for Germany after World War II, is also precedent for common law duties of researchers in the United States. Yet few American courts have even referred to the Nuremberg Code, much less applied it, in cases involving research subjects (3).

Revelations of unethical research and experimentation with human beings without their consent did not give rise to vigorous enforcement of common law principles, perhaps because common law rights must be enforced by lawsuits brought by those whose rights are violated. Instead, in the 1960s and the 1970s, such revelations inspired new federal guidelines and regulations (4–9). Federal laws requiring those who receive federal funding for research with human subjects to comply with specific regulations intended to protect subjects from such abuses have been harmonized into what is known as the Common Rule (10). The Common Rule applies to the Departments of Agriculture, Commerce, Defense, Education, Energy, Health and Human Services (including the Office of the Secretary, the Food and Drug Administration, the National Institutes of Health), Housing and Urban Development, Justice, Transportation, and Veterans Affairs, and the Consumer Product Safety Commission, Environmental Protection Agency, International Development Cooperation Agency (including the Agency for International Development), National Aeronautic and Space Administration, and National Science Foundation. Although the Common Rule incorporates basic elements of the common law into its regulatory provisions, the Common Rule has had more influence on the conduct of research than the common law (11).

Nevertheless, the common law remains the legal backstop to fill gaps left by regulations. More important, it provides research subjects with a legal remedy for injuries (which federal regulations do not). Thus the common law can be seen as describing the legal boundaries for lawful research and responding to the claims of injured research subjects, within which more specific statutes and regulations carve out particular additional duties.

Because common law is derived from judicial opinions deciding cases and controversies among parties to real disputes, it focuses on legal, not moral, rights and duties, and specifically on remedies for legal wrongs committed by one party against another or injuries to one party for which another party is legally responsible (12). Thus the common law does not enforce moral obligations or ethical principles, although many legal principles are based on moral theory and, in the case of research, codes of research ethics. While one's failure to act virtuously or adhere to moral principles may subject a person to moral opprobrium, the absence of virtue is not sufficient to warrant legal recourse.

Classification of Common Law Claims

Common law principles affecting research with human subjects are derived from more general principles applicable to all people and organizations. They are found primarily within the law of tort, or civil wrongs. Tort law assigns responsibility for certain duties, prescribes basic rules of conduct intended to prevent avoidable harms, and imposes penalties for unlawful conduct (13). The goals are typically described as deterring harm, compensating injury, and, sometimes, retribution for wrongs (14).

Tort law is enforced when a person brings a legal claim — called a cause of action — against another party (defendant) who has caused harm to the claimant (plaintiff) as a result of violating a legal obligation. The causes of action most relevant to research are intentional torts — including battery, fraud, intentional infliction of emotional distress, and invasion of privacy — and unintentional torts — specifically negligence, negligent infliction of emotional distress, breach of confidentiality, and products liability. Negligence is often thought to be the most likely basis for liability. However, most reported cases involve claims of battery (unauthorized touching), fraud, and misrepresentation — using people as research subjects without their knowledge or consent (15,16). Often several causes of action are brought in the same lawsuit, and these may be supplemented by actions for violations of statutes or constitutional rights. This article is limited to common law causes of action.

One can characterize the possible common law issues by (1) the type of legal claim (or cause of action) that might be brought by a research subject, (2) the type of defendant or entity claimed to be responsible for research harms, (3) the type of research subject, or (4) the nature of the research product or intervention, as summarized in Table 1. Elements in each of these categories can be combined in multiple variations to produce a staggering array of potential common law cases. There are more ways to classify the types of legal responsibility for research harms, however, than there are cases to use as examples. The full range of possibilities has not materialized in lawsuits, and there are relatively few published court decisions that address the common law duties and rights of researchers and subjects. Some consider this reassuring evidence of either the lack of harm caused by research or the negligible prospect of liability on the part of researchers. Others view the small number of cases as the tip of an iceberg of potential future litigation.

Table 1. Variables in Common Law Claims

1. Cause of action

> Intentional torts
> > Battery
> > Fraud, misrepresentation
> > Invasion of privacy
>
> Negligence
> > Research design
> > Research conduct
> > Informed consent
> > Failure to notify of later-discovered risks
> > Breach of confidentiality
> > Invasion of privacy
>
> Product liability
> > Design defects
> > Manufacturing errors
> > Failure to warn of risks
>
> Defenses
> > Charitable immunity
> > Sovereign immunity
> > Statute of limitations
> > Waiver, release
> > Preemption

2. Responsible party

> Individual investigator
> Research institution/employer
> > Institutional review board
>
> Research funder/sponsor
> > Private product sponsor
>
> Government

3. Research subjects

> Competent adults
> > Women
> > Fertile women
> > Pregnant women
>
> Vulnerable populations
> > Children
> > Fetuses, embryos
> > Incompetent adults
> > People with terminal illness
> > People in medical emergency
>
> Exploitable populations
> > Prisoners
> > Illiterate populations
> > Impoverished populations
> > Minority populations
> > Subordinates of investigators
> > Elderly populations

4. Product or intervention

> Pharmaceutical
> Medical device
> Biological product
> Diagnostic technique
> Surgical procedure
> Medical procedure
> Preventive intervention
> Health services
> Genetic therapy
> Tissue collection
> Psychiatric intervention
> Data collection
> > Epidemiological
> > Genetic
> > Sociodemographic

Injuries to Research Subjects

The number of research subjects to whom common law duties are owed is difficult to estimate. The U.S. General Accounting Office reports that the U.S. Department of Health and Human Services (DHHS) alone funds about 16,000 studies involving human subjects each year (for $5 billion) (17). Other federal and state agencies and private organizations that fund or conduct research significantly increase this number.

There is little empirical data on research injuries and even less on the proportion of research subjects who pursue legal remedies for their injuries. In 1976 the federal Department of Health, Education, and Welfare (HEW) Secretary's Task Force on the Compensation of Injured Research Subjects conducted a survey of researchers to estimate the number of injuries to subjects who had participated in research studies funded by the National Institutes of Health (NIH) and Alcohol, Drug Abuse, Mental Health Administration (ADAMHA) (18). That survey, which relied on telephone interviews with 331 investigators, reported 4957 injuries among 133,000 human subjects over the three preceding years (19). Research subjects who were injured represented just under 4 percent of all subjects who participated in such studies. The investigators characterized 3926 injuries as trivial, 974 as temporarily disabling, 14 as permanently disabling, and 43 deaths (fewer than 1 percent of subjects). The majority of injuries befell subjects who participated in so-called therapeutic research, which was defined as "an experimental program expected to benefit the research subject directly." Of the 39,216 subjects in these therapeutic research studies, 4246 (or 10.8 percent) were injured. In nontherapeutic studies, 711 (0.8 percent) out of 93,399 research subjects were injured. The survey was limited by reliance on reports by the principal investigators of the studies themselves and may understate actual injuries. The Harvard Medical Practice Study found that about 3.7 percent of hospitalized patients in 1984 were injured as a result of ordinary medical care (not research), which might be expected to result in fewer injuries than research (20). Without a similar study of research injuries, it is difficult to estimate the prevalence of research-related injuries that might justify claims of liability for personal injury suffered by research subjects.

There is no empirical study of the proportion of injured research subjects who seek legal redress for their injuries. As far as can be determined from published judicial decisions and anecdotal reports, nothing remotely approaching 4 percent of injured research subjects file common law claims against a third party. An Institute of Medicine committee reported that "the NIH Office of the General Counsel is only aware of three legal actions for research injuries where NIH was involved in the . . . twenty years" before 1994 (17). The Harvard Medical Practice Study found that only a tiny fraction (less than 2 percent) of patients injured as a result of negligent medical care (not research) actually file a claim for malpractice (21). The same is probably true for negligent research injuries.

There are several possible explanations for the small number of cases. Not all injuries are caused by unlawful conduct that gives rise to a legal claim. The nature of research often precludes clear findings of negligence or other unlawful conduct. Research is done because, by definition, the safety and efficacy of what is being studied is not known (22). (This is distinct from the standard of care used in carrying out the research.) If the Secretary's Task Force's findings hold true today, the majority of injuries involve subjects who are ill so that it is often difficult to sort out whether an injury was caused by underlying illness or by the research study. Subjects may not distinguish between problems with the experimental product or intervention, and problems with the way the research is conducted. Also subjects should be aware that research necessarily entails risks and may assume that they cannot sue. Some subjects may not be aware that they were injured as a result of research. History suggests that the most common legal wrong in research has been the failure to tell research subjects that they are involved in research. However, if subjects are not injured, there are few legal remedies available. Many subjects may not be inclined to sue in any circumstances, especially in light of the time and expense required. Lawyers who represent clients on a contingency basis (receiving a percentage of the money award only if the action succeeds) are unlikely to take cases that have poor or uncertain prospects of success or low potential awards. On the other hand, some cases may be settled voluntarily before any legal claim is brought or before the case is tried or otherwise finally decided in court. Some research institutions voluntarily provide remedial medical care, insurance, or other assistance to subjects who are injured as a result of research.

Perhaps all that can be said is that no one knows how many people have been injured as a result of participation in research, but injuries are likely to befall a relatively small percentage of all research subjects, and in all probability, only a small proportion of these research-related injuries will give rise to legal claims (23).

COMMON LAW CAUSES OF ACTION AGAINST INVESTIGATORS AND RESEARCH ORGANIZATIONS

All investigators and organizations that design, supervise, carry out, or report research are accountable to research subjects for violation of their common law obligations. These duties include (1) determining that the proposed research is properly designed and can be conducted without posing unnecessary, avoidable, or unreasonable risks of harm to research subjects, (2) ensuring that each research subject who participates in the study has voluntarily agreed to participate with full knowledge of the potential risks of participation, (3) ensuring that all investigators are qualified and competent to carry out the research, and (4) ensuring that the research is in fact implemented properly and that the subjects' safety and welfare is protected (7).

These duties are derived both from the growing body of ethical literature, including "codes" or declarations of research ethics that help to create the standard of practice among researchers, as well as more deeply entrenched common law principles of self-determination and reasonable care (7,24). Many of these common

law duties have been supplemented by the federal regulations that govern federally funded research (10,25). For example, the common law does not require an institutional review board (IRB) to review or approve research with human subjects, but federal regulations do. Arguably, IRB review and the threat of losing federal research funding may have improved compliance with common law duties as well as federal rules, or become the custom or standard of conduct for research institutions. In some states, violation of federal regulations can be considered negligence per se which eases a litigant's path to success. On the other hand, federal regulations do not grant research subjects a personal remedy against those who violate federal regulations. Although federal agencies may impose sanctions against violators, by withholding future funding to the researcher or organization, individual subjects are not compensated by such penalties. Thus the common law serves as the research subject's only source of rights to personal redress for harm.

The most common causes of action are described below.

Using Human Subjects Without Their Knowledge or Consent

The majority of litigated cases involve using human beings as research subjects without their knowledge or consent. The most horrific example — which generated the Nuremberg Code and sowed the seeds of codes of ethics in Western countries — was the Nazi physicians' medical experiments on concentration camp prisoners during World War II (26,27). Notorious American examples include the U.S. Public Health Service's Tuskegee study of syphilis in poor, black men, (5,28) federally sponsored cold war era studies of radiation exposure and poisoning, (29–31) and the Willowbrook study of hepatitis B in retarded children (4,32). However, less dramatic studies have also been conducted without consent (33).

More than a century ago, the U.S. Supreme Court said that "no right is more sacred, or is more carefully guarded, by the common law, than the right of every individual to the possession and control of his own person, free from all restraint or interference of others unless by clear and unquestionable authority of law" (34). The right to decide whether to permit anyone to violate one's own bodily integrity is a fundamental principle of common law (35). Informed consent is a "concept, fundamental in American jurisprudence, that the individual may control what shall be done with his own body" (36). Not even a physician who believes that medical care is necessary for a patient is permitted to act without the informed consent of the patient (37,38). Although the details of the legal cause of action have evolved over the past decades, the law has never permitted the involuntary treatment of a competent patient by a physician. Likewise no one is permitted to conduct research on any human being without that person's competent, voluntary, informed, and understanding consent (39).

Battery. At common law the failure to obtain consent is considered a battery — an unauthorized offensive touching (40). Researchers who use human beings as research subjects without telling them that they are conducting a study or giving the subjects an opportunity to refuse to participate commit a battery. Battery is a straightforward cause of action that requires a plaintiff to prove that some form of offensive, intentional contact took place without the plaintiff's consent or against his will (41). The touching need not cause any physical harm to be unlawful. The wrong lies in the offense to personal dignity caused by invading the inviolability of the person without permission. Successful plaintiffs can recover money damages for battery. These can range from nominal amounts (e.g., $1) for minor intrusions that cause no harm to substantial awards for highly offensive acts or those that cause serious physical injury.

The offense of battery may be more common in the research context than it is in the realm of medical care. This is because physicians who provide medical care to patients are more likely to obtain consent at least to the treatment given, so that disputes tend to focus on whether the patient was advised of the risks of treatment, as discussed below. Some research studies, on the other hand, have been concealed entirely from subjects.

A notorious example occurred at the Jewish Chronic Disease Hospital in Brooklyn, New York, in 1963 (42). A researcher wished to study whether the human immune system could be used to prevent cancer. It was known that healthy people had strong defenses against injections of cancer cells from other people, while cancer patients did not, but it was not known whether the cancer patients' response was due to their cancer or their overall debilitation. Drs. Southam and Levin of the Sloan–Kettering Institute for Cancer Research injected cancer cells into 22 Jewish Chronic Disease Hospital patients who had diseases other than cancer to see whether their immune systems would reject the cancer cells. The subjects were told only that they would receive a test of their immunity; they deliberately were not told that cancer cells were injected under their skin because researchers thought this might cause them to refuse to be in the study. Three young staff physicians refused to participate and informed a member of the hospital board of directors who in turn notified the New York State Department of Education and the Supreme Court of Brooklyn. The state's highest court allowed the hospital director to inspect the subjects' medical records as part of an investigation into illegal and improper experimentation on patients (43). The licenses of the investigators were suspended for one year, but the suspension was later replaced by one year's probation (4). The patients themselves were not parties to the lawsuit, but the hospital instituted a policy of requiring informed consent in future experiments.

In *Mink v. University of Chicago*, a federal court described why the failure to tell subjects that they are being used as research subjects is a battery (44). The University of Chicago and Eli Lilly & Company conducted an experiment from 1950 to 1952 to see whether diethylstilbestrol (DES) would prevent miscarriages. They gave DES to women who came for prenatal care without telling them of the experiment or that the pills were DES. The court noted that the cause of action was battery, not lack of informed consent, because the researchers performed an action to which the women did not consent

at all. The court compared their actions with performing unauthorized surgery. The administration of a drug without a person's knowledge fits within the meaning of offensive contact for purposes of battery. The fact that the women had consented to prenatal care did not mean that they had agreed to take experimental pills.

Instances of undisclosed research can give rise to other common law claims, including invasion of privacy and fraud and misrepresentation. Invasion of privacy—specifically, an unreasonable intrusion upon a person's seclusion or unreasonable publicity given to a person's private life or affairs—can occur when researchers seek out or disclose private information without the person's permission (45). Also possible is a cause of action for fraudulent misrepresentation, which includes a misrepresentation of fact or intention on the part of a researcher, the subject's reliance on the researcher's false statements, and resulting damages (46). Researchers may also be liable for failing to disclose a fact that would induce a person not to enter a study because researchers have a duty to provide all relevant information to prospective subjects. For example, in *Craft v. Vanderbilt University*, a group of women brought a class action claiming battery, fraudulent concealment, negligent misrepresentation, infliction of emotional distress, negligence, and invasion of privacy, as well as federal statutory claims, when they discovered, in 1993, that they had unknowingly been part of research studies decades earlier (47). In a study conducted from 1945 to 1947 by the university and the state of Tennessee, researchers sought to determine iron absorption in the uterus. Researchers gave 829 pregnant women patients in Vanderbilt's prenatal clinic a beverage containing the radioactive isotope Iron 59 but told them only that it was vitamins, "a cocktail" or "a sweet." The women were never told what the drink contained or that they were research subjects, even during a follow-up study in the 1960s to determine the long-term health effects of radiation exposure during pregnancy. That study found three deaths from cancer among children of the exposed mothers (and none among controls) that suggested a cause and effect relationship with the radioactive drink (48). The suit was settled for $10.3 million in 1998 (49).

Research conducted after World War II on the effects of radiation exposure has given rise to other lawsuits by surviving subjects or their families in the 1990s. Like most lawsuits, these assert multiple causes of action—slightly different legal wrongs—for the same injury. Where the subjects were not told that they were part of a research study or that they would be exposed to radiation, a claim of battery can be made. In one such experiment, the Massachusetts Institute of Technology fed Quaker Oats cereal with radioactive isotopes to children institutionalized at the Fernald School. A class action on their behalf was partially settled for $1.85 million in 1999 (50).

Where the research is misrepresented as medical care, a claim of fraud or misrepresentation is possible. For example, the plaintiff in *Stadt v. University of Rochester*, brought a claim of fraud, instead of battery, against the university hospital for injecting his 41-year-old mother with plutonium without her consent (51). Janet Stadt was used as a subject in an Army research study of radiation effects in 1946 but allegedly was told that she was being treated for scleroderma. In 1972 she underwent additional tests, again without being told they were part of the study, and died of cancer in 1975. The failure to disclose the research was sufficient to allow a lawsuit based on fraud, although, in this case, the court converted the state common law claims into a federal claim against the federal government for violating the subject's right to bodily integrity protected by the Fourteenth Amendment to the U.S. Constitution.

A more complicated example is the Human Radiation Experiments conducted at Cincinnati General Hospital between 1960 and 1972 (52). There, at least 87 patients with inoperable cancer were given radiation in doses ranging from 25 to 300 rads—the level expected to be experienced by military personnel exposed to nuclear attack—to study the effects of radiation on human beings. The study was part of the Defense Department's cold war efforts to prepare for possible nuclear war. The patients were told that the radiation was treatment for their cancer, although it shortened their lives and caused nausea, vomiting, burns, and other suffering. The plaintiffs claimed the experiments were intentionally concealed from them, which would give rise to causes of action for battery and fraud. Although a consent form was used beginning in 1965, it said only that the patients were participating in scientific experiments without indicating the nature or purpose of the experiment or the risks of the high-dose radiation. For subjects who received this form, a cause of action for lack of informed consent (discussed below) would be possible. In May 1999 a judge approved a settlement of the class action lawsuit for $5.4 million, which provided an average of $50,000 to each family (53).

A notable exception to these general principles lies in research conducted by the U.S. military using military personnel as subjects. The armed forces exposed soldiers to radiation during a nuclear explosion without telling them they were subjects of an experiment on radiation exposure (54). The Central Intelligence Agency gave LSD to servicemen without their knowledge to study the drug's effects (55,56). Although the Federal Tort Claims Act permits the federal government to be sued for certain acts of negligence and other torts (57), military personnel who are "injured in the course of activity incident to [military] service" are prohibited from suing the federal government for damages for personal injury (56,58). The courts justify excepting military personnel from the remedies available to civilians for the reason that allowing civil damage claims would intrude on military discipline. In addition the federal government has been protected from liability for what courts construe as military acts, including the use of investigational vaccines among soldiers serving in the Gulf War (59). The fact that the Nuremberg Code was, in effect, a common law decision written by U.S. military judges to apply to wartime experiments, has been uniformly ignored by American courts in these cases (3). Thus the common law has afforded no remedy to military personnel who are subjects of research conducted by military or quasimilitary officials.

Informed Consent. The common law doctrine of informed consent grew out of the principles underlying battery—bodily integrity, autonomy, and self-determination (7,39,40,60–62). While battery applies to cases in which no consent is given at all, informed consent applies to cases in which consent is given without sufficient information to render it meaningful (38,40). A person cannot make a meaningful decision in the absence of information about the benefits, risks, and consequences of the options. Ordinary lay people are not expected to have medical or scientific knowledge necessary to determine whether or not to participate in a research study. Thus the law imposes on researchers a duty to explain the research study and its potential risks and benefits to the prospective subject.

Explanation is especially important in research that uses patients as subjects. Individuals who seek medical care may fail to appreciate the uncertainty inherent in research or may assume that they are receiving proven medical care instead of an experimental technique (29,63,64). Subjects may find it difficult to keep in mind that the investigators do not assume the role of their personal physicians (65). In addition, unlike medical care, research necessarily entails unavoidable conflicts of interest for investigators who are obligated to protect the welfare of their research subjects but may also be eager to ensure the success of the study (66). This is particularly true when physicians act as investigators and use their patients as subjects (17).

The doctrine of informed consent emphasizes that every competent adult is completely free to accept or reject any medical or scientific intervention for any reason or for no reason at all. No competent adult can be forced to undergo medical treatment, even if it is certain to save his life (67,68). It should be clear, therefore, that no one can be forced to participate in research. There is almost never any justification for withholding information about a research study. (An exception may be permitted in certain behavioral experiments in which complete information will prejudice a subject's response during the experiment, but not the decision to participate, and where the subject will suffer no risk for lack of the information (69). In *Beno v. Shalala* (70), a family sued the Secretary of Health and Human Services seeking to invalidate her granting a waiver permitting California to conduct a work-incentive demonstration project to reduce welfare benefits by 1.3 percent and waive limits on income that beneficiaries could earn without losing their welfare benefits. The court found that the beneficiaries were human subjects of research but suggested that the benefit reduction was not large enough to require the informed consent of each beneficiary under specific federal law governing demonstration experiments.)

Although informed consent is required for both research and medical treatment, disclosure requirements have been more stringent for research than for medical care. As the *Belmont Report* concluded, informed consent standards applicable to patients in malpractice cases are not sufficient for research (7). At a minimum, disclosure must include (1) that fact that research is being conducted; (2) the purpose of the research, what will happen, and why; (3) the requirements of participation; (4) what experimental agents and techniques will be used; and (5) the potential risks, as well as inconveniences, to the subject of participation. A cause of action for failure to obtain informed consent is treated as a negligence action because the researchers have a professional duty to provide information sufficient to permit a prospective subject to make a voluntary, informed decision. To succeed in an action for failure to obtain informed consent, a plaintiff must prove each of the following four things:

1. The defendant had a duty to disclose certain information to the subject (usually information about the risks of participating in the study).
2. The defendant did not disclose that information.
3. The undisclosed risk or problem occurred and caused physical injury to the subject.
4. The failure to disclose the risk or problem was the proximate cause of the subject's injury because a reasonable person in the subject's circumstances would not have consented to participate in the research if he had known of the undisclosed risk.

For example, in *Halushka v. University of Saskatchewan*, a student was awarded $22,500 from a university that failed to obtain his fully informed consent to research (71). The student volunteered for a study of circulatory response and was told that a new drug would be used. He was not told that the drug was an untested anesthetic nor that a catheter would be inserted through his heart to his pulmonary artery. During the experiment, the subject's suffered cardiac arrest, was resuscitated, and remained unconscious for four days. Ultimately he dropped out of his university studies because of inability to concentrate. The appeals court held that subjects of experimentation are entitled to more information than patients must receive, including the full and frank disclosure of all facts, probabilities, and opinions that a reasonable person might be expected to consider.

The courts that have considered the issue agree that the subject of research that uses an experimental agent or procedure must be told that the agent or procedure is experimental (72–76). The cases endorsing this basic rule tend to involve patients whose physician used an experimental medical device or surgical procedure during the course of medical treatment without telling the patient that the device or procedure itself was experimental. Several cases concerned the implantation of investigational intraocular lenses. For example, in *Kus v. Sherman Hospital*, Richard Kus agreed to surgery to implant an intraocular lens in his eye (77). Dr. Vancil used an investigational implant without telling Kus that the device was experimental. The hospital's IRB had required all consent forms for the surgery to specify that the lens was experimental, but Vancil removed that information from the consent form he gave Kus and other patients. There was no question that the physician was liable for the injury to Kus's eye caused by the implant because he failed to tell Kus that the device was experimental, and Vancil settled with Kus out of court. The court found that the hospital might be liable for its own failure to obtain Kus's informed consent where the hospital, as a participant in a study governed by federal regulations, assumed a duty to ensure that informed consent was obtained for all subjects in the study.

The first implantation of a totally implantable mechanical heart in a human being, Haskell Karp, resulted in a lawsuit following Karp's death shortly after the experimental surgery (78). Karp's widow claimed that her husband had not given his informed consent to the use of the artificial heart. She contended that Dr. Cooley emphasized that he would surgically repair Karp's own diseased heart and had described the implantable artificial heart as like a heart and lung machine used to sustain life during open heart surgery. However, the court found that the Karp had actually consented to the experimental surgery and that the consent form, albeit only 179 words, noted that this device "has not been used to sustain a human being and that no assurance of success can be made." It is unlikely that courts today would be as willing to accept such a vague, abbreviated description as evidence of full disclosure for such a dramatic experiment (79).

The question of what risks to disclose can be difficult in research where not all risks can be known in advance. For example, in *Whitlock v. Duke University*, Whitlock volunteered for a deep sea diving experiment and suffered permanent organic brain damage as a result, even though the research was conducted properly (80). The court held that the investigator had a duty to warn subjects of all risks that were reasonably foreseeable when the research began. The plaintiff had signed a consent form that warned of the risks of death and other unknown risks, but did not specifically mention permanent organic brain damage. Expert evidence showed that brain damage was not a foreseeable risk and had not happened in the past. Thus the investigators could not have foreseen that specific harm and therefore had no obligation to mention it. The plaintiff had no basis for recovering damages.

Risks to the subject include psychological, social, and financial risks. For example, some types of research can expose subjects to discrimination if they are publicly identified as having socially undesirable conditions or traits, such as drug abuse, sexually transmitted diseases, or HIV infection (81). Other types of information about a subject, including drug abuse and child neglect, can lead to criminal prosecution. Research involving genetic analysis may identify susceptibility to genetic diseases, which can give rise to serious psychological and emotional concerns, as well as affecting the ability to obtain insurance or employment (82–84).

Beyond the risks to the subject himself or herself, there may be other aspects of the research that warrant disclosure. For example, the Public Health Service studied uranium miners to learn whether they were at increased risk of cancer without telling the miners of the cancer risk (85). In that case the reason the subjects were chosen was an important piece of information that should have been disclosed. Genetic research can produce probabilistic information about family members who do not want to participate in a study. Genetic research such as family pedigree studies to identify patterns of gene transmission may present risks to familial relationships.

Another type of information is the use to which information or tissue collected from the subjects will be put. In *Moore v. Regents of the University of California,* the investigator used cells extracted from a subject's spleen, removed as part of standard therapy for hairy cell leukemia, to produce a lucrative cell line later patented by the university (86). Moore sued the investigators for converting his property—his tissue—to their own profitable use without his consent. The court decided that the researcher had a duty to inform Moore of its intent to develop a cell line from his tissue but, in a controversial decision, did not allow Moore to share in the profits made from his cells, even though they had been taken for that purpose without his knowledge or consent. (The court reasoned that Moore had relinquished his ownership interest in his cells because he had not expected to keep possession of them after his spleen was removed.) The court described the physician's duty to disclose broadly to include "personal interests unrelated to the patient's health, whether research or economic, that may affect the physician's professional judgment." In the case at hand, the court presumably reasoned that henceforth a subject could refuse to participate in the research if he did not want to donate his cells to a commercial operation. Theoretically, subjects could refuse to consent to participate unless they receive a share of the profits that result. Those options may be unrealistic if the research is conducted in conjunction with a person's medical treatment, as was the case with Moore. Patients may find it difficult, financially or emotionally, to obtain treatment from another physician who is not engaged in research.

Many of the first lawsuits involving informed consent to research involved experiments intended, at least in part, to help an individual patient—often called innovative therapy—rather than an organized research project with many subjects. Until the late twentieth century, courts took a dim view of such experimentation, holding physicians accountable for injuries resulting from methods or procedures that were not generally accepted in the medical profession (39,87,88). Such experiments were considered deviations from standard medical practice (89,90). However, some experiments were tolerated if no accepted therapy worked, the patient knowingly agreed, and the physician was sufficiently skilled (91).

Today few physicians conduct isolated experiments on their own initiative. Most research is conducted more formally at large institutions. Although few lawsuits have resulted, courts no longer appear to summarily reject such research as a deviation from accepted medical practice (92). However, misrepresenting an experiment resulted in a jury verdict against several California physicians who treated AIDS patients with a drug, Viroxan, made by one of the physicians at home without FDA approval (93). The injected drug caused tissue necrosis and the patients became ill without receiving standard therapy like AZT, Bactrim, or Pentamidine. The physicians were found liable for intentionally misrepresenting the drug and other unorthodox practices as "new," safe therapy that was better than conventional treatment, with intent to defraud the patients, who were awarded $925,000 in compensatory damages. The medical center was also found liable as a co-conspirator for failing to remove the physicians from its staff after learning of their unusual remedies. This suggests that highly unorthodox methods can give rise to a cause of action for fraud, as well as lack of informed consent and medical malpractice, where the "innovation" is not fully disclosed and the patient relies on the misrepresentation to his detriment.

The 1960s and 1970s saw the growth of research as a more systematic endeavor with increasing methodological sophistication as the common law developed more rigorous protection of individual subjects' self-determination with respect to research. As a result subjects are more likely to be aware that they are involved in a research study than they were several decades ago. However, the common law still has a role to play. Research projects that use subjects who might benefit from the experimental methods or procedures—so-called therapeutic research—continue to have difficulties with ensuring that subjects are fully informed (94). Numerous groups and commentators have criticized the persistent confusion of research with treatment, among researchers as well as subjects (29,39,62,66,95). Thus the importance of ensuring that all subjects are fully informed and that their consent is competent and voluntary has not diminished.

Research With Subjects Who Cannot Consent. The general principle that no one can be used as a research subject without his or her consent assumes that individuals are legally capable or "competent" to make the decision. For example, the doctrine of informed consent applies only to legally competent adults—those who are capable of understanding their circumstances, the proposed research study and its potential risks, and making and communicating a decision to participate or not (35,36,96). (All states presume that every person over 18 years of age is legally competent, unless a court has adjudicated the person to be incompetent.) Some research, however, seeks to use subjects who are not legally competent, either temporarily or permanently. This broad category includes adults who are unable to make or communicate an informed decision because of lack of consciousness, medication, pain, developmental disabilities, or mental disorders, as well as children (97). Such research raises difficult legal, as well as ethical, questions about whether incompetent persons can participate in research at all and, if so, whether some form of surrogate authorization is necessary or sufficient (98).

There are many examples of research studies that, rightly or wrongly, used incompetent adults and children as subjects (4,29,39,97–99). The ethical justification for using such subjects is that it is impossible to discover the etiology or find a therapy for conditions that render people legally incompetent or that affect only those who lack competence—such as schizophrenia or childhood leukemia—unless the investigational modality is tested in that very group of people as research subjects (97,98). In such cases the need for research results directly conflicts with the general principle prohibiting the use of individuals as research subjects without their voluntary, competent and informed consent (100).

The states have the sovereign authority (the *parens patriae* power) to protect the safety and welfare of incompetent adults and children. Parents of children and legal guardians of incompetent adults also have the obligation to protect the safety and welfare of their children and wards and to act in their best interests. Whether such legal representatives have the legal authority to consent to the use of their children and wards as research subjects has not been finally decided in the common law.

There is ample precedent for the principle that incompetent adults have the common law right to *refuse* to participate in research, and that their legal representatives can refuse on their behalf. This assumption is based on an analogy to the right to refuse medical treatment (60). Virtually all state courts that have considered the issue have decided that incompetent adults have the same rights as competent adults to refuse *medical treatment*, even if the treatment would save the person's life (101,102). The right to refuse medical treatment is part of the common law right to bodily integrity and self-determination — the same right the allows competent individuals to refuse to participate in research. The U.S. Supreme Court has also assumed, without deciding, that the Due Process Clause of the federal Constitution protects the right to refuse treatment (67,68). The U.S. Supreme Court has also noted that individuals who are mentally ill — including those who are involuntarily committed to a mental hospital — retain a constitutionally protected liberty interested in avoiding unwanted medication (103,104). There is general agreement that a person who refuses or objects to participate in research must not be used as a research subject (97,98).

There is very little legal precedent for the prevalent assumption that a parent or guardian can *consent* to an incompetent person's participation in research. Legal guardians are obligated to act in the best interests of the incompetent person (105). Participation in research is rarely in the best interest of a subject. Where the research is intended solely to gain generalized knowledge, no benefit can be expected to accrue to the subjects. In the case of so-called therapeutic research, where an experimental technique offers the possibility of curing a disease suffered by the person and no standard or accepted treatment has proved effective for that person, one might argue that the person might benefit from participating in the research. However, the potential risks of the research must be weighed against any benefit. It could be argued that a surrogate decision maker should be able to consent to research that offers possible benefit to an incompetent individual as long as it carries little or no risk (106). It is more difficult to argue that a legal representative could consent to potentially beneficial research that also entails significant risks. Therefore it is possible that researchers can be liable for battery or failure to obtain informed consent if they use incompetent subjects in research, whether or not they obtain surrogate consent.

There is little case law concerning research with subjects who are legally incompetent to consent to participation. Most cases in which the courts have considered the problem of incompetence involved isolated experiments related to medical treatment for an individual patient rather than a research study with a defined study population and detailed protocol. Some courts have permitted experiments on an incompetent person if a responsible family member or guardian gives informed consent and the subject himself or herself does not appear to object (105).

One of the few court decisions describing common law as well as constitutional limitations on using incompetent individuals as research subjects involved a group of patients in hospitals and facilities licensed and operated by the New York State Office of Mental Health (OMH) (107). In *T.D. v. New York State Office of Mental Health*, the patients challenged OMH regulations that allowed them to be used as research subjects even though they were considered incapable of giving informed consent. The court found that the state regulations, which provided for surrogate consent, were unlawful because they had not been approved by the Commissioner of Health. The court also found that the regulations violated the common law and constitutional rights of the patients, but the Court of Appeals, New York state's highest court, noted that this part of the decision was unnecessary and therefore an inappropriate advisory opinion. Still it suggests what courts might decide in the future. The lower court recognized that the goal of achieving important medical advances might not always be compatible with the goal of protecting human rights:

> It may very well be that for some categories of greater than minimal risk nontherapeutic experiments, devised to achieve a future benefit, there is at present no constitutionally acceptable protocol for obtaining the participation of incapable individuals who have not, when previously competent, either given specific consent or designated a suitable surrogate from whom such consent may be obtained. The alternative of allowing such experiments to continue, without proper consent and in violation of the rights of the incapable individuals who participate, is clearly unacceptable (107).

The New York regulations analyzed in *T.D.* did not apply to federally funded research that complied with federal regulations. However, federal regulations do not yet address the participation of incompetent adults. Federal regulations do permit certain types of research with children under narrowly defined conditions (25). In 1996 the Food and Drug Administration (FDA) adopted a regulation to permit research using emergency medical technologies without the consent of individuals who are incompetent to consent to participate because they are temporarily unconscious or in pain as a result of an emergency medical crisis (108). Several states prohibit research with residents of mental institutions by statute. Others permit surrogate consent in carefully defined situations. In one case California's law permitting experimentation with incompetent persons was held to be limited to research studies and could not authorize a surrogate to allow a physician to use an experimental bone graft in the treatment of an incompetent patient (109).

Thus there remain significant gaps in the law with respect to whether and, if so, how children and adults who are incapable of giving informed consent may participate as subjects of research. Several government agencies have recognized these gaps and have recommended regulations permitting people who are not legally competent to participate as subjects of research in certain circumstances, often with the informed consent of legally authorized representatives (97,110,111). The recommendations offer slightly different answers to such questions as what types of research should be permitted, who should be eligible to give surrogate consent, and what procedures should be used to protect research subjects. Although this is an area where legislation or regulations may prove useful to clarify individual rights and duties, the proposals would apply to specific jurisdictions, such as federally funded research or research conducted within a particular state. The common law, which is less likely to permit incompetent individuals to participate in research, will still apply to research that is not governed by federal regulations or specific state laws.

Liability for the Design and Implementation of Research

Negligence. Researchers have common law duties to design and carry out research properly and can be liable for injuries suffered by research subjects that are caused by negligence in the design or implementation of the research study itself (112). The fact that a research subject has consented to participate in the study and has been informed of the foreseeable risks of participation does not mean that the subject assumes the risk of the researcher's own negligence (113).

The law of negligence requires individuals and organizations to conduct themselves as reasonably prudent persons so as to avoid causing harm to others (13). In ordinary circumstances reasonably prudent conduct can be judged by the ordinary citizen, as represented by the jury in a jury trial. The standard of conduct to be observed by professionals (e.g., physicians) in the conduct of their profession, however, is that of an expert in the field, as established by the profession itself, for ordinary citizens are not expected to be familiar with specialized knowledge and skills. Those who conduct research are legally responsible to subjects who suffer injury as a result of their negligence (39,112,113). Thus organizations that design and supervise research, as well as investigators who carry out studies with human subjects, can be liable for the injuries they cause when they fail to conform their conduct to professional standards of care. Researchers, like physicians, can also be liable for intentionally or negligently disclosing confidential personal information about research subjects or invading their privacy without their consent.

A research subject who claims injury as a result of negligence must prove the following four things:

1. The researcher had a duty to the research subject.
2. The researcher breached that duty.
3. The research subject suffered physical injury.
4. The subject's injury was caused by the researcher's breach of his duty to the subject.

Violation of a statute or regulation can sometimes be considered to be negligence per se if the injured person belongs to the class of people that the law intends to protect (114). Private individuals cannot sue researchers merely because the researchers have violated federal law governing federally funded research with human subjects (115). However, if the statute imposes specific requirements on the researcher's conduct, injured persons may sue for common law negligence and use the statute as evidence of the standard of care that the defendant should have followed (116).

Individual Investigators. The case of *Vodopest v. Mac-Gregor* illustrates negligence by an individual investigator (117). Patricia Vodopest suffered permanent brain damage from cerebral edema in a high altitude climbing study in the Himalayas in Nepal. The study was intended to test breathing methods as a way to prevent altitude sickness. Vodopest experienced symptoms of altitude sickness at 8700 feet, but MacGregor, the project leader who was also a nurse, discounted the symptoms and told Vodopest to "breathe away" the symptoms and continue climbing higher. By 11,300 feet, Vodopest had developed cerebral edema and had to be evacuated. Vodopest was able to prove all four elements of negligence in this case. As study leader, MacGregor had a duty to act responsibly to protect the safety and welfare of the research subjects. Because the study was designed to prevent altitude sickness, the investigators should have been able to recognize its symptoms and respond with medically appropriate care. MacGregor failed to recognize (or refused to acknowledge) obvious symptoms of altitude sickness (nausea, headache, dizziness, and mental confusion). She also failed to follow the standard of care for treatment, which required having the person descend. Vodopest suffered physical injury (permanent brain damage), and the evidence indicated that her injury was proximately caused by MacGregor's negligence.

Organizations. Organizations can be held responsible for negligent injury to research subjects in two ways. First, and most simply, they can be vicariously liable for negligent acts committed by their employees and agents (118). (The employees and agents themselves also remain personally liable for their own negligence. However, employees may have insufficient assets to satisfy a large judgment or their employers may agree to pay their liability awards directly or through liability insurance.) The doctrine of *respondeat superior* (from the Latin "let the master answer" for the wrongs of his servant) holds an employer legally accountable for the unlawful acts of its employees committed during the course of employment. The purpose of the rule is to encourage employers to supervise their employees to ensure that they act responsibly. For example, the plaintiffs in *Mink*, described above, sued both the University of Chicago, where the research took place, and Eli Lilly & Company, the DES manufacturer, for vicarious liability for the acts of their employees in failing to tell plaintiffs that they were given DES as part of a medical experiment (44). In *Schwartz v. Boston Hospital for Women*, a federal district court found that a hospital could be liable for a physician's failure to obtain a subject's informed consent to a study procedure where the hospital paid the physician as assistant project director of a study of diabetic pregnancies (119). Similarly organizations can be held liable for the negligence of those who act as agents on their behalf, where the organization controls and directs the actions of the agent.

The second basis for organizational liability is direct corporate liability for the organization's own negligent acts or violations of duty. Organizations have duties to use care in the selection, retention, and supervision of their research staff and can be responsible for injuries caused by individuals who are incompetent or should not have been hired or retained by the organization (119,120). Organizations also have a duty to maintain a safe environment, which requires them to keep their premises safe and equipment in good working order to avoid preventable injuries. Plaintiffs in the *Mink* case also claimed that the corporations breached their own duties to notify the women about the experiment when they learned that DES could cause an increased risk of cancer in the children of women who took DES (44). The court agreed that both corporations had a duty to notify the women as soon as they became aware or should have become aware of the relationship between DES and cancer. However, the women were unable to recover damages on the failure to notify claim because they could not show that they had suffered any physical injury as a result of the failure to notify.

Two subjects in a 1960s study were awarded $8 million, including $5 million in punitive damages, by a 1999 jury verdict against investigator Dr. William Sweet and Massachusetts General Hospital, where the study took place (121). The study was intended to determine whether radiation to treat brain tumors could be focused selectively on the tumor without destroying other brain tissue by using boron neutron capture to selectively attract radiation. Subjects who were terminally ill with brain cancer had a boron compound injected into their arteries to see if it collected in the tumor. The compound caused severe illness and premature death in many subjects. The investigator argued that the subjects could not be harmed because they were already terminally ill, but the jury found that the review process was negligent in allowing the study to proceed at all.

Media reports of studies inducing or allowing psychosis in patients with schizophrenia at almost a dozen medical schools in the 1980s and 1990s illustrate several possible grounds for negligence, although few lawsuits have been brought (122). A study of schizophrenic patients at the University of California at Los Angeles (UCLA) which began in the 1980s was intended to identify schizophrenic patients who could function without antipsychotic medication because long term use of certain drugs can cause tardive dyskinesia, a condition producing involuntary movements for which there is no known treatment. Antipsychotic medication was withdrawn from schizophrenic patients who had recovered from acute psychotic disorders. The patient-subjects were observed until they relapsed and exhibited severe psychiatric symptoms of "bizarre behavior, self-neglect, hostility, depressive mood [or] suicidability" (123).

One possible negligence claim might be that the researchers breached their duty of care by designing the study so as to induce subjects to experience severe psychiatric symptoms that had been controlled by their medication. Put more generally, the claim might be that the study design posed unreasonable risks and should not have been conducted in that manner or, possibly, at all. Another might be that researchers failed to adequately monitor or treat subjects who experienced symptoms during the study, allowing them to suffer unnecessarily and possibly risking irreversible deterioration. In addition it might be possible to claim failure to obtain informed consent if the subjects were not adequately informed about the nature of the study and the risks it posed. Although the subjects were asked to participate in a research study, the original consent form was ambiguous, allowing an inference that study was linked to their medical care, and did not make clear that, unlike regular patients, the subjects would not receive medication unless and until they had a severe relapse (123). By the late 1980s, 88 percent of the subjects had suffered a relapse.

After the parents of one subject complained, the federal Office for Protection from Research Risks (OPRR) in DHHS investigated and required UCLA to change some of its internal monitoring procedures and modify the consent form to point out the risks of participation and the fact that the study was not intended to meet the subject's own personal medical needs (123). OPRR also investigated psychiatric challenge and relapse studies at the University of Cincinnati, University of Maryland, Bronx Veterans Affairs Medical Center, New York State Psychiatric Institute, and the National Institutes of Mental Health, and found similar problems with researchers' informed consent practices (122,124).

The families of two subjects sued UCLA in 1992 for fraud, deceit, lack of informed consent, and civil rights violations. Gregory Aller dropped out of college, threatened his mother with a butcher knife to exorcise the devil he believed inhabited her, and tried to hitchhike to Washington, DC, to assassinate then President Bush. His parents asked the researchers to give Greg his medication, and claimed they did not do so for many months. Antonio LaMadrid jumped off a 12 story UCLA building and died about three months after participating in the study (125,126).

A study of schizophrenia subtypes took place at University Hospital in Cincinnati in the 1980s with similar results. According to newspaper reports, a patient with schizophrenia sought treatment to adjust her medication dosage to prevent a possible manic episode (126). Lacking the resources to pay for treatment, she was enrolled in a challenge study, withdrawn from her medication, given a different medication, and then placed in restraints when her manic and delusional behavior erupted. Like several other studies, this one used a medication to cause or exacerbate psychiatric symptoms. She sued the researchers, but the case was dismissed because it was not brought within the statutory time limit.

Relatively few lawsuits have focused on negligence in carrying out research, as compared with failure to obtain consent. It may be that problems are more likely to arise from expected risks or lack of consent than from poor implementation of a study. In some cases research subjects may be partly responsible for their own injuries if they deliberately or carelessly fail to follow instructions designed for their protection and are harmed as a result. The doctrines of contributory and comparative negligence reduce, and in some cases may eliminate, the damages to which an injured person would otherwise be entitled (127).

Increased publicity about research in the 1990s may encourage closer scrutiny of research design — and perhaps legal claims of negligence — in the future, especially where researchers are pushing the scientific envelope or stand to gain financially from the success of products they study. The research design was questioned in a gene therapy trial to test — for the first time in human beings — the safety of delivering genetic material missing in people with ornithine transcarbamylase (OTC) deficiency, a sometimes fatal genetic mutation that prevents the liver from breaking down ammonia (128). Among the review committee's concerns were the use of adenovirus as the vector for transmitting the OTC gene to the liver of subjects because of adenovirus's potential to cause liver damage and toxic, sometimes fatal, inflammatory reactions. There were also reservations about the method of infusing directly into the liver instead of into a distant blood vessel. In addition the study used healthy, asymptomatic patients with OTC deficiency as the first human subjects to receive the gene, instead of patients for whom an accepted regimen of diet and medications was not effective. In 1999 Jesse Gelsinger, an 18-year-old subject, died from an inflammatory response apparently caused by the adenovirus vector. Although there is no indication that any lawsuit will be brought, the circumstances illustrate the opportunity for claims of negligence against the investigators for using excessively risky methods in designing and carrying out the research and for exposing healthy volunteers to unreasonable risks.

Conflicts of interest among researchers who have financial interests in the products they investigate may trigger or provide supporting evidence for future legal claims (129,130). James Wilson, a researcher in the OTC gene therapy study, founded a company to sell the rights to his discoveries, including the liver-directed gene vector approach studied in the OTC gene therapy trial (128). The question is whether a researcher who stands to gain financially from the success of a product might neglect evidence of the product's risks, rush to use it in human subjects prematurely, or select a research design that poses unnecessary risks to human subjects. It is not clear whether disclosure of the researcher's financial interest is sufficient to preclude a claim that the researcher acted negligently or even fraudulently.

Duties to Third Parties. Research subjects may not be the only persons injured as a result of research. However, liability for negligence is predicated on violating a legal duty to the injured person, and in the absence of any duty to third parties, there is generally no liability to third parties. Indeed, excluding some groups, especially pregnant women, from participation in research studies was often thought to ensure freedom from liability to women and their offspring (131). This raises the question whether researchers have any duty to avoid harm to future generations. If so, may subjects waive any potential claim

that a future child might have for injury? Such questions are most likely to arise in connection with clinical research with fertile or pregnant women, and sometimes fertile men, using investigational drugs with teratogenic effects or genetic material that may affect a fetus or future child (132,133). Although tort law has gradually extended its application to cases of prenatal injury in which fetuses are injured by intentional or negligent conduct, the causes of action available to children remain limited and the cases rarely involve research (134). In one example, however, the University of Chicago settled a lawsuit brought by the daughters of women who had received DES as part of an experiment in the 1950s (135). The daughters claimed to have an increased risk of cancer as a result of exposure to DES in utero. Future research involving germ-line gene transfer may expose the next generation to the effects of research with today's subjects (136).

Recently it has been recognized that not using research subjects who are representative of the populations that will ultimately use or be affected by the research may also raise liability concerns if the investigational product or service is later marketed. For example, if research subjects do not include women, the research may fail to discover adverse effects unique to women. If women are injured from using the product when later marketed, they might argue that the manufacturer or researcher negligently caused their injuries by failing to use a reasonable research design to identify possible risks to women (131). Researchers have begun to respond to concerns about inappropriately underrepresenting specific groups in their study populations (132). Ethical principles for research design include the equitable selection of research subjects, and federal regulations may include more specific requirements. In the long run this may create a standard of care in research design that requires a reasonably representative study population. So far this concern has not given rise to any cause of action for individuals who are excluded from research. In the late 1990s, however, a family sued UCLA because their Asian-American child was not admitted to its experimental elementary school, claiming that the use of race as an admission criterion violated the equal protection clause of the U.S. Constitution. The students enrolled were research subjects protected by federal and state research guidelines, but the case was limited to the constitutional issue. The court found the use of race was justified in order to select a representative sample of local students to ensure the validity of the research, and that research itself served the state's compelling interest in improving urban public education (137). Just as there is no legal right to be a research subject, there is no duty on the part of researchers to include specific individuals in a particular research study. On the contrary, there may be a duty not to include vulnerable or incompetent persons in some circumstances (7,97,105,138).

Products Liability. Product manufacturers can also be liable for negligence in the manufacture, design, or distribution of their products. A cause of action for negligence parallels the four part format for negligence described above. A person who is injured by a product must prove that the manufacturer failed to adhere to the appropriate standard of care, which failure caused the person's injury (139). Manufacturers of specialized products, such as biotechnology products, are held to the standards of an expert in the field.

In the 1960s state courts began to adopt the doctrine of strict liability which holds manufacturers liable for injuries caused by defective products without requiring proof that the manufacturer acted negligently. The justification was partly that negligence could be presumed when a product turned out to be defective and partly because it was difficult for plaintiffs to obtain evidence of a manufacturer's internal manufacturing processes to prove negligence. In theory, strict liability focuses on the condition of the product, while negligence focuses on the conduct of the manufacturer. When a person is injured by a defective product (as opposed to the behavior of individuals), they often assert both negligence and strict liability claims. Courts, however, often analyze both types of claims under products liability, a special subset of tort law that borrows negligence principles from tort law as well as warranty principles from contract law (140). Manufacturers may also be liable for breach of express warranty or implied warranty of a product's fitness for a particular purpose or merchantability (141). These causes of action, although based in contract law, impose similar duties and are usually subsumed by products liability today.

Products liability law holds all manufacturers, sellers and distributors of products legally responsible under state common law for personal injuries caused by a defect in their products (140). The earlier *Restatement* summarized the rule as follows: "One who sells any product in a defective condition unreasonably dangerous to the user or consumer or to his property [is liable] for physical harm thereby caused to the ultimate user or consumer. . . ." (142). Product defects include (*1*) manufacturing defects or flaws, in which the manufacturing process contains an error that produces something different from the product intended by the manufacturer; (*2*) defects in product design, in which the product specifications themselves pose foreseeable risks of harm that could have been avoided or reduced; and (*3*) errors or omissions in directions or warnings that accompany the product (142–144).

Manufacturing defects, such as contamination, adulteration, or production errors, are rare in pharmaceutical and biotechnology products but have occurred less rarely in medical devices (145,146). A product design is considered defective if the product could have been designed in a different way that would reduce its inherent risks without significantly decreasing its utility or effectiveness (142–144). Whether a product could have been differently designed, however, is a technical question that depends on the state of scientific knowledge when the product was sold, the nature and results of product tests, and the feasibility of alternative designs (147). An experimental product can rarely, if ever, be accused of having a defective design while it is being studied precisely to determine whether the design is safe and effective. The fact of the research is ordinarily a defense against a claim of defect. In addition it is often difficult to sort out the

cause of an injury to a research subject, especially where the product has not been widely tested or the injury might result from other sources to which the research subject was exposed. The most plausible claim would be that the product was not ready for testing in human beings, and that additional laboratory or animal studies would have revealed dangers before human being were harmed, which is really a claim of negligence. Thus manufacturers are not liable for a defective design of an experimental product as long as they conduct reasonable studies and do not use a product design that is foreseeably and unnecessarily dangerous.

The recently issued *Restatement* of products liability law contains a somewhat narrower definition of design defect for prescription drugs and medical devices, which requires proof that the product produces no net benefits for *any* class of patients or that no reasonable manufacturer would produce it if it knew of the defect (148). This permits the marketing of drugs and devices with serious risks as long as they may benefit at least one class of patients. Outside the realm of ordinary consumer products and asbestos, claims of defective design have been directed primarily at lawfully marketed, nonexperimental medical devices, such as the Dalkon Shield, the Copper-7 IUD, the Bjork-Shiley heart valve, and silicon-gel breast implants (149).

Defects in product distribution include errors and omissions in product directions and warnings. Manufacturers are responsible for failing to provide to the user a warning of dangers inherent in the use of the product, or for providing an inadequate warning that failed to alert the user to the danger (150). The cause of action is similar to a cause of action against a physician or researcher for failure to obtain informed consent (151,152). The theory is that a consumer would not have used the product and suffered an injury had he or she been warned of the risk that caused the injury. Warning defects have been claimed in a substantial proportion of lawsuits against manufacturers of licensed pharmaceuticals and vaccines (153). Manufacturers of prescription drugs and vaccines that can be obtained only from a physician are not required to issue warnings directly to patients or subjects of research. The "learned intermediary" rule holds the manufacturer responsible only for providing adequate warnings to the physician, who in turn is responsible for judging the appropriateness of the medicine for an individual patient and informing the patient of its risks and benefits (154–158). In the future, however, courts may decline to apply the learned intermediary rule in the case of some prescription drugs that are marketed by direct-to-consumer advertising if the physician has only a minor role in determining whether the patient should use the drug (159,160).

Human subjects who participate in research involving investigational products should be made aware that they could be exposed to risks. If subjects are adequately informed, their consent to participate means that they assume responsibility for the risks that have been disclosed to them. The investigator's duty to obtain the voluntary informed consent of subjects also protects the product manufacturer.

Products liability law focuses on commercially marketed products—those lawfully for sale or distribution in the market. The new *Restatement* of products liability defines products as "tangible personal property distributed commercially for use or consumption" (161), and may not even apply to investigational products. (Although human blood and tissue qualify as products when they are sold or distributed commercially, most states have enacted blood shield statutes that limit liability for contaminated human blood and human tissue to liability for negligence.) It would not ordinarily apply to investigational products that are being tested in research studies before any commercial marketing. Such investigational products, like investigational new drugs, are not held out as safe or effective commercial products ready for marketing to consumers. Rather, they are being studied to determine their effectiveness and to identify possible defects and risks. Of course, some research studies commercial products that are already on the market. For example, two marketed products, such as diagnostic tests, may be compared for relative efficacy, or a product marketed for one purpose may be studied to determine whether it has another use, such as an "off-label" use of a drug that has been approved by the FDA for a different specific use.

In *Proctor v. Davis*, the Upjohn Company was found liable for failing to warn physicians of the risks of an off-label use of its corticosteroid suspension, Depo-Medrol, by injection near the eye to treat eye conditions (162). FDA had approved the drug for intramuscular, intrajoint, and intralesional use only. The court found that "Upjohn fostered and encouraged this unapproved use as experimentation on human beings" (162). The company gave ophthalmologists financial and technical assistance to test periocular use and write favorable reports. Although the company received reports of adverse reactions, including blindness, it did not issue any warning. The company's conduct was found to justify punitive damages of just over $6 million, as well as compensatory damages of over $3 million, to the plaintiff who lost his eye after the drug was injected in an off-label use.

Independent researchers — who are not employed by a manufacturer — are not likely to be subject to products liability unless they also qualify as commercial sellers or distributors who are engaged in the business of selling or distributing the products they study. Investigators ordinarily provide services, and services are not considered products. Courts are unanimous in refusing to categorize commercially provided services as products for purposes of strict products liability in tort. Thus strict products liability does not extend to professionally provided services, such as medical or legal help (163,164).

Most states have held that public policy should exempt hospitals, physicians, and dentists from liability for injury from products used to treat their patients, such as defective pacemakers, forceps, and syringes, even though similar product/service combinations have been the subject of liability (165). The justification for this exemption is that the need for medical care outweighs any need to hold providers strictly liable for the products they use or implant in medical treatment. However, courts do not ordinarily consider research to be part of medical care, so the public policy rationale may not apply to research. A few recent cases have found that liability might be imposed where a hospital or physician selected or sold the product (166,167). A few cases have held a hospital liable for injuries resulting from defective products that were not directly related to medical care, but neither did they relate to research (168,169).

Although most legal principles apply equally in the case of experimental and marketed products, few investigational products being tested in research studies are likely to qualify as commercial products or otherwise meet the criteria for products liability claims. Most new drugs, biologics, and medical devices are developed in compliance with FDA regulations governing investigational use, so the federal rules are the primary source of law governing such research with human subjects. A violation of federal regulations may be evidence of common law negligence in most states, although states vary with respect to whether such a violation should be considered negligence per se, evidence of a deviation of the applicable standard of care, or excluded from consideration as inflammatory and irrelevant to the issue of causation.

Institutional Review Boards. Most IRBs are created by hospitals, universities, and research organizations to comply with federal law as a condition of receiving federal funding for research. The institutions agree by contract (the "general assurance") with a federal agency (typically HHS, FDA, or DOE) to be bound by federal IRB regulations specifying general membership qualifications and obligations of an IRB (170). Thus IRB duties are defined in the first instance not by common law principles but by federal statutes and regulations — specifically the Common Rule — issued by virtue of the federal spending power (10).

Nonetheless, the institution retains significant discretion over IRB operations, determining procedures, funding and personnel (11). An IRB created by a hospital or university is typically part of that organization, and not a separate legal entity, so that the organization is legally responsible for IRB actions. A growing number of IRBs, however, are independent legal entities, not part of any hospital, university, or research organization, and therefore responsible for their own legal obligations. Some have been created by community research groups and by private commercial enterprises like pharmaceutical companies to review their own research studies. Others are independently organized and offer their services to any group that wishes to conduct clinical research, usually small commercial companies.

Two decades ago Robertson considered possible liability issues that IRBs and their parent organizations might face in theory, such as defamation of an investigator or termination of employment for conduct involving research (11). The rarity of reported claims against IRBs suggests that such concerns remain largely theoretical. There have been several publicized scandals but little litigation (171,172).

In theory, possible claims by injured subjects of research against an IRB might include IRB negligence in approving a study, failure to attach conditions to the study to protect subjects, failure to require adequate information or informed consent, negligent assessment of the study's risks and benefits, failure to review ongoing research that poses risks to subjects, failure to stop research when subjects are being harmed, and failure to notify subjects of a significant risk or harm during or after study. IRBs may have a common law duty to act with reasonable care. In addition a violation of federal regulations might be considered to be evidence of negligence in many states. To hold the IRB liable, however, a subject would have to prove that the injury would not have happened but for IRB misconduct. There are many steps between IRB approval and injuries to subjects that could negate IRB responsibility: An approved study might not be funded or carried out, important information might not have been available, investigators might disregard IRB requirements in conducting the study, investigators might have prevented the injury, or the subject might not act on warnings provided. As a Texas court noted, "Other than disapproving all or part of a study, the IRB does not and cannot control the direction, results, or use of the research" (173).

A 1996 U.S. General Accounting Office (GAO) report on IRBs serving NIH studies outlines some difficulties experienced by IRBs that could, in theory, prevent them from living up to relevant standards of conduct (17). IRB review is labor intensive and subject to considerable time pressure, with some IRBs devoting only a few minutes to reviewing a protocol. Members may rely on a single primary reviewer for their assessment of a proposed study and miss key issues in the protocol. Most IRBs studied by the GAO were composed of volunteer members and few had training in ethics or federal regulations governing the protection of human subjects. Lay volunteers may be reluctant to challenge the opinions of members with scientific backgrounds who may empathize more with other investigators than with potential subjects. Most institutions with IRBs derive substantial income from research grants and may put institutional pressure on the IRB to approve protocols, thereby creating a conflict of interest. Most IRBs were reported to spend much of their time reviewing the informed consent form rather than whether the research design posed unacceptable risks to potential subjects. In addition there was little time for continuing review of ongoing research, which might discover problems.

Despite these potential difficulties, the GAO report confirmed that there had been few complaints against institutions or IRBs that review research funded by the federal DHHS (17). FDA issued only 31 "Warning Letters" to institutions noting serious deficiencies in IRB oversight of drug research. Deficiencies included allowing researchers to participate in IRB review of their own research and false claims that research studies did not require IRB review. FDA had never disqualified an institution from submitting research studies. On the other hand, FDA issued 99 sanctions against 84 individual investigators between 1980 and 1995. Most violations were minor; serious violations included forging a subject's signature to a consent document, failing to obtain informed consent, fabricating data to make subjects appear eligible for a study, falsifying laboratory tests, and failing to report adverse reactions to investigational

drugs. However, a 1999 report of 1000 FDA spot checks found that 213 researchers failed to obtain necessary informed consent, 364 researchers failed to follow their research protocol, and 140 did not report adverse reactions experienced by research subjects (174).

These reports, as well as reports with similar findings by the Office of the Inspector General, suggest that IRBs may find it difficult to perform their legal duties adequately (175,176). If improvements are not made, they could be liable in the future if their wrongful actions cause harm. At the same time other factors argue against a significant upsurge in claims. Human subjects who are injured by research are more likely to have a cause of action against, and to sue, the investigator or the institution for research conduct that is unrelated to IRB actions. An IRB's contribution to any injury may be difficult to discover. Few nonmembers appear at IRB meetings, see its records, or are in a position to identify IRB misconduct.

Federal agencies that oversee IRBs are in a good position to discover problems and act directly to halt research that may threaten human subjects. In the late 1990s OPRR began to impose sanctions on research institutions that do not comply with federal regulations. In May 1999 OPRR suspended Duke University's authorization to conduct federally funded research with human subjects (177). The action was taken only after the university had been given an opportunity to improve its system and failed to respond adequately. OPRR has imposed similar suspensions at other institutions, including the University of Rochester, University of Southern Florida, University of Minnesota, Mt. Sinai, Rush-Presbyterian-St.Lukes Medical Center in Chicago, University of Illinois in Chicago, and University of Colorado (178,179). Reasons for suspensions varied but were consistent with the problems identified in the GAO report — failures to submit studies for IRB approval, inadequate consent procedures and records, failure to keep track of research studies after they began, and failure to document why studies were approved. Enforcement action by OPRR and FDA may herald more consistent and searching scrutiny of federally funded research. It may also encourage a re-evaluation of the entire system for protecting human subjects of research, including the use of IRBs themselves.

There have been a few lawsuits by individuals attempting to discover IRB records, often to use as evidence in their lawsuits against other parties. IRBs typically treat research protocols and investigators' reports on ongoing research as confidential and do not voluntarily release them outside the IRB, except to federal agencies that are entitled by statute to review IRB compliance with a general assurance. Federal regulations prohibit the disclosure of IRB records that identify individual research subjects. The common law, however, does not specifically protect IRB records from disclosure to investigators, subjects, or the public, or discovery in a lawsuit, although some state statutes may grant protection (180). Most state statutes that grant a privilege against discovery to hospital peer review committee proceedings and records do not apply to IRBs because IRBs are not peer review committees that oversee patient care (181). Nonetheless, even courts that have permitted disclosure of IRB records have kept confidential the names of subjects, as well as a company's proprietary information about its investigational products (181,182).

One might argue that there is a scientific or academic privilege protecting against discovery of IRB records (183,184). Research data, however, do not appear to be privileged. Data may be protected from discovery where they are not relevant or necessary to a lawsuit, where disclosure would be unduly burdensome on the researcher, or to protect a researcher's interest in completing a study or in publishing conclusions first in a peer-reviewed journal (185,186).

The fact that IRB records are not automatically protected from discovery does not necessarily mean that IRBs have any obligation to publicly disclose them. For example, an Iowa Supreme Court decision found that a public hospital (not an IRB) was not required to make public its summaries of nosocomial infection data that the hospital was required by statute to collect (187).

What counts as IRB records is not clear. They could be limited to minutes of the meetings, so that research protocols and monitoring reports would not be covered (188). A litigant might be able to obtain more information, including the research protocol, from the researcher who conducted the research.

Defenses Against Liability

Researchers charged with liability for injuries to research subjects can interpose several defenses, in addition to simply denying the facts alleged in the claim, to defeat a cause of action.

Release of Liability. Under the doctrine of informed consent, because researchers must disclose the risks of research to all subjects who agree to participate in a study, subjects assume the risk of being injured as a result of the risks that were disclosed to them. An analogous principle of contract law may allow human subjects to release researchers from liability for the foreseeable risks of participating in research, as long as those risks were disclosed and the research was properly carried out (189). However, the general rule is that human subjects cannot release researchers from tort liability for the researcher's own negligence (190). In addition federal regulations forbid the use of any document that waives or otherwise excuses, releases, or indemnifies researchers from liability for injuries to research subjects in federally funded research (191): "No informed consent, whether oral or written, may include any exculpatory language through which the subject or the representative is made to waive or appear to waive any of the subject's legal rights, or releases or appears to release the investigator, the sponsor, the institution, or its agents from liability for negligence." As a practical matter, most IRBs are thought to reject attempts to use any such releases (192).

In *Vodopest v. MacGregor*, described above, the research subject Vodopest had signed a form releasing the investigators from "all liability, claims and causes of action arising out of or in any way connected with my participation in this trek" (117). The Supreme Court of Washington found that the release was against public policy and therefore void and unenforceable because it would release researchers from their own negligence in the conduct of research. The Court noted that if Vodopest had fallen off a trail, the release might have been effective because falling is an expected risk of climbing mountains and was not necessarily part of the research. However, Vodopest's injury specifically related to the research—monitoring subjects for high altitude sickness—and therefore liability could not be released. The court concluded, "The public's interest in the safety of human subjects and the public's interest in the integrity of legitimate and necessary research militate against allowing researchers to negligently conduct research with impunity."

Sovereign Immunity. Neither the federal nor a state government can be sued without its consent. Sovereign immunity—based on the ancient notion that the king can do no wrong—can be a defense to tort claims brought against officials, employees, and agents of state and federal governmental institutions, such as state universities and city hospitals (13). The Federal Tort Claims Act permits certain tort claims to be brought against the federal government (57). All states have enacted statutes that permit certain claims to be made against the state, although some states limit the amount of damages for which a state can be held liable. For example, Massachusetts limits its own liability for personal injury to $100,000 per claim (193).

Researchers who are employees or agents of state entities may be protected by sovereign immunity if they are acting on behalf of the state. A Virginia court applied sovereign immunity to protect a researcher from liability for a research subject's death after an overdose of asthma medication (194). The researcher was employed as an allergy fellow by the University of Virginia Hospital, a state university protected by sovereign immunity. The court found that the hospital's employees and agents were also protected if they (1) are subject to the control of the state and have little or no control over the patients they see, and (2) participate in activities in which the state has a strong interest, especially where the activity is not readily available in the private sector. The court also noted that a third factor ordinarily required for sovereign immunity in Virginia—that the employees have duties that require them to exercise a substantial degree of judgment or discretion in the activity complained—is likely to exist in most medical research studies. In this case the court found that the research was important to the state. However, not all research is necessarily important to the state or anyone else, and much research can be performed as well or better by the private sector. Thus the importance of the research is not necessarily a reliable factor for predicting whether specific research will be protected by sovereign immunity elsewhere.

Physicians who are independent contractors—not employees or agents—are not protected by sovereign immunity. They may also be liable for the negligence of public hospital employees, including residents, if their supervision of those employees is negligent (195).

Sovereign immunity protected an assistant attorney general of the State of New York from liability for concealing the fact that a man's death from an injection of synthetic mescaline was the result of a covert experiment by the U.S. Army Chemical Corps to test the drugs as a chemical warfare agent (196). The assistant attorney general received absolute immunity because he acted as the lawyer for the state psychiatric institution in a lawsuit brought by the decedent's family. However, several federal attorneys representing the U.S. Army, which was not party to the lawsuit, were not entitled to immunity for their actions in fraudulently concealing the experiment and thereby depriving family members of their cause of action for battery.

Charitable Immunity. Some private, nonprofit, charitable organizations are also protected from liability for personal injury to patients and research subjects by the doctrine of charitable immunity. Massachusetts was the first state to adopt the doctrine, in 1876, and still retains it (197,198). However, charitable immunity has been repealed in most states, and states that still apply the doctrine have crafted so many exceptions that immunity is more the exception than the rule (120). The trend appears to be in favor of holding all private organizations liable for their own negligence. Reasons for ending charitable immunity include the fact that most research organizations that qualify as "charitable" for tax purposes are no longer charities in the historical sense, and also often purchase liability insurance to protect against depleting their assets (199). In addition the nineteenth-century assumption that patients who accept charity care must also accept the risk of harm is inconsistent with the premise that all patients and research subjects are entitled to the same legal safeguards (200). Nonprofit hospitals are among the few organizations that still qualify for the remnants of charitable immunity. Most private companies that conduct research in biotechnology are not charities and would not be protected.

Federal Preemption. Federal law sometimes supersedes (preempts) state common law, so individuals cannot use the state law as the basis for a lawsuit. In the absence of a federal statute explicitly preempting state law, however, there is a general presumption that federal law does not supersede state law, especially with respect to health concerns and common law claims of negligence and product liability (201). Even where a federal statute regulates an industry, the presumption against preemption of state common law actions remains (202,203). Although some products are subject to federal labeling requirements, few such laws specifically preempt state common law claims of failure to warn, inadequate directions or warnings, or fraudulent misrepresentation (204). Thus, federal preemption of state common law claims for personal injury is rare.

The Medical Devices Amendments of 1976 contain a provision that preempts state law safety or effectiveness requirements that differ from requirements for medical devices imposed by FDA under federal law (205). The Act states that "No State or political subdivision of a State may establish or continue in effect with respect to any device intended for human use any requirement — (1) which

is different from, or in addition to, any requirement applicable under this chapter to the device, and (2) which relates to the safety or effectiveness of the device or to any other matter included in a requirement applicable to the device under this chapter." (This specific provision in the statute does not apply to drugs.) In *Medtronic, Inc. v. Lohr*, the U.S. Supreme Court found that the federal law did not preempt common law claims for personal injury resulting from certain medical devices that are not subject to specific review and premarket approval by FDA (206). In that case, which did not involve research, Lora Lohr claimed that her pacemaker lead had a defect that caused a complete heart block requiring emergency surgery and also that the manufacturer had not warned of the device's tendency to fail. The device was sold pursuant to the federal law's §510(k) premarket notification procedure, which does not require FDA approval of the device or its specifications before marketing (207). The Medical Device Amendments may preempt state common law claims involving a Class III medical device, which is subject to premarket review and approval with specific FDA requirements, but the precise scope of preemption has not yet been fully addressed by the Supreme Court. Aside from its limited scope, this preemption provision may have little effect on research subjects because it is limited to claims about the medical device itself and does not necessarily preclude claims concerning the conduct of research.

The federal Biomaterials Access Assurance Act of 1998 limits the liability of suppliers of raw materials and component parts that are used in medical devices (208). Independent suppliers (who are not organizationally related to a medical device manufacturer) are liable to individuals who are injured by an implanted medical device only if the parts supplied to make the implant did not meet the supplier's or device manufacturer's specifications. Suppliers who sell raw materials and component parts that are used in many different products are not ordinarily in a position to conduct the research necessary to ensure that their materials are safe for implantation in human beings. Medical device manufacturers remain responsible for ensuring that the resulting product complies with FDA requirements and may be sued by injured users. However, the supplier may be sued for its own negligence or intentional conduct if the device manufacturer demonstrates that the supplier's actions caused an injury and the device manufacturer should not be held solely responsible or will not be able to pay the full amount of damages awarded to the injured person.

Statutes of Limitations. A defense that is available to all public and private defendants is that a claim has been brought after the time permitted by law to begin a lawsuit. All states have "statutes of limitation" or "repose" that bar the bringing of a legal action, typically two to six years after the claim arose. The purpose of such statutes is to limit a defendant's exposure to potential liability and to encourage claims to be brought while evidence is still fresh and available (13). In an unusual decision, a California trial court dismissed a suit brought against the University of California at Los Angeles for fraud, deceit, lack of informed consent and civil rights violations in a schizophrenia study because the plaintiffs had exceeded the five-year statute of limitations for bringing the case to trial after it was filed (209).

Different legal claims or causes of action often have different maximum time periods for bringing suit. Many states have shortened the statute of limitations for medical malpractice to two years, while the limit for ordinary negligence and other personal injury claims is typically three to six years. Several courts have held that the longer personal injury statutes of limitations (and not the medical malpractice statute) apply to cases involving research injuries, because research does not constitute medical care and does not create a physician–patient relationship—even if the research is conducted by physicians (47,52,210,211).

Most states apply a "discovery rule" that extends or "tolls" the time period in which a claim may be brought (13). The time does not begin to run until the date on which the plaintiff actually discovered or should have discovered that he or she had a legal claim (212). The discovery rule might be applied in cases of fraudulent concealment, in which investigators hid the fact of research from a research subject who suffered latent injuries, as in *Mink v. University of Chicago* or *Craft v. Vanderbilt University*, or if new information indicates a previously unsuspected causal relationship between an injury and earlier research (44,47). But the rule does not excuse a plaintiff with obvious injuries from acting on reasonably available information (74).

Because children are not ordinarily empowered by law to bring suit on their own behalf, many states toll the applicable statute of limitations until a child who has been injured reaches the age of majority, typically 18 years of age (13). Research sponsors and investigators who use young children as research subjects may be exposed to potential liability for many years. Some states have limited this exception by requiring that suit must be brought within a specific number of years following the injury, usually a longer period than the otherwise applicable time periods for adults.

CONCLUSION

The common law arguably offers the most comprehensive statement of legal principles for the protection of human subjects of research and redress for their injuries. Although the Common Rule is probably more familiar and influential, the common law is more comprehensive, in theory, than either federal regulations, which apply only to federally funded research, or state statutes, which address specific issues like fetal research. Moreover the common law is the only source of redress and compensation for human subjects injured as a result of research. Although the federal government can penalize investigators and researchers for violating federal rules, federal enforcement actions do not compensate research subjects personally.

Because they are the product of court decisions in particular disputes, common law principles are not easily summarized in a textbook. Caution must be used in extracting lessons from individual cases with particular facts. Reasonable people can disagree about precisely what is and is not required of researchers in specific circumstances. Moreover several issues have yet to be fully addressed in court decisions, including whether, and if so, when and how, individuals who are not legally competent might participate as human subjects of research; what, if any, duties investigators owe to people who are not subjects of their research; and whether, and if so, what kinds of conflicts of interest preclude researchers from conducting certain research.

Although the research enterprise—and the number of research subjects—is large and growing, there have been relatively few published court decisions addressing common law claims arising out of research. A majority of such lawsuits have complained of failure to tell subjects that research was being conducted at all. Few cases have been based on negligence in the conduct of research or invasion of privacy. In the absence of comprehensive empirical data on violations of legal obligations and injuries arising from research, it is hazardous to draw general conclusions about the incidence of particular problems from the few published court decisions and media reports.

Greater public visibility for research may focus new attention on the common law rights of research subjects and increase the potential for legal claims in the future. Publicity about cold war era radiation experiments conducted without the subjects' knowledge served to focus public awareness on the rights of human subjects in the 1990s and may have inspired lawsuits based on common law causes of action to compensate injured subjects or their families. A series of studies identifying problems and gaps in the current federal regulatory system may have encouraged federal agencies to initiate bolder enforcement of existing federal regulations governing federally funded research. Although these federal initiatives do not alter common law, they may lead to a more comprehensive restructuring of statutory legal protections for human subjects, which in turn may affect the scope of common law protections. The rise of biotechnology, new forms of financing research, innovative financial relationships between academe and private industry, and public–private partnerships between government and commercial enterprises all create new opportunities for research and for research subjects. These may foster a new era of public concern for protecting human subjects in research. Whatever form that concern takes, it is likely that the common law will remain the legal backstop for the rights of human subjects.

BIBLIOGRAPHY

1. G. Hughes, in A.B. Morrison, ed., *Fundamentals of American Law*, Oxford University Press, New York, 1996, pp. 9–33.

2. G.J. Annas and M.A. Grodin, eds., *The Nazi Doctors and the Nuremberg Code: Human Rights in Human Experimentation*, Oxford University Press, New York, 1992, pp. 94–104.

3. G.J. Annas, in G.J. Annas and M.A. Grodin, eds., *The Nazi Doctors and the Nuremberg Code: Human Rights in Human Experimentation*, Oxford University Press, New York, 1992, pp. 201–222.

4. J. Katz, *Experimentation with Human Beings*, Russell Sage Foundation, New York, 1972.

5. Tuskegee Syphilis Study Ad hoc Advisory Panel, Final Report, U.S. Department of Health, Education and Welfare, Washington, DC, 1973.

6. National Research Act of 1974, 88 Stat. 342, codified, as amended, at 42 U.S.C. 201-300aaa-13.

7. National Commission for the Protection of Human Subjects of Biomedical and Behavioral Research, *The Belmont Report: Ethical Principles and Guidelines for the Protection of Human Subjects of Research*, DHEW Publ. No. (OS) 78-0012, U.S. Government Printing Office, Washington, DC, 1978.

8. W.J. Curran, *Daedalus: Ethical Aspects Exp. Hum. Subj.* **98**, 542 (1969).

9. L.H. Glantz, in G.J. Annas and M.A. Grodin, eds., *The Nazi Doctors and the Nuremberg Code: Human Rights in Human Experimentation*, Oxford University Press, New York, 1992, pp. 183–200.

10. Office of Science and Technology Policy, Federal Policy for the Protection of Human Subjects, *Fed. Regist.* **56**, 28002–28032 (1991).

11. J.A. Robertson, *UCLA Law Rev.* **26**, 484–549 (1979).

12. O.W. Holmes, *The Common Law*, Little, Brown, Boston, MA, 1881 (1963 ed.).

13. W.P. Keeton and W.L. Prosser, eds., *Prosser and Keeton on the Law of Torts*, 5th ed., West Publication, St. Paul, MN, 1984.

14. G. Williams, *Curr. Leg. Probl.* **4**, 137–146 (1951).

15. H.K. Beecher, *N. Engl. J. Med.* **274**, 1354–1360 (1966).

16. D.J. Rothman, *N. Engl. J. Med.* **317**, 1195–1199 (1987).

17. U.S. General Accounting Office (USGAO), *Scientific Research: Continued Vigilance Critical to Protecting Human Research Subjects*, GAO/HEHS-96-72, USGAO, Washington, DC, 1996.

18. U.S. Department of Health, Education and Welfare (USDHEW), *Secretary's Task Force on the Compensation of Injured Research Subjects*, Report, USDHEW, Washington, DC, 1977.

19. P.V. Cardon, F.W. Dommel, Jr., and R.R. Trumble, *N. Engl. J. Med.* **295**, 650–654 (1976).

20. T.A. Brennan et al., Incidence of adverse events and negligence in hospitalized patients — Results of the Harvard Medical Practice Study I. *N. Engl. J. Med.* **324**, 370–376 (1991).

21. A.R. Localio et al., *N. Engl. J. Med.* **325**, 245–251 (1991).

22. B. Freedman, *N. Engl. J. Med.* **317**, 141–145 (1987).

23. President's Commission for the Study of Ethical Problems in Medicine and Biomedical and Behavioral Research, *Compensating for Research Injuries: The Ethical and Legal Implications of Programs to Redress Injured Subjects*, vol. 1, U.S. Government Printing Office, Washington, DC, 1982.

24. Council for International Organizations of Medical Sciences, World Health Organization, *International Ethical Guidelines for Biomedical Research Involving Human Subjects*, CIOMS, Geneva, 1993.

25. U.S. Department of Health and Human Services, Protection of human subjects, *Code of Federal Regulations*, Title 45, Part 46.

26. R.N. Proctor, *Racial Hygiene: Medicine under the Nazis*, Harvard University Press, Cambridge, MA, 1988.

27. R.J. Lifton, *Nazi Doctors*, Basic Books, New York, 1986.

28. J.H. Jones, *Bad Blood: The Tuskegee Syphilis Experiment*, Free Press, New York, 1981.

29. Advisory Committee on Human Radiation Experiments, *Final Report*, U.S. Government Printing Office, Washington, DC, 1996.

30. U.S. Congress, General Accounting Office, *Human Experimentation: An Overview on Cold War Era Programs*, Testimony Before the Legislation and National Security Subcommittee, Committee on Government Operations, House of Representatives, U.S. Congress, GAPO/T-NSIAD-94-266, U.S. Government Printing Office, Washington, DC, 1994.

31. U.S. Congress, House of Representatives, Committee on Energy and Commerce, Subcommittee on Energy Conservation and Power, *American Nuclear Guinea Pigs: Three Decades of Radiation Experiments on U.S. Citizens*, Report, U.S. Government Printing Office, Washington, DC, 1986.

32. R.M. Veatch, *The Patient as Partner: A Theory of Human Experimentation Ethics*, Indiana University Press, Bloomington, 1987.

33. H.K. Beecher, *J. Am. Med. Assoc.* **195**, 34–35 (1966).

34. *Union Pacific Railroad Co. v. Botsford*, 141 U.S. 250, 251 (1891).

35. *Scholendorff v. Society of New York Hospital*, 211 N.Y. 125 (1914).

36. *Canterbury v. Spence*, 464 F.2d 772, 780 (D.C. Cir.), *cert. denied*, 409 U.S. 1064 (1972).

37. *Natanson v. Kline*, 186 Kan. 393, 350 P.2d 1093 (1960).

38. *Cobbs v. Grant*, 8 Cal. 3d 229, 502 P.2d 1, 104 Cal. Rptr. 505 (1972).

39. G.J. Annas, L.H. Glantz, and B.F. Katz, *Informed Consent to Human Experimentation: The Subject's Dilemma*, Ballinger, Cambridge, MA, 1977.

40. R.R. Faden and T.L. Beauchamp, *A History and Theory of Informed Consent*, Oxford University Press, New York, 1986.

41. American Law Institute, *Restatement of the Law, Second, Torts*, §§13–20, ALI, Philadelphia, PA, 1965.

42. J.C. Fletcher, in K. Berg and K.E. Tranoy, eds., *Research Ethics*, Liss, New York, 1983, pp. 187–228.

43. *Hyman v. Jewish Chronic Disease Hospital*, 15 N.Y.2d 317 (1965), rev'g 21 A.D.2d 495, 251 N.Y.S.2d 818 (2d Dep't 1964).

44. *Mink v. University of Chicago*, 460 F. Supp. 713 (N. Dist. Ill. 1978).

45. American Law Institute, *Restatement of the Law, Second, Torts*, §§652A, 652B, 652D, 652I, ALI, Philadelphia, PA, 1977.

46. American Law Institute, *Restatement of the Law, Second, Torts*, §§304, 310, 525-530, 550, 551, 557A, ALI, Philadelphia, PA, 1977.

47. *Craft v. Vanderbilt University*, 18 F. Supp. 2d 786 (M.D. Tenn. 1998).

48. *Craft v. Vanderbilt University*, 174 F.R.D. 396 (M.D. Tenn. 1996).

49. *Boston Globe*, July 28, p. A5 (1998).

50. *Massachusetts Lawyers Weekly*, May 3, B6 (1999).

51. *Stadt v. University of Rochester*, 921 F. Supp. 1023 (W.D.N.Y. 1996).

52. *In re Cincinnati Radiation Litigation*, 874 F. Supp. 796 (S.D. Ohio 1995).

53. Associated Press, Judge Approves $5 Million Settlement in Radiation Case, May 5, 1999.

54. *Jaffee v. United States*, 663 F.2d 1226 (3d Cir. 1981), *cert. denied*, 456 U.S. 972 (1982).

55. *Central Intelligence Agency v. Sims*, 471 U.S. 159 (1985).

56. *United States v. Stanley*, 483 U.S. 669 (1987).

57. *Federal Tort Claims Act*, 28 U.S.C. 1348.

58. *Feres v. United States*, 340 U.S. 135 (1950).

59. *Doe v. Sullivan*, 938 F.2d 1370 (D.C. 1991).

60. President's Commission for the Study of Ethical Problems in Medicine and Biomedical and Behavioral Research, *Making Health Care Decisions*, vols. 1, 3, U.S. Government Printing Office, Washington, DC, 1982.

61. R.J. Levine, *Ethics and Regulation of Clinical Research*, 2nd ed., Urban & Schwarzenberg, Baltimore, MD, 1986.

62. A.M. Capron, *Univ. Penn. Law Rev.* **123**, 341–438 (1974).

63. J. Katz, *St. Louis Univ. Law J.* **38**, 7–54 (1980).

64. R.J. Levine, *Arch. Intern. Med.* **143**, 1229–1231 (1983).

65. G.J. Annas, *Standard of Care: The Law of American Bioethics*, Oxford University Press, New York, 1993.

66. P.S. Appelbaum et al., *Hastings Cent. Rep.* **17**, 20–24 (1987).

67. *Cruzan v. Director, Missouri Department of Health*, 497 U.S. 261 (1990).

68. *Vacco v. Quill*, 521 U.S. 793 (1997).

69. T.L. Beauchamp and J.F. Childress, *Principles of Biomedical Ethics*, 4th ed., Oxford University Press, New York, 1994.

70. *Beno v. Shalala*, 853 F. Supp. 1195 (E.D. Cal. 1993).

71. *Halushka v. University of Saskatchewan*, 1965 D.L.R2d. 436, 52 W.W.R. 608 (Sask. 1965).

72. *Estrada v. Jaques*, 70 N.C. App. 627, 321 S.E.2d 240 (N.C. App. 1984).

73. *Ahern v. Veterans Administration*, 537 F. 2d 1098 (10th Cir. 1978).

74. *Monroe v. Harper*, 164 Mont. 23, 518 P.2d 788 (1974).

75. *Clemens v. Regents of the University of California*, 87 Cal. Rptr. 108 (Cal. Ct. App. 1970).

76. *Wilson v. Scott*, 412 S.W. 2d 299 (Tex. 1967).

77. *Kus v. Sherman Hospital*, 268 Ill. App. 3d 771, 744 N.E.2d 1214 (Ill. App. 1995).

78. *Karp v. Cooley*, 493 F.2d 408 (5th Cir.), *cert. denied*, 419 U.S. 845 (1974).

79. G.J. Annas, Death and the magic machine: Consent to the artificial heart. In G.J. Annas, *Standard of Care: The Law of American Bioethics*, Oxford University Press, New York, 1993, pp. 198–210.

80. *Whitlock v. Duke University*, 637 F. Supp. 1463 (M.D.N.C. 1986), aff'd, 829 F.2d 1340 (4th Cir. 1987).

81. W.K. Mariner, *Eval. Rev.* **14**, 538–564 (1990).

82. M. Chapman, *Am. J. Hum. Gene.* **42**, 491–498 (1992).

83. K. Rothenberg, *J. Law, Med. Ethics* **23**, 312–319 (1995).

84. K. Rothenberg, *Science* **275**, 1755–1757 (1997).

85. *Begay v. United States*, 591 F. Supp. 991 (D. Ariz. 1984), aff'd, 768 F.2d 1059 (1985).

86. *Moore v. Regents of the University of California*, 51 Cal. 3d 120, 793 P.2d 479, 271 Cal. Rptr. 146 (1990).

87. *Carpenter v. Blake*, 60 Barb. 488 (N.Y. Sup. Ct. 1871).

88. D.J. Rothman, *Strangers at the Bedside: A History of How Law and Bioethics Transformed Medical Decision Making*, Basic Books, New York, 1991.

89. *Owens v. McCleary*, 281 S.W. 682 (Mo. 1926).

90. *Fortner v. Koch*, 261 N.W. 762 (Mich. 1935).

91. *Stammer v. Board of Regents*, 39 N.E.2d 913 (N.Y. 1942).

92. *In re Guess*, 393 S.E.2d 833 (N.C. 1990), *cert. denied*, 498 U.S. 1047 (1991).

93. *Looney v. AMI Medical Center of North Hollywood* (1994), *Jury Verdict Review Publications, Inc.*, Feb. 1995.

94. N.M.P. King and G. Henderson, *Mercer Law Rev.* **42**(3), 1007–1050 (1991).

95. N.M.P. King, *Hastings Cent. Rep.* **25**, 6–15 (1995).

96. T. Grisso and P.S. Appelbaum, *Assessing Competence to Consent to Treatment: A Guide for Physicians and Health Care Professionals*, Oxford University Press, New York, 1998.

97. National Bioethics Advisory Commission (NBAC), *Research Involving Persons with Mental Disorders That May Affect Decisionmaking Capacity*, vol. I, Available at: NBAC, Rockville, MD, 1998. *http://bioethics.gov/capacity/TOC.htm*

98. National Commission for the Protection of Human Subjects of Biomedical and Behavioral Research, *Research Involving Those Institutionalized as Mentally Infirm*, U.S. Department of Health, Education and Welfare, Washington, DC, 1978.

99. S.E. Lederer and M.A. Grodin, in M.A. Grodin and L.H. Glantz, eds., *Children as Research Subjects — Science, Ethics, and Law*, Oxford University Press, New York, 1994, pp. 3–25.

100. A.M. Capron, *N. Engl. J. Med.* **340**(18), 1430–1434 (1999).

101. *Matter of Quinlan*, 70 N.J. 10, 335 A.2d 647, *cert denied*, 429 U.S. 922 (1976).

102. *Superintendent of Belchertown State School v. Saikewicz*, 373 Mass. 728, 370 N.E.2d 417 (1977).

103. *Washington v. Harper*, 494 U.S. 210 (1990).

104. *Riggins v. Nevada*, 504 U.S. 127 (1992).

105. L.H. Glantz, in M.A. Grodin and L.H. Glantz, eds., *Children as Research Subjects — Science, Ethics, and Law*, Oxford University Press, New York, 1994, pp. 103–130.

106. R. Michels, *N. Engl. J. Med.* **340**(18), 1427–1430 (1999).

107. *T. D. v. New York State Office of Mental Health*, 228 A.D.2d 95, 650 N.Y.S.2d 173 (1st Dept 1996), *app. dism'd mem. opinion*, 89 N.Y.2d 1029 (1997).

108. 21 CFR §50.23.

109. *Trantafello v. Medical Center of Tarzana*, 227 Cal. Rptr. 84 (Cal. Ct. App. 1976).

110. Office of the Maryland Attorney General, *Final Report of the Attorney General's Research Working Group*, June 12, 1998.

111. State of New York Department of Health Advisory Work Group on Human Subject Research Involving Protected Classes, *Recommendations on the Oversight of Human Subject Research Involving Protected Classes*, State of New York, New York, 1998.

112. N. Hershey and R.D. Miller, *Human Experimentation and the Law*, Aspen Systems Corp., Germantown, MD, 1976.

113. I. Ladimer and R.W. Newman, eds., *Clinical Investigation in Medicine: Legal, Ethical, and Moral Aspects*, Law-Medicine Research Institute, Boston University, Boston, MA, 1963.

114. *Johnson v. Farmers & Merchants State Bank*, 320 N.W.2d 892 (Minn. 1992).

115. 21 U.S.C. §337(a).

116. *Femrite v. Abbott Northwestern Hospital*, 568 N.W.2d 535 (Minn. App. 1997).

117. *Vodopest v. MacGregor*, 128 Wash. 2d 840, 913 P.2d 779 (1996).

118. K.R. Wing, M.S. Jacobs, and P.C. Kuszler, *The Law and American Health Care*, Aspen Publishers, New York, 1998.

119. *Schwartz v. Boston Hospital for Women*, 422 F. Supp. 53 (S.D.N.Y. 1976).

120. *Darling v. Charlestown Memorial Hospital*, 33 Ill. 2d 326, 211 N.E.2d 253 (1965), *cert. denied*, 383 U.S. 946 (1966).

121. *Heinrich v. Sweet*, No. 97-12134-WGY (D. Mass. Oct. 15, 1999).

122. R. Whitaker and D. Kong, *Boston Globe*, November 15, p. 1 (1998).

123. J. Katz, *St. Louis Univ. Law J.* **38**(1), 7–54 (1993) (quoting the research protocol).

124. D. Kong, *Boston Globe*, February 9, p. A8 (1999).

125. J. Willwerth, *Time* **42**, 21–42 (1992).

126. D. Kong, *Boston Globe*, November 16, p. 1 (1998).

127. American Law Institute, *Restatement of the Law, Second, Torts*, §463, ALI, Philadelphia, PA, 1977.

128. D. Nelson and R. Weiss, *Washington Post*, November 21, p. A01 (1999).

129. K. Eichenwald and G. Kolata, *N.Y. Times*, November 30, pp. A1, C16 (1999).

130. K. Eichenwald and G. Kolata, *N.Y. Times*, May 16, p. A1 (1999).

131. E. Flannery and S.N. Greenberg, in A.C. Mastroianni, R. Faden, and D. Federman, eds., *Women and Health Research—Ethical and Legal Issues of Including Women in Clinical Studies*, vol. 2, National Academy Press, Washington, DC, 1994, pp. 91–102.

132. A.C. Mastroianni, R. Faden, and D. Federman, eds., *Women and Health Research—Ethical and Legal Issues of Including Women in Clinical Studies*, vol. 2, National Academy Press, Washington, DC, 1994.

133. L. Walters and J. Palmer, *The Ethics of Human Gene Therapy*, Oxford University Press, New York, 1997.

134. E.W. Clayton, in A.C. Mastroianni, R. Faden, and D. Federman, eds., *Women and Health Research—Ethical and Legal Issues of Including Women in Clinical Studies*, vol. 2, National Academy Press, Washington, DC, 1994, pp. 103–112.

135. *Wetherill v. University of Chicago*, 565 F. Supp. 1553 (N.D.Ill. 1983).

136. P. Kitcher, *The Lives to Come*, Simon & Schuster, New York, 1996.

137. *Hunter v. Regents of the University of California*, 971 F. Supp. 1316 (C.D.Cal. 1997).

138. *Bailey v. Lally*, 481 F. Supp. 203 (D.Md. 1979).

139. American Law Institute, *Restatement of the Law, Second, Torts*, §282, 388-390, 394-398, ALI, Philadelphia, PA, 1977.

140. American Law Institute, *Restatement of the Law, Third, Torts: Products Liability*, §1, Comment a, ALI, Philadelphia, PA, 1998.

141. *Mitchell v. Collagen Corporation*, 126 F2d 902 (7th Cir. 1997).

142. American Law Institute, *Restatement of the Law Second, Torts*, §402A, ALI, Philadelphia, PA, 1977.

143. American Law Institute, *Restatement of the Law, Third, Torts: Products Liability*, §2, ALI, Philadelphia, PA, 1998.

144. W.P. Keeton, *Mo. Law Rev.* **45**, 579–596 (1980).

145. Institute of Medicine, *Vaccine Supply and Innovation*, National Academy Press, Washington, DC, 1985.

146. *Martin v. Telectronics Pacing Systems, Inc.*, 105 F.3d 1090 (6th Cir. 1997).

147. J.A. Henderson, Jr. and A.D. Twerski, *Hofstra Law Rev.* **26**(11), 667–695, 672 (1998).

148. American Law Institute, *Restatement of the Law, Third, Torts: Products Liability*, §6, ALI, Philadelphia, PA, 1998. (No examples of experimental products are given.)

149. S. Garber, *Product Liability and the Economics of Pharmaceuticals and Medical Devices*, R-4285-ICJ, Rand Institute for Civil Justice, Santa Monica, CA, 1993.

150. P.D. Rheingold, *Rutgers Law Rev.* **18**, 947–1018 (1964).

151. J.A. Henderson, Jr. and A.D. Twerski, *Product Liability: Problems and Processes*, 3rd ed., Aspen Law & Business, New York, 1997, pp. 342–350.

152. M.A. Franklin and R.L. Rabin, *Cases and Materials on Tort Law and Alternatives*, 6th ed., Foundation Press, Westbury, NY, 1996, pp. 608–614.

153. W.K. Mariner, in Office of Technology Assessment, Congress of the United States, *Adverse Reactions to HIV Vaccines: Medical, Ethical, and Legal Issues*, OTA-BP-H-163, U.S. Government Printing Office, Washington, DC, 1995, pp. 79–159.

154. *Hill v. Searle Laboratories*, 884 F.2d 1064 (8th Cir. 1989).

155. *Kirk v. Michael Reese Hospital and Medical Center*, 117 Ill. 2d 507, 513 N.E.2d 387 (Ill. 1987), *cert. denied*, 485 U.S. 905 (1988).

156. *Plummer v. Lederle Laboratories*, 819 F.2d 349 (2d Cir.), *cert. denied*, 484 U.S. 898 (1987).

157. *Payne v. Soft Sheen Products, Inc.*, 486 A.2d 712 (D.C. 1985).

158. *In re Norplant Contraceptive Product Liability Litigation v. American Home Products Corp.*, 165 F.3d 374 (1999).

159. *Edwards v. Basel Pharmaceuticals*, 116 F.3d 1341 (10th Cir. 1997).

160. *Perez v. Wyeth Laboratories, Inc.*, 161 N.J. 1, 734 A.2d 1245 (1999).

161. American Law Institute, *Restatement of the Law, Third, Torts: Products Liability*, §19, ALI, Philadelphia, PA, 1998.

162. *Proctor v. Davis*, 291 Ill. App. 3d 265, 683 N.E.2d 1203, 1212 (1997).

163. American Law Institute, *Restatement of the Law, Third, Torts: Products Liability*, §19, Comment j, Philadelphia, PA, 1998.

164. *Hoven v. Kelble*, 256 N.W.2d 379 (Wis. 1977).

165. *Estate of Hannis v. Ashland State General Hospital*, 554 A.2d 574 (Pa. Cmwlth. Ct. 1989).

166. *Bell v. Poplar Bluff Physicians Group*, 879 S.W.2d 618 (Mo. Ct. App. 1994).

167. *Parker v. St. Vincent Hospital*, 919 P.2d 1104 (N.M. Ct. App. 1996).

168. *Thomas v. St. Joseph Hospital*, 618 S.W.2d 791 (Tex. Civ. App. 1981).

169. *Johnson v. Sears, Roebuck & Co.*, 355 F.Supp. 1065 (E.D. Wis. 1973).

170. National Institutes of Health (NIH), *Protecting Human Subjects: Institutional Review Board Guidebook*, NIH, Bethesda, MD, 1993.

171. *Science* **189**, 383–396 (1977).

172. R. Weiss, *Washington Post*, August 1, p. A1 (1998).

173. *Cason v. E.I. Dupont de Nemours & Co.*, 1997 Tex. App. LEXIS 811, *62 (Ct. App. Tex. 1997).

174. S. Kaplan and S. Brownlee, *U.S. News and World Report*, October 11, pp. 34–38 (1999).

175. Office of the Inspector General, Department of Health and Human Services, *Institutional Review Boards: A Time for*

Reform, Testimony before the Committee on Government Reform and Oversight, Subcommittee on Human Resources, U.S. House of Representatives, Washington, DC, 1998.

176. Office of Inspector General, Department of Health and Human Services, *Investigational Devices: Four Case Studies*, OEI-05-94-00100, USDHHS, Washington, DC, 1995.

177. R. Weiss, *Washington Post*, May 12, 1999 Available at: *http://www.washingtonpost.com/wp-srv/national/daily/may99/duke12.htm*

178. A. Scharader, *Denver Post*, November 1 (1999).

179. L. Song, *The Chicago Tribune*, October 19 (1999).

180. *Doe v. Illinois Masonic Medical Center*, 297 Ill. App. 3d 240, 696 N.E.2d 707 (Ill. App. Ct. 1998).

181. *Konrady v. Oesterling*, 149 F.R.D. 592 (D.Minn. 1993).

182. *Deitchman v. E.R. Squibb & Sons, Inc.*, 740 F.2d 556 (7th Cir. 1984).

183. *In re American Tobacco Company*, 880 F.2d 1520 (2d Cir. 1989).

184. *Plough Inc. v. National Academy of Sciences*, 530 F.2d 1152 (D.C. Cir. 1987).

185. *Burka v. U.S. Department of Health and Human Services*, 87 F.2d 508, 520 n.13, 318 U.S. App. D.C. 274 (D.C. Cir. 1996) (collecting cases).

186. *Dow Chemical Company v. Allen*, 672 F.2d 1262 (7th Cir. 1982).

187. *Burton v. University of Iowa Hospital & Clinics*, 566 N.W. 2d 183 (Iowa 1997).

188. *Washington Research Project, Inc. v. Department of Health, Education, and Welfare*, 504 F.2d 238 (1974), *cert. denied*, 421 U.D. 963 (1975).

189. *Colton v. New York Hospital*, 98 Misc. 2d 957, 414 N.Y.S.2d 866 (1979).

190. American Law Institute, *Restatement of the Law, Second, Contracts*, §195, ALI, Philadelphia, PA, 1997.

191. 45 CFR 46.116.

192. W.F. Bowker, Exculpatory language in consent forms (letter to editor). *IRB: Rev. Hum. Sub. Res.*, March, pp. 1, 9 (1982).

193. Massachusetts General Laws, c. 258, §2.

194. *Hicks v. Pollart*, 27 Va. Cir. 7 (Va. Cir. Ct. 1991).

195. *Dang v. St. Paul Ramsey Medical Center, Inc.*, 490 N.W.2d 653 (Minn. App. 1992).

196. *Barrett v. United States*, 798 F.2d 565 (2d Cir. 1986).

197. *McDonald v. Massachusetts General Hospital*, 120 Mass. 432 (1876).

198. Massachusetts General Laws, c. 231, §85K.

199. *Harv. Law Rev.* **100**, 1382–1399 (1987).

200. *Malloy v. Fong*, 37 Cal. 2d 356, 232 P.2d 241 (1951).

201. *Levesque v. Miles, Inc.*, 816 F. Supp. 61 (D.N.H. 1993) (collecting cases).

202. *Rice v. Santa Fe Elevator Corp.*, 331 U.S. 218, 230 (1947).

203. *New York Conference of Blue Cross & Blue Shield Plans v. Travelers Ins. Co.*, 514 U.S. 645 (1995).

204. *Cipollone v. Liggett Group, Inc.*, 505 U.S. 504 (1992).

205. 21 U.S.C. §360(k)(a).

206. *Medtronic, Inc. v. Lohr*, 518 U.S. 470 (1996).

207. D.A. Kessler, S.M. Pape, and D.N. Sundwall, The federal regulation of medical devices. *N. Engl. J. Med.* **317**, 357–366 (1987).

208. Biomaterials Access Assurance Act of 1998, 21 U.S.C. §§1601–1606.

209. D. Anderson, Judge dismisses lawsuit over schizophrenia patients. *Associated Press Report*, June 5, WL 4869407 (1997).

210. *Collins v. Thakkar*, 552 N.E.2d 507 (Ind. Ct. App. 1990).

211. *Payette v. Rockefeller University*, 220 A.D.2d 69, 643 N.Y.S.2d 79 (1996).

212. *Baird v. American Medical Optics*, 301 N.J. Super. 7, 693 A.2d 904 (1997).

See other HUMAN SUBJECTS RESEARCH entries.

HUMAN SUBJECTS RESEARCH, LAW, FDA RULES

BARUCH BRODY
Baylor College of Medicine
Houston, Texas

OUTLINE

INTRODUCTION

In the United States, as in most other countries, a new drug cannot be marketed unless a national regulatory agency has determined that it is safe in, and effective for, its intended use. In order to establish that a new drug is safe and effective, research must be carried out on human subjects. In the United States, as in most other countries, this preliminary research on human subjects must be carried out with the concurrence of a national regulatory agency. The Food and Drug Administration (FDA) is the national regulatory agency that carries out both of these functions (concurring with the research and approving the drug as safe and effective) in the United States; it has other functions as well. FDA has issued many important rules governing the performance of both functions. These rules, and the ethical and policy issues raised by them, are the focus of this article.

Each FDA rule raises its own set of ethical and policy issues. But there are two themes that run through FDA's treatment of rules and it is helpful to identify them in advance. The first is the recognition that complex ethical and policy issues are best resolved by recognizing the legitimacy of many values, even if they are sometimes conflicting values, and by attempting to formulate rules that properly balance these different values. The second is the recognition that participation in research can be both a benefit and a burden to subjects, and the rules governing research need to reflect this dual nature of participation.

BRIEF HISTORY OF FDA

A series of national scandals led to the legislation that created FDA as we know it today (1,2,3). In 1906, in response to a national outcry related to false labeling and marketing of patent medicines, Congress passed legislation creating an agency to deal with adulteration or mislabeling of drugs. In 1938, in response to the sale of a liquid form of sulfanilamide which turned out to be poisonous and caused the death of more than 100 children, Congress passed a law prohibiting the sale of drugs in interstate commerce until the seller had submitted to that agency a New Drug Application (NDA) that demonstrated that the drug was safe in its intended use. In 1962, in response to the worldwide outbreak of phocomelia in children whose mothers had taken thalidomide during pregnancy, Congress passed a law requiring (1) that no drug be tested in human subjects until its sponsors submitted to that agency an Investigational New Drug (IND) application and (2) that no drug be approved for sale until the seller demonstrated in its NDA that the drug is effective as well as safe. FDA, as it exists today, is largely the product of these three legislative acts. Because these legislative acts were passed in response to scandals arising out of new drugs hurting those who used them, their emphasis was on the value of protecting research subjects and the general public.

FDA is also the product of its own regulatory responses to public criticisms. In response to criticisms raised through the 1970s and early 1980s that the FDA process took too long and resulted in useful drugs being available elsewhere but not in the United States (the "drug lag" claim), FDA issued in 1985 and 1987 new regulations, the NDA Rewrite (4) and the IND Rewrite (5), that were designed to speed up the approval process. In response to the demand for quicker access to drugs by desperate Acquired Immune Deficiency Syndrome (AIDS) and cancer patients, FDA developed in the late 1980s and the early 1990s the Treatment IND program and the accelerated approval programs. In response to the claim that it was insensitive to the needs of special populations, it developed in the 1990s policies relating to emergency room patients, geriatric patients, pediatric patients, and patients who were women of childbearing potential. Many of these regulatory changes were incorporated into the Food and Drug Administration Modernization Act of 1997 (6). Because these regulatory responses, and the resulting legislation, arose in response to concerns about overregulation and overprotection, their emphasis was on

widening access of research subjects to the benefits of research and on speeding the availability of the results of research to the general public.

CURRENT PROCESS FOR NEW APPROVALS

The sponsor of a new drug begins the process by submitting an IND application to the FDA. The application must include information about the composition of the drug, information about preclinical testing of the drug (including animal studies), information about all proposed protocols for research on humans, information about the approval of those protocols by an independent Institutional Review Board (IRB), and information about the informed consent process proposed in such protocols. In certain special circumstances involving emergency research, that last requirement may be waived. The protocols may be for Phase I studies in a limited number of subjects designed primarily to study the effects of increasing dosages of the drug. For such studies, the FDA's focus in its review of the application will be on the safety of the proposed protocols. The protocols may be for Phase II or Phase III studies in larger numbers of patients designed to study effectiveness and the overall benefit–risk ratio of the drug. For such studies FDA's focus in its review of the application will also include the scientific quality of the studies to see whether they can generate the data that are sufficient to support an application for marketing approval.

Unless the FDA objects within 30 days (or a longer period if further information is requested of the sponsors), research on human subjects may commence. During that research period, the new drug is only to be provided to research subjects under the approved protocols. There has been a traditional exception for emergency use authorized by the FDA for an individual case. In recent years, in response to the AIDS crisis, the FDA has developed a Treatment IND Program under which whole classes of patients may receive a drug while large scale clinical trials are continuing; the details of that program are discussed below. In general, prior FDA approval is required to charge patients for drugs received in research protocols; in the case of drugs provided under the Treatment IND program, only notification of the FDA and nonobjection by the FDA is required. In either case, the price may not exceed the actual costs of manufacture, research and development, and handling of the investigational drug.

If the results of the research are satisfactory, the sponsors may file an NDA with FDA. This application must include information about the composition and production of the drug, about the proposed labeling for the drug, and about the research results which support the claim that the drug has a favorable benefit–risk ratio. FDA regulations provide an extensive description, to be analyzed below, of the research data that are required to support such a claim. Some of those requirements have been modified, in ways that will be described below, for the Accelerated Approval programs. FDA has 180 days to respond to a filing of an NDA, but that period is open to extension. In making a determination of the application's acceptability, FDA draws upon the expertise of advisory committees, but the final decision belongs to FDA. Congress has in the last few years authorized FDA to charge user fees to those filing NDAs. The user fees have supported the hiring of additional staff, which has helped speed the NDA review process. Once the FDA responds with its approval, the new drug can be marketed in accordance with the approved labeling.

The set of regulations just described was developed for the approval of new drugs. It also applies to the approval of already approved drugs for new indications. The approval regulations are somewhat parallel to those governing the approval of new medical devices (7). Approval does not apply to new surgical procedures (as opposed to new devices that they may employ), to new testing services including genetic testing services (as opposed to new testing kits that they may employ), or to new physician uses of approved drugs for nonapproved purposes (as opposed to manufacturer promotion of such "off-label" new uses). The 1997 Food and Drug Administration Modernization Act partially addressed the "off-label" use issue by codifying the conditions under which manufacturers can distribute scientific information related to "off-label" uses (8). Much controversy exists both about the merits of these exceptions to the FDA's regulatory authority and about the new legislation related to "off-label" promotion.

FDA RULES FOR IND STAGE

Rules Governing Human Subjects Research

When a sponsor submits an IND application to the FDA, it must submit documentation that the proposed research protocols have been reviewed and approved by an appropriate IRB and that the protocols contain provisions for obtaining the informed consent of research subjects or their representatives. These requirements are consonant with the internationally accepted consensus about the conditions required for ethical research on human subjects.

All such research protocols must be reviewed and approved by an IRB that is independent of both the sponsors and the investigators. In approving the research, the IRB must determine that the following very standard requirements have been met (9):

- Risks to the subjects have been minimized and the remaining risks are reasonable in relation to the anticipated benefits.
- Selection of the subjects is equitable, and potentially vulnerable subjects are provided with additional safeguard.
- Informed consent has been obtained and documented.
- Privacy of subjects and confidentiality of data are protected.
- Adequate provisions are made for ongoing safety monitoring.

The requirements for the information that must be provided as part of the informed consent process are also standard. Subjects must be informed of at least the following (10):

- Purpose of the research, its duration, and the experimental procedures involved.
- Risks and benefits of participation and of the alternatives to participation.
- Extent to which research records are confidential.
- Any compensation and/or treatment for research related injuries.
- Right not to participate and right to discontinue participation.

Comparison to Federal Common Rule

In 1991 most federal agencies that conduct or sponsor research on human subjects adopted regulations governing human subjects research. Among these agencies was the Department of Health and Human Services (HHS) (11). The commonly adopted regulations are referred to as the Federal Common Rule. FDA did not adopt the Federal Common Rule, because of certain special features of FDA's regulatory process. The above-described FDA regulations closely resemble the Federal Common Rule. In 1996 FDA issued a description of the differences between its regulations and the Federal Common Rule (12). The three most important to be noted are the following:

- FDA regulations apply to all research submitted to it in IND applications and NDA submissions, regardless of the funding source for the research, while the Federal Common Rule only covers research funded and/or conducted by the relevant federal agency.
- FDA has developed special exemptions from the requirement of informed consent for emergency research; these will be described below.
- The Federal Common Rule has an elaborate system of reports by IRBs (called the assurance mechanism) to ensure that they meet the standards in the Federal Common Rule. No such system is imposed under the FDA regulations.

International Application of These Rules

As noted above, the FDA regulations apply to all research submitted to it, whether or not the research is conducted in the United States (13). Because of the significance of FDA approval in allowing access to the U.S. market, this requirement impacts greatly on the conduct of human subjects research in other countries. There are those who see this as a form of ethical imperialism, of the United States imposing its ethical standards on other countries. There are others who see this as a U.S. reaffirmation of fundamental moral truths, which remain true even in other countries that have not yet recognized their value. Both views assume that the standards are unique to the United States, and it is this assumption that is in error. It has recently been demonstrated that the essentials of the FDA standards have been broadly adopted throughout the world, even if the precise details vary from country to country, and that this broad adoption is strong evidence of the moral validity of FDA standards (14).

FDA regulations require that research conducted on human subjects in foreign countries meet the standards of the Declaration of Helsinki, a declaration of the cross-national World Medical Association, rather than the standards of the FDA regulations. In fact the standards are very similar, reflecting the broadly accepted international consensus about human subjects research. Still, imposing the standards by reference to the Declaration of Helsinki clarifies that doing so is not a form of U.S. ethical imperialism. Moreover, if the country in which the research is conducted has its own more stringent standards on human subjects research, then FDA requires that those more stringent standards be met, showing respect for those stricter standards.

Exception for Emergency Research

The international consensus embodied in the FDA regulations requires that informed consent be obtained from all human subjects or from their representatives before the research is conducted. That requirement has often imposed great difficulties on those conducting vitally needed research on the management of medical emergencies. Such research often must be conducted as soon as possible after the patient-subject presents in the emergency room. The patient may be temporarily incompetent to give consent and no legally authorized individual may be present to give consent. Moreover the very short time frame for the effective use of the investigational intervention often precludes the possibility of accurately informing the competent subject and/or the available representative and giving them a meaningful choice about participation. So vitally needed research suffers from imposing that requirement of prospective informed consent. Even when there are enough subjects or representatives available and competent to give meaningful consent, so the research can go forward, that requirement has prevented many other noncompetent emergency patients from getting access to the most promising new interventions, and that seems inequitable. There are, then, important moral considerations favoring a waiver of the requirement of informed consent in these special circumstances (15).

There are, on the other hand, important moral considerations favoring retaining that requirement. To begin with, we should try to avoid violating the right of individuals not to be used as research subjects without their consent. Second, we need to protect these vulnerable subjects from being harmed by research to which they have not consented when (as often happens) the experimental interventions do not fulfill their promise.

Many proposals have been made as to how this value conflict should be resolved (16,17). One of the most promising is the regulations adopted by FDA in October 1996. FDA allows for an IRB to waive the requirement of prospective informed consent when (18):

- the subjects are in a life threatening condition and available treatments are unsatisfactory;
- animal and preclinical human studies support the likelihood of the intervention's being helpful;
- obtaining informed consent from the subjects or their surrogates is not feasible, so the clinical investigation cannot be carried out without the waiver; and

- additional procedural safeguards, as well as the usual IRB approval, are adopted, including attempting to find surrogates to obtain their consent and/or informing them and the subjects afterward, community consultation and notification, appointing an independent data safety monitoring board, and FDA approval.

If the values protected by the requirement of informed consent are absolute moral values, these regulations are inappropriate, since they allow for the waiver of the requirement of obtaining that consent. They are appropriate, however, if those values need to be balanced against the competing values of social need and potential subject benefit. As noted in the Introduction to this article, this type of balancing approach, which has recently won much favor among scholars, is characteristic of many of the FDA's newer rules.

Treatment INDs

In general, new drugs being tested under an IND may not be distributed for clinical use by patients not enrolled in research protocols. The one exception to this rule is the Treatment IND program which allows for the distribution outside of research protocols of some new drugs being tested for use in treating immediately life-threatening or serious illnesses (19).

The Treatment IND program, formally adopted by FDA in regulations in 1987, grew out of the experience with AZT for treating AIDS patients. In 1986, Phase II trials began for AZT. They were stopped after six months when it was clearly demonstrated that AZT was life-prolonging in the short run. An NDA was filed in December 1986 and approved in March 1987. From the time the drug was stopped until the application was finally approved, over 4000 patients received the drug outside any research protocol. This experience demonstrated the need for such a program in certain cases (20). From a moral perspective, the value of quicker access takes precedence in the relevant cases over the value of protecting desperate patients from not-yet-approved drugs. This is one more example of the need to balance values, rather than treating any one value as absolute, in research ethics.

Although the program was first introduced in response to the need for an AIDS drug not yet approved, it has been used to allow access to drugs for treating other conditions. In the first six years of the program, 28 Treatment INDs were issued. Nine were for AIDS-related drugs, 9 were for cancer-related drugs, and 10 were for drugs to treat other conditions (e.g., newborn respiratory distress syndrome and neurological conditions such as Alzheimer's dementia, MS, and Parkinson's disease) (21). Some Treatment INDs were issued after the clinical trials were completed but before the NDA was approved, while other INDs were issued while clinical trials were still being conducted.

Under the regulations the following are the requirements for issuing a Treatment IND (22):

- The drug is to be used for treating an immediately life-threatening condition (death will occur in a matter of months, or premature death will result from nontreatment) or a serious disease.

- There are no comparable or satisfactory treatments available.
- The drug is being investigated in controlled trials, or the trials have been completed and the sponsor is actively pursuing approval.
- For the drug to treat immediately life-threatening conditions, there must be a reasonable basis for believing that the drug may be effective and would not expose the patient to unreasonable additional risks. For the drug to treat a serious illness, there must be sufficient evidence of safety and effectiveness.

These provisions were adopted by Congress in the Food and Drug Administration Modernization Act of 1997 (23).

Special Rules for Genetic Research

Investigational biologic agents used for somatic cell and gene therapy must undergo the same process of FDA approval as other investigational agents (24). INDs must be secured for the research to begin, and NDAs must be approved before the agent can be distributed for general use. Until recently IRB and FDA approval of gene therapy protocols was not sufficient; a special NIH committee, the Recombinant DNA Advisory Committee (RAC), had to approve all federally funded research protocols. That requirement of additional approval has recently been eliminated, although RAC continues to review individual protocols, and it has been suggested that its authority to approve protocols be reinstituted.

Biologic agents pose special problems of product development which have received special attention by the FDA. Measures must be taken to control the biologic sources of the material, the production process, and the final product. These are necessary to deal with concerns both of safety and potency. The FDA issued in 1991 a *Points to Consider* document to clarify many of these technical issues (25).

In contrast to gene therapy, new forms of gene testing have not traditionally required FDA approval unless they use new testing kits; if they do, the kits, as opposed to the tests themselves, do require FDA approval. In recent years concern has been expressed about the proliferation of new genetic tests not employing new kits and not therefore being approved by any regulatory agency. Many feel that this allows for the marketing of tests that may not be analytically or clinically valid and useful, and that this can lead to a variety of problems for those being tested. Others have less concern, thinking that the decisions to use these tests should be left to the individuals in question, guided by their physicians or other health care providers. A recent report from a joint task force of the National Institutes of Health and the Department of Energy (26) supports the former approach, arguing that there is a need for review at a national level of new genetic tests before they enter into clinical practice and suggesting that the FDA does have the regulatory authority to conduct that review. It also points that if the FDA were to adopt this role, it would need to look at a broader set of issues, including the contribution of testing to producing better long-term outcomes, than the FDA usually does when reviewing lab tests. It remains to be seen whether this approach will be adopted.

FDA RULES FOR NDA STAGE

Traditional Rules for Adequate Evidence

The research conducted under an IND is designed to acquire data sufficient to support the approval of an NDA. In order to do that, the data must establish with a high enough degree of certainty that the drug has a sufficiently favorable risk–benefit ratio to justify its intended use. This section will review the FDA regulations related to the NDA approval process.

It is helpful in reviewing FDA regulations to understand that any drug approval process must answer two different, although related, questions, and that answers to these questions necessarily involve trade-offs among important values. The first of these questions is a *content* question: How should the values of effectiveness and safety be balanced in deciding whether the drug in question has a sufficiently favorable risk–benefit ratio? We want drugs that are effective. We also want drugs that are safe. But the most effective drugs often carry with them risks. The content question asks about how these two legitimate values should be balanced in deciding what is a sufficiently favorable risk–benefit ratio. The second of these questions is an *epistemic* question: How should the demand for adequate evidence and the demand for speedy approval be balanced in deciding that the evidence in question has established the drug's risk–benefit ratio? We want firm evidence that the risk–benefit ratio is sufficiently favorable. We want good drugs to be approved as soon as possible. But getting firm enough evidence often requires delays in the approval process. The epistemic question asks about how these two legitimate values should be balanced in deciding whether the evidence is firm enough so that the drug should be approved without further delay. The FDA regulations have much more to say about the epistemic question than the content question.

There is a special section of the NDA regulations that address the type of evidence required (27). The basic point of the section is that the FDA considers adequate and well controlled studies as the primary basis for supporting the claim that there is sufficient evidence to justify the approval of a NDA. The following characteristics of such studies are listed as essential:

- The study must involve a control group to compare with the treatment group in order for the drug's effect to be assessed quantitatively. Five types of control groups are identified. Four are concurrent control groups (placebo concurrent, dose-comparison concurrent, no treatment concurrent, and active treatment concurrent). For these types of control groups, subjects should be randomly assigned to the treatment group or the control group. The fifth type is a historical control group; in that type of study, randomization is not possible.
- Patients must be assigned to the treatment and control groups by a method that minimizes bias.

Randomization is the way to accomplish this in all concurrently controlled trials.

- Procedures such as blinding subjects and investigators must be adopted to minimize bias in the conduct or interpretation of the study.
- The protocol for the study must carefully define the population to be studied and the methods to ensure that the subjects are part of that population, the nature and duration of the treatment to be studied, the number of subjects to be studied, the methods to assess the response of the subjects, and the planned statistical analysis.

Based on its reading of the 1962 statute, FDA has traditionally insisted that two adequate and controlled studies provide sufficient evidence to justify the approval of an NDA. That is, of course, much more that just a point of statutory interpretation. It reflects the well-known scientific ideal that good scientific results are reproducible results. Nevertheless, such a requirement can often result in approval delays if the two trials are run sequentially, and additional costs even when the two are run concurrently. As part of its effort to develop a new balancing of values in response to the epistemic question, Congress, in the Food and Drug Administration Modernization Act of 1997, rejected that interpretation and left the need for a second trial to the discretion of the agency (28).

There are many important issues raised by this traditional set of rules. Probably the most controversial are those surrounding the choice of the control group, particularly when there already are available other drugs for treating the condition in question. From the perspective of easily getting well-established answers (the perspective of investigators, sponsors, and future patients), placebo or no treatment concurrently controlled trials are preferable because they allow for smaller and more easily interpretable studies. From the perspective of insuring access to at least some treatment (the perspective of the subjects and of their treating clinicians), dose-comparison or active treatment concurrently controlled trials are preferable. Both sets of values are legitimate, and the issue becomes one of how to balance the values. Several crucial official statements seem to support the use of only the latter types of control groups whenever an effective alternative treatment exists (29,30). This approach seems too absolute in stressing only the subject-centered values. FDA, by contrast, has stressed the scientific advantages of the first types of control groups, and has urged their use (with appropriate subject-protection mechanisms such as early rescue and minimization of study duration) except when existing treatments are life-prolonging (12). This approach seems to not sufficiently weigh the subject-centered values; it does not, for example, take into account the losses to subjects in the control group when existing therapies effectively limit morbidities and discomfort. What is needed in the case of each controlled trial is a careful look at all possible trial designs (different types of control groups and different types of protection mechanisms) and a determination of which best balances the research-centered values with the subject-centered values (31).

Accelerated Approval Rules

The accelerated approval rules, like the Treatment IND rules discussed above, arose in response to those suffering from AIDS demanding quicker access to promising drugs. They were joined in this demand by others, such as patients with cancer, who were dissatisfied with the available treatments and who wanted quicker access to promising drugs. Unlike the Treatment IND rules, the accelerated approval rules relate to final approval of the use of the drug, and not just to interim access while approval is being considered.

The accelerated approval rules apply to drugs being tested for the treatment of serious or life-threatening illnesses (32). The drugs in question must have the potential of providing meaningful therapeutic benefits to patients over existing treatments, either because patient response is improved with the new drugs or because patients are unresponsive to, or intolerant of, the existing treatments. The crucial provision of the rules is that such drugs can be approved on the basis of well-controlled trials that establish a favorable effect on surrogate endpoints (e.g., tumor shrinkage for cancer patients or reduced viral loads for AIDS patients), endpoints that are thought to be predictive of true clinical benefits (e.g., improved length of survival or decreased morbidity). This provision accelerates the approval because it is often possible to get data on surrogate endpoints quicker than data on true clinical benefits.

Critics have pointed to many examples where the use of surrogate endpoints has led to mistaken conclusions about the effectiveness of the new drugs as measured in terms of the true clinical benefits (33). They are certainly right to raise these concerns. Even after accelerated approval is given on the basis of data concerning surrogate endpoints, postmarketing studies of the effectiveness of the drug on true clinical endpoints need to be conducted when there is doubt about the effect on the true clinical endpoints. But this practice does not undercut the moral validity of the accelerated approval rules. There is rather a need to balance the demand for adequate evidence with the demand for speedier approval. The accelerated approval rules do so for drugs whose promise is supported by surrogate endpoint data but not fully established by true clinical endpoint data in cases where the patients are very sick and existing treatments are not helpful. The actual choice to use the new drugs will be made by patients with the advice of their doctors, but society will not stand in their way once promising surrogate endpoint evidence becomes available. When these provisions are coupled with the requirement of further studies as appropriate, they seem like a reasonable balancing of the competing values. The Food and Drug Administration Modernization Act of 1997 accepted this conclusion and incorporated the accelerated approval rules into the statute governing the FDA (34).

Rules Governing Special Populations

There are a number of populations that traditionally have been perceived as requiring special protection in the research setting. One such group is the elderly, who often live in conditions of dependency, a dependency that makes them vulnerable to being exploited, and who sometimes suffer from impaired intellectual functioning that may impair their capacity to protect themselves. A second group is children, whose immaturity often impairs their capacity to protect themselves and whose dependency on their parents makes it difficult for them to make their own independent decisions about participation in research projects. A third group is women of childbearing potential, whose potential fetuses may require special protection. One way to provide special protection for these subjects is to exclude them from research, and this ethical concern is one of several factors that have often led in the past to the exclusion of these groups from research projects.

While this value of protection of the vulnerable is deserving of respect, there are other values that need to be considered as well, especially the value of justice. Individual members of excluded groups may unfairly be denied the personal benefit of obtaining access to promising new treatments. The exclusion of entire groups may result in the nongeneralizability of the results of the research to the members of the group, so they and their treating physicians are unfairly denied the basis for making decisions as to whether to use new treatments whose use in the general population is supported by the research in question (35).

In the earlier discussions of research ethics, greater emphasis seems to have been placed on the value of protecting the vulnerable from the risks of research. In recent years, as the benefits from participating in research have become better understood, more emphasis is being placed on balancing that concern with the justice-based concern of not excluding these individuals and groups from obtaining the benefits of research. As part of that more recent understanding, FDA has made several major initiatives in its rules dealing with these populations.

In 1989, in response to concerns about the exclusion of the elderly from research, FDA issued guidelines that, while explicitly not introducing further requirements for drug approval, provided recommendations about the inclusion of the elderly in research on drugs that are likely to have significant use in the elderly (36). The crucial recommendation was that the elderly should be included in such trials in reasonable numbers so that the patients enrolled in the trial reflect the patients who will use the drug if it is approved on the basis of the research. Following this recommendation would result only in the detection of fairly large age-related differences, but it was thought that those are the only differences important enough to be of concern.

These 1989 guidelines concerning the elderly were part of a larger effort in 1988 and 1989 by FDA to ensure that the subjects in research protocols were more representative of the population that would use the drug if it were approved (37). This more general effort also attempted to address the issue of the inclusion of women of childbearing potential in clinical trials. There is some controversy as to whether this effort was successful. The Government Accounting Office (GAO) claimed that there was a continuing problem, especially in connection with cardiovascular trials, while the FDA argued that the remaining differences resulted from the need to include more younger men because they were more vulnerable than younger women to cardiovascular disease. Nevertheless, FDA issued further clarifying guidelines in 1993 (38):

- Enough members of both genders should be included in order to detect "clinically significant gender-related differences," and the total study population should "reflect the population that will receive the drug when it is marketed."
- The integrated analysis of data from the trials should contain an analysis of gender differences in terms of both safety and effectiveness; additional studies might be required if these differences are significant enough.
- The definitive trials to support approval should employ gender-differing dosages if pilot pharmacokinetic studies reveal gender differences.
- Research protocols, unless specifically designed to study the effects of a drug during pregnancy, should include measures to minimize the exposure of fetuses to the drug being tested; the most important of these measures are pregnancy testing before administering the drug being tested and proper counseling about the selection and use of reliable methods of contraception.

In 1997 FDA proposed still further regulations to ensure that women of childbearing potential were included in clinical trials (39). These proposed regulations would only apply to trials of drugs intended for treating life-threatening illnesses. Unlike the above-discussed regulations, which have a direct effect only on the full set of data needed to secure approval and do not affect specific clinical trials, these proposed regulations will put constraints on certain trials. Under the proposed regulations, FDA can put a hold on trials, preventing further enrollment, if women of childbearing potential are being excluded because of fears of reproductive or developmental toxicities. FDA's argument for these proposed regulations is that such exclusion unfairly denies these women the choice to participate because of risks that can be minimized by the measures to minimize fetal exposure described above. These proposed regulations represent one further step in FDA's balancing of values in this complex area.

A third population for which FDA has developed special rules is the pediatric population. A very high percentage of the drugs used in the treatment of children have not been adequately tested for use in that population. In 1992, for example, 79 percent of the new drugs approved by FDA that could be used in children were approved without labeling for use in children because they had not been sufficiently studied in pediatric subjects (40). Based on 1994 data, the FDA identified the 10 most commonly used drugs in the pediatric population for which there was no pediatric labeling. The list included: albuterol inhalation for asthma, ampicillin injections for infection, and Prozac for the treatment of depression (41). These are serious drugs for serious medical problems, and their pediatric use should be guided by better data. The lack of research data results in a fundamental ethical dilemma for pediatricians: Should they take the risk of using these new drugs without adequate information about their safety and efficacy at various dosages in children, or should they not use these drugs and risk denying to their patients valuable treatments? Clearly, policies that promote more testing of new drugs in children would help resolve this dilemma, and would constitute one more balancing of protecting subjects from research risks with insuring the access of patients to the benefits of research.

In 1994 FDA attempted to address this problem by adopting what can be called "the extrapolation" approach (42). According to that approach, FDA would allow labeling for pediatric use on the basis of clinical trials conducted with adults if there was sufficient pharmacokinetic and adverse reaction data from pediatric subjects to justify extrapolating the adult results of a favorable risk–benefit ratio to the pediatric population. That rule did not, however, require sponsors to conduct the studies required to justify the extrapolation. By 1997 FDA concluded that this approach had been inadequate to resolve the problem. It proposed supplementing that rule with an "including children" approach, similar to (although not exactly the same as) the approach it had adopted toward the elderly and toward women of childbearing potential. According to this proposed approach, the NDA application for drugs that represent a meaningful therapeutic advance and that are likely to be used in a substantial number of pediatric patients must include (unless a waiver is issued) data about safety, effectiveness, dosages, and mode of administration in the pediatric population. This proposal drew much criticism from those who were concerned with the cost of new drug development and from those who were concerned with the pace at which new drugs are approved for use in the adult population, but a version of it was adopted in 1998 (43).

CONCLUSION

As one reviews FDA's rules, two major themes clearly emerge. The first is the need to recognize that participation in research can be a benefit as well as a burden. It is this general recognition that has been central to the FDA's efforts to expand the types of subjects enrolled in clinical trials to include the elderly, children, and women of childbearing potential. The second is the need to balance values rather than to treat some as absolute. It is this general recognition that has led to FDA's rules on emergency research (which balances the need for informed consent with the individual and social need for emergency research), on Treatment INDs (which balances the need to complete clinical trials with the need for speedier access), and on accelerated approval (which balances the value of optimal evidence with the need for speedier access). These recognitions of the legitimacy of multiple values and of the need to balance them to arrive at optimal rules are a strength of FDA's rules governing human subjects research.

BIBLIOGRAPHY

1. H.G. Grabowski and J.M. Vernon, *The Regulation of Pharmaceuticals*, American Enterprise Institute, Washington, DC, 1983.
2. P. Temin, *Taking Your Medicine*, Harvard University Press, Cambridge, MA, 1980.
3. B. Brody, *Ethical Issues in Drug Testing Approval and Pricing*, Oxford University Press, New York, 1995.
4. 50 Fed. Reg., 7452–7519 (February 22, 1985).
5. 52 Fed. Reg., 8798–8847 (March 19, 1987).
6. Report 105-399 (November 9, 1997).
7. R. Merrill, "Regulation of Drugs and Devices: An Evolution" *Health Affairs* 47-69 (Summer, 1984).
8. PL105-115 Section 401.
9. 21 *Code Fed. Regulations*, 56.111.
10. 21 *Code Fed. Regulations*, 50.25.
11. 45 *Code Fed. Regulations*, 46.
12. FDA, *IRB Operations and Clinical Investigation Requirements*, Food and Drug Administration, Washington, DC, 1996, App. E.
13. 21 *Code Fed. Regulations*, 312.120.
14. B. Brody, *The Ethics of Biomedical Research*, Oxford University Press, New York, 1998.
15. B. Brody, *Hastings Center Rep.* **27**(1), 7–9 (1997).
16. N. Abramson and P. Safar, *An. Emergency Med.* **19**, 781–784 (1990).
17. P. Grim et al., *J. Am. Med. Assoc.* **262**, 252–255 (1989).
18. 21 *Code Fed. Regulations*, 50.24.
19. 21 *Code Fed. Regulations*, 312.34.
20. F. Young et al., *J. Am. Med. Assoc.* **259**, 2267–2270 (1988).
21. FDA, *Treatment Investigational New Drugs Allowed to Proceed*, Food and Drug Administration, Washington, DC, 1993.
22. 21 *Code Fed. Regulations*, 312.34.
23. 21 U.S.C., §360bbb.
24. M. Couts, *Kennedy Inst. Ethics J.* **4**, 63–83 (1994).
25. FDA, *Hum. Gene Ther.* **2**, 251–256 (1991).
26. Task Force on Genetic Testing, *Promoting Safe and Effective Genetic Testing in the United States*, NIH ELSI Program, Washington, DC, 1997.
27. 21 *Code Fed. Regulations*, 314.126.
28. PL105-115 Section 115.
29. World Medical Association, *Declaration of Helsinki*, Principle II.3.
30. *NCBHR Communique*, No. 7, pp. 20–21 (1996).
31. AMA, *CEJA Rep.* 2-A-96 (1996).
32. 21 *Code Fed. Regulations*, 314.150.
33. T. Fleming and D. DeMets, *An. Internal Medi.* **125**, 605–613 (1996).
34. 21 U.S.C., §356.
35. B. Brody, in J. Kahn, A. Mastroianni, and J. Sugarman, eds., *Beyond Consent*, Oxford University Press, New York, 1998, pp. 32–46.
36. FDA, *Guideline for the Study of Drugs Likely to Be Used in the Elderly*, Food and Drug Administration, Washington, DC, 1989.
37. FDA, *Guideline for the Format and Content of the Clinical and Statistical Sections of New Drug Applications*, Food and Drug Administration, Washington, DC, 1988.

38. FDA, *Fed. Reg.* (July 22, 1993), 39406–39416.

39. FDA, *Fed. Reg.* (September 24, 1997), 49946–49954.

40. Committee on Drugs, *Pediatrics* **95**, 286–294 (1995).

41. FDA, *Fed. Reg.* (August 15, 1997), 43900.

42. FDA, *Fed. Reg.* (December 13, 1994), 64240.

43. 21 *Code Fed. Regulations*, 201.23.

See other entries FDA REGULATION OF BIOTECHNOLOGY PRODUCTS FOR HUMAN USE; GENE THERAPY, LAW AND FDA ROLE IN REGULATION; see also HUMAN SUBJECTS RESEARCH entries.

HUMAN SUBJECTS RESEARCH, LAW, HHS RULES

CHARLES R. MCCARTHY
Kennedy Institute of Ethics
Georgetown University
Washington, District of Columbia

OUTLINE

INTRODUCTION

The U.S. Department of Health and Human Services (HHS) is the single largest supporter of biomedical research in the world. HHS conducts or supports more biomedical and behavioral research involving human subjects than all other federal agencies combined (1). It is not surprising, therefore, that HHS is charged by law to issue regulations for the protection of human research subjects. HHS has played and continues to play the leading role among U.S. departments and agencies in promulgating and implementing policies and protections for the rights and the welfare of human research subjects.

This article will present a brief history of the origins, development and astonishing growth of the HHS agencies, particularly that of the National Institutes of Health (NIH), and its laws, policies, programs, and regulations for the protection of human subjects involved in research.

ORIGINS OF FEDERAL SUPPORT FOR BIOMEDICAL RESEARCH

Marine Hospital Service

In 1798 the U.S. Congress authorized, and President John Adams signed into law, a bill establishing the federal Marine Hospital Service to care for sick and disabled seamen. No one living at that time could have foreseen that the Marine Hospital Service, a Division of the Department of the Treasury, would evolve into a complex network of federal departments and agencies that collectively support the world's largest health research enterprise.

Federal support for biomedical research did not begin for nearly a century after the health care service program for merchant seamen was initiated. The creation, in 1887, of the one-room Bacteriological Laboratory for investigation of cholera and other infectious diseases at the Marine Hospital on Staten Island, New York, initiated federal support of biomedical research dedicated to public health (2).

U.S. Public Health Service

The U.S. Congress gradually widened the responsibilities of the Marine Hospital Service, and assigned additional research responsibilities to its laboratory, designated as the Marine Hospital Service Hygienic Laboratory in 1891.

In 1902 the Marine Hospital Service was renamed the Public Health and Marine Hospital Service (MHS), and in 1912 the mission of the MHS was further broadened; the Service was reorganized and it was renamed the U.S. Public Health Service (PHS). Over the next decade, a small military unit called the Public Health Service Commissioned Corps was added to the PHS. The Corps was composed of trained medical and research personnel who dedicated themselves to the protection and promotion of the health of the U.S. population. PHS Commissioned Corps officers were placed under the command of the Surgeon General of the United States (SG). Members of the Corps were expected to accept assignment to areas of the country that were medically underserved, or assignment to areas of the country where an outbreak of disease created special demands for medical personnel. The officers of the Commissioned Corps frequently labored alongside their counterparts, civilian employees of the PHS agencies. From the earliest days of the Corps, some of its officers have been assigned to conduct or administer biomedical research programs.

The Chamberlain-Kahn Act of 1918 (40 Stat.L. 309) directed the SG to conduct a new research initiative into the causes, prevention, and cure of venereal diseases, and empowered the SG to initiate grants-in-aid to further such research. Accordingly in 1918, the SG authorized grants-in-aid-of-research to twenty five extramural (outside the

PHS) institutions. These grants were the first biomedical research awards involving human subjects made by the federal government to research institutions in the private sector. They mark the beginning of an unprecedented partnership involving academe, industry, and the general public (represented by the federal government).

Creation of Categorical Research Institutes

The Ransdell Act of 1930 (P.L. 71-251) reorganized, expanded and renamed the Hygienic Laboratory as the National Institute of Health. Over time the Ransdell Act served as a template for what eventually would become a cascade of legislation that created additional research institutes aimed at studying disease categories, or studying human organs that are subject to specific categories of disease. The National Institute of Health gradually became a federation of categorical research agencies called the National Institutes of Health (NIH).

The first of the new of categorical institutes (later called the National Institute of Mental Health) was authorized by Congress to carry out research into the causes, treatment, and prevention of mental and nervous disorders, and to conduct research into narcotics abuse (P.L. 71-357).

The Social Security Act of 1935 (P.L. 74-271) is often considered to be the high water mark of the pre–World War II Roosevelt administration because it entitles persons who qualify — for reasons of age or disability — to receive income from the federal government. However, few people recall that the Social Security Act also contained landmark provisions that authorized the PHS to advise and assist state and local health officials to offer services and conduct research to prevent the spread of disease within and across state lines, and to improve regional and local health programs. This responsibility has, for the most part, been assigned to the National Centers for Disease Control (CDC) in Atlanta, Georgia.

The Congress created the National Cancer Institute (NCI) in 1937 (P.L. 75-244). The new institute not only provided for grants-in-aid to study the causes, treatment, and prevention of the many varieties of cancer, but it provided for fellowships, personnel training, and cancer prevention and control programs within and across the several states. The legislation that established NCI served as a model for the subsequent establishment by statute of 24 additional institutes and centers that currently constitute NIH.

In 1939 PHS was transferred from the Treasury Department to the Federal Security Agency (FSA) (P.L. 76-19). In 1953 FSA was folded into the newly created Department of Health Education and Welfare (HEW) headed by a Secretary who is a senior cabinet officer. When the Department of Education was separated from HEW and established as a new Department in 1979, HEW was transformed into the Department of Health and Human Services (HHS) (3).

World War II: A Time of Change

The Second World War brought profound changes to health research in the United States. In 1941 President Franklin D. Roosevelt established a federal Committee for Medical Research (CMR) to address war-related disease and injury. The CMR carried out its responsibilities by issuing multiple contracts for targeted research. Spectacular results including, development of sulfanilimides, gamma globulin, adrenal steroids, cortisone, and a wide variety of new surgical techniques and treatment regimens, were credited to the CMR (4).

Enactment of the PHS Act with Section 301 Authority

The enthusiasm generated by successful wartime research efforts led by the CMR was channeled, following World War II, into unprecedented congressional support for biomedical research. Before World War II had come to a close, Congress responded to the success of wartime research by enacting the landmark PHS Act of 1944 (P.L. 78-410) that revised and consolidated into a single law, all of the authorities that governed the various PHS agencies. The PHS Act required major reorganization of PHS agencies. Most important, it gave the PHS unprecedented generic powers to support research "into [all of] the diseases and disabilities of man" (commonly known as "Section 301 authority"); and authorized the involvement of subjects in research conducted at PHS medical facilities. The significance of "301 authority" lay in the fact that the PHS agencies, especially NIH, were authorized to exploit promising research opportunities in basic research and in a wide variety of medical specialty fields without waiting for Congress to authorize such research. Thus, although the institutes of NIH were created with categorical missions, they were able to devote a large portion of their growing research budgets to basic research. The result was a spectacular expansion of the biological knowledge base. Similarly the institutes of the NIH, using "301 authority," were able to follow promising research leads without waiting for Congress to authorize and fund their efforts.

The NIH Clinical Center was erected in 1953 under Section 301 authority. Today Section 301 authority is seldom cited because Congress has created so many research programs targeted at specific diseases and disabilities that there is less need to invoke generic research authority. Furthermore Congress now recognizes that giving direction to specific research programs into diseases such as AIDS, heart disease, cancer, and diabetes is politically more attractive to the tax-paying public than giving nonspecific authority to support expansion of the biomedical research knowledge base. Much basic biological research is still supported by NIH, but it is called "cancer research" or "heart research" or "diabetes research," even though it focuses on basic biological structures and functions that may find application in many disease categories. Since the enactment of the PHS Act, virtually all federal legislated research initiatives have taken the form of amendments to the PHS Act that authorize or underwrite both intramural and extramural research programs.

Absence of Research Ethics Policies in Pre-war Years

The research agencies and programs of the U.S. government were created and repeatedly reshaped by the Congress during the first half of the twentieth century. It is interesting to note, however, that during that same period of time, the Congress wrote no laws, held no hearings, and created no policies for the protection of human subjects involved in research.

By failing to provide either substantive or procedural ethical rules for research funded by PHS, the Congress tacitly implied that responsibility for the ethical conduct of research should be left almost entirely to awardee institutions and to the conscience of each investigator who conducted research supported by the federal government (5).

The absence of ethics policy governing research involving human subjects conducted or supported by the federal government can be explained by several factors. In the first place, the Hippocratic tradition of medicine was the centerpiece of the prevailing ethics of medicine and research conveyed to physicians in training, including those who would function as clinical research investigators. That tradition was imparted by mentors, role models, and institutional traditions. Bioethics was not recognized as a distinct discipline that lent itself to a methodological approach and systematic teaching. Biomedical ethics, including research ethics, was not taught, nor was it even available in the library collections of most of the medical schools of that period.

Many medical ethical problems were addressed by moral theologians teaching in educational institutions established and operated by religious denominations. These theologians systematically addressed fundamental questions of medical ethics, but because theological deliberations were grounded in religious faith, traditions and practice — as well as human reason — neither the Congress nor the Executive Branch demonstrated a strong interest in incorporating the opinions and conclusions of moral theologians into public policy. Furthermore, although a few theologians enjoyed broad knowledge of medical practice, few had credentials in research. Most theologians lacked credibility with the research community (6).

Second, the distinction between medical care and biomedical research was seldom made — by the Congress, by the public, by subjects of research, or even by research investigators themselves. Consequently the ethics of medical research (dedicated to systematic development of generalizable medical knowledge) was not clearly distinguished from the ethics of medical practice (dedicated to the best interests of each individual patient). Research subjects often assumed that research investigators were serving their best interests, whereas in fact the primary concern of research investigators was to gain generalizable knowledge for the sake of the public health. (This error is sometimes referred to as the "therapeutic mistake.") In some cases the best interests of subjects coincided with the development of new knowledge in research, but in most cases the interests of the subjects were subjugated to the development of general knowledge. Today the distinction between medical practice and medical research is widely recognized. Prior to World War II and for many years thereafter, it was seldom cited or acknowledged, even by the strongest proponents of research.

Third, prior to World War II, research subjects had often been patients of physicians whose medical practice evolved beyond standard care of patients into innovative therapy, and finally into carefully constructed research projects. Neither medical practitioners who evolved into research investigators, nor their patients who evolved into research subjects, seemed to recognize the fundamental change in their relationship. Rothman notes that research subjects of this period had a high level of trust in research investigators because they usually thought of the investigators as their private physicians. Public confidence in the ethical integrity of physicians who provided care to patients was at an all time high. That confidence was easily extended to physicians who crossed the threshold from practice to research (5).

As the funding of research by the federal government expanded, research became a full-time career for many investigators. Because their research cohorts were made up mostly of subjects referred to them by physicians, it gradually became less common for research investigators to have prior trust relationships with their research subjects.

Postwar Development

After the close of World War II in 1945, a series of laws created the National Institute of Mental Health (P.L. 79-487), the National Heart Institute (P.L. 80-655), and the National Dental Institute (P.L. 80-755). The Hill-Burton Act of 1946 (P.L. 79-725) authorized grants to states for the construction of hospitals and public health centers, each of which included a research component. The Omnibus Medical Research Act of 1950 (P.L. 81-692) created the National Institute of Neurological Diseases and Blindness and the National Institute of Allergy and Infectious Diseases.

As research funding components of NIH multiplied following World War II, the biomedical research budgets of PHS agencies, particularly that of NIH, experienced meteoric rises that reflected unprecedented enthusiasm for the support of research by the tax-paying public. Between 1946 and 1949 NIH budgets leaped from $180 thousand to more than $800 million (7).

In 1953 two events of great importance to medical research in general and to research involving human subjects in particular occurred. First, PHS, including SG and PHS Commissioned Corps, was transferred to the new HEW. Second, the NIH Clinical Center, a state-of-the-art research hospital, opened its doors. The Korean conflict was in progress at the time, and young American men (including physicians and scientists) were subject to military draft. Physicians who qualified as research fellows at the CC at that time were given credit for military service. Consequently keen competition for NIH Fellowships developed. Fellows from the NIH CC program subsequently assumed leadership roles in the research programs of medical schools and in private industry. As a consequence of NIH's "doctor draft," biomedical research in America not only expanded, but the quality of such research improved dramatically.

It is not surprising, therefore, that HHS (and its predecessor, HEW) played and continue to play a leading role in the United States in providing protections for the rights and the welfare of human research subjects. That leadership was scarcely discernable after World War II, but since 1966 it has gradually expanded and has become increasingly prominent.

The HHS includes many agencies. The largest—measured in terms of disbursement of funds—are the Social Security Administration (SSA) and the Health Care Financing Administration (HCFA) that administers the Medicare program. Biomedical research supported by HHS is, with a few exceptions, conducted or supported by eight agencies that comprise PHS (9). (The eight agencies today include the Agency for Health Care Policy and Research, the Health Resources and Services Administration, NIH, the Indian Health Service, CDC, FDA the Substance Abuse and Mental Health Services Administration, and the Agency for Toxic Substances and Disease Registry.) The PHS agencies were, in the past, coordinated by the PHS administration, operating under the direction of the Assistant Secretary for Health (ASH) and SG (sometimes these positions have been held by a single person who was at both ASH and SG, at other times they were held by different persons). SG and ASH reported to the Secretary of HHS.

Beginning in 1998, the administrative Offices of ASH and SG have been absorbed into the offices of the Secretary of HHS. PHS agencies are expected to continue to report to the Secretary of HHS through ASH and SG, but an intermediate level of administration will largely disappear.

NIH is by far the largest biomedical research agency within HHS. NIH is currently composed of 25 institutes and centers. The fiscal year (FY) 1999 budget of NIH is $15.56 billion dollars (8).

In FY 1998 the CRISP data collection program showed that 27,782 research projects involving human research subjects were conducted or supported by HHS. The vast majority of these projects are supported by NIH. Because a study involving human subjects may involve as few as one or two subjects, and as many as 18,000 or more (e.g., the ongoing study on prevention of prostate cancer), it is nearly impossible to estimate how many subjects are involved in research conducted or supported by HHS.

First U.S. Policy for the Protection of Human Subjects

The most visible effect of the newfound public endorsement of research can be seen in completion of the NIH state-of-the-art Clinical Center (CC) in 1953. The CC is a 500-bed hospital dedicated exclusively to the conduct of research. Only subjects actively participating in research studies are eligible for admission to the Center. Even in the setting of an institution totally dedicated to research, it was tacitly assumed that research procedures involving Clinical Center "patients" should be conducted in accord with the traditional doctor-patient relationship—a relationship of trust in which the patient had reason to believe that the doctor was uniformly acting in the best interests of the patient. To require a formal process of informing subjects of the risks and benefits involved in research and eliciting informed consent was judged to be intrusive in the doctor-patient relationship. It was left to

each research investigator to decide what information, if any, would be conveyed to "patients," whether a consent process was to be employed, and how the process would be recorded.

However, the decision not to intrude on the doctor-patient relationship that was mistakenly thought to be the same as the physician-patient relationship could not be applied to "normal volunteers." Normal volunteers, recruited from colleges in nearby states, served to provide "control data" that was compared with data derived from "patients" in clinical studies. Since normal volunteers were expected to take medical risks and to receive no direct benefit from their participation in research, they could not be identified as "patients." A special policy was developed by the CC for normal volunteers (9). The policy stated: (1) that informed consent would be obtained from normal volunteers prior to their participation in any research project, and (2) that a research project involving normal volunteers could go forward only if it had been approved by a committee of scientists. Once a committee of scientists had approved a research protocol involving normal volunteers, and the normal volunteers had consented to participate, responsibility for the normal volunteers' safety and health was assigned to the principal investigator conducting the research. Dr. Philippe V. Cardon, M.D. was assigned to supervise the program for normal volunteers. The program would, in subsequent years, provide the first template for the PHS policy that was to come.

A Policy Vacuum

During the 1950s the United States was simultaneously recovering from the human and materiel expenditures of World War II and focusing its attention on the increasingly complex nature of the cold war. The United States was also enmeshed in the Korean conflict. American citizens were dissatisfied with the ambiguous stalemate that ended military conflict in Korea but produced no lasting peace. The U.S. economy was expanding, and America turned its attention away from military conflict to fighting battles against disease. Little attention was given to the ethics of research. Within NIH attention focused more on the development of a fair and just peer review process (employed to decide which research projects were worthy of public funding) than on the rights and welfare of human research subjects.

In 1959 Senator Estes Kefauver (D.TN) began a series of hearings focused on the impact of the rising cost of prescription drugs on the health care costs of patients. In the course of the Kefauver hearings, the media revealed that hundreds of infants born to mothers who had taken the drug thalidomide were grotesquely deformed. By this time many American homes had television sets that brought graphic pictures of the thalidomide tragedy into their living rooms. Most of the babies affected by thalidomide were born in the United Kingdom, in Europe, or in Canada. Tragedy had been averted in the United States only because one FDA employee, Dr. Frances Kelsey, had demanded more animal tests before she would agree to its being prescribed for use by pregnant women in the United States. Nevertheless, fear of a similar tragedy in this country preoccupied much of the nation.

Senator Kefauver quickly changed the focus of his hearings from the economics of drug testing to the safety of drug testing. He unveiled for the public the fact that drugs were commonly tested on patients without their knowledge. Physicians simply administered experimental pharmaceuticals without informing their patients that the drugs had not been tested for safety or efficacy. The Food and Drug Administration (FDA) had little authority to prevent the widespread practice, though it encouraged the presentation of animal data prior to licensing drugs for market use.

In 1962 Senator Kefauver, despite opposition from the pharmaceutical industry, was successful in persuading the Congress to enact the Kefauver-Harris amendments to the Food, Drug and Cosmetic Act (P.L. 87-781). The Act, as amended, required rigorous testing of drugs for safety and efficacy before they could be marketed. When the bill came to the floor of the Senate, Senators Jacob Javits and John Carrol introduced an amendment that required that subjects provide informed consent before being involved in the testing of investigational drugs.

The thalidomide episode seemed to prompt disclosure of other research problems. *The New York Times* carried a story of research involving the injection of live cancer cells into elderly, indigent, and possibly incompetent patients without their consent. These "charity" patients already were suffering from various forms of cancer, and the investigators wished to know whether their immune response systems would reject additional cancer cells. Drs. Southam and Mandel were found by the New York State Board of Regents to be "guilty of fraud or deceit and unprofessional conduct..." (10). Dr. Denton Cooley attempted transplantation of a sheep's heart into a dying human patient (11). He sought permission from no one. His act was soundly condemned as unethical by many of his colleagues in the medical community, but he did not violate any law or regulation. He defended himself by contending that he was acting in the best interests of his patient who died several days later.

Dr. James Shannon, the dynamic Director of NIH at the time, was alarmed by these incidents. He not only regarded them as violations of the rights of the research subjects, but he believed that they jeopardized the future of publicly supported medical research. Intense discussions began to take place in his office and across the NIH campus in Bethesda, MD, concerning appropriate professional ethical standards for the conduct of research. Questions of informed consent, minimizing research risks, excellence of research design, protection of subjects' privacy, and submission of research to review by other scientists were discussed. It was initially thought that NIH could not, without additional legal authority, impose standards of conduct on investigators who received support for their research from the agencies of PHS. Ethical behavior toward research subjects was considered to be a major responsibility of the research investigator.

Gradually the view emerged that federal agencies could place conditions, including ethical conditions, on the awards that they made to research institutions and their investigators. Dr. Shannon created a committee headed by Dr. Robert Livingstone to recommend a suitable set of protections for subjects involved in Public Health Service-supported research. Livingstone's report recommended making a careful assessment of "ethically responsible relationships" and an examination of the range and tenor of present professional ethical practices. The report concluded that "the NIH is not in a position to shape the educational foundations of medical ethics..." Shannon found this conclusion "wholly unsatisfactory" (12).

FORMULATION OF THE FIRST EXTRAMURAL FEDERAL POLICY

Concerns for the ethics of publicly supported research involving humans had been relatively dormant during the decade of the 1950s, at least until the thalidomide hearings raised serious questions about the conduct of research. The years from 1960 to 1972 provided a sharp contrast to the previous decade. A generation of young adults had grown to adulthood under the constant threat of nuclear destruction. Trust of science-based technology had gradually eroded. For a new generation of young Americans, suspicion of technology and fear of its misuse replaced the confidence and trust that had characterized the period following World War II. Virtually every kind of science-based technology was now viewed as a potential threat to health, well-being, or survival of the planet unless it was carefully controlled.

New federal regulatory agencies—the Environmental Protection Agency (EPA), the Occupational Safety and Health Administration (OSHA)—were created to make the environment and the workplace safer for all living things and especially for humans. Both old and new federal departments and agencies produced volumes of regulations to implement laws restricting perceived threats to the environment, public health, and public safety.

The new-found public suspicion of science and technology provided a dramatic and sharp contrast to the attitude of many in the research community who believed that "the scientific method rests [on] the integrity and independence of the research worker and his freedom from control, direction, regimentation, and outside interference" (13).

The director of NIH had few illusions about the difficulty of formulating an ethical framework that could serve as a guide for research involving human subjects. "To win general acceptance within, not only the medical research community, but also our society at large," Dr. Shannon wrote, "the final statement of principles should probably emerge from ... representatives of the whole ethical, moral, and legal interests of our society" (12, p. 152). Shannon knew that creating acceptable ethical guidance for research, and building a consensus for a statement of ethical principles, would be the work of many years. He also believed that the Congress would hold the PHS agencies, particularly NIH, responsible for failures to protect human subjects. He had the vision to see that failure to protect human research subjects was tantamount to a failure to protect the public support of biomedical research.

At Shannon's request Surgeon General William Stewart issued the first PHS extramural Policy for the Protection of Human Subjects. The policy was elegant, simple, and easy to understand. To this day, although it has been revised many times, the core of the policy remains essentially unchanged. Each awardee institution was required to file an Assurance of Compliance document with PHS stating that the institution would adhere to the following procedures.

> The awardee institution is to provide prior review of the judgment of the principal investigator or program director by a committee of his institutional associates. This review should assure an independent determination: (1) of the rights and welfare of the individual or the individuals involved [as subjects], (2) of the appropriateness of the methods used to secure informed consent, and (3) of the risks and potential medical benefits of the investigation (14).

It almost seemed that Henry K. Beecher, a prominent Harvard research investigator, wished to underline the importance of the policy when he published a shocking article in the *New England Journal of Medicine* in June 1966 (15). Beecher identified 22 research projects published in refereed research journals that, in the author's judgment, violated the rights of the research subjects involved. Because he enjoyed a prestigious position as a researcher in a leading academic institution, Beecher's article sent shock waves through much of the American research community. His article lent urgency and needed credibility to the new PHS policy.

Responsibility for implementing the PHS policy was assigned to the tiny Institutional Relations Branch of the Division of Research Grants within NIH. The process by which the policy was to be implemented required considerable clarification. It underwent minor revisions in the summer of 1966, and further revision in 1967. In 1969 the Surgeon General revised the policy to make it clear that the policy extended to behavioral and social science research as well as to biomedical research. The Institutional Relations Branch implemented the policy by negotiating Assurance of Compliance documents with each awardee institution. Few universities, clinics, and laboratories welcomed the policy, but after Beecher's article was published, most of them accepted the policy as inevitable.

Because the policy was directed toward extramural research, it did not apply directly to NIH's own intramural CC. It would be many years and many policy revisions later before the NIH intramural program came into full compliance with the policy that governed extramural institutions that received awards for research involving human subjects. Nevertheless, in 1966 Dr. Jack Masur, director of CC, appointed a committee headed by Dr. Nathaniel Berlin to update the CC policy of 1953. Masur was responding, in part, to the recently published PHS policy. Clinical Research Committees (CRCs) were created within the intramural programs of the categorical institutes (16). Consent of subjects (still referred to as "patients") was required only to the extent that the investigator was expected to make a note in each "patient's" chart that verbal consent had been obtained.

From the outset, the Institutional Relations Branch used education and negotiation as the primary tool of promoting compliance with the policy. Although the Institutional Relations Branch had authority to withhold awarded funds, for many years no sanctions were imposed on any institution for failure to comply.

In 1971 the PHS policy was revised by Dr. Donald S. Chalkley, director of the Institutional Relations Branch. The scope of the policy was expanded to cover all research supported by any agency, office, or unit within HEW (17). Consistent with the educational approach described above, the new HEW Policy for the Protection of Human Subjects (commonly described as the "Yellow Book" because of the color of the cover of the pamphlet in which it was published) not only set forth requirements that institutions were to meet (review by committee, informed consent, and evaluation of risks and benefits), but it provided a running commentary presenting reasons why these requirements were necessary.

In July 1972 the public press uncovered details of the infamous Tuskegee Syphilis Study involving 600 black males from Macon County, Alabama Started in 1932, the study systematically denied treatment for syphilis to approximately 400 men who were afflicted with the disease (the other 200 men were used as control subjects). Subjects were not informed of their diagnosis. They did not know they were involved in a research study, and even after penicillin became the drug of choice to treat syphilis, they were denied treatment.

The Tuskegee Study received national publicity because of hearings held in the Senate Health Subcommittee chaired by newly elected Senator Edward Kennedy (D. MA). The study was conducted by a series of PHS investigators over a period of more than 35 years. Assistant Secretary for Health, Dr. Monty DuVal convened a committee chaired by Dr. Jay Katz, M.D., J.D., of Yale University to review the study. At the recommendation of Dr. Katz, the study was closed. HEW subsequently paid millions of dollars to survivors and to families of those who did not survive to compensate them for harms caused by the study. In FY 1995 HHS paid $2.8 million and in FY 1996 HHS paid $1.88 million in compensation to survivors and heirs of participants in the Tuskegee Syphilis Study (18). On May 26, 1997, President Clinton, on behalf of the U.S. nation, apologized to survivors of the study and to their families. (Despite the fact that HEW/HHS has paid compensation to subjects in the Tuskegee Study, it has never adopted a policy of providing compensation to injured subjects in other situations.)

The Tuskegee Study was given wide publicity in the media, and it triggered public disclosure of a number of other alleged research abuses relating to psychosurgery, fetal experimentation, and illicit experimentation involving contraceptives. All of these matters were aired in the Senate Health Subcommittee hearings held periodically between 1972 and 1974.

In the meantime Dr. Robert Q. Marston, who succeeded Dr. Shannon as director of NIH, made a decision to promote protections for the rights and welfare of human research subjects. In an address to the College of Nursing at the University of Virginia, he declared an obligation on the part of society to carry out research aimed at improving the health of vulnerable populations and to protect the rights of vulnerable research subjects in the process (22).

Dr. Marston then created a PHS-wide drafting committee chaired by Dr. Ronald Lamont-Havers to revise and upgrade the HEW policy of 1971. The mandate to the PHS committee was based on the Marston talk at the University of Virginia. In 1972 Dr. Marston also upgraded the Institutional Relations Branch to a position within the Office of the Director, NIH. Marston changed the name of the office to the Office for Protection from Research Risks (OPRR). Dr. Chalkley, who had directed the Institutional Relations Branch, was named the first director of OPRR.

Largely as a result of the Kennedy hearings and the Tuskegee scandal, a series of legislative initiatives pertaining to research involving human subjects were introduced in Congress in 1973–74. These included a bill sponsored by Senator Kennedy to create a regulatory commission (similar to the Securities and Exchange Commission) to oversee research. Senator Walter Mondale (D. MN) introduced legislation to create a National Advisory Commission to study the impact of advances in science and technology on American society. Congressman Paul Rogers (D.FL) introduced a bill that called for a "National Advisory Commission for the Protection of Human Subjects of Biomedical and Behavioral Research" (20). Congressman Angelo D. Roncallo (D. NY) introduced legislation to ban federal support for human fetal research. Roncallo claimed, erroneously, that NIH had been conducting gruesome research involving live aborted fetuses ex utero (12). Senator Jacob Javits (R. NY) urged passage of legislation extending and strengthening the requirements for informed consent from research subjects. Congressman Rogers and Senator Kennedy cobbled together a bill that included some features of all of these legislative initiatives. Senator Kennedy indicated that he would support the compromise bill if HEW would publish regulations for the protection of human subjects. HEW then hurriedly transformed the HEW policy contained in the "Yellow Book" into regulatory form and published it as Regulations for the Protection of Human Subjects (21). The Congress then passed the pending compromise bill as Title II of the National Research Act (P.L. 93-348) signed July 12, 1974. The Act required: (1) that HEW promulgate regulations requiring institutions to ensure compliance with the regulations in a manner acceptable to the Secretary HEW (this requirement was completed prior to enactment of the law), (2) creation of a National Commission for the Protection of Human Subjects of Biomedical and Behavioral Research to study a wide range of health research issues, (3) a temporary moratorium on fetal research to remain in place until such time as the National Commission reviewed the matter of fetal research and made recommendations concerning continuation of the moratorium, and (4) a special study of the impact of science and technology. To prevent the recommendations of the new Commission from being ignored, the Act contained a "forcing clause," that is, the Secretary of HEW was required by law to accept recommendations of the National

Commission, or publish in the Federal Register reasons for not accepting them. The law also recommended that the commission be succeeded by a HEW Secretary's Ethics Advisory Board to address the question of federal funding of human fetal research and other matters that the Secretary might assign to it.

Between 1974 and 1978 hearings and deliberations of the National Commission, chaired by Dr. Kenneth Ryan of Harvard University, captured the attention of the research community and the media. The National Commission endorsed the approach taken by HEW in 45 CFR 46, and it incorporated much of the work of the PHS committee headed by Dr. Lamont-Havers. The commission recommended adding additional protections for pregnant women and human fetuses (1975), prisoner subjects of research (1977), children who are subjects of research (1978), and persons who are institutionalized because they are suffering mental infirmity. The National Commission deliberated for four years, and it published 9 reports. Perhaps its best known report is the *Belmont Report: Ethical Principles and Guidelines for the Protection of Human Subjects of Research* (22).

After the commission completed its work, the Secretary directed the Public Health Service to accept all of its recommendations and to implement them.

HEW Secretary's Ethics Advisory Board

In 1978 HEW Secretary Joseph Califano established the Secretary's Ethics Advisory Board (EAB) to review controversial biomedical ethical issues and to issue reports and recommendations concerning them. James Gaither, an attorney from San Francisco who had worked with Califano in the Johnson administration, was designated as chairperson. The first issue that the EAB addressed was whether HEW should fund human in vitro fertilization (IVF) research. Shortly after the EAB began its work, Louise Brown, the first "test tube" baby, was born in England as a result of the experimental work of Drs. Steptoe and Edwards. Enormous publicity followed the birth of Louise Brown, and considerable ethics debate both in England and the United States followed the publicity. In the United States several research investigators submitted applications for federal support for human IVF research. Suddenly the deliberations of EAB were followed by reporters from all the major media and from the tabloid press as well. In May 1974 EAB issued a Report and Conclusions: HEW Support of Research Involving Human In vitro Fertilization and Embryo Transfer. While the board stopped short of recommending to the Secretary that such research should be funded, it unanimously declared that such research, if it meets certain conditions, is acceptable from an ethical standpoint. On the basis of the report, Secretary Califano was preparing to authorize investigators seeking support for human IVF research to compete for awards when he became engaged in a controversy with Hamilton Jordan, assistant to President Carter. The President fired Califano and replaced him with Patricia Harris. Secretary Harris had little interest in the ethical questions pertaining to research and never took action on the recommendations of EAB. EAB issued reports on (1) fetoscopy (recommending that the department make funds available for the procedure), (2) nosocomial (hospital-induced) infections (recommending that the public be allowed to see data maintained by CDC showing the rate of hospital-induced infections), and (3) clinical trial Data (recommending that investigators be allowed to withhold release of preliminary clinical trial data under the provisions of the Freedom of Information Act).

In 1980 Secretary Harris failed to re-charter EAB, and its remaining budget was transferred to the newly created President's Commission for the Study of Medicine and Biomedical and Behavioral Research.

REVISION OF HHS RULES FOR THE PROTECTION OF HUMAN SUBJECTS

During the final two years of the National Commission for the Protection of Human Subjects of Biomedical and Behavioral Research and for several years thereafter (1976–1980), a PHS committee chaired by the director of OPRR scrutinized hundreds of public comments on the reports of the National Commission and, in the light of those reports and the public comment on them, revised and expanded the regulations for the Protection of Human Subjects (45 CFR 46) last issued in 1974.

The work was completed in the final months of 1980. The proposed regulations included (1) Subpart A: Generic Regulations Providing Protections for all Human Subjects participating in Research; (2) Subpart B: Additional Protections for Research Involving Pregnant Women and Human Fetuses; (3) Subpart C: Protections for Prisoners involved in Research; and (4) Subpart D. Protections for Children involved in Research. HHS Secretary Harris signed the regulations on January 19, 1981, and they were published in the Federal Register six days later.

In addition to the recommendations made by the National Commission, the PHS drafting committee created some categories of research that were exempt from the regulations, and other categories that would allow for expedited review by the chairperson of the institutional review board or a person designated by the chair. The exempt and expedited categories were created precisely to reduce the workload of IRBs that were already heavily burdened. They were welcomed by the research community that felt that a reasonable trade-off had occurred — greater protections for high-risk research and fewer safeguards for research involving negligible or minimal risks.

In 1978 Congress enacted legislation that created a new ethics Commission called the President's Commission for the Study of Ethical Problems in Medicine and Biomedical and Behavioral Research (23). Although the bulk of the work of the President's Commission was dedicated to ethical concerns in the delivery of health care, the commission issued two reports that dealt directly with the system of protections for human research subjects. It also issued a report on research involving genetic engineering (24).

From the point of view of federal policies for the protection of human subjects, the most important contribution of the President's Commission was the following set of recommendations: (1) All federal departments and agencies should adopt regulations of HHS, (2) the Secretary of HHS should establish an office to coordinate and monitor governmentwide implementation of the regulations, and (3) each federal agency should apply one set of rules for the protection of human subjects consistently to all research conducted or supported by the federal government (25).

The Secretary of HHS, through ASH, designated OPRR as the "lead" office to develop a common set of regulations across the government. However, OPRR was dealing with reduced budgets and severe restrictions on its outreach due to downsizing. Requests for personnel and budgets to carry out the Secretary's orders were denied. OPRR approached each agency in the federal system with a request for compliance with the recommendations of the President's Commission. Most of the agencies replied that they too were facing downsizing and could undertake no new initiatives. Nevertheless, OPRR was able to obtain some backing from the Office of Management and Budget on the ground that what was proposed was a simplification of a complex regulatory structure. The initial response from the agencies was disheartening. Each was willing to adopt the HHS rules — but with conditions attached. Each agency had different conditions so that, if all had been accepted, there would have been little left of the HHS Regulations for the Protection of Human Subjects that were used as a model for the proposed new governmentwide regulations. The tortuous process of winning agreement from all of the departments and agencies continued for more than seven years. Finally, the OPRR, after rebutting objections from the President's legal advisor, was able to gain support from the Office of the President's Science Advisor and from OMB. On June 18, 1991, final clearance was obtained. Sixteen departments and agencies, in addition to the HHS, simultaneously published the "Common Rule."

APPLICATION OF THE REGULATIONS TO WOMEN AND MINORITIES

As noted above, the Common Rule was based, in many ways, on the recommendations of the National Commission for the Protection of Human Subjects of Biomedical and Behavioral Research. That commission tended to regard biomedical research as risky and dangerous. It feared that persons who were sociologically disadvantaged would be exploited as research subjects. It was concerned that the burdens of research would fall on the poor, the uneducated, minority groups, and women. All of these were regarded as "vulnerable" populations. However, in the decade of the 1980s, a long period passed when there were few reports of research injuries or deaths. Furthermore, with the coming of the AIDS epidemic, most persons infected with HIV found that the best care, and the most advanced "treatment" for AIDS, could be obtained by participation in clinical trials. Consequently the protections that had been employed to prevent exploitation of the disadvantaged were now seen as discriminatory because they offered protections that served to discourage overinvolvement of these populations in research.

Furthermore, in the late 1970s and early 1980s the feminist movement identified many forms of discrimination against women. Women's rights became a major national political issue. Along with other concerns for women's rights came the awareness that many drugs administered to women had not been tested in women. Physicians, fearing malpractice suits, began to warn women that the drugs they were prescribing had not been tested in women and that women must be willing to take them at their own risk. Dr. Bernadine Healy, who became director of NIH in 1987, made repeated demands on the Congress for increased funding of research involving women. Dr. Healy created the Office of Women's Health Research at NIH and dedicated discretionary research funds to projects related to women's health (26).

Although AIDs research was gradually modified to include women, complaints that women, especially pregnant women whose offspring might be infected with HIV, were excluded from highly desirable research were given wide publicity. Clinical studies that a decade before had been considered to be a heavy burden for pregnant women were now considered to be a prized benefit (20).

CURRENT PROBLEMS

The history of federal protections for human subjects continues to develop. In 1994 President Clinton created the Advisory Committee on Human Radiation Experiments (ACHRE) that dealt with some 4000 studies sponsored by the federal government that exposed human subjects to radiation. Most of these studies were conducted prior to the existence of any federal policies or regulations. Although much of the information concerning these studies is incomplete and fragmentary, ACHRE identified some cases of clear abuse and recommended compensation for those who had suffered injury as a result of such studies. ACHRE further identified serious deficiencies in the current system for protecting the rights and the well-being of human subjects (27).

Partly in response to ACHRE's report, President Clinton issued an Executive Order establishing the National Bioethics Advisory Commission (NBAC) to address questions dealing with the adequacy of federal regulations, particularly regulations for the protection of the cognitively impaired. This commission has attempted to address the implementation of the Common Rule and has raised the question whether the location of OPRR within the NIH constitutes an apparent conflict of interest. Since NIH funds research, OPRR can easily be caught in a position of opposing its own agency if it decides that NIH-funded research is not being conducted in compliance with federal regulations. Partly as a consequence of the possible conflict of interest, a decision has been made to transfer OPRR to the Office of the Secretary of HHS.

NBAC also has issued reports on the ethics of research involving human cloning, and research involving persons who are cognitively impaired. The list of issues pertaining to human subjects that could be explored or revisited by NBAC is almost endless (28).

Research involving human subjects funded by the U.S. government is expanding each year. The United States is already the largest single source of such funding, and such research appears to be a growth industry. Slowly but surely such research has come under increasing public scrutiny and regulation. Some believe that much more

needs to be done. Others believe that overregulation may stifle the biological revolution that is extending human life, conquering disease, and providing better health not only to Americans but to humans in every part of the globe. The new millennium offers new and ever more complex challenges to the twin goals of advancing biomedical and behavioral science and protecting the rights and welfare of human research subjects.

BIBLIOGRAPHY

1. *NIH Almanac*, 1997, Available at: *http:www.nih/gov/welcome/alamanac97.htm*

2. Available at: *http//www.nih/gov/welcome/alamanac97/chapt1/legis.htm*

3. Department of Health and Human Services, October 1979, P.L. 96-88.

4. *Records of the Office of Scientific Research and Development*, Committee on Medical Research, Contractor Records, National Archives, Washington, DC Record Group 127.

5. D.J. Rothman, *Strangers at the Bedside*, Basic Books, New York, 1991.

6. A.R. Jonsen, *The Birth of Bioethics*, Oxford University Press, New York, 1988.

7. C.R. McCarthy, in *Encyclopedia of Bioethics*, vol. 4, Macmillan, New York, 1995, pp. 2285–2290.

8. Available at: *http://www.nih.govod/ofmbudget/totalbymechanism.htm*

9. *Group Consideration of Clinical Research Procedures Deviating from Accepted Medical Practice and Involving Unusual Hazard*, a policy statement issued by the NIH Clinical Center, 1953.

10. E. Langer, *Science* **143**, 551–553 (1964).

11. D.A. Cooley et al., *Am. J. Cardiol.* **22**, 804–810 (1968).

12. M. Frankel, *Public Policy Making for Biomedical Research: The Case of Human Experimentation, Dissertation*, George Washington University, Kansas City, MO, 1976.

13. C.J. Van Slyke, *Science* **104**, 569 (1946).

14. *PHS Policy and Procedure Order 129*, Feb. 8, 1966.

15. H.K. Beecher, *N. Engl. J. Med.* **274**, 1360–1364 (1966).

16. *NIH Clinical Center Policy*, issued in July 1966, Updated 1976, 1977.

17. *Institutional Guide to the Department of Health Education and Welfare Policy on Protection of Human Subjects*, Publication No. 72–102 HEW, Washington, DC, 1971.

18. Personal communication from the office for Protection from Research Risks to the author.

19. R.Q. Marston, *Medical Science, the Clinical Trial, and Society*, Presented at dedication of McLeod Nursing Building and Jordan Medical Education Building, University of Virginia, November 10, 1972.

20. C.R. McCarthy, in J. Kahn, A.C. Mastroianni, and J. Sugarman, eds., *Beyond Consent: Justice in Federal Research Policy*, Oxford University Press, New York, 1998, p. 20.

21. Title 45 Part 46, *Code of Federal Regulations*, May 30, 1974, *Regulations for the Protection of Human Subjects*.

22. *The Belmont Report: Ethical Principles and Guidelines for the Protection of Human Subjects of Research*, U.S. Government Printing Office, Washington, DC, 1979.

23. The President's Commission for the Study of Ethical Problems in Medicine and Biomedical and Behavioral Research, P.L. 95-622 Title III 1978, P.L. 93-377, 1982.

24. President's Comission for the Study of Ethical Problems in Medicine and Biomedical and Behavioral Research, *Splicing Life*, U.S. Government Printing Office, Washington, DC, 1983.

25. President's Commission for the Study of Ethical Problems in Medicine and Biomedical and Behavioral Research. U.S. Government Printing Office, *Implementing Human Research Regulations*, Washington, DC, 1981.

26. C.R. McCarthy, *Acad. Med.* **69**(9), 695–698 (1994).

27. Advisory Committee on Human Radiation Experimiments, *Final Report*, Oxford University Press, New York, 1996.

28. A. Capron, *Kennedy Inst. Ethics J.* **7**(1), 63–80 (1997).

See other HUMAN SUBJECTS RESEARCH entries.

HUMAN SUBJECTS RESEARCH, LAW, LAW OF INFORMED CONSENT

MARSHALL B. KAPP
Wright State University School of Medicine
Dayton, OH

OUTLINE

Introduction

Legal Regulation of Human Subjects Research

Specific Legal Issues in Human Gene Transfer Research

Conflation of Research and Therapy

National Institutes of Health (NIH) Oversight

Acknowledgment

Bibliography

INTRODUCTION

This article discusses the legal implications of research involving somatic cell gene transfer for the treatment of human disease. The genetic process is based on the transfer of normal genetic material (deoxyribonucleic acid, or recombinant DNA) from another organism into a diseased human being. Gene delivery can be achieved by either directly administrating the gene-containing viruses or DNA to the blood or tissues (i.e., in vivo) or indirectly introducing cells manipulated in the laboratory to harbor foreign DNA (i.e., in vitro). Gene therapy seeks to treat disease in an individual person by the administration of normal DNA rather than a drug (1). This therapy would be available at all stages of human development.

Somatic cell gene transfer is distinct from medical interventions that target the manipulation or engineering of germ-line cells (i.e., eggs and sperm) (2), although the ethical distinction has been questioned by some writers (3). Additionally therapeutic gene transfer is distinct from genetic manipulations done for personal enhancement. Although both forms of cell gene transfer attempt to "improve" the human body, therapeutic manipulations aim at normalizing a person with a genetic disease or abnormality, while genetic enhancement tries to optimize some designated area of a normal person's body or performance (e.g., athletic prowess) (2).

Recently there has developed much enthusiasm in the scientific and public communities about the therapeutic potential of genetic manipulation at the individual patient level (4). It has followed the successful efforts of biologists who in 1998 cultivated human embryonic stem cells. These embryonic stem cells are the primordial human cells from which an entire person may be created, via fertilized human eggs (ova) that are implanted in the uterus (5). As philosopher Daniel Callahan has stated it:

> [N]o excitement [in the biomedical arena] has quite matched that which genetic research has engendered. The claim in its behalf is sweeping and radical: genetic research and its clinical application promise to finally bring medicine to the root causes of disease.
> Once these casual, molecular mechanisms are understood, clinical medicine and medical technology will be in a superb position to eliminate many, if not most, of the deadliest diseases (6, p. 70).

According to a 1995 federal report:

> Somatic gene therapy is a logical and natural progression in the application of fundamental biomedical science to medicine and offers extraordinary potential, in the long-term, for the management and correction of human disease, including inherited and acquired disorders, cancer, and AIDS. The concept that gene transfer might be used to treat disease is founded on the remarkable advances of the past two decades in recombinant DNA technology (1, p. 1).

The positive potential of somatic cell gene transfer was recognized in a 1982 report of the President's Commission for the Study of Ethical Problems in Medicine and Biomedical and Behavioral Research (7) and in a 1984 report of the Office of Technology Assessment (OTA) (8).

LEGAL REGULATION OF HUMAN SUBJECTS RESEARCH

Research involving somatic cell gene transfer in human beings is governed by the same federal laws that apply to human subjects research generally, namely regulations codified at 45 Code of Federal Regulations Part 46. These 1981 regulations define research as "a systematic investigation, including research development, testing and evaluation, designed to develop or contribute to generalizable knowledge," 45 Code of Federal Regulations §46.102 (d). The requirements established by these regulations apply explicitly to all human subjects research that is federally funded, although most research-sponsoring institutions subject all of their human subjects protocols to these requirements. Compliance with federal regulations is overseen by the Office of Protection from Research Risks (OPRR), which Congress in 1999 placed in the Office of the Secretary, U.S. Department of Health and Human Services.

Under federal law, research protocols utilizing humans as research subjects must be reviewed and approved by a local, interdisciplinary Institutional Review Board (IRB) whose composition includes at least one public member (9). The IRB is supposed to evaluate the ethical implications of several aspects of a research protocol, including the following items, in terms of human subjects protection:

- Fairness or equity of subject selection
- Minimization of risks to subjects (including nonphysical risks, e.g., inconvenience and financial or loss of privacy costs), and whether any foreseeable risks are justified by reasonably expected direct, indirect, or social benefits
- Sufficiency of the informed consent process and its documentation
- Sufficiency of confidentiality safeguards

Questions have arisen regarding the propriety and degree of IRB involvement in evaluating the scientific merit of research proposals. The most cogent position is that usually the scientific merits of a research proposal cannot be disentangled from ethical considerations of subject protection. As a 1998 report (10, p. 32) asserted, "Any research involving humans entails some risk. Even for research in which the risk to subjects is minimal, the risk should not be taken unless the research has scientific merit."

In addition to federal requirements, a number of individual states have enacted their own statutes or promulgated regulations that apply to human subjects research generally. State requirements may be more, but not less, stringent in terms of human subjects protection than those created by federal law, and they apply with full force to research on somatic cell transfer.

One Canadian working group has suggested a detailed schema for IRB review of gene therapy/gene transfer protocols. This schema proposes that the IRB delve in depth into the following aspects of research protocols in this sphere: background and justification; research design; procedures; confidentiality; subject selection; risks, discomforts, and benefits; and information to subjects (11).

Another potential source of regulation is the civil justice system under which an individual human subject may sue investigators and/or the institutions and sponsors for injuries resulting from a negligently conducted experiment (12). The extent of possible liability exposure under traditional or novel tort theories remains uncharted at this time, but certainly investigators and their institutions will need to develop and implement appropriate risk management strategies to minimize that exposure (13).

SPECIFIC LEGAL ISSUES IN HUMAN GENE TRANSFER RESEARCH

Conflation of Research and Therapy

Background on the Confusion. The tremendous potential of somatic gene transfer technology to correct or manage many serious human diseases and abnormalities has been vastly oversold thus far, to both the scientific and general public communities. This excessively exuberant publicity has created, or at least fostered, a widespread confusion among members of the public who comprise the pool of potential human subjects for gene transfer research protocols regarding the important distinction between medical interventions that are properly categorized as therapy, on one hand, versus medical interventions more correctly considered research studies, on the other (14,15).

The terminology frequently used by many of those who participate in or report on this activity illustrates, and contributes to, the conflation of therapy and research in the public's mind. When terms such as "gene therapy research," "patient," "immunotherapy," "vaccine," "drug," and "cure" are repeatedly employed in this context, the public impression is established and strengthened that this form of medical intervention must be undertaken primarily for the benefit of the individual person receiving the recombinant DNA from another organism. As one set of authors has observed, "The call of activists, patients, researchers, and regulators for greater access to research protocols emerges as a symptom of the overselling of medical research as therapy." (14, p. 43) Although the commonly used terminology represents the long-range aspirations of researchers, it neglects the reality that interventions currently being studied in clinical trials are a significant way off from achieving the status of accepted therapies.

Properly understood, the practice of accepted therapy consists of professionally agreed-upon interventions that are expected to provide direct, personal benefit to an individual patient. By contrast, research has new, generalizable knowledge as its primary aim; any direct or indirect benefit enjoyed by specific human subjects, while certainly welcome, is quite secondary and incidental (16). Recognizing that overselling the current capabilities of gene cell transfer misleads potential human subjects into expecting personal benefit as a direct consequence of their participation, an NIH panel has advised more restraint by investigators and their sponsors in presenting information about this activity to the public (1).

"Informed" Element of Informed Consent. Federal regulations and common law doctrine (17) mandate that individuals may participate as human subjects in biomedical and behavioral research protocols only if and when they have given consent to such participation that is voluntary, competent, and informed. Widespread public failure to accurately distinguish between research and therapy in the gene cell transfer context is manifested by a tendency on the part of many people to already see this intervention as a benefit to which there is a right rather than as a hazard from which vulnerable people might need external protection (14). This popular attitude presents investigators and IRBs with a major challenge for ensuring that the process of informed consent operates in this sphere.

The 45 Code of Federal Regulations §46.116 (a) (1) requires that in seeking informed consent for research participation, each potential subject must be provided with "a statement that the study involves research, [and] an explanation of the purposes of the research ...." Fulfilling this requirement obligates the investigator to make clear to each potential subject that the protocol is intended to gather generalizable information that may help others in the future who have the same disease or abnormality as the subject, but that any immediate, personal benefit to that particular person would be unexpected and merely incidental to the research endeavor. In the written consent document that the possible participant is asked to sign, the educational effort about the research versus therapy distinction ought to be promoted by replacing misleading words like "treatment," "patient," and "therapy" with more accurate terms such as "potential subjects," "potential participants," "research," and "experiment." (15, p. 51).

Many human subjects in gene therapy research might benefit from long-term follow-up by the investigators. However, there often will be practical difficulties in the capacity of investigators and their institutions to conduct such follow-up over a sustained period of time. These limitations should be explicitly described to potential subjects as part of the informed consent process (18).

"Desperate Use" Exception. The widespread confusion regarding the research versus therapy status of gene cell transfer today is illustrated by the "desperate use" exception that has been officially recognized as a circumstance where gene cell transfer investigators are legally authorized to conduct transfer interventions on a person without first complying with the ordinary informed consent requirements. The "desperate use" exception for gene transfer research is modeled after earlier-created "emergency use" exceptions authorizing the conduct of limited types of research in the absence of timely informed consent from the participant (19,20). Although the "desperate use" exception has been defended on grounds of compassion for afflicted individuals who have no other available alternatives (an argument that assumes the likelihood of some direct benefit for that specific patient/subject), some commentators question the wisdom of this exception in light of the present dearth of scientific knowledge about the actual benefits of gene transfer (15, pp. 49–50). There also is concern that, despite the lack of any scientific basis, the "desperate use" exception was recognized nonetheless as a result of pressure brought on Congress by patients, and over the strong scientific objections of the NIH's Recombinant DNA Advisory Committee (RAC) (15, pp. 49–50).

Voluntariness When the Subject Is Also a Patient. Voluntariness is an essential element of informed consent for research participation. The 42 Code of Federal Regulations §46.116 provides, "An investigator shall seek such consent only under circumstances that provide the prospective subject or the representative [of a decisionally incapacitated subject] sufficient opportunity to consider whether or not to participate and that minimize the possibility of coercion or undue influence."

Voluntariness is jeopardized in the gene transfer research context when, as is ordinarily and logically the case, potential human subjects are recruited from the ranks of patients who are currently under a physician's care for the precise disease or abnormality that is the focus of the research protocol. Many lay persons (as well as

clinicians) have difficulty making the distinction between the patient seeking direct, immediate therapeutic effect from a medical intervention and the research subject altruistically contributing to generalizable knowledge of practical application only to others in the future. This confusion also impairs the "informed" element of informed consent. As a result patients may imagine, or accurately perceive, subtle or more blunt pressure to participate in research protocols as a way to continue access to whatever therapeutic treatment they are currently receiving, even though federal law expressly forbids investigators to conditionally link together research participation and the continued availability of standard treatment in that fashion.

As one commentator has noted, "Most of the persons enrolled in gene transfer research to date are a special class of research subjects—namely, sick persons who present special vulnerabilities and require special protections not relevant to healthy, 'normal' subjects." (15, p. 52) Clinical investigators and IRBs are legally and ethically challenged to establish an environment in which patients' perceived compulsion to "volunteer" for gene transfer research protocols solely in order to preserve their present patient status is eliminated or reduced as much as possible.

Conflicts of Interest. Investigators in gene transfer research involving human subjects may have actual or apparent conflicts of interest of two different sorts (21). First, since many potential human subjects for these research protocols are current patients receiving care for the exact disease or abnormality that the protocol intends to study, the clinical investigator may also occupy the professional role of attending physician currently caring therapeutically for the person (patient) that he or she is recruiting to now also become a research subject. This dual investigator/clinician position can easily contribute to the conflation of research and therapy in the minds of many patients/potential human subjects as just discussed, raising implications for both the voluntary and informed elements of informed consent to research participation. At the very least, the informed consent process—as enforced by IRBs—ought to require that potential human subjects be told in understandable lay terms about the dual investigator/clinician role, the distinction between the goals and expectations of research versus the goals and expectations of therapy, and the various kinds of professional and financial incentives that the treating clinician has to enroll his or her patients into research protocols for which that clinician also serves as investigator.

A second source of possible conflict of interest arises when the investigator in a gene transfer research protocol is also an investor in, or otherwise holds some financial stake (e.g., as a paid consultant), in a corporation that stands to profit financially from positive results of the particular study. A number of ethical commentators have argued that, as a matter of informed consent, financial interests of the investigator in the conduct or outcome of a study should be fully disclosed to potential human subjects at the time that their participation in that study is solicited (22–24).

Proxy Consent When the Subject Is a Child. As noted by one geneticist:

> Often the individual for whom DNA treatment [sic] is proposed is a very young child and thus unable to give consent except via proxy. In-depth counseling of the proxy — often the parents — should be provided to ensure full review of the state of knowledge concerning the risks of changing the DNA: what the known risks are, as well as what is unknown. The relative risks and benefits of alternative treatments and options should be fully discussed (25, p. 570).

The challenge of using minors as research subjects is exacerbated by the fact that many of them will achieve adulthood during the follow-up phase of the research.

> Thus, it may be necessary to obtain assent at the time of enrollment as well as consent at adulthood. This must be done in a way that both protects the rights of the patient [sic] to withdraw from research at any time and yet encourages follow-up, which may provide benefits to both the individual and society (26, p. 661).

National Institutes of Health (NIH) Oversight

In addition to the generic federal and state regulations applicable to human subjects research and common law principles of informed consent, gene transfer research involving human subjects that is funded by the federal government through the National Institutes of Health (NIH) also may be influenced by specific policies of that executive agency. Specifically, Section 402(b) (6) of the Public Health Service Act, as amended, codified at 42 United States Code §282(b) (6), established the NIH Recombinant DNA Advisory Committee (RAC). According to this Technical Committee's May 27, 1997, Charter, its function is to "advise the Director, NIH, concerning the current state of knowledge and technology regarding DNA recombinants, and recommend guidelines to be followed by investigators working with recombinant DNA." Review of specific protocols is conducted initially by the Human Gene Therapy Subcommittee (HGTS). The RAC is governed by the provisions of the Federal Advisory Committee Act, as amended, 5 United States Code Appendix 2.

ACKNOWLEDGMENT

Some of the writing of this entry took place while the author was the 1998–99 Dr. Arthur Grayson Memorial Distinguished Visiting Professor of Law and Medicine, Southern Illinois University School of Law, Carbondale, IL.

BIBLIOGRAPHY

1. S.H. Orkin and A.G. Motulsky, Co-chairs, *Panel to Assess the NIH Investment in Research on Gene Therapy, Report and Recommendation*, National Institutes of Health, Washington, DC, 1995.
2. T.H. Murray, *Int. J. Technol. Assess. Health Care* **10**, 573–582 (1994).
3. R. Moseley, *J. Med. Philos.* **16**, 641–647 (1991).
4. M.C. Coutts, *Kennedy Inst. Ethics J.* **4**, 63–83 (1994).
5. N. Wade, *N.Y. Times*, November 6, p. A1 (1998).
6. D. Callahan, *False Hopes*, Simon & Schuster, New York, 1998.
7. President's Commission for the Study of Ethical Problems in Medicine and Biomedical and Behavioral Research, *Splicing Life: A Report on the Social and Ethical Issues of Genetic Engineering With Human Beings*, President's Commission, Washington, DC, 1982.
8. U.S. Congress, Office of Technology Assessment (OTA), *Human Gene Therapy: Background Paper*, OTA, Washington, DC, 1984.
9. R. Levine, *Ethics and Regulation of Clinical Research*, 2nd ed., Urban & Schwartzenberg, Baltimore, MD, 1986.
10. N.A. Holtzman and M.S. Watson, eds., *Promoting Safe and Effective Genetic Testing in the United States: Final Report of the Task Force on Genetic Testing*, Johns Hopkins University Press, Baltimore, MD, 1998.
11. K.C. Glass et al., *IRB: Rev. Hum. Subj. Res.* **21**, 1–9 (1999).
12. J.G. Palmer, *Hum. Gene Ther.* **2**, 235–242 (1991).
13. J.G. Palmer, *Ann. N.Y. Acad. Sci.* **716**, 294–305 (1994).
14. L.R. Churchill et al., *J. Law, Med. Ethics* **26**, 38–47 (1998).
15. M.T. Lysaught, *J. Law, Med. Ethics* **26**, 48–54 (1998).
16. Office for Protection from Research Risks, *The Belmont Report*, OPRR Rep., ACHRE No. HHS-011795-A-2, U.S. Department of Health, Education and Welfare, Washington, DC, 1979.
17. Karp v. Cooley, 493 F2d 408 (5th Cir 1974).
18. F.D. Ledley, *Adv. Genet.* **32**, 1–16 (1995).
19. C. Marwick, *J. Am. Med. Assoc.* **269**, 83 (1993).
20. B.J. Crigger, *Hastings Cent. Rep.* **23**(3), 3 (1993).
21. R.G. Spece, Jr., D.S. Shimm, and A.E. Buchanan, eds., *Conflicts of Interest in Clinical Practice and Research*, Oxford University Press, New York, 1996.
22. J. LaPuma, C.B. Stocking, W.D. Rhoades, and C.M. Darling, *B. Med. J.* **310**, 1660–1661 (1995).
23. J. Moreno, A.L. Caplan, P.R. Wolpe, and Members of the Project on Informed Consent, Human Research Ethics Group, *J. Am. Med. Assoc.* **280**, 1951–1958 (1998).
24. E.J. Emanuel and D. Steiner, *N. Engl. J. Med.* **332**, 262–267 (1995).
25. P.A. Baird, *Perspect. Biol. Med.* **37**, 566–575 (1994).
26. F.D. Ledley, B. Brody, C.A. Kozinetz, and S.G. Mize, *Hum. Gene Ther.* **3**, 657–663 (1992).

See other HUMAN SUBJECTS RESEARCH entries.

INFORMING FEDERAL POLICY ON BIOTECHNOLOGY: EXECUTIVE BRANCH, DEPARTMENT OF ENERGY

DANIEL W. DRELL
U.S. Department of Energy
Washington, District of Columbia

OUTLINE

INTRODUCTION

The intention to establish an activity, linked to a major U.S. government science research program, directed at Ethical, Legal, and Social Issues (ELSI) was announced to an unsuspecting world on October 1, 1988, by Dr. James D. Watson, the co-discoverer with Francis Crick of the structure of the deoxyribonucleic acid (DNA) molecule. Watson was accepting the directorship of the newly created Office of Human Genome Research at the National Institutes of Health (NIH). Together with the Office of Health and Environmental Research (now the Office of Biological and Environmental Research) at the U.S. Department of Energy (DOE, which had been supporting genome research on a modest scale for the prior two years), these programs were preparing to map and sequence the human genome in its entirety, an estimated 3.2 billion nucleotides. In response to a question about dealing with the societal implications, Watson said that he thought that a small percentage of the research budget, initially 3 percent, should be used to study the implications of mapping and sequencing the human genome. Thus, even before the formal beginning of the Human Genome Project (HGP) in 1990, project managers, researchers, and lawmakers recognized that increasing knowledge about human biology and personal genetic information would raise a number of complex issues for individuals and society and that the agencies ultimately paying for the research needed to engage in anticipating its impacts. In response to Congressional mandates for identifying and defining such issues and developing effective policies to address them, DOE and NIH have devoted 3 to 5 percent of their annual HGP budgets, since the outset of HGP, to studies of the project's ELSI. [Much of this early history is well described in Cook-Deegan]. These expenditures for ELSI research by DOE and NIH now total nearly $85 million dollars.

DOE, which initiated HGP from roots in its radiation biology programs dating from the establishment of the Atomic Energy Commission (DOE's predecessor) in 1947, has for most of its 53-year history been an agency whose top priority was producing the materials necessary for the U.S. nuclear arsenal. This in turn made it an appropriate place to support studies of the biological effects of the radioactivity that characterizes the materials needed for nuclear weapons. Over the years, the greatest fear about radiation and radioactivity derived from its well-known capacity, first established by Herman Muller in the 1920s, to cause mutations and various forms of cancer. Mutations are alterations in the arrangement or sequence of the specific informational units in the genome, all made up of DNA. Just as a refrigerator magnet passed over a prerecorded cassette tape can alter the music on the tape, so too can various radiation exposures alter the information content of the DNA in the genome. To better devise tools and methods for assessing radiation effects, it is helpful to begin with knowledge of the original information prior to any exposures (e.g., the prerecorded music on the cassette *before* the magnet passes near it). This was DOE's rationale for initiating HGP. Additional reasons for DOE's activity included the availability of advanced technologies from the DOE National Laboratory system (e.g., for DNA fragment isolation, DNA sequencing, and computational analysis of sequences), better understandings of genetic contributions to workplace susceptibilities (given the particular nature of DOE facilities, characterized by mixed radioactive and toxic material waste), and the extensive experience of DOE with managing large interdisciplinary projects. For NIH, knowing the sequence of the human genome is the substrate for exploring what all the gene products do, how aberrations in them can lead to diseases, and how variations in individual genomes can lead to variation in individual humans. Thus both agencies had important and compelling mission-related reasons to participate in genome research. For the constituencies of both agencies, and for the larger general public, the genome project promised (and to a considerable degree has begun to deliver) many benefits in the way of new technologies, new industrial entities in the biotechnology sector of the economy, and new approaches to a better understanding of human biology in health and disease.

Still, there were worries about uses of genetic information, often of a very personal nature, that would not be welcomed by society. Early workshops enumerated several major ones. Among these were the implications of being able to predict future illnesses well before any symptoms or medical therapies existed; potential abuses of the privacy of genetic information by employers, insurers, direct marketers, banks, credit raters, law enforcement agencies, and many others; the availability of large amounts of genetic information in largely unprotected data banks; and the possible discriminatory misuse of genetic information. One hypothetical outcome, albeit perhaps an extreme one, of the wide use of genetic screening would be the creation of a new genetic underclass, leading to a host of new societal conflicts and exacerbating others of long standing. Additional concerns included the difficulties that physicians (many of whom had little genetics in their medical school curricula) would have in incorporating rapidly advancing genomic knowledge and technologies into practice, especially in an Internet enabled world where a patient often might know more about a particular condition than his or her physician. As biotechnology entered the business world, issues of whether gene sequences could (or should) be patented arose. In addition, since newly discovered genes often lead first to a test for the presence or absence of alleles associated with a disease, and the demand for a new test could be high, tension could be forseen between the natural economic imperative to sell a new test and the uncertainty of the significance of test results when the strength and nature of the putative disease association was not clear (e.g., BRCA1 and susceptibility to breast and ovarian cancer). As genetic advances rushed forward, there was concern that a variety of communities, from specific groups such as judges in the court system, to high school biology classes, to legislators, the press, MDs of various specialties, nurses, ultimately to the wider public, could not keep up. One group of concern were genetic counselors, who were appreciated as vital to helping patients and their families understand the implications (to the degree they were understood at all) of genetic test results; as a community, genetic counselors are recognized to be in drastically short supply.

The ELSI component of HGP began by emphasizing the privacy of genetic information, its safe and effective introduction into the clinical setting, fairness in its use, and professional and public education. In 1991, recognizing the breadth of this scope, DOE narrowed its ELSI focus to concentrate on genetic education, privacy and fair use of personal genetic information, the implications of intellectual property protection (e.g., patenting) of genes, and genetics and the workplace. As its portion of the total ELSI program, DOE has supported peer-reviewed studies on the uses, impacts, and implications of personal genetic information in various settings the ownership, access, and protection of genetic information in computerized databases, tissue and sample archives, and the commercialization of products of genome research. DOE also supports studies of ways in which society and its institutions deal with ELSI issues surrounding the complex (and common) multigenic conditions and disease susceptibilities most of us fear. One of DOE's goals is to ensure the wisest possible distribution of the knowledge from HGP to the general public as well as to the scientific, academic, minority, judicial, medical, educational, sociological, and political communities.

To avoid unnecessary duplication of effort by their two independent ELSI programs, DOE and NIH have collaborated on a number of activities and maintained close communications over the years. In addition to the NIH–DOE Joint ELSI Working Group, which periodically consulted with program staff and assisted in coordination of the ELSI programs, collaborations also involved several research projects, joint conferences and workshops, and programs supported with other agencies, organizations, and commercial companies. DOE and NIH collaborated in supporting the ELSI Research Planning and Evaluation Group (called ERPEG), a successor to the ELSI Working Group.

EDUCATION ACTIVITIES

In keeping with the long-standing commitment of DOE to education, the DOE ELSI program has emphasized the promotion of knowledge about the HGP and its ELSI implications to such groups as institutional review boards, medical professionals, genetic researchers, judges, policy makers, and the public. These efforts have included conferences, seminars, publications, videos, Web sites, and radio and television programs. A few examples are given below (a more complete listing is available at *http://www.ornl.gov/hgmis*) maintained by the Human Genome Management Information System (HGMIS) at the DOE Oak Ridge National Laboratory. Although not an ELSI program, HGMIS helps the DOE Human Genome Program fulfill its educational commitment by making information accessible to scientists, policy makers, and the public about the program's goals, funded research, implications, and applications. HGMIS carries out this mission by publishing and distributing documents and information both in print and via its Web site, which serves a wide and varied audience that includes the general public, scientific investigators, and medical professionals. One HGMIS product is the *Human Genome News*, the newsletter of the U.S. Human Genome Project. This Newsletter, originally jointly sponsored by both the DOE and the NIH Genome Programs, facilitates communication among genome researchers and informs the broader public about the project.

The DOE ELSI program has sponsored, in whole or in part, several widely distributed booklets on the genome project. These include *To Know Ourselves*, by Douglas Vaughan, an overview of the underlying science of the Human Genome Project; the *DOE Primer on Molecular Genetics*, by Denise Casey, an introduction to the basic science of genetics; *Your World, Our World*, a magazine dealing with the science of genomics and its ELSI implications designed for grades 7 through 10 by the Pennsylvania Biotechnology Association, in cooperation with the Alliance for Science Education; *Your Genes, Your Choices*, a book and video designed for low-literacy adults, by Catharine Baker and Maria Sosa of the American Association for the Advancement of Science (AAAS); and a comprehensive bibliographic database of publications related to ethical, legal, and social issues surrounding the genome project and including thousands of books and articles from 1990 to 1995, compiled by Michael Yesley of the DOE Los Alamos National Laboratory.

The DOE ELSI program has also assisted with video documentary projects including, "Medicine at the Crossroads," a four-part documentary jointly sponsored with NIH, produced by George Page and Stefan Moore at WNET/Thirteen in New York and shown around the country on public television in the spring of 1993. A book, *Medicine at the Crossroads: The Crisis in Health Care* by Melvin Konner, also resulted from this series. Another documentary that DOE ELSI supported was a WGBH series, "The Secret of Life," produced by Paula Apsell and Graham Chedd and shown in the fall of 1993. "A Question of Genes," a two-hour television documentary produced by Noel Schwerin, looked at a series of families and individuals challenged by the outcomes of genetic testing for inherited diseases. This was broadcast on the Public Broadcasting System in September 1997, and it received an Emmy nomination. "Seeking Truth, Finding Justice," a three-hour television documentary series, is now in production by Noel Schwerin now of Backbone Media. This series will explore the impact of cutting-edge science (e.g., genetic technology) on the courts and its profound effect on democratic institutions, people's relationships, and notions of truth, justice, and individual rights.

DOE ELSI has also supported two radio programs. For the Spanish-speaking public, DOE ELSI has supported the broadcast of 50 Spanish-language radio episodes within a nationally syndicated science and technology series, produced by The Self Reliance Foundation in Santa Fe, NM. "The DNA Files," a nationally syndicated series of radio programs on the social implications of human genome research, by SoundVision Productions in Berkeley, CA was broadcast over National Public Radio stations in November 1998.

Other DOE education activities have included a web site launched by the Shriver Center in Waltham, MA called "The Gene Letter," (now called GeneSage) a freely available, online quarterly ELSI newsletter for interested professionals and consumers (*http://www.genesage.com/professionals.html*). The San Francisco Exploratorium and the Smithsonian's Museum of American History each received DOE/ELSI support in 1995 for exhibits on genetics and the HGP. "Diving Into the Gene Pool" was on display at the Exploratorium from April to September 1995. Most recently, at Stanford University, a project is producing an interactive multimedia CD-ROM medical education course for physicians, most of whom have received little or no training in clinical applications of molecular genetics. It is expected that many other groups will benefit as well from such a resource.

Over the years, DOE ELSI has supported many workshops and conferences to explore and educate audiences about the genome project and ELSI. Among the more notable of these have been a May 1997 meeting at the University of Maryland in Baltimore to inform minority communities about HGP and to make the aspirations and interests of these communities better known to genome project scientists and policy makers. In September 1996 DOE and NIH jointly sponsored a major conference at Tuskegee University on "Plain Talk about the Human Genome Project," which addressed some of the HGP's implications for African-Americans. An updated compilation of all the conference talks was published in 1997. The Cold Spring Harbor DNA Learning Center on Long Island conducted several workshops early in the Genome Program to inform opinion leaders and public policy makers on genomics and the genome project's implications for society. In a three-year DOE and NIH project that began in 1990, the Baylor College of Medicine and the Texas Medical Center Institute of Religion organized two national conferences on genetics, religion, and ethics. This project led to a book, *On the New Frontiers of Genetics and Religion* (1995). In April 1998, a symposium attended by about 850 people on "The Human Genome Project: Science, Law, and Social Change in the twenty-first Century" was held at the Whitehead Institute for Biomedical Research. Following the symposium, which was supported in part by DOE, a CD-ROM containing the meeting syllabus, transcripts of all plenary talks, and links to relevant Web sites, were distributed to over 4000 people. A second conference in this series was held in Spring 2000.

Education for Professional Groups

The Einstein Institute for Science, Health, and the Courts has convened workshops around the country for more than 1500 federal and state judges. These workshops, conducted since 1994 and continuing through 2002 (with co-funding from the National Institue for Environmental Health Sciences of NIH), prepare judges for the expected onslaught of cases that will involve some aspect of genetics. In addition, the Summer 1997 issue of *The Judges' Journal of the American Bar Association*, with support from DOE, was devoted to "Genetics in the Courtroom." This project has a particular priority in the DOE ELSI program, since in our litigious society it can be anticipated that the courts will be the place of last resort for many issues. Many judges are touchingly honest about their very limited science education and their feeling of being ill-prepared since several recent decisions of the U.S. Supreme Court have assigned responsibility for deciding what scientific evidence is allowed into court to the trial judges and, for federal courts at least, altered the criteria by which scientific evidence must be judged before it is admitted. There are many potential issues for the courts, including numerous ones that are not rooted in medical practice. Issues from forensic uses, the penalty phase after a felony conviction, adoption cases, cases of the custody of a minor or of an incompetent parent, inheritance cases, liability, and many others may be influenced by assertions of genetic causation or involvement. Courts will have little alternative but to deal with cases arising from assertions that "my genes were responsible" cloaked in various guises.

Institutional review boards (IRBs) are responsible for overseeing clinical research procedures at such institutions as hospitals and research facilities, particularly protocols that might affect the rights and welfare of human research subjects. Increasingly complex ethical, regulatory, and scientific issues are proving challenging to these boards, which often are composed largely of nongenetic professionals. To assist IRBs in reviewing genomic protocols, expecially those involving human subjects and tissues, a project with the Public Responsibility in Medicine and Research is focusing on educating IRBs in the special language, technologies, and issues that typify these protocols.

Curricula

The DOE ELSI program has supported the production and dissemination of a number of curricula for high schools on the genome project and its implications. These have included a nationwide series of workshops created by the Cold Spring Harbor Laboratory in New York which introduce high school biology teachers to a laboratory-based unit on human DNA polymorphisms (genetic differences) and the ELSI aspects of the genome project. An innovative program initiated at the University of Washington, Seattle, allows high school students to perform DNA synthesis and sequencing in the classroom. Local teachers, as well as those from other states, attend a weeklong summer workshop in Seattle and receive continuing assistance with the experiments through equipment and technical advice after they return home. This program has allowed high school students to actually contribute DNA sequence to the human genome project database making them participants in the project.

The Biological Sciences Curriculum Study (BSCS) group in Colorado Springs has produced 4 high school modules for the DOE ELSI program, roughly every two years since 1993. The first was, "Mapping and Sequencing the Human Genome Program: Science, Ethics, and Public Policy," followed by "The Human Genome Project: Biology, Computers, and Privacy." The third module was "Nontraditional Inheritance and the Nature of Science." Most recently BSCS has released "Genes, Environment, and Human Behavior" which deals with complex traits and behavioral genetics. At California State University in Los Angeles, an ELSI-funded project has translated into Spanish the first BSCS module for use by students and their parents in selected high schools where the majority of the student body is Hispanic. In conjunction with other local and national organizations, the University of Kansas Medical Center trained over 175 high school science teachers annually as state "resources" in molecular genetics and the latest biotechnology methods. In a "trainer of trainers" model, they then prepared thousands of additional teachers, who are now reaching millions of students. At the Fred Hutchinson Cancer Research Center in Seattle, a researcher has established an electronic educational resource on the Web for republishing classic genetics literature (both papers and monographs). This project helps promote a foundation for understanding the new genetics and genome technology (*http://www.esp.org*). A more complete list of educational curricula developed with support from the DOE ELSI program is available at *http://www.ornl.gov/hgmis*.

Exploring Public Policy Issues

There are many public policy issues raised by advances in genome research and no ELSI program could hope to explore all of them. At the outset of the HGP, when the scope of potential ELSI issues was uncertain, DOE and NIH jointly supported a study by the Institute of Medicine to address a variety of issues raised by the rapid proliferation of predictive genetic tests in otherwise healthy individuals. This study led to the 1994 publication of *Assessing Genetic Risks: Implications for Health and Social Policy*, a report with recommendations for the use of genetic information in healthcare. Concentrating on Florida and Georgia, researchers at Morehouse School of Medicine and the University of Florida explored differences among state-supported programs for genetic testing, screening, and counseling. Jointly sponsored with NIH, this 1992 to 1994 project addressed the issue of confidentiality in a mobile society of broad ethnic diversity. A series of published papers resulted from these studies.

At the University at Albany in New York a project is examining confidentiality concerns raised by the possible misuse of DNA-based test results in the managed-care (MCO) setting. This setting presents unique ethical dilemmas because the MCO is both payer and provider and because physicians, and quite often, testing laboratory personnel are MCO employees. Additionally, a significant fraction of working Americans get their health care through MCOs. The uses they make of genetic information are a valid concern of ELSI.

Many disorders associated with mental retardation have genetic contributions, with Down syndrome and fragile X syndrome the most common, and new genetic findings from HGP pose very difficult ethical questions and legal and social concerns to those with disabilities and their family members. To address these concerns, The Arc of the United States developed and distributed a series of reports, fact sheets, and a workshop training package to all 1100 of the organization's chapters.

Threats to privacy based on genetic testing in various circumstances have been cited often as reasons for ELSI research and societal concern. The fair use of genetic information raises particularly difficult practical and philosophical problems related to access and disclosure. Third parties such as insurers, employers, adoption agencies, and educational institutions may feel they need access to genetic data having predictive or diagnostic value, while others feel that such access will lead to discrimination and decisions based too heavily on genetics. The DOE ELSI program has focused on a broad range of genetic privacy issues from the perspectives of several disciplines, including philosophy, social science, and law. Through grants and commissioned papers, the program has supported studies to consider such factors as attitudes toward genetic privacy in different populations, the need for appropriate measures to protect genetic information in various contexts, and evolving policies of private institutions and state, federal, and foreign governments in this area. These studies were designed to help increase the growing body of knowledge and to promote informed discourse leading to policy development.

A study by a leading social scientist at Columbia University related existing social science work on privacy to anticipated genetic-privacy issues. This study also examined current privacy-protection measures, debates over the need to update privacy protection, and implications for social and legal policies to deal with expected future genetic testing and applications of genetic data. The results will appear in book form. To strengthen the dialogue between the professional genetics community and federal policy makers, a congressional fellowship program was initiated in 1995. This program allows genetic professionals to spend a year as special legislative assistants on the staff of members of Congress or on congressional committees. The fourth fellow (of a planned five) began her fellowship at the beginning of 2000. Under a grant to the Library of Congress, Philip Kitcher, a philosopher at the University of California, San Diego, researched genome ELSI issues and wrote *The Lives to Come: The Genetic Revolution and Human Possibilities* (2). The book explores both the science and the ethical and moral dilemmas arising from the genome project and concludes by arguing that society should make active use of genetic testing to avoid the worst genetic conditions that presently are not amenable to medical therapies, but involve horrific suffering, for example conditions such as Tay-Sachs disease.

In a project jointly sponsored by NIH and DOE, researchers at the University of California, Berkeley, conducted a major study between 1992 and 1997 to illuminate the processes by which genetic screening and genetic concepts of health and illness were integrated into the health concerns of high-risk families. This study focused on issues of privacy, stigmatization, and discrimination and how these issues were managed within family and institutional networks in two contrasting communities: one in which the disorder was generally recognized as race-associated (sickle cell disease) and the other in which this association was not part of the public consciousness (cystic fibrosis). Interestingly, cases are being identified of individuals of the "wrong" community with these conditions, namely Cacausians with sickle cell disease and African-Americans with cystic fibrosis. A series of scholarly papers and book chapters have resulted.

DOE sponsored a workshop on "Medical Information and the Right to Privacy" in June 1994 in Washington, DC, which led to the book *Genetic Secrets*, edited by Mark Rothstein of the University of Houston (3). A compilation of chapters written by leading experts, the book explores the full range of issues related to genetic privacy, particularly focusing on issues arising from the possible use of genetics in the workplace, a particular focus of the DOE ELSI program.

Privacy Legislation

Focusing on privacy concerns, some proposed legislation has attempted to establish a legal framework of fair practices for health information and to regulate its access, disclosure, and use. A draft bill (the Genetic Privacy Act), was drawn up in 1995 by George Annas of the Boston University School of Public Health to assist legislators. The bill proposed that access to information in genetic data banks should be regulated during sample collection and when it is stored, disclosed, and used. Several state lawmakers used language and concepts from this draft bill in drawing up proposals for legislation in their own states. The Genetic Privacy Act and Commentary is on the Web (http://www.ornl.gov/hgmis). In 1994 a Shriver

Center study surveyed existing bills and laws with a view to drafting model legislation for protecting the privacy of personal genetic information. They found that state legislative efforts to regulate the use of such data have increased, particularly in employment and insurance, but major gaps and deficiencies in statutory coverage persist. From 1997 to 1998, Mark Rothstein determined the effects of a unique Minnesota law limiting employee medical records to job-related matters, with a view to using the law as a model for protecting genetic privacy in the workplace. He found that the Minnesota law had little impact because people were not aware of it. A journal article describing this study was published in 1998. At this writing, numerous state measures affecting genetic privacy have been passed, but each is somewhat different from the others, and no federal legislation focused on this issue has been passed by both the House of Representatives and the Senate of the United States.

Data Banks

Interest in forensic DNA data banks is growing, with all states except Massachusetts having laws that authorize the collection of samples from convicted felons. Two separate DOE studies carried out at the Shriver Center in Waltham, Massachusetts, focused on the growing practice of banking individuals' DNA or genetic data in forensic, academic, military, and commercial settings. These studies involved research on privacy in these settings and on developing and refining proposed policies and guidelines. One project, which reported widespread uncertainty about the types of sample releases that are legally or ethically prohibited, led to the production of a 28-minute video called *Banking Our Genes*. [Video: Fanlight Productions (800/937-4113)] Another study carried out a survey of life insurance companies and showed that they were more interested in obtaining existing genetic test information than in performing new tests on applicants. Company ratings based on genetic conditions reflect a considerable degree of subjectivity rather than actuarial data. How this will evolve as HGP nears completion and technologies make the acquisition of personal genetic profiles easier and less expensive is an ongoing ELSI challenge.

Patents

Since 1996 Rebecca Eisenberg at the University of Michigan Law School has explored the role of patents in transferring technology generated by the genome project to society at large. This issue, which was not expected to be a concern at the outset of HGP, exploded suddenly when, in 1991, NIH filed patent applications on ESTs (expressed sequence tags, which are short sequences of DNA from a gene that is known to be expressed in a cell). While this specific application was rejected by the Patent and Trademark Office in 1993 and NIH elected not to appeal the rejection, many other biotechnology companies have aggressively sought patent protection on genetic information. One company, Incyte Pharmaceuticals in California, has been awarded a patent on ESTs. Eisenberg now is conducting a study with Michael Heller on the trend toward privatization and patenting of the early stages of biomedical and microbial genome research. A series of influential publications has resulted from this work.

CONCLUSION

The continuing importance of ELSI studies is rooted in the basic fact that each person has a unique genome that both identifies her or him as an individual and has predictive implications for her or his future health. An "ideal" or "perfect" genome does not exist, even if such a concept could be defined. All genomes contain many small differences that could severely and adversely affect health under different circumstances or if not influenced or masked by other genes. This information has potential value to other people and groups who may have their own agendas. Thus, while HGP's potential benefits are enormous for medicine, bioremediation, agriculture, and many other socially and economically important areas, we must remain alert to the more problematic implications as well.

HGP is rapidly moving to its goal of obtaining the complete human reference DNA sequence by 2003 and a useful "working draft" by Summer of 2000, well ahead of the original schedule. All of us will therefore face, much sooner than anticipated, many questions surrounding differences in our individual sequences, uncertainties regarding their significance for health and longevity, and the implications of knowing about these subtle distinctions before the biological effects are understood. Although many ELSI issues are not novel to the genome project, they nonetheless remain challenging and need to be addressed. Only by dealing directly and openly with such issues through the collective best efforts of bioethicists, scientists, policy makers, and the public can the benefits of genome research be realized and the difficulties minimized.

The ELSI programs have had significant impacts, both in terms of the research and other activities they have supported, but also in other tangible ways. Within DOE, ELSI has led to the creation of parallel programs, particularly the BASIC (Bioremediation and its Societal Implications and Concerns) program element of the larger Natural and Accelerated Bioremediation Research (NABIR) program. NABIR is focused on exploring biological approaches to the legacies of radionuclide wastes created by the 50-year history of the nuclear weapons programs of DOE and its predecessor agencies. DOE also has an obligation to address the challenges of human subjects experimentation associated with this legacy. Outside of the DOE, ELSI has contributed to the creation of the National Bioethics Advisory Commission (NBAC) attached to the White House Office of Science and Technology Policy. NBAC has a charter to explore both human subjects issues and genetic information issues. The creation of NBAC also takes cognizance of the challenges that will result from the novel technologies arising from HGP, among them genechips and microarrays. These technologies, and others, will make it easier to acquire accurate and precise information about specific genetic variations present (or absent) from large numbers of people and to correlate them with clinical conditions.

Just as genome scientists need to become active participants in the discussion of these difficult issues, so too must the bioethicists, lawyers, social scientists, and other "ELSI scholars" learn the relevant genetic science so that these very important dialogues remain firmly grounded in scientific reality. This is an area where much remains to be done and where dialogue and crosstalk are less in evidence (and sometimes less enthusiastic) than could be wished. There is also a major challenge for the private sector that has aggressively entered genomics to a degree exceeding even the public sector, spurred by the promise of commercial yields from patents on human and other genes. What one company does can influence the way other companies are regarded and the way their products are received. The genomics revolution (it is nothing less) will have major impacts on our lives in the twenty-first century, and we must use all the wisdom and insights of our intellectual and political ancestors to inform us and to build on as we try to make wisely the many difficult and momentous decisions that lie ahead. These decisions will have great impacts on the lives of our children and our society years into the future.

INFORMATION SOURCES

Abstracts for current and past ELSI projects can be found in the DOE Human Genome Program reports and in Contractor-Grantee workshop reports all of which are available on the HGMIS Web site (*http://www.ornl.gov/hgmis*)

ACKNOWLEDGMENT

I want to acknowledge the valuable and helpful suggestions of my colleagues in the DOE Office of Biological and Environmental Research (OBER). I particularly want to acknowledge contributions from Ari Patrinos, Marvin Frazier, Mike Riches, and Betty Mansfield (Oak Ridge National Laboratory).

BIBLIOGRAPHY

1. R. Cook-Deegan, *The Gene Wars: Science Politics and the Human Genome*, Norton, New York, 1994.
2. P.Kitcher, *The Lives to Come: The Genetic Revolution and Human Possibilities*, Touchstone Books, 1997.
3. M. Rothstein, ed., *Genetic Secrets: Protecting Privacy and Confidentiality in the Genetic Era*, Yale University Press, New Haven, CT, 1997.

ADDITIONAL READINGS

The ELSI literature has grown extensively since the onset of the ELSI programs at the DOE and the NIH. The selection below is not representative and other contributors will make their own suggestions which can differ with these. I don't necessarily endorse or agree with everything that is in these volumes, but as a place to start, they offer good introductions to the history of the Human Genome Project and many of the issues that have arisen associated with it.

L.B. Andrews, *Medical Genetics: A Legal Frontier*, American Bar Foundation Press, Chicago, IL, 1987.

L.B. Andrews, J.E. Fullarton, N.A. Holtzman, and A.G. Motulsky, *Assessing Genetic Risks: Implications for Health and Social Policy*, National Academy Press, Washington, DC, 1994.

G. Annas and S. Elias, *Gene Mapping: Using Law and Ethics as Guides*, Oxford University Press, New York, 1992.

R. Cook-Deegan, *The Gene Wars: Science Politics and the Human Genome*, Norton, New York, 1994.

C. Cranor, ed., *Are Genes Us? The Social Consequences of the New Genetics*, Rutgers University Press, New Brunswick, NJ, 1994.

M.S. Frankel and A. Teich, eds., *The Genetic Frontier: Ethics, Law and Policy*, AAAS Press, Washington, DC, 1994.

S.J. Gould, *The Mismeasure of Man* (revised and expanded), Norton, New York, 1996.

D. Hamer and P. Copeland, *Living With Our Genes*, Random House/Doubleday, New York, 1998.

D.J. Kevles, *In the Name of Eugenics: Genetics and the Uses of Human Heredity*, University of California Press, Berkeley, 1985.

B.M. Knoppers, C.M. Laberge, and M. Hirtle, eds., *Human DNA: Law and Privacy, International and Comparative Perspectives*, Kluwer Law International, The Hague, The Netherlands, 1997.

C. Long, ed., *Genetic Testing and the Use of Genetic Information*, AEI Press, Washington, DC, 1999.

R. Plomin, *Nature and Nurture*, Brooks Cole, Pacific Grove, CA, 1990.

M.A. Rothstein, ed., *Genetic Secrets: Protecting Privacy and Confidentiality in the Genetic Era*, Yale University Press, New Haven, CT, 1997.

Y. Segal, *The Human Genome Project: Legal, Social, and Ethical Implications. Proceedings of an International Workshop, Jerusalem, 1995*, Israel Academy of Sciences, Jerusalem, 1997.

E. Smith and W. Sapp, eds., *Plain Talk about the Human Genome Project: A Tuskegee University Conference on its Promise and Perils ... and Matters of Race*, Tuskegee University Press, Tuskegee, AL, 1997.

L. Weir, S.C. Lawrence, and E. Fales, *Genes and Human Self Knowledge: Historical and Philosophical Reflections on Modern Genetics*, University of Iowa Press, Iowa City, 1994.

See other FEDERAL POLICY MAKING FOR BIOTECHNOLOGY entries.

INTERNATIONAL ASPECTS: NATIONAL PROFILES, FRANCE

FRANÇOISE TOURAINE-MOULIN
Immunology Laboratory
Hôpital Neurologique Pierre Wertheimer
Lyon, France

MARISSA VICARI
La Tour de Salvagny
Lyon, France

MARIA RUANO
CATHERINE CHABERT
Alain Bensoussan — Avocats
Lyon, France

OUTLINE

INTRODUCTION

The development of biotechnology is in France, as in other European countries, moving ahead strongly. Today the ethical debate, which has urged the reevaluation of new biotechnologies and commercial dynamics of the biotechnology industry, is also proceeding in full swing at the level of the general public, special interest group, government structures, and research facilities.

Although France is considered as being behind in comparison with the United States in industrial development and innovative entrepreneurship support, it is, without a doubt, competitive in life sciences research as well as for legal and ethical rules on biotechnology.

There has always been a high regard for life-sciences research in France, where pioneering advances were made in certain fields; the classic example is Louis Pasteur's contributions in microbiology, and later vaccinology. There were also major contributions by André Lwolff, François Jacob, and Jacques Monod (his achievement recognized internationally by the 1965 Nobel Prize in Medicine) to molecular biology in the discovery of the role of messenger ribonucleic acid (mRNA), and by Pr. Jean Dausset to knowledge in organ and tissue transplantation regarding the major histocompatibility complex (MHC). More recently in the field of virology a French team at the Pasteur Institute in Paris, directed by Pr Luc Montagnier, won acclaim with their discovery of HIV1 which is responsible for AIDS.

ACADEMIC INVOLVEMENT

In the academic sphere the research in biotechnology is directed by sizable institutions for research and technology. The Centre National de Recherche Scientifique (CNRS), National Center of Scientific Research comprises a large networks of university laboratories in which fundamental research in science and technology, including biotechnology, is carried out. In 1999, 25,400 people were employed in about 1300 CNRS research and service units throughout France and the French territories. The Institut National de la Santé et de la Recherche Médicale (INSERM), National Institute of Health and Medical Research is another such large organization dedicated to the "research, understanding and bettering of the human condition, with the main aim of promoting health for all." Created in 1964, INSERM includes 275 research laboratories and a community of over 10,000 scientific and medical professionals. Of special interest for bioethics INSERM houses a collection of 3000 books and 100 specialized journals on ethics in the Documentation Center on Ethics of Life Sciences and Health (CDEI). There are also the National Institute for Agricultural Research (INRA), the Centre d'Energie Atomique (Center for Atomic Energy) (CEA), and public universities which, like CNRS and INSERM, are under the direction of the Minister of Education, Research, and Technology.

INNOVATION AND ENTERPRISING IN BIOTECHNOLOGY

Before the law on innovation and research was passed on July 12, 1999, it was difficult for scientists working for public institutions to create their own companies to commercialize their innovative technologies. As a matter of fact, those scientists who are government employees cannot easily take part or become a partner in a private firm. The new law will allow French academic research to evolve into commercial enterprises.

Besides, the Ministry for National Education, Research and Technology (MENRT) a national database on biotechnologies, be set up, called "Biotechnologies/France," (1). The MENRT objective, among other things, is to show the wide range of French biotechnological activity and to promote the work abroad, as well as to inform and educate the public.

Biotechnology Companies

Until recently there were only around one hundred biotechnology companies in France. Their number has started to increase with the organization of an association called France Biotech and also the Syndicat National de l'Industrie Pharmaceutique (SNIP), the National Union of the Pharmaceutical Industry. The situation is helped by administrative efforts at the local level in the creation of regional "biopoles," or concentrations of biotechnology-related resources, and in facilitating the founding of biotechnological enterprises.

French biotech industry therefore may be said to be experiencing a full expansion, which is in keeping with a long tradition of quality scientific research. Health-related biotechnology companies in France employed in 1999 around 2000 people with a market of two billion French francs. Industrial and agricultural biotechnology has been strong in France since the 1970s within large pharmaceutical companies. Associated start-up companies could be seen by the 1980s, although without much development in technology transfer. Since 1990 the biotechnology industry in food and agriculture has seen another surge in the development of new companies.

Public Interest Groups

In French biotechnology research, associations and foundations, such the Centre d'Etudes de Polymorphisme Humain (CEPH), or (Center of Studies for Human Polymorphism), play an important role in upholding public interest. Their combined effort with the French government has led to expanded research and technological innovation in networking projects on the human genome (e.g., the first mapping of the human genome and in the robotics field, the newly designed "GenHomme" project) and on plant genomics (e.g., Genoplante, dedicated to transgenesis and genetically modified organisms).

The recent governmental projects aimed at fostering collaborations between various public and private laboratories have led to a concentration of start-up biotechnology companies in the area of genomics, and most of them are located in the Genopole (or "Genomic Valley") in Evry, near Paris.

THE "BIOETHICS LAWS"

History

Turning to the legislation on biotechnology, we find in France a historical consciousness of the ethical debate that has been cultivated and influenced by a strong collective memory of Nazi ideology and the disclosures of the Nuremberg proceedings (1947), by the Universal Declaration of Human Rights, and by the works of the World Medical Association, namely the Declaration of Helsinki. The translation of the human into legal terms in France has thus been difficult and attempted in consideration of issues in bioethics, specifically those that have arisen with the explosion of information coming from discoveries in molecular biology.

To rise to the challenge, France formulated the three Lois de Bioéthique, or "bioethics laws," of July 1994 (2), which specifically address medically assisted procreation, the protection of the embryo, and diagnostic medicine. These laws, which were among the first pieces of legislation adopted on the subject, were proceeded by numerous preparatory studies, beginning with the 1988 report by Guy Braibant, *Sciences de la vie, de l'éthique au droit* (Life sciences, from ethics to law) (3). Also intrinsic to the creation of the bioethics laws is the large work completed by Nöelle Lenoir in 1991, *Aux frontières de la vie* (4).

Une éthique biomédicale à la française (At the frontiers of life: French biomedical ethics) as well as *Les science de la vie et les droits de l'homme: bouleversement sans contrôle ou législation à la française? Questions clefs et réponses contradictoires*, (Life sciences and human rights: tumbling out of control or French legislation? Key questions and contradictory answers) by Senator F. Serusclat (1992) were also influential (5). The work continued in 1994 with a report by M.P.J.F. Mattéi, *La vie en questions:*

pour une éthique biomédicale (Life in questions: toward a biomedical ethic) (6). Punctuating these is a series of advisory reports issued by the Comité Consultatif National d'Ethique pour les sciences de la vie (CCNE, the National Consultative Ethics Committee for life sciences), created in 1983, the first committee of its kind (7). In addition, existing legislation such as the Caillavet law of 1976, which regulated organ donation, and regulations relative to the protection of people participating in biomedical research (1988), thus defining their legal status, were taken into consideration, namely the Law of Huriet Sérusclat, December 20, 1988 (modified July 1994).

Already present in the Code Civil were the principles of the inviolability and nonobjectification of the human body, which have been reaffirmed by the bioethics laws. The idea of the protection of the human body has been expanded as the necessary consideration of a persons' totality, including the genetic identity has become apparent.

It is to the credit of the bioethics laws that introduced into the Code Civil is a chapter on the respect of the human body. In this regard the law ensures the primacy of the person, prohibiting all diminution of their dignity and guaranteeing the respect for the human being from the beginning of his or her life. It is also interesting to note that the Conseil Constitutionnel (which is charged with controlling the conformity of the laws with the French Constitution) asked, with the formation of the bioethics laws, that the laws enunciate the set of principles designed to ensure the constitutional preservation of human dignity.

General Principles

The principles at the origin of the bioethics laws include (8):

- previous consent for all procedures that relate to the integrity of the body;
- the prohibition of all modification of heritable genetic material with eugenic ends and prohibition of all intervention aiming at affecting the descendants of the person concerned;
- noncommercialisation of the human body, therefore prohibiting the selling of organs and tissues;
- anonymity with regard to the donation of organs, elements, or products of the human body;
- the impossibility of contesting the familial relations or descent of a child for reasons related to medically assisted procreation;
- the regulation of the utilization of genetic tests and techniques of medically assisted reproduction;
- the regulation of information related to the donation of organs.

The bioethics laws thus offer very specific regulations on essential procedures that protect human dignity. The laws strive to consider the ethical issues brought up by each relevant technological step. They have arisen from the general apprehension of the ethical implications of biotechnology for the human and living organisms. The largely governmental initiative has materialized via the promulgation of numerous legislative reports and regulations, which leave little room for individual initiative. The ultimate aim of the heavy control of research in general, and biotechnology in particular, is to avoid at all costs, any legal lacunae. Nevertheless, the body of extremely complicated and rich legislation that has resulted is highly technical and, because of continual developments in this technology, eternally incomplete.

Provisions

The first of the bioethics laws is in the Code Civil, law 94-653 of July 29, 1994, which pertains showing to respect for human life. This law covers the study of the genetic characteristics of a person and the identification of a person via genetic fingerprinting, protection of the embryo, and filiation (legal status relations between one person and his or her descendants) in cases of medically assisted procreation.

On the respect for the human body, the law clearly prohibits any contravention from the beginning of life (Article L. 16), although the point at which life begins is not fully defined by French law. The human body is considered as inviolable (Article L. 16-1), and the integrity of the human body may only be challenged in cases where the intervention is therapeutic (Article L. 16-3). Any action challenging the integrity of the human species is prohibited, including all eugenic practices designed to select the sex of the embryo, and any modification of the genetic characteristics, with the exception of research directed at the prevention of genetic diseases (Article L. 16-4). No remuneration may be given to a person who agrees to submit to medical experimentation or to donate parts or products of his or her body (Article L. 16-6). Of interest is Article L. 16-7 which officially renders null and void any contract relevant to procreation or gestation regarding a third party. This is relevant to the donation of gametes and also to the issue of contracting surrogate mothers, which is prohibited. Any information on the identification of an organ donor, and the person who received the organ is strictly confidential (Article L. 16-8). Neither the donor's family nor the recipient may learn the identity of the other, and only in the case of medical necessity may the doctor of the patient receiving the donation have access to such information.

The second law (94-654) of July 29, 1994, relates to the donation and utilization of human fetal tissues, cells, and other parts, medically assisted reproduction and prenatal diagnosis. It is known as the Code de la Santé Publique (Public Health Code). This law also addresses organ donation from persons living and deceased, organ transplantation, the donation, use and conservation of fetal tissues, cells and products, medical assistance to procreation, and specific regulations regarding the donation and utilization of gametes. Notably this law prohibits in vitro fertilization (IVF) for the purpose of research or experimentation and has enacted a ban on all embryo experimentation, with certain exceptions for medical research. In line with the principle of noncommercialization of the human body, the law prohibits the creation and use of embryos for commercial or industrial purposes. In addition no remuneration may be received for an embryo. The same principle is applied to parts of the fetus, including tissues, cells and organs. Because of the large number of surplus embryos resulting from in IVF procedures, and the therapeutic possibilities associated with breakthroughs in cloning, the mandatory five-year legislative review of this law has been the subject of much public debate.

There is also an information protection law that has been modified by a third bioethic law; law 94-548 of January 1, 1994, on the confidentiality of personal data, amending law 78-17 of January 6, 1978, on information, medical records, and access, and the confidentiality of personal data for research purposes in the field of health.

Medically Assisted Reproduction

The extent to which the laws control certain technical processes, and their features related to the protection of human dignity within the letter of the law, are topics of notable interest. A look at the procedures of medically assisted reproduction can provide a good idea of the legislations' effect.

By the reproduction legislation and the decrees that have followed, France has instituted a highly organized structure to oversee medically assisted reproduction cases. These laws regulate public hospitals and private clinics alike. In particular, the Commission Nationale de Médicine et de Biologie de la Reproduction et du Diagnostic Prénatal (CNMBRDP, National Commission for Biological and Reproductive Medicine and Prenatal Diagnosis) was created by decree in April 1988. The members convene regularly, give advice, and authorize IVF practices in individual cases (which must also be further ratified by the Minister of Health). This committee has been responsible for the elaboration of several decrees on reproductive medicine and prenatal diagnosis. As a result the techniques pertaining to medically assisted reproduction are strictly codified to the point where practically every necessary act is the object of a specific decree.

Medically assisted reproduction procedures are in fact covered by La Sécurité Sociale, the French national health insurance. Thus economically the procedure is available to the general population. The whole process is therefore mitigated by the state via Sécurité Sociale, the organization that actually reimburses the expenses related to these procedures.

Under the law 94-654, medically assisted reproduction is restricted to couples whose infertility is due to a medically diagnosed pathology or who have the aim of avoiding the transmission of a serious disease to their offspring (Article L. 152-2). The couple must consist of a man and woman who are of reproductive age, and are either married or can prove that they have lived together for at least two years. Both members must be alive and consenting. An embryo may only be conceived in vitro for the purposes of medically assisted reproduction, and it must be conceived with the gametes from at least one member of the parental couple. In exceptional cases a couple may decide to donate, via written consent, an embryo to another couple who qualifies for medically assisted reproduction and for whom the technology has

not been successful (Articles L. 152-4 and 5). In this case the whole process is mediated by a judge. The donating and receiving couple remain anonymous; only for therapeutic reasons may the doctor have access to non-personal medical data regarding the donating couple and no payment may be received for the embryo (Article L. 152-5).

The excess embryos are kept frozen for a period of five years, and each year a letter is sent to the parent couple regarding the options of using the embryos for reproductive purposes and the continuation of embryo storage. The involvement of a third party as a gamete donor is only permitted in cases where the use of the couple's own gametes has not been successful, and the donation and utilization of gametes is also controlled by law. The donor must be part of a couple who has already procreated, and written consent for the donation is required of the donor and his or her partner (Article L. 673-1). Written consent is also required by both members of the receiving couple. Artificial insemination with fresh sperm or a mixture of sperm samples is prohibited, and the use of gametes coming from one donor is limited to the creation of five embryos (Articles L. 673-3 and 4).

The establishments practicing medically assisted reproduction technology must be authorized by decree. Authorization is accorded by the CNMBRDP and is for five years only. All establishments and laboratories authorized to practice medically assisted reproduction technology or prenatal diagnosis must present an annual written report of their activities to the Minister of Health, and must register the gametes and embryos that they hold in storage (Article L. 184-2).

The Code Pénal imposes a range of fines and prison terms, along with the temporary or permanent revocation of the right of the establishment to practice medically assisted reproduction or related activities. In addition any professional personally involved in a violation is subject to a maximum of 10 years suspension from professional activity. The nature of the sanctions include the taking of gametes from a living person without her consent, which is punishable by five years imprisonment and a fine of 500,000 FF. (Article L. 675-9). The same sanctions apply in obtaining gametes by payment of any kind, except that authorized to prepare and properly conserve the cells (Article L. 675-10).

Divulging information relative to the identity of a person or couple who have donated gametes and the couple who received them is punishable by two years imprisonment and a fine of 200,000FF. (Article L. 675-11). The same penalty is applied for the procurement of gametes from a living person without performing the required sanitary tests for transmissible diseases and proceeding with artificial insemination with fresh or mixed sperm samples.

Revision of "Bioethics Laws"

Taking into account the evolutionary nature of science, the bioethics laws include a legal exception requiring that they be reviewed every five years, thus underscoring the difficulty of reconciling the progress of the life sciences and respect for the essential values necessary for the preservation of human dignity.

The philosophical principles that form the basis of the laws were formulated with scientific progress in mind, and thus to guide the evolution of the bioethics laws along with the evolution of science. Consequently the bioethics laws were to be reviewed in 1999, and a certain number of modifications are in view to take effect during the year 2000.

Today several basic ethical questions still remain unresolved regarding for instance the legal status of the embryo. The debate in France today is at the level of whether an embryo should or should not, from the first step of development, be considered as a potential human being. While considered, this question was not answered in full with the creation of the bioethics laws, and it has been raised once again with the first five-year revision of the laws, notably concerning the embryo research issue, nowadays forbidden in France. On this issue the French Conseil d'Etat proposes, in a recent report of November 1999, to permit research on frozen embryos no longer intended for a parental project. In performing such the research, the embryo in question cannot be used later for reproductive purposes (9).

On human reproductive cloning, all proposals that have been submitted to the French government, for the revision of the bioethics laws, consider that existing French rules do not suffice to make clear the prohibition in France of such practices. In this context it is clear that new legislation will be taken to forbid human reproductive cloning activities. However, therapeutic cloning could, in certain conditions, be allowed.

COMITE NATIONAL CONSULTATIF D'ETHIQUE POUR LES SCIENCES DE LA VIE ET DE LA SANTE

A highly significant reference body for decisions in bioethics is the National Consultative Ethics Committee for Health and Life Sciences (CCNE) established in 1983 via presidential decree and enacted with the law of July 29, 1994. CCNE is an independent body with the aim of forming opinions and publishing recommendations on ethical issues arising with technological progress in the fields of biology, medicine, and health. The group us comprised of 40 individuals. The president and 5 members of the main philosophies and religious faiths are chosen by the president of the republic, 19 members are engaged due to their qualifications and interest in ethical issues, and 15 are research scientists. Topics may be referred to the CCNE by government officials, a university or establishment for higher education, a public institution, or foundation working in research, technology development or health issues, other persons, or committee members themselves. So far the committee has revisited topics as necessary, in keeping with scientific developments or social issues. CCNE opinions are considered by legislators both in the development and drafting of new laws (10).

Such has been the case with the CCNE Opinion 1 on *the sampling of human embryonic tissue for therapeutic, diagnostic, and scientific purposes.* CCNE has revisited it under various subheadings in several Opinions since

1984. This was shortly after its formation, and CCNE then expressed the Opinion that the human embryo must be considered a potential human person and therefore must never be subjected to in utero experimentation, CCNE further prohibited the commercial or industrial use of living embryos and any remuneration for embryo tissue samples. The same sentiments appeared in Opinion 8 (1986) *on the research and use of in vitro human embryos for scientific and medical purposes.* The emphasis again is on human dignity, with the conclusion that fertilization should not be done for research purposes alone and must exclude any form of industrial or commercial use. These conclusions, among others, are reflected in the bioethics laws of 1994.

Of further interest, the CCNE developed, in Opinion 1, the policy that the use of embryonic tissue for therapeutic purposes must be exceptional and, *considering the present state of scientific knowledge*, justified by certain criteria including the rarity of the disease to be treated, the absence of alternative, equally effective therapies, and a benefit to the recipient, such as survival. An important factor of this policy is the suggestion that the successful development of scientific knowledge may alter the ethical issues at hand. As exemplified by the bioethics laws, this emphasis on the ethical consideration of new technologies has been incorporated directly into legislation.

This policy was applied in a CCNE report in 1997 in which there was re-evaluated the use of embryos in research in light of new technology: the creation of embryonic stem (ES) cell lines from human blastocysts obtained by IVF. Although such research was earlier prohibited under the bioethics laws, with agreement with previous CCNE reports, the CCNE concluded that the therapeutic possibilities of developing this technology actually weighed in favor of a modification of law. This modified opinion is a part of the CCNE's report on the five-year evaluation of the bioethics laws.

CCNE has addressed many other ethical issues and through its Opinions has presented a rich discussion of national position on ethics in technology and health issues. Among the other topics discussed by CCNE are AIDS, drug abuse in the workplace, local ethics committees, human genome research, the utilization of placebos in therapeutic trials involving antidepressants, the care of autistic children in France, and xenotransplantation to name only a few (5).

RESEARCH ETHICS

The Use of human subjects in research is addressed by the law Huriet-Sérusclat of December 20, 1988, modified in July 1994, on the protection of persons participating in biomedical research. The two main objectives of this law are the regulation of biomedical research on humans and the creation of a system of regional Consultative Committees for the Protection of Medical Research Subjects (CCPPRBs). CCPPRBs are charged with reviewing proposals for biomedical research projects as regulated under the law Huriet-Sérusclat. Established by the Minister of Health according to regional needs, CCPPRBs are independent and have legal standing. Members are chosen by the regional government representative or the head regional committee from various specialities in the biomedical field to ensure the committee's independence with regard to ethical, social, psychological, and legal issues.

Under the law, the institution supporting the research project (whether public or private), called the sponsor, is responsible and held liable for any harm that comes to the subject as a result of the experiment. The investigator, the person directing or supervising the experiment, must be a medical doctor. Two kinds of research projects are distinguished under the law: those with direct therapeutic benefit to the subject and those without. The evaluative criteria for the two classes of research differ in order to better protect the subjects. In the case of research without direct therapeutic benefit to the individual, for example, there must not be any foreseeable serious risk to the participants health, and the research must be aimed at people of the same age group, or with the same disease or handicap as that of the of participant. There must further be no alternative way of performing the studies.

In research proposed to have direct therapeutic benefit to the subject, a limited risk to the subject is permissible. However, any risk must be weighed against the potential benefit.

CCPPRB, in their evaluation of research proposals, applies the ethical principle of protection of human dignity which are specifically codified in the law. CCPPRB may consider, for example, the recruitment methods used for the study and check that the methods respect the confidentiality of the subject. In keeping with noncommercialisation and nonobjectification of the human body, and noncoercion of the subject, there must be no volunteer remuneration, save the cost (or loss) to the subject in participating. The recruitment of certain subjects is restricted by law, such as pregnant women, people without health insurance, children, prisoners, the mentally ill, and people in critical medical conditions. Informed consent is required; volunteers must be informed by a medical doctor of the objective, methodology, and duration of the research and of the possible benefits or risks. Consent must be expressed clearly in writing.

CCPPRB also review the scientific validity of proposals. It adheres to three general principles: that all biomedical research on humans must be founded on the current state of scientific knowledge and sufficient preclinical experimentation, that the foreseeable risk of a project must be proportionate to the benefit to the subject or the research interest, and that biomedical research must be aimed at raising scientific understanding of the human being and the methods for ameliorating its condition. CCPPRB shall consult with local and institutional ethical committees if necessary.

REGULATION OF GENETICALLY MODIFIED ORGANISMS

The development of genetically modified organisms (GMOs) destined for food products and for pest control in agriculture has been a big topic of public debate in France over the past decade. The government's position has shifted concerning the different uses of GMOs and changed

several times. This is essentially due to very strong public protest against GMOs. Although administrative regulatory groups (Commission Nationale de classement des recombinaisons génétiques in vitro, later Commission de génie génétique) and specifical rules (decree of July 30, 1985, and AFNOR guidelines), have been in place since 1975 and 1985, respectively, the controversy in France, as in several other European countries, erupted around 1990 with the enactment of the European Council Directives 90/219/EEC on the Contained Use of GMOs and 90/220/EEC on the Deliberate Release into the Environment of GMOs. It should be mentioned that the contained use of GMOs refers to use within the laboratory, and this use has not been a subject of controversy. However, the release and marketing of GMOs affects a larger public which is concerned about the accidental release of such organisms into the environment and other risks associated with their use in food and agriculture.

The two European Directives provide the foundation for regulations regarding GMOs in France; they were transcribed into a single French law 92-654 July 13, 1992, related to the control, utilization, and release of GMOs. This regulation falls under the rubric of environmental protection and modifies a preexisting law 76-663 of July 19, 1976, related to classified installations for environmental protection.

Overall, the European Directives defined two major points. The first was the clear separation within the regulatory framework of the contained use of GMOs and their release into the environment, which includes marketing in the later stages. The second was the establishment of national authorities in all European member states to deal with application procedures as defined under the Directives. In France, although the two Directives have been transcribed into national law under one regulation, two separate national authorities have been established, one dealing with contained use and one dealing with release into the environment and marketing.

Contained Use

The national authority for the contained use of GMOs is the Commission on Genetic Engineering (CGG), which operates under the State Secretary of Research. The purpose of the CGG is to determine measures of confinement suitable to the risks of the use of GMOs, the processes used to obtain them, and the utilization of genetic engineering technologies.

CGG itself is comprised of experts in genetic engineering, public health, and environmental protection plus a member of the parliamentary office for evaluation of scientific and technological options (OPECST). It may also bring in any necessary experts.

Deliberate Release

The deliberate release of GMOs into the environment is mediated by the Commission du génie biomoléculaire (CGB, the Commission on genetically bioengineered organism field releases, or the Commission on Biomolecular Genetics). CGB is responsible for risk assessments related to the release of GMOs into the environment, and advises the Minister of Agriculture and the Permanent Technical Committee of the Selection of Plants and Cultivars (CTPS). CGB is composed of scientists in genetic engineering, public health, and environmental protection, a member of OPECST, consumer groups, and concerned professionals. It is required to produce an annual report.

The conditions for the release and marketing of genetically modified plants, seeds and seedlings, are given by decree 93-1177 of October 1993 and European Council Directive 94/15/EEC on the data needed for the application. The Directive was transcribed into French law by an Order of September 21, 1994.

The authorization of genetically modified plants must also come from the Minister of Environment and the Minister of Agriculture. Notification of releases are done at the local level, at the town hall. In addition, as with all plants, GMOs must be authorized and recorded as official varieties by CTPS. Marketing authorization comes from CTPS and generally requires two years of agronomic testing which must be accompanied by the opinion of the Superior Counsel of Public Hygiene.

Additional legislation relevant to GMOs includes decree 94-46 of January 6, 1994, which sets the conditions for the deliberate release of cleaning substances or products containing GMOs that may enter human or any animal diet through contaminated tools that come into contact with any drink or food substance. The planned release of pharmaceuticals composed of or containing GMOs is regulated under decree 94-359 of May 5, 1994, related to phytopharmaceuticals control. Finally, decree 95-487 of April 25, 1995, sets the conditions for the deliberate release of genetically modified animals.

Public Debate

In France, even with protective legislation in place, the introduction of GMOs into agriculture has met with much public resistance. The main problem is that there are still too many unknowns with regard to the safety of GMOs to humans, animals, and the environment.

For the French public a series of recent technology-related European biological disasters, namely the scandal of HIV contaminated blood and the "Mad Cow" disease problem, have brought home the importance of risk assessment and the precautionary principle. In addition the central place of the French cuisine in the food culture demands more vigilant regard of the process by which foods are produced and of the differences between industrial and nonindustrial produce. The tendency is to see genetically modified food as an aberration or degradation of the natural product. Besides the alarm about the safety of GMOs and the active movement by the agricultural community against GMOs, a real problem of acceptability based on food aesthetics exists in France.

The Precautionary Principle

Precaution is born of the concern the public, has about the risks that may exist in the uncertainty of scientific knowledge and the possibility of blunders creating grave and irreversible damage. New forms of risk are being considered starting with the environment and the effect

on food. The issues focus on how to anticipate the risks and resolve them. Human safety is considered a civil right; this logic is used in decision making to regulate conduct in situations of uncertainty and to guide the regulatory process with prudence.

The precautionary principle is therefore enscribed at the heart of democracy. It must not be taken as a matter of fear but rather as a basic right. Society exercises precaution as it learns from past experiences and uses that information in mitigating future risks. The realm of the precautionary principle is at the interface of science, politics, and law. The principle responds to a priority of our century: safety.

Covering the domain of the environment in French law is law 95-101 of February 2, 1995 (the Law Barnier) on environmental protection. The idea behind it is that in the absence of certainty, scientific knowledge and cutting-edge technology must not defer the adoption of effective and proportionate measures designed to prevent the risk of serious and irreversible damage to the environment at an economically acceptible cost. A report about the precautionary principle from Pr. Philippe Kourilsky and Pr. G. Viney, requested by the Prime Minister, is in progress.

Several structures have been put in place with law 98-535 of July 1, 1998, relative to the reinforcement of sanitary surveillance and the sanitary control of products destined for use by humans. The agencies replace the Agence du Médicament (January 1993) (Medicines Agency) and the Agence Nationale du Médicament Vétérinaire (the National Agency for Veterinary Medicine).

In the surveillance and sanitary control system we find the following groups: the Institut de Veille Sanitaire (the Institute for Sanitary Surveillance), and two new agencies charged with alerting the public when there appear any menace to public health, of any origin. These are the Agence Française de Sécurité Sanitaire des Produits de Santé (AFSSAPS, or French Agency for Sanitary Security of Health Products) and the Agence Française de Sécurité Sanitaire des Aliments (French Agency for Food Sanitary Security). Finally, an institution named the Comité National de la Sécurité Sanitaire (National Committee for Sanitary Security) governs the functioning of these structures.

Unlike the developments leading to the bioethics laws, the regulation of GMOs was not preceded by a large body of government papers and public debate. Rather, the government and public involvement has mostly occurred after the fact, and its effect can be seen in the changes in French policy regarding GM plants over the past decade. In 1995 authorization by the Commission du Génie Biomoléculaire made France the first country to propose the marketing of transgenic corn from Novartis, which received wide European approval in December 1996. The cultivation of Novartis corn in France was authorized in February 1998, and in September of that year the authorization to market Novartis corn cultivated on French soil was revoked by the Conseil d'Etat. In December 1998 the Conseil d'Etat brought the problem of the cultivation of Novartis corn before the European Court of Justice, and in June 1999 the French Minister of the

Environment joined with Ministers of the Environment of several other European member states to declare a moratorium on new authorizations for the marketing of GM plants.

Highlighting the recent public debate on GMOs, and government involvement, is the Conférence de Citoyens (or the Citizens Conference on the Utilization of GMOs in Food and Agriculture), a public discussion event orchestrated by the Parliamentary Office for Evaluation of Scientific and Technological Options (OPECST) in 1998. OPECST'S administrative structure is devoted to technology assessment, with the goal "to inform Parliament of scientific and technological options in order, specifically, to make its decisions clear," as established by Act 83-609 of July 8, 1983. OPECST is an independent organization to which only members of Parliament may refer matters for study. OPECST then "collects information, launches study programmes and carries out assessments." In the Conférence de Citoyens, a representative sample of 15 people was chosen through a survey to participate in a panel of citizens in a public discussion with a panel of experts and government officials with competence in GMOs issues related to food and agriculture. The citizens had two weekends of training prior to the conference, in order to furnish them with an understanding of the principles and issues at hand. The conference was a two-day public media affair, where the panel of citizens posed questions to and entered into discussion with the panel of experts. The conclusions of this panel of citizens is published in annex to the OPECST report on biotechnology in food and agriculture: *From Understanding Genes to Making Use of them.*

Other OPECST reports relevant to biotechnology include topics of biotechnology and the agro-food industry, biodiversity, the life sciences, human rights, and the landmark review of the Lois Bioéthiques, *Application of Law No. 94654 of 29 July 1994 concerning the Donation of Human Body Parts and Products, Medical Assistance with Reproduction, and Prenatal Diagnosis,* by Alain Claeys M.P. and Senator Claude Huriet, 1999 (11).

Despite the strong resistance to the marketing of GMOs, several releases, mostly in the form of field trials, are underway. According to a report published by the European Commission Joint Research Centre, there have been 485 summary notifications for the release of GMOs in France between October 21, 1991, and October 1, 1995. Since each notification covers either the release of one GMO at several sites or an ensemble of GMOs at one site, this number does not represent the total number of test sites in France since 1991, which is much larger. The tests have included 188 strains of transgenic corn, 106 strains of oilseed rape, and 61 strains of sugar beet, among several other species of plants, and they have been conducted by a number of international companies and included French companies and French public research institutions. So, while negative public opinion has had a large influence on the marketing and cultivation of GM plants in France, the number of summary notifications for the environmental release of GMOs in France for research purposes is actually the highest in Europe.

SUMMARY

As a summary we present a list of the principal official French regulatory and consultative groups in the field of biotechnology:

- Commission Nationale de Informatique et des Libertés (CNIL) (law of January 6, 1978)—National Data Protection Commission
- Comité Consultatif de Protection des Personnes dans la Recherche Biomédicale (CCPPRB) (law of December 20, 1988)—Consultative Committee for the Protection of People in Biomedical Research
- Commission de Génie Génétique (CGG) (decree of May 11, 1989)—Commission on Genetic Engineering.
- Commission d'étude de la dissémination des produits issus du génie biomoléculaire (decree of February 23, 1993)—Commission on genetically bioengineered organism field releases, or the Commission on Biomolecular Genetics
- Comité Consultatif National sur le traitement de l'information en matière de recherche dans le domaine de la santé (law of July 1, 1994, and decree of May 9, 1995) Consultative Committee on the treatment of data in research in the health sciences.
- Comité Consultatif National d'Ethique pour les sciences de la vie de la santé (CCNE) (decree of February 23, 1993)—National Consultative Ethics Committee for Health and Life Sciences.

CONCLUSION

To conclude this overview on Biotechnologies and Bioethics in France, we note that France has had 15 years of reflection concerning the ethical and legal issues generated by new technologies in the life sciences. By instituting the bioethics laws of 1994, the government took a step that brought the message to the Minister of Social Affairs, Simone Weil at the time, that the law established "the primacy of ethics over technique." The French politic has thus had the objective of prohibiting eugenic or commercial derivations and sharing with society the choices that the scientific world cannot assume alone. Must ethics be legislated? This is the route that France has chosen; it has preferred a strict legal framework to less constraining guidelines. However, France has inaugurated an unusual statute with this legislation, the provision of a five-year review; a wise decision that will hopefully help avoid conflicts between the progression of scientific research and the general ethic of our society, which is concerned with prevention of eugenics and the appropriation of human by human.

BIBLIOGRAPHY

1. Biotechnologies/France, Available at: *http//biotech.education.fr*
2. Code de la Santé Publique, Dalloz, 1999.

3. G. Braibant, *Sciences de la vie, de l'éthique au droit, Conseil d'Etat, Notes et études documentaires*, La Documentation française Ed, Paris, 1988.

4. N. Lenoir, *Aux frontières de la vie. Une éthique biomédicale à la française, rapports officiels*, La Documentation Française Ed, Paris, 1991.

5. F. Sérusclat, *Les sciences de la vie et les droits de l'homme: bouleversement sans contrôle ou législation à la française? Questions clefs et réponses contradictories, Economica*, Coll. Office parlementaire d'évaluation des choix scientifiques et technologiques, Paris, 1992.

6. J.F. Mattéi, *La vie en questions: pour une éthique biomédicale, Rapport au Premier ministre*, La Documentation française Ed, Rapports officiels, Paris, 1994.

7. Comite National Consultated d'Ethique pour les sciences de la Sante (CCNE), Available at: *www.ccne-ethique.org*

8. C. Chabert-Peltat, *Les Biotechnologies l'Ethique Biomédicale et le Droit, Mémento-guide Alain Bensoussan*, Hermes Ed, Paris, 1995; Alain Bensoussan, Informatique et Telecoms, Editions Francis Lefebvre 1997, reprinted in 1999.

9. A. Claeys and C. Huriet, *Rapport sur l'application de la loi du 29 juillet 1994 relative au don et á l'utilisation des éléments du corps humain, à l'assistance médicale à la procréation et au diagnostic prénatal*, OPECST, Paris, 1999.

10. *Cahiers du comité Consultatif National d'Ethique pour les sciences de la vie et de la santé* (CCNE), Biomed Ed, Paris, quarterly data.

11. J.-Y. Le Déaut, *De la connaissance des gènes à leur utiliisation, Première partie: L'utilisation des organisms énétiquement modifiés dans l'agriculture et dans l'alimentation*, vol. 1, Assemblée Nationale No. 1054, OPECST, Paris, 1998.

See other INTERNATIONAL ASPECTS entries; INTERNATIONAL INTELLECTUAL PROPERTY ISSUES FOR BIOTECHNOLOGY.

INTERNATIONAL ASPECTS: NATIONAL PROFILES, GERMANY

MATTHIAS KETTNER
Cultural Studies Center of Northrhein-Westfalia (KWI)
Essen, Germany

OUTLINE

INTRODUCTION

In Germany biotechnology is expected to be a prominent future industry ranking as high in perceived economic importance as today's information and communication technologies. However, public opinion is sharply divided between advocates of strong public sector involvement in this emergent field of cutting-edge industrial and scientific research and those who fear, for various reasons, that biotechnological "progress" means above all else overrated promises, technoeconomic determinism, and more "colonialization of the lifeworld" (Jürgen Habermas) at ever deeper levels of somatic existence. The resistance against biotechnology has for a long time been focusing on possible risks unique to genetic engineering. Since no strong evidence for this supposition has come forth, lately the "risk debate" has lost much of its momentum. The arguments that appeal to economic shareholders, and to patients and other stakeholders of medical and pharmaceutical progress, are winning as far as the public is concerned. Recent innovations in Germany's legal culture, partly owing to legislation on the level of the European Union (EU) of which Germany is a member state, support the trends on the scope, ease, and speed of patenting. In this sense, one could say that biotechnology in Germany is well on its way of becoming "normal" business.

GOVERNMENTAL SUPPORT OF BIOTECHNOLOGY

In Germany, in the recent past, advancement of biotechnology was a topic of strong political controversy. Many studies were made of public opinion in the 1980s and 1990s and showed that the public perception of biotechnology, specifically the perception of cutting edge genetic technology, tended to be more negatively biased in Germany than in other countries within the EU. At the same time, spokespersons from the economic sector and ministerial representatives of the federal government designated biotechnology as one of the "key technologies," as a field of promising research activity and economic interest for the future of Germany's economy. The federal government's perspective on biotechnology is richly documented in a 1985 official declaration by the Ministry for Research (1). (For a sceptical appraisal of this perspective from a sociological vantage point, see Ref. 2, for an optimistic one from the vantage point of a large life sciences enterprize, see Ref. 3.)

The tension that shaped the state administration's political discourse on biotechnology during the 16 years in which the Christian Democratic Party (CDU) was the ruling majority party in Germany did not ease after the Social Democrats (SPD) 1997 return to power. It also shaped the discourse on biotechnology of nongovernmental organizations that either promoted the advancement and acceptance of biotechnology or a control of developments in biotechnology which they perceived as politically dangerous or ethically unsound.

Political Economy of Biotechnology in Germany

Because of the long product development times in biotechnology, experts claim that the technical application of knowledge produced by the biological sciences will not become a palpable reality until sometime after the year 2000. Forecasting studies of the commercial use of biotechnology have estimated that in Germany it will reach around $2 billion by the year 2000 (compared to $1.25 billion in 1995), with a growth rate of up to 25 percent a year depending on specific products. For the state governments, such estimates serve mainly (1) to justify support measures for research and development in biotechnological sectors and (2) to provide hope against the prevalent impression that economic globalization will mean a growth of joblessness for Germany's economic system. Current estimates are that the biotechnological employment potential in Germany will be at well over 100,000 jobs in 2000. Commercial biotechnology alone is expected to account for up to 40,000 jobs (as compared to 19,000 in 1992), with up to 50,000 additional jobs in the supplier and services sector and another 20,000 jobs in universities or related to academic research. In 1999 roughly 500 firms with substantial or basic biotechnological profiles were counted in Germany (4).

The Federal Ministry for Education, Science, Research, and Technology (Bundesministerium für Bildung, Wissenschaft, Forschung, und Technologie, BMBF), has long supported the biotechnology industry in Germany. BMBF has consistently worked at improving the framework for biotechnological research and development in Germany by giving suitable incentives to industry and start-up companies, and by maintaining the quality of research institutes through state aid in appropriate core areas. In biotechnology the government has effectively assumed the role of a supervisor aiming systematically at creating productive economic alliances, among scientists and promoting a favorable political environment for such alliances. This policy was the focus of an important report (5) on future technology that was sponsored by BMBF and produced by the Karlsruhe-based Fraunhofer-Institut für Systemtechnik, a leading German institute for systems analysis.

One such remarkable initiative is the BioRegio competition (*BioRegio-Wettbewerb*) launched in 1995. This competition enables the best among the biotechnology-regions in Germany to come up with economic strategies to reach new goals. Incentives are given that focus on existing financing, funding, and investment potentials. Out of all proposals a jury selects the three regions with the most convincing biotechnology concepts. In the second phase these three regions receive priority in the appropriation of funds from the Biotechnology 2000 program of BMBF. Currently, the annual funds earmarked for the project are around $110 million. The BioRegio contest has become a way of establishing priorities that helps allocate BMBF funds for biotechnology. The total funding was about for the first phase starting in 1996 $0.6 billion, and this amounts to a clear signal of the priority given the development and implementation of new technologies in the biosciences and molecular medicine. The financial resources of BMBF are envisaged to support basic research, nonuniversity research institutions, and special federal government projects. In the first round, 17 German regions competed. In late 1996 a commendation was given to the Rhine-Neckar Triangel, Munich, and

the Rhineland, and the Bioregion of Jena received a special commendation from the jury for its BioInstruments initiative. The BioRegio competition has fueled the regions' desire to be able to boast as many company start-ups as possible. It has cleverly used the traditional federal nature of Germany and the independence of the regions to initiate an upswing in biotechnology in many locations simultaneously. Moreover the project grants provided by the BMBF's BioFuture program (with an allotted $8 million for the year 2000) are, in principle, available to all applicants regardless of region.

Another development of the government's BioRegio contest was the creation in 1996 of a "virtual enterprise" located geographically in the triangle of the university cities of Braunschweig, Göttingen, and Hanover. This has become the largest self-contained research region for natural and engineering sciences in Germany. This virtual enterprise, called BioRegioN, is designed as a network of dense communication and extensive cooperation among more than one hundred scientific institutes, industry-related facilities, and administrative offices (5). A special consulting Web site for handling issues of financing and siting is available to entrepreneurs who want to start up new biotech firms. Support is available on legal matters concerning the use of genetially modified plants and permits for genetic engineering work as well as for marketing and project organization. Moreover BioRegioN offers research institutes and companies in the region the opportunity to make presentations at German and international trade fairs.

A 1995 study by the federal Minister for Education, Science, Research, and Technology revealed that almost half of the surveyed small and medium-sized companies saw a lack of equity as an essential impediment to innovation. Germany seems to have much ground to cover in catching up in the use of venture capital. In 1995 only 6 of the more than 100 capital investment companies were fully venture capital companies specializing in nascent technological businesses. Partly, this poor showing is due to the absence of large institutional investors, for example, pension funds, in Germany. Recently this problem has been debated as due to Germany's underdeveloped shareholders' culture. However, in 2000 it is becoming easier for biotechnological start-up firms to attract venture capital because a number of banks have set up special services in response to this conspicuous problem.

In the legal realm the federal government has been urging a more flexible legal framework for biotechnology in Germany. In accordance with European directives, the federal government has moved in favor of standardizing the safety and application regulations that open up scope for more efficient methods, with more flexible structure and organization while maintaining risk protection. Germany also pioneered in the implementation and commercialization of biotechnological research and development by an initiative in 1996 that facilitates patenting procedures for research findings. The federal government has further regulatory responsibility for the German contribution to the Human Genome Project.

Since the late 1980s expenditure on research and development (R&D) has declined in Germany in relation to the gross domestic product (GDP). Between 1988 and 1994 R&D expenditure decreased from 2.88 to 2.33 percent. As a results there were considerable declines in research. In expenditure on R&D in the industrial sector declined even more, and public funds have not been able to compensate for the loss of funds. There has been much cause for political concern, particularly in view of the fact that Germany's industries depend heavily on R&D. Moreover, funding for R&D in the biotechnology sector has remained comparatively low in Germany in relation to corresponding activities in other European countries. As a result there is much concern about the efficiency and selectivity in the government's discourse on biotechnology. Funds and human capital need to be managed specifically in prominent future-oriented and pioneering aspects of R&D. The same applies to state aid. Correspondingly the bioindustrial sector has articulated concerns on how to improve the conditions for commercializing results of "pure" biological research in Germany.

Furthermore there are continual concerns about identifying the appropriate policies in order to correct institutional and financial inflexibility currently prevailing in research establishments. Increasingly academe, especially biology and chemistry departments, has had to confront criticism, voiced in the economic sector as well as by curricular commissions and higher education authorities, as to obsolete training and career structures, outdated conceptions of autonomous research ("freedom of science"), and lack of openness for teaching inter- and transdisciplinary skills and methods. An often-cited fact is that German universities have a share of less than 1 percent in biotechnological patent applications compared to about 15 percent in the United States.

To sum up, the record of the federal government's efforts reflects a fairly robust political majority consensus on creating an innovation-friendly framework for biotechnology. This policy encounters opposition for ideological reasons mainly from the "fundamentalist" wing within the Green politial party (DIE GRÜNEN). Apart from the somewhat special case of the Green party, what opposition there is to public policies in support of "biotechnological progress" is scattered across the other three major political parties and seems to be independent of party-line orientations. Owing to the majority political consensus, the government's budgetary preferences for R&D in the biosciences and biotechnology, despite falling overall rates for R&D funding, has not encountered serious political roadblocks.

Technology Assessment and Public Participation

In German law, administrative procedures for high-risk technologies generally involve an element of public participation. The 1990 Genetic Engineering Act, for example, made a public hearing (Erörterungstermin) a legal prerequisite for every license application to release genetically modified plants. The former public knowledge condition stipulated that the public notification of the project involve display of the submitted documents and provide for public opportunity to lodge objections and a public hearing. Anyone could make written objections, and these were then discussed at the hearing. In 1993 there were three hearings on applications for the release of genetically modified potatoes, sugar-beets, rape, and maize. The hearings each received between a few hundred and 20,000 written objections. The procedure, it turned out, frustrated both sides. For this reason, the provision of direct public participation was removed and substited by a much less demanding requirement for a written submission with the first amendment to the Genetic Engineering Act in 1994 (6).

Among the German people there is the feeling that industry and politics, and to some extent even the biological sciences, lack credibility, transparency, and democratic accountability in biotechnical matters. Neither concerted effort for citizen debate by the government nor various public relations campaigns by the biotech-industry have so far been able to dispell this attitude. Sociologists are apt to explain this as indeterminate suspicion caused by a knowledge gap. Nevertheless, no number of publications in every media form has translated into a corresponding degree of knowledge among the population. This knowledge gap itself requires more clarification. Political scientists concerned with technology assessment point to the fact that with certain exceptions, such as the citizens' forums in Baden-Württemberg (8) managed by one of Germany's leading institutes for technology asessment (Akademie für Technikfolgenabschätzung, Stuttgart), there has been too little open public discussion of the opportunities and risks of biotechnology. Pounding out information packages is one thing. Promoting attitudinal change through free and open public dialogue is quite a different thing. Government and industry have been strong on the former but weak on the latter communication policies.

Observations on cases where the general public was invited to take part in decision-related discussions (e.g., about sites for genetic engineering plants) indicate a number of recurring deficits and problems: (1) Those in positions of responsibility often fail to answer the questions that really concern critics and the public at large. This might be so because such questions tend to be of a fundamental nature. Frequently they concern consequences of technology on a broad cultural and societal scale. (2) Sometimes exaggerated expectations are aroused regarding the technical and economic potential of genetic engineering. The much-hyped gene therapies for cancer are a case in point regarding "red" (i.e., medically related) biotechnology. Other prominent examples are furnished by unrealistic pronouncements about expected positive impacts of biotechnology on employment and economic growth figures. (3) The depiction of biotechnology in the mass media is disturbingly polarized. Industry-financed advertising campaigns are apt to present progressive biotechnology as a safe panacea, whereas professional journalists writing in the mass media tend to capitalize on hypothetical risks of biotechnology (9,10). Consequently any balanced reporting that is both interesting and beyond obvious partisan interests is structurally disadvantaged and blotted out.

Between 1991 and 1993 a technology assessment (TA) of crop plants with genetically engineered herbicide resistance in agriculture was organized by the Wissenschaftszentrum Berlin, Germany's leading sociological research center (11). Herbicide resistance was chosen as the subject for TA because it seemed to be sufficiently relevant and controversial. It could be expected that a broad spectrum of developmental problems of modern biotechnology would be considered in the process, such as (1) the possible risks of transgenic plants, (2) the toxicological and ecological effects of the use of nonselective herbicides, (3) the future of genetic resources, (4) the advantages and disadvantages for farming, (5) the long-term safeguarding of world food supplies, and (6) the ethics of plant manipulation.

This pioneering TA project was not merely a forum of experts for evaluating the state of knowledge on the possible consequences of a technology; rather, it provided an arena in which the social conflicts related to the introduction of a new technology could be articulated and discussed in an exemplary manner. The procedural scheme was somewhat conventional in taking an emerging technology-induced development as its starting point for an analysis of possible desirable and undesirable consequences. The goal of technology-induced TA is to determine the political actions that might be necessary in order to cope with that technology. Critics of the Berlin herbicide resistance TA project called instead for a "problem-induced" approach. The starting point would shift to the social problem the technology purports to help solve (e.g., the agricultural problem of weed control). In problem-induced TA, various ways of tackling the problem would be compared (e.g., solutions by intensive industrial farming in comparison to solutions provided by ecological farming). Any comparisons would take questions of larger social context and fundamental political issues into account. The decisive question for transgenic herbicide resistance technology when considered from a "problem-induced" TA-perspective would have been whether the technology was needed and what kind of farming was socially desirable and ecologically acceptable. Problem-induced TA was found to be more demanding in terms of time, money, and other resources than the technology-induced approach. In retrospect, the Berlin project aimed at a broad and demanding interpretation of TA with resources appropriate only for dealing with narrow TA.

Participants from research institutions, industry, environmental groups and other NGOs, and governmental agencies (totaling 48 groups) reflected the interests and positions of the ongoing political concerns over biotechnology. Among them were outspoken advocates and critics. The debates that normally take place outside the TA procedure and only heat up as soon as results of "closed" TA are made public, were internalized in the procedure. TA became a social dialogue between representatives of opposing positions. After a series of meeting, the participants had to define a study framework, evaluate the results of expert reports, and discuss any conclusions. The idea was that a dialogical framework would promote a rational form of discussion. Whether this expectation was fulfilled remains an open question. As a matter of fact, shortly before the final meeting was to take place, the environmental associations announced their withdrawal.

The Berlin TA project reenergized the scientific and public debate in Germany on the proper form, scope, and aim of assessment of biotechnology and other new technologies. There is now emerging a consensus that a narrow TA is necessary but not sufficient for the democratic governance of technological development to flourish. The narrow TA is essentially an investigatory strategy aiming at the production of information where the validity of such information is conditioned by truth-claims and not by the factual acceptance by a majority of participants. The narrow TA is not to be conflated with purely political dialogues or consensus conferences where discussion focuses on goals and criteria of desirable development for society. The narrow TA contributes factual information about potential risks and expected advantages. Citizens have a right to identify the state of knowledge on politically controversial subjects. The TA procedures can at least advance answers as to whether publicly declared risks actually exist, whether claimed advantages exist, and so forth. Criticism of the methods, scientific and otherwise, through which the procedure arrives at its conclusions must also be submitted to the scrutiny of the public. Consequently technology-induced assessments, though narrow, must not be so narrow to involve only scientific experts. All stakeholders are to be considered in any political conflict over a new technology for a rational assessment of factual and other validity-claiming and procedural fairness.

LEGAL ASPECTS: SALIENT STRUCTURES IN THE REGULATION OF BIOTECHNOLOGY

Germany's Genetic Engineering Act (Gentechnikgesetz, GenTG) adopted in 1990 seeks to regulate the approval and registration procedures for genetic engineering facilities and for genetic engineering work geared to research and commercial purposes on microorganisms (viruses, bacteria, fungi, parasites) and macroorganisms (plants and animals). The GenTG does not cover reproductive medicine nor the use of somatic gene therapy.

The GenTG, in its amended 1993 version (12), is intended to safeguard the life and health of humans, animals, and plants; to protect the environment as an integrated system; to shield property from possible risks of genetic engineering methods and products; to prevent any such risks from emerging, and to create a legal framework for research, development, promotion and use of the scientific, technical and commercial possibilities of genetic engineering.

The GenTG legally defines a consensus-building processes within a network of agencies that share decisional power, political responsibility, and scientific competence concerning the issues that arise within the scope of the GenTG's regulatory framework. The network defined by the GenTG comprises four principal actors on the national level: (1) the Biological Federal Institute for Agriculture and Forestry (Biologische Bundesanstalt für Land- und Forstwirtschaft) that is situated within the Federal Minstry of Food, Agriculture and Forestry, (2) the Federal Environmental Protection Agency (Umweltbundesamt), in cases of animal applications, (3) the Federal Research Center for Virus Diseases in Animals (Bundesforschungsanstalt für Viruserkrankungen der Tiere).

In accordance with relevant European Community Directives, Germany's GenTG determines that a scientific body, namely (4) the Robert Koch–Institute (RKI) in Berlin, integrate the relevant consensus-building processes into resulting decisions. The RKI's Department of Genetics and Genetic Engineering serves as the institutional base of special advisory committee set up by the RKI, the Central Advisory Committe for Biological Safety (Zentrale Kommission für Biologische Sicherheit). This committee comprises a broad range of scientific and socially relevant points of view. Its 30 expert members come from micro- and cell biology, genetics, hygiene, virology, ecology, the trade unions, occupational safety, economics, research-promoting organizations and environmental protection organizations.

European Community Law and German Law

On the national level of law, the German GenTG implements directives of European law. Mainly three directives of the European Council are relevant: (1) Directive 90/219/EEC (13) stipulates joint measures to be implemented for the application of genetically engineered microorganisms in contained situations in order to safeguard human health and the environment. (2) Directive 90/220/EEC (14) on the intentional environmental release of genetically modified organisms (GMOs) serves both to promote the formation of the single European market for methods and products of genetic engineering and to constrain this process by the observance of suitable environmental and health protective considerations. (3) Directive 90/679/EEC (15) contains minimum requirements for the protection of employees in countries of the European Community against exposure to safety and health hazards caused by biological substances.

Unlike Germany's GenTG, the European Community directives include no special liability provisions. Approvals based on the directives focus on the genetic engineering product and not on the genetic engineering facilities, whereas the German GenTG prescribes extreme hazard liability for the operator of the facility concerned. In the European Community directives, hearings are only an optional feature. In the German GenTG, public hearings remain a legal requirement in approval procedures for operating sites, notwithstanding that the 1993 amendment of the GenTG narrows this requirement to commercial projects classified as "risky" or "very risky." (There are four legally recognized safety levels ranging from 0, no risk, to 4, very risky.) Despite these comparatively more demanding national requirements, the GenTG in its 1993 amendment fully exploits the scope for simplification and acceleration of approval procedures contained in the directives on the legal level of the European Union (EU).

In specific, the legal framework that regulates genetic engineering in Germany contains the following legal ordinances:

1. Genetic Engineering Safety Ordinance (*Gentechnik-Sicherheitsgesetz*) (16). This Ordinance is the most important legal ordinance. It consists of safety requirements for genetic engineering work, risk assessment of organisms, and safety measures in laboratory and production facilities, greenhouses, and animal facilities. Also covered are occupational safety, and sewage and water treatment, requirements that have to be met by project managers and biological safety engineers.

2. The amended Genetic Engineering Hearings Ordinance (*Gentechnik-Anhörungsverordnung*) (17). This Ordinance restricts public hearings to approval procedures for safety levels 3 and 4 of commercial projects. It regulates the formalities of hearings. It allows simplifications and modifications that favor the operator while still serving the purpose of protection.

3. Amended Genetic Engineering Records Ordinance (*Gentechnik-Aufzeichnungsgesetz*) (18). This amendment specifies the content, structures and time period for genetic engineering records. The amended Ordinance for the first time included environmental release in the range of requirements. The amendment introduced simplifications for safety level 1 projects. The safekeeping periods prescribed for records amount to 10 years for safety level 1 and 30 years for levels 2 to 4 and for environmental release. Laboratory logbooks are not an admissible substitute for proper records.

4. Amended Ordinance on Genetic Engineering Procedure (*Gentechnik-Verfahrensverordnung*) (19). This amendment regulates the formal documentation requirements to be met by facility operators in the registration and approval procedures for genetic engineering facilities, genetic engineering work, and environmental release and introduction of GMOs. The amendment has again enabled simplifications, clarifications, and accelerations. A distinction is made between genetically engineered "minor" and "major" plants.

5. Amended Ordinance on the Central Commision for Biological Safety (*Verordnung über die Zentrale Kommission für die biologische Sicherheit*, ZKBS) (20). This amendment describes the tasks, capabilities, and internal structure of the ZKBS.

6. Federal Cost Ordinance to the Genetic Engineering Act (*Bundeskostenverordnung zum Gentechnikgesetz*) of 1991. This Ordinance stipulates the fees and charges for offical acts performed by the Robert Koch–Institute as a senior federal authority. Fees for granting approval for environmental release range from $3,000 to $80,000, and fees for obtaining approval of introduction into circulation from $4,000 up to $170,000 for extremely expensive procedures.

7. Genetic Engineering Participation Ordinance (*Gentechnik-Beteiligungsverordnung*) (21). This Ordinance regulates participation of the European Council, the European Commission, and the authorities of the member states of the EU and those of the European Economic Area in the approval procedures for environmental release and introduction as well as in procedures for any supplementary measures based on the German Genetic Engineering Act. It is the

legal basis of the Robert Koch–Institute's obligation to observe set time limits in transferring received applications (for environmental release or introduction) to the European Commission. It also details the institute's right to information if such applications are submitted in other EU member states.

Intellectual Property Rights and Patenting

Industrial property rights are protected in Germany by the 1980 Patent Act (*Patentgesetz*, PatG) (22,23), by the 1973 Agreement on Granting European Patents (24) and the Species Protection Act (*Sortenschutzgesetz*) (25). PatG provides comprehensive regulation of all aspects regarding the patented item, such as patenting preconditions, effects of the patent, patent-granting procedures, revocation proceedings, patent infringements and their consequences. PatG is a major economic and political instrument for promoting innovation. The granting of exclusive rights over a limited time for the exploitation of inventions provides incentives. Inventors are rewarded, early revelation and dissemination of technical findings and know-how are safeguarded.

The German PatG, in keeping with European agreements, excludes from patenting any method invented exclusively for surgically or otherwise treating or diagnosing maladies of the human or animal body. Gene therapies, according to this criterion, are not eligible for patenting, whereas biopharmaceutical products produced by gene therapies are eligible. Ruled out from patentability are discoveries that are not inventions. By this criterion, nucleic acids, and more or less complete indications of sequences of nucleic acids, do not qualify for patent protection if the availability of the substance is merely based on a discovery. Likewise the PatG excludes "essentially biological" methods for breeding plants or animals, or creating new plant or animal species. Microbiological methods and their products, however, are eligible for patenting. Until recently this implied that there was no patent protection at all for generic creations like transgenic plants and animals. In comparison to the United States and Japan, this put the German bioindustry at serious disadvantage with regard to plant and animal biotechnological creations.

Things have changed somewhat with a recent amendment to PatG. The new rules of the General Agreement on Tariffs and Trade (GATT) took effect in Germany in 1995. All member countries must now modify their national intellectual property right laws so that they match with the new agreement on trade-related intellectual property rights (TRIPs). Since the issue of TRIPs and their discretionary leeways continue to be notoriously controversial, the GATT negotiations stipulated a review period in 1999 for the enforcement of the agreement on TRIPs. In September 1999 the GATT–TRIPs agreement was accepted by the European Patenting Organization in the format of the European Biopatents Directive, which in turn was enacted by the European Patent Office in Munich. Whereas former article 53b of the European Patent Agreement clearly ruled out the patenting of plant varieties and animal species, the recent Biopatents Directive allows for a much more patent-friendly legal interpretation on these crucial

points. However, the acceptance of the Biopatents Directive in Germany and the EU generally is so far only provisional, pending a ruling by the Supreme Chamber of the European Patent Office on a number of complaints about would-be patented genetically modified plants and, moreover, pending the European Supreme Court's decision on complaints against "the patenting of life" that have been filed by Italy and the Netherlands.

In the issue of deregulating European (and hence also German) patenting restrictions the parliamentary assembly of the European Council (*Europarat*) in Strassbourg is pitted against the bureaucracy of the European Commission (*Europäische Kommission*) in Brussels. In September 1999 the former organization, representing 42 European states, has recommended that patenting of genes, cells, tissue, and organs be prohibited altogether, whether they be of human beings, animals, or plants (the key argument is that these entities are discoverable but not inventable). Interestingly the former organization is perceived in the political public sphere as having far more democratic standing than the latter. This makes it easier for biotechnology critics to capitalize on the issue of patenting by interpreting it as an antagonism between a democratically incorporated humanistic ethos and the organized commercial interests of business corporations and their lobbies.

Consumer Sovereignty and "Novel Food"

Big agro- and food companies began promoting in the second half of the 1990s the prospect that within a few years a new generation of genetically modified novel food products would attract consumers through the appeal of both superior quality (longer lasting and tastier fruits and vegetables) and lower prices as compared to natural food. Commercially, this confident vision continues to prevail.

Public outrage at biotechnologically produced or modified "novel" food is reflected in German food legislation. This legislation took the form of the Food and Consumer Act (*Lebensmittel- und Bedarfsgegenständeverordnung*), numerous follow-up ordinances, and regulations pertaining to specific products (e.g., milk legislation), plus directly applicable EU law. The aim of this legal framework is to safeguard human health and to protect the population against fraud and deception. All provisions apply equally to standard products and products manufactured by genetic engineering methods. Specific differences, however, remain. The GenTG, with its special regulations on introducing products into circulation, refers to food products or food ingredients that contain or consist of GMOs. However, these regulations do not apply to food products or food ingredients that are manufactured from GMOs without containing any such organism. These foods remain subject to the "general food law."

Thus regulatory uncertainty exists about the proper implementation of the EU novel food regulation (NFR). As far as genetic engineering is concerned, NFR refers to (*1*) introducing novel food products and novel food ingredients into circulation within the EU; (*2*) food products and ingredients that are "novel" in the sense that they were previously not widely used for human consumption in member states; (*3*) products and ingredients that contain or consist of GMOs, or (4) products and ingredients that were manufactured from but do not contain GMOs.

Exempt from the regulation are food additives, flavorings and extraction solvents. They are subject to other special EU regulations. This exemption has recently become the focus of much public concern. In one interpretation at least, it creates a gap in the generally desirable principal goal of NFR, namely to ensure that novel food products and food ingredients do not pose a hazard for the consumer. According to NFR, approval decisions for novel food check for environmental compatability in order to prevent any harmful effects of GMOs on the nonhuman environment. In addition the Scientific Food Committee of the EU must give an opinion on all aspects of food that relate to public health.

One considered hazard is that novel food differs so greatly from comparable products and ingredients it intends to replace that normal consumption patterns would result in nutritional deficiency in people. However, the hazardous nutritional deficiency argument, though frequently marshaled by ecological consumer organizations and other critics of "Frankenfood," fails concerning mere additives, flavorings, and the like. Another critical argument — call it the consumer sovereignty argument — has better logic. It has also drawn more public support. It is a second explicit goal of NFR to ensure that nutritional improvements (biotechnological or otherwise) do not lead to consumer confusion. This goal is subverted, according to the consumer sovereignty argument, if consumers who (for whatever reasons) repudiate novel food completely lack the necessary information for making fully informed choices in following their convictions.

The appeal of the consumer sovereignty argument becomes lost if companies begin to discharge their requirement for full consumer information in a way that turns labels for food products into something like the patient information leaflets that accompany prescription drugs. Legally, how much leeway biotech-food companies have for their labelling policy is defined by conditions that are already quite demanding: (1) Food products containing or consisting of GMOs must be labelled under all circumstances. (2) They must also be labelled if the use of genetic engineering methods means that they are no longer equivalent to an existing food product or ingredient. "Not equivalent" means that a scientific assessment based on appropriate analysis of existing data indicates that the tested features differ from conventional food products or ingredients. Accepted limited values for natural fluctuations in these features are to be taken into account. If a product or ingredient is no longer seen as equivalent, the label must indicate the changed features or characteristics along with the method used to bring about this change (so-called method labeling). (3) The label must also indicate substances that are not present in existing equivalent food products and that could affect people with certain health problems like allergies or that could foreseeably violate ethical or religious dietary restrictions.

The preamble of NFR states that a label may declare that a food or ingredient is not a novel product in the sense of the regulation. A label "free of genetic engineering" or "without genetic engineering" is allowed, pending proper definition. Germany submitted a proposal to the EU in 1998 for defining this label, and still is awaiting approval. The proposal aims to require that labels state whether raw materials from transgenic plants, enzymes, or additives for flavoring, and the like, were derived from GMOs or used in food production. This would include animal feed and feed additives derived from transgenic organisms, even if the end product (e.g., milk from cows fed on transgenic soy) is chemically indistinguishable from the product of animals fed conventionally. What is decisive is that the labelled products must be garanteed not to have been in contact with genetic engineering. To give an example, though genetically modified barley, transgenic hops, or genetically modified yeast are not commercially available in Germany, most beers cannot be labelled "without genetic engineering" because conventional yeasts are fed GMOs during their fermentative reproduction.

ETHICAL ASPECTS: CATEGORIES, ASSESSMENTS, AND CONTROVERSIES

Genetic engineering and the underlying molecular biological research is viewed, in Germany as elsewhere, as cutting-edge biotechnology. In the public's mind genetic biotechnology comes in three colors: It is red if its application is mainly medically related (in the diagnosis and treatment of diseases), green if mainly related to agriculture and food, and gray if used for purposes of environmental conservation and the protection of natural resources, ranging from pollutant disposal to environment-friendly bio-mining processes.

If a 1995 survey of public opinion is any indication, the percentages of positive (+), neutral, and negative (−) attitudes toward certain applications of biotechnology are nearly the converse of each other in terms of the red and green ends of the spectrum of applications. The most positive results were for the diagnosis of incurable diseases, 75+ and 7− out of 100; negative attitudes increased for treatment of cell diseases, production of vaccines, use of modified bacteria to reduce oil pollution in the soil, modification of crop resistance against insects or plant diseases, breeding laboratory animals for pharmaceutical research. The most negative result was for the modification of the taste, keeping quality or appearance of food, 8+ and 76− (26,27). The basis for these attitudes is probably a mix of conceptions. Clearly, the big differences in the attitudes are to some extent due to the fact that the different practices to which biotechnology is applied are loaded with different ethical background beliefs. Biotechnology and genetic engineering in particular continues to be perceived in public opinion as something special resisting easy assimilation to "normal" new technologies (e.g., energy, information, and communication technologies). The new biotechnology is mysterious and evokes a pandora's box. The negative attitude persists despite an increasing evidence that genetic engineering biotechnology keeps dispelling all initially raised claims of associated high and unusual risks compared to other large-scale new technologies. Public risk perception, even if well informed, and moral public opinion are surprisingly independent of each other.

Recent analyses (28) of public debate about and public perception of genetic engineering in Germany point to the fact that it would be unrealistic to expect broad acceptance of such biotechnology to be achieved by media campaigns, improved distribution of information, inclusion in school-curricula, and the like. All that can be realistically expected is that a consensus on the acceptability of a limited number of specific applications where the application is, and is perceived to be, juridically regulated and politically controlled in ways that ensure that relevant ethical convictions (whether rooted in rational and secular views or in religious comitments) that are held by substantial groups of citizens are being respected.

In reaction to the strategically insurmountable momentum of ethical beliefs concerning biotechnology, the European Association for Bioindustries (EuropaBio), a business interest nongonvermental organization in Europe representing 45 multinational corporate members and 4 national associations that total around 600 small and medium-sized biotechnology enterprises, created an Advisory Group on Ethics (AGE) in 1997. The first task set for AGE was the crafting of a document on *Core Ethical Values*. This code of ethics has achieved adherence among EuropaBio members independent of their respective national legislation, and it contains commitments to ethical values and ethically important goals in all major fields of biotechnological applications. It is less specific and less stringent than the *Convention on Human Rights and Biomedicine* (29) that was crafted by the Council of Europe. The convention was officially adopted in 1997 and already ratified by six member states, though not by Germany. At present the hub of distinctively ethical normative regulation of biotechnology in Germany is informal in nature (by nonmandatory codes of ethics) rather than formal (by actual law).

The use of philosophical arguments of applied ethics in Germany is still very tentative. Four well established centers dominate the research: The Academy for Ethics in Medicine located in Göttingen, the Center for Ethics in the Sciences in Tübingen, the Center for Ethics and Science in Bonn, and the European Academy for the Study of Technology in Bad Neuen-Ahrweiler (30). Each of these major centers has invested a considerable part of its capacity into biotechnological research. The growing literature on bioethics generally in Germany is replete with attempts to appropriate arguments that have already come to enjoy some currency in contexts of applied ethics in England or the United States. This fact could attest to a level of similarity in the ethical problems raised by global progress in biotechnology independent of the cultural contexts. Such similarity could also reflect a level of homogeneous academic discourse layering itself over heterogeneous cultural and moral inclinations at an intuitive level. At the level of theory, it is noteworthy that one of the above-mentioned centers, the Center for Ethics in the Sciences, addresses social ethics. Unlike prevalent notions of principlism (e.g., the bioethics principles addressed at the Kennedy Institute in the United States), a social ethics approach emphasizes all reliable moral considerations that have already been worked out in cultural traditions and affect the target practices.

Moreover a social ethics approach works with what is, technically speaking, a "wide reflective equilibrium methodology" and takes into account a wide spectrum of religious, juridico-legal, and other relevant normative textures and theories in modern society, modern law, technological progress, risk asessment, and so on. As the qualifier "social" indicates, a social ethics approach has its integrative frame in taking seriously the impact that the foreseeable social change is likely to make a on moral conceptions that shape a society (30).

Gray, Green and Red

Gray Biotechnology. Although gray biotechnology is frequently hailed by the government as a dynamic and commercially rewarding field (32), there is at present hardly any ethical controversy about gray biotechnology. Speculating about the deeper reasons for the comparative moral inconspicuousness of gray biotechnology could be interesting, but it is safe to say is that a good reason is that no dramatic accident has occurred up to now. Absent something like gray biotechnology's Chernobyl, public attention will continue to focus on the more salient perplexities in red and green biotechnological developments (33). A recent development in public debate is an attempt to add ethical merit to gray biotechnology by pointing out its potential for political strategies of sustainable development. Such strategies enjoy much political enthusiasm in Germany and they have large support among the public. Interpreted optimistically, sustainable development presents itself as a viable response to most of today's pressing environmental problems.

Green Biotechnology. In public debate the following points provide most of the standard argument in favor of green biotechnology: Because of global population growth, and limited agriculturally utilizable terrain, the food problems of future generations must be solved by an environmentally friendly increase in yield and by the production of high-yield, low-cost food by means other than conventional methods in agriculture and in the foodstuff industry with new biotechnological methods and products. This argument is invoked for justifying genetic engineering based strategies for maximizing agro-industrial turnover as well as for associated strategies for minimizing the deployment of herbizides and insectizides in combination with suitably modified crops.

Despite efforts to provide an ethical underpinning for green biotechnology by linking it to moral responsibility for future generations, the vision of green biotechnology has not fared well in Germany (34). Many people have reservations or even reject its use in agriculture and in the food industry. Surveys indicate that the public sees no adequate benefits that could compensate for existing negative expectations and anxieties. For the individual consumer, it is obviously of major significance that products of green biotechnology have concrete and palpable advantages. By contrast, most advantages of biotechnologically "improved" food that have been touted by the biotechnology industry do not refer directly enough to the final product. Public relations campaigns designed to swing consumer attitudes in favor of bio-food are widely recognized as poorly concealing the fact that the would-be advantages are essentially commercial advantages on the side of bio- and agro-industrial entrepreneurs, and not on the side of the Consumer. At the moment, consumers do not perceive the show-cased advantages (e.g., the Flavrsavor tomato) as personally benefiting. Even tangible product improvements like longer shelf lives for fruit and vegetables do not automatically result in greater consumer appreciation in the face of the ample range of food products offered in Germany and in advanced capitalist countries generally. Also there is an as yet small, but in economic terms increasingly important, segment of the population that cherishes "natural" food products notwithstanding its comparative cost disadvantage.

General knowledge about green (genetic) biotechnology, especially about food stuff production, is not very good on the average in the population. Attitudes toward food are frequently characterized by culturally validated consumption patterns and a good deal of wishful thinking concerning the quality and the production process of preferred food. These attitudes are not shaped by rational choices in terms of cost–benefit calculations in everyday shopping. This, the green biotech-industry has been very slow in realizing. Many German consumers are afraid not only that gene technology is creeping surreptitiously into their food without their consent, but also that they are being misused as guinea pigs by agro-industrial corporations. Highly sensitive analytical techniques (e.g., PCR) can detect even minute quantities of transgenic ingredients, like soybeans used in the production of soya oil. Soya products are used in thousands of different foods. The fact, much publicized in the media, that Monsanto's "Roundup Ready" genes can now be detected even in tofu made from soya that was grown on controlled ecological farms engenders in consumers an insidious feeling of powerlessness. On the other hand, with the attractiveness of modern lifestyle preferences mounting, more people are becoming interested in consuming "functional" (e.g., high energy) foods. It will be interesting to observe how prevailing patterns of resistance to green biotechnology in Germany will change when vitamin enhanced vegetables and fruits become available in the supermarkets (35).

Red Biotechnology. Advocacy for red biotechnology in Germany mainly follows the conventional argument of medical utility: "Our rapid increase in understanding molecular genomics and genetic engineering has brought us to the brink of a health care revolution. Biotechnology brings the tools (gene therapy, recombinant proteins, and cellular therapies) for not only curing common diseases but also many rare diseases." The medical promises of biotechnology have materialized so far mainly in the pharmaceutical sector. For instance, in 2000 there were 42 (compared to 29 in 1998) medical drugs with genetically engineered components fully licensed in Germany (36).

Unsurprisingly, genetic engineering technologies and diagnostic techniques based on molecular genetics are generating most of the moral perplexity found in red biotechnology. In Germany as everywhere, great anxiety is being raised about the benefits of somatic cell gene therapy. Any original enthusiasm was dampened by the medical profession failing to deliver on the goods that had been promised by biotechnological advances. Ironically, somatic gene therapy has turned out to be technically more intractable than first thought but ethically less perplexing. At first arguments likening somatic gene therapy to other forms of transplant therapy won the day in the German debate. A trail-blazing government report on the chances and risks of gene-technology concluded as early as 1987 that "somatic gene therapy is a special form of transplant therapy" and that the "transfer of genes must be evaluated in the same way as the transfer of living material" (37). The German Association of Physicians (*Bundesärztekammer*) followed with similar policy statements in 1989. If successful therapies were available, many people would view the prospect of somatic gene therapy as less problematic than organ transplants.

Germ line gene therapy is ethically taboo in Germany, and it is legally prohibited by the 1990 Embryo Protection Act (*Embryonenschutzgesetz*) (38,39). This act prohibits (*1*) the sale, use, or acquisition of in vitro fertilized eggs for all purposes other than for an intended pregnancy. It also prohibits (*2*) the generation of more than three embryos per cycle in in vitro fertilization (IVF) procedures, (*3*) the selection of sex, (*4*) fertilization involving gametes from dead persons, (*5*) deliberate altering of the genomic information of gametes intended for procreation, (*6*) embryo cloning, and (*7*) the production of animal-human chimeras. These restrictions (with the exception of 2) in Germany's national law overlap more or less with relevant articles of the aforementioned European Convention on Bioethics (Articles 13, 14, and 18). However, the Convention has not yet been, and probably will never be, juridically implemented in Germany. The Convention has been politically attacked by relevant groups of stakeholders (e.g., by organizations representing the interests of disabled people) as being unduly liberal especially in matters foremost concerning research on subjects incapable of giving informed consent. The ethical arguments against germ-line gene therapy, cloning, and enhancement genetics that have gained most currency in Germany are (*1*) the violation of human dignity in identity-altering genomic manipulations and thus the integrity of human beings, and (*2*) the slippery slope of socially amplified genetic discrimination which recalls the outragious practices of Third Reich "eugenics." The metamorphosis of "healing" into "enhancement" and the interpretation of "eugenics" as "breeding" easily evokes the image of human beings treated as cattle. Many people in Germany associate eugenics not only with mass murder policies by totalitarian states but also with sexism and a patriarchal attitude toward women (40,41). This dramatic subtext of eugenics in Germany tends to distract from the disquieting fact that socially conditioned value-of-life judgments and selective decisions enter often into the "private" realm of parental responsibility as more and more prenatal diagnostics becomes a matter of course in "normal" pregnancies.

The German Research Association (*Deutsche Forschungsgemeinschaft*, DFG) cooperates with the BMBF in managing Germany's part within the global Human

Genome Project. The German Human Genome Project (DHGP), established in 1995, has a budget of about $30 million a year (42). Integral to DHGP is a program for the study of ethical, legal, and social issues. Between 1996 and 1999, during the first phase of this ELSI-program, seven interdisciplinary conferences were sponsored with a budget totaling about $0.3 million. Theses conferences ran the gamut of interesting topics, but very little overarching direction was achieved besides a set of general topics that function as preferences when new project grants are offered. Conferences were held on (1) Genetic Knowledge and Human Self-Understanding, (2) Talking Human Genetics: Verbal Communication, Knowledge and Genetic Make-Up (43), (3) The New Genetics: From Research into Health Care. Social and Ethical Implications for Users and Providers (44), (4) Predictive Genetic Tests, (5) Ethical and Legal Problems in the Patenting of Genetic Information, (6) Postgenomics? Historical, Techno-Epistemic and Cultural Aspects of Genome Projects, (7) The Human Analyzed (45). For the second phase of the DHGP's ELSI-program, these preferences are (1) ethical, legal, and social aspects of human genome research in practice (e.g., informed consent, protection of privacy for genome-related information, questions of patenting and of the commercialization of research), (2) the application of genetic testing (e.g., quality of genetic counseling and diagnostics, genetic testing beyond the confines of medical human genetics, aspects of health care economics), and (3) the cognitive apraisal and the social context of human genome research (e.g., social and cultural differences in the handling of genetic knowledge, impact of genetic knowledge on the concepts of prevention and malady, and sociological metaresearch on the human genome project itself).

The DHGP has not raised much protest in comparison with other contested technologies in Germany (like nuclear energy). What moral outcry there is applies indiscriminately to the global Human Genome Project. Prominent are arguments on the subversion of human dignity in critical debates about bioethics, and this is perhaps more pronounced in Germany than else where. There is a mixed coalition of small groups of diverse political orientation that oppose any research with a potential for furthering prenatal selection of human traits and other forms of what they perceive as genetic discrimination. Some of these groups are radically suspicious of the integrity of bioethics as a scholarly discipline (46). Besides such insignificant "fundamentalist" opposition against biotechnology, major nongovernmental organizations, such as *Gen-ethisches Netzwerk*, have developed a highly differentiated dissentious culture of counterexpertise and civic mobilization, and brought forth excellent journals (e.g., *Gen-ethischer Informationsdienst, Wechselwirkung, Politische Ökologie*) for critically monitoring developments in biotechnology (47).

Arguments Against "Patenting Life"

Patents, trademarks, and copyrights, are forms or intellectual property protection. The practical point of such protection is to ensure, for a certain period of time, that the individual or a corporation that rightfully claims to have invented that product or technology maintains an exclusive right to make, use, and sell a new product or technology. Any intellectual property rights regime can be construed as a purportedly legitimate compromise between industry's desire to capitalize on its investments in technological development and the prima facie justified claim of society to benefit from the knowledge and resources of its members in a terrain of activities that are only made possible by that society. However, embedded in the normative texture of the (amended) German patent law is a moral component that goes beyond those moral considerations of commutative justice that are pertinent to societally useful inventions and investments.

This moral amendment can best be understood by considering which exclusions from patentability of biotechnological inventions are claimed specifically on ethical grounds. As it turns out, not patentable in Germany (and generally in those EU member states that have implemented Article 6 of the Directive 98/44/EC) are processes (1) for cloning human beings, (2) for modifying human germ-line genetic identity; (3) for use of embryos in industrial or commercial capacities; (4) for modifying the genetic identity of animals that are likely to cause them suffering without any substantial benefit to human or animal, and (5) for creating new animals from genetic modifications.

The popular claim against permitting the patenting of genetic sequences in living organisms ("no patents on life") makes the following key arguments:

1. Patenting would blur the conceptual distinction between discovery and invention for the sake of vested private interests. It upsets the carefully established balance, controlled by that distinction, between a monopolistic commercial privilege and an associated benefit for the common good.

2. Patenting would regionally and globally threaten genetic diversity. For pharmaceutical, food, and seed companies, and the biotechnology firms behind them, the ability to scan, pick from, and patent the world's biological diversity harbors prospects of great new sources of revenue. But the emphasis on finding and isolating plants and other living matter with the most marketable traits leads to the decline of other plant species, since only the cultivation of those species that are required for the creation of new varieties becomes constantly reenforced. Tailoring property rights to the privatization of genetic resources that have been engineered and patented also promote crop monocultures. A study by the German Parliament's Office of Technology Assessment concludes that in comparison to conventional methods, no substantial risk of loss of biodiversity obtains for the use of genetic biotechnology in plant and crop cultivation (48).

3. Patenting of genetically modified seeds would cause farmers individually and in developing countries to incur extreme economic strain. The economic

incentive of royalties set by intellectual property rights would benefit the technically advanced countries (being the principal producers of economically valued biotechnologically modified seeds and crops).

4. Patenting would encourage "biocolonialism" and "biopiracy." Genetic resources should be treated as a common heritage of humankind, with the moral implication that any commercial use of a genetic resource whose origin can be determined to belong to a certain country must also benefit the people living in that country. (A recent example is the Indian Neem tree and its pharmaceutical exploitation by an English medical drugs corporation.) This argument overlaps with argument 2 in that the ability of business companies to gain monopolies over what were formerly freely available community resources (seeds, plants, and even microorganisms) is assumed to have devastating effects on both human communities and the protection of biodiversity.

5. Patenting would in the long run slow down the development of medical drugs; similarly it would stifle the advancement of very successful conventional techniques of animal breeding and plant cultivation.

In order to underscore the ethical meaning of these five issues, they can be summarized as concern over (1) greed, (2) loss of biodiversity, (3) exploitative commercialization, (4) biotechnological colonialism, and (5) misplaced utility. There is one further argument that is harder to categorize because it taps into a more general resentment against economic globalization. This is that (6) permissive biotechnological patenting engenders an ethically undesirable shift away from treating people as bearers of human dignity toward an image of people as biological material. This attitude is a possible outcome of colonialization of the nature by economic interests that turn genetic and cellular materials and human (and nonhuman) organisms into potential sources of revenue.

Interestingly, the last argument is not confined in its scope to genetic engineering. It can be applied to patenting in many areas of biotechnology. The argument comprises complex cultural consequences. The idea of life being something not of human making, or something beyond what human ingenuity may try to control, is deeply engrained in European culture. The outcry against patenting life captures a concern that, theologically speaking, scientists might be attempting to play God. Biotechnological patenting would give them license to play God. Yet this resistance to patenting could also be spelled out in less elevated terms. The more commonsense attitude is that whatever can be done is going to be done. This attitude which used to be a flippant remark on uncontrollable technical progress is now gradually being displaced by the realization that whatever pays off is going to be done. The anxiety over biotechnology patenting bespeaks widespread fears over uncontrollable market forces unleashed by economic globalization.

BIBLIOGRAPHY

1. Der Bundesminister für Forschung und Technologie, *Biotechnologie. Programm der Bundesregierung*, Bonn, 1985.

2. U. Dolata, *Politische Ökonomie der Gentechnik. Konzernstrategien, Forschungsprogramme, Technologiewettläufe*, Sigma Verlag, Berlin, 1996.

3. G. Vita, *Internationale Politik* **8**, 7–14 (1998).

4. Schitag Ernst & Young, *Germany's Biotechnology Takes Off in 1998*, Stuttgart, 1998. Survey available at: *www.dechema.de / deutsch / isb / firmen / sme.htm*

5. Available at: *www.gbf-braunschweig.de / bioregio / home.html*

6. H. Grupp, ed., *Technologie am Beginn des 21. Jahrhunderts*, Physica Verlag, Heidelberg, 1993.

7. A. Bora, *Law Policy* **20**(1), 113–133 (1998).

8. T. von Schell and H. Mohr, eds., *Biotechnologie — Gentechnik. Eine Chance für neue Industrien*, Springer, Heidelberg, 1995.

9. H.M. Keplinger, in P. Stadler and G. Kreysa, eds., *Potentiale und Grenzen der Konsensfindung zu Bio- und Gentechnik: vom 11. bis 14. März 1996 in der Evangelischen Akademie Schloss Tutzing am Starnberger See*, DECHEMA, Frankfurt, 1997, pp. 15–50.

10. K. Menrad, in P. Stadler and G. Kreysa, eds., *Potential und Grenzen der Konsensfindung zu Bio- und Gentechnik: vom 11. bis 14. M,,rz 1996 in der Evangelischen Akademie Schloá Tutzing am Starnberger See*, DECHEMA, Frankfurt, 1997, pp. 93–124.

11. W. an den Daele, A. Pühler, and H. Sukopp, *Transgenic Herbicide-Resistance Crops, A Participatory Technology Assessment, Summary Report* FS II 97-302, Wissenschaftszentrum Berlin (WZB), Berlin, 1997.

12. *Bundesgesetzblatt* (Federal Law Gazette), part I, December 16 (1993), p. 2033.

13. *Official Journal*, No. L 117/1 (1990).

14. *Official Journal*, No. L 117/15 (1990). European directive on release of GMOs, Available at: *www.bbba.de / gentech / gene-law.htm*

15. *Official Journal*, No. L 374/1 (1990).

16. *Bundesgesetzblatt* (Federal Law Gazette), part I (1995), p. 297.

17. *Bundesgesetzblatt* (Federal Law Gazette), part I (1996), p. 1649.

18. *Bundesgesetzblatt* (Federal Law Gazette), part I (1996), p. 1644.

19. *Bundesgesetzblatt* (Federal Law Gazette), part I (1996), p. 1657.

20. *Bundesgesetzblatt* (Federal Law Gazette), part I (1996), p. 1232.

21. *Bundesgesetzblatt* (Federal Law Gazette), part I (1995), p. 734.

22. *Bundesgesetzblatt* (Federal Law Gazette), part I (1981), p. 1. Comparison of Germany with 21 other countries available at: *www.oecd.org / ech / tradedoc / htm*

23. *Bundesgesetzblatt* (Federal Law Gazette), part I (1996), p. 1546.

24. *Bundesgesetzblatt* (Federal Law Gazette), part II (1976), p. 826.

25. *Bundesgesetzblatt* (Federal Law Gazette), part I (1985), p. 2170.

26. Akademie für Technikfolgenabschätzung in Baden-Württemberg, ed., *Bürgergutachten Biotechnologie/Gentechnik — Eine Chance für die Zukunft?*, Stuttgart, 1995.

27. J. Hampel et al., in J. Durant, M.W. Bauer, and G. Gaskell, eds., *Biotechnology in the Public Sphere*, Science Museum, London, 1997, pp. 63–76.

28. Akademie für Technikfolgenabschätzung in Baden-Württemberg, ed., *Verbundprojekt Chancen und Risiken der Gentechnik aus der Sicht der öffentlichkeit*, Stuttgart, 1998. Reports available at: *hampel@afta-bw.de*

29. Council of Europe, *Convention for the Protection of Human Rights and Dignity of the Human Being with Regard to the Application of Biology and Medicine*, European Treaties (ETS), No. 164, Oviedo, 1997.

30. Available at: *www.gwdg.de/~ukee; www.uni-tuebingen.de/-zew; http://ibm.rhrz.uni-bonn.de/iwe/iweframe.htm; www.-europaeische-akademie-aw.de*

31. D. Mieth, in T. vonSchell and H. Mohr, eds., *Biotechnologie — Gentechnik. Eine Chance für neue Industrien*, Springer, Berlin, 1995, pp. 505–530.

32. The Council for Research, Technology and Innovation (*Technologierat*), *Biotechnology, Genetic Engineering and Economic Innovation. Making Responsible Use of Existing Opportunities. Assessment and Recommendations*, Bonn, 1997. Available at: *www.technologierat.de/vdi/*

33. J. Radkau, in M. Bauer, ed., *Resistance to New Technology: Nuclear Power, Information Technology, and Biotechnology*, Cambridge University Press, Cambridge, UK, 1995, pp. 335–356.

34. W. Bender et al., eds., *Gentechnik in der Lebensmittelproduktion: Wege zum interaktiven Dialog, Dokumentation eines Workshops an der Technischen Hochschule Darmstadt*, Darmstadt, 1997.

35. W. Bender, K. Platzer, and K. Sinemus, *Sci. Eng. Ethics* **1**, 21–32 (1995).

36. European Agency for the Evolution of Medical Products, *www.eudra.org/en_home.htm*

37. Deutscher Bundestag, *Bericht der Enquete-Kommission 'Chancen und Risiken der Gentechnologie' des 10. Deutschen Bundestags*, Bonn, 1987, p. 83.

38. *Bundesgesetzblatt* (Federal Law Gazette), part I (1990), p. 2746.

39. E. Deutsch, in D. Evans, ed., *Conceiving the Embryo*, Kluwer, Dordrecht, The Netherlands, 1996, pp. 343–346.

40. U. Wessels, in A. Dyson and J. Harris, eds., *Ethics and Biotechnology*, Routledge, London, 1994, pp. 230–258.

41. S. Graumann, *Biomed. Ethics* **2**(1), 12–16 (1997).

42. German Research Association (Deutsche Forschungsgemeinschaft), *www.dhgp.de*

43. *med. Genet.* **4**, 618–624 (1997).

44. I. Nippert, H. Neitzel, and G. Wolf, eds., *The New Genetics: From Research into Health Care, Social and Ethical Implications for Users and Providers*, Springer, Berlin, 1999.

45. *Ann. Rev. Law Ethics* **7**, 3–340 (1999).

46. S. Boshammer et al., *J. Med. Philos.* **23**(3), 324–333 (1998).

47. Available at: *www.gen-ethisches-netzwerk.de*

48. R. Meyer, C. Revermann, and A. Sauter, Gentechnik, *Züchtung und Biodiversität*, TAB, Bonn, 1998, Available at: *www.tab.fzk.de*

See other INTERNATIONAL ASPECTS entries; INTERNATIONAL INTELLECTUAL PROPERTY ISSUES FOR BIOTECHNOLOGY.

INTERNATIONAL ASPECTS: NATIONAL PROFILES, JAPAN

DARRYL R.J. MACER
Institute of Biological Sciences, University of Tsukuba
Tsukuba Science City, Japan

OUTLINE

Introduction

Genetic Engineering and Biotechnology in Japan

Public Acceptance of Biotechnology

Japanese Law and Biotechnology

Japanese Culture, Biotechnology, and Bioethics

The Future and Trust

Bibliography

INTRODUCTION

Japan spends a high percentage gross domestic product (GDP) on research relative to other nations, and biotechnology spending is a high priority (1). The public in Japan is well educated, and is aware of biotechnology, perceiving both benefits and risks of most applications, and has a reasonable degree of bioethical maturity. Most of the regulation of modern biotechnology is through guidelines and directives issued from numerous government ministries. Among the many ethical issues that have been discussed, trust in authorities is one of the central public policy issues that must be dealt with in the future policy toward biotechnology in Japan.

GENETIC ENGINEERING AND BIOTECHNOLOGY IN JAPAN

Japan is one of the world leaders in modern biotechnology, producing about half of the world's antibiotics, building on a long history of fermentation technology (2). Biotechnology itself, as the use of living organisms to produce goods or services, has a history as long as the humans who have shaped the environment (3). While some may consider biotechnology to be a term more suited to genetic engineering and cell manipulation, to consider its ancient origins is important when we look at the ethical issues it raises and the legal approaches that have evolved. This is especially apparent when we look at the origin of bioethics in Asia, because the links to the past are more emphasized there than in Europe or North America, where terms like "gen-ethics" have arisen (4).

There have been several surveys of the progress of the policy decisions behind Japanese biotechnology research (5–7). The government and industry promoted biotechnology throughout the 1980s, and it was then predicted that by the year 2000 bioindustry would represent 10 percent financially of the Japanese economy (8), with 90 percent of this in traditional industries such as fermentation of food and drink. There have been some joint government and industry efforts to promote biotechnology, including the Bioindustry Development Centre (BIDEC), now called the Japan Bioindustry Association (JBA), a private think-tank of the Ministry of International Trade and Industry (MITI). The Science and Technology Agency (STA) also has invested in public acceptance of biotechnology. The prefix "bio" has been attached to many new words in the spoken Japanese language, like biocandy or biocosmetics, but perhaps not more so than in the argot of most other countries (9).

There is a very positive view of the contribution of science to improving the quality of life and economy. Research spending in Japan, at 3 percent of GDP, is the highest level in the world, with U.S.$120 billion being spent in the 1996 fiscal year (10). The United States spends about 2.5 percent of GDP of which 0.5 percent is earmarked to the defense industry. About 15 percent of the funding in Japan is from the government, which is close to that of the United States (16 percent) but less than that in the UK (23 percent) or France (37 percent) (11). Between 1996 and 2001 the Japanese government increased spending on research by almost 50 percent.

Agricultural applications were slow to develop, with few field releases of genetically modified organisms (GMOs) in Japan. Although the Ministry of Health and Welfare released guidelines to assess applications for foods and food additives made from GMOs in 1992 (12), at present most of the 30 foods accepted by the regulatory committee are from foreign imports. Despite the efforts to promote biotechnology, there appear to have been some bottlenecks caused by strict or bureaucratic regulations. Up to 1994 there had been 13 GMO field trials compared with over 1000 in the United States by that time (13). However, Japan leads the world effort in sequencing the rice genome.

The Human Genome Project (HGP) had its origins in Japan in the early 1980s, and the genetics programs have been on the rise but without much thought given to ethical, legal, or social impact (ELSI) issues (14). In the 1998 government budget, U.S.$149 million was allocated for genome research (15), and industry also made significant contributions as in other research areas. For example, Takeda Chemical Company was the leading patent claimer in a 1995 survey of world patents on human gene sequences (16), and it also obtained exclusive rights to use of the genetic database bought by SmithKline Beecham from the Institute for Genomic Research in 1992.

PUBLIC ACCEPTANCE OF BIOTECHNOLOGY

The public acceptance of biotechnology in Japan is reasonably high, somewhere in between the attitude in the United States and that in Germany (3,9,17). A number of studies on public acceptance since 1991 have described the bioethical concerns that different groups within Japan have toward biotechnological applications. It may be that the Japanese have the highest familiarity with the word "biotechnology" in the world. In 1991 two surveys found that 97 percent had heard of the word (9), and close results of 94 percent in 1993 (3) and 89 percent in 1995 were confirmed by Hoban (18). Clearly, there is at least high recognition of the word, since programs on genetics and biotechnology are seen on Japanese television almost daily and developments in the field are covered by most major

newspapers. There are many science magazines, though they are more in the style of the English language *Scientific American* than of the *New Scientist*.

The importance of medical care, agriculture, and aquaculture to human life is generally acknowledged among the peoples of large societies. The question, is, however, to what extent are the attitudes toward the use of organisms to provide these goods, universal, and the relationships with the organisms and ecosystems that provide the organisms universal, as well as the attitudes to the consumption of the products? To answer this question we need to consider a number of strategies. First, we can look at the use of organisms and new products by different groups with in a society and compare the results. For example, do the people eat meat? Although in Japan meat was not eaten widely a few decades ago, thought to be due to Buddhist influence, it is very difficult to find meals which are strictly vegetarian now. Do the people farm animals in open spaces or in factory farms? In Japan land constraints mean that most animals are in factory farm situations, except in Hokkaido during the warm seasons. We have further to standardize for environmental and economic conditions, and look at the religious traditions.

The religious traditions include guidance on ethical issues, answers to problems that are faced around the world. In one sense looking at the end result of choices, the adoption of science and technology products by consumers, is the best description of acceptance of science and technology. However, if we only look at the consumption statistics we may still not understand the reasons behind the choices, and whether, for example, there was really much choice for the consumer in the home environment and society. The ideal model would say that a consumer determines what products are best, but this may not be apparent in a world dominated by large commercial interests, trade groups and associations, and connections among producers, retailers, and regulators. Since Japan is not self-sufficient in any major species of food (even rice is imported for processed rice products), it is going to be dependent on exporters. However, purchasing power means selection among suppliers, and some new practices in air freight have been introduced in Japan that enable the import of live seafoods and many fresh fruits and vegetables. The market proportion of organic foods and pesticide-free foods has been increasing, though surveys find this to be more for reason of interest in health than for environmental concern (3). Some supermarkets provide nongenetically modified soybean products, like tofu (bean curd), and significant resistance to the products of GMOs emerged in late 1998 and continued through the year 2000.

Another strategy that is used to judge public acceptance is to look at cultural tradition in determining what could be adopted. Schmid in 1991 observed that public acceptance of biotechnology is high, "reflecting a high level of education and information within Japanese society, and the specific way of reaching decisions, which usually involve lengthy discussions with all groups" (7). However, if one asked the Japanese public if they had been involved in the decisions associated with the promotion of biotechnology, it is very doubtful that anyone would say yes. Decision-making in Japan tends to exclude public participation (13), although certain applications of biotechnology like organ transplantation technology have introduced the idea of individual choice in use of the technology (19), as will be discussed below. In Japan the strong public support of biotechnology is something that cannot be conclusively obtained from public opinion surveys. There is a predominant cultural attitude among the Japanese not to create friction with those they disagree with. Therefore protest movements and oppositions toward biotechnology are small and lack unity. The major outlet for dissent is the media, through discussion forums in magazines, newspapers, and television chat-shows.

The survey strategy that allows us to look at what individuals will accept, and their reasons for this, must be supplemented in Japan by topics covered in small group discussion forums. The survey results from Japan, and also from other Asia-Pacific countries (3), compared with the rest of the world, reveal an important distinction in the main concerns people have about biotechnology. They want to protect nature, not because of its value or property, but simply because it is there. The bioethics part of risk assessment, which elsewhere takes precedence over the analysis and prediction of risks, is combined with the value of avoiding harm and the benefits of doing good or beneficence. The assessment of risk in biotechnology involves both the potential to change something and the potential to do harm (20). The extent to which a change is judged to be a subjective harm depends on human values, whether nature should be "intransient" or modified. Open survey questions in all countries reveal that the major determinant of moral acceptability of a technology is whether it is perceived to be unnatural, or morally acceptable (3,9,17,21).

In the 1993 International Bioethics Survey (3), when asked about specific developments of technology, including in vitro fertilization (IVF), computers, pesticides, nuclear power, biotechnology, and genetic engineering, both benefits and risks were cited in open comments by many respondents in Japan as in Australia, Hong Kong, India, Israel, New Zealand, the Philippines, Russia, Singapore, and Thailand. In Japan 74 percent saw biotechnology as worthwhile, which was less than the 85 percent in the 1991 survey (9) but still a high positive response. In both years 37 percent said that they had not worried about its development. Thirty percent of those who cited a benefit in 1993 said it would help humanity, 19 percent said agriculture in general, and 15 percent said science. About a half did not describe any benefit or concern. In addition to those who saw it as unnatural, another common response was of human misuse.

In the 1997 telephone surveys (17), a single question on the perceived impact of seven areas of science and technology was used. Comparisons with the data from the European Commission, Eurobarometer 46.1, reveal that there is more optimism about solar energy, new materials, and space exploration in Japan (and New Zealand and Canada) but similar optimism toward computers, information technology, and telecommunication in the EU. There is less optimism about biotechnology and genetic engineering in Japan (with New Zealand being even lower). Only 62 percent in Japan thought that biotechnology would improve the way we live in the next 20 years, 12 percent thought biotechnology would make things worse, and 4 percent said they perceived no effect, with 22 percent saying they do not know. For genetic engineering 54 percent saw it as worthwhile, 12 percent as making things worse, and 7 percent as having no effect. In 1991, 76 percent in Japan said that genetic engineering would be a worthwhile area to explore, while 20 percent were extremely worried about the consequences (9). In 1993, 57 percent said that genetic engineering was worthwhile scientific research, while 15 percent still had a lot of worries about it. In Japan there does not appear general trend against genetic engineering over time, unlike the situations observed in Europe (22) or New Zealand (17).

When people were asked in 1997 to say what came to mind on hearing the term "biotechnology," 8 percent expressed a concern and 4 percent expressed a positive view of science, but most people just said "something technical" (17). In all Asian countries there is strong support for certain kinds of environmental release of GMOs (3). Plant genetic engineering is regarded more favorably than microbe, animal, or human genetic modifications (23), except for gene therapy for diseases like cancer, which is received very positively in Japan (24,25). Despite the concern expressed about genetic engineering, in 1997, 35 percent in Japan said they would buy genetically modified fruits if they tasted better, suggesting a positive image of the products. However, only 8 percent said current regulations are sufficient to protect people from any risks connected with modern biotechnology.

JAPANESE LAW AND BIOTECHNOLOGY

There are few specific laws on modern biotechnology in Japan but rather a series of regulations by the different Ministries. Many scientists and people in industry claim that these regulations have inhibited research development, which is a different view from the public as shown above.

The modern Japanese Constitution was drafted by the occupation forces after the Second World War, was reviewed by the Japanese government, and voted into force by the Japanese Diet (Parliament) in 1948. It has had almost no changes since then, reflecting a trend for laws to become fixed. It includes 31 Articles on the rights and duties of the people. The right of people to life, liberty, and the pursuit of happiness, to the extent that it does not interfere with public welfare, should be the supreme consideration of the law and government (Article 13). Two articles are relevant to access to biotechnology:

1. *Article 11.* The people shall not be prevented from enjoying any of the fundamental human rights. The fundamental human rights guaranteed to the people by this Constitution shall be conferred upon the people of this and future generations as eternal and inviolable rights.

2. *Article 25.* All people shall have the right to maintain the minimum standards of wholesome and cultured living. In all spheres of life, the State shall use its

endeavors for the promotion and extension of social welfare and security, and of public health.

Article 25 assumes a welfare state but does not have much legal meaning. It does not vest in each individual a concrete right that can be enforced by the judicial process, as such type of right comes into force only through implementing legislation. There are six major Codes in addition to the Constitution. Many medical procedures and medical protocol are regulated by specific legal Acts. The Civic Code is concerned with family and inheritance and was completely amended after the Second World War. The Criminal or Penal Code includes some relevant Articles. When there is a more specific law and it conflicts with a general code, the specific law usually takes precedence. Administrative guidance by government agencies and local authorities plays the more significant role in many of the biotechnology issues. The power in the guidance is that the Ministries have the power to grant licenses or permissions. There is little legislation on recent bioethical issues.

The basic philosophy of the Japanese health care system is universally mandated, government-provided health insurance coverage. There is little choice over which insurance scheme a person must join. Employees must join the one statutory plan offered by their employers, and self-employed persons must join the plan administered by the local government or by their trade associations (26). In the year 2000 a public long-term care insurance program will provide for some extra services for the elderly or chronically sick, such as home help, visiting nurses, or day care (27).

The Preventive Vaccination Law established a national program for influenza vaccination in 1976. The Law was weakened in 1987 by removing its obligatory nature, and was further weakened in 1993 with the broadening of exceptions and the removal of provisions that penalized parents who failed to have their children vaccinated. Influenza vaccination is performed annually in children aged 3 to 15 years in certain target groups. However, it has not been very effective, and has recently been recommended for young children only (28). Since 1987 it has been easy for parents to refuse influenza vaccinations, resulting in large differences among the various kindergarten, primary, and junior high schools. Between 1951 and 1965, 169 persons died because of reactions to vaccination. Articles 16 to 19-4 provides for a system of national compensation, under which there are set reimbursements for injuries. For example, in April 1992 a death was compensated by 20.5 million yen, with funeral costs of 140,000 yen. A pension after 18 years of age was 2,925,900 yen for the first category of disability and 1,910,500 yen for second class of disability (29).

A controversy erupted in 1993 over the high incidence (1 in 400) of side effects from a MMR (mumps, measles, rubella) vaccine made and used in Japan. It was withdrawn after the media released unpublicized government risk data. In 1994 no MMR vaccine was offered to children because the government refused to use the U.S. vaccine, which has a 20-year history of safe use with almost no side effects. The scandal reveals that the Japanese Ministry of Health and Welfare has been attempting to encourage Japanese industry by not using a foreign vaccine, while risking public health with a vaccine with 100 to 200 times more side effects. Parents who want their children vaccinated had to pay about U.S.$80 for a vaccine that was previously free (29).

Another sign of the support for the biotechnology industry is the system of drug reimbursement, and the overuse of antibiotics in Japan. One of the embarrassments of the Japanese health care system is the corruption that is implicit in the way drug prices are set and reimbursement is made, and the contributions from pharmaceutical companies to doctors who use their drugs. The Japanese are the world's highest spenders on prescription drugs (30). Almost all general practitioners and hospitals have their own pharmacies for outpatients. Every two years the Ministry of Health and Welfare sets the "official" prices for all drugs. These prices are used to determine the charges to patients and the national health insurance systems. However, pharmaceutical companies offer drugs to hospitals at a discount. The permitted discount is 10 percent, which means there is even official sanction of the scheme to have financial reimbursement for dispensing prescription drugs. In practice, the current discounts are 20 to 30 percent or more in competitive markets. This means that hospitals and doctors benefit from prescribing drugs, and it explains why the consumption of drugs is so high.

The average use of antibiotics is 3 times the U.S. average and 20 times the United Kingdom average (29). There is concern over methicillin resistant *Staphylcoccus aureus* (MRSA) infections and deaths, which are relatively common in Japan. In one old person's home in Chiba, Japan, 12 residents out of 80 were infected with MRSA in 1992. Many products are allowed to support the presence of a large local pharmaceutical industry.

Before a new drug is approved for use, the Central Pharmaceutical Affairs Council within the Ministry of Health and Welfare must examine the results of toxicity tests, animal tests, and clinical trials. The basic policy is outlined in the Notice No. 645 of 1967, on "Control of the manufacture of and trade in pharmaceutical products: approval of manufacture and import of medicaments." Drugs must pass three phases of clinical trials:

1. Phase I is to check a drug's safety in humans.
2. Phase II is to access its therapeutic index (the response rate and severity of side effects) in selected populations of patients for whom the drug is intended.
3. Phase III is to determine whether the new treatment is better than existing ones.

All these tests are done at universities or research organizations at the request of pharmaceutical companies, and there have been cases of bribery (29). There is pressure today for more of the trial results to be presented in scientifically refereed journals, but until now much of the data has been either in internal documents or in the drug companies' own journals. International standardization of tests in 1991 in the United States, Europe, and Japan avoided some duplication of tests and halved the length of long-term studies. In practice, when a drug has passed phase two trials, it may be widely prescribed in "trials," and there are some top-selling drugs that are said to be anticancer drugs which are only prescribed in Japan (30). The Phase III trials can become part of the pharmaceutical companies' marketing plan. There are also widespread issues of lack of informed consent in such clinical trials at the Phase II and Phase III level because most patients do not know what medicine they are being given.

The government can suspend production by a company if it fails to report details of clinical trials, especially if these involve deaths. On May 20, 1994, the chairman of Japan's fourth largest pharmaceutical wholesaler resigned to take public responsibility for insider trading among employees, and for the sale of sorivudine, which has led to the deaths of 18 people. The company had reached settlements by May 1994 with 10 of the 23 families of patients who died or had serious consequences as a result of the using drug. The penalty of a 105-day suspension of production at the Nippon Shoji Kaisha, Ltd. Okayama prefecture factory in response to the company failing to report the deaths of two persons during clinical trials of sorivudine occurred on September 2, 1994. This is the longest suspension of products that has ever occurred because of an infringement of the Drugs, Cosmetics, and Medical Instruments Law (31). It also gave time for reflection on the system of clinical trials in general (32).

The journals that publish the results are often the same journals that sponsor the trials or make the drug (32), and all major pharmaceutical companies are also involved in research on production of new drugs using modern biotechnology methods. Another feature of Japanese biotechnology is that most research is conducted in large established companies, usually multinational, rather than the small biotechnology companies that are a feature of North America. However, it could be that the sponsorship links are just more obvious than elsewhere.

Despite the blood donation system, which in 1991, saw 8,861,137 persons donate blood, 6.5 percent of the population, there is also a large import of blood, since Japan uses more blood per person than any other country in the world. The Product Liability Law passed unanimously by the National Diet on June 22, 1994, came into effect from July 1, 1995. It virtually excludes any liability on transfusion products. "Complications of blood transfusion such as those caused by contamination of viruses whose complete removal by existing technology is impossible cannot be considered as product defects" (33). There have been compensation claims paid to some of the blood transfusion victims of delayed implementation of heat treatment procedures to eliminate HIV from imported blood products (34) and some officials were sentenced to prison.

Patent claims on products in Western countries are recognized in Japanese patent cases, and there have recently been cases involving the use of recombinant DNA products. The approval of such products is independent of the patent claims. A Japanese court rejected a claim by Hoffman-Roche that a Japanese company infringed its patent on interferon, but the case will be appealed (29). The sales of interferon has rapidly been rising since it was approved for use against chronic hepatitis C. An Osaka local court in January 1991 decided that the company

Toyobo cannot market tissue plasminogen activator (TPA) because it conflicts with the Japanese patent given to Genentech in January 1991. Genentech licensed two other companies to sell TPA in Japan. TPA has been sold in Japan since May 1991. The case involving the rights to sales of erythropoietin (EPO) in Japan was solved out of court. EPO has also a very large market in Japan where kidney transplant rates are low.

Privacy of communication is guaranteed in the Constitution. Article 21 of the Constitution guarantees freedom of assembly and association as well as speech, press, and all other forms of expression. Censorship is prohibited, and secrecy of any means of communication must not be violated. There is the Law on the Protection of Computer Information on Individuals, which states that government agencies are prohibited from using the information on individuals for purposes other than the original purpose for which the files were compiled. Any person may require a government agency to disclose the information on themselves that is stored in the computer, and if necessary, demand its alteration. It could be interpreted to mean the truth of any health check information entered into a computer must be revealed following a person's request.

If someone informs others of the medical data of a person, for example, the result of genetic screening test to an employer, Section 134-1 of the penal code could apply. If the person who leaked the information is a national employee, he or she will be punished by the Law on Government Employees. There is still debate over how to control life insurance companies' questioning on the results of genetic tests, but other family history data and smoking are currently used in deciding policies.

For any medical intervention, physicians are required to obtain consent to medical treatment according to the Medical Practitioner's Act, Article 23. The Supreme Court, in 1949, (Decision 3.1) said that the obligation for treatment is based on assessing what can reasonably be expected in view of the knowledge and experience that ought to characterize the average physician. However, in practice and in court cases in the 1980s and 1990s, the doctrine of informed consent has yet to be fully recognized (35,36) as a right for patients to be told all information. Nevertheless, more doctors are starting to use informed consent, and truth telling in cancer cases has also been increasing (37).

Organ transplantation using cadaver donors, especially those that are determined to be dead by brain death criteria, is rare. A law permitting such transplants, and allowing whole brain death criteria to be used for determination of death if patients themselves have signed a donor card to that intent and if family members do not object to it, was passed in 1997. Until then, Law No. 64 (Article 4.17) enabled cornea transplantation since 1958, while in 1979 the Act Concerning the Transplantation of Cornea and Kidney was passed allowing kidney transplants. Kidney transplants since 1979 have been from both live and dead donors, with prior consent and family approval. In the mid-1990s there were few transplants from brain dead donors while the new organ transplant law was being debated. There was much discussion of the issue and whether it was related to any particular Japanese ethos or just suspension of the medical profession (14,19,38). From 1968, when the first heart transplant was performed, until 1997 there had been wide debate on the question in Japan, which was a rare, if not unique, occasion for extensive public debate on a biotechnology process. Live liver donations were possible but most patients had to seek heart and liver transplants overseas.

Until 1996 the 1948 Eugenic Protection Law governed the use of abortion services in Japan. The number of abortions conducted is declining, but it is still high among developed countries, and in 1996 the title was changed to Mother's Body Protection Law (39). After World War II, the Japanese government changed the population policy into "to stabilize and not to increase" from "to increase." How to popularize family planning became the primary policy in health care of postwar Japan. At the same time the Eugenic Protection Law was promulgated in 1948, and Japan became the second largest populated country after the Soviet Union in the liberalization of induced abortion.

The Eugenic Protection Law was a modification of the Preventive Law of Offspring with Hereditary Diseases (*das Gesetz zur Verhuetung erbkraanken Nachwuchses*), 1933 of Germany under Hitler, combined with a liberal view of induced abortion (39). In June 1996, however, some inappropriate parts of this law were amended by the omission or elimination of the eugenic articles, and the title of this law changed from Eugenic Protection Law to Mother's Body Protection Law.

Fetal diagnosis and selective abortion, while not explicitly allowed under these laws, is however, widely practiced when the pregnancy could cause psychological distress to the mother or economic hardship, both of these being acceptable reasons under the law for induced abortion. The marketing of genetic diagnosis and triple marker biochemical tests is governed under existing laws for pharmaceutical products and devices. Abortion is restricted to the period in which the fetus is not viable outside of the uterus, and this period is determined by the notification from the Ministry of Health and Welfare, currently being 22 weeks. The Japan Society of Human Genetics has voluntary guidelines on use of genetic screening (40).

IVF and assisted reproductive technology for married couples are guided by the voluntary guidelines of the Japan Society of Obstetrics and Gynecology (JSOG), and moves to introduce a law have been resisted by the medical community (41). The first baby was born after IVF and embryo transfer (ET) in 1983, and there was much media attention. There are significant differences in attitude among infertile couples, medical practitioners, and the general public on the technology (3,9,42). Most Japanese obstetricians belong to JSOG, and after the first baby was born they rushed to form an ethical committee concerning IVF ET. This committee consisted of 14 members of JSOG. While listening to the opinions of representatives from the mass media and highly educated laypersons, in October 1983, they wrote and announced a statement to the Society of Obstetrics and Gynecology. Even though it was a little too late, this was probably the first time that a Japanese

medical society made and announced ethical guidelines for its members. The statement is as follows (29,39):

1. The Method should only be used for women who are judged unable to become pregnant by any other medical method.

2. The individual implementing this Method must be a qualified doctor who has mastered a high standard of knowledge and technology in the field of reproductive medicine. Every procedure and treatment should be carried out with the utmost care. The procedures and expected results of the Method should be sufficiently explained to the applicants concerned prior to implementation of the Method. Upon obtaining consent from the applicants, and acknowledgment should be filled out and signed by the applicants and retained by the doctor.

3. The applicant receiving the Method should be married, have a strong desire for a child, and be in satisfactory mental and physical condition for pregnancy, delivery, and raising of a child. It must be possible to successfully conduct retrieval of mature ova, implantation into the uterus and maintenance of pregnancy.

4. The fertilized ovum should be carefully handled in respect to the basic moral values of life.

5. When implementing the Method, no gene manipulation is permitted.

6. The privacy of the couple and their delivered child should be respected and protected according to relevant laws and regulations.

7. Considering the importance of the Method, the organization using it should provide opportunities to hear opinions from individuals other than those directly concerned.

The number of babies born through IVF/ET has now reached an annual total over 10,000 since the first case in 1983 in Japan. Surrogacy, however, is not permitted, though foreign surrogacy agencies have been used by Japanese clients, and at least two agencies operate for U.S. surrogacy businesses in Japan (29).

Artificial insemination by donor sperm (AID) is conducted largely through the Obstetrics and Gynecology Department of Keio University, Tokyo. AID has no law to regulate it, and it started in 1948 (39). Now there are about 500 attempts at AID a year at Keio University, and about 250+ births per year. Each sperm donor is used for up to 15 pregnancies, and only married women are accepted. Keio University is the most public about its program. Other institutes do not admit having a program. The guidelines used are those of Keio University and Japan Society of Obstetrics and Gynecology. Since a conference discussion of the Japanese Association of Civil Law in 1953, many legal scholars have construed the law to allow the AID baby of a married woman to be a legitimate child of her husband, so long as the procedure was carried out according to the current practice, but there is still no specific law.

Preconception sex selection has been investigated in Japan, but in a 1993 survey, 76 percent said that if they had only one child they would want a girl, suggesting that traditional ideas of family inheritance are discounted by many people (29). The reason why more people wish to have a girl than a boy, which is in contrast to many other Asian countries, may be because girls are considered cuter, or better caregivers for elderly parents. Following its announcement in 1986, the Ethical Committee of JSOG, and at the same time, the newly founded Ethical Committee of the Japan Medical Association (JMA) came to about the same conclusion in September 1986. That is, it was decided that this procedure should only be adopted to prevent the creation of conceptuses with severe sex-linked recessive genetic disorders.

The details of the Ethical Committee of JSOG's statement are as follows (39):

1. Any individual implementing the Method must be a qualified physician who has mastered a high standard of knowledge and technology in the field of reproductive medicine.

2. A physician intending to implement the Method must be previously registered with the Society according to the specified format. It is also desirable that the results be reported to the Society.

3. Before application of the Method, the physician should sufficiently explain the procedures and expected results to the individual(s) concerned, and should obtain their written consent.

It is also against the guidelines of the Ministry of Health and Welfare to generally inform parents the sex of the fetus during routine prenatal ultrasound diagnosis.

The Ministry of Health and Welfare in Japan set up a special Ethics Committee to assess applications for gene therapy in 1994. The Ministry of Education also made guidelines and set up a separate committee (with seven overlapping members). In university hospitals, drugs already need the approval of both Ministries, and so does gene therapy. The first protocol was approved by Hokkaido University in 1994 allowing research on one child with ADA deficiency. Approval was given by both ministries. The guidelines are basically those of the National Institutes of Health (NIH) in the United States. The guidelines rule out germ-line therapy and until 1999 limited cases to terminal illnesses without effective therapy. However, they only require verbal informed consent, not the written consent that may be determined by local hospitals policies. Japanese scientists and public strongly support the use of gene therapy (24,25), but progress has been slow due to regulatory delays. There is also a lack of domestic vector production, and many trials that are considered are in collaboration with U.S. companies. Japanese people disapprove of use of enhancement genetics in surveys, unlike tendencies seen in China, India, or Thailand (25).

In Japan by 1994, DNA fingerprinting had been used in 180 criminal investigations, but it had only been used 12 times as evidence (29). It is more common to use blood typing and other methods, but it is being introduced in the same manner as other modern forensic techniques. By 1997 advertisements to fathers to check their real genetic relationships to children had appeared in popular magazines, without apparent regulation.

JAPANESE CULTURE, BIOTECHNOLOGY, AND BIOETHICS

The data from modern public opinion surveys needs to be interpreted in the context of the cultural heritage of Japan. The relationships of human beings within their society, within the biological community, and to nature and God are a fact of prehistory; therefore we cannot precisely define the origins of bioethics (3). One of the major elements that needs to be considered in Japanese bioethics is the history of polytheism and animism. However, during the expansion of agriculture and paddy fields over 500 years Japan has seen similar disregard for the environment as have Western countries, suggesting that religious belief does not overcome economic or self-interest (43). The decision to burn a forest and plant a crop is a bioethical decision, and we can see that almost all possible land has been utilized for agriculture, industry, or urban life, with wilderness area remaining only in those regions that proved difficult to exploit.

Japanese ethics is a mixture of Buddhist and Confucian influences combined with a later Shinto influence, and more recently Western influences. From the fifth and sixth centuries the medical profession was restricted to care for the privileged classes. With the centralization of government in the seventh and eighth centuries, there was established a bureau of medicine, and by the Yoro penal and civil codes, there was created an official physician class. The shortages of doctors opened the profession to others. After the Heian period (800–1200) the government-sponsored health service was replaced by a council of professional physicians. In the sixteenth century a code of practice was drawn up the physicians that is very similar to the Hippocratic code and is called the Seventeen Rules of Enjuin (44).

In all areas of public policy, committees of experts work together with bureaucrats to issue reports and guidances. There is a Ministry for the Environment that attempted to introduce a law to govern genetic engineering in 1992, but it was blocked in a power struggle between resistant academics and the Ministries for Agriculture, Forestry, and Fisheries, the Ministry of Education, Science, Culture, and Sports, the Ministry for International Trade and Industry, the Science and Technology Agency and the Ministry of Health and Welfare—all who had their own regulations and committees on biosafety and release of GMOs (9). The first two guidelines had been introduced in 1979 by the Ministry of Education, Science, Culture, and Sports, and the Science and Technology Agency. In 1986, following the OECD recommendations, the other three Ministries also introduced guidelines. Given the interministry division of duties, it is not surprising that the smaller, then Agency for the Environment, could not push through a law claiming it had jurisdiction for all GMO releases over the other Ministries. Each Ministry has revised its guidelines gradually but has kept control over its traditional areas, the same as other biotechnology applications and research.

For the medical discipline, in addition to the Ministry of Education, Science, Culture, and Sports and the Ministry of Health and Welfare, there is a Council of Medical Ethics established under the provisions of Article 25 of the Medical Act. It is an advisory body supervised by the Minister of Public Welfare, consisting of the presidents of the Japan Medical Association, the Japanese Dental Association, and scholars and staffs from related administrative departments. It functions to take administrative measures to eliminate physicians and dentists who commit acts of malpractice or unethical acts. If the media exposes a scandal, then usually top officials or Ministers must resign, an investigatory committee may be established, and then there is a proclamation of policy change. However, Japanese politics is dominated by long-term stability, as the ruling parties and coalitions have been in power except for a year since the Second World War.

Currently Japanese medical ethics is under change (35), recognizing that Japanese society contains people with a similar diversity of views to that of Western countries. The hesitant introduction of bioethics is more related to the structure of Japanese society than to any difference in a person's attitude in Japan or the Western countries (13,19). This fact emerged from opinion surveys where individuals were asked to give their reasons for their responses to bioethical issues regarding genetic manipulation or screening. There was at least as much variety expressed by members of the general public in Japan as there has been in other countries (3).

It must be noted that in terms of equity of access to biotechnology applications, apart from access to prenatal genetic screening and gene therapy, and novel treatments for rare diseases, most other applications of modern biotechnology are accessible to all under the universal medical coverage system. There have been criticisms of the health research system in Japan (45). Open debate is still not common, and this may be the greatest public policy need in biotechnology.

THE FUTURE AND TRUST

In summary, considering the two sides of bioethics, descriptive and prescriptive, a key issue is trust. If we describe the ethical issues that people think are associated with biotechnology in Japan, we find great diversity, the same as in other countries. One common feature, however, is a lack of trust in the process and the policy makers. The prescriptive ethics, or processes that can be used to make decisions and/or the range of decisions that can be made, has been influenced by the relativism that is perceived as correct in Japan. This means rather than one absolute view being right or wrong, we should respect the view of others and not challenge them. This is enshrined in the Constitution and also is in the spirit preserved in the choices given over acceptance of brain death for organ donors (though the family can override the decision of an individual).

While people should not judge, policy must be formulated. What is good for one person may not be good for the broader society, and the global nature of agricultural economics and environmental impact mean we have to think far beyond the small field trial of a GMO. Prescriptive bioethics not only calls for certain factors to be included in decision making but that certain groups of people be involved. Different groups of people may call for

different levels of risk assessment, and of what constitutes a significant risk. Therefore a central question is who can be trusted, and how the public in Japan can regain trust in authorities.

In the 1997 survey, when given a range of bodies, international organizations like the United Nations and WHO were considered the best placed bodies to regulate modern biotechnology by 62 percent in Japan compared to 34 percent in Europe (17), reflecting another cultural value that the opinions of those outside the country may be more trusted than the opinions of policy makers within the country. Europeans and North Americans prefer their local authorities. Therefore the legal tolerance limits of acceptable risk and harm as broadly outlined in international covenants such as the Declaration of Human Rights, the UNESCO Universal Declaration on the Human Genome and Human Rights, and international treaties on environmental protection, have been well accepted as a cultural norm and something to aspire to. Japan is less likely to break with world opinion than, for example, the United States, which regards national autonomy as a higher ideal.

In the same question, industry was more trusted to regulate modern biotechnology than public authorities, the reverse of other countries surveyed (17). This may reflect that by 1997 there was very little trust left in the government in Japan, not that industry is particularly well trusted compared to consumer organizations or university scientists. In the 1991 survey, when asked how do you think biotechnology should be regulated, 62 percent of the public chose the option standards and practices agreed upon jointly by industry and government, with 19 percent saying by the government alone, and 2 percent by industry alone, and 5 percent by individual researchers (9). There was more support for government expressed by scientists and high school teachers who were included in the same survey.

The distrust of authorities by the Japanese does not stem from lack of knowledge. Surveys of the different groups in Japan showed that educated people have as much concerns; in fact biology teachers considered there to be more risk from genetic engineering than the general public (9,17). The risk perceptions among scientists had a tendency to be more concrete than among the public, but all groups expressed a wide variety of concerns. In related questions on the risks of genetic engineering to animals and humans, only 16 percent expressed concerns that it meant interfering in nature (9). However, as in most countries among Japanese academics, industry scientists, and public authorities, there are still claims that increased knowledge is correlated to decreased perception of risk. This is not supported by empirical studies. The balancing of benefit and risk is necessary for bioethics, and it is the most effective indicator of the bioethical maturity of a society (46). The use of surveys can provide us with knowledge of the degree to which a society can make well-reasoned, "mature" judgment rather than offer impulsive, "childish" views based on immediate gain.

In late 1997 one of the government agencies, the Council for Science and Technology of the Science and Technology Agency, established a bioethics committee and a special committee to consider legislation on human cloning (47). The Agency, however, has been criticized for allowing meetings to be closed to public participation, despite claiming as its mission the universal response of all Ministries to a problem where "open and nationwide debate and public consensus is needed." Many Japanese people, of course, are aware that debates do not reach the public, so they rely on opinions expressed by selected experts. Several Ministries have started to open to the public meetings on bioethical issues like brain death and gene therapy, but otherwise most meetings stay closed.

Bioethical decision making involves recognition of the autonomy of all individuals to make free and informed choices provided that they do not prevent others from making informed choices. This is consistent with democratic principles, and the extent to which a society has accepted this is the one criteria of the success of bioethics. However, the structured paternalism of Japanese society is built on the idea that only the views of so-called experts (*sensei*) should be heard (13). It also means that their views should not be questioned, in accordance with the traditional paternalistic Confucian ethos. Medicine is "an Article of Jin," the expression of loving kindness (*Jin*) by the health care professional (34). The main theme of Confucianist ethics was the maintenance of moral discipline for the nation, society, and the home, and it was to the benefit of rulers and family leaders. Therefore it is not surprising that many of the authorities in Japanese society share this ideal because it means respect for them, and hence the rejection of an autonomous development of bioethics (13). They may promulgate the idea that Japanese are different as an attempt to prolong the Confucian ethic.

The bioethics debate may be the catalyst required to transform Japan from a paternalistic democracy. People of any country have at times resisted rapid change, and the globalization of ethics, ideals, and paradigms (may cause) ethnic and national identities to be changed, or lost, especially those of countries with unique cultural histories. How countries approach globalization is a fundamental question, but many individuals in countries with access to common news media have already answered the question by their converging lifestyles and values. To the extent that human rights and the environment are more respected, this trend is to be encouraged.

When Japan opened its doors to Western society in the nineteenth century, it was introduced to a newly emerged science and scientific paradigm that was only part of the fabric of Western society. Meanwhile Western society has continued to evolve, and bioethics has emerged. There has been a series of meetings on bioethics initiated in Japan, both through the Japan Association of Bioethics founded in 1987 and through international seminars on topics such as the Human Genome Project (48–51). There are several bioethics centers and university departments at which it is possible to do research in bioethics, but no degree course is specialized only in bioethics. An early part of this development has included importing and debating ethical approaches, but the current phase has opened up a multidisciplinary dialogue that has included the

public in the discussion and development of its diverse, indigenous ethical traditions. Modern biotechnology may be the stimulus to transform Japanese public policy to better encourage people's involvement in technology decisions.

BIBLIOGRAPHY

1. Committee for Scientific and Technological Policy, Working Party on Biotechnology, *Economic Aspects of Biotechnologies Related to Human Health*, OECD, Paris, 1998.

2. U.S. Congress Office of Technology Assessment, *Biotechnology in a Global Economy*, U.S. Government Printing Office, Washington, DC, 1991.

3. D.R.J. Macer, *Bioethics for the People by the People*, Eubios Ethics Institute, Christchurch, New Zealand, 1994.

4. D. Macer, *Nature* **365**, 102 (1993).

5. M.V. Brock, *Biotechnology in Japan*, Routledge, New York, 1989.

6. R.T. Yuan and M.D. Dibner, *Japanese Biotechnology: A Comprehensive Study of Government Policy, R&D and Industry*, Macmillan, London, 1990.

7. R.D. Schmid, *Biotechnology in Japan. A Comprehensive Guide*, Springer-Verlag, Berlin, 1991.

8. BIDEC, *Impact of Biotechnology on Industrial Structure in the Year 2000*, Japanese Fermentation Association, Tokyo, 1986 (in Japanese).

9. D.R.J. Macer, *Attitudes to Genetic Engineering: Japanese and International Comparisons*, Eubios Ethics Institute, Christchurch, New Zealand, 1992.

10. Editor, *Science* **278**, 1887 (1997).

11. J. Grant and G. Lewison, *Science* **278**, 878–879 (1997).

12. Ministry of Health and Welfare, *Guidelines for Foods and Food Additives Produced by Recombinant DNA Techniques*, Ministry of Health and Welfare, Tokyo, 1992.

13. J. Kinoshita, *Science* **266**, 11845–1185 (1994).

14. D. Macer, *Nature* **359**, 770 (1992).

15. D. Normile, *Science* **278**, 1700–1792 (1997).

16. S.M. Thomas et al., *Nature* **380**, 387–388 (1996).

17. D. Macer et al., *Eubios J. Asian Int. Bioethics* **6**, 137–151 (1996).

18. T.J. Hoban, *Nature Biotechnol.* **15**, 232–234 (1997).

19. D. Macer, *Politics Life Sci.* **13**, 89–90 (1994).

20. D.R.J. Macer in A. van Dommelen, ed., *Coping with Deliberate Release: The Limits of Risk Assesment*, International Centre for Human and Public Affairs, Tilburg, The Netherlands, 1996, pp. 227–245.

21. D.R.J. Macer, in D. Brauer, ed., *Modern Biotechnology: Legal, Economic and Social Dimensions, Biotechnology*, vol. 12, VCH, Weinheim, Germany, 1995, pp. 115–154.

22. Biotechnology and the European Public Concerted Action Group, *Nature* **387**, 845–847 (1997).

23. D.R.J. Macer, in K. Watanabe and E. Pehu, eds., *Plant Biotechnology and Plant Genetic Resources for Sustainability and Productivity*, R.G. Landes, Austin, TX, 1997, pp. 87–99.

24. D.R.J. Macer, *Hum. Gene Ther.* **3**, 511–518 (1992).

25. D.R.J. Macer et al., *Hum. Gene Ther.* **6**, 791–803 (1995).

26. N. Ikegami, *Science* **258**, 614–618 (1992).

27. N. Ikegami, *J. Am. Med. Assoc.* **278**, 1310–1314 (1997).

28. S. Morio et al., *Epidemiol. Community Health* **48**, 46–51 (1994).

29. D.R.J. Macer, in *International Health Care Law*, VistaMedia Corp: USA, & Center for International Legal Studies, Salzburg, Austria (for CD-ROM and print), in Press.

30. M. Fukushima, *Nature* **342**, 850–851 (1989).

31. C. Ross, *Lancet* **343**, 1418–1419 (1994).

32. M. Fukushima, *Nature Med.* **1**, 12–13 (1995).

33. M. Yawata, *Lancet* **344**, 120 (1994).

34. T. Hannay, *Nature Med.* **1**, 396 (1995).

35. R. Kimura with L. Bishop, in W.T. Reich, ed., *Encyclopedia of Bioethics*, rev. edn., Simon & Schuster Macmillan, New York, 1995, pp. 1496–1505.

36. N. Tanida, *Lancet* **345**, 1176 (1995).

37. H. Hattori et al., *Soc. Sci. Med.* **32**, 1007–1016 (1991).

38. M. Lock, *Cult. Med. Psychiatry* **19**, 1–38 (1995).

39. S.N. Shinagawa, *Eubios J. Asian Int. Bioethics* **6**, 158–160 (1996).

40. Japan Society of Human Genetics, *Eubios J. Asian Int. Bioethics* **6**, 137–139 (1996).

41. K. Bai, Y. Shirai, and M. Ishii, *Hastings Center Rep.* (special supp.), 18–20 (1987).

42. Y. Shirai, *J. Law, Med. Ethics* **21**, 43–53 (1993).

43. K. Friday, *Eubios J. Asian Int. Bioethics* **8**, 46 (1998).

44. J.M. Kitagawa, in W.T. Reich, ed., *Encyclopedia of Bioethics*, rev. edn., Simon & Schuster Macmillan, New York, 1995, pp. 1491–1496.

45. K. Imamura, *Lancet* **342**, 279–282 (1993).

46. D.R.J. Macer, *Soc. Sci. Med.* **38**, 23–33 (1994).

47. A. Saegusa, *Nature* **391**, 313 (1998).

48. N. Fujiki and D.R.J. Macer, eds., *Human Genome Research and Society*, Eubios Ethics Institute, Christchurch, New Zealand, 1992.

49. N. Fujiki and D.R.J. Macer, eds., *Intractable Neurological Disorders, Human Genome Research and Society*, Eubios Ethics Institute, Christchurch, New Zealand, 1994.

50. M. Okamoto, N. Fujiki, and D.R.J. Macer, eds., *Protection of the Human Genome and Scientific Responsibility*, Eubios Ethics Institute, Christchurch, New Zealand, 1996.

51. N. Fujiki and D.R.J. Macer, eds., *Bioethics in Asia*, Eubios Ethics Institute, Christchurch, New Zealand, 1998.

See other INTERNATIONAL ASPECTS entries; INTERNATIONAL INTELLECTUAL PROPERTY ISSUES FOR BIOTECHNOLOGY.

INTERNATIONAL ASPECTS: NATIONAL PROFILES, SCANDINAVIA

MATS G. HANSSON
Uppsala University
Uppsala, Sweden

OUTLINE

Introduction

Public Concerns: Why? or Why not?

Biobanks and the Dual Interests of the Citizen

The Cautious Legislative Approach

Space for Self-Regulation

Public Consensus Conferences in Denmark and Norway

The Bill on Medical Databases in Iceland

Ethical, Legal, and Social Aspects of Genome and Gene Technology Research in Sweden

    Public Perceptions and Values

    Genetic Medicine

    Genetic Engineering in Agriculture, Forestry, and Fishery

    Implementation of Gene Technology

Genetics in Dialogue with Other Disciplines

Closing Remark

Acknowledgment

Bibliography

INTRODUCTION

The development of biotechnology is intense in the Scandinavian countries, Denmark, Finland, Iceland, Norway, and Sweden. In a sense this development is a natural continuation of a long tradition originating with the Swedish botanist and professor of medicine, Carl von Linné (1707–1778). Linnaeus was eager to learn everything there is to know about nature; he and his disciples traveled the world to collect information about both natural and human resources, information that he then systematized and described in detail. Linnaeus lived at a time when the significant economic potential of science was first being recognized in Swedish political life, and he furthered this development. The Swedish historian of science, Tore Frängsmyr, has described this contribution as part of the Linnaeus heritage:

> He himself helped spread the new economic thinking and often maintained that natural history, his own science, formed the basis of any sound economy. Both agriculture and industry used nature's products, so knowledge of the three realms was fundamentally important. ...Linnaeus was very willing to put his branch of science at the service of the economy. At the request of the Riksdag, he made three long journeys in the 1740s to different provinces—to the islands of Öland and Gotland in the Baltic, to Västergötland in the west, and to Skåne in the southern Sweden. The aim was to inventory and list the "utilities," i.e., natural resources, ores and tree species, edible plants and berries, animals that were suitable as food or had good pelts, waterways that could be used to power mills, anything that could be put to economic use (1, pp. x–xi).

Linnaeus wrote of the wonders of wide open landscapes, and of plants, and insects that filled him with awe and religious inspiration, but he was able to combine his feelings of reverence with a commitment to putting these wonderful natural resources to use. Linnaeus was a biotechnologist of mind in a very modern sense. In a way he was also a forerunner of the contemporary interest in social aspects of the implementation of biotechnology in different sectors of society. He said that it was not enough to understand the wonders and utilities of natural resources, one must also understand the society in which the new science is going to be implemented. Accordingly

Linnaeus devoted a great deal of his time to describing the local customs and traditions he encountered on his journeys. He was interested in how people lived, what kind of food they ate, their housing conditions, and their cultural values.

The Linnaeus tradition of combining passionate interest and erudition in biology with concern for social and cultural values is not only a source of inspiration for Scandinavian scholars and scientists, but it has international significance. One might add the value of also acknowledging the limits of biological understanding as an explanation of complex human behavior and social phenomena. This, however, is not part of the Linnaeus heritage. He was too much of an optimist and an ardent adherent to the faith in progress that characterized science in the eighteenth century. The respect for the limits of science and circumspection with regard to what can be rightfully claimed as knowledge is Kantian. As Frängsmyr observes, Linnaeus was not a man of the Enlightenment in the philosophical sense (2).

Scandinavia is rich in natural resources, with long-standing traditions in fishing, forestry, and agriculture. As may be expected, plant and animal breeding have become focal points for strategic developments in biotechnology. Strategic research programs in many areas of biotechnology range from basic research in gene technology and protein engineering to the application of the new biological tools in agriculture, forestry, fishery, and medicine. As the international projects for the mapping of the human genome approach completion, major efforts are being made to develop technological platforms for understanding the function of proteins encoded by the discovered genes. Bioinformatic tools are being developed in order to help scientists at universities and in the biotech industry make use of the results in molecular biology and information technology.

Finland, Norway, and Sweden are well known for their large forests and their timber exports. Accordingly there is a special emphasis in these Scandinavian countries on the development of forest biotechnology. There is a chain of research groups in the universities and in the forest industry working on the biological production of forest biomass and tree fibers in cooperation with groups interested in the utilization of tree fibers for pulp and paper production. Plant biotechnology is the field in which the application of the research results most often meets the general public. The development of photosynthetic starch utilizes familiar products such as barley and potato. The use of transgenic tools in order to produce plant oils and fats that can replace fossil products are other examples generally accepted by the public.

Genetic research and the general use of biotechnologies have become intrinsic to many fields of medicine, in both clinical and nonclinical contexts. Genetic intervention has become a continually expanding concept referring to medical genetics as practiced by clinical geneticists, to laboratory pathology and clinical chemistry, and to many medical subspecialties in between. A number of initiatives in Scandinavia in genetic medicine with solid links to industry and academy have been taken in fields such as genetic diagnosis, gene therapy,

pharmacological genomics, drug development, and nucleic acid research. One goal has been to open new avenues for the development of vaccines against viral, bacterial, and parasitic infections and their immunopathological consequences.

For centuries the populations of Finland, Iceland, Norway, and Sweden have been relatively homogeneous. Until the latter half of the twentieth century, immigration to these countries was rather limited. As a result many parts of these countries may be characterized as genetically isolated. This fact, together with the long-standing tradition of keeping records of the citizens in church books, in national health registers, and through social security numbers, has provided a strong impetus to research in population genetics and genetic medicine.

Universities in Scandinavia are governmentally funded, but close collaboration with industry in joint research and development structures is well established. Many companies take part in cooperative efforts with scientists and scholars at the universities. The figures vary among the Scandinavian countries, but it is estimated that as much as 30 to 60 percent of research funding at the universities comes from nongovernmental sources. This situation has given rise to an ongoing ethical discussion in the scientific community with industry and public authorities on the conflict of values related to vital research in medical and pharmaceuticals. The scientific community firmly upholds a principle of openness, arguing that research results should be made public for two reasons. First, it should be available to other scientists for them to build upon in their work. Second, accessibility to the results ensures a critical scrutiny of methods and scientific claims. With regard to clinical research, however, the universities lack the commercial means necessary for bringing scientific results into medical application in the form of new drugs and new treatment. In the interest of patient seeking cures, therefore, it seems that the much-celebrated principle of openness must be reconciled with commercial interests related to the seeking of patents (or immaterial rights). Accordingly, it has become increasingly common for university departments to sign contracts with industry in which they agree to postpone publication of results in order to secure funding for clinical and nonclinical research in biotechnology. The issue is not settled, however, and from time to time heated debate occurs over how to balance public and commercial interests in an ethically acceptable manner.

PUBLIC CONCERNS: WHY? OR WHY NOT?

Several surveys in the Scandinavian countries have revealed a skeptical attitude toward biotechnology among the general public (3–5). The average Scandinavian citizen has lower expectations of positive effect on everyday life of biotechnology compared with the average citizen in Europe. Significantly, however, public perception varies depending on the kind of biotechnological application involved. Genetic medicine receives the highest support, followed by the use of biotechnology for food production and plant and animal breeding.

There may be several explanations for Scandinavian skepticism toward biotechnology. There is a long-standing tradition of public environmental concern that may explain public resistance, at least in part. The general public wants justification in terms of likely benefits for society at large, as well as assurances concerning safety issues, before they approve of new technology. There have been rather frequent calls for moratoria on implementation, pending investigations into the consequences of biotechnological applications for different sectors of society. Recently a moratorium was proposed in Sweden regarding xeno-transplantation. Proofs of estimated benefits and well-founded assurances of safety claims were demanded before a new technology could be accepted. This was what had happened with nuclear energy technology, and it seems to apply in biotechnology as well. While people in other regions of the world may be more accepting and open-minded, Scandinavians are inclined to ask why instead of why not, when confronted with this new technology. In the United States it might be the other way around (6).

The public surveys conducted in Scandinavia demonstrate that the common assumption that more information about biotechnology and its applications will result in a more favorable opinion toward this technology is false. Both scientists and industry often consider lack of knowledge to be the big factor behind low acceptance figures for biotechnology. A European survey from 1993, however, followed by a similar survey done in Norway in 1995, found high levels of knowledge together with low degrees of acceptance (3,7). More information, according to the analyses of these surveys, means more nuanced opinions among people, who are able then to make their own judgments about the various biotechnological applications. Applications within medicine are often highly regarded, whereas applications in the animal-breeding industry or in the production of genetically modified food products are met with much skepticism. According to the surveys, people have fundamental values that remain unchanged despite the availability of more information. Those who are skeptical from the start might well change their arguments, but they will not so easily change their attitudes. Nonetheless, it remains to be proved that there is no correlation between basic knowledge in biology and attitudes toward biotechnology. A common finding in the Scandinavian surveys is that within the population knowledge of biology, in general, and of molecular biology or genetics, in particular, is low. This may not be unique to the Scandinavian countries. Research and development in biotechnology challenges traditional biological concepts of the educated person. The skeptical attitudes toward biotechnology may be tied to old ideas about the biological world, ideas that have been made obsolete by new findings.

In Scandinavia opponents of biotechnology are often found among political and activist groups working with environmental issues. There are two issues that have galvanized resistance toward biotechnology and may be related to an insufficient knowledge of biology or to an outdated understanding of biology. The first deals with the view that natural ecological systems are fragile. The second takes up the idea of the sanctity of a species.

Value surveys among young people show a strong inclination toward environmental concerns (8). Fundamentally the concern for protection of the environment is connected with a resistance to biotechnology. Behind the resistance, there is a belief that the creation of transgenic plants and the release of genetically modified organisms will destroy the natural ecological balance, a balance that is said to be very fragile. Opposition to biotechnology is then an intrinsic part of a concern about changes to the natural ecological systems that must be based on a careful examination of likely consequences to the conditions of a specific ecosystem. This is not a controversial point. However, the more fragile the natural ecological system is, the more strictly this principle must be interpreted and applied. Anyone who has seen the local ecological effects of a discharge of an oil-tanker or who has seen how several plant and animal species have disappeared from the flora and fauna will notice that the ecological balance is very fragile. Against this background, the principle of caution must be applied in a very strict sense. But it is not clear that this is always true in the larger perspective. Seabirds die, and sea plants and sea microbes die from oil contamination, but the ecological system has recuperated rather quickly. Ecological systems may be vulnerable in one sense, but there is good evidence that they are robust even after devastating attacks on their balance (9). At the level of the ecosystem, it seems really not to be a matter so much of whether one species or another disappears but that the overall life-maintaining capacity must prevail. Ecological systems have further proved to be self-preserving in the sense that they do not easily admit new species to be created nor allow modified species to survive the complex steps of an organism's reproductive cycle. Against the background of this biological evidence, one can argue that nature as such is not fragile, and therefore the principle of caution should not be applied in a very strict sense. Depending on which stand one takes regarding the fragility of nature, one could have a different opinion regarding the release of genetically modified organisms in nature. If the biological knowledge of the dynamic capacity of ecological systems is communicated, acceptance figures related to biotechnology at large may improve, even if specific applications might be resisted for good biological reasons.

Respect for the natural boundaries among species seems to be the basis of the line of argument taken by opponents of the production and use of transgenic animals and plants for research, new food, or new medical substances. This view also sees nature as fragile and may be likewise based on a limited knowledge of biology. Aristotle was the first person to suggest that each species is determined by a specific idea. Aristotle's starting point was the Platonic theory of ideas, but he rejected the theory that ideas exist independently of the sense world. For Aristotle ideas are found in the phenomena of nature. They are present within the phenomena as teleological forces. Each species carries a fundamental purposiveness present as a formative power, a purposiveness that determines its characteristics and its relationships with other species. This fundamental purposiveness cannot be changed. Human beings must instead be attentive to this inner formative power of nature. Here one might recognize a certain Aristotelian inspiration in the opposition toward biotechnology that asks the scientists to keep the boundaries of species sacrosanct (10). Research in plant and animal biotechnology entailing the creation of new transgenic organisms does not respect natural borders such as species barriers.

The Aristotelian conception of a species, however, has long been obsolete. After Gregor Mendel, the essentialist concept of species was replaced by a purely statistical concept. A species is the sum of an arbitrary selection of characteristics. A species does not carry an inner purposiveness but has a basis in a statistical description of its biological or practical purpose (11). This biological fact about life challenges the popular view held by many Scandinavians that the differences among species are sacred and must not be transgressed.

However, the more prevalent view of nature is that the present state of balance must not be disturbed. Humanity is not entitled to manipulate the genetic basis of life in cells. Among the general public, the ethical arguments are seen largely to respond to the idea of a normativity in nature. Particularly among the younger population the belief is that there is a "natural order of things" to which humanity ought to adjust (12). The values of the younger population reflect concerns about animal welfare and frequently refer to animal rights. Animals are believed to have natural rights the same as human beings. This attitude represents a shift in the value system of Scandinavia. Skepticism about biotechnology can be explained therefore, at least in part, by reference to beliefs of the kind described. Surveys in other parts of Europe, however, indicate that distrust of the biotech industry is based on its unwillingness to assume responsibility for environmental and safety concerns (13,14). Consumer organizations have pointed to the reluctance of the food industry to provide information to consumers on genetically modified products. Added to that is the distrust among the general public of the capacity of regulatory authorities to monitor these developments.

BIOBANKS AND THE DUAL INTERESTS OF THE CITIZEN

The Scandinavian National Health Service has increased dramatically the use of registers. Health information has long been available in the form of cancer registers, disability registers, cause-of-death registers, and health service registers. With the aid of social security numbers, these registers have proved to be valuable in social medicine and in epidemiological research. Developments in DNA research have brought to the register concept now further health information in the form of tissue cultures, tissue sections, and blood samples. The rise in the number of biological information banks, or biobanks, has become the focus of several governmental investigations in the Scandinavian countries. Denmark has taken the lead in proposing legislation to protect sensitive information and the value of privacy that are at stake in this practice (15).

The Scandinavian PKU (phenylketonuria) registers and biobanks are familiar to the general public, since a blood sample is taken from every newborn baby. During a workshop organized by the Nordic Committee on Biobanks in 1997, Bent Nørgaard-Pedersen from Statens Seruminstitut in Copenhagen described the development

and potential of this kind of register (16). In Denmark, a nationwide screening of newborns for PKU has been carried out since 1975. Filter paper blood samples are taken 5 to 7 days after birth, and Guthrie blood tests are used for the analysis. The parents are informed about the sampling and the storage of the samples. They have the option of declining on behalf of their newborn babies, essentially to exercise an informed refusal. The agenda of the biobanks is to store laboratory and clinical data as well as the filter samples for future research. The biobank provides the valuable data needed for (1) diagnosis and treatment of phenylketonuria, (2) control of previously performed analyses, (3) quality assurance for diagnosis, (4) new analyses, in order to check for other diseases, and (5) research projects using biochemical, genetic, and environmental marker analyses. The register has been approved by the Health Ministry. All research projects using material or information from a biobank in Denmark must be approved both by a research ethics committee and, in special cases, by the national data surveillance authority.

Ethics committees play a central role in the regulation of the use of the biobanks for research purposes in all Scandinavian countries. They have a difficult task to strike a balance between the values that are at stake. To understand this function, a distinction must be made between two fundamental citizen concerns. On the one hand, the individual citizen as a potential patient has interest in the efficient storage and use of the biobanks as tools for new medical treatment and for the development of new drugs. From this viewpoint it is not good if one pharmaceutical company obtains exclusive rights to, and therefore a monopoly on, the information in a biobank. This would prevent other scientists and companies from working with the material and posing other scientific questions of potential interest for the health and well-being of the citizens. On the other hand, if a company is denied exclusive rights, it may not find it worthwhile to invest in the first place. Since the cost of research and development in the field of biotechnology greatly exceeds what most governments can afford to invest on their own, such a policy would not be beneficial to actual or future patients.

It is a difficult and delicate task for Scandinavian governments to find an ethically acceptable balance between these two extremes. The costs involved are also a challenge for the scientific community, health professionals, and the health authorities. They have yet to agree on efficient procedures for storing, sharing, and distributing biobank samples. Without rules and guidelines that take into consideration the interests of patients (actual and future), scientists, universities, and industry, there is a great risk that the doctors and scientists who control the freezers and the drawers containing the samples will not be able to cooperate and coordinate their efforts, both within and across national borders.

Against the interest of the citizen in efficient stewardship there is a no less important interest in protecting his or her integrity. The individual citizen must have sufficient safeguards to guarantee that the information contained in the samples is not used in a way that is harmful to him or her. A problem is that no one yet fully understands the potential for the abuse of knowledge yielded by the medical information contained in tissue samples combined with hereditary and environmental factors. A particular problem is that the information acquired and processed is of relevance not only for the individual who is the source of the sample but also for genetic relatives of this individual. A stumbling block in the ethical discussion of biobanks is the formulation of necessary and sufficient rules of informed consent.

Article 22 of the Convention on Human Rights and Biomedicine by the Council of Europe from November 1996 states: "When in the course of an intervention any part of a human body is removed, it may be stored and used for a purpose other than that for which it was removed, only if this is done in conformity with appropriate information and consent procedures." This decision of the Convention has been the focus of intense discussion in Scandinavia in the effort to create ethically responsible legislation on biobanks. If the rule of informed consent is taken too rigidly, much of the epidemiological research will be precluded, with the consequence that advances in health improvement dependent on that knowledge will be lost. On the other hand, it must be acknowledged that the information gathered about the individuals might violate their integrity.

There is reason to believe that Scandinavia's long experience in providing medical benefits for patients by using health information registers will facilitate the application of a nuanced rule of informed consent that is sensitive to the values at stake for all concerned parties. In cases where no issues of integrity are at stake, or are inconsequential to the individual who is the origin of the sample, and the risks of harm are negligible, the consent procedures may be conducted in accordance with health and well-being as the primary objectives. The Danish model of informed refusal might be appropriate for some research protocols. In other cases, when more is at stake for the individual from which the sample has been taken, stricter rules, including written informed consent procedures, might be appropriate. If information and consent procedures are formulated too rigidly, they may be detrimental to the individual that the convention seeks to protect. There is a need for different information and consent procedures for different research and medical practices (17).

THE CAUTIOUS LEGISLATIVE APPROACH

Legislation affecting biotechnology is not the same among the different Scandinavian countries. Norway has enacted strict regulation. Sweden appointed a Gene-Ethics Committee as early as 1982, and its task was to conduct an inquiry into ethical, humanitarian, and social issues arising from genetic engineering. The report was completed in 1984 (18). After several years of public discussion, a law restricting research on fertilized human eggs was passed in 1991 (19). It stated a time limit of up to 14 days when research can be done on fertilized human eggs, and it allows eggs to be frozen and stored for up to one year. The storage time limit was later extended to five years (20). With some variation, there are now similar regulations in the other Scandinavian countries. The Swedish law is unique in one important aspect. The 1991 Swedish law not only prohibits the implantation of a fertilized egg that has been subject to experimentation, it also rules out all kinds of experiments directed toward altering heritable characteristics. Germ line gene therapy cannot be practiced; neither can it be studied for a potential future application. The law has prohibited the development of this research, and as such it is unique both in Scandinavia and presumably the rest of the world.

Laws have also been passed which regulate plant biotechnology and the release of genetically modified organisms (GMOs). On the whole, however, there have been relatively few attempts to regulate the field through legislation. One is given the impression that Scandinavian parliamentarians and governments do not want to create special legislation for biotechnological research and development. They first try to apply existing regulation for the safety of patients within the health care Acts or, with regard to ecological concerns, in the environmental acts (21). The Scandinavian countries also have long traditions of ethical review provided by specially appointed scientific-ethical committees and of public ethical debates. The scientific-ethical committees are well regarded and they are often asked by the legislators to take on a large responsibility in analyzing and judging the acceptability of research. Finland has until very recently relied exclusively on a system of ethical committees, instituted on a voluntary basis by the scientific community (the Finnish proposition RP 229/1998 on medical research). Denmark has legislation that supports the ethical committee system; a characteristic is the large presence of ordinary citizen representatives on the committees, although the legislators do not direct the review process normatively. Both Norway and Sweden are fundamentally dependent on a voluntary system. In Sweden, however, the recent parliamentary commission on research ethics has proposed that there should be a law stating that research on human subjects or human tissue should be the object of examination and approval by a research ethics committee from the university in question (22). In summary in the Scandinavian countries, legislation is believed to be too blunt an instrument for use on biotechnology, since developments occur at a rapid pace and new scientific facts are produced in an unending flow.

SPACE FOR SELF-REGULATION

Now and then there is public demand for stricter regulation. It turns out that the balancing mechanism is most often more moderate legislation and self-regulation on the part of scientists and industry. Part of the explanation behind this mechanism in the Scandinavian countries is the readiness of scientists to go public and express their own moral concerns. Until very recently, there have been few animal rights activists engaged in these activities such as in other parts of the world. The American scientist Andrew Rowan (personal communication), a writer on the science and ethics of animal research, has suggested a plausible explanation for this difference (23). Rowan has noted that scientists in Scandinavia went public and expressed their moral concerns with regard to animal research almost from the start. They invited representatives of nongovernmental organizations to a public dialogue on the means and ends of scientific research. These animal ethics discussions led to review by specially appointed ethics committees that included a substantial number of lay members (24). The scientists expressed concern about animal welfare, but they could also explain the necessity of using animals as experimental models in order to provide cures for both humans and animals suffering from diseases. Thus the scientists achieved two goals: they brought ethical and policy issues related to research into the public debate, and they secured self-regulation for the scientific community.

The willingness of scientists and the biotech industry in Scandinavia to assume the moral responsibilities associated with a certain latitude for self-regulation can further be seen in the area of animal breeding and husbandry. Over many years, there has emerged a Nordic profile with regard to breeding goals in which the focus is not only on production traits but also on efforts to stabilize or improve the genetic level of functional traits related to animal welfare (25). According to this breeding ideology, increased birthweight of calves is not a value in itself, but must be related in a significant way to the health and well-being of the animals. Thus it comes as no surprise that the "Belgian Blue" (known as the monstrous bull because of its extremely large muscles) has met with great resistance not only from the public but also from the breeders in Scandinavia. Health, calving performance, quality of udder and teats, and fertility have long been vital breeding goals in addition to production traits.

Many scientists active in biotechnology research and development in Scandinavia seem to act in accordance with a maxim that has a certain Kantian ring to it (10). The maxim is general in nature, and does not specify any concrete goals of action. It is here suggested as the codification of a long tradition of moral thinking in bioscience and is also reflected in regulations created by the public authorities monitoring development in this field. Its role is as an aid in sorting out the value conflicts related to specific proceedings and applications of bioscience. The maxim may be formulated thus:

> Act in your biotechnology research and development so that you protect the health and well-being of human beings and animals; minimize their suffering; protect biological diversity; and make use of natural resources so that justice prevails and a contribution is made to a sustainable development.

PUBLIC CONSENSUS CONFERENCES IN DENMARK AND NORWAY

Among the Scandinavian countries Denmark and Norway have created their own ways of including public opinion in dialogues on biotechnology. The usual form a dialogue takes is for the bioscience community to engage the public in different activities related to popular science. These activities are important means for bridging the confidence gap between scientists and the public, but they are not enough. What is needed is a dialogue where the questions of the public are allowed to set the agenda and direct the discussion. This is what has been established in Denmark through the Danish Board of Technology for many years, and has recently emerged in Norway under the auspices of the National Committees for Research Ethics and the Norwegian Biotechnology Advisory Board. The early experiment of Denmark have been described and discussed at length within the European consensus conference context (26). These conferences are described as meetings that enable technology assessments to be made by an expert panel and a panel of concerned citizens (27). Issues like gene therapy, the handling of genetic information, and genetically modified food products have come under intensive discussion. The panel of citizens directs the proceedings and decides what questions to ask and what experts to engage. This panel then puts together a consensus report of their opinions regarding the questions discussed.

THE BILL ON MEDICAL DATABASES IN ICELAND

Iceland has passed an Act on a Health Sector Database and a Bill on Medical Databases that are particularly interesting for Scandinavia. The Icelandic population (270,000) is considered to be optimal for such a database for three main reasons. It is a homogenous population, health records are good and reliable, and the Icelanders are cooperative and positive toward research. These three features taken together are, to a great extent, shared by Finland, Norway, and Sweden.

The Health Sector Database, which draws on information from the entire Icelandic population, may be consulted for the purpose of discovering new drugs, developing new or improved methods for prognostic or diagnostic purposes and treatment of diseases, seeking the most effective solutions in the operation of health systems, or for medical reports or other comparable purposes in the health sector. The Bill elicited severe criticism from medical doctors, researchers, ethicists, and lawyers. The Ethics Council of the National Director of Health analyzed the implications of this Bill for the patient's right to privacy and for the relationship of confidentiality between doctors and patients. The Council has also worked on the ethical aspects of the intended bill on biosamples.

Public acceptance of biotechnology in Iceland is quite high. The criticism comes rather from professionals: the Bill on Medical Databases met with substantial resistance in the scientific community. A national survey in 1998 showed, however, that 75 percent of the Icelandic population is willing to have depersonalized information from their health records in a central database available for biotechnological research (V. Arnason, personal communication) (28).

ETHICAL, LEGAL, AND SOCIAL ASPECTS OF GENOME AND GENE TECHNOLOGY RESEARCH IN SWEDEN

A safe and wise implementation of genetic engineering and biotechnology in different sectors of society requires the cooperation of scholars working in well-developed multi-disciplinary research environments. In 1999 the Swedish Foundation for Strategic Research initiated a national research program with the aim of stimulating research focused on the ethical, legal and social implications of genome research and its implementation in different sectors of society (29). The following areas of research or fields of interest have been identified as examples of areas and fields that will be addressed in the program.

Public Perceptions and Values

It is not enough to be able to master the new gene technology tools. It is also important to understand the values and the formation of norms in the society where the new technology is integrated. As described at the beginning of this article, this perspective is part of the Linnaeus heritage. Values, attitudes, beliefs and worldviews reflect the larger impact of gene technology. Developments on genetic technology have challenged established concepts of health, illness, disease, and human integrity. Changes in concepts and changes in popular notions about these concepts are to be investigated. What are the underlying motives and views supporting certain beliefs about health and disease, about being human, or about the relationship between humans and animals? How do people classify their views? What are the general perceptions and evaluations of risks?

Genetic Medicine

Advances in molecular biology have provided medicine with powerful tools for diagnosis and for monitoring diseases and their treatment. Gene therapy may still be years away from becoming established medical practice, but several clinical trial protocols have been approved. The most urgent problems described in the research programs are related to genetic diagnosis and the handling of genetic information about individuals and families. The communication of risk and risk-related issues in relation to single-gene disorders is also an important area for research. Clinical practice is in need of the contributions of psychologists, anthropologist, ethnologists, and theologians who have the skills to map the wide spectrum of individual responses to the implications of genetic medicine. Polygenic and multifactorial conditions must be addressed besides the problems associated with the development of treatments for common genetic disorders. In this regard there are many questions about the rules and guidelines for the involvement of human subjects in clinical trials and related research, and for the storage and utilization of tissue samples in genetic research. As gene therapy gives way to cell therapy and the use of stem cell biology provides opportunities for the replacement of organs of the body, there will be calls for ethical and psychosocial interpretations of this development. If xeno-transplantation is considered, to what extent is implantation of an animal organ a violation of human dignity? What does it mean for human identity?

Genetic Engineering in Agriculture, Forestry, and Fishery

Agriculture, forestry, and fishery are areas in which gene technology will enable the development of new characteristics and breeding traits with a strong economic potential for big industry. The need is for research that focuses on practical problems encountered by scientists, industrialists, and policy makers in this area. Among the issues in need of attention there are many problems relating to four areas: (1) the intentional release of transgenic plants into the environment, (2) the introduction of genetically modified foods into the market place, (3) transgenic animal research, and (4) the impact of agriculture and plant biotechnology on developing countries.

From an ethical perspective, a society free of risks is neither possible nor desirable. Vital values in terms of survival, better food, health, and well-being are at stake, and these values have to be weighed and balanced against each other. The approach to risk assessment is expected to be normalization of risks (i.e., toxicological, allergenic, health hazards, ecological) through a comparison of risks associated with gene technology with the risks posed by conventional and future technologies for breeding and development.

In risk assessment there are six general ways of considering what can be expected. Preferably these questions will be taken together in regard to multidisciplinary projects.

1. What is the risk? (Potential risk identification)
2. How likely is the risk to occur? (Quantifying the probability of occurrence)
3. What is the severity and extent of the effect if it occurs? (Quantifying the effects)
4. What are the expected benefits?
5. Is the risk acceptable? (Normalization of the risk and risk–benefit analysis)
6. How important is risk assessment for different actors?

Implementation of Gene Technology

No one disputes the fact that gene technology will constitute an important economic sector in society. However, before the development of gene technology from science to commercialization can occur, several steps must, be taken. Even knowledge on how to go about this process is incomplete. On the one hand, gene technology could provide economic growth by stimulating the development of new industries. On the other hand, no companies have yet produced services and products of the anticipated quality. The gap between optimism and slow reality points to an equally important question on how companies that can harm the environment to create new products can achieve economic growth. Then there is still the problem of access to and use of genetic information, both commercially and noncommercially. All these concerns involve questions of intellectual property, questions of confidentiality and privacy, and questions of equity.

Within plant biotechnology and within pharmaceuticals, there has been a rapidly growing concentration of control in a few hands. Large corporations have bought out small companies. The issue of labeling is soon becoming a nonissue. Nongenetically modified soya is now more a thing of the past. The need is becoming urgent to clarify what is happening in biotechnology with regard to both the scientific developments and the economic regulation of corporations taking over the developments.

The funding of plant and animal biotechnology needs to be investigated, likewise the funding in the production of medical products and diagnostics. Pharmaceutical companies have made enormous investments where there are indications of possible breakthroughs. Human diseases are expected to be treated with products not yet foreseen. Who will pay the bill? Who will be able to afford the new treatments? What will be the prioritization scheme in the health care sector? What price will be paid for success in biological and genome research?

GENETICS IN DIALOGUE WITH OTHER DISCIPLINES

It has long been obvious that there is a genetic history to many human diseases. Recent research, however, indicates that there is also a genetic component in more complex human behavior. If it can be proved that language behavior and learning ability express hereditary variability, the consequences will be great for established theories within disciplines such as linguistics and educational research. The old question about nature and nurture that has been so vividly discussed in relation to evolutionary theory is reactivated. Of particular interest is the development in brain research in which complex phenomena such as memory function and emotions may soon be comprehensively described in chemical terms. How is human society going to survive in a culture in which everyone is aware of the genetic components of one another? No doubt, in these new emerging fields of research geneticists, evolutionary theorists, linguists, philosophers and scholars in educational theories need to collaborate on such perplexing issues.

CLOSING REMARK

Clearly, any nation hoping to compete in the international economy must bring itself to the cutting edge of research and development in biotechnology. Without an investigation of the ethical and social implications, there is the danger that political decisions and legislation will be premature and not based on good, strong knowledge. There is also the risk that vital values related to health and survival will be neglected. In all five Scandinavian countries, there is a growing interest in the ethical implications of new technology. A number of ethics committees have been formed at various levels. Questions about the ethical implications of new technology are also recurrent themes in the media and in parliamentary debates. An interesting phenomenon in these discussions is that saying no to new technology is often believed to be of greater moral significance than saying yes. "Better safe than sorry" and "Safety first" seem to be the primary guiding principles of the debate on biotechnology. To be sure, when important issues are at stake such as survival, health, and well-being, there is great moral responsibility assumed in saying no or yes. A society free of risks is neither possible nor desirable, since, when vital values are at stake, these values must be balanced against one another. A "no" to biotechnology may deprive society of enormous benefits from research in bioscience and biotechnology. There may be good reasons to say no to certain problematic innovations. However, whether one says yes or no, this should be done only after a careful risk assessment and a weighing of the pros and cons.

ACKNOWLEDGMENT

I am grateful to Peter Sandoe, Per Sandberg, and Vilhjalmur Arnason for providing valuable background information and texts regarding ethical, legal, and policy issues in their countries. I am also grateful to Knut Erik Trany and Sharon Rider for valuable comments on an earlier version of this article.

BIBLIOGRAPHY

1. T. Frängsmyr, ed., *Linnaeus. The Man and His Work*, Science History Publications/USA, Division Watson Publishing International, Canton, MA, 1994.

2. A. Jeffner, *Biology and Religion as Interpreting Patterns of Human Life, The Idreos Lectures*, Harris Manchester College, Oxford, 1999.

3. T. Hviid Nielsen, in S. Lundin and M. Ideland, eds., *Gene Technology and the Public. An Interdisciplinary Perspective*, Nordic Academic Press, Lund, Sweden, 1997.

4. B. Fjaestad, ed., *Public Perceptions of Science, Biotechnology and a New University*, Mid Sweden University, Östersund, 1996, p. 10.

5. J. Durant, M.W. Bauer, and G. Gaskell, *Biotechnology in the Public Sphere. A European Sourcebook*, Science Museum, London, 1998.

6. P. Sandoe, *K. Skogs- Lantbruksakade. Tidskr.* **136**(20) (1997).

7. B. Nygård, *Ny bioteknologi i Europa*, Rapp. 1/1995, Senter for Bygdeforskning, Trondheim, Norway, 1995.

8. P. Ester, L. Halman, and R. de Moor, eds., *The Individualizing Society*, Tilburg University Press, Tilburg, The Netherlands, 1994.

9. B.G. Norton, *Why Preserve Natural Variety?* Princeton University Press, Princeton, Nd., 1987.

10. M.G. Hansson, *Human Dignity and Animal Well-being. A Kantian Contribution to Biomedical Ethics*, vol. 12, Acta Universitatis Upsaliensis, Uppsala Stud. Soc. Ethics, Uppsala, Sweden, 1991.

11. E. Sober, *From a Biological Point of View. Essays in Evolutionary Philosophy*, Cambridge University Press, New York, 1994.

12. C.R. Bråkenhielm and K. Westerlund, in S. Lundin and M. Ideland, eds., *Gene Technology and the Public. An Interdisciplinary Perspective*, Nordic Academic Press, Lund, Sweden, 1997.

13. S. Lundin and M. Ideland, eds., *Gene Technology and the Public. An Interdisciplinary Perspective*, Nordic Academic Press, Lund, Sweden, 1997.

14. A. Mauron and J.-M. Thévoz, *J. Med. Philos.* **16**(6), 649–666 (1991).

15. *Health Science Information Banks—Biobanks*, The Danish Medical Research Council, the Danish Central Scientific-Ethical Committee and the Danish Council of Ethics, Copenhagen, 1996.

16. B. Nørgaard-Pedersen, in M. Sorsa and J. Eyfjörd, eds., *Human Biobanks — Ethical and Social Issues*, Nord 1997:9, The Nordic Committee on Bioethics, Helsinki, 1997, pp. 59–67.

17. M.G. Hansson, *J. Med. Ethics* **24**(3), 182–187 (1998).

18. SOU, 1984:88, *Genetisk integritet. (Genetic Integrity)*, Betänkande av Gen-etikkommittén, Stockholm, 1984, Swedish Governmental Report 8F.

19. SFS 1991:115, (The Swedish Code of Statutes).

20. SFS 1998:282, (The Swedish Code of Statutes).

21. T. Achen, *Den bioetiske udfordring (The Challenge of Bioethics)*, Linköping Studies in Arts and Science, Linköping University, Sweden, 1997.

22. SOU, 1999:4, *God sed i forskningen (Good Research Practice)*, Stockholm, 1999, (Swedish Govermental Report).

23. A. Rowan, *Mice, Models and Men: A Critical Evaluation of Animal Research*, State University of New York Press, Albany, 1984.

24. B. Forsman, *Djurförsök. Forskningsteik, politik, epistemologi (Animal Research, Research Policy, Politics and Epistemiology)*, Almqvist & Wiksell International, Gothenburg, Sweden, 1992.

25. L.G. Christensen, *Acta Agric. Scand., Sect. A, Anim. Sci. Suppl.* **29**, 77–89 (1998).

26. S. Joss and J. Durant, eds., *Public Participation in Science. The Role of Consensus Conferences in Europé*, Science Museum, London, Sweden, 1995.

27. J. Grundahl, in S. Joss and J. Durant, eds., *Public Participation in Science. The Role of Consensus Conferences in Europé*, Science Museum, London, 1995.

28. J.R. Gulcher and K. Stefansson, *Nature (London)* **400**, 307–308 (1999).

29. Swedish Foundation for Strategic Research, Available at: *www.bioethics.uu.se/elsa*

See other INTERNATIONAL ASPECTS entries; INTERNATIONAL INTELLECTUAL PROPERTY ISSUES FOR BIOTECHNOLOGY.

INTERNATIONAL ASPECTS: NATIONAL PROFILES, SWITZERLAND

FRANÇOISE BIERI
ORESTE GHISALBA
OTHMAR KÄPPELI
HERBERT REUTIMANN
Swiss Priority Programme Biotechnology of the Swiss National Science Foundation
Basel, Switzerland

OUTLINE

INTRODUCTION

The 1980s were a turning point for biotechnology. Prior to 1980, biotechnology applications in Switzerland were limited to the industrial sector and a few, pioneering academic institutions such as the Swiss Federal Institutes of Technology in Zurich (ETHZ) and Lausanne (EPFL). In the 1980s, more and more leading Swiss institutions and Small and Medium size enterprises (SMEs) began using molecular biology and genetic engineering techniques for all types of applications in the life sciences. It soon became clear that the potential of biotechnology to benefit society was immense, if provided with the proper environment for its development.

Various groups and organizations in Switzerland, such as the Swiss Academy of Technical Sciences, the Swiss Coordination Committee for Biotechnology, the Board of the Swiss Federal Institutes of Technology, and the State Secretary for Science and Education, launched several proposals in order to induce national efforts for the promotion and development of biotechnology. In 1989, the Swiss Science Council (SSC) mandated the Swiss Coordination Committee for Biotechnology to perform a comparative study on national and international biotechnology R&D programs, their goals and development strategies. Swiss science policy makers used this document (1) to lay the foundation for the first nationwide biotechnology program, subsequently approved by the Swiss Parliament and initiated in 1992. This was the beginning of the Swiss Priority Programme Biotechnology (SPP BioTech).

PAVING THE WAY FOR THE DEVELOPMENT OF SWISS BIOTECHNOLOGY

Organization and Goals of SPP BioTech

SPP BioTech is financed through the Swiss National Science Foundation (SNSF). Its goal for 1992 through

2001 is to ensure the international competitiveness of Swiss biotechnological research and development (2,3). The program is applications-oriented, and it aims to bring basic and applied research closer to the development stage by encouraging synergistic collaborations among universities, research institutes, and private industry. Fields of biotechnology where Switzerland already holds a strong position are strengthened, while fields that need encouragement are given an opportunity to fortify their bases, through the setting of relevant research priorities that ease technology transfer in Switzerland.

From among the broad range of modern biotechnology applications possible, it would be impossible for a small country like Switzerland to fund all activities equally. Therefore SPP BioTech has created a number of modules based on a thorough assessment of the national research capacity and was able to consolidate applied biotechnology research in Switzerland as listed in Table 1.

In its goal of strengthening biotechnological research in Switzerland, SPP BioTech has not neglected the peripheral activities necessary for bringing technology innovations into society. SPP BioTech supports scientific activities that use modern biotechnology to help achieve sustainable development and efficient use of resources in industrial processes and agricultural systems. The program also recognizes the important role of continuing education in biotechnology for young researchers, and funds are accordingly allocated for Ph.D.s, postdocs, visiting scholars, and junior group leaders. In addition the program includes a unit of study concerning biotech-related issues that have significant interest for the public.

The SPP BioTech program has prioritized addressing public concerns regarding applications in biotechnology in a timely and informative manner. The level of public acceptance for technology applications can determine the speed at which development proceeds in certain critical research areas. It is for this reason that the agencies BATS (Biosafety Research and Assessment of Technology Impacts), BICS (Biotechnology Information Center), and Unitectra (technology transfer) were created under the aegis of the SPP BioTech.

The research activities within the SPP BioTech gradually proceeded from ideas and goal-oriented basic research to practical applications of the achieved results. The program comprised three distinct phases with a total budget of approximately 100 million Swiss francs:

Table 1. Research Modules of the SPP BioTech

Processes for the production and purification of proteins for medical applications
Biotechnology: bioengineering and biocatalysis
Food biotechnology (started in 1996)
Bioelectronics and neuro-informatics
Biosafety research and development of biotechnology
- Biotechnology Information and Communication (BICS Agency)
- Biosafety Research and Technology Assessment (BATS Agency)
- Technology Transfer (Unitectra)
Biotechnology of higher plants
SPP BioTech education program

- *Buildup phase, 1992 to 1995.* Focus on applications-oriented R&D, by introducing collaborative ventures involving universities, research institutions, and industries; begin technology transfer activities in the transfer of products, methods, and services.
- *Consolidation and Extension of Collaboration with Industry, 1996 to 1999.* Continue applications-oriented research and concentrate on successful strategies; extend and intensify contacts with industry; motivate SMEs to join; speed up technology transfer (including the creation of new SMEs).
- *Harvest and termination (outphasing), 2000 to 2001.* Continue the most successful and productive projects; focus on development aspects and technology transfer in order to exploit the achievements.

Participation in SPP BioTech has also helped a large number of research teams find easy access to Framework IV Programs of the European Union (EU). The success rate for Swiss applicants (36 percent for the first call in 1995) was by far above the European average (26 percent).

Achievements and Impacts of the SPP BioTech

Through SPP BioTech there have been created centers of competence and nationwide networks for biotechnology research (see Figure 2). Biotechnological activities at ETHZ and EPFL have been strengthened. An Institute for Neuro-Informatics, is now jointly operated by the University of Zurich and the Federal Institute of Technology, Zurich.

Further SPP BioTech has provided support for bioelectronics research and applications of this technology for the development of biomedical equipment. There has been created a nationwide network for Swiss biosafety research on recombinant and "naturally" occurring organisms. Universities and government institutions are closely collaborating in the field of plant biotechnology in developing a more sustainable agriculture.

Since 1996, SPP BioTech has promoted innovative research food biotechnology for healthier and safer dairy products. Technology transfer between academe and industry has been facilitated and given a priority.

Research Network in Biotechnology

A significant part of modern biotechnology research at Swiss universities occurs outside of the SPP BioTech. A survey carried out by Unitectra in 1997 (4) revealed more than 300 research groups active in various fields. There has been estimated, overall, between 350 and 400 biotech-oriented academic research groups in Switzerland (see Table 2).

Funding of Research in Biotechnology

At present, there are three types of public funding for biotechnology research in Switzerland (see Fig. 1):

- Funding of basic research projects (individual projects) directly via the Swiss National Science Foundation (SNSF).

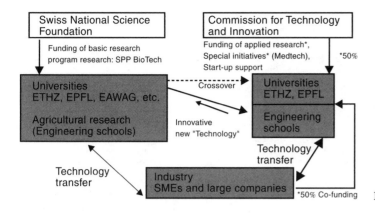

Figure 1. Funding, knowledge and technology flow.

Table 2. Main Areas of Interest of Research

Core area of nucleic acid technology (122)
Pharmaceutical biotechnology (for therapeutics and diagnostics) (69)
Agro/plant biotechnology (58)
Bioengineering, fermentation/reactor design (41)
Environmental biotechnology (27)
Bioinformatics (25)
Bioelectronics (23)
Biotransformation (22)
Biosafety (23)

Note: Number of research groups involved is given in parentheses.

- Funding of target-oriented program research (projects coordinated in units) via SNSF, with a strong emphasis on technology transfer at the precompetition level.
- Funding of applications-oriented R&D via the Commission for Technology and Innovation (CTI) (industry finances 50 percent of these projects).

In future developments of Swiss biotechnology, CTI will play a more important role. In 2001 SPP BioTech will be terminated. It can be assumed that a large number of SPP BioTech research teams will find new SNSF funding within the framework of the newly established National Centres of Competence in Research (NCCR), which are now in the evaluation phase. In the crucial attempts to organize smooth transitions, at present many researchers on SPP BioTech teams have already taken advantage of the extensive research network created by the SPP BioTech to access additional CTI and/or industrial funding.

USING RESEARCH RESULTS TO CREATE NEW JOBS

Technology Transfer — Universities Warming Up to Private Industry

The top researchers at Swiss academic institutions are a significant reservoir for of new inventions. Swiss academe thus provides opportunities for cooperative ventures with the private sector in the creation of start-up companies. This has been an important trend in the biotech industry worldwide, since many innovative concepts have emerged from the academic environment.

An example of the growing interest of companies in such ventures is the agreement signed in the fall of 1999 between Novartis, the Neuroscience Centre of the University of Zurich, and the Federal Institute of Technology Zurich. Under the agreement Novartis will fund research projects for up to 40 million Swiss Frances over a period of 10 years.

All Swiss universities are public. In recent years universities have further been given a high degree of administrative autonomy. In the course of these changes, ownership to all inventions resulting from research performed at the university has also been transferred from the state to the university. Corresponding laws are either in preparation or are enforced already, such as in Berne, Geneva, and Zurich.

Technology transfer has gained a lot of attention at universities in recent years and is strongly supported. Most universities in the meantime have established policies and technology transfer offices that support cooperative activities with the private sector, and provide support for faculty members on issues such as sponsored research agreements, the protection of intellectual property, licensing, and the creation of spin-off companies. Whereas technology transfer in the past was handled with mixed success by individual scientists, it is now being administered more professionally. Pragmatic and flexible guidelines for technology transfer are designed to facilitate interactions between academe and the private sector.

Increasingly academic researchers are considering the creation of spin-off companies as an interesting alternative or complement to their standard career paths. The high degree of entrepreneurial spirit among young academics is mainly due to the reduced job security in the large multinational companies that have all undergone significant restructuring in recent years due to the trends for globalization and the resulting mergers and acquisitions. Entrepreneurship at universities is even encouraged by various successful programs. A good example is Venture 98, a national business plan competition for university scientists organized by McKinsey & Company and the Federal Institute of Technology (ETH) in Zürich. More than 20 percent of the 215 projects submitted were in the field of biotechnology and life sciences, and they have spawned a number of companies. A similar subsequent program called Venture 2000 was launched in November 1999.

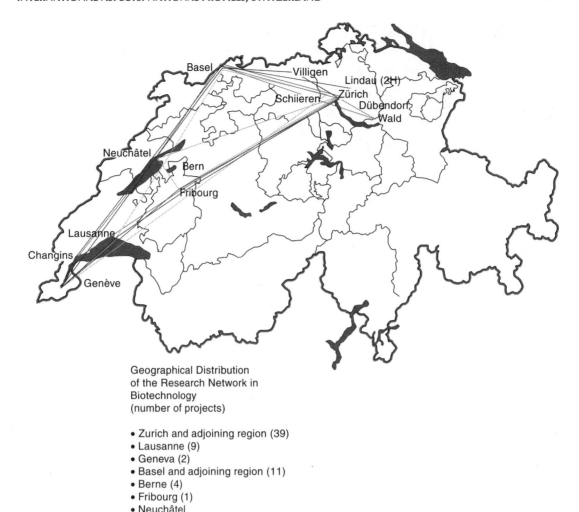

Geographical Distribution
of the Research Network in
Biotechnology
(number of projects)

- Zurich and adjoining region (39)
- Lausanne (9)
- Geneva (2)
- Basel and adjoining region (11)
- Berne (4)
- Fribourg (1)
- Neuchâtel

Figure 2. The Swiss Priority Programme. SPP BioTech was launched in 1992 with public funds. Six research areas in biotechnology and complementary activities in continuing education, information, communication, technology assessment, and technology transfer were designated to receive state support over a period of 10 years. The objective of the SPP BioTech is to consolidate strategic, applied biotechnology research in Switzerland.

Diversified and Rapidly Growing Swiss Biotech Industry

With a long tradition of economic strength in chemistry, Switzerland has added considerable expertise to bioscience in the past decades. The leading multinational drug companies Novartis, Roche, and Ares-Serono have, however, rather obscured the view of a very dynamic entrepreneurial bio-industry of small and medium-size enterprises. Unitectra, the technology transfer organization of the universities of Berne and Zurich and of the SPP BioTech, recently conducted the second comprehensive review of modern Swiss industrial biotechnology. This survey was published in September 1999 as *Biotechnology Industry Guide Switzerland* (5). It includes companies that meet the biotechnology definition of the European Federation of Biotechnology (EFB).

Biotechnology is the integration of natural sciences and engineering sciences in order to achieve the application of organisms, cells, parts thereof, and molecular analogues for products and services.

Overall, the new directory lists 234 companies compared to 177 in the first edition which was published in 1996. Half of the companies (117) are classified as biotech companies, that is their main business focus is on biotechnology. The other half (117) are "other companies," that is, enterprises where biotechnology represents only one segment of their activities.

Forty-five percent of the listed companies are manufacturers of biotech products in Switzerland, 30 percent are suppliers or distributors, and about 20 percent are service companies. The fields of activity of the various companies are listed in Table 3.

The majority of companies can be grouped in three geographical clusters. The Zurich area has 89 companies, the Basel area 74 companies, and the region around Lake Geneva has about 30 companies. The total number of biotechnology-related jobs in these companies is estimated at 6500 to 7000. This is more than three times as many as in the United States on a per capita basis. Behind these figures lies a typical feature of the Swiss biotechnology industry. In Basel companies are based around

Table 3. Number of Companies in Different Fields of Activity

Agriculture	6
Analytical services/quality control	10
Biomaterials	3
Bioreactors/equipment/engineering	31
Bioelectronics/bio-informatics	9
Bioseparations/down stream processing	13
Cell culture	11
Chemicals (specialty/commodity)	8
Consulting	18
Contract R&D/contract manufacturing	16
Cosmetics/health/beauty products	2
Diagnostics	25
Environmental treatment/waste disposal	9
Fermentation/production	4
Food	9
Laboratory equipment	59
Medical devices	4
Pharmaceuticals/therapeutics/vaccines	26
Platform technologies	16
Reagents/biochemicals	29
Veterinary	2

Note: Some companies are active in several fields.

the chemical and pharmaceutical multinationals. In the Zurich and the Lake Geneva area academe provides the main impetus.

The new survey reveals a sharp increase in the number of entrepreneurial spin-off and start-up biotech companies created over the past three years. Without considering the numerous consulting firms, more than 40 new biotech companies were created. Two-thirds of these start-ups have a strong R&D focus mainly in the pharmaceutical area; the rest can be divided evenly into engineering and service companies, respectively. About one-third of the newly formed companies are typical university spin-offs, whereas others are spin-offs from large pharmaceutical companies. The high number of university spin-offs, in relation to the population and the number of universities, reflects the new entrepreneurial spirit among young scientists in academe. Moreover it also is the result of the increased support provided by various start-up programs.

Some examples of recent innovative start-up companies are:

Actelion has its focus of research on the endothelium, which constitutes the innermost layer of blood vessels and plays a role in cardiovascular diseases, inflammation, asthma, and many types of cancer. The aim of the company is the discovery and development of innovative drugs.

Biolytix is a young company located close to Basel providing services in the area of molecular biological analyses. It is specialised in the field of real-time quantitative PCR using state-of-the-art technology.

Biospectra develops and manufactures novel on-line analytical equipment and state-of-the-art automation solutions for fast bioprocess development and bioprocess optimisation.

Cytos Biotechnology, a spin-off company of the ETH Zürich, is developing new process solutions in the area of cell culture technology to optimise protein production. It also develops innovative solutions in other areas e.g. vaccines.

Modex Thérapeutiques is a university spin-off company based in Lausanne. Its focus is on new approaches to cell therapy for chronic systemic diseases, such as anaemia or diabetes. It plans to soon start with clinical trials on its first treatment regimen.

Prionics specializes in the detection of prions which cause BSE ("Mad cow disease") in cattle, Traber's disease in sheep, and Creutzfeld–Jakob disease in humans. It markets a bovine spongiform encephalopathy test for slaughtered cattle and sheep. Future R&D efforts also aim at prevention and therapy of prion diseases. Prionics was created as a spin-off company from the University of Zürich.

Zeptosens has its core competencies in the areas of advanced optical sensor and array technologies, bio-interaction analysis and bioassay design and development. It develops analytical platforms for the detection of analytes at extremely low concentrations. Typical applications include gene expression analysis, investigation of drug-receptor interactions, identification of bioactive compounds, and so on.

The rapid growth and development of the industrial biotech sector is confirmed by the annual *European Life Sciences* report of Ernst & Young (6). Although the absolute numbers in this report differ from the Unitectra survey because of slightly different definitions and inclusion criteria, the recently published report lists about more than 90 entrepreneurial life sciences companies (ELISCOs). This corresponds to an increase of 600 percent in a period of only three years (7).

The Association of Swiss Biotech Companies (AESB) established in March 1998 has already more than 100 member companies, mostly small and medium-size enterprises. It promotes biotechnology in the country and actively represent its members' interests in political and other circles. One of the AESB's essential goals is the facilitation of technology transfer between its members and universities. AESB will also advise foreign biotech companies looking for development opportunities in Switzerland.

FINANCE

Over the past decade Swiss industry has attracted considerable media attention through a series of biotechnology acquisitions and partnerships in the United States. This has been particularly true of large chemical/pharmaceutical corporations.

The new drive in Swiss biotechnology is reflected in the country's financial community. The last few years have seen the creation of a number of funds focusing on private equity and venture capital financing. There are currently more than 60 different funds operating in Switzerland and many of them explicitly seek opportunities in the biotech field.

Investments in venture capital in Switzerland reached 215 million Eurodollars in 1998, an increase of almost

Table 4. Selection of Swiss Investment Funds and Companies

Life science funds in Switzerland

BB Bioventures LP
Clariden Biotechnology Equity Fund
CS Equity Fund Pharma
Global Life Science LP
Lombard Odier Immunology Fund
Lombard Odier Nutrition Fund
Novartis Venture Fund
Pharma wHealth
Pictet Global Sector Fund–Biotech
UBS (Lux) Equity Fund/Biotech
UBS (Lux) Equity Fund/Health Care

Life-sciences investment companies listed at SWX

BB Biotech
BB Medtech
Micro Value
New Venturetec
Pharma Vision

Life-sciences venture capital companies

Alta Berkeley Associates SA
Angel Capital
Apax Partners & Co.
Aventic AG
Castle Private Equity AG
Euroventure-Genevest
Friedli Corporate Finance
Dr. Jürg F. Geigy
Invecor AG
MiniCap Technology Investment AG
New Capital AG
New Medical Technologies
Nextech Venture
Private Equity Holding AG

300 percent over the previous year. About one-third of the 86 projects supported were in the seminal or early stage phase, another third in the expansion phase. Most of the investments went into the high-tech sector. A number of biotech and life sciences companies are listed on the Swiss stock exchange (Swiss Exchange, SWX). SWX has introduced a new market segment in summer 1999 especially designed to meet the needs of young companies (Table 4). This adds another exit opportunity for investors.

However, the main shortage currently is not in finance but in skilled management with experience in setting up and running high-tech start-up companies. A number of initiatives are set to improve this issue such as the CTI Start-up Program and the recent formation of a Swiss Business Angels Club.

SAFETY AND TRANSPARENCY WHEN INTRODUCING TECHNOLOGICAL INNOVATIONS INTO SOCIETY

Successful Launch of Swiss Biosafety Research Network

The safety of technological applications is a prerequisite to their introduction into society. Swiss policy makers have recognized the importance of a publicly funded

Table 5. Biosafety Research Projects

Viral recombination related to virus-resistant transgenic plants
Vertical gene flow
Biological containment for transgenes in plants
Ecological effects of transgenic plants
Horizontal gene transfer between plants and microorganisms, in aquatic systems and in the environment.
Fate of microorganisms in the soil
Projects related to health watch: monitoring for recombinant or pathogenic microorganisms, retroviruses, prions in water, food, and composts

program for carrying out biosafety research and technology assessment, and view this as a service to society. A special unit on Biosafety Research and Technology Development was created by the SPP BioTech to address the safety aspects of biotechnological applications. A national agency for biosafety research, the agency BATS, was also created to coordinate research projects on the biosafety of transgenic organisms, as well as hazardous, naturally occurring organisms (Table 5). The safety of transgenic plants has been an area of intense activity, for which there is considerable effort invested in safety research and the development of methodologies for the safety assessment of open biological systems (8–10).

Biosafety Research and Technology Assessment

Though important, safety is not the only criterion that is considered in evaluating technology applications. Decision makers in Switzerland rely on technology assessment (TA) for understanding the interrelationship of a technology or a product and society or its environment. In Switzerland, TA is coordinated by a central TA unit, founded in 1991, which was a time when the Federal Council and the Parliament decided that the Swiss Science Council (SSC) should develop a Swiss model for the assessment of the effects of technologies. In the field of biotechnology, the TA unit of SSC has coordinated several studies and organized a Publiforum, as the consensus conferences are called in this country. They are presently organizing a second Publiforum on xenotransplantation. The agency BATS is one of the partner institution of this federal TA program.

One example of a TA is a study that was carried out on the effects of potential widespread use of transgenic crops in Switzerland. This TA focused on the impact assessment in the ecological, toxicological, and economical dimensions (11,12). The culture of transgenic crops was compared to other agricultural alternatives, such as organic farming and integrated production, for each of the three impact dimensions mentioned. Another example of a TA project is a study of alternative agricultural production strategies for sustainability, based on ecological and economic indicators. From the information generated by this interdisciplinary effort, the existing scientific knowledge could be presented in a format that is useful to decision makers who define policy options related to transgenic crops.

Making Scientific Knowledge Available to Society

Access to reliable information is fundamental to good decision making on the personal and governmental level.

Members of the public require an adequate understanding of the meaning of new discoveries in order to make personal choices. Officials on all administrative levels need easily accessible knowledge and resources for the preparation of new legislation or for regulatory oversight.

The agencies BATS and BICS, granted by the SPP BioTech to provide information on all aspects of biotechnology, are both non profit and non lobby organizations. BICS publishes the unique Swiss quarterly review on Biotechnology, *BioTeCH forum*, available in a bilingual French/German edition. Other publications of BICS include facts sheets and brochures which are also available online and are a source of useful information not found in the media. The home page of the agency BICS allows the visitor an easy access to a vast selection of links covering all aspects of this field (13). The objectives of the Internet site developed by the agencies BATS and BICS are to (*1*) offer value-added knowledge on biotechnology impacts, and (*2*) pool and organize digital information for easy access (14). Contributors of information are research institutions, government agencies, and nongovernmental institutions. At this site the visitor can find information on a range of biotechnological applications. The information for the site is carefully gathered and checked for the quality of the source. In addition the retrieval of information is facilitated through a full-text retrieval system. Links are also given to other relevant sites. These sites are constantly being improved in order to serve the public better. A new feature of the *bioweb* site provides an interactive forum. Visitors to the site can discuss issues related to biotechnology with other citizens and a panel of scientists knowledgeable in the field.

REGULATORY FRAMEWORK AND PUBLIC DEBATE IN BIOTECHNOLOGY

Swiss Citizens Against a Ban on Genetic Engineering

Like their fellow European citizens, the Swiss are ambivalent about modern biotechnology. They have nevertheless acquired the distinction of being the first in the world to call for a national referendum, based on the complex technical and emotional issues surrounding genetic engineering. In June 1998 the Swiss were asked to vote on a constitutional prohibition of genetic engineering, therefore banning the use and patenting of transgenic animals and the deliberate release of transgenic animals, plants, or microorganisms into the environment. The political campaign leading to the vote lasted two and a half years and provided a unique opportunity for public and private organizations to hold informational meetings and public debates on the issues of genetic engineering. There was extensive media coverage of public debates, which helped to improve the overall public understanding of science, and to a significant extent, also the public acceptance for certain applications of genetic engineering. In the heat of the public discussion preceding the referendum, the Swiss Parliament also committed itself, in a motion called Gen-Lex, to enact a strict regulatory framework, in place of an all-out ban. On the day of the referendum, the Swiss people rejected a general ban on genetic engineering by a margin of 2 to 1.

At the time of writing, the Swiss government is still in the process of drafting the legislative framework regarding applications in genetic engineering. Nine existing laws pertaining to the various aspects of the use of genetically modified organisms (GMOs) and modern biotechnology are in the process of revision. Swiss legislation, based on the European Directives 90/219 on contained use of GMOs and 90/220 on deliberate release, has also introduced pioneering regulatory measures (14).

Prior to the drafting of the new legislation, a Federal Commission of Experts on Biosafety was created to oversee the use of GMOs. This was in the early 1980s, and the Commission followed the U.S. National Institutes of Health (NIH) guidelines on the safe use of GMOs. As a result, even in the absence of legislation, there have not been any abuses of GMO in Switzerland. In January 1997 a Swiss Expert Committee for Biosafety was created, as foreseen in the law on the protection of the environment and on epidemics. The role of this Expert Committee is to advise the administration regarding authorizations of field releases of transgenic organisms and on the drafting of the new legislation to encompass the newest knowledge in this field.

On November 1, 1999, three ordinances were enacted to regulate (*1*) the contained use of GMOs and pathogenic organisms, (*2*) safety at the workplace, and (*3*) the deliberate release of GMOs, including experimental field release or commercialization (15). Other aspects of genetic engineering applications are still not covered and are being hotly debated. In human medicine, these are genetic testing and xenotransplantation. Other contentious issues are liability insurance, intellectual property rights, and international trade, which must respect the guidelines of international agreements on trade and safety. The Swiss regulation for the labeling of foods derived from GMOs is similar to that of Europe. Labeling must be clear and unambiguous in Switzerland, with mention of the GMO origin, for example, in the list of ingredients. Chemically defined substances that are purified from GMOs and are free from traces of modified DNA or proteins do not have to be labeled. Switzerland was the first country in Europe to introduce, on July 1, 1999 (15,16), a threshold value of 1 percent for compulsory declaration. This means that any unintentional inclusion of a GMO equivalent during processing or transport of a product does not have to be declared, as long as the threshold is not surpassed. The threshold value of 0.5 percent for animal feed has also been legally accepted.

Although not legally prohibited, food derived from GMOs are deliberately kept off the shelves of department stores and groceries. No GMO-labeled product is currently sold in Switzerland because of the massive rejection by consumers of GMO-containing foods. The major food producers now ensure a sufficient stock from suppliers guaranteeing GMO-free crops. In addition environmental organizations act as watchdogs for the systematic monitoring of products suspected of containing unintended traces of GMOs. When a product tests positive for the presence of GMOs, it is withdrawn immediately from the shelves.

Swiss law does not prohibit deliberate releases, although a de facto moratorium currently exists in

Switzerland on the deliberate release of transgenic animals, plants, or microorganisms into the environment. In the spring of 1999 two experimental field releases were denied authorization by the Federal Administration despite the recommendations of the Commission of Experts on Biosafety to authorize the release, based on a thorough examination of safety. The final decision of the Federal Administration not to grant authorization was swayed by public opinion, which was against any type of deliberate releases of transgenic plants. At the time of writing, there has not been any field releases in Switzerland, experimental or commercial. Independently of this federal decision, the concluding opinion of a Publiforum on gene technology and food (June 1999) called for an official moratorium on the commercial cultivation of transgenic plants but not for experimental field releases. The discussion on how Switzerland will proceed in the future is still ongoing within the Federal Administration.

Ethics of Nonhuman Applications of Biotechnology

In Switzerland, public concern for the dignity of human and nonhuman organisms is taken very seriously. Switzerland is the only country in the world with a Federal Commission of Experts on Bioethics for the Nonhuman Applications of Biotechnology. The members of this Commission represent the various schools of thought in philosophy and ethics rather than lobby groups. This Commission works in conjunction with other Expert Commissions for biosafety, animal experimentation, and human applications of biotechnology (at the time of writing, this commission does not exist yet but is planned). The role of the Federal Commission on Bioethics is to advise the Swiss authorities and to provide them with criteria for a comprehensive evaluation of the ethical dimension of genetic engineering applications on nonhuman organisms. The elaboration of evaluation criteria in this new field is a fascinating and pioneering aspect of the Commission's work and will contribute to the legal recognition of the intrinsic value of animals and the environment.

Regulatory Framework for Biotechnology Applications on Humans

At the time of writing, several biotechnology applications on humans are close to being regulated and are hotly debated. Preimplantation diagnostics will most probably be banned in Switzerland by a law that is presently in preparation. In addition freezing of additional embryos, cloning, or any type of research work with human embryos or embryonic cells will most probably be banned, to avoid any abuse. Genetic testing is currently being debated because of its social implications beyond medical diagnosis. A public hearing has been organized for citizens to assess the various aspects of genetic diagnostics; Swiss legislators are currently consulting the report of the hearing.

Opinion on xenotransplantation is divided in Switzerland, as the biosafety and ethical aspects of xenotransplantation are highly controversial. Some people would like to see a restrictive legal framework, while others prefer a contingent ban. A Publiforum has been planned for citizens to discuss this issue.

The liability concerning transgenic plants or products derived from GMOs is currently under discussion. It is possible that the liability period for adverse effects will be extended from 10 to 30 years for transgenic organisms.

Technology, Law, and Society

The Swiss legislation that is under preparation attempts to accommodate the needs of the commercial sector, as well as public expectations on safety, information, and dialogue. During the years 1995 to 1998, the threat of a ban on genetic engineering applications generated a feeling of uncertainty that hampered business decision making, particularly in small and medium-size companies. Some scientists and entrepreneurs considered leaving Switzerland. Therefore the outcome of the vote in June 1998 has given the development of Swiss biotechnology a definite boost.

In the aftermath of the referendum in 1998, media coverage on bio- and gene technology issues remains passionate in Switzerland and is sustained by public opinion coming from abroad, such as on the Pusztai report of the possible adverse effects on rats after eating transgenic potatoes containing lectins or the adverse effects of Bt treated corn on the monarch butterfly. The experience of the political campaign prior to the referendum in 1998 has also propelled the scientists into the public debate.

Gen-Lex, the proposed legislative framework for biotechnology applications, is now undergoing the final step of approval in Switzerland. Applications such as experimentation with human embryos, xenotransplantation, sociopolitical consequences of gene diagnostics, and liability insurance on agriculture products derived from GMOs are the topics currently being debated in the Parliament and by federal agencies and the public.

Switzerland's first-rate university research and strong position in modern biotechnology have already produced important results in the areas of health, nutrition, environmental protection, raw materials, and specialty chemicals. In building firmly on its existing strengths, Switzerland is in a good position to keep pace internationally with future rapid developments in biotechnology. This will take place, however, with a firm commitment to the needs of society for safety, information, and innovation. The structure is now in place, through government policies, for promoting public dialogue at all stages of technology development.

ACKNOWLEDGMENTS

Dr. Lillian Auberson is gratefully acknowledged for editing the manuscript. The authors are financially supported by the Swiss National Science Foundation.

BIBLIOGRAPHY

1. O. Ghisalba and H. Vogel, *Frueherkennungstudie zur Biotechnologie*, Swiss Science Council, Bern, 1990, pp. 97, 154.

2. Swiss Priority Programme Biotechnology, *Paving the Way for the Development of Swiss Biotechnology, Swiss National Science Foundation*, Bern, 1999, p. 12.

3. *Prisma 99, Statistics on the Priority Programmes in Switzerland, Swiss National Science Foundation*, Bern, 1999, p. 160.

4. Biotechnology Research Compendium Switzerland, *SPP BioTech*, vol. 7, Unitectra, Basel, 1997.

5. *Biotechnology Industry Guide Switzerland*, 2nd ed., Unitectra and SPP BioTech, vol. 5, Unitectra, Basel, 1999.

6. *Ernst & Young's European Life Sciences 99, Communicating Value*, Ernst & Young International, London, 1999.

7. *Evolution, Ernst & Young Seventh Annual European Life Sciences Report 2000*, Ernst & Young International, London, 2000.

8. O. Käppeli and L. Auberson, *Trends Biotechnol.* **15**, 342–349 (1997).

9. O. Käppeli and L. Auberson, *Chimia* **52**, 137–142 (1998).

10. O. Käppeli and L. Auberson, *Trends Plant Sci.* **3**, 276–281 (1998).

11. E. Schulte and O. Käppeli, eds., *Gentechnisch veränderte krankheits- und schädlingsresistente Nutzpflanzen. Eine Option für die Landwirtschaft?* vol. 1, Swiss National Science Foundation, Bern, 1996.

12. E. Schulte and O. Käppeli, eds., *Gentechnisch veränderte krankheits- und schädlingsresistente Nutzpflanzen. Eine Option für die Landwirtschaft?* vol. 2, Swiss National Science Foundation, Bern, 1997.

13. *http://bics.ch*

14. *http://www.bioweb.ch*

15. *http://www.admin.ch/buwal*

16. *http://www.admin.ch/ch/f/rs/81.html* in french or *http://www.admin.ch/ch/d/sr/81.html* in German.

See other INTERNATIONAL ASPECTS entries; INTERNATIONAL INTELLECTUAL PROPERTY ISSUES FOR BIOTECHNOLOGY.

INTERNATIONAL ASPECTS: NATIONAL PROFILES, UNITED KINGDOM

RICHARD E. ASHCROFT
Imperial College
London, England

BENJAMIN CAPPS
RICHARD HUXTABLE
University of Bristol
Bristol, England

OUTLINE

INTRODUCTION

The term "biotechnology" covers "any technological application that uses biological systems, living organisms, or derivatives thereof, to make or modify products or processes for specific use." This definition appears in Article 2 of the Convention on Biological Diversity (1). Biotechnology touches our lives in many ways. It is instrumental in health care, food, the environment and agriculture, as well as such diverse fields as waste disposal, biomediation, and the promotion of more energy efficient, less polluting, and cheaper production processes. The diversity of biotechnologies, and their relevance to so many different spheres of human concern, is reflected in the breadth of different kinds of regulatory instrument and the range different administrative responsibilities and structures in place in the United Kingdom of Great Britain and Northern Ireland (UK) at the present time. Similarly the relative novelty of many biotechnological methods, and the ethical uncertainty surrounding many of them, is reflected in the mix of statute law, common law, advisory committees, and public policy in the UK at the present time. The British media and public have repeatedly indicated ambivalent concerns about much biotechnology, as is shown by ongoing debates over cloning, genetically modified (GM) foods, various reproductive medicine techniques, hunting and animal experimentation, to name but a few instances.

In order to understand the political and legislative structure of UK biotechnology regulation, we now briefly review UK law and policy making, and the relationship between this and European Community (EC) law and policy making.

UK Law

It is not strictly necessary to provide a detailed overview of the UK's system of law making, particularly as the UK regulatory scheme largely derives from two sources. For a general overview of the English legal system, see Ward (2). Those laws that possess relevance are not derived from the case law. Rather, in the context of biotechnology, various Acts of Parliament (or "statutes") and associated subordinated legislation govern the vast majority of the UK regulation. Essentially an Act of Parliament, once it has passed through Parliament (comprising the House of Commons and House of Lords) to become law, will bind those within the jurisdiction. Such Acts as apply to biotechnology often enable bodies other than Parliament to issue subordinate (or delegated) legislation. An example is the "statutory instrument," which enables a Minister to make a legally binding regulation. As we will witness below, there is a proliferation of such regulations within the scheme for regulating biotechnology.

UK Law and European Law

Much of the impetus for the regulation in the UK has been provided by European legislation in the area. For a general overview of the European impact on UK law, consult Ward (2). Two sources of European legislation are of particular relevance. First, the UK has been a member state of the Council of Europe since its inception in 1949. The Council of Europe strives to establish Europe-wide standards on a range of issues. Bolstered by various authorities and institutional machinery (e.g., the European Court of Human Rights), the Council of Europe itself comprises the Committee of Ministers and the Parliamentary Assembly. The former is the decision-making body; it may lay down binding legislation such as Conventions, or adopt recommendations to governments. The latter is the deliberative body; it too may make recommendations or resolutions. The Council of Europe has issued a number of legislative documents relating to biotechnology and the most important of these will be noted.

Second, having joined the EC in 1972, EC law is of prime importance to the UK's regulation of biotechnology. Since the Treaty of Maastricht, the EC is but one of three "pillars" of the European Union (EU). For the most part, the following discussion relates to the EC (see Ref. 3). The EC treaties set out broad frameworks that must be fleshed out by more specific measures. This task is performed by the various EC institutions: the Court of Justice (ECJ), the Council, the Commission, and the Parliament. These bodies provide three types of secondary legislation: regulations, directives, and decisions.

Regulations are directly applicable in all member states, and are binding in their entirety. Directives differ because these may be implemented by those means chosen by the member states to which they apply. There are a number of such regulations and directives that relate to biotechnology, either directly or indirectly. Finally, decisions are addressed to a specific person or persons: for example, the decisions of the ECJ. This latter source of EC law does not warrant exhaustive study, as the numerous regulations and directives suffice to provide the EC perspective.

One other, albeit non-statutory, area of European standards-setting warrants mention. The European Committee for Standardisation (CEN) is responsible for the planning, drafting and adoption of European technical standards (with the exception of those pertaining to the electrotechnology and the telecommunications sectors). In Europe, CEN works in partnership with CENELEC — the European Committee for Electrotechnical Standardization and ETSI — the European Telecommunications Standards Institute.

Such technical specifications ensure compatibility between products; guarantee appropriate levels for their safety, quality, or efficiency; and provide the test methods necessary to establish conformity. Once the need for a European standard has been firmly established, and nonduplication of work verified (CEN may also use an international standard), an experts Technical Committee is established. CEN's remit is to promote voluntary technical harmonisation in Europe in conjunction with international bodies and its partners in Europe. This harmonization is designed to diminish trade barriers, promote safety, allow interoperability of products, systems and services, and to promote common technical understanding.

In essence, then, legislation in a particular area of biotechnology is governed through Acts of Parliament, Regulations and the relevant legislation from Europe. By way of example, [this example is discussed in greater detail later in this article (contained use of GMOs and deliberate release of GMOs).], Genetically Modified Organisms (GMOs) are controlled in the UK by a number of regulations, including the GMO (Contained Use) Regulations 1992 and the GMO (Deliberate Release) Regulations 1992. These three pieces of legislation were created under the Health and Safety Act 1974 (contained use) and the Environment Act 1992 (deliberate release). Within Europe there are two major Directives that relate to the use of GMOs and are implemented by the UK regulations: Directive 90/219 (contained use) and Directive 90/220 (deliberate release). To assist in the implementation of these Directives, CEN is drafting appropriate safety standards that will apply to Europe in the use and release of GMOs; these standards should further serve to harmonize work in this area.

UK REGULATION OF BIOTECHNOLOGY

Introduction

In this section we will outline the legislative and regulatory framework that exists at present in relation to biotechnology in the UK. The wide-ranging scope of biotechnology means that the regulatory bodies — which are both governmental and nongovernmental — often have blurred remits. Numerous committees, subcommittees, and groups advise these bodies. Because of the rapid advancement in this field, the existing bodies are often inadequately equipped to deal with specific issues. To compensate for this there is a practice of creating new statutory and ad hoc committees to fill in the gaps that develop in the framework. Additionally, the government is also advised through independent bodies, such as the Royal Society, which periodically produce reports on matters of interest.

The regulatory and advisory framework, according to the UK government, has two distinct functions: to consider whether to grant approvals for individual products or processes (based on ethical, legal, regulatory, and scientific criteria), and to set a strategic framework for development of the technology in the UK (4). The main measures to provide safeguards against any real or hypothetical risks in biotechnological products are rigorous pre-market assessment of safety, research to improve scientific understanding of the particular product, and health surveillance to provide reassurance against any unexpected adverse effects. There is a concentrated effort to ensure that the governmental review of, and action on, issues relating to biotechnology are sufficiently transparent. Transparency is urged, in order to facilitate input from interested parties (such as the public and concerned industries) on matters of policy.

The regulation of biotechnology in the UK is divided into five areas of responsibility under a lead governmental departmental body. As observed, the scope of these areas is often blurred, with certain issues falling within the jurisdiction of one or more of the responsible organizations. The areas of legislative responsibility are divided among: the Department of Trade and Industry (DTI) (consumer safety, product liability, trading standards, and patents); the Health and Safety Executive (HSE) (health of biotechnology workers, the control of hazardous substances and the contained use of GMOs); the Department of the Environment, Transport and the Regions (DETR) (release and marketing of GMOs into the environment); the Ministry for Agriculture, Fisheries and Food (MAFF) (food safety and labeling, animal feed and veterinary medicines, and plant health and pesticides); and the Department of Health (DOH) (therapeutic medicinal products, medical devices, gene therapy, and medicines licensing). These same groups are responsible for specific guidance and advice, access to funding, and research and expert services. This additionally includes the input of research councils, universities and trade associations.

The various statutory and advisory committees that exist within these five areas regulate and provide advice on the safety and broader impact of biotechnology. In general, the issues fall within the remit of, on the one hand, food and agriculture and, on the other, medicines and therapeutics. Some bodies, however, such as the Advisory Committee on Genetic Modification and the Advisory Committee on Release into the Environment, involve themselves in each of the two areas. There are essentially two types of committee: those that are established (either statutorily or ad hoc) by the government to specifically address issues arising from developments in biotechnology, and those that are not biotechnology specific, but nevertheless undertake significant amounts of biotechnology based work. There are other groups, such as English Nature, that only occasionally touch on biotechnology as part of a much wider area of interest.

Eight of the existing committees have a statutory function to advise Ministers on the exercise of their powers under specific pieces of legislation. These committees are the Animal Procedures Committee (APC), the Veterinary Products Committee (VPC), the Human Fertilisation and Embryology Committee (HFEA), the Advisory Committee on Release into the Environment (ACRE), the Advisory Committee on Pesticides (ACP), the Food Advisory Committee (FAC), the Committee on Safety of Medicines (CSM) (which also advises the Medicines Control Agency, (MCA)), and English Nature (along with its equivalents in Scotland, Wales, and Northern Ireland). The relevant Ministers, and MCA in the case of CSM, are statutorily required to take into account the advice of three of these committees when taking decisions. Thus the relevant Secretary of State consults ACRE; the recommendations of ACP must be taken into account by Ministers at

the Ministry of Agriculture, Fisheries, and Food, the Department of the Environment, Transport, and the Regions, the Department of Health, and the Scottish and Welsh Offices; and MCA must consult the CSM.

In addition there are a number of nonstatutory committees. These offer advice on specific matters of interest or concern. These committees are the Advisory Committee on Novel Foods and Processes (ACNFP), the Farm Animal Welfare Council (FAWC), the Advisory Committee on Genetic Modification (ACGM), the Human Genetics Advisory Commission (HGAC), the UK Xenotransplantation Interim Regulatory Authority (UKXIRA), the Genetics and Insurance Committee (GAIC), the Gene Therapy Advisory Committee (GTAC), the Advisory Group on Scientific Advances in Genetics (AGSAG), and the Advisory Committee on Genetic Testing (ACGT).

The Department of Health

DOH is concerned with the health of humans, and it plays an active role in relation to biotechnology and the relevant UK legislation that governs the medical field and public health. DOH's main responsibilities concern the protection of public health from any possible hazards arising from the application of biotechnology. With this overarching objective in mind, DOH acts as a focal point in developing and coordinating policies, both national and international. The Department has legislative responsibility for developments in biotechnology in therapeutic medicinal products, medical devices and gene therapy also plays a role in the active encouragement of inward investment and sponsors the UK's biopharmaceutical industry. Research plays a large part within DOH and is divided into three main areas: the policy research program, NHS research and development strategy, and the research of nondepartmental public bodies. It is also involved in the European Community's Biomedicine and Health Research Programme.

DOH and the Office of Science and Technology (OST) jointly form the secretariat of the Human Genetics Advisory Commission (HGAC) which reports to and advises DOH and OST. HGAC remit is to "keep under review scientific progress at the frontiers of human genetics and related fields; to report on issues arising from new developments in human genetics that can be expected to have wider social, ethical and/or economic consequences, for example in relation to public health, insurance, patents and employment; and to advise on ways to build public confidence in, and understanding of, the new genetics" (5). It works alongside other committees that have an interest in human genetics. DOH also jointly forms the secretariat of the Advisory Committee on Novel Foods and Processes with the Ministry of Agriculture, Fisheries, and Food and provides members for the Inter-department Group on New Food Developments which covers feed intended for animal consumption.

The following discussion examines those various forms of biotechnology regulation which effectively operate under the auspices of DOH. Thus we address research, medicinal products, medical devices, and a variety of measures designed to survey the general field of human genetics. Before commencing the overview, two caveats must be noted. First, although efforts have been made to distinguish and categorize the areas of interest, in an effort to avoid unnecessary duplication, naturally, such duplication cannot be entirely avoided. Second, and related to this first point, we should note one issue that will not be rehearsed: contained use and deliberate release. Here DOH does not have a statutory responsibility, although it is closely involved with the relevant independent expert advisory committees (ACGM and ACRE) (6).

Research. Research in DOH is divided among the Public Health Laboratory Service (which develops and implements epidemiological typing methods and laboratory diagnostics), the National Institute for Biological Standards, the Centre for Microbiology Research, and the Edward Jenner Institute for Vaccine Research. The Chief Scientist Office in the Scottish Office Department of Health and the Welsh Office Health Department are the overseers of research in Scotland and Wales, respectively.

From the point of view of medical research on humans, the most important bodies are the local and multi-center research ethics committees. The first Local Research Ethics Committees (LRECs) began to be set up in the mid-1960s, in emulation of the U.S. Institutional Review Boards. Initially these LRECs were locally originated, and without official status, although their pattern of constitution and process was defined by the Royal College of Physicians guidance on the Research Ethics Committees in 1967. Gradually the number of committees grew and acquired Department of Health recognition. There are now over 230 LRECs in the UK, regulated since 1991 by official Department of Health guidelines, and responsible to the area health authorities. Since 1997 any research protocol involving five or more centers has been reviewed in the first instance by a Multi-centre Research Ethics Committee (MREC). There are 10 MRECs, one each for Scotland and Wales, and one for each of the eight English health Regions. The MRECs are responsible to the Regional Health Authorities, and to the Department of Health centrally. For information and documentation on RECs in the UK, see Smith (7).

LRECs and MRECs are responsible for any research on (National Health Service) NHS patients, the recently dead in NHS premises, fetal tissue research on NHS premises, access to the medical records of NHS patients, and any other research on human beings that takes place on NHS premises. Their remit is to protect the subjects of such research, and to facilitate useful research; when these two aims conflict, the presumption is supposed to be that protection of individual patients takes priority (as required by the Declaration of Helsinki).

The constitutions of LRECs and MRECs are similar. Such committees must have a chair and a vice-chair, one of whom must be a lay person, together with at least eight members drawn from a range of professional backgrounds and including at least one other lay person. The committees are consensus-forming committees, and are not intended to be voting committees or "representative" in any but the broadest political or social sense.

Research approved by the MREC must also be reviewed by the relevant LRECs, but the LRECs in this situation can only consider four factors: the suitability of the site for the research, the suitability of the local investigator to do this research, the suitability of the local population to take part in this research, and the usability and comprehensibility of the patient information sheet to patients in this locality.

LRECs and MRECs are expected to take their decisions in the light of a fair process of discussion, and in the light of the best available written guidance (including that from government advisory committees such as the ACGT, the medical Royal Colleges, and international agencies such as the World Medical Association, the Council of the International Organisations of Medical Sciences, and the International Committee on Harmonisation of Good Clinical Practice).

Medicinal Products. Therapeutic medicinal products are primarily controlled by the Medicines Control Agency (MCA). MCA acts on behalf of Health Ministers and the UK licensing authority to issue marketing authorizations for medicinal products for human use and other licenses governing manufacture, clinical trials, wholesale dealing and parallel imports. This is based on the product reaching required levels of safety, efficiency, and quality. Independent advisory committees support the MCA in its tasks: of chief importance are the Committee on the Safety of Medicines and the Medicines Commission. The MCA is responsible for enforcing compliance with those authorisation provisions issued under the *Medicines Act 1968* (and its associated legislation, such as the Medicines for Human Use (Marketing Authorisations etc.) Regulations 1994) (8), and is assisted in this by the Royal Society of Great Britain, the DOH and the Social Services in Northern Ireland.

The Committee on Safety of Medicines (CSM) is a statutory body that advises the UK Licensing Authority (part of DOH) and MCA on the quality, efficiency, and safety of medicines in order to ensure that appropriate public health standards are met and maintained. A final important feature of the regulation in this context operates at the international level. The National Institute for Biological Standards and Control (NIBSC) oversees international standards in medicinal products. This is the executive arm of the National Biological Standards Board, which essentially exercises certain controls on the purity and potency of biological substances. Specialist medicines inspectors are responsible for the inspection of biological — including biotechnological — manufacturing sites.

The *Medicines Act 1968* and its associated legislation gives effect to the European medicines legislation, as initially laid down in *Directive 65/65* (9). A subsequent Directive in 1975 began to flesh out this basic framework in, among other matters, providing the data requirements for testing (10). Another Directive from 1975 established the Committee for Proprietary Medicinal Products (CPMP) (11), which now forms part of the European Agency for the Evaluation of Medicinal Products (EMEA). This latter agency controls the European authorisation of UK medicinal products. *Council Regulation 2309/93* set up the EMEA and a centralized procedure for biotechnological, and other "high-tech," medicines (12). This allows an authorization, through the EMEA, which is valid throughout the Community. There is also a decentralized procedure for member states to mutually recognise each other's authorizations, which is also controlled through the EMEA.

Other Directives provide further amendment to, and elaboration on, the European legislative scheme relating to medicinal products. Those warranting particular mention are the laws governing specific products such as immunological products (13), radiopharmaceutical products (14) and products derived from human blood or human plasma (15), alongside more general measures dealing with good manufacturing practice (16), wholesale distribution (17), the classification of medicinal products (18), labels and leaflets (19), and advertising (20).

Medical Devices. Medical devices "are those diverse and extensive products, other than medicines, which are used in the healthcare field for the prevention, diagnosis, monitoring and treatment of disease and injury" (21). These are controlled by the Medical Devices Agency (MDA). This body is responsible for ensuring that medical devices and equipment for sale or use in the UK meet acceptable standards of safety, quality, and effectiveness and that these standards comply with the relevant EC Directives. At present, there are three directives that regulate the safety and marketing of medical devices throughout the EU. These are Directive 90/385 (governing active implantable medical devices) (22), Directive 93/42 (which covers all medical devices except those covered by Directive 90/385 and medical devices for in vitro diagnostics) (23), and Directive 98/79 (governing in vitro diagnostic medical devices) (24). The first of these Directives, which effectively covers such devices as heart pacemakers and cochlear implants, finds expression in UK law in the Active Implantable Medical Devices Regulations 1992 (25), as amended by the Active Implantable Medical Devices (Amendment and Transitional Provisions) Regulations 1995 (26). The second Directive, which has a broader scope, is implemented by the *Medical Devices Regulations 1994* (27). It appears that the third of these Directives has yet to be implemented.

A number of aims and themes can be identified in this legislation. Emphasis is placed on the requirements that devices must not compromise the health and safety of the patient, user or any other person, and that the risks associated with the device must remain compatible with the patients health and protection. In order to achieve these aims, a number of specific requirements must be met. Thus clinical investigations are to be carried out, adverse incidents must be reported, devices must be classified and controlled according to the degree of risk inherent in their application, and monitoring must occur in order to ensure compliance with the requirements.

Gene Therapy. The UK's approach to the control of gene therapy has been informed by the 1992 *Clothier Report* (28). The Clothier Committee observed that gene therapy should be regarded as research involving human subjects. It concluded, inter alia, that research in the area should be restricted to disorders that are life threatening or cause serious handicap and for which treatment is either unavailable or unsatisfactory. The Committee decided that, for the time being, no attempt should be made to intervene in germ line cells. Finally, the recommendation a national supervisory body be established to consider and advise on the acceptability of gene therapy protocols resulted in the establishment of the Gene Therapy Advisory Committee (GTAC). GTAC is responsible for both the case-by-case review of individual protocols and an assessment of more general issues relating to such therapy. In addition GTAC provides advice to Health Ministers on developments in this field and on their implications. As well as working closely with LRECs, GTAC works closely with the MCA. As previously noted under the *Medicines Act 1968* and *Directive 65/65* (as modified), the MCA has a responsibility for regulating the quality, safety, and efficiency of medical products and applications for their clinical trials. This possesses relevance for gene therapy. HSE and DETR are similarly involved when the scope of the gene therapy falls within their remit.

Genetic Testing. The Advisory Committee on Genetic Testing (ACGT) is a nonstatutory body that advises UK Health Ministers on developments on genetic testing, taking account of ethical, social, and scientific aspects. Established in 1996, its remit is to provide advice on developments in testing for genetic disorders and to establish requirements, especially in respect of efficiency and product information, to be met by manufactures and suppliers of genetic tests.

Xenotransplantation. The UK Xenotransplantation Interim Regulatory Authority (UKXIRA) is a nonstatutory body that provides the voluntary regulatory framework for biotechnology in the area of human genetics and xenotransplantation. At present there is no domestic legislation, although the need for primary legislation has been realised. Set up through the Advisory Group on the Ethics of Xenotransplantation, UKXIRA, with the assistance of the Committee on Dangerous Pathogens, reviews and assesses the safety and efficiency of xenotransplantation procedures. All treatments for patients in this field have to be approved by the Secretary of State (29). As for specific legislation, some aspects of xenotransplantation are covered by the Human Fertilisation and Embryology Act 1990 (including cell therapies and gene therapies involving viable tissue) (see also the section "Home Office").

Genetically modified (GM) animals created in the course of xenotransplantation research are disposed of (although some GM animals can be used as food) under the assistance of the ACNFP, FAC (for labeling issues) and ACRE. Any live GM animals used in containment will be subject to the GMO (Contained Use) Regulations 1992 (as amended 1996) and the GMO (Risk Assessments) (Records and Exemptions) Regulations 1996 with respect to environment risk assessment (see below). The control of animals under these regulations will be the responsibility of HSE and DETR, as advised by ACGM. Xenotransplantation protocols that involve animals are covered by the Animals (Scientific Procedures) Act 1986, and UKXIRA works closely with the Home Office in this matter. It is widely accepted that primates should not be used in such procedures, although the possibility has not been entirely ruled out (30). As of yet, no xenotranplantations from animals to human beings have taken place in the UK, although, while being cautious in its policy, the government has by no means excluded the possibility (31).

Infertility Treatment. The Human Fertilisation and Embryology Act 1990 established the Human Fertilisation and Embryology Authority (HFEA). Another important, albeit nonbiotechnology specific, piece of legislation in the field of infertility is the *Surrogacy Arrangements Act 1985* (which essentially prohibits commercial surrogacy arrangements). Its primary function is to licence and monitor centres providing treatment, research and care in this field. The Act and the Authority are therefore concerned with the use of gametes and embryos; as for HFEA's more general functions, advice will be disseminated on issues arising from developments in assisted conception and associated research. Key concerns include the need for safe, efficient, and ethical advances in the field.

The Health and Safety Executive

The Health and Safety Executive (HSE) is responsible for the health and safety of workers (and others) engaged in biotechnology in Great Britain. In Northern Ireland, the Health and Safety Inspectorate of the Department of Economic Development is responsible.

In Great Britain, the Health and Safety Commission (HSC) aids the HSE in its duties. HSC, whose members are appointed by the Secretary of State for the Environment, Transport, and the Regions, therefore considers, and also develops, health and safety policy. HSE advises the HSC on the shaping of policy and is responsible for implementing it. HSE specialist inspectors provide advice on the areas in question, specifically on risk assessment and containment. In relation to biotechnology, HSE's main responsibilities concern the regulation of the contained use of GMOs and the implementation of general health and safety legislation. (HSE works with the DETR with regard to the related issue of deliberate release when the release has implications on the health and safety of individuals. The COSHH regulations deal with certain deliberate releases of GMOs.)

Contained Use of GMOs. The former responsibility derives from those requirements laid down in the Genetic Manipulation Regulations 1989 (32), and the GMO (Contained Use) Regulations 1992 (as amended) (33). These regulations were made under the *Health and Safety at Work Act 1974*. The regulations revoke the regulations from 1989 (32) and replace them insofar as they relate to contained use. The latter contained use legislation assesses risks to humans and the environment, and essentially derives from EC Directive 90/219 (34). The 1996 amendments maintain this dual aim, as well as including various other requirements. These include the need to keep records; the need to establish a local

GM safety committee, the need to classify all activities and organisms used; the need to notify the HSE of first an intention to use premises for GM for the first time and second, an intention to engage in certain subsequent individual activities (and, in some cases, work must not begin without the HSE's prior consent); and, finally, the need to adopt adequate controls, including suitable containment measures. Since the amendment, DETR also plays a role—alongside the HSE—in the regulation of this area.

The contained use regulations are thus concerned with the health and safety of both workers involved in the contained use of genetic engineering and those members of the public who may come into contact with such work. HSE and HSC rely on advice from the Advisory Committee on Genetic Modification (ACGM). ACGM advises all relevant government departments on human health and the environmental aspects of the contained use of GMOs, including laboratory and industrial installations. It is not involved in product approval. ACGM is accordingly advised by a technical subcommittee formed to provide specialised technical advice on all aspects of the human and environmental safety of the contained use of GMOs. The Advisory Committee on Dangerous Pathogens (ACDP) and the Department of Health's Health Promotion Division Select Committee on Science and Technology (SCST) also have a role in the control of the contained use of GMOs. SCST is divided into three subcommittees: the Human Genetics Advisory Commission (nonstatutory advisory body that also advises OST), the National Screening Committee, and the Advisory Committee on Genetic Testing (nonstatutory advisory body). Other committees may or may not advise on specific issues of GMO contained use depending on their remit. Finally in relation to contained use, ACRE advises the HSE/HSC and any other bodies appropriate on the possible human consequences of releases into the environment.

General Health and Safety. The legislation focusing upon general health and safety includes the *Health and Safety at Work Act 1974* and the *Control of Substances Hazardous to Health (COSHH) Regulations 1999* (35). Here the risks to be assessed are those risks to humans. The *Health and Safety at Work Act 1974* applies to all persons at work in Great Britain, whether employees or self-employed. Its requirements cover biotechnology, including the application of genetic modification techniques. Under this Act, employers have a duty to ensure the health and safety of the employees and to ensure that the general public is not put at risk by the work. The COSHH regulations apply to biological agents, including those which have been genetically modified, that may cause an infection, allergy, toxicity, or otherwise cause a hazard to human health. The COSHH regulations implement those EC directives relating to the protection of workers from risks associated with biological agents (36). Employers must therefore assess the risk of working with certain biological agents, to adopt appropriate control measures, and to notify the HSE of work involving certain biological agents.

The Department of the Environment, Transport and Regions

The Department of Environment, Transport and the Regions (DETR) is responsible for the regulation of the deliberate release and marketing of GMOs in Great Britain, and in furtherance of its aims, it promotes an extensive research program into associated risks. Before analyzing the various bodies that work with DETR, the (distinct) position in Northern Ireland deserves mention. There the position is virtually identical to that in Great Britain, although the requisite notification must be made to the Department of Economic Development; it is enforced by its Health and Safety Inspectorate. The Department of the Environment for Northern Ireland controls legislation governing the release and marketing of GMOs.

Other bodies, however, also have a role to play in this context. In addition to DETR, the Ministry of Agriculture, Fisheries, and Food, the Scottish Office Agriculture, Environment and Fisheries Department, the Department of the Environment for Northern Ireland, and the Welsh Office Agriculture Department are implicated in regulating specific areas. In addition the Department of Health addresses those releases of GMOs that have an impact on human health. The Ministers are advised primarily by the Advisory Committee on Release to the Environment (ACRE) and the Advisory Committee on Genetic Modification (ACGM). With regard to food and the marketing of GMOs, the Food Advisory Committee (FAC) and the Advisory Committee on Novel Foods and Processes (ACNFP) have particular relevance. FAC advises Ministers on the exercise of powers in the *Food Safety Act* relating to the labeling, composition, and chemical safety of food. It also advises on general matters relating to food safety.

Turning to the particular role of DETR, the regulation of the release of GMOs is primarily covered by the Genetically Modified Organisms (Deliberate Release) Regulations 1992 (37), issued in accordance with the Environmental Protection Act 1990 (Part IV). Other domestic legislation that impacts upon biotechnology and the environment must also be noted (38). The latter Act sets out the offences and penalties which apply in the event of a breach of its requirements. Any release of GMOs, with a few specialized exceptions, into the environment must be approved by the Secretary of State for the Environment, Transport, and Regions (acting jointly with other appropriate Ministers). Its purpose is to minimize any damage to the environment or the public that might arise from the deliberate release or escape of GMOs. The Secretary of State is therefore empowered to revoke a consent and to take enforcement action. The *GMO (Deliberate Release) Regulations* are enforced jointly by DETR and HSC. The HSC, in turn, can direct the HSE inspectors to perform the delegated enforcement functions.

Both of the relevant pieces of legislation (i.e., the Environmental Protection Act 1990 and the GMO (Deliberate Release) Regulations 1992; note that parallel legislation exists in Northern Ireland) implement the EC *Directive 90/220* (39), which specifically addresses the deliberate release of GMOs. This Directive has been amended in the light of progress in relation to the (new, simplified) procedure for applications to release GM crop plants (40), and the technical progress made, for example, regarding the information requirements of GM higher plants (41). These amendments were implemented by the GMO (Deliberate Release) Regulations 1995 (Section 7.40). The EC has passed other legislation dealing with deliberate release, but this may best be dealt with in other contexts. It should also be noted that the Council of Europe is similarly committed to the safety of both humans and the environment, as evinced in a number of legislative documents (42).

Ministry of Agriculture, Fisheries and Food

The policy of the Ministry of Agriculture Fisheries and Food (MAFF) is to support the development and exploitation of biotechnology within the food and agriculture industries, while protecting people, livestock, crops, and the natural environment. Where applicable, MAFF is jointly responsible for regulations governing pesticides, plant health, veterinary medicines, food products and imports. The approach of MAFF is coordinated with other government departments. Thus, for example, it consults with ACNFP to provide guidance on, and to regulate the use of, GM food. ACNFP is also responsible for assessing all applications made under the EC regulations relating to novel food and novel food ingredients. HSE, DETR, and the DOH also consult MAFF on the contained use and deliberate release of GMOs and act jointly, where appropriate. Finally, in Northern Ireland, the Department of Agriculture for Northern Ireland coordinates its work with MAFF with regard to biotechnology as it applies to its jurisdiction.

Novel Foods. In the UK assessment of GM and other novel foods, Ministers are advised by the independent Advisory Committee on Novel Foods and Processes (ACNFP). This committee carries out safety assessments of individual novel foods as part of the pre-market approval scheme controlled by the EC. In carrying out such assessments, the ACNFP is assisted by other governmental advisory committees, such as the Committee on Medical Aspects of Food and Nutrition Policy, the Committee on Toxicity of Chemicals in Food, Consumer Products and the Environment, and the Food Advisory Committee on the labeling of GM foods.

MAFF operates under the Food Safety Act 1990 (which applies to England, Scotland, and Wales and is parallel to the Food Safety Act (NI) 1991), which controls food consumption in Northern Ireland. The legislation makes it an offence to render any food injurious to health by adding or using any article or substance, abstracting any constituent from the food, or by subjecting the food to any other processes or treatment. In the main, local authorities enforce those portions of the Act that relate to hygiene and health. Environmental Health Officers and Trading Standards Officers enforce the requirements governing labeling and composition. In the capital, the London borough and Metropolitan authorities carry out both of these sets of enforcement duties.

Evidently food safety is a key concern. This concern is also detectable at the more general European level, although competing interests have been cited. For example, Reports from the Council of Europe concede that risks should be assessed and minimised, but observe that advances — specifically in biotechnology — might increase yields and therefore prosperity (43).

Nevertheless, in the specific field of biotechnology, food safety is the pervasive theme. The primary piece of European legislation relating to novel foods and novel food ingredients is EC Regulation 258/97 (44). The Commission has taken many decisions in this context, too numerous to mention. These decisions concern such plants and vegetables as swede, maize, and soya, and an overriding concern is that the products will not adversely affect health. The UK has provided for its enforcement in The Novel Foods and Novel Food Ingredients Regulations 1997 (45). The EC regulation created a comprehensive EU-wide regulatory framework controlling all aspects of GM crops in Europe, from seed to final product. Accordingly, member states cannot introduce their own requirements in this area without the agreement of the other countries of the Commission, who are advised by the EC Scientific Committee for Food (SCF). (The SCF also re-evaluates any additives if they are prepared significantly differently from the original evaluation.) The regulation introduced a statutory pre-market clearance system for all novel foods, including those produced by genetic modification, and it is binding on all member states. Under this regulation the safety of individual GM foods is assessed by all member states, and any differences of scientific opinion are resolved by reference to a number of scientific committees within the EC.

The primary regulation has since been the subject of a recommendation concerning the scientific aspects and the presentation of information relevant to a safety assessment (46). Materials not originally covered by the regulation have also been brought within its procedures. Accordingly Regulation (EC) No 1813/97 (47), which generally concerned labeling requirements, dealt with genetically modified soya and maize, which was originally approved for food use under Directive 90/220 (supra), prior to the novel foods regulation. Detailed rules relating to the labeling stipulations contained in the later regulation have since been laid down in *Council Regulation (EC) No 1139/98* (48). These regulations governing the labeling of GM foods enter UK law via the Food Labelling (Amendment) Regulations 1999 (49).

Animal Feedingstuffs. The Agriculture Act 1970 (as amended) governs the marketing of animal feed in the UK. The Act makes it an offence to sell any material for use as feed that contains any ingredient that is deleterious to animals and, secondly, to human beings, who consume the products of an animal fed with the material. The Feeding Stuffs Regulations 1995 (as amended) (50) implement those EC Directives. The regulations cover a permitted list of single-cell proteins in feedingstuffs (51) (this Directive may be extended to encompass novel feed material; at present, a voluntary scheme for the approval of new feed material is in operation in the UK); assessment of "certain products" used in animal nutrition (52); and assessment of additives used in animal feedingstuffs (53). These Directives set out permitted additives that are

allowed to be used or present in animal feed. The Directives also laid down requirements governing the information that must be provided (54).

Veterinary Medicines. The manufacture, distribution, marketing, and administration of veterinary medicines are controlled by the Medicines Act 1968, in conjunction with the secondary legislation issued under it and other UK legislation implementing the apposite EC legislation. Veterinary medicines are also controlled by the Marketing Authorisations for Veterinary Medicinal Products Regulations 1994 (55). These, in implementing parts of Directive 81/851/EEC (as amended) (56) state that only veterinary medicinal products subject to a marketing authorisation valid in the UK may by placed on the UK market. The Agriculture Departments and the Department of Health and Social Security (Northern Ireland) enforce the provisions governing veterinary procedures, acting on behalf of the health and agriculture ministers. The Royal Pharmaceutical Society of Great Britain also undertakes responsibility for the enforcement of provisions relating to certain retail sales. The Veterinary Medicines Directorate (VMD) administers the control of veterinary medicines, on behalf of DOH and MAFF. These bodies are advised by the independent Veterinary Products Committee (VPC) (created under the Medicines Act 1968) on the safety, quality, and efficiency of veterinary medicines covered by the 1968 Act. The VMD monitor and regulate all veterinary products on the market, including postauthorization monitoring.

The Council's Regulation (EEC) 2309/93 establishes a European centralized authorization procedure for high technology products in veterinary medicine (57). Compliance with this procedure is obligatory for certain biotechnological products and for novel growth promoters. Under this regulation initial applications are made to the European Medicines Evaluation Agency. The application is then submitted to the EC Committee for Veterinary Medicinal Products, comprising representatives from all member states, which assesses the application for authorization throughout the Community.

Pesticides and Plant Health. The Control of Pesticides Regulations 1986 (COPR) (as amended) addresses the majority of pesticides (58). The responsibility for regulation in this area is divided between MAFF and HSE. MAFF deals with the approval of products for use in agriculture and horticulture and in food storage practice. HSE is concerned with products for use with regard to mainly nonagricultural and nonfood uses.

COPR, as it relates to the MAFF remit, is progressively being superseded by the body of legislation which implements EC law. MAFF is responsible for the domestic legislation that implements Directive 91/414/EC (59), which covers the placing on the market of plant protection products (broadly, agricultural pesticides) and the import of pesticides. Included within this legislation are the Plant Protection Products Regulations 1995 (as amended in 1996 and 1997) (60), the Plant Protection Products (Fees) Regulations 1995 (as amended in 1997) (61), and the Plant Protection Products (Basic Conditions) Regulations

1997 (62). The legislation also contains powers to control the import of pesticides.

Under this legislation, manufacturers seeking to gain approval for pesticide products must apply to the independent Advisory Committee on Pesticides (ACP), according to the Food and Environment Act 1985. It is expected that this legislation will be superseded by the recent EC Directive on biocidal products (63), which will have a larger scope than that presently under the control of the COPR.

The legislation is enforced by the HSE, local authorities, and agriculture departments. Local authorities are concerned with consumer aspects of the legislation (as overseen by Trading Standards Officers) and issues, including storage, which are not covered by HSE (as overseen by Environmental Health Officers). The agriculture departments enforce provisions relating to wildlife, including the impact of pesticides and of pesticide residues in the environment.

Finally with regard to plant protection measures, following European legislation (64), the Plant Health (Great Britain) Order 1993 (as amended) places restrictions on the import and movement within the EC, and keeping in Great Britain, of particular plant pests, including GM plant pests, plants and products (65). The Order further provides that no unauthorized person may engage in any activity that involves genetic modification of a plant pest without proper authorization. Licences to undertake such work are supplied by the Plant Health Division of MAFF and, in Scotland, by the Scottish Office Agriculture Environment and Fisheries Department.

Plant Breeders and Plant Varieties. Applications for plant breeders' rights and the National Listing of Varieties are governed by the Plant Varieties and Seeds Act 1964 (as amended), the Plant Breeders' Rights Regulations 1998 (66), and the Seeds (National Lists of Varieties) Regulations 1982 (as amended) (67).

Discussions are persisting on the most appropriate manner in which to embody the EC provisions in this context. Although a lengthy discussion of the European law in this context is unnecessary, a few points do warrant mention. Accordingly there are provisions relating to the marketing of GM material that require domestic adoption. The new EC Plant Varieties system (introduced on April 27, 1995) makes specific licensing provisions for essentially derived varieties (i.e., those produced from existing varieties using biotechnological techniques). A recent important Directive allows varieties to be marketed. Thus member states may authorize producers in their own territory to place GM materials on the market (68). Such authorization may be granted only if all appropriate measures have been taken to avoid adverse effects on human health and the environment (as determined in accordance with other Directives) (69). Other stipulations relate to such matters as the need for labels to identify GM products and the need to protect varieties threatened with genetic erosion. (The novel foods legislation is also taken into account.)

Animal Welfare. The Farm Animal Welfare Council (FAWC) advises MAFF Ministers on the welfare of farm animals on agricultural land, at market, in transit, and at place of slaughter. It can freely investigate, advise, and communicate with any outside body, including the European Commission and the public, on any legislative or any other changes that may be necessary in this context.

As for existing legislation in this context, the Animal Health Act 1981 provides for Ministers to control the spread of disease, and the Specified Animal Pathogens Order 1993 (SAPO) (70), prohibits the import of animal pathogens and carriers of pathogens except under licence. It appears that the prohibition essentially relates to Third World countries. Under the latter Order, GMOs require a licence regardless of their origin (i.e., whether or not they are from the Third World). The licence is granted by Agriculture Departments, as advised by the state veterinary service. With specific regard to fish, broadly similar requirements apply, as provided for in the Diseases of Fish Act 1983. (The approaches of the EC and the Council of Europe to animals (and specifically animal welfare), are discussed below.)

Department of Trade and Industry

The Department of Trade and Industry (DTI) is legislatively responsible for product liability, trading standards, and the Patent Office. DTI is also the lead sponsor department for biotechnology. DTI's Chemicals and Biotechnology Directorate works within the DTI and with other government departments on the regulation and general appraisal of issues surrounding competitiveness in biotechnology. DTI thus has the ultimate responsibility for championing the biotechnology industry in all aspects of governmental, European, and international policies that affect its competitiveness. For this reason DTI has strong links with both the industry and the regulatory bodies of the UK government. The Office of Science and Technology, part of DTI, is responsible for managing the science budget and coordinating government policies on science and technology.

Product Liability. The DTI regulates consumer safety through a number of specific Acts and Regulations, depending on the product. Among these are the General Product Safety Regulations 1994 (71), and those other measures that implement the relevant EC Directives. The 1994 regulations impose a general requirement for safety in all consumer products which have not already been comprehensively covered by extant, specific European product Directives, and British and European standards. Local Authority Trading Standards Officers enforce the regulations. There is also the Consumer Protection Act 1987, which implements the EC Directive on product liability.

Patenting. The central piece of legislation in the UK is the *Patents Act 1977*, which succeeded the prior European Patent Convention (72). In the context of biotechnology, it must be emphasized that animal or plant varieties and biological processes for the production of animals or plants not involving significant technical intervention *cannot* be patented. (However, certain interested parties are not left unprotected: for example, plant breeders; see the discussion below.) Nevertheless, as certain biotechnological innovations do remain subject to patenting, a basic overview of the UK system must be provided.

As with other inventions, the granting of a UK patent for a biotechnological invention depends on satisfying certain criteria, including those of novelty, inventiveness, and industrial applicability. In order to obtain a patent for any invention, an application that clearly and fully discloses the invention must be filed. It has been recognized that in some cases it may not be possible to describe a microorganism in words. In such cases a culture of the microorganism must be deposited in a culture collection not later than the date of the filing of the application. Indeed, the National Culture Collections offer a number of services, including the supply, identification, and safe deposit of cultures for patent purposes. Once the applicant has satisfied certain procedural requirements, a patent may be granted.

In the UK an applicant may deal with the Patent Office or the European Patent Office. The systems are broadly similar, although differences do exist. It should be noted that a UK grant of a patent is effective only in the UK. Patents for other countries generally have to be obtained locally. It is possible under the Patent Co-operation Treaty to obtain patents in a number of countries through a single initial application. It is also possible to obtain patents in up to 18 European countries, including the UK, by a single application to the European Patent Office.

We do not need to further probe the UK system governing patents because, with regard to Europe, the position on patenting is due to undergo some revision. It has been recognized that the biotechnology industries require a secure and effective intellectual property regime. The EC has answered the calls for such a regime in a 1998 Directive (73). This constitutes a significant piece of binding legislation, particularly for biotechnology. Member states will have until July 2000 in which to ensure that their national patent law is consistent with the requirements laid down by the Directive. The Directive maintains the general position with regard to patenting and biotechnology. Notable additional requirements include the denial of patents to inventions whose commercial use would be "immoral" (e.g., where suffering may be caused to a GM animal without particular gains for humans or animals). The Parliament and Council have also issued a regulation that impacts upon patenting biotechnology (74). This regulation enables certification, and strives to overcome some of the difficulties surrounding the granting of patents. Nevertheless, the 1998 Directive is undoubtedly the most important development in the area. Finally, and in contrast to the EC, the Council of Europe has appeared less willing to perceive patents as the panacea for the troublesome issue of rights in biotechnological advances. In that context, intellectual property rights are still being debated (75).

Home Office

In relation to the Home Office, a particular area of interest is the use of animals in scientific procedures (see the discussion below). The Animal Procedures Committee (APC) advises the Home Secretary on this issue, under the Animal (Scientific Procedures) Act 1986. The Committee is bound to have regard for both the legitimate requirements of science and industry and the protection of animals from avoidable suffering and unnecessary use in scientific procedures.

Similar themes are detectable in the relevant European legislative documents (for obvious reasons, not every document will be cited). The Council of Europe has had much to say on the status and treatment of animals (76). A key convention from 1986 regarding animals used in experimentation was, to a large extent inspired, by Directive 86/609 from the EC (77). It is notable that the contents of the two documents are broadly similar in their concern with ethical issues, such as the welfare of animals. Nevertheless, there are some distinctive differences. Thus, for example, whereas the Directive is primarily interested in the harmonization of national laws, in order to avoid any distortions in the internal market, provides better explanations of "alternative methods," and is supported by an Advisory Committee, the convention, by way of contrast, directs increased attention to the ethical issues, such as animal rights and humankind's needs.

In relation to biotechnology, two issues seem to have particular relevance to animals. These are cloning and genetic modification. As to the former issue, the EC has called for strict controls, with a particular view to ensuring that harm—to humans, animals and the environment—is minimized (78). As for the latter issue, both the EC and the Council of Europe have devoted some considerable efforts to assessing the permissibility of transgenesis; it is evident that each the ethical—primarily welfare-related—issues (79).

LOOKING TO THE FUTURE OF BIOTECHNOLOGY REGULATION IN THE UK

It is clear from the developments surveyed below that the regulatory philosophy of the UK government is unlikely to change much in the near future. It is unlikely, for instance, that a single National Bioethics Commission will be set up, partly because of the complexity of existing relationships within the administrative structure of the state (as shown), partly because of some scepticism regarding the merits of "bioethics" as an academic or policy discipline within the UK, partly because the experience of international debates with countries that have such a commission do not lead UK commentators to hope for much from such a commission, partly because of such a commission's enormous work load, and partly because there is no pressure for change at present. Bioethics advice is normally seen as the province of scientific experts, together with individual profession-based insights from key members of religious confessions (especially the established Church of England), the legal professions, and some social scientists and philosophers.

The credibility of some of the recent advice (e.g., on GM foods) has come under attack of late, in the main because many expert advisors have become seen to be out of step with the public mood, and in part also because some advisors have been seen as *parti pris*, in virtue of the intellectual or commercial links with the activities they regulate. This is a problem in particular for scientific advisors, as has been seen in the GM foods debate.

It is true that the present Blair administration, and the previous Major administration, public commitments have been made to "open government" and to accountability of governmental and quasi-governmental committees. Appointments to all advisory committees must now be opened to applications for membership, and stringent appointment procedures now apply. But many important groups (e.g., commissions of inquiry) remain committees appointed by the minister on advice of his or her civil servants. Accountability is secured by accountability to parliament, and to the press, rather than any more direct method (e.g., public meeting).

Moving on from processual considerations of how committees are set up, structured, and members appointed, it has been noted by some commentators that most UK committees apply some very specific kinds of ethical and regulatory considerations to the legitimation of research and technology. Most committees apply some sort of risk–benefit calculus, modified in the light of broad social or cultural concerns. The initial Warnock report on Human Fertilisation and Embryology was much criticized by philosophers and by members of conservative pressure groups for the way it tried to synthesize analytic philosophical argument with more intuitive ideas about right and wrong. The result was felt to be unsatisfactory both from a consequentialist point of view (which underlies the risk–benefit arguments) and from the mainstream religious viewpoints (because the risk–benefit arguments were felt to miss the point). Nonetheless, the Warnock committees recommendations were broadly taken up in the form of the Human Fertilisation and Embryology Act. The general methodology of Warnock is similar to that of all the chief regulatory committees; what differentiates them is the way they assess risk, the significance attached to particular risks, and the way in which socially expressed concerns are permitted to influence these considerations (80). This point of method has significant consequences for the politics of such decision making (Risk to whom? How much is tolerable and by whom? Which risks are considered?), not least in the essential informality of the way judgments about risk must be made. At least one chair of a major committee has indicated great dissatisfaction with this state of affairs, but it is unlikely that the philosophy of such committee decision making will change significantly unless a different but similarly mechanical means for taking decisions can be found (81).

One major change over the next few years will be the impact of research programs in bioethics and biolaw in the UK; recent funding initiatives by the Wellcome Trust and by the European Commission, among others, are likely to result in major research projects into both the sociology of public discourse on biotechnology, and the incorporation of public attitudes into bioethical decision making. It will be interesting to see how (and if) such research affects policy making in the more formal arena of central government. However, it is likely that the main source of advice will continue to be the scientific and medical communities, whether or not the public credibility of these sources of advice, qua impartial advice, rises or falls.

In the light of these considerations, then, the main structural changes to biotechnology regulation will be constitutional changes in the state, rather than in the organization of regulation, its deliberative processes, or its underlying philosophy. The two main changes in the next few years are increased devolution of power away from the UK government in London "down" toward the national and regional authorities in Scotland, Wales, Northern Ireland, London, and the English regions, and "up" toward the European Commission and the juristic functions of the European Courts.

Recent Changes: The Implications of Devolution

With the devolution to the Scottish, Welsh, and Northern Ireland Parliaments, certain changes will be made to the regulation of biotechnology in the UK and Northern Ireland. Assisted conception, genetics, and health and safety legislation will be reserved to the Westminster Parliament, and therefore, the makeup of the primary and secondary legislation-making process and the advisory structure will generally remain the same. All other biotechnology related fields will become the responsibility of the devolved parliaments. The Northern Ireland and Scotland Parliaments will gain the ability to make primary legislation under the new structure, while the Welsh Assembly will acquire a similar range of functions, including powers to make secondary legislation.

All the devolved administrations will have access to the existing committees, irrespective of whether legislation is devolved or not. Where legislation is devolved, the new administrations, once in place, will be required to decide whether they wish to use the existing committees or create an alternative. They will also possess powers with regard to appointments to existing committees that are effected by devolution. The FSA and the proposed HGC and AEBC will operate on a UK-wide basis, with the devolved administrations having a say in appointments.

Of the statutory committees, those that will remain reserved to Westminster will be HFEA, CSM (which primarily advises the MCA, which will itself have some of its remit devolved; the CSM itself will become a UK-wide committee), ACGM (the environmental aspect of this committee will become devolved), VPC, and APC. The nonstatutory committees will all remain reserved, with no changes to the ACGT, HGAC, AGSAG, and GTAC. The remaining four committees, GAIC, UKXIRA, FAWC, and ACAF, will be altered so as to report to all the UK administrations within a UK-wide remit. Four of the remaining five statutory groups, although operating under devolved legislation, will continue to advise Westminster, and the devolved administration, if requested to do so. English Nature will be replaced in its UK advisory role by the Joint Nature Conservation Committee.

Recent Changes to UK Biotechnology Regulation

In May 1999 a government report was published that reviewed the UK's regulatory and advisory structure in biotechnology (4). In the report the government confirmed its existing policies, which are to protect the health of the public and to protect the environment while ensuring that the potential benefits of this technology are not denied to the British people. Through consultation with the public, industry and experts, it was decided that there was a need to introduce a new regulatory framework. The new framework is designed to be less complex and more transparent, reflect the broader ethical and environmental questions and views of potential stakeholders, and is sufficiently forward-looking to encompass the rapid developments in this field.

The new comprehensive strategic advisory structure will be headed by the Food Standards Agency (FSA), and two new nondepartmental public bodies, the Human Genetics Commission (HGC) and the Agriculture and Environment Biotechnology Commission (AEBC). The remit of these three bodies is to advise relevant ministers on issues, receptively, on food safety, genetic technologies, and their impact on humans, as well as all other aspects of biotechnology without a direct impact on humans and food safety or food standards. Many of the present advisory committees' work has been taken on by the commissions, while other committees and technical bodies will be involved in cross-boundary issues between the three commissions. The commissions are not be involved in case-by-case review. This is still the responsibility of the remaining specialist technical bodies (84).

The ultimate responsibility in this area will lie with the appropriate Ministers. They will be responsible for giving guidelines to regulatory and technical bodies, changing the regulatory and advisory structure, and granting individual consent decisions. To aid the Ministers, the framework of independent expert regulatory and advisory committees will still exist under the remit of three nondepartmental public commissions. Direct advice to Ministers will come from these three strategic advisory commissions. The commissions will lease with the regulatory and technical bodies to give Ministers an overall picture of issues in biotechnology. In addition they advise on future strategy, changes in the guidelines for advisory/regulatory bodies, broader issues including ethics, and gaps in the framework.

Proposed Changes in to the European Legislation

There are a number of changes being proposed at the European level. Some of these are only entering the earliest debate and report stages, while others have progressed to encompass a draft Directive, which needs only to be debated. Due to the usual constraints, this discussion of potential innovations will have to be restricted. One of the more important proposals relates to those EC Directives governing deliberate release. In 1998 a draft Directive was drawn up (82), which is proposed to amend the central Directive 90/220 in a number of significant ways. These include strengthening the provisions on environmental risk assessment and monitoring, streamlining some procedures, ensuring greater transparency and improved labeling requirements, and providing for the consultation of EU ethics and scientific committees. Advances have also been proposed in the specific of novel foods, specifically in relation to additives (83).

CONCLUSION

The main advantages of the UK regulatory system are that it is flexible and responsive to the rapid changes in the biotechnology world; it is accountable readily to parliament and the courts; and some sort of workable balance is maintained, for the most part, between the politicians, scientists, technologists and industrialists, industrial consumers, and public consumers of biotechnological goods. For the most part the existing system has maintained public credibility, despite such recent disasters as the BSE/CJD affair and the GM foods affair. To some extent, at least, these crises are seen as failures of government rather than failures of advice. But underlying this assignment of responsibility is a growing sense of the politicization of science and the interested nature of scientific advice, and a response to this politicization that has not been seen since the nuclear power debates of the 1960s and 1970s. The weaknesses of the system of regulation are then (perceived to be) vulnerability to regulatory capture and lack of responsiveness to public debate. There is a sense abroad that "public debate" is perceived by government and the scientific community as necessarily ill-informed and misled, and therefore to be ignored or "educated." At the same time, certain kinds of public concern are taken note of (as in the cloning debate). It is not clear which sorts of public concern get to dominate scientific concerns and which are dominated by scientific and policy concerns. What is clear is that the political agenda, if not regulatory practice, will be increasingly shaped by the new single-issue agendas prompted by biotechnological change.

BIBLIOGRAPHY

1. Convention on Biological Diversity, *Off. J. Eur. Commun.* **L 309**, 3–20 (1993).

2. R. Ward, *Walker and Walker's English Legal System*, 8th ed., Butterworth, London, 1998.

3. P. Craig and G. de Burca, *EU Law: Text, Cases, and Materials*, 2nd ed., Oxford University Press, Oxford, UK, 1998, p. 3.

4. Cabinet Office: Office of Science and Technology, *The Advisory and Regulatory Framework for Biotechnology: Report from the Government's Review*, London, Cabinet Office, 1999, p. 4.

5. See HGAC at http://www.dti.gov.uk/hgac/

6. *GMOs (Contained Use) Regulations 1992* (see HSE, section 2.3.1) and the *GMOs (Deliberate Release) Regulations 1992, 1995* (see DETR, section 2.4).

7. T. Smith, *Ethics in Medical Research*, Cambridge University Press, Cambridge, UK, 1999.

8. SI 1994 No. 3144.

9. Council *Directive 65/65/EEC* of 26 January 1965 on the approximation of provisions laid down by Law, Regulation or Administrative Action relating to proprietary medicinal products [*Off. J. Eur. Commun.* **022**, 0369–0373 (1965)].

10. Council *Directive 75/318/EEC* of 20 May 1975 on the approximation of the laws of Member States relating to analytical, pharmaco-toxicological and clinical standards and protocols in respect of the testing of proprietary medicinal products [*Off. J. Eur. Commun.* **L 147**, 0001–0012 (1975)].

11. Second Council *Directive 75/319/EEC* of 20 May 1975 on the approximation of provisions laid down by Law, Regulation

or Administrative Action relating to proprietary medicinal products [*Off. J. Eur. Commun.* **L 147**, 0013–0022 (1975)].

12. Council *Regulation (EEC) No 2309/93* of 22 July 1993 laying down Community procedures for the authorisation and supervision of medicinal products for human and veterinary use and establishing a European Agency for the Evaluation of Medicinal Products [*Off. J. Eur. Commun.* **L 214**, 0001–0021 (1993)].

13. Council *Directive 89/342/EEC* of 3 May 1989 extending the scope of Directives 65/65/EEC and 75/319/EEC and laying down additional provisions for immunological medicinal products consisting of vaccines, toxins or serums and allergens [*Off. J. Eur. Commun.* **L 142**, 0014–0015 (1989)].

14. Council *Directive 89/343/EEC* of 3 May 1989 extending the scope of Directives 65/65/EEC and 75/319/EEC and laying down additional provisions for radiopharmaceuticals [*Off. J. Eur. Commun.* **L 142**, 0016–0018 (1989)].

15. Council *Directive 89/381/EEC* of 14 June 1989 extending the scope of Directives 65/65/EEC and 75/319/EEC on the approximation of provisions laid down by Law, Regulation or Administrative Action relating to proprietary medicinal products and laying down special provisions for medicinal products derived from human blood or human plasma [*Off. J. Eur. Commun.* **L 181**, 0044–0046 (1989)].

16. Commission *Directive 91/356/EEC* of 13 June 1991 laying down the principles and guidelines of good manufacturing practice for medicinal products for human use [*Off. J. Eur. Commun.* **L 193**, 0030–0033 (1991)].

17. Council *Directive 92/25/EEC* of 31 March 1992 on the wholesale distribution of medicinal products for human use [*Off. J. Eur. Commun.* **L 113**, 0001–0004 (1992)].

18. Council *Directive 92/26/EEC* of 31 March 1992 concerning the classification for the supply of medicinal products for human use [*Off. J. Eur. Commun.* **L 113**, 0005–0007 (1992)].

19. Council *Directive 92/27/EEC* of 31 March 1992 on the labelling of medicinal products for human use and on package leaflets [*Off. J. Eur. Commun.* **L 113**, 0008–0012 (1992)].

20. Council *Directive 92/28/EEC* of 31 March 1992 on the advertising of medicinal products for human use [*Off. J. Eur. Commun.* **L 113**, 0013–0018 (1992)].

21. D. Longley, *Med. Law Inter.* **3**(4), 319–345 (1998).

22. Council *Directive 90/385/EEC* of 20 June 1990 on the approximation of the laws of the Member States relating to active implantable medical devices [*Off. J. Eur. Commun.* **L 189**, 0017–0036 (1990)].

23. Council *Directive 93/42/EEC* of 14 June 1993 concerning medical devices [*Off. J. Eur. Commun.* **L 169**, 0001–0043 (1993)].

24. *Directive 98/79/EC* of the European Parliament and of the Council of 27 October 1998 on in vitro diagnostic medical devices [*Off. J. Eur. Commun.* **L 331**, 0001–0037 (1998)].

25. SI 1992 No. 3146.

26. SI 1995 No. 1671.

27. SI 1994 No. 3017. (Note that these Regulations are pursuant to the *Consumer Protection Act 1987.*)

28. *Report of the Committee on the Ethics of Gene Therapy*, H.M. Stationary Office, London, 1992, Cm 1788.

29. See Health Service Circular HSC 1998/126 (*Clinical Procedures involving Xenotransplantation*), which utilizes powers within the *National Health Service Act 1977* and the *Community Care Act 1990.*

30. J.K. Mason and R.A. McCall Smith, *Law and Medical Ethics*, Butterworth, London, 1999, p. 340.

31. *The Government's Response to Animal Tissue into Humans: The Report of the Advisory Group on the Ethics of Xenotransplantation*, London, Advisory Group, 1997.

32. SI 1989 No. 1810.

33. SI 1992 No. 3217, SI 1995 No. 2626, SI 1996 No. 967, and SI 1998 No. 1548.

34. Council *Directive 90/219/EEC* of 23 April 1990 on the contained use of genetically modified micro-organisms [*Off. J. Eur. Commun.* **L 117**, 0001–0004 (1990)].

35. SI 1999 No. 437 (revoking SI 1994 No. 3246), which implements a number of EC Directives, including some of those noted below (n. 39). In Northern Ireland, see *Control of Substances Hazardous to Health (COSHH) Regulations*, Northern Ireland, 1994.

36. The primary piece of European legislation is Council *Directive 90/679/EEC* of November 26, 1990 on the protection of workers from risks related to exposure to biological agents at work (seventh individual Directive within the meaning of Article 16 (1) of *Directive 89/391/EEC*) [*Off. J. Eur. Commun.* **L 374**, 0001–0012 (1990)]. This Directive has been amended since. See Council *Directive 93/88/EEC* [*Off. J. Eur. Commun.* **L 268**, 0071–0082 (1993)], Commission *Directive 95/30/EC* [*Off. J. Eur. Commun.* **L 155**, 0041–0042 (1995)], Commission *Directive 97/59/EC* [*Off. J. Eur. Commun.* **L 282**, 0033–0035 (1997)], and Commission *Directive 97/65/EC* [*Off. J. Eur. Commun.* **L 335**, 0017–0018 (1997)].

37. SI 1992 No. 3217. This has since been amended by SI 1993 No. 152 and SI 1995 No. 304. See also *Genetically Modified Organisms (Deliberate Release and Risk Assessment–Amendments) Regulations 1997* (SI 1997 No. 1900).

38. See the *Health and Safety at Work Act 1994, Wildlife and Countryside Act 1981* (s. 14) (as amended in 1985), the *Nature Conservancy Council Act 1973*, the *Countryside Act 1968*, the *Food and Environment Protection Act 1985*, and the *Food Safety Act 1990*. More specifically, see the *GMO (Risk Assessment/Records and Exemptions) Regulations 1996* (SI 1996 No. 1106, as amended by SI 1997 No. 1900) (restricting import and acquisition of GMOs) and the *Environmental Information Regulations 1992* (SI 1992 No. 3240, implementing Council *Directive 90/313/EEC* of 7 June 1990 on the freedom of access to information on the environment [*Off. J. Eur. Commun.* **L 158**, 0056–0058 (1990)].

39. Council *Directive 90/220/EEC* of 23 April 1990 on the deliberate release into the environment of genetically modified organisms [*Off. J. Eur. Commun.* **L 117**, 0002–0024 (1990)].

40. Commission *Decision 94/730/EC* of 4 November 1994 establishing simplified procedures concerning the deliberate release into the environment of genetically modified plants pursuant to Article 6.5 of Council Directive 90/220/EEC (*Official Journal of the European Communities* L 292 of 12/11/1994 pp. 0031–0034). See also the Written Questions Nos. 2394/96 to 2405/96 by Hiltrud Breyer to the Commission, on establishing simplified procedures concerning the deliberate release into the environment of genetically modified plants [*Off. J. Eur. Commun.* **C 96**, 0005–0009 (1997)].

41. See, first, Commission *Directive 94/15/EC* of 15 April 1994 adapting to technical progress for the first time Council *Directive 90/220/EEC* on the deliberate release into the environment of genetically modified organisms [*Off. J. Eur. Commun.* **L 103**, 0020–0027 (1994)]. See also Commission *Directive 97/35/EC* of 18 June 1997 adapting to technical progress for the second time Council *Directive 90/220/EEC* on the deliberate release into the environment of genetically

modified organisms [*Off. J. Eur. Commun.* **L 169**, 0072–0073 (1997)].

42. See *Convention on Civil Liability for Damage Resulting from Activities Dangerous to the Environment* (ETS No. 150) of 21 June 1993, *Recommendation R (92) 9 of the Committee of Ministers to Member States* on the potential ecological impact of the contained use and deliberate release of genetically modified organisms, and *Resolution 870 (1986)* on the biogenetic revolution in agriculture—a blessing or a curse? (as essentially adopted in Recommendation 1213 (1993) on developments in biotechnology and the consequences for agriculture).

43. See *Report (Doc. 7943)* of 7 October 1997 on food supply in the world and *Report (Doc. 8194)* of 14 September 1998 of the Committee on Agriculture and Rural Development on consumer safety and food quality.

44. European Parliament and the Council of 27 January 1997 [*Off. J. Eur. Commun.* **L 43**, 0001–0007 (1999)].

45. SI 1997 Nos. 1335 and 1336.

46. Commission *Recommendation 97/618/EC* of 29 July 1997 concerning the scientific aspects and the presentation of information necessary to support applications for the placing on the market of novel foods and novel food ingredients and the preparation of initial assessment reports under *Regulation (EC) No 258/97* of the European Parliament and of the Council [*Off. J. Eur. Commun.* **L 253**, 0001–0036 (1997)].

47. Commission *Regulation (EC) No. 1813/97* of 19 September 1997 concerning the compulsory indication on the labelling of certain foodstuffs produced from genetically modified organisms of particulars other than those called for in Directive 79/112/EC [*Off. J. Eur. Commun.* **L 257**, 0007–0008 (1997)].

48. Council *Regulation (EC) No. 1139/98* of 26 May 1998 concerning the compulsory indication of the labelling of certain foodstuffs produced from genetically modified organisms of particulars other than those provided for in *Directive 79/112/EEC* [*Off. J. Eur. Commun.* **L 159**, 0004–0007 (1998)].

49. SI 1999 No. 747.

50. SI 1995 No. 1412, as amended by SI 1996 No. 1260, SI 1998 No. 104, SI 1998 No. 2072, and SI 1999 No. 1528.

51. Council *Directive 82/471/EEC* of 30 June 1982 concerning certain products used in animal nutrition [*Off. J. Eur. Commun.* **L 213**, 0008–0014 (1982)].

52. Council *Directive 83/228/EEC* of 18 April 1983 on the fixing of guidelines for the assessment of certain products used in animal nutrition [*Off. J. Eur. Commun.* **L 126**, 0023–0027 (1983)].

53. Council *Directive 87/153/EEC* of 16 February 1987 fixing guidelines for the assessment of additives in animal nutrition [*Off. J. Eur. Commun.* **L 064**, 0019–0028 (1987)], as revised by Commission *Directive 94/40/EC* of 22 July 1994 amending Council *Directive 87/153/EEC* fixing guidelines for the assessment of additives in animal nutrition [*Off. J. Eur. Commun.* **L 208**, 0015–0019 (1994)] and Commission *Directive 95/11/EC* of 4 May 1995 amending Council Directive 87/153/EEC fixing guidelines for the assessment of additives in animal nutrition [*Off. J. Eur. Commun.* **L 106**, 0023–0024 (1995)].

54. See also Council *Directive 93/113/EC* of 14 December 1993 concerning the use and marketing of enzymes, microorganisms and their preparations in animal nutrition [*Off. J. Eur. Commun.* **L 334**, 0017–0023 (1993)].

55. SI 1994 No. 3142.

56. Council *Directive 81/851/EEC* of 28 September 1981 on the approximation of the laws of the Member States relating to veterinary medicinal products [*Off. J. Eur. Commun.* **L 317**, 0001–0015 (1981)].

57. Council *Regulation (EEC) No. 2309/93* of 22 July 1993 laying down Community procedures for the authorisation and supervision of medicinal products for human and veterinary use and establishing a European Agency for the Evaluation of Medicinal Products [*Off. J. Eur. Commun.* **L 214**, 0001–0021 (1993)].

58. SI 1986 No. 1510. This was made under the *Food and Environment Act 1985*. It has since been amended (see SI 1990 No. 2487, SI 1994 No. 3142, and SI 1997 No. 188). This regulatory scheme has been partially superseded by the *Plant Protection Regulations*, which provide that the *Control of Pesticides Regulations* do not apply to any "plant protection product" (as defined in SI 1995 No. 887).

59. Council *Directive 91/414/EEC* of 15 July 1991 concerning the placing of plant protection products on the market [*Off. J. Eur. Commun.* **L 230**, 0001–0032 (1991)].

60. SI 1995 No. 887, as amended by SI 1997 No. 7, SI 1997 No. 2499, and SI 1999 No. 1228. The latter of these revoked the amendments in SI 1996 No. 1940 and SI 1998 No. 2760).

61. SI 1995 No. 888.

62. SI 1997 No. 189.

63. *Directive 98/8/EC* of the European Parliament and of the Council of 16 February 1998 concerning the placing of biocidal products on the market [*Off. J. Eur. Commun.* **L 123**, 0001–0063 (1998)].

64. See Council *Directive 93/77/EEC* of 21 September 1993 relating to fruit juices and certain similar products [*Off. J. Eur. Commun.* **L 244**, 0023–0031 (1993)]; Commission *Directive 92/76/EEC* of 6 October 1992 recognizing protected zones exposed to particular plant health risks in the Community [*Off. J. Eur. Commun.* **L 305**, 0012–0015 (1998)]; Commission *Directive 95/44/EC* of 26 July 1995 establishing the conditions under which certain harmful organisms, plants, plant products and other objects listed in Annexes I to V to Council *Directive 77/93/EEC* may be introduced into or moved within the Community or certain protected zones thereof, for trial or scientific purposes and for work on varietal selections [*Off. J. Eur. Commun.* **L 184**, 0034–0046 (1995)]; Commission *Directive 98/22/EC* of 15 April 1998 laying down the minimum conditions for carrying out plant health checks in the Community, at inspection posts other than those at the place of destination, of plants, plant products or other objects coming from third countries [*Off. J. Eur. Commun.* **L 126**, 0026–0028 (1998)]; and Commission *Decision 98/109/EC* of 2 February 1998 authorising Member States temporarily to take emergency measures against the dissemination of Thrips palmi Karny as regards Thailand [*Off. J. Eur. Commun.* **L 027**, 0047–0048 (1998)].

65. SI 1993 No. 1320, made under the *Plant Health Act 1967* and *Agriculture (Miscellaneous Provisions) Act 1972*. The Order replaced the *Plant Health Order (Great Britain) 1987* (SI 1987 No. 1758). The Order has been amended by: SI 1993 No. 3213, SI 1995 No. 1358, SI 1995 No. 2929, SI 1996 No. 25, SI 1996 No. 1165, SI 1996 No. 3242, SI 1996 No. 1145, SI 1997 No. 2907, SI 1998 No. 349, SI 1998 No. 1121, and SI 1998 No. 2245.

66. SI 1998 No. 1027 (revoking SI 1978 No. 294).

67. SI 1982 No. 844, as amended by SI 1985 No. 1529, SI 1989 No. 1314, SI 1990 No. 1353, SI 1992 No. 1615, and SI 1998 No. 2726.

68. Council *Directive 98/95/EC* of 14 December 1998 amending, in respect of the consolidation of the internal market, genetically modified plant varieties and plant genetic resources, Directives 66/400/EEC, 66/401/EEC, 66/402/EEC, 66/403/EEC, 69/208/EEC, 70/457/EEC and 70/458/EEC on the marketing of beet seed, fodder plant seed, cereal seed, seed potatoes, seed of oil and fibre plants and vegetable seed and on the common catalogue of varieties of agricultural plant species [*Off. J. Eur. Commun.* **L 025**, 0001–0026 (1999)].

69. Such as *Directive 90/220* (see n. 49).

70. SI 1993 No. 3250, as amended by SI 1994 No. 3142 and SI 1994 No. 3144.

71. SI 1994 No. 2328, as amended by SI 1994 No. 3142 and SI 1994 No. 3144.

72. Convention on the Grant of European Patents (Munich, 5 October 1973; TS 20 (1983); Cmnd 7090).

73. *Directive 98/44/EC* of the European Parliament and of the Council of 6 July 1998 on the legal protection of biotechnological inventions [*Off. J. Eur. Commun.* **L 213**, 0013–0021 (1998)].

74. *Regulation (EC) No. 1610/96* of the European Parliament and of the Council of 23 July 1996 concerning the creation of a supplementary protection certificate for plant protection products [*Off. J. Eur. Commun.* **L 198**, 0030–0035 (1996)].

75. See, particularly, Report (Doc. 8459) of the Committee on Agriculture and Rural Development of 9 July 1999 on biotechnology and intellectual property.

76. See ETS Nos. 65, 87, 102, 125, 145 and Recommendations Nos. 1213 and 1289.

77. European Convention for the Protection of Vertebrates used for Experimental or Other Scientific Purposes (ETS No. 123) of 18 March 1986; Council *Directive 86/609/EEC* of 24 November 1986 on the approximation of laws, regulations and administrative provisions of the Member States regarding the protection of animals used for experimental and other scientific purposes [*Off. J. Eur. Commun.* **L 358**, 0001–0027 (1986)].

78. See, amongst other documents, the Resolution on cloning [*Off. J. Eur. Commun.* **C 115**, 0092 (1997)], which followed the reports of the successful cloning of a sheep ("Dolly").

79. See, for example, the Council of Europe documents (ETS Nos. 87 and 145). With regard to the EC, see Council *Decision 1999/167/EC* of 25 January 1999 adopting a specific programme for research, technological development and demonstration on quality of life and management of living resources (1998 to 2002) [*Off. J. Eur. Commun.* **L 064**, 0001–0019 (1999)]; Council *Directive 93/35/EEC* of 14 June 1993 amending for the sixth time *Directive 76/768/EEC* on the approximation of the laws of the Member States relating to cosmetic products [*Off. J. Eur. Commun.* **L 151**, 0032–0037 (1993)]; and *Study Concerning the Ethical Implications of Transgenic Animals (Legal Aspects)* SEC/8581/95: A report to the European Commission Group of Advisors on the Ethical Implications of Biotechnology, May 1995.

80. A. Maclean, *The Elimination of Morality*, Routledge, London, 1992.

81. M. Banner, *Christian Ethics and Contemporary Moral Problems*, Cambridge University Press, Cambridge, UK, 1999, pp. 204–224.

82. Proposal for a European Parliament and Council Directive amending *Directive 90/220/EEC* on the deliberate release into the environment of genetically modified organisms [COM (1998) 85 final of 23/02/1998].

83. Draft Directive III/5574/98.

84. http://www.hgc.gov.uk

See other INTERNATIONAL ASPECTS entries; INTERNATIONAL INTELLECTUAL PROPERTY ISSUES FOR BIOTECHNOLOGY.

INTERNATIONAL INTELLECTUAL PROPERTY ISSUES FOR BIOTECHNOLOGY

CYNTHIA M. HO
Loyola University of Chicago School of Law
Chicago, Illinois

OUTLINE

Overview of International Intellectual Property
   Introduction
   Types of Protection
   Examining International Issues
Defining the Parameters of Protection
   Patents Pursuant to TRIPS
   Plant Protection Under UPOV
Present Systems of Protection
   Europe
   Japan
Bibliography

OVERVIEW OF INTERNATIONAL INTELLECTUAL PROPERTY

Introduction

International intellectual property issues are becoming increasingly important as biotechnology is used and sold worldwide. This is because intellectual property rights are inextricably linked to the right to exclude others from use. The right to exclude can provide a competitive advantage or a barrier to entry into a commercial market. Several types of intellectual property provide some right to exclude others from an invention in the area of biotechnology—trade secrets, patents, and plant variety rights. Trade secrets enable their holder to prevent others from wrongfully appropriating valuable information for a potentially infinite time period. But they do not protect against independent invention, and they terminate once the information becomes public. A patent provides a right to exclude others, including independent inventors, from using the patented invention without consent, and only for a limited time. Plant variety rights function similarly to patents with respect to the ability to exclude others, but they are only available for plant "varieties."

Types of Protection

Trade Secrets. A trade secret typically consists of any information that is not generally known to others in the same business; it provides a competitive advantage to its owner. This is the easiest type of intellectual property to acquire, but it also provides the weakest protection. Unlike other types of intellectual property protection, formal procedural requirements are usually unnecessary to "obtain" a trade secret. Rather, the inventor of a trade secret merely needs to keep the information reasonably secret. The trade secret lasts as long as the information remains secret. However, once the information becomes publicly known, the trade secret ceases to exist. One way this can happen is if the information is independently developed and patented by another. Under this scenario, a trade secret would be exterminated because a patent on the same information would reveal the information to the public (since patents are public documents). Moreover, under this scenario, the patent owner could preclude the former trade secret owner from using the now patented invention because patent owners generally have rights to exclude all others from the patented invention. Although some countries soften this approach by providing those who used an invention prior to its patenting by another (prior users) with the right of continued use, there is no uniformity in such protection; for example, while some European countries allow a limited right, no analogous protection exists for biotechnology trade secrets in the United States.

Even without the complications of a superceding patent, trade secret protection offers minimal protection. To begin with trade secret protection does not allow exclusion of those who independently invent the identical "trade secret" therefore, two or more individuals or corporations could theoretically be using the same trade secret without infringing on each other's rights if they all did so independently. A trade secret does not confer any affirmative rights except as to those who misappropriate the information (e.g., an employee who leaves with confidential information). Even then, the "protection" provided is usually inadequate because monetary compensation for the trade secret misappropriation cannot restore information to trade secrecy status if it has been disclosed to the public.

Patents. Patents provide inventors a reward or incentive for publicly disclosing an invention, by providing the inventor with the right to exclude others from the patented invention for a limited term. The exclusivity provided by a patent is considered critical to stimulate ideas and lead to further advances. The requirements established for patentability are aimed at securing the goal of promoting innovation. Although the requirements vary somewhat between various countries, the typical requirements for obtaining a patent are that (1) the invention constitute patentable subject matter, namely constitute the type of subject matter that country wants to encourage innovation in, and (2) that the invention, as disclosed in a patent application, satisfies technical patentability requirements. The scope of patentable subject matter may include both products and processes in all areas of technology. Technical patentability requirements typically require that the invention be at least "new," "useful" (or have "industrial application"), "nonobvious" (or have an "inventive step"), and fully disclosed in a written document such that someone who was similarly technically competent could reproduce the invention. These requirements are intended to define inventive activity deserving of a patent. A national patent office typically examines patent applications to determine whether the patentability requirements are met. Additionally, in some countries, third parties are allowed to oppose the issuance of a patent or petition to revoke an existing patent for failing to meet the technical requirements.

Patents are generally considered the preferable type of intellectual property to protect biotechnology because they provide the most exclusive rights. Unlike trade secrets, patents protect against independent invention because the owner of a patent can exclude all others from using the patented invention. In addition, because a patent can entitle its owner to exclude others, including competitors, a patent or even a potential patent can justify the often high cost of research and development involved in biotechnology. The potential to exclude all others through patent protection is considered more valuable than attempting to maintain a trade secret indefinitely with no potential to affirmatively exclude others. Thus patents are the principal type of protection that is sought for biotechnology even though disclosure of the invention is required and patent protection is not a certainty.

Plant Variety Rights. A plant variety right also confers some exclusive rights. A plant variety right, which is also referred to as a "breeder's right" (because the right is provided to the breeder of a plant variety), functions analogously to patent rights — a relatively exclusive right is provided to breeders of new plant varieties to further the development of agriculture. As with patents, plant variety rights are not automatic; rather, they must be applied for and examined to determine whether they meet the requisite technical requirements. The requirements for plant variety rights are intended to function similarly to those for patent rights in that both are intended to provide protection to subject matter that is truly innovative. Some of the technical requirements for plant breeder rights parallel those for patent rights — a variety must be "new" (although the definition differs from that for patents, as will be discussed later), "distinct," "uniform," and "stable." These requirements are generally less onerous than those needed to meet patentability requirements. In addition, unlike patents, disclosure of the invention, or at least the method of making the invention, is not always required, which could be seen as an advantage. Although plant variety rights will be addressed in this article, the focus is on patent protection because patents provide coverage for more types of biotechnology and broadest protection.

Examining International Issues

Intellectual property rights are national rights provided by individual countries such that examining international protection requires examining the laws of individual countries. The need to examine what protection is available for biotechnology on an international scale is of obvious importance. However, national laws are currently evolving, particularly in the area of biotechnology. One way in which an international perspective of important issues can be obtained is by examining key international agreements, as well as illustrative national laws. Although no international agreement creates an international right, or even identical national rights, some agreements mandate minimum levels of protection by member countries, such that examining these agreements establishes what minimum levels of protection are globally available for biotechnology. The most important international agreement for this discussion is the Trade-Related Agreement on Intellectual Property (TRIPS) (1), which mandates minimum levels of protection for patents concerning all technology, as well as other types of intellectual property. In addition the International Convention for the Protection of New Varieties of Plants (UPOV) requires that its members provide minimum levels of protection for plant "varieties" (2,3). Both TRIPS and UPOV provide a useful framework for analyzing international protection as they establish a foundation of protection for all member countries. However, TRIPS and UPOV only provide a framework of minimum protection, rather than binding law. To determine global protection of intellectual property, national and regional laws will be discussed as illustrative of current approaches, as well as anticipated approaches in the near future. The laws applicable to Europe and Japan will be addressed as representative of current law impacting international protection of biotechnology (relevant U.S. law is addressed in a separate article of this encyclopedia). In addition to representing areas that are generally considered important for biotechnology, the laws of Europe and Japan highlight the application of both older and newer laws to biotechnology. Europe will be addressed as a single entity for this discussion because many European countries have adopted similar laws pursuant to international agreements other than TRIPS. The applicable agreements are the European Patent Convention (EPC), which applies to any European nation that signs the agreement and the European Union's Directive on the Legal Protection of Biotechnological Inventions (EU Directive), which applies to any country that is or becomes a member of the European Union (EU) (4,5).

Table 1 highlights membership in relevant international conventions based upon information obtained from the official website for each convention. However, it should be noted that this table only presents some members of these conventions — there are presently over 130 members of TRIPS and over 40 members of the UPOV and Budapest Conventions. For complete and membership regarding all conventions summarized in Table 1 (TRIPS, UPOV, and the Budapest Convention), the Web sites of those specific conventions should be consulted (6–11). In addition, for current membership to any of these agreements, official Web sites should be examined; this is particularly true for both TRIPS and the EU, as the scope of membership is expected to increase. For example, China is applying for membership to the WTO and if accepted, would be bound to comply with TRIPS. In addition countries that may be included within the EU include Hungary, Poland, and the Czech Republic (12).

Table 1 includes all members to EPC, as well as the EU (6–7). This table also includes important areas outside of Europe for which biotechnology protection is often considered important such as Australia, Canada,

Table 1. Membership in International Conventions Impacting Biotechnology

Country	EPC	EU	TRIPS	UPOV 1991	UPOV 1978	Budapest Convention
Australia			Y	Y		Y
Austria	Y	Y	Y		Y	Y
Belgium	Y	Y	Y		Y	Y
Canada			Y		Y	Y
Denmark	Y	Y	Y	Y		Y
Finland	Y	Y	Y		Y	Y
France	Y	Y	Y		Y	Y
Germany	Y	Y	Y	Y		Y
Greece		Y	Y	—	—	Y
Ireland	Y	Y	Y		Y	Y
Israel			Y	Y		Y
Italy	Y	Y	Y		Y	Y
Japan			Y	Y		Y
Korea			Y			Y
Liechtenstein	Y		Y	—	—	Y
Luxembourg	Y	Y	Y	—	—	
Monaco	Y			—	—	Y
Netherlands	Y	Y	Y	Y		Y
New Zealand			Y		Y	
Norway			Y		Y	Y
Portugal	Y	Y	Y		Y	Y
South Africa			Y		Y	Y
Spain	Y	Y	Y		Y	Y
Sweden	Y	Y	Y	Y		Y
Switzerland	Y		Y		Y	Y
United Kingdoms	Y	Y	Y	Y		Y

Japan, and New Zealand. Some areas where biotechnology protection is beginning to develop are also included for comparison such as South Korea and South Africa. For all these countries, Table 1 shows which are bound to the requirements of TRIPS and which provision of UPOV, if any, they comply with (as there are currently two versions of UPOV which govern) (8,9). Also countries that subscribe to the Budapest Convention, which governs deposits of biological material in relation to obtaining patent protection, are noted (10–11).

DEFINING THE PARAMETERS OF PROTECTION

Patents Pursuant to TRIPS

It is important to note that although TRIPS does not create a uniform worldwide patent law, it does establish a minimum floor below which no member may go without subjecting itself to potential retaliation by other countries in the form of trade sanctions. All countries that are members of the World Trade Organization (WTO) — an international organization designed to reduce trade barriers, including barriers to trade based on intellectual property — must comply with TRIPS. TRIPS binds a substantial number of countries — over 130 countries are presently members of WTO; these countries constitute 90 percent of world trade and include countries that utilize or expect to utilize biotechnology (2). Compliance with TRIPS requires complying with provisions of other international agreements that TRIPS incorporates, such as the Paris

Convention for the Protection of Industrial Property (13). TRIPS came into effect on January 1, 1995. However, the TRIPS requirements were not immediately binding; all members were provided time to enact laws to comply with TRIPS, with more time allotted for developing countries (many of which previously provided limited, or even no, patent rights whatsoever). The first date by which any WTO member had to comply with TRIPS was January 1996; at this time all developed countries were supposed to be in compliance with all provisions of TRIPS. In addition, as of January 2000, all developing countries (and those whose economies are in transition from a centrally planned economy into a free-market economy) should be in compliance with most portions of TRIPS. However, "least-developed" countries have until 2005 to provide patents for any subject matter that was previously deemed unpatentable. Moreover less-developed countries (LDCs) do not have to comply with any of the minimum patentability standards until 2005, with the possibility for a further extension of time.

Compliance with TRIPS requirements is aided by the potential of some type of retaliatory behavior by other WTO member nations. In particular, if a member state fails to comply with TRIPS, another member country may challenge the noncompliance. Members must initially try to resolve issues through a dispute resolution process pursuant to the Dispute Settlement Understanding (DSU) (1,14). If members fail to reach a mutually satisfactory resolution regarding any alleged noncompliance with TRIPS, a complaining member may request a WTO panel to examine the issue in an adjudicatory proceeding. The WTO panel must issue a decision within a relatively short timeframe, after which parties must comply or appeal to an appellate body of the WTO and subsequently comply with any decision by the appellate body. If a member fails to comply, sanctions may ultimately be imposed, including retaliatory action that includes withholding benefits under TRIPS or other WTO agreements such as the General Agreement on Tariffs and Trade (GATT). Although additional information on the DSU is beyond the scope of this chapter, the WTO Web site provides a current overview of active disputes; also additional information on dispute settlement is readily available from other sources (14–19). However, it should be noted here that the DSU rules have already been effectively applied to enforce protection of biotechnology. For example, India's compliance with the transitional provisions of TRIPS was challenged by the United States and the EU. After failing to convince both the WTO panel and Appellate Body that its laws were in compliance, India took action to amend its laws (20–21). The India dispute illustrates the effectiveness of TRIPS in effectuating real change to protection of biotechnology, as well as the relative rapidity under which such change occurs.

The TRIPS requirements relating to patents focus on three main issues: what must be patented, the scope of patent protection for issued patents, and the enforcement of patent rights. Although all three of these are important, this section will primarily focus on the first two issues since their impact is more particular to the field of biotechnology; information concerning enforcement of all

patent rights under TRIPS, including patent rights for biotechnology patents, is available from other sources (19). The TRIPS requirements are actually minimum standards of protection that must be satisfied by all members. Because of the substantial number of countries that must comply with TRIPS, the TRIPS requirements are very important.

Eligible Subject Matter. The requirements for what must be patented are set forth under Article 27 of TRIPS. In particular, TRIPS first sets forth that:

> 1. Subject to the provisions of paragraphs 2 and 3, [p]atents shall be available for any inventions, whether products or processes, in all fields of technology, provided that they are new, involve an inventive step and are capable of industrial application. Subject to paragraph 4 of Article 65, paragraph 8 of Article 70 and paragraph 3 of this Article, patents shall be available and patent rights enjoyable without discrimination as to the place of invention, the field of technology and whether products are imported or locally produced.

This provision imposes several requirements. First, it states that patents must be available for "inventions" subject to two limitations (in paragraphs 2 and 3) which exist for particular categories of subject matter as well as exceptional situations. Second, it states that a patentable invention must be "new," involve an "inventive step," and be "capable of industrial application." Third, it states that except for countries to whom a transitional period is allowed (under Articles 65 and 70 of TRIPS), patents and patent rights must be provided without discrimination; countries cannot discriminate in issuing patents or providing patent rights based on where the invention was made, the type of invention, and whether products are imported or locally produced. As the last requirement relates to both patentability and scope of patent rights, it will be discussed with respect to both issues separately.

Although these requirements are not presently binding on all WTO members, they must eventually be complied with. The TRIPS requirements of patentability indicate what types of biotechnology are presently patentable or will be in the future. TRIPS requires that patents be granted to all "inventions" that satisfy technical patentability requirements, subject to three specific exceptions. Table 2 summarizes these exceptions. As shown in this table, subject matter may be excluded if it is (1) not considered an "invention, or (2) specifically delineated within TRIPS as excludable, or (3) falls within the unspecified exclusion for inventions contrary to *ordre public* or "morality." Each one of these three requirements allows member states the opportunity to narrow the scope of patentability. The scope of patentable subject matter may be limited based on a narrow interpretation of undefined TRIPS terms; this table indicates such undefined terms by placing them all in quotes.

The requirement that patents be available for all "inventions" without regard to the type of technology involved and whether the invention is a product or a process is important for biotechnology. Prior to TRIPS more than 50 countries did not provide patent protection

Table 2. Permissible Exclusions of Patentable "Inventions" Pursuant to TRIPS[a]

Anything excluded from the definition of an "invention"[b]

Specific subject matter exclusions (Article 27(3))
- Methods of treatment of humans and animals
- Plants[c] and animals (other than microorganisms)
- "Essentially biological processes" for making plants and animals (other than microorganisms)

Unspecified exclusion for "*ordre public*" or "morality" (Article 27(2)) (subject matter may be excluded if preventing commercial exploitation is necessary to protect "*ordre public*" or "morality")[d]
- Protecting human, animal or plant life
- Protecting health
- Avoiding serious prejudice to the environment

[a]TRIPS *permits* members to exclude these categories from patentability, and members are free to provide patent protection for these categories.
[b]The word "invention," as well as other quoted words on this table, are undefined in TRIPS and may be subject to differing interpretations (which can affect the scope of patentable biotechnology).
[c]TRIPS does require that plant "varieties" (also undefined within TRIPS) be protected either under the patent system or a sui generis system.
[d]N/A where commercial exploitation is prohibited by local law.

for pharmaceutical and other products relating to health and medicine (22,23). In fact, the disparate treatment of such products was a prime consideration in enacting this language (24). Still it is unclear whether this provision mandates a general principle of patentability of "inventions" (25). In addition, even if a general presumption of patentability is established, the meaning of the word "invention," as well as the allowable exceptions from patentable "inventions," can substantially narrow what types of biotechnology are patentable (26).

Identifying what is an appropriate "invention" is the first point at which subject matter may be narrowed. TRIPS itself does not define an "invention." Although this is consistent with prior patent law in many countries, it leaves the scope of patentable subject matter particularly uncertain in the biotechnology area where there is not always a marked distinction between unpatentable "discoveries" and patentable "inventions." Countries generally consider discoveries and naturally occurring substances to be unpatentable inventions, but the genetic manipulation of naturally occurring substances has arguably blurred the line between discovery and invention. Some countries take a relatively expansive view of the term "invention" to include products that are found in nature, so long as they are isolated by an unnatural (biotechnological) process; isolated genes and gene sequences may be considered patentable under such a definition of invention. On the other hand, countries may elect to *exclude* isolated products of nature as noninventions (22,24). Because the term is undefined in TRIPS, countries are free to adopt either view. However, even if isolated biological material is considered a patentable invention, problems with other patentability requirements such as "industrial application" may exist (e.g., if there is no known use for the material other than as a general probe, as will be later discussed).

Specific Exclusions. The three types of subject matter that are specifically set forth as excludable subject matter under TRIPS all impact the scope of biotechnology that may be patented. TRIPS allows the following categories of subject matter to be excluded from patentability:

1. Methods of treating and diagnosing humans and animals
2. Plants and animals other than microorganisms
3. "Essentially biological processes" for creating plants and animals

It is important to note that these are *optional* exclusions from patentability. Accordingly, countries may adopt some, all, or none of these exceptions in their patent laws and still be in compliance with TRIPS. In addition, it should be noted that even if a country enables a broad range of subject matter to be considered an "invention," adoption of these exclusions, or a broad interpretation of these exclusions could significantly narrow the field of what is considered patentable subject matter. Moreover, even if none of these exclusions are adopted, what is actually patentable may nonetheless be a narrow range of inventions if the criteria of patentability are applied strictly, as will be explained later.

1. *Methods of Treatment and Diagnosis.* TRIPS allows methods of treating or diagnosing humans or animals to be excluded from patentability under Article 27(3)(a). This prohibition is similar to the language of many European and Latin American countries' patent laws (22). Application of similar prohibitions under national laws have shown that the scope of such an exclusion may be a function of the interpretation of individual terms, such as the meaning of "therapy," "method of treatment," or "diagnosis." (27–32). This exclusion is of great importance for biotechnology as the exclusion of "therapeutic and surgical methods" may exclude patents for gene therapy. In addition the bar on "diagnostic methods" may result in the exclusion of genetic testing or screening kits from patentability. Countries have taken varying stances in applying similar statutory exclusions to gene therapy; for example, some countries only exclude in vivo, but not ex vivo methods of treatment, whereas other countries exclude treatment involving humans but permit patenting of the identical treatment on animals.

2. *Animals, Plants, and Essentially Biological Processes.* TRIPS also states under Article 27(3)(b) that members may exclude from patentability "plants and animals other than micro-organisms, and essentially biological processes for the production of plants or animals other than non-biological and microbiological processes." However, protection of "plant varieties" must be provided either under the patent system or a sui generis system (1). This provision allows countries to disallow patents on transgenic plants and animals, even if they meet the technical requirements of patentability. Although genetically

modified plants may be entitled to some protection if considered a plant "variety" (which is not defined under TRIPS), exactly what protection is uncertain. While TRIPS requires an "effective" sui generis system, there is no definition of what this system requires. This may include one system for protecting plant varieties that existed at the time TRIPS was enacted—UPOV (24); however, not all commentators agree that UPOV is consistent with this provision (33). Moreover, even if applicable, UPOV does not provide the identical scope of protection as that required under TRIPS, as will be noted in the later discussion of UPOV. In addition it is uncertain what other system besides UPOV would be adequate. Moreover, TRIPS does not require any protection for transgenic animals. While there is the possibility that the process of creating a transgenic animal could be patented (assuming the technical patentability requirements are met), this provides less protection than if the animal itself were patentable. In addition patentability of a process for creating a transgenic animal is not certain as members need not protect "essentially biological processes" for the creation of animals. TRIPS does not define "essentially biological" other than to state that it does not include processes that are "nonbiological" or "microbiological." Accordingly member countries could define "essentially biological" to exclude any process that involved at least one biological step (i.e., one that does not involve genetic engineering). This exclusion will likely generate continued debate as it was controversial during the initial TRIPS negotiation and must be reviewed every four years after TRIPS enters into force (1,26,33).

Nonspecific Exclusion. In addition to the foregoing specific exclusions that implicate biotechnology, a potentially major hurdle for biotechnological inventions may be the unspecified exclusion under Article 27(2) that allows "inventions" to be barred from patentability if preventing "commercial exploitation" of the subject matter is necessary to protect "*ordre public* or morality." However, TRIPS also states that this exclusion is not satisfied by the mere fact that exploitation of an invention will be in violation of a national law. TRIPS does not provide additional guidance on the definition of *ordre public* or *morality* other than to indicate that it may involve the following broad categories: protecting human, animal or plant life, protecting health, and avoiding serious prejudice to the environment (1). Further, because this exclusion refers to "inventions" rather than categories of subject matter as in Article 27(3), it appears that it may need to be applied on a case-by-case basis (24).

This provision stands as a potentially large barrier to patenting new types of biotechnology as patenting biotechnology raises issues of morality even in countries where patenting of biotechnology is allowed. Similar provisions have been utilized, or noted as capable of being utilized, to deny patents on transgenic plants and animals as well as isolated gene sequences (27). However, interpretation of what constitutes ordre public or morality in the context of patenting biotechnology has been elusive.

This exception requires that the violation of "ordre public" or "morality" be the result of "commercial exploitation." This may pose some interpretative difficulty as commercial use or exploitation is not required to be disclosed in a patent application and may not be known. Accordingly, although a patent office is charged with determining whether commercial exploitation of an invention would violate these undefined terms, the necessary information may not be available at the time of examination to enable an educated decision.

The scope of this exception may be limited by prior interpretations of similar terms, albeit in contexts other than TRIPS, or even patent law. In particular, prior WTO dispute panels have examined and interpreted the meaning of "necessary" and "morality" under the General Agreement of Trade and Tariffs (GATT), which governs trade of goods (24,34–36). In that context, it has been found that the term "necessary" requires objectively justifiable measures and that there be no alternative measure available (24). Although prior WTO panels may be relied on by subsequent panels, it is unclear what weight a panel would give to a previously interpretation made in a different context (37). In addition it has been suggested that the definition of *ordre public* can be derived from case law by the European Court of Justice, the court that arbitrates issues among the EU countries (38). This may however be similarly unhelpful as there are no such cases interpreting *ordre public* in the context of patentability. Finally, countries or the WTO could look to prior interpretation of similar provisions, such as in Europe; although there are no decisions finding that inventions violate such a provision, suggestions as to the parameters of such a provision may be considered. For example, European courts have interpreted a similar provision under EPC article 53(a). The EU Biotechnology Directive also provides explicit examples of types of commercial exploitation that would violate an identically worded provision (5,27).

Illustrations of Patentable Subject Matter. Because TRIPS is a minimum standards framework with terms that are subject to differing interpretations, the scope of patentable subject matter may vary among different countries that comply with TRIPS. The diversity of approaches may be illustrated by a comparative view of specific types of patentable subject matter. A recently compiled report by the Working Party of the Trade Committee of the Organisation for Economic Cooperation and Development (OECD) provides a good source of such information. In response to an OECD questionnaire to individual countries concerning intellectual property practices in the area of biotechnology, information from patent offices of 22 OECD countries was obtained. Although the information was provided in varying degrees of detail and does not necessarily constitute official policy of individual patent offices, it is nonetheless useful for comparison.

Table 3 presents a summary of information submitted by individual patent offices to the OECD concerning the types of inventions in the area of biotechnology that are considered patentable subject matter. This information was submitted roughly in the time period 1998 to 1999. Accordingly, to the extent that laws have been amended

Table 3. Patentable Subject Matter

	Nucleic acid sequences	Amino acid sequences	Isolated materials	Living unicellular organisms	Plants, parts, varieties	Animals, organ varieties	Humans, organ, human-derived products	Methods of treatment	Methods of diagnosis
Australia	Y	Y	Y	Y	Y	Y	N	Y, except essentially biological processes that produce humans	Y except essentially biological processes that produce humans
Canada	Y	Y	Y	Y	N (but PVR exist)	N	Human-derived only	No, except ex vivo	Y, unless surgery or therapy
Japan	Y	Y	Y	Y	Y	Y	Human-derived; human organs if not immoral	Only animals or ex vivo	Animals only
Korea	Y	Y	Y	Y	N, except asexual plants	Y	Human-derived only	Animals only	Animals only
New Zealand	Y	Y	Y	Y	Y	Y	Human organs; human-derived (if process to create not immoral)	Animals only	Y, unless surgery involved
Norway	Y	Y	Y	Y	N, except plant parts that can't differentiate to whole plants	Animal organs only	N, except isolated elements	Ex vivo only	Ex vivo only
EPC/EPO	Y	Y	Y	Y	Y, except "varieties"	Y, except "varieties"	Human-derived only	N	N
EU	Y	Y	Y	Y	Y, if not technically limited to specific variety	Y, unless animal modified to cause suffering without substantial medical benefit	Human-derived if isolated or produced by "technical process"	N/A (national/EPC law to apply)	N/A (national/EPC law to apply)
United States	Y	Y	Y	Y	Y	Y	Human-derived only	Y	Y

767

since that time, this information would not reflect that. The table includes countries in which patents are often sought such as Japan, Canada, and Australia. In addition, the EPO is included because it issues EP patents that can become transformed into national patents. Information from individual countries of EPC are not included in this table as their laws largely mirror that of the European Patent Office (EPO), the governing office of EPC. However, some of these countries did provide separate information to the OECD survey, and this information is available from the OECD Web site (40). In addition some countries are included for comparison such as South Korea, New Zealand, Norway, and the United States. For information concerning all 22 countries that responded to the OECD questionnaire, the actual report should be consulted, as well as the responses of individual countries (39–41).

As noted on Table 3, there are several categories of biotechnological matter that are universally considered patentable subject matter. In particular, chemical structures composed of nucleic acid sequences and gene sequences are considered patentable subject matter in all responding countries; this is true regardless of whether the sequences correspond in part or whole to information found in living organisms. Material isolated from living organisms, other than such sequences, are all considered patentable subject matter as well as living unicellular organisms. It should be noted, however, that despite the uniform patent eligibility of such inventions, actual patents nonetheless may not issue because of how the technical requirements of patentability are applied.

More variation in patentable subject matter exists with respect to multicellular organisms and subparts of those organisms. The only common denominator among approaches is that patents on humans are universally rejected. However, "human-derived products," including cell lines, genes, and nucleic or amino acid sequences are considered patentable, for example, in Canada, Japan, and the United States. However, even to the extent that multicellular organisms and their parts are not categorically excluded from protection, they may nonetheless be denied protection on grounds similar to TRIPS 27(2) — namely they may violate a provision of the patent act that precludes patents on inventions that are unethical or immoral. Exclusions on the basis of morality could be a factor in the denial of patents for inventions by the EPO as well as by the patent offices of Japan, and Canada as all of them have such a provision.

Similarly there is wide variation with respect to what types of methods of treatment or diagnosis would be considered patentable. For the purposes of comparison, the methods of treatment column in Table 3 represents responses to both whether methods of treatment are patentable and whether methods involving genetic engineering for purposes other than surgery, therapy or diagnosis (e.g., for experimentation or research) are patentable. Some countries appear to take a fairly strict interpretation of statutory provisions against methods of treatment and diagnosis such that gene therapy is excluded. Others will allow methods of treatment or diagnosis only with respect to animals, or with respect to methods occurring outside the body

Table 4. Technical Patentability Requirements

A patent application must disclose an "invention" (see Table 2) that is:
- "new"
- "industrial application" or "useful"
- "inventive step" or "nonobvious"

Patent application (Article 29) must disclose invention adequately for duplication by one "skilled in the art"

(ex vivo). Finally, just as with patents on multicellular organisms, patentability of biotechnological methods are also impacted by provisions in patent acts that preclude patents that are immoral or contrary to *ordre public*.

Patentability Criteria. There are two basic requirements for patentability, as shown in Table 4. An application must establish that an invention meets technical requirements of patentability. The application must further sufficiently disclose the invention such that someone who is similarly skilled could replicate the invention. Article 27(1) of TRIPS requires that technical requirements include that an invention be new (novel), nonobvious (or have an inventive step), and be useful (or be industrially applicable) (1). Although these requirements are ones that are common to most patent laws, countries have differed in their definition of these terms, as well as in the application of these terms to biotechnology (22,24). For example, there is no consensus on whether patents on gene sequences (that are considered "inventions") may be denied for lack of "industrial applicability" if the function of gene sequences is unknown other than as a probe. Similarly there is no consensus on whether isolated biological matter is "new." While application of patentability requirements to biotechnology are developing with respect to all the requirements, novelty and adequate description will be focused on here because application of these standards to biotechnology raises unique issues. As was seen with patentable subject matter, the undefined patentability criteria allows wide variation among countries; each requirement is an additional juncture at which the scope of patentable subject matter may be further limited.

Novelty. The requirement that an invention be new is fundamental to the patent system policy of providing an incentive to produce things that would not otherwise be known to the public. An invention is generally new or novel if it was not known or available to others prior to the application of the patent; it is often stated that an invention is new if it is not previously known in the "prior art." Prior art may include descriptions in printed publications (e.g., foreign patents and published patent applications) as well as oral information and actual use. In all cases, to constitute prior art, the information must be such that the invention is essentially known to those in the same field (i.e., the description must be adequate to "teach" the invention to one in the field).

In determining novelty, it is important to note that pursuant to TRIPS, all WTO members must now recognize a "right of priority" for filing of patent applications (1,13,42). This means that if a patent application is filed in a member country, that date of filing may be relied on for subsequent filings of applications for the same invention in other member countries, so long as subsequent filings are within 12 months of the original application (1,13,42). This right of priority is very important to determining novelty because, without such a right, an applicant's own application could bar a patent in another country unless patent applications were simultaneously filed in every country in which protection were desired. Therefore recognition of priority date is important as it gives an applicant 12 months within which to decide where to file and actually complete the formal filing requirements.

There are different interpretations of the novelty requirement taken by countries to determine whether an invention is sufficiently deserving on patent protection. The different approaches reflect different perspectives of how new something must be before a patent is awarded. Under a "strict" approach to novelty, an invention is not novel if it was known in any way, in any country, prior to the date of the patent application — regardless of whether the applicant was aware of it, or if the applicant could have reasonably been aware of it; this strict approach is often referred to as "absolute novelty." However, another approach is to reward an applicant for bringing some invention to the particular territory in which the patent is sought that likely would not otherwise have been known. Under this "relative novelty" approach, nonwritten knowledge or use of an invention outside the territory is not considered prior art for purposes of novelty; this is the approach taken by the United States. The rationale for this scheme is that a printed publication is accessible to all even if printed in another country, whereas a use of an invention is much more difficult to ascertain and thus a patent should be granted to one who brings the invention to the public's attention.

In the biotechnology area, the distinction between the absolute and relative novelty standards are demonstrated by some biotechnology methods. For example, new uses of previously known products (often referred to as "second use" or "second medical use") may be disallowed under a strict novelty criterion. Although such uses are considered "new" under both United States law as well as the EPC, TRIPS certainly does not mandate such an interpretation. Similarly new products that result from previously known processes may be disallowed under a strict novelty criterion. Again, the United States provides patent protection for such processes.

It is possible to also provide a grace period in terms of novelty. In particular, a grace period may allow an applicant to disclose an invention to the public prior to filing the patent application without sacrificing patent rights. It is common for the grace period to last only six months from the triggering event and to be no longer than a year; for example, the United States has a one-year grace period, while Canada and EPC provide a more limited period of six months. The rationale behind such a system is that it enables speedier public access without removing the incentive of patent protection. Countries vary with respect to what activities are allowable during the grace period. For example, in the United States a grace period is provided for not only publications but also prior use, sale, and offers for sale. However, in other countries, a grace period may only be available for activity of an inventor, and the activity may be limited to disseminating information only at specified conferences; some illustrative subscribers include Canada, Australia, and EPC.

Adequate Description. TRIPS requires that for a patent to issue, an application must first disclose the invention "in a manner sufficiently clear and complete for the invention to be carried out by a person skilled in the art." (1). Once again, the TRIPS agreement has only mandated the result but has not dictated how countries must meet it. Countries vary with respect to whether a deposit of a sample of a biotechnological invention is required, as well as when and where this must be done (4,5,39). The Budapest Convention provides some guidance to its members (many of whom include WTO members) on when deposits are required, and the relevant procedures that must be followed (10).

Patent Rights

Patent Term. TRIPS requires that members provide a minimum patent term of 20 years, calculated from the date of filing of the patent application (1). An applicant generally has no rights during the application process. No extension of the term is required by TRIPS even if the effective term has been substantially reduced by the patent examination process; this is problematic for patents that involve biotechnology as they often take longer to examine than more traditional inventions. In addition no extension of term is provided for delays in the sale of a patented invention that often result from regulatory approval for pharmaceuticals. Accordingly the effective term of pharmaceutical patents may be considerably shorter than that provided to other subject matter. Although many developed countries, including the United States, and more recently Japan, provide an extended term of protection for such patents, TRIPS does not require it. Accordingly, although TRIPS mandates patent protection for all inventions and nondiscrimination in terms of patent rights, it does not mandate an effectively equal patent term.

Scope of Protection. Consistent with prior patent law doctrine, patents granted pursuant to TRIPS do not provide an affirmative right of use to the patent owner. Rather, patents only provide a right to exclude others. To be entitled to use the patented invention, the owner must determine if there are additional laws with which compliance is necessary. For example, a newly patented pharmaceutical typically cannot be sold without governmental approval and separate applications are required to obtain patent rights and the right to sell. Moreover the patent owner may need to determine if permission from another patent owner is required to avoid infringement. For example, the owner of a patent on a new use of a previously known and patented compound would need permission from the owner of the patented compound to actually use the newly patented invention (as the owner of the patented compound has the right to exclude others from making the compound).

TRIPS provides that a patent confers on its owner "exclusive rights" to prevent unauthorized persons from certain activities. The activities that may be excluded differ with regard to whether the subject matter of the patent is a product or process. If the patented subject matter is a product, the owner can exclude others from "making, selling, offering for sale, selling, or importing, the patented product. On the other hand, if the patented subject matter is a process, the owner is entitled to exclude others from using the patented process as well as offering for sale or selling the process; in addition the owner can exclude others from importing the product obtained "directly" from the patented process. Most of the exclusive rights established under TRIPS have been generally recognized in industrialized countries. It should be noted that many industrial countries also allow rights against persons who contribute or induce these activities, although TRIPS does not require this; the United States, Japan, and certain European countries allow such rights.

Although TRIPS specifies that certain activity is within the patent owner's "exclusive rights," the nature of that exclusivity may be limited pursuant to other articles of TRIPS (as will be discussed in the next section), as well as by a nation's interpretation of these provisions. For example, what constitutes a patented product or process must first be determined in order to determine what may be excluded. Generally, patents include one or more claims, which are sentences at the end of a patent that define the scope of the invention, and accordingly define the scope of patent rights. In addition patents generally can include claims to both products and processes; some countries allow claims to not only chemical structures but also claims to function and products made by certain processes. To determine what constitutes the patented invention, the claims most likely will be examined (often referred to as "claim interpretation"). However, countries differ vastly in their approaches to claim interpretation. Countries may look not only at the patent claims but also the rest of the patent, its history, and other extrinsic evidence to determine the "true meaning" of the claims. Some countries are more liberal in interpreting patent claims, and they find patent infringement even when a defendant is not precisely within the literal scope of patent claims (sometimes referred to as infringement under the "doctrine of equivalents" as the defendant's actions are considered equivalent as a matter of equity, to what is literally stated in the claims). Equivalents are considered either at the time the application is filed or at the time of the infringement. All of the above variations allow member countries substantial leeway in the extent of protection granted.

Regardless of how a country interprets the patent, TRIPS provides a higher threshold of protection for patented processes than was previously recognized in most countries. In particular, TRIPS provides the right to exclude unauthorized persons from products obtained "directly" from a patented process in addition to excluding persons from the patented process itself. The additional protection against products "directly" obtained by a patented process is intended to provide protection in the case where the patented process is used in a country where no protection exists, and the resulting product is imported into the country in which patent protection exists (24). The extension of protection to direct products also aids owners of such patented processes to establish

unauthorized use of a patented process as access to an infringing process is often impossible. In addition TRIPS now provides that judicial authorities may require the alleged infringer to rebut a presumption of infringement in the case of patented process if the infringer's product is identical to that produced by the patented process (1). However, it is only required in one of two circumstances: (1) where the product obtained by the patented process is "new," or (2) when there is a "substantial likelihood that the identical product was made by the process and the owner of the patent has been unable, through reasonable efforts, to determine the process actually used" (1).

Although the scope of the patent owner's exclusive rights is fairly clear, one exception is the scope of the importation right. In particular, TRIPS specifies that for patents on both products and processes, the owner has the right to exclude others from importing the patented product or the product obtained directly from the patented process. The right to preclude importation of a patented product is a right that was not generally recognized in industrialized countries prior to TRIPS. However, TRIPS does not clarify what this new right should include; there is a footnote after the importation right that cross-references an earlier article of TRIPS that states that for purposes of dispute settlement, TRIPS is considered not to address the issue of exhaustion of patent rights. The principle of "exhausting" a patent right is that if a patented product is legitimately sold or otherwise conveyed, no further patent rights exist with regard to that article. To the extent that there is an argument that imports of patented products fail to impinge on a patent owner's exclusive rights because of the principle of international exhaustion, such arguments will not be recognized in official dispute settlement proceedings.

Exceptions. Although the previously discussed provision of TRIPS provides "exclusive rights" to the patent owner, TRIPS also clearly contemplates that the "exclusive right" may be more circumscribed. Two separate articles of TRIPS provide exceptions to the "exclusive right" that is provided in Article 28. In particular, Article 30 provides a "limited exception" to the exclusive right, while Article 31 contemplates that member countries may allow unauthorized "other use" of a patent. As these exceptions are critical to defining the full scope of patent rights—just as possible exclusions to patentable subject matter were important to determining what inventions are patentable—the exceptions to the patent owner's exclusive rights will be discussed here.

"Limited" Exceptions. Although TRIPS provides that members may provide "limited exceptions" to the patent right, it does not clearly define what constitutes a "limited exception." In particular, TRIPS Article 30 states that "[m]embers may provide limited exceptions to the exclusive rights conferred by a patent, provided that such exceptions do not unreasonably conflict with a normal exploitation of the patent and do not unreasonably prejudice the legitimate interests of the patent owner, taking account of the legitimate interests of third parties" (1).

Although the scope of this exception is not explicity clear, some countries have tried to incorporate the exception into national law. Some countries have incorporated the language of TRIPS Article 30 wholesale into their national laws, without attempting to define any terms. Other countries, while not using the literal language in Article 30, have adopted language that is equally vague. For example, Argentina has amended its patent laws to state that the patent office is permitted to establish limited exceptions in "sectors of vital interest to the socio-economic and technological development of the country" (43). It remains to be seen whether countries with established patent systems, including exceptions to patent rights, are in fact TRIPS-compliant; this cannot be clearly established in the absence of challenge by another country. However, even in the absence of an official challenge, countries are becoming more cognizant of TRIPS requirements and the possibility of a challenge. In fact there has been some discussion in the United States concerning whether a provision of the U.S. patent laws that precludes patent owners from obtaining relief against infringing doctors of medical procedure patents either violates the TRIPS rights or is exempt under this provision (44–46).

The proper interpretation of the "limited exception" provision will probably be a major issue in international patent protection. One panel recently addressed WTO whether Article 30 permits (1) the manufacture of patented product for the purpose of obtaining regulatory approval prior to the patent expiration or (2) the manufacture and storage ("stockpiling") of generic drugs such that they can be sold the day the patent expires (15). In particular, Canada's patent laws were alleged to infringe on patent rights during the patent term (15,47). Canada argued that Article 30 provides it with an exception to what would otherwise technically be a violation of patent rights mandated by TRIPS (48). In particular, Canada believed that its actions were allowable as a "limited exception" in the case of patented pharmaceuticals where generic drug companies need to infringe a patented drug to obtain regulatory approval prior to the expiry of the patent (48). The rationale for such an exception was that without it, a patent owner would effectively be given an extension of patent term due to the generic manufacturer's need to wait for regulatory approval; allowing limited infringement by the generic manufacturer was considered to be in the public interest because it enables cheaper drugs to be provided to the public sooner. The WTO found that the regulatory approval exception was an allowable exception to the usual patent owner rights pursuant to article 30. However, the WTO panel found that the stockpiling provision was not a "limited exception" in accordance with article 30. Although the decision is not binding on other parties, it is important to the many nations which have regulatory approval exceptions similar to Canada's such as the United States and Japan; these countries are now more assured that their patent law exceptions are TRIPS-compliant. The WTO decision is also important beyond the specific subject matter of regulatory approvals. In particular, the WTO panel decision explained how article 30 should be interpreted in relation to patent rights under article 28 (47). The WTO panel underscored that a patent owner is entitled to the full panoply of patent rights listed under article 28 (and not just the right to sell as Canada had

previously asserted) (47). In addition, the WTO panel clarified that to be entitled to an exception under article 30, three requirements must be met:

1. there must be a "limited exception" to the exclusive rights;
2. the exception must not "unreasonably conflict" with the "normal exploitation" of a patent; and
3. the exception must not "unreasonably prejudice" the "legitimate interests" of the patent owner, taking into account the "legitimate interests" of "third parties."

The WTO panel decision itself may be consulted for the detailed explanation of each of these requirements (47). Even without a complete explanation of the panel's opinion, however, it should now be appearent that future disputes concerning exceptions to patent rights will be carefully evaluated by WTO panels in accordance with the specific facts of each case.

In addition, what other exceptions to a patent owner's exclusive rights may exist remains an issue. Some countries and commentators have assumed that certain activity is covered by Article 30 based on prior drafts of Article 30 (22,24). For example, activity that has been assumed to be allowable includes private noncommercial use, use of the invention for research, experimental use to test or improve the invention, use for teaching purposes, preparation of medicine, prior use of the invention by a third party before the date of application of the patent (22,26). However, whether such uses are actually consistent with article 30 is unknown. Although a WTO panel recently had occasion to address this issue, it specifically declined to comment on whether national patent laws providing exceptions for experimental use were consistent with TRIPS ariticle 30 (47).

Compulsory Licensing. Compulsory licenses are an important issue as they substantially limit a patent right. The term "compulsory licensing" refers to a license of the patent owner's invention that is involuntary, or compulsory. Compulsory licensing is important to biotechnology patents because even if patents are appropriately granted, the patent right is substantially diluted if compulsory licensing is allowed. Moreover, prior to TRIPS, it was not uncommon, particularly in developing countries, to require compulsory licensing of pharmaceutical patents, or other patents relating to health on the ground that it was necessary in the interest of public welfare. In addition compulsory licenses have been granted in some countries where the patent owner is not using the invention in the country (usually referred to as not "working" the invention).

Article 27 prohibits certain types of "discrimination"; in particular, it requires that all patents, once issued, be entitled to the same right to exclude without regard to the subject matter of the patented technology; in particular, it states that there should be no discrimination based on whether the invention is "imported or locally produced." The negotiating history of this provision indicates that developed countries intended this to exclude compulsory licenses for nonworking as previously allowed under the Paris Convention (24,49). Some interpretations of this provision interpret it as eliminating such a requirement (50–51). However, although this may have been the intent of some countries in drafting this provision, the correct interpretation is unclear. For example, Brazil explicitly provides compulsory licenses for failure to exploit an invention locally (52–54).

An important issue under TRIPS is determining what types of compulsory licensing are permissible. Several articles of TRIPS may relate to compulsory licensing. Article 31, which places procedural limitations on licenses, is often presumed to apply to compulsory licenses; indeed, although Article 31 presently refers to "other use" that is unauthorized by the patent owner, prior drafts of Article 31 used the term "compulsory licenses"(24). Two additional articles of TRIPS recognize that intellectual property rights, including patent rights, are not absolute. In particular, Article 7 states that:

> protection and enforcement of intellectual property rights should contribute to the promotion of technological innovation ... to the mutual advantage of producers and users of technological knowledge and in a manner conductive to social and economic welfare, and to the balance of rights and obligations.

Article 8 provides that:

> 1. Members may ... adopt measures necessary to protect public health and nutrition, and to promote the public interest in sectors of vital importance to their socio-economic and technological development, provided that such measures are consistent with the provisions of this agreement.
>
> 2. Appropriate measures ... may be needed to prevent the abuse of intellectual property rights by right holders or the resort to practices which unreasonably restrain trade or adversely affect the international transfer of technology.

It is unclear whether only situations described in Article 31 are subject to compulsory licensing, or if Article 31 imposes requirements on compulsory licenses of any subject matter. TRIPS implicitly provides certain ground for issuing compulsory licenses; in particular, compulsory licenses are deemed proper for a national emergency, noncommercial use, anticompetitive practice, if a patent owner has refused to license, and where necessary to practice another patented invention (1). However, TRIPS does not indicate what other grounds would be permissible for compulsory licensing, or whether this is the only provision that relates to such licensing. Although all compulsory licenses are an imposition on a patent right, at least those provided under Article 31 provide the owner with reasonable remuneration and are not to be granted for entire classes of inventions. However, if another provision of TRIPS is interpreted as justifying compulsory licensing — such as Article 8 — the protection provided to patent owners under Article 31 would essentially fail to exist. Interpreting Article 8 to establish an independent basis for compulsory licenses has been suggested by commentators, usually in the context of trying to accommodate developing countries who are attempting to comply with TRIPS (55–56).

Determining whether Article 31 is the only applicable provision to compulsory licensing is important because of the many limitations placed on licensing under Article 31. In particular, Article 31 requires that authorization be "considered on its individual merits." This seems to suggest that compulsory licensing of all patents within a certain category would be impermissible. In addition Article 31 requires that compulsory licensing should only be permitted if a patent owner has first denied a request; although the attempt to obtain permission is waived for a national emergency or public noncommercial use, even in these situations, the patent owner must be informed promptly after the use has begun. Also Article 31 places limitations on the scope and duration of the use; it must be nonexclusive, nonassignable, preferably for a domestic market, and subject to termination once the conditions that necessitated the unauthorized use cease. The decision to grant a license must be subject to judicial review, and the holder of the patent right must be entitled to "adequate remuneration," which takes into account the "economic value of the authorization," The holder of such a right is also entitled to judicial review of any decision regarding remuneration.

All of the above requirements raise issues of interpretation. For example, it is unclear whether compulsory licenses must be granted on an individual basis, or if categories of inventions may be considered together. TRIPS also does not specify any criteria to determine whether remuneration is "adequate" other than to state that it must take into account the "economic value of the authorization." Countries could continue to apply the average royalty rate paid in voluntary licenses within a given industry rather than consider the royalty that would have been paid with regard to the particular invention.

Transitioning to the Future. Although developing countries and "economies in transition" should have in compliance been with most TRIPS requirements as of January 2000, they need not provide product patents until the year 2005 in any areas that they had previously not patented. This is important to international biotechnology because many countries do not provide product protection of pharmaceutical and/or agricultural products. However, TRIPS does provide some protection for such products during this intervening period.

For countries that did not provide patents for pharmaceutical and agricultural chemical products on terms consistent with Article 27 as of January 1995, a system for filing what has been referred to as "mailbox" applications must exist as of that date (1,20–21). The mailbox applications consist of patent applications that will not be examined until the country must comply with the remainder of TRIPS provisions; however, the mailbox system provides patent applicants with a "date of application" on the day the application is deposited for purposes of novelty and priority (1,20). Patents will issue based on mailbox applications after the expiration of the transitional period.

For products that are subject to mailbox applications, an exclusive marketing right (EMR) must be granted to provide patentlike protection in the interim period before mailbox applications are examined. An EMR is available where (1) a patent on the same invention was granted in another WTO country as of January, 1995 and (2) marketing approval for the invention was granted in the same country (1,20). The EMR must last for the shorter of (1) five years after the EMR is granted or (2) the grant or rejection of the patent (1,20). Thus an EMR enables exclusion of commercial use even in the absence of a patent. However, an EMR provides no possible recourse to oppose noncommercial use.

Even after the transitional period expires, issues are likely to linger concerning the TRIPS requirements. Not only are some present requirements subject to interpretation, but there are also suggestions to modify the TRIPS requirements. For example, it has been suggested, mostly by developing countries, that the scope of items that may be excluded from patent protection should be broadened to include microorganisms, microbiological processes, gene sequences and essential drugs listed by WHO (55). Similarly African countries and Venezuela have argued that compulsory licensing should be allowed for essential drugs. On the other hand, developed countries want to expand patentable subject matter by eliminating the permissible exclusion of plants and animals. Moreover there is some desire to provide protection for holders of "traditional knowledge" or to disallow patents on items available to the public; this desire derives from the belief that TRIPS is not consistent with the Convention on Biological Diversity and International Undertaking on Plant Genetic Resources (32). However, while this issue continues to be studied, the TRIPS agreement is unlikely to be radically changed in the near future as any actual reconciliation with other international agreements is likely to require substantial time and negotiation. While all these issues will likely be discussed either within the context of TRIPS or some new international agreement, TRIPS is likely to be the relevant framework for the near future, considering its long path to enactment and the difficulty in establishing enough consensus to change the agreement.

Plant Protection Under UPOV

Overview. Intellectual property protection of plants should be examined outside the TRIPS framework because TRIPS expressly allows members to exclude plants from patentability, so long as an effective sui generis system exists to provide protection. Although TRIPS does not specify what such a system would require, patentlike rights have been provided to those who develop new plant varieties. In particular, plant breeders have been provided certain exclusive rights of exploitation when they create new varieties of plants; such rights have been referred to as "plant breeders' rights" (PBR) or "plant variety rights" (PVR). The rationale for such a system is very similar to the rationale for the patent system, albeit limited to one particular area. Namely breeders' rights are justified as providing an incentive to develop new plant varieties that further the development of agriculture; also the PVR enables the breeder to obtain a reasonable return on developing a new variety.

International agreement on the importance of such a right has existed since 1961 when UPOV was first signed. UPOV represents a union of member states that have agreed to provide a breeder's right in accordance with the terms of UPOV. Each member country enacts its own national laws to implement at least the minimum standards stated under UPOV. Accordingly, as with patents, an applicant seeking a PBR must seek protection from each country where protection is desired.

Although UPOV was revised in 1972, 1978, and 1991, there are primarily two versions under which countries are currently operating—the 1978 and 1991 versions of UPOV. Thirty countries operate under the 1978 UPOV, while only 12 countries use the 1991 version, which generally provides greater rights to breeders. Countries who were members of the 1978 UPOV had the option, but were not required, to sign the 1991 Convention. As of May 2000 the following countries had signed the 1991 UPOV: Bulgaria, Denmark, Germany, Israel, Japan, the Netherlands, Republic of Maldive, Russian Federation, Slovenia, Sweden, the United Kingdoms and the United States Although the EU is not a formal member of UPOV, it has enacted legislation consistent with the 1991 UPOV (100–101). As countries are currently operating under both systems, the highlights of both will be discussed here, as well as the critical distinctions; additional information is available elsewhere (57–59). Table 5 provides a summary and comparison of the requirements of the 1978 and 1991 UPOV; the distinctions between the two are noted in italics.

Eligible Subject Matter. The fundamental premise of all versions of UPOV is that breeders should be entitled to some exclusionary right with respect to the varieties they create; however, the different versions of UPOV vary in the type of right that must be provided. Under both the 1978 and 1991 UPOV, members may protect a particular variety through either patent protection or a breeder's right consistent with the provisions of UPOV. However, an important distinction is that under the 1978 UPOV—the version that is most widely followed at present—only one type of protection is available for the same genus or species such that dual protection under the patent system and a UPOV-type system is prohibited (2, Article 2(1)). However, there is a grandfather clause that allows members who had previously provided double protection to continue to do so. This provision was included chiefly to accommodate the United States which provided dual protection prior to the 1978 UPOV and wanted to continue do so (2, Article 37). In contrast, under the 1991 UPOV, dual protection is explicitly allowed, namely members may grant the same plant variety a patent as well as a breeder's right (3, Article 40). Accordingly, whereas dual protection was prohibited under the 1978 UPOV, it is now permissible under the 1991 UPOV but still not mandatory.

The scope of varieties that a member must protect also differs between the 1978 and 1991 versions of UPOV. The 1978 UPOV only requires 24 genera or species to be granted protection, and only after eight years of joining UPOV. On the other hand, the 1991 UPOV requires that *all* genera and species eventually be protected; any

Table 5. Plant Protection Under UPOV (non-patent protection)

	1978 UPOV	1991 UPOV
Available protection	Plant *or* UPOV protection *only*; no concurrent protection allowed except for grandfather clause	Plant *and* UPOV *possible*; members can elect for concurrent protection
Scope of varieties entitled to protection	*Requires* that *24* genera or species be granted protection within 8 years of member joining UPOV; *allows* protection for *all* genera or species	*Requires* protection for *all* genera or species within 5 years of signing 1991 UPOV if already a member, or within 10 years if new member
Technical requirements	• Distinct • Uniform • Stable • Commercially novel; grace period *optional*	• Distinct • Uniform • Stable • Commercially novel; grace period *required*
Excluded activities	• Production of the propagating material for purposes of commercial marketing • Offering the propagating material for sale • Marketing the propagating material • Repeated use of the plant variety to commercially produce another variety • Commercial use of ornamental plants as propagating material in the production of ornamental plans or cut flowers	• Production or reproduction • Conditioning for the purpose of • Offering for sale • Selling or other marketing • Exporting • Importing • Stocking for any of the above purposes
Excluded subject matter	Plant variety and its propogating material	• Protected variety • Varieties that are not "clearly distinguishable • Essentially derived varieties • Variety whose production requires repeated use of a protected variety
Exceptions to UPOV right	• Noncommercial use • Experimental use • Use for further breeding • Farmers saving seed from the harvest of a protected variety (farmer's exception)	• Noncommercial use • Experimental use • Use for further breeding *unless it creates an essentially derived variety* (members *may* provide a farmer's exception subject to the interest of the breeder; otherwise, activity will constitute infringement)
Rights during pendency of application	Members *may* provide provisional protection against the "abusive rights of third parties"	Members *must* provide provisional remedies during this time; at a minimum the breeder must obtain equitable remuneration for any activities within the right to exclude
Term of protection	15 years from grant of breeders right; 18 years from grant for vines, fruit trees, forest trees, and ornamental trees	20 years from grant of breeders right; 25 years for trees and vines
Sample countries operating under the convention	Australia, Austria, Belgium, Canada, Finland, France, Ireland, Italy, New Zealand, Norway, Portugal, South Africa, Spain, Switzerland	Denmark, Germany, Israel, Japan, Netherlands, Sweden, United Kingdom, United States

member who was already a member to the 1978 UPOV must do so within five years of signing the 1991 UPOV whereas those who are joining UPOV for the first time have an additional five years to do so (3, Article 3).

A UPOV right can only be obtained after an examination in a member state that determines whether the application satisfies the UPOV technical criteria. UPOV requires that breeders be entitled to a right of priority similar to that under TRIPS; the breeder may utilize the date of the first application in a UPOV member state in determining novelty of subsequent applications for the same variety in other member states. However, UPOV does not mandate how member states must perform the examination. Nonetheless, it does contemplate

that the breeder applicant provide information, including propagating material and seed (2, Article 7). In addition the 1991 UPOV suggests that the examination include growing the actual variety (3, Article 12). Accordingly, some countries compare the grown variety with the closest reference variety in the applicant's submitted description as well as a standard benchmark variety (26). The formalities of the application, including whether a description of the breeding process is necessary and whether it needs to be publicly disclosed, are left to individual member countries to determine.

Requirements for UPOV Protection. UPOV rights are provided to (*1*) a plant "variety" in accordance with the

UPOV definition that (2) meets the technical criteria under UPOV of being distinct, uniform, stable, and commercially novel (2, Article 6; 3, Article 5). Although the technical criteria are essentially the same under the 1978 and 1991 versions, the 1991 UPOV provides a definition of a "variety," whereas the 1978 UPOV did not. In particular, the 1991 UPOV explicitly defines a qualifying variety as "a plant grouping within a single botanical taxon of the lowest known rank which grouping can be defined by features characterizing a given genotype or combination of genotypes, and is distinguished from any other plant grouping by the expression of at least one of the said characteristics." (3, Article 2(i)).

Both Acts require that the variety be sufficiently distinct. They both provide that a variety meets this requirement if it is "clearly distinguishable" from another variety that is "common knowledge" at the time (2, Article 6; 3, Article 7). A variety is likely to be "clearly distinguishable from another" if the variety is different from one or more morphological characteristics (e.g., leaf shape and flower color) or physiological characteristics (e.g., disease resistance and hardiness) of other varieties known at the time of the application (26). The more difficult issue is to determine what was known at the time of the application. What is common knowledge appears similar to the concept of "prior art" under patent laws. Just as TRIPS does not define what should be considered prior art in determining novelty, UPOV does not define "common knowledge." However, both Acts include applications filed in the definition of "common knowledge" although such applications are not publicly available (2, Article 6(1); 3, Article 7). In addition the 1978 UPOV provides that common knowledge may include description in prior publications or commercial sale (2, Article 6(1)(a)).

Besides being distinctive, the new variety must be both "uniform" and "stable." The variety must be uniform with respect to features of its reproduction or propagation (26). The variety must maintain its essential characteristics after repeated reproduction or propagation, or if the breeder has defined a limited cycle or reproduction, it must remain true at the end of each cycle (2, Article 6(d); 3, Article 9). The amount of stability required is a function of the species and variety. However, it is generally assumed that a variety is stable if it exhibits a reasonable level of uniformity in its essential characteristics for a minimum of two successive growing seasons (26).

Finally, the breeder must establish "novelty" of the variety. Under the 1978 UPOV, novelty exists as long as the breeder does not authorize the variety to be offered for sale or market in the state where the breeder seeks protection prior to application. Novelty may also exist even after the breeder authorizes the variety to be offered for sale or marketed in a state where protection is sought if that state provides a one-year grace period after the commercial activity to file an application; states may provide a grace period of up to four years for commercial activity that occurred in other states. In contrast, the 1991 UPOV provides that members must provide a one-year grace period.

Commercial activity that affects novelty differs between the 1978 and 1991 versions of UPOV. Under the 1978

UPOV, the only commercial activity that defeats novelty is activity pursuant to the authority of the breeder (2, Article 6(1)(b)). However, under the 1991 UPOV, protection is barred if the "propagating or harvested material" of the variety was "sold or otherwise disposed of to others … for the purpose of exploitation of the variety"; as this no longer mentions consent of the breeder, it includes all commercial activity regardless of whether the breeder consents (3, Article 6).

UPOV Right

Scope of Protection. Although member states can always provide additional protection, at a minimum they must enable the breeder to exclude certain activities by others. Under the 1978 UPOV (Article 5(1)), the breeder can exclude others from the following:

1. Production of the propagating material for purposes of commercial marketing,
2. Offering the propagating material for sale,
3. Marketing the propagating material,
4. Repeated use of the plant variety to commercially produce another variety,
5. Commercial use of ornamental plants as propagating material in the production of ornamental plants or cut flowers.

Accordingly, under the 1978 UPOV, the reproductive material of a protected variety may be used to produce another variety as long as it is not for the purpose of selling the protected variety and as long as there is no repeated use of the material of the protected variety for the commercial production of the new variety. Thus a second breeder could generally breed and commercialize a new variety without providing any compensation to the initial breeder of the protected variety.

The activities that the breeder can exclude under the 1991 Act are even broader than those under the 1978 Act. The 1991 Act provides seven acts of exploitation for which authorization from the breeder is required (3, Article 14):

1. Production or reproduction
2. Conditioning for the purpose of propagation
3. Offering for sale
4. Selling or other marketing
5. Exporting
6. Importing
7. Stocking for any of the above purposes

The breeder must authorize these seven activities with respect to propagating material of the protected variety as well as varieties that are not clearly distinguishable from the protected variety (in accordance with the UPOV definition of distinctiveness). In addition the breeder must authorize these activities with regard to harvested material (including entire plants or parts of plants), if the harvested material has been obtained through the unauthorized use of propagating material and the breeder has had no reasonable opportunity to exercise his right in relation to the propagating material (3, Article 14(1)–(2)). The 1991 UPOV permits, but does not require, that member states provide the breeder with the right to exclude products made directly from harvested material of a protected variety through unauthorized use of harvested material where the breeder has had no reasonable opportunity to exercise his right in relation to the harvested material (3, Article 14(3)); a member state that adopts this provision would preclude farmers from saving harvested seeds. However, member states can instead elect to provide a farmer's right similar to that which is available under the 1978 Act. If the breeder is not provided rights with regard to farm-saved seeds (if a farmer's exception is provided), the breadth of any such farmer's exception is within a member state's discretion as long as it protects the breeder's "legitimate interests" (3, Article 15(2)). For example, the breeder's interests may be safeguarded by only providing a farmer's right to small farmers, or farmers of certain crops.

Under both the 1978 and 1991 UPOV, provisional protection may be granted to the breeder during the pendency of the UPOV application. Under the 1978 UPOV such protection may, but need not, be provided against "abusive rights of third parties" during the pendency of the application (2, Article 7(3)). Under the 1991 UPOV, however, provisional remedies during this time period are required. In particular, members to the 1991 UPOV *must* provide measures to safeguard rights during pendency so that at a minimum the breeder obtains equitable remuneration for any activities within the right to exclude. However, member states can reduce the impact of this provision by only applying it against those who knew of the application (3, Article 13).

Exceptions. A major distinction between the 1978 and 1991 UPOV is that under the 1978 Act the breeder cannot prevent farmers from saving products of their harvest to replant, namely farmers who save seed from the harvest of the protected variety do not infringe when they later replant the seed. However, under the 1991 Act, although a member state *may* provide a farmer's exception subject to the interests of the breeder, it can also allow such activity to constitute infringement—something that was not a possibility under the 1978 Act.

Other than the farmer's privilege, the exceptions to the breeder's right under the 1978 and 1991 Acts are similar. Both provide for three exceptions to the breeder's rights: private noncommercial use, experimental use, and use of the protected variety for further breeding (3, Article 1(i)–(ii)). However, with respect to use for further breeding, the breeder has more rights under the 1991 UPOV. In particular, the breeder has a right against those who use the protected variety to breed additional varieties if the new variety is "essentially derived" from the protected variety (3, Article 15(1)(iii)). A variety is "essentially derived," and subject to the breeder's right to exclude, when it is predominately derived from the initial variety and, except for differences that result from the act of derivation, it displays the same essential characteristics that result from the genotype, yet is clearly distinguishable from the initial variety (3, Article 7(5)(b)). This new requirement is intended to remove the unfairness

that operates under the 1978 Act, whereby a genetic modification of a protected variety could enable a second breeder to obtain rights without providing any recognition or compensation to the original breeder. UPOV explicitly provides examples of what should constitute essentially derived varieties: those obtained by selection of a mutant, of a somaclonal variant, or of a variant individual from plants of the initial variety, backcrossing, or transformation by genetic engineering (3, Article 14(5)(c)). Thus, although protected varieties may continue to be used as a source of initial variation, if the resulting variety falls within the definition of an essentially derived variety, authorization from the initial breeder is required.

Two final limitations to the PBR under UPOV are compulsory licensing and exhaustion of the PBR. Compulsory licensing is permissible under both versions of UPOV. However, it is only possible if it is in the public interest and equitable remuneration is provided to the breeder (2, Article 9; 3, Article 17). Only the 1991 UPOV provides for a principle of exhaustion. The UPOV exhaustion principle is analogous to that previously discussed with regard to patent protection; in particular, it provides that the breeder's right will not extend to acts concerning material of the protected variety (or essentially derived variety) if sold or otherwise marketed by the breeder unless further propagation is involved or export of the variety is involved (3, Article 16).

Comparison of UPOV and Patent Protection. It is important to note the distinctions between UPOV and patent protection, particularly for countries where only one type of protection is allowed such that an informed decision on the optimal type of protection may be made. First, it should be noted that whereas patents protect all "inventions" (whether the inventions are products or processes) UPOV seeks to protect one type of product—plant varieties. UPOV can only protect the variety itself, whereas patent protection, if available, can also cover a transformed gene or process of making such a gene. Second, rights against imports are only available under the 1991 UPOV, whereas any country that is a WTO member can prevent imports of patented products. Third, both UPOV Acts explicitly allow use of a protected variety for not only noncommercial purposes but also for the purpose of breeding other varieties. Although there is the possibility that such use could be permissible as a "limited exception," to patent rights under Article 30 of TRIPS, there is at least uncertainty as to whether that is possible, unlike the explicit exception provided under UPOV.

In one respect, the 1991 UPOV may provide more extensive protection than patent rights because of the requirement that states provide protection during the pendency of the UPOV application. Although WTO members may provide similar protection, TRIPS does not require any rights be provided prior to the issuance of a patent. In addition the requirements necessary to obtain UPOV rights potentially require less disclosure than a patent application and potentially no disclosure to the public. Whereas TRIPS mandates disclosure of the invention to obtain patent protection, UPOV does not require a description of the "invention," and does not require any disclosure that would allow replication by another. There is no requirement of either a deposit or of public access to the deposit. Although a member state may impose such requirements on an applicant, UPOV itself does not.

The term of protection may also be different if patent protection is chosen versus UPOV protection. Notably TRIPS calculates the term of patent from filing date, whereas UPOV calculates it from the date of grant of the right. While TRIPS requires a minimum period of 20 years from the filing of an application, the term is 15 years from the grant of the breeder right under the 78 UPOV (18 for vines, fruit trees, forest trees, and ornamental trees) or 20 years from grant of the breeder right under the 91 UPOV (and 25 years for trees and vines).

PRESENT SYSTEMS OF PROTECTION

Europe

Relevant Laws. Most European patents are sought from the EPO rather than from patent offices of individual countries. EPO law includes (*1*) the EPC, which is an agreement among European countries (24), (*2*) regulations implementing the EPC, and (*3*) judicial interpretation of the EPC by EPO courts (60,61). EPC provides a streamlined process for obtaining patents in Europe — a single application to the EPO may lead to national patents in every member state; EPC thus provides a short-cut to obtaining patent protection in multiple European countries.

Eighteen months after an application to EPO, a European Patent (EP) application is published. Subsequently EPO will determine whether to grant a patent (an EP patent). If EPO decides to grant a patent, the EP patent is published and available for opposition on almost any ground on which EPO could have denied patentability. If the EP patent survives the opposition period (is not revoked in its entirety), it will become transformed into individual national patents once certain procedural requirements are fulfilled. Once those requirements are met, EPC ceases to govern the EP patent, and national laws control how the national patents are enforced in each country (4,27).

In addition to the EPO law on patentability, patentability requirements in the European Union (EU) will be discussed. The EU is a union of presently 15 countries. It was founded to further political, economic, and social cooperation; formerly, the EU was known as the European Community (EC) or the European Economic Community (EEC). Unlike EPC, the EU is an organization that governs many facets of national law through a governmental structure including a Parliament, Council, and Commission that roughly represent branches of an executive government. These EU bodies can require member states to take actions through the issuance of regulations (immediately and directly binding on member states) or directives (binding as to result only and usually not immediately effective). In the area of biotechnology, the EU has enacted a directive concerning patent protection for biotechnology that includes both patentability and enforcement of such

inventions (5). Member states are to be in compliance with this by July 20, 2000 (5). It is unlikely that all members will actually be in compliance by this time since the EU directive is presently subject to legal challenge by several member countries (62,63). However, it is perhaps more important to note that the EPO has issued regulations that adopt many of the articles of the EU Biotechnology Directive. Thus, even if EU member states have not yet altered their national laws, that may be of minor import because most patent applicants utilize the EPO's streamlined application system, rather than applying to individual national patent offices. This is particularly true because all of the current EU member states are also members of the EPC. Most importantly, all of the current EU member states are also members of EPC. Accordingly, because the EPO has made its laws consistent with the EU Directive, the laws and policies of the EPO are presently of paramount importance and will be the focus of this section. More information on patent laws of individual European countries is available from other sources (64–67).

Patentable Subject Matter. The first hurdle to patenting biotechnology is to establish that there is a patentable "invention." Natural discoveries are excluded from the scope of patentable inventions under EPC. Accordingly an issue that has been raised is whether certain types of biotechnology, including isolated natural substances, are natural discoveries and thus unpatentable. Under current EPO policy, "biological material," which is defined as "any material containing genetic information and capable of reproducing itself or being reproduced in a biological system" may be patentable regardless of whether it previously occurred in nature if it is "isolated from its natural environment or produced by means of a technical process" (5, art. 2; 40–41). Isolated elements may be patentable even if the elements are isolated from the human body such as gene sequences or partial gene sequences (5, art. 3(2); 29). However, the mere discovery of the elements of the human body, including embryonic stages of the human body is not considered to be patentable (5, art. 5; 40–41,61).

Specific Exclusions

1. Methods of treatment. The patentability of gene therapy and associated technology is an issue due to a provision of EPC that defines methods of treatment and diagnosis of either humans or animals to lack "industrial applicability," therefore making such methods unpatentable. In particular, Gene therapy can be considered a method of treatment; diagnostic kits utilizing genetic engineering can be considered as methods of diagnosis. However, EPO courts could interpret this exclusion narrowly to allow some methods to be patentable. For example, although the EPC excludes diagnostic methods, the EPO has interpreted this exclusion narrowly to only exclude diagnostic methods whose results can be immediately used to decide on a course of medical treatment; if the method provides interim results in the course of making a diagnosis, the method is patentable (67). Methods of diagnoses

are only excluded if actually carried out on the body; accordingly diagnosis of body tissues or fluids removed from the body may be patentable (67).

Even without relying on judicial construction of the method exclusion, some patent protection may be available pursuant to a related part of the EPC statute. In particular, the EPC explicitly states that the medical treatment prohibition "shall not apply to *products*, in particular substances or compositions, for use in any of these [prohibited] methods." (68, Article 52). Therefore, even though patent protection is excluded for the actual method of treatment, products used for such treatment may still be patentable subject matter. Thus, although both somatic and germ-line therapies are considered unpatentable methods of medical treatment, the EPO has indicated that products such as genetically modified cells intended for use in somatic gene therapy are patentable and would not be automatically barred (39). However, patentability may be ultimately precluded pursuant to another provision of the EPC or the EU directive that bars patenting of inventions that are contrary to *ordre public* or morality. Products for use in germ-line gene therapy, in particular, may be found in violation of such a provision (39).

2. Plant and animal varieties. EPC Article 53(b) declares unpatentable "plant or animal varieties or essentially biological processes for the production of plants or animals; this provision does not apply to microbiological processes or the products thereof." This provision does not bar the patenting of transgenic animal (65,80–81). However, other provisions of EPC, in particular, the exclusion of inventions that raise morality concerns could pose a problem even if this provision does not (4, art. 53(a); 80–81). The patentability of plants, on the other hand, is less certain.

The patentability of plant varieties is presently unclear because of conflicting EPO case law and the yet to be implemented EU Directive. The EU Directive as well as recent amendments to the regulations implementing EPC provide that plants are patentable "if the technical feasibility ... is not confined to the particular plant or animal variety" (7,61). This would suggest that transgenic plants could be patented and patent claims could cover varieties so long as it covered more than a single variety. Not all EPO case law is consistent with this — recent cases have held that so long as a claim to a transgenic plant may cover a plant varieties, it should be precluded from patentability (27,69–73). However, in the most recent EPO case T1054/96, In re Novartis, the Enlarged Board of Appeal declared that a claim to a transgenic plant could be acceptable even though it may include specific plant varieties (74). Although this would appear to make EPO practice consistent with the EU Directive, whether this will continue remains to be seen.

It should also be noted that article 53(b) states that although processes for creating the ambiguous "varieties" are unpatentable, other types of biotechnological processes may be patentable. In particular, microbiological processes and the resulting products are patentable — so long as they do not include the improper "varieties" and also meet the technical requirements of patentability. A microbiological process consists of "any process involving or performed UPOV or resulting in microbiological material" (7,61). A process for the production of plants or animals is deemed to be "essentially biological" if it "consists entirely of natural phenomena such as crossing or selection" (7,61). Accordingly only traditional breeding of plants and animals are excluded "essentially biological" processes under Article 53(b). Potentially patentable microbiological processes could thus include methods of obtaining, transforming and using microorganisms such as viruses and bacteria.

Nonspecific Statutory Exclusion. Inventions, whose publication or exploitation would be contrary to *ordre public* and morality, are excluded from patentability even if they otherwise constitute an invention and meet the technical patentability requirements. The important query here is the meaning of *ordre public* and morality. One EPO board has considered the term *ordre public* to include the protection of public security and the physical integrity of individuals within the society, including protection of the environment (71). However, there is no single unitary concept of either *ordre public* or morality in all members of EPC. Moreover it is unclear whether the accepted standards of conduct to which the invention is to be compared are those within one member state or all member states. An invention could be deemed to be lacking morality only if it was contrary to the accepted standards for all member countries, or, an invention could be deemed to be lacking morality if it was contrary to any one country. In addition it has been noted by certain EPO courts that there must be an "overwhelming consensus" of opinion before Article 53(a) will bar patentability (76,77). In any event, there have been instances where this has been raised as a ground for denying a patent on genetic engineering inventions. In particular, patents on transgenic animals, plants, and isolated gene sequences have been opposed based on this provision (76,77–78).

Although not an absolute bar to patenting transgenic animals, the *ordre public* and morality exclusion may preclude patents on some animals. To determine whether a transgenic animal is barred by this provision the EPO has previously used a balancing test that takes into account the suffering of animals and possible dangers to the environment, on one hand, with the potential benefits to humans, on the other. Accordingly, in the first and most famous case where this was applied, the Harvard OncoMouse application, it was held that because a mouse genetically engineered to be predisposed to cancer had such substantial utility to humankind, that it outweighed the suffering of the individual animal (81). However, not all transgenic animals have been held to meet this test. For instance, the EPO has denied applications of transgenic animals whose utility is to study baldness (27).

In the future a modified balancing test may be applied as the EU Directive, as well as the amended EPO regulations, state a differently worded test. In particular, patents on processes for creating transgenic animals as well as the resulting animals are precluded if it is "likely to cause them suffering without any substantial medical benefit to man or animal" (7,61). It literally seems to require some showing of "*substantial* medical benefit" where a process is "likely" to cause animal suffering. However, it is unclear what would constitute a "medical" benefit, let alone a "substantial one." For example, cows that are genetically engineered to produce more milk would appear to be of questionable "medical" benefit. Moreover it is unclear what "suffering" must consist of. Some might even argue that the cow is "suffering" by having its genes altered. On the other hand, unlike the OncoMouse, the altered cow is not programmed to a hastened death.

The appropriate test to apply to plants is unclear. It has been noted that the OncoMouse balancing test is not the only way to evaluate violation of Article 53(a) (71). However, in the absence of a balancing test, the vaguer standard of *ordre public* and morality must be applied. This appears difficult to establish. For example, in a case where a transgenic plant engineered to be pesticide-resistant was opposed as violating Article 53 (a) (among other provisions), the EPO court dismissed alleged danger to the environment on the ground that surveys of the general public showing opposition to genetic engineering were inadequate; actual danger, rather than the "mere possibility" of danger was stated to be necessary and infringement of environmental regulations alone was not considered adequate (71).

Although the interpretation of the morality requirement is still unclear, the EU Directive and the EPO regulations provide a noninclusive list of what will per se violate morality or *ordre public*. Current examples of biotechnology that are unpatentable because of this include processes for cloning humans, processes for germline gene therapy, and use of human embryos for industrial or commercial purposes (7). Processes for modifying the genetic identity of animals which are likely to cause them suffering without any substantial medical benefit to human or animal, and animals resulting from such processes" (7). Because EPO has amended the regulations to EPC to essentially adopt the EU Directive wholesale, the categorical bars to patentability stated in the Directive presumably apply (61). In addition, although EPO has noted that "the provisions of Article 27(2) of TRIPS will be considered" (39), it is presently unclear how that will affect the analysis, if at all, as there are notable similarities in the language, as indicated on Table 6. In fact, it may be that the EPO case law is drawn upon by the TRIPS council and member countries as the most pertinent and largest body of law in interpreting what is otherwise vague language.

Patentability Requirements. As the focus of this article is on issues relating to patenting biotechnology, major issues regarding technical requirements will be noted. However, more detailed information on the specifics of each technical requirement under EPC alone is discussed elsewhere (27,64).

Novelty. EPO takes an "absolute novelty" approach to determining whether an invention is patentable.

Table 6. Exclusions from Patentability

	EPC/EU[a]	TRIPS
Categorical exclusions of specific subject matter	• Methods for treatment or diagnosis of human/animal body • Plant and animal "varieties" • "Essentially biological processes" for the production of plants or animals except "microbiological" or other technical processes • "Human body, at the various stages of its formation and development"	• Diagnostic, therapeutic and surgical methods for the treatment of humans or animals; • Plant and animals other than micro-organisms, • "Essentially biological processes" for the production of plants or animals other than "nonbiological" and "microbiological processes"
Nonspecific exclusions	Inventions whose commercial exploitation would be contrary to "*ordre public* or morality" Listed examples: • Processes for cloning humans • Processes for germ line gene therapy of humans • Commercial uses of human embryos • Methods of genetically modifying animals that are "likely to cause them suffering without any substantial medical benefit to man or animal," and any animals resulting from such methods	Inventions whose commercial exploitation must be prevented to protect "ordre public or morality," including to protect human, animal or plant life or health or to avoid serious prejudice to the environment

[a]Mandatory exclusions.
[b]Permissive exclusions.

Accordingly relevant prior art includes any written publications, abstracts, or drawings; prior use anywhere in the world; and prior, though later published, patent applications (3, Article 54). However, EPO does recognize a very limited grace period of six months before the filing of the application for disclosure made by the applicant at an official, or officially recognized, international exhibition (3, Article 55). In addition there has been some discussion concerning whether a more expansive grace period should be provided.

Novelty may be satisfied for isolated products found in nature so long as it is done through a technical process. Thus a new substance discovered as being produced from a microorganism, or a previously unknown gene or protein, would be considered novel and patentable if the other patentability requirements were satisfied (40).

Novelty may be established in cases of newly discovered uses of previously patented compounds. In particular, a compound may be patented with a use limitation even if the compound itself was previously known as long as its use for treatment or diagnosis was not previously known (first medical use). A second patent can be obtained for the newly discovered use of a previously known compound, even if the compound was previously known to have a use, if it is newly discovered to be useful for additional therapeutic purposes (second medical use). In either event, it should be recalled that only the compound, and not the process of using the compound, can be patented so that the medical treatment prohibition is avoided. This often results in just a technical distinction of how the invention is claimed — a claim reciting a "method of treating X using substance Y" would be considered unpatentable, whereas a first medical use claim reciting "substance Y for use as an active pharmaceutical substance" would be acceptable, as would a second medical use claim stating "use of substance Y for the preparation of a pharmaceutical compound for the treatment of X."

Industrial Applicability and Inventive Step. The invention must also have "industrial applicability," which requires that the invention be useful in "any kind of industry, including agriculture." (3, Article 57). In general, this is a fairly easy requirement to meet. However, industrial applicability may be an issue with some nucleic acid sequences. The EPO has noted that it is "questionable" as to whether sequences that have no known use other than as probes satisfy the industrial applicability requirement (40). Similarly most inventions will be found to have met the "inventive step" requirement unless the invention is "obvious to try" by a person skilled in the art with a reasonable chance of success. However, in the area of biotechnology, the application of the inventive step requirement is uncertain as different tests have been applied by EPO and national courts (82–83).

Description and Deposit. The disclosure requirement merely requires sufficient disclosure such that the invention can be carried out by a skilled person for the claimed subject matter using common general knowledge and the information provided in the application. No specific examples are required in the application if a skilled person could carry out the invention without undue experimentation. In addition, pursuant to the EU Directive, the patent application for a sequenced or partially sequenced gene must be disclosed; however, that alone may not be sufficient — there may be other issues with patentability requirements for expressed sequence tags (ESTs) pursuant to Articles 52(1) and 56 (40).

A deposit is required if necessary to enable reproduction of the invention claimed in an application. A deposit is required, for example, if a specific microorganism cannot be reproduced based on the application alone. However, if a protein or nucleic acid is sufficiently defined in the application such that it may be synthesized by one of skill in the art, it need not be deposited.

If a deposit is required, it must be made no later than the filing date. The deposit may be made to any institution recognized under the Budapest Treaty as well as other institutions as noted in the Official Journal of EPO. The deposit may become available to others after the patent is published. However, if the applicant informs EPO prior to the publication of the application of a preference to limit access, there will be no public access until an EPO patent is granted, or 20 years after the filing date if the application is refused or withdrawn. If public access is not specifically requested to be restricted, the deposit will be available after the application is published to any person who requests it on the condition that the person does not make the material available to a third party and if the person is using the deposited material for experimental uses only.

Patent Rights

Challenging an EP Patent. Even after an EP patent has been issued, there is some uncertainty involved because the patent is subject to a period of opposition during which anyone, including competitors or special interest groups, may file an opposition against the patent. Oppositions may be filed within nine months after the patent grant is published. There are many grounds for contesting a published EPO patent, including lack of novelty (Articles 54–55), lack of inventive step (Article 56), lack of industrial application (Article 57), improper subject matter (Article 53), noninvention (Article 52) and inadequate disclosure (Articles 100(b) and 83). Opposition may result in amendment of the patent scope or complete revocation of the patent.

It should be noted that opposition proceedings are conducted by EPO before the EP patent becomes a collection of individual national patents. Accordingly successful opposition of an EP patent is a much cheaper and more efficient to challenging parallel national patents in different countries. However, even if a patent withstands opposition in EPO, it may still be subject to revocation in national proceedings. An EPO decision in opposition proceedings has no binding effect, even on the same parties in later national proceedings for which national law, rather than EPC, will govern.

There may be inconsistency between the EPO system and national courts as well as inconsistency within courts of same nations. For example, in the recent biotechnology case of *Biogen v. Medeva*, several different courts reached different decisions on whether the Biogen patent was valid, and even courts that reached the same result had different reasoning (82–87). The Biogen patent was first subject to opposition within EPO, after which the EPO Technical Board of Appeal maintained the patent (84). However, once the Biogen EP patent was transformed into national patents, trouble began. Biogen first sued Medeva for using its patented composition. In its defense, Medeva asserted that Biogen's patent was invalid on a number of grounds. In the initial court, Biogen's patent was considered valid on the same grounds as the EPO had found (85). However, the next court to hear the case, the High Court of Justice, came to the opposition conclusion — namely that Biogen's patent was invalid for all asserted grounds — that it was insufficient, obvious, and possibly not even an "invention" (86). Finally, the House of Lords also declared Biogen's patent invalid, but for different grounds than the High Court of Justice; in particular, the House of Lords declared the critical issue was one that had not been formally raised in any of the decisions of the prior courts (87). The House of Lords determined that the claimed invention was too broad and held that the patent from which priority was claimed could not support the patent at issue (87).

Scope of Protection. Granted EP patents have the same scope of protection within a given country as do patents granted by the individual country's patent office (4, Article 2(2)). EPC generally does not govern activity after the opposition period with two exceptions. EPC provides that there should be uniform protection conferred by European patents pursuant to Article 69. EPC provides that there should be protection for the "direct product" of a patented process under Article 62(2). However, because national courts may interpret identical provisions differently, there is bound to be some inconsistency despite the requirement of "uniform protection."

The scope of protection for biotechnological matter is still under development. Although EPO issued regulations in June 1999 implementing some of the provisions of the EU Directive, it did not include any of the EU Directive provisions concerning the scope of protection of biotechnological inventions (61). However, the EU Directive will be considered here as all EU members are EPC members. The EU Directive provides that patents on biological material provide protection on any biological material derived from the initial patented material through propagation or multiplication; accordingly offspring of a patented transgenic animal would be within the scope of protection (7). In addition, for patented processes that produce biological material with specific characteristics as a result of the invention, protection is extended to any biological material "directly obtained" from the patented process as well as any biological material "derived from the directly obtained" material through propagation or multiplication (7).

The EU Directive provides limitations on the scope of protection for certain biological material. In particular, it is stated that no protection is provided for material obtained from material placed on the market by the holder of the patent with his consent (7). In other words, once the patent owner sells the patented product, rights for subsequent products are extinguished. In addition the Directive provides a farmer's rights similar to the rights provided under UPOV as well as a corollary right for animal breeders (7). Member states have control over how much of an exception to provide with respect to the newly created animal breeder's right (7). Finally, compulsory licensing is also specifically provided for, although most of the specifics of such licensing are left up to individual countries (7).

Another limitation on infringement that has been established in some countries is an exception for experimental use. Many EU and EPC countries exempt from infringement "acts done for experimental purpose relating to the subject matter of the patented invention" (64,88); this is often referred to as an experimental use exception. It is unclear whether experiments for market approval constitute infringement and whether there should be any difference when testing is done by manufacturers in preparation for the sale of generic drugs (89–90). However, the most recent case on this issue in Germany found that clinical trials on a patented compound to determine its properties and effects were protected by the experimental use exception even if conducted with the goal of acquiring marketing approval; the only limit to the exclusion was stated to be if the tests were solely directed at determining commercial facts such as market needs and price acceptance (91). Even if national courts uniformly interpret a broad exception, such an interpretation could be challenged as violating TRIPS and not protected by the "limited exception" provision of Article 30 of TRIPS.

Enforcement Issues. An important enforcement issue in Europe is that in recent years, some courts have issued "cross-border" or "pan-European" injunctions against defendants in patent infringement actions (92,93). An injunction is an order issued by the court; in the context of a patent infringement action, an injunction is often issued against a defendant who is infringing. The injunction orders a defendant to stop infringing or be subject to court sanctions. What is notable about cross-border injunctions is that typically courts only issue injunctions within their territorial boundaries, that is, within the nation in which the court sits. However, cross-border injunctions are injunctions in which the court orders a defendant not to infringe in other states or countries (across its own borders).

The availability of cross-border injunctions are particularly useful to patent owners desirous of relatively inexpensive means of stopping a defendant from infringing in multiple nations. While a patent owner attempting to enforce an EP patent would traditionally have to pursue litigation in multiple jurisdictions with potentially different verdicts, a cross-border injunction enables a single litigation to potentially enjoin a defendant's action in all European countries.

The legal basis for cross-border injunctions is based on a provision of the Treaty of Brussels (94) that allows a case to be brought before the court of any defendant's residence when there are multiple courts. Through a liberal interpretation of this provision, courts, and particularly the district court of the Hague, have found themselves competent to hear cases against defendants in patent cases as long as one defendant was Dutch (92–94,97). Jurisdictions that have issued cross-border injunctions include the Netherlands, England, Germany, Belgium, and France, although the majority of such injunctions have been issued by Dutch courts (92–94,97). Although subsequent case law has limited application of this principle in cases against foreign defendants (95), the potential for such injunctions still exists in the absence of an effective means to enforce patents issued by EPO. Indeed, commentary on cross-border injunctions has noted that the only long-term solution is a more uniform enforcement mechanism (97). Such a possibility is in fact a potential reality.

Additional Protection

Pharmaceuticals. In addition to patent protection, manufacturers of patents on pharmaceuticals in the EU are entitled to some patentlike exclusionary rights after the term of a patent via a supplementary protection certificate (SPC) (98–99). A SPC allows a limited exclusivity right after the ordinary 20-year patent term expires. The term of SPC is calculated as the time elapsed between the date of filing the application for the patent and the date of the first marketing authorizing minus five years, up to a maximum of five years. There are four requirements that must be met to obtain a SPC: (*1*) The product sought to be protected by an SPC must be protected by a basic patent in force, (*2*) the product must not have been granted marketing authorization, (*3*) the product must not have already been the subject of an SPC, and (*4*) the marketing authorization is the first to place the product on the market (Article 3). Only the patentee, not a licensee, is entitled to apply for a SPC.

A SPC allows its owner the right to exclude others from the commercial sale of covered goods. However, because a SPC is not a patent, the owner of SPC cannot exclude others, such as generic drug manufacturers, from testing the drug that is no longer patented. SPC effectively achieves the same result as an infringement exception provided to manufacturers of generic drugs under the U.S. system.

Plants. Regardless of the status of patenting plant "varieties" under EPC or the EU Directive, an alternative means for protection exists both under the EU and under national laws of individual countries. Plant Variety Rights within the Community are governed by an EU regulation that provides a communitywide right; this right is additional to any rights available under national regimes. This means that for plants, protection is potentially available through EPO (subject to the bar on plant varieties), through national patent systems of EPC or EU countries; in addition PVR may be available either from individual countries or from the EU. The benefit of obtaining rights under the EU system is that the breeder obtains an exclusionary right that applies throughout the entire EU system with only one application. Plant variety protection pursuant to the Community regulation is consistent with protection available under the 1991 UPOV (100–101).

For patented plant-related inventions, additional protection is available in the form of a patentlike right along the same lines as for pharmaceuticals. Since 1996, an SPC has also existed for "plant protection products" that

are protected by a patent (102). The term and procedural requirements of an SPC for such products is identical to the one provided for pharmaceuticals. What differs is the product that is protected. In relation to plant products, those that qualify for SPC protection are defined as "active substances and preparations containing one or more active substances ... intended to": (*1*) protect plants against harmful organisms, (*2*) influence the life processes of plants, such as a growth regulator, (*3*) preserve plant products, (*4*) destroy undesirable plants, or (*5*) destroy parts of plants or otherwise minimize undesirable growth (102).

Future Developments. Patent protection for biotechnology in Europe is likely to continue to be in a state of flux in the near future due to the existence of different systems for obtaining and enforcing patents. A uniform system for obtaining as well as enforcing patent rights throughout the EU member states has been previously envisioned, although it has not yet become a reality.

In particular, although the Community Patent Convention (CPC) provides for such a system, it has not been ratified by all EU members. Because of political and constitutional reasons, CPC has not come into force, although it was first signed in 1975 (65). At this point, it is anticipated that the EU will cease efforts toward effectuating CPC, and instead work toward enacting a regulation to create a unitary EU patent (103); the EU Commission is planning to propose such a regulation in the year 2000 but no such regulation has been proposed as of June 2000. The EU has been clear that such a system would be largely consistent with the EPC provisions and at least for a transitional period, coexist with the present two-tiered system of national and EP patents (103). However, even if such a regulation were adopted, the protection of biotechnology is likely to continue to be remain unsettled for quite some time.

Japan

Japanese patent law has changed markedly in the last 10 years, both because of the TRIPS agreement and because of international pressure to conform its laws to those of Europe and the United States. Some doctrines that have been established for decades in the United States have only recently been established in Japan. It is likely that amendments and clarifications to its patent laws will continue. Current Japanese Patent Office (JPO) practice with respect to biotechnology can be determined by examining the JPO Guidelines, as well as the JPO's response to some hypothetical biotechnology patent examples, all of which are publicly accessible through the JPO web site (104,105). Because of the availability of this information, as well as the fact that Japanese patent law will likely continue to evolve, this discussion will provide an overview of recent changes to Japanese patent law that affect biotechnology rather than attempt to provide a comprehensive description of all laws concerning biotechnology.

Patentability. The JPO has clarified that biotechnology inventions are not per se precluded from patentability (39–40,105). Therefore the usual patentability standards apply to such inventions. However, application of those standards has posed some difficulties in the area of biotechnology. For example, it is sometimes difficult to determine when genetic engineering inventions are novel or have an inventive step. Nonetheless, the JPO has indicated a wide variety of biotechnological innovations to be considered patentable subject matter, so long as they can still meet the technical patentability requirements. In particular, living unicellular organisms, animals and animal parts, plants, and plant parts are all considered patentable subject matter. As a "general rule," human-derived products are said to be eligible for patentability, although humans themselves may be barred under a morality-based exclusion (40). New uses of known compound are also considered patentable (40).

However, "industrially applicable" has been statutorily defined to exclude certain subject matter that narrows the scope of patentable biotechnology. As in EPC certain types of "medical activity" are considered not industrially applicable. Of particular relevance to the pharmaceutical area is that methods of treating humans by surgery or therapy, as well as diagnostic methods practiced on the human body are all excluded as being not industrially applicable; however, pharmaceutical compositions for use in the unpatentable methods can be industrially applicable. In addition methods of treating animals are patentable so long as there is no attempt to patent treatment on the human body. The relevance of these exclusions to biotechnology is that gene therapies are not patentable because they are considered a method of treatment within this exception. However, pharmaceutical products produced by gene therapy techniques may be eligible to be patented (40).

Another bar to patenting biotechnology is that inventions "liable to contravene public order, morality or public health" may be considered unpatentable under Section 32 of the Japanese Patent Laws (106). This language has been the basis for denial of patents on new medical treatments as well as methods of breeding new plants or animals. In addition JPO has indicated that this may also preclude patents on humans as well as human organs (40).

While not a bar to patentability, it is important to be aware of deposit requirements for biotechnology. For inventions concerning microorganisms, the microorganism must generally be deposited with an institution designated by JPO or international depositary authorities. This requirement can be met simultaneously with application but is improper after filing. There are certain situations where microorganisms need not be deposited, for example, if it can be created by a person skilled in the art based on the specification description alone.

Patent Rights

Opposition and Patent Term. Prior to 1996 the JPO published applications for opposition by third parties after examination, but prior to the issuance of an actual patent and tied the patent term to this opposition period. The pre-1996 patent term was 15 years from the date of publication of the postexamination (and pre-grant) application, with a maximum duration of 20 years from filing date (106, Article 67(1)). This scheme has been substituted with an opposition procedure *after* the patent grant, similar to the situation described with regard to EPO (106, Article 113). Now, any third party can file an opposition, and there is a uniform patent term of 20 years from the date of filing of the patent application.

Patent terms of pharmaceutical or agricultural inventions that are subject to pre-marketing administrative approval may be extended for up to an additional five years. However, there are certain limitations to obtaining such an extension in addition to qualifying subject matter. For instance, the request must be submitted within three months of regulatory approval of the patented invention, and no request may be filed within six months from the end of the patent term. The request must show that the patent could not be commercialized for at least two years after issuance due to delay necessitated by regulatory approval (from either the day the patent issues, or the day approval is first sought, whichever is *later*).

Scope of Protection. Infringement of a patent can occur through both direct and indirect means. However, the infringement must be commercial and public, unlike infringement under U.S. law. It should be noted that pursuant to TRIPS requirements, the rights of patent owners were recently extended to also include the ability to exclude unauthorized offers for sale of the patented invention. In particular, literal infringement occurs when an authorized party:

1. commercially makes, uses, sells, offers to sell, or imports a patented product;
2. commercially uses a patented process; or
3. commercially makes, uses sells, offers to sell or imports a product made by a patented process.

Under a recent revision to Japanese patent laws, unless an alleged infringer (the defendant in a patent infringement suit) shows proof to the contrary, if a defendant's product is identical to one obtained by the patented process, it is presumed to be manufactured by such process (107).

Infringement under the doctrine of equivalents (DOE) has been a relatively recent phenomenon in Japan. Although courts had applied the DOE sporadically in recent years, it was not until 1998 that the Supreme Court affirmatively embraced the doctrine and set forth clear standards for its application (108–109). The Supreme Court clarified that the DOE analysis should be considered *at the time of the infringement* (rather than at the time of filing as was previously held by lower courts) (108); this change is consistent with U.S. law and favors patentees and particularly pioneering patents that can now potentially cover related advances that develop after a patent issues. Although courts were initially hesitant to find infringement under DOE, at least one district court has done so as of 1999 (110).

Under the DOE law in Japan, infringement may be found even if the patent has an element that does not cover the accused product, if the accused product is regarded as being equivalent. Equivalence is determined based on a multifactor test. The essential inquiry in finding infringement via DOE is whether the accused product or process contains elements identical or equivalent to each claimed element of the patented invention; again, this central inquiry echoes current U.S. law requiring that DOE be conducted on an element-by-element analysis. Infringement under DOE may be found where:

1. portions are not essential to the patentability of the patented invention;

2. objectives and effects of the patented invention are still attained after replacing the portions with their counterparts in the accused product;

3. the replacement would have been obvious to those skilled in the art at the time of manufacturing the accused product;

4. the accused product was neither anticipated nor obvious to those skilled in the art as of the filing date of the application for the patented invention; and

5. no special conditions exist, such as the accused product being intentionally excluded from the scope of claims during the prosecution of the application for the patented invention (no prosecution history estoppel).

There are several exceptions to the basic patent right to exclude others from commercial use of the invention that may affect biotechnological inventions. No patent rights are available with respect to products that existed in Japan prior to the filing of the patent application (106). Also no patent rights exist for acts of preparing medicines in accordance with medical prescriptions (106). The Japanese Patent Act has long since established that there is no right to exclude those who use a patented invention for "purposes of experiment or research" (105, Article 69(1)). However, this has been recently interpreted to cover the manufacture of patented products for pre-marketing approval by companies seeking to manufacture generic versions of patented drugs (110). This judicial interpretation brings Japanese patent law in line with practice under U.S. law. However, as noted with the similar provision in the EU, it is unclear whether this interpretation is consistent with TRIPS.

Compulsory licenses of patented inventions are also available in certain circumstances under Japanese law. In particular, one who has independently created a patented invention and has been commercially using it or made preparation to do so, can obtain a nonexclusive license (Article 79). JPO has noted that compulsory licenses are available in the cases of nonworking of a patent for at least three years (Article 83), dependent patents (Article 92), and public interest (Article 93). However, it is unclear whether the provision of such compulsory licenses are entirely compatible with TRIPS.

Enforcement Issues. Finally, damages are important in examining patent protection of biotechnology in Japan because damages enable a patent owner to effectively enforce rights and preclude infringement. An examination of damages under Japanese Patent law is particularly relevant because of recent changes that substantially improve, recovery for prevailing plaintiffs (i.e., patent owners). Previously only limited damage awards were available (at least in comparison to U.S. awards). Plaintiffs generally did not even attempt to seek lost profits as

courts rarely granted them; the primary reason for this was that plaintiffs could not establish causation between infringement and damages because plaintiffs had no access to confidential information of the defendant and Japanese patent law did not provide for any inferential causation of lost profits. However, under the present law, which became effective as of January 1, 1999, the types of damages available have been explicitly broadened in the patent statute to include, for the first time, lost profits as a measure of patent infringement damages.

Under the new section, the plaintiff is relieved of the causation burden as the statute provides a presumptory amount that the defendant must then rebut. The presumptory amount of lost profits consists of the infringer's sales volume multiplied by the patentee's profit rate, as long as such amount does not exceed the amount a patentee would be able to obtain based on its own manufacturing capacity. The burden is on the infringer to establish that the presumed damages are incorrect; the infringer would need to show that the actual number of infringing products was lower.

In addition to providing for lost profits, the amended patent law also allows the potential for increased awards based on a reasonable royalty calculation. Previously damage awards were calculated based on industry-standard royalty rates and rates for licensed government-owned patents; this resulted in royalty rates ranging from 2 to 4 percent, which was in stark contrast to a more typical rate of 8 to 10 percent in Europe (112). This stringent calculation was in part derived from the fact that the word "normally" qualified the term "reasonable royalty." The amended law removes the qualification "normally," such that courts should be able to grant higher amounts than previously. The cap for damages available from an infringing corporation has been raised from 5 million yen ($36,500) to 150 million yen ($1.09 million) (106, Article 201).

There are additional provisions in the Japanese patent laws that may assist plaintiffs in patent infringement actions to recover more damages. In the past plaintiffs were forced to extrapolate sales of defendants infringing products based solely on public documents because there was no requirement that parties produce documents other than those that it intends to rely on at trial. However, under a new provision of the patent laws, if a party so requests, a court may order the opposing party to produce documentation necessary to determine damage caused by an infringement (barring some legitimate reasons for failing to produce such documents) (106, Article 105). Moreover a new provision of the Civil Procedure Code, provides judges with discretion to determine the appropriate amount of damages where "it is extremely difficult to prove the amount of damages from the nature of such damage" (114, Article 248). Although this provision was intended to assist in calculating intangible damages such as emotional distress, it has been relied upon to discount a defendant's calculation of lost profits and to instead adopt a calculation closer to the plaintiffs when the defendant's internal documentation was not made available (112,115).

In the future, it is possible that even more monetary compensation may be available for plaintiffs, which would bring Japanese patent law closer to the situation in the United States In 1998 the JPO proposed to provide punitive damages and partial attorney fees for successful plaintiffs — both of which are available remedies to prevailing patent owners in the United States and serve as a deterrent effect against infringement. Although no such legislation was passed in 1998, JPO, as well as the Ministry of International Trade and Industry, continue to lobby for such changes.

BIBLIOGRAPHY

1. Agreement on Trade-Related Aspects of Intellectual Property Rights (April 15, 1994), Marakesh Agreement Establishing the World Trade Organization, Annex C, Legal Instruments — Results of the Uruguay Round, 33 I.L.M. 81, Available at: *http://www.wto.org*

2. International Convention for the Protection of New Varieties of Plants (1978 Act), Available at: *www.upov.int/eng/convntns*

3. International Convention for the Protection of New Varieties of Plants (1991 Act), Available at: *www.upov.int/eng/convntns*

4. Convention on the Grant of European Patents (October 5, 1973), 13 I.L.M. 270.

5. Directive 98/44/EC of the European Parliament and of the Council on the Legal Protection of Biotechnological Inventions (July 6, 1998), OJ (L 213) 13.

6. EPO Member States, Available at: *http://www.european-patent-office.org/epo/members.htm*

7. Information on the European Union, Available at: *http://www.cali.co.uk/bis/europe.htm*

8. About the WTO, Available at: *http://www.wto.org/wto/about/organsn.htm*

9. States Party to the International Convention, Available at: *http://www.upov.int/eng/ratif/index.htm*

10. Budapest Treaty on the International Recognition of the Deposit of Microorganisms for the Purposes of Patent Procedure (April 28, 1977), 32 U.S.T. 124, Available at: *http://www.wipo.int/eng/iplex/index.htm*

11. Status of Member States to Budapest Treaty on the International Recognition of the Deposit of Microorganisms for the Purpose of Patent Procedure, Available at: *http://www.wipo.org/eng/ratific/q-budpst.htm*

12. Europa, Key Issues: Enlarging the EU, Available at: *http://europea.euint/geninfo/key-en.ht*

13. Paris Convention for the Protection of Industrial Property (March 20, 1883), *as last revised*, Stockholm, July 14, 1967, arts. 1-12, 21 U.S.T. 1583, 828 U.N.T.S. 305, Available at: *http://www.wipo.int/eng/iplex/index.htm*

14. Understanding On Rules and Procedures Governing the Settlement of Disputes (April 15, 1994), WTO Agreement, Annex 2, 33 I.L.M. 1226.

15. Overview of the State-of-Play of WTO Disputes (November 26, 1999), Available at: *http://www.wto.org/wto/dispute/bulletin.html*

16. E.-U. Petersmann, in E.-U. Petersmann, ed., *International Trade Law and the GATT/WTO Dispute Settlement System*, Kluwer Law International, London, 1998.

17. T.P. Stewart and M.M. Burr, *N.C. J. Int. Law Com. Reg.* **23**, 431 (1998).

18. J.H. Reichman, *Virginia J. Int. Law* **37**, 334, 337 (1997).

19. R.C. Dreyfuss and A.F. Lowenfeld, *Virginia J. Int. Law* **37**, 275 (1997).

20. India, Patent Protection for Pharmaceutical and Agricultural Chemical Products (December 19, 1997), AB 1997-5, W.T./DS50/AB/R.

21. C. MacDonald-Brown and L. Ferera, *Eur. Intell. Prop. Rev.* **2**, 69 (1998).

22. C. Correa, in C. Correa and A. Yusuf, eds., *Intellectual Property and International Trade: The TRIPS Agreement*, Kluwer Law International, Brentford, UK, 1998.

23. World Intellectual Property Organization, *Existence, Scope and Form of Generally Internationally Accepted and Applied Standards/Norms For the Protection of Intellectual Property*, Doc. No. MTN.GNG/NG11/W/24/Rev.1, Annex II (September 1988).

24. D. Gervais, *The TRIPS Agreement: Drafting History and Analysis*, Sweet & Maxwell, London, 1998.

25. C. Correa, *Eur. Intell. Prop. Rev.* **8**, 435, 437–438, 441 (1997).

26. C.M. Correa, *J. World. Intell. Prop.* **1**, 75–98 (1998).

27. E.S. Van de Graf, *Patent Law and Modern Biotechnology*, Sanders, Rotterdam, 1997.

28. T820/92, *Contraceptive Method*, 1995 Off. J. Eur. Pat. Off. 113; 26 Int. Rev. Indus. Prop. & Copy. L. 543 (1995).

29. 290/86 (*Cleaning Plaque/ICI*), 1992 Off. J. Eur. Pat. Off. 414; 23 Int. Rev. Indus. Prop. & Copy. L 815 (1991).

30. T 144/83 (*Appetite Suppressant*), 1986 Off. J. Eur. Pat. Off. 301; 18 Int. Rev. Indus. Prop. & Copy. L 258 (1987).

31. *T 74/93, Contraceptive Method/British Technology Group*, 1995 Off. J. Eur. Pat. Off. 712, *reprinted in* 27 Int. Rev. Indus. Prop. & Copy. L. 99 (1996).

32. R. Moufang, *Methods of Medical Treatment under Patent Law*, 24 Int. Rev. Indus. Prop. & Copy. L. 18-49 (1993).

33. P. Cullet, *J. World Intell. Prop.* **2**, 617 (1999).

34. S. Charnovitz, *Virginia J. Int. Law* **38**(6), 639–744 (1998).

35. C.T. Fedderson, *Minn. J. Global Trade* **7**, 75–122 (1998).

36. *Analytical Index of the GATT*, pp. 562–596 (1994).

37. D. Palmeter and P. Mavroidis, *Am. J. Int. Law* **92**, 398 (1996).

38. T. Ackerman, *Tex. Int. Law J.* **32**, 489–510 (1997).

39. Organisation for Economic Cooperation and Development, *Intellectual Property Practices in the field of Biotechnology*, TD/TC/WP(88) 15/final (1999), Available at: *www.oecd.org*

40. Organisation for Economic Cooperation and Development, *Responses from Member Countries to Working Part of the Trade Committee of the OECD Questionnaire on Intellectual Property Practices in the Field of Biotechnology* (1999), Available at: *www.oecd.org*

41. Responses from EPO and JPO, Available at: *http://appli1.oecd.org/ech/BiotechIPR.nsf*

42. G.H.C. Bodenhausen, *Guide to the Application of the Paris Convention for the Protection of Industrial Property as Revised at Stockholm in 1967*, United International Bureau for the Protection of Intellectual Property, Geneva, 1968.

43. Argentina Patent Law art. 44.

44. *Cong. Rec.* **142**, S11843–11844 (September 26, 1996).

45. American Bar Association, Section of Intellectual Property Law, *Annual Report 1996–1997*, ABA Sect. I.P. Law Rep. 70, 103 (1997).

46. C. Ho, *U.C. Davis Law Rev.* **3** (2000).

47. WTO, Canada-Patent Protection of Pharmaceutical Products Report of the Panel, WT/DS114/R (March 17, 2000).

48. Canada Department of Foreign Affairs and International Trade, *Trade Negotiations and Agreements — Panel Cases to which Canada is a Party* (1999), Available at: *http://www.dfait-maeciigc.ca/tna-nac/summary-e.as*

49. Draft of July 23, 1990 (Doc. MTN.GNG/NG11/W/76), reprinted in D. Gervais, *The TRIPS Agreement: Drafting History and Analysis*, Sweet & Maxwell, London, 1998.

50. M. Halewood, *Osgoode Hall Law J.* **35**, 243, 264–265 (1997).

51. R. Weissman, *Univ. Ponn. J. Int. Econ. Law* **17**(4), 1069–1125 (1996).

52. Brazilian Industrial Property Law (May 14, 1996).

53. J. Giust, *Hastings Int. Comp. Law Rev.* **21**, 597–637 (1998).

54. I. Ahler, *Int. Rev. Indus. Prop. Copy. Law.* **28**(5), 632–671 (1997).

55. E. Duran and C. Michalopoulos, *J. World Intell. Prop.* **2**, 865–873 (1999).

56. J. Watal, *J. World Intell. Prop.* 281–307 (1999).

57. International Convention for the Protection of New Varieties of Plants (1961 Act), Available at: *www.upov.int/eng/convntns*

58. S.K. Verma, *Eur. Intell. Prop. Rev.* **6**, 281–289 (1995).

59. B. Greengrass, *Eur. Intell. Prop. Rev.* **12**, 466–472 (1991).

60. EPO Guidelines for Substantive Examination, Available at: *www.european-patent-office.org/guidelines/english/*

61. Decision of the Administrative Council Amending the Implementing Regulations to the European Patent Convention, 1999 Off. J. Eur. Pat. Off. 43; 29 Int. Rev. Indus. Prop. & Copy. L. (1998).

62. *Netherlands v. Parliament*, Case C-377-78, 1998 O.J. C 378/13.

63. A. Scott, *Eur. Intell. Prop. Rev.* **4**, 212–215 (1999).

64. I. Muir, M. Brandi-Dohrn, and S. Gruber, *European Patent Law: Law and Procedure under the EPC and PCT*, Oxford University Press, Oxford, UK, 1999.

65. G. Van Overwalle, *The Legal Protection of Biotechnological Inventions in Europe and in the United States: Current Framework and Future Developments, Technical and Ethical Approaches*, Leuven University Press, Leuven, Belgium, 1997.

66. K. Goldbach, H. Vogelsang-Wenke, and F. Zimmer, *Protection of Biotechnological Matter under European and German Law: A Handbook for Applicants*, VCH Publishers, New York, 1997.

67. T 385/86 (Noninvasive measurement/Bruker), 1988 Off. J. Eur. Pat. Off. 308 (September 25, 1987); 1988 Eur. Pat. Off. Rep. 357.

68. T 143/94, *Trigonelline*, 1996 Off. J. Eur. Pat. Off. 430, reprinted in 28 Int. Rev. Indus. Prop. & Copy. L. 95 (1997).

69. T 49/83, Propagating material/Ciba Geigy, 1984 Off. J. Eur. Pat. Off. 112; [1979-85] Eur. Pat. Off. Rep. vol. C. 758.

70. T 320/87, Hybrid plants/Lubrizol, 1990 Off. J. Eur. Pat. Off. 71.

71. T 356/93, Plant Cells/ Plant Genetics Systems, 1995 Off. J. Eur. Pat. Off. 545.

72. R. Nott, *Eur. Intell. Prop. Rev.* **21**, 33 (1999).

73. I. Voelker, *Europe Won't Reverse Controversial EPO Ruling*, IP Worldwide, July/August 1997, Available at: *http://www.ljx.com/patents/7-8europe.html*

74. T 1054/96, In re Novartis, 1998 Off. J. Eur. Pat. Off. 511 (Tech. Bd. Appeal 1997); 30 Int. Rev. Industrial Property and Copyright Law 78 (Tech. Bd. App. 1997).

75. G. Van Overwalle, *J. L. & Tech.* **39**(2), 143–194 (1999).

76. G 1/98, Novartis, 2000 Off. J. Eur. Pat. Off. 111–171 (Enlarged Board of Appeal, December 20, 1999), Available at: *http://www.european-patent-office.org/epo/pubs/oj000/3_00/index_e.htm*

77. Relaxin, 1995 Off. J. Eur. Pat Off. 388 (Exam. & Opp. Div. 1994).

78. *Greenpeace UK v. Plant Genetic Systems NV*, 24 Int. Rev. Indus. Prop. & Copy. L. 618–625 (Opp. Div. 1992).

79. H.R. Jaenichen and A. Schrell, *Eur. Intell. Prop. Rev.* **9**, 345 (1993).

80. Harvard/Onco-mouse, 1990 Eur. Pat. Off. Rep. 501 (Tech. Bd. App.).

81. Harvard/OncoMouse, 1991 Eur. Pat. Off. Rep. 525 (Examining Division).

82. I. Karet, *Eur. Intell. Prop. Rev.* **1**, 21–26 (1997).

83. M. Spence, *Law Q. Rev.* **113**, 368–374 (1997).

84. *Biogen v. Medeva*, 1977 R.P.C. 1 (House of Lords).

85. *Biogen v. Medeva*, 1995 R.P.C. 25 (Patent Court).

86. *Biogen v. Medeva*, 1995 R.P.C. 68 (Ct of Appeals); 27 Int. Rev. Indus. Prop. & Copy. L. 856 (1996).

87. R. Crespi, *Int. Rev. Indus. Prop. Copy. Law* **28**(5), 603–622 (1997).

88. Community Patent Convention, Art. 27.

89. ICI/Medicopharma, Sup. Ct, December 18, 1992, The Netherlands.

90. Ars/Organon, 1997 Int. Rev. Indus. Prop. & Copy. L 702 (Ct of Appeal 1994), affd 1998 Int. Rev. Indus. Prop. & Copy. L 702 (Sup Ct 1996).

91. Clinical Trials II (kirin Amgen/Boehringer Mannheim), 1998 RPC 423 (April 1997).

92. W. Meibom and J. Pitz, *Eur. Intellec. Prop. Rev.* **8**, 469–478 (1997).

93. G. Gauci, *Eur. Intellec. Prop. Rev.* **9**, 361–362 (1998).

94. J. Thomas, in K. Hill and T. Takenaka, eds., *International Perspectives on Intellectual Property: The Doctrine of Equivalents, Enforcement in East Asia and Other Issues*, CASRIP, Seattle, WA, 1997, pp. 189–233.

95. Treaty of Brussels par. 6(1).

96. *EGP v. Boston Scientific* (April 24, 1998).

97. P. Coletti, *J. Pat. Off. Soc.* **81**, 351 (1999).

98. Council Regulation No. 1768/92 of 18 June 1992, Concerning the Creation of A Supplementary Protection Certificate for Medicinal Products, 1992 OJ (L 182) 1-5.

99. H. Von Morze and P. Hanna, *J. Pat. Off. Soc.* **77**, 479 (1995).

100. Council Regulation No. 2100/94 of 27 July 1994 on Community Plant Variety Rights, 1994 OJ (L 227) 1-30.

101. P.A.C.E. van der Kooj, *Introduction to the EC Regulation on Plant Variety Protection*, Kluwer Law International Brentford, UK, 1997.

102. Council Regulation No. 1610/96 of the European Parliament and the Council on the Creation of a Supplemental Protection Certificate for Plant Protection Products, (July 23, 1996), OJ (L 198) 3-35.

103. EU Green Paper on Patents, IP/97/558; Memo 97/65 (June 1997).

104. Trilateral Project B3, Available at: *http//www.european-patent-office.org/tsw/sr-3-b3b.html*

105. Implementing Guidelines for Inventions in Specific Fields: Biological Inventions (Ch. 2), Available at: *http://222.*

jpo-miti.go.jp/guidee/sisine.html (published February 27, 1997).

106. Japan Patent Law reprinted by J.P. Sinnott, *World Patent Law and Practice*, Matthew Bender, New York, 1968.

107. *SmithKline Beecham v. Fujimoto*, 30 Int. Rev. Indus. Prop. & Copy. L. 457 (Tokyo Dist. Ct. October 12, 1998).

108. THK v. Tsubakimoto ("Ball-Spine case"), No. 3292, March 29, 1996.

109. S. Yamamoto and J.A. Tessensohn, *J. Pat. Off. Soc.* **81**, 483 (1999).

110. *Pharmacia & Upjohn Aktiebolag v. Nippon Eli Lilly* (Osaka District Ct., May 27, 1999), Available at: *http://courtdo mino.courts.go.jp/chiteki.nsf/caa027de696a3bd3492567950 07b825/7da56fc63308e18a492567a20013a 5c8? Open document.*

111. Clinical Trial III, 30 Int. Rev. Indus. Prop. & Copy. L. 448 (Japan Sup. Ct. 1999).

112. R. Gunther and D. Yoshida, *Pro-Patent Trend Signaled in Japan, N.Y. Law J.*, S3 (July 26, 1999), citing Kogyo Shoyuken Shingikai [Industrial Property Advisory Committee (IPAC)], Tokkyoho nado no kaisei ni kansuru toshin [Report on Reforms to the Patent Law] (December 16, 1997) [1997 IPAC Report].

113. A. Hirai, *IP Worldwide* 8 (July/August 1998).

114. Japan Civil Procedure Code 220.

115. K. Hagberg, E. Doi, and J. Ishiguro, *Nat. Law J.* C7, (September 14, 1998).

See other INTERNATIONAL ASPECTS entries; MEDICAL BIOTECHNOLOGY, UNITED STATES POLICIES INFLUENCING ITS DEVELOPMENT; OWNERSHIP OF HUMAN BIOLOGICAL MATERIAL; see also PATENTS AND LICENSING entries; SCIENTIFIC RESEARCH, POLICY, TAX TREATMENT OF RESEARCH AND DEVELOPMENT.

M

MEDIA COVERAGE OF BIOTECHNOLOGY

DOROTHY NELKIN
New York University
New York, New York

OUTLINE

INTRODUCTION

In 1924 Edwin E. Slosson, editor of the first science writing syndicate in America, described his view of science journalism. "The public that we are trying to reach is in the cultural stage when three-headed cows, Siamese twins and bearded ladies draw the crowds to the side shows." That is why, he explained, science is usually reported in short paragraphs ending in "-est." "The fastest or the slowest, the hottest or the coldest, the biggest or the smallest, and in any case, the newest thing in the world" (1).

In some respects little has changed in the world of science reporting. In the 1990s journalists still play up the hottest scientific discoveries, the riskiest technologies, and the latest miracle cures. The media coverage of rapid advances in biotechnology and especially in human genetics illustrates these tendencies: extravagant reports about wonder therapies, genetically engineered pigs, and cloned sheep attract readers and sell magazines. "The gene of the week" is "hot" news even though the significance of particular discoveries has often been questioned. The controversial theories emerging from research in behavioral genetics that purport to explain human behavior and social differences are especially newsworthy though their validity is often in doubt.

The media reporting on biotechnology and genetics is mainly promotional, directly reflecting the influence of scientific sources of information. But it is also polarized, swinging from enthusiastic promises to warnings of peril. Journalists are especially attracted to controversies, and new developments in biotechnology have created many provocative and disputed issues. Biotechnology products, intended to control pests or plant diseases and to increase agricultural yields have been controversial because of potential risks to environmental quality or human health. The bioengineering of animals and the creation of new transgenic foods has raised moral and aesthetic objections as well as concerns about risk.

Media coverage of science and technology provides a useful window on public attitudes, but science reporting also has an important influence on public perceptions. The way people perceive research in human genetics or developments in agricultural biotechnology — the way they interpret their costs and benefits — may be influenced less by the details of scientific evidence than by the repeated messages conveyed in the popular press. These media messages help to create the beliefs and assumptions that underlie personal decisions, social policies, and institutional practices.

Following a brief overview of the history of the media coverage of biotechnology, this article will illustrate its most important characteristics by describing five areas that have received considerable media attention. The first widely reported development in biotechnology was the discovery and synthesis of interferon. The media's presentation of news about this therapeutic product was volatile and polarized, a style that has since characterized many reports about biotechnology events. The coverage of research on gene therapy suggests the important influence of scientific sources in shaping the content and tone of science news. The coverage of cloning demonstrates the appeal of drama, myth, and image to the journalists reporting on science events. The media attraction to behavioral genetics suggests the appeal of scientific theories that conform to popular social stereotypes or support prevailing policy agendas. Finally, reports on biotechnology risks, and the problems that may follow from genetic predictions illustrate the media's attraction to policy disputes. A persistent theme pervades this media coverage of biotechnology issues — a concern about the social implications of the growing ties between science and commercial interests.

HISTORICAL OVERVIEW

"After years of being a dowdy old lady, biology has become belle of the ball." Its revolutionary potential has attracted researchers "in droves," and "bankers [are] in hot pursuit." To the media in the early 1980s, biotechnology was expected to become the next economic miracle. But, ironically, only a few years earlier the reports on biotechnology had been more about risks than revolutions. In the mid-1970s molecular biologists held an international meeting at the Asilomar conference center in California to assess the potential risks of recombinant DNA research (2). This was mainly a technical discussion, but the press evoked images of Frankenstein monsters and Andromedalike strains spreading like an incurable disease. Some reporters worried about "warping the genetic endowment of the human race"; others about "biological holocaust." The message? Runaway science needs to be controlled (3).

789

Yet, only a few years later, questions of safety ceased to be news. Journalists dropped the subject, turning their attention to promises and applications. The discovery of ways to synthesize interferon brought reports of miracle cures. Techniques of gene splicing, once represented as dangerous, became "a mundane tool," news and headlines began to tout the potential applications of biotechnology research as "miracles," as the key to better health. In the space of only a few years, media attitudes had markedly changed. In 1977, for example, *Time* magazine had run a cover story called "The DNA Furor: Tinkering with Life" Only three years later a *Time* cover story was called "DNA: New Miracle." Similarly in 1976 the *New York Times* magazine section published an article called "New Strains of Life or Death?" Later a 1980 article in the same section of the *New York Times* was called "Gene Splicing: The Race towards Better Human Health" (4).

By the early 1990s the "runaway science of genetic engineering" had become a "technological frontier." Stories welcomed the patenting of genetically engineered products, the implications of these products for resolving medical, agricultural, and industrial problems, and the proliferation of genetics R&D firms. Local papers touted the importance of these firms to the regional economy. Journalists described geneticists as pioneers, "unlocking the basic laws of nature," discovering the "secrets of life," solving the problems of devastating disease. The biotechnologists working on "high tech veggies" will "do wonders" to help meet nutritional requirements and to enhance the economy. Genomic researchers are "riding the DNA trail." Geneticists are "relentless hunters of genes," involved in a "race" to find the markers for disease. Media accounts are reverent—almost religious: A 1994 cover of *Time* depicts a figure on a pedestal, his arms extended in a Christlike pose, his torso inscribed with a double helix. The caption reads: "Genetics—The Future Is Now." The image, of course, is the Ascension.

Encouraged by the enthusiastic response to research in medical genetics, behavioral psychologists in the 1990s began to publish and to publicize their long controversial studies on the genetic basis of behavioral conditions and personality traits. Always seeking provocative copy, journalists reported these claims, focusing especially on the most controversial—those concerning predisposition to antisocial behavior. And, often uncritically, they drew conclusions about the implications of this research for social policy (5).

Despite the general media enthusiasm about biotechnology, media reports are frequently tempered by doubts. Many journalists, for example, have been critical of the growing links between the biotechnology industry and universities for their affect on open research and they have called attention to the conflicts of interest that are endemic to this field where profits and ethics collide. The media have amplified the critical views of activist Jeremy Rifkin, a persistent and media-savvy biotechnology gadfly. They have extensively reported the protests of animal rights groups against the creation of transgenic animals, and the concerns of religious groups about genetic engineering. In response to the rapid development of genetic tests, media attention turned to the issue of genetic discrimination as tests reveal information about individual predispositions that could influence access to insurance or to jobs (6). And the old Frankenstein metaphors have reappeared in reports about genetically modified foods.

Reporting on biotechnology—from the development of interferon to the creation of clones—the press has batted readers back and forth from biotechnology miracles to visions of apocalypse, from celebrations of progress to warnings of peril, from optimism to doubt. These dramatic shifts in media reports take place as journalists respond to the promotional hyperbole of scientists and technical institutions, but also to protestors and changing popular fashions. This fickle and volatile style of reporting serves the interests of the media in their endless quest for newsworthy and dramatic material. The discovery and synthesis of interferon provided the first opportunity for this style of science journalism.

INTERFERON

Interferon, a protein manufactured in the body when a virus invades a cell, was discovered in 1953 as a natural therapeutic agent, a so-called interfering protein that inhibits infection. The possibility of isolating the protein raised hopes for eventually developing a cancer cure, and this caught media attention. However, the scarce supply of the agent at that time limited scientific progress and clinical possibilities and journalists lost interest until 1975 when Mathilde Krim, a politically astute geneticist, organized a conference intended to publicize the potential of interferon and to win public support for research (7).

Krim's persistent efforts and the growing interest of the American Cancer Society (ACS), which began to sponsor clinical trials, brought a deluge of media coverage in the late 1970s. The research articles in scientific journals explicitly qualified the promises of interferon, indicating the tentative nature of existing studies, the high cost of isolating the protein, and its therapeutic limits. In the popular press, however, interferon became a "magic bullet," a miracle cure for everything from cancer to the common cold.

In 1980 Biogen, a new biotechnology firm, developed a DNA clone for this protein, opening the possibility of producing large quantities of the product at low cost. Uncritically accepting promotional information provided by the company at a press conference, journalists welcomed this new technological development as still another miracle. "Like the genie in a fairy tale," the *Detroit Free Press* told its readers, "science came up with the key to the magic potion." *Reader's Digest* wrote about a "wonder therapy," *Newsweek* about "cancer weapons" and "the making of a miracle drug." Business journalists focused on interferon as a profitable commodity, calling attention to the dramatic increase in stock prices of biotechnology firms. *Business Week* described the efforts to synthesize the substance as a "race" to capture the market: "We have just passed the quarter mile pole and all the horses are in a bunch." *Time* wrote described interferon as a "gold mine" for patients and for biotechnology firms.

The *New York Times* science writer Harold Schmeck, however, broke this promotional pattern. Writing cautious reports, he suggested that while interferon was promising, there was no definitive evidence of its effectiveness. He also reported on possible harmful effects, suggesting that "the seemingly ideal weapon" was less of a panacea than anticipated, and he reported on interferon studies that "put cancer use in doubt" (8). Emphasizing the "modest, controversial, and even negative results of research," Schmeck observed that the promise of a scientific advance was raising research money, but also raising false hopes. In response to this article four scientists from the Sloan Kettering Institute for Cancer Research wrote a letter to the *New York Times*, expressing concern that such qualified reporting could undermine public support of interferon research (9).

The difficulties of using interferon as a therapeutic agent became public in 1982 when four patients in France died after they were treated with the drug. Abruptly, the tone of reporting changed from exaggerated optimism to disillusionment: "From wonder drug to wall-flower." The wonder drug was demoted from a magic bullet to a "research tool." Newspapers and magazine articles assessed the situation pessimistically: "Jury's out on interferon as a cancer cure"; "Studies cast doubt on cancer drug"; "It's a hard row to hoe." Research continued, but little more appeared in the press until a series of patent disputes turned media attention to the question of proprietary interests in commercially promising biotechnology products.

These reports on interferon research demonstrate several themes that have since characterized media reports on biotechnology. First, the content is limited. Little appeared in the press coverage of interferon about the actual nature of the research; instead, most articles appealed to public concerns about cancer and the hopes of soon finding a cure for this dread disease. While interferon's short-term usefulness as a therapeutic agent was problematic, the research did yield significant scientific understanding of basic biological concepts (e.g., the control of gene expression in mammalian cells and the regulation of immunity) that in the long term have affected the practice of medicine. But those readers who followed the interferon story learned little about such developments.

Second, the media coverage of interferon placed great emphasis on scientific and technological competition. Scientists and the firms developing interferon were in a "race" for breakthroughs, for solutions to a dread disease. The gradual accumulation of information that is inherent to the research process was not considered news. Whether the goal is to discover a new genetic marker or to clone a sheep, the media are attracted to the competition in science, the race to be the first to get results.

The media reports about interferon were also volatile. Readers were mainly treated to hyperbole—to a promotional coverage designed to raise their expectations and whet their interest. Scientists played an important role in shaping this coverage. Far from being neutral sources of information, they actively sought a favorable press, equating public interest with research support, and journalists for the most part were inclined to accept the claims of scientific sources. However, when predictions about interferon's curative powers failed to materialize, unqualified optimism in the press quickly shifted to the opposite extreme. This pattern of premature promotion followed by a bitter backlash when optimistic promises fail has continued to be a striking characteristic of the media coverage of gene therapy.

PROMISES OF GENE THERAPY

The media have welcomed claims about gene therapy with extravagant headlines and promotional hype as reporters convey—often uncritically—an array of futuristic scenarios presented by enthusiastic scientists. In the future, said a geneticist to *Discover*, "present methods of treating depression will seem as crude as former pneumonia treatments seem now" (10). In the future, said another scientist, food companies will sell infirmity related breakfast cereals targeted to aid those with genetic predispositions to particular diseases. "Computer models in the home will provide consumers with a diet customized to fit their genetic individuality, which will have been predetermined by simple diagnostic tests" (11). In the future, geneticist French Anderson told a *Time* Reporter, "Physicians will simply treat patients by injecting a snippet of DNA and send them home cured" (12).

To the media, the National Institutes of Health (NIH) Human Genome Initiative has replaced the NASA Space program as a new frontier, the cutting edge of high-technology exploration, and just as in the heyday of NASA, journalists are receptive to scientific enthusiasm. On their part, scientists seek media coverage as a form of public relations, regarding public visibility as a means to attract funds for their research. Thus scientists and the press offices of their institutions are inclined to turn preliminary experimental findings into magic bullets. For example, in 1995 the Scripps Research Institute issued a press release announcing that researchers had found a cure for cancer through a small injection of a protein that would cut off the blood supply from tumors and cause them to shrink while leaving normal tissue intact. The announcement about a cancer cure, designed to attract media attention, was, it turned out, only about a laboratory observation; there had been no experimental trials testing the relevance of the observation to human pathology. But as anticipated, the press release was covered as welcome news in the press (13).

Traditionally working in a context where success is measured by the judgment of peers, scientists have long assumed that a record of accomplishment is sufficient to maintain research support. Thus information, the scientist's "stock-in-trade," has been directed primarily toward professional colleagues. Most scientists have not been interested in public visibility; on the contrary, they have feared it could result in external controls on their work. And "visible scientists"—those who seek media attention—have often been marginalized or disdained (14). But attitudes in the scientific community have changed. Dependent on direct congressional appropriations or, increasingly, on corporate support of research, many scientists now believe that scholarly communication is no longer sufficient to maintain their enterprise. They see gaining national visibility through the mass media as crucial to securing the financial support required to run major research facilities and to assuring favorable public policies toward science and technology.

Geneticists seeking to maintain public support have become skilled in rhetorical strategies designed to attract the media. In media interviews they have described the genes as "master molecules:" We are but "readouts" of our genes. They describe the body in deterministic terms — as a set of instructions, a blueprint, a map, or a program that is transmitted from one generation to another. They suggest that by deciphering the text, classifying the markers on the map, and reading the instructions, they will unlock the key to human ailments and human nature, revealing the secrets of human life. Molecular biologist Walter Gilbert, introduces his public lectures on gene sequencing by pulling a compact disk from his pocket and announcing to his audience: "This is you." Scientists, he claims, will "provide ultimate answers to the commandment, know thyself" (15).

Geneticists also emphasize the predictive powers of their science by calling the gene "a Delphic oracle," "a time machine," "a trip into the future," "a medical crystal ball." James Watson, the first director of the Human Genome Project, has announced in frequent media interviews that "our fate is in our genes." The metaphors scientists use to describe their work convey several messages about the meaning of genetics that have been widely disseminated through a receptive press: a definition of the gene as an essentialist and deterministic entity, and a promise that genetic research will enable the prediction of future behavior and disease and thus lead to therapeutic solutions.

The biotechnology industry has further encouraged media hype about gene therapy. The industry has made a major financial commitment to gene therapy, expecting this will be the basis of future medicine. Over 60 percent of gene therapy studies are directly financed by industry in anticipation of a profitable market in the near future. Corporate advertisements announce "a great leap in the treatment of disease," and promise "a healthy future one gene at a time." Immediately following the discovery of the mutation in the BRCA-1 breast cancer gene, one pharmaceutical company announced in a newspaper advertisement that they had made progress in finding a "breakthrough . . . new treatments and ultimately the cure for breast cancer." Another advertisement appearing in sports magazines, said it all: "'Bad Genetics?' Use Optigenetics — 'The first genetic optimizer.'"

The development of pharmaceutical products and the proliferation of clinical trials on new therapeutic procedures have encouraged the tendency towards technological optimism in the press. For example, the introduction of therapeutic molecules, especially TPA (tissue plasminogen activator) for dissolving clots, and the use of human growth hormone to treat dwarfism became newsworthy issues. Then the first FDA-approved gene therapy experiment in 1990 — the injection of cells containing ADA genes in a child with an immune system malfunction — became a major news event. "The long awaited era of genetic therapy has at last arrived" said a writer in *The Sciences* (16).

Discover called gene therapy, "The Ultimate Medicine." Writing on gene-transfer techniques, the reporter proclaimed, "Genetic surgeons can now go into your cell and fix those genes with an unlikely scalpel: a virus." Interviewed for this article, molecular biologist Richard Mulligan declared that "We can use gene transfer to make a cell do whatever we want. . . .We can play God in that cell." Similarly *US News* told its readers that gene therapy is the medicine of the future. "No disease has given up more of its secrets to genetic sleuths than cancer." Genetics, promised the writer, will allow doctors to "do something" about the disease. The isolation of the colon cancer gene in 1993 prompted an enthusiastic scientist to tell a *New York Times* reporter of its implications: "Deaths are entirely preventable" (17).

When they report on complex scientific issues, journalists rely heavily on press releases, often responding with uncritical enthusiasm to promotional hype. "Genetic Research Leaves Doctors Hopeful for Cures," "New Hope for Victims of Disease," "Genetics, the war on aging . . . [is] the medical story of the century . . . Genetic technologies will dramatically curtail heart disease, aging, and much more" (18). In a story called "the Age of Genes," *US News* reported that "advances bring closer the day when parents can endow children not only with health but also with genes for height, good balance, or lofty intelligence" (19).

While promises, backed by scientific authority, raise hopes of instant cures, aside from controlling reproduction there is little that can be done to cure genetic disease. The gap between identifying a predisposition to a genetic condition and finding a therapeutic solution is very wide. The problems of clinical application follow in part from the complexity of genetic diseases. Some are caused by the absence of the activity of a gene product, others by altered proteins that disrupt cellular function, and still others by alterations of chromosomal structure. The early expectations about a successful therapy for cystic fibrosis, reported widely in the press, were confounded by the fact that this disease has many more mutations than originally anticipated.

There are also problems in finding safe vectors capable of transporting genes into targeted cells. Most gene therapies use viruses as the carrier mechanism, for this is the most feasible way of targeting appropriate cells. But there are risks. Inserting a gene in the wrong place along a strand of DNA could cause an undesirable effect. And the immune system may attack cells treated with gene therapy, responding to them as foreign or infected. Gene transfer experiments on monkeys were found to cause malignant T-cell Lymphoma. Clinical trials of Genentech's promising drug called Pulmozyme, developed to treat cystic fibrosis, were halted when they found significant mortality rates among treated patients.

Discovering risks and side effects is, of course, the purpose of animal research and clinical trials. But in an area hyped by both scientists and the media as "the medicine of the future," failures become more than routine science; they also become a newspeg for journalists who seldom convey to their readers the fact that a failed clinical trial can itself yield useful information. The death of a gene therapy patient in 1999 brought an abrupt end to the gene therapy hype.

The growing realization of the practical difficulties of extending laboratory studies to clinical applications is puncturing inflated expectations, and this is reflected in skeptical media reports. Describing the research on the unusual frequency of a mutation in the BRCA1 gene among Ashkenazy Jewish women, reporters commented again and again on the absence of effective therapies. "Does it make sense to screen healthy women for the defect given that there is no good therapy to offer those in whom it is found?"

Some media reports associate experiments in gene therapy with genetic engineering or "tampering" with genes. "Lurking behind every genetic dream come true is a possible Brave New World nightmare," says a *Time* reporter. "To unlock the secrets hidden in the chromosomes is to open up the question of who should play God with man's genes." An accompanying image portrayed scientists balancing on a tightrope of coiled DNA (20). And an illustration for a *New York Times* article on gene therapy and the potential of genetic engineering featured a drawing imitative of the famous Edvard Munch painting, "The Scream." A figure stands, horrified, mouth ajar, eyes wide open, her hair a mass of coiled DNA (21). Such images proliferated in the reports about the creation of Dolly, the first cloned sheep.

CLONING

This section "Cloning" has been adapted from Ref. 22. The consequences of cloning have long captured the popular imagination: cloning has been a major theme in horror novels and science fiction films. Genetic engineering research that has been associated with cloning has met a critical press. For example, in 1993 scientists at a George Washington University laboratory conducted a genetic engineering experiment that "twinned" a nonviable human embryo. The purpose was find a way to create additional embryos for in vitro fertilization, but major newspapers, popular magazines and talk shows covered the experiment as if it had actually yielded a cloning technology for the mass production of human beings. The media response was remarkable and diverse. The *Los Angeles Times* announced the glorious news that "infertility, virginity and menopause are no longer bars to pregnancy" (23). But also envisioned were embryo and selective breeding factories, cloning on consumer demand, breeding of children as organ donors, a cloning industry for selling multiples of human beings, and even a "freezer section of the biomarket" (24). *Time*, wrote of the "Brave New World of cookie cutter humans" (25). And repeatedly, scientists in media reports were accused of "playing God."

Then, in February 1997, scientists at the Roslin Institute in Edinburgh cloned Dolly, after 276 attempts, from the genetic material of a six-year-old sheep. The media response to the production of a sheep by cloning a cultured cell line reflected futuristic fantasies and Frankenstein fears about science and especially about genetic engineering. The meaning of Dolly that was conveyed by the media reflected a pervasive assumption of "genetic essentialism" — that human beings in all their complexity are simply readouts of a molecular text, that

human identity is contained entirely in the sequences of DNA in the human genome (26). Thus, speculated journalists, why not clone great athletes like Michael Jordan, or great scientists like Albert Einstein, or popular politicians like Tony Blair, or less popular politicians like Newt Gingrich, or wealthy entrepreneurs like Bill Gates. Some reporters lauded cloning as a way to assure immortality. Again and again, media stories predicted that cloning will allow the resurrection of the dead (e.g., bereaved parents might clone a beloved deceased child). Or the technology could provide life everlasting for the deserving (narcissists could arrange to have themselves cloned).

But there were also anxious scenarios developed in the press, including futuristic stories about making new Frankenstein monsters, or creating Adolph Hitler clones, or producing "organ donors" only to harvest their (fully compatible) viscera (27). While cloning could theoretically make both sexes irrelevant to reproduction, the technology appeared as a threat to the male of the species — men would no longer be necessary! It also held a promise of creating perfect cows, sheep, and chickens, or perhaps even perfect people. If sperm banks (as portrayed in some women's magazines) were a place to "shop for Mr. Good genes," why not, asked reporters, use cloning to produce and reproduce perfect babies?

Journalists elicited views from people in various professions about the implications of cloning for their fields. A divorce lawyer predicted the doubling of his business. Historians wondered if the founding fathers could be cloned for display in a "living history" exhibit in a theme park: They suggested that the park might be called "Clonial Williamsburg." Some facetious policy commentators announced that cloning experiments could be developed to solve social problems: The race problem could be resolved by manipulating the balance between melanin and IQ genes. The age-old nature-nurture dispute could be definitively settled by creating clones and raising them systematically in different environments.

News articles covered religious perspectives on cloning (28). One writer quipped that cloning offered a "second chance for the soul." If you sin the first time, try again. But a theologian, Rabbi Mosher Tendler, a professor of medical ethics at Yeshiva University in New York City, warned in a news interview that "whenever man has shown mastery over man, it has always meant the enslavement of man." Other theologians, long concerned about the implications of genetic engineering, worried that the scientists who experimented with cloning were "playing God" and "tampering with God's creation." Articles in evangelical magazines such as *Christianity Today* or *The Plain Truth* have regular articles opposing genetic engineering as "tampering" with genes." Pope John Paul II has taken a position on genetic manipulation, arguing that: "All interference in the genome must be done in a way that absolutely respects the specific nature of the human species, the transcendental vocation of every being and his incomparable dignity..." (29).

In his scientific paper itself, Dr. Wilmut called attention to the problem of whether "a differentiated adult nucleus can be fully reprogrammed." He called the lamb in question

6LL3 rather than Dolly, and made it clear, in diagrams and illustrations of gels, that there is some question about the precise genetic relationship between Dolly and the "donor" (30). Somatic DNA, which was the source of Dolly's genes, is constantly mutating. Dolly, in fact, may not be genetically identical in every way to her "mother," a point that is of some importance for the possible agricultural applications of cloning techniques.

For the media, however, such technical details were less important than symbolic associations. The cloning of a lamb was immediately set in a context of other fears about genetics and genetic manipulation, and even in a context of more general fears about science and its applications. One journalist compared cloning to weapons development. Another worried that the shortage of organs for transplantation would be resolved by cloning anencephalic babies (who are born without a brain but are otherwise normal) so that their organs could be harvested for patients in need. And a writer for *Newsweek* related the creation of Dolly to broader concerns about food biotechnology by speculating about "cloned chops" (31).

Dolly also evoked an amazing range of media humor. A *New York Times* journalist interviewed writer Wendy Wasserman who wondered what you would say to your shrink if you are your own mother (32). A cartoonist in the London *Guardian* depicted a women comforting a cab driver who had just run over her husband: "That's alright, I have another one upstairs." A writer predicted a new action movie called "Speed Sheep" in which thousands of cloned sheep clogged Interstate 95. Headlines of cloning stories revelled in puns: "An udder way of making lambs," "Send in the clones," "Little Lamb, who made thee?" "Will there ever be another ewe?" and "Getting stranger in the manger." And inevitably there was the anticipation of "Double Trouble."

More pointed jokes — as well as serious editorial commentaries — expressed the growing tensions over commercial control of biotechnology and its implications for the commodification of the body. Just as the GWU experiment evoked images of a cloning industry and breeding factories, so Dolly evoked cynical references to "test tube capitalists," and sardonic queries about a market for genetic "factory seconds" and "irregulars." Meanwhile *Business Week* anticipated "The Biotech Century" in which cloning animals is just the beginning: "It's all happening faster than anyone expected" (33).

As in the media coverage of gene therapy, the messages evoked by Dolly have ranged from promises of miracles to portents of disaster. Editorial appeals called for regulation and for a moratorium on cloning experiments. As political and social pressures began to grow, scientists responded, defending the importance of the work. Media images were "selling science short." The calls for regulations and restrictions, some argued, ignored the medical benefits that could follow from cloning experiments and their potential contribution to the development of life-saving treatments and the testing of new drugs. We are not interested in playing God, said James Geraghty, president of the biotechnology firm, Genzyme, but in "playing doctor" (34). Mammalian cloning could help to generate tissue for organ transplantation and encourage

transgenics experimentation. And certainly research using cloning would enhance scientific knowledge about cell differentiation. The politicians who sought a ban on cloning research, said the scientists, were "shooting from the hip."

But media coverage continued to reflect mistrust of this kind of science, concern that commercial interests would ignore social considerations, and fear that the outrageous possibilities suggested by a cloned sheep will eventually, perhaps inevitably, be realized. News reports and media headlines suggested that "Science fiction has become a social reality." "Whatever's Next?" And, inevitably, "Pandora's Box."

Dolly, for a very brief period had reinforced media myths about science — evoking both euphoric fantasies and horrible nightmares and eliciting a fear of science out-of-control. Yet, only a few months after the media blitz, Dolly and the problems of cloning ceased to be news. By June 1997, when the National Bioethics Advisory Commission appointed by President Clinton reported its recommendations to continue the moratorium on federal research funding in this area, and to consider federal legislation, the media had lost interest and paid little attention to the report. In July 1997 the British scientists who had cloned Dolly cloned Polly and three other lambs, this time from fetal rather than adult cells. Polly, a transgenic lamb, carries a human gene in her cells. But cloning was already old news, and this event was reported as merely one more technical advance — the fusion of a fibroblast cell from a fetus to an egg cell. For the media, cloning, even when it involved a human gene, was accepted as routine.

Dolly in the media had been more than a biological entity; she became, ever so briefly, a cultural icon, a symbol, a way to define the meaning of personhood and to express concerns about the forces shaping our lives. She provided a window on popular beliefs about human nature, on public fears of science and its power in society, and on concerns about the human future in the corporate-driven climate of the biotechnology age. A more lasting preoccupation for the media has been research on the genetics of human behavior, research that purports to explain age-old questions about human differences and offers tantalizing, if problematic, prospects for developing technical solutions to social problems.

GENETICS OF HUMAN BEHAVIOR

The language of biological determinism is pervasive in the press. A media survey found references — some more plausible than others — to the genetic basis of shyness, directional ability, aggressive personality, caring tendencies, exhibitionism, homosexuality, dyslexia, job success, arson, traditionalism, preferred styles of dressing, tendencies to tease, political leanings, religiosity, criminality, intelligence, social potency, and zest for life (26). The media refers to selfish genes, pleasure-seeking genes, criminal genes, celebrity genes, homosexual genes, couch potato genes, depression genes, genes for genius, genes for saving, and even genes for sinning. They are presented in the media as if they are simple Mendelian disorders, directly inherited like brown hair or blue eyes.

Theories of behavioral genetics seek to explain human differences, and like other theories purporting to explain race or gender differences, they have enjoyed a very active press. The idea of biological determinism had attracted considerable news coverage following the controversy over Jensen's claims about the relationship between race and IQ. It later reappeared as sociobiology in media reports that were less concerned with substance than with provocative images, social implications, and policy applications. In selecting this subject for extensive coverage, journalists in effect have used a controversial theory to legitimize a particular point of view about the importance of biological determinism.

Sociobiology is a field devoted to the systematic study of the biological basis of social behavior. Its premise is that behavior is shaped primarily by genetic factors, selected over thousands of years for their survival value. Its most vocal proponent, E.O. Wilson from Harvard University, contends that genes create predispositions for certain types of behavior and that a full understanding of these genetic constraints is essential to intelligent social policy. He believes that sociobiology is "a new synthesis," offering a unified theory of human behavior. "The genes hold culture on a leash," he wrote in his book *On Human Nature*. "The leash is very long but inevitably values will be constrained in accordance with their effects on the human gene pool" (35).

During the early 1980s Wilson's arguments about human behavior, extrapolated from his research on insect behavior, were attacked by other scientists for their purported justification of racism and sexism, their lack of scientific support, and their simplistic presentation of the complex interaction of biological and social influences on human behavior (36). But the 1980 publication of Wilson's first book on the subject, *Sociobiology, A New Synthesis*, was welcomed in the *New York Times* as a "long awaited definitive book." Subsequently sociobiological concepts appeared in newspaper and magazine articles about the most diverse aspects of human behavior. They were used, for example, to explain:

- Child abuse. "The love of a parent has its roots in the fact that the child will reproduce the parent's genes." (*Family Week*)
- Machismo. "Machismo is biologically based and says in effect: 'I have good genes, let me mate.'" (*Time*)
- Intelligence. "On the towel rack that we call our anatomy, nature appears to have hung his-and-hers brains." (*Boston Globe*)
- Promiscuity. "If you get caught fooling around, don't say the devil made you do it. It's your DNA." (*Playboy*)
- Selfishness. "Built into our genes to insure their individual reproduction." (*Psychology Today*)
- Rape. "Genetically programmed into male behavior." (*Science Digest*)
- Aggression. "Men are more genetically aggressive because they are more indispensable." (*Newsweek*)

The press has been especially attracted to sociobiology's controversial implications for understanding stereotyped sex differences. The theory, we are told, directly challenges women's demands for equal rights, for differences between the sexes are innate. *Time*, for example, tells its readers that "Male displays and bravado, from antlers in deer and feather-ruffling in birds, to chest thumping in apes and humans, evolved as a reproductive strategy to impress females" (37). And a *Cosmopolitan* reporter, citing the "weight of scientific opinion" to legitimize his bias, writes: "Recent research has established beyond a doubt that males and females are born with a different set of instructions built into their genetic code" (38).

The media have readily picked up on every research project suggesting there might be a genetic basis of sex differences. In 1980, for example, two psychologists, Camille Benbow and Julian Stanley, published a research paper in *Science* on the differences between boys and girls in mathematical reasoning (39). Their study, examining the relation between Scholastic Aptitude Test scores and classroom work, found that differences in the classroom preparation of boys and girls were not responsible for differences in their test performance. The *Science* article qualified the implication of male superiority in mathematics: "It is probably an expression of a combination of both endogenous and exogenous variables. We recognize, however, that our data are consistent with numerous alternative hypotheses." But the popular press was less qualified, writing up the research as a strong confirmation of biological differences and a definitive challenge to the idea that differences in mathematical test scores are caused by social and cultural factors. The newspeg was not the research but its implications.

The authors themselves encouraged this perspective in their interviews with reporters, where they were less cautious than in their scientific writing. Indeed, they used the press to push their ideas as a useful basis for public policy. According to the *New York Times*, they "urged educators to accept the possibility that something more than social factors may be responsible. ...You can't brush the differences under the rug and ignore them" (40). The media were receptive. *Discover* reported that male superiority is so pronounced that "to some extent, it must be inborn" (41). *Time*, writing in 1980 about the "gender factor in math," summarized the findings: "Males might be naturally abler than females" (42).

The most striking feature of the media articles on sociobiology was how easily reporters slid from noting a provocative theory to citing it as fact, even when they knew that the supporting evidence was flimsy. A remarkable article called "A Genetic Defense of the Free Market" that appeared in *Business Week* clearly illustrates this slide. While conceding that "there is no hard evidence to support the theory," the author wrote: "For better or worse, self-interest is a driving force in the economy because it is engrained in each individual's genes. ...Government programs that force individuals to be less competitive and less selfish than they are genetically programmed to be are preordained to fail." The application of sociobiology that he calls "bio-economics" is controversial, he admitted; nevertheless, it is "a powerful defense of Adam Smith's laissez-faire views" (43).

The journalists who write about the genetics of behavior recognize, indeed rely on, the existence of controversy to enliven their stories. Yet most articles convey a point of view by giving space to advocates and marginalizing critics — often described as "few in number but vociferous," or people who are "unwilling to accept the truth." In 1976, for example, *Newsweek* suggested that Wilson was a victim like Galileo: "The critics are trying to suppress his views because they contradict contemporary orthodoxies" (44). In 1982 *Science Digest* compared the criticism of sociobiology to the attacks by religious fundamentalists on the theory of evolution: "Like the theory of evolution, sociobiology is often attacked and misinterpreted" (45). This a comparison places sociobiology's scientific critics, such as Stephen J. Gould and Richard Lewontin of Harvard University, in the same league as William Jennings Bryan.

In a seamless transition, the sociobiological ideas that appealed in the early 1980s have drifted into genetic explanations of social behavior and human differences. During the 1990s the media have disseminated the ideas about the inherited basis of behavior that were generated by studies of identical twins reared apart. These controversial studies gained both media attention and public credibility as part of the growing popular interest in genetics. Offering a simple explanation of complex behavior and a reinforcement of prevailing stereotypes, these ideas appealed to the media that began to attribute an extraordinary range of behaviors to "the genes." *US News and World Report* published an authoritative-looking table providing precise percentages of how much personality traits were determined by heredity rather than culture: extroversion 61 percent, conformity 60 percent, worry 55 percent, creativity 55 percent, optimism 54 percent, and so on (47). In the same issue, *US News* published an article called "How Genes Shape Personality," claiming that "solid evidence demonstrates that our very character is molded by heredity." It suggested that the future of Baby M, the child in a controversial surrogacy dispute, may not rest on which family got her, but in her genes (46). In 1992 *Time* once again offered an explanation of sex differences: "Nature is more important than nurture" and it is just a matter of time until scientists will prove it (47).

The concept of genetic predisposition has appeared to explain a range of personality characteristics. A 1993 *New York Times* article was called: "Want a room with a view? Idea may be in the genes" (48). But especially attractive to the media are explanations associating aggression and violent behavior with biological predisposition. Throughout the coverage of behavioral genetics are references to "bad seeds," "criminal genes" and "alcohol genes." To a *New York Times* writer "evil is embedded in the coils of chromosomes that our parents pass to us at conception" (49). And in a news report of a murder involving the arrest of a 14-year-old high school boy from a "good home," the *New York Times* interpreted the event as a key piece of evidence in "the debate over whether children misbehave because they had bad childhoods or because they are just bad seeds." The reporter used the power of inheritance to explain the incident: "Raising Children Right Isn't Always Enough"

read the headline. The implications? There are simply "bad seeds" (50).

The acceptance, indeed promotion of genetic explanations of behavior, reflects in part the media's idealization of science as an ultimate authority. But it also reflects the tendencies of science to justify social stereotypes and popular policy agendas. By its selection of what theories to champion, the press in effect uses the imprimatur of science to support a particular world view. It does so, however, with little attention to the substance of science, its slow accumulative process, and the limits of these theories as an adequate explanation of complex human behavior, shaped by multiple genetic and environmental influences.

THE REPORTING ON BIOTECHNOLOGY RISK

The media coverage of biotechnology has been, in large part, enthusiastic, optimistic, and, indeed, promotional. Yet there remains a persistent and pervasive ambivalence about the implications of this rapidly developing field. Though journalists raise few questions about the ultimate benefits of genetics research or the credibility of research claims, they frequently question the potential abuses of genetic information. We read of the importance of genetic explanations of disease, and then are warned that the ability to identify genetic predisposition is far ahead of therapeutic possibilities. Stories extol the benefits of genetic research, but then decry the risk of gathering genetic information. We are told that this is the dawn of a new genetic era, and then cautioned about an impending eugenic nightmare.

Journalistic attention has often focused on the risks of biotechnology, especially in the area of agriculture and food production where biotechnology applications have been a target for public demonstrations. Some of this reporting is futuristic — abstract speculations about the possible harm of bioengineered products that are yet to appear. But journalists have mainly reported on existing controversies. One of the earliest biotechnology disputes focused on the field testing of Ice Minus, the genetically altered microbes that were developed to inhibit ice crystallization so as to protect strawberries and other fragile crops from injury from frost. Environmental groups, concerned about health hazards, were strongly opposed to these tests. Attracted to a growing controversy and to public demonstrations, news reports about the Ice Minus field tests included striking and provocative photographs of the workers who were spraying the fields, wearing protective clothing that resembled the moon suits associated with the cleanup of toxic chemicals and nuclear wastes (51).

Opposition to the bioengineered Flavr Savr tomato also gained considerable media attention. As in the case of cloning, the issue appealed to journalists as much for the irresistible potential for puns as for evidence of real risk. The genetically engineered tomato, introduced by the biotechnology firm Calgene in late 1991, was initially welcomed in the press as a fruit that would not rot on the way to the market. The product generated media stories on the "wonders" of high technology foods — leaner meat, celery sticks without strings, crisper and sweeter vegetables — and the press supported Calgene's effort to classify its product as a food rather than a drug subject to FDA regulations. But then, as critics of biotechnology moved in, skepticism became fashionable, and journalists began to write about the tomato as a "frankenfood," a "killer tomato." There was a "tomato war" and a "tomatogate" (52). The idea of injecting mouse genes into food, the spectacle of restaurant chefs boycotting a tomato, the concern about "safe soup," attracted reporters who covered this product as an example of the risks of biotechnology. The business press responded by denouncing "crackpots and scaremongers" who hold back the "wheels of progress" by playing on public fears.

The bioengineering of transgenic animals was also controversial especially among animal rights activists whose colorful antics have often attracted the attention of reporters (53). The media uncritically followed the antics of animal rights groups who projected images of composite cattle, "geeps" (half goat, half sheep) and grossly oversized, distorted pigs. Reporters also addressed the concerns of small farmers who believed that costly applications of biotechnology advances would give further economic advantage to agribusiness, and the opposition of religious groups who worried about the meaning of scientists tampering with nature and "playing God." Here again, media-savvy Jeremy Rifkin was able to use the press to attract publicity for his antibiotechnology campaign. But, as we saw in the evolution of media coverage of cloning disputes, reporting on transgenic foods and animals is usually focused on the newest or most dramatic events. The media, for example, only briefly mentioned Polly, the ultimate transgenic animal who had been cloned in 1997 with human genes.

In some striking ways the images pervading the media coverage of biotechnology risks are remarkably similar to those that were projected during the nuclear power controversy—Frankenstein monsters, mutant animals, mad scientists, and consumers without choices who are captive to an industry portrayed as out of control. Reporting in both of these areas has suggested that risk in the media is often a surrogate issue. Fears of biotechnology are linked to ethical and religious issues, to concerns about economic inequities, and to deep mistrust of a commercially driven science. And media reporting reflects the sensitive question of consumer choices about food and environmental quality—controversial issues of considerable interest to newspaper readers.

CONCLUSION

The media serve, in effect, as brokers between science and the public, framing social reality for their readers and shaping the public consciousness about science-related events. They are for most people the only accessible source of information about important scientific and technical choices. Through their selection of news about science and technology, the media help set the agenda for public policy. Through the information they convey about biotechnology risks, they may affect stock market prices in a volatile industry and influence product sales. And through their presentation of science news, they shape personal attitudes and public actions.

Thus scientists and the companies involved in biotechnology research have been extremely sensitive to their image in the press. And like advocates in any field, they are prone to overestimate the benefit of their work and minimize its risks. Hoping to shape that image, they have become adept at packaging information for journalists. It was not journalists, but scientists who initially employed attention-seeking metaphors to describe the genome as a "blueprint of life," a "Book of Man," a "medical crystal ball." Geneticists themselves have promoted the "gene of the week" and touted the latest therapeutic possibilities. And agricultural researchers were the first to promote the economic benefits of cloning. Courting media attention, those engaged in biotechnology and genetics research have also helped to evoke premature enthusiasm and optimistic expectations. The media have not created the science news; they have mainly amplified and disseminated the messages conveyed by scientists themselves.

Most journalists have limited knowledge about science, and they are vulnerable to manipulation by their sources of information. But as the above review suggests, the media are conveying mixed messages about the costs and benefits of biotechnology. Journalists seem to welcome the notion of biological determinism—simple, startling, and easy to convey—yet many writers remind their readers of the history of eugenics and place current research in this threatening historical context. While journalists report with enthusiasm and wonder the promises of gene therapy, they also warn about potential abuses of genetic manipulation. The media have welcomed agricultural innovations, but they have also warned about potential health or environmental risks.

Perhaps the most striking feature of the media reporting on biotechnology is a pervasive concern about the social and economic context in which this field is developing. The ties between the science of genetics and its commercial applications have invited widespread media cynicism. Reporters have repeatedly called attention to the nonscientific interests—the investments and profits in an intensely competitive field—that are driving biotechnology and its clinical and agricultural applications. Science journalists have long maintained an image of academic science as a pure and unsullied profession, a neutral source of authority, and an objective judge of truth (4). They are skeptical of corporate driven science. These days, according to some disillusioned journalists, scientists working in biotechnology and genetics are "greedy entrepreneurs" or "molecular millionaires" driven by economic interests that threaten their objectivity and override concerns about abuse. Political cartoons portray geneticists in less than flattering terms as bumbling, naive, and unaware of the social implications of their discoveries. And news reports and editorials repeatedly call attention to troubling aspects of the growing links between science and industry, the conflicts of interest that are inevitable when profits and ethics collide.

The context of science, especially in the fields of biotechnology and genetics, has radically changed in recent years. And media coverage, appropriately, is beginning to reflect the implications of these changes.

The media have retained their history of technological optimism, but they are also expressing a growing concern that the expanding commercial interests in these profitable areas of science will overide important social considerations.

BIBLIOGRAPHY

1. E. Slosson, quoted in D.J. Rhees, Master's Thesis, University of North Carolina, Raleigh, 1979.
2. S. Krimsky, *Genetic Alchemy*, MIT Press, Cambridge, MA, 1982; M. Rogers, *Biohazard*, Knopf, New York, 1977.
3. M. Altimore, *Sci. Technol. Hum. Values* Fall, 24–31 (1982).
4. D. Nelkin, *Selling Science: How the Press Covers Science and Technology*, 2nd ed., Freeman, New York, 1995.
5. D. Nelkin, in M. Rothstein, ed., *Biology and Culture*, Johns Hopkins University Press, Baltimore, MD, 1998.
6. D. Nelkin and L. Tancredi, *Dangerous Diagnostics*, 2nd ed., University of Chicago Press, Chicago, IL, 1994.
7. S. Panem, *The Interferon Crusade*, Brookings, Washington, DC, 1984.
8. *N.Y. Times*, May 27, 1980.
9. *N.Y. Times*, Letter, June 17, 1980.
10. L. Wingerson, *Discover*, February, 60–64 (1982).
11. F. Clydesdale, *Food Technol.*, September, 134–136 (1989).
12. *Time*, January 17, 1994.
13. L. Altman, *N.Y. Times*, January 10, 1995.
14. R. Goodell, *The Visible Scientists*, Little, Brown, Boston, MA, 1977.
15. W. Gilbert, *Current State of the HGI*, Harvard University Dibner Lecture, Harvard University, Cambridge, MA, 1990.
16. K.W. Culver, *Sciences* **1**, 18–24 (1991).
17. N. Angier, *N.Y. Times*, December 3, 1993.
18. A. Rosenfeld, *Longevity*, May, 42–53 (1992).
19. S. Brownlee and J. Silberner, *U.S. News and World Rep.*, November 4, 64 (1991).
20. P.E. Dewitt, *Time*, March 20, 70 (1989).
21. Cartoon by S. Goldenberg, *N.Y. Times*, September 16, 1990.
22. Adapted from D. Nelkin and S.M. Lindee, *Cambridge Rev.* (1998).
23. *Los Angeles Times*, October 27, 1993.
24. *Review of Media Coverage of Human Embryo Cloning*, Newsletter, January 1, Center for Biotechnology Policy and Ethics, Texas A&M University, College Station, 1994.
25. *Time*, November 8, 1993.
26. D. Nelkin and S. Lindee, *The DNA Mystique: The Gene as Cultural Icon*, Freeman, New York, 1995.
27. R. Langreth, *Wall Street. J.*, February 24, 1997.
28. G. Niebuhr, *N.Y. Times*, March 1, 1997.
29. Address to the Pontifical Academy of Sciences, October 8, 1994, quoted in *Family Resource Center News*, Winter (1996).
30. I. Wilmut et al., *Nature (London)* **35**, 810–812 (1997).
31. L. Reibstein and G. Beals, *Newsweek*, March 10, 58–59 (1997).
32. Quoted in *N.Y. Times*, February 27, 1997.
33. Cover story of *Business Week*, March 10, 1997.
34. J. Geraghty, quoted in *Gene. Eng. News*, April 1, 10 (1997).
35. E.O. Wilson, *On Human Nature*, Harvard University Press, Cambridge, MA, 1980.
36. U. Segerstrale, *Biol. Philos.* **1**, 53–87 (1986).
37. *Time*, August 1, 1977.
38. *Cosmopolitan*, March 1982.
39. C. Benbow and J. Stanley, *Science* **210**, 1262–1264 (1980).
40. *N.Y. Times*, December 7, 1980.
41. P. Weintraub, *Discover*, April, 15–20 (1981).
42. *Time*, December 15, 1980.
43. *Business Week*, April 10, 1978.
44. *Newsweek*, April 12, 1976.
45. *Science Digest*, March 1982.
46. *US News and World Report*, April 13, 1987.
47. *Time*, January 20, 42 (1992).
48. *N.Y. Times*, November 30, 1993.
49. D. Franklin, *N.Y. Times Mag.*, September 3, 36 (1989).
50. M. Newman, *N.Y. Times*, December 22, 1991.
51. S. Krimsky, *Biotechnology and Society*, Praeger, New York, 1990.
52. *Barrons*, May 31, 1993, and the *N.Y. Times Mag.*, November 21, 24 (1993).
53. J. Jasper and D. Nelkin, *Animal Rights Crusade*, Free Press, New York, 1992.

See other entries EDUCATION AND TRAINING, PUBLIC EDUCATION ABOUT GENETIC TECHNOLOGY; INTERNATIONAL INTELLECTUAL PROPERTY ISSUES FOR BIOTECHNOLOGY; PUBLIC PERCEPTIONS: SURVEYS OF ATTITUDES TOWARD BIOTECHNOLOGY.

MEDICAL BIOTECHNOLOGY, UNITED STATES POLICIES INFLUENCING ITS DEVELOPMENT

ROBERT MULLAN COOK-DEEGAN
Georgetown University
Annapolis, Maryland

OUTLINE

Introduction

Definitions

Factors Affecting the Commercial Development of Biotechnology

Capital Availability

Funding for Research and Development

Patents

University–Industry Relations

Regulation of Products and Services

Public Perception and Political Process

Bibliography

INTRODUCTION

National policies, including but not restricted to government rules and actions, have profoundly influenced the pace and direction of biotechnology. Medical products and services were among the first and most significant applications of biotechnology. The economic impact of new biotechnologies was felt first in pharmaceuticals, even before other sectors that saw early applications, such as agriculture and environmental applications. Many nations have pursued policies to cultivate the growth of biotechnology. Many of the first applications were developed in the United States. This was not because the United States had a coherent policy to promote biotechnology. It did not. The United States does, however, have a set of policies to promote health research, and in particular a political structure that has supported the consistent growth of the National Institutes of Health (NIH) since the end of World War II. Those health research policies spawned medical biotechnology.

Development of diagnostic and therapeutic products depends to an unusual extent on government and private funding for research, intellectual property protection, norms governing academic science, product regulation by government, and historical and cultural factors that influence how national governments frame issues arising from medical biotechnology. This entry reviews how national policies influenced the development of medical biotechnology, using the United States as a case example.

DEFINITIONS

Biotechnology is the practical use of living things. At this level of generality, however, agriculture, forestry, fishing, ranching, and many other activities would be included, whereas most intend to refer to the practical applications of molecular biology, and in particular the structural analysis of DNA and proteins. In its most common sense, biotechnology was a term used first by stock analysts to describe a set of companies that began to form late in the 1970s to exploit recombinant DNA, cell fusion, and other methods of molecular biology (1).

One of the first and most influential reports on biotechnology, *Commercial Biotechnology, An International Analysis*, was completed in January 1984 by the Congressional Office of Technology Assessment (OTA) (2). That report distinguished so-called new biotechnology from the old, and focused mainly on "dedicated biotechnology firms," those largely or solely devoted to using the new molecular biological techniques. OTA's definition, modified to accommodate other new techniques of molecular biology, remains useful and formed the basis for another OTA report eight years later, *Biotechnology in a Global Economy* (3).

Technologies to analyze the structure of DNA and proteins have evolved rapidly, so the methods that are new change each year, but the term biotechnology has continued to refer mainly to academic and commercial activities that depend on the structural analysis of DNA and proteins. This definition also includes some activities of major pharmaceutical firms, companies, and research groups that develop instruments used in biology. Activities with direct practical relevance to use of information from cellular and molecular biology have come into being since biotechnology became a widely used term, such as computer analysis of DNA and protein structure and large-scale genetic analysis of organisms (genomics). Medical biotechnology, as used here, refers to the use of molecular biological techniques to develop drugs and diagnostic

services in established firms as well as those founded for this purpose. Biological instrumentation and informatics firms (or activities) are also often counted as biotechnology, but their focus is generally on markets to supply research tools. They are generally excluded here, except in a section on patent policies for research tools.

FACTORS AFFECTING THE COMMERCIAL DEVELOPMENT OF BIOTECHNOLOGY

The 1984 and 1992 OTA reports analyzed several industrial sectors for which biotechnology was relevant. Among the areas considered, only the pharmaceutical sector was mainly focused on human medical applications. OTA identified 10 factors that influenced the commercial development of biotechnology (Fig. 1). The first nine are listed in descending importance as judged by OTA. OTA judged the tenth factor, public perception, more variable and unpredictable, at times playing a major role in policy and at other times taking a back seat to the other factors:

1. Financing and tax incentives for firms
2. Government funding of basic and applied research
3. Personnel availability and training
4. Health, safety, and environmental regulation
5. Intellectual property law
6. University-industry relationships
7. Antitrust law
8. International technology transfer, investment, and trade
9. Targeting policies in biotechnology
10. Public perception

All 10 factors are significantly affected by social values and government policies.

CAPITAL AVAILABILITY

The practical applications of molecular biology, especially recombinant DNA techniques, took the form of dedicated biotechnology companies with explicitly commercial aspirations. This was particularly true in the United States. The geographic origins in the United States are best explained by a combination of three factors: availability of capital to form new companies, public funding for biomedical research, and strong university–industry ties.

The availability of capital to form companies to exploit technological opportunities has proved crucial in most high-technology sectors. After World War II several methods of raising capital were developed. The first venture capital firm was established in conjunction with MIT scientists and Boston bankers (4). Eventually the San Francisco Bay Area became an even more active center for venture capital (5). By the 1970s and the dawn of the new biotecholgy, venture capital firms that had grown up to fund computers, software, and telecommunications were open to help finance the launch of commercial biotechnology. The impetus to found Genentech, for example, came from venture capitalist Robert Swanson

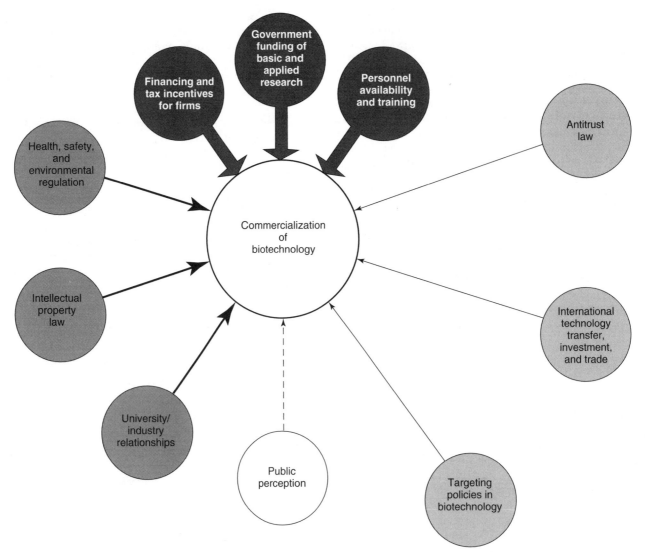

Figure 1. Relative importance of factors affecting the commercialization of biotechnology. *Source: Office of Technology Assessment.*

who approached scientist Howard Boyer of the University of California, San Francisco (6). The financier approached the scientist, and not vice versa. Once the seed was planted and companies had formed, some scientists and university administrators began to approach venture capitalists, rather then the reverse, but the origin of the modern biotechnology sector clearly arose from a member of the investment community. The national and regional environment for capital formation and access to investors willing to fund new technological ventures were critical to the emergence of biotechnology.

The need for early investment to found new companies has attracted attention in Europe and Asia, leading to government incentives and private efforts to create markets along the lines of those started in postwar Boston. Venture capital firms have become a worldwide phenomenon.

In addition to formal venture capital firms, there is another larger but less conspicuous and less easily characterized capital market. Individuals or small groups of wealthy individuals who invest several hundred thousand or millions of dollars in new ventures are often called "angels." They function as an informal market to finance new ventures and often fill gaps in the venture capital markets because they can move more swiftly. They typically close deals with a handshake. The angel market is harder to study, but is comparable in size and at least as important as the formal venture capital market (7).

The angel and venture capital markets enticed university and government scientists to found companies that were privately held, with most equity shared among the investors, scientists, and founding managers. Privately held firms often later become corporations with publicly traded stock. Once a firm's stock was publicly traded, investors could more readily trade their equity for cash.

The availability of venture capital and angel markets depended on a diverse set of government policies, including antitrust, tax (local, state, and national), and other domains of public policy. As a general rule, however, public policy governing individual investments and venture capital was far more subject to private sector actions than deliberate government policy fostering innovation. Indeed, formal government policies intended to encourage startups and availability of risk capital has tended to be local and late in the game, and sometimes even been impediments, rather than a coherent national policy helpful early in the process of spawning a new industry (4,5). The policies that most influence the availability of startup capital have been in financial institutions and among individual investors rather than the product of deliberate government action. In contrast, government policy has been absolutely crucial in the other most important factor influencing the genesis of medical biotechnology: publicly funded biomedical research.

FUNDING FOR RESEARCH AND DEVELOPMENT

Biotechnology companies were founded to exploit a technological base that grew from substantial and sustained public investment in biomedical research. The term molecular biology first referred to a grants program funded by the private Rockefeller Foundation in the 1930s (6). As Rockefeller Foundation administrator Warren Weaver first used the term, molecular biology addressed scientific problems in the life sciences by importing techniques and scientists from the physical sciences, especially physics and chemistry. Molecular biology, and its spin-offs into commercial biotechnology, is the child of federally funded research but the grandchild of policies first developed in private philanthropy.

Before World War II, government funding for biomedical research was relatively sparse throughout the world. Academic medicine, of which research was a component, had a strong tradition in Germany, France, Great Britain, and other countries in Europe and Asia. Most research was conducted "on the side" in hospitals, or funded through private philanthropy. In the United States, the federal government funded less than private sources — the Rockefeller Foundation, the Foundation for Infantile Paralysis (later the March of Dimes), the Carnegie Corporation, universities and hospitals, and other private philanthropies.

In the years leading up to World War II, Mary Lasker and the American Cancer Foundation (later the American Cancer Society) began to pursue a new strategy for biomedical research funding that focused on inducing federal investment through the political process. After the war, this strategy caused an explosive growth of the NIH budget (9,10). The movement already underway to fund cancer research, and then heart disease research, merged with a consensus favoring federal investment in basic research most conspicuously articulated by President Roosevelt's wartime science advisor, Vannevar Bush (11). This consensus did not become embodied in the form of Bush's proposed National Research Foundation, which would have administered biomedical, military, and general science under a single roof. Instead, the Navy established the Office of Naval Research in 1946, and the other armed services then created their own research and development (R&D) organizations. The Atomic Energy Commission was created to support nuclear physics and to apply it to both military and civilian uses. Most relevant to biotechnology, NIH began to take shape as the nation's foremost patron of biomedical research.

Mary Lasker, then-Senator Claude Pepper, and a succession of NIH directors formed an "iron triangle" to expand federal funding (the vertices of the triangle were nongovernment advocates for medical research, NIH administrators, and congressional champions). They chose to focus resources on university-based research, funded through disease-oriented institutes. Through the late 1960s, Lasker forged strong ties to the chair of the House appropriations subcommittee that funded NIH, Representative John Fogarty, and to his Senate counterpart, Senator Lister Hill. Supported strongly on the inside by NIH Director James Shannon, these congressional patrons boosted biomedical research funding substantially year after year (Fig. 2) (12).

The iron triangle was reconstructed with different players after the death of Fogarty in 1967 and retirement of Hill and Shannon in 1968, and the tactics were replicated by groups wanting "their" institute (for heart disease, for neurological illnesses, for eye diseases, etc.). The direct access that disease advocacy groups had to Congress drove the rise of NIH's budget. Health research grew consistently for five decades, more consistently and more substantially than other federal science programs (Fig. 3).

NIH grew into the world's largest source of support for biomedical research. In a survey of articles from 1973 to 1980, for example, the National Cancer Institute alone accounted for 40 percent of all cancer research publications in 275 medical journals (13). Other U.S. sources — including government (mainly other NIH institutes), private nonprofit, and for-profit firms — accounted for roughly another third, with the remainder unknown or funded by an institution outside the United States. The level of public support for health research in the United States has been a critical factor in biotechnology, and it helps explain biotechnology's geographical origins. Medical applications of new technologies appear likely to remain dependent on science for the foreseeable future, so biomedical research policy will continue to be a decisive factor in the development of medical biotechnology.

The basis for this consistent and substantial growth in health research was not a desire for economic growth but for the conquest of disease. The relevance of medical

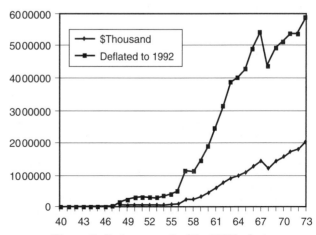

Figure 2. Postwar growth of the NIH budget.

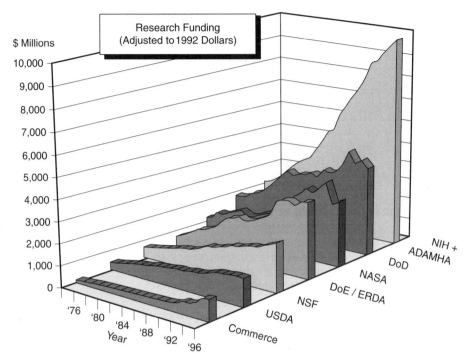

Figure 3. Rise of NIH among U.S. R&D agencies. ADAMHA = Alcohol, Drug Abuse, and Mental Health Administration, DoD = Department of Defense, NASA = National Aeronautics and Space Administration, DOE/ERDA = Department of Energy (and Energy Research and Development Administration for relevant years), NSF = National Science Foundation, USDA = U.S. Department of Agriculture, Commerce = U.S. Department of Commerce.

research to the private commercial development of drugs, devices, and services was foreseeable and foreseen, but promoting such uses was a subsidiary goal, not the primary goal.

The most important two institutions in the birth of biotechnology were universities and start-up firms, with the start-up firms seeded by university scientists. As biomedical research budgets rose and NIH institutes proliferated, the science base grew enormously. It grew most in universities and private research institutes. While Vannevar Bush's vision of one central R&D agency did not become policy, another important feature of his vision did — the emphasis on federal funding for university-based research, as opposed to government owned and operated laboratories more common in other nations and prewar U.S. R&D. (Agricultural research, the grandfather of federal R&D, was typically done by federal employees in government laboratories; military research until World War II waxed and waned with military commitments, and scientists were brought directly into the military in times of war (14).) NIH did conduct some research on its own campus, and in 1953 added a research hospital. During the Vietnam War, the intramural program of NIH grew dramatically in both size and international prestige, in part because working at NIH was a means to avoid induction into the military, but even during this era, most biomedical research dollars flowed to universities and private academic research centers affiliated with them. Those universities were in turn the birthplace of biotechnology.

Coupled with a powerful biomedical research lobby, the decentralized structure of U.S. science enabled a disproportionate growth of the life sciences, particularly medical research. One reason that health research enjoyed growth rates in excess of those in physics, chemistry, engineering, and other fields is the political popularity of

health research. Improving health through research is an accepted federal responsibility in both political parties. Boosting health research budgets, moreover, does not mean having to reduce other research budgets because the United States lacks a central research ministry or ministries. With the exception of the National Science Foundation, research budgets are part of budgets for mission agencies, health and defense being the two largest, but also including space, agriculture, energy, and environment. Budgets for research are decided by congressional appropriation subcommittees that do not have to trade off a reduction in defense research that might affect computing, for example, to obtain more dollars for health research. That is, whereas most countries have a research ministry that must set priorities, no unitary science budget exists in the United States. While this is irrational to the degree that research accounts, which are intended as investments to secure future benefits, are mixed with other accounts that focus on current consumption spending, it has enabled different research fields to expand and contract independently, and health research has expanded consistently, in part because of the perceived scientific opportunities, but also because of political popularity and citizen advocacy. In some countries the science and technology ministries are independent, in others, part of a larger body with responsibility for education (e.g., Japan) or industry (e.g., the United Kingdom). The unique separation of powers and the delegation of appropriation authority to Congress under the U.S. system has led to a uniquely rapid growth of health research that would have been far less likely under Bush's unitary National Research Foundation, and it has not been replicated in other developed economies.

The United States funds both a larger total R&D budget and a far higher share devoted to health within that than other countries (see Table 1). If items not directly

Table 1. Share of R&D for Health from NSF S&E Indicators

Objective	United States (1994)	Japan (1994)	Germany (1993)	France (1993)	United Kingdom (1994)	Italy (1993)	Canada (1992)
Total R&D ($U.S. billion, 1995)	68.3	18.1	15.0	13.7	8.7	5.2	3.4
Percent R&D for health	16.5	3.0	3.3	4.5	7.2	6.1	7.8

Sources: From *Science and Engineering Indicators, 1996*, National Science Board (National Science Foundation), table 4–32; Office of Economic Cooperation and Development, Main Science and Technology Indicators database, Paris, June 1995.

related to creation of new knowledge and technology are taken out of the U.S. R&D budget, as recommended by the National Academy of Sciences (NSF), the total U.S. 'federal science and technology' budget is a smaller total, but the fraction devoted to health in 1994 rises to 28 percent (15), more than twice as concentrated on health as the next closest country. Some nations (especially Japan) categorize a substantially higher fraction of science funding as "advancement of knowledge;" there general university funds cover research that in the United States would appear in NIH "health" research accounts. Michaud and Murray have estimated health R&D as a fraction of gross domestic product (GDP), arguably the most pertinent statistic, and by their estimates the United States remains first, but by a far smaller margin (16). The real disparities are therefore not as large as the figures of the Office of Economic Cooperation and Development (OECD) figures suggest, but the U.S. research system clearly tilts much more heavily toward life sciences and health than do other countries.

Anemic public support for research is an important factor explaining Japan's relatively small role in biotechnology (2,3) for example, and the rich base of public and nonprofit support in Europe, especially the United Kingdom, goes a long way to explain Europe's large role in biotechnology. While government funding for R&D has not been generous in the UK compared to other developed countries, the R&D fraction devoted to health is high, and private philanthropies play an unusually large role, funding biomedical research at levels comparable to government (the Wellcome Trust and the Imperial Cancer Research Fund, among others). Countries with a strong pharmaceutical base, such as the U.K., Switzerland, Germany, France, and Denmark are major powers in biotechnology.

In addition to the overall level of support, the responsiveness of the R&D system is also important. Investments in promising new scientific fields and in emerging technologies must be available to sustain innovation. A system that funds specific projects case-by-case through a system of peer review is far more adaptable than one that allocates most funding through institutions (15). Examples at each end of the peer review spectrum are investigator-initiated grants funded by NIH and NSF, on one hand, and the system of institutional funding of the former Soviet Academy of Science, on the other. While institutional funding achieves many notable successes, and peer-reviewed grants can fund poor science,

as a general heuristic, systems that channel a substantial fraction of their funds through competitive peer review appear to produce better work and respond better to new opportunities in the long run.

OTA's third crucial factor, the availability of trained experts, is closely tied to academic science, and thence to publicly funded health research. This is especially true in the emergent phase of a science-dependent technological sector, when the only source of training may be university laboratories. In many countries, academic training is funded through educational ministries, and the size and flexibility of those ministries, and their attentiveness to needs of an emerging field such as biotechnology, is highly dependent on policy. In the United States, most academic training in the life sciences is covered by NIH research budgets, and so the growth of NIH has created an ample supply of labor. Indeed, indicators suggest the increase of available graduate students and postdoctoral fellows has exceeded even the growth of NIH research budgets, creating a surfeit of young, trained personnel who may turn to careers in biotechnology (17).

PATENTS

Patents are more important in pharmaceuticals and biotechnology than in most other economic sectors. Patents are grounded in national laws and international treaties, and therefore heavily dependent on government policy. Governments (or organizations delegated authority by governments, e.g., the European Patent Office) convey patent rights to private parties, enabling them to exclude others from making, using, or selling an invention. The period of exclusivity generally extends from the date a patent is issued until 20 years after the patent application is filed, although patent terms can be extended in the United States under certain circumstances. Enforcement of patent rights ultimately falls to the government through administrative procedures and litigation in national court systems.

Several different kinds of patents are relevant to medical biotechnology. The most valuable are patents covering both the composition of matter (that is, the protein or chemical) and its method of manufacture. The path to current biotechnology patents leads through several seminal events. In 1980, the U.S. Supreme Court permitted patenting of a microorganism, in the case of *Diamond v. Chakrabarty* (18). That same year, the U.S.

Patent and Trademark Office issued the first of three landmark patents covering recombinant DNA to Stanford and the University of California (19–21). These Cohen-Boyer patents ultimately produced over $200 million in revenues for the two universities until the patent expired in December 1997 (22). Thus began a succession of patents for genes, DNA fragments, methods of producing useful gene products, and methods for making and characterizing DNA. Between 1980 and 1993 the U.S. Patent and Trademark Office issued over 1000 DNA-based patents, and many thousands more have been issued since (23). The rationale for patenting DNA was that while genes were found in nature, isolating and making those genes into a useful form required substantial inventive activity (24–26). A DNA or other biotechnology patent must meet the same criteria as other inventions. Three principal criteria are used worldwide, with some variation in details of interpretation. An invention must be (1) new (or "novel"); (2) inventive (or "nonobvious"), and (3) useful (or have "utility"). In return for the exclusion rights conferred by the patent, the inventors must describe the invention in sufficient detail to enable others to make it work ("enablement").

Therapeutic pharmaceuticals constitute the largest and most financially rewarding applications for medical biotechnology. There is considerable argument about the cost of discovering and marketing a new therapeutic pharmaceutical. The average cost is estimated in the hundreds of millions of dollars to introduce a new product (27). This average is not meaningful in any particular case because the cost varies tremendously. Through the pharmaceutical sector as whole, however, and hence for policy purposes, it is clear that R&D costs are high, and highest for therapeutics. New therapeutic products thus face the highest costs but also yield the highest payoffs. Patent protection is most important for therapeutic drugs, medical biotechnology's most profitable subsector.

Medical biotechnology encompasses many other products and services with diverse product cycle times. Some biotechnology instruments may resemble short computer or telecommunication device product cycles, measured in years rather than decades, and biotechnology informatics firms face the rapid cycle times of software, with product cycles as short as a year or two. Medical diagnostic products typically reach the market faster than therapeutics because they face a less elaborate regulatory gauntlet. Diagnostics do require FDA approval, and so face regulatory hurdles higher than for most nonmedical goods and services. The capital and time needed to secure market approval for diagnostic products, therefore, are generally lower than for therapeutics but higher than for most nonmedical goods and services. Where product cycle times are short, lead-time advantages, trade secrets, and other factors tend to be more important competitive factors than patents, although patents may still be important in negotiating licensing and technology-sharing arrangements among firms, and in raising capital for new firms. The importance of patents in therapeutic pharmaceuticals builds on a century of industrial history.

Modern biotechnology is the extension of a long trend in which pharmaceutical discovery and development has become increasingly dependent on scientific and technical advance. The modern pharmaceutical industry grew by applying science to medicine (28). Early products were often discovered initially as folk remedies or through clinical observation, but modern pharmaceutical research grew to prominence by making the hunt for new products more systematic. A century ago acetylsalicylic acid (aspirin), was identified as more potent for pain relief than the salicylic acid in bark extract, and then manufactured on a massive scale by Bayer. Companies began to synthesize and screen large numbers of compounds for medical effects. They turned to the emerging methods of organic chemistry to create the compounds and to manufacture them. Firms also used the growing power of microbiology and physiology to discover and screen potential drugs. Antibiotics were first found as products of soil fungi that used them as defenses against bacterial attack, and the industrial innovation was to identify and purify the active agents, and devise ways to produce them cheaply on a grand scale.

In the pharmaceutical business, drug discovery is only one determinant of success. Efficient conduct of clinical testing and manufacture are just as important. Pharmaceutical firms must also manage large and complex distribution systems, manufacturing plants, and they devote enormous sums to marketing. The end markets in health care are financially complex and heavily regulated, requiring expertise and management. At root, however, success follows introduction of new drugs, and innovation through drug discovery where patent protection is paramount.

The time horizon for investment in pharmaceuticals is quite long compared to other industries (29). More than a decade typically passes between discovery of a lead compound and market introduction of a therapeutic pharmaceutical (27). The level of R&D investment is also unusually high in pharmaceuticals, rivaled only by software and a few other high-technology sectors. The pharmaceutical industry invests an estimated 19 percent of revenues in R&D (30). Firms invest heavily in R&D because drug discovery is a major basis for competition. Since the early 1980s private pharmaceutical R&D investment has grown even faster than the NIH budget.

The strength of patent protection in pharmaceuticals is one major reason such substantial, long-term private R&D investments are possible. Patents increase the price that can be charged on drugs that make it all the way through the pipeline, producing revenues to fund drug discovery and development of future products. Gambardella's econometric monograph on pharmaceutical innovation concluded that managing the process of drug discovery was a powerful predictor of financial growth (31). In several other studies of pharmaceutical innovation, economists have used patents as indicators of success among pharmaceutical competitors (32,33), empirically corroborating this message.

The general criteria for patents are shared worldwide, but the criteria can be applied differently in different countries. The interpretation of patent claims is left to litigation when one party sues another for infringing its patents. Such litigation is costly and takes place long after the R&D results have been disclosed in the patent application. Patent litigation over recombinant DNA-derived insulin cost over $30 million and consumed half a decade (34,35), and the case was decided in 1997, two decades after the crucial events disclosed in the relevant patents. Battles raged over most first generation protein therapeutics — insulin, growth hormone, tissue plasminogen activator, interferons, interleukins, and others. Such litigation is not only costly and slow, but its outcome is uncertain and may differ from country to country. The same drug may be patented in one jurisdiction and not another (e.g., Genentech's tissue plasminogen activator patent is vadid in most countries, but not the United Kingdom). This can occur with traditional drugs as well, but uncertainty is higher in biotechnology because the interpretation of patent criteria is less well settled. Patent policies may differ among countries in three areas relevant to biotechnology: animal patents, gene patents, and research tools.

If patent policies diverge, it is the patent policies in the largest markets or the first markets that will most influence the development of biotechnology. An invention made in a country can be patented in any other country, and the patents that matter most are those that cover inventions in the most lucrative markets, where prices are highest or more units can be sold. While an inventor tends to file the first patent application in his or her home country, this is not necessarily the case, and many biotechnology patents have been filed first in the largest expected markets, rather than in home countries. The largest national pharmaceutical market is the United States, followed by Japan, which has higher per-capita consumption but generally lower prices. Europe is, in aggregate, roughly comparable in size to the United States. Asian and other markets are growing in size, and recent international harmonization of patent law should lead to policies in developing economies coming to resemble those in the developed economies. Because the United States is among the few countries with no pharmaceutical price controls, its drug prices are higher, and for many drugs, the U.S. market alone accounts for most of the global profit.

UNIVERSITY–INDUSTRY RELATIONS

The pharmaceutical sector in general, and biotechnology in particular, is uniquely dependent on academic research. The late Edwin Mansfield queried industrial executives about the degree to which their product stream depended on academic science (36,37). Pharmaceuticals stood out. Executives estimated that twice as many products either would not have been developed at all or would have been substantially delayed without academic research, compared even to other high-tech sectors. The survey results are corroborated by patent data in this unusually patent-dependent sector. Pharmaceutical and biotechnology patents are much more likely to cite academic research than others (38,39). In most patent classes, academic institutions hold only a few percent of patents, but the fraction

is much higher for pharmaceuticals and rises to almost one-third of all DNA-based patents in the period 1980 to 1993 (23). The number of academically owned patents has risen steeply in pharmaceuticals, and unlike a drop-off of citations to academic patents in other fields, citation of academic patents in pharmaceutical and medical patent classes has risen over the past two decades (40).

The patent story is but a small part of a larger story of academic-industrial mutualism in biotechnology. The impetus to create biotechnology firms did come first from private investors, as noted above, but it was the commercial potential of academic science that they recognized. To an even greater extent than the already science-dependent pharmaceutical business, biotechnology has emerged as a hybrid academic-industrial enterprise.

The national policies most influencing academic science are public funding decisions for health research, covered above. The policies of universities and private firms, both individually and in aggregate, also influence the development of biotechnology. Biotechnology companies develop around scientists who first use the technologies and take action to apply them commercially (41). Biotechnology companies first appeared close to universities and centers of biomedical research excellence, especially near (1) Stanford and the University of California campuses in San Francisco, Berkeley, Los Angeles, and San Diego; (2) MIT and Harvard; and (3) Johns Hopkins University and the NIH campus. The highest concentrations of biotechnology firms thus arose in California, Massachusetts, and Maryland, not because of a national targeting policy but because that is where the science was based.

The academic research base is necessary, but not sufficient. If federal funding for academic science were the sole determinant, then biotechnology investment would more closely parallel federal spending for health research in specific regions. While California does lead the nation in NIH grant funding, if the intramural program in Maryland is taken into account, California receives less total NIH funding than Maryland. New York is third in NIH funding and Massachusetts fourth, but Massachusetts is second in number of firms, ahead of Maryland and New York. Moreover the fraction of firms based in California and their capitalization is far more heavily weighted to California than the funding would predict. A combination of academic policies conductive to industrial collaboration at Stanford and the University of California, the availability of venture capital, and the history of specific technologies, especially recombinant DNA techniques, led to biotechnology taking root around San Francisco Bay in California.

Policies on the industrial side of the equation are clearly important in addition to factors on the academic side. The formation of small biotechnology companies is in part a signal that established firms in the relevant markets are not fully exploiting an emerging technology. Established firms have been important in introducing many of the first protein therapeutic products, but the initial discoveries generally took place either in academic laboratories or in dedicated biotechnology firms collaborating closely with university scientists. Pharmaceutical firms that adopt the R&D ethos of academic science — encouraging open publication, forming numerous collaborations with academic groups, and allowing scientists to influence R&D rather than exclusively dictating objectives from the top down—tend to be more innovative (31,32). While it has not been similarly corroborated by empirical studies of biotechnology firms, the "academic ethos" in dedicated biotechnology firms is widely perceived to be even more important than among established pharmaceutical firms.

The importance of federally funded academic research is not unique to biotechnology, but it is more important than for most other sectors. An historical study of computing and software research notes the critical importance of federal funding and academic research in the commercial development of those fields (42). Publicly funded research at academic institutions appears to be important for several reasons. First, it is directly linked to training those who will further develop the science and technology, whether in academic science, industrial R&D, or in management. Second, the results of academic science are generally published openly, and are thus in theory available to all potential beneficiaries. This enables information to flow readily not only among academic scientists, but also among disparate industrial users. Third, the federal government can invest in research that is expected to produce broad social benefit over the long term. Private firms are not similarly motivated, except in unusual cases of monopoly or extensive market dominance (e.g., AT&T in telecommunications until the 1970s or IBM in computing until the 1980s), because no one firm can expect to capture the benefits of its R&D investment. Federally funded academic research is thus a tide that raises all ships in the biotechnology and pharmaceutical sector.

The pattern of geographic clustering is also shared with other high-technology sectors, such as computer manufacture, software engineering, and telecommunications. The region along Route 128 outside Boston and the San Francisco Bay Area have been particularly well studied (4,5,43). The reliance of industrial innovation on academe is similar to other industries whose early history depended critically on science and technical innovation (44), and the academic–industrial nexus is particularly salient in medicine (45).

REGULATION OF PRODUCTS AND SERVICES

The modern pharmaceutical industry started by purifying and manufacturing agents using chemical methods. A century ago, "patent medicines" were as likely to be peddled using grandiose, unsubstantiated medical claims as they were to be legitimate and effective medications. A system of regulation grew up to combat charlatans and quacks, and that system of regulation became most explicit and elaborate where the most money changed hands and where the health stakes of misuse were highest, that is, for products marketed as therapeutic pharmaceuticals.

For therapeutic pharmaceuticals, that is, chemical compounds that are claimed to alter a body function, firms developing the drug must test the compound for safety and efficacy. Clinical testing is the most expensive and prolonged component in the pharmaceutical product development cycle. It entails several phases of clinical trials. The usual development process entails testing the compound for toxicity in a small number of health individual volunteers (phase I), based on evidence from laboratory and animal experiments that suggest the compound might be clinically useful. If the compound proves safe to administer to humans, phase I is followed by use in a larger number of individuals (usually tens or hundreds) to establish dosage and preliminary evidence of efficacy (phase II), and then larger trials (involving hundreds or thousands of individuals) to establish efficacy and to monitor adverse outcomes and side effects (phase III). If a compound proves safe and effective, it is approved for market. A phase IV of clinical trials for new indications or for other reasons (e.g., a requirement for further data by the regulator) may take place once the drug is approved. Approval to market may be withdrawn if concerns about safety come to light.

Drugs and devices are regulated somewhat differently and details of such regulation vary among countries. In general, however, devices that affect essential body functions or raise potential safety concerns, such as heart valves or respirators, face standards of evidence about safety and efficacy similar to therapeutic pharmaceuticals, entailing a series of clinical trials. Products that pose less direct threats to safety or that merely modify well-studied devices, including most diagnostics, face lower regulatory thresholds but still require premarket approval. Some products of biotechnology, such as analytical software or research tools and medical procedures that do not entail introduction of a drug or other product into the body, may not require regulatory approval before entering the market.

The main impact of product regulation on the emergence of medical biotechnology has been to increase the barriers to entry into pharmaceutical markets by substantially increasing both the R&D costs and the time from discovery to market. The need to raise tens to hundreds of millions of dollars to cover clinical testing has deeply affected the development of biotechnology. A few dedicated biotechnology firms that quickly discovered protein therapeutics of substantial value have grown to rival the smaller established pharmaceutical firms. Most biotechnology firms, however, have found the cost of independent development prohibitive, and have forged strategic alliances with larger firms or have been wholly or partially purchased by established firms. Established pharmaceutical firms have created in-house R&D efforts that do molecular biological research quite similar to that done by dedicated biotechnology firms. The relationship between the established firms and dedicated biotechnology firms has become highly complex, and varies so much case-by-case that generalization is perilous.

PUBLIC PERCEPTION AND POLITICAL PROCESS

The economic, R&D, patent, and regulatory policies analyzed above fail to capture some factors that have influenced the speed and direction of biotechnology development in different nations. OTA refers to these, for want of a better term, as "public perception," lumping together disparate factors that influence industrial development in different ways and to different degrees. Historical and political factors were the most conspicuous elements OTA addressed in this category. Religious views have also played a major role in national debates about select areas of biotechnology policy, such as reproductive technologies, gene transfer, stem cell and embryo research and their applications. Religious demographics of different countries vary so much that the U.S. case may not be suitable grounds for generalization.

The influence of such factors is not surprising or unusual, and is far from unique to biotechnology. Indeed, all areas of policy are influenced by religious, moral, historical, and political factors. It is impossible to capture all the factors that influence public perceptions, but many of them share a common element—they increase or mitigate public fears about biotechnology and its applications. A few factors that have been particularly important in the development of medical biotechnology deserve special mention. Two historical events, the eugenics movement in the first half of the twentieth century and the recombinant DNA debate of the 1970s, had a particularly strong influence on biotechnology policy.

The rise of national bioethics commissions is a new feature of political process in many countries and international organizations. Such commissions have developed as an attempt to grapple with moral and religious pluralism as well as the desire for explicit analysis of ethical implications of rapid scientific and technical advances. Such commissions typically cover more than biotechnology, often addressing topics in health care, end-of-life issues, reproduction, and other areas. The focus here is on bioethics and genetics, where "public bioethics" has an especially rich history.

Eugenics in the first half of the century cast a dark shadow over genetics in the second half. Eugenics as a term refers to many different notions with varying levels of political coercion, having in common directed inheritance intended to improve human populations. Eugenics first became a political movement in England and the United States (46). The ideology of eugenics spread worldwide and took its most extreme expression in association with racial hygiene, part of the ideology underlying the National Socialist Holocaust (47,48). Eugenics and racial hygiene as political movements became associated with senior academics in anthropology and human genetics (49), and this association carried over into postwar German genetics. Applications from molecular genetics were regarded as suspect in the German Green political party, and in German-speaking Europe, public distrust of biotechnology emerged in both agriculture and medicine. The first applications of biotechnology were controversial in many countries, but the controversies were more protracted and pervasive in Europe. The causal association between the history of eugenics and different perceptions of biotechnology is weak and circumstantial. It is historically plausible, but far from demonstrated. At the least, however, the resurgence of scholarship about eugenics, and particularly about the coercive social policies associated with its most florid expressions on both sides of the Atlantic, has contributed to greater vigilance about

untoward consequences of social policies that intrude on choices about marriage, immigration, and reproduction, and also about discrimination based on disability or illness.

The recombinant DNA controversy of the mid-1970s also left a legacy for several decades. The recombinant DNA debate began with scientific concerns about biohazards from gene-splicing experiments. The central concern was that bacteria carrying spliced genes could spread uncontrollably and cause harm to individuals subsequently infected. Molecular biologists declared a moratorium on experiments involving recombinant DNA, which was lifted after a famous meeting at Asilomar in California, when the moratorium gave way to federal guidelines for such experiments devised and effected by NIH (50). The debate activated several social activists, who turned their energies to opposing applications that emerged from recombinant DNA and other techniques of molecular biology, and hence biotechnology. Jeremy Rifkin and his Foundation on Economic Trends became the most prominent antibiotechnology activist in the United States; and similar sentiments entered social movements commingled with Green politics in Europe and Asia. At the root of public distrust of biotechnology lay concern that the speedy advance and power of the new technologies were outstripping government and other social mechanisms to ensure they were applied fairly and safely. As with patent policy, capital formation, and public support for academic science, biotechnology emerged first in the most permissive environment, the United States. While this was where the recombinant DNA controversy first arose, it was also where guidelines governing the research were relaxed first and most extensively among developed economies.

National bioethics commissions and biotechnology share a common origin in policies governing the application of molecular biology. The first national bioethics commission, the U.S. National Commission for the Protection of Human Subjects of Biomedical and Behavioral Research (National Commission), was the result of almost a decade of congressional debate. The first hearings on such a commission were convened by Senator Walter Mondale in the late 1960s, out of concern that technical advances in biology were proceeding in advance of policies to ensure wise use of the technologies in genetics, reproduction, and medicine. In 1971 attention shifted to human cloning, but there was insufficient congressional support for a bioethics commission until a series of scandals about the conduct of medical research involving human participants led to hearings by Senator Edward Kennedy. The National Commission was established in 1975 and operated until 1978. The legacy of its origins in concern about advances in biomedical research was a single report (51). As the National Commission faded out of existence, it recommended that Congress establish a new national bioethics commission with a mandate beyond protection of human participants in research. Congress did so in the form of the President's Commission for the Study of Ethical Problems in Medicine and Biomedical and Behavioral Research

(President's Commission), whose title was admirably self-explanatory if lamentably long. The President's Commission operated from 1980 through March 1983. Two of its reports dealt directly with genetics, on genetic testing and screening and on human uses of recombinant DNA, particularly human gene therapy (52,53). These reports came out just as the term biotechnology was coming into general use. As it neared the end of statutory term, the President's Commission recommended that Congress establish a more permanent body to deliberate and make recommendations about ethical issues in medicine, in particular about rapidly advancing technologies that touched areas of controversy and moral uncertainty, including reproduction and genetics (54). In 1985 Congress acted on this recommendation by establishing a new congressional advisory agency modeled on OTA, the Biomedical Ethics Advisory Commission. That Commission took three years to get underway, largely because of delays in congressional process, and in the end never issued a report. In effect, there was a hiatus from 1983 until 1995, when President Clinton established the National Bioethics Advisory Commission by executive order (55).

During the interregnum of bioethics commissions in the United States, many countries established national bioethics commissions and several international organizations also did so. A November 1992 OTA survey of international bioethics bodies found 31 countries and 5 international organizations had some form of advisory apparatus for bioethics (56). Many countries had more than one body, such as Denmark, France, Australia, and Canada.

The growth of national bioethics initiatives and international organizations with bioethics consultative capacity grew from an effort to link explicit analysis of ethical, social, and legal implications of medical science and technology. In many cases they were an attempt to clarify values and to sample disparate religious and social perspectives on medical practices and emerging medical technologies. Most bioethics bodies included attention to advances in genetics or biotechnology or both within their remit. Several of these bodies, particularly those in Europe and Canada, have proven influential in government policy about biotechnology. In Denmark, Sweden, and Germany, for example, commissions recommended proscribing germ line genetic alterations (introduction of DNA into a person so that genetic changes are inherited), and the national parliaments did so. In many cases bioethics commissions are serving to find common ethical arguments that resonate throughout their respective national cultures, grappling with moral pluralism in the face of advancing technologies and pressing problems arising from practical applications, both immediate and in prospect.

Medical biotechnology differs from other high technologies in that its development directly depends on the participation of people in research, as its applications are intended to influence human physiology. This means that the advance of biotechnology is more immediately relevant to the relief of human suffering than most other technologies, but it also more directly threatens fundamental social values connected to reproduction and inheritance.

One response has been an effort to anticipate the social implications of biotechnology through bioethics commissions and other mechanisms. It remains too early to judge the success or failure of these responses.

BIBLIOGRAPHY

1. R. Teitelman, *Gene Dreams: Wall Street, Academia, and the Rise of Biotechnology*, 1989, New York: Basic Books.

2. U.S. Congress, *Commercial Biotechnology: An International Analysis*, Office of Technology Assessment, Washington, DC, 1984.

3. U.S. Congress, *Biotechnology in a Global Economy*, Office of Technology Assessment, Washington, DC, 1991.

4. S. Rosegrant and D.R. Lampe, *Route 128: Lessons from Boston's High-Tech Community*, Basic Books, New York, 1992.

5. A. Saxenian, *Regional Advantage: Culture and Competition in Silicon Valley and Route 128*, Harvard University Press, Cambridge, MA, 1994.

6. R.E. Kohler, *Partners in Science: Foundations and Natural Scientists 1900–1945*, University of Chicago Press, Chicago, 1991, pp. 299–300.

7. S.S. Hall, *Invisible Frontiers: The Race to Synthesize a Human Gene*, Atlantic Monthly Books, New York, 1987.

8. U.S. Congress, *Innovation and Commercialization of Emerging Technologies*, Office of Technology Assessment: Washington, DC, 1995.

9. J.T. Patterson, *The Dread Disease: Cancer and Modern American Culture*, Harvard University Press, Cambridge, MA, 1987.

10. S.P. Strickland, *Politics, Science, and Dread Disease: A Short History of United States Medical Research Policy*, Harvard University Press, Cambridge, MA, 1972.

11. V. Bush, *Science: The Endless Frontier*, Office of Scientific Research and Development, Washington, DC, 1945.

12. R.E. Miles, *The Department of Health, Education, and Welfare*, Praeger, New York, 1974.

13. H.H. Gee and F. Narin, *NIH Program Evaluation Report: An Analysis of Research Publications Supported by NIH 1973–1976 and 1977–1980*. NIH & Computer Horizons, Inc., Washington, DC, 1986.

14. A.H. Dupree, *Science and the Federal Government*, Johns Hopkins University Press, Baltimore, MD, 1985.

15. National Academy of Sciences, *Allocating Federal Funds for Science and Technology*, National Academy Press, Washington, DC, 1995, Available at: *http://www.nap.edu/readingroom/books/fedfunds/*

16. C. Michaud and C.J.L. Murray, *Resources for Health Research and Development in 1992: A Global Overview*, World Health Organization, Geneva, Switzerland, 1996.

17. National Research Council, *Trends in the Early Career of Life Scientists*, National Academy Press, Washington, DC, 1998.

18. *Diamond, Commission of Patents and Trademarks, v. Chakrabarty*, Supreme Court of the United States, Washington, DC, 1980.

19. S.N. Cohen and H.W. Boyer, U.S. Pat. 4,740,470 (1988), (to University of California and Stanford University).

20. U.S. Pat. 4, 237,224 (1980), (to S.N. Cohen and H.W. Boyer, University of California and Stanford University).

21. U.S. Pat. 4, 468,464 (1984), (to S.N. Cohen and H.W. Boyer, University of California and Stanford University).

22. National Research Council, *Intellectual Property Rights and Research Tools in Molecular Biology*. National Academy Press, Washington, DC, 1997.

23. S. McCormack, *Personal Communication, DNA Patent Database*, Kennedy Institute of Ethics, Georgetown University, & Foundation for Genetic Medicine, Manassas, VA, 1998, Available at: *http://www.geneticmedicine.org/*

24. R.S. Eisenberg, *Emory Law J.* **39**, 721–745 (1990).

25. R.S. Eisenberg, *Science* **257**, 903–908 (1992).

26. R.S. Eisenberg, *Nature genet.* **15**, 125–130 (1997).

27. U.S. Congress, *Pharmaceutical R&D: Costs, Risks, and Rewards*, Office of Technology Assessment, Washington, DC, 1993.

28. J.P. Swann, *Academic Scientists and the Pharmaceutical Industry*, Johns Hopkins University Press, Baltimore, MD, 1988.

29. National Academy of Engineering, *Time Horizons and Technology Investments*, National Academy Press, Washington, DC, 1992.

30. Pharmaceutical Research and Manufacturers Association, *Annual Survey*, Pharmaceutical Research and Manufacturers Association, Washington, DC, 1997.

31. A. Gambardella, *Science and Innovation: The U.S. Pharmaceutical Industry during the 1980s*, Cambridge University Press, New York, 1995.

32. I. Cockburn and R. Henderson, *Proc. Nat. Acad. Sci.* **93**, 12725–12730 (1996).

33. R. Henderson, A. Jaffe, and M. Trajtenberg, *Universities as a Source of Commercial Technology: A Detailed Analysis of University Patenting, 1965–1988*. National Bureau of Economic Research, Cambridge, MA, 1994.

34. E. Marshall, *Science* **277**, 1029 (1997).

35. E. Marshall, *Science* **277**, 1028–1030 (1997).

36. E. Mansfield, *Res. Policy* **20**, 1–12 (1991).

37. E. Mansfield, *Academic Research Underlying Industrial Innovations: Sources, Characteristics, and Financing. Rev. Econ. Statist.* **77**, 55–65 (1995).

38. F. Narin and R.P. Rozek, *Res. Policy* **17**, 139–154 (1988).

39. F. Narin and D. Olivastro, *Res. Policy* **21**, 237–249 (1992).

40. R. Henderson, A.B. Jaffe, and M. Trajtenberg, *Review of Economics and Statistics* **80**, 119–127 (1998).

41. L.G. Zucker and M.R. Darby, *Proc. Nat. Acad. Sci.* **93**, 12709–12716 (1996).

42. National Research Council, *Funding a Revolution: Government Support for Computing Research*, National Academy Press, Washington, DC, 1999.

43. E. Moscovitch et al., *MIT: The Impact of Innovation*, BankBoston, Boston, MA, 1997.

44. R.R. Nelson, ed., *National Innovation Systems: A Comparative Analysis*, Oxford University Press, New York, 1993.

45. N. Rosenberg, A.C. Gelijns, and H. Dawkins, *Sources of Medical Technology: Universities and Industry*, National Academy Press, Washington, DC, 1995.

46. D.J. Kevles, *In the Name of Eugenics*, University of California Press, Berkeley, CA, 1985.

47. R.N. Proctor, *Racial Hygiene: Medicine under the Nazis*, Harvard University Press, Cambridge, MA, 1988.

48. M.B. Adams, ed., *The Wellborn Science: Eugenics in Germany, France, Brazil, and Russia*, Oxford University Press, New York, 1990.

49. B. Müller-Hill, *Murderous Science*, Oxford University Press, New York, 1988.

50. S. Krimsky, *Genetic Alchemy: The Social History of the Recombinant DNA Controversy*, MIT Press, Cambridge, MA, 1982.

51. National Commission for the Protection of Human Subjects of Biomedical and Behavioral Research, U.S. Department of Health, Education and Welfare, *Special Study: Implications of Advances in Biomedical and Behavioral Research*, U.S. Government Printing Office, Washington, DC, 1978.

52. President's Commission for the Study of Ethical Problems in Medicine and Biomedical and Behavioral Research, *Screening and Counseling for Genetic Conditions*, U.S. Government Printing Office, Washington, DC, 1983.

53. President's Commission for the Study of Ethical Problems in Medicine and Biomedical and Behavioral Research, *Splicing Life: The Social and Ethical Issues of Genetic Engineering with Human Beings*, U.S. Government Printing Office, Washington, DC, 1982.

54. President's Commission for the Study of Ethical Problems in Medicine and Biomedical and Behavioral Research, *Summing Up*, U.S. Government Printing Office, Washington, DC, 1983.

55. W.J. Clinton, *Executive Order 12975: Protection of Human Research Subjects and Establishment of a National Bioethics Advisory Commission*, Executive Office of the President, Washington, DC, 1995.

56. U.S. Congress, *Biomedical Ethics in U.S. Public Policy*, Office of Technology Assessment, Washington, DC, 1993.

See other entries FDA REGULATION OF BIOTECHNOLOGY PRODUCTS FOR HUMAN USE; FEDERAL REGULATION OF BIOTECHNOLOGY PRODUCTS FOR HUMAN USE, FDA, ORPHAN DRUG ACT; see also PATENTS AND LICENSING entries; STRATEGIC ALLIANCES AND TECHNOLOGY LICENSING IN BIOTECHNOLOGY; TRANSFERRING INNOVATIONS FROM ACADEMIC RESEARCH INSTITUTIONS TO INDUSTRY: OVERVIEW; UNIVERSITY-INDUSTRY RESEARCH RELATIONSHIPS, ETHICS, CONFLICT OF INTEREST.

OWNERSHIP OF HUMAN BIOLOGICAL MATERIAL

LISA M. GEHRKE
Nilles & Nilles, S.C.
Milwaukee, Wisconsin

OUTLINE

INTRODUCTION

In recent years biological materials have become increasingly important in biomedical research. These materials, which were once considered waste materials, have become essential to research in many promising areas. As a result the demand and potential commercial value for these materials continues to rise. The legal status of these materials depends primarily on the source of origin of the materials. If the material is obtained from a nonhuman source, the law will generally consider it as property and therefore, a commodity that may be freely bought and sold in the market place. However, if the material is obtained from a human source, the law generally fails to recognize it as property.

OVERVIEW OF THE USE OF HUMAN BIOLOGICAL MATERIALS IN BIOTECHNOLOGICAL RESEARCH

The study of the human body, body parts, and biological materials has traditionally played an important role in the advancement of the biomedical and pharmaceutical sciences. The observation and study of postmortem organs, tumors, and other biological materials obtained during surgery has provided physicians and researchers with an understanding of the nature and function of the human body and human disease processes.

The introduction and refinement of biotechnological tools such as recombinant DNA, cell fusion, cloning, and bioprocessing techniques in the early 1970s dramatically changed the role of human biological materials in biomedical research and development. These techniques enabled researchers to go beyond merely observing tissues and cultivating cell lines to actually incorporating the material into new organisms to create life forms not previously known in nature. Recognizing the tremendous economic potential of these products, researchers and venture capitalists began investing in the development of biotechnologically based products and the private commercial biotech industry was born. As a result, the demand for the raw materials, namely biological materials, dramatically increased.

In 1980 the United States Supreme Court significantly changed the U.S. patent law by expanding the scope of patentable subject matter to include to artificial living organism in the landmark decision *Diamond v. Chakrabarty* (1). Prior to *Chakrabarty*, U.S. patent law considered all plants, animals and microorganisms products of nature and not products of invention. Therefore they did not qualify as patentable subject matter. *Chakrabarty* recognized that a living organism not found in nature and created by the intervention of humans, is a product of invention and not a product of nature. Consequentially artificial living organisms, qualified as patentable subject matter under the U.S. patent law.

In 1988 the United States government allocated three billion dollars to fund the human genome mapping project commonly known as the Human Genome Project (HGP). Concurrently, the Human Genome Organization (HUGO) was formed in Europe to coordinate research and foster collaboration between scientists (2). The information obtained from HGP has been a tremendous source in developing understanding of the genetic basis of human function and the genetic roots of disease. This information serves as a foundation for research and development in biopharmaceuticals and therapeutics around the world.

The biotech industry continues to grow as investors recognize the economic incentive created by the availability of patent protection for artificial living organisms combined with the value of the information obtained from HGP. This growth and increased interest in gene-based therapies has also created a corresponding increase in the demand for both human and nonhuman biological materials.

Human biological materials such as human tissues, cells, and other materials previously viewed as biological waste products are now seen as having significant economic value. As a result some groups support legally recognizing human biological materials as property. Other groups, however, opposed any legal recognition of property rights in human biological material because it commodifies the human body, which is believed to be contrary to the U.S. legal tradition, public policy, and human dignity.

PROPERTY, OWNERSHIP, AND PROPERTY RIGHTS

Ownership is a term arising out of common law that refers to an individual's right to exercise control or dominion over property. Ownership allows an individual to use, control and transfer the owned property to the exclusion of others.

Property is something having recognized value that may be owned by an individual. Property is either real or personal. Real property, such as land and items affixed to the land, is immovable. Personal property, however, is movable property. Personal property may be tangible items, such as cars, jewelry, and furniture, or intangible concepts, such as stocks, annuities, and intellectual property. For example, nonhuman biological materials such as microorganisms are considered personal property because they are tangible materials. A patent claiming a process of making the microorganism, however, is considered intangible property.

A property owner has a bundle of rights known as property rights that are created and recognized by the state. The definition of property and the nature and extent of an owner's rights in that property are determined by the state through either legislation or common law. In addition the state defines the remedies available to a property owner if the property owner's rights are violated.

Many states currently recognize nonhuman biological material as property. For example, the statutes of Illinois (3) and Maryland (4) recognize "samples, cultures, microorganisms and specimens" as property. If it is misused, the owner of the property may sue under a theory such as trespass to chattel, to recover damages for any injury to the property resulting from unauthorized use or misuse of the property. For example, an individual may own a specific bacterial culture. If a culture is misused, harmed, or destroyed by another, the owner has legal standing to sue the third party to obtain restitution for the damage to the property.

Human biological material, unlike biological material from a nonhuman source, is not currently recognized as property by any state within the United States. Several legal theories, however, have been suggested as a basis for recognizing human biological materials as property. The legal recognition of human biological material as property would create a unique situation in which the property owner is also the property subject to ownership. In this situation an injury to the person would not only create standing to sue the harming individual for injury to person but would create standing to sue for injury to property as well. Legal scholars have looked to various theories to support the recognition of human biological materials as property.

HUMAN BIOLOGICAL MATERIAL

The term "human biological material" is used to broadly refer to any replenishable or nonreplenishable substance obtained from a human body. Replenishable or regenerative materials include blood, skin, bone marrow, hair, urine, perspiration, saliva, milk, semen, and tears. Nonreplenishable or nonregenerative substances however, include body parts such as oocytes and organs, whether vital or nonvital (5).

Research facilities generally obtain human biological materials from patients, research subjects, paid donors or repository collections. The individual providing the material may be either identified or unidentified and the material sample may be used directly in research or saved in a repository collection for later use. The laws dealing with the use and transfer of these materials vary greatly depending upon the specific type of human biological material, and the source from which it was obtained.

HUMAN BIOLOGICAL MATERIAL AS PROPERTY

The recognition of human biological material as property would allow an individual to sell or transfer his or her biological material for valuable consideration such as cash. This change would impact society and the tissue recipients and as well as the tissue donors. Therefore all of their interests should be considered in this decision.

Society's Interest

U.S. public policy generally opposes recognizing a property right vested in the donor of human biological material. This position arises out of society's responsibility to ensure the safety and welfare of its citizens, and to promote scientific advancement.

Over the past 40 years, the U.S. government and the global research communities have made great advances in enacting legislation and developing ethical standards to ensure the protection of human subjects in biomedical research. A central principle common to all the laws, codes, and standards is respect for the human dignity of each person. In other words, society is generally prohibited from permitting any treatment or activity involving a patient, research subject or tissue donor that would be contrary to the dignity and integrity of that individual. Opponents of recognizing human biological material as property argue that the commodification of the human body or any of its parts is equivalent to the commodification of the human person. Since society has determined that it is a violation of human dignity to treat human beings as commodities, it is also a violation to treat the human body or any of its parts as commodities. In addition, opponents argue that recognizing a property interest in human biological materials will encourage individuals to take unnecessary risks with their health for a short-term economic gain. For example, an individual in need of money may feel compelled to sell tissues or an organ to obtain compensation even though the donation may seriously compromise his or her health.

Proponents of recognizing human biological material as property argue that any imposition of regulations restricting the sale and purchase of body parts, tissues, or materials is a paternalistic infringement upon an individual's freedom. Under this approach, if an individual is provided all of the facts surrounding the risks and benefits of the donation, he or she should be able to sell his or her biological material for cash or other valuable consideration. Any law or regulation restricting such a transaction is considered an abuse of the state's authority. Furthermore the sale of human biological material is not equivalent to the sale of a human being.

Material Donor's Interest

It is clear that the value of human biological material has increased within the past 30 years. Research involving the use of human biological materials has produced products with great economic potential. In some cases the discovery or the invention is directly related to a specific trait found only in the biological material of a specific individual or group of individuals. Consequentially proponents of legally recognizing human biological material as property assert that the individual or group of individuals who provided the essential biological material should share in the financial gain generated by the product arising out of their unique trait. For example, if an individual exhibits a specific genetic trait that is essential to the formation of a new pharmaceutical composition, he or she should be entitled to a portion of the profits realized by the sale of that product. Alternatively, since human biological material has a recognized value, proponents argue that an individual should be entitled to sell his or her biological material in the same manner as a farmer sells a crop.

Opponents to recognizing human biological material as property assert that the treatment of human biological material as property is contrary to human dignity and integrity in the same manner as discussed above.

Material Recipient's Interest

Researcher are generally opposed to the recognition of a property interest in human biological material for several reason including fear of increased research costs, fear of future liability for the use of biological material obtained in a wrongful manner. Since a large portion of the research involving human biological materials is directed toward the production of biopharmaceuticals and biotherapeutics, researchers argue that a creating any property interest in human biological materials would only serve to stifle research and should therefore be contrary to public policy. Alternatively, some researchers advocate creating a property interest in human biological materials because they believe the tying an economic incentive to material donation would increase the overall supply of biological materials available for research.

POTENTIAL SOURCES OF PROPERTY RIGHTS

In 1987 the U.S. Office of Technology Assessment issued a report entitled *New Developments in Biotechnology; Ownership of Human Biological Tissues and Cells-Special Report* (5), which addressed the various ethical legal and public policy concerns regarding legally recognizing human biological material as property. The report identified several areas of law that may support creating a property interest in human biological material. The most relevant of these areas are discussed below.

Patent Law

A patent is broadly defined as a grant made by a government to an inventor which conveys and secures in him the exclusive right to exclude others from making, using, selling or offering for sale his invention for a term of years (6) Governments grant patent rights to their citizens as economic and social incentives to publicly disclose scientific and technological innovations. Patent protection is intended to provide economic and social incentives for inventive efforts made in the advancement of sciences and for the benefit of all. The promise of high financial returns provides the economic incentive for inventors to publicly disclose the fruits of their research, rather than concealing it from competitors. Therefore society benefits from the public disclosure of scientific advancements.

Under U.S. patent law, a patentable invention must qualify as patentable subject matter and be novel, useful and nonobvious to one skill in the technology of interest. As discussed in a previous section, in 1980 the United States Supreme Court expanded the scope of patentable subject matter to include artificial living organisms in the case of *Diamond v. Chakrabarty* (1). Since that time, inventions directed to biotechnologically modified organisms are commonly the subject matter of U.S. European, and Japanese patents.

Although biological materials, such as human biological materials, are commonly used in the development of the invention claimed in the patent, the patent is directed only to the invention formed of the human biological materials and not the biological materials themselves.

For example, an inventor may use a tissue expressing a particular genetic trait as raw material for new pharmaceutical. Although the tissue may have been essential to the development of the new pharmaceutical, the actual invention is the new pharmaceutical and not the tissue or raw material used in its creation. Since patentable subject matter must be the product of invention, only the new pharmaceutical and not the tissue used to create the new pharmaceutical would qualify for patent protection.

Although patent law may provide protection for inventions formed out of human biological materials, it does not provide a basis for creating a property right in human biological material.

Laws Relating to the Donation and Transfer of Organs Tissues and Other Biological Materials

The law relating to the donation and transfer of human biological materials looks to see if the biological material is replenishable or nonreplenishable. No federal or state law prohibits the sale of replenishable human biological materials such as blood, plasma, and semen. As a practical matter, however, state laws categorize this type of transfer as a service rather than a sale of a commodity to avoid any potential liability, such as product liability, created by the sale of an imperfect or harmful good. If the transfer of human biological material were treated as a sale of goods, the implied warranties would be subject to implied warranty of merchantability and the implied warranty of fitness under the Uniform Commercial Code (UCC) (7).

The Implied Warranty of Merchantability would require the human biological material to be of "fair and average quality" as described by the seller. It is unclear how this would apply to the transfer of human biological materials. If the materials, such as blood or semen, were infected with a contagion, it would create significant liability for the merchants who distribute the material obtained from donors (8).

Similarly the Implied Warranty of Fitness requires the human biological material described to be suitable for the buyer's purpose (9). The law dealing with the transfer of nonreplenishable materials generally falls within the Uniform Anatomical Gift Act (UAGA) (10) and the National Organ Transplant Act (NOTA) (11).

The UAGA was finally approved in 1968 and has been adopted by all 50 states and the District of Columbia. The Act allows any competent adult to donate any or all of his or her body to medical education, research and transplantation at the time of death.

The NOTA was enacted by Congress in 1984. It prohibits the sale of any human organs such as kidneys, hearts, livers, eyes, and bone marrow. It does not, however, prohibit the sale of human tissues and cells for research, nontransplantation or commercial purposes.

In summary, neither federal nor state law explicitly prohibits the sale of either replenishable or nonreplenishable human biological material if the sale is for research, commercial or nontransplantation related purposes.

Law Trespass to Chattel and Conversion

Under civil law, any intentional interference with the personal property of another is considered "trespass to chattel." Trespass is a civil action that allows an injured party to seek restitution for any injury done to his or her person or property. A trespass to chattel action refers more specifically to any injury done to one's personal property.

If the personal property has simply been taken, the usual remedy is to have the trespassing party return the personal property and to provide monetary compensation for any diminution of the value of the property. If, however, the personal property is converted into another form rendering it useless to the owner, the law allows the injured party to seek monetary compensation under a conversion action, equivalent to the value of the property that was converted. In determining whether a conversion has taken place, the court considers the following factors: the extent and duration of the actor's exercise of dominion or control over a chattel, the actor's intent to assert a right inconsistent with the owner's right of control, the actor's good faith, the extent and duration of the resulting interference to the owner's right of control, the harm done to the chattel, and the inconvenience and expense caused to the owner.

If human biological material is recognized as property, a material donor could assert a property interest in that material. Therefore any unauthorized use of that material would constitute a misappropriation of the donated material and the donor could bring a conversion action for compensation for the unauthorized use.

The Case of John Moore's Spleen

In 1990 the Supreme Court of California considered the legal theories discussed above in the case of *John Moore v. The Regents of the University of California* (12). In this case, the Court determined whether the plaintiff, John Moore, a patient with hairy cell leukemia, stated a cause of action against his treating physician, David Golde, for using his cells in potentially lucrative medical research without his permission. Moore alleged that Golde failed to disclose his preexisting research and economic interests in the cells before obtaining his consent to extract the cells. The Court determined that Moore did state a cause of action for breach of a physician's disclosure obligations, but the Court did not for a conversion action.

Statement of the Facts. On October 5, 1976, Moore visited the UCLA Medical Center after having learned that he had hairy cell leukemia. Moore was hospitalized and extensive amounts of blood, bone marrow aspirate, and other substances were removed. Golde confirmed the diagnosis. At that time the defendant was allegedly aware that certain blood products and blood components were of great commercial value and that these substances, which were found in Moore's biological materials, would provide certain competitive, commercial, and scientific advantages.

On October 8, 1976, Golde recommended the Moore's spleen be removed. Golde informed Moore that the splenectomy was necessary to slow down the progress of his disease. Moore signed a written consent form authorizing the surgery.

Before the surgery, Golde had made arrangements to obtain a portion of Moore's spleen for a special research project following its removal. Moore was never informed of Golde's plan.

Between November 1976 and September 1983, Moore traveled from his home in Seattle to the UCLA Medical Center for treatment at Golde's direction and based on the representations that the visits were necessary for his health and well-being. On each visit, Golde withdrew samples of blood, blood serum, skin bone marrow aspirate, and sperm. Moore was told that these procedures could only be performed there and under Golde's direction.

During the course of these treatments, Golde was actively involved in research on Moore's cells and planned to benefit financially and competitively from the exclusive access to these cells through the physician–patient relationship.

Prior to August 1979, Golde established a cell line from Moore's T-lymphocytes (Mo Cell line). On January 30, 1981, the Regents applied for a patent on the cell line listing Golde as one of the inventors. The patent, U.S. patent number 4,438,032, was issued on March 20, 1984. The Regents and Golde entered into several agreements intended to commercialize the Mo cell line.

Based on the allegations discussed above, Moore filed suit stating 13 causes of action including conversion, lack of informed consent, and breach of fiduciary duty. The superior court only recognized the cause of action for conversion.

Majority Opinion

The California Supreme Court addressed the issues of Golde's alleged breach of fiduciary duty to his patient Moore, and Moore's action for conversion of his biological materials. The Court ruled that Moore stated a cause of action for breach of fiduciary duty but did not state a cause of action for conversion.

Breach of Fiduciary Duty. Under the law of California and all U.S. jurisdictions, a physician has a duty to inform a patient of all the information material to a proposed treatment or procedure before the patient consents to the procedure. This allows the patient to make an informed decision about undergoing or proceeding with a treatment or procedure. For example, a physician has a duty to inform a patient of the risks that are believed to be relevant to the patient's decision such as the mortality rate and success rate of the treatment or procedure, and the possible side effects associated it.

The Court ruled that a physician's economic interest in a patient's treatment should be disclosed to the patient because it may be relevant to that patient in his or her decision to proceed with the treatment. Golde therefore had a duty to disclose his economic interest and the research involving Moore's biological materials to Moore before Moore then underwent the splenectomy. In addition Golde has a continuing duty to disclose his financial interest in Moore's follow-up care. Since Golde did not inform Moore of this information, the Court rule that Moore did state a cause of action against Golde for breach of fiduciary duty to a patient for failing to obtain informed consent. The suggestion is that if Moore would have been aware of Golde's financial interest in Moore's biological materials, Moore may not have undergone the treatment. This situation raises a question of fact that would be answered by a jury if the case proceeded to trial.

Conversion. The California Supreme Court did not allow Moore's claim for conversion. To state a cause of action for conversion, a party must show that he or she has a property interest in the material converted. A property interest is only present if the material is legally recognized as property. California law, however, does not recognize human biological materials as property. Therefore Moore cannot have a property interest in his biological materials and his action for conversion could not stand.

The Court discussed the public policy concerns in favor of and in opposition to recognizing human biological materials as property. Ultimately the court determined that human biological material is necessary to the advancement of biomedical research, and any impediments to the advancement of this research, such as any uncertainty of title, would be against public policy. The Court determined that a legal recognition of biological material as property should be done through legislation.

Justice Arabian's Concurrence

Justice Arabian believes the majority's opinion correctly concludes that a tissue donor does not have a property interest in his or her donated tissue. Arabian's opinion, however, stems from moral rather than economic concerns. It is feared that a recognition and enforcement of a property interest in human body parts would have a highly negative effect on human dignity. Arabian notes that the plaintiff is not left without a remedy. Moore may pursue an action under the breach of fiduciary duty theory.

Justice Broussard's Dissent

Contrary to the majority opinion, Justice Broussand believes that Moore does, in fact state a conversion cause of action because he alleged that Golde wrongfully interfered with Moore's right to determine how his spleen would be used. Unlike the majority, justice Broussard asserts that conversion is not only applicable when personal property is taken but also when it is used without authorization. This position may be supported by the Uniform Anatomical Gifts Act.

The UAGA allows a donor to control the express manner in which the biological materials will be used after his or her death. Accordingly Moore arguably had a similar right to determine how his biological materials would or would not have been used after their removal.

Justice Mosk's Dissent

Justice Mosk believes the majority's opinion is incorrect because it fails to recognize a difference between the use of materials used for nonprofit research purposes and commercial use. Mosk believes that Golde's failure to disclose his economic interest in Moore's biological materials amounted to commercial exploitation of Moore. Furthermore Mosk concluded that Moore should have a property interest in his tissues that would allow him to enter into contracts for the exploitation of his tissues if he so desires.

Impact of the Moore Decision

Following the Moore case, both public and private parties having interest in the use of human biological materials began to reevaluate their methods and policies regarding the procurement of these materials. In view of the majority opinion the focus of the discussion shifted from concern over the creation of a property interest to reviewing and revising the standards of informed used in obtaining human biological materials from patients and/ or research subjects.

ROLE OF INFORMED CONSENT IN THE PROCUREMENT OF HUMAN BIOLOGICAL MATERIALS

Since the Moore decision, society has developed a deeper understanding of the value connected to human biological material and the potential harms and benefits that may arise out of its commercialization in research. As a result the ethical principle of respect for persons has expanded to require disclosure of any potential harm or benefit resulting from the intended harvesting or use of the material, to the potential donor.

The ethical principle of *respect for persons* recognizes an individaul as an autonomous person capable of deliberation about personal goals who acts under that deliberation. Therefore, to respect the autonomy of the individual is to give weight to his considered opinions and choices while refraining from obstructing his actions unless they are harmful to others. To show lack of respect for an autonomous person is to repudiate the person's considered judgement when there is no compelling reason to do so (13).

Respect for persons requires that each research subject be given the opportunity to choose what will or will not happen to his or her body or biological materials. This opportunity is provided when adequate standards for informed consent are satisfied (13). To accomplish this, an individual must be given all the information regarding the nature of the procedure, the potential risks and benefits arising out of it and all other factors, such as economic considerations, that may bias the situation.

The requirement of informed consent was originally applied to a traditional biomedical research setting in which the subject received an experimental treatment. The requirement was intended to ensure that the research subject knew the risks involved in the procedure to be performed. Traditionally the doctrine of informed consent was not extended to require a disclosure of any future use of the human biological materials taken from a person. In recent years the informed consent requirement has been extended to apply to the procurement and use of any human biological materials obtained from a known donor as in the Moore case.

On December 13, 1995, the *Journal of the American Medical Association* published the results of an NIH workshop charged with formulating informed consent requirements and their application to the gathering of human biological materials that may be used in genetic studies, and with identifying when further consent for the use of material samples already in the possession of researchers should be obtained (14). The participants concluded informed consent is required for all genetic research using samples that may be linked to an identifiable person (14). The participants further concluded that the disclosure of genetic information could have medical, psychological, and economic implications on the material donor. Therefore individuals have the right to put limits on the use of their biological materials. For example, they may specify that their tissues shall only be used by a noncommercial entity.

In August 1999 the U.S. National Bioethics Advisory Commission (NBAC) issued its report entitled Research Involving *Human Biological Materials: Ethical Issues and Policy Guidance* (15). As in the 1995 NIH report, NBAC discusses the importance of human biological materials in the advancement of medical research and recommend obtaining informed consent from all identifiable tissue donors including disclosure of any economic interests of the parties and all potential physical, social, and psychological harms they may arise from its procurement or use.

PRACTICAL APPLICATIONS RECOMMENDATION

As a practical matter, it is essential that any individual, company or organization that is involved in the procurement and use of human biological materials, obtain full informed consent from all potential material donors. More specifically, the economic interests of the researchers, or procuring organization, the proposed use of the materials, and any physical, social, psychological or economic harms that may arises from the use of the materials, must be disclosed to the potential donor prior to the donation. In addition, it is advisable to have the potential donor explicitly transfer any title or property interest to the procuring individual or organization before the donation occurs. Although a donor's property interests in their human biological materials are not currently recognized by any state in the United States, it is important that the potential donor is fully informed about the implications and potential economic ramifications of the transfer of title before it is granted. The biological material may be transferred to a jurisdiction that recognizes such a property interest in the future. Therefore, it is important to take every precaution to ensure the title to the human biological material in question is clear.

BIBLIOGRAPHY

1. *Diamond V. Chakrabarty*, 444 U.S. 397 (1980).
2. Comment, *Am. Univ. Law Rev.* **44**, 2433, 2435 (1995).
3. 38 Illinois Annotated Statute 15-1 (1979).
4. 3A Maryland Code, Article 27, §340(h)(10).
5. U.S. Congress, Office of Technology Assessment, *New Developments in Biotechnology: Ownership of Human Biological Tissues and Cells — Special Report*, U.S. Congress, Office of technology Assessment, Washington, DC, 1987, p. 24.
6. 35 U.S.C. §271 (1996).
7. American Law Institute and National Conference on Uniform State Laws, Uniform Commercial Code (1978).
8. Uniform Commercial Code §2-314.
9. Uniform Commercial Code §2-315.
10. Uniform Anatomical Giftt Act, 8 A.U.L.A. 15 (1985).
11. National Organ Transplantation Act: Public Law 98-507 (1984).
12. *Moore v. Regents of the University of California*, 51 Cal. 3d. 120 (1990).
13. Office of the Secretary, The National Commission for the Protection of Human Subjects of Biomedical and Behavioral Research, *The Belmont Report: Ethical Principles and Guidelines for the Protection of Human Subjects of Research*, (April 18, 1979).
14. E.W. Clayton et al., *J. Am. Med. Assoc.* **274**, 1786 (1995).
15. National Bioethics Advisory Commission; *Report and Recommendations: Report Involving Human Biological Materials: Ethical Issues and Policy Guidance*, vol. 1, NBAC, Washington, DC, August 1999.

See also PATENTS AND LICENSING entries.

P

PATENTS AND LICENSING, ETHICS, INTERNATIONAL CONTROVERSIES

GÖRAN HERMERÉN
Lund University
Lund, Sweden

OUTLINE

BACKGROUND

The view, in general, on what biotechnological inventions may be patented is more restrictive in Europe than in the United States, although the differences have diminished over the last few decades. Moreover there have been important variations in the United States, Canada, and the European Union (EU) positions, as well as among their peoples over time, as a brief glimpse at their historical backgrounds will indicate.

The 1961 UPOV Convention (Union internationale pour la protection des obtentions végétales) has continued to be endorsed by many European countries and also by the United States and Japan. It provides protection for new plant varieties in a different way than the patent system does. It means that a plant breeder and plant improver can obtain exclusive right to use a particular brand of plant according to certain conditions that are specified in the convention. This convention was revised in 1978 and later in 1991. The new UPOV convention provides a somewhat stronger protection of the rights of the plant breeders.

A new instrument of International patent cooperation, the Eurasian Patent Convention, was signed in Moscow on September 9, 1994, and entered into force on August 12, 1995.

United States

In biotechnology a number of difficulties have been encountered in the attempts to patent living materials. Products of nature are not patentable under U.S. law. Nevertheless, before the rapid expansion of modern commercial biotechnology, protection by patents had been granted on materials derived from natural sources through human intervention, such as substances obtained by purification of naturally occurring products.

In 1930 the Plant Patent Act made it possible to extend protection to new and distinct asexually propagated plant varieties. To be sure, the protection was subject to certain limitations: a research exemption that allows use of protected varieties to develop new varieties, and a farmer's exemption that allows farmers to save protected seed for use on their farms or for sale to other farmers.

Before 1980 patents could be obtained on processes using bacterial strains for commercially valuable purposes such as in producing antibiotics. But bacterial strains per se could not be patented in the United States. However, the United States Supreme Court decided in 1980 that a genetically modified living single-celled bacterium, transformed to give it the capacity to break down multiple components of crude oil, could be patented. The Patent and Trademark Office (PTO) soon extended the categories eligible for patent protection, to such products as corn plants (1985), polyploid oysters (1987), as well as to "nonnaturally occurring nonhuman, multicellular living organisms, including animals."

The first patent on a transgenic animal was issued in April 1988 to Harvard University for the development of a mouse with a human onco-gene that makes it susceptible to cancer, the OncoMouse. This extension to animals has generated considerable public controversy, however. Restrictive legislation, including a moratorium on animal patenting, has been proposed (1–3).

The present American patent law is the Utility Patent Act, in Title 35 of United States Code, which among many other things states that "whoever invents or discovers any new and useful process, machine, manufacture, or composition of matter, or any new and useful improvement thereof, may obtain a patent therefor" (35 US Code 101).

Canada

In Canada, patent legislation governing drugs evolved through a series of amendments to the Patent Act. From 1923 to 1993 Canada operated a system of "compulsory licensing," described in some detail by Mathews (4), allowing generic copies of patented medicines to be manufactured within Canada and, by 1969, to be imported.

In 1987 the Act was amended. A compulsory license could be issued on patented medicines only after a fixed period of market protection. Moreover a price review board was created to monitor and control prices charged. Brand-name drug companies promised to invest a growing percentage of sales revenue in research and development in Canada, in exchange for patent protection.

A further amendment to the Patent Act in 1993 fundamentally changed the legislation by abolishing the system of compulsory licensing and applying general patent regulations to medicines. In that way Canadian law was brought into line with that of its trading partners, particularly the United States. It is now illegal to sell a copy of the drug until the patent has expired.

Europe

The European Patent Convention was signed in Munich on October 5, 1973. The European policy on patentability is in general stricter than the one in the United States, even though there have been variations among the national patent offices and the European Patent Office (EPO) in Munich. Before 1998 the existing legal framework did not allow the patentability of living organisms in the European Community countries, notwithstanding some decisions of the European Patent Office (which still are contested).

This changed with the Directive 98/44/EC of the European Parliament and the Council issued on July 6, 1998. This directive grants free circulation of patented biotechnological products, guaranteeing compliance with the European Patent Convention from 1973, the trade-related aspects of Intellectual Property Rights agreement of April 15, 1994, and the Rio de Janeiro Convention on Biological Diversity of June 5, 1992.

The European Patent Convention also states that European patents will not be issued for plant or animal varieties and essentially biological processes for the production of plants or animals, with the exception of microbiological processes or the products thereof. Nevertheless, EPO in 1991 issued a European patent to Harvard University on its transgenic mouse, after having refused this application a few years earlier.

According to Schatz (1998) in Europe, only some 15,000 European patent applications have been filed in biotechnology and over 2000 patents have been filed for DNA sequences isolated from the human genome used to develop therapies and medicines.

CONTROVERSIES

There are controversies within and among countries as to what can be patented, the conditions of patentability, how these conditions have been interpreted and applied, as well as concerning the exceptions to these conditions. There are also conflicting views on the entire system of patenting forms of life, in particular animals and plants. By implication, these controversies have a bearing on the reasons accepted for and against the system of patenting

different forms of life. Even though there are variations in views between the developing and developed countries, the latter on the whole take a more negative view of the present system, which they feel has been exploited.

The competition over transgenic technology rights has financial implications. Those individuals who have the intellectual property rights can protect their discoveries by patents and earn money by getting royalties from others or by exploiting the intention themselves. It is then hardly surprising that there are objections to such rights and that they have sometimes resulted in court cases (5–7). But the basic issues concern what can be patented and on what conditions.

Conditions of Patentability

The conditions of patentability include utility, novelty, nonobviousness. Some of the clauses in patent laws and patent conventions contain difficult but important expressions, like "utility," "novelty," and "inventive step" as well as "living organism." Each of these conditions can be interpreted and applied in different ways (8,9). Controversies as to how they are to be interpreted and applied in practice are to be expected.

What Is Excluded? And Why?

What is excluded from patenting, and how are these exclusion clauses to be interpreted? In the European Patent Convention (EPC) a broad concept of invention is being used. There are three express exceptions:

1. Methods for treatment or therapy.
2. Essentially biological processes for producing plants and animals (and the results of these processes).
3. Inventions contrary to public order and morality (in Europe at least).

The terms "plant variety," "animal variety," "essentially biological processes," "invention," and "public order and morality" can be interpreted in different ways, since the praxis of interpretation has evolved over time. The novelty requirement presupposes that there is a distinction between earlier existing plant varieties and a new one. In this context the definition and criteria of identity of "plant variety" becomes relevant.

Klett (10), Moser (11), and Thomsen (12) have discussed the interpretation of various clauses of exemption in the legislation and practice of several different countries (Germany, Switzerland, Denmark). Other legal scholars Kern (13) and Schatz (14) have elucidated the views on patentability and the patenting practice in the United Kingdom and EPO. The various changes in the Belgian attitude toward the patenting of biological material have been described and analyzed by van Overwalle (15).

Lugagnani (16) proposes that in setting practical limits to patentability via ethical considerations, the moral judgment should move from exploitation of the invention to the nature and/or objectives of the research and development (R&D) projects that produced it. In his view, unethical R&D activities should not be rewarded by granting intellectual property rights. The crucial question then becomes: What is the basis of these moral judgments? And who has the privilege of interpretation?

Lugagnani suggests that ethical guidance be derived from the 1996 Council of Europe Convention on human rights and biomedicine as well as from the Directive 98/44/EC of the European Parliament. According to Article 7 in the directive, the European Commission (EC) states that its European Group on Ethics in Science evaluates all ethical aspects of biotechnology. The group has already discussed several aspects in earlier reports, and several others are on their way.

Several tasks ought to be separated in the discussion of these clauses of conditions and exceptions. Clearly, there is the empirical task of finding out how, in fact, they have been interpreted by particular patent offices. In the literature, this has been discussed extensively. But, since the clauses, like a musical score, are in general open to many interpretations, there is also a question with normative implications: How are they to be interpreted? In a way the question can be understood as a means-end question: How are the clauses to be interpreted if one wants to achieve (or avoid) certain ends? Obviously an author can at the same time be interested in discussing and answering all these questions.

What Can Be Patented?

Today's candidates are microorganisms, plants, animals, and genetic material from humans as well as the methods or processes used in biotechnology to modify them. While attempts to manipulate and patent human life, also before birth (embryo, fetus), is controversial and forbidden in some countries, the situation is different when it comes to microorganisms.

If gene technology and patents only served good purposes, for example, in the production of medicines, vaccines, diagnostic methods, or protection of environment (e.g., producing bacteria that break down crude oil), objections would be more difficult to obtain and sustain. A great deal of criticism has to do with the use of human growth hormone to get salmon to grow quicker, for example, and a range of similar enhancement purposes. With regard to animals, the public generally considers it unethical to produce and patent animals with properties that cause them to suffer. It has been argued that plants, animals, and large organisms should not ever be patentable.

The general idea in the July 1998 Directive 98/44/EC of the European Parliament and the Council is that discoveries per se are not considered patentable, nor are plants and animal varieties per se, nor essentially biological procedures for the production of plants and animals. Also methods of surgical and therapeutic treatment and diagnostic methods applied to animal bodies are not considered to be inventions of the sort that may be patented. But plants and animals with newly introduced genetic traits, like the famous OncoMouse, may be protected by patents. The same goes for biological materials and material isolated from its natural environment and isolated elements of the human body with technical processes.

In Europe, inventions contrary to law and order or public morality may not be patented. This includes processes for human cloning for reproductive purposes — banned by a protocol to the Council of Europe Convention on human rights and biomedicine (17). The same holds for processes for modifying the germ-line identity of human beings and the use of human embryos. Processes for modifying the genetic identity of animals with substantial medical benefit for humans, because they are useful for treating serious diseases such as cancer or AIDS, may be protected by patents. If this is not the case, they are not considered to be patentable.

Computer-aided molecular design techniques have recently been used along with more traditional methods to design new peptides possessing specific properties, such as bactericidal activity. Patel and Colleagues (18) have argued that the ability to protect the intellectual property rights associated with the discovery of the new molecules is a key issue for commercial utilization of such peptides, and the authors have described and proposed an extension of established patenting practice.

Moreover Green (19) argues that combinatorial libraries are patentable as long as the library meets the statutory criteria of utility, novelty, and nonobviousness, and the application meets the standards of enablement, best mode, and written description. Licensing and alternatives to patenting are also considered, along with potential problems unique to combinatorial chemistry agreements. A particularly interesting illustration of the diversity of views on the ethical aspects of gene patents, reflecting deep disagreements as to both conclusions and arguments, concerns patenting of disease genes.

Patenting of Disease Genes?

The possibility and desirability of patenting disease genes is a much-debated area where heated controversies have attracted a great deal of attention around the turn of the twenty-first century. For example, McGee (20) has argued that gene patents can be ethical. He examines some arguments against patents in this area and concludes that patenting on methods for detecting the presence of a genetic correlation with disease-related (and other) phenotypes can be appropriate.

The main reason offered for this position by McGee is that there is a subtle distinction to be made between "observing DNA" and constructing a DNA-based product for diagnosis of some disease or phenotype. He examines two arguments that oppose his view: (1) genetic information is part of nature and (2) gene patents create a "toll bridge" barring research using patented genes.

Taking the first point, McGee argues that disease gene patents are more an innovation of scientists than a discovery. "Finding DNA is discovery. Correlating it with human life for the purpose of creating a diagnostic process is innovation" (20).

The second position poses a problem: How is one to avoid the issue that genetic research will be hindered, that genetic tests will not be developed, if researchers in the early stages of work are not able to pay the tolls necessary to start the research? McGee claims that this can be prevented by "correctly framed and issued gene patents."

He argues that infringement of the patent occurs not when methods are used for further research but only when the actual innovation is put to use in the same sense that the method of detecting the gene's or allele's correlation with a particular phenotype is applied to a patient.

Kimbrell (21) is an important early work that has played a central role in the debates on engineering and marketing of life. Various positions in the subsequent debate have been advocated by several researchers (22–27).

Caplan and Merz (23) have instructively compared the identification of disease genes to the land rush. They agree that the debate is timely, for a number of reasons, but contend that McGee's position is untenable. Caplan (26) argues laying claim to own what we find in the biological world cannot be done without an arrogance or hubris about our creativity, that patenting genes is profane — it is "commercializing something that ought to remain outside the realm of commerce," and that "allowing patents ignores that it is our common genetic glue that holds us together as a species, and that is something all of humanity has as its common legacy." These metaphysical concerns are especially at the heart of religious-based objections to patents, according to Caplan.

Merz and Cho (27) argue that McGee is wrong and that methods of screening for Alzheimer's disease should not be patentable. Obviously they criticize that a U.S. patent on such methods has been granted by PTO. More specifically, they argue that there are two false premises in McGee's argument:

1. The difficulty and effort involved in making a discovery makes the discovery patentable.
2. Patentability is based on the usefulness of the discovery.

They argue that market potential is not a necessary, much less a satisfactory, condition to determining whether something comprises patentable subject matter. They point out that gold and diamonds, though valuable, are not patentable, regardless of who first discovered gold and diamonds and how difficult this discovery was.

Referring, for example, to Eichenwald (28), they conclude by arguing that broad diagnostic patents may led to the abandonment of methods of prenatal or genetic testing that long have been the standard of care: "This is simply unacceptable. These patents are contrary to good medical practice, and must be prohibited" (27).

Also Tribble (29) argues against exclusive licenses with a single commercial developer for all uses of the invention: "This limits both the potential benefits for human health and the potential for greater financial returns to the academic institution that could stem from their research" (29). Therefore he proposes a kind of compromise position. He argues that scientists could protect their intellectual property rights by granting access to the research tools for research use via nonexclusive licenses to scientists in both academic and industrial laboratories, and simultaneously reap financial returns and realize the commercial potential of the patented technology. He points out that "refusal to grant non-exclusive licenses ... could have a chilling effect on biomedical research and would be contrary to public interest" (29).

Magnus (30) criticizes the approach of PTO in the United States, in particular, the constructivist interpretation of genetic testing to avoid the "product of nature" doctrine. (If something is a product of nature, it does not satisfy the earlier mentioned conditions of novelty and inventive step.) He argues that the strategy of the defenders like McGee will result in an intolerable dilemma for physicians.

ALTERNATIVES AND CONSEQUENCES

Some important questions in this debate are as follows:

1. Which are the available alternatives: protection by patents, with or without specified exemptions, trade secrets, protection of intellectual property rights in other ways, such as by the UPOV-convention, compulsory licensing or no protection at all?
2. What is the range of application, if any, of these methods of protection? Each method could have a different area of application (one for microorganisms, another for plants, still another for animals, etc.), they could be partly overlapping, or totally overlapping.
3. What are the actual or likely consequences of these alternatives for the various stakeholders, including the pharmaceutical and agrobiotechnological industry, farmers and animal breeders, the researchers, the universities, the developing countries, and the entire population of the earth, expected to double in less than 50 years?
4. What ethical considerations should be used in evaluating these consequences: (which) deontological considerations, utilitarian theories, contractualist ethics, and the like? For obvious reasons, it will be impossible to take a stand on the latter issue here. What is important is only to make clear that the choice of ethical framework will have practical consequences; hence transparency and explicitness is important.

The first of these questions deserves some elaboration. Patents provide protection of intellectual property rights. Patents also expire after a number of years and are limited to a geographic area. A patent granted by a national patent office is valid in that country, a patent granted by EPO is valid only in Europe, and so forth. Moreover it takes time and is costly to obtain a patent, and the patentee has to check that no infringements on his or her rights are made. If infringements are made, the patentee has to prove that he or she has suffered damage. Besides, it is always somewhat uncertain what the rulings of the courts will be. What is certain, however, is that the system provides patent lawyers with a great deal of work; it is good for them.

Therefore alternatives to the patenting system have been considered, and the consequences of each alternative, within each particular area (microorganisms, plants, animals, etc.) ought to be evaluated in some detail. An alternative is to keep the invention secret in order to use and develop trade secrets. This may work (as it has for Coca Cola) if the product is difficult to copy. Further alternatives include inventing other forms of protection, like the UPOV convention for plants, or to carry on without protection of intellectual property rights at all.

REASONS FOR THE PATENTING SYSTEM

The controversies over protection of intellectual property rights via patents also concern the foundations of this protective system and the reasons offered for and against patenting. The main arguments in favor of patenting biotechnological inventions center on justice, economic growth, and openness.

Justice

The standard reasons for patenting include considerations of justice. The idea is simply that the inventor should not be without financial reward while others make a fortune on the basis of his or her invention. The inventor is entitled to economic compensation. The underlying conception of justice in this context is justice according to desert. This means, for instance, that universities can get royalties on biotechnological inventions made by their researchers. For big universities like Harvard or Stanford in the United States these royalties amount to very large sums every year. If these inventions had not been protected by patents, big companies could use these inventions free of charge to increase their own profits.

Economic Growth

The second main line of argument is concern for economic growth—the desire to stimulate the biotechnology industry by the possibility of financial rewards. Society needs research and development, and patent laws are designed to encourage this. Protection of intellectual property rights encourages investments by the private sector in agrobiotechnology simply because it increases the chances of financial reward. For instance, Dompe (31) argues that patents are a useful legal instrument to protect intellectual property rights, allowing society to promote and to support the innovation and development of useful inventions.

Openness

The disclosure requirement in patent laws prevents secrecy and attempts by researchers to preserve exclusivity in access to their discoveries. Thus patent laws work in favor of openness. To obtain a patent, the applicant has to send in a detailed description of what the applicant wants to protect. This description should make it possible for anyone in the field to repeat the process or to manufacture the product in the future. These descriptions can be bought at a very reasonable price from the patent office.

It could be objected that the patent system slow's down scientific publications. But in many countries this problem is taken care of by a clause in the patent laws. In order not to discourage early publication of research results because the inventor is considering to apply for a patent on his discoveries, patent laws in the United States and Canada contain a clause about a "grace period," during which time the author may wait to apply for a patent. In fact this encourages early publication, since the author who publishes first also has the right to (apply for) a patent. But to say this is not to say or suggest that the U.S. patent law requires filing before publication.

REASONS AGAINST THE PATENTING SYSTEM

The standard arguments offered against protection of biotechnological inventions by patenting could be classed into two broad categories: (1) objections to the underlying technology itself, rather than to its protection under patent laws and (2) objections to the actual or anticipated effects of patenting itself on research and health care, for example (32). Here focus will be on the latter ones. But on a more general note, Hoffmaster (33) has argued that patenting of life forms promotes an irreverent materialistic conception of life. It may affect our attitudes toward life negatively when life forms are treated as commodities to be bought and sold in the market (34), particularly when the organisms to be patented have been modified by gene technology so that they possess human genes (2).

Effects on Research

It has been feared that the patent system will distort the research agenda in favor of potentially lucrative projects rather than the traditional mission of expanding knowledge regardless of potential financial gain. It is clear that there is at least a potential tension between commercial imperatives and traditional scientific norms, such as is evident in Merton's oft-quoted CUDOS norms (35):

- Communism, according to which there is no private ownership of knowledge
- Universalism: hypotheses, objections, and criticism should be treated equally, regardless of the age, sex, race, and nationality of the author
- Disinterestedness: impartiality is essential and researchers should not be influenced by their own economic, religious, political and other ideological interests
- Organized Scepticism: everything may be questioned, and this ought to be done systematically

Heller and Eisenberg (36) take the metaphor of the "tragedy of the commons" as their starting point. This metaphor helps to explain why people overuse shared resources, like water and air. The authors argue that the recent proliferation of intellectual property rights in biomedical research may lead to a different tragedy, a tragedy of the "anticommons," in which people underuse scarce resources because too many owners can block one another. The authors conclude that privatization of biomedical research must be more carefully deployed to sustain both research and product development. Otherwise, more intellectual property rights may lead to fewer useful products for improving human health. This article has provoked several comments and responses (37).

Along similar lines Merz (38) has argued that the rapidly growing number of disease gene patents — patents protecting all methods for diagnosing a particular genetic condition — will threaten the ability of physicians to provide optimal medical care to their patients. Disease gene tests are being monopolized by a small number of providers. This, according to Mertz, threatens to restrict research activities, creates unacceptable conflicts of interest, may reduce patient access to testing, and may lead to inequitable extensions of patent terms on tests and related discoveries. It will make it possible for patentees to dictate the standard of care for testing, and in that way to interfere with the practice of medicine. In view of these risks, Merz suggests an amendment of the patent law to require compulsory licensing of physicians providing medical services. In the United States, a National Institutes of Health (NIH) proposal, discussed by Flores (39), will restrict the licensing of federally funded biomedical research tools for commercial gain.

Effects on Developing Countries

The development of genetically modified plants, fruits, and animals may undermine the export of products (plants, fruits, sugar, raw material, etc.) from developing countries; the products will no longer be competitive. Since the biotechnological industry is situated in the developed countries, this could increase the gap between rich and poor countries. The rich will be made richer, while the poor will become poorer.

It has been feared that patents can be used as an instrument to exploit the third world. A double exploitation has been evoked. Researchers from rich countries, where advanced biotechnology industries are located would harvest and refine desired biological material from the developing countries. The derived products now protected by patents or the UPOV convention, would be sold back to farmers in the countries from which they originate, and the seeds made sterile so that they can only be used for one crop.

Concern has also been expressed in the *Bulletin* of the World Health Organization (WHO) that recent global developments in the regulation of trade and intellectual property rights threaten to hinder the access of populations in developing countries to essential drugs. Velasquez and Boulet (40) have argued for state intervention in the health and pharmaceutical markets in order to guarantee equitable access to these essential products.

Serageldin (41) has argued that biotechnology can contribute to food security if it benefits sustainable small-farm agriculture in developing countries. The agrobiotechnology focus has been on ethical, safety, and intellectual property rights issues. An effective system of protection of intellectual property rights would encourage investments by the private sector in agrobiotechnology by increasing the chances of financial gain. But, as Serageldin argues, in developing countries the needs of small farmers and environmental conservation are unlikely to attract

private funds. Public investment will be needed, and new and imaginative public–private collaboration could make the gene revolution beneficial to developing countries as well.

Thus there is particular cause of concern for the effects of licensing and patents on farming and agriculture in the developing countries. Traditionally these countries have had two important sources of income. The first is cheap labor, which is less important today than it was, because of the rapid development and widespread use of industrial robots and information technology. The second important source of income is their raw material and agricultural products, which the developed countries do not need today to the same extent as before, because of the rapid progress of biotechnology.

Effects on Environment

The major part of the world's collected genetic resources are to be found in tropic or subtropic areas, that is to say, mainly in the developing countries. Will genetically modified plants yielding better crops in the future replace traditional domestic plants? Will this lead to a reduction of biological diversity? And is this to be deplored? For what reasons? Here empirical and value questions are intertwined. There is clearly a risk that biological diversity is threatened by present developments in plant breeding.

Will patenting increase the risk for environmental problems? Will patenting of animals make the situations of animals worse? To view the problems in a proper perspective, however, it is important to realize that environments can be endangered also by release of products that are not protected by patents. This is usually the case in air and water pollution. Thus the pressing problems of preservation of the environment need to be addressed independently of the problems discussed in this article.

A precautionary principle has been launched in the international debate, in the wake of the Bovine Spongiform Encephalopathy (BSE) ("mad cow disease") and various threats to our environment: Irreversible processes in nature should not be started, such as by using genetically modified and patented seed (or salmons) that may spread in nature in uncontrollable ways. Ethical arguments in this area had better be regarded as preliminary and provisional, and if mistaken judgments are made, such mistakes should preferably have been done for a good purpose. This, of course, presupposes that ethical analyses are made both of the research that eventually leads to a patent, in the application for a patent, and in the use of a patented product.

Effects on Industry, Farming, and Agriculture in the Developing Countries

The impact of patents on demographical structure of society may alter the settlement patterns of populations. Sparsely populated regions may become more sparsely populated due to migrations. The direct and indirect impacts of migration on various economic interests is another cause of concern and another reason for criticism of the patent system.

It has been argued, for example, that the higher cost of patented livestock or seed will restructure farming. Small farms will be put out of business, and there will be even more concentration of corporate agriculture. Agricultural patenting also undervalues the contributions of less developed countries to biological diversity (42).

Further, with the increasing globalization of the world economy, the biotechnology industry may be restructured worldwide. Knowledge and economic resources may be concentrated in a few big corporations. Already large corporations have merged, and small companies are finding it difficult to survive in the market place, particularly those in the developing countries.

VALUE CONFLICTS AND FURTHER ETHICAL ASPECTS

What are the possible benefits? Which are the risks? To make this explicit, we must look to identify some of the value conflicts. The possible economic benefits are as follows:

- Consumer values, in terms of better-tasting grapes, firmer tomatoes, bigger oranges
- Producer values, in terms of increased crops for farmers, pigs with faster or increased muscle growth
- Company values, increasing revenue, such as by developing and marketing plants resistant to pests afflicting specific species
- Environmental values, such as in developing plants resistant to herbicides so that fewer herbicides are used by farmers

The monopolies achieved by patents will stimulate technological inventions, but they may clash with ideals of free competition. A patent will give an air of legitimacy to an invention, but this is sometimes what is challenged. Free competition is a value, and so is reward for beneficial work.

These values may clash or be combined, according to each particular case. They have to be compared to other positive and negative values wherever the values clash: increasing unemployment among the population, immiserizing growth in the developing countries, health risks for patients (if patents and royalties change the standard practice of medicine so that tests cannot be performed because of expensive royalties), health risks for large segments of the population (if genetically modified products are not safe to consume), environmental hazards (if genetically modified organisms are released into nature and cause ecological problems).

So what economic, social, psychological, environmental, and medical values are at risk, and whose values are these? Further, which values are favored or promoted by the various alternative methods of protecting, or not protecting, intellectual property rights?

Is it the prosperity of one group against the prosperity of another? The prosperity of one group against the health of another? The prosperity of one group against the welfare or quality of life of another? If two families of values clash, and both are equally legitimate, we are facing an optimization problem. If one is arguably more legitimate than the other, the situation is, of course, different.

A Strategy of Analysis

The questions are difficult, strong economic and other interests are at stake, and the terminology invites talking at crosspurposes. Therefore, it is essential to be explicit and open, try to separate different kinds of disagreements, make clear what we know as well as in what areas there are considerable gaps in our knowledge, and to identify the alternatives, the stakeholders and their interests, and the principles used in assessing these interests.

Consequences for human and animal welfare, broadly defined, are clearly important considerations in this debate. Consequentialists argue that they are the only important reasons. Moral philosophers of other persuasions want to supplement consequentialist arguments with deontological, contractual, egalitarian, and other considerations, to use appeals to human rights as a basis for claims as to what human beings and animals are entitled to, or as negative restrictions for what others may do to them.

Many other ethical problems than those discussed here are raised by recent developments in biotechnology, such as issues of informed consent and remuneration to those from whom genetic material has been collected. The celebrated case Moore case (*Moore vs. University of California*) illustrates this. The court ruled that Moore was not entitled to a share in the profit made by the company that had used tissues he donated, but the court criticized the information Moore had received as inadequate (43). The special problems raised by human tissue banks (collection, preservation, access, purpose, etc) are also relevant but will not be discussed here.

Underlying Assumptions

The controversies described here are important and interesting because they bring to the foreground a number of underlying and culture-dependent assumptions on what the human being is, what the place of humans is in nature, whether there is a radical and sharp distinction between humans and other living organisms and what constitutes this difference. That is to say, what differences, if any, exist between humans and the so-called undignified biomass, between what is considered natural or unnatural? To what extent is everything living an end in itself, or are humans entitled to exploit animals and plants as well as human DNA for their own purposes, and so forth? Views on these matters may vary somewhat among cultures, and also within a culture over time. There are also indications of interesting recent value changes in the relations between humans and nature. Here is an interesting area for research.

Who owns the genetic resources, in particular the human genome? Among the alternative views are the following:

1. They are the common heritage of humankind. This is a position undermining the whole system of protecting intellectual property rights in this area

via patents: It implies no patents on disease genes, no royalties on biotechnological inventions involving the human genome.

2. The genetic resources belong to the country where it is found. This could be taken to mean that royalties for patents of biotechnological inventions based on genetic material found between the borders of a country should go to the country of origin. This is a more general point, and it is applicable outside the field of biotechnology and genetic resources as well.

3. The invention belongs to the company or the researcher who developed it, applied for, and was granted a patent — the inventor is entitled to royalties. As already mentioned, at the base of this position is an argument of justice, but a different one from that used to support the previous position.

Since a large share of the world's collected genetic resources is to be found in tropic or subtropic areas, consisting mainly of the developing countries, the second alternative might favor developing countries, and this has been their line of argument for the most part. Alternatives discussed at the Rio Convention have also been tried, calling for the transfer of technology from developed to developing countries and royalties on patents developed from genetic material found in the country, in return for the right to harvest and refine genetically the desired material from a developing country.

These basic views could be backed up by religious, social, economic, and legal reasons. Like the underlying assumptions mentioned above, they point to the cultural context in which this debate is taking place, and where there are considerable differences as to views and traditions both between and within developing as well as developed countries.

Appel (44) argues in his analysis of the human body and patent law that human life must not be allowed to be reduced to "undignified biomass" and that protection should be guaranteed from the beginning of life, from embryonic stage. Can patent offices be in a position of assessing the risk factor of a genetic engineering invention? Is there justification for excluding therapeutic treatment from European patent law? Appel argues that this has an extremely inhibiting effect on research.

As is often true in difficult cases, there are good arguments on both sides and many difficulties in assessing the relevance and tenability of the reasons given. Often one has to take a close look at each separate case before making a judgment.

CONCLUDING REMARKS

In view of the strong commercial interests, and the rapid development of biotechnology, future trends are hard to predict. But it is essential that the ethical aspects of patenting and licensing are not neglected. Different constructions of licensing and protection of intellectual property rights, whether by patents or in other ways, will create winners and losers. Who will be the winners, and who the losers, in the short and the long run? And who should decide this? For careful consideration of these

issues, the many underlying conflicts of interests and value clashes need to be made explicit, as well as the principles used in weighing the conflicting interests.

The ethical issues underlying licensing and patents are relevant from the start of the research that eventually leads to the patent, to the examination of the patent application and the decision whether or not to grant a patent, and to the actual use of the patent. This means that all biotechnological research project should be decided under the aegis of an ethical review board, and not just certain medical projects involving patients.

The various patent offices sometimes express frustration when they have to deal with ethical issues. This is not something they are prepared for. It may not be the role of EPO to engage in ethical debates and take a position (45). Other national and international fora exist for the discussion and analysis of the social, ethical, and other extrajuridical aspects of patenting and patentability. One solution may be to suggest different fora and to make a clear division in the discussion and analysis between legal and ethical aspects of patent applications.

Ideally, then, a patent will not be granted if the application offends public order; the issue of morality will be relevant, and controversial applications will be contested. Lawyers will review each application and provide examples of what they consider to be compatible or not compatible with this or that related clause. An attempt could then be made to interpret the clauses in the light of this praxis or to describe the changing praxis of various patent offices.

Since patent applications are certain to come in an increasing extent from different countries, international cooperation and harmonization of patent laws will be important in the future. Comparative research in the present situation of globalization is obviously essential. What are the similarities and differences between the approaches in Europe and the United States concerning plant protection? Concerning protection of biotechnological inventions in general? With focus both on the current framework and future developments (8,46)? The studies should include countries not mentioned expressly here, such as Japan and China. By noting similarities and differences, problems as well as possibilities can be identified.

BIBLIOGRAPHY

1. R.G. Adler, 1988. *Harvard J. Law Techn.* **1**, 1–61 (1988).
2. R. Dresser, *Science* **243**, 1002–1003 (1988).
3. T.T. Moga, *J. Pat. Trademark Office Soc.* **76**, 511–545 (1994).
4. J.H. Matthews, *Clin. Invest. Med.* **19**(6), 470–478 (1996).
5. S. Hensley, *Mod. Healthcare* **27**(23), 52 (1997).
6. J.L. Fox, *Nat. Biotechnol.* **15**(13), 1332 (1997).
7. R. Dalton, *Nature* **400**, 601 (1999).
8. G. Van Overwalle, *The Legal Protection of Biotechnological Inventions in Europe and in the United States. Current Framework and Future Developments*, Leuven University Press, Leuven, Belgium, 1997.
9. R. Rogge, *Int. Rev. Indust. Property Copyright Law* **28**(4), 443–466 (1997).
10. K. Klett, *Int. Rev. Indust. Property Copyright Law* **28**(6), 845–850 (1997).

11. W. Moser, *Int. Rev. Indust. Property Copyright Law* **28**(6), 845–850 (1997).

12. H.C. Thomsen, *Int. Rev. Indust. Property Copyright Law* **28**(6), 845–850 (1997).

13. M. Kern, *Int. Rev. Indust. Property Copyright Law* **29**(3), 247–282 (1998).

14. U. Schatz, *Int. Rev. Indust. Property Copyright Law* **1**, 2–16 (1998).

15. G. Van Overwalle, *Int. Rev. Indust. Property Copyright Law* 1999 (in press).

16. V. Lugagnani, *Forum (Geneva)* **9**(3, Suppl. 3), 93–98 (1999).

17. Council of Europe, Convention for the Protection of Human Rights and Dignity of the Human Being with Regard to the Application of Biology and Medicine, Strasbourg, France, 1996.

18. S. Patel et al., *J. Comput. Aided Mol. Des.* **12**(6), 543–556.

19. G.D. Green, *Mol. Divers.* **3**(4), 247–252 (1998).

20. G. McGee, *Camb. Q. Healthcare Ethics* **7**(4), 417–421 (1998).

21. A. Kimbrell, *The Human Body Shop: The Engineering and Marketing of Life*, Free Press, New York, 1993.

22. G. Poste, *Nature* **378**, 534–536 (1995).

23. A.L. Caplan and J. Merz, 1996. Patenting Gene Sequences. *Br. Med. J.* **312**, 926 (1996).

24. J.F. Merz et al., *Mol. Diagnosis* **2**(4), 299–304 (1997).

25. E. Marshall, *Science* **275**, 780–781 (1997).

26. A. Caplan, *Camb. Q. Healthcare Ethics* **7**(4), 422–424 (1998).

27. J. Merz and M.K. Cho, *Camb. Q. Healthcare Ethics* **7**(4), 425–428 (1998).

28. K. Eichenwald, *N.Y. Times*, May 23, A1 (1997).

29. J.L. Tribble, *Camb. Q. Healthcare Ethics* **7**(4), 429–432 (1998).

30. D. Magnus, *Camb. Q. Healthcare Ethics* **7**(4), 433–435 (1998).

31. S. Dompe, *Forum (Geneva)* **9**(3, Suppl. 3), 113–117 (1999).

32. R.S. Eisenberg, *Yale Law J.* **97**, 177–231 (1987).

33. B. Hoffmaster, *Intellect. Property J.* **4**, 1–24 (1989).

34. T.H. Murray, *Univ. Michigan J. Law Ref.* **20**(4), 1055–1088 (1987).

35. R. Merton, in *The Sociology of Science*, University of Chicago Press, Chicago, IL, 1973.

36. M.A. Heller and R.S. Eisenberg, *Science* **280**, 698–701 (1998).

37. A.H. Fried et al., *Science* **281**, 517 (1998).

38. J.F. Merz, *Clin. Chem. Mar.* **45**(3), 324–330 (1999).

39. M.A. Flores, *Nat. Biotechnol.* **17**(8), 819–820 (1999).

40. G. Velasquez et al., *Bull. World Health Organ.* **77**(3), 288–292 (1999).

41. I. Serageldin, *Science* **285**, 387–389 (1999).

42. J. McGowan, *Cult. Survival Q.* **15**(3), 20 (1991).

43. B.M. Dickens, *Transpl. Proc.* **24**(5), 2118–2119 (1992).

44. B. Appel, *Der menschliche Körper im Patentrecht*, Carl Heymann, Munich, 1995.

45. G. Van Overwalle, in *Biotechnology, Patents and Morality*, Aldershot, Ashgate, UK, 1997, pp. 139–148.

46. G. Van Overwalle, *IDEA-J. Law Technol. (USA)* **39**, 143–194 (1997).

See other entries FEDERAL REGULATION OF BIOTECHNOLOGY PRODUCTS FOR HUMAN USE, FDA, ORPHAN DRUG ACT; HUMAN GENOME DIVERSITY PROJECT; OWNERSHIP OF HUMAN BIOLOGICAL MATERIAL; see also PATENTS AND LICENSING entries.

PATENTS AND LICENSING, ETHICS, MORAL STATUS OF HUMAN TISSUES: SALE, ABANDONMENT OR GIFT

COURTNEY S. CAMPBELL
Oregon State University
Corvallis, Oregon

OUTLINE

Introduction

Body and Self

Attitudes Toward the Body

Conflicts: Bodily Integrity and Disintegration

The Mechanistic Body

Understandings of Body Tissue

Metaphors of Body Tissue

The Moore Case

Transfer and Acquisition

Modes of Transfer

The Donation Paradigm

Contributions

Abandonment

Sales

Conclusion

Bibliography

INTRODUCTION

Almost every person who enters a doctor's office for a routine examination can be sure to leave some of their body tissue behind. Typically requests for a blood or a urine sample are a minimally invasive part of the standard routine of clinical practice, enabling physician diagnoses of ailments, and more commonly, of good health. In nonclinical settings, such as research studies, samples of human body tissues may be collected for purposes of assessing disease prevalence or susceptibility. DNA analysis of waste tissue, such as saliva or hair, can reveal very detailed genetic information about a person. Persons may also part with their body tissue with the intent to benefit others: Gametic tissue, for example, may be donated or sold to enhance the prospects of fertility treatment of another person in otherwise good health. In acute crisis settings, donation of an organ or tissue, such as bone marrow, may be a life-saving therapy.

There are, in short, many purposes in clinical medicine and biomedical research that involve retrieval and sampling of human body tissues. The procurement and use of human tissue in diagnostic, therapeutic, research, and educational purposes illustrates how contemporary scientific study understands the human body as a source of medical information. The National Bioethics Advisory Commission recently estimated there are over 282 million body tissue specimens collected throughout the United States, a figure that is increasing by an additional 20 million specimens each year (1). This practice has been accelerated significantly by research on the human genome and by the advent of genetic technologies in medicine, which can provide valuable diagnostic and prognostic information about a person, family or ethnic community.

This increasing use of body tissue as an information resource in the grand projects of contemporary biomedicine presupposes views about the moral status of the human body, of its parts and tissues individually and in aggregate, and of conceptions of self-identity. These biomedical assumptions may be compatible with, or in conflict with, philosophical and religious perspectives, as well as cultural and personal attitudes. This article will reflect on some major issues regarding the moral status of body tissues, beginning with an examination of the relation between the self, the body, and excised body tissue. Analysis and ethical reflection on the status of body tissue requires attention to prominent metaphorical understandings in scientific and research discourse, and provides a basis for ethical transfers of body tissue from patient to researcher. However, important questions must be addressed regarding the limits of use of body tissue in order to avoid abuse.

BODY AND SELF

Attitudes Toward the Body

Moral reflection on the status of body tissue must first begin by considering tissue in relation to the organic bodily whole and to the self. Bioethicist William F. May has distinguished five philosophical attitudes on the status of the body and its parts that are useful to review in the context of retrieval of human tissue (2):

Dualism — Gnosticism — Holism — Materialism — Idealism

May differentiates these attitudes based on questions regarding whether the body has a phenomenal reality, is ontologically good, and is intrinsic or incidental to personal identity.

Dualism. Dualism affirms the phenomenal reality of the body but denies its goodness because of the body's association with flesh and matter. Dualism portrays the body as "at war" with the self, its literal mortal enemy. Parts of the body, disposal of the body, and persons who come into contact with the corpse are denigrated, stigmatized, and considered taboo and sources of pollution or uncleanness.

Idealism. At the opposite end of the continuum, idealism denies that bodily life has any ultimate significance. The body, disease, and death are constructs of the mind that can be transcended through identification with a separate realm of the spiritual. The self seeks its true home in this noncorporeal realm of spirit. Idealism supports practices of spiritual healing rather than medical ministrations in response to disease. Body parts, or study of the body, are viewed as existentially indifferent within the idealistic worldview.

Gnosticism. The gnostic attitude seeks liberation of the true self from the body through knowledge. The true self resides in a disembodied mind or consciousness, while the body is a prison of the soul. Liberation entails overcoming the burdens of mortality, including finitude, disease and death. Body parts have no significant value, and may even have disvalue, insofar as a preoccupation with bodily life can hinder the quest for liberation.

In contemporary Western thought, May argues that the gnostic attitude has assumed the form of philosophical Cartesianism. Philosophical and moral emphasis is placed on the mind or consciousness, distinct from the body. The body (and body tissue) is a possession and instrument of the rational will, useful as a tool to achieve the self's own goals and aims, but assumes no independent value of its own. Except in those circumstances where the body may impinge on and frustrate the plans and desires of the rational self (e.g., in illness), the body and constituent body tissues can be assessed as morally indifferent.

The Cartesian separation of self (mind) and body (matter) is embedded within the ideology of biomedical research; the body may be perceived merely as a resource for obtaining raw biological materials from which information and knowledge can be retrieved. In turn, this knowledge can be converted into technologies, therapies, and pharmaceutics that enable mastery over nature (3). The separation of self and body would also seem to imply there is no moral necessity for informed consent to the removal of bodily tissue, or subsequent research and manipulation. Even though bodily space is violated by such procedures, since a body is incidental to personal identity, the retrieval of an organ, tissue, or cell cannot be said to violate the person and their integrity. In this regard informed consent appears to be a rule without a rationale. However, the increasing importance of informed consent procedures in the procurement of human tissue for research (4) is a signal of a fundamental inadequacy of the Cartesian/gnostic perspective.

Materialism. In contrast to a knowledge-based quest for philosophic liberation and biomedical mastery of nature, the materialistic attitude to the body is informed by an ideology that human life is subject to, and at the mercy of, powers in nature that are arbitrary, abusive, and destructive. This can lead to two different and conflicting perspectives, avoidance and denial (of aging, death, etc.), or resistance, both of which resonate in contemporary Western cultures, including biomedicine. The perspective of resistance is enacted primarily through the practice of medicine and its war on death and human disease (5). In the medical effort to resist the inevitable natural course of bodily morbidity and mortality, the body assumes the role of primary "battleground." Patient consent to invasive procedures (an "invasion" not only of the body, but against nature) still leaves the patient as principally a passive observer to the battle plan carried out by physicians. A successful war often requires excision of body parts and removal of tissue. This body tissue is not only surplus but also is a valuable diagnostic and prognostic information source.

Holism. Within the typology constructed by May, the dualistic, gnostic, and materialistic attitudes are compatible with the interests of biomedical research in human tissue. The idealistic attitude, by contrast, finds in medical research a misguided attempt at medicalization of the metaphysical. This implies that scientific proposals for using human tissue samples are an *ethical* issue primarily in those traditions — philosophical, religious, and medical — that affirm a holistic attitude of human embodment. Holism expresses an intrinsic relationship of body and self, in contrast to the instrumental value embedded in Cartesian and materialist thought. Holistic traditions claim for the body an ontological reality of the body (in contrast to idealism) and an intrinsic moral goodness (in contrast to dualism). Moreover the traditions of holism affirm the intrinsic nature of the body to personal identity: Human beings are embodied selves, not simply a soul or a mind housed within a corporeal prison.

The holistic attitude thereby supports a presumption in favor of bodily integrity and intactness (even beyond death in some cultural and religious traditions). However, this presumption can be overridden for purposes of preserving the health of the person or for providing direct therapeutic benefits to others, as in transplantation. The reality and goodness of the body entail the use of medical procedures to restore and heal. The fundamental connection of body and self makes consent of the person a moral mandate with respect to invasive medical procedures and removal of bodily tissue, healthy or diseased. Given the integration of body and self, holism places significant emphasis on accountability for uses of the body and an orientation of uses of excised body tissue toward the common good. Ethical positions and liturgical rituals, for example, justify sharing of the body and its constituent organs and tissues as a form of altruistic service to others. Thus an understanding of body tissues as "gifts" is prominent in discourse with respect to organ transplants or transfusions of vital tissues, such as blood or bone marrow (6,7).

A more circumspect attitude is characteristic of holism toward research or educational uses of body tissue. Until the onset of the era of genetic and molecular medicine in the mid-1980s, holistic discourse largely followed the lead of technological capacities and examined the status of human organs and tissues with respect to donation for therapeutic goals. Despite ongoing collection and analysis of pathological tissue specimens, very little ethical discussion was devoted to nontherapeutic research uses of human tissue, other than questions pertaining to autopsy procedures.

Conflicts: Bodily Integrity and Disintegration

Although these five attitudes about the body are compelling as basic or ideal types, none can fully accommodate the complete range of bodily experience nor the new requests for body tissue precipitated by emerging biomedical technologies. Holism requires modification to accommodate scientific interest in body tissue, particularly those tissues that are not core to the conception of self. Similarly attitudes congruent with scientific interests — dualism, gnosticism, materialism — must be adapted to account for human experience that displays that the self is more than the mind, and that the body, and at least some bodily tissues, are not mere instruments but are morally and symbolically significant.

For example, empirical studies suggest that it is not only the body as a whole but also body parts, organs, and tissues that can be formative of self-identity (8). Visible parts of the body, such as skin, genitals, fingers, hands, legs, and eyes, as well as the heart, have a strong correlation with a sense of self. With the exception of the heart, nonvisible organs and tissues are not as strongly incorporated into a sense of self. Thus not all body parts possess equal ontic status, or are equally important to self-identity. It is possible, as Belk (8,9) suggests, that the less an organ or tissue is connected with a sense of self-identity, the more willing a person will be to donate it for use by others or to have it retrieved for purposes of a scientific research study.

Despite this possible convergence of scientific and nonscientific attitudes on the body regarding the specific question of the status of body tissue, an important question about the starting point for inquiry is still unresolved. Philosophical or religious claims about the ethical significance of bodily integrity can conflict with scientific and research interest in parts of the body, that is, the body as a disintegrated entity.

While the human body as an organic totality has long been the subject of philosophical reflection and a symbol of political (Plato's "Republic") or religious ("the body of Christ") communities, the interest of medical science often begins with a micro unit as its point of inquiry, the "building blocks of life," or DNA. DNA constitutes the fundamental components of life, with stem cells and somatic cells, tissues, organs, bodily systems, and the like, viewed as more complex, functional expressions of basic genetic materials. The scientific value of the body as a totality is thereby a means to the goal of deciphering the codes, messages, and functions of the genetic building blocks that contain valuable information.

In some circumstances these different starting points will generate quite different ethical conclusions about appropriate interventions, manipulations, and excisions of the body and its tissues. Legal philosopher E. Richard Gold cites the "disparate claims of scientific investigation and religious belief" as the exemplary case of incommensurate values regarding the body. According to Gold, "The body, from a scientific viewpoint, is a source of knowledge of physical development, aging, and disease. From a religious perspective, the body is understood as a sacred object, being created in the image of God. . . .The scientist values the body instrumentally, as a means to acquire knowledge; the believer values the body intrinsically, for being an image of God" (10).

The Mechanistic Body

Perhaps the most prominent discursive scientific understanding of the body, which has its roots in the intellectual Enlightenment, the industrial revolution, and Cartesian science, is that of a "machine." Among others of his generation, Thomas Jefferson clearly expressed this understanding toward the end of his life: "But our machines have now been running seventy or eighty years, and we must expect that, worn as they are, here a pivot, there a wheel, now a pinion, next a spring, will be giving way; and however we may tinker them up for a while, all will at length surcease motion" (10). The role of biomedicine becomes very clear in the mechanistic understanding of the body: to "fix" the worn-out parts through technical acumen and knowledge, perhaps to replace others through transplantation, and to improve on still others through research and engineering. Research on body tissue becomes an important gauge by which to determine the various developmental and functional processes of cells and tissues, as well as providing a means to identify what has gone wrong in case of malfunction.

Clearly, a valuation of the body that is mechanistic and instrumentalist will permit more research interventions and manipulations than will a position that affirms the intrinsic value of the body and its inextricable intertwining with the self. Moreover tissues retrieved from a mechanized body may elicit sentiments of revulsion rather than respect. The distinction between the status of organs, tissues, and fluids when they are incorporated within the body rather than outside and separate from the body is displayed in a memorable illustration by psychologist Gordon Allport: "Think first of swallowing the saliva in your mouth, or do so. Then imagine expectorating it into a tumbler and drinking it! What seemed natural and 'mine' suddenly becomes disgusting and alien. . . . What I perceive as separate from my body becomes, in the twinkling of an eye, cold and foreign" (12).

The emphasis on bodily integrity in the western religious faiths has culminated in the development of stigmas and taboos regarding certain bodily tissues when they are external rather than internal to the body (12). Characteristically the dis-incorporation of bodily tissues is assessed by religious thought with reference to issues of purity and cleanliness. A very prominent historical illustration of purity, which has permeated secular culture and has not been entirely overcome in contemporary religious communities, are stigmas and taboos surrounding menstruation (13).

There are then different ways of assessing the moral status of the body and of body organs and tissues depending on the "place" of the organs or tissues, that is (1) intrinsic to self-identity (e.g., heart) or incidental, (2) visible (eyes, skin) or hidden (kidney), and (3) integrated (circulating blood) or dis-incorporated (bodily excretions). In general, it may be claimed that the more an organ, tissue, or fluid possesses intrinsic connection, visibility, and integration, the more its retrieval and use for biomedical research purposes may present ethical questions. Put another way, philosophical and religious thought on the body begins with a strong presumption that the status of the body as a whole is greater than the sum of its parts. Body organs and tissues contain potent symbolic significance when considered as part of the bodily whole. Yet, when considered in isolation from the rest of the body, organs and tissue may generate revulsion and stigmas. This ambivalence can be heightened or ameliorated by scientific research that seeks to use body tissue for any number of research, therapeutic, or commercial purposes.

These perspectives and conflicts must be faced directly in the face of, and in anticipation of, technological advances in the realms of genetics and reproduction. Such advances have vastly expanded the biomedical gaze and the scope of scientific and medical interest in the body and its tissues. There is now virtually no part of the body that may not be the subject of scientific scrutiny. Genetic analysis in particular offers the prospect of gaining information about human character traits and behaviors, including susceptibilities to illness and bodily responses to disease, through study, analysis, and understanding DNA. Such information is not otherwise possible through an examination of the bodily totality. An assessment of body tissue as it is related to, and separated from, the body is inescapable in the context of biomedicine and biotechnology.

UNDERSTANDINGS OF BODY TISSUE

Metaphors of Body Tissue

Assessing the moral status of disincorporated body tissue can be illuminated by analysis of important metaphors that emerge in scientific and ethical discourse. These metaphors function to provide an understanding of something relatively unfamiliar, such as the significance of DNA or body cells, in terms of something more familiar and concrete.

Library. A central metaphor in discourse about the human genome project, for example, is that of a "book" or "library." This metaphor highlights the "information" stored in genes and cells, which scientists seek to retrieve and interpret through genetic analysis. Although information itself is often said to be value free, the process of determining what counts as information and what is "junk DNA" is clearly value laden. Indeed, as Rosner and Johnson observe, "By the choices [the genome project] makes — the choices of what books to include in the library and in what condition — the Project will determine what is 'correct,' what is 'real.' It will necessarily set standards, defining and cataloging what it means to be human, limiting what range of diversity is acceptable" (14).

The library metaphor highlights the extraction of information as the primary feature in research on body tissue. This implies that bodily tissue has at best instrumental moral status. What is concealed by the metaphor, however, is that unlike the information contained in a library, which is generally publicly accessible, the information that is retrieved from genomic studies is highly personal and sensitive. The information stored in any DNA library can be interpreted in ways that hold out implications for personal identity, including comparisons to a collective group standard or prospects of personal risks of disease susceptibility.

The information itself is of moral significance and should thus be treated in a manner equivalent to other forms of medical information. As with any information that is personally identifiable, there are legitimate concerns about the uses and potential for abuses of genetic information. Thus, in some jurisdictions, regulations and laws have been passed to ensure privacy and confidentiality of genomic information and to protect persons from discrimination.

For example, the Genetic Privacy Act passed in 1995 in Oregon affirms: "The DNA molecule contains information about an individual's probable medical future. . . .Genetic information is uniquely private and personal information that should not be collected, retained, or disclosed without the individual's authorization." While employers or insurance companies may, with consent, have access to a person's genetic information, the law prohibits the use of such information for purposes of hiring or setting insurance rates (15).

Resource. A second prominent metaphorical understanding of human body tissue is that of a "resource," most commonly, a resource that has been removed from its "natural" integrated state in the body and is thus accessible for a variety of scientific manipulations and purposes. More specifically, body tissue is portrayed in terms of a valuable mineral resource: Genetic analysis is described as a "gold rush," umbilical cords are referred to as "clinical gold," a patient's body as a "gold mine," and the researchers themselves assume the role of "gene prospectors" (16).

Ultimately the image behind these metaphors is that of the body as the "new frontier" encountered by cutting-edge science, a wilderness that biomedical science seeks to map and mine and ultimately to master and domesticate. Exploring the frontier requires and rewards individual initiative, thus it is not surprising that aspects of the resource metaphor move in the direction of commercial and entrepreneurial exploitation of body tissue. As Andrews and Nelkin express it, "Body parts are extracted like a mineral, harvested like a crop, or mined like a resource" (16).

The frontier or wilderness is also a realm beyond moral control; thus, as the frontier begins to be explored, and resources extracted, rules must be developed to facilitate a coordinated discovery effort. If the metaphor follows historical patterns, however, then there is ethical reason for concern: If the human body of the early twenty-first century is treated in terms analogous to the land and mineral rushes of the nineteenth century, short-term benefits may be obtained at the expense of harms that befall future generations. When the bodily frontier is viewed as an exploitable natural resource of immense scientific interest, a question arises about the extent to which researchers will take steps that respect and ensure the integrity of the whole.

The resource metaphor of body tissue is distinct from but certainly compatible with the book or library metaphor. Indeed, body tissue is a resource *precisely* because it contains information. The purpose of procuring tissues for research is to generate generalizable knowledge, for example, advancing researchers' understanding of the genetic dimensions of human diseases. However, this research has yet to yield significant promise of therapeutic benefits to persons afflicted with those diseases. Similarly investors in a biotech company engaged in genetic research may have to stay with their investment over the long haul until the research bears fruit in a commercially marketable product. This distinction between basic research that provides information or enhanced diagnostic capabilities in the short term but defers therapeutic (and entrepreneurial) benefits to the distant future is one point of criticism in ethical assessments of genetics research (17).

Property. Both the library and the resource metaphor are inadequate in at least one morally essential respect: They overlook authorization from the person from whom the body tissues are retrieved. The genetic library is not a public library, but contains very intimate, private information, that can accessed only by a select few, and only then upon consent. Similarly a person's body tissue is not a resource commons. Thus, lest scientific explorations on the frontier of the body become unnecessarily invasive of privacy and trespass integrity, or body tissue be subject to the research equivalent of a public land grab, a third metaphor of body tissue has been promoted to address issues of control and authorization. This is the model of the body and of body tissues as "property" (18,19). The property understanding stems from a claim of self-ownership, and it seeks to authorize the individual person, patient, or research subject, rather than the scientific prospector, with control over the use and disposition of the body and of body parts.

Not unlike some of the scientific attitudes about the body discussed previously, the property metaphor tends to treat the body as incidental rather than intrinsic to personal identity. The philosophical assumption is that the body even as a totality is distinct from the self, and body organs and tissues can therefore be transferred or alienated to others without compromising the integrity of the self. Thus, with the proviso that informed consent is obtained from the person, these assumptions make the property perspective very conducive to scientific interest in body tissue. However, conflict can arise when, for example, a patient and a researcher assert competing claims or "property rights" to excised body tissues, or to cell lines that have been isolated from retrieved tissue.

The Moore Case

The most prominent example of such conflict involved the case of John Moore, who sought treatment for a rare disease, hairy-cell leukemia (20). Moore's physician recommended a splenectomy, which was performed with Moore's consent. However, the physician recognized early into the course of treatment that Moore's body was overproducing lymphokines, an important component of the human immune system. Samples of Moore's blood, skin, bone marrow, and sperm were then retrieved in follow-up visits. Cells from these tissues were cultured and altered to create an immortal cell line, which was patented and commercially developed, without Moore's knowledge. Upon learning of this activity, Moore brought a lawsuit, alleging that he retained ownership of his body tissues and cells and that the research had wrongfully used his "property" to benefit others. Moore's proprietary interests in the removed body tissue and cells were rejected by the Supreme Court of California, although the court did find he had a compensatory claim if he had not been informed by the physician of the prospective research use and commercial development of the tissue. The majority opinion rejected Moore's claim on the basis that granting Moore property rights would impede scientific progress. A decision in favor of Moore, it was claimed, would open the door to actionable claims by patients any time body tissue was retrieved for research purposes. Moreover pharmaceutical companies would be deterred from investing capital and energy in new product development for fear of litigation.

The Moore case throws into sharp relief many questions about ownership and control embedded in any situation where body tissue is removed from a patient. There is no dispute that, when the tissues and cells are functioning as integrated parts of the intact bodily organism, they are under the authority and control of the patient. However, does the patient relinquish any and all claims and interests to the tissue once it is excised from the body? Do physicians or researchers gain rights of ownership to the tissues that are removed by their labor? Such questions necessitate attention to various modes of transfer and acquisition of body tissue.

TRANSFER AND ACQUISITION

The property, library, and resource metaphors assume the validity of the use of body tissue, so long as consent is voluntary and informed. Consent provides authorization for the transfer of body tissue from a patient or subject to a researcher. However, consent is only a necessary, rather than a sufficient, condition of morally justifiable transfers. Questions must still be addressed concerning the limits to be placed on such transfers and on avoiding abuse in research on acquired tissue.

Modes of Transfer

An analogy can usefully illustrate the moral context of these issues. I will adapt Belk's observation that "the house is a symbolic body for the family" (8,9), and construct an analogy wherein household goods take the place of human tissue samples. The relevant question is the manner of transferring household goods. I want to focus on four modes of transfer, all of which have correlates in methods of transfer of bodily tissue (21).

Donation. One transfer method is to donate certain goods such as clothing to a community goodwill program. This presents an example of a gift or an altruistic action designed to benefit others and to enhance a recipient's quality of life. Similarly various life-saving organs and tissues, such as blood, have been referred to as "gifts of life" in popular discourse.

Contribution. A second form of transfer concerns those household goods whose designed use has been depleted by members of the household, for example, food products that come in plastic or cardboard containers or newspapers. These containers and other materials could still be used for some household functions, such as for storage or for

starting a fire; they are not without household utility. It is also possible, however, to contribute these materials for a greater good, the benefit of the community through recycling. An important difference between donation and a contribution is that the donated goods are typically transferred and used in their original form, while a contributed good, be it a household container or body tissue, undergoes some alteration before re-use.

The recycling approach requires communal support of organizations with the knowledge and expertise to convert the recycled materials into valued products. Similarly a biomedical research institution may retrieve certain tissues contributed for the greater good of the advancement of scientific knowledge. Researchers can then subject the tissue to a research process in which cells are isolated to yield a valuable immortal cell line or genes analyzed to provide information.

Abandonment. A different set of household materials are those goods that have been consumed completely, and whose benefits have been exhausted by household members. The packaging protecting these materials is now "waste" and is commonly discarded through a community utility such as a trash collection service. Household refuse has no personal meaning to the discarder, who is typically quite willing to abandon the refuse and even pay a fee to have the items removed. This does not, of course, preclude the possibility that this refuse might have value to someone else who is willing to take the time to sort through the materials. Similarly at least some bodily tissues are referred to as "waste" or "garbage" in biomedical literature (22,23), even though research processes may transform the tissues into a beneficial and profitable product.

Property Sale. A fourth form involves transferring the household goods to others through an informal market transaction, the "garage" or "yard" sale (or through classified advertising). In these situations the original owner hopes to obtain some financial gain through the transfer, although the expectation is that neither the purchase price, nor the quality of the goods, will be as high as if the purchaser had obtained a similar item through an established business. These informal transactions occur parallel to, but outside of, formal and regulated market mechanisms and controls; once the transaction has taken place, there is no expectation of a continued buyer–seller relationship. Procurement of some body tissue does occur through buying and selling, although typically the seller remains anonymous.

It may be the case that all of these modes of transfer, and their biomedical analogies, have a role in considering ethical issues in research acquisition of body tissues. As suggested, there are precedents already for tissue as a donated gift, as a research recyclable, as abandoned refuse, and as a commodity. However, certain tissue may be considered more or less appropriate to a particular form of transfer.

This differentiation may, in part, rely on a sense of connectedness to self-identity delineated above. Some body parts, such as the heart, eyes, or blood, may have such symbolic significance and connection to personal identity that their donation is the moral equivalent of a gift of self. Donation or contribution imagery may seem more fitting with the transfer of such tissues than abandonment or selling. Other body tissues, for example, urine or hair clippings, may have such minimal value to a sense of self that they are considered "waste" and routinely abandoned. Still other organs and tissues, such as a pancreas, liver, spleen, or marrow, may fall in between these examples, not as central to personal identity as the heart or eyes but not as incidental as urine either. Still other tissue, such as gametes or blood, may be considered so vital to the creation of life or to the maintenance of life that society is willing to permit a limited market to function to procure these tissues.

Of the above modes of transfer, gifts or donations have a prominence in discourse relative to contribution, abandonment, or sales. The discussion below will explain this prominence through examining the features of what may be designated the "donation paradigm," and it then will assess the extent to which methods of contribution, abandonment, and selling retain or violate these features.

The Donation Paradigm

Ethical thought on the body and its use within medicine has presupposed a context within which organs and tissues are donated for therapeutic purposes of healing, restoring, or saving life. This moral ideal is emphasized through the language of "gift," "altruism," "sacrifice," and so on, on the part of the donor and that of "benefits" for recipients. The donation paradigm seems constructed by four principal features:

Altruistic Intent. The intent of the donor of an organ or tissue is structured by gift-giving to beneficiaries or recipients, such as persons on a waiting list for a transplant (although the identity of such persons may be veiled from the donor).

Therapeutic Expectation. The expected outcome of a gift of the body is that it will be used in such a way as to offer a pronounced therapeutic prospect for the recipient. These prospects encompass both enhancing quality of life and preserving life.

Generative Re-incorporation. Body tissue that has been retrieved from the donor, or dis-incorporated, should in most circumstances be re-incorporated within the body of the recipient. As noted above, tissue that is dis-incorporated may evoke sentiments of revulsion and practices of stigma and taboo. Some religious practices and rituals require burial of removed body parts, or re-incorporation in the earth. This is particularly the case with body parts that have an identifiable human form: In Jewish thought body parts such as limbs, composed of "flesh, sinew, and bones," should under most circumstances be buried. Roman Catholic tradition distinguishes major from minor parts of the body in a similar manner. Major parts of the body are those, such as a limb, that retain their "human quality" following excision

and should be buried (24). Such concerns reflect in part the importance of these visible body parts for self-identity.

Re-incorporation of body organs or tissues in a human recipient has a generative power in that it offers the prospect of new or renewed life to the recipient. In general, then, the donation paradigm prioritizes practices in which body tissue remains with a body (even if transferred and transplanted to the body of another person) and thus symbolizes the significance of bodily integrity and holism.

Recipient Responsibilities. The gift of body tissue carries with it certain responsibilities on the part of the recipient, which are embedded in everyday practices of sharing and gift-giving (25,26). These responsibilities include a sentiment of gratitude toward the gift-giver, or toward the institutional structure that mediates the gift transfer. Gratitude may also be enacted in the actions and conduct of the recipient by which he or she makes grateful use of the gift. In addition a gift induces a responsibility of reciprocity. Reciprocity does not necessarily mean the continuation of the gift relationship between the initial giver and recipient; rather, a recipient of donated blood, for example, may at some time in the future become a donor for other strangers.

The donation paradigm provides a moral justification for medical use of human body tissue. It is especially prominent in bioethics discourse because it highlights simultaneously moral virtue and altruism, and medical success, the restoration of life to the nearly dead. The paradigm is limited, however, for the most part to medical practices of transplantation or transfusion, that is, those practices that promise some form of therapeutic outcome from the gift. However, different models seem necessary to accommodate biomedical retrieval and use of body tissue in circumstances where nontherapeutic uses of body tissue are contemplated (e.g., tissue contributed for research and educational purposes) or where altruistic intent is absent (e.g., abandonment and sales).

Contributions

The idea of body tissue as a contribution shares with the donation paradigm the intention of gift-giving or altruism. However, the anticipated outcomes are rather different. The goal in research or educational uses of human tissue is to advance scientific learning and to generate knowledge that is generalizable, not to furnish a prospective therapy for an individual recipient (although certainly some research progresses in the direction of therapeutic outcomes). This is not an incidental issue, for some cultural and religious traditions that are generally supportive of donation for therapeutic goals find it much more difficult to provide a justification for biomedical research undertaken without therapeutic intent. In examining the ethical status of tissue contributions, we will rely on the fourfold features of the donation paradigm to illuminate distinctions and comparisons.

Intent. A person who places their recyclable household materials at the curbside is engaging in nonobligatory gift-giving in an important respect because they could well retain the materials for their own purposes, such as storage, collections, or fire-starters. While they are not making a personalized gift to a specific individual, they do contribute to a cause that is larger than themselves and the benefits they might provide in a direct or mediated relationship with another person in need. The cause in the domestic case may be "environmental preservation"; in the case of body tissue, it may be designated as "scientific discovery" or "medical progress." The contribution in both cases is one of nonspecific generosity; it is "nonspecific" in that a "cause," rather than a person, is the intended beneficiary. The contribution is also a generous act in that a person is participating in the advancement of the larger cause when they could just as easily place the recyclable material in the refuse bin without moral blame, or request abandonment or return of the body tissue.

Contributor Expectation. Unlike the therapeutic expectation embedded in the donation paradigm, a contribution of body tissue for research or educational purposes does not bring direct and immediate benefits to a specific or designated individual. To be sure, there is an expectation of benefits on the part of the contributor but these are much more diffuse in space and remote over time. The recipient of the tissue contribution, a science researcher or medical educator, may develop protocols or pedagogies that over time will accrue benefits to the larger good of society, or at least to those persons with a stake in biomedical research or education. Perhaps the tissue will turn out to have "immortal" features, either through its cell progeny or through laboratory learning that is transferred to clinical practice. Still the objective in contributions is advancing knowledge, not therapy; learning, not curing. The good of the greater cause, not an individual in need, predominates.

This difference in expectation marks a distinction between donations and contributions of body tissue. There are also marked differences between contribution and abandonment or sales. A willingness to contribute does not imply that the contribution has minimal or no value to the contributor. A plastic milk container that can be recycled is equally serviceable as a water jug. Similarly a person may attribute value to many tissue specimens, including blood, reproductive tissues, skin or hair that have been retrieved or excised from the body. The prime difference of contributions lies in the fact that something of value is contributed to a person or organization through whose work the society realizes a benefit that is greater than would have been the case had the contributor decided to retain the material. A contribution thereby intends a benefit for the common good of all.

Symbolic Re-incorporation. The donation paradigm involves re-incorporating removed tissue or organs into another body, whether an organ transplant or blood transfusion into a person, or as is common among some religious rituals, burial in the earth. Re-incorporation practices are not literally possible in contributions, because the research and educational purposes necessitate analysis or study of body tissue in isolation from the body totality. However, a form of symbolic re-incorporation is possible with contributions of body tissue. For example, just as recycling contributes to the good of the larger communal whole,

the contribution of body tissues for research can provide information that can then be integrated within, or require revisions to, a larger, symbolic whole, designated as the "body of scientific knowledge."

Recipient Responsibilities. Contributions in general are acknowledged in some form by the recipient, but implementing an acknowledgment is more difficult with respect to contributions of bodily tissue. It is important that those who seek informed consent for nontherapeutic uses of body tissue not presume that a patient or subject will simply "sign off" to any and all uses made of retrieved body tissues.

However, since the moral point of the contribution is to facilitate the achievement of some cause greater than the interests of either tissue source or recipient, the researcher should assume a role of trustee and steward for the community. This entails a research responsibility to use the contribution of body tissue for the common good. At a minimum this requires treating the information generated by tissue research with safeguards that ensure protection of the contributor against discrimination or harm.

The appeal to the "common good" does not preclude recourse to the private sector to carry out research; in some cases, as with domestic recyclables, the good of all can be more efficiently and effectively achieved through private sector initiatives. However, in a contribution approach, profit interests must be subordinated to the common good and the greater cause that the contribution is designed to advance. In short, retrieved body tissue is a source of good, and not merely a resource for financial gain. Body tissue that is contributed for purposes of the general cause of advancing science should not be viewed as merely an economic asset. Such limitations may not apply, however, in situations of acquisition of body tissue through abandonment or sale.

Abandonment

Many of the 282 million specimens of human biological materials stored in labs in the United States have been acquired following abandonment by the patient or research subject. A person who has undergone a surgical procedure during which certain body tissue has been removed is characteristically not interested in retaining the tissue themselves; the removed tissue assumes the status of "surplus" or bodily "refuse" that the patient is quite willing to discard, especially if the tissue is implicated in a disease condition. In some cases (though not all) the person may sign a general consent form permitting research or educational uses of their tissue. This tissue is a resource for scientific study and exploitation, but this permission is an attenuated form of donative or contributory intent.

In other circumstances, as with the placenta or umbilical cord, the removed tissue may not have been a source of disease but a sustainer of life that is no longer biologically needed. Nonetheless, for some persons, as well as cultural traditions such as Native American, these tissues are profound symbols of life and relationships, and the patient may retain them. For many others, however, they are surplus membranes that can be discarded following birth; even if research consent is given, the primary intent is to be rid of the tissue.

This suggests departures in several respects from the features of the moral model that inform the donation paradigm, and with some modifications, contributions of tissue. If a donative intent is present at all, it is relatively insignificant and possesses secondary importance. There is also likely to be no expectation of a therapeutic outcome and minimal interest in any research or educational uses. Are efforts at re-incorporation attempted? As Murray writes, in most cases of the removal of a diseased organ, "the tissue is disposed of, usually by incineration." Yet this does not relieve pathologists or researchers of recipient responsibilities, including respect and dignity appropriate to "the fact that the tissue is from a human" (7). Moreover, since research analysis could establish linkage to the tissue source, considerations of privacy and confidentiality continue to apply.

Sales

Certain body tissues, though not solid organs, are allowed to be bought and sold. These include blood, skin, and reproductive tissue. Financial transactions for body tissue stretch the language of "donor" beyond all reasonable usage. The language of tissue "source," "seller," or "vendor" is more appropriate. In sales of body tissue the altruistic intent of the donation paradigm has been replaced by the commercial prospects of property (27). The property understanding of the body, and its correlate, an understanding that excised or retrieved body tissue is, potentially, a "commodity," recognizes the ownership rights of the person to his or her own body and body tissue. As with any other form of property, these rights can be transferred to, and acquired by others, including researchers or tissue banks through contractual agreement.

A contract that transfers property rights in body tissue means that a person's interests in their body tissue are effectively ended with the transaction. The tissue source may harbor expectations for a successful fertility outcome for the buyer, or some therapeutic developments by researchers, but that is not the primary rationale for the commodity exchange of tissue for money. If it were, then the tissue source should be equally satisfied with a donation or contribution of body tissue. Rather, the point is to use a shared procedure, the market mechanisms that function to transfer goods and services in various social realms, to facilitate mutual self-interest. Once the transaction is completed, the new property holder possesses rights to use and dispose of the commodity at his or her discretion.

Within the transaction, the social identity of the researcher-recipient of body tissue marks a striking contrast to that found in the donation paradigm or in contribution. The latter approaches present the recipient as a trustee and steward of what are communal resources, with a responsibility to use the resources as consonant with the common good. No such strictures apply in a commodity exchange of body tissue. The researcher functions more along the lines of a private entrepreneur, seeking to protect personal property as both a scientific and an economic investment. In some cases this protection is secured through patenting of an innovation, such as an immortal cell line derived from body tissue. For example, in the case

of John Moore, the physician entered into contracts with pharmaceutical companies to enable successful exploitation of the estimated $3 billion potential of the cell line. A patent on the cell line and its derived products was secured and transferred to the research university (9).

The property paradigm thus has a very attenuated view of recipient responsibilities; those responsibilities are primarily directed toward advancing the interests of the recipient rather than shared societal interests. The researcher is not bound by responsibilities of gratitude or reciprocity because the work of the scientific community has made possible the retrieval of the tissue in the first place. Were it not for the initial labor of physicians and researchers, there would be neither donations nor sales of body tissue. The main responsibility of the researcher-recipient thus appears to be ensuring that informed consent takes place.

While the donation paradigm has few critics, other than those who may fault it for providing an inadequate supply of organs and body tissue, the property paradigm of the body, and the idea of tissue as a purchasable commodity, has received criticism similar to that directed at (so far, unsuccessful) proposals for a commercial market in solid organs. These criticisms include questions about the impact of financial incentives on voluntary, informed consent; "quality assurance" mechanisms to ensure safety; and the potential for the property paradigm to compete with, and ultimately erode, communal support for donation. Underlying these procedural and consequentialist concerns is a substantive criticism that commodification of body tissue, and understanding the body as personal property, ultimately is an affront to human dignity. Proponents of the property paradigm have argued in response that it provides assurance of personal control over the disposal of body tissue and freedom of choice for the tissue source. A commercial market, it is claimed, can and should supplement gifts of the body; after all, property can be sold as well as donated. Moreover a commercial market could increase supplies of scarce body tissue.

One common concern is that the effort to increase supplies of tissue for either therapy and research through permitting sales risks undue influence and coercion of the decision maker. Thus laws prohibit women from selling tissue from their aborted fetuses, and regulations are currently being discussed that would impose similar restrictions on couples in vitro fertilization (IVF) programs who are informed about the possibility of research on stem cells derived from surplus embryos. In other circumstances, however, such as procurement of gametes, financial incentives are not necessarily deemed as compromising the voluntariness of a choice. With respect to gamete procurement, the voluntary guidelines promulgated by various professional societies that discourage excessive financial inducements seem to have gained general acceptance.

CONCLUSION

Innovative biomedical research, especially in the field of molecular genetics, has heightened scientific interest in tissue samples, and thereby posed new questions about the moral status of body tissue in research and educational settings. In a diverse and pluralistic culture, wherein many traditions, cultures, and communities vest body tissue with tremendous symbolic significance, the moral burden of proof falls on the scientific community and researchers to justify acquisition of body tissue. Researchers seeking access to body tissue must make a compelling case for its use, ensure that their scientific design will likely generate significant results, determine that there are no other alternatives to achieve the research objectives, facilitate a process of informed consent with tissue sources, and provide guarantees of confidentiality and anonymity as appropriate. In this way body tissue can legitimately be used while minimizing abuses.

BIBLIOGRAPHY

1. National Bioethics Advisory Commission, *The Use of Human Biological Materials in Research: Ethical Issues and Policy Guidance*, Rockville, MD, 1999.

2. W.F. May, *The Patient's Ordeal*, Indiana University Press, Bloomington, IN, 1991.

3. C.S. Campbell, in L.S. Cahill and M.A. Farley, eds., *Embodiment, Morality and Medicine*, Kluwer Academic Publishers, Dordrecht, The Netherlands, 1995, pp. 169–183.

4. National Bioethics Advisory Commission, *The Use of Human Biological Materials in Research: Ethical Issues and Policy Guidance*, Rockville, MD, 1999.

5. J.F. Childress, *Practical Reasoning in Bioethics*, Indiana University Press, Bloomington, IN, 1997.

6. W.F. May, *Hastings Cent. Rep.* **15**, 38–42 (1985).

7. T.H. Murray, *IRB* **8**, 1–5 (1986).

8. R.W. Belk, in J. Shanteau and R.J. Harris, eds., *Organ Donation and Transplantation: Psychological and Behavioral Factors*, American Psychological Association, Washington, DC, 1990, pp. 139–149.

9. R.W. Belk, *J. Consumer Res.* **15**, 139–168 (1988).

10. E.R. Gold, *Body Parts: Property Rights and the Ownership of Human Biological Materials*, Georgetown University Press, Washington, DC, 1996.

11. S. Nuland, *How We Die: Reflections on Life's Final Chapter*, Knopf, New York, 1994.

12. G. Allport, *Becoming*, Yale University Press, New Haven, CT, 1955.

13. J. Lee and J. Sasser-Coen, *Blood Stories: Menarche and the Politics of the Female Body in Contemporary U.S. Society*, Routledge, New York, 1996.

14. M. Rosner and T.R. Johnson, *Hypatia* **10**, 104–129 (1995).

15. Legislative Assembly, State of Oregon, *1995 Oregon Revised Statutes*, vol. 10, Salem, OR, 1995 (ORS 659.700).

16. L. Andrews and D. Nelkin, *The Lancet* **351**, 53–57 (1998).

17. H. Bouma et al., *Christian Faith, Health, and Medical Practice*, W.B. Eerdmans Publishing, Grand Rapids, MI, 1989.

18. R. Scott, *The Body as Property*, Viking Press, New York, 1981.

19. L. Andrews, *Hastings Cent. Rep.* **16**, 28–38 (1986).

20. *Moore v. Regents of the University of California* (51 Cal. 3d 120 [1990]); Cf. E.R. Gold, *Body Parts: Property Rights and the Ownership of Human Biological Materials*, Georgetown University Press, Washington, DC, 1996.

21. C.S. Campbell, *Kennedy Inst. Ethics J.* **8**, 275–305 (1998).

22. G.J. Annas, *Hastings Cent. Rep.* **18**, 37–39 (1988).

23. National Bioethics Advisory Commission, *The Use of Human Biological Materials in Research: Ethical Issues and Policy Guidance*, Rockville, MD, 1999.

24. J.F. Childress, in D.A. Knight and P.J. Paris, eds., *Justice and the Holy*, Scholars Press, Atlanta, GA, 1989, pp. 215–240.

25. P. Camenisch, *J. Relig. Ethics* **9**, 1–34 (1981).

26. T.H. Murray, *Hastings Cent. Rep.* **17**, 30–38 (1987).

27. C.S. Campbell, *Hastings Cent. Rep.* **22**, 34–42 (1992).

See other entries GENETIC INFORMATION, ETHICS, ETHICAL ISSUES IN TISSUE BANKING AND HUMAN SUBJECT RESEARCH IN STORED TISSUES; HUMAN GENOME DIVERSITY PROJECT; OWNERSHIP OF HUMAN BIOLOGICAL MATERIAL; see also PATENTS AND LICENSING entries.

PATENTS AND LICENSING, ETHICS, ORGANIZATIONS WITH PROMINENT POSITIONS ON GENE PATENTING

R. COLE-TURNER
Pittsburgh Theological Seminary
Pittsburgh, Pennsylvania

OUTLINE

Introduction

Religious Organizations

Indigenous Peoples and their Coalitions

Environmental Groups

Public Interest Science Groups

Scientific Organizations

Industry Organizations

Bibliography

INTRODUCTION

One might think that the question of the patentablity of DNA and other biological components is narrow and technical, of interest only to lawyers, directors of research, venture capitalists, and a handful of scientists. In fact it is a question of great interest and concern to a wide range of organizations from every continent that for some reason feel compelled to express an opinion on this subject. And while the technical matters of patenting law and trade policy are often lost in the public debate, it is clear from the wide range of organizations with positions and from the intense passion with which they sometimes express their opinions that deeply held convictions are at stake in this question of the ownership and commercial use of biological components. Because patent law and trade policy are created through political processes, the conflict among organizations on biological patenting may have wide significance for the biotechnology industry for years to come.

The organizations with positions on patenting range across the entire spectrum of institutional styles and continuities of identity, from religious organizations, with thousands of years of unbroken institutional identity, to conferences that meet briefly but leave a lasting impact. What counts as an organization? For the purposes of this summary, transient organizations, such as coalitions and conferences, are included, not only because their impact often outlives their structure but because of their galvanizing effect on their complex constituencies and on public consciousness. Some of these "organizations" are virtual; that is, they exist primarily if not exclusively on the Internet. Internet technology, so important to genome research and sequence publication, and thus partly determinative of what is patentable and what is not, is also used by individuals and communities worldwide to organize their opposition to biotechnology, to commercial use of genetics, and specifically to biological patenting. The impact of such "virtual organizations" and of Internet-based community organization should not be underestimated. Furthermore, precisely because of the fluid boundaries of organizations, "virtual organizations" can be remarkably successful in swaying the position of traditional organizations, such as religious bodies or governments or the United Nations.

The positions of these organizations on biological patenting also span the whole spectrum of possible views. Some of the organizations included here (mostly the environmental and indigenous peoples groups) oppose not only patenting but biotechnology itself and see opposition to patenting as a strategy toward their broader end. Others (mostly the religious organizations) support biotechnology, at least for many purposes, but oppose many if not all forms of biological patenting. Many others (e.g., scientific organizations) support biotechnology and therefore oppose certain forms of patenting on the grounds that they impede progress in the field. Others organizations (largely the industry groups) support broad patenting of genetic and biological discoveries, including (for some) short DNA sequences of unknown function.

The organizations that have take public positions on biological patenting may be grouped into seven categories, namely religious, indigenous peoples, environmentalist, public interest science, scientific, and industrial. Each category will be considered in the sections that follow.

RELIGIOUS ORGANIZATIONS

Of all the public statements made on biological patenting, perhaps the most widely reported is the May 1995 statement, the *Joint Appeal against Human and Animal Patenting*. The text of the statement, which was endorsed by the heads of over 80 religious organizations in nearly all faith traditions, consists of three sentences: "We, the undersigned religious leaders, oppose the patenting of human and animal life forms. We are disturbed by the U.S. Patent Office's recent decision to patent human body parts and several genetically engineered animals. We believe that humans and animals are creations of God, not humans, and as such should not be patented as human inventions" (1).

The Joint Appeal was organized by the General Board of Church and Society of the United Methodist Church and promoted in cooperation with the Foundation for Economic Trends, whose leader, Jeremy Rifkin, is a well-known activist in opposition to genetic technology. In a press statement that accompanied the release of the Appeal, the Rev. Kenneth L. Carder, Resident Bishop of the Nashville, Tennessee, Area, the United Methodist Church, argued for the Appeal on religious grounds: "The issue is the commodification of life and the reduction of life to its commercial value and marketability.... Life is a sacred gift from God the Creator. As a gift from God, life has intrinsic value. The patenting of genes, the building blocks of life, tends to reduce it to its economic worth.... Patenting further identifies life mechanistically and blurs the distinction between the animate and the inanimate.... The conflict is between reverence for life and exploitation of life, life valued for its marketability and life valued as an intrinsic gift. It is a conflict between utilitarianism and transcendent meaning. It is conflict between control based on economic profit and access based upon shared need" (2).

The 1995 Appeal was not the first time that religious leaders had joined to speak publicly on this issue. In 1987 a *Statement of Religious Leaders against Animal Patenting* was signed by seven national leaders of mainstream Protestantism and of Judaism and issued in response to U.S. Patent Office decision to allow patenting of transgenic animals. The statement objected to the decision, saying: "No decisions about the patenting of genetically altered animals should be made without the most careful examination of all possible consequences." The statement urges "support for congressional measures to halt such actions by the Patent Office, and to enable a careful consideration of these questions by the Congress and the public. The gift of life from God, in all its forms and species, should not be regard solely as if it were a chemical product subject to genetic alteration and patentable for economic benefit. Moral, social, and spiritual issues deserve far more serious consideration before binding decision are made in this area" (3). It should be noted that most of the religious leaders who signed the 1987 or the 1995 statements did so without their religious body having discussed a policy on patenting, much less having taken an official stance.

In these two statements, theological opposition to biological patenting is grounded on the belief that life is a gift from God, its creator and (one might say) inventor, who alone should possess any intellectual property rights in biological phenomena. The sheer act of patenting, the claim to possess a property right in the design of a biological organism, is an affront to this theological conviction. Furthermore patenting is an act that reduces an organism to mere matter and denies the special status that living things enjoy in relation to their creator. These two theological principles are the twin pillars upon which religious opposition to patenting is based. The first, that God is the creator, giver, and rightful source and definition of the value of the creation, leads in many of these statements immediately into the second principle, that God has so ordered creation that living things enjoy a special status that must not be violated or ignored. The first of these principles warns against the danger of anthropocentrism or the view that we (or any other creature) can define the value or purpose of creatures. The second principle warns against reductionism, the view that all creation is nothing but matter, valueless apart from its usefulness to us.

It is the second of these twin principles that comes to the fore in positions taken by the Church of Scotland and found in a report of the European Ecumenical Commission for Church and Society (EECCS), a commission created by the Protestant, Anglican, and some Orthodox Churches to relate to the European Union and the Council of Europe. (It should be noted that the Church of Scotland has been especially strong in supporting the development of the positions taken by the EECCS.) The 1998 EECCS report, entitled "EECCS and Bioethics," holds that, "[c]oncerning patenting, it is a matter of ethical concern that commercial demands are now tending to abuse the normal distinctions between what is alive and what is not, and what is discovery and what is human invention. The mere knowledge of a gene should not be patentable in itself, nor should an entire transgenic organism — animal or plant — when it is only a tiny modified gene sequence that is novel. Moreover, there seems to be a real danger that genetically modified organisms are looked upon only as commodities in a global market place" (4, p. 5). The theological significance of the distinction between the living and the nonliving is also asserted in a major position paper of the World Council of Churches (WCC), whose participants come from every continent and nearly every tradition within Christianity. Among the concerns raised about biological patenting, the report states: "The US Supreme Court decision on patenting of life forms rested upon a specific, highly reductive conception of life, which sought to remove any distinction between living and non-living matter that could serve as an obstacle to the patenting of living but unnatural organisms. It would be easy to underestimate the philosophical, moral and ideological importance of the abolition of such a distinction, precisely because it allows a shift in accepted ideas as to what may be done to living things" (5, p. 18). It must be noted, however, that what is being opposed here is biological patenting, not biotechnology. In fact the statements are clear that they do not intend to prohibit biotechnology, even if they do raise concerns about some of its uses. But it is not clear in the statements why the living/nonliving distinction would prohibit patenting but would permit biotechnology. Nor is it clear just where, in biological phenomena, one should find the distinction between the living and the nonliving, or whether theology can offer any help in locating that distinction.

Religious opposition to some aspects of biological patenting is of course grounded in other theological and ethical concerns beyond the two central principles. Indeed, the WCC 1982 report lists "five causes for concern," of which the living/nonliving distinction is only one and which does not put "life as gift" on the list. Among the other four concerns, the report notes that "some argue that patent protection in this form will allow existing and emergent corporations to make excessive profits out of human suffering, to sustain an unhealthy worldwide dependence on pharmaceutical products, and to ignore real health needs because they offer lesser economic returns." In addition, due to inconsistencies and imprecision in

patent law, there is a concern about "extremely costly patent litigation. This prospect is likely to cause major distortions in the process of innovation and to restrict the range of products made and the kinds of organizations making them" (5).

The report notes the concern that the patent system will lead "to a curtailment of communication and greater reluctance to exchange new research materials and to engage in collaborative projects." The final concern noted in the 1982 WCC report is that in view of the fact that "much of the research leading to commercially valuable knowledge was supported by public funds," patents leading to private profit will cheat the public twice. "Not only does this deplete the public exchequer but the needs of the community might be better served by arranging for the commercial exploitation of publicly funded research by publicly owned corporations." In particular, "churches and other institutions should press for alternative ways in which the labours of science are transmuted into the maximum human good." In light of these concerns, however, the report does not call for sweeping opposition to patenting but merely "[t]hat the WCC maintain a watching brief on the issues arising from the patenting of micro-organisms" (5, p. 26). The general thrust of the argument is that excessive patenting will lead to injustices in profits and to social inequities in the progress of research for the benefit of all.

The WCC 1982 report was followed in 1989 with a report from the WCC Subunit on Church and Society, entitled *Biotechnology: Its Challenges to the Churches and the World*. Like the 1982 report, the 1989 document included a major section on intellectual property, specifically on "The Patenting of Life." The report notes the economic impact of biotechnology patents on farmers and on scientific research, where patenting "is leading to a curtailment of communication and sharing of resources in the scientific community." The 1989 report notes, of course, that the alternative to patenting might be "complete research secrecy." But of all the economic impacts of patenting, "perhaps the greatest concern in the patenting of life is that the lure of patenting can cause a misappropriation of Third World genetic resources by corporations looking for patentable, genetic products. Since a significant amount of useful genetic material is found in the tropical and subtropical countries, First World patenting of life could increasingly exploit the collective resources and germplasm of Third World countries and peoples." This is not an insurmountable barrier, however, because already there are examples of "responsible sharing of information and benefits" (6, p. 22).

While the 1989 report observes that patenting may itself contribute to environmental problems, it moves quickly to the "Ethical and Theological Impacts" of patenting, where its strongest concerns are expressed. Here again, we find the living/nonliving distinction clearly articulated as the leading theological concern: "The patenting of life encodes into law a reductionist conception of life which seeks to remove any distinction between living and non-living things.... This mechanistic view directly contradicts the sacramental, interrelated view of life intrinsic to a theology of the integrity of creation. As expressed by Arie Brouwer, former General Secretary of the National Council of Churches of the USA, 'Reverence for all life ... may be eroded by subtle economic pressures to view life as if it were an industrial product invented and manufactured by humans.... The gift of life from God, in all its forms and species, should not be regarded solely as if it were a chemical product subject to genetic alteration and patentable for economic benefit.' The churches should question all those technologies, whether traditional or modern, whose only stance toward creation is one of exploitation and profit, ignoring the biblical call to 'tend the garden and keep it'" (6, pp. 22–23). The final sentence is instructive in that the target of questioning is not patenting but technology itself; if technology is reductionist and blurs the distinction between the living and the nonliving, it should be challenged and rejected.

The report then enumerates (with cautious endorsement) a list of concerns about patenting: (1) First World control of Third World genetic resources by corporations looking for patentable genetic products, (2) encoding a reductionist conception of life by removing the distinction between living and nonliving things, (3) creating the loss of genetic diversity, (4) animal patenting creating the profit incentive for cross-species genetic transfers that leads to great animal suffering, (5) threatening environmental destruction by encouraging the deliberate release of genetically engineered organisms, (6) severe effects of patenting life on small farmers and producers, and (7) continued and growing confusion in the legal community over precise and acceptable definitions of patentable living matter (6). In light of these concerns, the report concludes with the recommendation from the Subunit on Church and Society that "The World Council of Churches believes that animal life-forms should not be patented and calls for further study of the profound moral and social implications of patenting life forms" (6, p. 23). In contrast to the preceding discussion, which is wide-ranging in scope, the final recommendation is limited to opposition to patenting of transgenic animals and leaves aside many other areas of biological and genetic patenting.

Somewhat in structural parallel to WCC, but operating on the national level within the United States, the National Council of the Churches of Christ (NCC) has also engaged in studies of the religious and ethical implications of biotechnology and has specifically addressed biological patents. A 1986 NCC report, entitled *Genetic Science for Human Benefit*, notes that "[s]cientists, investors and managers who provide the knowledge and capital necessary for biotechnological development and marketing deserve fair compensation for their ingenuity, work and willingness to incur economic risks." Patenting, however, can lead to "threats to science itself.... More serious still is the admonition against monopolistic ownership of genetically modified organisms or substances which are known to be essential to human life or for nourishment and health" (7, p. 14).

In contrast to this somewhat "pro-patenting" stance, more recent church positions have tended to view patenting with greater suspicion. For example, one of the largest Protestant denominations in the United States, the United Methodist Church, adopted a resolution at its highest national gathering, the 1992 General Conference. The resolution states that "[t]he position taken by the church ... is consistent with our understanding of the sanctity of God's creation and God's ownership of life. Therefore, exclusive ownership rights of genes as a means of making genetic technologies accessible raises serious theological concerns. While patents on organisms themselves are opposed, process patents, wherein the method for engineering a new organism is patented, provide a means of economic return on investment while avoiding exclusive ownership of the organism and can be supported.... We urge that genes and genetically modified organisms (human, plant, animal) be held as common resources and not be exclusively controlled, or patented. We support improvements in the procedures for granting patents on processes and techniques as a way to reward new developments in this area" (8, pp. 332–333). It was on the basis of this position that United Methodist officials went on to organize the May 1995 Joint Appeal, which was signed by many leaders of other religious bodies.

The largest U.S. Protestant denomination, the Southern Baptist Convention, adopted a resolution "On the Patenting of Animal and Human Genes" at its June 1995 national gathering, declaring that "the scriptures of both the Old and New Testaments plainly teach that God alone is creator and owner of all he has made ... [and] humans legitimately may own individual or groups of animals of a given species, but not an entire species and its progeny ... [Therefore,] we, the messengers of the Southern Baptist Convention ... do hereby affirm our conviction that God alone is Creator and owner of all creation ... we call upon the President, the Congress, the National Institutes of Health, and the United States Patent Office to place an immediate moratorium on the patenting of animal and human tissues and genetic sequences until a full and complete discussion has occurred" (9, pp. 7–8). The United Methodist and the Southern Baptist statements both stress the first theological principle noted earlier, namely, that God is the creator and owner of life. While the living/nonliving distinction is not unimportant, it is less important that the principle that God is creator and therefore owner of all living things.

The principle that God is owner of life and therefore we cannot claim ownership could be understood in a way that makes God appear possessive and petty. On the other hand, it can be interpreted as pointing to something important about the creation and its transcendent (as opposed to anthropocentric) ground of value. In this view, "as Creator, God reserves the right to determine how the knowledge of organic development is to be used. In a way, it is as if God the Creator were the first to patent genes, not to exclude us from using the knowledge, but to exclude us from excluding others.... It is as if anyone who tries to patent a gene should have the application rejected because it encroaches upon a prior patent whose holder, God, has given free license to everyone. As Creator, God has and reserves the right to define how we may use the knowledge of the creation, including the knowledge of DNA sequences. God exercises this right by saying, in effect, that this knowledge is to be used without exclusion, that is, without patent protection" (10, p. 154).

The second key principle, that of the importance of the distinction between the living and the nonliving creation, is more commonly asserted than the first in the theological statements opposing patenting. The fullest development of this distinction is found in the report noted earlier of the EECCS Bioethics Working Group. The report declares: "We object to the patenting of living organisms, to the patenting of human genetic material of any sort, and to non-human genetic material which has not undergone a major change by inventive means" (11, p. 38). This is because there is an important conceptual distinction that must be maintained between what is living and what is not. "Boundaries need to be drawn to make this distinction [between animate and inanimate matter] clear, to avoid reducing life conceptually to being merely an economic commodity, and then treating it as such" (11, p. 43).

It should be noted, however, that alongside the living/nonliving distinction, another boundary, this time between the human and the nonhuman, is also asserted in this report and elsewhere in church statements. For example, the report declares that "[p]atenting any part of the human genome is ethically abhorrent, in principle" (11, p. 44). In this context we should note that Roman Catholic concerns about patenting are not as sweeping or as urgent as those voiced by Protestants, in part because of the relatively greater emphasis Catholics place on the human/nonhuman distinction than do Protestants. For example, in one of the few Catholic statements on the subject, Pope John Paul II said to the Pontifical Academy of Sciences in 1994 that "...we rejoice that numerous researchers have refused to allow discoveries made about the [human] genome to be patented. Since the human body is not an object that can be disposed of at will, the results of research should be made available to the whole scientific community and cannot be the property of a small group" (12, p. 3). The opposition to patenting is both mild and limited to human DNA.

Generally speaking, Protestant statements rely on the living/nonliving distinction more than on the human/nonhuman distinction to define the limits of acceptable use of patenting as the means of intellectual property protection. According to the statement of EECCS, patenting is simply unacceptable for living things, in part because patenting is so firmly associated in the public mind with mechanical inventions that its application to living things conveys the wrong attitude toward life, one that is prejudicial in regarding living things as mere matter, mere mechanisms. In contrast to a mechanistic view, "[a]nimate material presents a radical discontinuity from mechanical and chemical inventions, which requires a different way of thinking about intellectual property.... Living organisms of any kind should not be patentable" (11, p. 42). For jurists or corporations to patent living things is an inappropriate use of governmental and financial power to the rule on a moral and theological question. In the strongest terms, the statement holds that "[w]e deplore the implications of various court decisions regarding patenting of living organisms.... This represents an unacceptable paradigm shift in how life forms are regarded.... This view sees nature entirely in anthropocentric terms of its utility to humans, as tools and products, and has lost the sense

of respect for animal or plant as of value in itself. This perception runs contrary to Christian understanding that all of creation owes its existence to God, and its significance is first of all what it is before God, irrespective of any use to which human beings might think of putting it.... An animal, plant or microorganism owes its creation ultimately to God, not human endeavour. It cannot be interpreted as an invention or a process, in the normal sense of either word" (11, p. 43). But this does not necessarily mean that biotechnology should not be given *some form* of intellectual property protection. If "patenting" connotes "mechanism," surely some other legal category could be created that protects inventive value without demeaning natural value. And so the report suggests "that consideration should be given to developing an alternative form of intellectual property for biological material" (11, p. 38). In other words, a new form of intellectual property protection that encourages research but does not contribute to reductionism would be acceptable.

It should be expected that religious organizations, perhaps including those of other major faith traditions beside Christianity, will continue to express opinions on biological patents. The question of patenting can easily become the lightening rod for general concerns about technology, the human role in nature, and the meaning of life itself. "While patent law may seem to be a technical and arcane subject and therefore of little relevance to the religious community, patenting biotechnology raises important questions traditionally within the domain of religious ethics. These include such issues as the nature and status of life, the relationship between the Creator and the creation, and the protection of human dignity. Placing biology within the jurisdiction of intellectual property law also has important symbolic implications relating to whether humanity can or should claim ownership of forms of life" (13, p. 8).

INDIGENOUS PEOPLES AND THEIR COALITIONS

At the 1995 United Nations conference on women in Beijing, women representing indigenous peoples met for extended discussion resulting in the *Declaration of Indigenous Women*, which addressed many concerns, among them attitudes toward nature and the question of biological patenting. The Declaration affirms a view of the earth as "our mother. From her we get our life, and our ability to live. It is our responsibility to care for our mother and in caring for our mother, we care for ourselves. Women, all females, are the manifestation of Mother Earth in human form" (14). From this consciousness arose a concern about technology, global trade, and the right of corporations to patent biological components. Recent global trade agreements are criticized as "new instruments for the appropriation and privatisation of our community intellectual rights through the introduction of the trade-related intellectual property rights (TRIPs). This facilitates and legitimises the piracy of our biological, cultural, and intellectual resources and heritage by transnational corporations." These global agreements threaten traditional social values: "Our indigenous values and practice of sharing knowledge among ourselves, and mutual exchange will become things of the past because we are being forced to play by the rules of the market." Furthermore global trade agreements legitimize the theft of traditional knowledge by incorporating it into patentable technological knowledge. "Bio-prospecting, which is nothing but the alienation of our invaluable intellectual and cultural heritage through scientific collection missions and ethnobotanical research, is another feature of recolonisation.... Their bid for the patenting of life forms is the ultimate colonisation and commodification of everything we hold sacred." The threat envisioned here is threefold: patenting endangers the value of traditional knowledge of plants or animals, cultural values such as sharing, and ultimately the survival of the people as a distinct identity. "It won't matter any more that we will disappear because we will be 'immortalised' as 'isolates of historic interest' by the Human Genetic Diversity Project." In highly passionate language, the Declaration urges resistance: "It is an imperative for us, as Indigenous Peoples, to stand in their way, because it means more ethnocide and genocide for us. It will lead to the disappearance of the diverse biological and cultural resources in this world which we have sustained. It will cause the further erosion and destruction of our indigenous knowledge, spirituality, and culture. It will exacerbate the conflicts occurring on our lands and communities and our displacement from our ancestral territories" (14).

The most immediate and specific form of resistance, according to the Declaration, lies in refusing to accept the universal validity of global trade agreements and their assertion of biological patenting. "We demand that the Western concept and practice of intellectual property rights as defined by the TRIPs in GATT, not be applied to indigenous peoples' communities and territories. We demand that the World Trade Organisation recognise our intellectual and cultural rights and not allow the domain of private intellectual rights and corporate monopolies to violate these.... We call for a stop to the patenting of all life forms. This to us, is the ultimate commodification of life which we hold sacred" (14). The Human Genetic Diversity project is particularly offensive, and the idea of any patent applications derived from these samples is especially abhorrent.

An incident that served to galvanize global opposition by indigenous peoples to biological patenting was the attempt by the United States Department of Commerce to file a patent application on a cell line derived from a Guaymi Indian woman from Panama. The Guaymi tribal president was widely quoted as saying, "I never imagined people would patent plants and animals. It's fundamentally immoral, contrary to the Guaymi view of nature, and our place in it. To patent human material ... to take human DNA and patent its products ... that violates the integrity of life itself, and our deepest sense of morality" (15, p. B5). Not only did this event and comment inspire opposition among indigenous people; they evoked a response far beyond indigenous peoples and served to mobilize opposition to patenting within Christianity and other religions.

So important is the issue of intellectual property to indigenous peoples that international conferences have been convened on this subject alone. For instance, 150 delegates from as many as 14 countries met in 1993 in New Zealand and issued the *Mataatua Declaration on Cultural and Intellectual Property Rights of Indigenous Peoples*. In the view of those gathered, "[w]e declare that Indigenous Peoples of the world have the right to self determination: and in exercising that right must be recognised as the exclusive owners of their cultural and intellectual property." The conference issued a call to indigenous peoples to develop protection strategies to resist corporations and governments in their unilateral pursuit of traditional knowledge or intellectual property protection. "Commercialisation of any traditional plants and medicines of Indigenous Peoples, must be managed by the indigenous peoples who have inherited such knowledge" (16) In the same year the *Intellectual Property Rights and Biodiversity* statement was developed by the Coordinating Body of Indigenous Organisations of the Amazon Basin (COICA). The COICA statement voices generalized suspicions about international intellectual property accords, which are seen as a new form of colonialist exploitation. The issue then becomes a matter of self-determination and of territorial sovereignty, especially in controlling access to traditional knowledge of plants within the traditional territory. "Biodiversity and the culture and intellectual property of a people are concepts that mean indigenous territoriality.... For members of indigenous people, knowledge and determination of the use of resources are collective and intergenerational. No indigenous population, whether of individuals or communities, nor the Government, can sell or transfer ownership of resources which are the property of the people and which each generation has an obligation to safeguard for the next" (17). Here again, the issues of traditional knowledge and of cultural values are both seen as threatened by global trade agreements. In particular, indigenous people are resisting the idea that they lack intellectual property or that they have no system to protect and regulate its value, and that therefore a intellectual property system can be asserted over their heads, so to speak. "Adjusting indigenous systems to the prevailing intellectual property systems (as a world-wide concept and practice) changes the indigenous regulatory systems themselves." Precisely because intellectual property systems are already in place, the assertion of alien Western-style "patents and other intellectual property rights to forms of life are unacceptable to indigenous people" (17).

In part through the assistance of the United Nations, indigenous peoples from around the world have come together to address common concerns, such as intellectual property. A gathering in Geneva in July 1999 resulted in *No to Patenting of Life! Indigenous Peoples' Statement on the Trade-Related Aspects of Intellectual Property Rights (Trips) of the WTO Agreement*, which begins with language that echoes the 1995 Beijing Women's Declaration. "*We, Indigenous Peoples* from around the world, believe that nobody can own what exists in nature except nature herself. A human being cannot own its own mother. Humankind is part of Mother Nature, we have created nothing and so we can in no way claim to be owners of what does not belong to us" (18). The concern that is raised in the statement is that by asserting a global regime, trade agreements force global and therefore local consensus in favor of a western and modern worldview. What looks to be a conflict of law and economy is ultimately a clash of worldviews, as "western legal property regimes have been imposed on us, contradicting our own cosmologies and values" (18). The idea of owning nature "goes against the very essence of indigenous spirituality which regards all creation as sacred." This concern is remarkably similar to that expressed by EECCS, which protested that courts, by ruling on patenting, were in effect ruling on theology, on worldview, and deciding in favor of a mechanistic reductionism and against traditional theological or spiritual views of nature and life.

In the 1999 No to Patenting of Life! statement, as in the Beijing Declaration, the claim is made that traditional cultures have rich legacies of intellectual property, but that the intellectual value is commonly held because it is the result of common effort. These two systems of intellectual property are bound to clash, and indigenous peoples will inevitably lose in the ensuing struggle. This conviction is clearly asserted in the closing statement of the Consultation on the Protection and Conservation of Indigenous Knowledge, meeting in East Malaysia in 1995, which declared: "The intellectual property rights system is in favour of the industrialized countries of the North who have the resources to claim patent and copyright, resulting in the continuous exploitation and appropriation of genetic resources, indigenous knowledge and culture of the indigenous peoples for commercial purposes" (19). The problem is both structural and conceptual. As relatively weaker partners in political and economic discussions, indigenous peoples are unable to assert claims and to protect them. But the problem is also deeper in that the very concept of intellectual property advanced by the international trade agreements is at odds with the cultural foundations and worldview of many indigenous peoples. "The intellectual property rights system totally ignores the close inter-relationship between indigenous peoples, their knowledge, genetic resources and their environment" (19). Or as stated elsewhere: "The inherent conflict between these two knowledge systems and the manner in which they are protected and used will cause further disintegration of our communal values and practices. It can also lead to infighting between indigenous communities over who has ownership over a particular knowledge or innovation" (19). It must be clearly recognized that in these statements, indigenous peoples are making the claim that they possess intellectual property in the form of their traditional knowledge of nature. But their claim, they believe, is wholly incommensurate with Western legal protections and indefensible within global trade accords. If the accords win, their claims will inevitably lose. This is because the accords reflect one view of nature and ownership, one that favors technical over traditional knowledge. Furthermore it should be understood that it is not the idea of the value of knowledge of biology that is disputed in these statements. At dispute is something far more profound, namely, how *bios,* life itself, is to be regarded and how value is to be owned.

ENVIRONMENTAL GROUPS

Environmental groups have been active in voicing opposition to biological patenting, largely out of the conviction that widespread patenting will likely reinforce certain tendencies in the development of genetics and biotechnology that will have especially detrimental effects on the environment. Groups such as Greenpeace have organized and issued press releases to try to persuade the European Parliament (EP) to pass legislation that prohibits or sharply limits biological patents. In a press release in response to a 1997 vote by the EP, Greenpeace states that "[g]enes, and living organisms are not inventions and therefore they should not be patentable. Patent law was introduced to protect technical inventions. It has been enshrined in law that patents should only be granted if inventions are novel, are not discoveries and can be manufactured. Genes, plants and animals clearly do not fit with patent protection. They exist in nature so are discovered and not novel. Because they reproduce biologically, they are not man made. Allowing patents on life is highly immoral because it would give monopoly control over life to private interests for profit alone" (20). The argument here is based on Greenpeace's view that biological patents fail to meet the technical requirement of novelty. That argument expanded on a press statement issued just two weeks earlier: "Greenpeace believes it is immoral to claim genes, cells and living organisms are inventions of man to be used and controlled by commercial interests.... Greenpeace believes that by allowing patents on animals they will simply be seen as machines to be treated as people wish. Their dignity will be sacrificed for flimsy supposed benefits to humans." Of course, the core conviction for Greenpeace or any broad-based environmental group is that biological patents favor biotechnology, which threatens the environment. "Greenpeace also believes genetically engineered plants bring threats to the environment which are irreversible and unpredictable in nature" (21).

This environmental focus is echoed in a 1997 *Statement on Life and Evolution* drafted during the 1997 State of the World Forum (November 4–9) in San Francisco by environmentalists and posted on the Internet for others to sign. "Life is an intimate web of relations that evolves in its own right, interfacing and integrating its myriad of diverse elements. The complexity and interdependence of all forms of life have the consequence that the process of evolution cannot be controlled, though it can be influenced. It involves an unpredictable creative unfolding that calls for sensitive participation from all the players, particularly from the youngest, most recent arrivals, human beings." The statement makes a broader claim when it asserts that "[l]ife must not be treated as a commodity that can be owned, in whole or in part, by anyone, including those who wish to manipulate it in order to design new life forms for human convenience and profit. There should be no patents on organisms or their parts." The statement then returns to the core theme of threat to the environment, which of course leads not just to opposition to patenting but to a rejection of biotechnology: "We must also recognise the potential dangers of genetic engineering to health and biodiversity, and the ethical problems it poses for our responsibilities to life. We propose a moratorium on commercial releases of genetically engineered products and a comprehensive public enquiry into the legitimate and safe uses of genetic engineering" (22). In contrast to the statements of religious and indigenous peoples groups, the statements of environmentalist groups about patenting are limited in the depth of their argument.

PUBLIC INTEREST SCIENCE GROUPS

Also limited in depth are the brief but influential statements of organizations like the Council for Responsible Genetics (CRG), which argues that "[n]o individual, institution or corporation should be able to claim ownership over species or varieties of living organisms. Nor should they be able to hold patents on organs, cells, genes or proteins, whether naturally occurring, genetically altered or otherwise modified." Their reasons are largely prudential or consequentialist, namely, that patent protection will restrict the practice of good science. The CRG statement adds that "[p]atenting organisms and their DNA promotes the concept that life is a commodity and the view that living beings are gene machines to be exploited for profit" (23). In an effort to build support for its position, CRG circulates a petition on the Web entitled *No Patents on Life!*

A similar strategy is found in the *World Scientists' Statement Calling for a Moratorium on GM Crops and Ban on Patents* (1999), issued by a group that is largely self-identifying and uses the Internet to recruit supporters. Its statement contains the following call: "Ban patents on living organisms, cell lines and genes." The supporting argument echoes the concerns raised by indigenous peoples: "The patenting of living organisms, cell lines and genes under the Trade Related Intellectual Property Rights agreement are sanctioning acts of piracy of intellectual and genetic resources from Third World nations, and at the same time, increasing corporate monopoly on food production and distribution. Small farmers all over the world are being marginalized, threatening long term food security for all" (24).

Another organization, the Crucible Group, met in 1993 and claims to "represent the widest cross section of sociopolitical perspectives and agricultural experience that may have ever been assembled...." Perhaps because of the diverse perspectives, the gathering produced a report but did not claim that it is a "consensus document." Notably they did not achieve an agreement on biological patenting except to conclude that the matter is urgent, divisive, and requires global attention. "Sensing, on the one hand, a certain uncertainty and lack of understanding related to intellectual property regimes and, on the other hand, the opportunity to create a new covenant in support of wider innovative processes, the Crucible Group recommends that the United Nations convene an international conference on society and innovation. Now, and at this conference, policymakers must bear in mind that some people, countries, and cultures have deep ethical concerns about biotechnology and the concept of life patenting" (25).

SCIENTIFIC ORGANIZATIONS

The Human Genome Organization (HUGO), which is an international association of leading researchers in human genome research, has issued a *Statement on Patenting of DNA Sequences*. While the statement does not oppose gene patenting altogether, it notes that "[i]t would be ironic and unfortunate if the patent system were to reward the routine while discouraging the innovative. Yet that could be the result of offering broad patent rights to those who undertake massive but routine sequencing efforts—whether for ESTs or for full genes—while granting more limited rights or no rights to those who make the far more difficult and significant discoveries of underlying biological functions. A second, equally unfortunate outcome would arise if a partial sequence publication or submission to a database precluded patenting of innovative disease gene discoveries leading to improved medical diagnostics and therapeutics. This could lead to inhibition of contributions to databases and lack of investment protection for the innovative. We hope that the system will find some way to adjust to the changing realities in this field to promote and protect this important and ongoing process of discovery in the public interest" (26, p. 7).

This statement was followed by HUGO in 1997 with a one-page comment that updates and clarifies HUGO's position, namely, that "*reaffirms* ... that HUGO does not oppose patenting of useful benefits derived from genetic information, but does explicitly oppose the patenting of short sequences from randomly isolated portions of genes encoding proteins of uncertain functions ... *regrets* the decision of some patent offices, such as the US PTO, to grant patents on ESTs based on their utility 'as probes to identify specific DNA sequences,' urging these offices to rescind these decisions and, pending this, to strictly limit their claims to specified uses, since it would be untenable to make all subsequent innovation in which EST sequence would be involved in one way or other dependent upon such patents ... [and] *urges* all large-scale sequencing centres and their funding agencies to adopt the policy of immediate release, without privileged access for any party, of all human genome sequence information in order to secure an optimal functioning of the international network, as well as to avoid unfair distortions of the system" (27).

HUGO's position is quite similar to the view put forward by the American Society of Human Genetics (ASHG) in its *Position Paper on Patenting of Expressed Sequence Tags* of November 1991. ASHG states that it has "taken the position that the issuing of patents for ESTs is likely to do far more harm than good...." Once again, this partial objection to patenting is not to be construed as general opposition, for "ASHG has not opposed patenting of genetic information when that information had utility." The fear is that the patenting of short sequences of unknown function will inhibit research in genetic science, and therefore "[t]he ASHG does not support the concept of patenting a short sequence from a randomly isolated portion of a gene encoding a protein of unknown function" (28).

As should be expected, the opposition of science organizations to patenting will be based on and limited to the negative effects of patenting upon scientific research.

Likewise, physicians might be expected to object to patenting that would limit the prompt application of research to medicine, and in fact the American College of Medical Genetics (ACMG) has voiced this concern. In its *Position Statement on Gene Patents and Accessibility of Gene Testing*, ACMG asserts the belief that gene testing "must remain widely accessible and affordable, and that the development and improvement of safe and effective genetic tests should not be hindered. The decision of the Patent and Trademark Office (PTO) to permit the patenting of naturally occurring genes and disease-causing mutations has produced numerous difficulties. While the ACMG disagrees with the PTO over this fundamental issue, we have further concerns over current patterns of enforcement of patents on genes that are important in the diagnosis, management and risk assessment of human disease." It should be noted that the scope of ACMG's objection is broader than that of the genetic science organizations, precisely because ACMG is concerned not merely for research but for the transfer of the benefits of research to the clinical setting. With that in view, ACMG complains of "...exorbitant up-front fees and per-test fees, and licensing agreements that seek proportions of reimbursement from testing services. These limit the accessibility of competitively priced genetic testing services and hinder test-specific development of national programs for quality assurance. They also limit the number of knowledgeable individuals who can assist physicians, laboratory geneticists and counselors in the diagnosis, management and care of at-risk patients." Patent protection and excessive licensing fees unduly restrain the field of clinical genetics, and "restricting the availability of gene testing has long-term implications beyond patient care. It affects the training of the next generation of medical and laboratory geneticists, physicians, and scientists in the area enveloped by the patent or license. It also retards the usually very rapid improvement of a test that occurs through the addition of new mutations or the use of new techniques by numerous laboratories that have accumulated samples from affected individuals over many years" (29).

As a result of this analysis, which is argued entirely on grounds that patenting will inhibit clinical genetics, the statement concludes that "it is the ACMG's position that ... [g]enes and their mutations are naturally occurring substances that should not be patented ... [p]atents on genes with clinical implications must be very broadly licensed ... [and] [l]icensing agreements should not limit access through excessive royalties and other unreasonable terms" (29).

Most scientific organizations draw back from such broad opposition to patenting and limit their concerns to specific misuses of the patenting process, especially the patenting of short sequences of unknown function. In a statement issued by the British Society for Human Genetics (BSHG) in 1997, the general principle of the patentability of genetic knowledge is affirmed: "Patenting is a valuable means of protecting intellectual property and promoting investment in developing products for the diagnosis and treatment of genetic disease." However, "[t]he discovery of gene sequence has for some little time

been a well understood process. There is nothing novel or inventive about this in principle, and as such new gene sequences should not be patentable, even where a straightforward utility e.g. diagnostic testing has been specified, unless there has been real progress towards the design of a specific commercial product." Patent offices must address, in greater precision, questions of usefulness or novelty, which "cannot reside in the mere description of a nucleotide sequence. It must rest in either novel methodologies for discovering the sequence or a novel use or application of the sequence. Conventional technology, conventionally applied, should not result in patents on newly isolated sequences." In a similar way, the standard for a claim for usefulness must be clear and fairly high: "A claim for utility must describe a utility specific to the sequence in question, and not simply rehearse those possible applications of any known gene sequence which are part of the general public state of the art." Specifically, claims of utility should be denied if they are based upon "use for isolating the full gene sequence ... use for detecting mutations in the gene ... [or] use for studying expression or function of the gene." Stated positively, the criterion of utility applicable to gene patenting ought to be "some meaningful indication that the sequence being patented has a reasonable prospect of being developed into a marketable product (which may be a diagnostic test) ... [or] a proposed specific use — for example, diagnosis of mutations in people with a specified clinical indication." The BSHG statement concludes with a call to patent offices to limit the scope of gene patents to the "specified applications which meet the novelty and utility criteria" (30).

It was noted earlier that the possible patenting of discoveries linked to the Human Genome Diversity Project is especially offensive in the view of the statements of indigenous peoples. To respond to this concern, the North American Regional Committee of the Human Genome Diversity Project has drafted a model protocol for collecting DNA samples. The model contains this description of the position of the HGDP regarding intellectual property rights: "The HGDP has no position on questions of patentability, although individuals participating in the HGDP hold a variety of positions. The HGDP does, however, hold clear positions about the commercial use of its samples and of the information derived from them." There are two principles to which all HGDP researchers are required to give assent: "First, it [HGDP] has resolved that it will not profit from any commercial uses of samples it gathers or knowledge derived from those samples. Second, it has vowed to ensure that, should commercial products be developed as a result of the HGDP's collections, a fair share of the financial rewards shall return to the sampled populations" (31, p. 1466). Enforcement mechanisms are not specified, nor is it clear that this will satisfy the concerns of indigenous peoples.

INDUSTRY ORGANIZATIONS

In response to the May 18, 1995, Joint Appeal of religious leaders that called for sweeping bans on biological patenting, Gerald J. Mossinghoff of the Pharmaceutical

Research and Manufacturers of America (PhRMA) stated that "PhRMA believes that it would be immoral for the pharmaceutical and biotechnology industries to walk away from new technologies that could stop pain, suffering and hunger. Because patents on animals, cell lines, genes, and their products are necessary to foster such scientific enterprise, PhRMA believes it is a moral imperative that the patenting of these types of inventions be maintained and encouraged" (32). This position is more fully developed in a PhRMA Policy Paper, *Strong Patent Protection Is Essential*, which argues that patent protection is necessary if research is to be funded and new pharmaceutical products brought to market. "Without strong patent protection, there simply would be no research-based pharmaceutical industry, which discovers and develops virtually all new medicines. As a result, few new life-saving, cost-effective medicines would be developed, and improvements in the quality of health care would be sharply curtailed" (33). The view is echoed by the major firms engaged in this area of research, such as SmithKline Beecham, whose 1997 "Patenting Statement" states: "Patent protection offers the only effective incentive for bringing to market the many commercial and industrial applications for which genetic inventions may be used.... SmithKline Beecham supports the patentability of any inventions which meet the federal patentability requirements of subject matter, utility, novelty, and non-obviousness" (34).

Any suggestion that genes or biological components are *living* and therefore inherently unpatentable is of course rejected in industry statements. The position of the Biotechnology Industry Organization (BIO) is expressed by Alan Goldhammer in these words: "BIO supports continuation of the current law to allow patenting of human genes when the applicant meets the necessary criteria for securing any patent: the invention must be novel, nonobvious and useful. When these standard criteria are met, patents should be issued irrespective of the nature of the invention. No exception should be made for patents on genes, life forms, or any other subject matter" (35). Without any doubt, BIO's position is fundamentally at odds with the views put forward by the organizations of indigenous persons and by some of the religious groups.

While the position of industry does not change significantly from country to country, it is interesting to note that regional industry groups advance the additional pro-patenting argument that national opposition to patenting will undermine a nation's biotechnology industry and therefore the nation's (or the region's) entire economy. For example, the Forum for European Bio-Industry Coordination (FEBC), in a Directive on the legal protection of biotechnological inventions, argues that a positive environment on biological patenting will encourage regional reinvestment in research and development. "Without patents there would be less investment in research. Patents are the foundation on which the development of new products like pharmaceuticals ... depend." At a time when Europe was considering a sharp limitation on biological patents, the industry warning was clear: "FEBC believes that any weakening of the draft would put Europe at a further disadvantage and will shift the emphasis of research

in biotechnology further towards the USA and Japan." The statement documents the warning with evidence that appeared to show that biotechnology investment in Europe was falling behind that in the United States and Japan, and stated that "[p]art of the reason for this gap is the lack of harmonised patent practice. The message is clear: although European investment has increased, the competitive gap between Europe and the US and Japan is still increasing" (36).

Many in industry and in government recognize the urgency of the problem posed by intellectual property protection for advances in knowledge in genetics and biology. The Human Genome Project, which has established an informal agreement that raw sequence data will be posted daily on the Web rather than patented, finds itself in competition with private efforts to sequence the genome in order to gain proprietary advantage. Speaking of the differences between the publically funded National Human Genome Research Institute (NHGRI) and private efforts such as the joint Perkin-Elmer-TIGR genome project, Francis Collins, the Director of NHGRI, told Congress that the "release of sequence data from the Perkin-Elmer-TIGR effort will occur quarterly, rather than daily. The policy of daily release of DNA sequence data by publicly-funded efforts was arrived at because of the great interest in the scientific community in gaining access to this highly valuable information. Any delay can result in wasted effort in research" (37). Deep differences of opinion and philosophy exist even here between scientists and laboratories that are engaged in essentially the same research.

BIBLIOGRAPHY

1. General Board of Church and Society of the United Methodist Church, Press Conference, Washington, DC, May 18, 1995.

2. Rev. Kenneth L. Carder, Statement to the press, Washington, DC, May 18, 1995.

3. Rev. Wesley Granberg-Michaelson, Statement of Religious Leaders against Animal Patenting, August 6, 1987; appended to testimony before the Subcommittee on Courts, Civil Liberties and the Administration of Justice of the Committee on the Judiciary, U.S. House of Representatives, Concerning Animal Patenting, November 5, 1987.

4. European Ecumenical Commission for Church and Society, Working Group on Bioethics, *EECCS and Bioethics*, EECCS, Strasbourg, France, 1998.

5. World Council of Churches, *Manipulating Life: Ethical Issues in Genetic Engineering*, WCC, Geneva, 1982.

6. World Council of Churches, Subunit on Church and Society *Biotechnology: Its Challenges to the Churches and the World*, WCC, Geneva, Switzerland, August 1989.

7. National Council of the Churches of Christ in the U.S.A., *Genetic Science for Human Benefit*, NCC, New York, 1986.

8. United Methodist Church, *Book of Resolutions, 1992*, Nashville, 1992.

9. Southern Baptist Convention, *SBC Bulletin*, Nashville, 1995, pp. 7–8.

10. R. Cole-Turner, in A.R. Chapman, ed., *Perspectives on Genetic Patenting: Religion, Science, and Industry in Dialogue*, American Association for the Advancement of Science, Washington, DC, 1999, pp. 149–165.

11. Ecumenical Commission on Church and Society, Bioethics Working Group, *Critique of the Draft EU Patenting Directive, a Submission to the European Parliament and the European Commission*, EECCS, Strasbourg, France September 1996; printed *Drawing the Line—The Ethics of Biotechnology*, Occasional Paper No. 5, by the Ecumenical Association for Church and Society, 1997, pp. 37–50.

12. Pope John Paul II, *L'Osservatore Romano*, November 9, 3 (1994).

13. A.R. Chapman, in A.R. Chapman, ed., *Perspectives on Genetic Patenting: Religion, Science, and Industry in Dialogue*, American Association for the Advancement of Science, Washington, DC, 1999, pp. 7–40.

14. *Declaration of Indigenous Women*, Beijing, 1995, Available at: *http://www.southside.org.sg/souths/twn/title/indig-cn. htm*

15. P.L. Bereano, *Seattle Times, Sunday*, August 27, B5 (1995).

16. *The Mataatua Declaration on Cultural and Intellectual Property Rights of Indigenous Peoples*, Aotearoa New Zealand, June 1993, Available at: *http://users.ox.ac.uk/~wgttr/ mataatua.htm*

17. Coordinating Body of Indigenous Organisations of the Amazon Basin (COICA), *The COICA Statement: Intellectual Property Rights and Biodiversity*, Santa Cruz de la Sierra, Bolivia, September 1994, Available at: *http://users.ox.ac. uk/~wgtrr/coica.htm*

18. "No to Patenting of Life! Indigenous Peoples' Statement on the Trade-Related Aspects of Intellectual Property Rights (TRIPS) of the WTO Agreement," signed at the United Nations, Geneva, Switzerland, on July 25, 1999, Available at: *http://www.alphacdc.com/ien/intellectual property.html*

19. United Nations Development Programme, *Consultation on the Protection and Conservation of Indigenous Knowledge*, Sabah, East Malaysia, February 24–27, 1995, Available at: *http://users.ox.ac.uk/~wgtrr/sabah.htm*

20. Greenpeace, *Greenpeace Deplores EP Vote on Bio Patent Directive; Leverage Buyout of Nature—Ethics under Corporate Control?* Strasbourg and Brussels, July 16, 1997.

21. Greenpeace, *Patents on Life: Immoral, Driven by Greed and Damaging to Medical Research*, Strasbourg and Brussels, July 2, 1997.

22. *Statement on Life and Evolution*, San Francisco, November 1997, Available at: *http://www.psrast.org/swf97reg.htm*

23. Council for Responsible Genetics, *No Patents on Life! DNA Patents Create Corporate Monopolies on Living Organisms*, Cambridge, MA, 1994, Available at *http://www.genewatch.org/*

24. *World Scientists' Statement Calling for a Moratorium on GM Crops and Ban on Patents* (1999), Available at: *http://www.isis.dircon.co.uk/home.htm*

25. Crucible Group, *Statement*, 1993, Available at: *http:// www.idrc.ca/books/725/preface.html*

26. Human Genome Organization, *HUGO Statement on Patenting of DNA Sequences*, Bethesda, MD, January 1995.

27. Human Genome Organization, Bethesda, MD, 1997.

28. American Society of Human Genetics, *Position Paper on Patenting of Expressed Sequence Tags*, November 1991.

29. American College of Medical Genetics, *Position Statement on Gene Patents and Accessibility of Gene Testing*, August 2, 1999.

30. British Society for Human Genetics, *Patenting of Human Gene Sequences and the EU Draft Directive*, September 1997, Available at: *http://www.bham.ac.uk/BSHG/patent eu.htm*

31. North American Regional Committee of the Human Genome Diversity Project, Model Protocol: Proposed Model Ethical Protocol for Collecting DNA Samples, *Houston Law Rev.* **33**(5), 1431–1473 (1997).

32. Pharmaceutical Research and Manufactures of America (PhRMA), News Release, Washington, DC, May 18, 1995.

33. Pharmaceutical Research and Manufacturers of America (PhRMA), Policy Paper, 1995.

34. SmithKline Beecham, Patenting statement, 1997.

35. A. Goldhammer, Director of Technical Affairs, Biotechnology Industry Organization (BIO), Testimony before the Subcommittee on Technology of the House of Representatives, Committee on Science, September 17, 1996.

36. Forum for European Bio-Industry Coordination (FEBC), *FEBC Views on the Directive on the Legal Protection of Biotechnological Inventions*, Brussels, 1995.

37. Statement of Francis S. Collins, M.D., Ph.D., Director, National Human Genome Research Institute on the Human Genome Project, Statement before the Subcommittee on Energy and the Environment Committee on Science, United States House of Representatives, June 17, 1998.

See other entries HUMAN GENOME DIVERSITY PROJECT; OWNERSHIP OF HUMAN BIOLOGICAL MATERIAL; see also PATENTS AND LICENSING entries.

PATENTS AND LICENSING, ETHICS, OWNERSHIP OF ANIMAL, AND PLANT GENES

MICHELE SVATOS
Iowa State University
Ames, Iowa

OUTLINE

INTRODUCTION

Current biotechnology patenting practices raise major ethical, legal, and economic controversies. Patenting "life," including plants, animals, and their genes, has been criticized as unethical and inimical to social justice. The Council for Responsible Genetics circulated "The No Patents on Life! Petition," which argues:

> The plants, animals and microorganisms comprising life on earth are part of the natural world into which we are all born. The conversion of these species, their molecules or parts into corporate property through patent monopolies is counter to the interests of the peoples of this country and of the world. No individual, institution, or corporation should be able to claim ownership over species or varieties of living organisms. Nor should they be able to hold patents on organs, cells, genes or proteins, whether naturally occurring, genetically altered or otherwise modified (1).

Patent advocates retort that of course life cannot be patented — merely genes and genetically modified organisms (GMOs) (2). However, it is unlikely that patent critics actually misunderstand that point, as is evident in No Patents on Life! petition; indeed, many of its supporters are scientists. Life itself, as some élan vitale, surely cannot be patented, although one can patent living organisms and genes; that alone is bad enough according to critics. Let us consider the arguments for and against animal and plant patents. After a brief review of the legal status of life patenting and the relation of patent law to morality, we will look at arguments specifically addressed at animal patenting first, before considering arguments that apply more generally to both animal and plant patents.

PATENT LAW AND MORALITY

Patents grant the right to exclude others from making, using, or selling an invention for a period of 20 years. A patentable invention must be novel, useful, and nonobvious. Ideas, theories, mathematical algorithms, laws of nature, and the like, cannot be patented; processes (e.g., production methods, special techniques, and diagnostic methods), products (e.g., microorganisms, enzymes, plasmids, cell lines, and DNA and RNA sequences), and new uses of an existing product can be patented.

Plants were the first living organisms to receive explicit patentlike protection. The Plant Patent Act of 1930 provided protection of asexually reproduced plants, and it was mainly applied to flowers and certain fruits. The broader 1970 Plant Variety Protection Act (PVPA) provides patentlike protection for new varieties of plants, although it differs from a patent in lacking the utility requirement. Plant breeders use both plant variety protection and patents to protect genetically modified plants and plant genes.

Since the landmark 1980 Supreme Court decision *Diamond v. Chakrabarty*, living organisms — "anything under the sun that is made by man" — can be patented in the United States (3). However, the *Chakrabarty* decision focused on a microorganism, and did not explicitly address higher life forms (4). In 1987 the Commissioner of the U.S.

Patent and Trademark Office (PTO) formally affirmed the patentability of "nonnaturally occurring non-human multi-cellular living organisms, including animals." It clarified that "Products found in nature will not be considered to be patentable subject matter ... unless given a new form, quality, properties or combination not present in the original article existing in nature in accordance with existing law" (5).

In 1988 the first transgenic animal was patented, the "Harvard mouse" or "Oncomouse" (6), genetically modified for hypersensitivity to carcinogens for medical research. Animals are genetically modified for medical research, agriculture, and as pharmaceutical biofactories. Animal, plant, and human genes may be patented, including cloned, unmodified genes. Most animal patents thus far are for mice and rats.

Morality comes into play at various levels in the legal practice of patenting (7). Normally the extension of the patent system to a new technological discipline is automatic and may be assumed. However, unique features of new technologies can result in difficult questions of interpretation for patent law. In modern biotechnology, the distinction between discovery and invention is becoming blurred. Moreover genetically modified organisms (GMOs) are unique as inventions. Not only are some of them alive, but also they are able to reproduce on their own, and are not well standardized, easily described, and so on. If they are released into the environment, they will interact with it unpredictably. These are good reasons to pause, rather than extending or excluding patent protection automatically. Old definitions and criteria must be reconsidered and sometimes redrawn.

Although the legal question of the patentability of life forms has been settled in the United States and Europe, some popular sentiment against it remains in these countries. Moreover it remains a lively issue within the international community, as developed nations pressure less developed countries (LDCs) to adopt Western intellectual property laws. LDCs are rich in genetic resources, and many object to intellectual property laws and alleged "biopiracy" on grounds of morality and social justice. Thus the moral debate over life patenting has become increasingly important to discussions of international trade relations and international justice. Are there any special reasons to believe that animal and plant patents are more or less justified than other types of patents, that they should be subject to special conditions or exclusions, or that they raise special problems of ethics and social justice?

ANIMAL PATENTS

Animal patents are criticized on the grounds that they:

- Encourage increased production of transgenic animals, many of which suffer greatly
- Violate animals' rights
- Encourage the reduction of animals to mere inventions
- Engender additionally most of the criticisms that apply to plant patents (see the discussion below)

Several philosophers and legal scholars, such as Baruch Brody (9), Robert Merges (10), and Rebecca Dresser (11), have considered the arguments against animal and plant patenting in depth. They have argued that the criticisms are without merit or point, to problems that are not fundamentally about patenting. Let us consider the arguments for and against animal patenting in greater detail.

Animal Welfare

Animal welfare advocates warn that animal patents encourage more research that, on balance, tends to increase animal suffering. With animal patents, claims Michael Fox of the U.S. Humane Society, "the wholesale industrialized exploitation of the animal kingdom will be sanctioned, protected, and intensified" (8). Although genetic engineering may sometimes decrease animal suffering, critics are concerned that its record thus far is not good, and that this trend will continue. In fact genetic engineering poses special problems for animal welfare, above and beyond normal concerns about the treatment of farm and research animals. While many traditionally bred animals are routinely subjected to inhumane treatment, transgenic animals may be engineered to inevitably suffer even under the best conditions, for example, cancer-prone or hairless mice (12). The U.S. Animal Welfare Act does not apply to farm animals, birds, or rodents, making critics skeptical that genetic engineers will be subject to limits on cruelty (9). Insofar as genetic engineering is likely to cause more suffering for animals than it prevents, and insofar as patents promote the development and commercialization of genetically engineered animals, animal welfare advocates attack animal patenting as an economic incentive for much morally problematic research.

Patent advocates respond that animal welfare advocates should attack the problem of cruelty to animals directly, rather than through patenting (9,11–13). Banning such patents would discourage even research designed to alleviate animal suffering, and which may help many humans. Even if some transgenic animals suffer more, their suffering might be outweighed if it is possible to use fewer but more efficient animals. Moreover patents are perfectly compatible with strong animal welfare regulations, and in themselves are morally neutral.

A few animal welfare advocates have admitted that if sufficient animal protection regulations were in place, they would have less or no opposition to animal patenting. This makes sense for utilitarian animal welfare advocates, many inspired by utilitarian philosopher Peter Singer's book *Animal Welfare*. Singer argues that animal suffering is morally relevant but may be overridden by the likelihood of greater benefit to others (14). Thus some research on animals could be justified—if there were no adequate alternatives, if the benefit were great enough, and if the suffering were small enough. How much animal research would remain is a matter of debate; many utilitarians argue that it would be very little. Most of animal agriculture is not justified on a utilitarian animal welfare view, since vegetarianism is usually a feasible alternative. Thus utilitarian animal welfare advocates, as critics of most of animal agriculture and research, are natural critics of animal patents, but mostly because and insofar as it is likely to increase the suffering of animals. The key assumption is that patented transgenic animals are likely to suffer in order to more efficiently promote our medical and agricultural ends. While this initially may seem extremely pessimistic, the charge warrants examination. For utilitarian animal welfare advocates, criticisms of animal patenting would naturally be combined with a more direct approach toward alleviating animal suffering and strengthening animal welfare laws—and indeed that is often the case, as with Michael Fox.

Animal Rights

A more radical view of animals and morality, inspired by philosopher Tom Regan, holds that animal suffering is morally relevant but that animals also have moral *rights* and may not be used as a mere means to our medical, agricultural, or recreational ends—regardless of the benefits (15). Animal rights advocates are likely to be categorically opposed to animal patenting, on the ground that it involves the use of animals as property and inventions, mere means to our ends. They are likely to accept genetic engineering of animals only where it might be justified on human infants, done on behalf of the child or animal rather than directly or indirectly in service of adult goals. Thus disease-resistant transgenic chickens would probably not be justified even if the animals suffered less, in the animal rights view. Even where genetic engineering might be justified, animal rights advocates could say that it is unjust to patent the genetically engineered animal, since doing so amounts to treating that variety of animal as mere property rather than as a living entity with rights and its own ends.

Thus, even radical revisions in animal welfare laws are unlikely to move animal *rights* advocates (as opposed to utilitarian animal welfare advocates). Patent defenders sometimes are confused and frustrated by animal rights advocates' criticism of patenting. We treat animals as property all the time. Most people do not consider animals to have rights, so animal welfare concerns could be more directly addressed through animal welfare regulations (10). However, all this makes sense given a certain philosophical grounding for concern for animals, namely an animal rights view. Not only could animal patents work to perpetuate or promote animal suffering, the mere practice of patenting animals violates animal rights by treating them as means to our ends. Patent advocates may respond that public policy should not be molded to accord with one particular minority view of animals and morality. However, for animal rights advocates, the situation is analogous to slavery. If all humans have moral rights, then slavery should be abolished, regardless of whether the majority believes in it and benefits from it. Animal patenting could be compared to patenting of genetically engineered slaves: although slavery is the main practice that should be abolished, the logical implication is that patenting of slaves should be abolished as well, especially if it eases the commercialization of slaves. If the main practice seems unlikely to disappear in the near future, then attacking the newer, more controversial patenting practice makes sense, particularly if there is a real possibility of banning patenting.

Extending an animal rights view further, biotechnology critics such as Jeremy Rifkin argue that animals have a right to species integrity violated by genetic engineering (16). Defenders reply that genetic material can move between species in nature, and we have been genetically manipulating plants and animals for generations through traditional breeding. Moreover the notion of species' rights is extremely controversial, even compared to the concept of animal rights. (4,10). Finally, this criticism does seem directed more toward genetic engineering than patenting.

Reductionism

Animal and sometimes plant patents are also criticized on grounds that are less directly related to animal welfare. Critics argue that patents are reductionistic and encourage and intensify the commodification and objectification of living organisms, particularly animals. Characterizing GMOs as human "inventions" attaches too much importance to the contribution of scientists and too little importance to what God or nature has given us. It demonstrates an attitude of hubris or arrogance that is also apparent in our release of these organisms into the environment when we cannot fully understand or control the risks. Let us consider these objections, which take both secular and theological forms, in more detail.

Objectification. Patent critics argue that the language used to describe transgenic animals reveals a mechanistic view of them. They are bioreactors, biofactories, disease "models," and so on. While objectification occurs without animal patents, patenting takes the objectification to a new level by reconceptualizing animals as human-made utility inventions and "compositions of matter" (17,18). Critics argue that God or nature creates life, and it is absurd and arrogant to patent living organisms as human inventions. Even GMOs usually have only one or a couple of transgenes, which are merely transferred from another living being rather than being created de novo. At most we are moving around a few parts of God's or nature's creations (19,20).

Patent advocates reply that a patent does not mean that one has created an invention from nothing; patents are allowed for improvements on preexisting things. Of course, humans did not invent the mouse, and they do not patent mice, the entire species. Instead, they patent mice that have been genetically altered in a small but significant way. This does not detract from nature or God's handiwork, it merely adds to it. Scientists would be the first to admit that they are far from being able to create an entirely new animal out of nothing. Moreover genetic engineering, patenting, and the use of terms such as "biofactory," are compatible with proper respect for life in other contexts. Although humans are technically "compositions of matter," they are also much more, and so are other living organisms (21).

Patent critics may be skeptical of this response. It might be compared to the argument that treating people as sex objects is compatible with respecting them in other

contexts; feminists are skeptical that compartmentalizing really works that well. Objectifying animals helps us view their plight as morally irrelevant, just as objectifying women helps rationalize violence against them, and objectifying the enemy in war helps soldiers commit acts of violence against them (18,22). While it is not clear that this argument holds merit, it is also not clear that patent advocates are correct that viewing animals and plants as inventions has no larger implications for our treatment of them and our conception of humanity's relationship to the rest of the living world. There may be grounds for making a distinction between the ownership of particular animals, and the ownership of, say, all mammals with a certain gene. There is, one might argue, a difference in attitude (which is simply a fact of our society) between the owner of a cow or dog, and the owner of a technology/invention. This takes the mechanistic view of nature even further than it has previously been extended. We may think that we have finally refuted and rejected Descartes's view that animals are simply machines—but wait! They are, after all, but our own technologies, our own inventions. To say that we own an animal is different than saying that we invented them or that they are a technical solution to a technical problem (7). It is a mistake to completely attribute this debate to a misunderstanding of the practical, patent law distinction between discoveries (unpatentable) and inventions (patentable). Indeed, some molecular biologists agree with the public on this point, hardly out of ignorance of science or patent law. Science and technology, discovery and invention, are becoming increasingly blurred in biotechnology. If a gene, its function, and its structure are scientific discoveries, how can the purified and isolated gene become an invention (23)? The public is perplexed and accuses patent lawyers of playing word games. Critics may charge that patent law has focused too much on technical definitions at the expense of the commonsense "fact" that human, animal, and plant genes are discoveries, not inventions, and that even genetically modified plants and animals still do not qualify as inventions (24,25). We will raise this issue again in the final section.

Commodification. The Council for Responsible Genetics argues that "Patenting organisms and their DNA transforms them into commodities for profit and promotes the view that living beings are little more than 'gene machines'" (26). One may well not be opposed to the research itself but rather to certain forms of its commercialization made possible by patenting. Most people support genetic research that would lead to new forms of human gene therapy. They might even be in favor of commercializing gene therapy, to the extent of making it a private for-profit venture and allowing many aspects of the research (e.g., new processes) to be patented. Opponents of patenting genes argue that genes should and in some sense do belong to everyone and no one. Every researcher should be free to work on them and try to use them to benefit humankind, without having to get a license and pay a royalty. Of course, if companies are able to patent only the process they use but not the gene product, it will be difficult for them to control profits, but to stake a claim on genes themselves is to go too far. Opponents of animal patents may view the primary problem as patenting itself rather than, or above and beyond, the suffering that occurs. Groups such as Global Action in the Interest of Animals (GAIA) object to the patenting of animals even where the animals do not suffer, or actually benefit, from a genetic modification (17).

Patent advocates reply that we have owned and bred animals for centuries. The fact that a genetically engineered line of mice is patented is unlikely to affect our treatment of those animals, and animal welfare concerns should be separated from patenting (4,9). Moreover patents are needed in order for companies to have an incentive to invest in and commercialize socially useful research. However, critics, including many scientists, are skeptical that this is actually necessary, and argue that it may actually hinder research (27–29).

Why object to patenting genes and GMOs, apart from concerns about genetic engineering itself? This view should not be confused with a criticism of research into a particular area. It is rather an argument for limiting the subjects of product patents, regardless of whether we wish to limit the research itself, and regardless of the fact that patenting genes probably does meet the criteria of current patent law and is explicitly supported by the *Chakrabarty* decision.

HUBRIS AND RISKS

"From an ethical perspective, the patenting of animals reflects a human arrogance toward other living creatures that is contrary to the concept of the inherent sanctity of every unique being and the recognition of the ecological and spiritual interconnectedness of all life" (30). Michael Fox argues that patenting will accelerate the "transformation of life and of the creative process to serve purely human ends, and as many see it, the end of the natural world" (8). Such critics also often question the wisdom of moving genes around. They worry that we do not adequately understand the risks, that we are transgressing natural barriers that might be best respected, and that we often harm transgenic animals. In our hubris we consider GMOs and even unmodified genes to be our inventions, and we further assume that these creations are so under our control that the risks of introducing them are minimal (31). Critics see patenting living organisms and genes as arrogant and unwise. They worry that the upshot of patenting living organisms may be the devaluation of all life, including human life (18,31).

Patent advocates respond that once again, critics should attack the real "culprit," rather than patenting. Concerns about the risks of genetic engineering should be addressed by the U.S. Environmental Protection Agency (EPA), the Food and Drug Administration (FDA), or other regulatory agencies, rather than the patent office. The mere fact that one has patented a genetically engineered organism provides no assurance that one will actually be able to use and market it; it may not receive regulatory approval for release (32). If regulatory processes for genetically engineered organisms are seen as inadequate, then they should be criticized, rather than patenting. If genetic engineering is unwise or unnatural (either in general,

or in particular cases), then criticize genetic engineering itself rather than patenting (9).

Rather than being a *symptom of* worries about risks, the hubris objection to patenting "life" is sometimes more accurately classified as *parallel to* concerns about risks. Biotechnology critics hold that our arrogance and greed leads us to engage in risky behavior, and at the same time leads us to consider living organisms and genetic resources as mere inventions to be patented. Another part of the same arrogance and greed is our treatment of animals, which is poor to begin with and often only intensified by genetic engineering. We hold ourselves up as masters and creators of the natural world rather than a humble part of it. This view can take theistic or secular forms, and it may be accompanied by more or less skeptical attitudes toward the possibility of wise, safe genetic engineering. Some critics see genetic engineering as entirely unnatural or immoral, and others do not reject it out of hand but only want it to be done and applied with wisdom and justice.

PATENTING, FARMERS, AND "BIOSERFDOM"

Some farmers' advocates oppose plant and animal patenting. One major concern is that patenting will further encourage factory farms and the consolidation of agribusiness at the expense of small farms and farming communities. Critics charge that agribusiness is becoming an oligopoly. As companies merge and become vertically integrated, they control more of agriculture, and independent animal and plant breeders and dealers can scarcely compete (33). With large multinational corporations heavily invested in genetic engineering, they have to heavily promote their products to recoup research costs. Large corporations are better equipped to handle patent applications and infringement cases. If there are few competitors left, then there will eventually be few affordable, nongenetically engineered varieties available to small farmers, who are so squeezed for profits that they can scarcely afford to make more expensive or less efficient purchasing decisions than their corporate competitors. Practically speaking, they will have little alternative to purchasing expensive patented varieties, both because little else will be offered for sale and because their larger competitors will also be doing it. The genetically engineered varieties are more likely to be well suited to large corporate farms, with their high technology inputs and large economy of scale.

Patents may further mean that farmers are prohibited from breeding animals and seeds for their own use. Patent critics and even sympathizers have argued that at minimum, there must be an exemption for breeding by farmers, particularly small- and medium-sized farmers. Currently U.S. law does not have such an exemption, although European law has an exemption for "small" farmers (33). In the absence of such an exemption, patent advocate Robert Merges argues there are two practical problems. First, considerable record keeping would probably be required of farmers to prove that they did not breed any patented animals without paying the proper royalty. Second, enforcement procedures might be difficult, and might require farmers to open their farms' animals to inspections to ensure that no patent law violations had taken place (10). Even if farmers are allowed to breed patented varieties without paying royalties, genetic drift assures that in a few generations, most will want to renew their stock from commercial breeders to retain desired traits.

These practical problems also raise moral issues, namely the reduction in the autonomy and privacy of farmers. Critics charge that even Western farmers are becoming poorly paid employees with no applicable minimum wage laws and little autonomy. "As the life industry dictates more and more of the farm-level management decisions, the farmer becomes little more then a 'renter' of proprietary germplasm and information, a step in the food/industrial manufacturing process. Farmers and consumers thus increasingly lose control over what products they grow and consume, and which food production processes they chose to support" (34, p. 5). Some critics have denounced this as bioserfdom. They worry that animal and plant patenting is merely another step on the way to bioserfdom and corporate agriculture, with poor conditions for farmers and animals alike. Let us consider the role of licensing restrictions and the so-called terminator seed in promoting this situation. As we will see, an exemption or derogation for farmers might not be sufficient to prevent it.

Seed Saving and Licensing Restrictions

Farmers traditionally have saved seeds from part of their crop for breeding, trading, and planting the next season. This still occurs to a certain extent in developed countries, particularly for non-hybrid crops such as cotton and rice. The practice is more common among poor farmers in less developed countries (LDCs). "Between 15 and 20 percent of the world's food supply is grown by poor farmers who save their seed. These farmers feed at least 1.4 billion people" (35). Seed companies view seed saving as a problem in expanding their market and protecting their intellectual property, namely proprietary seeds. An analogy may be made to computer software; farmers buy the right to plant the seeds but not to make "pirated copies" by planting second generation seeds. "From an industrial perspective, plant varieties are to be used for growing a crop. All other unauthorized uses of varieties lower returns on investment and therefore must be eliminated" (33, p. 139). Constantly improving seeds is one way to encourage farmers to buy seeds every growing season. However, seed research is expensive and most improvements are small, so this alone may not be sufficient, particularly for nonhybrid crops. So seed companies look to new ways to protect their intellectual property from what they view as the piracy of farmers.

Restrictive licensing agreements attempt to ensure that seeds will not be saved and replanted without payment of royalties. Monsanto has been criticized for its practice of requiring such contracts, which include a rather unpopular provision that Monsanto's agents — sometimes dubbed "gene police" — may enter the farm and test for Monsanto genes for a period of three years following the initial purchase of seeds (36). Monsanto even provides a toll-free number for farmers to report suspected violations and

has publicized the names of violators on radio stations, in addition to suing for compensation. Of course, no one is *required* to buy seeds from Monsanto, so if farmers don't like this policy, they may take their business elsewhere, and avoid trading seed with Monsanto customers.

Such a policy of policing intellectual property is criticized as an example of bioserfdom. It is obviously not popular with many farmers and requires considerable expenditures for enforcement and litigation. In LDCs, matters are even more difficult; enforcement costs might be higher, and legal support dubious or lacking. Indeed, proprietary seeds could make it to a developing country without the company's knowledge, and many LDCs are notorious for failing to protect intellectual property. Such licensing agreements may turn out to be an unpopular and not especially effective strategy of protecting intellectual property.

Technology Protection System: "Terminator" Seeds

A further step in protecting seed intellectual property from farmers is to build a technological "fence" around proprietary seeds, comparable to software copy protection schemes (37). Indeed, such a fence is in the development stages: TPS (technology protection system) seed is being developed by the U.S. Department of Agriculture (USDA) and Delta and Pine Land Co., dubbed "the terminator seed," "suicide seeds," or "traitor technology" by Rurual Advancement Foundation International (RAFI) and others. TPS seeds are genetically modified with an extra gene that seed companies can "turn on" before selling the seeds (38) so that second generation seeds are sterile. "The Terminator provides a built-in biological 'patent,' enforced by engineered genes" (35, p. 1). Farmers would buy TPS seeds, which offer farmers no direct benefit in themselves, because the TPS trait would be paired with other valuable traits. They would then have no alternative but to purchase new seeds each growing season, or to buy whatever seed varieties do not include TPS. Of course, critics worry that agribusiness industry is shaping up in such a way that there will be few choices.

The media, developing countries, and several nongovernmental organizations (NGOs) have expressed outrage over TPS. NGOs such as RAFI have called for a global ban, claiming that:

> It is a global threat to farmers, biodiversity, and food security. The seed-sterilizing technology threatens to eliminate the age-old right of farmers to save seed from their harvest and it jeopardizes the food security of 1.4 billion people — resource-poor farmers in the South — who depend on farm-saved seed. . . .If the Terminator technology is widely utilized, it will give the multinational seed and agrochemical industry an unprecedented and extremely dangerous capacity to control the world's food supply (39).

Similarly the Center of Education and Technology in Chile warns, "This is an immoral technique that robs farming communities of their age-old right to save seed and their role as plant breeders. Farmers and governments everywhere should declare the use of the technology as contrary to public order and national security. This is the neutron bomb of agriculture" (40). Influential groups such as CGIAR (Consultive Group on International Agricultural Research) have condemned TPS, and several countries have banned it.

In response to these sorts of criticisms, the USDA released a *Fact Sheet: Why USDA's Technology Protection System (a.k.a. "Terminator") Benefits Agriculture—A Discovery to Spur New Crop Improvement*. The fact sheet claims:

> Because of [farmers'] seed-saving practice, companies are often reluctant to make research investments in many crops because they cannot recoup their multiyear investment in developing improved varieties through sales in one year (38).

The fact sheet goes on to assert that farmers as well as the environment will benefit from new seed varieties that seed companies will be inspired to develop, given the reassurance that their intellectual property will be protected. In particular, "small farmers may benefit greatly if the invention stimulates the extension of biotechnology to 'minor crops' such as tomatoes," since the seed industry will see greater potential for return on their research investment into these minor crops (38). Advocates argue that farmers in developing countries will still be able to save their traditional and public seed varieties. Rather than posing a threat to world food security, it will do the opposite, as seed companies will feel more secure in releasing genetically improved varieties in developing countries with poor intellectual property protection. This will help even the playing field between U.S. farmers who abide by intellectual property laws, and LDC farmers who often do not. In other words, it requires everyone to pay their fair share of the seed research that they benefit from. The market for terminator seeds will not extend beyond its benefits to farmers since no one is forced to buy TPS seeds. Moreover it is a safe technology that poses no risks to the environment, and that may indeed help prevent the unintentional spread of other genetically modified traits. Finally, TPS is part of a larger research program that will allow the controlled expression of many traits in plants, and that has the potential to enormously benefit agriculture.

This is scarcely the place to resolve the controversy, which has received little serious academic attention (41). However, it is clear that TPS is appropriately named Technology Protection System. The debate must be understood as part of the debate over plant patents, especially restrictive licensing agreements. Although TPS is several years away from commercial release and may never prove viable in its current form, expect further attempts at "fencing" plant—and eventually animal—intellectual property.

"Bioserfdom," Western Farmers, and LDCs

In sum, if patents tend to work more to the advantage of large corporations, and if increasing control of agribusiness by large corporations tends to work to the advantage of corporate factory farms, then small farmers view animal and plant patents as another stake being driven through the heart of the Jeffersonian ideal of the yeoman farmer. Critics believe that patents are a

government incentive that works mostly in the favor of corporate agribusiness, and contrary to the purposes of preserving the independent family farm. If the government does not want to support family farms, they should at least not provide incentives for the corporate competition that already enjoys many advantages. Again, this is not to say that individual patents might not help small farmers, but on balance the practice of biotech patenting is seen as a tool for agribusiness and corporate farming. Corporate breeders want farmers to keep coming back to them every time rather than breeding their own sometimes, and patents and licensing agreements help ensure that this happens. Many farmers are skeptical that this ultimately will be to their advantage, even if it does encourage more investment in research. The agricultural technology treadmill may help early adopters of new technologies, but given current low prices and overproduction, in the end these new technologies will only drive down food prices further—which might be nice for the consumer but not for most farmers. A strong system of patents in agriculture is, critics say, likely to promote corporate dependence rather than sustainable farming.

If this is true, will patents at least benefit farmers and consumers in countries where overproduction is scarcely a problem? Patent critics argue that these countries are most in need of sustainable solutions rather than increased dependency on Western corporate agriculture. The poorest farmers in those countries are subsistence farmers who cannot afford genetically engineered animals and seeds, plus the capital-intensive farming methods that tend to go with them. Critics worry that like the Green Revolution, new genetic technologies will simply perpetuate or even worsen inequalities of wealth by benefiting large, wealthy farms in those countries (60). Moreover, even if the result is cheaper food, this does not benefit subsistence families. Furthermore many of the crops grown in those countries are actually for export, and do not help feed the poor in that country. Multinational corporations argue that patents are good for American farmers because it is unfair that American farmers pay royalties while farmers in LDCs do not. However, advocates of LDCs argue that they can hardly afford to pay such royalties; how much are U.S. farmers really disadvantaged when compared to farmers in LDCs?

Patent advocates respond that "the economic forces driving a family farm into liquidation, or an academic institution into embracing some corporate suitor, operate quite independently of patents ... [patenting] is essentially neutral as to oligopolistic trends that may be at work in the present economy. ...Think twice before turning to the patent system for a means to alter their course" (32, p. 9). "If the government wants to avoid any negative impact of animal patents on the family farm, the appropriate approach is to create mechanisms to enable all farmers to gain access to this new technological development through agricultural extension services and special subsidies" (4). Although other approaches may be necessary to address many of these concerns, patent critics may still be justified in their concern that extending patent law to living organisms is likely to provide incentives that help structure the market in ways that are more favorable to corporate agriculture.

PATENTING AND BIODIVERSITY

If farmers are not allowed to breed plants and animals themselves, biotechnology critics such as Vandana Shiva charge that this increases the centralization of breeding and promotes genetic uniformity rather than biodiversity (19,43). As farmers relinquish their independence and are compelled to forgo most breeding, and as independent breeders go out of business or are bought out, there are fewer options offered and less difference among the options. Monoculture is already a dangerous trend in agriculture, and patenting will just encourage the marketing of varieties with a great deal of genetic uniformity.

Patent advocates respond that monoculture is indeed a worry, however, patents do not contribute to it. Indeed, allowing the patenting of genetically engineered plants and animals may make it more profitable for researchers to innovate to avoid the dangers of genetic uniformity, and to preserve genetic resources for use as raw materials (32). Shiva and others reply that genetic engineering may alter traits or create new ones, but generally this involves only one or a few genes. Thus, even if genetic engineering provides new varieties, they may differ little from other varieties in ways other than the engineered trait (19). Moreover regardless of how new the trait is, if millions of "copies" are sold, there is considerable genetic uniformity. Insofar as patents and licensing agreements limit the ability of farmers to engage in independent breeding, these practices in themselves do contribute to monoculture and the loss of biodiversity.

PATENTING, BIOPIRACY, TRIPS, AND NATIONAL SELF-DETERMINATION

In response to concerns about patenting and bioserfdom, patent advocates assert that "the patent office is not the place to structure a morally appropriate program for the international economic order" (9). However, in the age of information technology, developed countries want and demand strong intellectual property protection worldwide, and consider the absence of such laws to be an unfair trade barrier; developed countries use the patent office to promote their own goals including fairness. Patent advocates often miss the fact that plant and animal patenting raises issues of national self-determination, at least in the era of TRIPS (Trade Related Intellectual Property Rights). "With the advent of TRIPS, virtually all the world's nations have lost their right to determine the balance of private and public benefits designed to meet national goals. Instead, they must comply with a single international standard designed to open their markets to transnational corporate interests" (44, p. 1). Although many LDCs are opposed to plant and animal patenting (if not all patenting), they are being required to adopt patent or patentlike protection for plants and animals, or else suffer serious economic consequences.

According to patent critics, TRIPS are particularly egregious, considering the fact that developed countries have taken or "pirated" the genetic resources of the South for centuries; each side accuses the other of piracy and attempts to take the moral high road. While the North is economically and technologically rich, the less developed South is genetically rich. Plant and animal varieties have been exported from LDCs for centuries, with no compensation to the countries where they were domesticated and made more valuable over many generations of work by indigenous peoples. Indeed, they are treated as and referred to as unimproved genetic resources, rather than the collective property of indigenous communities. International seed banks have been established to save these genetic resources, and many Northern researchers come looking for genetic "gold" to bring back to their laboratories, often with indigenous people as their free "tour guide" (19,45). Recently some researchers have offered compensation to the country of origin if a successful product is created, but much so-called biopiracy occurs without compensation. Moreover compensation is usually small, and often the indigenous community has no control over how the money is used. If it is devoted to preserving the country's genetic resources, this may be rather self-serving on the part of researchers who want to mine further genetic treasures (33).

Many LDCs see genetic resources as the common heritage of humankind which they have no desire to privatize, much less have privatized by others (16,46). They are outraged to discover that medicinals or crops first developed in their country are patented by multinational corporations, or even by the U.S. government. Indeed, the U.S. National Institutes of Health (NIH) has gone so far as to apply for a patent on cultured tissue samples taken by indigenous peoples themselves (which it later dropped). Critics are aghast that not only the country's genetic resources, but also the tissues from its people, are being privatized and commercialized. Patents on some of India's genetic resources have been extremely controversial. India accuses corporations of pirating the sacred neem tree, traditional medicinal uses of turmeric, and basmati rice. While some of these patents have been dropped or overturned, countries such as India continue to worry about biopiracy, and object to international pressure to adopt strong intellectual property protection, particularly over living organisms and genes.

This debate raises issues of international justice, national self-determination, and what should remain in the commons rather than being privatized (47). Many patent critics view genetic resources as a commons, to be shared for the benefit of all (46). Some patent advocates reply that if LDCs are concerned about biopiracy, they should privatize their own genetic resources. If research companies come and find these resources unclaimed, the indigenous peoples have no one to blame but themselves. Simply privatize and fence in the genetic commons, and corporations will have to buy the genetic resources they need from other countries (48). Critics view this response as impractical and beside the point. Even if the practical problems could be surmounted, the moral objections to privatizing the commons would remain. Part of the concern is that developed countries are taking unfair advantage of poor countries, but another part of the concern is that privatizing genetic resources is simply not appropriate; they should belong to no one and everyone, for the mutual good of all.

CONCLUSION

The debate over patenting the products of biotechnology is often polarized. Critics are characterized as being anti-biotechnology, or at least anti-genetic engineering, and quite possibly technophobic. Often the objections to patenting, which to a great degree are interrelated, are characterized as attacking the wrong practice, perhaps to increase public criticism of biotechnology in any way possible, or perhaps because the critics themselves are confused. However, sometimes the attack of patenting is perfectly logical, even if it may rest on the attack of other practices.

Moral Justification of the Patent System

Patenting is characterized by its advocates as morally neutral, and even most critics would agree that we do not want the patent office to act as the arbiter of technology and morality. On the other hand, patenting is supposed to advance morally significant goals. Patents were established to promote useful inventions and thereby benefit society — a form of utilitarian justification. Alternatively, they were established to protect the natural rights of inventors to the fruits of their labors (50). Thus, in a sense, the patent system depends on moral argument. What is the appropriate scope of patent law in biotechnology? It depends on what we take to be the real philosophical justification of the patent system. If patents are justified by the extent to which they encourage research, would this extend to the patenting of living organisms and even human genes? Or if the justification is the protection of the natural rights of inventors to the fruits of their labors, can this be extended to living organisms and especially human genes (27)?

If we choose the former rationale, then it is reasonable to question the extent to which plant and animal patents are likely to benefit society as a whole, particularly in an era when the Western patent system is being imposed internationally against the wishes of numerous countries. Many Westerners simply assume that patents are effective and beneficial. However, it is instructive to examine this assumption, which some argue is unwarranted (27,49). Patents historically were grants of monopoly protection for those who imported, not invented, technologies. Patents initially played a greater role in technology transfer and excluding competitors than in promoting research. They are now being used to keep developing countries from the "piracy" European countries initially designed them to promote when it was in their own interests. In nineteenth-century Europe, many philosophers and economists opposed patents on the ground that they were forms of corporate protectionism that prevented the efficient operation of the free market (50). Now most people assume the opposite, but is the assumption justified? Perhaps not. Patents have played an interesting role in the transformation of agriculture into agribusiness, and it is instructive to study the history of agribusiness opposition to and later support of patents (33,45,51). Indeed, most of the debate over "patenting life" is explicitly tied to larger questions and assumptions about the structure of agriculture and the effects of globalization.

If, on the other hand, the protection of the natural rights of inventors is the primary justification for patents, then it is perfectly reasonable to question the extent of these rights. In particular, it makes sense to consider what belongs in the genetic commons as discoveries and the natural heritage of humankind rather than industrial or government property. Patent defenders, even philosophers, often ignore arguments about the genetic commons. Yet this is the most unique and independent part of the debate. This is a very philosophical issue about the nature of property rights and the commons, and does not rest on larger concerns about animal welfare, family farms, and the like. As such, and given its role in the international debate over TRIPS, this issue deserves greater attention, especially from philosophers.

Novelty, Nonobviousness, Breadth, and Patenting Discoveries

Many responses to criticism of plant and animal patenting ignore problems of the breadth, novelty, and utility of some of the patents being issued. Such problems are not necessarily new or unique, but they may be worse when applied to GMOs. Questions about patenting living organisms have been raised by many scholars who are not necessarily biotechnology critics, such as Louis Guenin (52), Philippe Ducor (53), and Brian Cannon (54). It is instructive to see how the moral arguments against this patenting practice fits within this more academic debate. Moreover some of these scholars have specific proposals that might make patent law more consistent and logical while at the same time addressing some of the moral concerns regarding the patenting of living organisms and genes (although such proposals are beyond the scope of this article). Let us briefly consider some of these criticisms.

Charges of biopiracy may be recast as concerns about the way that the patent law criterion of novelty is being applied. Patent applications must identify any prior art that would be relevant to the patent. The fact that indigenous peoples may have cultivated and improved a variety over generations, or developed a particular use for a plant, is not reflected in prior art, which consists primarily of Western scientific writings. Thus something invented by indigenous peoples could easily be patented as a new invention — even if it had not been modified at all. On appeal, the patent might or might not be rejected — assuming that someone has the knowledge and resources to appeal it. Our Western patent system does very little to acknowledge our transfer of genetic resources and indigenous knowledge. Thus the charge of biopiracy may be seen as at least partially a concern about how the criteria of novelty and nonobviousness can be understood within an international, cross-cultural context.

Plant and animal patents are also criticized for being overly broad (16). Patents are often characterized as a trade-off between the interests of inventors to recoup their costs and the interests of society in having inventions widely and cheaply distributed. Patent advocates argue that this trade-off works to the best interests of society in the long run. However, overly broad patents may strike the wrong balance in this trade-off, and allow patent holders too much control over the development of further research. Since *Diamond v. Chakrabarty*, "anything under the sun that is made by man" can be patented in the United States. Unfortunately, there is a tendency for patent applications to claim not just anything but rather everything under the sun. Allowing broad patents may not be in the best interests of developing nations, some of which already depend on genetically engineered rice or other crops to feed their growing populations and perhaps produce a small amount for export. Even within one country, very broad patents may benefit one firm at the expense of other firms and often also the public interest (27).

Finally, patenting unmodified genes (human, animal, or plant) is often criticized, even by those engage in biotechnology research. The U.S. PTO allows patenting of cloned, unmodified genes. Cloned genes are not the same as their counterparts in nature, and cloning involves an inventive step. Thus, when scientists discovered and cloned a human breast cancer gene, they patented it. Likewise animal and plant genes can be patented, even if they have not been genetically modified. Patent lawyers explain that GMOs and genes that have been isolated and purified (cloned) are a technical solution to a technical problem, and do not occur naturally. Contrary to the No Patents on Life! petition, they are not "part of the natural world" and do not exist without considerable human innovation.

To patent critics, this is the height of absurdity. Inventions are patentable, but not discoveries. We say that scientists *discover* a gene, not that they *invent* it (49). Patent critics, some of them scientists who understand what goes into cloning a gene, remain unconvinced by the answers of patent lawyers. The general public may not have the knowledge of patenting criteria to express their concern properly, but they are appealing to an intuitive form of the idea that patents must meet the criteria of novelty, utility, and nonobviousness. A cloned gene seems to them to fail the criterion of novelty and/or nonobviousness. To say that scientists *invented* an animal hormone or a breast cancer gene seems inaccurate and arrogant, even though most of these critics would applaud breast cancer research. Although the location and sequence of the gene might be nonobvious, once a gene is discovered, is the process of cloning it nonobvious and the resulting cloned gene novel?

Finally, many critics of plant, animal, and gene patents are perfectly happy to accept process patents. It is a mistake to attribute all of these criticisms to concerns about the research itself, or concerns about animal welfare (7). Rather, they are often genuine concerns about what should constitute a valid patent, what constitutes a criterion such as novelty, and whether it makes sense to say that an entire line or species of animals or plants could be owned (as opposed to owning individual animals and plants). Patent advocates often oversimplify the views of critics, whose spokespersons often are not philosophers or legal scholars and whose arguments are sometimes admittedly not fully developed. However, there are indeed serious issues here. The patenting of animals and plants raises many moral and social issues that have not yet been resolved and that will only become more important in an increasingly globalized information economy.

ACKNOWLEDGMENT

This article was made possible in part by an Iowa State University Science, Technology, and Society grant. Some of the initial research was made possible by a research leave at the Social Philosophy and Policy Center. I would also like to thank my capable and well-organized research assistants, Stacy Bastian and Katherine Dresser, my secretary Janet Krengel, and my husband Richard for playing single dad while I hid in my office.

BIBLIOGRAPHY

1. Council for Responsible Genetics, Fall 1998.
2. S.R. Crespi, in S. Sterckx, ed., *Biotechnology, Patents and Morality*, Ashgate, Brookfield, VT, 1997, pp. 220–221.
3. *Diamond v. Chakrabarty*, 447 U.S. 303, 308, 309 (1980).
4. R. Dresser, *Jurimetrics J.* **28**, 399–435 (1989).
5. U.S. PTO Commissioner D.J. Quigg, Ex parte Allen, April 7, 1987.
6. U.S. Pat. 4,736,866 (April 12, 1988).
7. M. Svatos, in S. Sterckx, ed., *Biotechnology, Patents, and Morality*, Ashgate, Brookfield, VT, 1997.
8. M. Fox, *Atlanta J.* (1987), cited by S. Bent, in W.H. Lesser, ed., *Animal Patents: The Legal, Economic, and Social Issues*, Macmillan, New York, 1989.
9. B.A. Brody, in W.H. Lesser, ed., *Animal Patents: The Legal, Economic, and Social Issues*, Macmillan, New York, 1989.
10. R. Merges, *Maryland Law Rev.* **47**, 1051–1075 (1988).
11. R. Dresser, *Jurimetrics J.* **28**, 399–435 (1989).
12. B. Rollin, *The Frankenstein Syndrome*, Cambridge University Press, New York, 1995.
13. L. Walters, Committee on the Judiciary, 1988.
14. P. Singer, *Animal Liberation*, Avon Books, New York, 1990.
15. T. Regan, *The Case for Animal Rights*, University of California Press, 1983.
16. J. Rifkin, *The Biotech Century*, Putnam, New York, 1998.
17. M. Vandenbosch, in S. Sterckx, ed., *Biotechnology, Patents and Morality*, Ashgate, Brookfield, VT, 1997.
18. B. Belcher and G. Hawtin, *A Patent on Life: Ownership of Plant and Animal Research*, International Development Research Centre, Ottawa, 1991.
19. V. Shiva, in *Biopolitics*, V. Shiva and I. Moser, eds., Zed Books, London, 1995.
20. J. Densberger, *J. Bioethics* 91–115 (1984).
21. M.J. Hanson, *Hastings Center Rep.* (November, December), 1–18 (1997).
22. A. Garry, *Soc. Theory Practice* **4**, 395–442 (1998).
23. L. Kluver, in S. Sterckx, ed., *Biotechnology, Patents and Morality*, Ashgate, Brookfield, VT, 1997.
24. B. Mitchell, in *Patenting of Biological Entities, Proceedings of the ITEST Workshop*, ITEST Faith/Science Press, St. Louis, MO, 1997.

25. N. Hettinger, *BC Environ. Affairs Law Rev.* **22**, 267–305 (1995).

26. Council for Responsible Genetics, *Synthesis/Regeneration* **17**, (Fall 1998).

27. M. Svatos, *Social Philosophy and Policy* **13**(2), 113–143 (1996).

28. B. Weutrich, *Sci. News* **144**, 154–157 (1993).

29. D. Nelkin, *Science as Intellectual Property*, Macmillan, New York, 1984.

30. Committee on the Judiciary, 1988, pp. 64–65, quoted in B.A. Brody, in W.H. Lesser, eds., *Animal Patents: The Legal, Economic, and Social Issues*, Macmillan, New York, 1989.

31. L. Kass, *Toward a More Natural Science*, Free Press, New York, 1985.

32. S. Bent, in W.H. Lesser, eds., *Animal Patents: The Legal, Economic, and Social Issues*, Macmillan, New York, 1989.

33. R. Pistorius and J. van Wijk, *The Exploitation of Plant Genetic Information*, Academisch Proefschrift, Amsterdam, 1999.

34. RAFI Communique, *Bioserfdom: Technology, Intellectual Property and the Erosion of Farmers' Rights in the Industrialized World*, March, April, 1997.

35. R.A. Steinbrecher and P.R. Mooney, *Ecologist* **28**, 4 (1988).

36. R. Weiss, *Washington Post*, February 3, p. A01 (1999).

37. T.G. Palmer, *Hamline Law Rev.* **12** (1989).

38. U.S. Department of Agriculture, *Fact Sheet Why USDA's Technology Protection System (a.k.a. "Terminator") Benefits Agriculture—A Discovery to Spur New Crop Improvement*, Government Printing Office, Washington, DC, 1998.

39. RAFI Communique, *The Terminator Technology*, March, April, 1998.

40. C. Montecions, Centro de Education y Tecnologia, Chile.

41. M. Svatos, Ethical and social issues raised by the 'terminator' seed, and technological fences for intellectual property: Comparing the 'terminator' seed to software copy protection unpublished manuscripts, 1999.

42. M. Crouch, *J. Agri. Environ. Ethics* **4**, 151–158 (1991).

43. J. Doyle, *Altered Harvest*, Viking Press, New York, 1985.

44. K. Dawkins, *Gene Watch* **12**(4), 1 (1999).

45. C. Fowler and P. Mooney, *Shattering: Food, Politics, and the Loss of Genetic Diversity*, Tucson: University of Arizona Press, 1990.

46. H. Shand, *J. Agri. Environ. Ethics* **4**(2), 131–142 (1991).

47. P. Drahos, *A Philosophy of Intellectual Property*, Dartmouth, Brookfield, VT, 1996.

48. J. Vogel, *Genes for Sale*, Oxford University Press, New York, 1994.

49. E.R. Gold, *Body Parts*, Georgetown University Press, Washington, DC, 1996.

50. F. Machlup and E. Penrose, *J. Econ. Hist.* **10**(1), 279–314 (1950).

51. J. Kloppenburg, *First the Seed: The Political Economy of Plant Biotechnology, 1492–2000*, Cambridge University Press, New York, 1988.

52. L. Guenin, *Theoret. Med.* **17** (1996).

53. P. Ducor, *Patenting the Recombinant Products of Biotechnology and Other Molecules*, Kluwer, Boston, 1998.

54. B. Cannon, *Cornell Law Rev.* **79**, 735–765 (1994).

See other entries MEDICAL BIOTECHNOLOGY, UNITED STATES POLICIES INFLUENCING ITS DEVELOPMENT; see also PATENTS AND LICENSING entries.

PATENTS AND LICENSING, POLICY, PATENTING OF INVENTIONS DEVELOPED WITH PUBLIC FUNDS

DANA KATZ
JON F. MERZ
University of Pennsylvania
Philadelphia, Pennsylvania

OUTLINE

Introduction

Early Innovation

Historical Milestones

Industrialization

Bureaucratization

Development of Technology Transfer

Publicly Funded Research and Biotechnology

Bibliography

INTRODUCTION

Should governments fund scientific research? If so, what types of research, basic or applied? How should taxpayers' money be allocated? What priorities should the government adopt, and how should those priorities be set? If government research yields commercializable products (other than solely for military purposes), how should those developments be moved into the market? Should private parties be permitted to have exclusive rights to publicly funded inventions? And, ultimately, who are, and who ought to be, the beneficiaries of publicly funded research?

To analyze these questions, we look here at the development of research funding policy in the United States. Because of its free-market roots, the United States has approached these questions with trepidation. Over the last 200 years the country has moved incrementally from funding only research having military importance to current policies that provide substantial support for basic biomedical sciences, among many others, and permit private commercialization of and profit from the resulting intellectual property. This intellectual property is heralded as the means to economic prosperity both domestically and in the global marketplace. Though this can be considered standard free-market industrial policy, the public that invests in scientific innovation conceptualizes its benefits in ways other than economic, such as through the development of life-saving medical technology or an increase in general knowledge. Although there are cases where both economic and social benefits of science can be achieved concurrently, technological and industrial growth, particularly in the area of biotechnology, has increased the potential for these goals to compete. Today the government promotes scientific innovation with marketplace incentives which, while serving to promote technological advance and the development of public goods, raises numerous ethical concerns, such as its perpetuation of inaccessible health care for many individuals and the privatization of basic research knowledge.

EARLY INNOVATION

Although in 1776 there was little tangible evidence of the benefits of scientific advancement, the founding fathers intuitively recognized the importance of innovation for a developing nation. It was argued during the Constitutional Convention that the wording of the constitution ought to reflect the new republic's "duty and ability to encourage progress in the arts and sciences" (1). To this end, some argued the need for constitutional provision for technical schools, societies, seminaries, and a national university. Those who saw pecuniary incentives as the cornerstone of innovation sought constitutional provisions for patenting rights, as well as rewards, prizes, and direct subsidies for citizens who endeavored creatively to promote agriculture, commerce, and other social goods.

Ultimately innovation was afforded very limited constitutional protection for fear that broader commitments would strengthen the central government and thereby increase the potential for later abuse of power. The only explicit constitutional provision for innovation grants Congress the power to:

> ...promote the Progress of Science and useful Arts, by securing for limited Times to Authors and Inventors the exclusive Right to their respective Writings and Discoveries (2).

A. Hunter Dupree, in his seminal work *Science in the Federal Government*, interprets this clause as sanctifying the creation of national scientific institutions (1). Prior to its first comma, the clause states the intention to promote advancement of the arts and sciences. The intention is qualified by the English practice of affording inventors exclusive ownership of their work as an incentive for further innovation and development. This qualification, Dupree argues, is neither a prohibition of, nor suggestion for, publicly funded research because the Constitution does not address the concept directly. For the century following, however, opponents of government support of research argued that Congress was permitted to do only what the Constitution explicitly authorizes so that the absence of a provision permitting research funding prohibits Congress from doing so.

Although the language of the Constitution falls short of reflecting its framers' high regard for innovation, the government was constructed to rely on scientific expertise. From its inception, the basic operations of the federal government have relied on technical skill, such as surveying new territory and establishing coinage, weights, and measures. However integral to its function, the role of science in the federal government was informal and would remain so for generations.

According to Dupree, the nation's inherent regard for science stems from the invocation of natural law used to deliver its founders from the political and clerical despotism of England (1). With the same rationalist approach used to create the new republic, the founding fathers sought to expand natural history, philosophy, political theory, and science in the New World (1). The widespread dislike of bureaucratic and oppressive rule, however, led the federal government to call on scientists only when needed rather than to expand its organization to employ them full-time. As a result the pursuit of knowledge was slated to occur largely outside of the government domain.

The nation's early scientists were upper-class, professional males who pursued scientific inquiry as a hobby. Though their science was amateurish, their powerful positions in society allowed them to effectively introduce scientific concepts and traditions into American society. Their observations rarely concentrated on one aspect of the natural world, such as chemistry, physics, or botany, and this lack of specialization limited the potential for scientific accomplishment (1). With little distinction between pure and applied research, "science" became the catch-all category for products of systematic investigation that ranged from philosophical theories of electricity to the construction of ornamental clocks. Although it was recognized that science offered potential benefits, the most significant innovations of the era were the results of trial and error by those unacquainted with the scientific method (1).

In 1791 the federal government received its first request to fund scientific research. In seeking a patent for navigational calculations, a surveyor requested that Congress finance an arctic voyage to test his theories. Though there was some question as to whether the matter should be deferred to state legislators, the larger ethical question of whether the government ought to use public funds for such uncertain ends went unasked. Many supported the concept, seeing the issue only in terms of the nation's need to improve geography and navigation. Others, while reluctant to prematurely reject a potentially beneficial enterprise, advised against the hasty support of philosophical patent applications (1). Ultimately the request was refused on account of the poor success rate of arctic voyages.

A federal patenting board had already been instituted in 1790 to process what was fast becoming an overwhelming number of patent applications. Arguments that either a panel of experts or ordinary citizens ought to compose the board were defeated on the belief that a few politicians, under the guidance of Thomas Jefferson, would offer inventors the best protection. Authorized to withhold patent rights for inventions lacking novelty, utility, or importance, the nation's first patent board demonstrated how government regulation could be used to promote the general welfare.

Jefferson understood that the fundamental task of a patenting board is to advance society by delicately balancing two often-competing goals: the desire to encourage innovation versus the need to prevent destructive monopolies. Patents essentially function as a government-endorsed monopoly in that they confer exclusive rights to use, produce, and sell an invention. The provision of exclusivity serves as an incentive for individuals to generate useful, novel, and important inventions and develop them into public goods. By eliminating the threat of competition for a fixed period of time, innovative and enterprising individuals are able to recoup their investment in product development and reap the rewards of their ingenuity.

The patent system benefits society by providing a mechanism for new, technologically advanced goods to enter the marketplace and spur economic growth.

However, if patents were issued indiscriminately, the market could become overrun with monopolies that would impede the competition necessary for promoting technological and economic growth. In the absence of a competition-driven marketplace, the distribution of wealth would be inequitable, the prices of goods would remain artificially inflated, and there would be little incentive to improve existing products. To prevent rampant exclusivity, it is necessary to impose criteria for patentability so that incentives remain for collaborative efforts to expand general knowledge. Otherwise, technology, and in turn society, is unlikely to advance as rapidly. Jefferson sought to protect the public from monopolization that would impede innovation by formalizing the distinction between pure and applied research.

On the belief that exclusive ownership of scientific theories would hinder rather than promote innovation, Jefferson declared theoretical discoveries, or pure science, as ineligible for patent protection. While far from its intention, Jefferson's stipulation transformed the congressional sentiment toward scientific research from one of relative support to complete disinterest in any form of innovation that was not reducible to a "machine of potential cash value" (1). Congress, consistent with its aversion to bureaucratized government, used Jefferson's patent policy to free itself from any formal involvement in scientific research. It contended that the constitutional provision empowering Congress to issue patents limited the congressional role in innovation to that function. The federal government thereby distanced itself from basic science by declaring the subsidization of basic research unconstitutional and its findings ineligible for patent protection.

Jefferson, in contrast, believed Congress ought to exercise its constitutional power to encourage innovation by subsidizing the advancement of pure science. He understood that applied science would prosper in the hands of private enterprise because the potential for profit would attract investors. Pure science, however, would be less likely to prosper through private support because it offers a public good that has little profit potential. It was therefore in the nation's interest to invest pubic funds into pure scientific research because innovation ultimately stems from the progress of both pure and applied sciences.

HISTORICAL MILESTONES

Congress never failed to see the potential benefits of the scientific enterprise. To avert bureaucratization, it was simply preferable to fund science on an "as needed" basis rather than formalize an expensive, open-ended commitment. The occasional research project was conducted within the traditional units of a federal government that was focused almost exclusively on military and industrial needs. As a result federally funded research began through the Departments of Commerce and War.

Jefferson, when he was elected to the presidency in 1801, paved the way for military research by securing congressional support for the Lewis and Clark expedition. While convincing Congress that a technical exploration of the west was essential for military and commercial gain, he disclosed to the Spanish and French, who controlled western trade, that the mission was for scientific discovery only. The explorers, trained by the nation's scholars in topography, botany, zoology, agriculture, geology, astronomy, paleontology, and ethnology, returned with a body of knowledge so great that it was used to establish an American corridor to the Pacific and a trading presence in the northwest. The success of the expedition inspired Congress to fund, under the auspices of the War Department, further expeditions and requisite technical training. Arguably the earliest arm of the federal government receptive to scientific research, the early nineteenth-century American military gave rise to a national weather service, the first accurate observations of digestion, and research studies of diet and nutrition. (It is through the military that the National Academy of the Sciences was established in 1863, during the Civil War, and the National Research Council was formed in 1916, during World War I.) The research, however, was neither explicitly nor entirely publicly funded. While the government provided scientists with materials and laboratory space, direct funding was considered unconstitutional. As a result the processes of data analysis and publication depended on private support.

This mix of publicly and privately supported research did not confront the issue of ownership rights, for the scientists of the time, though increasingly specialized and respected, "had no aptitude for applying their knowledge to a downstream product (1)." Nevertheless, the potential relationship between the science reported in academic journals and the duty of the patent office was recognized. Henry L. Ellsworth, appointed the nation's first Commissioner of Patents in 1836, expanded the patent office into a "central depository" for innovation in the belief that the eventual merger of science and invention would serve the nation's economic interests (1).

Ellsworth collected patent application fees and used the revenue to acquire patented mechanical models as well as unpatented models, specimens, manufactures, minerals, seeds, and scientific findings culled from various collected publications. He drew from his library various solutions to the nation's agricultural crisis, becoming the first to use statistical analysis in large-scale problem assessment. In his promotion of a recently published German study calling for the use of chemistry to remedy soil exhaustion, Ellsworth recognized what was in fact "the first direct application of pure scientific research to agricultural success" (1). As well as devoting his career to promoting the utility of science, his efforts on behalf of the congressionally neglected farming community laid the foundation for the Department of Agriculture. In his striving to develop a great scientific bureau, the first Commissioner of Patents exceeded the traditional scope of both the patent office and the federal government.

Meanwhile, in 1838, S.F.B. Morse had presented Congress with his invention of the electric telegraph, another striking example of the convergence of pure and applied scientific advancement. Without a niche in the nation's economy and few precedents regarding the development and ownership of his invention, Morse lobbied for four years to receive a $30,000 grant for construction of a telegraph line connecting Baltimore to Washington, DC (Congress justified the expense under the constitutional provisions for commerce and postal service). In 1845 Morse issued a proposal to extend the line to New York City, but it was voted down in Congress because the southern contingency thought it unconstitutional. Eleven years later the same fate was suffered by the electric motor; after an initial research grant, federal support was withdrawn despite extremely successful results.

Congress justified its recoil from research support with the contention that laying any groundwork for a research bureau would open a Pandora's box of new and unwanted federal responsibilities. It was argued that federally subsidized research would inappropriately risk public funds, while it forced the government into the role of examining agent for both the scientific enterprise and any devices it produced. In addition the deluge of research proposals requiring evaluation and, to exclude charlatans, investigation, would be too expensive and expansive to take on. Thus it came to be that the mid-nineteenth century would see Congress neglecting many opportunities to formally engage in the "industrial development of devices born from scientific discovery" (1).

At the same time Congress had stepped up its support of basic science through its technical explorations of the North American continent. Increasingly basic scientists were becoming government employees, but they were hired only for specific tasks and not for the construction or maintenance of a permanent research bureau. In 1834, when the Coast Survey, traditionally overseen by the Secretary of the Treasury, was reassigned to the Navy Department for cost-containment reasons, the scientists revolted by refusing to share their work with nonscientists. Two years later the Coast Survey was returned to the Treasury; the scientists succeeded in securing pay commensurate with their abilities while establishing civilian control of research.

As the power and prestige of the scientific community grew, congressional sentiment toward their research evolved accordingly. By 1842 Congress had even assumed responsibility for publishing the findings of a military expedition. However, the publication process was a fiasco and the finished product ridden with mistakes. It was clear that congressional committees were incapable of effectively administrating the highly technical enterprise that science had become. It is ironic that by the time Congress came to explicitly appreciate the utility of science, a government of nonscientists was no longer intellectually equipped to oversee its advancement (1).

INDUSTRIALIZATION

Despite the federal government's express regard for science, it had no intention of expanding its narrowly conceived administrative interests. By the end of the nineteenth century, industrialization, urbanization, and mobility had imposed vast and unfamiliar problems on a small-town, agrarian nation that had seen its traditional

institutional framework erode (3). While Congress saw issues of social welfare as falling outside its sphere of responsibility, the early industrialists acted to address the issues that were a side effect of their success. Beholden to the Protestant work ethic, they developed philanthropic foundations in the belief that as stewards of their wealth they had a moral obligation to manage it in a socially redeeming way. Technology and standardization were the tools of choice for the industrialists-turned-philanthropists who strove to enhance social welfare in a manner that would serve their industrial interests.

Through addressing the educational and medical needs of the masses, the industrialists established a national infrastructure that would, for years to come, foster industrial and technological growth. Private universities, by educating middle class men in science and engineering, were able to equip an emerging professional stratum with the skills necessary to manage the future of industry. To advance the fields of science and engineering, universities and foundations funded the research of pure scientists as a means to further their respective disciplines.

The expansion of the scientific enterprise proved particularly beneficial to the field of medicine. Standardized and cost-effective, the application of science to medicine, called biomedicine, employed a reductionist approach to disease that complemented the political and economic agenda of industrial capitalism. While diverting attention from the environmental and social roots of working class illness, biomedical practitioners were able to restore health to the level of functionality necessary to labor toward capital accumulation (4). Consequently biomedicine was afforded tremendous industrialist support, for while it sustained a productive work force, its approach to research dovetailed nicely with the broader trend of scientific development.

By the 1920s biomedical research conducted throughout the nation's universities and foundations required coordination in order to make significant progress. Leading industrialists, as part of a larger trend of tying the nation's technologically driven "capitalist infrastructure to government function," requested that Congress act to centrally organize the nation's efforts in medical research (3). The government was open to the proposition. Despite its traditional reluctance to expand in organizational structure or function, it had been taking incremental, yet affirmative, steps toward establishing a formal commitment to national health for over a century.

The government's first intervention in public health came in response to the late eighteenth century scourges of Yellow Fever in the nation's port cities, for it had impeded federal revenue collection. The correlation between the presence of trading ships, communicable disease, and the effect on revenue, properly classified the matter as one of interstate commerce and hence one of the principle responsibilities of the federal government. The Department of Commerce enacted quarantine laws and established hospitals initially intended to care only for merchant seamen, and shortly thereafter for the military as well. By the twentieth century the government had organized a national, comprehensive medical research effort intended to investigate "the origin and causes of epidemic diseases" and established the Hygienic

Laboratory for bacteriological research in its Staten Island Marine Hospital (5). In 1912 the nation's Public Health Service was established to coordinate the government's involvement in issues of public health that quickly expanded to include numerous research studies on various facets of communicable and noncommunicable disease.

With the backing of powerful industrialists (who had already committed to large donations upon the success of the effort), scientific societies, the American Medical Association, and the life insurance industry, Louisiana Senator Joseph E. Ransdell, sought to establish out of the Public Health Service a central office to coordinate the nation's medical research. The 1930 enabling act of the National Institute of Health (NIH) changed the name of the Hygienic Laboratory to NIH and relocated it to a new building. Significantly, the act authorized the implementation of a graduate-level fellowship program to train young scientists in biomedical research. The Secretary of the Treasury was permitted to accept the philanthropic donations that would support these newly organized efforts in medical research. The congressional record of the enabling act, filled with rhetoric about doing God's work, is a reflection of the traditional interests and priorities of the federal government and the influence of industry in its decision making (6). In addition to arguing that science and medicine are intrinsic goods worthy of support, the industrialists convinced Congress of the relationship between health status and productivity to explain how health maintenance is in the nation's economic interest. In order to maintain national health, public and private efforts in research and data collection needed to be coordinated on a national scale. Further, with the nation still recovering from World War I, the fact that German progress in biomedical research appeared to be outpacing the United States served as an added incentive to engage more American scientists in government endeavors. Chemists, in particular, were understood as valuable to military and industrial research, but without an expansion in employment opportunities, they would be unable to apply their expertise to biomedicine. Fellowships, instead, would provide opportunities for this specialized graduate training. Last, perhaps the most attractive aspect of the Ransdell Act was that the NIH was expected to advance biomedicine ostensibly through private funds, with periodic support from the public treasury as needed (6).

BUREAUCRATIZATION

By the 1930s the promise of both public and private support of medical innovation made its potential appear boundless. Dupree's observation of the early republic, however, would remain applicable centuries later, for science is "a group activity carried on by limited and fallible men, and much of their effectiveness stems from their organization and the continuity and flexibility of their institutional arrangements" (1). The twentieth century would not only foster the growth of scientific research, but of corresponding organizational structures that would inevitably dilute its social and economic benefits. This process of bureaucratization permeated

America's scientific enterprise whether conducted through foundations, the government, universities, industry, or combinations thereof.

While the foundations of the 1920s and 1930s funded scientific research, the scientists themselves were responsible for resource allocation. Over time, however, a stratum of science-administrators developed who, according to Barry Karl, were "not content in the function as mindless vehicles for the products of the academic mind" and so aimed to create their own professional identity (3). This pseudoprofessionalization of the scientific bureaucracy was counterproductive to advancement. Pure science was no longer conceptualized as simply a public good but as a means to enhance the careers and reputation of the administrators and their foundations. By the 1950s the foundation bureaucrats had taken to "reshaping the substance and form" of research proposals and placing undue demands on scientists who, due to a tight university job market, had few alternatives for employment (3).

During this period of the 1920s through the 1950s, the agenda of the federal government evolved to rely on both pure and applied scientific advancement. While an era of progressivism helped to focus congressional attention on public health needs, an effusive international agenda led to massive scientific undertakings, particularly in the areas of aeronautics and atomic physics. Starting with the National Advisory Committee for Aeronautics in 1920 and continuing through the Atomic Energy Commission and the National Aeronautics and Space Administration, new scientific agencies were established to exploit the quasi-military technology that would have remained uncultivated if development were left to the existing federal agencies and the private sector (7). It is clear that this organizational expansion of scientific support was initially justified by the technological needs of both World Wars and the cold war. Beyond assisting the military, however, the new agencies served the nation's economic interests and, in doing so, demonstrated that pure research is a viable investment in the nation's economic future.

The size and complexity of the government's scientific endeavors, particularly since World War II, required it to extend beyond its own laboratories to cull specialized expertise from the nation's universities, foundations, and industries. No longer was there pervasive opposition to such government expansion. Post–World War II organizational theorists heralded bureaucratization as a product of modernization, "a rational management technique that, while striving for effectiveness and efficiency, would leave the sources of intellect and power under popular control" (3).

By the 1960s the federal government became the nation's primary supporter of pure scientific research. A welcome employment alternative to the politicized foundations, scientists were eager to both receive government grants and organize the process of resource allocation. Like the foundations, however, the government produced a stratum of science-administrators whose professional aspirations would come to intrude on the enterprise itself. The function of government bureaucrats was to reconcile the goals of the scientific community with the "intentions of Congress and sensitivities of its committees" (3). This differentiated them from the foundation bureaucrats who, at the very core, remained guided by the mission to serve humanity. Government bureaucrats had no such moral agenda but instead endeavored to engage in political maneuvering as a means to climb a burgeoning bureaucratic hierarchy. Karl attributes to T.S. Kuhn the observation that science-administrators are essentially "manipulated into producing results that are determined elsewhere; they have no interest in the process beyond its efficient progress toward its stated goal" (3). Karl explains that the role of such administrators is not intended to interfere with knowledge development, for an administrator's knowledge of the discipline he manages may be limited to that which is directly related to his job function.

This concept of isolating the administrator from knowledge development proved problematic for academic institutions whose researchers were the recipients of the government's expanded support of science. A need developed for a "scholar-administrator" to mediate between the university's internal needs and interests and the increasing external demands placed upon it. Karl explains that university faculty members discovered quickly that "even a relatively brief career as an administrator could jeopardize their scholastic careers" (3). Academic research science was faced with the challenge of finding administrators who, in addition to having basic institutional affection and technical skills, had a professional understanding of the discipline to be managed (3). The typical manager of academic research was either a "critical compromise" between the faculty and university administration, a rejected tenure candidate, a doctoral candidate who never completed a dissertation, or a legal or business professional (3).

As the practice of federally funded, privately managed research grew, it introduced an unprecedented level of government intrusion into university matters (e.g., investigations for equal opportunity and audits for accountability in the use of public and private resources). Left with the very nebulous distinction between public and private affairs, the university relied on its bureaucracy to protect the once independent, private academic system from imposition by the government. Similar to the experience of the foundation and the government, the expansion of academic research institutionalized the bureaucrat into intellectual life (3). Over time these bureaucrats became permanent stakeholders in the American scientific enterprise and became the individuals largely responsible for the mechanics of technology transfer.

DEVELOPMENT OF TECHNOLOGY TRANSFER

In his 1968 book *The Government of Science*, Harvey Brooks discusses public concern over the priorities of the federal government (7). During the 1960s the government was perceived to have "deprived the civilian economy of its sources of technological innovation" with its cold-war preoccupation with military and space technology, while neglecting to apply this innovation to the improvement of social welfare (7). Brooks attributes the government's reluctance to resolve social concerns, such as pollution, urban transportation, and disease, in part, to its inability to reach consensus regarding: (*1*) the degrees to which issues are federal responsibility, and (*2*) the means and ends of proposed solutions to civilian problems. He believes that the feasibility of general agreement on the process and goals of research and development determines how the government uses technological progress to address social needs (7).

Brooks argues that direct subsidization of research and development is an effective means for developing highly technical areas, such as defense and space, since, in reality, their advancement rests on the consensus of only a small number of technocrats. In contrast, civilian technology requires numerous parties, often with divergent interests, to strike a compromise on who, where, when, and what is to be sacrificed in the interest of the general good. To illustrate his point, Brooks describes what the installation of an urban rapid transit system would entail if completed exclusively on government subsidy. Test results of the technology and system would constantly have to be weighed against public opinion. Not only would the process be exceedingly time-, labor-, and cost-intensive, it would be wasteful as well if the result was a system people elected not to use. Brooks questions whether "technical-economic analysis is sufficiently refined to justify large gambles with public funds" (7). Ironically a response to this question requires a public consensus that is, for the most part, impossible to reach.

Many believe that the marketplace is the most efficient and effective tool with which to measure mass opinion of a public good. In addition to turning out the most workable solutions to public demand, it produces the incentives necessary for the demand to be satisfied through private investment. It is widely contended therefore that the government need only provide a "framework of information, incentives, and underlying general technology" for an entrepreneur who, by responding to the marketplace, can accomplish the goals of public policy without risking public funds (7). This indirect subsidization of research and development is often heralded as the means through which innovative technology born from public funds ought to return to the public in the form of social goods. In order for it to be an effective method of technology transfer, however, the government must offer sufficient incentives so private developers will invest in bringing federally funded inventions to market.

The federal government's World War II expansion of research support led to an investigation of the ways in which it could promote private development of its discoveries while, at the same time, retaining an unrestricted right to their use. The Roosevelt administration observed that while the practice of publication protects the government's rights to discoveries arising from its research, those rights would be better secured through patent protection. A 1945 advisory report by the National Patent Planning Commission recommended that the government patent its inventions, but retain exclusive ownership of the titles only in cases where private ownership would be detrimental to national welfare. The commission advised the government to make its titles generally available, on a nonexclusive basis, to anyone wishing to develop inventions into public goods. It recognized, however, that certain circumstances require the government to offer exclusive ownership rights in order to "induce private manufacturers to commercialize an invention" (8). Roosevelt's commission therefore recommended that exclusive licenses be issued in cases where it was reasonable to assume that the invention would otherwise remain idle.

Rebecca Eisenberg, in her review and analysis of the U.S. history of technology transfer policy, explains that the aim of transferring title ownership from the government to the private sector is to promote economic prosperity by "stimulating innovation, new products and new jobs" (8). Many believe that in order to effectively promote industrial growth, the government must have the freedom to grant private developers exclusive ownership of inventions made at public expense. This is controversial, however, for while exclusive private ownership may be an effective method of putting publicly funded technology to practical use, its method of empowering industry for the promotion of large-scale economic growth often entails the sacrifice of other social goods, such as the sharing of scientific knowledge. On the other hand, the absence of a provision for exclusive licensure would sacrifice potential technological and industrial advancement. A technology transfer policy based solely on nonexclusive licensure would dissuade private developers from investing to develop government inventions and discoveries because, in some cases, their competitors could "copy successful inventions without having shared in the initial cost and risk of making them" (8). What would likely happen is that each developer would generate a portfolio of improvement patents covering specific applications of the basic government-funded invention. The threat of competition in the race to develop marketable products, however, increases the risk of development and thus lowers the value of the basic invention. As a result the nation's best firms might refrain from involvement in government innovation and many publicly funded inventions would remain undeveloped.

In 1947 the U.S. Attorney General, Robert H. Jackson, issued a recommendation for technology transfer that underscored the drawbacks of conferring exclusive ownership rights to private parties. His report called for the government to retain, with few exceptions, exclusive ownership rights to all inventions funded in part, or in whole, with public funds. To encourage the development of these inventions, he recommended that rights to government inventions be licensed to private parties on a nonexclusive basis only. In cases where development hinged on the provision of exclusive rights, Jackson contended that the government itself should finance product development rather than endorse the monopolization of a publicly funded invention. He further advised the government against charging royalties for the use of this technology.

The recommendation of the Attorney General was intended to encourage the commercialization of government-held inventions while, at the same time, protecting the public's equitable claim to the technology created at their expense. When a private enterprise has exclusive rights to a publicly funded invention, the public is required to "pay twice for the same invention — once through taxes to support the research that yielded the invention, and then again through higher monopoly prices and restricted supply when the invention reaches the market" (8). This hurts both consumers and small business because it concentrates innovative technology, and its attending economic power, in the hands of beneficiaries of "government favoritism" (8). The practice of issuing exclusive licenses demands that the government undergo a certain level of bureaucratic expansion in order to orchestrate an application process, "select a licensee, police its operations, and detect and prosecute patent infringement" (8). If the licensees are continually large technology firms, entrepreneurs of limited means are prevented from competing in an increasingly technologically driven global market. As a result the practice of doling out exclusive rights to government inventions may contribute to a progressive concentration and centralization of power. Nonexclusive licensure, on the other hand, would allow publicly funded inventions to be used by many firms, thereby introducing them to a competitive, as opposed to monopolistic, marketplace. This would somewhat level the playing field for small business and benefit consumers through reduced product costs.

The congressional response to the 1945 and 1947 advisory reports for technology transfer policy was to refrain from enacting any governmentwide policy for over 30 years. From the 1940s to the 1980s, the federal agencies involved in research were broadly encouraged to license their inventions to private developers. Actual policies, however, were instituted only in response to particular agency-specific issues. It was believed that the tremendous disparities between federal agency missions, collaborator agendas, and the type and commercializability of federally funded inventions, made a standardized policy both "unfeasible and undesirable" (9). As a result much of the technology transfer legislation during this time was directed toward authorizing federal agency heads to manage collaborative research and development efforts in whatever manner best suited their agency's operations. The perception that disparities in agency needs and practices precluded the institution of a uniform policy was supported by a 1965 study commissioned by the Committee on Government Patent Policy. The study found collaborator decisions regarding whether or not to invest resources in government research and development to rest primarily on the commercial potential of specific research endeavors and inventions, as opposed to the particulars of licensing agreements. In fact, until the 1980s, overall commercial utilization of government-sponsored inventions was "very low, regardless of who held the title" (8). The 1965 study, while reluctant to issue a blanket recommendation, asserted that in some cases the provision of exclusive rights would promote the development of inventions better than acquisition of title by the government.

The Kennedy administration moved to standardize technology transfer policy by issuing a memorandum outlining those circumstances in which the federal government would retain ownership rights to patent titles, and those where rights should be licensed to private developers. The situations in which the government was to retain title ownership included (*1*) when the products of research were intended for the public's health, welfare, or commercial use; (*2*) when the contractor was coordinating a government owned facility or operation; and (*3*) where the government was the principle developer or leading authority in the field of expertise and granting exclusive rights to a contractor would designate that contractor as the dominant figure in the market. Eisenberg explains that the contractor, in turn, is to acquire ownership rights to an invention when exclusive rights are essential for development, and

> ...where the contract research is to build upon existing technology to develop information, products or processes for use by the government, and the contractor has acquired technical competence and established a nongovernmental commercial position in the field, the contractor would normally acquire title, subject to a non-exclusive, royalty-free license in the government (*8*).

Although Kennedy, and later Nixon, took affirmative steps to improve and standardize technology transfer policy, their attempts toward establishing governmentwide uniformity were negated by designating agency heads, with their disparate policies, to administer the practice. Nevertheless, these efforts served to improve government assessment and oversight of federal patenting and licensing practices. Significantly, exclusive owners of publicly funded inventions were required to issue progress reports on their commercialization process. If, after three years, they failed to take reasonable steps to bring their invention to practical application, the government was entitled to terminate the contractor's right to exclusivity.

The tragic flaw of the technology transfer policy from the 1940s to 1980 was not that it varied from agency to agency but that it failed to provide adequate incentives for government contractors and grantees to pursue research with commercial potential. In their paper, *Technology Transfer Laws Governing Federally Funded Research and Development*, James V. Lacy, Bradford C. Brown, and Michael R. Rubin attribute the large numbers of government-owned, unlicensed patents to a "lack of statutory basis for royalty sharing" (*9*). They argue that the absence of a legal provision that entitled government collaborators to a portion of the profits generated by their invention, denied researchers any incentive to create "commercially viable technology" (*9*). Whereas private sector researchers were motivated by goal structures and profit-oriented management techniques, government researchers and grantees were motivated by salary alone (*9*). Consequently one of the key problems with technology transfer during this period was not simply the low commercial potential of many government-held inventions, but the lack of incentive for government employee-inventors to transfer any inventions to the private sector for development.

In the late 1970s Congress, in an attempt to improve the nation's low economic productivity, set out to resolve the deficiencies of technology transfer. A "series of bipartisan initiatives" were enacted to "revise government patent policy, reduce legal and bureaucratic barriers, and create incentives to improve federal technology transfer to the private sector" (*9*). At last, Congress understood that "it was not enough to fund, invent, and patent inventions. The government had to actually make its way into the market in order [for technology transfer policy] to produce positive economic results" (*9*).

In passing the 1980 Bayh-Dole Act, Congress sought to improve the practice of technology transfer by aligning federal research policy with the nation's economic needs. The Act entitled small business firms and nonprofit organizations collaborating in government research to retain ownership rights in subsequent inventions. According to Eisenberg, the conspicuous omission of large firms was a reflection of the Carter administration's "strategy for improving the industrial competitiveness of the nation" (*8*). She explains how many of Carter's supporters believed small business to be "innovative, adaptive, risk-taking, entrepreneurial and competitive, yet [inequitably] burdened by the practice of obtaining case-by-case waivers of title from sponsoring agencies" who were traditionally reluctant to grant research funds and patent rights to small businesses (*8*). This made it difficult to compete with large firms that during this time period, Eisenberg recounts, were often painted as "short-sighted, risk-averse, and predatory — more likely to suppress new technologies than to adopt them, yet savvy and powerful in their dealings with government agencies" (*8*). It was a hallmark of the Carter presidency to blame the nation's large firms for the decline in the global position of U.S. industry. Thus a policy for technology transfer that promoted small business growth as a means to enhance the American marketplace was a reflection of the Carter era. By 1984, however, the Reagan administration, with its markedly different economic agenda, extended the provisions of the Bayh-Dole Act. This revision, which remains in effect today, entitles *all* private enterprisers in government collaboration, including large firms, to own any inventions generated in whole or in part with public funds.

Through the Bayh-Dole Act, and related legislation throughout the 1980s, Congress was seeking to improve the competitive position of the United States in world markets. The ideal policy for technology transfer would see to it that every dollar invested in scientific research would, in essence, be a dollar invested in national economic prosperity. In order for the policy to meet this objective, Congress modified the earlier system of technology transfer in three key areas. The first was discussed previously: Policies were instituted so that private parties contributing to publicly funded research would retain the right to develop any subsequent inventions. The second was the establishment of an incentive system to motivate the employees of government-owned, government-operated laboratories to make and license commercializable inventions. Third, a legal basis was provided for favoring American over foreign industry in conferring ownership rights to publicly funded technology.

The policy motivating government agencies and employees to invent and license technology was established as part of the 1986 Federal Technology Transfer Act, which made technology transfer a top priority for agencies involved in research. Employee-inventors are now required to actively seek licensees for their inventions and are evaluated on their ability to do so. In certain cases the inventors are permitted to assume ownership rights and pursue commercialization. A system for royalty-sharing was created that gives the agencies and employees of government-owned, government-operated laboratories a financial stake in the inventions they create. Royalties from commercialized inventions are collected by the sponsoring agency and shared with the employee-inventors. A portion of the remaining revenue is put toward the inventing laboratory's budget for the next year and the rest used for activities that encourage technology transfer within the agency.

Particular provisions of the Bayh-Dole Act ensure that the economic benefits of federally funded research are enjoyed primarily by the United States. Agencies are to favor U.S. industry when a developer (1) is not located in the United States, (2) does not have a place of business in the United States, or (3) is subject to control of a foreign government. Further, in order for a developer to assume exclusive rights to an invention, the "products embodying the subject invention or produced through use of the invention" must be manufactured substantially in the United States (9). Exceptions are made, however, if domestic manufacture is either patently infeasible or not possible at the time. Violation of these terms entitles the government to terminate the licensing agreement.

The Bayh-Dole Act also provides the government with residual rights to all publicly funded inventions to ensure its access to the technology in certain circumstances. The government retains the freedom to employ a licensed invention, royalty-free, for its own use or on behalf of a foreign organization and federal agencies are permitted to retain additional rights. To maintain fairness in the marketplace, the technology transfer policy aims to minimize the monopolization of publicly funded inventions by encouraging the use of nonexclusive licenses. Agencies are only permitted to issue exclusive licenses when they are proved to be in the best interest of the public. Exclusivity is beneficial only when it is a necessary incentive for development, and does not threaten competition or concentrate a particular technology in a specific geographic area. To further protect fair competition, when federal agencies issue licenses, they are required to give first preference to small businesses who have adequate resources for successful commercialization. In addition the government is entitled to exercise "march-in" rights and terminate the exclusivity of a contract if (1) The licensee has taken, or is not expected to take in a reasonable amount of time, effective steps toward developing an invention, (2) requirements for public use specified by federal regulations are not being reasonably satisfied by the licensee, or (3) action is necessary to alleviate health or safety needs that are not reasonably satisfied by the licensee. These march-in rights have never been exercised.

PUBLICLY FUNDED RESEARCH AND BIOTECHNOLOGY

The current policy of technology transfer is a reflection of the federal government's traditional conceptualization of scientific research. Although it offers an array of societal benefits, the primary reason science receives extensive federal support is that its advancement has become vital to national military and economic progress. Thus biotechnology, which is believed to hold great potential for the field of medicine, is primarily conceptualized by the government as a means to promote industrial growth. This is reflected in the 1989 argument by Michael A. Andrews before the House of Representatives in a plea to secure future funding for the Human Genome Project:

> The United States has a soaring trade deficit. We are slowly awakening to a growing weakness in international competition.... The Japanese are developing automated sequencing devices. The English have almost completed the mapping of the roundworm genome. The West Germans and the French have set up international reference data banks to collect the results of genome research. International competition has often spurred the United States into action on major scientific endeavors: Sputnik caused us to put a man on the moon, World War II brought about the Manhattan project.... I believe that international competition will shore up a commitment of the United States to the Human Genome Project more than any other single factor.... One by one, we have watched the pillars of our economy fall: the steel industry, the auto industry, the electronics industry, and the energy industry. Biotechnology is one area where the U.S. can have a clear lead (10).

The government apparently heeded Andrews' advice. Not only was extensive funding secured for the Human Genome Project, but by 1994, federal laboratories were the nation's leading inventors and enablers of new technologies for the biotechnology and pharmaceutical industries (11). The Public Health Service continues to lead the nation as the organization with the largest number of "therapeutics in active development, both in terms of those licensed out and those being developed internally" (11). This vast federal support of biotechnology raises special concerns because discoveries in this field hold value beyond their contribution to industrial growth. Biotechnological innovation carries the potential to improve the lives of the public, who, through their investment of tax money, and by serving as clinical research subjects, make medical research possible. Arguably, because the public is so uniquely invested in the products of medical research, particular aspects of the technology transfer policy ought to be reconsidered.

The absence of price controls, for example, allows private developers to set the price of publicly funded medical technology beyond what portions of the population can afford. In other words, current technology transfer policy confers on private parties the right to ration publicly funded therapies according to what the market is willing to pay. The government refrains from exercising its right to "march-in" and terminate exclusive licenses on behalf of the public health because the ultimate goal of technology transfer is not necessarily to improve social well-being but to serve the nation's economic interests.

The circumstances surrounding the implementation and repeal of the "reasonable price clause" illustrate this point. In 1984, when HIV was identified as the virus that causes AIDS, a screening program was initiated where drug companies submitted shelved drugs to NIH for testing against the retrovirus. Burroughs-Wellcome submitted a drug called AZT that was invented in 1964 by the National Cancer Institute (NCI). In 1985 AZT was deemed effective against HIV in vitro and was therefore worthy of clinical investigation. With NCI supplying the thymidine necessary for AZT production, Burroughs-Wellcome provided NIH with AZT to run the clinical trials necessary for FDA approval. A year later AZT, having successfully prolonged the survival of AIDS sufferers, was approved by FDA. To promote its development, Burroughs-Wellcome was granted a seven-year exclusive marketing privilege and patent rights until 2005 for its use in the treatment of HIV. AZT was introduced to the public in 1987, at the exorbitant price of $10,000 to $12,000 per patient/per year (12).

Outraged, the public demanded to know why the price of AZT was so high when both the initial discovery and later recognition and research of its modern application were publicly sponsored. In his recounting of the AZT controversy, Baruch Brody asks whether "the public's need for the drugs [is] being served by allowing drug companies to charge that much for drugs?" (12). A plausible argument can be made that the needs of the public should take precedence over the promotion of technology transfer. In response, some would argue that technology transfer never actually takes precedence over public needs because it serves those needs in the long term. Allowing companies to set high prices is society's way of rewarding them for transferring the public's scientific research investment into important public goods. The question arises, however, whether a public good is provided when a product is largely inaccessible to its sponsors.

In 1989 the government sought to improve access to products of technology transfer by promulgating a "reasonable price clause" in exclusive licenses arising from Cooperative Research and Development Agreements between government and industry. The clause mandated that the price of inventions must reasonably reflect the health and safety needs of the public and their investment in the product (13). In his explication of the reasonable price clause, Brody outlines its numerous presuppositions. For one, it assumes that the government is entitled to a financial return on the intellectual property rights conferred to private developers. Reducing product prices according to the degree of public subsidization will result in Medicare and Medicaid savings. In addition the clause assumes that access to the products of technology transfer should not be determined by price alone but weighed against public health and safety needs. Last, the reasonable price clause implicitly assumes that its price control measures would not deter private developers from investing in publicly funded inventions. Brody believes that the assumptions contained in the reasonable price clause are problematic in that a pricing policy based on the degree of public funding may yield an entirely different price than a policy focused on accessibility. Furthermore,

it is presumptuous to assume that either approach offers the return on investment necessary for private developers to engage in technology transfer (12).

In 1995 the reasonable price clause was repealed, for numerous reasons. NIH claimed that it was not only difficult to enforce but that it had a chilling effect on industry–government research collaboration. The repeal was clearly influenced by the pharmaceutical/biotechnology lobbyists who were vehemently opposed to price controls, by the patient advocacy groups who were incensed over a possible delay of new products, and by a Congress that was quick to abolish regulations without instituting safeguards to secure the public's equitable claim to the products of technology transfer.

The resulting policy omission exemplifies the federal government's prioritization of economic interests over any genuine commitment to improve the fundamental problems of the nation's (and the world's) access to health care. As patients, providers, and payers alike struggle with the burgeoning costs of medicine, the current technology transfer policy propagates the belief that "the proliferation of new technology developed at public expense is an unqualified good" (14). William Sage questions what he has dubbed "the conventional wisdom" of the policy in light of its contribution to health care inflation (14). He points out that most new products introduced by the biotechnology and pharmaceutical companies neither prevent nor immediately cure illness. They tend, rather, to "palliate suffering and prolong life" which, coupled with their high price tags, serves to funnel a significant portion of increasingly scarce health care dollars into industry pocketbooks (14). The government incurs the escalating expense of medical technology through Medicare and Medicaid. To contain costs, it chooses not to question the cost-effectiveness of new technology but rather to reduce provider reimbursement rates and tighten Medicaid eligibility criteria. Whereas it could fund cost-effectiveness studies with the royalties collected from technology transfer, the government has little incentive to do so because any policy that restricts the market for new technology would be detrimental to the pharmaceutical/biotechnology companies that are regarded as key to national economic prosperity. Private insurers, on the other hand, and the employers and individuals who pay insurance premiums, are increasingly reluctant to cover expensive technology when cost-effectiveness has not been demonstrated. At this point, however, programs to evaluate cost-effectiveness are too expensive and controversial to permit their widespread implementation.

For the 44 million Americans without health insurance, the industrialization and bureaucratization of health care has rendered rudimentary care, let alone technologically advanced treatment, virtually inaccessible. Sage argues that to retard the trend of using scarce health care dollars to purchase products that, in the long run, are a cost rather than a benefit to the system, new medical technologies, particularly those arising from public funds, should be brought to market only after determined to be cost-effective (14). Whether an invention's cost-effectiveness is sufficient for it to be brought to market ought to be based on a societal consensus to purchase the technology for all, regardless of their ability to pay. To increase access, royalties could be used to subsidize the cost of technology for the indigent. Additionally the government could require private patent holders to provide their product to the disadvantaged gratuitously or at a discount. Sage concedes, however, that it is unlikely these suggestions would adequately offset "the incentives for unbridled innovation and consequent cost pressures created by the current technology transfer policy" (14).

Whereas royalties can potentially patch some of the holes in the system, some believe they create more problems than they solve. It is unclear, for example, why the government collects royalties from licensed inventions if the objective of technology transfer is to promote the private development of public inventions. Royalties make publicly funded technology a less desirable investment because they function essentially like a tax on development that reduces the overall profit potential of government-funded technology and increases the risk of the investment. In the end it is the consumer who bears the cost of royalty agreements because product prices must offset royalty expenditure, and higher prices reduce the accessibility of public goods. Again, the public pays twice for inventions.

The Bayh-Dole Act encouraged university ownership as a means to promote academe–industry collaboration. Underlying this decision was the perception that universities are well suited to determine which research results ought to be patented and developed, and which would best serve the interests of science as part of the public domain. University–industry collaboration appeared to optimize the potential for scientific advancement. Industry would provide university researchers with incentives to generate inventions that would benefit society in the form of public goods and economic productivity. However, in order to simplify their administrative burden, universities tend to prefer to grant exclusive rather than nonexclusive licences, which undermines the rights of the public to relatively unrestrained access to publicly sponsored inventions and discoveries (15).

The financial return on commercialized inventions would provide future funding for innovative academic research. The collection of royalties by universities therefore may provide a much-needed source of revenue for sustaining research initiatives and other institutional needs. It would be unwise, however, for all universities to rely on such revenue, since it is only a minority that can generate a sufficient amount of royalty revenue to reliably sustain institutional functions. Further there is some debate over whether the patents to publicly funded inventions should have been transferred to universities in the first place because they offer no advantage over the government in that both are unable to develop inventions into public goods.

The university–industry research collaboration has changed, and it is likely to continue to change the culture of academic science. It is a tradition in academic research to promptly share information with the scientific community with the incentives for doing so largely taking the form of professional accolades for furthering the advancement of knowledge. Commercial sponsorship threatens this tradition by motivating researchers to privatize the knowledge they generate in order to ensure both patent eligibility and gain a competitive edge. This trend is reflected in a study conducted by David Blumenthal that found biotechnology faculty with industry sponsors to be more than four times as likely as colleagues without commercial support to report that they had kept research results a secret in order to protect their proprietary value (16). His study also found these researchers to be five times as likely to report than they had conducted research the results of which were the property of private sponsors and could not be published without the sponsor's consent (16), reflecting a serious sellout of academic freedom. The incentive to patent research results increases the likelihood that researchers will engage in the lengthy patent application process, during which time they may be reluctant, or even contractually forbidden, to share information regarding their inventions. Thus the industrial support of academic research imposes barriers that hinder the sharing of knowledge among colleagues. Yet it is this cooperation that has traditionally advanced science.

In biotechnology patentability is still a gray area, and publicizing information, such as sequence data, may render an invention "obvious" and thus ineligible for patent protection. When patents are granted in young fields like biotechnology, earlier inventions are often granted broader patent coverage than those that follow because "patent claims are drafted to encompass not simply what the inventor has done, but the idea which underlies the specific detail. Sometimes, in the absence of much detail, patent rights may be granted which many regard as excessively broad" (17). The uncertainty that surrounds biotechnology patents is exemplified by the 1988 patenting of the Harvard OncoMouse, a transgenic mouse produced for carcinogenicity testing (18). The OncoMouse patent raises numerous questions, including whether the patent was limited to transgenic mice, or whether it extended to transgenic rodents, or even to transgenic mammals in general (19). When the boundaries of intellectual property are nebulous, the stage is set for complicated turf battles that will likely send potential developers in search of investment opportunities with less potential for complication.

Broad patent claims, when applied to basic research tools like the OncoMouse, may chill entire areas of research. Licenses to use patented research tools may be unaffordable to some institutions, or researchers may be unwilling to purchase licenses that would essentially allow their research to be shaped by the interests of the patent holder. By owning and exploiting the rights to research tools, industry can gain considerable control over the nation's research agenda. This was demonstrated in the controversy over Cre-lox (20), a recombinant technology owned by DuPont that was used freely in NIH genetic research for years (17). In what was thought to be a prudent business maneuver, DuPont began to require researchers to purchase licenses for the use of Cre-lox. Recognizing that these licenses would be unaffordable for some institutions, DuPont permitted researchers to use Cre-lox with the understanding that DuPont would retain ownership of all their inventions that either incorporated the technology or used it somewhere along the line. Researchers were prohibited from transferring Cre-lox technology to unlicensed colleagues and were required to send all Cre-lox-related papers to DuPont for review prior to publication. Ultimately the research community struck a less restrictive agreement with DuPont for the use of Cre-lox, but the circumstances reflect that the strong presence of industry, and its ownership of technology, can wield such power over research that the government has to intervene, a heretofore unprecedented action.

Companies are involved in the development of products when they are still in the early stages of research. Researchers, including those working under federal grants, often receive private compensation for their ideas in the form of income, equity interest in the developing company, a seat on its board of directors, or a percentage of future sales (14). Admittedly, scientists are entitled to benefit financially from their inventions, and the technology transfer system is constructed to tolerate personal gain at taxpayer expense in order to encourage the development of public goods. However, when universities and individual faculty members have a direct profit motive to invent a product, it "represents a clear departure from past practices, and creates real risks for universities and the functions they are designed to serve in society" (14). When academic research begins to take on market characteristics, competition may tempt scientists "to circulate misinformation about a project's likelihood of success or even to commit outright fraud" (14). In the past it appeared harmless to allow researchers to earn royalties from their inventions because this occurred only in exceptional cases and after a prolonged period of research and development. Today a financial stake has been driven through the lab bench by companies who pay universities and researchers large sums of money up-front for the long-term rights to their inventions. Sage explains that "when the enrichment of scientists is directly related to the success of the scientific endeavor, society runs the risk that researchers will knowingly influence the outcome of neutral scientific inquiries" (14). The compromise of scientific integrity for greater reward violates, particularly in the case of medical research, the scientist's fiduciary duty to the public who entrusts the research community with their health and safety (14).

Royalty incentives, much like researcher equity positions in end products, comprise a facial conflict of interest. The federal government, however, relies on these incentives to drive technology transfer (21) and so excludes them from federal conflict-of-interest regulation. The use of royalty incentives in the absence of safeguards to protect the public from unsound science questions whether the government is using the public's funds responsibly and in the interest of their welfare. The pervasive use of royalty incentives raises the additional concern that if the scientific community is motivated by financial reward, it may cause basic research, because its products lack immediate commercial potential, to become undervalued and underdeveloped. Basic science, however, is essential to societal advancement. It has been understood since the nation's inception that the greatest breakthroughs often occur out of investigations into uncertainty. The information derived from basic research is the foundation for applied research which, in contrast, is defined by certainty in that it usually requires a set of "unambiguous facts" and specific targets toward which to work (22). There is a prevailing fear that as the industrialization of science progresses, the breadth of scientific inquiry may narrow to areas with foreseeable market potential. Thomas Jefferson believed that the government is obliged to promote innovation, and it ought to do so by subsidizing basic science because the goods it produces, and the incentives it requires, cannot be adequately sustained by the private sector. In that case the driving of publicly funded research toward the invention of commercializable goods is not what the government's role in science should be. Without the comprehensive support of basic science, the socioeconomic structure of the nation may suffer as it will be denied the benefit of scientific uncertainty and its resulting discovery. Arguably, the government tilts the nation's research agenda away from basic science through the policy of technology transfer.

If the government funds basic research, then what happens to the results should depend on the commercial potential and usefulness of the results. For example, some space research might hold military importance, some long-term strategic importance (e.g., claiming bases on Mars for future uses), and even some potential commercial utility (e.g., for commercial satellite, communication, or other types of businesses). Other research might be pure science performed in an attempt to understand the universe. Arguably, if the public sponsors the former, then some mechanism should be found to give a commercial preference to domestic companies, which will, in turn, return direct benefits to the country by employing citizens and paying taxes. In the latter case, however, the intellectual contribution of understanding black holes may not and should not be considered property but knowledge. Knowledge should be more freely shared with all. The government, through its policy for technology transfer, promotes the patenting and development of technology that provides society with public goods and economic productivity, but it does so at the expense of the collaboration necessary for overall scientific advancement (23). This is consistent with the federal government's conceptualization of science. Because it is valued as a means to spur economic growth, the measures used to promote this growth may threaten the proliferation of knowledge is not such a pressing concern. This view is rather short-sighted because it is indeed the proliferation of knowledge that ultimately gives rise to overall societal advancement.

Aspects of the technology transfer policy run counter to the ideal of a "more perfect union" where the federal government fulfills its constitutional duty to "establish justice," "ensure domestic tranquility" and "promote the general welfare" (2, preamble). The investment of public funds in the life-enhancing field of medical research, without ensuring the public's equitable claim to resulting therapy, is arguably unjust, despite the aggregate economic benefits. An expensive, bureaucratized, and largely inaccessible health care system is not in the interest of the general welfare, yet aspects of technology transfer perpetuate these systemic flaws.

As this review shows, the evolution of technology transfer policy reflects a difficult balancing between promoting development and use of government-funded inventions and discoveries, and protecting public welfare. One constant throughout this historical account is the military or economic justification for government support of research. A second constant is that as science becomes increasingly complex, it is becoming industrialized and, as a result, bureaucratized. All the while, however, it is not necessarily becoming more accessible to the public. A third constant is the concern about monopoly: Exclusive rights can benefit the powerful at the expense of the public, but exclusive rights sometimes are the best way to move inventions off the shelf and into the market.

BIBLIOGRAPHY

1. A.H. Dupree, *Science in the Federal Government: A History of Policies and Activities to 1940*, The Belknap Press of Harvard University Press, Cambridge, MA, 1957.
2. United States Constitution, Art. 2, §8, cl. 8.
3. B.D. Karl, *Daedalus* **105**(4), 129–149 (1976).
4. H. Baer, *Soc. Sci. Med.* **28**(11), 1103–1112 (1989).
5. *NIH Almanac*, September 1998.
6. Congressional Record H 9180-9183 (daily ed. May 19, 1930), Congressional Record S 9258-9265 (daily ed. May 21, 1930), Congressional Record S 10358-10362 (daily ed. June 10, 1930), Congressional Record H 10950-10952 (daily ed. June 16, 1930).
7. H. Brooks, *The Government of Science*, MIT Press, Cambridge, MA, 1968.
8. R. Eisenberg, *Virginia Law Rev.* **82**, 1663–1726 (1996).
9. J.V. Lacy, B.C. Brown, and M.R. Rubin, *Pepperdine Law Rev.* **19**, 1–28 (1991).
10. 101 Congressional Record H, E1418 (1989).
11. NIH Fall 1994 press release/study. Originally released 8/22/94; revised 9/8/94 including new data for 43 NIH CRADAs. <http://www.bioinfo.com/fed_biotech_transfer_study.html>
12. B. Brody, *Hastings Center Report* **26**(2), 5–11 (1996).
13. *NIH Backgrounder*, April 11 (1995).
14. W. Sage, *Virginia Law Rev.* **82**, 1737–1752 (1996).
15. A. Schissel, J.F. Merz, and M.K. Cho, *Nature* **402**, 118 (1999).
16. D. Blumenthal, *J. Am. Med. Assoc.* **268**, 3344–3346 (1992).
17. D. Wood, *Whose Life is it Anyway*, InterPharma (1996). <http://www.interpharma.co.uk/iss3.htm>
18. U.S. Patent No. 4,959,317 (1988) Site-specific recombination of DNA in eukaryotic cells.
19. E. Russo, *Scientist* **12**(18), 1 (1998).
20. U.S. Patent No. 4,736,866, Transgenic nonhuman mammals.
21. M. Witt and L. Gostin *J. Am. Med. Assoc.* **271**(7), 547–551 (1994).
22. T. Lewis, *The Lives of a Cell: Notes of a Biology Watcher*, Viking Press, New York, 1974.
23. *The Organization of the National Research Council*, Available at: <www4.nationalacademies.org/arc.nsf>

See other entries HUMAN GENOME DIVERSITY PROJECT; MEDICAL BIOTECHNOLOGY, UNITED STATES POLICIES INFLUENCING ITS DEVELOPMENT; OWNERSHIP OF HUMAN BIOLOGICAL MATERIAL; see also PATENTS AND LICENSING entries.

PATENTS, ETHICS, HUMAN LIFE FORMS

LOUIS M. GUENIN
Harvard Medical School
Boston, Massachusetts

OUTLINE

INTRODUCTION

It might be supposed that morality operates as a side constraint on patentability. On this view, even though a process or device might meet conceptual and scientific criteria for recognition as an invention, moral considerations might override so as to deny a patent. Or again, it might be held that morality, genetics, and biotechnology so intertwine that whenever we construct criteria of patentability with respect to "genetic inventions," we perforce impose some moral view.

EXPLOITATION AND MONOPOLY OF CLONED DNA SEQUENCES

Whereas U.S. patent authorities formerly declined to issue patents on gambling devices and phony medicines, the U.S. Patent Code of 1952 dissociates law and morality. It leaves to other laws the matter of restraining the use of inventions. Efficiency alone commends this division of labor, since many patents are never exploited. For example, Pasteur obtained a U.S. patent in 1873 on a yeast for making beer. But, so far as we know, he never developed a commercial product (1). If immorality of use does not count as an objection to a patent application in general—even if the proffered device be mischievous—should immorality of use count as an objection to patents on life forms?

We might answer this with another question. If a patent on a gene or other human life form confers ownership over something human, and if, on moral grounds, we reject claims to own humans, are not patents on genes and human substances illegitimate? This question probes not an invention's use but the appropriateness of the patent privilege itself. A patent lawyer will reply, reprovingly, that a patent does not confer ownership, that a patent merely grants for a term of twenty years the privilege to exclude others from making, using, or selling an invention. This reply does not end the discussion. For various circumstantial reasons any policy on biological patents brings moral controversy in its train. In the first instance, allowing commercial entities to wield even limited monopolies on things human will seem morally problematic to many observers. Some will regard such privileges as threats to the autonomy of persons (as discussed below for clinical settings). Others will point to various economic consequences of wielding patents, among them high prices and restricted output of end products. When a DNA sequence patent issues but the patentee fails or declines to introduce a product predicated on the sequence, the only benefit of the patent, if one may call it that, is to prevent the patentee's competitors from exploiting the sequence. It may be granted that for some the welfare loss of squandering an opportunity to improve beer production, especially for a mere scientific career, is cause for lament. But if a patentee shelves a human gene patent and denies society an opportunity to develop beneficial drugs or to perform gene therapy, the cost may be human suffering. As we shall see, good reasons obtain to resist the generalization that biological patents enhance aggregate welfare. In respect of the foregoing concerns, one hears not merely the voices of patent examiners and courts—unlikely arbiters of morality in any event—but a variety of moral views held among citizens to whom accountability for governmental decisions is owed.

Because the decision to award a patent may be publicly perceived as at least implicitly a decision to condone any and all uses of the invention, it may behoove us first to resolve objections concerning morally problematic uses of certain biotechnological innovations before we attempt a consensus on monopoly of the innovations. If prudence commends this two-part agenda in the United States, the European patent system demands it. The European Patent Convention of 1963, whose criteria of patentability are otherwise roughly coincident with the American, proscribes patents on inventions whose commercial exploitation would be "contrary to *l'ordre publique* or morality." This phrase was long considered so vague as to lack teeth. But as adopted in 1998, the Directive on the Legal Protection of Biotechnological Inventions of the European Parliament (the "European Directive," or "ED") declares unpatentable, on the ground that their commercial exploitation would be contrary to *l'ordre publique* or morality, the following: human germ line intervention, "cloning" humans, commercial use of embryos, and both somatic and germ line genetic intervention in animals that is "likely to cause suffering without any substantial medical benefit to man or animal" (2). To this the ED curiously adds, "exploitation shall not be deemed to be so contrary merely because it is prohibited by law or regulation."

Anticipated Benefits of Transgenesis

Mankind has bred plants and animals for millennia. Since Mendel, breeders have exploited knowledge of dominant and recessive alleles. Moral controversy about genetic engineering stems not from manipulation by breeding, but from recombinant DNA. It is not that recombinant techniques clearly violate any moral view in particular. Rather it is the case that recombinant DNA technology poses questions not previously raised within any traditional moral theory.

Transgenesis consists in isolating a gene of one living being and inserting the gene at an early embryonic stage, before somatic and germ cells separate, in such a way that the gene enters the germ line of another living being of a different species. The insertion may be accomplished by introducing the foreign gene into (i) a retrovirus that infects an embryo, (ii) a plasmid microinjected into the pronucleus of a zygote, or (iii) cultured embryonic stem cells injected into the cavity of a blastocyst. The inserted genes are usually few and manifest themselves in only a small subset of an animal's phenotype. Transgenesis enables improvements in the growth, heartiness, and yields of animals and plants as sources of food, vaccines, and other compounds, affords models for study of diseases, the immune system, and gene regulation and expression, holds promise for direct therapeutic use in humans, allows the "pharming" of animal organs so that, upon transplant to humans, they will not be rejected, and allows production of cotton, plastics, and other industrially valuable compounds. A vaccine-enriched transgenic banana holds promise as a vehicle for surmounting economic and practical obstacles to vaccine delivery in many regions of the world.

Reservations Concerning Transgenesis

As encouraging as these prospects may be, they are not without their detractors. Objections to transgenesis include the following. Even if genes insert at a targeted locus, in animals the effect of transgenesis may be suffering, a theme frequently rehearsed in European discussions. The usual defense of animal experimentation (as in the ED) adverts to collective human benefit. A net increase in aggregate human preference satisfaction is all that need occur to satisfy a utilitarian; the second form of Kant's categorical imperative permits using even humans as means so long as they are not used solely as means. But risk-averse humans worry about their own welfare in eating transgenic plants and animals—even assuming full disclosure in the grocery store. To introduce a vaccine into a banana crop raises questions about imposed risk-taking and paternalism when informed consent may not be feasible. Risk of human suffering is sometimes cited as a consideration against human gene therapy. Risks about where and in how many copies genes insert and whether a procedure will otherwise work are chanced by any single recipient of somatic cell therapy; to this germ line intervention adds the risk that an untoward result may burden future generations who, it may be said, have no voice in what is done to their ancestors' genomes. Even when gene therapy achieves an intended result, the long-term effect may be a population less diverse, a gene pool that is diminished. A suite of controllable genetic characteristics may eventually generate a canon. By reference to that canon, persons lacking certain traits may be treated by others as inferior. Perhaps indeed we shall cavort down a slippery slope from disease-related therapies to frivolous enhancements. To engage in germ line intervention, it is decried, is to "play God."

Defense of Human Germ Line Intervention

A defense of human germ line intervention might run as follows. Genetic engineering may be a way to improve man's contribution as co-creator in God's work (3). One might argue that God would wish caring physicians to use it. Gene therapy will not invent discrimination, a practice already thriving with respect to many traits. We should not count even against enhancement that someone will be born with a trait more desirable than another trait. Instead a temptation to be invidious should remind us, as do Kant and many religions, to recognize the dignity of each person. As for an effect on descendants, that is not a new category of moral responsibility. Future generations are already affected by innumerable influences on one's germ cells of how one lives. One may imagine a complaint of wrongful life by someone born with a genome adversely affected by something that went wrong in a gene therapy procedure, but in another case, the same tort might be committed by failing to attempt gene therapy. It would appear dubious to bar a physician from using an available method of averting disease in a consenting patient's offspring if, for example, the odds of the method's success exceed those of any treatment after birth. It is also difficult to expect a family or society to forgo eradicating a lethal gene if that be possible. A few decades from now, germ line intervention may be considered routine, its provision the duty of a competent physician, its inclusion a requirement of the health care a just society ought to provide.

The slippery slope to enhancement is not fairly ascribed to gene therapy (though perhaps to recombinant DNA). Recombinant human growth hormone, for example, is already dispensed. At least at present, the imagined efficacy of gene therapy is limited to diseases involving single genes, and among those, to diseases mediated by

recessive genes, because an inserted gene's locus may not be controllable and any dominant defective gene remains in the patient. On the other hand, someday prospective parents may be speaking of a "designer child" polymath 9′ basketball player. In any event a distinction between "therapy" and "enhancement" may not be sharp or necessary. If a slippery slope connects therapy and enhancement, the transition from first violation of any ban on genetic enhancement to widespread violation thereof may be an avalanche. A treaty presented for adoption by members of the European Union bans enhancement (4). As of this writing, human germ line intervention for enhancement purposes is not feasible, but upon its availability, one may expect the following. Though a government may ban genetic enhancement, as soon as one person manages to procure an enhancement, others acting rationally will likely rush to procure enhancement in order to remain competitive (5). These aspirants will either violate the ban or migrate to a sovereignty that lacks one. Sovereignties will likely behave the same way, rushing to follow the first innovator for fear of being dominated by superiors. Unless one contemplates intrusive "audits" comparing parental and progeny DNA, a ban on enhancement seems unenforceable.

The foregoing brief account reveals moral concerns about uses that insinuate themselves into discussions whether to approve monopolies of uses. We may now turn to moral concerns about a patent privilege itself.

Autonomy and Patent Claims Against Parents and Children

The effect that a patent might exert on individual autonomy may be studied through a dramatic example. This first requires that we explain the rationale for patents on transgenics.

Patented Transgenic Organisms. Although Pasteur's yeast long before gained a patent, modern recognition of a nonplant life form as patentable occurred in a decision with respect to a bacterium into which were introduced plasmids rendering the bacterium capable of decomposing oil (6). Patentability was later confirmed for multicellular organisms (polyploid oysters) (7). A furor ensued over ethical concerns, and for a time, the U.S. Patent and Trademark Office ("PTO") imposed a moratorium on animal patents. Thereafter the PTO issued a patent on the Harvard mouse (8).

Designed for the study of cancer, the Harvard mouse contains DNA sequences, comprising an oncogene such as *myc* and a promoter, that effect a high proclivity to form tumors when the mouse encounters carcinogens. The introduction and expression of the *myc* gene in the mouse was innovationary. One could not have presumed that a zygote acquiring the oncogene would survive its insertion and expression, or if the zygote did survive and a mouse were born, that the offspring would be fertile. The Harvard mouse invention was exhibited for patent purposes by deposit of DNA in a plasmid. But the patent extends not merely to the inserted DNA, not merely to the oncogene's introduction and expression, but to the whole "oncomouse."

Why should a patent embrace an animal? Two arguments might be mustered. First, there has endured throughout the history of patents the notion that patents should not be available on nature's extant treasures — in a phrase attributed to Thomas Jefferson, on any "product of nature" — but should be available only on what humans manufacture. The discovery of uranium garnered no patent, but the PTO issued a patent to Glenn Seaborg on curium and isotopes of americium, transuranic elements believed to exist on earth only in a cyclotron or reactor as a result of human efforts. Of course such elements may be abundant in stars. A precise statement of patent eligibility would not exclude from patentability every naturally occurring thing. Rather we may state patent eligibility by the following proposition, which we may call the "unpatentability of nature": to be eligible for a patent, a thing must be such that there obtains only a very low probability that, without human intervention, the thing exists near the surface of the earth or on other astronomical bodies to which humans travel. What constitutes a "very low" probability requires specification. Since patent lore speaks confidently of things that "exist" or "do not exist," no guidance on probability may be found within it.

It is logically possible for there to evolve an organism whose genome is identical to some modern transgenic organism. Exchange of genes across species occurs and any mutation is possible. Yet the probability may be extremely low that a creature will contain genes of two given organisms that do not mate. In such case a patent examiner may treat a transgenic genome as if it does not naturally occur. An animal possessing such genome is then seen not as an unpatentable product of nature but as a patentable "manufacture" or "composition of matter" (9). With respect to the European Patent Convention, it would be said that such an animal is not an unpatentable "variety" (10). In general, the fruits of breeding programs are considered varieties, but transgenic animals are not considered varieties because transgenesis was unknown in 1973 when the European concept of a variety was introduced.

Second, when introduced into a recipient's germ cells, transgenes pass to descendants. A transgene will not be expressed in all offspring of the first generation, but those in which it is expressed will be selected for further breeding. Were a patent to cover only cells expressing a trait, it would not capture the invention, which, by virtue of being genomic, appears in every cell. Were a patent to cover only an inventor's process of introduction and expression, anyone who purchased one transgenic animal could breed others without infringement; natural reproduction is not the same process as laboratory transgenesis. A purchaser of transgenic agricultural livestock could breed the livestock through unlimited generations. Hence inventors are accorded the protection of a patent on the transgenic genome, which is effectively a patent on the animal. Breeding descendants of a patented transgenic animal without license would just as clearly constitute infringement as would duplicating a patented laboratory process for inserting transgenes into an embryo. The patent system assimilates reproduction, whether natural or artificially aided, to "making" a duplicate.

The Human Qua Infringement. A moment's reflection reveals that if the foregoing two grounds (a claim to originate a manufacture; self-reproducibility of a recipient) entitle an architect of transgenesis to a patent on recipient and progeny, then in the case of human germ line intervention, infringement claims will lie against the birth and existence of humans. To call human birth or life a "patent infringement" seems perverse. But on what principled grounds should we reject such claims? Even in the somatic cell case, as scientists perfect the manufacture of yet more human enzymes and other proteins, as they progress to substantial tissues, should society continue to grant patents on human "parts"? Manufacture of a liver or other major organ, or someday even of a brain, may confound previous thinking.

Abjuring the Human Qua Infringement. To resolve the solecism of the human qua infringement, we may reason as follows. We do not imagine infringement claims against any plant or animal. Instead we recognize claims against people who control breeding. We do so because we recognize a farmer's ownership of plants and animals. When the "designer" of a transgenic organism applies for a patent, the contest concerns only which humans (or corporations they represent) own property in the nonhuman species. When the issue is which of two humans owns a human, we say that humans own themselves. They do not own each other. Human births, we hold, are not analogous to breeding, to manipulation by owners of mating subjects. Hence we may decline to recognize property rights in humans.

The premise that humans do not own each other, that we each enjoy a "bodyright" (11), is not categorically held in all societies, and given that the common law describes an adolescent's maturity as "emancipation," perhaps it is not unequivocally held anywhere. Defense of the premise often comes round to some distinction between humans and animals. According to a Cartesian distinction, man is a singular creature possessed of reason. According to Kant, only man and angels are capable of reason. To say that a human being's existence infringes property rights would seem inconsistent with the second and third forms of the Kantian categorical imperative, which together enjoin that we treat each person not solely as a means but as an end-in-himself in a kingdom of ends. If we allow an ownership claim on a person, we condone treating the person solely as a means. We condone interfering with the person's autonomy. Were someone to assert an ownership claim that purportedly extended to only part of a person, the claim would appear indistinguishable from a claim on the whole. Bodily parts are integrated. For the same reasons, conception and birth, the instantiation of human nature, may be held immune from claims of others. We may also say that conception and birth are private.

Distributive Justice and Patent Claims on Extracorporeal Compounds

Although we may thus deny the permissibility of exerting dominion over, impairing the autonomy of, or disrespecting an individual, a different case, actual in biotechnology, is the following. There is adduced a substance that is human in the sense of being found in the human species, but which has been made outside the human body and is not ascribable to any individual. Were a patent to issue on that substance, the patent would not appear to interfere with any individual.

To this it must be added that it is not easy to steer clear of the DNA sequences that distinguish an individual or that make any individual akin to another. Only about three million of the three billion base pairs in the human genome account for individual differences, but the genetic code is redundant, the most interesting traits are polygenic and beyond present understanding, and mutation never ceases. For now, individual identity, to the extent it is genetic, is genomic. We have not demarcated a nonindividuating subset of the genome that we may cede. We do know that individuality is greatly affected by a relatively small number of regulatory sequences that control which genes are expressed. Such sequences are indeed used in biotechnology manufacturing unless a bacterium's or other host's regulatory sequences effect expression.

Let us assume for the moment that it is possible to grant monopolies on proteins and DNA sequences without there resulting any interference with the autonomy of any individual. The PTO effectively allowed as much when it began to grant patents on human DNA sequences despite its earlier declaration that a patent on a human would violate the prohibition of slavery in the Thirteenth Amendment to the Constitution of the United States. It contravenes common usage to say that a nonpossessory interest in a protein or gene constitutes slavery.

Were one confident that a system of limited monopolies would lead to advances that prevent or alleviate human suffering, one might decide that the conceptual coherence of the patent system should give way to the promotion of aggregate welfare. If patents on genes contravene the unpatentability of nature, so much the worse for that premise. It seems that one perforce resorts to that stance for the defense of patents on plant antibiotics. One might go so far as to say that there should issue any patent, even if the patenting process is purely piecemeal, that results in net aggregate welfare gains.

Whether compromise with intellectual purity be systematic or piecemeal, and even assuming any contingent results that an advocate of such conceptual indulgence predicts, this talk of welfare effects presupposes a criterion for discerning improvements in welfare. That in turn implicates some version of a social welfare function. A social welfare function is a function that yields or induces a positioning of possible resource allocations on which one may predicate a claim such as "α is a welfare improvement over β." The specification of a social welfare function is the main problem of distributive justice. For this reason, what begins as a moral problem concerning respect for personal autonomy, which arguably is tractable by virtue of the ability to eschew interfering with any individual, endures as a challenging problem of collective morality.

To the extent that patents are distributive mechanisms, this problem arises within an economy in respect of any patent. But concerns abound with respect to the welfare effects of biotechnology patents. Some may espy

unacceptable burdens and risks for the human species as a whole from various patents on molecular or structural human life forms. For instance, when a patent owner sets what seems an exorbitant price for a vital drug, one observes an arguably undesirable effect of market power conferred by an unqualified government-created monopoly.

PRODUCT PATENTS ON HUMAN DNA SEQUENCES

Supposing that the prospects of collective benefit or some other morally persuasive consideration have justified the alteration and use of life forms, why confer exclusive control on one party? The orthodox *quid pro quo* of a patent is that, instead of keeping an invention a trade secret, the patent teaches the details. When a patent expires, the invention will be in the public domain, and even during its term, what others learn from its teaching may foster other innovations. Whether the patent's revelations are in fact valuable will depend on whether one may easily infer the invention by reverse engineering (12). An alternative and more familiar rationale asserts that a patent provides an incentive that fosters ingenuity and effort. Or as Bentham put it, "He who has no hope that he shall reap will not take the trouble to sow" (13).

Isolation-Purification Rationale for DNA Product Patents

Organic compounds found in humans are, *ipso facto*, naturally occurring. Suppose that an organic chemist discovers a way to synthesize a protein in a purified form not found in humans. If the protein appears extractable from another organism, then perhaps we should not regard the protein as distinctively human. But in fact the human version of a given protein is unlikely to be identical with that of another organism. Through mutations in duplicate genes, species have evolved a variety of genes coding for different versions of proteins that we call by single generic names. The notion of isolation and purification (a creature of case law, not statute) was popularized by product patents on inorganic chemicals. (In patent parlance, a "product patent" is a patent on a thing as opposed to a patent on a process.) The notion was then borrowed in support of patents on the products of biological processes, including purified human adrenalin, prostaglandins, vitamin B_{12}, and, most recently, human DNA sequences. For the last, investigators' counsel have persuaded patent examiners that investigators have "isolated and purified," which is to say cloned, human genes.

There is reason for scepticism whether a patent must be available in order to induce a given result. "The large amount of research that has already occurred when no researcher had sure knowledge that patent protection would be available," noted the Supreme Court of the United States in affirming the patent on the oil-eating bacterium, "suggests that legislative or judicial fiat as to patentability will not deter the scientific mind from probing into the unknown any more than Canute could command the tides" (6). There arrived for filing a spate of plant patent applications in Europe, many presumably from European companies, quickly after the first plant patent was allowed there in 1989, which suggests that the research had long before been done. And rivals face effort and expense to follow a first entrant into the market. In the United States the lead time that an imitator of a drug or medical device would need to obtain approval from the Food and Drug Administration ("FDA") for selling the product provides a period of *de facto* exclusivity to the product's originator once the originator obtains approval. There may also be observed a tendency after a favorable experience for physicians to continue prescribing, and consumers to purchase, the first drug of a genre. Assertions about the necessity of incentives can be facile, but evidence is lacking.

Inventions concerning nonliving phenomena make use of materials that mankind has long exploited with an aplomb perhaps attributable to the mistaken belief that we cannot alter earth's vastness. Biological inventions obviously effect alterations of nature. If we approve the engineering of some protein, we might view the circumstance that it is found in humans as a reason against monopoly. Hence the isolation–purification rationale originated for chemistry in general cannot be assumed to carry the day for human compounds. One might add that to adopt such construct for humans would not follow the model of chemistry faithfully enough. Patents have been granted to those first to synthesize chemicals, but courts tend to find evidence that chemical patents have been infringed only insofar as a patentee's process has been copied.

The probable benefits of recombinant innovations may in a given case outweigh the acknowledged detriments. But each wave of innovation evokes a new comparison of risks, costs, and welfare gains.

Unpatentability of Nature and DNA Patents

An enduring challenge for the molecular biologist is to understand a disease or bodily function, to identify a protein related to it, to ascertain the nucleotide sequence of a gene coding for the protein and the protein's amino acid sequence, to locate the gene on a chromosome, and to explain the gene's regulation and expression. Once sequenced, a cloned gene may be preserved as complementary DNA ("cDNA") in a vector. When vectors transform and infect, not only do they multiply an inserted gene, but the gene can integrate into the transformant genome, causing such host to produce the protein for which the gene codes. It is by growing such transformants under suitable conditions that a biotechnology manufacturer may produce a protein in high volume.

A typical patent claims at least three inventions: (a) an isolated and purified DNA sequence encoding some protein, (b) any vector that contains that sequence and any transformed host possessing that sequence, and (c) one or more processes. The patent system indulges the notion that the cloning of genes produces "inventions" that do not naturally occur. By contrast, a detailed examination of what occurs in the laboratory, though providing ample evidence to confirm our admiration for scientific achievements, reveals no *entity* that mankind creates (14).

A gene encoding any human protein exists in nature. It is embodied in a chromosome. Its transcript also exists in mature messenger RNA ("mRNA"), a single strand of DNA-complementary nucleotides. Transcription of DNA into mRNA, followed by splicing that eliminates introns, is nature's own "isolation" of the coding sequence (with uracil in place of the thymine of DNA). This alone seems to tell against the argument that only an inventor has achieved isolation. Is "purification" then the inventor's trump over nature? The process of making cDNA is not thought to occur naturally in humans (though many viruses that infect humans make DNA from RNA). But once a gene is known, the laboratory process of making cDNA can be routine. Perhaps then vector and host deserve credit as the ingenious embodiments of purification? Where a host and a donor of foreign DNA are members of species that do not mate, asserted Stanley N. Cohen and Herbert W. Boyer in teaching the first recombinant process, a recombinant host "could not exist in nature" (15). This imagined impossibility of course is an exaggeration. Any mutation is possible. As Bernard D. Davis observed in debating the hazards of recombinant DNA research, bacteria absorb foreign DNA from lysed cells of their host, including the human gut. The rate of bacterial absorption of foreign DNA is low, but the number of bacteria in the gut is enormous, and bacteria have thrived for millions of years. There is some probability for any given human gene that it has already occurred in a bacterium. We may not observe that gene today, since the gene may have conferred no selective advantage on the bacterium, and the strain vanished (16). On the other hand, what is the likelihood that a plasmid or bacterium naturally contains a given human gene? That probability may approximate that of unicorns existing. Cohen and Boyer evoked a sense of mythological improbability when they called an altered plasmid a "chimera."

When a patent claims a chimera and host, the two are usually notable only in one respect: they contain a DNA sequence that the applicant purports to have invented. The plasmid and host are effectively the sequence's housing and factory. Once an investigator has selected a sequence as described above, the process of cloning it, and hence the "invention" of the sequence in a vector and transformed host, is mechanistic. Advanced techniques for sequencing proteins may make straightforward the selection of probes and primers, and hence the discovery of a known protein's gene. The "invention" of a protein variant may also follow straightforwardly from a variant, experimentally achieved, of a gene sequence. An investigator who finds naturally occurring genes and proteins merits accolades. But it remains to be shown why a patent should issue, and if so, on what.

What Should Suffice for a Biotechnology Product Patent?

As necessary conditions for award of a product patent, consider the following: (*1*) a claimed invention is such that it is highly improbable that we shall find it as such on earth or on any other astronomical body to which human travel is possible, and (*2*) the invention is ingenious. The "as such" phrase in (*1*) would allow some particularly convenient forms (e.g., a vector with a foreign DNA insert) to gain recognition apart from a natural form. In (2), "ingenious," which shares an etymological root with "genetics," is a placeholder for what constitutes an invention, of which more later. Sequences fail (1) if they are found in chromosomes and mRNA. Chimeras, transformed hosts, and cDNA meet (1) but fail (2) when they are mere mechanistic steps from discovery of a sequence. The vectors and hosts of microbiology are not as fastidious as the species united in mythological chimeras: they accept DNA inserts regardless what creature originates them.

An "artificial gene" may satisfy both (1) and (2). In theory such a gene may be constructed of any codon-containing sequence one likes; in practice the amino acid sequence of a protein may inspire the sequence. Some caution may be needed in characterizing a sequence as "artificial" because if a sequence codes for a human protein, then either that sequence, another differing only by substitution of alternative codons for the same amino acids, or yet another that is insignificantly different exists somewhere in the genome. Because of the phenomenon of overlapping genes, the sequence may also be part of another gene that its discover has not even envisioned. On the other hand, it may be that an "artificial gene" meets (1) and that its gene product is in some sense superior. (The artificial gene product might, for example, lack contaminants usually found in the natural gene product.) Not every such sequence will be ingenious. Courts have often declared DNA sequences inferred from protein sequences to be obvious (17, p. 50).

In view of patents on algorithms — a departure from previous conventional wisdom that ideas are not patentable — one might appeal to the notion of patents on information as a defense of DNA patents. But this defense would seem to fail insofar as any information encoded in cDNA is encoded in naturally occurring DNA and mRNA.

Adverse Welfare Effects of DNA Patents

If the autonomy of no one in particular is threatened by a product patent, the autonomy of everyone together might be.

The PTO in 1987 granted a product patent on isolated and purified natural erythropoietin. Merely four months later it granted a second patent to another party relating to a recombinant DNA technique for making the protein. The second patentee had cloned the gene after screening a genomic DNA bank with two sets of probes. It then produced the protein in transformed hamster ovary cells. The first patent blocked the invention of the second. This portended that patients would be deprived of a recombinant method of producing erythropoietin in high volume at low cost. As a group, patients were saddled with the first patentee's production method (extracting extremely low yields of the protein from thousands of gallons of urine). When, four years later, the first patent was invalidated on unrelated procedural grounds, the second became a barrier against any better recombinant process employing the claimed sequence (18). Similarly did a biotechnology firm discover the human gene for factor VIII:C by probing a human cDNA bank, inserting the gene in plasmids, transforming hamster kidney cells with the plasmids, and producing factor VIII:C. This recombinant advance was blocked by an earlier patent on factor VIII:C itself (19). The patentee's process not only required enormous amounts of donated blood plasma for a small yield, but in contrast with the recombinant method, it risked contamination. Contamination was a critical risk because many hemophiliacs who received contaminated factor VIII:C died of AIDS. The recombinant's manufacturer protested unsuccessfully that the patentee had not invented factor VIII:C (though, given the chance, the manufacturer might have argued for its own invention of the recombinant). The erythropoietin and factor VIII:C episodes illustrate how product patents may frustrate society's interest in encouraging, at the same rapid pace at which biomedical research is otherwise moving, helpful innovations in processes for making therapeutically valuable human compounds.

Farmers raise the specter that, burdened by the high cost of patented animals and crops, they may turn to cheap unpatented strains, that crops will become less diverse, and that more crops will succumb to pests. In transgenesis, often an investigator cannot control the place within a genome at which a foreign gene inserts or the number of copies that insert, or in the case of plants, the weeds or other unwanted plants to which a transgene may migrate via airborne pollen. We also have reason to rue "blind promotion of technological innovation" (20). Agriculture, after all, is an industry afflicted with overproduction. A patent granted by the European Patent Office on all manner of genetically engineered soybeans has been criticized on the ground that soybeans are among the world's most important crops and monopoly of soybeans will threaten "world food security." Suppose that a patent on a critical crop, organism, or substance has been conferred on an enterprise that becomes bankrupt. Or suppose that the patent is acquired by some foreign entity that is involved in international intrigue, that uses the patent as leverage for some disreputable purpose, or that otherwise seems to control output contrary to the common good. As exemplified by experience with the anti-AIDS drug zidovudine (or "AZT"), the price of a patented product is a monopolist's price.

Scientists have become acutely aware that availability of patents on DNA sequences may be generating a patent race that misallocates resources and delays publication of results. This would run contrary to the hope underlying the Human Genome Project that disseminating chromosomal mapping and sequence data will foster growth in collective knowledge. It took four years after a gene implicated in breast cancer, *BRCA1*, was mapped to chromosome 17 before one of twelve rival collaborations found the gene, a feat they all recognized as a "discovery" (21). Yet the winner immediately sought a patent on *BRCA1* and related diagnostic processes. About a year later, one of the competing groups contributed to a public database the sequence of a large portion of chromosome 13 where *BRCA2* was thought to repose. "It will not be helpful to medicine," the scientist John Sulston was quoted as saying, "if, by the year 2003, control of every single gene is tied up by one company or another for twenty years. That would be an enormous ball and chain. ...[F]or the good of humanity, we should try to keep these things in the publicly exploitable domain" (22). The group contributing the chromosome 13 sequence data urged that DNA sequences be public information. Within a month thereafter, *BRCA2* was found (23). This seemed to exemplify the rapidity of progress when results are shared. Thereupon the discoverers of *BRCA1* filed for a patent on *BRCA2*, launching a dispute over who found *BRCA2* first. Seemingly ignored was the untenability of claiming to invent parts of nature's storehouse.

One cannot dismiss objections to product patents as the outpouring of any single, disputed moral view. Even without an appeal to morality, it may be argued on exclusively scientific and economic grounds that patents on human DNA sequences violate the unpatentability of nature. Many moral views assign significance to the aggregate welfare consequences of that violation. In such case the moral case against human DNA sequence patents reprises the scientific.

ALTERNATIVE INCENTIVES FOR BIOTECHNOLOGICAL INNOVATION

Measures for Holding onto the Availability of Product Patents

The legal criteria for patentability are that a "process," "manufacture," or "composition of matter" be "new," "useful," and "nonobvious" (24). According to a conservative article of faith espoused by patent practitioners, these criteria possess such protean qualities as to suffice for the resolution of all questions that arise from time to time. The criteria need only be interpreted by the courts. In reply to this, it must be said that, under prevailing interpretations, "new" and "useful" erect only minimal thresholds. "New" eliminates from patentability only what has already been published. "Useful" eliminates from patentability only the utterly useless, a rare creature among proffered inventions anyway. (Scientists at the National Institutes of Health, NIH, dramatized the weakness of the "utility" requirement in 1991–1992 when they ostensibly satisfied the criterion by citing a seemingly trivial use for parts, "expressed sequence tags," of cDNA sequences. The applicants conceded ignorance about the feature of usual biological interest, viz., what the sequences encode or regulate, and ventured only that the tags could be useful as genetic markers, primers, or probes in diagnostic kits for unnamed diseases. But any DNA sequence may be a marker in genomic mapping.)

To resolve a question of patentability, two sobriquets, "manufacture" and "nonobviousness," must carry most of the load. In fact nonobviousness must do all the work. For it is considered settled in U.S. patent law that cloned DNA sequences, as fruits of "isolation and purification," constitute a patent eligible genre. Being a "product of nature" is now seen as no impediment to patent eligibility; the question is whether a sequence constitutes a new, useful, and nonobvious manufacture or composition of matter (25). Thus stood on its head is Jefferson's use of "product of nature" for the unpatentable. But the point is only semantic: since every extant thing's ingredients are naturally occurring raw materials, every extant thing may be called a "product of nature" in some sense. The

semantic point entails no practical consequence if some other provision insures the unpatentability of nature. (As defined earlier, the unpatentability of nature is the premise that to be eligible for a patent, a thing must be such that there obtains only a very low probability that without human intervention, the thing exists near the surface of the earth or on other astronomical bodies to which humans travel.) We might think that the statutory term "invent" secures the unpatentability of nature. But instead for the domain of biotechnology though not for others, we observe patent examiners and courts effectively either rejecting the unpatentability of nature or exhibiting remarkable restraint as they construe the premise. The DNA sequences that they pronounce patentable are sequences that *chromosomes of living beings contain*. The only apparent way to reconcile this with some version of the unpatentability of nature is to emphasize that a given cDNA sequence corresponding to a chromosomal sequence differs from the chromosomal sequence insofar as the chromosomal sequence is littered with introns. Still it must be said that the chromosomal sequence *includes* the cDNA sequence. That is to say nothing of the chromosomal sequence's uninterrupted transcript in the form of mRNA. Courts and patent examiners keep faith with only a weak version of the unpatentability of nature.

Given that DNA sequences are recognized as a patentable genre, whether a given sequence garners a patent turns on whether the sequence is deemed obvious. When courts first struggled with arguments about recombinant DNA technology, it seemed obvious that what was obvious was not obvious. As courts came to recognize recombinant techniques as commonplace, they bent over backwards to conclude that newly discovered cDNA sequences were nonobvious. If the prior art did not enable a method of finding a proffered sequence "with a reasonable prospect of success," a court would pronounce the sequence nonobvious. (This move vindicated a patent on the sequence encoding erythropoietin, a sequence found by screening a genomic DNA bank with two fully degenerate sets of oligonucleotide probes.) As further reasons to sustain a verdict of nonobviousness, courts have even recognized circumstances extraneous to the intellectual process of discovery and invention, including commercial success, long-felt need, failure of others, unexpected results, and the scepticism of rivals (17, p. 19). As critics would have it, one influential judicial decision saves the day for cDNA patents only by tortuously construing "obvious" so as effectively to declare patentable *per se* any DNA sequence found to encode a protein (26). According to one observer, the obviousness of many purported cDNA inventions is betrayed "in the very attitude of the persons skilled in the field. Today, if a researcher discovers a new protein and its probable properties, he usually does not publicize the information until he has found the corresponding gene. How to explain this in a community whose motto is 'publish or perish' save that it would be obvious to another research team to pick up the information, and clone the gene?" (17, p. 90). In hopes of securing future DNA patents against a tide of progress that may render ever fewer cDNA sequences nonobvious, it has been suggested that the nonobviousness requirement, to the extent not already emasculated by the aforementioned judicial decision, might be weakened. If the steeplechase jump proves too high for the average contestant, lower the bar. Where a nucleotide sequence is itself a drug, as with anti-sense RNA or the use of a DNA sequence to achieve expression without integration into the genome, it has been suggested that obviousness might be replaced with superiority over prior art in therapeutic efficacy (17, pp. 141–143, 148). Without some such move, it is urged, future application of the obviousness standard may thwart the availability of product patents on the expectation of which the biotechnological industry arose. Here an appeal is made to the biotechnologist's familiar prediction that a world without DNA patents will be a world without therapeutic innovation. But rehearsing that prediction does not provide evidence for it.

The issue remains whether, all things considered, more DNA product patents should issue. By virtue of considerations mentioned earlier, the answer may be in the negative. In such case, what might replace such patents?

Categorical Prohibitions

In the Biotechnology Patent Protection Act of 1995 (27), procured at the behest of the biotechnology industry to allow a biotechnology process claim to piggyback on a product claim, precedent was set for legislation that speaks in biological parlance about patents concerning molecular biology. Despite the supposed protean generality of the patent conceptual scheme, the door has been thrown open to discipline-specific rules. What most commonly seems to flow through that door is a stream of *ad hoc* prohibitions, usually categorical and often embracing uses as well as monopolies on uses. One favorite in the U.S. Congress and the European Parliament is a moratorium on a given sort of patent or research. Sometimes the rationale for a prohibition will appear to assimilate a property interest in an extracorporeal molecule to invasion of personal autonomy, which earlier we found reason to distinguish. As the portion of the human genome claimed by patents expands, motivation arises to prohibit any more such patents. Were a ban imposed, the control experiment of life without patents would run in real time. Biotechnology firms would compete with no intellectual property save for trade secret protection of whatever they managed to keep secret. Such competition might produce salutary results. It might also diminish aggregate welfare unless some mechanism replaces at least some of the incentives fostered by product patents.

Compulsory Publication of DNA Sequence Data to Thwart Patents

When without first filing for a patent, a scientist publishes a DNA sequence, no one may obtain a patent on the sequence. Mere citation of that publication as prior art will spike anyone's claim that the sequence is "new." Mindful of this, some scientists acting of their own volition and other scientists acting in compliance with funding mandates have promptly and systematically released DNA sequence data as discovered. A concerted effort so to publish could

thwart most new DNA patent applications. Thereupon it becomes open season for any and all to explore therapies predicated on all unpatented portions of the genome. The benefits of expanding the universe of potential investigators would seem apparent. To the extent that research motivated by profit may contribute applications that might not flow from academic laboratories, incentives must now be sought elsewhere.

Subsidies

When a public good is underprovided, as is familiarly the case in perfect competition (e.g., as to education and national defense), government may step in to provide it. Valuable public ideas are intellectual public goods. Suppose then that no further patents issue on DNA sequences other than artificial sequences. Instead, the government systematically subsidizes biotechnological research. Subsidies are awarded not only to academic institutions but to nonprofit biotechnology research centers. Specifically organized for the pursuit of applied as well as basic research, these centers tackle applications that might not be pursued, or pursued with less zeal, in academic laboratories. This scheme could implement coordinated decisions, reached with benefit of expert extramural advice, concerning which fields of fundamental biomedical and biotechnological research should be pursued and to what extent. The scheme entails substantial expenditures and may importune taxes earmarked for research (20). But the subsidies assure that society gains the benefit of valuable innovations.

Were it widgets that society sought to encourage, subsidies for institutional laboratory research might not succeed in coaxing the same innovations as would market incentives for entrepreneurs experimenting in their shops. When the desired innovations are biotechnological, it happens that academic laboratories constitute society's most fertile source of ideas. What academic laboratories do not pursue by way of applications may be pursued in the research centers. Were a share of sales revenues promised to any laboratory originating an end product, market incentives could also be brought to bear within both academic laboratories and research centers.

In the marketplace, with valuable discoveries being contributed to the public domain and available for exploitation, firms would now compete less on the basis of their discoveries and more on their efficiency in production. As with any subsidy, it may be difficult to ascertain whether the extent of biotechnological innovation induced is optimal. But one could at least compare the extent of technological innovation during the present era of product patent availability with the extent of technological innovation under a new regime.

Exclusive FDA Approval for a Term of Years

A government may also engraft an incentive mechanism upon the process by which, with a view to public safety, the government grants approval for the sale of medical products and devices. The Orphan Drug Act (28) affords a model for such an incentive scheme. According to that statute, if the FDA grants a manufacturer approval to sell a drug targeted at a disease that affects fewer than 200,000 persons in the United States, or whose likely sales cannot reasonably be expected to recoup the costs of development, the agency must refrain, for seven years after such approval, from approving sale of the drug by anyone else for use against that disease. Routine delay in obtaining FDA approval for any drug affords to the first party who gains FDA approval some period of *de facto* postapproval protection against imitators; in respect of an orphan drug, the first party to gain approval enjoys seven years of *de jure* postapproval exclusivity. The orphan drug scheme is not without its complications. For purposes of identifying which compounds are blocked for seven years by an approved orphan drug, it has been necessary to define what constitutes "the same drug." The FDA defines a new drug to be the same as a previously approved orphan drug if the new drug has the same "principal molecular features"—unless the new drug is "clinically superior" to the approved orphan drug (29). This seems to rehearse, though with variations, a judicial patent doctrine that a claimed invention is obvious if the prior art includes a structurally similar compound—unless the claimed invention possesses an unexpected property (17, pp. 145–148).

This incentive scheme could be extended. From orphan drugs it could be extended to any genre of products that seem likely to serve the public interest—indeed to any and all biotechnology products. To specify the genre of products for exclusive approval, the government could rely on advice from extramural scientific panels. Such a scheme would spare the costs, burdens, and uncertainties of patents. It would reward the development of valuable products without tying up the human genome with property claims. It would respect the unpatentability of nature. The number of years and other terms of the exclusive sale privilege are of course variable. One might also replicate the provison of the Orphan Drug Act that allows the FDA to approve sale of an orphan drug by a second applicant if the original manufacturer "cannot assure the availability of sufficient quantities of the drug to meet the needs of persons with the disease or condition" (30).

Human Methods Patent

Ambivalence between patents on products *vis-à-vis* patents on processes has been evident since recombinant DNA technology began. The Cohen-Boyer patent protects a process. It was followed, as the technology developed, by many process as well as product patents (31). The Cohen-Boyer application also sought claims on recombinants, but no product patents issued until 1984 (on plasmids) and 1988 (on plasmid-transformed hosts). Stanford University's licensing of the process patent thrived beginning in 1981 before Stanford acquired any product patent (32).

A possible resolution of the ambivalence would be to require hereafter that to secure a patent pertaining to a DNA sequence, one must invent some new process that can be performed in respect of the sequence rather than claim to have invented the sequence or its gene product. A new form of patent predicated on this principle has been proposed (14) in the following statutory phrasing:

There shall be allowed a patent pertaining to a human life form (a "human methods patent" or "*HMP*"), the scope of which patent shall not exceed the least inclusive description of an ingenious process. Such a process may consist in the production, use, alteration, amplification, or attenuation of human life forms outside the human body. An *HMP* may include an additional claim on nonhuman reproduction of any transgenic and its progeny if and only if (a) the ingenious process produces such transgenic, (b) such transgenic produces a human life form, and (c) without reference to such human life form, the process would not be patentable.

No product patent shall be allowed on a human life form or anything in which it is included. The foregoing shall not preclude a patent on a synthesized, fully explicated nucleotide sequence or protein that is not present, consecutively or otherwise, in the human body.

Research in a nonprofit institution for nonprofit purposes shall be exempt from any claim of infringement.

The significance of an *HMP* may be made more clear by the following observations.

(i) Interspecies homology is only similarity to a degree, not identity of nucleotide sequences. Absent evidence of identity with a nonhuman form in a given case, "human life form" may be assumed distinct.

(ii) The confinement of an *HMP* to the least inclusive description of an invention protects against the detriment of overbreadth as illustrated by experience with erythropoietin and factor VIII:C. Such limitation would depart from the law's tendency to allow contributors a claim on a whole—as when a farmer obtains a claim on grain in an elevator with which the farmer's is commingled, or a security interest in a part attaches to a mass in which the part is commingled or assembled. Reasons for parsimony obtain concerning the human genome.

(iii) Suppose that an *HMP* claims "a method for obtaining DNA sequence $b_1, b_2, \ldots b_n$ from genomic DNA as follows: ...," and describes the ingenious method by which the sequence was discovered. Without more, such a patent would afford little protection. Everyone may now read the sequence disclosed by the patent. Free of infringement, anyone may then obtain the sequence by employing any process, including the polymerase chain reaction, other than the patented process. To avoid this vulnerability, the discoverer might seek to claim "cloning of the sequence in vector v and transformation by v of host h that results in production of protein p_i as follows: ...." Perhaps this investigator has ingeniously devised a way to use a new v to produce p_i in some mammalian h never before used to produce human substances. In general, it will not be ingenious to clone an identified sequence, nor to produce a protein by means of a known gene. The principle of least inclusiveness allows a claim on only so much of the process as is ingenious. The discoverer's successors may find it unnecessary to use the process first used to discover the thing, and may proceed to "event around" the process. This is true about the discovery of any natural thing. Successors may also invent methods by which to use, alter, or promote or attenuate the effect of the thing.

Process patent opportunities still await—in protein chemistry, insertion of foreign DNA, transformation and infection, gene expression, and protein-manufacturing techniques. One process might describe a technique for making a protein, another how to use it. Or a firm might use knowledge of a gene not to produce but to curtail the effect of a given protein, including a newly discovered protein.

(iv) It may be possible to state certain minimal conditions for work to be ingenious. If a claimed method is predicated on a human life form, it may be unlikely to exhibit an advance over present knowledge unless the life form is fully explicated. "Fully explicated" entails, in the case of DNA, that a specific nucleotide sequence (the "explicated sequence") is identified, including all regulatory sequences necessary for any exons in the explicated sequence to be transcribed into RNA and for a gene to be expressed, that all such sequences have been inserted into a vector or maintained in some stable form, that it is known what the explicated sequence encodes or regulates (or perhaps only that it is implicated in the etiology of a disease), and that the process succeeds in expressing or preventing expression of such gene. For a protein, full explication would embrace biological function, amino acid sequence, and encoding gene sequence.

(v) One could circumvent a patent thus far described if, for example, one were to pay a royalty in order to perform a patented transgenic process of producing a human hormone in a pig, and then, without paying any more royalties, one were to breed a line of pigs. Thereby one could obtain copious amounts of the hormone. Natural breeding of course produces naturally occurring progeny, and, except for plants in the United States, such progeny would seem unpatentable. To prevent the foregoing circumvention, which would defeat an inventor's reasonable property expectations, the *HMP* allows a claim on growing or nonhuman breeding of a transgenic if the transgenic is a result of the invented process and the transgenic produces some human life form without reference to which the process would not be patentable. The additional claim may be defended as a claim on reproduction of an "unnatural" organism, one not likely to be found in nature. Such scheme resolves the predicament to which the self-reproducibility argument for the Harvard mouse patent is directed. It allows no claim on a human life form itself. Nor may the additional claim encompass human reproduction. Should the invented process happen to be one of artificial human reproduction, remedies may be provided (as discussed in the next section) against infringing providers, but never against a parent or child as such.

Whether an "artificial gene" or the protein it encodes will qualify for an *HMP* is contingent on how close a variant or equivalent the gene may be to what is found in the human genome. Will this contingency discourage fruitful research on the genome? Significant disincentives seem unlikely unless firms so greatly prefer product to process patents that they choose to pursue the more difficult task of sequencing proteins rather than finding naturally occurring genes encoding for proteins. Where proteins may be sequenced automatically, a disincentive may occur. But if the therapeutic value of an artificial gene product is not sufficient, it will not be an appealing product no matter what the patent availability. At least the products of naturally occurring genes have known worth.

Objections to the *HMP* and replies thereto include the following. Industrialists preferring product patents often contend that recombinants are more potent and free of contaminants, that recombinants thus differ from natural isolates and from each other, and hence that product patents will not prevent new advances from reaching the market. This conjecture seems belied by the history of erythropoietin and factor VIII:C in particular, and in general by the hegemony of any product patent over improvements. Whichever industrialist happens to be first in time will often hoist another on the petard of contradiction. In scientific publications and in advertising, sellers of recombinants are wont to describe their products as virtually identical to the corresponding natural isolates. But when forced to defend against a claim of patent infringement, the same sellers may be heard invoking the "reverse doctrine of equivalents," which, under U.S. patent law, excuses some literal infringements if the accused product displays differences in specific activity and purity from the patented product. (The doctrine of which this is called the "reverse" sustains a claim of infringement against an accused protein somehow differing from a patented protein if no functional differences obtain between them.)

A more orthodox industrialist objection to the *HMP* would be to say that without product patents, businesses will not invest the millions of dollars needed to find a gene and to produce a protein by recombinant methods. This bluff is handy because it is counterfactual. As earlier indicated, when one looks at the relatively scant evidence of inventive behavior without patents, and then conjectures about what happens if only process and not product patents are available, one may be sceptical about the claim that biotechnology cannot thrive without product patents. The effective protection afforded by process patents depends on how easy it is to design around a process. Large, complex proteins found in humans may be more difficult to design than, say, pharmaceuticals. Biotechnology patents are replete with process claims. It appears that firms have found ways to protect their intellectual property even though patent examiners vary in their view of product *vis-à-vis* process claims, and even though, given how often courts invalidate them, the status of any product patent is contingent. It must be granted that process patents are often less convenient to enforce because a patentee must show what transpired in a rival's plant. Even so, if a patent has been issued on a recombinant process, ordinarily the recombinant result has only a very low probability of naturally occurring. The patent holder may invoke this probability to refute a defendant's claim to have bred transgenics without using a patented process and without using offspring of the patented process.

One previous motivation for a U.S. product patent is now obviated. When the Harvard mouse emerged, anyone could avoid infringement of a U.S. process patent by performing the patented process in a foreign country outside the reach of U.S. law and then importing the end product into the United States; no such move would defeat a product patent. A statutory amendment changed this by declaring that any such importation is an infringement of the process patent (33). By virtue of the Biotechnology Patent Protection Act, one may obtain a process patent on a recombinant process that uses or makes a patented product, although this piggyback rule will be moot if product patents become unavailable. Instead of this piggyback rule, it might better be declared that a patent is available on an invented process if what the process uses or produces would be patentable but for the fact that the product is a human life form. Such is the effect of the *HMP*. It allows a process claim to be predicated upon a human life form while allowing no claim on the life form itself.

It remains necessary to show an ingenious process. An industrialist may object that there seem to be few new processes to invent, that current biotechnology employs standard processes that differ only by genes expressed. Mere substitution of a different gene in a known process may indeed be perfunctory. It would not seem to state an argument for product patents to say that innovation is difficult. Opportunities for process innovations abound. The Cohen-Boyer patents expired in 1997. It may simply be that the challenge of finding genes commands more attention at present.

The *HMP*, subsidies, and a period of exclusive FDA drug approval could be implemented separately or together.

ANCILLARY MECHANISMS

Compulsory Licensing

A patent subjects society to the vagaries of a monopolist's choices and fortunes. A possible protection against such risk with respect to biological patents is compulsory licensing according to which anyone may use a patented process upon payment of no more than some reasonable royalty. Another protection is ceilings on the prices of goods made by patented processes. As early as the federally supported Cohen-Boyer research, the NIH considered seeking patents on funded innovations. NIH asserted patent rights to AZT based on the research contributions of NIH intramural scientists, all with the declared purpose of restraining prices of products. This prompts the suggestion that a government agency other than the patent office be empowered to determine what events trigger, and the royalty rate of, a compulsory license established as a condition of any biotechnological patent. A further condition might empower the agency to set maximum prices on goods produced and processes performed in the practice of the patent. An ideal scheme would foster commercial incentives and allow a reasonable return on investment while preventing exorbitant prices.

Such a scheme, it may immediately be objected, would interfere with markets. The industrialist might contend that governments should not restrain returns on genetic inventions since they do not restrain prices of patented artificial hearts or organ transplants. One might reply that when a government grants the privilege of selling a drug or medical product, or of enjoying a monopoly on anything importantly related to human health, the public interest may justify conditioning the privilege on end product price restraint. Compulsory licensing would also protect against disasters with respect to things other than price. As earlier noted, the patentee of the sole therapy for a serious disease could become bankrupt or for other reasons decline to practice or license the invention. The common weal may demand that the invention be available. The industrialist's appeal to the case of an organ transplant does not provide a persuasive counterexample against a compulsory license because organs are donated and recipients pay only for services. An artificial organ is not perfectly analogous to a gene since the organ lacks person-defining genetic information. In any case there may be good reasons to interfere concerning any commerce in human parts.

Expert Guidance

A U.S. patent is only presumptively valid. Since courts often invalidate patents, no one knows for sure that a patent is valid until and unless it is upheld in court. Consider how numerous are the courts within the sovereignties that comprise the international biotechnology market. Trial courts decide only questions placed before them by a flow of cases that is nearly stochastic. The same is true for appellate courts on which depend the prospects of resolving conflicts among trial courts. In contrast to scientists for whom dialogue is a way of life, judges of different courts do not, as a matter of decorum, communicate with each other on pending cases. The science on which they rule is also limited to that practiced a few years, if not a decade, before trial. This obtains because time of invention is the reference point for what is obvious. Hence judicial decisions provide uncertain guidance about patentability of today's scientific processes. Moral issues, as we earlier saw, are not even tackled.

It seems improbable that any one word such as "nonobviousness" or "ingenuity" can bear the load of defining what is a sufficient feat to merit a monopoly. For instance, a claimed invention might be a *tour de force* of genetic engineering, even though the investigator knows neither a sequence's chromosomal locus nor the sequence's coding or regulatory function, if the investigator correctly infers that the sequence is involved, by homology or otherwise, in the etiology of a disease. To transform "ingenious" from placeholder to admission ticket, we may have to settle for a notion of family resemblance. For if ingenuity were to admit of precise definition, would anything be ingenious?

To meet the difficulty of recognizing ingenuity as science progresses, to overcome the lag between research and adjudication, and to improve upon the limited expertise brought to bear in patent adjudication, a mechanism could be confirmed for introducing scientific expertise. A government agency, otherwise involved in scientific research, could exercise authority continually to revise published standards for patenting life forms in reliance on recommendations of expert scientific panels. For purposes of judicial review, the law could preserve the practice of judging a patent by the standards in effect at the time of alleged invention. From such expertly framed standards, the biotechnology industry could obtain guidance more current and systematic than case law or statute is likely ever to be.

IMPLICATIONS OF PATENTS IN THE CLINIC

Introduction of Human Substances Outside the Germ Line

Ex vivo somatic intervention involves removing patient cells (e.g., tissue-infiltrating lymphocytes or bone marrow stem cells), growing them in culture, transferring genes into them using nonvirulent retroviruses or otherwise, and reintroducing the cells into the patient's body, not necessarily at the site of their effect. *In vivo* intervention is exemplified by the introduction of retrovirus vectors containing human genes at the site of the condition to be overcome. The ED, which would allow patents on substances isolated from the human body, would permit, while the *HMP* would deny, a patent on such cultured cells or vectors. They are ineligible for an *HMP* because they are or contain human life forms. Indeed the cultured cells are grown from the patient's. Except for attempted enhancement, the cultured cells would be unlikely candidates for "inventions" anyway. They are not intended to be innovations. The goal of therapy is to insert a normal gene. The ultimate achievement is homologous recombination. Thereby a normal gene replaces a defective one rather than entering the genome at an indeterminate locus.

Somatic interventions involve medical procedures on patients. Medical treatment, surgery, and diagnosis are not patentable in Europe (34). Their eligibility for U.S. patents has been dubious since 1862 when a patent was sought on the use of ether. It has seemed to many that it would be wrong to discourage physicians, on pain of infringement, from deploying in the relief of human suffering the most efficacious procedures they can muster under the exigencies they face. Hence one might conclude that the only processes of somatic intervention that may qualify for an *HMP* are ancillary laboratory processes. Similarly might patents be confined to laboratory processes with respect to tissues or organs grown in cell culture — especially if, as may be typical to avoid rejection, the cultured cells are grown from the patient's. Opportunities for process innovations would appear abundant when one apprises present difficulties in somatic cell therapy and the challenge of growing tissues and organs.

A contrary moral view might be that the foregoing is too generous. Suppose that one opposed patents on reproduction of any sort. One might assimilate the culturing of cells to reproduction, thereby reversing the patent law's assimilation of reproduction to manufacture. One might add that a laboratory process ancillary to a medical treatment is indistinguishable for these purposes from the treatment. The contention that human reproduction cannot be an infringement does not entail any claim about what is human reproduction. One might conclude that a patent on growing cells outside the human body does not threaten any patient's autonomy so long as there is no claim on the cells themselves. The difficulty of developing successful methods of somatic cell therapy,

and of cultivating tissues and organs, suggests the benefit of patent incentives. One need not claim that laboratory processes ancillary to medical procedures are in general nonmedical. One need only allow some of them to be patentable.

Human Germ Line Intervention

Germ line intervention affects reproduction in two ways. (a) It alters the genome, an offspring's complement of genes that appear in all cells including the gametes. (b) In order to achieve (a), it is performed before germ and somatic cells of an individual differentiate, i.e., on zygotes and early stage embryos. A moral objection might be lodged against a patent on any such method because of these links to reproduction. As noted, the ED would allow no patent on any method of human germ line intervention. Again a reply may be that collective benefit could result from creating patent incentives on certain laboratory processes. It is also noteworthy that a patent on gene therapy would not be a patent on *in vitro* fertilization. Therapy is subsequent to fertilization. The choice to conceive may be seen as a different choice than the choice whether to intervene genetically for the health of a child whose conception has been chosen, even if the former is contingent on the latter. On the other hand, the opposite may be the case if eggs fertilized *in vitro* are screened for genetic defects or traits, thereby exercising a choice of which shall live. Two *in vivo* methods also merit mention. One consists in altering an embryo *in utero* by retroviral infection. Another consists in causing adult testes or ovaries to produce genetically engineered gametes (35). A requested European patent on the latter technique was criticized as contrary to *l'ordre publique* or morality (36). For these also one may ask whether the prospect of collective benefit suffices to warrant property claims on ancillary laboratory processes of medical procedures.

If government grants patents on any germ line interventionary process, does that comport with the stance that human reproduction cannot be infringement? The answer lies in stipulating that no remedy will lie against a parent or child as such. Damages and preconception injunctive relief could be made available against unlicensed providers of patented processes. If Mr. and Mrs. Thurston, learning of Mendipulate Inc.'s patented technique for germ line manipulation, arrange with their physician for the technique but no one pays the royalty, a damage remedy may lie against the providers. We can scarcely imagine a suit by Mendipulate against Mrs. Thurston, her daughter or granddaughter, or their physicians or hospitals, complaining of the conception of a child, not to mention injunctive relief, i.e., an order for an abortion. Mere pragmatism makes clear that Mendipulate's interests require no remedy against a patient. Drug manufacturers do not sue patients who infringe by "using" an infringing drug. They sue rival manufacturers and distributors who "make" and "sell" the drug in quantity.

Mendipulate may protest that if it cannot obtain a product patent, every Thurston descendant will benefit from Mendipulate's invention without paying for it. Mendipulate is correct that the *HMP* allows claims on reproducing the progeny of transgenesis only for nonhuman reproduction. But consider that Mendipulate will advertise a patented process of germ line therapy as a method to remedy a genetic defect. It cannot tenably assert that if it had a product patent, many Thurston descendants would become good-paying customers when they inherit the defect! Moreover, whether the process is therapy or enhancement, Mendipulate's twenty years of monopoly will run before any transgenic Thurston reaches adulthood. Mendipulate may still complain that if Mr. or Mrs. Thurston undergoes a patented Mendipulate process that causes them to produce genetically engineered gametes, no more compensation will be gotten by Mendipulate if the Thurstons have a dozen children than if they have one. This of course overlooks the difference between having children and copying a patented contraption for profit. People are not motivated to have children because they can copy a gene for free. Mendipulate may anticipate fecundity when it prices the royalty for its laboratory process.

Since interventions will be performed by physicians, enforcement of a process patent will require showing what happened in the doctor's office. To Mendipulate this will seem inconvenient. It would prefer a product patent whose infringement it could establish by comparison of parental and progeny DNA. Such a comparison would be peculiar, to say the least, as it would be mustered in support of a complaint that a child is healthy or possessed of some enhancement. It should suffice to protect Mendipulate that licensed specialists may generally be expected to pay royalties on patented processes. What would be troublesome would be the enterprising move of a patient who sells gametes that contain altered genes. This concern may be minimized for the moment by realizing that only enhancement genes, not corrected disease-causing genes, would be likely to be marketable.

Society might deem the collective benefit of enhancement to be insufficient for allowing a patent. If concerns about playing God and discrimination prevail, refusing patents on enhancement would be a means to discourage the practice. A contrary view might be that if we demarcate certain interventions to be outside the physician's armamentarium for maintaining health, no public policy will be disserved by a patent.

There remains possible a product patent on a synthetic gene nowhere found in humans. To use such a gene might depart from the present vision of installing normal in lieu of defective genes. The prospect of such departures no doubt explains the habitual mention of Frankenstein when observers discuss germ line intervention. Regardless, the immunity of parents and children as such from claims of infringement would control. The inventor of a human genetic intervention surrenders the product of the process for integration into an unownable being. If a process alters an early stage embryo, integration occurs into a human in gestation. If alterations are made in gametes or the means of their production, integration occurs into the body of the patient.

CONSISTENCY OF POLICY FOR PLANTS AND ANIMALS

Unless policies about forms of life evince a consistent understanding of innovation and reflect generalizable moral principles, a stable consensus seems unattainable. Conditions (*1*) and (*2*) above stated for a biotechnology product patent — low likelihood of finding the claimed invention in nature, and ingenuity — appear applicable to any life form patent. Some transgenic plants and animals may be improbable of natural occurrence and recognizable as the products of ingenuity. Others may possess transgenes from members of their own species for which the odds of acquisition by mutation are better than trivial, or as to which the process of transgenesis is not ingenious. Where a product patent would be unwarranted, a process patent could be available. As may an *HMP*, a process patent could claim a process by reference to an identified plant or animal life form. It could add a claim on the breeding of any plant or animal that the patented process produces and without which the process would not be patentable. Such an additional claim would obviate the self-reproducibility rationale for a transgenic product patent.

One may argue for bounding a patent's enforceability by operation of a "farmers' privilege," a derogation imposed for plants in the United States and often proposed for animals there and in the ED. This permits a farmer to breed patented animals to the extent needed to replenish stocks on the farm, or to plant seeds generated by transgenic plants grown on the farm. A farmers' privilege would avail a typical farmer who does not seek to compete with breeders in selling varieties as such but who wishes to sell what is raised on the farm. The derogation would entail that, as Mendipulate must do concerning the Thurstons, commercial breeders must collect their royalties on the first generation.

ACKNOWLEDGMENT

This article is adapted from the author's *Norms for Patents Concerning Human and other Life Forms* (14).

BIBLIOGRAPHY

1. P.J. Federico, *Science* **86**, 327 (1937).
2. Directive 98/44/EC of the European Parliament and of the Council of 6 July 1998 on the Legal Protection of Biotechnological Inventions, *Off. J. Eur. Commun.* **L213**, 13–21, Article 6.1 (1998).
3. M.A. Ryan, *J. Med. Philos.* **20**, 419–438 (1995).
4. Council of Europe, Convention for the Protection of Human Rights and Dignity of the Human Being with Regard to the Application of Biology and Medicine, Eur. Treaty Ser. 164, Article 13.
5. W. Gardner, *J. Med. Philos.* **20**, 65–84 (1995).
6. *Diamond v. Chakrabarty*, 447 U.S. 303 (1980).
7. *Ex parte Allen*, 2 U.S.P.Q. 2d 1425 (1987).

8. U.S. Pat. 4,736,866 (April 12, 1988), P. Leder (to the President and Fellows of Harvard College).

9. 35 U.S.C. §101.

10. European Patent Convention Article 53.

11. M. Tallacchini, *Bibl. Della Liberta* **33**(147), 21–50 (1998).

12. M.S. Greenfield, 44 *Stanford Law Rev.* 1051–1094, 1058 (1992).

13. J. Bentham, *A Manual of Political Economy* in J. Bowring, ed., *The Works of Jeremy Bentham*, vol. 3, Russell and Russell, New York, 1962, p. 71.

14. L.M. Guenin, *Theor. Med.* **17**, 279–314 (1996).

15. U.S. Pat. 4,237,224 (December 2, 1980), S. Cohen and H. Boyer (to Trustees of Leland Stanford Jr. University).

16. B.D. Davis, *Science* **193**, 442 (1976); *Am. Sci.* **65**, 547–555 (1977); *The Genetic Revolution*, Johns Hopkins University Press, Baltimore, MD, 1991, pp. 21, 243.

17. P.G. Ducor, *Patenting the Recombinant Products of Biotechnology*, Kluwer, London, 1998.

18. *Amgen v. Chugai Pharmaceutical Co., Ltd.*, 706 F. Supp. 94 (D. Mass. 1989); 13 U.S.P.Q. 2d 1737 (D. Mass. 1989); *aff'd in part and reversed in part*, 927 F. 2d 1200 (Fed. Ct. 1991), *cert. denied*, 112 U.S. 169 (1991).

19. *Scripps Clinic and Research Foundation v. Genentech, Inc.*, 666 F. Supp. 1379 (N. D. Cal. 1987); 678 F. Supp. 1429 (N. D. Cal. 1988); 707 F. Supp. 1547 (N. D. Cal. 1989); *aff'd in part and reversed in part*, 927 F. 2d 1565 (Fed. Cir. 1991).

20. M. Svatos, in S. Sterckx, ed., *Biotechnology, Patents and Morality*, Ashgate, Aldershot, 1997, pp. 291–305.

21. Y. Miki et al., *Science* **266**, 66–71 (1994).

22. D. Dickson, *Nature* **378**, 424–425 (1995).

23. R. Wooster et al., *Nature* **378**, 789–792 (1995).

24. 35 U.S.C. §101–103; European Patent Convention Article 52 (expressing the last two criteria as "industrial application" and "inventive step," and, corresponding to the notion of "manufacture," excluding "discoveries" of nature).

25. *Merck & Co., Inc. v. Olin Mathieson Chemical Corp.*, 116 U.S.P.Q. 484 (4th Cir. 1958).

26. *In re Deuel*, 34 U.S.P.Q. 2d 1210 (Fed. Cir. 1995), discussed at length in (17), pp. 70–90, 141–142.

27. P.L. 104-41, adding 35 U.S.C. §103(b).

28. 21 U.S.C. §§360aa–360cc.

29. 21 C.F.R. §316.3 (13).

30. 21 U.S.C. §360cc.

31. D.W. Plant, N.J. Reimers, and N.D. Zinder, eds., *Patenting Life Forms*, Cold Spring Harbor Lab., Cold Spring Harbor, NY, 1982, pp. 75–78, 95–106.

32. U.S. Pat. 4,468,464 (August 28, 1984) and 4,740,470 (April 26, 1988), S. Cohen and H. Boyer (to Trustees of Leland Stanford Jr. University). See N. Reimers, *J. Assoc. Univ. Technol. Managers* **7**, 25–47 (1995).

33. 35 U.S.C. §271 (g) added by P.L. 100-418.

34. European Patent Convention Article 52 (4).

35. R. Brinster and M.R. Avarbock, *Proc. Natl. Acad. Sci.* **91**, 11303–11307 (1994).

36. *Nature* **368**, 572 (1994).

See other entries HUMAN GENOME DIVERSITY PROJECT; MEDICAL BIOTECHNOLOGY, UNITED STATES POLICIES INFLUENCING ITS DEVELOPMENT; OWNERSHIP OF HUMAN BIOLOGICAL MATERIAL; see also PATENTS AND LICENSING entries.

PHARMACOGENETICS

MICHAEL M. SHI
MICHAEL R. BLEAVINS
Warner-Lambert Company
University of Michigan Medical School
Ann Arbor, Michigan

OUTLINE

INTRODUCTION

Pharmacogenetics is a new and quickly evolving scientific discipline that studies how genetic differences determine an individual's response to therapeutics. Recent advances have shown that many drug-metabolizing enzymes have genetic variations in expression and regulation. When these genetic variations affect the enzyme's function significantly, different clinical outcomes can occur among people exposed to a particular drug. Pharmacogenetics can help to individualize dosing regimens, thereby maximizing a drug's therapeutic effect and minimizing toxicity. The rapid development of genotyping as a molecular diagnostic tool nevertheless raises ethical, legal and policy issues, as have been the case with other DNA-based testing. Although pharmacogenetics shares concerns with other genetic research in clinical practice, this discipline also has unique objectives and goals. It therefore is important to address the issues associated with this emerging discipline. The purpose of this article is to describe the concepts and advances in pharmacogenetics, discuss social implications of this field and provide recommendations in this area.

PHARMACOGENETICS

Genetic Variations in Drug Response

Significant variations in drug response exist among both populations and individuals. These variations can be due to genetic and/or environmental factors. Age, gender, body size, diet, alcohol or tobacco consumption, pregnancy, kidney or liver dysfunction, concurrent disease states, and drug interactions can all modify the bioavailability, distribution, protein binding, metabolism and excretion of drugs (1). Interindividual variations in drug response also result from genetically based differences in drug metabolism. Individual differences in the absorption, distribution, metabolism, and excretion of therapeutics can alter the effects of a given dose, leading to a spectrum of responses ranging from clinical benefit to adverse effects or therapeutic failure. When taken up by the human body, foreign compounds typically undergo metabolism via Phase I and II enzyme reactions (2). Phase I reactions involve hydrolysis, reduction, and oxidation. These enzymes introduce a functional group (i.e., hydroxyl, amino, sulfhydrol, or carboxyl) to the compound, usually increasing water solubility. Within the category of Phase I enzymes are the cytochorome P450 multigene family, AND(P)H quinone oxidoreductase, and aldehyde dehydrogenase. Phase II metabolism includes glucuronidation, sulfation, acetylation, methylation, glutathione conjugation, and amino acid conjugations. Most Phase II reactions dramatically increase hydrophilicity, thus greatly enhancing the excretion of foreign compounds. Phase II enzymes include UDP-glucuronosyl transferases, glutathione-S-transferases (GSTs), sulfotransferases, catechol-o-methyltransferases, phenol-o-methyltransferase, and thiol methyltransferases (3). Recent research has identified functionally important variations in most Phase I and II enzymes, which can lead to different metabolic profiles among individuals.

Genetic Polymorphisms of Drug-Metabolizing Enzymes

Genetic polymorphism refers to distinct traits derived from a single gene that exists in more than one form. These polymorphisms are transmitted from generation to generation, sometimes with striking differences in allele distributions among different ethnic groups (4). A majority of the genetic polymorphisms do not affect protein function and therefore have no phenotypic importance. Other polymorphic genes encode drug-metabolizing enzymes with dysfunctional or nonfunctional activity. As a result the subgroup of the population with the genotype(s) will metabolize drugs that are eliminated via this pathway differently from individuals with the normal (wild-type) genotype. A classic example of genetic polymorphism is

class 2 aldehyde dehydrogenase (ALDH2). Approximately 50 percent of the Asian population has a single amino acid change of Glu^{487} to Lys^{487} in this enzyme, causing impaired acetaldehyde metabolism (5). These individuals can rapidly convert ethanol to acetaldehyde but only slowly metabolize acetaldehyde to acetic acid. Affected people experience of flushing syndrome after alcohol consumption due to the release of catecholamines triggered by the sustained high blood acetaldehyde levels, which does not normally occur in people with the fully functional form of the enzyme. Pharmacogenetics emerged as a discipline to study genetic variations in drug-metabolizing enzymes that may determine an individual's responsiveness to therapeutic agents (6,7). Advances in this area have important clinical implications and practical values for the design of dosing regimens. It is important to recognize that the genetic variations in drug metabolism can lead to significantly different therapeutic responses, including either low or exaggerated pharmacological effects or side effects.

An increasing number of drug-metabolizing enzyme polymorphisms have been identified in recent years (8). Examples of Phase I and II enzymes with functionally important polymorphisms are listed in Table 1 (9–22). A well-characterized drug metabolizing enzyme with functionally important variants, cytochrome P450 2D6 (CYP2D6, debrisoquine hydroxylase), will be highlighted. The genetic polymorphisms of CYP2D6 are perhaps the most well-established alterations with known clinical significance. This microsomal isozyme is responsible for the oxidative metabolism of approximately 50 clinically important drugs of varying therapeutic classes (2). Its substrates include widely used antiarrhythmics, tricyclic antidepressants, β-adrenergic blocking agents, neuroleptics, and other classes. These drugs frequently have narrow therapeutic windows, meaning that slightly lower than targeted plasma concentrations will not have the desirable therapeutic effect while only somewhat higher concentrations cause toxicity.

CYP2D6 has three clinically distinct phenotypes: (1) the normal (or extensive) metabolizers, (2) slow (or poor) metabolizers, and (3) fast (or ultraextensive) metabolizers. The same dose of a drug metabolized via CYP2D6 will result in plasma concentrations that vary greatly among these individuals. In normal metabolizers, steady-state plasma concentrations will normally fall within the desired therapeutic range and toxic effects will be non-existent or minimal. In fast metabolizer individuals, steady-state drug levels will be below therapeutic concentration and this group of patients is unlikely to respond to standard treatment regimens. It has recently been reported that the cholesterol-lowering drug simvastatin did not reduce plasma lipid levels in CYP2D6 fast metabolizers at standard doses (23). While it may be possible to increase the dose given to these patients and achieve the same therapeutic effect as normal metabolizers, this approach also increases the potential for undesirable side effects, particularly those not related to CYP2D6 metabolism. In slow metabolizing individuals, plasma drug concentrations will be significantly above therapeutic levels when

Table 1. Example Polymorphic Enzymes in Drug Metabolism and/or Disease Susceptibility

Enzymes	Enzyme Reaction	Phenotype
CYP1A1	PAH oxidation	FM associated with lung cancer in smokers (9)
CYP2C9	Tolbutamide hydroxylation	SM for tolbutamide (10)
CYP2C19	S-Mephenytoin hydroxylation	SM for mephenytoin and other drugs (10)
CYP2D6	Debrisoquine hydroxylation	SM and FM for over 50 clinically important drugs (11)
CYP2E1	Chlorzoxazone hydroxylation	Associated with lung cancer (12)
ADH2	Ethanol metabolism	SM for alcohol metabolism (5)
NQO1	Quinone reduction	Associated with urological (13) and lung cancer (14)
GSTM1	Conjugation of epoxide	Gene deletion associated with lung (15) and bladder cancer (16)
NAT2	Acetylation	SM and FM for isoniazid and other drugs (17) associated with bladder and colon cancer (18)
UDPGT1A1	Bilirubin conjugation.	Deficiencies in Crigler-Najjar (19,20) and Gilbert syndrome (21)
TPMT	Methylation	Deficiency associated with mercaptopurine and azathioprine toxicity (22)

Abbreviations: FM, fast metabolizer; GSTM1, Glutathione S-transferase M1; NAT2, N-acetyltransferase 2; NQO1, NADPH-quinone oxidoreductase 1; PAH, polycyclic aromatic hydrocarbons; SM, slow metabolizer; TPMT, Thiopurine S-methyltransferase; UDPGT, UDP-glucuronosyltransferase.

conventional doses are used. In this case undesired toxicity can proceed or mask the desired pharmacological effects, and these individuals are likely to suffer adverse side effects. This is particularly true for many antipsychotic and antidepressant agents (11). In other instances, the parent drug requires biotransformation to an active form. For example, the analgesic effect of codeine largely depends on its conversion to morphine through o-demethylation, and so adequate analgesic effect cannot be achieved in CYP2D6 slow metabolizers (24). Therefore clinical practice in the future may benefit from dose individualization to avoid toxicity or achieve optimal therapeutic benefit.

The human CYP2D6 gene consists of nine exons and has been mapped to chromosome 22 (25). After transcription, the premature mRNA undergoes splicing and only the exonal region encodes the protein synthesis. The poor metabolizer phenotype occurs in 7 to 10 percent of the Caucasian population (11) and results from autosomal recessive inheritance of nonfunctional alleles (7). In addition to the wild-type gene (CYP2D6* 1), over 20 different alleles of CYP2D6 are associated with deficient, reduced, or increased enzyme activity (26). The most frequent inactivating mutation among Caucasians is the CYP2D6* 2 genotype, a splice–site mutation involving $G_{1934}A$ transition in the 3'-end of intron 3, leading to a mis-splicing of the premature transcript, and loss of enzyme activity (27,28). The CYP2D6* 3 mutation is a 1-bp A_{2637} deletion in exon 5 leading to a frame–shift change in the translation of CYP2D6 mRNA (29). The CYP2D6* 5 mutation is a deletion of the entire CYP2D6 gene (30). The CYP2D6* 2 mutation constitutes about 75 percent of all mutant alleles, with the CYP2D6* 5 mutation responsible for 14 percent and the CYP2D6* 3 mutation for 5 percent. Together these three polymorphisms account for approximately 95 percent of the slow metabolizer genotypes (29).

Detection of Genetic Polymorphisms

Standard procedure for evaluating metabolic capacity involves administration of a probe compound and measuring the ratio between the parent drug and its metabolite in urine and/or plasma. This procedure involves analytical techniques and often requires a week or more before a conclusion can be drawn. Metabolic phenotyping has additional disadvantages in that results can be influenced by sample stability and that conversion of the drug can be affected by external factors, including age, nutrition, general health, and other medications. Furthermore some poor metabolizers experience unpleasant side effects of the probe drugs (11). Therefore metabolic phenotyping is not widely used in clinical practice. These limitations can be circumvented for many enzymes by genotyping the patient at a centralized laboratory.

Genotyping is relatively easy to perform and generally requires only a sample of peripheral blood from patients. Therefore it is potentially less invasive than phenotyping and is not influenced by drug–drug or drug–food interactions. If a polymorphic site changes the DNA recognition sequence of a restriction enzyme, or if the genetic polymorphism involves a large deletion or insertion, the genetic polymorphism can be identified using polymerase chain reaction (PCR) coupled with restriction fragment length polymorphism (PCR–RFLP) analysis. In this approach, DNA sequences containing the polymorphic site are amplified by PCR, followed by restriction digestion and gel electrophoresis. PCR-RFLP tests have been developed to detect most of the CYP2D6 mutations (26). Other genotyping methods include allele-specific PCR (11), fluorescent dye-based high throughput genotyping (31,32), and the recent gene chip technology (33,34). Genotyping for genetic polymorphisms in drug metabolism can help explain drug toxicity or therapeutic failures, and help predict potential drug interactions. The addition of pharmacogenetic testing will help clinicians to manage drug therapy, especially for drugs with low therapeutic indices.

Prescriptive Medicine

Interindividual variability in uptake and metabolism of many drugs make clear dose–response relationships for these compounds difficult to predict and toxic side effects a real possibility. A dose that produces the desired therapeutic response in one individual may be toxic or subtherapeutic for another person. Therefore it would be valuable to know in advance the dose of medication to prescribe based on each individual's metabolic capabilities. Pharmacogenetic testing can provide a powerful tool for optimizing therapeutic efficacy and reducing drug toxicity for those compounds known to be metabolized via pathways with functionally important genetic polymorphisms. It has been estimated that a typical marketed drug is efficacious in approximately 20 to 40 percent of patients (35). The same substance is likely to have no therapeutic benefit for 20 to 40 percent of patients, and to cause significant side effects in 20 percent of patients (35). Adverse drug effects negatively impact the health of patients and their quality of life, as well as adding substantial financial burden to society in terms of costs and acceptance of the compound to treat disease. More effectively predicting drug efficacy and toxicity offers significant benefits to society and may be described as "prescriptive medicine." Genotyping for relevant DNA markers may help physicians prescribe the most efficacious drugs for a given individual and disease, while minimizing adverse drug side effects. Patients also could be provided an opportunity to choose among therapeutic agents, both prescribed and over the counter, to obtain suitable medication and avoid undesired side effects. If genotyping and phenotyping can predict which patients are most likely to benefit from a specific treatment, only those people need be exposed, with individuals likely to experience adverse or ineffective responses not administered the drug. In addition health care costs can be reduced, especially for drugs that must be taken for extended periods, such as hypocholesteremic, antihypertensive, and antidiabetic drugs. Once validated scientifically and proved to be cost effective, pharmacogenetics will provide significant benefits to both patients and society. In the future physicians may have the opportunity to prescribe the most effective and safest drug based on the patients' genetic blueprint.

Pharmacogenetics may also help drug makers to design therapeutic agents that specifically target patient subpopulations. A recent study (36) found that a drug given to 400 Alzheimer's patients had no statistically significant effect. However, when patients were stratified according to ApoE subtype, a clinically significant response was demonstrated. Due to the heterogeneity of human populations, genetic stratification can be the difference between a drug's success and failure. Pharmacogenetics thus can assist in identifying safer and more efficacious drugs by targeting subpopulations of optimal responders and predetermining those most at risk of undesirable reactions. It also holds great potential for accelerating the drug discovery process by providing clearer answers, as well as reducing the length and cost of clinical trials.

Ecogenetics and Preventive Toxicology

Ecogenetics is a broader definition of interindividual differences in response to environmental toxic chemicals (37). Just as in the metabolism of therapeutics, some slow metabolizers might detoxify environmental or occupational agents significantly slower than normal populations. By the same token, fast metabolizers may more readily activate some foreign agents to their toxic intermediates. Therefore certain allelic forms of drug-metabolizing enzymes could render an individual either more sensitive or more resistant to the toxic effects of specific classes of foreign compounds. For example, molecular epidemiology studies have identified associations between specific genotypes of CYP1A1, CYP2E1, and GST-M1 with a variety of cancers (9,12,38). By defining these susceptibility genes, those people at increased risk can be advised to avoid certain exposures and safer standards established for workers and the public. An informed decision to avoid exposure to some occupational hazardous materials could be made by individuals whose genotypes had been associated with cancer or diseases. Clearly, there are also uncertainties and controversies in the use of these genotyping and association studies. For example, an association between CYP2E1 DraI genetic polymorphism with increased lung cancer susceptibility was suggested in a Japanese population (12) but was not observed Caucasian populations (39,40). This discrepancy may be due to a significantly low frequency of CYP2E1 DraI polymorphism in Caucasians (41). Clearly, more critical research has to be done to effectively use the research and epidemiological data emerging from this area of study.

ETHICAL, LEGAL, AND POLICY ISSUES ASSOCIATED WITH PHARMACOGENETICS RESEARCH

Most genetic research and tests share similar ethical, legal, and policy issues. The most common concerns involve risks of psychological distress, loss of insurance or employment, as well as confidentiality of genetic information. Since the objectives and goals of pharmacogenetics are different from other genetic research, it is important to discuss these issues separately. Up until now, only a few ethical issues have been briefly touched upon (42). There are potential controversial aspects such as informed consent, confidentiality of genetic information, sample and data ownership, potential discrimination against people identified as genetically "deficient," and access to human genetic materials and information. This section will review the current status in the genetic field and provide some recommendations for pharmacogenetics.

Informed Consent

As pharmacogenetic research requires population-based sampling for genetic variation and gene–environmental interaction studies, it frequently involves collection of large numbers of volunteer and patient samples. In order to obtain the testing materials, researchers need the informed consent of test subjects. Informed consent involves a process of education and counseling that facilitates voluntary, reasoned decision making. The prospective participants or patients must understand the purpose and the nature of the study or test, understand his/her role in that study, and be cognizant of the benefits and risks that may result from the study. The document should be comprehensive, easily understandable, and serve as the means to protect both patient and care provider.

There is a growing belief that genetic information is particularly sensitive and that some people may not want genetic information about them obtained, even in therapeutic indications. These concerns must be addressed in order to protect individual's rights, pursue important research, fulfill medical ethics, satisfy regulatory requirements, and benefit society. Significant controversy also surrounds the ethical issues associated with archived blood and tissue samples used for molecular genetic testing for either basic research or clinical parameters (43). With PCR technology, even material from archived paraffin-embedded tissue or frozen blood can yield sufficient DNA for genetic analysis. These materials have tremendous value for pharmacogenetic research, particularly when a new genetic polymorphism is identified. Researchers then have the possibility of retrospectively identifing banked samples and conducting genotyping that establishes whether or not a genotype/phenotype correlation exists. This raises concerns about invasion of privacy, loss of individual autonomy, and stigmatization if test results were released. At the same time, however, a balance needs to be achieved that allows researchers access to these human samples for improving disease treatment and test validation. Another compelling reason to use archived DNA is that some historical samples are extremely valuable in that it may not be possible to reconduct the same study or collect new samples. In addition, with the continuing progress in human genetics, these specimens may have significant value in furthering medical discoveries beyond even those currently envisioned. A recent article offers the suggestion to treat the test samples, however obtained, as anonymous by keeping the patient-specific portion of molecular genetic test results confidential from even those scientists conducting the evaluation (44). A general guideline on informed consent for genetic research has been provided by the American Society of Human Genetics (45).

Recommendations. When obtaining informed consent for research and clinical testing, it is imperative to clarify the following:

- Purpose. limitation, and potential outcomes of the research
- Methods for maintaining confidentiality of results
- Anticipated use of testing samples
- Duration of storage and disposal of the materials
- Potential for research to lead to new clinical diagnostic tests
- Final publication or disclosure of study results

When it is not possible to give informed consent to an incomprehensive patient, a legally authorized guardian or appropriate decision maker may be substituted. In addition, a general notification of potential future use of the samples should be included. If the patient does not object and if samples are coded or remain anonymous, the DNA may be used for research not specifically defined in the informed consent statement (46). The collection and storage of DNA used for genotyping should follow established guidelines for DNA databanking (47).

Confidentiality of the Research and Test Results

Confidentiality of test results has been and remains a major focus of the ethical, legal, and policy issues related to genetic testing. Pharmacogenetics provides an opportunity to observe a person's molecular genetic makeup independently of the visible characteristics. The genotyping data may reveal asymptomatic conditions that would only manifest with age, or upon exposure to specific drugs or compounds. Genetic testing therefore may allow better diagnosis for disease risk at earlier stages of life.

Pharmacogenetic testing promises to provide value in making diagnostic decisions and assessing medication risk. As described earlier, recent epidemiological studies and animal models also have identified a strong association between some metabolic enzyme genetic polymorphisms and cancer. Although specific associations exist, these studies identify risks, not certainties. In reality the development of cancer is a complicated process and depends on multiple gene–environmental interactions. Obviously more critical research has to be done to clearly establish the biochemical pathways important in carcinogenesis and the role of genetics in susceptibility. On the positive side, knowing this information may help patients prepare for the risk and adjust their work or lifestyle to minimize potential hazardous occupational or environmental agent exposure that could trigger disease. One major concern in this area revolves around the possibility that insurers and employers might regard an increased genetic risk as the final concrete outcome and use this information to establish policies that discriminate among individuals. For example, employers might propose to identify workers with lower genetic risk for toxicity or malignancy from exposure to particular occupational agents and select only those individuals to perform "high-risk" jobs. Genetic information also could be used by employers to predict health care costs or an employee's productivity. These predictions might influence hiring, retention or promotion decisions.

Similarly genotypic information could be used for insurance purposes to weed out individuals at highest disease risk. This information might be used to justify higher premiums or cancellation of policies. It therefore is very important to establish guidelines that prevent abuse of genotyping information. Researchers and clinicians must carefully consider the risks and benefits and the potential impact of genetic information for participants and others. Protection of patient confidentiality may require further protection through legislation (48).

It is advised that patients or volunteers be informed of pertinent aspects related to acquisition, storage, and use of data, as well as the degree to which third parties can obtain access. Researchers should adhere to the principle of least-intrusive disclosure, in which the data are stored using identifiers such that patient identity and sensitive personal factors connections are not possible, or where the fewest number of investigators necessary to achieve the research goal is maintained (49). A security infrastructure should be in place to ensure the confidentiality of research information, including access control, audit trails, disaster recovery, and encryption of patient-identifiable data before transmission on networks (48).

Recommendations

- Patients should be informed about storage and access to test results
- Study center should safeguard genetic information
- Disclosure of information and access to DNA samples should respect principles of privacy

CLINICAL APPLICATIONS AND ASSOCIATED ETHICAL, LEGAL, AND POLICY ISSUES

Benefits and Risks of Clinical Pharmacogenetic Testing

Recently genetic testing to predict linkage to late-onset diseases such as ApoE in Alzheimer's disease, or BRCA1 and 2 mutations in breast cancer, have resulted in substantial public debate (50). Genetic testing is complicated by uncertainties in predicting and diagnosing these diseases, and more importantly, by the social, ethical, and legal implications of disclosing genotype results.

Unlike the BRCA and ApoE genotyping, where no current therapeutic intervention is available, pharmacogenetic tests may be more acceptable to the public. The reason for this is that there are intervention strategies available, namely either withdrawal of the drug or switching to another compound belonging to a different chemical class. Based on a patient's drug metabolizing genotype, physicians also may be able to adjust the standard doses, thus achieving therapeutic value and avoiding toxicity. In addition to satisfying patients' physical well-being, substantial financial benefits may be achieved by utilizing the data obtained by pharmacogenetic testing. Incentives from managed care organizations and insurers to control health care costs may strongly support these tests. If a pharmacogenetic test accurately predicts that a commonly prescribed drug will be ineffective, or has serious adverse effects for a relative large percentage of patients, these individuals could be given preselected medications to avoid the lack of efficacy or the severe toxicity.

Predictive Value and Limitations of the Tests

Unlike molecular biology tests for pathogens, definitive or absolute results cannot be easily achieved in current pharmacogenetic testing. For example, CYP2D6 has more than 20 polymorphisms leading to altered enzyme activity. Until testing is performed for all polymorphisms, many of which are extremely rare, it is not possible to have an absolutely accurate prediction about every individual's phenotype. It also must be kept in mind that other functional polymorphisms may exist that have not yet been identified. Since pharmacogenetics is a relatively

new area, the prediction of results often must be qualified. There are only a handful of pharmacogenetic markers currently available commercially for genotyping, such as CYP2D6, CYP2C19 and ApoE (8). When used properly, they can be valuable for decisions regarding dosing, counseling and prognosis. Some general principles can be expected for clinical testing in pharmacogenetic testing:

- Testing is useful for detecting functionally important polymorphisms of drug metabolizing enzymes
- Techniques used for genotyping are relative simple and noninvasive
- Testing can be expected to benefit the patient and society
- Pretreatment genotyping might be desirable for drugs with low therapeutic indices and high toxicities
- Positive testing results will be more informative than negative ones

Standards in Clinical Testing

Due to the complexity of interpreting genotyping results and the requirement of expensive and complex laboratory equipment, pharmacogenetics is likely to be conducted primarily through service laboratories. Currently the clinical laboratory is required to establish analytic validity and Clinical Laboratory Improvement Amendments (CLIA) certification (51). Yet the rapid progress in human genome research and development in genetic testing technologies have outpaced the quality assurance and quality control for molecular diagnosis. A major concern is that a complete set of specific standards to assure proficiency for genetic testing in the clinical laboratory has not yet been developed, even though genotyping procedures are classified as being of high complexity. A recent report from the NIH–DOE Task Force on Genetic Testing recommended that clinical validity, such as test sensitivity and the predictive value of a positive test result, as well as institutional review board (IRB) approval of the protocol, should be part of the prerequirement of CLIA certification (52). In addition the Task Force suggested an external review before the genetic test can be commercially offered even after CLIA certification (53).

In the future pharmacogenetic testing is likely to fall within the competency testing currently imposed on clinical pathology laboratories as described under CLIA and Commission on Office Laboratory Accreditation (COLA) regulations. In lieu of the current situation, clinical pharmacogenetic tests should be done by laboratories that meet accepted standards of general laboratory quality assurance, including patient test management, personnel requirements, specimen handling, quality control, test validation, and confidentiality. In addition to CLIA requirements, an inspection of the laboratory's competence in performing the genetic test should be conducted by certification agencies. Interlaboratory comparisons of reference samples might be necessary to assure the quality control mechanism, a system already in place for more routine clinical biochemistry and hematology procedures.

Indications and Demands in Pharmacogenetics Testing

Although pharmacogenetic testing could have predictive value in clinical situations, it is not routinely invoked unless patients have previously had severe toxicity and the primary care providers have a knowledge of pharmacogenetics. Predictive tests, such as the association with cancer risk or late-onset genetic disorders, should not be routinely performed until additional research supports a beneficial outcome and effective treatment.

Several important questions should be asked before initiating pharmacogenetic testing. What are the benefits of conducting these tests? What is the most cost-effective way to do this? Are these tests accurate enough to predict the clinical outcome? At present, it is questionable which variant alleles should be routinely genotyped to allow a sufficiently reliable, but still practical, estimation of a person's metabolic capacity. With rising health care costs and the scarcity of resources, it is not acceptable to adopt expensive testing that does not add any value in patient care or new technologies of which the benefits are still unclear (54).

A genotyping test will only be of value to consumers when it can provide predictive value and a reasonably precise answer regarding individual risk. Currently very few pharmacogenetic tests are available that meet these standards. Although some genotyping tests are commercially available, they primarily provide services for research studies and have not been integrated into routine clinical practice. Another limitation is that pharmacogenetics as a discipline has not been integrated into the medical training curriculum and few physicians are familiar with the underlying concepts, benefits, and applications. Giving the progress in pharmacogenetics, it is easily foreseeable that it will soon become another subspecialty in medicine and pharmacy. One complexity is that there can be many genotypes that produce the same phenotype. Attempting to identify all polymorphisms associated with a defective phenotype will significantly increase testing complexity and expense. The practical approach at this time is to screen for the most common genotypes leading to altered enzyme activities. With the rapid development of modern technology in genetic diagnosis, it will be possible to detect multiple polymorphisms in a single test. For example, there is a report indicating a human P450 DNA chip could identify all the currently known polymorphisms of human CYP2D6 and CYP2C19, but its clinical usefulness has yet to be identified (55). However, these chips only detect those variants specifically programmed onto them and modifications to include new polymorphisms are expensive and time-consuming. The progress in fluorescent-based high throughput genotyping and DNA chip technology will definitely add significant value to pharmacogenetic diagnostics. Genotyping tests for the major drug metabolizing enzyme polymorphisms will soon be as easy as a routine blood test.

Cost Effectiveness of Pharmacogenetic Testing

Pharmacogenetics has the potential to be cost-effective in the managed care community. A simple diagnostic genetic test will enable the drug to be selectively prescribed to those patients for whom a drug would be safe and effective. This would provide cost savings to the health care providers by increasing drug efficacy, reduce follow-up and doctor visits, eliminate costly ineffective drugs, and reduce possible drug toxicity at "normal" doses in slow metabolizers. For example, a patient who metabolizes drugs more rapidly than other patients will not respond to the drug treatment under standard dosing. Identifying these rapid metabolizers of the drug could help these patients to either increase the dose or to use other appropriate drugs without undergoing three or four months' trial and error. This will be cost saving for many expensive prescription drugs for treatment of chronic disease such as dislipidemia, diabetes, and Alzheimer's disease.

Avoiding adverse drug effects alone may bring significant savings to society. It is estimated that approximately 3 to 5 percent of all hospitalizations result from adverse drug reactions, and as many as 30 percent of patients hospitalized for other reasons may have an adverse drug effect during hospitalization (56). The cost of treating drug reactions in the United States alone is estimated at approximately $3 billion annually (57).

Current genotyping generally costs up to two to three hundred dollars per test depending on specific assays. With the rapid development of automated high-throughput genotyping technologies, pharmacogenetic testing will likely become a relatively low-cost/high-volume service, just like a routine blood test. It is reasonable to develop cost-efficient pharmacogenetic tests by using multiplex PCR or non-PCR based genotyping technologies. Therefore pharmacogenetics holds great promise for prescriptive medicine, and it is expected that monitoring of pharmacogenetic markers will be routinely used clinically, especially for patients receiving drugs with low therapeutic indexes.

NEW DIRECTIONS OF PHARMACOGENETICS

Genomic Approach in Pharmacogenetic Studies

Studies of families with disease are informative for identifying highly penetrant gene variants. However, other approaches are needed to study less penetrant alleles, which may not be easily identified in family members. This is particularly true for environmental susceptibility genes and drug response genes. Such alleles may identify those people at risk, but who otherwise would only be observable in an exposed population. While some drug toxicities have been identified, the genetic polymorphisms associated with the effects of most drugs have not been characterized. Targeting these genes will be another goal of pharmacogenetics. The new concept of pharmacogenomics will utilize high-density markers to conduct genome scans to better predict drug efficacy and toxicity. In contrast to the candidate gene method, the strength of this system is the ability to scan the entire genome (58). It is becoming increasingly popular to use single nucleotide polymorphism (SNP) for association and linkage analysis, since they are the most frequent DNA sequence variations found in the human genome. Researchers will be able to conduct whole genome scans for identification of critical drug–response genes in nonfamilial studies. Creation of high-density SNP maps is feasible using high-throughput DNA sequencing (59) and chip hybridization (60). Cataloging common variants in human genes is moving very rapidly (61). It will be necessary for the scientists to prove the technology can work in the real world and transfer the research to clinical practice. Again, with the progress in this area, and as more genes have been identified, additional and increasingly complicated social, ethical, and policy issues will be encountered. Because of potential social consequences, researchers have been encouraged to pursue anonymous testing whenever possible and to ensure that the results of genetic testing are separated from an individual's record.

Education for Clinical Practitioners

The advances in pharmacogenetics will have a significant impact on the practice of diagnostic and preventive medicine. Currently there are only a limited number of medical practitioners familiar with and conversant in this area. It is important for clinicians to understand the concepts and applications of pharmacogenetics, since they will be explaining genetic tests and implications to their patients, determining when testing is appropriate, selecting specific tests, and interpreting the results. It is therefore necessary to develop training programs, that efficiently transfer a working knowledge of this field. Until such programs are available, physicians and pharmacists are advised to contact specialists or consult with colleagues having expertise in pharmacogenetic testing. As pharmacogenetic testing becomes more commonly offered by clinical pathology and reference laboratories, these facilities also will be called upon to provide expertise and appropriate indications.

Patient Stratification in Clinical Trials

For drugs prescribed on a limited basis due to a high incidence of adverse effects, pharmacogenetics may provide means to identify those most likely to benefit therapeutically without serious side effects. By targeting a specific subpopulation, pharmacogenetics offers the possibility of wider and safer drug use by creating a clear prescription path. The use of patient stratification also offers the potential to rescue an existing drug with great promise but undesirable problems, or to provide the data needed to withdraw a dangerous compound earlier. Pharmacogenetics data similarly could be used to logically design clinical trials and to increase the amount of information obtained from these studies. The greatest, but yet unproved, promise of pharmacogenetics is to alter trial-and-error application of a new medication into prescriptive medicine. Differentiating patient groups to improve the risk–benefit ratio of a new drug is already common practice. Therefore the concept of this predictive medicine approach looks attractive and can build on existing principles. But clearly, any tests associated with drug toxicity must be rigorously reviewed before a conclusion is made.

With the emerging global economy, pharmaceutical companies need to market new drugs in multiple countries. Due to the differential distribution of some drug metabolizing enzyme genetic polymorphisms among populations, a well-developed and extensively tested drug might not be suitable for patients with different ethnic backgrounds. In utilizing data obtained from genotyping both ethnic groups, prediction of drug efficacy and toxicity in a different population group become possible, as well as potentially reducing the need to conduct pivotal clinical trials in multiple countries. The most important step for pharmacogenomics now is proof of principle. It is critical to clearly show that pharmacogenetic concepts will yield improved and more predictive results in clinical trials. This process is actively underway.

CONCLUSIONS

The introduction of pharmacogenetic testing into clinical medicine has great promise for affecting the future of prescriptive and preventive medicine. Physicians may be able to prescribe drugs based on genotype, as well as allowing pharmacists to check for potential drug interactions and side effects. It will be important to educate medical practitioners and patients on both the concepts and clinical practice of pharmacogenetic testing. The upfront discussion of social, legal, and policy issues should not be used to block the collection of genetic data, but should serve as a safeguard to benefit and protect patient rights. With the advances from the Human Genome Project and functional genomics, massive increases are taking place in the information available on individual genes and functionally important polymorphisms. These differences hold the potential to improve effectiveness and limit toxicities of the available drugs while providing an understanding of gene/environmental interactions. Consideration of the ethical, social, legal, and policy aspects of accurate genetic prediction, and the design of more specific and safer drugs to meet individual's needs, will be important considerations as we enter the new millennium.

BIBLIOGRAPHY

1. A.S. Nies and S.P. Spielberg, in J.G. Hardman and L.E. Limbird, eds., *Goodman & Gilman's the Pharmacological Basis of Therapeutics*, 9th ed., McGraw-Hill, New York, 1996, pp. 43–62.

2. L.Z. Benet et al., eds., *Goodman & Gilman's the Pharmacological Basis of Therapeutics*, 9th ed., McGraw-Hill, New York, 1996, pp. 3–27.

3. A. Parkinson, in K.C.D. Klaassen, ed., *Casarett & Doull's Toxicology*, 5th ed., McGraw-Hill, New York, 1996, pp. 113–186.

4. U.A. Meyer et al., *Adv. Drug Res.* **19**, 197–214 (1990).

5. H.W. Goedde and D.P. Agarwal, in W. Kalow, ed., *Pharmacogenetics of Drug Metabolism*, Pergamon, New York, 1992, pp. 281–311.

6. W. Kalow, ed., *Pharmacogenetics of Drug Metabolism. International Encyclopedia of Pharmacology and Therapeutics*, Pergamon Press, Oxford, UK, 1992.

7. F.J. Gonzalez and J.R. Idle, *Clin. Pharmacokinet.* **26**, 59–70 (1994).

8. W.E. Evans and M.V. Relling, *Science* **286**, 487–491 (1999).

9. K. Kawajiri et al., *FEBS Lett.* **263**, 131–133 (1990).

10. J.A. Goldstein and S.M. de Morais, *Pharmacogenetics* **4**, 285–299 (1994).

11. M.W. Linder, R.A. Prouch, and R. Valeds, Jr., *Clin. Chem.* **43**, 254–266 (1997).

12. F. Uematsu et al., *Jpn. J. Cancer Res.* **82**, 254–256 (1991).

13. A.S. Wolfgan et al., *Pharmacogenetics* **7**, 235–239 (1997).

14. J.K. Wiencke et al., *Cancer Epidemiol, Biomarkers & Prev.* **6**, 87–92 (1997).

15. J.E. McWilliams et al., *Cancer Epidemiol. Biomarkers & Prev.* **4**, 589–594 (1995).

16. A. Hirvonen, *J. Occ. Env. Med.* **37**, 37–43 (1995).

17. S.P. Spielberg, *J. Pharmacokinet. Biopharmaceut.* **24**, 509–519 (1996).

18. D.A.P. Evans, in W. Kalow, ed., *Pharmacogenetics of Drug Metabolism*, Pergamon, New York, 1992, pp. 95–178.

19. P.J. Bosma et al., *FASEB J.* **6**, 2859–2863 (1992).

20. J.K. Ritter et al., *J. Clin. Invest.* **90**, 150–155 (1992).

21. P.J. Bosma et al., *N. Engl. J. Med.* **333**, 1171–1175 (1995).

22. D. Otterness et al., *Clin. Pharmacol. Ther.* **62**, 60–73 (1997).

23. C. Nordin et al., *Lancet* **350**, 29–30 (1997).

24. I. Johansson et al., *Eur. J. Clin. Pharmacol.* **40**, 553–556 (1991).

25. F.J. Gonzalez et al., *Genomics* **2**, 174–179 (1988).

26. C. Sachse et al., *Am. J. Hum. Genet.* **60**, 284–295 (1997).

27. N. Hanioka et al., *Am. J. Hum. Genet.* **47**, 994–1001 (1990).

28. A.C. Gough et al., *Nature* **347**, 773–776 (1990).

29. R. Saxena et al., *Hum. Molec. Genet.* **3**, 923–926 (1994).

30. A. Gaedigk et al., *Am. J. Hum. Genet.* **48**, 943–950 (1991).

31. M.M. Shi et al., *J. Clin. Pathol: Mol. Pathol.* **52**, 295–299 (1999).

32. M.M. Shi et al., *Res. Commun. Mol. Pathol. Pharmacol.* **103**, 3–15 (1999).

33. M. Chee et al., *Science* **274**, 610–614 (1996).

34. M.M. Shi, M.R. Bleavins, and F.A. de la Iglesia, *Mol. Diagnosis* **4**, 343–351 (1999).

35. V. Glaser, *Genetic Engineer. News* **18**, 1998.

36. F. Richard et al., *Lancet* **349**, 539 (1997).

37. E.J. Calabrese, *Ecogenetics*, J. Wiley, New York, 1983.

38. K.T. Kelsey et al., *Am. J. Indus. Med.* **31**, 274–279 (1997).

39. S. Kato, *Cancer Res.* **52**, 6712–6715 (1992).

40. A. Hivonen et al., *Carcinogenesis* **14**, 85–88 (1993).

41. J.Y. Hong and C.S. Yang, *Environ. Health Prospect.* **105**, 759–762 (1997).

42. D.W. Nebert, *Am. J. Hum. Genet.* **60**, 265–271 (1997).

43. W.W. Grody, *Diag. Mol. Pathol.* **4**, 155–157 (1995).

44. E.W. Clayton, *J. Am. Med. Assoc.* **274**, 1786–1792 (1995).

45. ASHG Report, *Am. J. Hum. Genet.* **59**, 471–474 (1996).

46. J. Stephenson, *J. Am. Med. Assoc.* **279**, 184 (1998).

47. G.J. Annas, *J. Am. Med. Assoc.* **270**, 2346–2350 (1993).

48. L. Gostin, *Ann. Intern. Med.* **127**, 683–690 (1997).

49. M.J. Mehlman et al., *Am. J. Hum Genet.* **58**, 393–397 (1996).

50. W.C. McKinnon et al., *J. Am. Med. Assoc.* **278**, 1217–1219 (1997).

51. Public Law 100-578, Clinical Laboratory Improvement Amendments of 1988, 42 USC 263a et seq.

52. N.A. Holtzman et al., *Science* **278**, 602–605 (1997).

53. NIH–DOE Task Force on Genetic Testing, Available at: *www.med.jhu.edu/tfgtelsi*

54. S. Gevers, *Med. Law* **15**, 407–411 (1996).

55. M. Cronin, in J. Schlegel, ed., *Pharmacogenetics: Bridging the Gap Between Basic Sciences and Clinical Application*, IBC, Southborough, MA, 1996, pp. 5.1.1–5.1.18.

56. H.J. Jick, *Allergy Clin. Immunol.* **74**, 555–557 (1984).

57. H.C. Slavkin, *JADA* **128**, 1157–1160 (1997).

58. L.M. Fisher, *N.Y. Times*, February 25 (1998).

59. E. Lai et al., *Genomics* **54**, 31–38 (1998).

60. D.G. Wang et al., *Science* **280**, 1077–1082 (1998).

61. M. Cargill et al., *Nat. Genet.* **22**, 231–238 (1999).

PROFESSIONAL POWER AND THE CULTURAL MEANINGS OF BIOTECHNOLOGY

ROBERT M. NELSON
University of Pennsylvania
Philadelphia, Pennsylvania

PAUL BRODWIN
University of Wisconsin
Milwaukee, Wisconsin

OUTLINE

Introduction

Prolonged Ventilation for Children with Neuromuscular Disease

Meaning of Tears

Knowledge and Communication

Knowledge and Power

Difference Between Starting and Stopping

Is Technology Value-Neutral?

Contesting Power Over Technology

Boundary of Body

Concluding Remarks

Acknowledgment

Bibliography

INTRODUCTION

Contemporary bioethics arose in the 1960s in the wake of innovations in dialysis, transplants, artificial organs, and assisted reproduction. These biotechnologies sparked debates about the allocation of scarce resources and the quality and limits of life. In response, ethicists developed a set of abstract normative principles — autonomy, beneficence, and distributive justice — which structure professional debates to this day. These core principles of American bioethics take a generic concept of the person and make it the basis for a universal morality. This is, of course, the autonomous individual of Western liberalism: the sovereign individual who acts freely according to a self-chosen plan. However, the very technologies that sparked early bioethics unsettle this tacit understanding of the person. They have created new ways of exerting one's will and gauging one's present identity and future fate. For example, predictive genetic testing alters the way people calculate their life prospects—the likely mixture of happiness and suffering they will encounter—and it can erase or intensify certain aspects of their identity. New strategies to assess people's subjective experience are demanded by transplantation and mechanical ventilation. Such technologies set in motion profound transformations in our cultural model of personhood and in the ways we experience and enact moral agency. These transformations, as much as the conflict between abstract principles, motivate our deepest ethical concerns.

This article takes up a vexing tendency in our current use of biotechnologies: the replacement of moral discourse by technical expertise. This tendency is magnified when health care professionals must make educated guesses about the subjectivity of technologically altered individuals. If we regard biotechnology as a simple collection of devices—a morally neutral means to the ultimate good of prolonging life—certain procedures become a standard, unquestionable component of care (1). This, in turn, justifies technical discourse as the sole guide for treatment decisions. It provides certainty for medical workers, but it also rules out other ways to understand the experience of, for example, chronically ventilated children and our ethical obligations toward them. This article examines the conflict of interpretation over the subjective experience of the mechanically ventilated child. Too often, the authority to read it one way or the other remains solely in the hands of medical professionals, illustrating how rational, technical expertise can foreclose genuine moral debate.

PROLONGED VENTILATION FOR CHILDREN WITH NEUROMUSCULAR DISEASE

An infant or young child with a neuromuscular disorder, such as nemaline rod myopathy or spinal muscular atrophy, usually presents with generalized poor muscle tone, which then progresses to respiratory failure and the need for assisted ventilation. Often a muscle biopsy is necessary to make a diagnosis, and the first biopsy may be inconclusive as it takes time for the characteristic pathological findings to develop. Nemaline rod myopathy, selected as an illustration of the issues raised in this entry, is a rare and slowly progressive neuromuscular disease that renders a person immobile (2,3). Unable to move and unable to breath, such an infant may undergo a tracheostomy procedure so that the airway is secure and the infant can be ventilated more easily. A tracheostomy, which is a plastic tube inserted into the windpipe through a surgical incision in the front of the neck, is done when the plan is to provide long-term ventilation, perhaps including sending the patient home on a ventilator. In addition a so-called gastrostomy tube may be placed into the infant's stomach through the abdominal wall so that the infant can be fed without placing a temporary feeding tube through the nose. Once these procedures have been performed, the infant is often transferred from the intensive care unit to a "step-down" unit designed for long-term care and is placed on a simpler breathing machine whose primary purpose is for use in home ventilation.

We do not know what causes nemaline rod myopathy, nor do we have any treatment for it other than putting someone on a ventilator. It only affects the skeletal muscle, so the other muscles of the body such as the heart and gut work fine. In its most severe form, the child cannot move, cannot swallow, cannot breathe, and may not be able to move his eyes or close his eyelids. Eyes open and barely moving, an affected infant or child will stare at you without expression, unable to move the muscles of the face to show pain or pleasure, unable to smile or frown, unable to laugh or cry, unable to communicate at all—the face a frozen, expressionless mask. The disease, however, does not affect the brain. An affected infant or child is alert and aware of everything going on around him.

At times, conflict may arise over the continued use of mechanical ventilation, with a parent insisting on the right to remove the ventilator and let the child die, and the professional staff (both nurses and physicians) resisting or disputing this claim. Usually this conflict is addressed in such terms as the child's quality of life, the "value" of living with a severe disability, the child's "best interest," the authority of the physician/state in determining "medical neglect," the parent's authority to make decisions concerning the child's medical care, and so forth. All of these approaches assume that the technology itself is morally neutral: It sets the stage for ethical conflict but does not influence the outcome. In this article we question that assumption. We suggest that once biotechnology is introduced into patient care, it constrains our subsequent moral choices, and this belies the claim that the technology itself is morally neutral. We examine, in particular, tracheostomy and mechanical ventilation, techniques that undercut parental (or nonprofessional) control over the medical care of children suffering from a neuromuscular disease such as nemaline rod myopathy. Because medical professionals control both how to use these technologies and how to interpret their effects upon individual subjectivity, ethical conflicts become prematurely translated into matters of technical expertise.

To examine these issues in the case of a ventilator dependent child, we need to understand how the technology affects the child, how the ventilator defines and redefines the boundary of body, and how the ventilator produces a subject who is both body and machine. Rather than a neutral technology designed to achieve goals that are selected for non-technical reasons, the ventilator seems to impose its own agenda and values. However, we should not reify this technology in a way that obscures and thus privileges the agency of the medical professionals who control it—as in the Wizard of Oz when we are told "don't pay attention to that man behind the curtain." Moreover, we need to pay attention to the organizational and cultural context within which professionals operate. Only then can we understand how professional power infiltrates both the use of technology and the interpretation of its effects.

MEANING OF TEARS

An infant with nemaline rod myopathy often has tears in his eyes, lending support to the belief (usually by a parent or family member) that this constant tearing indicates emotional and physical distress. The professional staff, however, may interpret these tears as a simple result of an inability to close the eyelids, rather than a reminder of an infant's suffering. Such an infant becomes the locus of a contested interpretation. The parents view the infant as too fragile and unable to tolerate activities such as being propped up in a wheelchair or taken out of the hospital on field trips. The apparent suffering of the infant often motivates their desire to stop the ventilator at some point in the future. The staff regards the infant as able to take pleasure in simple things and to sit contentedly in a chair for hours. On occasion, the staff may blame a parent for the very episodes that the parent interprets as fragility, arguing that the infant would cry when his parent arrived and "suffocated" him during a brief visit. This conjures up the image of an overprotective parent who fails to appreciate the strength and ability of her or his child; however, the use of the metaphor of suffocation takes on a more literal and provocative meaning given the staff's suspicion of the parent's (implicit or explicit) desire to stop the ventilator.

Fundamentally the infant's parent continues to see the ventilator as something other than the infant: as a threat, as an invasion of his body, as something foreign. Removing it in order not to prolong the infant's suffering simply returns the infant to a more natural state. In the eyes of the staff, the threat to the infant is not the ventilator but the parent. The staff often regards such an infant not as a body on a machine, but as both body and machine—that is, a machine-human/ human-machine cyborg (cybernetic organism) whose body and machine components are mutually interdependent (4). Therefore, to ask the staff to participate in turning off the ventilator is to ask them to amputate part of the infant's body. Is there a fact of the matter that could settle this dispute about removing the ventilator? Is there a third point of view, acceptable to both parties, from which the relationship between tears and suffering could be "objectively" determined? Glossing the difference between pain and suffering, the absence of other signs of pain such as sweating and a rapid heart rate may support the staff's argument. This is an argument that a parent can only lose; the best one could hope for is for everyone to agree to disagree.

There often is no apparent disagreement over the moral principle that mechanical ventilation can be stopped if the burden of such treatment outweighs the potential benefit (5). What is in dispute is the description of the child as either suffering or simply unable to close his eyelids. The staff is able to substitute a physiologic argument about the presence of pain for the existential question of whether the child is suffering—in effect, shifting a moral argument about the worth of living life on a ventilator to a technical dispute about the interpretation of a physical sign. In addition the staff has the political power to threaten a parent with a charge of failing to provide necessary medical treatment, that is, "medical neglect,"

thus throwing the matter into court. Consequently the moral and political questions about what to do are transformed into a rational technical discourse that in the minds of the staff is unambiguous. As such, the dependence of knowledge on subject position or "point of view" is implicitly denied; the power to determine objectivity is invisibly exercised. Rather than moral and political discourse about the conflicting visions of the infant's experience serving as the "paradigm of rational discourse," the professional technical discourse determines the political stakes and reduces moral discourse to the vanishing point (6, p. 194).

KNOWLEDGE AND COMMUNICATION

An appeal by the parent to the shared "sense experience" of the child's tears usually will not persuade the staff that the child is suffering. In addition the staff will probably fail to convince a parent that the child's tears simply mean that he cannot close his eyelids. The staff may additionally dismiss the parent's claim to know that the child is suffering as mere subjective opinion. The staff's insistence that the child is not suffering clearly reinforces their professional interest in continuing treatment.

In answer to the question—"What protects knowledge from being [either] the arbitrary expression of subjective desires [on the part of a parent] or the tool of social and personal interests [on the part of the medical and nursing staff]?"—Helen Longino, a philosopher of science at the University of Minnesota, offers an approach she refers to as "contextual empiricism." Longino, as do other philosophers in the pragmatic tradition, grounds "objectivity" or the truth of a statement concerning a sense experience in an "inter-subjective" or "social" process that should ensure "the inclusion of all socially relevant perspectives in the community engaged in the critical construction of knowledge" (7, pp. 200, 202–203). A necessary part of this communicative process is a critical examination of the implicit assumptions that establish the relevance and interpretation of observational or empirical data. The natural world cannot impose one single interpretation, that is, the empirical observation of the presence of tears does not establish the truth of one or the other interpretation. However, differences of power in this social or communicative process may limit the plurality of interpretations to the one that is consistent with the dominant discourse.

This is the outcome, for example, when a particular powerful group or individual constrains the freedom of expression and diversity of legitimate knowledge, or restricts the community of discourse in such a way as to predetermine which interpretation is accepted. This process involved (fragment) discounting the parent's interpretation of a physical sign such as an infant's tears. Parents may also be isolated from outside family and community supports and effectively alone with the medical staff during conversations in the hospital about the care of their child. In this setting medical professionals fail to establish a meaningful community of inquiry concerning the question of a child's suffering.

In discussing contemporary policy debates about technology in general, Langdon Winner observes that this lack of a coherent community of discourse "contributes to two distinctive features ... (1) futile rituals of expert advice and (2) interminable disagreements about which choices are morally justified" (8, p. 75). The moral uncertainty involved in the application of ventilator technology to the indefinite support of patients cannot be resolved by an appeal to the technical advice and expertise of the physician. Such a "futile" appeal to expert advice will not achieve a consensus. Moreover the lack of an appropriate community of discourse and the resulting disagreement over what choices are morally justified privileges the physician's technical expertise and thus interpretation of the patient's experience. While this does not avoid a conflict of interpretations, it guarantees that the conflict gets resolved in ways that favor the power and interests of the physician.

KNOWLEDGE AND POWER

In response to this professional prerogative which constrains the available choices, how do we empower a parent to make decisions concerning her child's medical care? The notion of personal autonomy or self-rule has resulted in a significant shift of power from the physician to the patient over the past two decades. However, once we abandoned the concept of the child as property, the notion of parental autonomy as a justification for the right of a parent to direct a child's medical care became problematic. Each one of us may have an absolute right to determine our own medical care. A parent has, at most, a prima facie right that is limited by the child's right to life and freedom from serious bodily injury or disability (9). Within this constraint we expect that a parent will make decisions that benefit the child or, in other words, are in the child's "best interest." Thus the parent's vision of the good is imposed on or becomes the child's vision — an imposition we accept given the diversity and, at times, incompatibility of competing visions of the good within our society. This creates a disturbing paradox. On the one hand, we expect a parent to express a decision concerning his or her child not as "what is good for the parent" but rather as "what is good for the child." On the other hand, the only possible way to give "voice to the voiceless" is by articulating adult values and projecting them upon the child (10). An infant with nemaline rod myopathy cannot speak; so when we speak for such an infant, we ask: "If I (the adult) were in this condition, what I would want?"

If we seek to escape this paradox and avoid this imposition of adult values by supporting the child until he is capable of self-expression, we inadvertently reinforce the physician's tendency for the relentless application of life-sustaining technology. Consequently the concept of a child's "best interest" appears to be the arena for an unavoidable expression of adult power on the part of either the physician or the parent. The stakes are high, for if the parent understands the child's "best interest" in such a way as to refuse what the physician otherwise believes to be necessary medical care, the parent may find him- or herself in court facing a charge of medical neglect.

The past two decades have seen a lively debate in the bioethics literature and the courts concerning the withholding and withdrawal of life-sustaining technology (11). Some have argued that removing a person from a ventilator is to choose death based on the judgment that the anticipated quality of life is not worth living. Others, concerned about the potential abuse of quality-of-life judgments, have argued that such decisions are better understood as the choice of how to live while dying (12) or as simply the decision to remove technology that is no longer medically indicated (11,13). The first argument, that of how to live while dying, makes the decision to remove a ventilator dependent on a prior determination that the patient is dying — a determination that the technology itself makes more difficult. A child with nemaline rod myopathy who is on a ventilator may not die for years in the absence of an intervening complication. Thus, once you put him on the ventilator, you cannot remove it unless he is dying, and he is not dying unless you remove the ventilator. The second argument, that technology can be removed when it is no longer medically indicated, either restricts the removal of technology to those situations where more narrow technical goals cannot be achieved or obscures the physician's own determination of an acceptable quality of life behind the veil of professional technical competence. A ventilator is medically indicated when a patient has respiratory failure; it is not indicated when either the patient recovers or the ventilator fails to correct the respiratory failure. Thus, in most cases, the ventilator for an infant with nemaline rod myopathy is medically indicated. If a physician argues that the ventilator is not medically indicated, since correcting the patient's respiratory failure does not contribute to the overall good of the patient, we necessarily must engage the question of what is or is not in the patient's "best interest" — a discussion that cannot avoid questions of the patient's quality of life. The problem then of trying to avoid an explicit discussion of a child's anticipated quality of life is that the physician's power and authority is inadvertently reinforced.

Physicians impose their power by establishing what counts as legitimate and credible knowledge, rather than by forcing a choice for one of either two credible options. In asking whether a child on a ventilator is suffering, a parent and the health care team may disagree over the description of the child's life, not over the moral evaluation of an agreed upon description (14). It is simply not credible to the medical staff that the child is suffering. In discussing the problem of technology as ideology, Robert Pippin asks whether we have "been so influenced by technical instruments ... that our basic sense of the natural world has changed ... so fundamentally that ... possibilities for social existence are seen only ... in terms of such technical imperatives." The physician's reliance on technology "reaches a point where what ought to be understood as contingent, an option among others, open to political discussion, is instead falsely understood as necessary; what serves particular interests is seen, without reflection, as of universal interest; what is a contingent, historical experience is regarded as natural" (15, p. 46). Physicians appear to have lost any sense of the natural or the contingent as a moral category. Rather the natural serves to mark that domain that resists the physician's intervention, as in "let nature take its course." The natural becomes that which cannot be technically overcome, rather than that which should not be overcome. The natural is subservient to the technical, which in turn resists the explicit introduction of moral and political questions.

DIFFERENCE BETWEEN STARTING AND STOPPING

The belief that technology is a neutral means to whatever ends are selected on moral, political, or more narrow physiologic grounds is a fundamental conviction and ideology of medical practice. For example, the decision to perform a tracheostomy may not be intended as a decision for long-term home ventilation but may be seen as consistent with a desire to defer any decision to limit or withdraw support given any remaining uncertainty about a child's diagnosis and prognosis. A parent may be told that a decision to perform a tracheostomy does not preclude a decision at some point in the future to remove a child from the ventilator — "what is done can be undone." Such a statement is consistent with the widely endorsed bioethical teaching that there is no significant moral or legal difference between withholding and withdrawing treatment (11).

There are a number of important assumptions behind the use of this bioethical maxim. First, it assumes a symmetry in the application and removal of medical technology consistent with the prejudice that technological means are value-neutral. It also assumes a symmetry between an endotracheal tube and a tracheostomy by reducing each to its essential function of establishing an airway for the purpose of mechanical ventilation. However, as an endotracheal tube is inserted either through the mouth or nose, the tape required to hold it in place covers a major portion of the face. A tracheostomy surgically inserted through the front of the neck results in the entire face being visible and thus capable of expression. Second, the maxim appears to ignore any relevant differences that may occur between the moments of application and removal of the technology, apart from any changes in the medical indications. Third, and related to this historically naive stance, is the view that the organizational context in which these decisions are being made is apparently unimportant. After a tracheostomy a child may be transferred out of the intensive care unit and to the ward that houses patients in the home ventilation program. One wonders whether the use of this bioethical maxim that there exists no significant moral or legal difference between withholding and withdrawing treatment is based on a reasoned ethical stance, or used as a rhetorical device to postpone the discussion of more difficult ethical issues to a later date. The latter interpretation is confirmed, for example, when a receiving physician in conflict with a parent over the removal of a ventilator is unable to find any other physician willing to assume the child's ongoing medical care (and thus agree to withdraw support), including any of the physicians who have previously cared for the child prior to the tracheostomy. An appreciation of the value-laden nature of a tracheostomy, along with the importance of time and

context, counsel against a premature surgical procedure and then transfer to a home ventilation program.

IS TECHNOLOGY VALUE-NEUTRAL?

The bias that our medical technology is simply a "collection of devices" emphasizes the functional aspects of technology and obscures its social context. As a result of this dichotomy between function and context, our technology appears value-neutral, while only the application of that technology becomes morally problematic. Andrew Feenberg, in an article reflecting on the relationship between technology and power, points out that this "dichotomy of goal [function] and meaning [context] is a [contingent] product of functionalist professional culture" rather than a necessary component of technology (16, p. 9). Echoing criticisms of the common view of science as value-free, Feenberg asserts that this ideology of technology as value-neutral reinforces the dominant forms of power that compose the cultural horizon and social meaning of technology. Feenberg refers to this as the "bias of technology" by which "apparently neutral, functional rationality is enlisted in support of a hegemony," that is, the professional dominance of those who control the technology. The professional claims to exclusive control of technology are strengthened insofar as its associated values and presuppositions drop out of sight (16, p. 12).

This professional control is also reinforced by the perception that technology should always be used when it can be used: the so-called technological imperative. Barbara Koenig suggests that the technological imperative acquires a certain moral force as the technology becomes habitual or routine (1). Her field research focused on therapeutic plasma exchange, a procedure that involves the removal and then replacement of blood plasma. Koenig identified three steps in the process by which plasma exchange became a routine therapy for certain conditions. The first step was a transformation in roles and responsibilities. The physician–nurse relationship shifted from egalitarian to hierarchical while, at the same time, the physicians moved from being closely involved to delegating many of the routine tasks to nurses (1). Similarly the physicians involved in a home ventilation program maintain close control while shifting many of the routine tasks from in-hospital nurses and respiratory therapists to parents and visiting home nurses. The second step was the use of treatment rituals that appeared to reduce uncertainty, anxiety, and disorder and thus established the meaning of the technology as standard therapy for both patients and staff (1). When a patient is placed on a home ventilator after a tracheostomy, there is an orderly and nearly invariable process of parental training, arranging for nursing services, equipment purchase, and so forth, that must take place prior to discharge from the hospital. Any deviation from this process generally results in uncertainty, inefficiency, omissions, and the like. The third step that Koenig identified was the generation of research data. Noting the enthusiasm with which the physicians engaged in plasma exchange collected clinical data as part of their ongoing research, she speculated that the machine's capability of producing research data supported the physician's tendency to use the technology (1). Although this may be true with therapeutic plasma exchange, it does not appear that the development of home ventilation programs was driven by a research imperative. The physician's use of home ventilator technology is more likely driven by such factors as the need to find alternative placements for children who otherwise would survive intensive care but remain dependent on ventilator technology. Also the immediate efficacy of the ventilator when compared to plasma exchange is obvious, for otherwise the child would die. Despite these differences Koenig's conclusion remains essentially correct. The technological imperative is transformed into a moral imperative through the development of a "sense of social certainty experienced by health professionals" (1, pp. 485–486). The technology simply begins to feel routine, and hence both appropriate and necessary.

The decision to perform a tracheostomy and then to transfer a child to a unit where the use of chronic home ventilation is considered routine is governed by a similar moral imperative. The unit is organized so that home ventilator technology is accepted as standard therapy. Within this social context, it becomes difficult, if not impossible, to question whether this technical standard of care ought to be used for any particular child. The moral question of what is in a particular child's "best interest" thus receives an axiomatic answer applicable to all children: "Given these circumstances, we should provide the standard technology." The moral meaning of our medical technology is thus created and sustained by the professional culture of the hospital. Since the statutory definition of medical neglect in Wisconsin, for example, is simply failure to provide necessary medical care, the technological and moral imperative experienced by medical and nursing professionals clearly has "the potential to wrest control of decisions about the use of technology" from parents and patients (1, p. 489).

Andrew Feenberg proposes that one of the assumptions behind our modern image of technology is that social institutions must adapt to the technological imperative. Noting that "the economic significance of technical change often pales besides its wider human implications in framing a way of life," Feenberg encourages us to study the "social role of the technical object and the lifestyles it makes possible" through defining "major portions of the social environment, such as ... medical activities and expectations" (16, p. 9, 16). This assumption that we must adapt to technology is readily apparent over the past two decades with the development of home care programs for so-called technology-dependent children (17). The family is explicitly expected to change in response to the demands of caring for a child who is to be discharged from the hospital on a home ventilator. The only other available option is foster care, which is problematic for two reasons. While the child is in foster care, a parent may lose control over any decisions to either withhold additional medical treatment or withdraw existing medical treatment. In addition there is often an unspoken assumption that to choose foster care reflects poorly on the ability of a parent to provide for his or her child. Although many parents

choose to take their ventilator-dependent child home out of a sincere concern for their continued life and well-being, the normative pressures against choosing otherwise are enormous once the child is within the context of the home ventilation program. This assumption then that social institutions such as the family must adapt to the technological imperative is another manifestation of the extension of professional power implicit in the ideology of value-free technology.

The apparent inevitability of the technological imperative is rejected by both Koenig and Feenberg. Consistent with Koenig's thesis, Feenberg asserts that "technology is just another dependent social variable" and the "scene of social struggle" (16, p. 8). Contrary to the claim that technology itself requires professional control, Feenberg argues that technology has been used to block the extension of public or democratic control to "technically mediated domains of social life" (16, p. 20). Thus the professional medical culture seeks to reinforce the image of technology as both value-neutral and complex in order to maintain control despite the "routinization" process of placing that same technology into the home.

CONTESTING POWER OVER TECHNOLOGY

To insist that technology is "socially constructed" may give the impression that people ultimately hold complete power over its meanings and uses. The typical circumstance of one group having more control and another group having less control over technology results, in this view, solely from social and political considerations independent of the constraints of particular devices. To deny that technology is "socially constructed" may imply the opposite extreme: that we have no power over technology and that our moral and cultural response is determined by its concrete and independent reality. We argue against both extremes. To assume that technology is neutral—the core of the anticonstructionist position—reinforces the professional dominance of physicians. By failing to recognize the extent to which technical knowledge is constructed by and for the interests of a particular community, we are likely to ratify this group's power and authority. At the same time the strong constructionist position ignores the material effects of this technology and the way it constrains moral deliberation. By use of this technology, a child's breathing becomes, ineradicably and by definition, assisted breathing. As a result the object of clinical decision making has become altered. It is no longer the child as such but the hybrid object of the ventilator/child. Once this massive technological intervention has taken place, it is not clear who gives life to whom: the ventilator to the child, or the child to the ventilator? In the face of this ambiguity, medical workers substitute technical rationales for action (algorithms and expertise about ventilator use) for the search for mutual understanding about the child's experience and, possibly, suffering. In the end this process makes humans subservient to things (18). However, this result is produced by both the social power of the profession and the particular way this technology transforms the very objects of clinical decision-making.

Cleaving to either the strong constructionist or the strong anticonstructionist position misses this complex result.

Controlling the technical mediation of social activities such as medical care is a major source of public power within our society. The ability to manage or expand this technical mediation results in the concentration of power in an elite group of experts, the narrowing of acceptable options for public discussion, and an increase in the extent of administrative or professional control over aspects of daily life (15). Changes in the way medical technology is delivered or applied to a particular problem will require a shift in this expert control of technology. As Feenberg writes: "If authoritarian social hierarchy is truly a contingent dimension of technical progress, ... and not a technical necessity, then there must be an alternative way of rationalizing society that democratizes rather then centralizes control" (16, p. 5). Is the link between the physician's social role and the control of medical technology necessary or contingent? For example, one approach to the issue of physician-assisted suicide is to allow for assisted suicide while preserving the traditional social role of the physician by making available to the general public the technical tools that to-date remain under the physician's prescriptive authority.

If we move the control of medical technology into the public domain, we will need to create an appropriate community of discourse to monitor development and application. Such a task may be difficult given the diversity of our current communities. Although the reform of technology is a better option than simply resistance, it is not clear that those (e.g., nurses and physicians) who have been socialized in the modern medical ethos could resist attempting to impose new forms of professional control (15). The creation of a community for the reform of medical technology should include those who anticipate needing or who may resist medical technology and thus will require abandoning the notion of professional expertise. In addition such a community of discourse must begin by questioning the assumption that technology is a value-free instrument—an assumption that serves to reinforce professional control and hinder rational debate. Or should we simply recognize the legitimate existence of disparate communities and thus reframe the question of the appropriate application of medical technology as a choice of which community to belong to?

BOUNDARY OF BODY

Let us now return to the question of whether the medical and nursing staff simply see a mechanically ventilated child with a tracheostomy differently than he is seen by his parent(s). As the disease progresses, a child with nemaline rod myopathy cannot move, cannot breath, cannot express emotion, indeed cannot make any facial expressions; communication at best may occur through the movement of an eye in response to a question. Consequently it may be difficult if not impossible to get any indication of what a physical sign such as "tearing" meant to the child. As the passive object of our application of ventilator technology, the child is reduced to either a resource for our instrumentalist projects or a mask for

our dominant interest in maintaining control (6). Modern medical technology, as we have seen, clearly includes the feature of the technical control of some human beings by others. Donna Haraway attributes this modern tendency towards technical domination to the dualism between objective nature and subjective culture so that the projects or interests that shape our determination of natural objects are hidden from view. As an alternative, she offers us a view of "objectivity as positioned rationality" (6). To capture a notion of the object as active and not passive, Haraway asserts that "bodies as objects of knowledge ... materialize in social interaction. Boundaries are drawn by mapping practices; 'objects' do not pre-exist as such" (6, p. 200–201). The issue then is the various positions from which each one of us, including the child's parents, view the ventilated child—a question that necessarily draws us back into an explicit discussion of the power of professional "mapping practices" in determining the boundaries of the ventilated child as the object of our attention.

How then are we to understand who the ventilated child is, this body attached to a ventilator? Through an exploration of the "semiotic use of the body" among the Kayapo of the Brazilian Amazon, Terence Turner illustrates how "the body is at once a material object and a living and acting organism possessing rudimentary forms of subjectivity that becomes, through a process of social appropriation, both a social identity and a cultural subject" (19, p. 145). For example, the Kayapo use various modifications of their body surface to define and redefine their social identity, as in the use of ear piercing to indicate age cohort, marital status, and other social identities. The individual Kayapo, as both a social body and an embodied subject, assumes the dual role as product and producer (19). In our case the body of the ventilated child as a material object of our technical interventions takes on the social identity of a patient in the home ventilation program. Although his parent(s) may try to resist this medical appropriation of the child's body, the tracheostomy and attached ventilator tubing are key modifications of his body that produce the child's social identity as a patient in the home ventilation program (19). The ventilator-infant as embodied subject appears to be the socially patterned product of our technical activity, rather than the producer of its own activity. Similar to the ideological consequences of the view of technology as value-neutral, the misrepresentation of the "cultural subject" of the ventilator-infant as an "objective (natural) feature existing independently" of our social production further reinforces the dominant power of the physician (19). In infancy, it is unclear that there is any content to the notion of the subject existing prior to and independently of the social production of the embodied subject by others. In other words, what meaning can we give to the notion of the "best interest" of the child apart from the specific interests of a particular embodied social subject? Once a child undergoes a tracheostomy and is placed on a chronic home ventilator, he is and will remain a patient in the home ventilation program. This much is visibly announced on his body. Thus we come full circle to the notion of the ventilator-infant as cyborg, the machine-human as "embodied subjectivity" rather than the machine as external to the body. The social identity of the ventilator-infant/infant-ventilator is a product of being a machine-human hybrid, that is, the ventilator gives life to the body and the body gives life to the ventilator. To contemplate taking the patient off of the ventilator would be to contemplate amputation — a request that the medical and nursing staff cannot and will not honor.

CONCLUDING REMARKS

What have we learned from this story of the social production of the ventilator-infant as a patient in a home ventilation program? We have come to doubt the universality of the classic teaching of the symmetry between withholding and withdrawing technology. We have a renewed understanding of the insight that our medical technology is not value-neutral, and that it often serves to reinforce the professional dominance of physicians. While we acknowledge that specific hospital units have different cultures, the general impact of the organizational context on the ability of patients and parents to control the application of medical technology is greater, on reflection, than previously appreciated. This impact occurs not primarily through the imposition of a different set of moral values, but through fundamental shifts in our point of view, and thus how we see and come to know our patients. The fundamental conflict between medical staff and the parents of young, chronically ventilated children does not turn on a choice between competing ethical principles. As we have seen, medical staff often do not advance any explicit ethical principle in support of their action (i.e., refusing to withdraw the ventilator). The conflict turns rather on what counts as the proper object of concern: the child, the ventilated child, or the hybrid "ventilator-child." In this conflict, medical workers enjoy enormous power to make authoritative readings of the child's subjective experience and, more generally, what is admitted as knowledge in the medical setting.

Using the example of ventilator technology, this article demonstrates how the use of biotechnology constrains our subsequent moral choices concerning the application of that technology in a manner that belies the claim that the technology itself is morally neutral. Physicians and other medical staff are thus "technicians" in the following sense: They translate moral and political issues surrounding the application of biotechnology into the dominant technical discourse. To accomplish this, these technicians may constrain the freedom of expression and diversity of legitimate knowledge, or they may structure the community of discourse in such a way as to reinforce their own power and interests. Whatever strategy they ultimately follow, the ideology of technology as value-neutral reinforces their dominance in the clinical encounter and their authority to establish the local meanings of technology. The belief that biotechnology is a value-neutral means to certain ends selected according to entirely different criteria (moral, political, or physiological) thus perpetuates professional dominance over patients and their families. Moreover this belief obscures that the process is happening at all. Clinical

actions, such as withdrawing or continuing ventilation, as well as knowledge claims about the child at the center of attention, are always underdetermined by available physiological evidence. Ethics and politics, even if hidden, play the crucial role in the outcome of conflicts between medical staff and parents. Elucidating that role, and restoring moral discourse where it has been banished, demands that we abandon the model of morally neutral biotechnology.

ACKNOWLEDGMENT

Portions of this article have been previously published in R.M. Nelson, The Ventilator-Infant as Cyborg: A Case Study in Technology and Medical Ethics, and P. Brodwin, Introduction, in P. Brodwin, ed., *Biotechnology, Culture and Body*, Indiana University Press, Bloomington, IN, 2000.

BIBLIOGRAPHY

1. B.A. Koenig, in M. Lock and D. Gordon, eds., *Biomedicine Examined*, Kluwer Academic Publishers, Dordrecht, The Netherlands, 1988, pp. 466–490.

2. S.T. Iannacone and T. Guilfoile, *J. Child Neurol.* **3**, 30–32 (1988).

3. M.B. Connolly, E.H. Roland, and A. Hill, *Pediatric Neurol.* **8**(4), 285–288 (1992).

4. G.L. Downey, J. Dumit, and S. Williams, *Cult. Anthropol.* **10**(2), 264–269 (1995).

5. L.J. Nelson and R.M. Nelson, *Crit. Care Med.* **20**(3), 427–433 (1992).

6. D.J. Haraway, in *Simians, Cyborgs, and Women: The Reinvention of Nature*, Routledge, New York, 1991, pp. 193–201.

7. H.E. Longino, in A. Feenberg and A. Hannay, eds., *Technology and the Politics of Knowledge*, Indiana University Press, Bloomington, IN, 1995, pp. 200–203.

8. L. Winner, in A. Feenberg and A. Hannay, eds., *Technology and the Politics of Knowledge*, Indiana University Press, Bloomington, IN, 1995, pp. 75–77.

9. Committee on Bioethics, American Academy of Pediatrics, *Pediatrics* **99**(2), 279–281 (1997).

10. D. Archard, *Children: Rights and Childhood*, Routledge, New York, 1993.

11. President's Commission for the Study of Ethical Problems in Medicine and Biomedical and Behavioral Research, *Deciding to Forego Life-Sustaining Treatment: A Report on the Ethical, Medical, and Legal Issues in Treatment Decisions*, U.S. Government Printing Office, Washington, DC, 1983, pp. 60–90.

12. A.J. Dyck, in S.J. Reiser, A.J. Dyck, and William J. Curran, eds., *Ethics in Medicine: Historical Perspectives and Contemporary Concerns*, MIT Press, Cambridge, MA, 1977, pp. 529–535.

13. P. Ramsey, *Ethics at the Edges of Life: Medical and Legal Intersections*, Yale University Press, New Haven, CT, 1978, pp. 145–188.

14. S. Hauerwas and D.B. Burrell, in S. Hauerwas, *Truthfulness and Tragedy: Further Investigations in Christian Ethics*, University of Notre Dame, Press, Notre Dame, IN, 1977, pp. 15–39.

15. R.B. Pippin, in A. Feenberg and A. Hannay, eds., *Technology and the Politics of Knowledge*, Indiana University Press, Bloomington, IN, 1995, pp. 43–51.

16. A. Feenberg, in A. Feenberg and A. Hannay, eds., *Technology and the Politics of Knowledge*, Indiana University Press, Bloomington, IN, 1995, pp. 5–20.

17. J. Arras, ed., *Bringing the Hospital Home: Ethical and Social Implications of High-Tech Home Care*, Johns Hopkins University Press, Baltimore, MD, 1995.

18. S. Vogel, in A. Feenberg and A. Hannay, eds., *Technology and the Politics of Knowledge*, Indiana University Press, Bloomington, IN, 1995, pp. 23–37.

19. T. Turner, *Cult. Anthropol.* **10**(2), 145–169 (1995).

See also GENETIC DETERMINISM, GENETIC REDUCTIONISM, AND GENETIC ESSENTIALISM.

PUBLIC PERCEPTIONS: SURVEYS OF ATTITUDES TOWARD BIOTECHNOLOGY

DOROTHY C. WERTZ
The Shriver Center
Waltham, Massachusetts

OUTLINE

INTRODUCTION

Surveys can provide a useful baseline for policy planning, educational development, marketing, and ethical discussions. Surveys may include the general public or may seek to identify the views of various stakeholders, such as researchers, physicians, genetic counselors, or patients. Surveys also have limitations. This article will present the rationale for doing surveys, outline major types of survey methods, and describe some results of major national and international surveys of public and stakeholder views on biotechnology, including genetic testing, screening in the workplace, gene therapy (including germ-line gene therapy), and enhancement of average human characteristics.

RATIONALE FOR CONDUCTING SURVEYS

Uses of Surveys

Surveys can be useful in providing a baseline for public policy debates or ethical discussions, devising educational programs, marketing genetic services, and identifying existing or potential abuses of genetics. In the public policy arena, for example, it makes little sense to outlaw a test or procedure that most people would either use themselves or think others should have a right to use. Policies related to abortion for fetal "defects" are one example. The General Social Surveys (GSS) conducted by the University of Chicago's National Opinion Research Center on a random sample of the United States adult public have indicated that about 80 percent believe a woman should be able to get an abortion if there is a "serious defect" in the fetus. This percentage has varied by only a few points between 1971 (two years before *Roe v. Wade*, the Supreme Court decision legalizing abortion) and 1998 (1). Reports such as this can be useful to lawmakers, courts, advocacy organizations, and lobbyists. Surveys have the advantage of allowing a wide variety of people to express their views, including many who would not otherwise come forward to comment on proposed policies, regulations, or ethical guidelines.

In the arena of education, surveys can identify areas of ignorance that may require special educational programs. For example, a 1986–87 survey of 1473 physicians conducted by the U.S. Congress Office of Technology Assessment (OTA) found that 63 percent would reject a sperm donor with a family history of Huntington's disease (an autosomal dominant disorder with severe effects on the nervous system, transmitted to 50 percent of the offspring and not presymptomatically diagnosable in the donor at the time), while 61 and 49 percent respectively would reject a healthy donor with a family history of Duchenne muscular dystrophy or hemophilia A, both of which are transmissible only by females (2). The survey results indicated need for better education of physicians, greater oversight by sperm banks, and new guidelines from professional societies. A 1995 survey of 499 U.S. primary care physicians found that substantial percents held inaccurate knowledge about the life expectancy and functioning of people with common genetic disorders such as cystic fibrosis or Down syndrome (3).

Surveys also inform ethical discussion. A 1985 survey of 683 geneticists reported that 62 percent in the United States would either perform prenatal diagnosis at parental request solely to select the sex of the fetus (34 percent) or would offer a referral (28 percent) (4). Subsequent discussions among bioethicists and medical practitioners led to various international ethics guidelines rejecting this practice, notably those of the World Health Organization (WHO) (5). These discussions appear to have had little effect on attitudes of professionals in the United States; a 1995 survey of 1085 genetics professionals found 72 percent willing to do prenatal diagnosis for sex selection or refer (5).

Market researchers have always used surveys to identify markets for new tests and treatments. Many of these researchers work for nonprofit institutions or health departments, trying to predict demand for new tests or treatments. Surveys of potential uptake of presymptomatic tests for Huntington's disease (7) and carrier status for cystic fibrosis (8) were conducted before such tests become available, and results (indicating that most people at risk would take such tests) affected allocation of government and institutional funds. More recently, studies of willingness to take tests for mutations in BRCA1 and BRCA2 genes (partially predictive of breast cancer) have led to commercial outlays of plant and equipment (9). More general studies of public attitudes toward genetic testing in general and gene therapy (10,11) have enabled researchers, companies, and policy makers to identify trends toward increased public acceptance of genetic technologies, at least for some purposes. Surveys can also identify the extent of possible abuses of new technologies, such as "genetic discrimination" by insurers or employers (12–14), the testing of asymptomatic children for genetic mutations for adult-onset disorders (15), or perceptions of minority populations that they are being used as guinea pigs (16).

In sum, surveys of the public or stakeholders, including patients, professionals, or others (e.g., religious organizations) can help to prevent making policy decisions or ethical guidelines in a vacuum. Surveys provide a window on the opinions of those who will ultimately benefit or lose from research or policy. Surveys also identify differences among the various stakeholders. The public may think differently from the members of bioethics commissions; patients may (and do) think differently from the majority group (18); women may think differently from men (19). Surveys provide a means — albeit imperfect — for gauging some of these differences.

Limitations of Surveys

Most surveys involve hypothetical questions ("What would you do if...?"). In genetics, many surveys involve tests, treatments, or products not yet available. The reliability and validity of responses to hypothetical questions are always debatable. Reliability means that the respondent would answer the question the same way if given the survey instrument a second time. In other words, the survey measures the attitudes or intentions it was designed to measure. In real life it is almost impossible

to go back to the same respondents and re-ask the questions, especially if a survey was anonymous. In ethics, particularly, the reliability of surveys is always in doubt. The more difficult the ethical, social, or emotional implications of a question, the more likely it is that respondents themselves will report that they might answer the same question differently another time. One possible response to the reliability problem is to present analogous situations (perhaps in different case vignettes) in different portions of a questionnaire and to measure whether there is a meaningful cluster of responses pointing to a possibly stable underlying attitude (one statistical procedure used is called *factor analysis*). Another response is to conduct longitudinal surveys on the same population (e.g., the public) and to look for wide swings of opinion in the short term. Underlying opinions on most ethical issues do not change rapidly unless there is some well-publicized new discovery or other precipitating event, so a major change in 6 to 12 months suggests that the initial survey may have been unreliable.

The problem of validity (predictive value) is perhaps more important. Validity means that a respondent will actually do what he or she says, if presented with the "real" situation. One limitation of surveys is that the questions posed usually cannot describe accurately the complex characteristics of a hypothetical genetic test or predict whether it will actually provide benefits once it becomes available. Time and again research has predicted an active uptake of new genetic tests, yet when such tests actually became available, few used them (9). There are two major reasons for this lower-than-predicted uptake: (1) the uncertainty inherent in many of the tests, and (2) the absence of proven or acceptable prevention or treatment (20). For example, BRCA mutation testing provides only a "risk" of breast cancer (about 80 percent for women with a family history of BRCA-related cancer, somewhat lower for women without such a history), not a certainty. "Treatment" involves removal of breasts and ovaries and even then does not provide absolute certainty of freedom from the disease. For Huntington's disease there is no prevention or treatment. For autosomal recessive disorders, such as sickle cell anemia or cystic fibrosis, carrier testing may lead to difficult reproductive decisions, including decisions about prenatal diagnosis and abortion, but does not lead to treatment. People who say in surveys that they will take future genetic tests perhaps implicitly believe that treatment will become available at the same time as testing (11). When no treatment materializes, they do not take the tests. Although the validity of surveys may be improved by detailed explanations of what is and is not possible, lengthy explanations tend to reduce response rates and are difficult to apply to technologies still under development. Future technologies often turn out differently than presented in surveys. For example, no one could have predicted that there would be over 800 different mutations for cystic fibrosis or that a few individuals could have two mutations without the classical symptoms (21). The most controversial future technologies — germ-line gene therapy, genetic enhancement, and human cloning — will undoubtedly present as-yet-unforeseen scientific and technical possibilities posing new ethical problems.

SURVEY METHODS

Sampling

Most large-scale surveys use one of two methods to select samples: random probability sampling or quota sampling (22). Random probability sampling selects a proportion of the population to be sampled at random, to avoid researcher bias, and requires a comprehensive, accurate list of the elements to be sampled (census tracts, blocks, house numbers in the case of door-to-door or mail samples, telephone numbers in the case of random digit dialing). Usually a table of random numbers is used to select the first element in the survey, and interviewers count down the list and select elements at regular intervals, depending on the proportion of the population to be sampled. (For a 1 in 10 sample, every tenth telephone or house number is selected.) Interviewers continue to make visits or calls to the numbers selected until they contact someone. In budgeting for surveys, researchers usually allow for at least three or four visits or calls per selected household in order to obtain a response.

Stratified sampling is a modification of random probability sampling. It is used when researchers are especially interested in comparing several groups, for example, African-American and white women with and without a family history of breast cancer, in regard to their attitudes toward BRCA testing. In a stratified sample, the population would be organized into four groups (African-American women without a family history, African-American women with a family history, etc.) and an appropriate number selected at random from each group. Stratification ensures that each group of interest is included but limits the numbers in each group, thereby reducing the statistical "power" of finding real differences between groups, especially if these differences are small. Power always depends on the smallest group in a comparison. Usually this group must include at least 200 in order to provide an 80 percent likelihood of identifying a difference of 10 percent between groups.

Quota sampling differs from probability sampling. Quota sampling begins with a matrix based on relevant characteristics of the population to be sampled. Researchers first need to know the overall characteristics of the population, for example, sex, age, race, geographical area, income, or education. The population is divided into cells (e.g., white women aged 25–30, living in southern suburban areas, of Catholic background, with a college education and income of $30,000–40,000), and each cell is given a weight according to its percentage in the overall population. Interviewers seek to fill an assigned "quota" of respondents for each cell. They do not attempt repeated visits or calls to randomly preassigned addresses or telephone numbers.

The random probability sample is the "gold standard" of survey research, and is the method preferred by U.S. government agencies and used by the University of Chicago National Opinion Research Center's General Social Surveys (1), the U.S. Congress Office of Technology Assessment (10), and a 1995 survey of primary care physicians (3). The method is expensive, however, since a sampling frame (list of "units" — persons, house numbers, telephone numbers — to be sampled) must be assembled and interviewers have to make repeated attempts at contact. Quota sampling usually costs less than one-quarter the price of random probability sampling and can be completed much more quickly. Most political campaign polls use quota sampling. Some survey firms combine probability and quota sampling. Households are selected randomly, but individuals within each household are selected by quota. This is called *modified probability sampling*. At least one public survey of ethical views in genetics has also used modified probability sampling (23).

Other sampling methods include so-called convenience sampling (e.g., stopping people on a street corner or giving out questionnaires in a college cafeteria) and "snowball sampling" (letting the initial interviewees direct the researcher to other possible interviewees, who in turn suggest further interviewees). For precise statistical reporting of population views, neither method is adequate, though college students frequently use both. These methods can, however, be useful in anthropological studies of communities and in identifying questions for further large-scale research. The snowball method, by providing access to a cross section of willing interviewees sharing particular social characteristics, has enabled researchers to sample in depth the emotional lives of particular groups, for example working-class women (24). The convenience sample is sometimes the only sample readily available, and the researcher faces the choice of using this or not doing the research at all. Scholarly journals have published results of surveys using these methods, providing that authors acknowledge their limitations.

Survey Approaches: Interview, Telephone, Mailed Questionnaire

The various approaches are approximately equal in terms of reaching study populations. Most people now have telephones (although in some of the poorest rural or innercity areas, about 5 percent do not). People with unlisted telephones can be included in surveys by dialing from random number tables instead of phone books. People appear willing to answer phone questionnaires up to about 15 minutes in length (about the same length as in-person interviews) though some surveys have lasted up to half an hour. Mailed questionnaires have the advantage of allowing for greater length and complexity, but they require more tasks from the respondents and must be accompanied by stamped return envelopes. Sometimes written questionnaires are delivered by in-person interviewers, who wait while the respondent completes them. Many people feel more comfortable with methods that promise anonymity, such as mailed questionnaires or randomly dialed calls, especially if questions touch on deeply held personal beliefs (25). In-person and telephone methods now provide similar response rates. Questionnaires may produce lower response rates, but they can convey more complex questions and can use case vignettes.

Requirements for Consent

Participation in anonymous surveys does not require written informed consent of the type usually required in medical research. Anonymous studies usually fall under the educational exemption of the Federal Office for Protection from Research Risks (OPRR) rules, though researchers funded by the federal government or at institutions that receive federal funds must request this exemption from their Institutional Review Board. Answering the survey constitutes consent. Political pollsters and market researchers usually give a very brief description of the survey ("I'm going to ask you a few questions about..., which will take about ... minutes") and ask if the person is willing to participate. Questionnaires usually have a brief introduction describing the purpose of the survey, naming the individual researchers and institution conducting the research and stating that participation is voluntary. If sensitive questions are involved, the introduction gives a warning to this effect (sometimes repeated at intervals in the questionnaire) and tells people that they do not have to answer questions that make them feel uncomfortable. (Similar warnings can be given in voice interviews.) The most sensitive question of all—even in ethics questionnaires—is usually about income. Many people are reluctant to disclose how much money they make. Repeating the word "optional" before this question is advisable.

Anonymity can be preserved even in "before-and-after" surveys by placing matching numbers on sets of questionnaires. In studies where respondents receive payments after completing sets of questionnaires, anonymity can still be preserved by preventing names from reaching researchers and by destroying code numbers once payment is made. Surveys where researchers can match individuals' names to questionnaires or interviews require a signed informed consent document, provided that the survey is sponsored by an agency of the U.S. government or conducted at an institution receiving federal funds. Most medical and educational institutions receive federal funds.

Developing Survey Instruments

Focus Groups. Focus groups are frequently used to help develop questionnaires or interviews. A focus group is somewhat like a group interview focused on a certain topic (26). Focus groups usually consist of 6 to 12 people, often selected by research companies specializing in focus groups. There is no attempt at random selection. Focus groups are not surveys. Ideally participants will be as different from each other as possible. Over an approximately two-hour period, a facilitator gently leads the group through a list of areas that the organizers wish covered and also coaxes out the quieter members of the group. Comments by one member may reverberate in others, leading to a fuller range of responses than would occur in a one-on-one interview. The group is tape-recorded and transcribed. Researchers continue to conduct new focus groups until new viewpoints cease to arise. Focus groups are not "town meetings" of people with special agendas, nor are they neighborhood gatherings.

Participants are selected so as not to know others in the group, and usually do not include the most vocal activists. The transcripts help identify concerns and wordings that should be included in the survey instruments. Focus groups are also widely used by market researchers (to develop advertising) and politicians (to develop campaign strategies).

Wording. In developing the survey instruments, researchers will need to consult with some of the major interest groups, whether or not they have used focus groups. For example, in developing questions on relationships between biotechnology and people with disabilities, persons from groups in the disability community should be consulted. It is important to try to avoid language that some people may regard as insulting, such as "burden," "affected child," "defect," or phrases that put diseases ahead of persons, such as "Tay-Sachs baby" or Down syndrome case." The disability community prefers "person-first" language, such as "persons with mental retardation."

Choice of words can have strong effects on responses. For example, most people do not want to be "genetically engineered" or to eat food that is "genetically engineered." People are more receptive to "biotechnology" than to "genetic engineering" (27). Technical terms such as "germ-line gene therapy" are confusing to most people. An explanation, such as "correcting genes that would carry the disease to future generations" (11) may be better. Even "enhancement" is a questionable word for inclusion in a public survey. It is usually better to give an example, such as "increase athletic abilities" or "improve performance in school."

Sometimes researchers skew questions toward a particular type of response that they hope to find. For example, a researcher who hopes that people will agree to participate in biomedical research may say "Do you agree or disagree with the following statements: A. The benefits of the proposed research outweigh the risks. B. This study may lead to an important treatment. I would want to participate in this research myself." Both statements emphasize the benefits and ignore the risks of the research. The vast majority of people will agree with them. In most cases the surveyor constructs skewed questions inadvertently, rather than by design. Even bioethicists construct skewed questions. The usual methodological approach to counteract this is to construct an equal number of questions that may elicit the opposite response. For every question that may produce a positive response to biomedical research, a question phrased in a negative light (stressing risks or uncertainties) should be included. A balanced set of questions not only contributes to internal reliability of responses but is also useful in the data analysis.

Ideally surveys should be short. Exploring complex issues can require more lengthy instruments, however. Projects should not be rejected on the ground that "the public will never understand the issues." Researchers in developing nations have found that most people, including those with no formal education, understand complex concepts (including risk and the placebo effect) if these are explained adequately (28). Simplistic global questions, such as "Do you think the Human Genome Project poses greater risks than benefits?" tend to produce answers of limited usefulness. Questions about self-perceived knowledge may measure self-confidence rather than actual knowledge, unless accompanied by actual knowledge tests.

Cross-Cultural Surveys. Survey research can be conducted crossculturally (29). Biotechnology will affect populations differently, but human beings face similar issues. For federally funded surveys requiring translation, two steps are ordinarily required. First, the survey is translated and tested on several members of the population who will receive it. Second, the translation is "back translated" into English by an independent translator who has never seen the original and compared with the original English version so that mistakes and nuances of language can be corrected.

Pilot Testing. All survey instruments are ordinarily field-tested on several people, revised for clarity, and then pilot-tested on a larger group. Observers watch people completing questionnaires or interviews and then debrief them regarding comprehensiveness, emphasis, and clarity. Usually an instrument undergoes several rounds of pilot testing.

Response Rates

Some authorities consider 50 percent an acceptable response rate (22). Some journals, such as the *American Journal of Public Health*, have preferred 60 percent but are willing to publish surveys with lower response rates if they include an especially hard-to-reach group, such as physicians. A 70 percent response rate is good by anybody's standards (22). In-person quota surveys usually have very high (over 95 percent) response rates. Questionnaire surveys usually have low response rates (about 30 percent) on the initial mailing, but these are increased by successive mailings and telephone reminders. Most responses to the initial mailing arrive within two weeks. At that time, a second mailing, sometimes in the form of a postcard reminder, is sent, followed by a third and final mailing (sometimes a complete questionnaire in case the first one has been lost) in a week or two. Usually successive mailings produce about half the number of responses received in the previous mailing.

Physicians are a notoriously difficult group to survey and are difficult to reach even for a telephone reminder. Some survey firms specialize in physicians, charging approximately $100 per completed interview. A small payment of $25 for completing a questionnaire or interview can increase response rates dramatically, even for physicians.

Including Members of Minority Groups

Members of minority groups may hold considerably different views from those of the majority with regard to the ethical conduct of medicine and research, in view of past experiences where they were used as guinea pigs without their knowledge (31,32). Unfortunately, most survey research on ethical, legal, and social issues

in genetics, like most medical research (33), does not adequately reflect the views of minorities. Two possible reasons are (1) underrepresentation of minorities in some groups, notably service providers, and (2) extra costs of including adequate numbers of minorities in public and patient surveys, costs that may exceed a funding agency's customary limits.

According to the American Society of Human Genetics, of 4810 members based in the United States, 4031 (84 percent) are white, 604 (13 percent) are Asian, 57 (1 percent) are African-American, 34 (0.7 percent) are Hispanic, 30 (0.6 percent) are Native American, and 54 (1 percent) are "other." These percentages apply to the entire membership, which includes researchers as well as service providers and also includes many people who are not board certified. An examination of names (where sex can be attributed) suggests that about half of board-certified members in the United States may be women. The National Society of Genetic Counselors (the professional association for Master's level counselors) reports that 96 percent of its members are women, 93 percent are white, 4 percent are Asian, 1 percent are Hispanic, 1 percent are African-American, and 1 percent are "other" (34). Women are better represented in genetics than in some other medical specialties; according to American Medical Association data, 44 percent of pediatricians, 29 percent of obstetricians, and 23 percent of family practitioners are women (35). According to U.S. Census Bureau data, 21 percent of all physicians were women in 1990 (36). Some minorities, however, are underrepresented in genetics when compared with other medical specialties and with their proportions of the U.S. population. According to 1990 U.S. Census data on 587,675 physicians, 80 percent were non-Hispanic white, 11 percent were Asian, 5 percent were Hispanic, 4 percent were African-American, and 0.1 percent were Native American (37). In the 1990 census, African-Americans constituted 12 percent of the U.S. population, and Hispanics constituted 9 percent. Today Hispanics are almost 12 percent of the population, excluding Puerto Rico (38). These groups are underrepresented in medicine generally, but especially underrepresented in genetics. There are no national data on race, ethnicity, or sex of individuals or families receiving genetics services.

Surveys of minority groups may cost several times the fee for surveys of the general population. For example, because African-Americans represent only 12 percent of the adult U.S. population, it is necessary to contact successfully and to screen more than 4000 households in order to locate 500 that include an eligible African-American respondent for a door-to-door or telephone survey. Surveying Hispanic populations requires translation and use of bilingual interviewers. Spanish-speaking groups originating from different areas (e.g., Puerto Rico, Mexico, Cuba, Peru) hold different views and their responses cannot be lumped together under the general label "Hispanic" or "Latino." Each group must be surveyed separately. Reaching members of "low-incidence" groups such as Southeast Asians or Native Americans requires construction of special sampling frames, something that most survey research organizations are unable or unwilling to do.

It is not sufficient simply to conduct a general population survey and hope that minorities will be adequately represented. According to Roper Starch Worldwide, a survey research firm, the average English-speaking population sample achieved is 80 percent white, 10 percent African American, and 10 percent other or unknown. Telephone surveys may result in inclusion of even fewer African-Americans than door-to-door surveys, since fewer households have phones. In order to represent minority views, each group must be "oversampled." In survey research language, "oversampling" means that to get an adequate sample of a minority group, which represents its proportion of the population, it is necessary to solicit participation from proportionately more members of the minority group then are represented in the population as a whole. The low incidence and difficulty of reaching minority groups may quadruple the cost of a survey. For example, a survey of 1000 members of the U.S. public may cost around $25,000, while a comparable survey of only 500 African-Americans may cost a minimum of $50,000. A survey of Native Americans or Asians will probably be unobtainable from most survey research firms. Some researchers consider a survey of 500 persons inadequate to represent the views of an entire minority. The sampling error is plus or minus four percentage points, and many persons will never be reached, either door-to-door or by telephone. To represent all of America's various minorities in adequate numbers for analysis could be economically prohibitive by the usual funding standards for survey research.

Data Analysis and Presentation of Survey Results

Survey data are usually coded (put into numeric form) and entered into statistical programs in computers. The most common statistical packages are SAS (Statistical Analysis System) and SPSS (Statistical Package for the Social Sciences). Qualitative data (people's "write-in comments" on questionnaires or verbal responses to open-ended interviews that ask them to reply in their own words) can be quantified for purposes of statistical analysis. This requires development of categories to organize the responses. Each piece of numerical data is entered by two independent operators, whose work is compared for corrections, a procedure known as "punch and verify."

Survey results are usually expressed as one figure. However, if another randomly selected group of similar size were to be sampled, another figure might emerge. The results of separate surveys would fall on a bell curve, with 95 percent of these results falling within a certain range. Thus, if 80 percent of 1000 people surveyed give answer A, a statistician would say that the true figure is in the range of 77 to 83 percent with 95 percent confidence. In other words, we are likely to be right that the true value falls in this range 95 percent of the time, but there is a 1 in 20 chance that the answer is outside this range. The range is sometimes reported in tables as a "confidence interval"; the 95 percent is the "confidence level" for this interval.

VIEWS ON BIOTECHNOLOGY

Major Surveys

The most careful and comprehensive surveys were conducted for the U.S. Congress Office of Technology Assessment, which is no longer extant. These include a mail survey of the 500 largest U.S. Corporations and 11 labor unions, reported in *The Role of Genetic Testing in the Prevention of Occupational Disease* (1983); a survey of surrogate mother matching services, reported in *Infertility: The Medical and Social Choices* (1988) (40); a survey of 1575 physicians, 1213 fertility specialists, and 30 commercial sperm banks, reported in *Artificial Insemination: Practice in the United States* (1988) (2), and, most important, a 1000-member telephone Survey of Public Attitudes toward Biotechnology, Science, and Engineering, conducted in 1986 and reported in *Public Perceptions of Biotechnology* (1987) (10). The National Center for Human Genome Resources (NCGR), a private, nonprofit research organization in Santa Fe, New Mexico, commissioned a partial repeat of the OTA Survey of Public Attitudes in 1996, by the same survey researchers, in order to examine possible trends (11). In addition, the NCGR study surveyed 521 primary care physicians, 100 leaders of patient organizations, 102 research and development directors from the biotechnology industry, 76 genetic researchers, 50 science journalists, 50 religious leaders, 70 medical directors of insurance companies, and 79 federal and state policy makers. In the United States, Singer (41) has conducted public surveys on the ethics of utilizing prenatal diagnosis, including complex issues such as sex selection, and has examined the views of labor unions and religious groups.

In Canada, the Royal Commission on New Reproductive Technologies surveyed over 9000 people by mail question-naire and telephone, regarding opinions about use of fetal tissue and treatments for infertility (42).

In Europe, the Eurobarometer surveys in 1991, 1993, and 1996, commissioned by the European Union (EU) and conducted by INRA, a European network of market and public opinion research agencies, are the most comprehensive (27). In the most recent, Eurobarometer 46.1 (1996) public surveys were conducted in all 15 member states of the EU, plus Norway and Switzerland, using random probability sampling, for a total of 16,246 face-to-face interviews about public perceptions of risks and benefits of various aspects of biotechnology.

A series of large-scale surveys of portions of the Japanese public by Macer also looks at global attitudes, but response rates were about 24 percent, well below standards of acceptability in the West, although Macer claims it is average for Japan.

Two worldwide surveys of genetics professionals by Wertz and Fletcher, a 19-nation survey of 683 in 1985 (4) and a 36-nation survey of 2901 in 1995 (44) together with surveys of 499 primary care physicians, 476 genetics patients, and 988 members of the general public in the United States, concentrate largely on ethical issues in genetics services, rather than research, but some questions touch on new technologies.

Overall Views on Biotechnology

In the United States, the public's self-perceptions of adequacy of knowledge increased between 1986 and 1996 (11). The percent who thought they knew what a gene is increased from 85 to 91 percent; those who thought they understood the meaning of "human gene therapy" increased from 29 to 49 percent. In 1996, 68 percent knew that scientists were trying to map the human genome, 48 percent thought this effort would have at least a moderate effect on themselves or their families, and 53 percent knew that genetic tests were available. About 7 in 10 approved of gene mapping. Among the various leadership groups, 87 percent approved mapping the human genome, and 92 percent expected improvement in the early diagnosis of disease in the next ten years; 88 percent of patient group leaders, 78 percent of industry representatives, and 68 percent of scientists thought there would be at least moderate improvement in treatment of chronic diseases, though notably fewer scientists (13 percent) thought there would be "a lot" of improvement, as compared to 42 percent of patient organizations and 39 percent of policy makers. A majority of all leaders (62 percent) said they followed the scientific literature fairly closely, but only 17 percent followed it very closely, including 17 percent of patient organizations, 8 percent of policy makers, 7 percent of insurers, and 0 percent of the media. All groups thought that funding would present the biggest frustration for scientists in biotechnology, though 27 percent of the biotech industry also cited government regulations. Overall, few leaders saw research confidentiality (3 percent), patient confidentiality (2 percent), or discrimination (1 percent) as the biggest source of frustration. Four out of five (81 percent) in the leadership sample expected current efforts to identify markers of genetic disease to have at least a moderate effect on society. Nine out of 10 (92 percent) expected society to benefit from medical applications of biotechnology in the next 10 years, including 51 percent who expected "a lot" of benefit. Notably fewer religious leaders (20 percent) expected a lot of benefit than patient organizations, scientists, and industry (65–67 percent). About two-thirds (65 percent) thought that medical applications of biotechnology would pose some risk to society, ranging from 86 percent of religious leaders and 80 percent of policy makers to 49 percent of scientists and 53 percent of industry. Only 15 percent thought there would be "a lot" of risk. Most (85 percent) thought the benefits of biotechnology would outweigh the risks, while 8 percent thought risks would outweigh benefits. Leadership groups were divided with regard to government regulation. About 40 percent would like to see regulations left as they are now, including 57 percent of scientists, 47 percent of industry, and 27 percent of insurers; 27 percent would like more stringent regulation, including 44 percent of religious leaders, 37 percent of policy makers, 30 percent of insurers, and 17 percent of scientists and industry; 18 percent would like less stringent regulation, including 32 percent of industry but only 5 percent of policy makers.

In the 1996 Eurobarometer survey (27), 50 percent thought that biotechnology would improve life, with Finland the most optimistic and Greece, Norway, and Germany the least optimistic; 15 percent thought it would make things worse, with Austria, Norway, and the Netherlands the most pessimistic. When questions used "genetic engineering" instead of biotechnology, optimism fell to 39 percent and pessimism rose to 27 percent. Although 54 percent thought biotechnology would lead to cures for most genetic diseases in the next 20 years, 70 percent thought it would create dangerous new diseases. Overall optimism about telecommunications, computers, solar energy, and new materials exceeded optimism about biotechnology. Optimism about biotechnology decreased slightly, but significantly, between 1991 and 1996, from 51 to 48 percent. About half of respondents had heard about and talked with someone about biotechnology in the three months preceding the survey. "Textbook knowledge" did not increase significantly between 1993 and 1996; there was wide variation among countries, and generally greater pessimism in countries with lower levels of textbook knowledge, such as Austria. There was great national variation in beliefs about heritability of human characteristics. In all, 62 percent, ranging from 78 percent in Ireland and 77 percent in the Netherlands to 39 percent in France, thought that human intelligence was mainly inherited; 17 percent, ranging from 32 percent in Ireland to 9 percent in Switzerland, believed that attitudes toward work were mainly inherited; 15 percent, ranging from 24 percent in Italy, and 23 percent in Austria, to 10 percent in Denmark and 9 percent in Sweden, thought criminality was mainly inherited; 25 percent, ranging from 39 percent in the Netherlands and 34 percent in Germany to 13 percent in France and 17 percent in Portugal, thought homosexual tendencies were mainly inherited. There are no simple cultural explanations for these findings.

Europeans thought that international organizations such as the WHO (35 percent), or scientific organizations (22 percent) were better able to regulate biotechnology than national governments (17 percent) or the EU (6 percent) (27). Only 24 percent thought that existing regulations were sufficient. Respondents placed the greatest confidence in consumer organizations to tell the truth about biotechnology, followed by schools and universities. Public authorities were way down the list. Over half of respondents (54 percent) thought that "irrespective of regulations, biotechnologists will do what they like" especially in Denmark (71 percent), Switzerland (65 percent), France, Germany, and the United Kingdom (all 60 percent). The Japanese surveys, which were also distributed in New Zealand for comparison, used the words "genetic engineering," "genetic manipulation," and "genetically modified organisms," and the framing of questions makes comparison with European and U.S. surveys difficult. Nevertheless, they suggest a considerable amount of fear about and concern over new developments in biotechnology, as well as need for more education (43).

Views on Genetic Testing

In 1996, 53 percent of the United States public said that they were aware that genetic tests were available, including 42 percent of high school graduates and 68 percent of college graduates (11). More than 9 out of 10 (93 percent) approved of the use of genetic information for early diagnosis of disease; 85 percent approved of presymptomatic genetic testing for diseases that occur later in life, including 43 percent who approved strongly. About three-quarters (73 percent) thought it was "a good thing for a healthy person to be able to find out how likely they are to get a serious disease in the future," while 17 percent thought it was "bad" to know. More women (21 percent) than men (13 percent) thought it was bad to know. Those who thought knowledge was good cited the possibility of prevention (57 percent), changing lifestyles (11 percent), finding a cure (10 percent), making decisions about having children (7 percent), or early treatment (6 percent), but 23 percent simply said they wanted to know in advance. Those who thought foreknowledge was bad cited worry (34 percent), negative influence on life plans (27 percent), depression (11 percent), fatalism (no need to know) (13 percent), negative self-image, or emotional effects (12 percent). Perspectives on why knowledge is good or bad may explain one of the most persistent problems in survey research on genetics: why the majority of people say that they would take a test and then do not take it once it becomes available. People who think foreknowledge is good implicitly associate the knowledge gained through testing with prevention or eventual cure. If a test becomes available without acceptable means of prevention or cure, most people are not interested (20).

Nine out of 10 (88 percent) approved the use of genetic tests to find out whether future children are likely to have a serious genetic disease. The 1996 NCGR survey did not specify what kind of testing this meant (prenatal diagnostic or parental carrier testing). A 1995 survey of 476 genetics patients (44) (mostly working-class Catholic white women bringing children in for evaluation) found that 64 percent agreed that "before marriage, responsible people should find out whether they could pass on serious diseases or disabilities to their children" (64 percent also agreed in a 1994 public survey); 81 percent thought that "a woman should have tests on the unborn baby if she is at risk of having a child with a serious disease or disability" (62 percent agreed in the public survey); and 80 percent thought that "tests on unborn babies should be available to all women who request them." Fewer patients (21 percent) thought that "a woman should have an abortion if tests say the unborn baby has a serious disease or disability"; 43 percent were neutral on this question.

In Europe, 81 percent thought it was "useful for society to use genetic testing to detect diseases that we might have inherited from our parents," and 72 percent thought that people should be encouraged to take such tests (11). In Japan, most thought prenatal testing (76 percent) and presymptomatic testing (73 percent) should be available under national health insurance (43).

The level of certainty provided by a test might be expected to affect public acceptance of it (20). This was not the case in the 1996 NCGR survey. Most people were willing to accept some uncertainty (11). Only 27 percent said tests should be made available only if they predicted

with certainty that someone would develop a disease. Most of the rest were willing to settle for tests that could indicate "a high risk." ("High" was not defined in the survey.) Treatability of the condition had less effect than might be expected on the degree of risk that people would tolerate. For serious, untreatable conditions 27 percent thought tests should be available even if they indicate only a slightly increased risk, 20 percent would require a moderately increased risk, and 40 percent thought tests should be available only if they indicate highly increased risk. Slightly higher percents would approve tests for slightly (36 percent) or moderately (26 percent) increased risk if a condition were treatable. Most (94 percent) thought doctors should "advise" people with a family history of cancer to take a genetic test for cancer, and 48 percent thought doctors should advise everybody to take such a test.

Reports of personal willingness to take genetic tests are of dubious validity, tending to exaggerate greatly the numbers who would take such tests. In 1996, 65 percent in the NCGR survey (11) said they would take a test indicating whether they would develop a fatal disease later in life, almost the same percent as in 1986. There was a statistically significant decline between 1986 and 1996 from (83 to 76 percent) in the number who said they would take a genetic test, before having children, that would indicate whether their children would inherit a fatal genetic disease. Nevertheless, the percent who would be tested on behalf of future children exceeds those who would be tested on their own behalf. A review of recent similar studies of willingness to be tested indicated high percentages of acceptance of BRCA testing, as in earlier surveys, but little evidence that people were actually requesting tests (45).

In Japan, 63 percent of the public, when asked the same question, said they would have tests before conceiving children; 53 percent would have presymptomatic tests for themselves, and 76 percent would have prenatal diagnosis (43). Percents were not substantially different among students, general public, university staff, and scientists.

Prenatal Testing

A survey of labor unions and Protestant religious groups found that most either supported genetic testing and prenatal diagnosis or (in the case of unions) had no position (41). A survey of public attitudes toward using prenatal diagnosis to select the sex of the child found that only 5 percent approved this use; however, when presented with the "hard-luck case" of a couple with four children of the same sex, a substantial minority (38 percent) supported sex selection (41). In a 1995 survey, 72 percent of U.S. genetics professionals, 68 percent of U.S. primary care physicians, 59 percent of patients, and 38 percent of the general public thought the doctor should do prenatal diagnosis in this case (percentages for professionals include those who would offer a referral) (44). This points to another potential weakness of surveys: People may answer one way on a general question and another way in response to a more concrete situation. In medical practice the exigencies of the concrete situation usually win. This is why some ethics surveys employ case vignettes.

Surveys of women having prenatal diagnosis for genetic disorders generally document increased anxiety, need for sympathetic and accurate counseling, reduction in anxiety after receiving favorable results, and willingness to undergo the procedure in a future pregnancy (46). For women whose tests indicate presence of a genetic condition, studies document the difficulty of decisions (46). Women who already have a child with a genetic condition are often reluctant to abort a fetus with the same condition (47). Most women who choose abortion recover psychologically within months, but a small minority continue to have strongly negative emotions (48). In general, Americans appear reluctant to use selective abortion. In a 1994 public survey describing eight fetal conditions, the majority would not abort for any condition listed, including severe mental retardation, with the child "unable to speak or understand" combined with death in the first few months of life; 48 percent would abort in this situation, 47 percent would abort for severe mental retardation accompanied by a nearly normal lifespan, and 41 percent would abort for paralysis from the neck down, with no retardation and a normal lifespan. Smaller percents would abort for mild retardation ("child could live independently") (17 percent), severe incurable mental disease appearing at age 40 (21 percent), moderate retardation ("could communicate but not live independently") (22 percent), "gross overweight" (16 percent), and "child not of the sex desired" (7 percent). For all of these conditions, except sex selection, majorities believed that abortion should be legal for others.

Testing Children

The genetic testing of asymptomatic children for disorders that occur later in life has occasioned much discussion. A survey of U.S. laboratories found that most had no comprehensive policies and many had performed such tests (15). Other surveys found that geneticists in English-speaking nations and Northern/Western Europe rejected the practice, but majorities of those in other parts of the world would test children for mutations for Huntington's disease or Alzheimer's disease at parents' request (49).

Perspectives on Confidentiality

Access to an individual's genetic information by spouses and genetically related family members is one of the most controversial areas in bioethics. Although surveys of genetics professionals indicate that most believe that spouses should not have access to information without the individual's consent (44), surveys suggest patients and the public think otherwise. In one set of surveys, 20 percent of the public thought spouses should have general access without consent; among patients, substantial minorities thought spouses should have access to their partner's genetic information without consent if there were a risk that a child could inherit mental retardation (43 percent), if the spouse/partner had mutations predisposing to mental illness (31 percent), or alcoholism (23 percent), or if the spouse/partner

was a carrier of cystic fibrosis (30 percent). In the NCGR Survey, 75 percent thought spouses should have access, but the question did not specify whether this was with or without consent (11). There was far less support (9 to 12 percent) in any group for access for blood relatives, at least in general questions. When the situation was presented as a case vignette describing Huntington's disease, however, 75 percent of patients and 38 percent of U.S. genetics professionals thought the doctor should tell the relatives, against the patient's wishes (50). These results point to divisions within the professional community and differences among professionals, patients, and the public. Beliefs about privacy are different outside English-speaking nations and Northern/Western Europe, where the unit of privacy is the individual. In most of the world, the unit of privacy is the family (50–52). Surveys point to an overwhelming consensus against access for employers and insurers without a person's consent; many respondents believed that these institutions should have no access at all, even with consent.

Eugenics

An international survey of geneticists found little support anywhere for state-mandated testing or sterilization (53). However, substantial percents of geneticists in developing nations (especially China and India) and Eastern Europe believed that "reducing the number of deleterious genes in the population" was "an important goal of genetics." Except in the English-speaking world, majorities would offer purposely pessimistically slanted information after prenatal diagnosis, so that people would abort without the professional suggesting it directly. Overall, genetics professionals held a pessimistic view about disability (53).

Gene Therapy

The NCGR survey suggests a trend toward public acceptance of gene therapy. In 1996, 87 percent approved of "correcting genes that cause serious illness" (11). The percentage who believed that "changing the genetic makeup of human cells is morally wrong" decreased from 42 percent in 1986 to 22 percent in 1996. In 1986, 83 percent approved of "changing the makeup of human cells to cure a usually fatal genetic disease"; in 1996, 85 percent approved (not a statistically significant increase). There was a significant increase (77 to 84 percent) in those who approved changing human cells "to reduce the risk of developing a fatal disease later in life." There has been a significant decline in approval of gene therapy for enhancement ("to improve the physical characteristics children would inherit.") In 1986, 44 percent approved this use; in 1996, 35 percent approved this use. Approval was highest among those with less than a high school degree (61 percent), lowest among college graduates (28 percent). Among leadership groups, 25 percent of patient organizations but only 12 percent of scientists approved this use.

The majority of the public approved both somatic cell and germ-line gene therapy. In 1996, 68 percent thought doctors should be allowed to correct both the gene affecting the disease in the patient, *and* the gene that would carry the disease to future generations, an increase from 62 percent who approved in 1986. Four of five (83 percent) primary care physicians thought that doctors should be allowed to correct genes carrying disease to future generations. Majorities in all leadership groups, including 79 percent of patient organizations, 74 percent of religious leaders, 79 percent of insurers, 72 percent of policy makers, 66 percent of industry, and 55 percent of scientists (who may be more aware of the risks) approved of both somatic cell and germ-line gene therapy.

Genetics in the Workplace

The OTA survey of industry suggested that most companies were not using genetic testing, though some had used it in the past (mainly for sickle cell trait, which is irrelevant to occupationally-related disease) and some expected to use it in the future (39). One study found that while employers supported genetic testing to identify presumably susceptible workers and move them to less hazardous jobs before damage occurs, unions preferred genetic monitoring, which means testing to see whether cellular or molecular damage is actually occurring (54). The unions' argument was that genetic testing could have low predictive value for disease and would lead to unfair discrimination against many workers. In a 1995 international survey there was agreement among genetics professionals everywhere that testing should be voluntary, the worker should have access to the results, and no one else should have access without the worker's consent (44). Half of U.S. patients, however, thought that testing should be required, apparently because they thought testing would protect the worker. In the United States, the Americans with Disabilities Act (ADA) now prevents employers from refusing to hire because of family history or presymptomatic tests, provided that the person is able to do the job.

Genetic Discrimination

Although many people believe firmly that insurers and employers are using genetic information to deny insurance or employment, surveys have found little evidence that companies are singling out genetic information for special treatment. What surveys have found is considerable fear of discrimination. A survey of educated members of genetic consumer groups found that most were afraid that information would be used against them (12). The majority of Americans thought that health insurers (85 percent) and employers (59 percent) will probably ask applicants in the future to take genetic tests (11). In Europe, 41 percent of the public thought insurance companies would use genetic tests to set premiums within the next 20 years (43). In the United States, surveys of medical directors of insurance companies and state insurance commissioners suggest that insurance companies intend to rely primarily on family histories, as they have always done (55–57). In an attempt to assess the prevalence of genetic discrimination, one group of researchers sent out over 30,000 questionnaires to members of support groups for families with Huntington's disease, hemochromatosis, and sickle cell anemia in anticipation of identifying

asymptomatic people who had been discriminated against solely on the basis of genotype (13). The response rate was about 3 percent, of whom about half reported some form of discrimination. In another set of surveys (14), 1084 genetics services providers reported 76 clients refused employment and 474 refused life insurance because of genetic predisposition or carrier status, for a total of 550 persons refused employment or insurance. The 499 primary care physicians in the survey, with a median of 14 years in practice and 100 to 150 patients per week, reported 29 patients refused health insurance on the basis of genetic predisposition. Patients were asked "because of a genetic disability or disease, have you or a member of your family been refused employment or health or life insurance?" Two percent reported being denied or let go from a job, 3 percent were refused health insurance, 7 percent were refused coverage for some services, 5 percent were refused life insurance, and 1 percent were refused school admission. Most patient descriptions fell within the scope of employment and insurance practice generally and were only indirectly related to genetics and not at all to genetic testing.

Views of Minority Groups

A recent two-wave survey of 500 African-Americans and 500 members of the public before and after President Clinton's May 16, 1998, apology for the unethical treatment of African-Americans in the Tuskegee Syphilis Study (31) showed dramatic differences between African-Americans and the general public on many questions related to medical research in general (16). For example, almost three-quarters of African-Americans thought they were very likely (36 percent) or somewhat likely (38 percent) "to be used as guinea pigs without their consent," as compared with 16 and 34 percent of the general public (16). A 1995 survey on ethics and genetics, while reaching too few African-Americans for proper analysis, found significant differences in responses to half the ethical questions (18).

Forensics

Survey responses have shown overwhelming approval of DNA identification, especially for law enforcement agencies. Over 90 percent of genetics professionals and 95 percent of patients believed that persons convicted of serious crimes (not only sex crimes) should be required to have DNA fingerprinting and that the DNA should be kept on permanent file, like regular fingerprinting (44). Majorities of both professionals (59 percent) and patients (72 percent) would also require DNA fingerprinting for persons charged with, but not convicted of, serious crimes. Most professionals (80 percent) and patients (86 percent) favored DNA fingerprinting of members of the armed forces, to identify casualties. Most patients (73 percent), but fewer professionals (37 percent), would DNA fingerprint newborns to prevent mixups in the nursery. Half the patients, but only 20 percent of professionals, would require DNA fingerprinting for passport applicants. Almost half the patients (47 percent) would require it for people receiving welfare, to prevent fraud, and 34 percent would require it for credit applicants. In the Eurobarometer surveys, 70 percent thought that genetic fingerprinting would lead to solving more crimes in the next 20 years (27).

Genetically Modified Organisms

The largest surveys of public attitudes have been done in Europe (27). In all, there is greater support for introducing human genes into bacteria to produce medicines or vaccines and for introducing genes into crop plants to make them more resistant to insect pests than there is for using modern biotechnology in food production, for developing genetically modified animals for cancer research, or for xenotransplantation (27). Majorities regarded all these technologies as useful for society, but majorities also found food production (60 percent), genetically modified animals for laboratory research (52 percent), and xenotransplantation (60 percent) risky. Only minorities of respondents believed it morally acceptable to develop genetically modified animals for laboratory research (44 percent), to use biotechnology in food production (48 percent), or to produce animals for xenotransplantation (35 percent). Japanese surveys indicated considerable public concern about the health effects of eating genetically modified foods (43).

CONCLUSION

Surveys are useful in gauging public optimism or pessimism about biotechnology, identifying sources of concern, and pointing to differences of opinion among stakeholder groups. Most surveys have shown that publics, especially in the United States, are generally optimistic about biotechnology and believe that benefits outweigh risks. Majorities support genetic research, testing and gene therapy, including germ-line gene therapy. In the United States, fears center on possible misuses of information by insurers and employers. There appear to be some substantial differences among views of professionals, patients, and the general public. The views of minority groups on issues specific to biotechnology remain largely unknown. In some areas of ethical concern, such as views on research uses of biological samples, potentially useful surveys are still lacking.

BIBLIOGRAPHY

1. National Opinion Research Center, *General Social Survey, 1972–1997, Cumulative Codebook*, University of Chicago, Chicago, IL, 1998.

2. United States Congress, Office of Technology Assessment, *Artificial Insemination: Practice in the United States: Summary of a 1987 Survey—Background Paper*, No. OTA-BP-BA-48, U.S. Government Printing Office, Washington, DC, 1988.

3. D.C. Wertz, Primary Care Physicians' Knowledge of Genetics, *Am. J. Hum. Genet.* **61**(4), A 56 (1997).

4. D.C. Wertz and J.C. Fletcher, *Ethics and Human Genetics: A Cross-cultural Perspective*, Springer-Verlag, New York, 1989.

5. World Health Organization, Hereditary Diseases Programme, *Proposed International Guidelines on Ethical Issues in Medical Genetics and the Provision of Genetics Services*, WHO, Geneva, Switzerland, 1998.

6. D.C. Wertz and J.C. Fletcher, *Soc. Sci. Med.* **46**(2), 255–273 (1998).

7. G.J. Meissen and R.L. Berchek, *Am. J. Med. Genet.* **26**, 283–293 (1987).

8. M.M. Kaback, in D. Lawson, ed., *Cystic Fibrosis: Horizons*, Wiley, New York, 1984.

9. *Science* **275**, 782 (1997).

10. United States Congress, Office of Technology Assessment, *New Developments in Biotechnology. 2: Public Perceptions of Biotechnology-Background Paper*, No. OTA-BP-BA-350, U.S. Government Printing Office, Washington, DC, 1987.

11. J.M. Boyle, *National Survey of Public and Stakeholders' Attitudes and Awareness of Genetic Issues*, National Center for Genome Research, Santa Fe, NM, 1997.

12. E.V. Lapham, C. Kozma, and J. Weiss, *Science* **274**, 621–624 (1996).

13. L.N. Geller et al., *Sci. Eng. Ethics* **2**, 71–78 (1996).

14. D.C. Wertz, *Health Law Rev.* **7**(3), 7–8 (1999).

15. D.C. Wertz and P.R. Reilly, *Am. J. Hum. Genet.* **61**, 1163–1168 (1997).

16. S.B. Thomas, *N.Y. Times*, April 27 (1997).

17. D.C. Wertz, *Health Law J.* **6**, 1–42 (1998).

18. D.C. Wertz, *Comm. Genet.* **1**(3), 175–179 (1998).

19. D.C. Wertz, *J. Am. Med. Wom. Assn.* **52**(1), 33–38 (1997).

20. S. Shiloh et al., *Psychol. Health* **13**, 1071–1086 (1998).

21. United States Department of Health and Human Services Secretary's Advisory Committee on Genetic Testing, *A Public Consultation on Oversight of Genetic Tests*, U.S. Government Printing Office, Washington, DC, 1999.

22. E.R. Babbie, *The Practice of Social Research, Second Edition*. Wadsworth, Belmont, CA, 1979.

23. D.C. Wertz, *Ethical Issues in Genetics Study*, Roper-Starch Worldwide, New York, 1994.

24. L.B. Rubin, *Worlds of Pain: Life in the Working-Class Family*, Basic Books, New York, 1976.

25. R.M. Lee, *Doing Research on Sensitive Topics*, Sage Publications, Thousand Oaks, CA, 1993.

26. D.L. Morgan, *Planning Focus Groups*, Sage Publications, Thousand Oaks, CA, 1998.

27. G. Gaskell, M.W. Bauer, and J. Durant, in G. Gaskell, M.W. Bauer, and J. Durant, eds., *Biotechnology in the Public Sphere: A European Sourcebook*, Science Museum, London, 1998, pp. 189–216, 231–298.

28. Nuffield Council on Bioethics, *The Ethics of Clinical Research in Developing Countries*, The Nuffield Council, London, 1999.

29. D.C. Wertz, in R. DeVries and J. Subedi, eds., *Bioethics and Society: Constructing the Cultural Enterprise*, Prentice-Hall, Upper Saddle River, NJ, 1998, pp. 145–164.

30. E.S. Tambor et al., *Am. J. Pub. Health* **83**(11), 1599–1603 (1993).

31. J.H. Jones, *Bad Blood: The Tuskegee Syphilis Experiment—A Tragedy of Race and Medicine*, Free Press, New York, 1981.

32. P.S. Eichstaedt, *If You Poison Us: Uranium and Native Americans*, Red Crane Books, Santa Fe, NM, 1994.

33. Advisory Committee on Human Radiation Experiments, *The Human Radiation Experiments*, Oxford University Press, New York, 1996.

34. S. Goering, in A. Dula and S. Goering, eds., *It Just Ain't Fair: The Ethics of Health Care for African Americans*, Praeger, Westport, CT, 1994, pp. 182–192.

35. K.A. Schneider and K.J. Kalkbrenner, *Perspect. Genet. Couns* **20**, S1–S8 (1998).

36. American Medical Association, *Physician Characteristics and Distribution in the United States*, 1996, American Medical Association, Chicago, IL, 1996.

37. United States, Department of Commerce, Economics and Statistics Administration, *Bureau of the Census: 1990 Census Population, Supplementary Reports, Detailed Occupation and Other Characteristics from EEO File for United States*, U.S. Government Printing Office, Washington, DC, 1992, p. 2.

38. United States, Bureau of the Census, *Estimates and Projections 1998*, U.S. Government Printing Office, Washington, DC, 1998.

39. United States Congress, Office of Technology Assessment, *The Role of Genetic Testing in the Prevention of Occupational Disease* U.S. Government Printing Office, Washington, DC, 1983.

40. United States Congress, Office of Technology Assessment, *Infertility: Medical and Social Choices*, No. OTA-BA-358, U.S. Government Printing Office, Washington, DC, 1988.

41. E. Singer, *Pop. Res. Pol. Rev.* **10**, 235–255 (1991).

42. Royal Commission on New Reproductive Technologies, *Social Values and Attitudes Surrounding New Reproductive Technologies*, Final Report, vol. 2, Royal Commission, Ottawa, Canada, 1993.

43. D.R.J. Macer, *Attitudes to Genetic Engineering: Japanese and International Comparisons*. Eubios Ethics Institute, Tskuba, Japan, 1992.

See other entries BEHAVIORAL GENETICS, HUMAN; EDUCATION AND TRAINING, PUBLIC EDUCATION ABOUT GENETIC TECHNOLOGY; GENETIC DETERMINISM, GENETIC REDUCTIONISM, AND GENETIC ESSENTIALISM; MEDIA COVERAGE OF BIOTECHNOLOGY.

R

RECOMBINANT DNA, POLICY, ASILOMAR CONFERENCE

CHARLES WEINER
Massachusetts Institute of Technology
Cambridge, Massachusetts

OUTLINE

INTRODUCTION

The development and laboratory use of recombinant DNA research techniques in the early 1970s made it possible to manipulate genes, opening the path to genetic engineering. The research and its potential applications were embroiled in controversy from the beginning and became one of the major ethical and public policy issues in the 1970s and 1980s. Unresolved questions persist through the present and the earlier experiences of scientists, citizens and policy makers continue to influence perceptions and actions today. In the 1970s the concern was about the potential health and environmental hazards of laboratory use of these novel research techniques. The researchers, their institutions, and their funding agencies developed a system of self-regulation to avoid hazards and to forestall legislative control. They focused on the means not the ends, on the tools of genetic engineering rather than on the moral limits. Rapid and pervasive commercialization of academic research in the field in the 1980s provoked continuing controversies on the role and nature of universities, the effects of corporate ties on education and on research goals and communication processes, and the threat of potential conflicts of interest of commercially involved academic researchers whose expert advice was sought on ethical and public policy issues. Commercial applications of recombinant DNA and related genetic engineering techniques continue to raise public concern about environmental and health hazards in agriculture and medicine. And the moral limits to applications of human genetic engineering are deeply controversial (1,2).

SCIENTISTS' ORIGINAL CONCERN ABOUT HAZARDS

At the Gordon Research Conference on Nucleic Acids in July 1973, invited specialists on DNA research heard fascinating reports of new techniques for manipulating and moving genetic material. The use of the newly discovered restriction enzymes made it possible to cut strands of DNA at specified precise points and to insert them into the DNA of other organisms, combining the hereditary material of animals and bacteria. These recombinant organisms could be replicated in billions of copies through cloning. It was apparent to the involved scientists that they now had a tool for studying the structure and functions of genes and to probe the details of DNA and its transcription in cells of higher organisms. Biologists recognized that this would open up a new field of work, enabling the posing of fundamental research questions that would not have been feasible before. They expected that the answers to these questions would help solve problems at the forefront of knowledge with important applications.

Amid the excitement about the potential of the new recombinant DNA technique some of the conference participants were alarmed over its possible immediate hazards in their own laboratories. They were concerned that using some of these hybrid DNA molecules might cause unforeseen hazards to human health and the environment. There was a possibility that harmless microbes could be unintentionally changed into human pathogens through introduction of antibiotic resistance, which was part of the technique; through the production of dangerous toxins, which was a possible outcome; or through the transformation into cancer-causing agents of materials that previously were benign. In this relatively new field there was a great deal of uncertainty and little information about the hazards.

The Gordon Conference participants asked for a special discussion of these larger questions. At that brief special session, they decided to write a letter to ask the National Academy of Sciences to study the potential hazards and to devise a plan to do something about them. They voted by a large majority to compose the letter and they approved the content of it. They also voted, this time by a slim majority, to send a copy of the letter to be published in *Science*. The reluctance of many of the participating scientists to call public attention to the problem was an indication of a continuing conflict. They were concerned about a possible public health problem, and yet they feared that talking about it publicly might bring intrusion, as they saw it, into the scientific process. The Gordon Conference letter, replete with technical language, was intended for other scientists (1, pp. 70–80). It was published in *Science* in 1973 and did not generate much public attention (3).

THE BERG COMMITTEE LETTER

The National Academy asked Paul Berg, a distinguished biochemist and a principal researcher in the field, to organize a group of scientists to consider the issues. They met at MIT on April 17, 1974, and planned a conference for February 1975 to evaluate the hazards of the research and ways of dealing with them. Feeling a sense of urgency, they also drafted a letter to alert the larger community of biologists. Two months after the MIT meeting, Berg described these actions and the group's motivations in a letter to a colleague in England:

> We met at MIT for a day and settled on the idea of calling a conference next February of those scientists working on methods of joining DNA molecules and particularly those involved in constructing hybrid DNAs. It was our plan that one of the major purposes of the Conference, besides a report on the scientific progress, would be a wide ranging discussion of potential hazards growing out of these types of experiments. Were there any experiments that should not be done? How could such a moratorium be proposed or enforced? In short, we expected a frank and searching review of what people were doing or wanted to do, particularly from the point of view of whether they should be done. But as we talked we realized that the pace of events might not wait for February and that some of the experiments many people would agree could be hazardous would be done by then (e.g., attempts to fuse portions of Herpes DNA to appropriate plasmids for cloning in E. coli were imminent). Since the technology for constructing hybrids has become ridiculously simple, that fear was well founded.
>
> Consequently we decided to devise a letter to be submitted to Science and Nature calling on scientists to defer certain kinds of experiments until these potential hazards could be better evaluated and certainly until there was an opportunity to discuss the issues at the February meeting (4).

Drafts of the Berg committee letter were circulated privately among the relevant scientists, and in July 1974 the final version was published in Science and in Nature (5). Why did the letter go public? Because the committee felt it was the quickest way to bring the potential hazards to the attention of the community of researchers who would be likely to use the new recombinant DNA techniques. They felt that the situation was urgent, because of pending experiments and because the power and fruitfulness of these research tools rapidly would attract many scientists to the field who were not experienced in handling pathogenic organisms. The letter called for a voluntary moratorium, a temporary deferral of those experiments which at the time were thought to be potentially hazardous. This appeal for self-restraint was linked to an end point, the conference scheduled for February 1975.

The response to the letter by the relevant scientific community was generally favorable. When the draft was read at the Cold Spring Harbor meeting in June, 12 of the European scientists in attendance immediately drafted a letter to John Kendrew, director of the European Molecular Biology Organization, requesting urgent consideration of the matter. They felt that it was essential that research utilizing the new recombinant DNA techniques be made possible in Europe by providing appropriate special risk laboratories.

When the Berg et al. "moratorium letter" was published in Nature, it was followed by responses from leading British biologists. Michael Stoker, head of the Imperial Cancer Research Fund wrote:

> No doubt a good many dirty tricks have been attempted and discarded by nature in the course of evolution, but the disquiet arises from the utterly novel associations of genetic material which are now possible. The potential benefits should, therefore, be delayed, not for ever, but until consequences can be assessed, and preliminary experiments carried out under conditions of maximum security. . . .[I]t is encouraging that the very leaders in the field have taken the initiative and have been supported by the [National Academy of Sciences]. It is now to be hoped that academics and learned societies in other countries will add their weight, and that international organizations such as the European Molecular Biology Organization will lend support. . . .For many it will be a test of self denial and social responsibility in the face of strong intellectual temptation (6, p. 278).

Kenneth Murray of the University of Edinburgh stated:

> The NAS request is both reasonable and responsible and deserves to be universally respected. It recognizes both the difficulty in evaluating real or potential hazards that may be involved in such work, as well as the obvious criticism that these will remain obscure in the absence of experimental study; urgent consideration of the latter is explicitly recommended. Fears that the proposed limitations to experiments will seriously obstruct research in vital areas of biology seem unfounded. The NAS initiative, by focusing attention on the hazards involved, could well promote rather than hinder work on in vitro recombination in animal viral systems, an area believed by many to hold the key to gene therapy in its broadest terms. . . .[I]f we follow the moderate tone set by the NAS we shall be careful not to oversell the social benefits devolving from recent experiments (7, p. 279).

THE 1975 ASILOMAR CONFERENCE

The February 1975 meeting at the Asilomar Conference Center in California evaluated knowledge in the field and its potential for research. It was the equivalent of an international review conference which ordinarily would be held well into the development of a research field and not at such a very early stage. The detailed review enabled the conference participants, who were the researchers and the potential researchers in the field, to consider the potential risks and ways to control them. The motive from the start was to avoid public interference and to demonstrate that scientists on their own could protect laboratory workers, the public, and the environment. Of course, there is that contradiction again: They were dealing with a public health issue and simultaneously attempting to keep the public out of it.

Initially it was not clear whether any media representatives would be allowed to attend the publically funded conference, but later the organizers decided to limit press attendance to eight invited reporters. A deal was struck with 16 journalists, most of whom were invited, that they would not report on the conference until it was over, because things would be too much in flux. That pleased the reporters because they did not have to call in stories to their editors every day. Instead, the telephone booths were jammed with scientists calling their laboratories in Europe and the United States about the need to tool up for this very exciting new research. The conference gave them an opportunity to learn as much as possible about the recombinant DNA techniques, and it stimulated the growth of the field while producing a framework for pursuing it safely.

Several technical working groups met independently over a period of months in preparation for the conference. The most active was the Plasmid Working Group, focusing on the circular pieces of DNA which were the main tools for this new technique. They scoured the literature and their own knowledge, talked with other people in the field, and produced a detailed technical document (8). Reports of the working groups were presented and discussed at the meeting and one session was devoted to presentations of lawyers on policy and liability issues. Participants paid special attention to their legal responsibility for damage resulting from their laboratory work.

The narrow technical focus of the conference was evident in the opening remarks of David Baltimore, one of the organizers. He first acknowledged that the techniques that were developed could have applications in a number of areas, including biological warfare, and that it had larger societal implications, but that such issues would be excluded, since there was a full agenda of technical issues:

> The issue that ... [brings] us here is that a new technique of molecular biology appears to have allowed us to outdo the standard events of evolution by making combinations of genes which could be immediate natural history. These pose special potential hazards while they offer enormous benefits. We are here in a sense to balance the benefits and hazards right now and to design a strategy which will maximize the benefits and minimize the hazards for the future (9).

What happened at Asilomar? The recombinant DNA issue was defined as a technical problem to be solved by technical means, a technical fix. Larger ethical issues regarding the purposes and the long-term goals of the research were excluded, despite the rich discussions that had occurred among geneticists and other biologists in the 1960s about where to draw the line when it became possible to do genetic engineering. The 1960s discussions led to congressional proposals for anticipatory study of the ethical limits of genetic engineering, which were resisted as premature by several leading biologists (10). Instead of those longer-term issues, the focus at Asilomar in 1975 was on safety of the newly developed technical tools for genetic engineering, on the means not the ends.

The Asilomar participants adopted provisional safety guidelines based on a two-part system of physical and biological containment of potentially hazardous recombinant organisms (11). The extent of physical containment was graded according to the anticipated level of hazard an organism might present if it escaped the laboratory, ranging from good laboratory technique for those experiments deemed to be of low hazard, to hooded glove boxes, negative pressure, showers and clothes changes for laboratory workers dealing with organisms thought to be especially dangerous. Biological containment would introduce mutations in the organisms that were to be used in the experiments so that if they escaped they could not survive in the environment beyond the laboratory.

RECOMBINANT DNA ADVISORY COMMITTEE

In November 1974 the NIH had established the Recombinant DNA Advisory Committee (RAC), advisory to the director of NIH. The first meeting was held immediately after the Asilomar conference at the end of February 1975. RAC appointees were knowledgable researchers in the field, who were asked to develop and extend the Asilomar provisional safety guidelines to control all recombinant DNA work at institutions receiving NIH funding of any kind. They were designing safety protocols that had the potential for restricting their own work. These controls were to be administered by the NIH, which funded and encouraged the research and therefore was itself in a position of conflict of interest. NIH officials acknowledged the potential conflict, and maintained that although NIH was not a regulatory agency, it had the best expertise in the field and needed to act in the absence of any other government group playing a role. Similar efforts were also underway in other countries.

During 1975 and 1976 scientists on the RAC argued about whether the proposed guidelines were too strict or too permissive, and the document went through many drafts (12). All of this occurred in the absence of risk assessment experiments. At the same time scientists at laboratories throughout the country were tooling up to use the new technique and were impatiently waiting for the green light that would allow them to proceed as rapidly as possible. They exerted a great deal of pressure on RAC and NIH. The process of establishing safety rules involved a series of compromises aimed at achieving a consensus within that portion of the scientific community affected by the guidelines while providing assurances to the public that they would be protected from possible hazards.

LOCAL AND NATIONAL POLITICAL RESPONSES

The long expected NIH safety guidelines for recombinant DNA were approved by the director of NIH on June 23, 1976. On that day when the green light flashed, an extraordinary event took place in Cambridge, Massachusetts. Scientists from MIT and Harvard and representatives of NIH appeared at a special City Council hearing. They had been invited to explain to the citizens of Cambridge why the scientists themselves had been arguing about the safety of recombinant DNA and whether the guidelines were adequate to protect the communities in which the research was to be done. Was there any danger to citizens? Who was going to monitor and enforce the safety standards? Could the scientists and their universities be trusted to regulate themselves? Testimony by several biologists that recombinant DNA techniques posed few risks and that they could be contained by the new guidelines was countered by testimony from other biologists who argued that the guidelines were inadequate and that they were formulated by self-interested advocates of the research. After a second hearing in July 1976 the City Council established a citizens' review board to examine the problem and, pending the outcome of the board's deliberations, placed a temporary ban on experiments classified in the guidelines as posing moderate to major hazards.

The nine member Cambridge Experimentation Review Board met twice weekly for a total of more than 100 hours over a four month period. About one half of the time was used for testimony by scientists on both sides of the issue. The board presented its report to the city council on January 5, 1977, recommending the creation of a city biohazards committee to oversee adherence to the NIH guidelines for all recombinant DNA work in the city whether funded by NIH or not, and several additional safeguards on experimental procedures, containment and testing of organisms (1, pp. 302–307; 13). These community confidence building measures were incorporated in a city council ordinance passed in February 1977, which was the first recombinant DNA legislation in the United States and was interpreted as a qualified public endorsement of the NIH guidelines (14).

A major fear of the recombinant DNA scientists was that their own early concern about laboratory safety had initiated public scrutiny of the new research. This was emphasized by the events in Cambridge and in other communities such as Ann Arbor, Cambridge, San Diego, New Haven, and Princeton where academic biologists were tooling up to use recombinant DNA techniques. By 1978, 16 separate bills had been introduced in Congress to regulate recombinant DNA safety standards by making the NIH guidelines mandatory for both publically and privately funded research and providing enforcement and punishment provisions for any violations. Research universities and scientific organizations saw this local and national activity as public "overreaction" threatening their control of laboratory safety procedures and their research funding. They vigorously lobbied to oppose or influence legislation. Several prominent biologists who had shared the early concern about possible safety hazards of the research publicly recanted, and a resolution to Congress signed by most of the participants in a 1977 Gordon Conference stated that they previously had overstated the risks and now could provide reassurance that the work was safe (15). In the end no legislation was passed by Congress.

EVOLUTION OF RESEARCH GUIDELINES

By 1979 the NIH Recombinant DNA Guidelines had been made far more permissive than the original 1976 version. More than 90 percent of U.S. research in the field was either no longer covered by the guidelines or was subject to only minimal controls equivalent to standard laboratory practice. By 1982 most experiments subject to the guidelines were controlled at the local level through institutional biosafety committees and RAC reviewed only research that had the potential for special safety problems. No demonstrated harm had been caused by the research as conducted under the guidelines. A limited amount of risk assessment research had been done during that period, and several small consensus workshops of scientists in the field were held to review existing knowledge and to refute the earlier concerns (1,2). NIH's approach to the guidelines was that they would be flexible enough to respond to new scientific knowledge. That also opened them up to flexible response to pressures from researchers and their

interests, pressures from industry, and pressures from national policy priorities and political interests.

BEGINNINGS OF COMMERCIALIZATION

Downgrading of the guidelines coincided with rapid commercialization of the field and the involvement of academic scientists in biotechnology companies. In November 1974 during the moratorium period, a patent for the recombinant DNA technique was filed by Stanford University and the University of California on behalf of two of the scientists who developed the technique. The patent was granted in 1980 after the Supreme Court decision allowing patenting of human-made organisms (16). Biologists and their universities became involved in what soon became almost a complete commercialization of the work. In the 1980s political climate of deregulation the U.S. biotechnology industry was promoted as a national priority. Emphasis was on government, industry, and media claims of medical, practical, and economic benefits of the research and the need to develop the industry. Critical questions about the health and environmental safety of research techniques and products were met by arguments that if the United States did not move forward rapidly in biotechnology, the country would lose out in international competition. The "gene gap" argument was deployed to resist special regulation of the field.

ENVIRONMENTAL RELEASE OF MODIFIED ORGANISMS

As the guidelines faded away for most laboratory work, attention shifted from the accidental escape of genetically engineered microorganisms to the intentional release of these organisms into the environment for agricultural purposes. The U.S. Department of Agriculture (USDA) and the U.S. Environmental Protection Agency (EPA), the agencies who would ordinarily become involved, initially claimed that they did not yet have the expertise to evaluate the possible hazards and they urged NIH to provide safety oversight for these applications through the RAC. Evaluation by the RAC seemed like a very comfortable approach for scientists and companies who had been working with it. In the absence of federal legislation for recombinant DNA, industry had been in voluntary compliance with the NIH guidelines. It was not until 1984 that EPA issued an interim policy statement on field testing of genetically engineered microbial pesticides. By that time NIH had approved proposals for small scale field testing of a genetically modified organism that was to be sprayed on strawberry and potato plants to prevent frost damage. The "ice-minus" controversy of the mid-1980s involved approvals by NIH, EPA, and California agencies, legal challenges by genetic engineering critic Jeremy Rifkin, congressional hearings, and protests and demonstrations by citizens in the community where field testing was to occur. As in Cambridge several years earlier the citizens asked, "Why are we the last to know?" The test plot was definitely in their backyard, but they were not informed of its exact location. They were also concerned about unresolved safety questions raised by ecologists. By the time the tests were finally conducted in 1987, RAC's role in approval of environmental release of genetically modified organisms had been superceded by EPA (17).

HUMAN GENE TRANSFER EXPERIMENTS

The RAC also played a transitional role in the oversight of experiments in human gene transfer, generally referred to as gene "therapy" to reflect the as yet unrealized hopes of its advocates. In 1983 the RAC responded to the report of the President's Commission on Bioethics' study of genetic engineering which considered several approaches to the oversight of future human genetic engineering. The commission's study was initiated after the leaders of the three major U.S. religious groups wrote a letter to the President stimulated by the 1980 Supreme Court decision permitting patenting of genetically engineered organisms. They called for study of the ethical issues associated with genetic engineering and observed that "no government agency or committee was currently exercising adequate oversight or control, nor addressing the fundamental ethical issues in a major way" (18). RAC's response to the Commission's report was to establish a Working Group to consider whether it would review proposals for human gene transfer. In 1985 RAC's "Points to Consider in the Design and Submission of Human Somatic Cell Therapy Protocols" was issued by NIH. The RAC said it would be willing to review proposals for human gene transfer protocols for somatic cells but would not "at present entertain proposals for germ line alterations" (19,20). When pressed by a public interest group, the Council for Responsible Genetics, to specifically ban human germ-line engineering, the committee refused. Leroy Walters, the bioethicist who had been for many years a member of RAC and was the head of its human gene therapy subcommittee, subsequently argued that in his view voluntary programs of germ-line genetic intervention were "ethically acceptable in principle" (19).

Gene therapy became the primary task of the group in the late 1980s, and since then it has dealt with the scientific validity of proposals as well as risks for human subjects, the adequacy of informed consent, the role of local institutional review boards, and the liability of researchers. RAC nurtured the development of human gene therapy by applying the clinical standards of biomedical ethics, but bypassed the larger ethical issue of whether it should be done at all. The role of RAC remained as advisory to the director of NIH. In 1995 the Food and Drug Administration (FDA) became the regulatory oversight agency for human genetic engineering, with RAC playing an advisory role in reviewing proposals involving novel techniques or applications (21–24). The adequacy of this approach was questioned by Congress and government agencies when revelations and allegations about violations of regulations were reported in the media in September 1999. These included abuses of informed consent procedures, failure to report adverse effects and harm to human subjects, and possible commercial conflicts of interest. Several clinical trials were shut down by FDA and the situation was under intense review in 2000.

LIMITATIONS OF SELF-REGULATION

Throughout RAC's history—from its creation in 1974 to deal with initial concerns about laboratory safety to its current role in human gene transfer experiments—it has been friendly to researchers and dominated by their interests. At the same time the work of the committee has been relatively open and visible. NIH made efforts to create a full public record of RAC deliberations and documents in addition to the announcements of meetings, proposed changes in the guidelines and decisions required to be published in the Federal Register. However, very few citizens read that relatively inaccessible, small print publication. Nor do many people have the opportunity to travel to Bethesda, Maryland, to sit in on committee meetings. The RAC minutes list the noncommittee members who attended the meetings. As the commercialization of genetic engineering increased from 1980 on, the record shows that representatives of companies were consistently present to follow the deliberations and look after their interests.

Public participation on the RAC was broadened in 1978 in response to complaints that it was dominated by self-interested researchers. Yet there were built-in limits and constraints to this participation because most of the issues placed before the committee were technical and often beyond the expertise of the nonscientists. Another problem was that RAC was increasingly asked to review industry proposals. Biotechnology companies were in voluntary compliance with the guidelines and sought NIH approval for their recombinant DNA work with the condition that proprietary information would be kept confidential, as was the practice with federal regulatory agencies, even though NIH was a research-supporting agency. As a result, public representatives on the committee frequently were not able to report to the public about information relevant to environmental and public health.

The development of genetic engineering clearly involves more than the safety issues that have been the major focus of RAC's mandate and activities. The larger ethical concerns about where to draw the line in applications of genetic engineering were occasionally discussed when raised by some members of the committee or at the request of outside groups. RAC, however, resisted taking a stand against the use of recombinant DNA techniques for biological warfare and refused to recommend an unambiguous ban on the review of proposals for human germ line intervention (25). Instead, RAC's emphasis was to develop safe procedures for the research, focusing on *how* to do it rather than *whether* it should be done. As Leon Kass observed in 1997, "the piecemeal formation of public policy tends to grind down large questions of morals into small questions of procedure" (26). Recombinant DNA research was safer as a result of the NIH guidelines developed by RAC. The biologists at Asilomar in 1975 and the subsequent generations of RAC members raised important safety issues and set standards for good laboratory practice.

Despite the success in improving the safety of research, the quasi self-regulation model developed in the recombinant DNA controversy is not adequate for expressing and enforcing societal and moral limits for potential genetic engineering applications such as human cloning or human germ-line interventions. These potential applications are not inevitable, and they raise profound issues beyond laboratory and environmental safety and patients' rights. They occur in a context of increasing genetic determinism, pervasive commercialization, and aggressive efforts to sell genetic intervention as a cure-all for medical and even social problems. Separation of the technical issues from the ethical issues, and the narrowing of ethical concerns to clinical biomedical ethics, limits meaningful public involvement and obscures the larger picture.

ACKNOWLEDGMENT

This historical summary draws on my observations and documentation of the recombinant DNA controversy from 1975 to the present, utilizing archival materials and interviews collected from 1975 to 1979 in a project under my direction and deposited in the Recombinant DNA History Collection available for study at the Institute Archives and Special Collections, Massachusetts Institute of Technology. Portions of this account are included in C. Cranor, ed., *Are Genes Us? The Social Consequences of the New Genetics*, 1994, pp. 31–51 and C. Weiner, *Health Matrix* **9**, 289–302 (1999).

BIBLIOGRAPHY

1. S. Krimsky, *Genetic Alchemy: The Social History of the Recombinant DNA Controversy*, MIT Press, Cambridge, MA, 1982.

2. S. Wright, *Molecular Politics: Developing American and British Regulatory Policy for Genetic Engineering, 1972–1982*, University of Chicago Press, Chicago, IL, 1994.

3. M. Singer and D. Soll, *Science* **181**, 1114 (1973).

4. P. Berg to H. Kornberg, June 18, 1974, Recombinant DNA History Collection, Massachusetts Institute of Technology, Cambridge, MA.

5. P. Berg et al., *Science* **185**, 303 (1974); *Nature* **250**, 175 (1974).

6. M. Stoker, *Nature* **250**, 278 (1974).

7. K. Murray, *Nature* **250**, 279 (1974).

8. Plasmid Working Group, *Proposed Guidelines on Potential Hazards Associated with Experiments Involving Genetically Altered Microorganisms*, Asilomar Conference, February 24, 1975. Recombinant DNA History Collection, Massachusetts Institute of Technology, Cambridge, MA.

9. Audiotape of the International Conference on Recombinant DNA Molecules, Asilomar, February 24, 1975. Recombinant DNA History Collection, Massachusetts Institute of Technology, Cambridge, MA.

10. U.S. Senate, 1968. *Hearings before the Subcommittee on Government Research of the Committee on Government Operations*, 90th Congr., 2nd Sess., on S.J. Res. 145, a joint resolution for the establishment of the National Commission on Health, Science & Society, March 7, 8, 21, 22, 27, 28 and April 2, 1968, p. 5.

11. P. Berg et al., *Science* **188**, 994 (1975).

12. National Institutes of Health, Office of Recombinant DNA Activities, *Guidelines for Research Involving Recombinant DNA Molecules*, National Institutes of Health, Bethesda, MD, 1976, in press.

13. Cambridge Experimentation Review Board, *Guidelines for the Use of Recombinant DNA Molecule Technology in the City of Cambridge*, January 5, 1977.

14. R. Goodell, *Sci. Technol. Hum. Values* (Spring), 36–43 (1979).

15. W. Gilbert, *Science* **197**, 208 (1977).

16. *Diamond, Commissioner of Patents and Trademarks v. Chakrabarty*, No. 79–136. Argued March 17, 1980, decided June 16, 1980.

17. S. Krimsky and A. Plough, *Environmental Hazards: Communicating Risks as a Social Process*, Auburn House, Dover, MA, 1988, pp. 75–121.

18. U.S. President's Commission for the Study of Ethical Problems in Medicine and Biomedical and Behavioral Research, *Splicing Life: A Report on the Social and Ethical Issues of Genetic Engineering with Human Beings*, U.S. Government Printing Office, Washington, DC, 1982, pp. 95–96.

19. L. Walters and J.G. Palmer, *The Ethics of Human Gene Therapy*, Oxford University Press, New York, NY, 1997, pp. 148–151.

20. I.H. Carmen, *Am. J. Hum. Genet.* **50**, 245–260 (1992).

21. H. Miller, *Wall Street J.* A18 (May 10, 1994).

22. L.E. Post, *Hum. Gene Therapy* **5**, 1311–1312 (1994).

23. S. Krimsky, *Hum. Gene Therapy* **5**, 1313–1314 (1994).

24. D.T. Zallen, *Hum. Gene Therapy* **7**, 795–797 (1996).

25. National Institutes of Health, *Recombinant DNA Advisory Committee, Minutes of Meeting, January 30, 1989*, National Institutes of Health, Bethesda, MD, 1989, pp. 19–37.

26. L. Kass, *New Republic* No. 18, 17–26, (June 2, 1997).

See other GENE THERAPY entries.

RELIGIOUS VIEWS ON BIOTECHNOLOGY, BUDDHISM

RONALD Y. NAKASONE
Graduate Theological Union
Berkeley, California

OUTLINE

INTRODUCTION

The modifications and improvements of living organisms, including human beings, and the development of microorganisms through biotechnology to produce or modify products to improve plants or animals challenge Buddhists to reexamine their doctrines, sharpen their interpretative insights, and expand their moral imagination. Buddhist views on biotechnology require an understanding of the spiritual goals that govern the Buddhist life, its doctrines, and the practical demonstration of its ideals. This article begins with a review of Buddhism's origins and spread, continues with a description of the major doctrines, and surveys selected Buddhist responses to biotechnology. Since modern science and technology and population increases have resulted in problems neither confronted nor anticipated by the Buddha and, until very recently, by Buddhist thinkers, this article also describes the Buddhist attitude toward change and new knowledge. Buddhist attitudes toward scientific and technological manipulation of human beings and nature, however, are often tempered by indigenous traditions such as Confucianism and Shinto, by modernization, and by Western culture.

Buddhist interest in biotechnology is relatively recent, and its impact has been felt most keenly in technically advanced countries, such as Japan, and technologically advanced regions of China and other countries. Buddhist thinkers of Third World countries and their devotees who struggle to survive from day to day are concerned with agricultural yields, clean water, political stability, and basic medical care. The dearth of reflections on biotechnology from developing countries such as Cambodia, Laos, and Vietnam indicate their preoccupation with meeting basic survival needs. For these Buddhists exotic high tech innovations, such as mapping of the human DNA or organ transplants, are inconceivable luxuries. While Buddhist thinkers in Japan are reflecting on biotechnology, their thinking is dominated by modern Western bioethical paradigms. They are, however, beginning to reshape questions into Buddhist categories and envisioning Buddhist solutions. The speeds at which change is taking place and new possibilities are emerging, outpace, at least for the moment, religious and moral thinking.

ORIGIN AND SPREAD

Siddhārtha Gautama (563–483 B.C.E.) (1) founded the Buddhist faith and community in what is now northeast India and Nepal. The community remained a single unit until about a hundred years after the founder's death. Disagreements centering on the status of the *arhat*, the archetype personality, the status of Śākyamuni, the historical Buddha, and the *vinaya*, the monastic rules of conduct, split the early community into the Sthavira and the more liberal Mahāsaṅghika. The Sthavira argued for the perfectibility of human-nature in the guise of the *arhat*, the humanity of the historical Buddha, and strict observance of the *vinaya* outlined by the Buddha. Present-day Theravāda, an offshoot of the Sthavira, claims to observe the faith established by its founder. The Mahāsaṅghika, on the other hand, believed that the *arhat* is not completely free from impurities. Its devotees understood Śākyamuni to be a manifestation of a transcendental Buddha who is pure, infinite, and eternal. The tradition stressed the spirit of the *vinaya*, not its letter. These initial disagreements led to further splintering. Buddhist documents mention the existence of as many as 34 monastic sects between the second and fourth centuries after the Buddha's passing.

The origins of Mahāyāna (great vehicle) Buddhism is obscure. Many scholars believe that Mahāyāna (2) emerged during the two centuries between 100 B.C.E. and 100 C.E. from ideas advanced by the Mahāsaṅgika and related sects, and as a reaction against the aloofness of the monastic sects. Others, notably Akira Hirakawa, a Japanese Buddhologist, argue that Mahāyāna began as a lay movement that appeared immediately after the death of the historical Buddha. These devotees who honored the memory of the Buddha at the *stūpas*, memorials that housed his relics, evolved over time their own liturgies, doctrines, and institutions (3). Still others argue that Buddhism's encounter with non-Indian peoples and ideas and the increased influence of the laity stimulated the monastic tradition to redefine their goals, iconography, and doctrines to be more inclusive. These early Mahāyānists referred to themselves as *bodhisattvas*, "beings who aspire for wisdom." The bodhisattva, an outgrowth of the idealization of the historical Buddha, vows to save all beings before he or she achieves full enlightenment.

After subduing his adversaries, the Gupta king, Aśoka (circa 274–236 B.C.E.) embraced the Buddhist faith and sent missions throughout India, Sri Lanka, North Africa, Macedonia, and Central and South Asia. These missions initiated the gradual spread of Buddhist and Indian culture. Sri Lankan chronicles report the establishment of Theravāda Buddhism in the later half of the third century B.C.E. By the eighth century C.E. Indian culture, including Buddhism, stretched from the east coast of the Indian subcontinent all the way to Vietnam, and Bali in the Indonesian archipelago. Theravāda eclipsed Mahāyāna and remains dominant today. Buddhist missions did not leave a lasting impact in North Africa and the Near East. Meanwhile, to the north, Buddhism fared better. It had a well-established presence in Central Asia by the second century B.C.E. Among the many schools, Mahāyāna and the Sarvāstivāda, an influential branch of the Sthavira linage, were the most strongly represented. Buddhist culture established itself in Khotan, Kucha, Turfan, and other city states that straddled the caravan routes to China. Buddhism entered China sometime during the first century B.C.E. and the beginning of the common era (4), and gradually became an integral part of the national life. Distinct Chinese forms of Buddhism emerged between 500 and 800, its most prosperous and creative period. Buddhism continued its eastward advance into Manchuria, Korea, and Japan. It officially entered Korea in 372, and by about 525 had penetrated the entire peninsula. Buddhism arrived in Japan in 552. After six hundred years, the Japanese evolved forms of Buddhism that suited and reflected their temperament.

Mahāyāna in the form of Vajrayāna, later called Tantric Buddhism, traveled to Tibet in the seventh century.

It conceived the Buddha as a cosmic body and as the substance of all things and all beings. By harnessing the forces that pervade the universe, the devotee can achieve Buddhahood in this very life and this very body. Tibetan monks carried their faith to Mongolia in 1261 and again in 1577. The fourteenth Dalai Lama (1935–), the spiritual and secular leader of Tibet, fled his occupied country in 1959 with tens of thousands of other Tibetans. The Chinese claim that Tibet has always been part of China.

Since the mid-nineteenth century when Chinese laborers joined the California gold rush, Asian immigrants have carried their Buddhist faith to Hawaii, North and South America, and other parts of the globe. The 1893 World's Parliament of Religions in Chicago introduced Buddhism and other non-Christian faith traditions to the West. The United States and other Western nations today have sizable ethnic Buddhist communities. D.T. Suzuki (1870–1966) and Alan Watts (1915–1973) popularized Zen Buddhism at midcentury and spawned a still small, but vital American–Buddhist community. Political refugees from Southeast Asia introduced Theravāda Buddhism during the latter half of the century. At present, Theravāda Buddhism is dominant in Sri Lanka and the Southeast Asian countries of Burma, Kampuchea, Laos, and Thailand. Mahāyāna exists in North and East Asia. Tibet, Mongolia, China, and Japan have substantial Buddhist populations. In a once Buddhist country, South Korean Buddhists constitute about one-fourth of the current population of 50 million. The social reformer and former Minister of Law, Bhimrao Ramji Ambedkar's (1891–1956) conversion to Buddhism in 1956 generated a Buddhist rival in India after 700 years of its disappearance from the land of its origin.

BELIEFS AND DOCTRINES

Siddhārtha Gautama began his spiritual journey with the question of human suffering that accompanies old age, sickness, and death. After six years of spiritual exercises Gautama realized the Dharma, the truth of *pratītyasamutpāda* (dependent co-arising or interdependence) and became the Buddha, "the Enlightened One." The Buddha awakened to the truth that all things and all beings are mutually related and mutually dependent. *Pratītyasamutpāda* represents the ideological content of the enlightenment and is the common theme throughout Buddhist thought and practice (5). The history of Buddhist thought can be understood to be simply an unfolding of the implications inherent in this central idea.

This overview begins with a discussion of the temporal and relational aspects intrinsic to *pratītyasamutpāda*. It proceeds to explain the notions of *karma*, *samsara*, *nirvana*, *anātman*, and other key Buddhist ideas, and the Four Noble Truths within the context of *pratītyasamutpāda*. The following section, The Buddhist Posture, discusses the implications for our interest in biotechnology.

Pratītyasamutpāda can be understood to be an extension of karma, the law of cause and effect. The idea of karma, literally "action," appeared approximately two or three centuries before the birth of Siddhārtha

Gautama and is closely associated with the notions of samsara, literally "passage," and personal responsibility. There are three classes of karma: good, bad, or morally neutral. An individual's present station in life has been determined by the moral quality of action, or karma, generated in the past. Deeds committed in the present life affects one's status in the next. Buddhists divide karma into three categories: mental, verbal, and physical. Early Buddhists exerted considerable effort debating whether the essential nature of karma is mental or physical. Theravāda concluded that the mental is the essence of karma. Volition, which is mental activity, generates verbal and physical action. It is thus essential, if one wishes to realize nirvana or spiritual peace, to quicken thoughts that generate behavior that lead to that end. The notion that it is not possible to escape the consequences of one's deeds is intrenched in present day Theravāda Buddhist cultures. Sri Lankan Buddhists, for example, explain the death of an impaired infant as the result of the working of karma (6). Present-day Thai women considering abortion of a HIV-afflicted fetus or an unwanted pregnancy from rape or forced prostitution weigh the consequences of an unfavorable rebirth from their poverty as they struggle to be faithful to Buddhist teachings (7). While the idea of karmic retribution is also very strong among Mahāyāna devotees, the inexorable consequence of karmic action is mitigated by the compassion of Buddhas and Bodhisattvas.

The ideas of personal karmic retribution and successive lives are part of the fabric of popular Indian thought and played a key role in the development of Indian Buddhist doctrine. Rebirth, however, is not a necessary tenet of the Buddha's teaching and was not central to the development of East Asian Buddhist thought. Chinese, Korean, and Japanese beliefs in spirits and soul were not based on rebirth (8). Moreover the Buddha maintained that claims of rebirth, like questions of whether life continues after death, are not empirically verifiable and cautioned against such speculations. Does the Enlightened One exist after death? Or not exist? are two of 10 questions that the Buddha refused edification. Hakuin, (1685–1768), the Japanese cleric, offered a similar response when a wealthy parishioner queried about the nature of death: "Why ask me?" The parishioner replied, "Because you are a Buddhist monk." Hakuin retorted, "But not a dead one." Rather than engaging in endless speculation, the Buddha proposed that we deal directly with those problems that will ease human suffering and lead to spiritual ease.

In addition to the temporal understanding of karma, *pratītyasamutpāda* also describes the simultaneous presence of cause and effect. The individual threads of the warp and woof of a piece of fabric illustrate this expanded notion of karma. Individual threads constitute the entire fabric; the fabric in turn defines each thread in relation to all other threads. The metaphor illustrates the mutual dependency of cause and result and, by extension, the mutuality of all things. In a mutually dependent universe each individual does not simply exist in the world. By being involved in the world, he or she helps to create the world through the manner in which he or she thinks, speaks, and lives. This understanding of *pratītyasamutpāda* provides a vision of identity and responsibility to all beings and all things, and quickens a sense of gratitude for all things and beings. *pratītyasamutpāda* dissolves the preoccupation with the self and gives rise to sentiments of compassion and service to others.

Mahāyāna documents that appeared during the first century of the common era interpret *pratītyasamutpāda* to be compassionate and morally purposeful. The *Larger Sukhāvattvyūha sūtra* casts the doctrine of *pratītyasamutpāda* in the myth of the Bodhisattva Dharmākara, a spiritual hero who vows to forgo supreme enlightenment until all beings enter the Pure Land, the realm of spiritual ease. Dharmākara and other spiritual heroes accomplish this monumental task by transferring the vast store of merits they have accumulated over innumerable eons to the spiritually impoverished. *Parināma*, or the transference of merit, is a soteriological idea based on the belief that an individual's life is irrevocably linked with all beings and things. Bodhisattvas and Buddhas do not literally withdraw merits from their merit repositories and deposit them in another's. Merits are "transferred" in the sense that we benefit from their spiritual exercises. *Parināma* softens the harsh and uncompromising individualism of karma. In contrast, the rigidly individualistic view of karma dissolved society into isolated individuals, and it fails to acknowledge the mutuality among all beings and the complexity of the human experience. Also in an interdependent world, countless karmic forces intersect to often thwart our noblest intentions and propel us to violate our deepest instincts.

Though Theravāda and Mahāyāna Buddhism emphasize differing aspects of *pratītyasamutpāda*, they agree that change is the nature of reality, that suffering is endemic to the human condition, and that *nirvana* or spiritual bliss is a transcendent reality. For Theravāda, *nirvana* means the transcendence of *samsara*, the realm of suffering that is associated with change. Mahāyāna, on the other hand, identifies *nirvana* with *samsara* and thus speaks of spiritual release in the world of change. Both traditions assert the doctrine of *anātman* (non-self) which allows a person to identify and empathize with others. Early Buddhists explained *anātman* by analyzing the body and mind through five aggregates: form, sensation, perception, mental formations, and consciousness. A person, composed of these five constantly changing and mutually dependent aggregates, is devoid of a substantial and abiding self. Further investigation by the early Buddhist thinkers expanded these 5 categories into 75 dharmas or elements of existence. Eventually form was interpreted to include all material things. Mahāyāna accepted the view of the non-self but advocated a more radical view that dharmas themselves are without substantial and enduring reality. The Mahāyāna view is supported by the *prajñapāramitā hrdaya sūtra* or Heart Sutra which proclaims: "Form is emptiness [*śūnyatā*] and emptiness [*śūnyatā*] form." Later Nāgārjuna (circa 150–250), a Mahāyāna thinker, asserted "*Pratītyasamutpāda* is *śūnyatā* [emptiness]." The empirical person, the result of the coming together of countless dharmic elements and conditions correlates with "form." "Emptiness" of the self in early Buddhism evolved to designate the intrinsic reality of an individual devoid of all accidental characteristics or *śūnyatā*. *Śūnyatā* thus refers to the "suchness" or "thusness" of the person.

The Buddhist notion of *anātman* came under stiff attack. Critics asked: "If there is no-self, how does one account for the need for rebirth? Who is the agent responsible for action and change? Who becomes enlightened?" These questions arose from a misunderstanding of *anātman*. While the Buddha spoke against the psychological and nonrelational reality of an independent self, he never denied the ontological self (9). The *anātman* doctrine describes a relationship whereby any given person or thing derives its being and meaning, not from itself but from its relations with others. Simply, each individual is defined by his or her role in society and is affirmed by his or her interactions with others. Partly as a reaction against its critics and to affirm the great reverence for the "empty" and ontological self, Buddhists postulated such notions as Buddha-nature, *ālayavijñana* (storehouse consciousness), and *tathāgathagarbha* (womb of the *Tathāgata*, one who has touched the shore of *nirvana*) that posit an underlying reality on which the affectations occur. The significance of an ontological self is further emphasized by the observation in the prologue to the *Tri-śarana-gamana* or Threefold Refugees recited daily by Buddhists, that the appearance of an individual in the world is a rare event. The Japanese cleric, Dōgen (1200–1253), and the aesthetician and art critic, Yanagi Sōetsu (1898–1961) expanded the notion of the intrinsic value sentient beings to include things. Dōgen asserted that even inanimate objects are Buddha-nature (10). Yanagi spoke of enlightened things (11).

While the doctrine of *pratītyasamutpāda* describes the Buddhist understanding of reality and is the rationale for karmic interaction, the Four Noble Truths crystallize this doctrine's existential import. According to Buddhist lore, the lesson of the Four Noble Truths is the first the Buddha shared after the enlightenment. It relates directly to the doctrine of *pratītyasamutpāda* as a moral principle based on a reworking of the law of karma. The Four Noble Truths, an empirical–rational methodology that is closely associated with ancient Aryadevic medicine, parallels the steps — diagnosis, etiology, recovery, and therapeutics — that summarize the medical treatment of a disease. The Four Truths profile the condition of our lives, explain the cause of suffering, and the means by which we, residing in a samsaric world, can extract ourselves and realize an abiding spiritual reality. The Four Truths are (*1*) the Noble Truth of Suffering, (*2*) the Noble Truth of the cause of suffering is illusion and desire, (*3*) the Noble Truth of Nirvana, a realm free from suffering, and (*4*) the Truth of the Noble Eightfold Path is the way to enlightenment or nirvana. The Eightfold Path consists of Right View, Right Thought, Right Speech, Right Action, Right Livelihood, Right Effort, Right Mindfulness, and Right Meditation. In the First Truth, the Buddha acknowledges that spiritual suffering, though the most serious, was just one of many ills. The cause or etiology of this suffering, the Second Truth, stems from illusion and desire. Illusion is the belief in a substantial self and an unchanging world, and desire refers to wishing for unattainable things. The Fourth Truth is the Eightfold Noble Path that releases the individual from ignorance and delusion. The Eightfold

Path is the medicine or spiritual therapy that leads to the Third Truth, wisdom or nirvana.

The Four Noble Truths outlines a method to transcend, not escape suffering through understanding. In the *Vissudhimagga*, Buddhaghosa (circa 400 C.E.), the great Theravāda commentator, correlated the four phased method, for spiritual health systematized in the Four Noble Truths, with the treatment of a disease. "The truth of suffering is like a disease, the truth of origin is like the cause of the disease, the truth of cessation is like the cure of the disease, the truth of the path is like the medicine" (12). Modern Buddhists have applied the four-step method outlined in the Four Truths to remedy social and economic problems. The Sarvodaya Shramadana Movement, a rural self-help program initiated by Ahangamage Tutor Ariyaratnes of Sri Lanka, is the most celebrated example. Since the 1960s Ariyaratnes and his method have empowered poor villages to recognize their problems, discern their causes, envision solutions, and devise remedies.

THE BUDDHIST POSTURE

A vision of an interdependent world affirms the reality that we live in a complex, ever-changing web of interrelationships. This vision allows for important implications when thinking about ethical, legal, and policy issues concerning biotechnology. This brief section begins with a sketch of the ideal relationship between the whole and part explicated by *pratttyasamutpāda*, proceeds to discuss some of its inherent difficulties, and concludes with a some thoughts of the virtues of deliberation in the context of an interdependent world.

Ethical, legal, and policy deliberation and exercise that posit an interdependent world consider a microscopic view along side a macroscopic one. The symphony offers an analogy. To grasp the beauty in a Beethoven symphony, for example, it is not enough to listen to the individual instruments sequentially. One must hear the instruments together, as each musician modulates his or her instrument and timing in response to the other instruments. Each instrument participates in creating a greater whole. The whole in turn gives value to the sounds from each instrument. The microscopic and the macroscopic resonate together.

Like a symphonic composition, in the Buddhist vision of the world, we are parts of a larger whole. The whole in turn gives each individual value and worth. The idea of interdependence links our individual lives to each thing and each being in the universe. To act according to this vision is to work to nurture the lives and relationships that enrich and sustain the life of individuals and the whole universe (13). Such an interpretation of life dictates the various virtues and ends that Buddhists should consider when reflecting on ethical questions, legal issues, and formulating policies. A sense of fair play, compassion, gratitude, humility, and patience are some virtues to nurture and consider when reflecting on such questions as health care, allocation of resources, relating to the most vulnerable in our society, the environment, biodiversity, and the manipulation of life.

This harmonious ideal is difficult and almost always impossible to realize. In a world where lives are inexorably intertwined, like ice and water, our concern should be extended to all beings. In a practical sense, our energies should focus where we can make a difference, before extending our energies to embrace all lives and relationships. The Japanese cleric, Shinran (1173–1262) reiterates this attitude in his fourteenth letter of the *Mattōshō*, "the one who first attains nirvana vows without fail to save those who were close to him first and leads those with whom he is karmically bound, his relatives, and his friends" (14). Additionally our life experiences we are often beset by conflicting demands and responsibilities. Our wishes are continually frustrated by the demands of others and by events beyond our control. While some forces nurture our lives, others demean. In the Thirteenth Chapter of the *Tan'nishō*, Shinran underscores the reality that we are often swept up in events that thwart our best intentions. In his conversation with Yuienbō, a fellow devotee, Shinran says, "It is not that you keep from killing because you are good. A person may wish not to harm anyone and yet end up killing a hundred or even a thousand people" (15). Since our lives are intimately linked with the karmic tide of others, to society, and even the whims of nature, we may be propelled to violate our deepest moral instincts. Under those circumstances we yield — mournfully and perhaps, even justifiably — to the dictates of more powerful karmic forces. In such a world the best that we can hope for, according to Thich Nhat Hahn, the Vietnamese monk, is to be determined to go in the direction of compassion and try to reduce suffering to a minimum (16). The exercise of compassionate aspirations, no matter how insignificant, is based on the belief that an act of kindness resonates throughout the farthest reaches of the universe.

A vision of an interdependent world recognizes the complexity of even the most common event. The quiet unfolding of the morning glory is supported by the entire universe. Fa tsang (643–712) articulates the complexity of this singular event in his Ten Subtle Principles of the Unobstructed Fusion of *Pratttyasmutpāda* (Shih-hsüan-yüan-ch'i-mu-ai-fa-mên). He reasoned that in an interdependent world no dharma (thing or event) is independently established and thus all dharmas are mutually supportive and mutually dependent within the *dharmadhātu*, the realm of dharmas; each dharma is thus of equal importance. However, when a dharma is arbitrarily singled out for consideration, that particular dharma becomes the principal dharma and the remaining dharmas take on a secondary role. Each dharma has the potential of alternately assuming the principal role or a secondary role. The role a dharma assumes is determined by what is weighed to be important at any given moment. Moreover each cause and condition offers a different perspective of how a thing or event arises. No thing or event is ever the locus of attention for everyone. The construction of a much needed new bridge, for example, requires the approval of many public agencies and private interests. The necessity of efficient and reliable thoroughfares must balance commercial, environmental, engineering, aesthetic, and other needs of the community. The environmentalists concern for biodiversity relegates commercial interests to a secondary concern. The engineer is concerned first with structural integrity, rather than aesthetics. Knowing that we live in an interdependent world means that we may never resolve issues to everyone's satisfaction.

Predicaments also arise from competing interests as well as conflicting perspectives. We may never know for certain how a specific event transpired, as Akira Kurosawa (1910–1998) dramatizes in *Rashamon*. In this murder mystery we hear the testimonies of the bandit, the police agent, the woman who was raped, and through a shaman, the murdered husband. Each individual relates a slightly different version of the circumstances that surround the murder. The film ends without a resolution. Often we must deliberate, make decisions, and act knowing we are unable to reconcile or understand completely differing perspectives. Appreciating the complexity of an issue permits us to see many points of view and is perhaps the most productive way of ethical deliberation and action. We nurture humility and patience knowing that others may not approve or follow our example. Living with and appreciating alternative points of view is reminiscent of a Cubist painter who renders an object from different viewpoints. In contrast, perspective, a visual rendering technique perfected during the European Renaissance, renders an object from a single-fixed point.

Acknowledging the validity of other points of view, the Buddha urged his would be followers not to accept any of his teachings without first critically examining them. Only if any of his teachings lead to spiritual ease, should they be observed and accepted (17). Other paths may be more suitable for one's particular temperament. The Buddha insisted that he was a guide, not an authority. The Buddha's critical attitude toward religious authority, even his own, is seen in Thich Nhat Hahn, a Vietnamese monk whose experiences of the Vietnam War forced him to reinterpret and condense the traditional Buddhist precepts into Fourteen Precepts for Engaged Buddhism. The first Precept reads: Do not be idolatrous about our bound to any doctrine, theory, or ideology, even Buddhist ones. Buddhist systems of thought are guiding means; they are not absolute truth. Explaining this precept, Thich Nhat Hanh writes that clinging to our views can cause us to lose the opportunity to a higher and more profound view of reality. By being open to other points of view, we expand the frontiers of our knowledge and our understanding of the world (16).

BUDDHIST MEDICINE AND HEALTH CARE

Health and health care are metaphors common to Buddhist thought and practice. The Buddha, the great physician, dispenses the Dharma, the medicine that heals humanity's suffering and brings spiritual ease. To the terminally ill he administers the teaching of impermanence and to others meditative exercises. Though spiritual suffering was of paramount concern, the Buddha understood that spiritual well-being necessarily involved physical health, which in turn, is dependent on a

wholesome community and its sound management of economic and other resources. The Buddha and his devotees attended to their spiritual ills through self-cultivation, and served as nurses to the sick by dispensing medicine and compassionate deeds. While illness testifies to the frailness and transiency of the human condition, caring for the sick is also an opportunity for spiritual quickening. The sick, in turn, have an opportunity for abundant giving. Spiritual health means to realize and to live with gratitude and responsibility to all things and beings. This section begins with an overview of health and health care within the context of the early Buddhist theory of medicine. It then proceeds to discuss caregiving and the relationship between the caregiver and the patient.

Medical Theory

Grounded in the belief of an interdependent world, Buddhist medical theory understands mind and body to be a single unit. Illness of the body is illness of the mind, and mental illness is directly related to the illness of the body. Health requires balance and reciprocity among all the four elements: earth or the solid element, water or the wet element, fire or the hot element, and wind or the mobile element, and the three peccant humors of wind, phlegm, and bile that constitute the human body. Illness arises when one or more of these elements experience an abnormal augmentation or diminution. Medicine and medical therapies provide the means to restore and maintain a healthy physical balance. By contrast, present biomedical diagnostics understand the body to be made up of distinct divisible parts, whose organs and functions can be isolated and treated. Modern etiology seeks the sources of disorder from external pathogens, rather than internal disorders. While Buddhist physicians traced the etiology of disease to empirical causes, karma or past action is also a category of medical etiology. Past deeds relate to present mental suffering caused by greed, hatred, and doubt.

In addition to the disequilibrium among the four elements and the three humors, Buddhist medical theorists understood that external and societal conditions affected the internal working of the body and cause of disease. Diet, daily regimens, alteration of the seasons, stress from unusual physical activities, and past actions affect one's physical and mental well-being. In keeping with the belief that prevention is the best guarantee against illness and disease, the Buddha urged moderation in spiritual exercises and in all life activities. His monastic rules emphasized personal hygiene and public health. Straining water served to purify it and to prevent consuming water-dwelling organisms. Living quarters and privies were to be kept clean. Even today in Zen monasteries in Japan, certain days are set aside for washing and mending.

The Buddha traced much of human illnesses to poverty. The poor, he reasoned, had limited access to material and nonmaterial resources that ensured basic necessities: food, clothing, shelter, medicine, and education (i.e., spiritual development). A Buddhist state would have the responsibility to provide a wholesome living environment by safeguarding the natural environment, by ensuring the equitable distribution of resources, and protecting the poor and dispossessed against exploitation. These measures would optimize psychological and physical well-being, the basis on which to nurture spiritual development. The Buddha placed a high value on physical fitness and freedom from illness as a basis for mental and spiritual development. "When sentient beings have sick bodies, their minds cannot be at peace ... the bodhisattva who would cultivate awakening should first minister to the illness of the body" (18). This holistic approach to health and healing resonated within Taoist- and Confucian-based Chinese medicine. Both Buddhist and Chinese medicine do not distinguish between mind and body. In addition to the yin and yang theory which is based on harmony of the universe, the Chinese included the connections between a well-functioning social system and a healthy body.

Caregiving

Medicine and caregiving were integral parts of the early Buddhist community. The model for caregiving is the Buddha himself. On an occasion, the Buddha chanced on an unattended sick monk wallowing in his own excrement. He said, "O monks, if you do not nurse one another, whoever will nurse you?" Thereupon the Buddha bathed the monk, changed his garments, and laid a bed for his ailing comrade. This experience led the Buddha to declare, "Anyone who wishes to make offerings to me, let him make offerings to the sick." Initially the early community focused medical treatment and nursing activity on the care of monks and nuns by fellow cenobites or by pious lay devotees. Caring for their compatriots became part of the monastic code. From around the mid-third century B.C.E. medical care was extended to the population at large. Later the medical arts became part of the curriculum at Nalanda and other monastic universities.

Buddhist documents abound with medical injunctions and prohibitions. The *Aṅguttara Nikāya* lists the five qualities of those who tend to the sick. It also lists the five faults of a patient that impede his or her recovery. The competent caregiver (1) possesses knowledge of medicaments and their application, (2) tends to the sick with amity of mind and without thought of personal nourishment and profit, (3) is not lazy, (4) or prone to annoyances and does not loathe removing excrement, urine, sweat, or vomit, and (5) since the early Buddhists linked illness with mental states gone awry, the competent caregiver should delight in sharing the Buddha's Dharma and conversing with the sick (19). Conversing requires the art of listening which is personified in Avalokiteśvara, the Bodhisattva of Compassion. "Avalokiteśvara" or "Kanzeon," the Japanese rendering, clearly captures the Bodhisattva's special talent. "Kan" means "to hear," "ze" means "world," and "on" means "sound." Avalokiteśvara who hears the pleas of the world, is the ideal caregiver. Listening to the outbursts of anger, despair, and hurt, the Bodhisattva attends to the journeying spirit, not the momentary stammer. When we engage in conversation and allow another to speak, we act as a midwife who helps to bring new life into the world. By listening, we allow for self-discovery that gives birth to a new being.

Patient Responsibility

The vision of an interdependent world requires a patient's involvement in his or her care. The *Aṅguttara Nikāya* lists five faults of the patient that discourage healing. The patient impedes his or her healing (1) by not being selective of what he or she eats or drink, (2) fails to take nourishment at the proper times, (3) refuses to take medicaments, (4) abandons him or herself to melancholy, merriment, and annoyance, and (5) is pitiless toward the sick nurse (20). In addition to cooperating with nurses to hasten health, one can use his or her illness as the occasion for spiritual exploration and abundant giving. Vimalakīrti, the most famous invalid in the Buddhist canon, describes the responsibility of the sick:

> Through one's own experience, a Bodhisattva should have sympathy with the sick person. Let them know of the pains they suffered from the infinite past, but encourage them by advising them to endeavor and become the Buddha, the Great King of Medicine and cure the illness of all people ... (21).
>
> Though the body may be in the world of delusion and diseased, if one gives abundantly and tirelessly, that is called expedient means. The body may not be freed from the disease and the disease may not be freed from the body, but if the disease and body are seen neither as new nor old, that is called wisdom. Though the body may be diseased, if one does not forsake this world and does not intend to enter Nirvana, that is called expedient means (22).

Should we choose to heed Vimalakīrti's injunction, illness is an occasion for service and spiritual quickening. Through their experience of illness, caregivers can sympathize with others who are ill. They can urge the sick not to succumb to their pains, but encourage them to relieve the suffering of all beings. Even if one is ill, one can give abundantly. "Expedient means," a rendering of *upāya*, ordinarily refers to the wisdom to convey the Dharma according to the needs and capacities of the listener. But here, Vimalakīrti defines *upāya* to mean the efforts the devotees of the Dharma should expend to relieve suffering.

Abundant giving takes on many forms. The invalid and sick often inspire and instill faith in the human experience to those who attend to them. "*Dana*" or selfless giving is the first of the Six Paramitas or Perfections a Bodhisattva observes. The selfless gift is "*Dana* ... most profound among all joys ... that which is found through witnessing and experiencing the joy of others [so that] ... the joy of others becomes one's own joy" (23).

In an interdependent and ever-changing world, health and healing are possible with the advent of new knowledge and with new relationships. Change assures that illusion can be transformed into enlightenment, illness can be cured, and social decay arrested.

MANIPULATION OF LIFE

The discussion of the manipulation of life includes the environment and attended concerns of ecology, biodiversity, and agriculture, and modifications of living organisms that include human beings, microorganism, and animals. While Buddhist documents offer insight into dealing with the environment, we can only extrapolate what might be the Buddha's attitude toward the manipulation of life, organ transplants, and cloning.

Environmental Issues

Buddhist responses to ecological issues, biodiversity, and agricultural development are rooted in the sense of responsibility and gratitude, intrinsic to the doctrine of *pratttyasmutpāda*. As it was noted above, for the Buddha, personal health is established within the context of a wholesome society and environment. The *Abhidharmakośa śāstra*, an influential primer composed by Vasubandhu (circa 400 C.E.), defines the world to include both sentient beings and the container world, the realm that supports sentient life (24). Just as mind and body are considered to be a single unit, Buddhism understands the individual, society, and the natural world to constitute a single whole. The idea of "one is all and all is one" articulated in the *Avataṃsaka sūtra*, and the elaboration of this idea in such doctrines as the Ten Subtle Principles of Unobstructed Fusion of *Pratttyasmutpāda* and the Six Principles of the Causal Aspects of *Pratttyasmutpāda* (Yüan-ch'i-yin-mên-lu-i-fa) by Fa-tsang and other Hua-yen masters, articulate the ideological rationale for environmental concerns. These doctrines explain that human, plants, animals, and material entities do not simply exist in and by themselves. Their individual and separate existences are affirmed by and made possible through their relationship with others (25). Our individual well-being is dependent on the health of the world we live in. Our activities should thus be conducted with a sense of respect and reverence for all life. Conservation, harmonious coexistence, the need to care for and to restore the land (26), not exploitation, should be the hallmark of a Buddhist devotee. The destruction of the Amazon rain forests affects my well-being and the very life of the world. The Buddhist devotee should quicken feelings of gratitude to animals and plant life to whom he or she must depend on for life. Donald Swearer sums the modern Thai monk Buddhadasa's (1906–1993) rationale toward the environment:

> One cares for the forest because one empathizes with the forest, just as one cares for people ... [Empathy] is fundamentally linked with non-attachment or liberation from preoccupation with self, which is so central to Buddhasada's thought ... Caring in this deeper sense ... goes beyond the well-publicized strategies of the conservation monks to protect and conserve the forests ... [Empathy] translates as having at the very core of one's being the quality of caring for all things in the world and their natural conditions; that is to say, caring for them as they are in themselves rather than as I might benefit from them or as I might like them to be (27).

The modern environmental ethic that advocates minimum exploitation of and optimum utilization of natural resources and maximizes recycling finds much in common with the Buddhist idea of *pratttyasmutpāda*. Modern Buddhists actively work to preserve the environment and promote biodiversity. The Buddhist Peace Fellowship founded in 1978 under the leadership of Robert Aitken and others in Hawaii have galvanized Buddhists worldwide to work to protect the environment and issues of justice. In November 1997 villagers, Buddhist monks, and environmentalists gathered in Sai Yok National Park to protest the Thai government's decision to allow a pipeline to run through the park. To protest the deforestation and its animal inhabitants, and displacement of people, Buddhist monks ordained the trees (28). The voices of Buddhists thinkers and activists appeared in *Buddhism and Ecology, the Interconnection of Dharma and Deeds*, a collection of papers presented at the 1996 proceedings of Earth Charter, a project that set forth a vision of ethical principles for the twentyfirst century (29).

In the Buddhist countries of Asia, modernized agriculture resulted in increased yields, but it also brought about the large-scale depletion of natural resources. Deforestation for agricultural uses has destroyed much of the wild life, and use of chemical fertilizers and insecticides has killed off mudfishes and edible frogs that once thrived in the rice fields and served as a rich source of food (30). In response to the "destruction of human communities and nature in the name of globalization, of multinational corporations, governments, and local allies" more than 400 social activists and leaders from the Asia–Pacific region gathered in Kathmandu in 1996. The international gathering, "People's Convergence, Shaping Our Future" was the third event of the People's Plan for the twenty-first Century that began in 1989 in Japan to bring attention to environmental pollution. The Minamata Declaration called for a grassroots transborder participatory democracy to change the global structure (31). These grassroots nongovernmental organizations, including Buddhists and Buddhist organizations, attempt to live out the Buddha's insistence that material and environmental needs of the people be respected.

Brain Death and Organ Transplants

The question of brain death and the appropriateness of organ transplants generated great concern and a range of opinions in the Buddhist community. The controversy lies in part in the meaning of life and death, personal identity, and the belief in the inseparability of mind and body. In the United States and other countries where transplants are routine, death is defined as the absence of brain activity, which often occurs before a heart stops beating. Legally defining a patient, whose brain has ceased to function, dead is crucial in harvesting organs. Organs quickly deteriorate once the heart stops beating. Brain death is not satisfactory for Buddhists who subscribe to the traditional cardiopulmonary definition of death. More substantially, others object to the brain-death criterion of death because Buddhists have always associated life with sentience (32), which in its broadest sense includes feeling. Though the brain may have ceased to function, the individual with a beating heart may be pained by being cut, and having his or her organs removed. Doctrinally, death is defined as the dissolution of mind and body. Death dissolves the fortuitous interactions of karmic events that gave birth to and nurtured an individual. The separation of the mind from the body, however, is not death of the person, as we will presently see. Curiously, even with death defined as cessation of higher-brain functions, we commonly begin funeral preparations when the heart has been removed or has stopped beating, and not when the brain is dead. The medical and legal definitions of brain death often conflict with social notions of death.

Another pervasive attitude against organ transplants is the assumption that life is impermanent. Since life is transient and death inevitable, there is no meaning to artificially extend life by receiving the organs of another. The extension of life by organ transplant disrupts the natural karmic life span. Rather than extending life through heroic measures, humane end-of-life care would be more in keeping with the spirit of the Dharma. Further, organ transplants are possible only at the expense of another's life, a violation of the precept to abstain from taking life. Consequently some Buddhists advocate the development and use of artificial organs. However, those who favor transplants argue that the gift of life is the greatest gift an individual can give. The body is, after all, transient and ultimately worthless (33). Buddhist lore is replete with legends that relate the sacrifice of limb and life by the Buddha. In Sri Lanka, Hudson Silva has used a legend of the Buddha with great success to persuade people to donate their eyes for corneal transplants (34). The legend even makes mention of an eye transplant.

The Buddhist misgiving toward defining death as the cessation of brain activity may be another reason for the ambivalence of the Japanese toward organ, especially heart transplants. After more than 30 years of debate, the Japanese Parliament on June 17, 1997, passed a law that allows a person, whose brain has stopped functioning, to be defined as dead; in cases where a patient has agreed to this definition of death, he or she can request that his or her heart and lungs be donated for transplants. The bill does not provide a legal definition of death; it does nonetheless allow the brain-death standard to be used for donors of hearts and lungs. Current Japanese law defines death as the moment the heart stops beating. The bill does not give the donor an absolute right to ask physicians to decide whether he or she is considered brain-dead. The donor's family has the ultimate right to veto the doctors' diagnosis of brain death and the patient's wishes. A patient's rights, as noted above, are not absolute. Ironically the new Japanese law allows for a greater measure of individual autonomy. Physicians are given permission by the prospective donor to declare himself or herself brain-dead for the purposes of organ donation. In contrast, in the United States brain death is defined by law, and physicians do not need the patient's permission to declare an individual brain-dead.

On February 28, 1999, 21 months after the approval of the new law, Japanese doctors performed their first heart transplant since Dr. Juro Wada attempted a heart transplant 32 years ago. Dr. Wada's patient died and his operation and motives are clouded with legal controversy.

The difference in the importance of individual autonomy differentiates the U.S. and Japanese approach to organ transplants. As we noted, in Japan an individual does not have an absolute right to self-determination; the Japanese Parliament allowed the family to void any prior directive a brain-dead individual may have made concerning the disposition of his organs. The Japanese approach to brain death reflects the manner in which a Buddhist would approach the problem. In an interdependent world where lives are intertwined with countless others, individuals do not have exclusive claim on their lives. We may have separate lives, but we live in resonance with others.

The hesitation of organ transplants among East Asian Buddhists can also be traced in part to the Chinese Confucian notion of filial piety. The opening lines of the *Hsiao Ching*, or Classic on Filial Piety, states, "Filial piety is the basis of virtue and the source of our teachings. We receive our body, our hair, and skin from our parents, and we dare not destroy them" (35). When Buddhism first entered China, the Chinese appealed to this passage to argue against their sons and daughters shaving their hair when entering the Buddhist order. A person should be buried with every part of his or her body. The donation of one's organs would thus constitute a most unfilial act. This attitude has prevented wide acceptance of organ transplants. While the Chinese value keeping the body intact after death, there is a countervailing attitude that the use of organs from executed prisoners can benefit social and public good (36). This attitude toward the asocial elements has its roots in imperial China when social order and individual health were closely linked (37). Korean Buddhists' wariness of organ transplants stems from a strong Confucian imprint. The indigenous shamanic belief that a person who is not buried with all of his or her body will suffer in the next phase of life also contributes to their hesitation.

The importance of family lineage and the reciprocity between the living and dead account for the reluctance of organ transplants among the Chinese, Koreans, and Japanese. In traditional East Asia, death of the physical body is not the death of the person. Incorporating this belief, Buddhist mortuary rites mark the transformation of the person from a physical to a spiritual being. The person matures or proceeds to ancestorhood with the aid of memorial observances sponsored by the living descendants. In return, the ancestor ensures health and prosperity for the family. This accounts for the complex and lengthy memorial cycle. The Japanese Buddhist mortuary rites are especially long. The memorial cycle begins immediately after death and continues for at least 33 years. The 49th day, 100th day, 1st year, and the 3rd, 7th, 13th, 17th, 25th, 33rd year observances are especially important. On the island of Okinawa, a living repository of Japanese culture and language, the 33rd year memorial service marks the complete transition of the individual to an ancestral spirit. After the completion of the service, the individual's memorial tablet is burned. Services are no longer dedicated to the memory of the deceased and the individual is honored collectively as an ancestor with all other ancestors. While the long memorial cycle ritualistically marks the transformation of a person's identity, it in fact reveals something of the nature of our memories. As years pass, our recollections of the deceased become less and less distinct and he or she gradually loses his or her individuality. Korean Buddhist mortuary rituals continue for up until three years, and thereafter the deceased is honored at an annual memorial service for ancestors. Thai and Burmese Buddhists' memorial rituals are seven years (38). In accordance with Confucian sentiments, Chinese Buddhists mourn for three years.

Cloning

In late February 1997, when Ian Wilmut, a scientific researcher, announced the first successful cloning of a sheep named "Dolly," the prospect of cloning a human being prompted celebration, caution, and concern. Buddhists have not raised objections to this scientific breakthrough, but they are concerned with the ends of and motivations for human cloning. The creation of new life should not be seen as a product but an end or value in itself. Such reverence appeals to the idea of Buddha-nature and that the appearance of a clone is a rare event. Cloning a human being to produce organs for use in transplantation, however, would be repugnant (39). In contrast, some appeal to the principles inherent in the doctrine of *pratītyasmutpāda* to celebrate the reality of change. The technique of cloning is a tool that further expands human scientific and technological promises (40) and new moral possibilities (41). Since change is the nature of reality, the present challenge of the cloning question is how to accommodate change, expand our notions of humanity, and our moral parameters.

Organ transplants and genetic manipulation raise questions of family continuity in those East Asian cultures where family lineage is valued. If a person is considered to be the unique repository of prior generations, receiving an organ from another person raises the question of identity. If genes are manipulated what is the relationship between ancestor and descendent?

The manipulation of life through genetic engineering, like other human-generated innovations, is consistent with the Buddhist belief in change. Countless causes and conditions propel us to this present moment, providing for new achievements. Our present thinking and activities interact with current concerns and with the natural order. As active participants and an integral part of the process of interdependence in the life of the world, human beings have the capacity to affect the subsequent course of events. This should give us pause to reflect on our responsibilities and present and future action. Often we are unaware of the consequences of our achievements. For example, an August 1998 article in *Nature* reports that climatologists have linked pollution emitted by factories and automobiles with rainy weekends (42).

CONCLUDING REMARKS

Buddhists, caught up in rapid scientific and technological changes, have been slow to reflect on the ethical, legal, and policy issues generated by recent advances in biotechnology. Since much of these changes occurred in the United States and Europe, technically the most advanced countries, and where the issues surrounding modern biotechnology initially appeared, Christian theologians have reflected on and activists voiced concerns over these advances. By contrast, Buddhism, which is dominant in the countries of South and East Asia where biotechnology has begun to have an impact, has only recently confronted these problems. Japan is the most notable exception. Since the successful cloning of twin calves in July 1998 by Japanese scientists under the direction of Yukio Tsunoda of Kinki University, the Japanese Ministry of Agriculture and Forestry reported that as of March 31, 1999, 57 calves have been cloned from somatic cells and another 461 from nuclear embryo transplantation (43). Buddhist thinkers in that country have been reflecting on cloning.

The notion of *pratītyasmutpāda* holds the key to the Buddhist approach to implications of biotechnological advances. *Pratītyasmutpāda* offers an understanding of change, humanity's place in the process of change, and a vision of human responsibility to all things and all beings and to the world. Nothing in the Buddhist documents suggests halting changes that new knowledge generates. Change is a cardinal Buddhist presupposition. While change may be the opportunity for expanding Buddhists' moral imagination, the idea that all things and all beings are mutually and irrevocably interdependent instills a sense of humility that is necessary for ensuring that all species and all things are accorded respect. Further the vision of an interdependent world quickens concerns for the safety of food from cloned animals and plants, and the long-term consequences of gene manipulation on the environment and all sentient life. These and other ramifications of biotechnological advances and policy decisions must carefully consider all aspects of suffering that change generates. The karmic energies of a single individual have wide repercussions. "A wise man should do things that are beneficial to living-beings" (44).

BIBLIOGRAPHY

1. The most widely accepted dates, 563–483 B.C.E., are based on the Sri Lankan historical chronicles, *Dīpavaṃsa* and *Mahāvaṃsa*. Hakuju Ui (1882–1963), a prominent Japanese scholar, disputed these dates based on Northern or Mahāyāna sources. Thus, according to Ui, the Buddha's dates are 466–386 B.C.E.

2. Followers of Mahāyāna (great vehicle) Buddhism coined the expression, "Hīnayāna" (lesser vehicle) as a pejorative for those who did not accept their documents and their doctrines. Mahāyānists accused Hīnayānists and their *arhat* ideal that was reserved for the select few. The Mahāyāna ferries all beings across the sea of samsara to nirvana, while the Hīnayāna transports only a few. No Buddhist group referred to itself as Hīnayāna. The devotees of Theravāda, a non-Mahāyāna tradition, object to being labeled "Hīnayānaists."

3. K. Mizuno et al., *Butten kaidai jiten [A Bibliographical Dictionary of Buddhist Texts]*, 2nd ed., Shunshu sha, Tokyo, 1978, pp. 15–16.

4. S. Kamata, proposed alternative routes and dates for the eastward advance of Buddhism in L. Lancaster and C.S. Yu, eds., *Introduction of Buddhism to Korea, New Cultural Patterns*, Asia Humanities Press, Berkeley, CA, 1989. See *The Transmission of Paekche Buddhism to Japan*, pp. 143–160.

5. R. Fukuhara, *Bukkyōgairon [Outline of Buddhism]*, Nagata shodo, Kyoto, Japan, 1995, p. 145.

6. K.N.S. Subramanian, *Hastings Cent. Rep.* **16**(8), 20–22 (1986).

7. P. Ratanakul, *Bridges* **3**, 4–5 (1997).

8. A. Hirakawa, *A History of Indian Buddhism, from Śākyamuni to Early Mahāyāna*, University of Hawaii Press, Honolulu, 1990, p. 6.

9. H. Nakamura, *Jiga to muga [Self and Non-self]*, Heirakuji, Kyoto, 1976, pp. 55–59, also J. Pérez-Remón, *Self and Non-Self in Early Buddhism*, Mouton, The Hague, Japan, 1980, pp. 301–305.

10. Dōgen, *"Busshō" [Buddha-nature] in Shōbōgenzō*, Iwanami, Tokyo, 1970, p. 49.

11. S. Yanagi, *The Eastern Buddhist, New Ser.* **13**(2), 16 (1978).

12. Buddhaghosa, *Visuddhimagga*, R. Semage, Colombo, Sri Lanka, 1956, p. 586 (trans. by Bhikkhu Nanamoli as *The Path of Purification*).

13. R. Nakasone, in Watanabe Takaō kyōju kanreki kinen ronshū, ed., *Bukkyōshisō bunkashi ronsō [Essays on Buddhist Thought and Cultural History]*, Nagata shodo, Kyoto, Japan, 1997, pp. 302–330.

14. Shinran, *Mattōshō*, Nagata shodo, Kyoto, 1978, p. 47 (trans. by Yoshifumi Ueda as *Letters of Shinran, a translation of Mattōshō*).

15. Shinran, *Tan'nisho*, Ryukoku University, Kyoto, Japan, 1982, pp. 32–35 (trans. by D. Hirota as *Tan'nisho, a Primer*).

16. T. Nh'ât Hahn, *Interbeing, Commentaries on the Tiep Hien Precepts*, Parallax Press, Berkeley, CA, 1987, p. 27.

17. Saṅghabhadra, *Samantapāsādikā*, Bhandarkar Oriental Research Institute, Poona, India, 1970, pp. 171–172 (trans. by P.V. Bapat and A. Hirakawa).

18. J. Takakusu, ed., *Taishō Shinshū Daizōkyō hangyōkai, Tokyo* **10**(293), 710c–711a (1925–1931).

19. J. Takakusu, ed., *Taishō Shinshū Daizōkyō hangyōkai, Tokyo* **2**(125), 680c (1925–1931).

20. J. Takakusu, ed., *Taishō Shinshū Daizōkyō hangyōkai, Tokyo* **2**(125), 680b (1925–1931).

21. J. Takakusu, ed., *Taishō Shinshū Daizōkyō hangyōkai, Tokyo* **14**(475), 544c (1925–1931) (English trans. in *Buddha-Dharma*, Numata Center for Translation and Research, Berkeley, CA, 1984, pp. 319–320).

22. J. Takakusu, ed., *Taishō Shinshū Daizōkyō hangyōkai, Tokyo* **14**(475), 545b (1925–1931) (English trans. in *Buddha-Dharma*, Numata Center for Translation and Research, 1984, p. 320).

23. S. Makino, *The White Way* **34**(5), 3 (1990), quoted by J.C. Takamura, *Generations*, Fall/Winter (1991).

24. Vasubandhu, in J. Takakusu, ed., *Taishō Shinshō Daizōkyō hangyōkai, Tokyo* **30**(555), 57a (1925–1931).

25. Y. Matsunaga, *Bukkyō* [Buddhism] **6**, 121–130 (1991).

26. J. Takakusu, ed., *Taishō Shinshō Daizōkyō hangyōkai, Tokyo* **24**(1488), 1061a (1925–1931).

27. D.K. Swearer, in S. Sivaraksa, ed., *The Quest for a New Society*, Santi Pracha Dhamma Institute, Bangkok, Thailand, 1994, p. 7; quoted by R. Aitken in *Turning Wheel*, Summer, 1998, pp. 28–30.

28. *Turning Wheel* Winter, 7, 8 (1998).

29. M.E. Tucker and D.R. Williams, eds., *Buddhism and Ecology, the Interconnection of Dharma and Deeds*, Harvard University Center for the Study of World Religions, Cambridge, MA, 1997.

30. R. Bobilin, *Revolution from Below, Buddhist and Christian Movements for Justice in Asia*, University Press of America, Lanham, MD, 1988, p. 119.

31. I. Muto, *ANPO: Jpn.-Asia Q. Rev.* **27**(2), 40–45 (1996).

32. *Samantapāsādikā*, Bhandarkar Oriental Research Institute, 1970, p. 319 (trans. by P.V. Bapat and A. Hirakawa).

33. H. Masaki, *Seimei rinri* [Bioethics] **1**, 48–52 (1993).

34. M.V. van Andel, *Doc. Ophthalmol.* **74**(2), 141–150 (1990).

35. *Hsiao Ching*, St. John's University Press, New York, 1975 (trans. by Mary Lelia Makra as *The Hsiao Ching*, p. 4).

36. R. Qui and D. Jin, in B.A. Lustig, ed., *Bioethics Yearbook, Regional Developments in Bioethics: 1991–1993*, Kluwer Academic, Dordrecht, The Netherlands, 1995, pp. 339–365.

37. P.U. Unschuld, *Medical Ethics in Imperial China, a Study in Historical Anthropology*, University of California Press, Berkeley, CA, 1979, p. 27.

38. E. Namihira, in Kazumasa Hoshino, ed., *Japanese and Western Bioethics, Studies in Moral Diversity*, Kluwer Academic, Dordrecht, The Netherlands, 1997, pp. 61–69.

39. D. Keown, *Reflections*, Spec. Ed., May, 1997, p. 8.

40. Z. Kumatani, *Bukkyō* **42**(January), 94–106 (1998).

41. R. Nakasone, *Reflections*, Spec. Ed., May 1997, pp. 6–7.

42. R.S. Cerveny and R.C. Balling, *Nature (London)*, August 6, 561–563 (1998).

43. *Asahi Shinbun*, April 23 (1999).

44. *Samantapāsādikā*, Bhandarkar Oriental Research Institute, 1970, p. 331.

See other RELIGIOUS VIEWS ON BIOTECHNOLOGY entries.

RELIGIOUS VIEWS ON BIOTECHNOLOGY, JEWISH

ELLIOT N. DORFF
University of Judaism
Bel Air, California

OUTLINE

INTRODUCTION

Judaism, a religion tracing its roots to Abraham close to 4000 years ago and continuing through the Bible and rabbinic interpretations to our own day, has sought since its inception to use the world productively while yet preserving it, both seen as God's commands. This article describes the theological foundations for Judaism's activist, and yet respectful, stance toward the world. It then describes how this stance is articulated in issues at the beginning and end of life and in environmental matters.

OVERALL CONTEXT OF JEWISH BELIEFS AND PRACTICES RELEVANT TO TECHNOLOGY

Adam and Eve are told in the Garden of Eden "to work it and to preserve it" (Genesis 2:15). Judaism has ever since tried to strike a *balance* between using the world for human purposes while still safeguarding and sustaining it. We are not supposed to desist from changing the world altogether: "Six days shall you do your work" is as much a commandment as "and on the seventh day you shall rest [literally, desist]" (Exodus 23:12).

In changing the world to accomplish our ends, though, we must take care to preserve the environment, whether we are practicing medicine, farming, traveling, or doing anything else. This balance is demanded because, in the end, we do not own the world; God does (1). We are but tenants in God's world, with a lease on life and on the world.

During the duration of that lease, we may and should act as God's agents to improve it. God, in fact, intended that we function in that way. This is probably most starkly stated in a rabbinic comment about, of all things, circumcision. If God wanted all Jewish boys circumcised, the rabbis ask, why did He not create them that way? The answer, according to the rabbis, is that God deliberately created the world in need of fixing so that human beings would have a divinely ordained task in life, thus giving human life purpose and meaning (2). We are, then, not only permitted but mandated to find ways to bend God's world to our purposes—as long, again, as we preserve God's world in the process.

Thus technology, in and of itself, is not good or bad: it depends on how we use it. If we employ it to assist us in bending the world to our ends while yet preserving the world, our use of technology is theologically approved and morally good; if we disregard our duty to preserve the world when using technological tools, we are engaged in a theologically and morally bad act.

FUNDAMENTAL BELIEFS RELATING TO HEALTH CARE

Three underlying principles regarding Judaism's positions on issues in health care emerge from Jewish sources:

1. *The body belongs to God.* Since God owns everything in the world (3), our bodies do not belong to us. Rather, God loans our bodies to us for the duration of our lives, and they are returned to God when we die.

The immediate implication of this principle is that neither men nor women have the right to govern their bodies as they will. Since God created our bodies and owns them, God can and does assert the right to govern the care and use of our bodies. Thus Jewish law requires us to safeguard our health and life (4), and, conversely, to avoid danger and injury (5). So, for example, Conservative, Reform, and some Orthodox authorities have prohibited smoking as an unacceptable risk to our God-owned bodies (6). Ultimately human beings do not, according to Judaism, have the right to dispose of their bodies at will (i.e., commit suicide), for that would be a total obliteration of that which does not belong to them but rather belongs to God (7).

2. *The body is morally neutral and potentially good.* For Judaism the body is as much the creation of God as the mind, the will, and the emotions are. Its energies, like those of our other faculties, are morally neutral, but they can and should be used for divine purposes as defined by Jewish law and tradition. Within that structure, the body's pleasures are God-given and are not to be shunned, for that would be an act of ingratitude toward our Creator (8). The body, in other words, can and should give us pleasure to the extent that that fits within its overriding purpose of enabling us to live a life of holiness.

The Jewish mode for attaining holiness is to use all of our faculties, including our bodily energies, to perform God's commandments. Maimonides states this well:

> He who regulates his life in accordance with the laws of medicine with the sole motive of maintaining a sound and vigorous physique and begetting children to do his work and labor for his benefit is not following the right course. A man should aim to maintain physical health and vigor in order that his soul may be upright, in a condition to know God. ... Whoever throughout his life follows this course will be continually serving God, even while engaged in business and even during cohabitation, because his purpose in all that he does will be to satisfy his needs so as to have a sound body with which to serve God. Even when he sleeps and seeks repose to calm his mind and rest his body so as not to fall sick and be incapacitated from serving God, his sleep is service of the Almighty (9).

The medical and technological implications of this are clear. Jews have the obligation to maintain health not only to care for God's property but also so that they can accomplish their purpose in life, namely to live a life of holiness. Moreover, since pain is not perceived as a method of attaining holiness but is rather an impediment to acting according to God's law, it is our duty to relieve it. Thus perhaps the most pervasive corollary of Judaism's insistence on the divine source of our bodies is its positive attitude toward the body and medicine.

3. *Human beings are not only permitted but obliged to try to heal.* God's ownership of our bodies is also behind our obligation to help other people escape sickness, injury, and death (10). God is our ultimate healer, as the Bible asserts in many places (11), but God both authorizes us and commands us to aid in that process (12). In fact the duty of saving a life (*pikkuah nefesh*) takes precedence over all but three of the commandments in the Torah (13).

The Talmud reflects some ambivalence about the level of expertise of physicians of its time (most explicitly in comments like "The best of physicians deserves to go to Hell!"), and some later Jewish authorities were particularly wary of physicians' abilities to practice internal medicine (in contrast to surgery and healing external wounds and diseases). In the end, though, the Talmud prohibits Jews from living in a community in which there is no physician (14). Here this third principle wraps back into the first, for if we were not within easy reach of a doctor, we could not as effectively carry out our fiduciary obligation to God to take care of our bodies.

Medical experts, in turn, have special obligations because of their expertise. Thus Rabbi Joseph Caro (1488–1575), the author of one of the most important Jewish codes, says this:

> The Torah gave permission to the physician to heal; moreover, this is a religious precept and is included in the category of saving life, and if the physician withholds his services, it is considered as shedding blood (15).

The following rabbinic story indicates that the rabbis recognized the theological issue involved in medical care and in the use of technology generally, but it also indicates the clear assertion of the Jewish tradition that the use of technology to assist in good purposes like producing food and preserving health is legitimate and, in fact, obligatory:

> It once happened that Rabbi Ishmael and Rabbi Akiva were strolling in the streets of Jerusalem accompanied by another person. They were met by a sick person. He said to them, "My masters, tell me by what means I may be cured." They told him, "Do thus and so until you are cured." The sick man asked them, "And who afflicted me?" They replied, "The Holy One, blessed be He." The sick man responded, "You have entered into a matter which does not pertain to you. God has afflicted, and you seek to cure! Are you not transgressing His will?"
>
> Rabbi Akiva and Rabbi Ishmael asked him, "What is your occupation?" The sick man answered, "I am a tiller of the soil, and here is the sickle in my hand." They asked him, "Who created the vineyard?" "The Holy One, blessed be He," he answered. Rabbi Akiva and Rabbi Ishmael said to him, "You enter into a matter which does not pertain to you! God created the vineyard, and you cut fruits from it."
>
> He said to them, "Do you not see the sickle in my hand? If I did not plow, sow, fertilize, and weed, nothing would sprout."
>
> Rabbi Akiva and Rabbi Ishmael said to him, "Foolish man!... Just as if one does not weed, fertilize, and plow, the trees will not produce fruit, and if fruit is produced but is not watered or fertilized, it will not live but die, so with regard to the body. Drugs and medicaments are the fertilizer, and the physician is the tiller of the soil (16).

The rabbis quite explicitly, then, understand God to depend upon us to aid in the process of healing. We are, in the talmudic phrase, God's partners in the ongoing act of creation (17).

TECHNOLOGY AFFECTING THE BEGINNING OF LIFE

Underlying Principles Regarding Family, Sexuality, and Procreation (18)

Marriage and children are the epitome of blessing in the Jewish view. "It is not good for man to live alone,"
the Torah declares, and so one goal of marriage is companionship, sexual and otherwise (19).

The second goal of marriage is procreation. Children figure prominently in the Bible's descriptions of life's chief goods (20), and so God's blessings of the Patriarchs promise numerous children (21). Procreation is not only a blessing; it is a commandment. Indeed, the very first commandment in the Bible is "Be fruitful and multiply" (Genesis 1:28). In rabbinic interpretation, for exegetical and probably economic reasons, it is the man who bears the responsibility to propagate, even though men obviously cannot do so without women. A man, then, fulfills the obligation to propagate when he fathers two children, and since we are supposed to model ourselves after God, the ideal is to have both a boy and a girl, thus creating both male and female, just as God did (Genesis 1:27) (22).

The family is important in Judaism not only because it is in that context that adults gain sexual fulfillment and the next generation is produced; it is also important because it is in the family that the tradition is passed on. Parents have a biblical obligation to teach the tradition to their children (23), and even after schools were established in the first century, parents remained ultimately responsible for the education of their children.

Preventing Conception

Contraception. With the importance of marriage and children in mind, one can understand that traditional Judaism looked askance at interruptions in the process of conception and birth. Normally one was supposed to marry and have children. Birth control, sterilization, and abortion were, both physically and ideologically, counterproductive.

Until very recently the use of birth control or even abortion for family planning purposes, so common in our day, was simply unknown to the tradition. Methods of birth control—either a cloth inserted in the vaginal cavity or a "cup of roots" taken orally—were unreliable, and abortion posed a major threat to the life of the woman. Moreover, if a couple wanted to have two or three children survive to adulthood, they had to produce six or seven. We must keep in mind this major distinction in context and purpose, then, when we examine and evaluate traditional Jewish sources on methods of preventing conception.

The rabbis state that the methods of contraception they had are permitted and even required under certain circumstances. Because the tradition understands the command to propagate to be the obligation of the male, male forms of contraception are generally forbidden. The specific conditions under which female contraception is permitted (and, in some cases, even required) depend on one's interpretation of a second-century rabbinic ruling describing three classes of women who "use" contraceptives—namely a minor (less than 12 years of age), a pregnant woman, and a nursing woman. The present tense of the verb is ambiguous in Hebrew, as it is in English. If it means that these women *must* use contraceptives to protect their life or health or that of their nursing infant, women in other circumstances then *may* use contraceptives. On the other hand, if these three categories of women may use contraceptives only to preserve life and health, then when that is not a factor, women *may not* use contraceptives (24). In any case, because Judaism restricts the legitimacy of abortion to cases where the life or health of the mother is at stake, modern forms of contraception that prevent conception in the first place (e.g., the pill, the diaphragm) are preferred over those that abort the fertilized egg cell after the fact (e.g., RU486).

Sterilization. The same concerns govern the issue of sterilization, but there another issue arises, namely the prohibition against a person mutilating his or her body in light of the fact that the body is really God's property. Although the procedures are rather new, there are a few rabbinic rulings (responsa) available on the issues of vasectomies and tubal ligations. Both traditional and liberal respondents forbid male sterilization on the basis of the rabbinic interpretation and extension of Deuteronomy 23:2 ("No one whose testes are crushed ... shall be admitted into the congregation of the Lord") (25), or Leviticus 22:24 ("That which is mauled or crushed or torn or cut you shall not offer unto the Lord; nor shall you do this in your land") (26). They are more permissive about female sterilization, both because a woman does not come under those prohibitions and also because she is not legally obligated to procreate (27).

All sources agree, however, that even male sterilization is permitted and perhaps even required if the man's life or health makes it necessary as, for example, if he contracts testicular cancer. Moreover, even though I am not aware of any written opinion that would allow a vasectomy, I could imagine an argument consistent with Jewish law and principles that would permit a vasectomy when pregnancy would entail a severe risk to the man's wife. After all, that procedure is far easier and safer than tying a woman's tubes, and saving a person's life takes precedence over both the commandment to procreate and the prohibition of injuring oneself. Moreover, a vasectomy does not amount to castration or to crushing the testes, and so the biblical verses cited above are not directly violated by the operation. The question, though, would be whether pregnancy could be effectively prevented by other means that would not endanger the woman and would not even possibly violate the verses cited. If so, then such means would undoubtedly be preferable.

Most often, though, men contemplate vasectomies simply because they do not want to father any more children. In light of the strong bias of the Jewish tradition for having children, and in light of the major demographic crisis facing the Jewish community that I will describe below, rabbis have not endorsed vasectomies for family planning purposes, seeing it as a violation of Jewish law and values and a threat to the continuity of the Jewish people.

Abortion. There is a clear bias for life within the Jewish tradition. Indeed, it is considered sacred. Consequently, although abortion is permitted in some circumstances and actually required in others, it is not viewed as a morally neutral matter of individual desire or an acceptable form of *post facto* birth control. Contrary to what many contemporary Jews think, Judaism restricts the legitimacy of abortion to a narrow range of cases; it does not permit abortion at will.

Judaism does not see all abortion as murder, as Catholicism does, because biblical and rabbinic sources understand the process of gestation developmentally. Thus Exodus 21:22–25 makes a clear distinction between an assailant who causes miscarriage of a fetus, when only monetary fines are imposed, as opposed to one who causes the death of the mother, when the rule is "life for life." According to the Talmud, within the first 40 days after conception the zygote is "simply water" (28). Another talmudic source distinguishes the first trimester from the remainder of gestation (29). It is not a theory of ensoulment that determines these marking points; it is rather the physical development of the fetus.

The effect of these demarcations is to make abortion during the early periods permitted for more reasons than during the rest of pregnancy (30). Classifying the first 40 days of gestation as "simply water," though, does not amount to a blanket permission to abort. Thus the RU486 pill, advertised as a "morning after pill" for those couples who simply do not want to have a baby, would be forbidden as a *post facto* contraceptive. On the other hand, if the woman's life or health would be threatened by pregnancy, then use of the RU486 pill would be preferable to a later-term abortion, both because it poses less risk for the woman and because the fetus is further from becoming a full human being.

The fetus does not attain the full rights and protections of a human being until birth, specifically when the forehead emerges or, if it is a breech birth, when most of the body emerges (31). The mother, of course, has full human status. Consequently, if the fetus threatens the life or health of the mother, then it may and in some cases must be aborted, as the following Mishnah graphically stipulates:

> If a woman has (life-threatening) difficulty in childbirth, one dismembers the embryo in her, limb by limb, because her life takes precedence over its life. Once its head (or its "greater part") has emerged, it may not be touched, for we do not set aside one life for another (32).

While all Jewish sources would permit and even require abortion in order to preserve the life or organs of the mother (33), authorities differ widely on how much of a threat to a woman's health the fetus must pose to justify or require an abortion. Based on a responsum by Rabbi Israel Meir Mizrahi in the late seventeenth century, many modern authorities also permit an abortion to preserve the mother's mental health, and this has been variously construed in narrow or lenient terms in modern times (34). To the extent that Jewish law makes special provision for an unusually young or old mother, an unmarried mother, the victim of a rape, or the participant in an adulterous union, abortion is construed to preserve the mother's mental health (35).

There is no justification in the traditional sources for aborting a fetus for reasons having to do with the health of the fetus; only the mother's health is a consideration. As a result some people object to performing an amniocentesis at all, even when the intent is to determine whether to abort a malformed fetus (36). Others reason in precisely the opposite direction. They point out that the sources could not have contemplated abortions due to the condition of the fetus because nobody could know anything about that until very recently through technologies like amniocentesis and sonograms. Now that we have those tools, most rabbis justify using them to aid in the delivery of a healthy baby. Moreover, when those technologies reveal fetal abnormalities, many rabbis justify abortion on the basis of preserving the mother's mental health where it is clear that the mother is not able to cope with the prospect of bearing or raising such a child (37).

Many Conservative and Reform rabbis, and even a few contemporary Orthodox rabbis, have handled the matter in a completely different way. Our new medical knowledge of the status of the fetus, they say, ought to establish the fetus' health as an independent consideration in determining when abortion is justified (38).

In practice much of this discussion is moot, for Jews engage in abortion as if it were a matter of individual choice. That is a particularly problematic phenomenon for the contemporary Jewish community because Jews constitute only 0.2 percent of the world's population (while Christians make up a full 33 percent). To make matters worse, Jews are barely reproducing themselves in Israel and are falling far short of that in North America, where the Jewish reproductive rate is approximately 1.6 or 1.7 children per couple. Consequently, even those rabbis who are liberal in their interpretation of Jewish abortion law are also calling for Jews to marry and to have children so that the Jewish people and Judaism can survive.

Generating Conception

Artificial Insemination. Since Judaism prizes children so much, it is no wonder that rabbinic authorities have permitted unusual ways of having them for couples who cannot have them otherwise. Nevertheless, there are objections, or at least precautions, connected to some of the procedures.

Rabbis have not objected to uniting a man's sperm with his wife's ovum artificially, whether through artificial insemination or through in vitro fertilization (IVF) (39). Because of Judaism's appreciation of medicine as an aid to God, there is no abhorrence of such means merely because they are artificial.

The matter becomes more complicated when the donor is not the husband. Some rabbis object to such procedures on grounds of adultery. For many, however, adultery takes place only when the penis of the man enters the vaginal cavity of the woman, and that is clearly not the case when insemination takes place artificially. Not only is the physical contact missing; the intent to have an illicit relationship is also absent (40).

More commonly the objection to donor insemination is based on the possibility of unintentional incest in the next generation — specifically, if the product of the artificial insemination later happens to fall in love with a person of the opposite sex who is the child of the semen donor conceived with his wife. Since their biological father is the same man, these two people would be each other's natural half-brother or half-sister. That is problematic for some because it represents a violation of the Torah's laws against incest. Even for those who would invoke the lack

of intent to excuse the couple from those laws, there still remains a critical health concern — namely the increased likelihood among consanguineous unions of genetic diseases transferring from one generation to the next.

This issue dissolves if the semen donor is known or if the donor would not likely be a marital partner for someone in the Jewish community. It was on the latter basis that a prominent Orthodox rabbi, Rabbi Moshe Feinstein, ruled that donor insemination would be permissible if the donor were not Jewish, for in his community intermarriage between Jews and non-Jews was rare. Those Orthodox Jews who will use donor insemination will therefore often require that the donor be a non-Jew.

The Conservative Movement's Committee on Jewish Law and Standards has approved by rabbinic ruling, according to which donor insemination is permissible if either the identity of the donor is known or, lacking that, that enough is known about him so that the child can avoid unintentional incest in his or her sexual partners (married or not) and so that the child can know as much as possible about his or her family traits, both medically and characterologically. In view, however, of the psychological problems that may ensue for the child, the donor, and/or the parents who raise the child (the "social parents"), all parties to the insemination should seek and receive appropriate counseling (41).

Egg Donation. The considerations described above with regard to donor insemination apply as well to egg donation. If the identity of the egg donor remains confidential, the same problems arise with regard to possible unintentional incest in the next generation, and the same solutions by the various rabbinic authorities apply. Specifically, either the egg donor's identity should be shared with the couple who will raise the child and ultimately with the child him/herself, or the woman should be a non-Jew, or enough about the biological mother must be shared with the couple and child to enable the child to avoid such unintentional incest. The donor, in my view, must also share enough information about her talents and traits to help the child understand him/herself. Finally, psychological counseling is appropriate for all concerned both before the procedure and afterward.

Egg donation, though, raises some additional problems. Semen donors incur virtually no medical risks, but that is not true of egg donors. In order to procure as many eggs as possible during each attempt, the donor must be hyperovulated with drugs, and there is some evidence that repeated hyperovulation increases the risk of ovarian cancer (42). This is especially troubling since the donor herself will not, by hypothesis, be gaining a child of her own but will rather be helping another couple have a child. For all that Jewish law prizes procreation, it values the life and health of those already born even more. Consequently, while healthy women may undergo the procedure to donate eggs once or twice, they may not do so much more than that, unless new studies allay the fear of increased cancer risk.

Normally, a child is defined as Jewish in traditional Jewish law if born to a Jewish woman. In cases of egg donation, however, some rabbis have maintained that it is the donor of the gametes who is the legal mother. Most, though, have ruled that even if the egg comes from some other woman, as it does in egg donation, it is the bearing mother whose religion determines whether the child is Jewish or not, and the Conservative Movement's Committee on Jewish Law and Standards has adopted that view (43).

In vitro Fertilization (IVF), Gamete Intrauterine Fallopian Transfer (GIFT), Zygote Intrauterine Fallopian Transfer (ZIFT), etc. When a couple cannot conceive a fetus through sexual intercourse, even when assisted by timing their intercourse, by stimulating ovulation, or by surgery to correct a problem in either the man or the woman, and when the couple prefers to use their own gametes rather than those of donors, they may try any of a number of new techniques, some of which are listed in the title of this subsection. Since the Jewish tradition does not frown upon the use of artificial means to enable people to attain permissible ends, much less sanctified ones like having a child, the mechanical nature of these techniques is not an issue. On the contrary, the important thing to note in recent Jewish rulings is that infertile couples are *not obligated* to use these means to fulfill the man's duty to procreate, even though they *may* (44).

When a woman is impregnated with more than three fetuses, either naturally or artificially, an abortion may be indicated in order to preserve both the life of the mother and the viability and health of the remaining fetuses. For that purpose, such abortions are permitted (and possibly even required). When it can be determined through genetic testing that some of the fetuses have a greater chance to survive and to be healthy than others do, then it is permissible selectively to abort those less likely to survive. This is the same criterion to be used for triage decisions made at the end of life. If all of the fetuses are equally viable, the abortions must be done on a random basis. To avoid the necessity of selective abortion as much as possible, the Conservative Movement's Committee on Jewish Law and Standards has ruled that only three zygotes should be implanted at one time (45).

Surrogate Motherhood. This is really two different forms of overcoming infertility: "traditional surrogacy" or "ovum-surrogacy," in which the surrogate mother's own egg is fertilized by the sperm of the man in the couple who are trying to have a baby (presumably not the husband of the surrogate), and "gestational surrogacy," in which both the egg and the sperm are those of the couple, and the surrogate mother's womb is used to carry and deliver the baby.

From a Jewish perspective, this method of overcoming infertility, or at least something much akin to it, is among the oldest ways recorded in the Jewish tradition. Sarai (later, Sarah), after all, gives her handmaid, Hagar, to Abram (later, Abraham) specifically to conceive a son who would be attributed to Sarai, and Rachel and Leah likewise have their handmaids conceive children with their husband, Jacob. Leah, in fact, had already borne four sons by the time that she uses a surrogate mother because "she stopped bearing"—although she herself was later to bear him two more sons and a daughter (46). These are

all, in modern terminology, ovum-surrogates, and even so, because the handmaid belonged to the man's wife, the Bible attributes the child to the wife rather than the surrogate.

These precedents notwithstanding, though, surrogate motherhood raises difficult emotional and legal problems—although not technological problems beyond those of artificial insemination (in ovum surrogacy) or IVF (in gestational surrogacy). Thus rabbis raise some concerns about the way in which a surrogacy arrangement should be handled, but they do not ultimately prohibit it. Specifically, the couple must abide by civil law in their region and, in light of the recency of this matter in most systems of law, the couple must be informed of the possibility of legal challenges. Furthermore Jewish law would require that steps be taken to ensure that the surrogate mother has full and informed intent to abide by the agreement—perhaps, in ovum-surrogacy, at least, by giving her a period of time (usually 30 days) after birth to cancel the agreement. The surrogate mother must not have physical or other conditions that would make pregnancy dangerous for her beyond the risks normally associated with pregnancy. In ovum-surrogacy the child must either be told the identity of the woman whose gametes he or she inherited or at least be given enough information to be able to avoid incest in his or her own sexual relations and to know about his or her physical and characterological background. Within these parameters, the few rabbis who have written about this have generally permitted surrogacy (47).

Prenatal Diagnosis and Treatment. Both for their own good and for that of their fetuses, pregnant women should seek and get prenatal care. They should also take the preventive measures that modern medicine prescribes to ensure a healthy baby, including restrictions on alcohol, smoking, and some prescription drugs; avoidance of toxins (e.g., in paints) and people with diseases which have been shown to cause fetal damage (e.g., German measles); and adoption of generally health-promoting habits of eating, hygiene, exercise, and sleep.

If the age or genetic background of a couple puts the child at risk for a degenerative, fatal genetic disease (e.g., Tay-Sachs) or for being seriously malformed, the mother may—but not must—undergo prenatal testing, even though that puts the fetus at some risk. Moreover, if the tests reveal that the fetus suffers from such maladies, the mother may choose to abort it. If, however, techniques exist that can cure the child in utero or once born, she may, and probably should, choose to employ those techniques rather than abort the fetus.

According to all interpreters of Jewish law, it is generally not permissible to screen specifically for gender just because one wants a boy or a girl or to screen for any characteristic other than disease (e.g., height, intelligence). Similarly the new sperm-splitting machine (a flow cytometer) to enable couples to choose either a boy or girl would generally violate Judaism's appreciation of people of both genders as equally created by God in the divine image (48). At the same time, Jewish law, as noted earlier, requires a man to father at least two children, specifically a boy and a girl. While that could not be used to justify aborting a fetus of the same gender as those already born, it might justify using the flow cytometer in families who have produced three or more children of one gender and none of the other.

Gene Therapy, Genetic Engineering, and Cloning. Gene therapy is very new and only available in limited areas, and genetic engineering is still only a theoretical possibility. For example, techniques of genetic therapy are already being used to cure hydrocephalus while the fetus is still within the womb of the mother. The Human Genome Project has already discovered the genetic roots of many diseases, and that holds out the hope that someday soon those diseases may be cured. Indeed, on April 27, 2000, scientists in France reported the first success of gene therapy, using it to save three babies with severe combined immune deficiency (SCID) (48a).

There is already general agreement among rabbis that the legitimacy of human intervention to effect cure extends to procedures within the womb as well (49). When used in this therapeutic way, genetic engineering is an unmitigated blessing. Some rabbis have reservations about changing the stem cells themselves and thus all future generations, claiming that our divine mandate is to heal individuals who are ill but not to alter the nature of future human beings. That, in their view, would be to arrogate too much power to ourselves. Thus for them only therapeutic changes in the somatic cells of a diseased individuals would be permitted. Others, though, maintain that if we can root out a genetic disease not only from those who now have it but from descendants of such people as well, we should definitely do so, for our religious mandate is not only to cure diseases but to prevent them, if possible. With respect to degenerative genetic diseases like Tay-Sachs, I myself fall within the latter group.

Genetic engineering also, though, holds out the possibility that someday we will be able to change the nature of the human being so as to avoid the diseases that kill most of us today — heart disease, cancer, and the like. That already raises the question of whether we are effectively trying to reverse God's decision in the Garden of Eden to prohibit us from becoming immortal. Presumably, even if the currently deadly diseases are cured, we will eventually die of something else. So the Jewish justification for human beings to engage in curing may also be applied to genetically engineering ourselves to resist heart disease and cancer.

When we gain the ability to do these arguably permissible (and maybe even mandatory) things, we will also have the ability to change other things about ourselves. In fact, once we can change not only the genes of a particular fetus but even its germ line, we will be able to screen out traits that are not manifestations of a disease at all but merely characteristics that are deemed undesirable by certain individuals or groups. Abortion to eliminate defective fetuses poses the danger of the slippery slope where the definition of "defective" is broadened to the point of allowing only "perfect" children to be born, thus creating a master race. For example, we might change the genetic traits of shortness, merely average intelligence, a particular skin color, a propensity to alcoholism, and,

perhaps, homosexuality. Moreover genetic engineering will create a new organism, and that poses real risks to human beings and to the environment.

There are thus some uses of genetic engineering that are clearly legitimate or illegitimate, but there are many where it is, and will be, difficult to tell. How do we determine when we are using genetic engineering appropriately to aid God in ongoing, divine acts of cure and creation and when, on the other hand, we are usurping the proper prerogatives of God to determine the nature of creation? More bluntly, when do we cease to act as the servants of God and pretend instead to be God?

Although cloning has been much more thoroughly discussed in the media, it actually presents fewer moral problems for Jews than genetic engineering does. Cloning, after all, does not introduce into the environment any new organism; it just replicates an organism that already exists, thus posing lesser risks. If cloning is used to overcome infertility, to aid in the research of diseases, or, in plants and animals, to produce food for starving people, it will be a very positive thing. On the other hand, cloning to avoid the intimacy of sexual intercourse, to gain immortality (as if that were possible through this technique), or to replicate oneself without any admixture of someone else's genes would be illegitimate uses of the technique. They smack of self-idolization and of the denial of human mortality; they thus make the moral and theological error of confusing human beings for God.

Our moral doubts about genetic engineering and cloning do not mean that research into these techniques should stop; the potential benefits to our life and health are enormous. They should prompt us, however, to exercise care in how we use our new capabilities. The problems are not just medical and technological; they are moral and theological, requiring us to reexamine the very ways we understand ourselves as human beings, our relationships to others and to God, and the limits inherent in being human.

Care of Severely Handicapped Newborns. Once a child is born, the child is a full-fledged human being and is to be treated in its health care like all other human beings. That is true for disabled newborns (or adults, for that matter) just as much as it is for those with no disabilities. The image of God in each one of us does not depend on one's abilities or skills; in this way the Jewish way of evaluating life is distinctly at odds with the utilitarian view common in Western societies.

If the child is born with severe disabilities that threaten his or her life, however, heroic measures need not be employed to keep the child alive. Here the same rules that govern the withholding and removal of life-support systems of any human being apply to newborns, with all of the diversity of opinion among rabbis noted in that section below. Some rabbis, however, are more lenient with respect to the treatment of newborns than they are regarding people dying later on in life because of the possibility, noted in Jewish law, that the child was born prematurely. Specifically, until the child is 30 days old, he or she is not considered to be a person whose life is confirmed (a *bar kayyma*). Therefore, while we may not do anything actively to hasten the child's death, we may, according to these authorities, do less to sustain it than we would be called upon to do with regard to people who had lived beyond 30 days. Thus some who would insist on artificial nutrition and hydration for most dying people would not require it for life-imperiled infants less than 30 days old — except of course if the intervention holds out significant promise of curing the infant of the disease or condition. Some would require incubators, but most would not require surgery or medications beyond those necessary to relieve the child of pain (50).

Stem Cell Research

Jewish Views of Genetic Materials. Since human embryonic stem cells can be procured from aborted fetuses, the status of abortion within Judaism immediately arises. As we have seen, sometimes abortion is required by Jewish law and sometimes it is permitted, but mostly it is forbidden. The upshot of the Jewish stance on abortion, then, is that *if* a fetus was aborted for legitimate reasons under Jewish law, the aborted fetus may be used to advance our efforts to preserve the life and health of others.

In general, when a person dies, we must show honor to God's body by burying it as soon after death as possible. To benefit the lives of others, though, autopsies may be performed when the cause of death is not fully understood, and organ transplants are allowed to enable other people to live (51). The fetus, though, does not have the status of a full-fledged human being. Therefore, if we can use the bodies of human beings to enable others to live, how much the more so may we use a part of a body — in this case, the "water" or "thigh" that constitutes the fetus — for that purpose. This all presumes, though, that the fetus was aborted for good and sufficient reason within the parameters of Jewish law.

Stem cells for research purposes can also be procured from donated sperm and eggs mixed together in a petri dish and cultured there. Genetic materials outside the uterus have no legal status in Jewish law, for they are not even a part of a human being until implanted in a woman's womb, and even then, as we have noted, during the first 40 days of gestation their status is "as if they were simply water" (52). Abortion is still prohibited during that time except for therapeutic purposes, for in the uterus such gametes have the potential of growing into a human being, but outside the womb, at least as of now, they have no such potential. As a result frozen embryos may be discarded or used for reasonable purposes, and so may stem cells procured from them.

Other Factors in Stem Cell Research. Given that the materials for stem cell research can be procured in permissible ways, the technology itself is morally neutral. It gains its moral valence on the basis of what we do with it. The question, then, reduces to a risk-benefit analysis of stem cell research. The articles in a recent *Hastings Center Report* (53) raise some questions to be considered in such an analysis, and I will not rehearse them here. I want to note only two things about them from a Jewish perspective.

First, the Jewish tradition sees the provision of health care as a communal responsibility, and so the justice arguments in the *Hastings Center Report* have a special resonance for me as a Jew. Especially since much of the basic science in this area was funded by the government, the government has the right to require private companies to provide their applications of that science to those who cannot afford them at reduced rates or, if necessary, even for free. At the same time, the Jewish tradition does not demand socialism, and for many good reasons we, in the United States, have adopted a modified, capitalistic system of economics. The trick, then, will be to balance access to applications of the new technology with the legitimate right of a private company to make a profit on its efforts to develop and market applications of stem cell research.

Second, the potential of stem cell research for creating organs for transplant and cures for diseases is, at least in theory, both awesome and hopeful. Indeed, in light of our divine mandate to seek to maintain life and health, one might even argue that from a Jewish perspective we have a *duty* to proceed with that research. As difficult as it may be, though, we must draw a clear line between uses of this or any other technology for cure, which are to be applauded, as against uses of this technology for enhancement, which must be approached with extreme caution. Jews have been the brunt of campaigns of positive eugenics both here, in the United States, and in Nazi Germany (54), and so we are especially sensitive to creating a model human being that is to be replicated through the genetic engineering that stem cell applications will involve. Moreover, when Jews see a disabled human being, we are not to recoil from the *disability* or count our blessings for not being disabled in that way; we are rather commanded to recite a blessing thanking God for making people different (55). In light, then, of the Jewish view that all human beings are created in the image of God, regardless of their levels of ability or disability, it is imperative from a Jewish perspective that the applications of stem cell research be used for cure and not for enhancement.

We thus should take the steps necessary to advance stem cell research and its applications in an effort to take advantage of its great potential for good. We should do so, though, with restrictions to enable access to its applications to all Americans who need it and to prohibit applications intended to make all human beings into any particular model of human excellence. Instead, through this technology and all others, we should seek to cure diseases while simultaneously retaining our appreciation for the variety of God's creatures.

TECHNOLOGY AT THE END OF LIFE

Care of the Dying

General Concepts and Categories. Judaism prohibits murder, and it views all forms of active euthanasia as the equivalent of murder (56). That is true even if the patient asks to be killed. Because each person's body belongs to God, the patient does not have the right either to commit suicide or to enlist the aid of others in the act. Those who assist someone in a suicide violate Jewish law; the specific nature and severity of the violation depend on how the aid is proffered. No human being has the right to destroy or even damage God's property (57).

The patient does have the right, however, to pray to God to permit death to come (58), for God, unlike human beings, has the right to destroy His own property. Moreover Judaism does permit passive euthanasia in specific circumstances.

Point When Passive Euthanasia Is Permissible. When does the Jewish obligation to cure end, and when does the permission (or, according to some, the obligation) to let nature take its course begin?

Authorities differ. All agree that one may allow nature to take its course once the person becomes a *goses*, a moribund person. But when does that state begin? The most restrictive position is that of Rabbi J. David Bleich, who limits it to situations when all possible medical means are being used in an effort to save the patient and the physicians assume that he or she will nevertheless die within 72 hours (59). Others define the state of *goses* more flexibly and therefore apply the permission to withhold or withdraw machines and medications during that time more broadly, in some cases up to a year or more (60).

In a rabbinic ruling approved by the Conservative Movement's Committee on Jewish Law and Standards (61), I noted that classical Jewish sources describe a *goses* as if the person were "a flickering candle," so that he or she may not even be moved for fear of inducing death. That description and that medical therapy only apply to people within the last hours of life (not even the last three days). Consequently, I argued, the appropriate Jewish legal category to describe people with terminal, incurable diseases, who may live for months and even years, is, instead, *terefah*. Permission to withhold or withdraw medications and machines would then apply to people as soon as they are in the state of being a *terefah*, that is, as soon as they are diagnosed with a terminal, incurable illness. In judging a disease to be incurable, we are not responsible for knowing whether a cure is imminent, for we are not God; the attending physicians must just use their best judgment.

Artificial Nutrition and Hydration. While intravenous cannulation to provide nutrition and fluids is appropriately used in people where there is reasonable hope for recovery, where no such hope exists, may one remove such tubes? Some rabbis have said no, reasoning that since artificial nutrition and hydration supply the liquids and nutrients that all of us need to survive, they cannot be classified as medications, which are used only when specific people need them (62). Others, however, noting that the Talmud specifically defines "food" as that which is ingested through the mouth and swallowed, classify artificial nutrition and hydration as medicine and permit removing or withholding them when recovery is not anticipated. Those attending the patient, though, must still go through the motions of bringing in a normal food tray at regular meal times in fulfillment of our duty to feed the starving (63).

Curing the Patient, Not the Disease. The important thing to note, however, is that there is general agreement that a Jew need not use heroic measures to maintain his or her life but only those medicines and procedures that are commonly available in the person's time and place. We are, after all, commanded to *cure* based on the verse in Exodus 21:19, "and he shall surely cure him." We are not commanded to sustain life *per se* (64). Thus, on the one hand, as long as there is some hope of cure, heroic measures and untested drugs *may* be employed, even though this involves an elevated level of risk. On the other hand, physicians, patients, and families who are making such critical care decisions are *not* duty-bound by Jewish law to invoke such therapies and may instead follow a course of hospice care. Indeed, because hospice care involves the support of family and friends and subjects the patient to the least amount of physical invasion possible, it often is preferable to more technologically sophisticated forms of treatment.

Pain medication may be administered as needed. Even in the last stages of life, when the dosage needed may actually hasten the patient's death, it may be used so long as the intent is not to kill the person but rather to alleviate his or her pain (65).

Moreover our duty to cure the patient rather than any specific disease means that if a person who is suffering from multiple, incurable, terminal illnesses develops pneumonia, doctors may refrain from treating the pneumonia if that will enable the patient to die less painfully. This would be in line with the strain in Jewish law that does not automatically and mechanically assume that preservation of life trumps all other considerations but rather judges according to the best interests of the patient (66). That principle does not extend so far as to permit mercy killing (active euthanasia), but it does make it permissible to refrain from administering the antibiotic so that the patient can die of his or her other diseases.

A person may volunteer to undergo an experimental procedure that holds out no hope to improve his/her own health but may increase medical knowledge and thereby help others only if it subjects the person to minimal or no risk. One's duty to preserve one's own life takes precedence over one's obligation to help other people preserve theirs.

Care of the Deceased: Autopsies and Organ Transplants

General Principles. The treatment of these topics in Jewish law depends on two primary principles. The general tenet that governs treatment of the body after death is *kavod ha'met*, namely, that we should render honor to the dead body as a sign of respect for both the deceased person and for God's property. Honor of the corpse, then, underlies Jewish burial customs.

The other principle that affects the topics of this section is that of *pikkuah nefesh*. When interpreting Leviticus 18:5, which says that we should obey God's commands "and live by them," the rabbis deduce that this means that we should not die as a result of observing them. The tenet that emerges is *pikkuah nefesh*, the obligation to save people's lives. This tenet is so deeply embedded in Jewish law that, according to the rabbis, it takes precedence over all other commandments except murder, idolatry, and incestuous or adulterous sexual intercourse (67).

Jews are commanded not only to do virtually anything necessary to save their own lives; they are also bound by the positive obligation to take steps to save the lives of others. The imperative to do so is derived from the biblical command, "Do not stand idly by the blood of your neighbor" (Leviticus 19:16). This means, for example, that if you see someone drowning, you may not ignore him or her but must do what you can to save that person's life (68).

What happens, when you can only save your life or someone else's? Whose life takes precedence? Since the Torah says that one should not exact interest from a fellow Jew "so that your brother may live *with you*" (Leviticus 25:36), and since you must therefore be alive at the time that you care for your brother, "your life takes precedence (*hayyekha kodemim*)" (69).

Autopsies. The two procedures that may interrupt the normal Jewish burial process are autopsies and organ transplants. Even though autopsies require invading the body of the deceased, in 1949 Israeli Chief Rabbi Isaac Herzog enunciated what has come to be the generally accepted position among Jews—namely, that while autopsies may not be done routinely, they are permissible if required by civil law, if the cause of death cannot otherwise be ascertained, if three physicians attest that the autopsy might help save the lives of others suffering from an illness similar to that from which the patient had died, or if a hereditary illness was involved so that performing the autopsy might safeguard surviving relatives. In all these cases we honor the dead by using the body to save lives.

Organ Transplantation. The overriding principles of honoring the dead (*kavod ha-met*) and saving people's lives (*pikkuah nefesh*) also work in tandem in organ transplantation. So the default assumption is that a person would be honored to help another live through organ donation.

Living Donors. Because one's own life takes precedence over helping someone else live, contemporary rabbis have generally permitted, but not required, donations from living donors when their life or health is not thereby subjected to major risk (70). If a family member suffers from leukemia and no appropriate bone marrow match is available, a married couple may seek to have another child in an attempt to find such a match, but only if they will not abort the child even if it becomes clear that the child is not the match they seek.

Cadaveric Donors. Since a dead person incurs no health risk, cadaveric donations are not only generally held to be permissible but, according to a responsum approved by the Conservative Movement's Committee on Jewish Law and Standards, actually a positive obligation so as to prevent the need of living persons incurring such risks while also saving the recipient's life. While traditional Jewish sources define the moment of death as the cessation of heartbeat and breath (71), even the chief rabbinate of the State of Israel in 1987 approved heart transplants, thereby accepting evidence of full brain death (including the brain stem) as fulfilling those requirements (72).

If a fetus has been aborted for reasons approved by Jewish law—namely to save the life or health of the mother or because the fetus suffers from Tay-Sachs or some similar fatal illness—the fetus may be used for purposes of transplant or experimentation.

Use of Animal or Artificial Organs; Animal Experimentation. While Judaism seeks to minimize pain to animals (73), it permits their use for food, for work, and, certainly, for saving a life. This would include medical research based on animal trials and the use of animal parts for transplantation, if that proves successful.

ENVIRONMENTAL USES OF TECHNOLOGY

The Jewish tradition, from the Torah on, was concerned with preserving God's world, leading to a series of ecological laws (74). Although classical Jewish law could not contemplate all the opportunities and problems produced by modern technology, it already prohibits wasting natural resources, even if one owns them, and it makes people responsible for the air and water pollution they cause. Judaism's appreciation of the world as belonging to God would additionally require us, in modern times, to create less waste than we moderns do, especially in technologically sophisticated societies, to recycle, and to use our new technology to reduce and, if possible, prevent pollution.

One application of biotechnology that, on the face of it, might cause special problems for Jews is the use of technology to produce new foods. Jewish dietary laws (*kashrut*, or "keeping kosher") restrict the fish, fowl, and animals that Jews may eat and the way that they are killed and their meat prepared and served. The Torah also forbids mixing seeds (*kilayim*) (75), but if non-Jews, who are not subject to this law, create hybrids, Jews may use them. An established principle in Jewish law, though, is that if a substance is chemically changed so that it cannot be reconstituted in its original form, it is "a new thing" (*davar hadash*) and, as such, loses any characteristics of its origins (76). Therefore bioengineered foods, such as cloned tomatoes, may certainly be eaten in accordance with Jewish dietary laws if the original substances are kosher and, if there is sufficient chemical change, even if the original substances are not kosher. Similarly Jews may engage in bioengineering new foods without violating the laws against mixing seeds if all (or all but one of) the materials to be combined are already so chemically changed as to constitute a new substance.

BIBLIOGRAPHY

1. See, for example, Deuteronomy 10:14; Psalms 24:1. See also Genesis 14:19, 22 (where the Hebrew word for "Creator" [*koneh*] also means "Possessor," and where "heaven and earth" is a merism for those and everything in between); Exodus 20:11; Leviticus 25:23, 42, 55; Deuteronomy 4:35, 39; 32:6.

2. *Genesis Rabbah* 11:6; *Pesikta Rabbati* 22:4.

3. See, for example, Deuteronomy 10:14; Psalms 24:1. See also Genesis 14:19, 22; Exodus 20:11; Leviticus 25:23, 42, 55; Deuteronomy 4:35, 39; 32:6. For these three and four other foundational principles of Jewish medical ethics, see E.N. Dorff, *Matters of Life and Death: A Jewish Approach to Modern Medical Ethics*, Jewish Publication Society, Philadelphia, PA, 1998, Ch. Two.

4. Thus, for example, bathing is a commandment, according to Hillel: *Leviticus Rabbah* 34:3. Maimonides includes rules requiring proper care of the body in his code of Jewish law as *a positive obligation* (not just advice for feeling good or living a long life), parallel to the positive duty to aid the poor: Maimonides' *Mishneh Torah* (1177 C.E.) *Laws of Ethics (De'ot)*, Chs. 3–5.

5. Babylonian Talmud (edited circa 500 C.E.) *Shabbat* 32a; Babylonian Talmud (edited circa 500 C.E.) *Bava Kamma* 15b, 80a, 91b; Maimonides' *Mishneh Torah* (1177 C.E.) *Laws of Murder* 11:4–5; Joseph Karo's *Shulhan Arukh* (1565 C.E.) with the glosses of Moses Isserles. *Yoreh De'ah* 116:5 gloss; Joseph Karo's *Shulhan Arukh* (1565 C.E.) with the glosses of Moses Isserles. *Hoshen Mishpat* 427:8–10. Jewish law views endangering one's health as worse than violating a ritual prohibition: Babylonian Talmud (edited circa 500 C.E.) *Hullin* 10a; Joseph Karo's *Shulhan Arukh* (1565 C.E.) with the glosses of Moses Isserles. *Orah Hayyim* 173:2; Joseph Karo's *Shulhan Arukh* (1565 C.E.) with the glosses of Moses Isserles. *Yoreh De'ah* 116:5 gloss.

6. See J.D. Bleich, *Tradition* **16**(4), 130–133 (1977).

   S. Freehof, *Reform Responsa for Our Time*, Hebrew Union College Press, Cincinnati, OH, 1977, Ch. 11; *Proc. Rabbinical Assembly* **44**, 182 (1983). All of the above are reprinted in E.N. Dorff and A. Rosett, *A Living Tree: The Roots and Growth of Jewish Law*, State University of New York Press, Albany, 1988, pp. 337–362.

7. Genesis 9:5; Mishnah (edited circa 200 C.E.); *Semahot* 2:2; Babylonian Talmud (edited circa 500 C.E.) *Bava Kamma* 91b; *Genesis Rabbah* 34:19 states that the ban against suicide includes not only cases where blood was shed but also self-inflicted death through strangulation, and the like; Maimonides' *Mishneh Torah* (1177 C.E.) *Laws of Murder* 2:3; Maimonides' *Mishneh Torah* (1177 C.E.) *Laws of Injury and Damage* 5:1; Joseph Karo's *Shulhan Arukh* (1565 C.E.) with the glosses of Moses Isserles. *Yoreh De'ah* 345:1–3. Cf. J.D. Bleich, *Judaism and Healing*, Ktav, New York, 1981, Ch. 26.

8. The rabbis note that the Nazarite, who takes an oath to avoid such pleasures, must, according to Numbers 6:11, bring a sin offering after the time specified by his oath, and they derive from that law that abstinence is prohibited: Babylonian Talmud (edited circa 500 C.E.) *Ta'anit* 11a. Cf. also Maimonides' *Mishneh Torah* (1177 C.E.) *Laws of Ethics (De'ot)* 3:1.

9. Maimonides' *Mishneh Torah* (1177 C.E.) *Laws of Ethics (De'ot)* 3:3.

10. *Sifra* on Leviticus 19:16; Babylonian Talmud (edited circa 500 C.E.) *Sanhedrin* 73a; Maimonides' *Mishneh Torah* (1177 C.E.) *Laws of Murder* 1:14; Joseph Karo's *Shulhan Arukh* (1565 C.E.) with the glosses of Moses Isserles. *Hoshen Mishpat* 426.

11. E.g., Exodus 15:26; Deuteronomy 32:39; Isaiah 19:22; 57:18–19; Jeremiah 30:17; 33:6; Hosea 6:1; Psalms 103:2–3; 107:20; Job 5:18.

12. The permission and duty to heal: Babylonian Talmud (edited circa 500 C.E.) *Bava Kamma* 85a, 81b; Babylonian Talmud (edited circa 500 C.E.) *Sanhedrin* 73a, 84b (with Rashi's commentary there). See also *Sifrei Deuteronomy* on Deuteronomy 22:2 and *Leviticus Rabbah* 34:3. Nahmanides, *Kitvei Haramban*, Bernard Chavel, ed., (Jerusalem: Mosad Harav Kook, 1963 [Hebrew]), vol. 2, p. 43, bases the duty for the community to provide health care on Leviticus 19:18, "You shall love your neighbor as yourself"; this passage comes from Nahmanides' *Torat Ha'adam (The Instruction of Man)*, *Sh'ar Sakkanah (Section on Danger)* on Babylonian

Talmud (edited circa 500 C.E.) *Bava Kamma*, Ch. 8, and is cited by Joseph Karo in his commentary to the *Tur*, *Bet Yosef*, *Yoreh De'ah* 336. Nahmanides bases himself on similar reasoning in Babylonian Talmud (edited circa 500 C.E.) *Sanhedrin* 84b.

13. Babylonian Talmud (edited circa 500 C.E.) *Sanhedrin* 74a.

14. The best of physicians deserves to go to Hell: Babylonian Talmud (edited circa 500 C.E.) *Kiddushin* 82a. Abraham ibn Ezra, Bahya ibn Pakuda, and Jonathan Eybeschuetz all restricted the physician's mandate to external injuries: See Ibn Ezra's commentary on Exodus 21:19 and cf. his comments on Exodus 15:26 and 23:25, where he cites Job 5:18 and II Chronicles 16:12 in support of his view; Bahya's commentary on Exodus 21:19; and Eybeschuetz, *Kereti U'pleti* on Joseph Karo's *Shulhan Arukh* (1565 C.E.) with the glosses of Moses Isserles. *Yoreh De'ah* 188:5. See I. Jakobovits, *Jewish Medical Ethics*, Bloch, New York, 1959, 1975, 5–6. That a Jew may not live in a city without a physician: Jerusalem Talmud (edited circa 400 C.E.) *Kiddushin* 66d; see also Babylonian Talmud (edited circa 500 C.E.) *Sanhedrin* 17b, where this requirement is applied only to "the students of the Sages."

15. Joseph Karo's *Shulhan Arukh* (1565 C.E.) with the glosses of Moses Isserles. *Yoreh De'ah* 336:1.

16. *Midrash Temurrah* as cited in J.D. Eisenstein, ed., *Otzar Midrashim*, vol. 2, New York, 1915, pp. 580–581. Cf. also Babylonian Talmud (edited circa 500 C.E.) *Avodah Zarah* 40b, a story in which rabbi expresses appreciation for foods that can cure. Although circumcision is not justified in the Jewish tradition in medical terms, it is instructive that the rabbis maintained that Jewish boys were not born circumcised specifically because God created the world such that it would need human fixing, a similar idea to the one articulated here on behalf of physicians' activity despite God's rule; see note 2 above.

17. Babylonian Talmud (edited circa 500 C.E.) *Shabbat* 10a, 119b. In the first of those passages, it is the judge who judges justly who is called God's partner; in the second, it is anyone who recites Genesis 2:1–3 (about God resting on the seventh day) on Friday night who thereby participates in God's ongoing act of creation. The Talmud in Babylonian Talmud (edited circa 500 C.E.) *Sanhedrin* 38a specifically wanted the Sadducees *not* to be able to say that angels or any being other than humans participate with God in creation.

18. On this entire matter, the Rabbinical Assembly, the rabbinic body of the Conservative Movement within Judaism, has created a rabbinic letter designed for use with adults and with teenagers discussing the concepts, values, and laws of Judaism governing intimate relations, marriage, nonmarital sex, and homosexuality. See E.N. Dorff, *This Is My Beloved, This Is My Friend: A Rabbinic Letter on Intimate Relations*, Rabbinical Assembly, New York, 1996. See also E.N. Dorff, *Matters of Life and Death*, Jewish Publication Society, Philadelphia, PA, 1998, Ch. 3–6.

19. Genesis 2:18; cf. *Midrash Psalms* on Psalms 59:2. Exodus 21:10 prescribes that a woman has conjugal rights in marriage, just as a man does, and the rabbis then spell out exactly how often a man must offer to have sex with his wife and how long she can refuse his advances without losing part of her settlement in a divorce; see Mishnah (edited circa 200 C.E.); *Ketubbot* 5:6–7. Note that he may never force himself upon her.

20. For example, Deuteronomy 7:13–14; 28:4, 11; Psalms 128:6.

21. Genesis 15:5; 17:3–6, 15–21; 18:18; 28:14; 32:13.

22. Mishnah (edited circa 200 C.E.); *Yevamot* 6:6 (61b); Babylonian Talmud (edited circa 500 C.E.) *Yevamot* 65b–66a; Maimonides' *Mishneh Torah* (1177 C.E.) *Laws of Marriage* 15:2; Joseph Karo's *Shulhan Arukh* (1565 C.E.) with the glosses of Moses Isserles. *Even Ha'ezer* 1:1, 13.

23. Deuteronomy 6:5.

24. Babylonian Talmud (edited circa 500 C.E.) *Yevamot* 12b. For a thorough discussion of this, see D.M. Feldman, *Birth Control in Jewish Law*, New York University Press, New York, 1968; reprinted as *Marital Relations, Abortion, and Birth Control in Jewish Law*, Schocken, New York, 1973.

25. Cf. Maimonides' *Mishneh Torah* (1177 C.E.) *Laws of Forbidden Intercourse* 16:2, 6; Joseph Karo's *Shulhan Arukh* (1565 C.E.) with the glosses of Moses Isserles. *Even Ha'ezer* 5:2.

26. Babylonian Talmud (edited circa 500 C.E.) *Shabbat* 110b.

27. Cf. J.D. Bleich, *Judaism and Healing*, Ktav, New York, 1981, p. 65.

    D. Feldman and F. Rosner, eds., *Compendium on Medical Ethics*, New York, 1984, pp. 46–47.

    S. Freehof, *Reform Responsa*, Hebrew Union College Press, Cincinnati, OH, 1960, pp. 206–208.

28. Babylonian Talmud (edited circa 500 C.E.) *Yevamot* 69b. Rabbi I. Jakobovits [*Jewish Medical Ethics*, Bloch, New York, 1959, 1975, p. 275] notes that "40 days" in talmudic terms may mean just under two months in our modern way of calculating gestation due to improved methods of determining the date of conception.

29. Babylonian Talmud (edited circa 500 C.E.) *Niddah* 17a.

30. Cf. D.M. Feldman, *Birth Control in Jewish Law*, New York University Press, New York, 1968, pp. 265–266 and Ch. 15.

31. Mishnah (edited circa 200 C.E.) *Niddah* 3:5.

32. Mishnah (edited circa 200 C.E.) *Oholot* 7:6. There are variant versions of this. Like our Mishnah, Jerusalem Talmud (edited circa 400 C.E.) *Shabbat* 14:4 reads "its greater part;" Tosefta (edited circa 200 C.E.) *Yevamot* 9:9 and Babylonian Talmud (edited circa 500 C.E.) *Sanhedrin* 72b have "its head;" and Jerusalem Talmud (edited circa 400 C.E.) *Sanhedrin* 8, end, has "its head or its greater part."

33. Cf. I. Jakobovits, *Jewish Medical Ethics*, Bloch, New York, 1975, pp. 186–187 and No. 173 on pp. 378–379.

34. Cf. D.M. Feldman, *Birth Control in Jewish Law*, New York University Press, New York, 1968, pp. 284–294.

    M.H. Spero, *Judaism and Psychology: Halakhic Perspectives*, Ktav, New York, 1980, Ch. 12.

35. I. Jakobovits, *Jewish Medical Ethics*, Bloch, New York, pp. 189–190.

36. J.D. Bleich, *Contemporary Halakhic Problems*, Ktav, New York, 1977, pp. 112–115.

    J.D. Bleich, in F. Rosner and J.D. Bleich, eds., *Jewish Bioethics*, Sanhedrin Press, New York, 1979, Ch. 9, esp. pp. 161 and 175, No. 97.

37. Cf. D.M. Feldman, *Birth Control in Jewish Law*, New York University Press, New York, 1968, pp. 284–294.

38. E. Waldenberg, *Responsa Tzitz Eliezer*, 9:51 (1967) and 13:102 (1978) [Hebrew]; S. Israeli, *Amud Hayemini*, No. 35 cited in *No'am*, 16 (K.H.) 27 (note) [Hebrew]; L. Grossnass, *Responsa Lev Aryeh* 2:205 [Hebrew]; A.J. Goldman, *Judaism Confronts Contemporary Issues*, Ktav, New York, 1978, Ch. 3, esp. pp. 52–62.

39. I. Jakobovits, *Jewish Medical Ethics*, Bloch, New York, 1975, p. 264.

    J.D. Bleich, *Judaism and Healing: Halakhic Perspectives*, Ktav, New York, 1981, pp. 82–84.

40. J.D. Bleich [*Judaism and Healing*, Ktav, New York, 1981, pp. 80–84] cites all of the following authorities (all of whom wrote in Hebrew) as requiring physical contact of the genital organs for adultery to occur: R. Shalom Mordecai Schwadron, *Teshuvot Zekan Aharon*, II, No. 97; R. Yehoshua Baumol, *Teshuvot Emek Halakhah*, No. 68; R. Ben Zion Uziel, *Mishpetei Uziel, Even Ha-Ezer*, I, No. 19; R. Moshe Feinstein, *Iggrot Moshe, Even Ha-Ezer*, I, No. 10; and R. Eliyahu Meir Bloch, *Ha-Pardes*, Sivan 5713. Nevertheless, as he points out, these Orthodox rabbis would prohibit donor insemination on the grounds of potential incest in the next generation, as discussed in the next paragraph. Moreover some Orthodox rabbis Bleich cites (R. Yehudah Leib Zirelson, R. Ovadiah Hadaya, R. Eliezer Waldenberg) maintain that violating the prohibition against adultery does not require genital contact, and so they would object to donor insemination on that ground as well.

41. E.N. Dorff, Artificial insemination, egg donation, and adoption, *Conservative Judaism* **49**(1), 3–60 (1996). This rabbinic ruling, approved in 1994 by the Conservative Movement's Committee on Jewish Law and Standards, also discusses several ancillary concerns, such as the identity of the father for various purposes in Jewish law; the psychological issues raised by the asymmetry in the situation — namely that the child will be the biological product of the woman but not the man who will be raising him/her; and the psychological need of the child to know his/her genetic roots. This ruling was reprinted in E.N. Dorff, *Matters of Life and Death*, Jewish Publication Society, Philadelphia, PA, 1998, pp. 66–115.

42. R. Spirtas, S.C. Kaufman, and N.J. Alexander, *Fertility and Sterility* [J. Am. Fertil. Soc.] **59**(2), 291–292 (1993). I want to thank my friend, Dr. Michael Grodin, for sharing this article with me. The 1988 Congressional report also reported a number of other possible complications caused by commonly used drugs to stimulate the ovaries, including early pregnancy loss, multiple gestations (fetuses), ectopic pregnancies, headache, hair loss, pleuropulmonary fibrosis, increased blood viscosity, and hypertension, stroke, and myocardial infarction; see U.S. Congress, Office of Technology Assessment, *Infertility: Medical and Social Choices*, OTA-BA-358, U.S. Government Printing Office, (Washington, DC, 1988), pp. 128–129. Once again, the demonstrated risks are not so great as to make stimulation of the ovaries for egg donation prohibited as a violation of the Jewish command to guard our health, but they are sufficient to demand that caution be taken and that the number of times a woman donates eggs be limited.

43. Rabbi Aaron Mackler, *In vitro Fertilization*, Draft No. 3, November, 1995, adopted by the Committee on Jewish Law and Standards in December, 1995, (unpublished); see p. 12 of the typescript.

44. See, for example, J.D. Bleich, *Judaism and Healing: Halakhic Perspectives*, Ktav, New York, 1981, pp. 85–91.

    E.N. Dorff, *Conservative Judaism* **49**(1), 17–18, 47–48 (1996), with regard to donor insemination and egg donation, but the same considerations, although sometimes in different forms, apply to IVF, GIFT, and ZIFT. See also E.N. Dorff, *Matters of Life and Death*, Jewish Publication Society, Philadelphia, PA, 1998, Ch. 3 and 4.

45. E.N. Dorff, Artificial insemination, *Conservative Judaism* **49**(1), 47 (1996); *Matters of Life and Death*, Jewish Publication Society, Philadelphia, PA, 1998, pp. 56–57, 101–102, 129–130.

46. Genesis 16:2 uses a play on words in Hebrew when Sarai says to Abram, "Look, the Lord has kept me from bearing. Consort with my maid [Hagar]; perhaps I shall have a son [also, I shall be built up] through her." This indicates that Ishmael, the son that resulted from this union, was not only to be considered Abram's son, but Sarai's as well. Similarly Rachel tells Jacob, "Here is my maid Bilhah. Consort with her, that she may bear on my knees and that through her I too may have children" (Genesis 30:3). At that time, Rachel was infertile (she gave birth to Joseph and Benjamin only later), but Leah, who had already had four sons, also gave her handmaid, Zilpah, to Jacob, and when Zilpah bore two sons to Jacob, Leah says " 'What fortune!' meaning, 'Women will deem me fortunate' " (Genesis 30:13)—indicating that those two sons were ascribed to Leah as well. That Leah had stopped bearing and therefore resorts to the use of her handmaid Zilpah: Genesis 30:9. That she later bears three more children: Genesis 30:14–21.

47. See, for example, J.D. Bleich, *Judaism and Healing*, Ktav, New York, 1981, pp. 92–95. While Rabbi Bleich generally prohibits or limits the use of new medical procedures, here he specifically argues against Rabbi Jakobovits' claim that surrogacy is inherently immoral and spends most of his discussion on the question of the Jewish identity of the child. See also M. Gold, *And Hannah Wept: Infertility, Adoption, and the Jewish Couple*, Jewish Publication Society, Philadelphia, PA, 1988, pp. 120–127; and E. Spitz, Through her I too shall bear a child: Birth surrogates in Jewish law. *J. Religious Ethics* **24**(1), 65–97 (1996), a slightly different version of which was approved by the Conservative Movement's Committee on Jewish Law and Standards in September, 1997. At that same meeting the Committee also approved a rabbinic ruling, as yet unpublished, by Rabbi Aaron Mackler that ultimately permits surrogacy but with more disclaimers and restrictions.

48. For a good, popular article describing some of the issues involved, see L. Belkin, Getting the girl. *N.Y. Times Mag.*, July 25, 1999, pp. 26–31, 38, 54–55.

48a. *N.Y. Times*, April 28, pp. A1–A16 (2000).

49. E.g., Bleich, *Judaism and Healing*, Bloch, New York, 1981, p. 106.

50. Rabbi Avram Reisner reasons this way in a responsum entitled *Peri- and Neo-Natology*, approved by the Conservative Movement's Committee on Jewish Law and Standards in Fall, 1995 (unpublished). That a child's life is not confirmed until 30 days of age: Babylonian Talmud (edited circa 500 C.E.) *Shabbat* 135b.

51. For classical sources on this, see E.N. Dorff, *Matters of Life and Death*, Jewish Publication Society, Philadelphia, PA, 1998, Ch. 9.

52. Babylonian Talmud (edited circa 500 C.E.) *Yevamot* 69b. Rabbi Immanuel Jakobovits notes that "40 days" in talmudic terms may mean just under two months in our modern way of calculating gestation, since the Rabbis counted from the time of the first missed menstrual flow while we count from the time of conception, approximately two weeks earlier. See I. Jakobovits, *Jewish Medical Ethics*, Bloch, New York, 1959, 1975, p. 275.

53. *Hastings Center Report*, March–April, 1999, pp. 30–48.

54. See S.J. Gould, *The Mismeasure of Man*, Norton, New York, 1996, and G.J. Annas and M.A. Grodin, *The Nazi Doctors and the Nuremberg Code: Human Rights in Human Experimentation*, Oxford University Press, New York, 1992.

55. For a thorough discussion of this blessing and concept in Jewish tradition, see C. Astor, "*... Who makes people different*": *Jewish Perspectives on the Disabled*, United Synagogue of America, New York, 1985.

56. Mishnah (edited circa 200 C.E.); *Semahot* 1:1–2; Mishnah (edited circa 200 C.E.); *Shabbat* 23:5 and Babylonian Talmud (edited circa 500 C.E.) *Shabbat* 151b; Babylonian Talmud (edited circa 500 C.E.) *Sanhedrin* 78a; Maimonides' *Mishneh Torah* (1177 C.E.) *Laws of Murder* 2:7; Joseph Karo's *Shulhan Arukh* (1565 C.E.) with the glosses of Moses Isserles. *Yoreh De'ah* 339:2 and the comments of the Shakh and Rama there.

57. This includes even inanimate property that "belongs" to us, for God is the ultimate owner. Cf. Deuteronomy 20:19; Babylonian Talmud (edited circa 500 C.E.) *Bava Kamma* 8:6, 7; Babylonian Talmud (edited circa 500 C.E.) *Bava Kamma* 92a, 93a; Joseph Karo's *Shulhan Arukh* (1565 C.E.) with the glosses of Moses Isserles. *Hoshen Mishpat* 420:1, 31.

58. Cf. RaN, Babylonian Talmud (edited circa 500 C.E.) *Nedarim* 40a. The Talmud records such prayers: Babylonian Talmud (edited circa 500 C.E.) *Ketubbot* 104a, Babylonian Talmud (edited circa 500 C.E.) *Bava Mezia* 84a, and Babylonian Talmud (edited circa 500 C.E.) *Ta'anit* 23a. Note that this is not a form of passive euthanasia: in that, people refrain from acting, but here God is asked to act.

59. J.D. Bleich, *Judaism and Healing*, Ktav, New York, 1981, pp. 141–142.

60. E.g., I. Jakobovits, *Jewish Medical Ethics*, Bloch, New York, 1975, p. 124 and No. 46.
A.I. Reisner, *Conservative Judaism* **43**(3), 52–89, esp. pp. 56–58 (1991).

61. E.N. Dorff, *Conservative Judaism* **43**(3), 3–51, esp. pp. 19–26 (1991); *United Synagogue Rev.* **44**(1, Fall), 21–22 (1991).

62. E.g., Rabbi Avram Israel Reisner, holds this view; see *Conservative Judaism* **43**(3), 62–64 (1991).

63. E.N. Dorff, *Conservative Judaism* **43**(3), 34–39 (1991).

64. Thus the Talmud specifically says, "We do not worry about mere hours of life" (Babylonian Talmud (edited circa 500 C.E.) *Avodah Zarah* 27b). The Talmud also says, however, that we may desecrate the Sabbath even if the chances are that it will only save mere hours of life (Babylonian Talmud (edited circa 500 C.E.) *Yoma* 85a). The latter source has led some Orthodox rabbis to insist in medical situations that every moment of life is holy and that therefore every medical therapy must be used to save even moments of life; e.g., see J.D. Bleich, *Judaism and Healing*, Ktav, New York, 1981, pp. 118–119, 134–145. The only exception is when a person is a *goses*, which Rabbi Bleich defines as within 72 hours of death, at which time passive, but not active, euthanasia may be practiced. He then uses the source in *Avodah Zarah* only to permit hazardous therapies that may hasten death if they do not succeed in lengthening life. Rabbi Bleich's position is *not*, however, necessitated by the sources. On the contrary, they specifically allow us (or, on some readings, command us) not to inhibit the process of dying when we can no longer cure, even long before 72 hours before death (however that is predicted).

65. Rabbi Reisner does not accept this "double-effect" argument, but he would agree that pain should be alleviated as much as possible up to, but not including, the dosage that would have the inevitable effect of hastening the person's death, even if not intended for that purpose. See A.I. Reisner, *Conservative Judaism* **43**(3), 66, 83–85, Notes 50–52 (1991); and see, in contrast, E.N. Dorff, *Conservative Judaism* **43**(3), 17–19, 43–45, Notes 24–27 (1991). See also Rabbi Reisner's summary of the differences between the Dorff and Reisner positions, *Conservative Judaism* **43**(3), 91 (1991).

66. Tosafot, Babylonian Talmud (edited circa 500 C.E.) *Avodah Zarah* 27b, s.v., *lehayyei sha'ah lo hyyshenan.* See E.N. Dorff, *Conservative Judaism* **43**(3), 15–17, 43, No. 22 (1991). For a contrasting interpretation of this source, see A.I. Reisner, *Conservative Judaism* **43**(3), 56–57, 72, No. 21.

67. Babylonian Talmud (edited circa 500 C.E.) *Yoma* 85a–b (with Rashi there); Babylonian Talmud (edited circa 500 C.E.) *Sanhedrin* 74a–b; *Mekhilta on Exodus* 31:13; and for a general discussion of this topic, see I. Jakobovits, *Jewish Medical Ethics*, Bloch, New York, 1975, pp. 45–98.

68. Babylonian Talmud (edited circa 500 C.E.) *Sanhedrin* 73a.

69. Babylonian Talmud (edited circa 500 C.E.) *Bava Metzia* 62a.

70. I. Jakobovits, *Jewish Medical Ethics*, Bloch, New York, 1975, p. 291; cf. also pp. 96–98. That this is the generally held opinion regarding living donors is true not only for Orthodox rabbis, some of whom he cites but also for Conservative and Reform rabbis. For Orthodox opinions, see Moshe Feinstein, *Igrot Moshe*, Yoreh De'ah 229 and 230 [Hebrew]; E. Waldenberg, *Tzitz Eliezer*, vol. 9, No. 45; vol. 10, No. 25 [Hebrew]; Obadiah Yosef, *Dinei Yisrael*, vol. 7 [Hebrew]. For a Conservative position (the only one I know of to date on living donors), see E.N. Dorff, *Choose Life: A Jewish Perspective on Medical Ethics*, University of Judaism, Los Angeles, CA, 1985, p. 23. For Reform positions, see S.B. Freehof, *New Reform Responsa*, Hebrew Union College, Cincinnati, OH, 1980, p. 62ff.

S.B. Freehof, *Current Reform Responsa*, Hebrew Union College Press, Cincinnati, OH, 1969, pp. 118–125.

W. Jacob, *Contemporary American Reform Responsa*, Central Conference of American Rabbis, New York, 1987, pp. 128–133.

71. Babylonian Talmud (edited circa 500 C.E.) *Yoma* 85a; Pirkei de-Rabbi Eliezer, Ch. 52; *Yalkut Shim'oni*, "Lekh Lekha," No. 72.

72. Yoel Jakobovits, [Brain death and] heart transplants: The [Israeli] chief rabbinate's directives, *Tradition* 24:4 (Summer, 1989), pp. 1–14. For Conservative positions, see S. Siegel, Updating the criteria of death, *Conservative Judaism* **30**(2), (Winter), 23–30 (1976) D. Goldfarb, The definition of death, *Ibid.*, pp. 10–22; the Rabbinical Assembly resolution urging organ donation *Proc. Rabbinical Assembly* **52**, 279 (1990), and the 1996 responsum by Joseph Prousser making organ donation a positive obligation (unpublished). The Reform Movement officially adopted the Harvard criteria (presumably, as modified by the medical community) in 1980.

See W. Jacob, ed., *American Reform Responsa*, Central Conference of American Rabbis, New York, 1983, pp. 273–274.

73. See "Animals, cruelty to," *Encyclopedia Judaica* 3:5–7.

74. E.g., see R. Gordis, *Judaic Ethics for a Lawless World*, Jewish Theological Seminary of America, New York, 1986, pp. 113–122; and E. Schwartz and B.D. Cytron, *Who Renews Creation*, United Synagogue of Conservative Judaism, New York, 1993.

75. The choice of animals: Leviticus 11; Deuteronomy 14. The dietary laws additionally require a specific mode of slaughter to minimize pain to the animal (based on Deuteronomy 12:21), that the blood be drained from the meat (Genesis 9:4; Deuteronomy 12:23–25), and that meat and dairy meals be separated. The prohibition of mixing seeds: Leviticus 19:19; Deuteronomy 22:9.

76. See K. Abelson and M. Rabinowitz, Definition of a *Davar Hadash. In Proceedings of the Committee on Jewish Law and Standards of the Conservative Movement, 1980–1985*, Rabbinical Assembly, New York, 1988, pp. 187–190.

ADDITIONAL READINGS

J.D. Bleich, *Contemporary Halakhic Problems*, Ktav, New York, 1977.

J.D. Bleich, *Judaism and Healing*, Ktav, New York, 1981.

E.N. Dorff, *Choose Life: A Jewish Perspective on Medical Ethics*, University of Judaism, Los Angeles, CA, 1985.

E.N. Dorff, A Jewish approach to end-stage medical care, *Conservative Judaism* **43**(3), 3–51 (1991).

E.N. Dorff, A time to live and a time to die, *United Synagogue Review* **44**(1), 21–22 (1991).

E.N. Dorff, Artificial insemination, egg donation, and adoption, *Conservative Judaism* **49**(1), 3–60 (1996).

E.N. Dorff, *This Is My Beloved, This Is My Friend: A Rabbinic Letter on Intimate Relations*, Rabbinical Assembly, New York, 1996.

E.N. Dorff, *Matters of Life and Death: A Jewish Approach to Modern Medical Ethics*, Jewish Publication Society, Philadelphia, PA, 1998.

E.N. Dorff and A. Rosett, *A Living Tree: The Roots and Growth of Jewish Law*, State University of New York Press, Albany, 1988.

D.M. Feldman, *Birth Control in Jewish Law*, New York University Press, New York, 1968; reprinted as *Marital Relations, Abortion, and Birth Control in Jewish Law*, Schocken, New York, 1973.

D.M. Feldman and F. Rosner, eds., *Compendium on Medical Ethics*, New York, 1984.

S. Freehof, *Reform Responsa*, Hebrew Union College Press, Cincinnati, OH, 1960.

S. Freehof, *Current Reform Responsa*, Hebrew Union College Press, Cincinnati, OH, 1969.

S. Freehof, *Reform Responsa for Our Time*, Hebrew Union College Press, Cincinnati, OH, 1977.

S. Freehof, *New Reform Responsa*, Hebrew Union College, Cincinnati, OH, 1980.

M. Gold, *And Hannah Wept: Infertility, Adoption, and the Jewish Couple*, Jewish Publication Society, Philadelphia, PA, 1988.

D. Goldfarb, The definition of death *Conservative Judaism* **30**(2), 10–22 (1976).

A.J. Goldman, *Judaism Confronts Contemporary Issues*, Ktav, New York, 1978.

R. Gordis, *Judaic Ethics for a Lawless World*, Jewish Theological Seminary of America, New York, 1986.

W. Jacob, ed., *American Reform Responsa*, Central Conference of American Rabbis, New York, 1983.

W. Jacob, ed., *Contemporary American Reform Responsa*, Central Conference of American Rabbis, New York, 1987.

I.Jakobovits, *Jewish Medical Ethics*, Bloch, New York, 1959, 1975.

Y. Jakobovits, [Brain Death and] Heart Transplants: The [Israeli] Chief Rabbinate's Directives, *Tradition* **24**(4), 1–14 (1989).

A. Mackler, *In Vitro Fertilization*, Draft No.3, November, 1995, adopted by the Conservative Movement's Committee on Jewish Law and Standards in December, 1995 (unpublished).

Proceedings of the Committee on Jewish Law and Standards of the Conservative Movement, 1980–1985, Rabbinical Assembly, New York, 1988.

J. Prousser, *The Obligation to Preserve Life and the Question of Post-Mortem Tissue Donation*, Rabbinic ruling adopted by the Conservative Movement's Committee on Jewish Law and Standards, 1996, (unpublished).

A.I. Reisner, A Halakhic ethic of care for the terminally ill *Conservative Judaism* **43**(3), 52–89 (1991).

A.I. Reisner, *Peri- and Neo-Natology*, Rabbinic ruling approved by the Conservative Movement's Committee on Jewish Law and Standards in Fall, 1995 (unpublished).

F. Rosner and J.D. Bleich, eds., *Jewish Bioethics*, Sanhedrin Press, New York, 1979.

E. Schwartz and B.D. Cytron, *Who Renews Creation*, United Synagogue of Conservative Judaism, New York, 1993.

S. Siegel, Updating the criteria of death, *Conservative Judaism* **30**(2), 23–30 (1976).

M.H. Spero, *Judaism and Psychology: Halakhic Perspectives*, Ktav, New York, 1980.

R. Spirtas, S.C. Kaufman, and N.J. Alexander, *Fertil. Steril.* **59**(2), 292–293 (1993).

E. Spitz, Through her I too shall bear a child: Birth surrogates in Jewish law, *J. Religious Ethics* **24**(1), 65–97 (1996).

U.S. Congress, Office of Technology Assessment, *Infertility: Medical and Social Choices*, OTA-BA-358, U.S. Government Printing Office, Washington, DC, 1988.

See other Religious Views on Biotechnology entries.

RELIGIOUS VIEWS ON BIOTECHNOLOGY, PROTESTANT

Courtney S. Campbell
Oregon State University
Corvallis, Oregon

OUTLINE

INTRODUCTION

Protestant thought has played a critical and ironic role in the development of the ethical justifications and practices of biotechnology. The influential theologians of the Protestant Reformation, such as Martin Luther (1483–1546) and John Calvin (1509–1564), made a sharp distinction between God and nature (including human nature). All value is derivative of the divine being; nature and human beings, having experienced the dramatic and enduring consequences of the human fall from divine grace, have been divested of ultimate value and stand in need of redemption and reconciliation. One result of these profound theological claims was, to borrow from Max Weber's memorable phrase, the "disenchantment of the world" (1), that is, nature loses its reverential hold on human attitudes and actions, and comes to be seen as a realm open to the many manipulations of science and technology. Ironically, these manipulations over time would culminate in skepticism over the existence of the God that Protestant theology wanted to bring to the forefront of human consciousness. For some contemporary Protestants, this unanticipated result of a basic theological claim leads to religious and moral criticism of biotechnology.

It is nonetheless the case that Protestant perspectives on biotechnology are multiple and complex, and connected with deeply rooted theological claims about God, human beings, and nature. This article will provide an overview of these perspectives, beginning with attention to some of the formative views that characterize a position as "Protestant."

PROTESTANT DISTINCTIVENESS

Protestantism is first of all a "religion," rather than an "ethic," in which the primary questions concern the nature of God, God's relationship to humanity, and human nature and salvation. Nonetheless, themes that are embedded in these fundamentally theological issues have implications for ethical actions in the world and for perspectives on biotechnology.

Authority of Scripture

Protestant reformers and theologians are united in affirming the primary authority of the Bible as revealing the word of God to human beings about God, our relationship to God, and our prospects for redemption and reconciliation. God is revealed as Creator, Sustainer, and Redeemer. Human beings are mirrors or "images" of God in the world but have, through participation in sin and evil, fallen short of divine glory. Human reconciliation with God is made possible through the suffering of Jesus Christ for the sins of the world.

The authority of Scripture in the realm of salvation carries over, with some qualifications, in the realm of ethics. In Protestant thought, salvation is a matter of grace and mercy, which means the ethical life expresses salvation rather than being an instrument to salvation, as it is formulated in those religious traditions that emphasize salvation primarily by works or actions. Therefore the question of why human beings should be moral is for Protestants placed within a context of gratitude and gracious response to the divine gift of salvation, and a recognition of humility and human dependency upon powers beyond our control. One issue for Protestants in examining the ethical questions of biotechnology is whether scientific research and its application retains this motivational sense of gratitude and humility or effaces it by emphasizing human accomplishment.

There is debate within the various Protestant traditions as to the sufficiency of Scripture for moral conduct. While the Bible has moral authority within Protestantism, it may not be *the only* authority. Theologian James M. Gustafson has argued that four resources are required for a comprehensive and adequate Christian (including Protestant) ethics:

1. Scripture and its interpretation through the historical tradition of Christianity
2. Philosophical insights, methods, and principles
3. Scientific methods and information
4. Interpretations of human experience (2)

In this understanding, Scripture illuminates the context of ethical issues, but it is not possible to move directly from a biblical passage to a moral conclusion about a current controversy in an area such as biotechnology. While such a position has been very influential among prominent Protestant theologians, it is not necessarily shared by evangelical and fundamentalist Protestants and their communities (3).

One implication of the moral authority of Scripture for Protestants is a corresponding non-normativeness of nature or the natural order. Indeed, some Protestant thought affirms the ongoing "ordering" of creation, rather than the created "order," precisely to emphasize the fluidity and dynamic interactions of nature, including those changes introduced by human beings. Nature is created good but is now currently disordered and no less in need of redemption. According to Protestant theology, Scripture reveals Jesus as healer of the disorder in nature, and provides a pattern for human beings to emulate. This model of scriptural interpretation is invoked by one influential scholar, Ronald Cole-Turner, to support the genetic alteration of plants in order to enhance their disease resistance. By enhancing the usefulness of plants, and diminishing environmental damage, human beings participate in the divine workings of redemption (4).

Thus, even though Protestant perspectives on biotechnology are profoundly influenced by interpretations of nature and the non-normativeness of the natural world, this does not imply that nature is without normative significance. Under the governance of divine providence, nature is susceptible to human intervention; indeed, human labor and ingenuity can use nature in the service of preserving and enhancing human (and animal) life. In some views, moreover, human beings are called to restore fallen nature to an original, properly ordered condition, given constraints of human finitude and fallibility. Given that Scripture does not speak directly to concrete issues in biotechnology, and that nature does not possess normative status but does possess normative significance for human action, it is possible to establish a Protestant theological presumption in favor of biotechnology. Biotechnology should be directed and constrained by norms of love, freedom, and stewardship.

Christian Freedom

As illustrated by the preceding discussion, dissent or "protest" is itself a characteristic Protestant theological perspective. This internal disagreement is manifested in many areas in Protestant thought, including the nature of ecclesiastical authority and its relation to biblical authority and the authority of personal conscience, as well as the role of sacraments mediated by the Church and their relation to saving grace. Given these disputes on matters of profound ecclesiastical importance, it comes as little surprise that Protestant history has been marked by continual reformations and the founding of new and diverse churches. In addition to these ecclesiastical implications, important ethical issues are embedded within the principle of "Christian freedom."

Freedom and Choice. Protestantism expresses a pronounced commitment to the primary of personal freedom and choice. Christian freedom should not be conflated with the secular norm of autonomy, for freedom is directed and constrained by love and ultimate accountability before God. However, this emphasis does give Protestants significant personal discretion in moral action without necessary reliance on a structure of specific moral rules.

Moral Pluralism. These theological and ethical commitments inevitably give rise to moral pluralism within Protestantism, as different interpretations are offered of the requirements of love and freedom. On virtually any moral question of consequence, a range of Protestant perspectives can be identified, without any ecclesiastical teaching authority available to provide a definitive conclusion. This is no less true of approaches to biotechnology, where assessments span the spectrum from hostility to biotechnology as an arrogant intrusion upon God's created order to a celebration of biotechnology as a beneficial means of partnership with God in continuing creation. These models will be discussed more fully below.

The Image of God. The primary Protestant claim about human nature is derived from the biblical account of the earth's creation; in this narrative, human beings are created in "the image of God," a theological description with profound normative implications. By this designation, human beings are given a status that distinguishes them from both God and nature. First, human beings are not God; the creator is sovereign over the created being. In particular, human beings lack the abilities to predict the results of action, control actions once they are initiated, or to adequately evaluate outcomes. Human distinctness from God is manifest in limitations such as finitude and fallibility. Persons do not have the capacities of omniscience and omnipotence attributed to the divine. When human aspirations exceed human capabilities, the created being runs the risk of the sin of pride or hubris, and of "playing God." This theological anthropology means for Protestant thought that the general theological presumption supporting biotechnological interventions characteristically will be constrained by concern with unforeseen consequences, slippery slopes, and admonitions of caution.

Notwithstanding these limitations, human beings "image" God in the world, a status that distinguishes them from animals, plants, and other manifestations of creation. Human beings are given a mandate of dominion to (1) care for and (2) cultivate creation, a mandate that within Protestant thought has been subject to numerous interpretations, including models discussed below such as passivity, anthropocentric domination, stewardship, and co-creatorship. These various interpretations, however, all affirm a common theme that responsible dominion involves a covenant with God that the human person will exercise freedom and love toward others—these "others" include human beings, the earth, and its creatures—in a way that benefits the common good. Responsible dominion also entails a recognition that ultimate accountability must be rendered to God for one's actions.

Much of the ambivalence and caution in Protestant thought concerning biotechnology is generated from the intrinsic tensions of the dual mandates of dominion, those of care and of cultivation. The principle of "caring" implies practices of conservation and preservation of our current conditions; the principle of "cultivation" implies reliance on human creative potential in using natural resources to bring about improvements and progress. Theologies of human dominion will tend to give greater emphasis to one or the other of these basic principles, or hold them in some kind of balance or tension, which in turn shapes perspectives on biotechnology.

Since nature does not have normative status within Protestant thought, the use of living organisms in biotechnological methods for making or modifying products, such as pharmaceuticals, enhancing strains of plants and animals, or manipulating genes to provide therapies for humans does not seem intrinsically wrong on the grounds that, for example, such practices "violate nature" or constitute "playing God." This is significant since a study of public attitudes to biotechnology indicates opposition to biotechnology is frequently grounded in claims that it is "not natural," or is "against God's will" (5). Although such claims are frequently presented as religious arguments, it is not clear they are well-ground in Protestant theological discussion: While human beings are not God in Protestant views, humans are responsible to and accountable before God in bringing about the full fruition of creation and in redeeming it from its current disordered condition. This supports the normative theological presumption in favor of biotechnology, unless it can be shown that a particular application of the technology will violate norms of love, freedom, or responsibility. The following section develops a general typology of Protestant perspectives that seek to negate, limit, or enact this presumption.

TYPOLOGY OF PROTESTANT THEOLOGICAL PERSPECTIVES

Nonintervention

One strand of Protestant thought emphasizes a noninterventionist or passive posture regarding the use of biotechnology. This perspective gives primacy to the principle of "caring," while attributing diminished importance to the principle of "cultivating." The noninterventionist understanding does not claim that nature is intrinsically good, and should therefore be left alone. Like all creation, nature has fallen from its pristine paradisiacal state. Rather, the problem is that human beings, because of their finitude and fallibility, are through technological interventions likely to make matters worse rather than better. Thus one Protestant writer urges caution and warns of the potential dangers from genetic manipulations because of the limited understanding and knowledge of human beings regarding the genome (6).

Within this understanding, moral priority is given to a norm of nonharm (as derived from love) rather than seeking to provide benefits through technological progress. Nonetheless, it is open to criticism on several grounds. It must first address what is particularly distinctive about biotechnology relative to other forms of human interventions. In many cases the position of nonintervention does not portray anything intrinsically wrong or misguided about biotechnology but expresses instead concern, rooted in the anthropology of finitude, about human prospects for control of biotechnological applications. This minimized capacity for control increases the risks of harm from biotechnology, both in terms of probability of occurrence and severity of the harm, to substantial and even unacceptable levels. Since it is impossible to empirically establish a risk assessment without proceeding ahead with biotechnology, and since prevention of harms have moral primacy over promotion of possible benefits, the course of ethical wisdom on this view lies in forgoing biotechnology.

The noninterventionist or passive perspective may also be challenged on the grounds that by forgoing benefits, harm is inevitably caused. It is possible to engage in a thought experiment and readily determine what benefits humanity currently enjoys from medical and biotechnological interventions that would have been forgone had the noninterventionist perspective prevailed throughout history. Thus this strand of Protestant thinking appears to convey a "free-rider" approach, willing to receive and make use of the benefits bestowed from the scientific legacy of prior generations but unwilling to develop these capabilities still further to benefit future generations.

Finally, nonintervention seems theologically suspect because of its neglect of the mandate of "cultivation." Responsible dominion involves a judicious balancing of cultivation and care, for cultivation provides justification for human intervention, through biotechnology or other means, in the natural world, while care sets limits on the scope and extent of that intervention. That is, both are necessary principles of dominion, and neither is by itself sufficient. It is part of Christian freedom and responsibility to work out the practical implications of these principles when, as in some instances of biotechnology, they come into conflict.

Anthropocentric Domination

If nonintervention presents one pole of a continuum in which caring assumes primacy over cultivation, the other end of the role is represented by perspective of anthropocentric domination, which emphasizes the mandate of cultivation to the neglect of caring. Moreover the distinctive status of "image of God" of humans means that human beings should be the primary beneficiaries of cultivation of nature. In this understanding, ontology implies moral superiority: Humans are not only the culmination of creation but also its measure and purpose. Thus human beings are held to receive divine permission to use the resources at their disposal, including the natural resources of the earth and their own intellectual and creative potential to improve human welfare.

The position of anthropocentric domination shares with that of nonintervention the view of a fallen world and nature, but it differs in two important respects. A fallen world invites improvement, which in some views may support efforts of biotechnology and medicine to restore conditions similar to those of an Edenic paradise; in other interpretations, while paradise may be ineradicably lost, there nonetheless is a mandate for cultivating more humane and beneficial conditions for living. Nor is human finitude and fallibility as paralyzing on the domination account as it is within noninterventionist accounts. Indeed, the record of human history, particularly within medicine over the past century, shows dramatic improvements in health and welfare through sustained investigation, understanding, and manipulation of nature. Thus there is confidence that human interventions will culminate in greater benefits than harms, and that the risks of subsequent interventions, such as through biotechnological methods, can be controlled and minimized.

This position has been very influential in Protestant thought, particularly since the dawn of the scientific and industrial revolutions, and it is no surprise that it has also been the recipient of the most sustained philosophical criticism. In a very significant essay, Lynn White, Jr. laid the blame for the current "ecological crisis" precisely at the door of anthropocentric domination. This version of Christianity, in White's view, "not only established a dualism of man and nature but also insisted that it is God's will that man exploit nature for his proper ends" (7). And, such attitudes can easily be reflected in and perpetuated by contemporary biotechnology.

While White's thesis has been very controversial, and has been challenged on several grounds, it is important to differentiate descriptive and normative implications. Descriptively it is the case that (1) anthropocentric domination has been present in some Protestant understandings of nature and technology and (2) it can be rendered as compatible with at least part of the mandate of human dominion, that of cultivation. Normatively, however, cultivation is not unlimited; it should be directed and constrained by the principle of care, with its emphasis on protection, preservation, and conservation. Moreover cultivation is also limited by responsibility and accountability before God. This is to say that an anthropocentric account of dominion—which is the primary focus of White's critique—may not possess the fidelity to Protestant Scripture and teaching as other interpretations (8).

Stewardship

A third perspective that seeks a balanced response to the mandates of care and cultivation has historically been designated as *stewardship* or *trusteeship*. An ethic of stewardship can itself be articulated in terms of the Protestant themes of authority, agency, and accountability. Human beings have been given divine authority over nature, as well as the moral freedom to make choices regarding the use of natural resources. However, the content of such choices should reflect a concern for the common good (which is not limited to what is good for human beings) and persons are held to assume accountability before God for their choices. In short, the "dominion" of human beings is much more inclusive of other creatures than implied by the anthropocentric interpretation, and as stewards of the earth, human beings are in the service of God to render service to others, with "others" defined holistically rather than anthropocentrically.

Neither care nor cultivation receives moral primacy in this ethic. Rather, the stewardship ethic tries to maintain a responsible balance that both justifies human interventions on the grounds of improving the world and human welfare and limiting those interventions when they overreach these goals. Thus there is recognition that both benefits and harms can occur through human technologies, and making decisions about their use under the human conditions of finitude and fallibility is complex and permeated by genuine ethical uncertainties and dilemmas. Thus, while biotechnology can be justified, good reasons must be offered in its support, and constraints must be acknowledged and adhered to.

There is a depth of kinship between humans and the earth and its creatures present in the stewardship ethic. Human beings are "earth creatures," created by God, to be sure, but of the dust of the earth. Indeed, "human" and "earth" share a common etymological root, "humus." This commonality brings awareness of a sense of interdependence, mutuality, and humility that precludes the attitude of anthropocentric conquest present in the domination perspective.

While certainly very influential in the history of Christian and Protestant thought, to the point that some interpreters have conflated anthropocentrism with stewardship, the stewardship perspective is also not immune from moral critique. Since it tries to hold two principles in some kind of equitable balance, it often is found limited with respect to practical issues and controversies, where some choice about whether to give priority to care or to cultivate is not a theological abstraction but a practical necessity. In addition it has been argued that stewardship is an abstract ideal that is not embedded in cultural practices, which historically have reflected a domination perspective. Thus, on both counts, the practical relevance of this perspective may be much less compelling than its theoretical appeal.

Partnership

While the stewardship perspective affirms that human beings are authorized agents of God in the world, the purposes for which the earth should be cared for and cultivated are by and large given by divine design. The partnership perspective emphasizes by contrast a much more interactive and engaged role of human beings in shaping these ultimate purposes. It presumes that creation did not end at some prehistorical period, but that creation is a dynamic, ongoing process that human beings participate in as created co-creators with God, with human and bioecological destiny not predetermined but to be shaped contextually. Thus "human work, especially our technology, may be seen as a partnership with God in the continuing work of creation. "[O]ur genetic engineering has the potential for being an extension of the work of God" (4).

This perspective has emerged fairly recently within Protestant thought; it has been given thoughtful exposition in the context of genetics and ecology by such prominent writers as Ronald Cole-Turner, Philip Hefner (9), and Ted Peters (10), and will certainly be increasingly influential in attempts to facilitate dialogue between religious, scientific, and biotechnological interests. Partnership, or created co-creator, models rely more heavily on scientific understandings of cosmology that do their Protestant predecessors, which are largely formulated within biblical cosmologies. The analogical and substantive features of partnership, and by inference its differences from stewardship, are eloquently articulated by British theologian Arthur Peacocke:

> It is as if man has the possibility of acting as a participant in creation, as it were the leader of the orchestra in the performance which is God's continuing composition. ... [M]an now has, at his present stage of intellectual, cultural, and social evolution, the opportunity of consciously becoming *co-creator* and *co-worker* with God in his work on Earth, and perhaps even a little beyond Earth (11).

Indeed, Peacocke places human beings within a cosmos whose designs and purposes are known neither by human beings or by God; thus persons are even *"co-explorers"* with God. Such an understanding is certainly very compatible with, and gives theological justification for, biotechnology as, to continue the orchestral metaphor, the lead violin in the creative and explorative composition; technology is a metaphor and symbol for divine creative activity.

Certainly the partnership perspective is not without its theological detractors and these objections seem twofold. First, the position presumes that "the future of creation is uncertain, because God has not guaranteed its outcome" (4). Yet, precisely because the future is indefinite, and to be partly shaped and directed by human creativity, the perspective does not seem capable of generating clear limits or constraints. The necessity for some purpose or goal is an issue of acute importance in ethical evaluations of biotechnology, lest the technology create its own justificatory role. Second, the position seems to overstate human capacities to the point of pretension to playing God. While Peters has argued that the theological anthropology embedded in partnership is to "play human" fully and authentically, the long-standing concerns of Protestant thought about sin, human finitude, and fallibility are not as prominent in expositions of

partnership. And, as illustrated above, it is precisely those concerns that stand behind many of the limits or cautions some strands in Protestant ethics wish to impose on biotechnology.

LIMIT CIRCUMSTANCES

Certainly many of the innovations of biotechnology, particularly with respect to products that bring about improvements in human health (eg, genetically engineered human insulin) have been welcomed by Protestant thinkers and communities. In this respect, the efforts of biotechnology to alleviate or cure disease are commonly set within a context of divine creativity and redemption, working through the imaginative instrumentality of human beings. Thus, with very few exceptions, biotechnology per se seems to not be theologically suspect; especially on the partnership perspective, biotechnology can be theologically praiseworthy and even morally required. Nonetheless, given the background theological commitments delineated above, Protestant thinkers do raise questions about the purposes and the procedural controls of biotechnology. Some scholars have raised the possibility that "control" itself is fundamental to the enterprise of biotechnology, by seeking to diminish human vulnerabilities to the capriciousness of the natural world. Thus the theme of human control is an issue that cuts across both substantive and procedural questions. These two questions can be more carefully examined by considering selected innovations in biotechnology that have raised concern among a range of Protestant thinkers and traditions.

Gene Therapy

One of the most significant innovations in biotechnology is the possibility to alter the genetic makeup of a person. Protestant denominations, under the auspices of the National Council of Churches in Christ, joined in the late 1970s with Roman Catholic and Jewish ecclesiastical bodies to raise questions about the risks of genetic manipulations, as well as a perceived arrogance of human control and mastery presupposed by such manipulations. In 1982 theologian Roger L. Shinn articulated in congressional hearings five base points for Protestant reflection on genetic interventions:

1. A bias for the sacredness of human life requires minimization of risk to the patient or subject.
2. A sense of "human inviolability" both permits interventions and limits their scope.
3. Efforts to eliminate genetic-based diseases are justifiable.
4. Genetic enhancements cross the boundary of inviolability and are dangerous.
5. Equity and justice should guide the distribution of benefits and burdens in genetics research (12).

In the intervening years, Protestant thought on human gene therapy has tended to reflect the considerations delineated by Shinn, and skepticism has gradually given way to a cautious endorsement of gene therapy in some circumstances. Some questions have been amenable to resolution through increasing knowledge of scientific and technical issues, as well as through procedural safeguards of public oversight and monitoring. And Protestant commentators have characteristically insisted upon a specific moral rationale for genetic manipulations, that is, that the designed intervention have therapeutic potential for alleviating disease.

These concerns have tended to direct Protestant thought into ethical positions that some times converge and some times diverge with other ethical traditions, religious or secular. As one point of convergence, the Protestant commitment to human equality and the dignity of individual persons has required respect for genetic diversity and correlative opposition to efforts at genetic enhancements and positive eugenics. However, an issue of controversy has occurred over the validity of the line drawn in many bioethics discussions between somatic cell therapy and germ-line therapy. Protestant thought has argued for a greater continuity between somatic cell and germ line therapy insofar as the rationale for either form of genetic intervention is disease-based. A policy document of the National Council of Churches developed in 1986, even before attempts at somatic cell therapy had been conducted, asserted that while germ-line therapy needed "stringent control," it could not be precluded because of prospects of substantial benefits in alleviating disease (13). This view has evolved into a "wait-and-see" position presented in a document of the United Methodist Church: "We oppose therapy that results in changes that can be passed to offspring (germ-line therapy) until its safety and the certainty of its effects can be demonstrated and until risks to human life can be demonstrated to be minimal" (14). Such conditional opposition, or what Shinn describes as "cautious openness" (15), suggests that there is nothing intrinsically theologically objectionable with germ-line therapy, but it does place the burden of proof on those who wish to proceed with such interventions, in the sense that safety and efficacy must be demonstrated, rather than those who wish to prohibit germ-line therapy because of possible risks.

However, there are Protestant dissenters to cautious openness to germ-line therapy, and they are instead characterized by Shinn as adopting a position of "emphatic rejection." Some conservative Protestants have argued that as germ-line therapy will likely involve manipulation of the cells of human embryos, this method constitutes unethical experimentation on the unborn, and ultimately erodes the sanctity of human life. Indeed, some Protestants are concerned about somatic cell therapy, not on its own merits but because the underlying rationale of cure of disease raises a new set of concerns about responsibility to the vulnerable who cannot consent when considered in the context of germ-line manipulations.

A distinguished group of Protestant scholars has, while generally supporting development of gene therapies, raised some additional questions that revolve around the risk of humans losing control of the technology. One concern focuses on the relativity and elasticity of the concept of "disease." It may be difficult to uphold a firm line distinguishing legitimate uses of genetic therapies on the basis of whether they are directed to healing diseases or not because the very concept of disease is so fluid. In addition successes with gene therapy may encourage society and medical researchers to develop innovative uses for genetic manipulation that depart from the disease-based rationale. "[P]roper therapy also directs control to the goods of life and health, to the goal of healing genetic disease. If people simply celebrate genetic control itself, ... we fear that they will lose the capacity to direct and limit this new power to therapeutic uses" (16). An attitude of celebrate without caution, or a practice of cultivation without caring, may open the door to designing human descendants after our own preferred image and characteristics. Thus the slippery slope of most concern to Protestants may not be that of somatic to germ-line therapy but from therapy to enhancement, for enhancement is considered a response to "cosmetic purposes or social advantage" (14). In each case of concern about genetic manipulations — germ-line therapy, the concept of disease, and genetic enhancements — some common ethical perspectives are reinforced in Protestant thought by convictions about human finitude, fallibility, and pretensions to arrogance and pride.

Transgenic Research

For some Protestants, manipulation of the genome of human beings is not the only theologically problematic dimension of biotechnology. While animals do not generally have the theological and moral status of human beings within Protestantism, as creatures, animals do fall under the domain of responsible human stewardship or partnership. Thus, on some accounts, the insertion of genes from one species into another species to produce a transgenic organism or animal raises questions and objections. One kind of argument stems in part from an imperative rooted in the biblical creation narratives to plant and animal species to "reproduce after their own kind" (6). Transgenic biotechnological research may thus be understood to use genetic information for purposes not intended in the origins of plant, animal, or human life. Even though scientific research possesses the power and capability to bring about such genetic alterations, use of that power violates a normative ideal of species integrity and perpetuation.

A second argument regarding transgenic research is that it may compromise human distinctiveness. The claim in this instance is that by eroding distinctions between species, including animal and human species, through the creation of transgenic organisms, it then seems arbitrary to draw a line that would allow for such research on animals but not on humans. Protestant opposition to transgenic animal research in the 1980s led one researcher to reply, "I don't know what they [Protestant opposition] mean when they talk about the integrity of species. ... Much of all genetic material is the same, from worms to humans." This appeal to the commonality of genetic information between species leads Protestant theologian Andrew Linzey to question whether, if researchers are really convinced of this point, what grounds they would then have for opposing transgenic and eugenic research with humans (17).

Theological arguments against transgenic research do not always invoke abstract possibilities; they commonly cite examples of actual research projects that have culminated in harms rather than benefits. The creation of transgenic pigs in the 1980s through insertion of the human growth hormone gene has been cited by Protestant writers as a case in point of the risks and moral mistakes implicit in transgenic research. The point of the research, which was sponsored by the U.S. Department of Research, was to develop pigs that had greater muscle mass and leaner fat content, thus making the animals more commercially desirable. However, the pigs turned out to be excessively hairy, arthritic, impotent, and lethargic.

Significantly, theological critiques of this project argued that transgenic research was proceeding without ethical sensitivity to the harms experienced by the animals, rather than, as would be expected within a framework of anthropocentric domination, to human beings. It is important to note, however, that theological evaluations that assess biotechnology relative to its benefits and harms primarily focus on scientific and technical considerations, rather than on intrinsic theological issues. Such arguments may not be so compelling when the harms are uncertain and unforeseen, as in more recent projects that have created a chicken-quail hybrid as a prelude to understanding brain disorders in human. And, as noted above, other Protestant scholars have supported transgenic plant research on the grounds of increased productivity and diminished harms.

Of particular concern for some evangelical Protestants is maintaining the integrity or purity of the human genome. This follows from the special status of human beings in contrast to animals. Thus, on one account, "genetic information from any other organism which does or did not exist in the human genome should not be placed within humans" (6). However, this position raises questions over which Protestants would differ. First, a prohibition of mixing genetic information from other species with humans seems to suggest that human distinctiveness is constituted by genetic differences, rather than, for example, all that is embedded in the concept of the "image of God." The preservation of human genetic integrity seems bought at a price of genetic essentialism and theological reductionism, or what one theologian criticizes as the "gene myth" (10). Second, such a position presumes that there is something distinct or unique about genetic information, even though there are many other ways by which human beings might absorb or consume animals or plants or their products. While it is clear that not all Protestants hold similar views on transgenic research and organisms, it is clear that this form of biotechnology raises some widely shared questions (18).

Issues of Ownership: Patenting and Distributive Justice

The tools of biotechnology have raised important questions of ownership, particularly with regard to human cells that have been retrieved, modified, and immortalized, or of genetically engineered animals and plants. One issue of debate has focused on the legitimacy of patenting genetically modified life forms, while another has concentrated on ownership of the economic gains from commercial development of biotechnological products.

Protestant scholars have engaged both of these debates within their own faith communities and in public discourse.

Scholars and clergy from both conservative and mainstream wings of Protestantism have been in the forefront of challenging efforts to patent genetic material, derived either from animals or (of greater concern) from human beings. A report on biotechnology issued by the World Council of Churches in 1989 opposed patenting of genetically altered animal life forms, asserting that "the patenting of life encodes into law a reductionist conception of life which seeks to remove any distinction between living and non-living things" (19). More recently, in 1995, a group of 186 religious leaders, primarily from Protestant churches, issued a "Joint Appeal against Human and Animal Patenting": "We ... oppose the patenting of human and animal life forms. We are disturbed by the U.S. Patent Office's recent decision to patent human body parts and several genetically engineered animals. We believe that humans and animals are creations of God, not humans, and as such should not be patented as human inventions" (20,21). The "Joint Appeal" provoked a storm of controversy, not only between religious and scientific communities but within the Protestant community itself, with some theologians arguing that it was misguided and reactionary. However, gene patenting may simply be the issue that crystallizes concerns of many religious communities about biotechnology and the new genetics, concerns that have a deeper and broader significance. Indeed, religious objections to patenting the results of biotechnology appear to stem from a diversity of rationales, including (1) symbolism about life, (2) scientific reductionism, (3) human artifice, and (4) anthropocentrism.

As a statement about biology, one would be mistaken to infer from approval of a patent for a gene that "life" is thereby under the realm of the patent office, as is suggested by the language of the WCC report and by comments of individual religious clergy who supported the 1995 "Joint Appeal." However, critics who dismiss this objection as misinformed about science or naive as to the patenting process and public policy may overlook the theological symbolism at stake. The claim is best understood as seeking to resonate at a symbolic rather than literal level of interpretation. In a very probing study, Dorothy Nelkin and Susan Lindee have illustrated how a gene sequence, or the double-helix structure, is invariably used in scientific, academic, and popular literature as a symbol for life as a whole; the gene has become invested with a spiritual or sacral significance historically attributed to the soul (22). Thus a gene is not simply an object for scientific study and manipulation, it has become embedded with a complex matrix of cultural, ethical, and religious meanings. In this respect the willingness to proceed with gene patenting signifies an effort to extricate science from these embedded social meanings, and thereby is interpreted by some opponents as diminishing the value of life.

The value of life is underscored by a second rationale for objections to genetic patenting, a concern about scientific reductionism, an ideology embedded in contemporary biomedicine that is viewed as blurring the boundaries between life and nonlife. If patenting is an instance of such a reductionist ideology, then some theological questions assume a greater legitimacy. In particular, scientific research and technological applications of parts of the body, such as tissues, cells, or genes, can conflict with religious values about bodily integrity that are central to the Protestant understanding of the "image of God" present in human beings. E. Richard Gold draws on these themes to contrast scientific instrumentalism with theological intrinsicalism: "The body, from a scientific viewpoint, is a source of knowledge of physical development, aging, and disease. From a religious perspective, the body is understood as a sacred object, being created in the image of God. ... The scientist values the body instrumentally, as a means to acquire knowledge; the believer values the body intrinsically, for being an image of God" (23).

Put another way, Protestant (and Western) religious thought begins with a strong presumption that the status of the body as a whole is greater than the sum of its parts. The interest of medical science in the body stems from the prospect of gaining information about human character traits and behaviors, including susceptibilities to illness and bodily responses to disease, through study and understanding of the basic components of life, such as genes. The scientific value of the body as a totality is instrumental to the goal of deciphering the codes, messages, and functions of the fundamental components of the body that contain valuable genetic information. In this respect, gene patenting may reflect scientific reductionism in that (1) genes are viewed as scientifically more significant than the organic body totality and (2) the value of a cell or gene resides primarily in the information it provides researchers rather than as a symbol of life's dynamism and processes.

The third Protestant objection raised to gene patenting seeks to maintain a distinction between the realm of divine creation and the realm of human invention. The argument suggests that "creation" is the work of God, and within the Protestant Christian context, God creates *ex nihilo*, or "out of nothing." This is then differentiated from "invention" as a human affair, in which human beings re-organize, or more to the point with genetics, re-combine, already existing material elements to produce a new life form, such as the Harvard OncoMouse.

Given this distinction, however, it is difficult to see just how a patent application is any more an encroachment on divine creativity than the original research that inserted the human gene into the mouse. The latter might be objectionable with regard to the issues delineated concerning transgenic organisms, but it would then seem that the theological line needs to be drawn on that issue, not over the legal rights and restrictions granted by a patent. As with the objection that equates patenting DNA with patenting life, the creativity–inventiveness objection also seems to function as a symbol for objections that are important but not fully articulated.

Perhaps the strongest argument presented against biotechnology and patenting has been put forward by theologian Andrew Linzey who, in a broad-ranging theological critique of human use of animals more generally, sees in patenting the culmination of a departure from Christian stewardship and an embrace of anthropocentric domination. Linzey contends that "biotechnology in animal farming represents the apotheosis of human domination" (17). This development is the technological, if not logical, end point of the anthropocentric perspective that animals (despite also being created by God) belong to and exist for the benefit of human beings, and have no intrinsic value. While Linzey is aware that the human species has always made use of animal species, he nonetheless claims something distinctive is present in the application of biotechnology and genetic engineering to animals, namely it employs "the technological means of absolutely subjugating the nature of animals so that they become totally and completely human property" (17). In this interpretation, genetic engineering and patenting of animals is the moral equivalent of human slavery.

The Christian theology of responsible stewardship entails for Linzey maintenance and promotion of the good that already exists. The "artificial creation" of animals with disease-bearing characteristics, such as the OncoMouse, simply violates the integrity and design of creation. Moreover acceptance of patenting of genetically engineered animals, by which legal recognition is given to human property claims over animals, symbolizes biotechnological enslavement. The granting of patents over animals will "reduce their status to no more than human inventions, and signifies the effective abdication of that special God-given responsibility that all humans have towards the well-being and autonomy of sentient species" (17). Thus, on Linzey's account, animal patenting is a form of "idolatry" because it supplants God with human beings as owners of creation and "represents an attempt to perpetuate, to institutionalize, and to commercialize, suffering to animals" (17).

In explicating these Protestant reservations about genetic and animal patenting, it is important to acknowledge that other Protestant scholars have not found these reservations at all compelling. Ronald Cole-Turner, one of the most influential Protestant scholars at the intersection of theology and genetics, has maintained that because "there is no metaphysical difference between DNA and other complex chemicals, ... there is no distinctly religious ground for objecting to patenting DNA" (21). Cole-Turner does not contend that there cannot be legitimate objections to patenting, but only that there are no specifically religious grounds for those objections. This claim enables Cole-Turner to encourage the initiation of dialogue between religious communities and scientific researchers, for "religion gives science its purpose, and science gives religion its eyes and its hands." It also provides Cole-Turner a basis for interpreting genetic engineering not in terms of anthropocentric domination, but rather as participation in divine creative activity.

One feature of the Protestant debate over gene patenting raises a more general concern in Protestant discourse about biotechnology, the issue of "commodification." This question may take two different but related forms. First, objections may be raised against biotechnology on the grounds that it transforms what is found in the world, such as genetic material, into commodities for commercial development in accord with market values. This implies an understanding and valuation of genetic material that may be considered theologically objectionable (24). Thus, in its generally cautious but favorable appraisal of biotechnology, the World Council of Church admonishes, "the integrity of creation is damaged if biotechnology is utilized by commercial pressures to manufacture new life forms that are valued only as economic commodities" (19). Part of the concern embedded in this claim reiterates the Protestant concern with authoritarian control (i.e., commercial pressures) over biotechnology.

A second kind of argument may accept that certain things are legitimately classified as commodities, but protests against biotechnology on the ground that certain peoples and nations are better-positioned to participation in the biomarket and gain access to the benefits of biotechnology, while other peoples and nations will be excluded for economic reasons. In particular, the fruits of the biotechnological revolution are likely to be harvested by first-world peoples, while third-world countries may find themselves on the margins of the technologies. This runs contrary to the norm of distributive justice and the example of the ministry of Jesus, wherein Christians are encouraged to pay special attention to lifting the burdens and meeting the needs of the poor, the vulnerable, and the outcast. Some Protestant scholars believe the community of the vulnerable, and the moral primacy of responding to their needs, must be broadened from the "near" neighbor to the "stranger." Even though Protestant discussion about patenting of genetically engineered life forms is still in its embryonic stages, it seems clear that it will engage Protestants in dialogue and criticism with each other, and with the scientific and biotechnological communities.

Human Cloning

Recent scientific reports on successful mammalian cloning through the process of somatic cell nuclear transfer, which have in turn raised the prospects of human cloning in the near future, have revealed the pluralism of Protestant ethics perhaps more than any other question in biotechnology. Protestant theologians were invited to testify before the National Bioethics Advisory Commission established by President Clinton to recommend public policy on human cloning, and Protestant scholars have begun to contribute to the emerging ethics literature on this question (25). Yet this biotechnological development of the late 1990s is not a new question within Protestantism. Protestant theologians Joseph Fletcher and Paul Ramsey participated in influential academic and scientific forums in the 1960s and 1970s when cloning was first proposed as a scientific solution to many of the ills of the world. Fletcher and Ramsey staked out diametrically opposed positions and envisioned a world of human cloning that is remarkably prescient given the state of current discussion.

Fletcher advocated expansion of human freedom (self-determination) and control over human reproduction. He portrayed human cloning as one among a variety of present and prospective reproductive options that could be ethically justified under circumstances of overriding societal benefit. Indeed, for Fletcher, human cloning was a preferable method of reproduction relative to the "genetic roulette" of sexual reproduction: laboratory reproduction was "radically human" because it was deliberate, designed, chosen, and willed (26).

By contrast, Paul Ramsey portrayed cloning as a "borderline" or moral boundary for medicine and society that could be crossed only at risk of compromise to humanity and to procreation. He identified three "horizontal" (person–person) and two "vertical" (person–God) border-crossings of cloning. (1) Clonal reproduction would require dictated or managed breeding to serve the scientific ends of a controlled gene pool. (2) Cloning would involve non-therapeutic experimentation on the unborn. (3) Cloning would assault the meaning of parenthood by transforming "procreation" into "reproduction" and by severing the unitive and the procreative ends of human sexual expression. Theologically, cloning represented (4) the sins of pride or hubris, and (5) of self-creation in which human beings aspire to become a man-God (27). The legacy of the themes identified by Fletcher and Ramsey concerning human cloning have been perpetuated in both recent religious and secular reflection on cloning. Within the Protestant religious communities, these debates have revolved around several contested themes.

Sanctity of Life. Protestant evangelicals have appealed to the sanctity of human life to argue against human cloning. The process of somatic cell nuclear transfer for the purpose of making a new human being, at least as illustrated in the current animal studies, would inevitably result in loss of human embryonic life. In addition evangelical positions claim that contemporary societal disregard for the sanctity of human life could possibly lead to a re-definition of humanity, such that the clone may be treated as a repository for spare organs and tissues (28).

Parenthood. Conservative and evangelical Protestants also object to human cloning on the basis of an intrinsic connection between the unitive and procreative purposes of sexuality as embedded in the *Genesis* creation story. Sexuality is understood to be a divine gift with the twin purposes of uniting the partners through a physical expression of their love and for bringing offspring into the world. In this context human cloning runs contrary to critical biological, emotional, and symbolic connections between spouses and between parent and child. In particular, the idea of a child as a "gift" is effaced as the child becomes both a project and a projection of the self (29). This argument interprets human cloning to diminish humanity to "raw material" out of which an artifice can be designed and constructed in our image, rather than the "image of God," thus leading to power over other humans rather than enhanced choices.

The Image of God. Conservative and evangelical Protestant scholars maintain that as bearers of God's image, human beings gain insight into self-understanding and human uniqueness and receive a distinctive status relative to the rest of creation. Cloning risks devaluing this image of the person by suggesting genetics is the essence of personhood, or by valuing the clone because of its replication of valued characteristics of another person.

Some mainstream Protestant theologians have argued, by contrast, that human cloning can express the creative dimensions of the *imago Dei* insofar as the new genetics promotes human dignity and welfare (30). Moreover the Christian vocation of freedom warrants the pursuit of scientific knowledge, when coupled with the obligation of accountability delineated above. Even though the reality of sin will manifest itself in an ongoing disparity between a designed future and its reality, this position holds that Christians are given permission to "sin bravely" in the pursuit of progress. Thus, if further research on human cloning can establish a reasonable expectation of benefits, and ensure human dignity, then both research and eventually human cloning seem warranted.

CONCLUSION

Protestant thought can celebrate biotechnology because of the prospects of revealing more about God's creation and applying that knowledge for human betterment, and the betterment of life on this planet. Simultaneously Protestant thought characteristically urges caution about the biotechnological revolution, lest use be transformed into abuse. The powers that human beings can wield through biotechnology must be acknowledged as limited and beyond our capacity to fully control, but Protestant thought has historically been concerned not simply with the external action but what such action reflects or expresses about a person's moral character. In this regard Protestant theological ethics forces the question of what kind of persons we need to be in order to wield such powers for good rather than ill.

BIBLIOGRAPHY

1. M. Weber, *The Protestant Ethic and the Spirit of Capitalism*, Scribner's, New York, 1958.

2. J.M. Gustafson, *Protestant and Roman Catholic Ethics: Prospects for Rapprochement*, University of Chicago Press, Chicago, IL, 1978.

3. A.D. Verhey, in J.F. Childress and J. Macquarrie, eds., *The Westminster Dictionary of Christian Ethics*, Westminster Press, Philadelphia, PA, 1986.

4. R. Cole-Turner, *The New Genesis: Theology and the Genetic Revolution*, Westminster/John Knox Press, Louisville, KY, 1993.

5. T.J. Hoban and P.A. Kendall, *Consumer Attitudes about the Use of Biotechnology in Agriculture and Food Production*, North Carolina State University Press, Raleigh, 1992.

6. K.P. Wise, in R.D. Land and L.A. Moore, eds., *Life At Risk: The Crises in Medical Ethics*, Broadman & Holman, Nashville, TN, 1995.

7. L. White, *Science* **155**, 1203–1207 (1967).

8. J.B. Cobb, Jr., *Second Opinion* **18**, 11–21 (1992).

9. P. Hefner, in T. Peters, ed., *The Cosmos as Creation: Theology and Science in Consonance*, Abingdon Press, Nashville, TN, 1989.

10. T. Peters, *Playing God? Genetic Discrimination and Human Freedom*, Routledge, New York, 1997.

11. A.R. Peacocke, *Creation and the World of Science*, Clarendon Press, Oxford, UK, 1979.

12. J.R. Nelson, *Hum. Gene Ther.* **1**, 43–48 (1990).

13. National Council of Churches, *Genetic Science for Human Benefit*, National Council of Churches of Christ in the U.S.A., New York, 1986.

14. United Methodist Church, *Genetic Science*, General Board of Church and Society, Washington, DC, 1992.

15. R.L. Shinn, *The New Genetics*, Moyer Bell, Wakefield, RI, 1996.

16. H. Bouma et al., *Christian Faith, Health, and Medical Practice*, William B. Eerdmans Publishing, Grand Rapids, MI, 1989.

17. A. Linzey, *Animal Theology*, University of Illinois Press, Chicago, IL, 1994.

18. J.R. Nelson, *On The New Frontiers of Genetics and Religion*, William B. Eerdmans Publishing, Grand Rapids, MI, 1994.

19. World Council of Churches, *Biotechnology: The Challenge to the Churches and to the World*, World Council of Churches, Geneva, 1989.

20. Call for a Moratorium on Gene Patenting, *The Christian Century* **112**(20), 633 (1995).

21. R. Cole-Turner, *Science* **270**, 52 (1995).

22. D. Nelkin and M.S. Lindee, *The DNA Mystique: The Gene as a Cultural Icon*, Freeman, New York, 1995.

23. E.R. Gold, *Body Parts: Property Rights and the Ownership of Human Biological Materials*, Georgetown University Press, Washington, DC, 1996.

24. M.J. Hanson, *Hastings Cent. Rep.* **27**(6), 1–22 (1997).

25. R. Cole-Turner, *Human Cloning: Religious Responses*, Westminster/John Knox Press, Philadelphia, PA, 1997.

26. J. Fletcher, *The Ethics of Genetic Control*, Anchor Press, Garden City, NY, 1974.

27. P. Ramsey, *Fabricated Man: The Ethics of Genetic Control*, Yale University Press, New Haven, CT, 1970.

28. J.K. Anderson, *Genetic Engineering*, Zondeman Publishing House, Grand Rapids, MI, 1982.

29. G. Meilander, *Bio Law* **II**(S), 114–118 (1997).

30. C.S. Campbell, in *Cloning Human Beings*, vol. 2, National Bioethics Advisory Commission, Rockville, MD, 1997, pp. D1–D64.

See other RELIGIOUS VIEWS ON BIOTECHNOLOGY entries.

REPRODUCTION, ETHICS, MORAL STATUS OF THE FETUS

BONNIE STEINBOCK
SUNY Albany
Albany, New York

OUTLINE

INTRODUCTION

The moral status of the fetus is not only central to the abortion debate but is also relevant to reflection on such issues as assisted reproductive technology (ART), the moral status of extracorporeal and frozen embryos, prenatal genetic testing, fetal and embryo research, and fetal therapy. This article discusses the meaning of moral status, and its relation to moral value and moral rights. It goes on to discuss the moral status of the fetus in the context of the abortion debate, outlining the main positions that have been taken. These include genetic humanity, brain birth, viability, personhood, potential personhood, the possession of a "future like ours," sentience, and a multi-criterial approach.

WHAT IS MORAL STATUS?

Mary Anne Warren succinctly explains the concept of moral status:

> To have moral status is to be morally considerable, or to have moral standing. It is to be an entity towards which moral agents have, or can have, moral obligations. If an entity has moral status, then we may not treat it in just any way we please; we are morally obliged to give weight in our deliberations to its needs, interests, or well-being. Furthermore, we are morally obliged to do this not merely because protecting it may benefit ourselves or other persons, but because its needs have moral importance in their own right (1).

Some entities clearly have moral status: for example, people. It is difficult even to imagine a moral view that did not require us to consider the interests and well-being of people. Indeed, this requirement may define "the moral point of view" (2). Just as clearly, some entities do not have moral status: for example, ordinary rocks. They are just things, of no particular value or importance. The hard questions fall in between people and more things. What about nonhuman animals, permanently unconscious humans, plants, species, and the environment? A coherent, nonarbitrary answer to the question of whether these beings have moral status requires a general theory of moral status.

Moral Status and Moral Value

While some "mere things" have little or no value, others are valuable in various ways. Some things have commercial value, some aesthetic value, some scientific value, and so on. All of these ways of being valuable can be distinguished from being morally valuable or having moral value. What, then, is it for something to have moral value? A plausible suggestion is that something has moral value if there are moral reasons for valuing it. Consider, for example, the flag of the United States. Viewed solely as a piece of material, the flag has relatively little value. But it is more than that. It is the emblem of our country. School children pledge allegiance to the flag, it is flown over government buildings, lowered to half-mast when important people die, and so forth. Many people have strong feelings about the flag: They respect it, revere it, even love it. Of course it is not the flag itself that inspires these feelings, but what it stands for. This explains why it is held to be wrong to fly the flag in tattered condition or to allow it to touch the ground. It also explains the outrage felt by many patriots when the flag is burned in a political protest. The point is not that flag burning is necessarily wrong. In fact burning is the recommended way of disposing of a worn-out flag. In addition flag-burning could be a legitimate form of political protest. The point is rather that it is a not a matter of moral indifference what one does to a flag. There are moral reasons to treat flags in certain ways. This suggests that the flag, because of its symbolic significance, has moral value.

Bonnie Steinbock has suggested that moral status be distinguished from moral value (3). The difference lies in the reasons why certain treatment is regarded as morally wrong. In the case of the flag, the reason why it is held to be wrong to burn a flag in protest is that this shows disrespect for the flag and the country for which it stands. The reason is not a "golden rule" type reason, along the lines of "how would you like it if you were a flag and someone burned you?" In making this distinction, we recognize that it does not matter to a flag what is done with it, and this differentiates flags from, say, people or animals. To accord something moral status is not merely to consider it important or valuable, and worthy of protection; it is to take its perspective into account when making moral decisions. In this view, flags might have moral value, but they lack moral status. This is because flags do not have a point of view. They cannot have a point of view because they lack awareness of any kind. Mere things can be destroyed, but whatever is done to them cannot matter to them. They do not have a stake in their preservation or well-being (if mere things can be said to *have* a well-being). They have no interests because only beings who can care about how they are treated or what is done to them can have interests of their own. Lacking interests, mere things lack moral status, but it does not follow that it is morally permissible to treat them any way you like.

Many environmentalists object to the view that only sentiment beings can have moral status: the "sentience only" view. For "deep ecologists," like Aldo Leopold, moral status is not limited to sentient beings. Natural plant and animal species, populations, and habitats can all have moral status, just as much as sentient beings. As Mary Anne Warren expresses the point, "To many environmentalists, a theory which allows us to have moral obligations *regarding* the nonsentient elements of the natural world but never *to* them, seems just as inadequate as the Kantian theory, which allows us to have duties regarding animals, but never to them" (1, p. 72). While this debate is important for environmental ethics, it is not directly relevant to the moral status of the fetus, as those who maintain that the fetus has moral status do not usually base its status on its being an element of the natural world but rather on such features as its being genetically human, potentially a person, or the kind of future it will have.

Moral Status and Moral Rights

All beings that have moral rights have moral status. Indeed, if someone has a right to something, this imposes obligations on others to behave in certain ways. Your moral rights limit my freedom with respect to how I am permitted morally to treat you. This implies that from a moral point of view, someone who has a right counts or matters, which is the same as ascribing to it moral status. However, the reverse does not hold. It does not follow that every being that has moral status has moral rights. It is even possible that there are no moral rights, as some utilitarians maintain. There can be an account of morality that does not include moral rights, but there cannot be an account of morality that does not include a theory of moral status.

MORAL STATUS AND ABORTION

We start with the problem of abortion, where the moral status of the fetus has been a central and contentious issue. First, a word about terminology. During the first week of its existence, the fertilized egg is known as a conceptus. The term "embryo" refers to the entity between the second and eighth weeks. From the eighth week until birth, it is a fetus. However, the term "fetus" is often used to refer generally to the unborn throughout pregnancy. This article follows that usage, except where the different phases of gestation have a bearing on moral status and need to be distinguished. Those who are "pro-life" (often referred to as "conservatives" on abortion) argue that fetuses, throughout gestation, have the same moral status as born human beings, and therefore killing fetuses is seriously wrong, as wrong as killing born human beings. Those who are "pro-choice" (often referred to as "liberals" on abortion) usually argue that fetuses differ in morally important ways from born human beings, and for this reason, lack full moral status and in particular a right to life. Abortion, while not desirable, is not seriously wrong, in this view, and is certainly not equivalent to killing a born human being.

A number of writers have pointed out that the abortion issue does not turn solely on the moral status of the fetus. For example, Thomas Murray argues that "Fetal personhood is only one strand in the web holding our moral judgments about abortion" (4, p. 146). Sociologist Kristin Luker argues that it is a mistake to think that the views of pro-choice and pro-life activists about abortion are determined by their philosophical or religious views on the moral status of the fetus (5). Rather, their views on the morality of abortion stem from their differing views on the meaning and value of sexuality, motherhood, and the proper role of women. How they view the fetus is determined by how they regard abortion, not the reverse. Legal philosopher Ronald Dworkin argues that the abortion debate is not really about the moral status of the fetus, despite the rhetoric on both sides. According to Dworkin, even those vehemently opposed to abortion do not actually believe that a fetus is from the moment of its conception a full moral person with rights and interests equal in importance to those of any other member of the moral community. Nor do those committed to protecting a woman's "right to choose" think of the developing fetus as just a part of the pregnant woman's body. "The disagreement that actually divides people is a markedly less polar disagreement about how best to respect a fundamental idea we almost all share in some form: That individual human life is sacred" (6, p. 13). Judith Thomson maintains that it is a mistake to think that the abortion debate is over if the premise that the fetus is a person, with a right to life, is accepted. For the abortion debate also raises the question of whether the fetus has a right to the use of the pregnant woman's body, and how much of a sacrifice to sustain its life she is required to make. Thomson thinks that even if fetuses are persons, at least some abortions could be justified (7). Finally, some feminists regard the inquiry into the status of the fetus as irrelevant to the problem of abortion (8). They view abstract inquiries into the moral status of the fetus as distracting from the real issues, which have to do with creating the social conditions that permit women to make genuine reproductive choices.

All of these voices have deepened the abortion debate, but it is doubtful that any has proved the irrelevance of moral status. Even if Luker is right about the origin of people's views about the moral status of the fetus, that does not address the correctness or plausibility of their views. Feminists who focus solely on the interests of women can be fairly charged with simply avoiding the question of fetal moral status. Even if sexism, racism, poverty, and other bad social conditions were eliminated, there would undoubtedly still be unwanted pregnancies, and women would still want abortions. The question is whether abortion is a morally permissible choice. It is hard to see how this question can be answered without considering the arguments of those who claim that killing fetuses is seriously wrong. Finally, while Dworkin's view is ingenious, it is doubtful that most pro-lifers at least would accept his reconceptualization. For them, it is not a matter of how best to respect the sanctity of life, but rather a matter of preventing the murder of innocents. Their opposition to abortion cannot be taken seriously if this claim is not addressed.

THEORIES OF MORAL STATUS

Genetic Humanity

According to Roman Catholic teaching, abortion is murder. It is murder because it is "the deliberate and direct killing ... of a human being in the initial phase of his or her existence . . ." (9,10). In other words, there can be no doubt that a human fetus has the moral status of any other human being, because "from the time that the ovum is fertilized, a life is begun which is neither that of the father nor the mother; it is rather the life of a new human being with his own growth. *It would never be made human if it were not human already*" (9, p. 22, emphasis added). Despite the fact that a fertilized egg does not look like the human beings we know, and has none of the characteristic attributes of born human beings, its possession of its own genetic code makes it a separate, unique, and individual human person. It is worth noting that this argument, though religious in origin, is not religious in nature. It makes no appeal to religious notions, like God or the soul. Rather, the humanity of the fetus is based on its being biologically or genetically human. John Noonan makes the same point when he says, "The positive argument for conception as the decisive moment of humanization is that at conception the new being receives the genetic code. . . . A being with a human genetic code is man" (11, p. 264). All and only genetic human beings have full moral status, and this full or human moral status begins at conception.

Birth

Conception as the point at which a human being acquires moral status is often defended by showing the arbitrariness of any other stage. Consider, for example, birth. Live birth is often taken to be a significant landmark both in Anglo-American law and in religious traditions, such as Judaism. At birth, the fetus becomes an infant, separate from its mother and no longer physiologically dependent on her. Moreover, birth is a precise moment, noted on birth certificates. However, conservatives argue that the difference between a newborn moments after birth and a fetus moments before birth is insignificant. How can location alone determine moral status?

Viability

Moving back into gestation, some regard viability—the stage at which the fetus can survive outside the womb, albeit with artificial aid—as having moral significance. In *Roe* v. *Wade*, the Supreme Court chose viability as the point at which there was no longer a constitutional right to an abortion:

> With respect to the State's important and legitimate interest in potential life, the "compelling" point is at viability . . . State regulation protective of fetal life after viability thus has both logical and biological justifications. If the State is interested in protecting fetal life after viability, it may go so far as to proscribe abortion during that period except when it is necessary to preserve the life or health of the mother . . . (12).

Of course, the Court's decision was not intended to mark the point at which the fetus acquires *moral* status, since that is not a legal issue. Nor did the Court stipulate that a fetus undergoes a change in *legal* status at viability. *Roe* v. *Wade* did not hold that a fetus becomes a legal person, with a right to life, at viability. Rather, viability marks the point at which the *state's* interest in protecting fetal life becomes "compelling," namely the point at which the state can

outweigh the woman's privacy right to make the abortion decision. However, unless viability has moral significance, it would be a completely arbitrary point at which to allow states to ban abortion. It is clear that the Court did not regard it as an arbitrary point, but rather considered viability as a dividing line with "logical and biological" justification. But what exactly is this justification, and what does it have to do with fetal moral status?

The Court noted that at viability the fetus has the capacity for meaningful life outside the mother's womb. However, as several commentators have noted, this is not a justification for extending protection to it, but rather an explanation of what viability means. The question is why the capacity for meaningful, independent existence bears on moral status. Nancy Rhoden has suggested that the Court focused on viability because, especially in 1973, the capacity for independent existence was connected with late gestation and fetal development (13). In other words, by the time a fetus becomes viable, it shares many characteristics with infants, who are entitled to the law's protection. A late-gestation fetus looks like a baby, it can probably feel and hear, it sucks its thumb. It is sufficiently like a newborn that all of our protective feelings for babies "kick in" and incline us to extend the same protection to viable fetuses.

None of this is persuasive to the conservative, who asks, "Why should the dependence of the nonviable fetus on its mother deprive it of human moral status?" Moreover, a viable fetus has a chance at survival if removed from the uterus. The fact that a nonviable fetus cannot survive outside the womb is all the more reason not to eject it. Thus the "logical and biological justification" for permitting abortion of previable fetuses is puzzling. In any event, the conservative does not regard any part of fetal development as having special moral significance. If allowed to grow and develop, the fetus will acquire a nervous system, organs, a brain. It will begin to move and look like a born human being. Eventually it will have sensations and hear sounds. None of these stages makes the fetus a human being. It is already human, and each landmark is nothing more than a stage in its development.

Brain Birth

Most conservatives on abortion base moral status on being human, but not all believe that human life begins at conception. For example, Baruch Brody argues that a functioning brain is essential for being human (14). His position is based on a parallel with the end of human life. If a human being dies and goes out of existence when the brain irrevocably and completely stops functioning, then human life can be said to start when the fetal brain begins to function, or when brain waves can be detected, somewhere around six weeks after fertilization. It is not simply the parallel with the current criteria for death that motivates Brody's focus on brain birth. In addition brain function is the biological basis for consciousness, thought, and feeling. The importance of this for Brody is indicated when he says, "One of the characteristics essential to a human being is the capacity for conscious experience, at least at a primitive level. Before the sixth week, as far as we know, the fetus does not have this capacity. Thereafter, as the electroencephalographic evidence indicates, it does. Consequently, that is the time at which the fetus becomes a human being" (14, p. 83). However, from the fact that brain waves can be detected in a six-week old fetus, it does not follow that it can feel. While there is some disagreement on this point (15), and it is difficult to pinpoint the exact time at which the fetus becomes aware of its surroundings, most scientists believe that more than brain waves are necessary before the fetus has the capacity for conscious experience. It seems likely that sentience does not occur until well into the second trimester (probably between 22 and 24 weeks) as prior to that time the neural pathways are not sufficiently developed to be able to transmit pain (or any experiential) messages to the fetal cortex (16,17). So, if it is conscious experience that marks the beginning of human life, then this very likely does not occur until the second trimester of pregnancy. The occurrence of brain waves is certainly a necessary part of this development, but it is not clear why the beginning of brain function should have particular moral significance. Certainly most pro-lifers do not view "brain birth" as having any more significance than any other developmental stage in the life of the fetus.

Personhood

A different approach to the moral status of the fetus focuses on the fact that fetuses, while certainly genetically human, are not persons. Persons are essentially defined or recognized, not by their biology, but by certain psychological features, such as consciousness or an awareness of their surroundings, sentience, self-consciousness, thought, and the use of language. Fetuses, especially at the beginning of pregnancy, when 90 percent of abortions occur, have none of these characteristics. How, proponents of a personhood view ask, can a clump of cells be compared to an individual who feels, thinks, worries, enjoys, and so forth? It is persons who have a special moral status and a right to life. The mistake conservatives make is to equate being human with being a person.

This is a natural enough mistake, since all the persons we know are human beings. In contrast, personhood proponents argue that the two concepts are distinct in theory, however much they coincide in our experience. To see this, think about encountering an intelligent alien like *E.T.* in the movie of that name. Certainly E.T. is not human in a genetic or biological sense. He is not a member of the species *Homo sapiens*. Nevertheless, if we should meet someone like E.T., we would have no hesitation in according him the moral status of a person because, whatever his species membership, he resembles us in morally significant ways. That is, he is conscious, sentient, rational (indeed, far more advanced than human beings), self-conscious, uses language, and is a moral agent. Mary Anne Warren (18) claims that it is these features that have moral significance, not membership in a particular species. Why should a merely biological category make a moral difference? Or, as Don Marquis (an opponent of abortion who nevertheless criticizes the species criterion of moral status) puts the point, "Why . . . is it any more reasonable to base a moral conclusion on the number of chromosomes in one's cells than on the color of one's skin?" (19, pp. 26–27). On the other hand, if "human being" is taken to be a *moral* category, then the fetus's claim to be a human being cannot be a *premise* in the antiabortion argument, for that is precisely what is to be proved. The conservative argument appears either to be based on an arbitary, morally irrelevant category (species membership) or to beg the question. To say this is not to say that the genetic humanity criterion has been shown to be wrong, only that it has not been adequately defended. Perhaps membership in the human species can be shown to have moral significance so that the conservative argument can be rendered noncircular.

What are the implications of a personhood view of moral status for the fetus? Warren does not claim that possession of *all* the person-making characteristics is necessary to be a person. Her point is rather that a being who possessed *none* of the relevant characteristics would not be a person. Indeed, she thinks that anyone who claimed that a being who possessed none of the person-making traits was a person all the same "would thereby demonstrate that he had no notion at all of what a person is—perhaps because he had confused the concept of a person with that of genetic humanity" (18, p. 68). On Warren's view, early-gestation fetuses are certainly not persons, as they lack all of the characteristics of persons. But what are we to say about more developed fetuses, who have some of the characteristics of persons, such as sentience and a rudimentary form of consciousness? Warren notes that even late-gestation fetuses are not fully conscious, in the way that an infant of a few months is, and that it cannot reason or use language, and so forth. "Thus, in the *relevant respects*, a fetus, even a fully developed one, is considerably less personlike than is the average mature mammal, indeed the average fish" (18, p. 69). Warren concludes that a fetus at any stage cannot have more of a right to life than a newborn guppy "and that a right of that magnitude could never override a woman's right to obtain an abortion, at any stage of her pregnancy" (18, p. 69).

There are two central problems with a personhood criterion of moral status. The first is that it applies not only to fetuses but to other human beings, notably, newborns. If abortion is not seriously morally wrong, is infanticide equally morally neutral, something best left to parents to decide? Warren's response to this is to offer consequentialist reasons for keeping infanticide illegal and to acknowledge that infanticide need not be seriously wrong in a society that cannot care for all the infants who are born. Whatever the merits of this argument, the personhood criterion seems to leave out many individuals most of us thought were full-fledged members of the moral community: elderly senile people, for example, and people with severe mental disabilities. Are we to say that their deficits in rationality, language usage, and so forth, deprive them of "human moral status" and that therefore killing them is not seriously morally wrong? One way to avoid this unpleasant conclusion is to set the requirements for personhood relatively low. Perhaps consciousness and sentience will do. However, that will mean that a great many animals will qualify as persons whom it is presumably wrong to kill. This is a possible moral view, but one that requires radical revision of ordinary moral thought and practice. On the other hand, if the requirements for personhood are set relatively high, and rationality and language use are necessary conditions, that will keep out the animals, but at the cost of denying full moral status to many human beings, and not just fetuses.

The other problem with Warren's personhood view is that it suffers from the same defect she criticized in the conservative argument (20). According to Warren, the conservative confuses genetic humanity with moral humanity (or moral personhood). This confusion leads the conservative to think, wrongly, that just because the fetus is genetically human, it must be morally human, that is, have the moral status that human persons have, including a right to life. What is missing in the conservative's argument is an explanation of why species membership should endow a being with a particular moral status. However, Warren can be criticized for making the same mistake, for she does not explain what it is about personhood that endows a being with moral status. As Marquis explains the criticism:

> The principle "Only persons have the right to life" also suffers from an ambiguity. The term 'person' is typically defined in terms of psychological characteristics, although there will certainly be disagreement concerning which characteristics are most important. Supposing that this matter can be settled, the pro-choicer is left with the problem of explaining why *psychological* characteristics should make a *moral* difference (19, p. 27).

Warren says that it is "self-evident" that descriptive persons are moral persons, but that equation is no more self-evident than the conservative's claim that genetic humans are moral humans (i.e., entitled to be treated in certain ways). This is not to say that Warren's claim cannot be defended, but rather to say that it is not self-evident, and an argument connecting the psychological properties of personhood and moral status is needed. Even if such an argument can be given, it does not follow that moral personhood is limited to descriptive persons. The possibility remains open that it is justifiable to confer *normative personhood* (including a right to life) on human beings, such as babies and those with severe mental deficits, who lack the psychological properties of descriptive persons.

Being a Person and Having a Right to Life

Like Warren, Michael Tooley thinks that personhood is essential to moral status, but unlike Warren, he provides an argument to show why only descriptive persons can have a right to life (21). Tooley starts with Joel Feinberg's analysis of right-bearers as beings who can have interests (22). Interests are necessary for rights because the function of rights is to protect interests; if a being had no interests, there would be nothing to protect and the ascription of a right meaningless. Tooley then takes Feinberg's view one step further, arguing that particular rights are connected with specific sorts of interests. According to the "particular-interests principle," an individual cannot have a right to R unless it is capable

of having an interest in R. The question then is what is required for a being to have an interest in R. This depends, Tooley maintains, on the sort of thing that R is. Consider the interest in not being subjected to painful stimuli. To have an interest in that requires only that the being be capable of experiencing pain as a disagreeable sensation. Thus we could intelligibly ascribe a right not to be subjected to painful stimuli to any sentient being. Other rights require more conceptual abilities. It would be absurd to ascribe a right of freedom of expression to a cow, because cows have no interests that can be furthered by such a right. Now what about life? It might be argued that life is in the interest of all living things, but this is just what Tooley wants to deny. He maintains that the right to life protects the interest in one's own continued existence, and that therefore only beings who can have a concept of their own continued existence can have a meaningful right to life. Tooley's view is much more stringent than Warren's. It limits a right to life to those who, first, are able to think of themselves as continuing to exist into the future and, second, have desires about that future, in particular, that it exist. This rules out not only human fetuses and most animals, but also babies and young children. It is not clear when children obtain a conception of themselves as existing in the future. It is probably not before they become language users, somewhere around the age of two.

The practical implications of Tooley's view (that it makes infanticide morally permissible) are not the only objection to it. In addition it can be objected that a right to life can be meaningfully ascribed to any being whose life is a good to it now, even if it lacks the capacity to envisage, and have desires about, a future existence. Animals, babies, and severely retarded adult human beings can enjoy their lives; why then cannot we preserve their lives *for their own sake?* If the reason for not killing them is that their lives are a good *to them*, this suggests that they have an interest in living, which can be the basis for ascribing to them a right to life.

Potential Personhood

Some pro-lifers do not regard genetic humanity as intrinsically significant, but rather as an indication that the being will become a descriptive person. An embryo does not now have any of the properties of a person. It is not even sentient or conscious, much less capable of communicating or relating to other persons. However, even an embryo is potentially just like us. If left alone (i.e., not aborted), it will grow and develop into a human person. Therefore we ought not to thwart its natural development. "On its strongest interpretation," Stephen Buckle explains, "the argument is thought to establish that we should treat a potential human subject as if it were already an actual human subject" (23, p. 227).

Is it in fact true that embryos, if not deliberately aborted, will develop into persons? A great deal has been learned about the rate of miscarriage in the last 20 years. When John Noonan was writing in 1970, he claimed that only 20 percent of pregnancies ended in spontaneous abortion. This figure still holds for pregnancies that are physiologically recognized. However, it is now thought that up to 75 percent of all human conceptions are aborted spontaneously (24). Many of these spontaneous abortions occur before the woman realizes she is pregnant. With such a high rate of pregnancy loss, can it be maintained that every embryo is a potential person? The vast majority do not develop into persons, and would not develop into persons, even if allowed to develop naturally. This is one reason why implantation might be chosen by potentiality theorists as the moment at which the conceptus attains full moral status. After implantation, the prospects for live birth improve considerably.

There are other problems with arguments based on potentiality. The first is known as "the logical problem with potentiality." It is directed at the strongest version of the potentiality argument, which claims that the potential of the fetus to become a person gives it now the rights of a person. But, it may be objected, why should mere potential convey actual rights? As Stanley Benn has put it, "A potential president of the United States is not on that account Commander-in-Chief" (25, p. 143).

The logical problem can be avoided if the claim is weakened. It is not that the embryo or fetus *now* has the right to life. Rather, the claim is that because we think that the lives of persons are valuable and deserving of protection, so too we ought to recognize the value of entities that have the potential to become persons. While this might not accord to fetuses a full-fledged right to life, it would at least require that the reasons for killing a fetus be substantial ones.

A serious problem for potentiality arguments is that they seem vulnerable to a *reductio ad absurdum*. If the objection to abortion is that it kills potential person, why cannot the same complaint be made of contraceptive techniques that kill sperm, or prevent sperm and egg from joining? Why isn't an unfertilized ovum also a potential person? John Harris makes the point this way (although he refers to "human being" rather than "person"):

> To say that a fertilized egg is potentially a human being is just to say that if certain things happen to it (like implantation) and certain other things do not (like spontaneous abortion), it will eventually become a human being. But the same is true of the unfertilized egg and the sperm. If certain things happen to the egg (like meeting a sperm) and certain things happen to the sperm (like meeting an egg) and thereafter certain other things do not (like meeting a contraceptive), then they will eventually become a new human being (26, pp. 11–12).

If abortion is seriously wrong because it kills a potential person, then using Delfen foam (a spermicide) is mass murder! Indeed, even abstinence would have to be justified, since failing to have intercourse, at least during a woman's fertile period, would prevent the development of a new human being. Since virtually all potentiality theorists wish to differentiate between contraception (which they regard as morally neutral) and abortion (which they regard as seriously wrong), they must explain why an embryo is a potential person in a way that a gamete is not.

Some theorists look to probability to defend the claim that an embryo is a potential person, but a gamete is not. Even if a fertilized egg has only a 25 percent chance of becoming a person, this does not compare with a sperm's chance: about one in 200 million. A given ovum has a better chance to become a person than a given sperm, but still much less chance than that of a fertilized egg. However, it is not clear that potential should be understood in terms of probability. As Steinbock points out, "Is not every entrant in a lottery a potential winner, even if the odds of winning are extremely low?" (3, p. 63)

Others regard potentiality not in terms of the odds of success but rather in terms of natural development. Many fertilized eggs do grow and develop into embryos, fetuses, and babies. A gamete, on the other hand, is not growing or developing into anything (27,28). However, basing potential personhood on what fertilized eggs become in the natural course of events has notable consequences for extracorporeal embryos. Since they cannot develop into persons without human intervention (i.e., being placed in a uterus), extracorporeal embryos are not potential persons, and presumably, it is not seriously wrong to kill them. This conflicts with the opinion of many pro-lifers and the Catholic Church that extracorporeal embryos have the same moral status as embryos in a uterus. Indeed, some have regarded frozen embryos as "pre-born children" (29).

Some defenders of a potentiality principle try to distinguish between a gamete and a zygote by saying that prior to fertilization, no particular individual exists. Once the complete human genome is present, there is a new human being, the same individual that will be born, grow up, and die. However, as Mary Anne Warren notes, this claim can be disputed on empirical grounds:

> It is not clear that the zygote is the same organism or proto-organism as the embryo that will later develop from it. During the first few days of its existence, the conceptus subdivides into a set of virtually identical cells, each of which is "totipotent"—capable of giving rise to an embryo. Spontaneous division of the conceptus during this period can lead to the birth of genetically identical twins or triplets. Moreover, it is thought that two originally distinct zygotes sometimes merge, giving rise to a single and otherwise normal embryo. These facts lead some bioethicists to conclude that there is no individuated human organism prior to about fourteen days after fertilization, when the 'primitive streak' that will become the spinal cord of the embryo begins to form (1, pp. 203–204).

In this view, an implanted embryo is a potential person, while neither a gamete nor a newly fertilized ovum is. But some theorists argue that while the implanted embryo may be identified with some *particular* person, in a way that neither the zygote nor the constituent gametes are, nevertheless gametes and zygotes are potential persons. Potentiality is one thing; uniqueness or identity another (30).

A Future Like Ours

A variation on the potentiality principle is offered by Don Marquis in "Why Abortion Is Immoral" (19). According to Marquis, the reason why killing people is generally wrong is that killing deprives the individual of a valuable future, a future like ours (FLO). Killing a fetus by having an abortion deprives the fetus of its valuable future, and therefore abortion is (usually) seriously morally wrong. Marquis notes two ways in which his view differs from traditional accounts. First, it is not "species-ist." Moral status does not depend on being genetically human but rather on having a valuable future like ours. If there are members of other species elsewhere in the universe who have valuable futures like ours, then it would be seriously wrong to kill them. For that matter, it is possible that some nonhuman animals have FLO, and that killing them is seriously morally wrong. Marquis leaves indeterminate precisely what FLO consists in, but presumably he has in mind the kinds of capacities that make us persons: rationality, self-consciousness, the ability to have relationships with others, and so forth. Therefore, although his account does not refer explicitly to the wrongness of killing persons or potential persons, this is implicit in his account, since the beings it is seriously wrong to kill (i.e., those who have FLO) are persons or potential persons. If a being is neither a person, nor capable of developing into a descriptive person, then presumably it does not have a valuable future like ours. Second, Marquis's view differs from a sanctity of life approach that holds that killing people is always wrong. According to Marquis, killing someone who does not have a valuable future is not necessarily wrong. Thus Marquis's account is compatible with voluntary euthanasia and even nonvoluntary euthanasia of infants and fetuses whose lives will be filled with suffering or empty of the things that make life worth living.

An objection to Marquis's view is that it is vulnerable to the same *reductio* as other potentiality views. If it is wrong to kill a fetus because of its FLO, why is it not equally wrong to kill gametes? Why do not gametes have a valuable future? Of course, it is true that most gametes do not have much of a future at all. They pass out of or are reabsorbed into the body. However, use of a contraceptive prevents those few gametes that might develop into persons from doing so. It would seem that contraception should be somewhat morally problematic, in Marquis's view. He, however, denies this. We can ascribe a valuable future to a fetus because the fetus is identified with the person it becomes. It has the *same* future as the born individual. However, the born individual does not share a future with the egg and sperm that conjoined to form it. Neither the egg nor the sperm is the person; therefore neither has his or her future. As Marquis expresses it:

> If I were the same individual as a sperm and also the same individual as an ovum, then a particular sperm and a particular ovum were the same individual. This is obviously false. It follows that any argument that I was once a sperm (or an unfertilized ovum) is unsatisfactory. Therefore, the contraception objection fails (30).

However, why should the fact that I am not identical with either the sperm or the egg entail that they do not have valuable futures, of which they would be denied if they were killed or otherwise prevented from conjoining? Marquis thinks that neither gamete can have a future because it is only when the two conjoin that there is anything that has a valuable future. The valuable future is that of the new being, and it is not identical with either of its two components. However, it is not clear why only the new being can be said to have a valuable future. To be sure, neither gamete can have a valuable future all by itself, but why does that matter? It might be objected that Marquis has confused identity with having a valuable future.

One of the interesting implications of Marquis's account is the possibility that pre-implantation embryos lack FLO, and therefore lack moral status. Since a pre-implantation embryo might turn out to be two or three people, it cannot be identified with any particular individual. If having FLO depends on identification with a particular future person, then pre-implantation embryos do not have FLO. This means that killing a pre-embryo using a very early abortifacient, such as the morning-after pill, is not seriously wrong. In addition Marquis presumably would not oppose creating embryos either for research or possible implantation, and then discarding them. Only implanted embryos, which have the primitive streak and can no longer become twins, can be identified with the subsequent person, have FLO and are seriously wrong to kill. The fact that most conservatives would be unwilling to accept these implications is not an argument against the FLO theory. However, it remains an open question whether it, like potentiality theories generally, are vulnerable to the contraception objection.

Sentience-Based Views

Sentience is the ability to experience pain or pleasure. It is relevant to moral status because we normally assume that it is wrong to inflict pain without a good reason. Many pro-lifers seem to base their objection to abortion on the premise that abortion causes the fetus to suffer. One pro-lifer was quoted as saying that abortion is "mean." Since most methods of abortion require the embryo or fetus to be ripped from the uterine wall, and often cause it to be torn apart, the idea that abortion hurts the fetus is not surprising. However, as we have seen, it is extremely unlikely that a fetus in the first trimester can experience pain or anything else.

Rejection of unnecessary infliction of pain suggests that sentience is a sufficient condition of moral status. However, the infliction of pain is not the only kind of action with moral importance. Killing is also something that often requires justification, even when done painlessly. We need a deeper reason for thinking that sentience is a necessary, as well as sufficient, condition of moral status. The deeper reason comes from the connection between moral status and interests, and the connection between interests and sentience.

To say that a being has moral status is to say that it has moral claims against us. This in turn suggests that we should do, or forbear from doing, certain things for its sake. We are required to consider its welfare in deciding what to do. The next question is what kinds of beings can have a sake or a welfare. For some philosophers, anything that can be protected or preserved can have a welfare. Others limit a welfare to beings to whom it can matter how they are treated. In this view, it may be morally wrong to destroy the environment, but it is not a wrong *to* the environment because the environment does not, indeed, cannot, care what is done to it. The environment has no stake in what happens to it. Because nothing can matter to it, it has no interests. If it has no interests, then its interests cannot be considered, and therefore it lacks moral status.

The conceptual connection between moral status and the possession of interests seems self-evident. To accord something moral status is to require that its interests be considered. If a being has no interests, no sake, and no welfare, it is hard to see how it could have moral claims on us. More controversial is the claim that sentience is necessary for a being to have interests. What is it about the ability to experience pain and pleasure that enables a being to have interests? The intuitive idea is that sentient beings care about what happens to them, while nonsentient beings do not. However, perhaps it is not sentience that enables beings to have interests but rather simply consciousness: the ability to have experiences. Mary Anne Warren suggests this when she writes:

> One can imagine a being that has conscious experiences of many sorts, but that never experiences pleasure or pain, or any other positive or negative feeling, mood, or emotion. Such a being would be conscious, but it would not be sentient. Data, the brilliant and personable android of the television series *Star Trek: The Next Generation*, is described by himself and other characters as such a being. Although he is conscious, rational, morally responsible, and highly self-aware, his programming includes no capacity to experience pain, pleasure, or emotion ... such a being would have strong moral status by virtue of its moral agency; but it could not have any moral status that is contingent upon sentience (3, p. 56).

The ability to experience physical pain or pleasure does not seem essential to having interests. There are human beings who lack the ability to feel pain; they still have all kinds of interests, including an interest in continuing to live. Sentience, understood in purely physical terms, does not seem to be necessary for moral status. However, Data is alleged to have no feelings or emotions at all. This being the case, does it matter to him what happens to him? Does he care if he is destroyed or protected? And if he is indifferent to what happens to him, can preserving him be said to be in *his* interest? It might be argued that the connection between sentience (broadly understood to include emotions and feelings) and interests is not so easily broken.

Tom Regan argues that the belief that sentience is necessary for the possession of interests stems from a failure to distinguish between two senses of "interest." One sense of "interest" refers to what individuals *take* an interest in; what they are interested in or care about, what matters to them. It seems clear that nonconscious, nonsentient beings cannot have interests in this sense. However, there is another sense of "interest" that refers to what is *in* a being's interest. That these are not the same is easily shown. It can be *in* someone's interest to give up smoking, exercise regularly, and eat moderately, and yet the person in question might have no interest in doing so. Conversely, people often have an interest in things that are not, on balance, in their interest. The question is whether it makes sense to talk about what is *in* the interest of beings that do not *take* an interest in anything. Certainly we do sometimes talk this way, recommending that certain actions be taken "in the interest of preserving the environment." But it does not follow that the interests in question are those of the environment, nor that an inanimate object can have a good or welfare of its own in the same sense as a being that cares, if only in the most rudimentary sense, about what is done to it.

Some sentience-based theorists regard sentience alone as having moral significance. For example, Peter Singer is a "sentience-only" proponent, whose theory of moral status is inspired by that of Bentham and Sidgwick. According to Singer, the comparable interests of all sentient beings should be given equal weight in our moral deliberations (31). This does not entail treating all sentient beings alike, since their needs and interests will differ. It does not even mean valuing the lives of all sentient beings equally. Singer acknowledges that the life of a human — that is, a rational, self-conscious, morally autonomous agent — may properly be considered more important than the life of a mouse. However, he thinks that there is no justification for valuing the pain experienced by a person over the (comparable) pain experienced by a mouse. Pain is pain, no matter who feels it.

Not all sentience-based theorists accept the principle of equal consideration of interests. For example, L.W. Sumner, who thinks that sentience is a necessary and sufficient condition of moral status, argues that both sentience and moral status come in degrees (32). The moral status of a being is proportional to its degree of sentience. This is supposed to explain the intuitive view that the interests of people count for more than the interests of mice. However, it is not clear why rationality, self-consciousness, and moral agency should be conceived of as degrees of sentience. Nor does it seem impossible that a being lacking in those morally relevant features could nevertheless be intensely sensitive to pain and pleasure. Bonnie Steinbock does not accept the degrees of sentience view advocated by Sumner; she thinks that there can be features aside from sentience that are relevant to moral status. Thus she argues that while the possession of interests is a necessary condition of moral status, and potential personhood by itself does not endow beings who lack interests with moral status, the potential of a sentient being to become a person must be regarded as enhancing its moral status. Both human infants and nonhuman animals have minimal moral status as sentient beings, but the degree of moral status possessed by human infants is greater than that of nonhuman animals because the infants are potential persons. Another factor is relationships with others. Some humans, due to brain defects, will not develop into descriptive persons, capable of language, reasoning, and moral responsibility. Nevertheless, they remain normative persons. In part, this is because of the place they occupy in a network of affections. A retarded child does not cease to be someone's son or daughter, or loved the less because of that. Their moral status does not change because they will not fulfill normal human potential. And even if the parents reject the child, society should take on this caring role, out of kindness and compassion for the helpless child who needs our care.

The Multi-Criterial View

Steinbock's suggestion that there might be features besides sentience relevant to the degree of moral status is taken up explicitly by Mary Anne Warren, who argues for a "multi-criterial" view (1). She rejects uni-criterial views, whether based on life, sentience, or personhood, as simplistic and inconsistent with elements of commonsense morality that we cannot reasonably be expected to jettison. Her multi-criterial approach consists of seven principles that she regards as implicit elements of commonsense morality. The principles that might be thought relevant to fetal moral status are respect for life and the transitivity of respect. (The anticruelty principle applies only to beings who can be treated cruelly, that is, sentient beings. It does not apply to early-gestation fetuses.) The respect for life principle accords some moral status to fetuses, but not enough for full moral status. Moreover, the reasons why women seek abortions are sufficiently compelling to justify the destruction of a living thing that is not yet sentient and not yet a member of a human social community. The transitivity of respect principle requires us, within the limits of the other principles and to the extent that it is morally feasible, to respect other people's attributions of moral status. Since some people do regard fetuses as having full moral status, this justifies regarding fetuses as having some moral status. However, the transitivity of respect principle is limited by the basic moral rights of moral agents. Warren concludes that "although the fetus gains in moral status as it becomes increasingly likely to be capable of sentience, until it has been born it cannot be accorded a fully equal moral and legal status without endangering women's basic rights to life and liberty" (1, p. 222).

BIBLIOGRAPHY

1. M.A. Warren, *Moral Status: Obligations to Persons and Other Living Things*, Clarendon Press, Oxford, UK, 1997.

2. K. Baier, *The Moral Point of View*, Random House, New York, 1965.

3. B. Steinbock, *Life Before Birth: The Moral and Legal Status of Embryos and Fetuses*, Oxford University Press, New York, 1992.

4. T. Murray, *The Worth of a Child*, University of California Press, Berkeley, 1996.

5. K. Luker, *Abortion and the Politics of Motherhood*, University of California Press, Berkeley, 1984.

6. R. Dworkin, *Life's Dominion: An Argument about Abortion, Euthanasia, and Individual Freedom*, Knopf, New York, 1993.

7. J.J. Thomson, *Philos. Public Affairs* **1**(1), 47–66 (1971).

8. For example, S. Harding, in S. Callahan and D. Callahan, eds., *Abortion: Understanding Differences*, Plenum, New York, 1984.

9. John Paul II, *Evangelium Vitae*, Encyclical Letter, August 16, 1993. Copyright Libreria Editrice Vaticana, Rome, 1993, reprinted in Dwyer and Feinberg (10, pp. 21–23). Page references are to this edition.

10. S. Dwyer and J. Feinberg, eds., *The Problem of Abortion*, 3rd ed., Wadsworth, Belmont, CA, 1997.

11. J.T. Noonan, Jr., in J.T. Noonan, Jr., ed., *The Morality of Abortion: Legal and Historical Perspectives*, Harvard University Press, Cambridge, MA, 1970, reprinted in J.D. Arras and N.K. Rhoden, eds., *Ethical Issues in Modern Medicine*, 3rd ed., Mayfield Publishing, Mountain View, CA, 1989, pp. 261–265. Page references are to this edition.

12. *Roe v. Wade*, 410 U.S. 113, 163 (1973).

13. N.K. Rhoden, *Yale Law J.* **95**(4), 639–697 (1986).

14. B. Brody, *Abortion and the Sanctity of Life*, MIT Press, Cambridge, MA, 1975.

15. P. McCullagh, *The Foetus as Transplant Donor: Scientific, Social and Ethical Perspectives*, Wiley, Chichester, 1987.

16. K.J.S. Anand and P.R. Hickey, *N. Engl. J. Med.* **317**(21) (1987).

17. M.J. Flower, *J. Med. Philos.* **10**(7), 237–251 (1985).

18. M.A. Warren, *The Monist* **57**, 1973, reprinted in Dwyer and Feinberg (10, pp. 59–74). Page references are to this edition.

19. D. Marquis, *J. Philos.* **86**(4), 183–202 (1989), reprinted in Dwyer and Feinberg (10, pp. 24–39). Page references are to this edition.

20. J. Feinberg, in T. Regan, ed., *Matters of Life and Death: New Introductory Essays in Moral Philosophy*, 2nd ed., Random House, New York, 1986, pp. 256–293.

21. M. Tooley, *Abortion and Infanticide*, Clarendon Press, Oxford, UK, 1983.

22. J. Feinberg, in W.T. Blackstone, *Philosophy & Environmental Crisis*, University of Georgia Press, Athens, 1974.

23. S. Buckle, *Bioethics* **2**(3), 227–253 (1988).

24. J.W. Knight and J.C. Callahan, *Preventing Birth: Contemporary Methods and Related Moral Controversies*, University of Utah Press, Salt Lake City, 1989.

25. S.I. Benn, in J. Feinberg, *The Problem of Abortion*, 2nd ed., Wadsworth, Belmont, CA, 1984, pp. 135–144.

26. J. Harris, *The Value of Life: An Introduction to Medical Ethics*, Routledge & Kegan Paul, London, 1985.

27. R. Hursthouse, *Beginning Lives*, Basil Blackwell, Oxford, UK, 1987.

28. R. Warner, *Soc. Theory Pract.* **3**(4) (1974), revised and reprinted in R.A. Wasserstrom, ed., *Today's Moral Problems*, 2nd ed., Macmillan, New York, 1979.

29. P. Singer and K. Dawson, *Philos. Public Affairs* **17**(2), 87–104 (1988).

30. D. Marquis, in R. Edwards and E.E. Bittar, eds., *Bioethics for Medical Education*, JAI Press, Greenwich, CT, 1999.

31. P. Singer, *Practical Ethics*, Cambridge University Press, Cambridge, UK, 1979.

32. L.W. Sumner, *Abortion and Moral Theory*, Princeton University Press, Princeton, NJ, 1981.

33. P. Singer, *Animal Liberation*, Avon Books, New York, 1975.

34. T. Regan, *The South. J. Philos.* **14**, 485–498 (1976).

See other entries GENE THERAPY, ETHICS, GENE THERAPY FOR FETUSES AND EMBRYOS; see also REPRODUCTION, ETHICS entries; REPRODUCTION, LAW, REGULATION OF REPRODUCTIVE TECHNOLOGIES.

REPRODUCTION, ETHICS, PRENATAL TESTING, AND THE DISABILITY RIGHTS CRITIQUE

ERIK PARENS
The Hastings Center
Garrison, New York

ADRIENNE ASCH
Wellesley College
Wellesley, Massachusetts

CYNTHIA POWELL
University of North Carolina
Chapel Hill, North Carolina

OUTLINE

TECHNOLOGICAL CONTEXT

History of Prenatal Testing

The oldest reference of prenatal diagnosis is for anencephaly (absent skull and brain) diagnosed by X rays in 1917 by James T. Case. The first amniocentesis is attributed to Schatz in 1883 for the purpose of treating hydramnios (excessive amniotic fluid) (1). Amniocentesis to detect erythroblastosis fetalis (complication of Rh incompatibility) began in the 1950s. Prenatal testing with amniocentesis for genetic disorders began in 1955 when it was discovered that the sex of a human fetus could be predicted by analysis of fetal cells in amniotic fluid (2–5). Initially the testing was used to determine fetal sex when a woman was at risk for having a child with an X-linked condition such as hemophilia (6). In this situation she would have a fifty percent risk of having an affected male infant. If the baby were female, she would have a fifty percent risk of carrying the gene, but would be clinically unaffected. The sex chromatin body (Barr body) could be identified in nondividing amniotic fluid cells. A male fetus (absent Barr body) could be identified and the pregnancy terminated despite a 50 percent probability that it would be unaffected. The first report of the procedure being done for this purpose was from Denmark (7).

It was reported in 1959 that Down syndrome is due to an extra chromosome 21 and use of amniocentesis to detect fetal chromosome abnormalities began in the 1960s (8–11). Widespread use of amniocentesis for increased maternal age in the United States is partly attributed to lawsuit settlements in the late 1970s in cases where the patients had not been referred for testing and gave birth to children with disabilities. In 1983 the American Academy of Pediatrics and the American College of Obstetricians and Gynecologists recommended that all women over the age of 35 be offered amniocentesis; thus amniocentesis became a routine part of obstetric care (12). Other contributing factors to the rapid increase in demand for prenatal diagnostic services include the liberalization of abortion statutes, changes in cultural attitudes toward family size, and extensive media coverage and publicity given to prenatal diagnosis (13).

Chorionic villus sampling (CVS), aspiration of tissue that will become the placenta, was first done in Copenhagen in the late 1960s (14,15). Because of various technical problems it did not come into common use until the 1980s (16). This test has the advantage of being performed in the first trimester of pregnancy (9–13 weeks) as compared to amniocentesis, which is done in the second trimester (15–18 weeks). However, the spontaneous abortion rate following the procedure is approximately 1 percent, which is higher than for amniocentesis (1 in 300–400). There is also a higher rate of mosaicism (two or more cell lines with different chromosomal constitution) with CVS as compared to amniocentesis (0.5–1 percent for cultured CVS cells vs. 0.2–0.3 percent for amniocytes) (17). In most cases the abnormal cell line is confined to the placenta, so a follow-up amniocentesis is often needed to look for the possibility of true mosaicism in the fetus. CVS performed earlier than nine weeks may be associated with an increased risk of limb and facial malformations. And different from amniocentesis, it is not possible to measure alpha-fetoprotein levels with CVS; as a result CVS cannot be used to detect neural tube impairments. Moreover, because CVS is a technically more difficult procedure than is amniocentesis, fewer obstetricians offer CVS in their office. Not all tertiary care facilities or major medical centers offer it.

Screening for neural tube impairments was first demonstrated in 1972 by determining amniotic fluid concentrations of alpha-fetoprotein (AFP). Elevated levels were associated with spina bifida and anencephaly in the fetus (18). Elevated levels of AFP were also found in maternal blood serum when the fetus had an open neural tube (19). By the 1980s most women were being offered serum AFP screening during pregnancy. Maternal serum AFP (MSAFP) screening, ultrasound and amniocentesis are capable of detecting 100 percent of cases of anencephaly and 80 to 90 percent of cases of spina bifida (20).

In 1984 it was reported that AFP values in maternal serum were lower than expected when the fetus had Down syndrome or trisomy 18 (21). AFP measurement along with age of the mother were used to detect pregnancies with an increased likelihood of a chromosome abnormality in women under the age of 35. This technique could detect 20 percent of cases of Down syndrome in this population who otherwise would not be identified as having an increased risk. Other biochemical markers were identified which varied from normal when a woman was carrying a fetus with Down syndrome or trisomy 18. These include unconjugated estriol and human chorionic gonadotropin. Use of these three markers to screen pregnancies is commonly referred to as the "triple screen" and is generally done at the sixteenth to eighteenth week of pregnancy. Abnormal values are followed by ultrasound to confirm dating. If dating is correct, amniocentesis is offered for definitive testing for chromosome abnormalities. Detection rates with use of the triple screen have been reported as 67 to 75 percent, with a false positive rate of 4 to 5 percent (22).

Ultrasound was first developed in World War I as sonar, to look for underwater submarines. Using it to look at fetuses began in the 1960s (23). Ultrasound is used for determining gestational age, locating structures prior to invasive testing procedures, and identifying structural abnormalities in the fetus. Conditions such as anencephaly, spina bifida, congenital heart defects, hydrocephalus, and kidney abnormalities are detectable with ultrasound. Most women in this country have at least one ultrasound during their pregnancy. Studies have not shown a risk to the fetus as a result of ultrasound, but pregnancy outcomes have not been shown to improve through routine ultrasound use.

Fetoscopy, the direct visualization of the fetus by using an optical instrument, was first used in 1954 but is now rarely done except for fetal skin biopsy. Fetoscopy was used in the past to obtain fetal blood samples, but percutaneous umbilical blood sampling (PUBS) — also known as cordocentesis is currently the method of choice. In this procedure ultrasound guides placement of a needle inserted through the maternal abdomen into the umbilical vein. PUBS is used for diagnosis of chromosome abnormalities, fetal infections, coagulation defects, hemoglobin and red cell disorders, metabolic and immunologic diseases.

The first reported prenatal diagnosis established by molecular genetic techniques was for alpha-thalassemia by means of linkage analysis in 1976, and for sickle cell anemia by analysis of gene mutation in 1978.

Today's Prenatal Tests

Prenatal diagnosis is available for hundreds of genetic conditions including chromosome abnormalities, inborn errors of metabolism, neural tube impairments, and single gene disorders. Ultrasound detects many different structural defects including hydrocephalus, congenital heart impairments, limb anomalies, skeletal dysplasias, and diaphragmatic hernias. In the most recent edition of Catalog of Prenatally Diagnosed Conditions there are 940 conditions listed that have been diagnosed prenatally. These include chromosome abnormalities, congenital malformations, dermatologic disorders, fetal infections, hematologic disorders, inborn errors of metabolism, tumors and cysts, and multiple congenital anomalies of unknown etiology.

Methods of prenatal diagnosis include maternal serum screening for biochemical markers to look for neural tube impairments as well as chromosomal aneuploidies (i.e., abnormal numbers of chromosomes). The triple screen detects up to 75 percent of fetuses with Down syndrome. However, these are screening tests and must be followed by additional procedures such as ultrasound in the case of neural tube impairments and amniocentesis or chorionic villus sampling for chromosome abnormalities.

Data from the Council of Regional Networks for Genetic Services (CORN) for 1989 estimated that 50 percent of pregnancies were being screened for MSAFP, this number is most likely now increased. The 1989 data from this group showed that increased maternal age was the most common indication for prenatal tests (62 percent), with abnormal MSAFP accounting for 14 percent, positive family history in 7 percent, previous spontaneous abortion or stillbirth 1 percent, abnormal ultrasound 1 percent, parental concern or anxiety 1 percent. "Other" and "unknown or unrecorded" accounted for an additional 11 percent (24). In a survey of a university-based cytogenetics laboratory over a five month period in 1997, of 476 amniotic fluid samples, 52 percent were obtained for increased maternal age, 34 percent were for abnormal triple screen, 7 percent were for abnormal ultrasound, 3 percent were for a family history of a genetic disorder, 1 percent were for DNA testing, and the remaining 3 percent were for elevated MSAFP, multiple miscarriages, and maternal anxiety. The DNA diagnostic testing included testing for achondroplasia, Rh, Kell antibody, sickle cell, and X-linked hydrocephalus.

Who Is Offering Tests

Most prenatal genetic testing is obtained through obstetricians in private practices or public health clinics. The remainder is obtained through tertiary referral centers such as university medical centers or private genetic centers. In a U.S. study that looked at discussions between obstetricians or nurse-midwives and their patients during their first prenatal visit, it was found that time devoted to discussion of genetic testing averaged 3.7 minutes ±3.9 minutes (range 0–25.3 minutes). A comprehensive family history was not taken in any of the visits. Discussion of topics such as abortion or continuation of pregnancy if an anomaly were detected, or a description of the disorders for which testing was offered occurred in a minority of visits (24). Guidelines for Perinatal Care (1997) from the American Academy of Pediatrics and American College of Obstetrics and Gynecology does not recommend referral for genetic counseling for women of advanced maternal age because it states that the primary care physicians can explain the risks (25).

When an abnormality is detected through prenatal testing, information and counseling usually comes from an obstetrician, a genetic counselor, or a nurse. The information given varies depending upon the knowledge and experience the professional has of the specific condition. Many health professionals who offer prenatal genetic counseling have not had direct contact with children and adults with developmental disabilities and genetic disorders. Obstetricians often have had little contact with such patients since their medical school training, and even then this contact may have been minimal. Few genetic counselors work in both prenatal and pediatric genetics. Few genetic counseling training programs give students an opportunity to work with developmentally disabled children and adults. Therefore, most women who receive information about a specific chromosome abnormality or genetic impairment in their fetus receive this information from health care providers without personal knowledge of the natural history and outcomes of the condition.

SOCIAL CONTEXT OF THE DISABILITY RIGHTS CRITIQUE

Outright Discrimination Against and Unexamined Attitudes About People With Disabilities

The history of discrimination against people with disabilities, including episodes of infanticide and compulsory sterilization, is long, ugly, and well documented (26–28). Even with such important steps as the passage of the Americans with Disabilities Act (ADA) and the Individuals with Disabilities Education Act (IDEA), discrimination is far from over. People with disabilities are still often treated as inferior to nondisabled people. As disability studies scholar Lennard Davis has observed, even the most educated of Americans, professors who make a living by writing about the nature of discriminatory practices and who decry discrimination against women, people of color, and other minorities, leave their attitudes toward people with disabilities largely unexamined. According to Davis, in the writings of these literary theorists, while "others" whose bodies are normal become vivid, others whose bodies are abnormal remain invisible (29).

It is not just practitioners of fashionable literary theory who sometimes harbor unexamined and discriminatory attitudes toward people with disabilities. The bioethics and medical literatures of the last decade too reveal misinformation and stereotypic thinking about what disability means for individuals, families, and society. Many clinicians and bioethicists take it for granted that health status is mostly responsible for the reduced life chances of people with a disability, largely ignoring the role of societal factors such as educational and employment discrimination. Furthermore, these clinicians and bioethicists often discount data indicating that people with disabilities and their families do not view their lives in solely or even predominantly negative terms (30); instead, they may insist that such data reflect a denial of reality or an exceptional ability to cope with problems (31–32).

People who make policy concerning the dissemination of genetic information have reached a consensus that the purpose of prenatal testing is to enhance reproductive choice for women and families—not to decrease the number of children with disabilities who are born. Some have acknowledged, however, that there is a tension between the goals of enhancing reproductive choice and preventing the births of children who would have disabilities. Writing about screening programs for cystic fibrosis in the pages of the *American Journal of Human Genetics*, medical geneticist A.L. Beaudet observed: "Although some would argue that the success of the program should be judged solely by the effectiveness of the educational programs (that is, whether screenees understood the information), it is clear that prevention of CF is also, at some level, a measure of a screening program, since few would advocate expanding the substantial resources involved if very few families wish to avoid the disease (33, p. 603). Beaudet acknowledges that, in tension with the genetic professional's stated goal of educating individuals (without any investment in the particular decision those individuals might reach), those who pay for such education do so in part with a view to reducing the number of—and costs associated with—children born with cystic fibrosis.

The profession of genetic counseling is based on a deep commitment to helping clients discover what course of action, upon reflection, is best for them. Some evidence suggests, however, that when disabilities are involved, both trained genetic counselors and others who deliver genetic information do not always live up to that commitment. A recent study designed to understand the experience of mothers who received a prenatal diagnosis of Down syndrome and chose to continue the pregnancy found problematic attitudes toward people with disabilities, evidenced in the way that medical professionals spoke to those prospective mothers. According to David T. Helm, one of the mothers who received a diagnosis of Down syndrome reported the following exchange:

> Obstetrician: *You have to move quickly. There is a doctor at [Hospital X] who does late-term abortions.* Mother: *No, I told you I'm not going to have an abortion.* Obstetrician: *Talk to your husband. You might want to think about it* (34, p. 57).

Because Helm only provides this portion of a longer exchange, the reader cannot confidently interpret the exchange he reports. Advising a patient to discuss a major life decision with her spouse is not prima facie problematic, much less discriminatory. According to Helm's interpretation (and the interpretation of the Disability Rights Community), however, these words reveal the physician's unwillingness or inability to respect this woman's already stated decision to continue the pregnancy with the fetus carrying a disabling trait. The reported exchange provides no evidence that this obstetrician understands the ways in which many families welcome and nourish—and are nourished by—children with Down syndrome.

Research has shown that obstetricians may be more likely than genetic counselors to urge particular actions upon their patients (35,36). Helm's study also reports, however, that some genetic counselors reacted negatively to women who intended to bear and raise children with Down syndrome. A woman who was told that the fetus she was carrying would have Down syndrome reported the following: "[The genetic counselor] treated me as though I couldn't accept this news, although I told her I could. She asked, 'What are you going to say to people when they ask you how you could bring a child like this into the world?'" (34, p. 57) Those words suggest that this counselor has not thought deeply about what disabilities mean for individuals who live with them and for their families. At least from what we learn of her from Helm, she does not seem to appreciate that welcoming a child with Down syndrome into a family is not a decision that needs to be defended; she does not seem to appreciate that parental attitudes differ, that traits that matter a great deal to one couple may seem inconsequential to another. Such exchanges are probably not rare exceptions; similar examples can be found in other discussions of genetic counseling practices in the prenatal testing situation (37,38).

Nonetheless, many genetic counselors and physicians work extremely hard to live up to the central genetic counseling values of informed consent and nondirectiveness, and many of them are not only aware of but share the concerns voiced by the disability rights community. For example, at the New England Medical Center, women whose fetuses are diagnosed with Down syndrome are routinely scheduled to meet with a pediatric medical geneticist and a nurse clinician who specializes in the care of pediatric genetic patients. These women are scheduled to meet with pediatricians who specialize in genetics rather than obstetricians because pediatric geneticists understand better how Down syndrome influences the lives of children and their families. According to Dr. Diana Bianchi, who practices at the New England Medical Center, every attempt is made to introduce the pregnant woman and her partner to families who are raising infants, children, and/or young adults with Down syndrome. She reports that in her practice, only 62 percent of women who discover they are carrying a fetus with Down syndrome decide to have abortions. That rate of abortion upon a positive finding is believed to be relatively low. Disability critics point to such facts to suggest that when prospective parents obtain more accurate information about what life with disability is like, many realize that parenting a child who has a disability can be as gratifying as parenting a child who does not.

The disability critique proceeds from the view that discrimination results when people in one group fail to imagine that people in some "other" group lead lives as rich and complex as their own. The disability rights critics believe that everyone from literary theorists to bioethicists to obstetricians and genetic counselors are susceptible to such failures of imagination. Moreover they think that the desire of prospective parents to avoid raising children with disabilities may depend on that same failure.

Plurality of Disabling Traits and Plurality of Attitudes Toward Prenatal Diagnosis

In thinking about the meaning of using prenatal diagnosis to detect disabling traits, it is important to notice that the class of "disabling traits" is exceedingly heterogeneous. Prenatal diagnosis can now detect conditions as different as Lesch-Nyhan syndrome and ectrodactyly (a trait involving a partial fusion of the bones of the fingers and toes). Further, not only are the traits heterogeneous, but so are perceptions of their significance and/or seriousness. Nancy Press's research reveals that some generalizations can be made about what people take to be "serious": for example, mothers considering prenatal testing are most fearful of conditions like Lesch-Nyhan, which results in early and painful death (39). But as the infamous Bree Walker Lampley case indicates, there is debate about

the seriousness of ectrodactyly. In 1991 Bree Walker Lampley, a television news woman in Los Angeles who had ectrodactyly, discovered that the fetus she was carrying had the trait and, when asked, made it known that she had no interest in terminating for such a trait; some suggested that it was "irresponsible" to bring a child into the world with such a serious trait (40). Indeed, the research of Dorothy Wertz and colleagues suggests that even genetics professionals have very different ideas of what is and what is not "serious" (41). In one of Wertz's surveys, cleft lip/palate, neurofibromatosis, hereditary deafness, insulin-dependent diabetes, Huntington's disease, cystic fibrosis, sickle cell anemia, Down syndrome, and manic depression were deemed serious by some professionals and not serious by others (42).

A similar plurality of views exists within the disability community. Many groups representing people with disabilities, such as the National Down Syndrome Congress and Little People of America, have position statements affirming the value of life with disability for individuals and families (43,44). However, there is nuance and disagreement among groups, and in fact within some groups. This complexity is suggested by attitudes within the membership of Little People of America. Many of those who live with achondroplasia are concerned that prenatal testing, which can identify heterozygotes (i.e., fetuses that will develop into long-lived people with achondroplasia) will be used to obliterate the Little People of America community. In fact some members of that community might use the technology to select for the trait. Nevertheless, many couples who are heterozygous for achondroplasia would like to use prenatal testing to identify fetuses that are homozygous for the allele associated with achondroplasia. Homozygous achondroplasia is a uniformly fatal condition, and they would like to spare themselves the experience of bearing a child who will soon die. Adding to the complexity, some people with disabilities would use prenatal testing to selectively abort a fetus with the trait they themselves carry—and some people who would not abort a fetus carrying their own disability might abort a fetus if it carried a trait incompatible with their own understanding of a life they want for themselves and their child.

A similar diversity of views toward prenatal testing and abortion can be found among parents raising a child with a disability. Many such parents do not use prenatal diagnosis to determine whether their present fetus is affected (45). The reasons for this are many; to some, the trait has come to be unimportant or irrelevant. Some may refuse it on the ground that using the technology would say something hurtful to or about their existing child. Other parents of children with disabilities decide to use these technologies.

The point about the plurality of traits and attitudes toward testing is not to suggest that the terrain is too complex to be amenable to policy response. The point is simply that people committed to ending discrimination and improving life for people who have disabilities are not monolithic on the prenatal testing issue, any more than all feminists are monolithic on a host of "women's issues" or than members of racial minorities are monolithic in their stance toward affirmative action or other practices that affect them. Comprehensive evaluations of prenatal genetic testing will have to take such complexities into account.

Reproductive Liberty Premise

The proliferation of prenatal genetic testing occurs against the background of the controversy about abortion. Prenatal testing for genetic disability elicits unexpected responses from both sides of the abortion debate: Many of those who are uneasy with abortion based on a prenatal finding of a disabling trait are pro-choice. And many who, in general, are against the right to abortion nonetheless approve of abortions performed on a fetus carrying a disabling trait.

Virtually all the major work in the disability critique of prenatal testing emerges from those who are also committed to a prochoice, feminist agenda: Adrienne Asch, Marsha Saxton, Anne Finger, and Deborah Kaplan, for example (46–49). Other pro-choice feminists, including Ruth Hubbard, Abby Lippman, Carole Browner, and Nancy Press, draw on the disability critique to question the impact of prenatal testing (50–53). The shared premise of these scholars is that women (and men) have the right to determine when and how many children they will have; within the first two trimesters of pregnancy, abortion is a legally and morally defensible means of exercising that right.

What is new about prenatal testing is that it enables prospective parents to some extent to determine not only when and how many but also what kind of children they will have. With the exception of revealing the sex of the fetus, current prenatal testing is used to detect traits considered medically disabling—characteristics deemed undesirable or departures from species-typical functioning. In the future it may be increasingly possible to select for traits that we do value. That, however, is not the possibility that has motivated the disability critique; the motivation for the disability critique is the reality of using prenatal testing and selective abortion to avoid bringing to term fetuses that carry disabling traits.

ETHICAL ARGUMENTS OF THE DISABILITY RIGHTS CRITIQUE

As mentioned above, the number and variety of conditions for which prenatal genetic tests are available grows almost daily (54). Today we test for one trait at a time. In the future, however, with advances in diagnostic technology, it will be possible to test simultaneously for as many traits as one would like. In principle, it will be able to test for any trait one wishes that has been associated with any given allele. Not only will the cost of such testing likely decrease as the diagnostic technology advances, but advances in the technology will make it possible to do the testing earlier in the pregnancy. As mentioned earlier, one such technology will isolate the very small number of fetal cells that circulate in the maternal blood. Insofar as these earlier tests will be performed on fetal cells obtained from the mother's blood (rather than from the amniotic sac or chorionic villi) they will be minimally invasive. Thus it will be possible to do many more tests, at once, and with less cost to the pregnant woman in time, inconvenience, risk, or dollars, than is now the case (55).

As the ease of testing increases, so does the perception within both the medical and broader communities that prenatal testing is a logical extension of good prenatal care: The idea is that prenatal testing helps prospective parents have healthy babies. On the one hand, this perception is quite reasonable. Although no researcher has yet even attempted to correct a genetic impairment with in utero gene therapy, increasingly there are nongenetic approaches to such impairments. At the time of this writing, more than 50 fetuses have undergone in utero surgery to repair neural tube impairments (myleomeningoceles) (56). Moreover, negative (or reassuring) prenatal test results will reduce the anxiety felt by many prospective parents, and this in itself can be construed as part of good prenatal care. On the other hand, as long as in utero interventions remain relatively rare, and as long as the number of people seeking prenatal genetic information to prepare for the birth of a child with a disability remains small, prospective parents will use positive prenatal test results primarily as the basis of a decision to abort fetuses that carry mutations associated with disease and/or disability. Thus there is a sense in which prenatal testing is not simply a logical extension of the idea of good prenatal care.

Logical extension or no, using prenatal tests to prevent the birth of babies with disabilities seems to be self-evidently good to many people. Even if the testing will not help bring a healthy baby to term this time, it gives prospective parents a chance to try again to conceive. To others, however, prenatal testing looks rather different. A moment's reflection about the history of our society's treatment of people with disabilities makes it easy to appreciate why people identified with the disability rights movement might regard such testing as dangerous. Critics contend that prenatal diagnosis reinforces the medical model that disability itself, not societal discrimination against people with disabilities, is the problem to be solved. The charge that such testing is dangerous is supported by two, broad lines of argument. The first is that prenatal testing followed by selective abortion is *morally problematic*. The second line of argument is that the desire to undertake prenatal testing is based on *misinformation* about what disability is like for people with disabilities and for their families.

Prenatal Testing Is Morally Problematic

The disability critique holds that selective abortion after prenatal diagnosis is morally problematic, and for two reasons. First, selective abortion expresses negative or discriminatory attitudes not merely about a disabling trait, but about those who carry it. Second, it signals an intolerance of diversity not merely in the society but in the family, and ultimately it could harm parental attitudes toward children.

The Expressivist Argument. The argument that selective abortion expresses discriminatory attitudes has been called the *expressivist* argument (57). Its central claim is that prenatal tests to select against disabling traits express a hurtful attitude about and send a hurtful message to people who live with those same traits. In the late 1980s Adrienne Asch put the concern this way: "Do not disparage the lives of existing and future disabled people by trying to screen for and prevent the birth of babies with their characteristics" (58, p. 81). More recently, she has clarified what the hurtful or disparaging message is:

> As with discrimination more generally, with prenatal diagnosis, a single trait stands in for the whole, the trait obliterates the whole. With both discrimination and prenatal diagnosis, nobody finds out about the rest. The tests send the message that there's no need to find out about the rest (59).

Indeed, many people with disabilities, who daily experience being seen past because of some single trait they bear, worry that prenatal testing repeats and reinforces that same tendency toward letting the part stand in for the whole. Prenatal testing seems to be more of the discriminatory same: a single trait stands in for the whole (potential) person. Knowledge of the single trait is enough to warrant the abortion of an otherwise wanted fetus. On Asch's more recent formulation, the test sends the hurtful message that people are reducible to a single, perceived-to-be-undesirable trait.

This observation about letting the part stand in for the whole is surely enormously important. In everyday life, traits do often stand in for the whole, people do get looked past because of them. Indeed, one form of the expressivist argument has been regarded rather highly in another context. Many people who are concerned to support women's rights, have argued that prenatal sex selection is morally problematic because it embodies and reinforces discriminatory attitudes toward women (60). The sex trait is allowed to obliterate the whole, as if the parents were saying, "We don't want to find out about 'the rest' of this fetus; we don't want a girl."

Marsha Saxton has put the expressivist argument this way:

> The message at the heart of widespread selective abortion on the basis of prenatal diagnosis is the greatest insult: some of us are "too flawed" in our very DNA to exists; we are unworthy of being born. ...[F]ighting for this issue, our right and worthiness to be born, is the fundamental challenge to disability oppression; it underpins our most basic claim to justice and equality—we are indeed worthy of being born, worth the help and expense, and we know it (61).

And as Nancy Press has argued, by developing and offering tests to detect some characteristics and not others, the professional community is expressing the view that some characteristics, but not all, warrant the attention of prospective parents (62).

For several reasons, however, there is disagreement about the merit of the expressivist argument as a basis for any public policy regarding prenatal diagnosis of disability. Individual women and families have a host of motives and reasons for seeking out genetic information, and as James Lindemann Nelson and Eva Feder Kittay argue, it is impossible to conclude just what "message"

is being sent by any one decision to obtain prenatal testing (63,64). Acts (and the messages they convey) rarely have either a single motivation or meaning.

Some prospective parents no doubt have wholly negative attitudes toward what they imagine a life with a disability would be like for them and their child; others may believe that life could be rich for the child, but suspect that their own lives would be compromised. Others who have disabilities perhaps see passing on their disabling trait as passing on a part of life that for them has been negative. Parents of one child with a disability may believe that they don't have the emotional or financial resources for another. The point is that the meaning of prenatal testing for would-be parents is not clear or singular. In any case, those sympathetic to at least some forms of prenatal testing point out that prospective parents do not decide about testing to hurt existing disabled people but to implement their own familial goals. In that sense, there is no "message" being sent at all.

To many in the disability rights movement, however, regardless of the parental motive to avoid the birth of a child who will have a disability, the parent may still be letting a part stand in for the whole. That prospective parents do not intend to send a hurtful message does not speak to the fact that many people with disabilities receive such a message and are pained by it.

A second criticism of the expressivist argument is that it calls into question the morality of virtually all abortions. The argument presumes that we can distinguish between aborting "any" fetus and a "particular" fetus that has a disability—what Adrienne Asch has called the *any-particular distinction*. According to Asch, most abortions reflect a decision not to bring any fetus to term at this time; selective abortions involve a decision not to bring this particular fetus to term because of its traits. Pro-choice individuals within and outside the disability community agree that it is morally defensible for a woman to decide, for example, that she doesn't want any child at a given time because she thinks she's too young to mother well, or because it would thwart her life plan, or because she has all the children she wants to raise. The question is whether that decision is morally different from a decision to abort an otherwise-wanted fetus.

But it is not clear that the distinction is adequate. Sometimes the decision to abort "any" fetus can be recast as a decision to abort a "particular" fetus. James Lindemann Nelson, for example, argues that if parents of three children chose to end a pregnancy that would have produced a fourth child, such parents would not be making a statement about the worthwhileness of other families with four children, or about the worth of fourth-born children as human beings (64). Rather, they would be deciding what would be right for their particular situation. If, as Asch and others have argued, prenatal testing is morally suspect because it lets a trait stand in for the whole potential person, precisely the same argument would apply to aborting a fetus because it was the fourth child. The trait of being fourth-born makes the prospective parents ignore every other respect in which that fetus could become a child that would be a blessing to its family and community. Nelson's example of the potential fourth-born child suggests one reason to doubt the merit of the any-particular distinction; he thinks that the disability critics have failed to explain why traits like being fourth-born could be a legitimate basis for an abortion while disabling traits could not.

A third criticism of the expressivist argument is that it presumes that selective abortion based on prenatal testing is morally problematic in a way that other means of preventing disability are not. Such other means include, for example, taking folic acid to reduce the likelihood of spina bifida, or eschewing medication that is known to stunt the growth or harm the organs or limbs of a developing fetus. Such acts (or refraining from such acts) on the part of the pregnant woman are designed to protect the health of the developing fetus.

Disability critics hold, however, that abortion does not protect the developing fetus from anything. It prevents disability by simply killing the fetus. Proponents of this disability critique hold a strong prochoice position. Their objection is only to a certain way of using abortion.

But those from the mainstream prochoice community think of selective abortion in different terms. They do not see an important moral difference between selective abortion and other modes of preventing disability in large part because they do see an important moral distinction between a born child with a disabling trait and an embryo or fetus with a disabling trait. They argue that parents of all born children have an obligation to love and care for those children — regardless of their traits. They also argue, however, that the pregnant woman (and her partner) are not "parents" before the child is born. Just as a woman or couple may decide during the first two trimesters of any pregnancy that becoming a parent to a first child, or to any child, is not in accord with their life plans, so may they make the same decision on the grounds that the fetus has disabling traits. The woman may terminate the pregnancy and try again to become pregnant with a fetus that has not been identified as carrying a disabling trait. On this view, if it is reasonable to prevent disability in a developing child by adhering to a particular lifestyle, taking specified medications or refraining from taking others, it is equally acceptable to opt for abortion to prevent the birth of a child with a significant disability (65).

Even if expressivist arguments will not dissuade all people from using tests in making reproductive decisions for their own lives, there is widespread agreement that policies that would in any way penalize those who continue pregnancies despite knowing that their child will live with a disabling trait must be avoided. That is, there is widespread agreement that prospective parents who either forgo prenatal testing or decide that they want to continue a pregnancy despite the detection of a disabling trait should not have to contend with losing medical services or benefits for their child, nor feel obliged to justify their decisions.

The Parental Attitude Argument. The second argument that prenatal testing is morally problematic may be called the *parental attitude* argument. According to it, using prenatal tests to select against some traits indicates a problematic conception of and attitude toward parenthood. Part of the argument is that prenatal testing is rooted in

a "fantasy and fallacy" that "parents can guarantee or create perfection" for their children (58). If parents were to understand what they really should seek in parenting, then they would see how relatively unimportant are the particular traits of their children.

The parental attitude argument also involves the thought that in the context of prenatal testing, a part, a disability, stands in for the whole, a person. The prospective parent who wants to avoid raising a child with a diagnosable disability forgets that along with the disabling trait come other traits, many of which are likely to be as enjoyable, pride-giving, positive (and as problematic, annoying, and complicated), as any other child's traits. If prospective parents imagine that disability precludes everything else that could be wonderful about the child, they are likely acting on misinformation and stereotype.

According to the parental attitude argument, prospective parents should keep in mind that the disabling trait is only one of a fetus' characteristics. The activity of appreciating and nurturing the particular child one has is what the critics of selection view as the essence of good parenting. Loving and nurturing a child entails appreciating, enjoying, and developing as best one can the characteristics of the child one has, not turning the child into someone she is not or lamenting what she is not. If we were to notice that it is a fantasy and fallacy to think that parents can guarantee or create perfection for their child, if we were to recognize what is really important about the experience of parenting, we would see that we should be concerned with certain attitudes toward parenting, not with "disabling" traits in our children. Good parents will care about raising whatever child they receive and about the relationship they will develop, not about the traits the child bears. In short, what bothers those wary of prenatal diagnosis is what might be called "the selective mentality." The attention to particular traits indicates a morally troubling conception of parenthood, a preoccupation with what is trivial and an ignorance of what is profound.

Those who connect acceptance of disability to what is desirable in any parent–child relationship worry that our attitudes toward parenthood and ultimately toward each other are changing as a result of technologies like prenatal diagnosis (66,67). Do these technologies lead us, one might ask, toward the commodification of children, toward thinking about them and treating them as products rather than as "gifts" or "ends in themselves"? Is it making us as a society less resilient in the face of the inevitable risks that our children face, and less willing to acknowledge the essential fragility of our species? When members of our society are confronted with, for example, sex selection or with the possibility of selecting for non-health-related traits like sexual orientation, they often raise concerns about the selective mentality. Indeed, those who want to reject the parental attitude argument in the context of disabling traits should recognize that they are criticizing an argument that they themselves may sometimes use in the context of non-health-related traits. Certainly many worry about the cumulative effect of individual choices, about the technologization of reproduction, and about a decreasing cultural ability or willingness to accept the reality of uncontrollable events. These concerns trouble even those who profess to be comfortable with genetic testing and selective abortion.

Nonetheless, many find significant problems with the parental attitude argument. One of the most important is that it makes what William Ruddick has called the "maternalist assumption," namely that "a woman who wants a child should want any child she gets" (68). Ruddick acknowledges that many women do hold "maternalist" conceptions of pregnancy and motherhood, out of which that assumption grows. But he argues that there are other legitimate conceptions of pregnancy and motherhood that do not depend on or give rise to the same assumption. He suggests that some prospective parents may legitimately adopt a "projectivist" or "familial" conception of parenthood, and that either of these views is compatible with trying to ensure that any child they raise has characteristics that accord with these parental goals. In the projectivist parent's understanding of child rearing, the child is a part of her parental projects, and within limits, parents may legitimately undertake to ensure that a child starts out with the requisites for fulfilling these parental hopes and aims. Ruddick is not claiming that projectivist parents could ignore a child's manifested commitments to things beyond the parents' life plans, but he is saying, for example, that, the parent passionate about music may legitimately select against a future child whose deafness would make a love of some forms of music impossible. If a hearing child turns out to be tone deaf and enthusiastic about rock collecting and bird watching but not music, and if the parent views these activities as inimical to her parental values or projects, she need not support them, or (within limits) allow other people to do so.

According to Ruddick, the "familial" conception of parenthood highlights a parent's vision of her child as herself a parent, sibling—a participant in a nuclear and extended family that gives central meaning to life. For example, parents whose dreams of child rearing include envisioning their own child as a parent would be acting consistently with their conception of parenthood if they decided not to raise a boy with cystic fibrosis, whose sterility and shortened life span might preclude either biological or adoptive parenthood. A child of such a parent might, of course, reject family life in favor of solitude or communal adult companionship, but in using available technology to avoid raising a child who would never be able to fulfill a deeply cherished parental dream, the parent is acting in accordance with a legitimate conception of parenthood.

Though many share the disability community's concern that prenatal testing may threaten our attitudes toward children, parenthood, and ultimately ourselves, arguments such as Ruddick's and the others mentioned above make it unlikely that such concerns can undergird specific policies regarding prenatal testing for disabling traits.

Prenatal Testing Is Based on Misinformation

The second major claim of the disability critique is that prenatal testing depends on a misunderstanding of what life with disability is like for children with disabilities and their families. Connected with this claim is the question whether disability is one more form of "neutral" human variation, or whether it is different from variations usually thought of as nondisabling traits, such as eye color, skin color, or musicality.

There are many widely accepted beliefs about what life with disability is like for children and their families. Most of these beliefs are not based on data. They include assumptions that people with disabilities lead lives of relentless agony and frustration and that most marriages break up under the strain of having a child with a disability. Recent studies suggest, for example, that many members of the health professions view childhood disability as predominantly negative for children and their families, in contrast to what research on the life satisfaction of people with disabilities and their families has actually shown (69–70). For example, disability researchers Philip Ferguson, Alan Gartner, and Dorothy Lipsky have reviewed empirical data on the impact of children with disabilities on families (71), and have concluded that the adaptational profiles of families that have a child with a disability basically resemble those of families that do not.

According to Ferguson, Gartner, and Lipsky's reading of the data, families that include disabled children fare on average no better or worse than families in general. Some families founder, others flourish. Ferguson, Gartner, and Lipsky do not deny that families are often distressed upon first learning that their child has a disability. And they acknowledge that families with children who evince significantly challenging behavior experience more disruption than do other families. But recent research on raising a child with a disability offers happier news for families than many in our society have been led to expect. The Ferguson, Gartner, and Lipsky review of scores of studies about family life where children have significant cognitive, physical, and sensory disabilities, behavioral and health problems, suggests that on average, families with and without disabled children fare about the same on such measures as parental stress, marital satisfaction, and family functioning.

Although families of children with a variety of conditions have been studied, families of children with Down syndrome have received the most extensive examination. Nonetheless, all of those studies concerned children whose conditions are incontrovertibly disabilities of some consequence. The findings indicate that challenging behavior of a child is much more likely to disrupt families and cause negative consequences than significant intellectual disability or health problems. While studies differ in methodology, population studied, questions pursued, and types of conclusions, according to Ferguson, Gartner, and Lipsky's interpretation, what most reviewers find is that the mild-severe continuum is not the important one in terms of family outcomes. Behavior of the child is a much stronger predictor of negative consequences than is intellectual or physical impairment.

Studies of family adaptation, too, have begun to recognize the prevalence of positive outcomes in many families (72,73). Indeed, one recent study found that parents of disabled adolescents reported more positive perceptions of their children than do parents of nondisabled adolescents (74).

In a 1995 study intended to learn how a child's disability affected the work lives of dual career families, the authors found that the needs and concerns of families with and without children with disabilities were "strikingly similar." They did, however, observe:

> What seems to distinguish families of children with disabilities from other working families is the intensity and complexity of the arrangements required to balance work and home responsibilities successfully. For example, parents of children with disabilities, particularly those with serious medical or behavioral problems, find it more difficult to locate appropriate, affordable child care. ...Similarly, these families are more dependent upon health insurance policies with comprehensive coverage (75, p. 511).

This same study also suggests, however, that a child's disability may sometimes alter the customary parent–child life cycle, in which parents gradually relinquish daily guidance and caretaking and — if they are fortunate — see their children take on adult productive and caretaking roles. Depending on the impairment and on the social arrangements that parents help a growing child construct, some people with disabilities may require their parents' help through adulthood in securing shelter, social support, and safety. Increasingly, adults with disabilities such as muscular dystrophy, spina bifida, cystic fibrosis, Down syndrome, and other conditions do not stay "eternal children," as they were once thought to do. Nonetheless, some, albeit small, portion of the population of disabled people will be more vulnerable for longer than others, and more in need of what Kittay (borrowing from Sara Ruddick) describes as "attentive, protective love" (76).

While it is important to demolish the myth that disability entails relentless agony for the child and family, there is still considerable disagreement about what conclusions to draw from the literature on the family impact of a child with disability. In the view of the disability community, this literature suggests that prenatal testing to select against disabling traits is misguided in the sense that it is based on misinformation. That is, if prospective parents could see that families with children who have disabilities fare much better than the myth would have it, then parents would be less enthusiastic about the technology.

However, recognizing that there are erroneous beliefs that need to be dispelled may not show that the desire for prenatal testing stems from misinformation alone. The first problem with the argument from misinformation has to do with the difference between retrospective and prospective judgments. It is one thing to look back on a stressful but ultimately rewarding experience and say, I'm glad I did that. It is another to look forward to the possibility of a stressful and perhaps ultimately rewarding experience and say, I'm glad to give it a try. To appreciate that many families respond well to stress does not commit one to thinking that it would be a mistake for families to try to avoid it. It may be true that, as one of the studies of working families points out, the concerns of working parents with disabled children very much resemble the concerns of any working parent — ensuring that children are safe, happy, stimulated, and well cared for at home, at school, and in after-school activities. But that study also acknowledges that working parents of children with special medical or behavioral needs find that meeting those needs takes more time, ingenuity, and energy than they think would have to be spent on the needs of nondisabled children. To appreciate that many families emerge stronger, wiser, and even better as a result of such an experience may not suggest that it is unreasonable or morally problematic to try to avert it.

Disability in Society. One of the most difficult issues that emerges in the argument from misinformation concerns what having a disability is "really" like for people themselves and for their families. Just how much of the problem of disability is socially constructed? Is it reasonable to say that in a differently constructed social environment, what are now disabling traits would become "neutral" characteristics?

Undoubtedly, more of the problem of disability is socially constructed than many people generally believe. But does that imply that having a characteristic like cystic fibrosis or spina bifida is of no more consequence than being left-handed or being a man who is five feet, three inches tall? According to the disability rights critique of prenatal testing, if people with disabilities were fully integrated into society, then there would be no need for the testing. In the world they seek to create, if a given health status turned out to be a handicap, that would be because of societal, not personal, characteristics; the appropriate response would be to change society so that the person could live a full life with a range of talents, capacities, and difficulties that exist for everyone. In a society that welcomed the disabled as well as the nondisabled, there would be no reason to prevent the births of people with traits now called disabling.

Those sympathetic to at least some forms of prenatal testing are struck by the fact that, for reasons that seem to be complex, members of the disability community speak at different times in different modes about the nature of disability. Sometimes, members of that community are clear about the fact that disabling traits have a "biological reality" or are not neutral. Adrienne Asch writes, "The inability to move without mechanical aid, to see, to hear, or to learn is not inherently neutral. Disability itself limits some options" (58, p. 73). At other times, however, and this is the mode usually emphasized in critiques of prenatal testing, those in the disability rights movement speak as if those traits indeed are inherently neutral. Thus Deborah Kent writes: "I premised my life on the conviction that blindness was a neutral characteristic (77). In this other mode, the disability community argument often is that, different from what prospective parents imagine, these so-called disabling traits are not, to coin a term, "disvaluable" in themselves; they are disvaluable because of the way they are socially constructed.

Nora Groce's work illustrates the point about how social arrangements shape whether a characteristic is disabling (78). In Martha's Vineyard in the nineteenth century, Groce argues, being unable to hear was not disabling because everyone spoke sign language. Groce's work establishes that much of what is difficult about having a disability stems from manifold facets of society, from architecture to education to aesthetic preferences. In choosing how to construct our societies, we do, as Allen Buchanan puts it, "choose who will be disabled" (79). We could choose differently than we have, and if we were to choose differently, what's disabling about what we now call disabilities would be largely eliminated. Plainly, then, the social constructionist argument is powerful. The objection concerns, rather, what appears to be a correlative claim of the disability position: that so-called disabling traits are neither disabling nor "disvaluable," but neutral.

Again, adherents of the disability critique acknowledge that some characteristics now labeled disabilities are easier to incorporate into today's society, or into a reconstructed society, than are others. Thus no one would deny that disabling traits—departures from species-typical functioning—foreclose some options, or that some disabilities foreclose more options than others. A child with Down syndrome may never climb Mount Rainier because his strength, agility, and stamina may preclude it; he may also never read philosophy because he does not have the skills to decipher abstract material. Granting that people who can climb mountains and read abstract papers derive enjoyment and meaning from such activities, then being foreclosed from them, not by one's own choice, is regrettable. The lack of possibility is widely seen as disvaluable. In addition these lacks of capacity stem from the characteristics of the individual who is not strong enough or agile enough to climb, or who is unable by any teaching now known to us to grasp complex abstract discourse. In that sense, disability community critics acknowledge that these facets of some disabilities are "real," inherent in the characteristic itself and not an artifact of any interaction with the environment. Even if all traits are to some extent "socially constructed," that is irrelevant to the fact that the existence of these traits forecloses for those who have them the opportunity to engage in some highly desirable and valuable activities; not being able to engage in those activities is disvaluable. To the extent that spina bifida, Down syndrome, blindness, or cystic fibrosis currently preclude people from undertaking some parts of life that people who do not have those traits might experience, the disability critique acknowledges that disability puts some limits on the "open future" (80,81) people seek for themselves and their children.

As Bonnie Steinbock argues, if we really thought disability "neutral," we would not work as we do to maintain, restore, and promote health in ourselves and others. We use medicine in the hope that it will cure or ameliorate illness and disability. We urge pregnant women to refrain from activities that risk harming the fetus. If we thought that disabilities were "neutral," then we could tell women who smoke or drink during pregnancy to rest easy, for developmental delay, low birth weight, and fetal alcohol syndrome would all be just "neutral variations," of no consequence to the future child (82).

While disability community critics acknowledge that some disabilities foreclose some opportunities, they also hold that calling attention to the foreclosure obscures two important points. The first is that rather than dwell on the extent to which opportunities to engage in some activities are truncated, we should concentrate on finding ways for people with disabilities to enjoy alternative modes of those same activities. Philip Ferguson puts it this way:

> The point is not so much whether ... a blind person cannot enjoy a Rembrandt ... but whether social arrangements can be imagined that allow blind people to have intense aesthetic experiences. ...People in wheelchairs may not be able to climb mountains, but how hard is it to create a society where the barriers are removed to their experiences of physical exhilaration? ...Someone with Down syndrome may not be able to experience the exquisite joy of reading bioethics papers and debating ethical theory, but ... that person can experience the joy of thinking hard about something and reflecting on what he or she really believes. ...The challenge is to create the society that will allow as many different paths as possible to the qualities of life that make us all part of the human community (83).

The second fundamental point is that rather than concentrate on the truncation or loss of some opportunities, our society generally—and prospective parents in particular—should concentrate on the nearly infinite range of remaining opportunities. Every life course necessarily closes off some opportunities in the pursuit of others. Thus, while the disability critics of prenatal diagnosis acknowledge that disability is likely to entail some amount of physical, psychological, social, and economic hardship, they hold that when viewed alongside any other life, on balance, life is no worse for people who have disabilities than it is for people who do not. No parent should assume that disability assures a worse life for a child, one with more suffering and less quality, than will be had by those children with whom she or he will grow up.

The claim then is that overall, there is no more stress in raising a child with a disability than in raising any other child, even if at some times there is more stress, or different stress. In that sense the disability community claims that disability is on balance neutral. Even here, however, many find that the terms "neutral" and "normal" are either inaccurate characterizations of disability or are being used in confusing ways. Specifically some worry that these terms are used sometimes only to describe or evaluate traits and at other times to describe or evaluate persons.

Evaluations of Traits Versus Evaluations of Persons. As already mentioned, the disability community itself sometimes speaks about the descriptive and evaluative senses in which disabling traits are not neutral, not normal. Legislation like the ADA could not exist without a recognition that in some sense disabling traits are neither neutral nor normal. Indeed, the societal provision of special resources and services to people with disabilities depends on noticing the descriptive and evaluative senses in which disabling traits are not neutral, and how the needs of the people who live with them are, descriptively speaking, not normal. Yet the recognition of the obligation to provide those special resources is rooted in a commitment to the fundamental idea that the people living with those traits are, morally speaking, "normal"; the people bearing the traits are evaluatively normal in the sense of deserving the normal respect due equally to all persons. Unequal or special funding expresses a commitment to moral equality. Recognizing the nonneutrality of the trait and the "abnormality" of the person's needs is necessary for expressing the commitment to moral equality and equal opportunity. There is nothing paradoxical about appreciating the descriptive sense in which people with disabling traits are abnormal while also appreciating the evaluative or moral sense in which they are normal.

Some who are sympathetic to prenatal testing worry that people in the disability community (as well as others) often conflate descriptive claims about traits and evaluative or moral claims about persons. For example, Deborah Kent, who is blind, writes:

When I was growing up people called my parents "wonderful." They were praised for raising me "like a normal child." As far as I could tell, they were like most of the other parents in my neighborhood, sometimes wonderful and sometimes very annoying. And from my point of view I wasn't like a normal child — I was normal (77).

What does Kent mean when she says that she "was normal"? As a descriptive claim, it is not reasonable to say that the trait of blindness is normal. Statistically speaking, it is not. Also, as an evaluative claim, insofar as the trait can make it impossible to enjoy some wonderful opportunities, it does not seem reasonable to say that the trait is neutral. The trait may indeed seem neutral and insignificant when viewed in the context of the whole person, but that is a claim about the person, not the trait. On the view of those sympathetic to testing, the descriptive and evaluative claims about the trait do not bear a necessary logical relation to evaluative claims about the person who bears it. As an evaluative or moral claim about the person, it makes perfect sense to say that a person who is blind is normal; she is normal in the sense that she deserves the normal, usual, equal respect that all human beings deserve.

But if it is easy to notice the difference between the descriptive and evaluative claims about traits and the evaluative claims about persons, why do people in the disability community (and others) keep slipping between the two? Erik Parens has suggested that there may be an important reason for this seemingly imprecise slipping (84). Discrimination against people with disabilities often involves a tendency to allow the part to stand in for the whole; perhaps members of the disability community sometimes succumb to a similar, equally problematic error. It could be that as the majority community sometimes uses the trait to deny the moral significance of the person, the disability community sometimes uses the moral significance of the person to deny the significance of the trait. The majority community slips from an observation about a trait to a claim about a person; the disability community slips from an observation about a person to a claim about a trait. At important moments, both groups fail to distinguish evaluations of traits from evaluations of persons. While such slippage may be easily committed in both communities, and

particularly understandable on the part of the disability community, it may be equally counterproductive in both.

Regardless of whether one is or is not persuaded by the disability community arguments regarding prenatal testing, it is important to remember that the disability community arguments are not intended to justify wholesale restrictions on prenatal testing for genetic disability. Rather, they are intended to make prospective parents pause and think about what they are doing, and to challenge professionals to help parents better examine their decisions. They are intended to help make the decisions of prospective parents thoughtful and informed, as opposed to thoughtless and automatic. As the prenatal testing technology marches forward, the need for thoughtful private and public conversations about its uses will become increasingly great. The disability rights arguments will be an invaluable resource in the promotion of those conversations.

BIBLIOGRAPHY

1. D.D. Weaver, *Catalog of Prenatally Diagnosed Conditions*, 3rd ed., Johns Hopkins University Press, Baltimore, MD, 1999.

2. D.M. Serr, L. Sachs, and M. Danon, *Bull. Res. Council Israel* **5B**, 137 (1955).

3. F. Fuchs and P. Riss, *Nature* **177**, 330 (1956).

4. E.L. Makowski, K.A. Prem, and I.H. Kaiser, *Science* **123**, 542 (1956).

5. L.B. Shettles, *Am. J. Obstet. Gynecol.* **71**, 834 (1956).

6. R. Schwartz Cowan, *Fetal Diagnostic Ther.* **8**(suppl. 1), 10–17 (1993).

7. P. Riis and F. Fuchs, *Lancet* **2**, 180 (1960).

8. J. LeJeune, M. Gauthier, and R. Turpin, *C. R. Acad. Sci.* **248**, 1721–1722 (1959).

9. M.W. Steele and W.R. Breg, *Lancet* **1**, 383 (1966).

10. H.L. Nadler, *Pediatrics* **42**, 912 (1968).

11. C.B. Jacobson and R.H. Barter, "Intrauterine diagnosis and management of genetic defects," *Am. J. Obstet. Gynecol.* **99**, 796–807 (1967).

12. American Academy of Pediatrics, American College of Obstetricians and Gynecologists, *Guidelines for Perinatal Care*, Washington, DC, 1983.

13. M.M. Kaback, *N. Engl. J. Med.* **289**, 1090–1091 (1973).

14. J. Mohr, *Acta Pathol. Microbiol. Scand.* **73**, 73–77 (1968).

15. N. Hahnemann, *Clin. Genet.* **6**, 294–306 (1974).

16. Z. Kazy, I.S. Rosovsky, and V.A. Bakharev, *Prenat. Diag.* **2**, 39–45 (1982).

17. Canadian collaborative CVS-Amniocentesis Trial Group, *Lancet* **1**, 1–7 (1989).

18. D.J. Brock and R.G. Sutcliffe, *Lancet* **2**, 197 (1972).

19. D.J. Brock, A.E. Bolton, and J.M. Monaghan, *Lancet* **2**, 923–924 (1973).

20. A. Milunsky, ed., *Genetic Disorders and the Fetus Diagnosis, Prevention, and Treatment*, 4th ed., Johns Hopkins University Press, Baltimore, MD, 1998.

21. R. Merkatz et al., *Am. J. Obstet. Gynecol.* **184**, 896 (1984).

22. A. Kolker and B.M. Burke, *Prenatal Testing, A Sociological Perspective*, Bergin and Garvey, Westport, CT, 1994.

23. F.J. Meaney, S.M. Riggle, and G.C. Cunningham, *Fetal Diagnosis Ther.* **8**(suppl.), 18–27 (1993).

24. B.A. Bernhardt et al., *Obstet. Gynecol.* **91**, 649–655 (1998).

25. American Academy of Pediatrics and American College of Obstetrics and Gynecology, *Guidelines for Perinatal Care*, 1997.

26. A. Gartner and J. Tome, eds., *Images of the Disabled: Disabling Images*, Praeger, New York, 1987.

27. J. Shapiro, *No Pity*, Times Books, New York, 1992.

28. J. West, ed., *The Americans with Disabilities Act: From Policy to Practice*, Milbank Memorial Fund, New York, 1991.

29. L.J. Davis, *Enforcing Normalcy: Disability, Deafness, and the Body*, Verso, London, 1995.

30. P. Ferguson, A. Gartner, and D. Lipsky, in E. Parens and A. Asch, eds., *Prenatal Testing and Disability Rights*, Georgetown University Press, Washington, DC, 2000.

31. J.E. Tyson and R.S. Broyles, *J. Am. Med. Assoc.* **276**, 492–493 (1996).

32. National Organization on Disability, *Survey of Americans with Disabilities*, 1998.

33. A.L. Beaudet, *Am. J. Hum. Genet.* **47**(4), 603–605 (1990).

34. D.T. Helm, S. Miranda, and N.A. Chedd, *Mental Retard.* **36**(1), 55–61 (1998).

35. B.A. Bernhardt et al., *Obstet. Gynecol.* **91**, 648–655 (1998).

36. T.M. Marteau, J. Kidd, and M. Plenicar, *J. Reprod. Infant Psychol.* **11**, 3–10 (1993).

37. C. Dunne and C. Warren, *Issues Law Med.* **14**(2), 165–202 (1998).

38. T. Marteau, H. Drake, and M. Bobrow, *J. Med. Genet.* **31**, 864–867 (1994).

39. N. Press et al., in S. Franklin and H. Ragone, eds., *Reproducing Reproduction: Kinship, Power, and Technological Innovation*, University of Pennsylvania Press, Philadelphia, PA, 1998.

40. A. Kolker and B.M. Burke, *Prenatal Testing: A Sociological Persepective*, Bergin and Garvey, Westport, CT, 1994.

41. B.M. Knoppers et al., *Am. J. Hum. Genet.* **57**(4, suppl.), A296, 1723 (1995).

42. D. Wertz, Abstract, National Society of Genetic Counselors meetings, October 1998.

43. National Down Syndrome Congress, *Position Statement on Prenatal Testing and Eugeneics: Families' Rights and Needs*, Available at: *http://members.carol.net/ndsc/eugenics.html*

44. Little People of America, *Position Statement on Genetic Discoveries in Dwarfism*, Available at: *http://www2.shore.net/~dkennedy/dwarfism-genetics.html*

45. D.C. Wertz, in H. Holmes, ed., *Issues in Reproductive Technologies I*, Garland Publishers, New York, 1992.

46. M. Fine and A. Asch, *Reprod. Rights Newslett.* **4**(3), 19–20 (1982).

47. M. Saxton, in E.H. Barucch, A.E. D'Amado, and J. Seager, eds., *Embryos, Ethics and Women's Rights*, Haworth Press, New York, 1988.

48. A. Finger, *Past Due: Disability, Pregnancy, and Birth*, Seal Press, Seattle, WA, 1987.

49. D. Kaplan, in K.H. Rothenberg and E.J. Thomson, eds., *Women and Prenatal Testing: Facing the Challenges of Genetic Testing*, Ohio State University Press, Columbus, OH, 1994.

50. R. Hubbard, *The Politics of Women's Biology*, Rutgers University Press, New Brunswick, 1990.

51. A. Lippman, *Am. J. Law Med.* **17**(1–2), 15–50 (1991).

52. C. Browner and N. Press, *Med. Anthropol. Q.* **10**(2), 141–156 (1996).

53. M.A. Field, *Harvard Women's Law J.* **16**, 79–138 (1993).

54. C.M. Powell, in E. Parens and A. Asch, eds., *Prenatal Testing and the Disability Rights*, Georgetown University Press, Washington, DC, 2000.

55. T.H. Murray, *The Worth of a Child*, University of California Press, Berkeley, CA, 1996.

56. D.W. Bianchi, T.M. Crombleholme, and M. D'Alton, *Fetology: Diagnosis and Management of the Fetal Patient*, McGraw-Hill, New York, forthcoming.

57. A.E. Buchanan, *Soc. Philos. Policy* **13**, 18–46 (1996).

58. A. Asch, "Reproductive Technology and Disability," p. 81.

59. A. Asch, in E. Parens and A. Asch, eds., *Prenatal Testing and the Disability Rights*, Georgetown University Press, Washington, DC, 2000.

60. D.C. Wertz and J.C. Fletcher, in H.B. Holms and L.M. Purdy, eds., *Feminist Perspectives in Medical Ethics*, Indiana University Press, Bloomington, IN, 1992.

61. M. Saxton, in R. Solinger, ed., *Abortion Wars: A Half Century of Struggle, 1950–2000*, University of California Press, Berkeley, CA, 1997.

62. N. Press, in E. Parens and A. Asch, eds., *Prenatal Testing and the Disability Rights*, Georgetown University Press, Washington, DC, 2000.

63. E.F. Kittay and L. Kittay, in E. Parens and A. Asch, eds., *Prenatal Testing and the Disability Rights*, Georgetown University Press, Washington, DC, 2000.

64. J.L. Nelson, in E. Parens and A. Asch, eds., *Prenatal Testing and the Disability Rights*, Georgetown University Press, Washington, DC, 2000.

65. B. Steinbock, in E. Parens and A. Asch, eds., *Prenatal Testing and the Disability Rights*, Georgetown University Press, Washington, DC, 2000.

66. T.H. Murray, *Worth of a Child*, University of California Press, Berkeley, CA, 1996.

67. A. Asch and G. Geller, in S.M. Wolf, ed., *Feminism and Bioethics: Beyond Reproduction*, Oxford University Press, New York, 1996.

68. W. Ruddick, in E. Parens and A. Asch, eds., *Prenatal Testing and the Disability Rights*, Georgetown University Press, Washington, DC, 2000.

69. J.A. Blier Blaymore et al., *Clin. Pediat.* **35**(3), 113–117 (1996).

70. M.L. Wollraich, G.N. Siperstein, and P. O'Keefe, *Pediatrics* **80**(5), 643–649 (1987).

71. P. Ferguson, A. Gartner, and D. Lipsky, in E. Parens and A. Asch, eds., *Prenatal Testing and the Disability Rights*, Georgetown University Press, Washington, DC, 2000.

72. D.A. Abbott and W.H. Meredith, *Family Rel.* **35**, 371–375 (1986).

73. A.P. Turnbull et al., eds., *Cognitive Coping: Families and Disability*, Paul H. Brookes, Baltimore, MD, 1993.

74. J.P. Lehman and K. Roberto, *Mental Retard.* **34**, 27–38 (1996).

75. R.I. Freedman, L. Litchfield, and M.E. Warfield, *Families in Society: J. Contemp. Hum. Services* (October), 507–514 (1995).

76. S. Ruddick, *Maternal Thinking: Toward a Politics of Peace*, Beacon Press, Boston, MA, 1989.

77. D. Kent, in E. Parens and A. Asch, eds., *Prenatal Testing and the Disability Rights*, Georgetown University Press, Washington, DC, 2000.

78. N.E. Groce, *Everyone Here Spoke Sign Language: Hereditary Deafness on Martha's Vineyard*, Harvard University Press, Cambridge, MA, 1985.

79. A. Buchanan, *Soc. Philos. Policy* **13**(2), 18–46 (1996).

80. D.S. Davis, *Hastings Center Rep.* **24**(6), 15–21 (1997).

81. R. Green, *J. Law, Med. Ethics* **25**(1), 5–16 (1997).

82. B. Steinbock, in E. Parens and A. Asch, eds., *Prenatal Testing and the Disability Rights*, Georgetown University Press, Washington, DC, 2000.

83. P. Ferguson, Personal communication.

84. E. Parens and A. Asch, *Hastings Center Rep.* **29**(5, special suppl.), S1–S22 (1999).

See other entries DISABILITY AND BIOTECHNOLOGY; GENETIC COUNSELING; see also REPRODUCTION, ETHICS entries; REPRODUCTION, LAW, IS INFERTILITY A DISABILITY?; REPRODUCTION, LAW, REGULATION OF REPRODUCTIVE TECHNOLOGIES; REPRODUCTION, LAW, WRONGFUL BIRTH, AND WRONGFUL LIFE ACTIONS.

REPRODUCTION, ETHICS, SEX SELECTION

BONNIE S. LEROY
University of Minnesota, Institute of Human Genetics
Minneapolis, Minnesota

DIANNE M. BARTELS
University of Minnesota, Center for Bioethics
Minneapolis, Minnesota

OUTLINE

Introduction

Background

Sperm Separation

Preimplantation Diagnosis

Fetal Sex Determination

Attitudes Toward Sex Selection

Case Study

Ethical Arguments for and Against Sex Selection

Summary

Acknowledgments

Bibliography

INTRODUCTION

There are basically two ways to approach the issue of sex selection. The first includes an assortment of methods with variable results that can be used to enhance the odds of achieving a conception of one sex over the other. None of these methods ensure that a pregnancy will result in the desired sex, rather they are thought to increase the chances. This approach is often referred to as *sex preselection* as the methods concerned are employed prior to conception. The second approach involves techniques used to determine the sex of an embryo or fetus by directly analyzing the genetic material. Both approaches elicit considerable ethical controversy, but the second is more controversial since a conception has already been achieved. The following discussion will first review the history of sex selection and then examine present and possible future technologies. We will summarize a variety of attitudes regarding sex selection, present a case study, and finally, consider a number of ethical arguments that have been advanced toward this subject.

BACKGROUND

Myriad arguments have been extended in support of and in opposition to sex selection. Prior to discussing the issues surrounding sex selection, it is important to look at the history and understand present and possible future technological interventions. Attempts to influence the conception of a desired sex extend far back in our history. Gledhill, in one paper, and Reubinoff and Schenker, in another publication, present useful portrayals of the early history of this practice (1,2). In early Greece one theory held that males developed on the right side of the uterus while females developed on the left. This led to the belief that the sex of an offspring could be controlled by the position of the woman during intercourse. Another belief was that the right testis produced male sperm while the left produced female sperm. This theory sustained for quite some time and in the eighteenth century led to the procedure of removing the left testicle so that one supposedly would be guaranteed a boy. Even Aristotle also had an opinion on this subject. He argued that the partner who was most active during intercourse would determine the sex of a child.

The Talmud suggests that one can influence the sex of offspring by one partner having an orgasm before the other. If a woman has an orgasm first, the child was said to be a male and if a man was first to have an orgasm, the child would be female. Diet was also, at times, thought to play a role. In the Middle Ages, wine and lion's blood and later in the early twentieth century, bitter and sour foods and a diet rich in red meat consumed by the mother were thought to enhance conception of a male. In 1917 it was reported that the right ovary contained male eggs and the left contained female eggs. Also it was thought that ovulation occurred in an alternate fashion, releasing a male egg from the right ovary one month and a female from the left the next. Therefore it was believed to be possible to predict the sex of a conception by counting the months since the last child.

In the 1950s reports began to emerge that supported a belief that one could influence the conception of a child of the desired sex with the timing of intercourse. This theory maintains that the male determining sperm swim faster but have a shorter survival time than the female determining sperm. Therefore intercourse close to the time of ovulation would result in a male, while intercourse days prior to ovulation would more likely result in a female. Many researchers later followed with reports on the effects that timing of conception had on controlling the determination of sex. Some research also included enhancement techniques such as variations in vaginal penetration and douching with an alkaline solution prior to intercourse. Most of these methods have been alternatively supported and disputed with many attempts to reproduce the results.

Although today much skepticism surrounds these methods, it is clear from this history that considerable effort has been expended over centuries in the attempt to gain the ability to control the sex of offspring. This history provides some insight into the substantial role that sex of our offspring plays in reproductive issues and the personal identity of parents today.

Current technology involves three distinct approaches: the separation of a sperm sample prior to insemination, evaluation of the genetic material of an early embryo prior to implantation in the uterus, and evaluation of the sex of a fetus in an established pregnancy. Each approach will be discussed in the following text. Before examining the methods, it necessary to know that sex in humans is determined by the genetic constitution of the sperm cell that fertilizes the egg. Females normally have two chromosomes designated as Xs and males normally have one X chromosome and one Y. The combination of a normal X chromosome and a normal Y chromosome in a fertilized egg will result in the development of a male fetus and the presence of two normal X chromosomes will result in the development of a female. Females contribute an X chromosome in each oocyte while males produce sperm with either an X chromosome or a Y. A sample of sperm will normally, on average, be composed of equal numbers of X and Y bearing sperm. Given no abnormalities in the genetic material, the odds of conceiving a male versus a female are equal.

SPERM SEPARATION

Efforts to separate X from Y bearing sperm prior to insemination have revolved around various methods that utilize differences such as size, shape, density, charge, swimming characteristics, and DNA content (1–3). Most of these methods fail to produce consistent results and the research conclusions reported conflict with one another (4–6). Some researchers claim success rates of over 80 percent, while others are unable to replicate these results (7–10). However, some recent research appears to hold promise. Investigators using flow cytometry, a method utilizing the 2.9 percent difference in DNA content between an X bearing sperm and a Y bearing sperm, assert to be able to produce a sperm separation of 80 to 90 percent for X enriched samples and 65 to 70 percent for Y enriched samples (2,11,12). Because this method greatly reduces the number of sperm in the sample, it is necessary to couple it with in vitro fertilization (IVF) technology to achieve pregnancy. This makes the procedure complicated, time-consuming and expensive. As investigators become more proficient with this and other methods in the near future, it will likely be possible to effectively separate X from Y bearing sperm.

PREIMPLANTATION DIAGNOSIS

A second technique is preimplantation diagnosis, which is the evaluation of genetic material in an early embryo prior to implantation in the uterus (2,13). Eggs are harvested from the mother and IVF is performed. One or two cells are then taken from the early developing embryo for genetic analysis. This analysis can include chromosome studies to evaluate the number and structure of all of the chromosomes including those determining sex. In addition direct gene analysis can be performed for a limited number of genetic diseases. This technique is also complicated and expensive. Current use focuses mostly on the determination of the sex of an embryo to prevent a pregnancy with a sex-linked genetic disease, or to detect a known single gene disorder (14).

FETAL SEX DETERMINATION

The final group of techniques that can be used for sex selection involves the determination of fetal sex in an established pregnancy (15). Chorionic villus sampling (CVS) entails harvesting a small amount of early placental tissue containing fetal cells at about 10 weeks into the pregnancy. Chromosome analysis, as well as other biochemical and genetic tests, can be performed on these cells, and the sex of the fetus can be determined. Genetic amniocentesis is a procedure that has been used for quite some time to diagnose a pregnancy affected with a chromosomal or inherited disorder. Amniotic fluid is collected at about 15 to 16 weeks into a pregnancy, and this fluid contains fetal cells that can be analyzed. As in CVS, the sex of a fetus can be learned through chromosomal analysis from fetal cells present in the amniotic fluid.

Since CVS and amniocentesis directly analyze the genetic material that, among other things, determines the sex of the fetus, the accuracy of the testing is extremely high. In addition current prenatal ultrasonographic equipment provides the resolution to ascertain fetal sex with a high degree of precision. Ultrasound uses sound waves to provide a picture of the developing fetus. CVS, amniocentesis, and ultrasound are the most widely used procedures to determine the sex of a fetus, since they currently provide the greatest testing accuracy at the least cost.

ATTITUDES TOWARD SEX SELECTION

A number of studies describe the attitudes of specific groups toward sex selection by examining views of a number of diverse cultures, religions, and professional providers. Most studies separated the use of sex selection for prevention of the birth of a child with a sex-linked genetic disease, from sex selection for the sole purpose of choosing the desired sex. A majority of groups surveyed were supportive of sex preselection and sex determination followed by abortion as options for the prevention of a sex-linked genetic disease in offspring. The following discussion will describe attitudes toward sex selection for nonmedical reasons, that is, for the sole purpose of choosing the desired sex in offspring.

In exploring preferences for a child of one sex over another, reports describe a decided cultural difference between Western views and those of Asia, India, and some other less industrialized societies. In the United States, studies showed that couples preferred a son as their firstborn and a daughter for their second child. The inclination was clearly toward a balanced sex composition in the family, although the majority of women in the United States do not approve of controlling the sex of their offspring (16). In the early 1960s unmarried college students were asked about sex preference if they were to have only one child. At that time, 91 percent of men and 66 percent of women stated that they would prefer a boy (16). In 1972 and 1987 similar surveys showed that the preference for boys had dropped. In 1972, 55 percent of all students would prefer their only child to be a male, and in 1987 that number was down to 52 percent in favor of males (17). In Great Britain one study asked over two thousand pregnant women if they preferred one sex to the other for their child. The majority, 58 percent, stated no preference while only 18 percent leaned toward a boy and 25 percent toward a girl. When asked if they wanted to know the sex of their baby before birth, 62 percent said no and 20 percent said they were unsure. In a small number of Afro-Caribbean and Asian respondents, there was a slight bias toward wanting boys (18). Another study in Great Britain looked at academic and nonacademic men and women between the ages of 18 and 20 years and found that over 75 percent of all respondents did not support the idea of choosing a baby's sex. No differences were found in the responses based either on sex or between the academic and nonacademic populations (19).

In 1993 the London Gender Clinic opened in London, England. Data gathered on couples who attended the clinic during the first 18 months of service provides some insight into those who would use this technology (20). The ethnic distribution among clients was 57.8 percent East Indian, 32 percent European, 3.6 percent Chinese, and 6.8 percent designated as other. Of all the couples participating in the clinic, 80.6 percent stated that they would have had another baby even if sex preselection were not an option. As expected, Asian and East Indian couples overwhelmingly wanted a boy, while European couples stated a slight preference (62.9 percent) for girls. The main reason given for wanting a girl was the desire of the mother to have a daughter. Couples from the Indian community repeatedly stated the need for a boy to carry on the family name for religious and social reasons. Most interestingly, a major reason for seeking these services involved wanting to avoid having a large family in order to get a son. This was important since 54 percent of the Indian couples in the clinic population already had 3 or more girls in their family and 94 percent had not yet had a son. The authors concluded from their experience that those interested in sex selection are mostly couples with two to three children of the same sex who want one last child of the opposite sex to complete their families. A study based in New York City considered which populations were using sex preselection technology and why. Out of 178 couples studied, 58 were from other countries. All non-American couples in the study wanted a boy, while the American couples wanted boys and girls with equal frequency depending on the sex of the children they already had. They expressed a desire to balance their family. In the non-American couples,

reasons for wanting a boy included the custom in their country for a son to support their parents at old age, the belief that it was essential for a male to run a family business, for inheritance purposes, the belief that males are more intelligent, and cultural pressures where males are preferred (21).

Canada assembled the Royal Commission on New Reproductive Technologies to make recommendations to the Canadian federal government by the end of 1993 (22). Before making their recommendations, they held public hearings and conducted random surveys involving more than 40,000 people. Decisions were made in light of research findings, public input, and a set of guiding principles including autonomy, equality, and noncommercialization of reproduction. In addressing the issue of sex selection for nonmedical reasons, they concluded, "The commission viewed sex selection for preference as contrary to its guiding principles, and to generally held Canadian values. Policies are needed to ensure that the values of citizens are respected." In Japan a similar national commission, the Japanese Medical Ethics Advisory Board, set guidelines limiting the availability of sex selection due to the concern that couples would overwhelmingly choose boys if reliable methods of sex preselection were available. However, it was noted that physicians are not legally bound by these guidelines (16). In the Netherlands in 1995, attempts to open the first private clinic to offer sex selection for nonmedical reasons failed due to opposition from doctors and politicians (23). The Dutch health secretary stated that he would take legal measures to ban it because, "the clinic's claims are ethically unjust." Physicians were mostly said to be opposed because, "it crosses the border of what is ethically acceptable in the Netherlands." The Royal Dutch Medical Association said that doctors should not cooperate with the clinic.

In India, son preference is so strong that the use of sex selection technologies followed by abortion is widely accepted. Between the years of 1978 and 1983, about 78,000 female fetuses were aborted following the use of sex selection technologies in India (24). Khanna published a study about the practice of sex selection in Shahargaon, a small village in north India (24). It dramatically demonstrates the effects of wide usage of sex selection procedures on a society. The sex composition of the children in this community for ages 0 to 5 years was found to be 691 females to 1000 males. Khanna, the author of this study, reports, "In this society, the birth of a son is considered an economic and political asset associated with the honor of a family, whereas a daughter is born as an expense and as a moral burden." Lobbying groups brought their concerns surrounding this practice to the Indian government and in response the government has attempted to regulate the use of prenatal diagnostic technologies. The results of these regulations has been an increase in cost for services by the clinics to offset the risks of practice, and an increase in the number of illegal "unregistered" clinics. One of many concerns in India is that a sex-selection industry is rapidly developing because of the great potential for profit.

Attitudes toward sex selection vary among different religions as well. Grazi, Wolowelsky, and Jewelewicz compared the position of traditional Jewish law with that of Roman Catholicism on the subject of assisted reproduction that included sex selection as one of the issues (25). In the Roman Catholic document *The Instruction on Respect for Human Life and the Dignity of Procreation*, reproductive technologies and IVF specifically are considered to be morally and absolutely illicit practices. This position includes the attempt to conceive a baby of a desired sex for any reason. The Catholic church's position articulates a belief held by many that a fetus has a right to life from the moment of conception. Termination for sex selection would be a most egregious violation of that right.

In Jewish law, gametes and unimplanted embryos have no standing. This would suggest that IVF, if used in treatment of infertility followed by implantation of some of the embryos, the selection for a characteristic such as sex may be allowed. However, the position on IVF for the sole purpose of sex selection is not at all clear. Rabbinic authorities who allow IVF, and presumably other technologies, are doing so in support of the religious obligation to procreate. Some authorities forbid IVF completely, and others preserve the halakhic imperative, which is to maintain natural marital relations. The attempt to conceive a child of the desired sex has been described as, "simply too frivolous a halakhic concern" (25). Neither religion supports sex selection for the sole purpose of choosing the sex of offspring.

Several studies have focused on the perspectives of professional care providers, especially on those who provide clinical genetics services. One study evaluated the attitudes toward sex selection of members of the American Society of Human Genetics, the International Fetal Medicine and Surgery Society, the Society of Perinatal Obstetricians, and selected ethicists and clergy with experience in biomedical issues (26). The majority of respondents in all groups considered sex selection ethically unacceptable. Agreement on this position was stronger regarding the use of sex selection in the second and third trimesters than in the first. The authors of this study indicate that one reason for opposition is a belief that gender is not a disease, and therefore this practice is a form of eugenics and ought not to be a part of health care.

Wertz and Fletcher used hypothetical cases to assess ethical decision making by medical geneticists (27). One case involved a choice about whether to perform prenatal diagnosis for sex selection. The responses were fairly equally split three ways among those who would offer the procedure, those who would refuse the procedure, and those who would refer the patient to another facility that does offer the procedure. Respect for patient autonomy was the reason given by 68 percent of those who stated they would offer the procedure. One interesting finding was that women were twice as likely as men to state that they would perform the procedure in respect for patient autonomy. Burke interviewed genetic counselors to determine attitudes towards fetal sex identification and abortion (28). All but one of the 32 genetic counselors who responded strongly opposed the use of prenatal diagnosis for sex selection. Burke noted that this position imposes stress upon the genetic counselors who also support patient autonomy through the ideal of nondirective

counseling and almost universally uphold a patient's right to an abortion. In a similar study of genetic counselor attitudes, Pencarinha, Bell, Edwards, and Best found comparable results (29). When presented with a hypothetical case, 38.3 percent of genetic counselors responding would perform genetic counseling for sex selection while 18.3 percent would refuse to be involved in a such a case and 43.4 percent would refer the couple to another center. Many of the respondents who would refuse the request for these services defended their position, "it is not a medical indication for testing and because prenatal diagnostic services are a limited resource." Genetic counselors who would offer the procedure maintain that the patient has the right to choose and support the patient's decision out of their duty to respect patient autonomy.

Wertz compared the views of geneticists in the United States with those in other European countries (30). As a group, geneticists in the United States were more willing to perform prenatal diagnosis for sex selection, or offer a referral for such services, than geneticists from any other country. Only 4 percent of the participants saw sex selection as having social consequences. Wertz reported that the participants focused on the particular family involved in the case, not on society as a whole. In an extension of this study using the same hypothetical cases, attitudes of genetics service providers from 30 provinces in China were studied (31). The majority, 89 percent, of the participants supported the Chinese laws on termination of pregnancy for genetic abnormalities and for population control and family planning considerations. However, more than half opposed the use of prenatal diagnosis for sex selection.

In summary, most cultural and religious groups surveyed, as well as health care professionals, opposed the use of reproductive technologies for the sole purposes of selecting the desired sex. The opposition was most strong when the result was the abortion of a fetus of the undesired sex.

CASE STUDY

It is sometimes helpful to use a real case to begin to think about how to develop an ethical position involved in a particular issue. The following is a true case that occurred at the University of Minnesota perinatal clinic in the mid-1980s.

AJ was a 35-year-old woman 15 weeks into her fourth pregnancy. She came to clinic with her husband seeking prenatal testing. They were both East Indian and had resided in the United States for about three years. An ultrasound study performed previous to this visit revealed that this was a twin pregnancy. AJ and her husband stated that they have had three previous pregnancies resulting in two normal healthy girls and one son, born in India, who died of a heart defect and many other birth defects. The cause of the anomalies was reported to be unknown and medical records were not available. The couple expressed the concern that if they had another son, he would also be affected because their girls were born healthy. They were also concerned about their age-related risk for having a child with a chromosome abnormality such as Down syndrome.

The genetic counseling session involved a discussion of age-related risks for chromosome abnormalities, the amniocentesis procedure, and the risks and limitations of the testing. Also discussed was the possibility that their son may have had a chromosome abnormality which, if it were to recur, would be detected with the amniocentesis. He also could have had an undefined genetic birth defect that could recur, but the ability to make a prenatal diagnosis would be limited to what could be seen by a level II ultrasound study.

The couple expressed a strong interest in the amniocentesis and the level II ultrasound study. The results of these studies were normal for both fetuses. When results were called to the couple, they stated that they wanted to know the sex of the fetuses. Both fetuses were female. These results were given to the couple at about 18 weeks gestation. Although they both previously expressed concern if one or both babies were male, they did not appear to be relieved with the test results, and the conversation was short.

About three weeks following the results discussion, AJ called the genetic counselor again. She was calling from a local abortion clinic. She was clearly distressed, and she wanted to know if the results of the chromosome studies could possibly incorrect. We discussed the fact that it was possible but very unlikely. Laboratory errors are rare, and the ultrasound study at the time of the amniocentesis agreed with the chromosome analysis regarding the sex of the fetuses and no abnormalities were seen.

AJ then revealed that she had no real choice in her decision to terminate the pregnancy. Although she very much wanted these babies, she had to terminate this pregnancy. She feared that her husband would leave her and her two daughters if she decided to continue the pregnancy. She had no formal education, no money, and no skills necessary for making a living. She had no way to support herself or her children without her husband. She had struggled with the possibility of having to make this decision since the time she learned of her pregnancy. She then confessed that there had never been a son with birth defects. They had invented this story to explain their interest in the sex of the fetuses, fearing that the clinic would not supply them with this information. Her role in the family was to provide a son and so far she had failed. In reviewing this case we wondered, had she and her husband come to clinic requesting prenatal diagnosis solely for sex selection, what would be the ethical arguments that would help to make a choice for or against providing this service?

ETHICAL ARGUMENTS FOR AND AGAINST SEX SELECTION

Many arguments have been offered in opposition to the use of reproductive technologies for the sole purpose of having a child of a desired sex. A majority point to the injustice of sex discrimination and the value of women. Other arguments in opposition include, but are not limited to, treating children as commodities, inappropriate use of medical technology, setting gender in the same category

as disease, the unbalancing of the sex ratio, and lack of respect for human life.

A few organizations have taken a position on this subject. In 1996 the committee on ethics of the American College of Obstetricians and Gynecologists published a position paper specifically on sex selection (32). The committee approved of the use of sex selection for the prevention of sex-linked genetic disorders but strongly rejected the practice of sex selection on demand for the sole purpose of having a child of the desired sex. The main argument given in defense of this position was that they felt this practice, "may reflect and encourage sex discrimination." The committee was concerned that physicians meeting these requests, "may ultimately support sexist practices." The Council on Ethical and Judicial Affairs of the American Medical Association in their position paper on ethical issues related to prenatal genetic testing also opposed the use of sex selection except when it is employed to prevent or treat genetic disease (33). The Council considered selection for sex as, "the most evident example of the discriminatory potential of selection for benign genetic traits." They go on to say that this procedure encourages the value of one sex over the other and places sex in the same category as disease. They argue that the practice of sex selection may result in harms to society including discrimination and the view that children are products. They view sex selection as a form of eugenics. The Turkish General Directorate of Mother and Child Health and Family Planning, analyzed the technical and ethical issues of sex selection (34). This report was explicit in that it emphasized that a baby should never be considered as a technological product. It stated that, "parents should not or any other authorities should not have a right to choose any physical or behavioural features of the baby unless an associated medical problem exists." The Ethical Committee of the Turkish Medical Association agreed that gender should never be treated as a disorder.

In 1985 H.B. Holmes wrote an extensive review of the available technology and ethical arguments and came to the conclusion that sex selection is the practice of eugenics (35). She is careful to acknowledge that for women in countries where females are not valued, the decision to have male children may be a correct moral choice. Given the present social practices, women in these countries are attempting to, "maximize their own and their family's happiness and minimize the suffering of little girls." What is needed, she argues, is social change so that women are valued. She concludes that when people design their children through choosing particular characteristics, they practice eugenics. She states that, "No human is wise enough to choose the kinds of people who ought to perpetuate our species."

Grazi and Wolowelsky examine the issue of sex selection in relation to contemporary Jewish law and ethics (36). They conclude that although new reproductive technology represents an opportunity for alleviating pain and suffering, it should not be used for "frivolous considerations." Rabbinic authorities, not the couple, reserve the right to decide under which situations these technologies can be used. Choosing the desired sex would likely be considered frivolous by most authorities.

Shrivastav writes on this subject from the perspective of the United Arab Emirates (37). It is noted that most citizens of the United Arab Emirates are of the Muslim faith, and since abortions are unacceptable to Muslims, sex selection followed by abortion would not be allowed according to the faith. However, gender preselection may be acceptable to followers of Islam as it would not contravene the Sharia law where IVF and embryo biopsy to rule out disorders are permitted. Despite this, Shirvastav considers that the fact that male offspring are preferred and writes that, "as far as society in the United Arab Emirates is concerned, potential availability of techniques for selecting the gender of offspring will encourage couples to alter the sex ratio of their offspring in favour of boys. Without these techniques, they would probably accept whatever nature has in store for them!"

The issue of unbalancing the normal gender distribution in society by allowing couples to choose the desired sex is a major concern. Seibel, Seibel, and Zilberberstein address this issue and offer a unique solution (38). They first state that they consider using prenatal diagnosis followed by abortion on the basis of sex to be morally unacceptable and that this practice could ultimately lead to an unbalanced sex ratio in society. They go on to say that preimplantation diagnosis for the sole purpose of sex selection, where the unused embryos are destroyed, appears to be an inappropriate use of technology. However, they feel that this practice could be used in an ethically acceptable manner by taking the embryos of the undesired sex and donating them to infertile couples. They argue that if couples were synchronized, this would result in gender distribution. Shenfield, in another letter to the same journal, refutes a previously held position that nature will soon redress the balance of the sex ratio (39). The problem with this position, Shenfield contends, is that it implies the acceptance of the superiority of one sex over the other. Shenfield goes on to assert that the practice of sex selection, "would be detrimental to both sexes to be brought up in a society which acknowledged, by a selective practice, that personal freedom may be obtained at the cost of one's gender identity being constantly assaulted by the implicit disapproval entailed when it becomes a serious handicap worthy of termination."

Steinbacher (40) describes the advantages of the firstborn as being more intelligent, achievement oriented, and successful than second born. Since those who are already privileged will be the ones to utilize sex-selection techniques, supporting selection means enhancing the disparity between men and women globally. "Fewer firstborn females, a higher [male] sex ratio at birth, more poor women in developed countries and elimination of women in the third world are indeed devastating outcomes of sex preselection for women" (40, p. 190). To prevent women's lives from being controlled by technologies, she says that women must first have "voice and vote" when policies are made about all methods of preselection.

Baird's ethical evaluation separates the three approaches: sex-selective abortion, sex-selective implantation, and sex-selective insemination (41). Baird submits that all of these approaches raise ethical concerns about the values of a society of our choice. The concern about sex-selective abortion is lack of respect for human life and dignity. Sex-selective implantation is an invasive and expensive procedure posing risk to women and using medical resources to prevent something that is not a disease. This too, she says, demonstrates a lack of respect for human life. Baird argues that sex-selective insemination does not violate the respect for human life as a life does not yet exist. However, she reasons that this practice reinforces the belief that the sex of the child is important and that families with children of only one sex are less than ideal. Additional concerns about all of these technologies are the possibility that lack of regulation may lead to commercialization of reproduction, the exploitation of the public, and the transformation of children into commodities. Baird argues for regulations and policies to address these concerns.

Botkin examined the broad subject of prenatal screening with respect to policies that would limit parental choice and included the subject of sex selection in his analysis (42). Botkin argues that it would not be justifiable to require that a patient defend her reason for abortion after prenatal screening when abortion is available on demand. However, he does not suggest that parents should have the right to request the use of prenatal screening for any and all purposes, nor that all physicians are obliged to provide all services that patients request. Botkin argues that broad policies limiting parental choice are not workable without a social consensus on the relative values involved, but rather, "physicians should be strongly encouraged to establish and uphold personal moral standards with respect to prenatal screening," thereby respecting the autonomy of both patients and physicians. He believes that, "limits to parental choice may be more appropriately applied through the moral values of individual physicians in their provision of diagnostic services."

Arguments supporting the use of sex selection mostly address parental rights and freedom of choice. Other arguments in support include, but are not limited to, sex selection is merely an extension of other assisted reproductive technologies, it is the least harmful option in some countries, preselection would reduce the number of gender-based abortions and the incidence of infanticide, it would slow population growth, it would eventually result in increased value of women, and it would reduce the number of unwanted children.

In a chapter in *Biomedical Ethics Reviews*, Warren writes in support of the practice of sex selection by refuting the position of Holmes in the same reference (43). She disputes the position that sex selection is a sexist practice. She notes that although some people will only want a child of one sex, many would choose to have a child who is the sex opposite of the child(ren) they already have. Also, in societies where the preference for a son is strong, Warren believes that accusing women of sexism is commensurate with blaming the victim. She reasons that it is not considered wrong to condemn a couple who do not want a child because they are unable to afford to care for it, and so it is also wrong to condemn women in certain societies who decide not to have daughters. She also maintains that many of the arguments made in opposition are based on speculation about the possible long-term consequences or about how people might behave if this practice was widely accepted. She says that it is wrong to condemn something based on such speculation. In a later publication, she states that sex selection is not always a form of gendercide in that, "if it were inherently wrong to alter sex ratios, then it would be wrong to seek cures for heart disease, breast cancer, and other lethal illnesses which afflict primarily members of one sex" (44). She also holds that, "sexism and its potential for harm are very much a function of how it is done, why it is done , and the social context."Warren argues it would be wrong to condemn a practice outside of its social context.

Anand Kumar offers this social context from the perspective of the culture in India (45). In India, sons are considered an asset, while daughters are considered a great burden. The son preference is so strong that female infanticide is a prevalent occurrence, and although abortion and infanticide are illegal, legislative measures have failed to produce any change. Kumar submits that the real ethical choice lies between the prevention and the perpetuation of feticide, infanticide, and homicide of females. Kumar notes that social change is a long process and asks the question, "Can we afford to wait until these social changes occur and in the meantime silently witness female deaths at all stages of life?" In light of what is now taking place, one argument is that reliable methods of sex preselection would offer the least harm to this society.

In response to the position that that sex selection will lead to an altered sex ratio, Lilford points out that in many countries there is no real preference for one sex over the other but rather a preference for a balanced family (46). He then goes on to say that in countries where there is a strong preference for boys, sex selection may slow the population growth, and ultimately, the demand for girls would increase thereby eventually changing the direction of preference. In the same publication, Lilford refutes the claim that sex selection is a form of discrimination against women but rather a preference for a particular sex may be the result of discrimination. He states that, "peoples' choice for a particular sex is a mirror of their society."

Mahoney separates sex selection followed by abortion from sex preselection. He rationalizes that availability of sex preselection would decrease the number of gender-based abortions performed and the incidence of infanticide (47). Mahoney points out that in Great Britain abortion is legal due to serious social pressure, so it is reasonable that people have access to it for any reason including selecting the sex of the child. This argument has been used by others including Egozcue who wrote, "Sex selection: why not?" in the journal, *Human Reproduction* (48). He discusses the use of sex selection on embryos where those embryos of the unwanted sex are discarded. Egozcue submits that this should not be a problem in those countries where abortion is available on demand as this would be an extension of other assisted reproductive technologies. Also Egozcue states that planned parenthood organizations have always supported the view that every child should be a wanted child, and therefore a child of the desired sex very much is a wanted child. Smith, in a letter to the *British Medical Journal*, also discusses the issue of wanted children (49).

He points out that in many families unwanted children are abused. Smith argues that even if sex selection results in an altered sex ratio, the scarcer sex would be valued over time, and since fewer children would be born, it would slow the population growth. In the same letter, Smith discusses the justice problems in attempting to regulate the practice of sex selection. He says, "the rich and connected can usually gain access to any technological innovation that they want."

Pennings addressed sex selection for balancing families and proposed ethical guidelines (50). He proposed that sex selection should not be allowed for the first child nor when there is already a balance in a family. In response to the Pennings proposal, Dawson and Trounson ask the question, "Who will enforce these guidelines?" Although they do not support sex selection, they do point out that the United Nations Declaration of Human Rights states that each individual has the right and freedom to form a family. The application of this declaration to the issue of sex selection is not clear. However, they feel that the Pennings proposal represents, "a violation of the right to freely form a family given in the Declaration of Human rights and, given that the appropriate technology is available." In the end, however, they argue that sex selection is not a responsible use of technology (51).

Lilford challenges the argument that technology should only be used for medical purposes (46). He submits that it is difficult to determine the difference between a medical and a nonmedical mission. He states that, "the important factors in human life are those of suffering and happiness, and the eradication of disease is merely a means towards these ends. If medical technology can produce these ends without eradicating a disease, then it is equally worthwhile."

One final major argument in support of sex selection is that of respect for individual autonomy. Kaye and LaPuma emphasize this position from the perspective of the clinical geneticist (52). They assert that although clinical geneticists are not themselves ethically neutral on this subject, "the best interests of the patient, not of society or the human race, should determine diagnosis and treatment." They strongly maintain that the overriding ethical principle is that of beneficence for the individual patient. Shulman, of the Genetics and IVF Institute in Fairfax Virginia, concludes that it is fundamental to free societies for responsible individuals to have the freedom to differ and to make personal choices based on their own convictions (53). Stephens demonstrates the strong belief that patient autonomy must be the guiding principle when he writes, "It is my opinion that the only issue that is the sole responsibility of physicians really is support of the patient's (and it is usually women who are burdened with this responsibility) right to exercise her (or, in the instance of a couple in a counseling situation, their) reproductive options, regardless of the indication, regardless of the personal moral or ethics standpoint held by the practicing physician" (54). Such wholehearted support of patient autonomy is most prevalent among geneticists in the United States (27).

Finally, an examination of the arguments surrounding sex selection would be incomplete without looking at the work of John C. Fletcher who has written extensively on this topic and revised his ethical position over time. His initial arguments opposed sex selection because (1) sex is not a disease, (2) abortion for sex choice could contribute to social inequality between the sexes because of a preference for male offspring, (3) sex choice is a "frivolous" and indefensible reason for abortion, and (4) amniocentesis is a scarce resource in light of the total number of at-risk pregnancies (55). In reviewing his writings over the years, it is interesting to see how he has re-evaluated his position and his reasons behind his position many times. This speaks to the extreme complexity of this issue.

In this 1980 essay, Fletcher assumes that the main reason for discouraging sex selection is the belief of physicians that performing prenatal diagnosis that such abortions are morally unjustifiable (55). Although this is still his personal view, he believes that the legal rules on abortion defined by the U.S. Supreme Court supersede the clinician's personal moral views. Since a woman need not state reasons for abortion under any circumstances, sex selection ought not be subjected to public scrutiny. He concludes that, "it is inconsistent to support an abortion law that protects the absolute right of women to decide, and at the same time to block access to information about the fetus because one thinks that an abortion may be foolishly sought on the basis of that information" (55, p. 16). Fletcher continues to believe that physicians have a right to state their own moral views and to describe risk factors of prenatal testing, including "an unknown risk of insult to other numbers of the family and to wider society." However, if a couple continues to request the information for sex selection, Fletcher stated that the physician may not legally or morally refuse to provide it if he or she wants to "keep faith" with the moral intent of the law.

In a later re-evaluation of this issue, Fletcher, along with Wertz, considered all of the arguments given in support of sex selection and came to the conclusion that the medical profession as a whole has not demonstrated any responsibility in this arena. They state, "We hold that a very strong normative case exists against sex selection that transcends cultural boundaries, especially based on claims of equal worth of both sexes and justice in social life" (56). They strongly suggest that the medical community take a stand against sex selection. Fletcher and Wertz believe that by doing this, the medical community will ultimately be protecting important reproductive choices by demonstrating that they are able to set the standards for practice. With these standards in place, government intervention would not be necessary and reproductive choices involving medical decisions would not be lost (56). In a more recent study, Fletcher, along with Wertz, evaluated the attitudes of medical geneticists about sex selection in 19 nations. They found that in many nations women do not have access to prenatal testing to detect birth defects either because of the cost or because of the scarcity of such medical technology. They conclude that it is unfair to use these limited resources for nonmedical reasons. They state a concern that sex selection could be, "the first step on a 'slippery slope' toward cosmetic choices for height, weight, eye or hair color." Again, they call for the medical profession to abandon its nonjudgmental stance and set a standard of care with regard to sex selection (57).

SUMMARY

With few exceptions, positions of professional societies and governmental agencies oppose the practice of sex selection. Yet, sex selection continues to occur in practice, and the debate about the morality of the practice continues among ethicists and practitioners. Why is that?

There are two major reasons. One is that sex selection is only one of the possible traits that one can select prenatally. Most prenatal testing is accepted today because it serves the interests of people who want to make decisions based on health information. Even for many who are comfortable with prenatal testing and autonomous choice, however, there is concern about the intrinsic value of human life as it is created. Choosing the sex of their offspring represents the first real and available choice for parents who want to select a child with traits that fit their vision of an ideal family. The specter of sex selection could be the first step down a slippery slope to the "brave new world" of designer children. We worry about whether allowing choice of the sex of the child will open the door to the use of genetic technologies to select other traits for more "trivial" reasons than avoiding disease (55).

Second, sex selection also is a paradigm case for considering what values really ought to guide health care policy. The same tensions, between issues of justice for many and respect for individual autonomy, exist in determining the appropriate use and distribution of health care technologies more generally. A majority of those opposing sex selection address broad societal justice concerns. Some say that using such measures will continue, and even enhance, gender discrimination. Others believe that health technologies ought to serve the needs of improving health, and ought not be squandered in support of individual or societal determination of what human traits are valuable and worthy.

On the other side of the argument, are those who strongly support the individual's right to self-governance and the professionals' obligation to respect that autonomy. Individual (or couple) autonomy undergirds most arguments for allowing sex selection, since prohibiting the practice would be limiting autonomy in reproductive decision making. Genetic and reproductive technologies, particularly sex selection, make the professional obligation to respect autonomy more complex, however. Because decisions about how to use them are intrinsically about families they raise the question of who is the patient, and thus whose interests ought to be served by clinicians and policy makers. Is the primary obligation to the mother, the father, both parents, the child or potential child, existing children, or to the societies in which reproductive and genetic applications are made available?

Both clinicians and policy makers must ponder the right and appropriate use of health care technologies and from the framework that guide how these decisions are made. Sex selection is but one example of this challenge. Examining how positions on sex selection are cast may provide some insight into how other ethical challenges in health care will be addressed as well.

ACKNOWLEDGMENTS

The authors would like to thank the Center for Bioethics at the University of Minnesota for its support. In addition, we would like to extend special appreciation to Robert Koepp, a Center research assistant, for his assistance.

BIBLIOGRAPHY

1. B.L. Gledhill, *Semin. Reprod. Endocrinol.* **6**(4), 385–395 (1988).
2. B.E. Reubinoff and J.G. Schenker, *Fertil. Steril.* **66**(3), 343–350 (1996).
3. A. Botchan et al., *J. Androl.* **18**(2), 107–108 (1997).
4. S.A. Carson, *Fertil. Steril.* **50**(1), 16–19 (1988).
5. R.J. Ericsson, *Fertil. Steril.* **51**(2), 368–369 (1989).
6. J.H. Check, D. Katsoff, and A. Bollendorf, *Fertil. Steril.* **61**(6), 1181–1182 (1994).
7. J.H. Check and D. Katsoff, *Hum. Reprod.* **8**(2), 211–214 (1993).
8. J.J. Beernik, W.P. Dmowski, and R.J. Ericsson, *Fertil. Steril.* **59**(2), 382–386 (1993).
9. S.P. Flaherty et al., *Hum. Reprod.* **12**(5), 938–942 (1997).
10. M. Greier, J.L. Young, and D. Kessler, *Fertil. Steril.* **53**(6), 1111–1113 (1990).
11. R.G. Edwards and H.K. Beard, *Hum. Reprod.* **10**(4), 977–978 (1995).
12. S.P. Flaherty and C.D. Matthews, *Mol. Hum. Reprod.* **2**(12), 937–942 (1996).
13. I. Findlay et al., *J. Assist. Reprod. Genet.* **13**(2), 96–103 (1996).
14. J.C. Harper, *J. Assist. Reprod. Genet.* **13**(2), 90–95 (1996).
15. J.T. Queenan, *Semin. Perinatol.* **11**(3), 264–267 (1987).
16. C. Ruegsegger Veit and R. Jewelewicz, *Fertil. Steril.* **49**(6), 937–940 (1988).
17. T.M. Martequ, *Br. Med. J.* **306**, 1704–1705 (1993).
18. H. Statham et al., *Lancet* **341**, 564–565 (1993).
19. A. Green, C. Wray, and B. Balsuch, *Psychol. Rep.* **73**(1), 169–170 (1993).
20. P. Liu and G.A. Rose, *Hum. Reprod.* **10**(4), 968–971 (1995).
21. M.A. Khatamee et al., *Int. J. Fertil.* **34**(5), 353–354 (1989).
22. P. Baird, *J. Assist. Reprod. Genet.* **12**(8), 491–498 (1995).
23. T. Sheldon, *Br. Med. J.* **311**, 10–11 (1995).
24. S.K. Khanna, *Soc. Sci. Med.* **44**(2), 171–180 (1997).
25. R.V. Grazi, B. Wolowelsky, and R. Jewelewicz, *Gynecol. Obstet. Invest.* **37**, 217–225 (1994).
26. M.I. Evans et al., *Am. J. Obstet. Gynecol.* **164**(4), 1092–1099 (1991).
27. D.C. Wertz and J.C. Fletcher, *Am. J. Med. Genet.* **29**, 815–827 (1988).
28. B.M. Burke, *Soc. Sci. Med.* **43**(11), 1263–1269 (1992).
29. D.F. Pencarinha et al., *J. Genet. Couns.* **1**(1), 19–30 (1992).
30. D.C. Wertz, *Clin. Obstet. Gynecol.* **36**(3), 521–531 (1993).
31. X. Mao and D.C. Wertz, *Clin. Genet.* **52**(2), 100–109 (1997).
32. American College of Obstetricians and Gynecologists (ACOG), *Int. J. Gynecol. Obstet.* **56**, 199–202 (1997).
33. Council on Ethical and Judicial Affairs, American Medical Association, *Arch. Fam. Med.* **3**, 633–642 (1994).
34. Kalaca and A. Akin, *Hum. Reprod.* **10**(7), 1631–1632 (1995).
35. H.B. Holmes, in J.M. Humber and R.F. Almeder, eds., *Biomedical Ethics Reviews*, Humana Press, Clifton, NJ, 1985, pp. 39–71.
36. R.V. Grazi and J.B. Wolowelsky, *J. Assist. Reprod. Genet.* **9**(4), 318–322 (1992).
37. P. Shrivastav, *Hum. Reprod.* **10**(5), 1319–1320 (1995).
38. M.M. Seibel, S.G. Seibel, and M. Zilberstein, *Hum. Reprod.* **9**(4), 569–570 (1994).
39. Shenfield, *Hum. Reprod.* **9**(4), 569–570 (1994).
40. R. Steinbacher, in H.B. Holmes, B.B. Haskins, and M. Gross, eds., *The Custom-Made Child? Women Centered Perspectives*, Humana Press, Clifton, NJ, 1981, pp. 187–191.
41. P. Baird, *Annu. Rev. Med.* **47**, 107–116 (1996).
42. J.R. Botkin, *Obstet. Gynecol.* **75**(5), 875–880 (1990).
43. M.A. Warren, in J.M. Humber and R.F. Almeder, eds., *Biomedical Ethics Reviews*, Humana Press, Clifton, NJ, 1985, pp. 73–89.
44. M.A. Warren, *Bioethics* **1**(2), 189–198 (1987).
45. C. Anand Kumar, *Hum. Reprod.* **10**(5), 1319 (1995).
46. R.J. Lilford, *Hum. Reprod.* **10**(4), 762–764 (1995).
47. J. Mahoney, in P. Byrne, ed., *Medicine, Medical Ethics and the Value of Life*, Wiley, New York, 1990, pp. 141–157.
48. J. Egozcue, *Hum. Reprod.* **8**(11), 1777 (1993).
49. T. Smith, *Br. Med. J.* **307**, 451 (1993).
50. G. Pennings, *Hum. Reprod.* **11**(11), 2339–2343 (1996).
51. K. Dawson and A. Trounson, *Hum. Reprod.* **11**(12), 2577–2578 (1996).
52. C.I. Kaye and J. LaPuma, *Hastings Cent. Rep.* **20**(4), 40–41 (1990).
53. J.D. Shulman, *Hum. Reprod.* **8**(10), 1541 (1993).
54. J. Stephens, *Am. J. Obstet. Gynecol.* **166**(3), 1024–1025 (1992).
55. J.C. Fletcher, *Hastings Cent. Rep.* **10**(1), 15–20 (1980).
56. D.C. Wertz and J.C. Fletcher, *Hastings Cent. Rep.* **19**(3), 21–27 (1989).
57. D.C. Wertz and J.C. Fletcher, *Soc. Sci. Med.* **37**(11), 1359–1366 (1993).

See other entries GENETIC COUNSELING; see also REPRODUCTION, ETHICS entries; REPRODUCTION, LAW, REGULATION OF REPRODUCTIVE TECHNOLOGIES.

REPRODUCTION, ETHICS, THE ETHICS OF REPRODUCTIVE GENETIC COUNSELING: NONDIRECTIVENESS

BARBARA BOWLES BIESECKER
National Human Genome Research Institute
Bethesda, Maryland

OUTLINE

INTRODUCTION

In reproductive genetic counseling, nondirectiveness may refer to an ethic of practice or to the process itself. Different aspects of genetic counseling have been described as nondirective; the communication style, the offering of genetic testing, or the counseling interaction. These various interpretations of the term *nondirectiveness* have lead to confusion about the goals and practice of reproductive genetic counseling. As well, it has diluted conversation about important issues surrounding the personal nature of reproductive choice involving genetic risk. As an ethical principle, nondirectiveness suggests that pregnant women and their partners ought to be supported to make autonomous decisions about prenatal testing and their reproductive outcomes without the direct influence of the counselor. The personal autonomy of the client facing the genetic reproductive decision is paramount. Nondirectiveness should be used exclusively to describe an ethical principle in reproductive genetic counseling. Although it is not evident always how this principle translates into the practice of genetic counseling, the process may be discussed as a dialogue of client-centered counseling that is guided by nondirectiveness.

NONDIRECTIVENESS IN GENETIC COUNSELING

Nondirectiveness describes components of a young medically related professional service, called *genetic counseling*. This psychoeducational practice assists people who have concerns about birth defects, genetic conditions, or genetic risk (1–3). Throughout its short history, genetic counseling has been consistently described as nondirective, as opposed to advice giving. Genetic counseling may be the only medically related practice intended to be nondirective. The term has been used to describe not only the ethic of practice but also the goal of genetic counseling, the process, and an outcome. The literature discusses nondirectiveness assuming one of these practice components but frequently fails to distinguish its meaning. For those who strive to understand, to investigate, or to use genetic counseling services, it is unfortunate that the concept is inconsistently portrayed. Even those who practice genetic counseling have confused the meaning and interpretations (4,5).

Several scholarly articles have appeared to address the confusion in the meaning of nondirectiveness (4,6,7). The literature has begun to distinguish the various uses of the term in an effort to achieve some consensus on the goals and process of genetic counseling. Since the literature on nondirectiveness is discrepant, this chapter will delineate uses of the term and compare their implications. The success of the practice of genetic counseling depends on continued efforts to define and strive towards nondirectiveness assuming the profession can agree on what it is, that it is central to the process and that it can be achieved.

HISTORY OF NONDIRECTIVENESS

The original introduction of nondirectiveness into the genetic counseling literature remains elusive. Sheldon Reed, a medical geneticist who coined the phrase "genetic counseling" in 1947, spoke of a nondirectivelike practice but only used the term later in his writings after it had appeared in the literature (8). Reed described a genetic social worklike practice of explaining genetic concepts and supporting clients who use the information to make reproductive decisions. In this case the concept of nondirectiveness describes the process of genetic counseling more so than the overarching ethical principle. Some authors claim that nondirectiveness in genetic counseling arose in opposition to the eugenics movement. Resta points out that many of the medical geneticists writing about the process of genetic counseling in the 1950s used the term nondirective but then also described eugenic ideas about the practice (9,10). It is evident from the literature that certain supporters of nondirectiveness were not opposed to eugenic practices. Thus, such claims about nondirectiveness may be unfounded (4,11).

The general source of the term nondirectiveness predated genetic counseling by about two decades. Dr. Carl Rogers, a prominent psychologist, used the word to describe his theory of psychotherapy (12). By 1951, however, Rogers had come to describe his theory and practice as client-centered. This clarification in his terminology acknowledged the presence of directive components to the therapeutic relationship, yet emphasized the focus on the client's expressed needs rather than the explicit direction of the counselor. Rogerian psychotherapy developed prior to the existence of genetic counseling and in parallel to, not in reaction against, the eugenics movement in the United States. It is intriguing to consider why the profession of genetic counseling adopted as its mantra a term that was rejected early on by the field of psychotherapy. Since its introduction into genetic counseling, nondirectiveness has lead researchers, academics, and practitioners astray.

Genetic counseling has sustained the use of the term nondirectiveness despite its ambiguity. Client-centered theory and practice have offered one useful framework (within limits) of thinking about and practicing genetic counseling. Nondirectiveness has been used effectively to describe a client-centered counseling style not unlike a Rogerian approach. Since genetic counseling has evolved as a clinical and atheoretical practice, it has borrowed ideas from its theoretical neighbors. In a different sense, genetic counseling has long recognized the lack of desire or ability to make reproductive decisions for others. It has emphasized autonomy and voluntariness (13,14). Genetic counseling embraces a certain hands-off approach to sensitive issues of life and death that are entwined in reproductive decision making. This has proved to be a more comfortable stance for genetic counselors than entering into the difficult and sticky terrain of directing people in their childbearing decisions that involve genetic risk. Rather than as a reaction against eugenics, perhaps nondirectiveness has been sustained by an abhorrence of eugenic practices. Some would argue that it might also serve to shield practitioners from confronting difficult aspects of reproductive genetic counseling.

INTERPRETATIONS AND IMPLICATIONS OF NONDIRECTIVENESS

Nondirectiveness as a Guiding Ethical Principle

There have been at least four different, yet overlapping, meanings of nondirectiveness expressed in the literature. Most often, nondirectiveness has been used to mean a desire to uphold the personal nature of reproductive decision making. Nondirectiveness in this sense represents an underlying value or ethical principle of the profession. Genetic counselors in the United States have emphasized the principle of nondirectiveness conceptionally in their code of ethics: "Genetic counselors strive to enable their clients to make informed independent decisions, free of coercion, by providing or illuminating the necessary facts and clarifying the alternatives and anticipated consequences (15, p. 41). Yet as a value it does not readily translate into a way of practice or a specific goal. It is difficult to assess whether an individual or couple has made a "good" personal decision. How does a genetic counselor promote the decision-making process within an ethical framework of nondirectiveness? Clients experience many influences on their reproductive decisions. Exclusively personal or autonomous decision making is difficult, and not necessarily uniformly desirable, to achieve. Yet it is important that providers not assert undue influence on the reproductive outcomes of their clients. This is a blatantly eugenic goal and contradicts the desires of most geneticists and genetic counselors internationally (16). Kessler points out that even when there is an explicit goal to discourage certain reproductive outcomes (e.g., in a country that supports such practice), a significant number of clients ignore the advice (17). It is unclear that it is necessary for professionals to completely withhold advice from clients. Yet much of the international genetics community, and in particular, genetic counselors in the United States, Canada, and the United Kingdom, finds the notion of advising people directly on their reproductive choices to be loathsome. It is difficult to know or to appreciate the values, resources, thought-processes, and ideas of another person sufficiently to provide advice about having or not having children who may be affected with a certain genetic condition. The truth is, most people struggle to understand what choices they would make for themselves, let alone know better for another.

Genetic counselors need to be exquisitely self-aware and not harbor personal opinions of what constitutes a life worth living. If they do, they must be honest with themselves and disclose to clients that they may hold beliefs that children affected with certain genetic conditions should not be born. This differs from stated goals to enhance personal choice for clients. Yet it is more honest than undisclosed potential agendas. Most counselors, who also work with children and families affected with genetic conditions, serve as advocates for those with special needs as children or who are disabled as adults. As a profession they value diversity and often enter the field of genetic counseling concerned about genetic conditions and how society views disability.

Medical genetics services, such as triple screening for neural tube defects and carrier screening for recessive or sex-linked genetic conditions, may have the more or less explicit goal of reducing the number of individuals affected with genetic conditions (18–20). Cost–benefit analyses to justify such programs may be based on an assumption that a significant number of affected pregnancies will be aborted. In this case the genetic counseling that accompanies such practices may have values that are in conflict with the intention to reduce the incidence of genetic conditions. Genetic counseling may strive to help the individual make the best personal decision, yet the goal of the program may be to abort affected fetuses. Genetic counselors may find themselves caught in a dilemma between professional values that emphasize personal autonomy and programs that are justified by social policy to improve the health or well-being of the populace. If counselors uphold a nondirective ethic, then they should not paradoxically endorse genetics services that have a goal of preventing the birth of individuals who will be affected with a genetic condition. Genetic counselors should and do endorse services that emphasize informed and autonomous choice in reproductive decision making. An example is the choice about whether to undergo amniocentesis to determine the chromosomal status of a fetus. Nevertheless, aspects of service provision (e.g., assuming the outcome of the decision to undergo testing by scheduling the amniocentesis to follow the counseling session) do not always promote the genetic counselor's role to ensure personal choice about testing.

A challenging aspect of an ethic of nondirectiveness is not so much the goal to refrain from explicit influence on reproductive decisions, as it is the potential for more subtle and unintended (even unconscious) influence. Such practice may occur when a genetic counselor harbors a belief that a certain reproductive outcome is most desirable for a person or couple. But rather than state the bias outright, the counselor's approach is influenced by her or his beliefs. This would be an ethically directive approach even if the counselor did not intend to provide direction to the client.

When counselors successfully manage to facilitate the client's decision making without influencing the outcome, the process is flexible and difficult to operationalize. Counseling is inherently directive, as is providing education to ensure understanding about the options. Genetic counselors have no standard of practice to consistently uphold an ethic of nondirectiveness. Counselors recognize that the type of information they provide and how they present it may influence decisions (21,22). An ethical principle should translate into an effective mode of practice. White has proposed a counselor–client dialogue as a working description of the process (7). The practice is to facilitate client centered reproductive decision making, within an ethical framework of nondirectiveness. As a guiding principle this ethic would suggest a process of genetic counseling that emphasizes the values and beliefs of the client, but that tolerates the direction offered by a competent counselor who does not preconceive a decision for the client. In order to further clarify the underlying ethical principles of genetic counseling, the field may need to differentiate itself from other genetic services whose goals (such as abortion of affected pregnancies) are inconsistent with the values of the profession.

Nondirectiveness as a Guiding Policy on Genetic Testing

Nondirectiveness also has been used to describe the concept of not denying access to genetic testing. This definition relates to genetics health policy and access to services. It is a practical one, although it has overlap with the previous definition in its intention to uphold reproductive freedoms. In this case genetic counselors are reticent to deny access to any genetic test that an individual or couple may request (even one that puts a pregnancy at risk) provided that there is understanding about the risks and benefits of the test and its potential outcomes.

Historically much of genetic counseling has addressed risk for serious conditions, with the exception of certain mild birth differences (such as a cleft lip) and sex chromosome "anomalies" (such as Turner syndrome). Counselors offer genetic testing to determine whether a fetus will be affected (prenatal) or whether a couple may be at increased risk for having an affected child (carrier). One survey has suggested that the majority of genetic counselors, internationally, offer prenatal diagnosis (or a direct referral) for sex selection (23). This is worrisome when one considers that genetic counseling originated from a desire to help people grapple with difficult dilemmas about serious genetic conditions or birth defects. Genetic counselors, as represented by the U.S. professional society (NSGC), uphold a moral right to reproductive freedom (24). The majority of practitioners believe that if a woman (or couple) has consented to prenatal testing by considering the relative balance of risks and benefits, she should be offered the opportunity to determine the sex of the fetus, even if she desires to abort a fetus of undesired sex. Such a finding bodes poorly for the future of genetic testing, when prenatal tests may be offered for physical or personality traits. Will genetic counselors, in the name of nondirectiveness, offer prenatal testing for anything a couple desires as long as they are informed?

In this regard, the meaning of nondirectiveness has caused the profession of genetic counselors to be passive about taking a stand on what tests ought to be offered. There have not yet been professional guidelines written by U.S. genetic counselors discouraging certain prenatal testing. Within NSGC there are position statements and a resolution on genetic testing or screening, for instance, one exists on prenatal and childhood testing for adult-onset disorders (25). It states that while such testing is discouraged, each case should be considered individually and counselors should decide whether to offer parents testing of their children on a case by case basis. This leaves the judgment of reasons, fitness, and values of the client up to the counselor. While inherently flexible and accounting for individual differences, it neglects to take a clear stand and puts counselors in the position of practicing inconsistently. It leads to confusion for the profession. In response, members of the genetic counseling community published a substantial position statement on the genetic testing of children for adult onset conditions to more clearly state a testing policy (26). While it is unlikely that there are many moral absolutes in reproductive decision making, an insistence on nondirectiveness has stymied the process of policy making in prenatal genetic testing. With the promised onslaught of new genetic tests, reproductive counselors seem to be prepared to offer testing for any indication. In the name of nondirectiveness, genetic counselors have avoided their professional and moral obligations to take a stand on the appropriateness of certain types of prenatal testing.

The approach to reproductive genetic testing, "anything goes as long as the individual has had pre-test counseling," predicts that counselors will play less of a role in establishing genetics health policy. Yet genetic counselors may be one of the most important groups of professionals to be involved in helping to establish guidelines or polices about what testing may not be an appropriate use of resources or may be morally reprehensible (27). Do genetic counselors advocate for the use of prenatal testing to potentially abort fetuses found to be at somewhat increased risk for adult-onset cancer, for instance? Worse yet, for a slightly lower projected adult height? The role of testing gatekeeper may be an important one for genetic counselors in the future. Yet nondirectiveness has been misinterpreted to imply that any genetic tests that are technically feasible should be offered. In the name of nondirectiveness, counselors refrain from judging the choices of their clients. In doing so, genetic counselors may be washing their hands of the responsibility to offer morally, not to mention economically, responsible reproductive testing options.

Rather than interpreting that nondirectiveness holds no opinion on genetic testing, reproductive genetic counselors ought to offer genetic testing only for serious conditions that may significantly impede an individual's ability to achieve quality of life. While there is no consensus on what constitutes a serious genetic condition (28), this should not dissuade genetic counselors and other providers from establishing responsible genetic testing services and genetics health policy (29,30). This misunderstanding of nondirectiveness has led to a significant lost opportunity and an ongoing need for the professional practice of reproductive counseling.

Nondirectiveness as a Style of Communication

In contrast, nondirectiveness has been construed as a style of communication within the practice of genetic counseling (31,32). Genetic counseling has been described as a value-neutral encounter despite awareness that any human relationship is value laden. The mis-notion of value neutrality has further confused the issue of nondirectiveness (33). In communicating genetic information within genetic counseling, there are many directive components. In an educational relationship, the person with the information has more power and there is an inequality to the relationship (1). The way the information is conveyed and the amount of information given may be quite directive. Genetic counselors lead, guide, and even advise their clients. Each of these is a directive process. Genetic counselors strive to give complete and balanced information, but it is human nature to be inconsistent and influenced by individual experiences. This might be described as directive practice as well.

This interpretation of nondirectiveness implies information should be conveyed in a nonleading way. Studies that have been conducted to assess use of directive language have concluded that the process is directive (30). While such outcome studies are necessary and valuable for determining what happens in genetic counseling, they seek to document a nondirective psychoeducational practice. It is an unattainable paradox.

In a desire to use nondirective language to communicate, genetic counselors may seek to use words that are ambiguous. Such avoidance of direct language may not be useful to clients who are often seeking not only information but also advice on how to use it or how to make meaning of it. The irony of the use of vague language is that expert communication of complex genetics information is often heralded as a prominent goal of genetic counseling. A nondirective intent has guided counselors into inexplicit use of language that could otherwise make genetic concepts and their implications more obvious to clients. This use of nondirective communication has lead to process studies that have shed light on this perplexing notion of genetic counseling (31,34). Conclusions have been drawn that counselors are directive in a manner that implies they are undermining a guiding ethical principle. In fact they are merely communicating as professionals do, using language that is often directive and in a manner that may be directive. While there is merit in research toward understanding how an ethical principle such as nondirectiveness translates into practice, the mode of communication is only one component of a complex dialogue within a relationship of influence.

Without guidance on the adaptation of an ethic of nondirectiveness into practice, counselors have assumed a nonjudgmental approach that also involves noncommittal or evasive communication around difficult issues. This minimization of an ethical principle has led to one outcome counselors seek: clients who have not been explicitly directed in their reproductive decisions. But it has also lead to not providing clear messages about the implications of the information, and perhaps even not facilitating "good" reproductive decisions. For instance, Wertz and her colleagues found that in the majority of prenatal genetic counseling sessions they studied, abortion was not mentioned (35). Since it is the only intervention a couple could choose to take for the vast majority of conditions tested for, it should be prominent in discussion about the potential usefulness of prenatal testing. If nondirectiveness had been uniformly adopted as an ethical principle that supports a client-centered counseling process, rather than a communication style, word choice and tone would be considered less significant than the components of dialogue within a therapeutic relationship.

Nondirectiveness as a Theoretical Basis of Counseling Practice

This point segues into a further interpretation of nondirectiveness, the intent to provide client-centered counseling. This definition is not dissimilar to that of Rogers's theory of psychotherapy. In this regard, nondirectiveness represents a reasonable and responsible goal for genetic counseling. It heralds the role of the client as central and as a goal, can be achieved (1,36). In this venue nondirectiveness provides a model for genetic counseling that can uphold an ethic of personal reproductive decision making. But nondirectiveness should not be used to describe both the process of counseling as well as the underlying ethical framework or the existing confusion will pervade.

In describing genetic counseling as a psychoeducational process, the psychological or therapeutic goal is to explore the meaning that the genetic information has for clients. This is a client-centered approach that focuses on client values, beliefs, ideas, and desires. The process by which it is achieved varies but the client's agenda and needs are paramount. As Rogers previously discovered, the term nondirectiveness in this case compounds the confusion, since many strategies used by the counselor might be described as directive. Yet they are executed with the client in mind. For instance, the counselor may help the client to set an agenda to explore implications of the information in a way that is personal, useful, and lends itself to decision making. The counselor may be directive in helping a client determine what may be reasonable to try and accomplish in one or two sessions. While these behaviors are directive, they do not override the needs of the client. Rather, they represent the counselor's expertise that may be used to enhance the effectiveness of genetic counseling. The client's needs are the ones addressed, but the client is not left to talk randomly without focus on the problem or issue at hand. Without such structure, a session would never become therapeutic. This is only one example of directive practices of counselors that do not undermine a client-centered approach.

In this more appropriate use of nondirectiveness the term remains problematic and should be replaced with client-centered genetic counseling. In much the same way nondirectiveness did not accurately depict the therapeutic approach proposed by Rogers, it has lead genetic counselors to largely ignore the need to engage actively with clients in order to address their concerns. Transcripts of genetic counseling sessions indicate that counselors practice inadequate counseling skills to accomplish even a minimally client-centered approach (1,5,31). The mantra of nondirectiveness may have caused counselors to hesitate over using their own best judgment about people's ability to make good decisions for themselves, to grow from difficult experiences, and to cope and adjust. In the name of nondirectiveness, many counseling opportunities have been lost in genetic counseling. An active dialogue about options, alternatives, resources, strengths, and outcomes within a therapeutic relationship may best help clients (3,6,7). Such a dialogue is likely to have many directive statements in it but does not direct the client toward a certain outcome in a coercive or even persuasive way.

Although the various uses of the term nondirectiveness have been problematic for the profession of genetic counseling, the pervasiveness of the concept of personal autonomy in genetic reproductive decision making sets the practice apart from the majority of medical services. The ethical principles of autonomy and beneficence in reproductive decision making, the need for thorough informed consent for genetic testing, and the value of human diversity and the lives of those affected with disabilities are crucial to reproductive genetic counseling. And the few outcome studies that have been conducted suggest that reproductive genetic counseling clients are satisfied with the service. They like their genetic counselor and are grateful for the time spent teaching them genetic principles (37). From a process and outcomes perspective, clients are likely to be best served by a psychoeducational approach that includes a client-centered or cognitive, theoretically based practice. This therapeutic process and its desired outcomes of self-determination, feelings of personal control, and restored self-esteem have yet to be studied (38). To achieve them, a counselor may be directive and to conduct process studies that investigate how directive she or he is seems counterproductive. Studies are needed on the most effective therapeutic approaches in genetic counseling, to observe how successful they are in achieving desired outcomes. Research would be facilitated if genetic counselors embraced such a therapeutic approach, and if consensus could be achieved on the goal of restoring feelings of personal power to clients and on outcomes of the process that can be systematically measured.

NONDIRECTIVENESS IN PRACTICE

In the most common reproductive genetic counseling example of a couple learning that their fetus is affected with Down syndrome (due to an extra chromosome 21), genetic counseling is the process through which the couple can determine what the condition may mean for their lives. Down syndrome cannot be "repaired," although some of the symptoms can be treated. The child will be mentally retarded, although to what degree is unknown. The couple may continue the pregnancy as planned or have an abortion. This is an agonizing decision even for couples who initially feel confident about what they would do in such circumstances. In facing the situation couples often take into account their expectations of parenthood, family life, economic resources, previous experiences with persons who have Down syndrome, opinions of family and friends, spiritual beliefs, social influence and expectations, and so on. Decisions about a pregnancy are complex, deeply personal, and irreversible. Important aspects of the decision may even be intangible or elusive to the couple themselves.

A genetic counselor in this situation seeks to establish an empathic connection or a therapeutic bond with the couple in order to help them make personal meaning of the information about Down syndrome. The counselor may strive to identify resources useful to the couple in making the decision so that they can live with their decision (one way or the other) in the years to come. A therapeutic approach focuses on enhancing self-determination and perceived personal control. Couples are helped to recognize that they have the strength to make such a difficult decision and that they have made other decisions successfully in the past. The counselor works toward facilitating the decision-making power of the couple in addressing their needs and concerns. This process may be described by some as nondirective counseling. Yet it is more appropriately described as client-centered and personally empowering.

The example of a client or couple asking the counselor what he or she would do in the same circumstances is often used to illustrate nondirectiveness in genetic counseling. Common responses by genetic counselors may be:

- "I am not in your situation so I couldn't possibly know what I would do."
- "Other people in your situation have chosen to continue the pregnancy, while others have had an abortion."
- "There are no right answers, I am here to help you make the best decision for yourself."
- "I will support any decision you make."

Evasive answers such as these do not address the concerns of the client. The client is asking for advice because she has not received the help she needs to make her decision. It is unlikely that she is literally handing over responsibility for the decision to the counselor (although a minority of clients may do so). Nor is she likely to mimic the choice of the counselor in order to solve her dilemma. However, all too often counselors neglect to work toward exploring and understanding where the client's anxiety and concerns come from in an effort to best help her with the decision. Kessler reminds genetic counselors that if the clients are frequently asking this question, there is something fundamentally flawed about the process (4). There are many respectful and considerate ways to address this question without abandoning the client in a time of great need. They challenge genetic counselors to fully experience with clients some of the hardest decisions of their lives. It takes a lot of hard work and *direction* on behalf of the counselor. A nondirective mode of practice misinterpreted is a missed counseling opportunity and at its worse an abandonment of a client in need of help.

SUMMARY

Nondirectiveness is a term to be reserved for an ethical principle of practice in reproductive genetic counseling. It emphasizes the importance of autonomy in genetic reproductive decision making. As a guiding principle, nondirectiveness provides a moral framework for providing client-centered counseling. Reduced to merely how a counselor communicates or to a lack of health policy on the use of genetic tests, nondirectiveness is an ineffectual concept. Its counterpart, direction, is an essential aspect of effective client centered counseling that supports informed reproductive choices involving genetic risk.

BIBLIOGRAPHY

1. S. Kessler, *J. Genet. Counsel.* **6**(2), 287–295 (1997).

2. R. Kenen and A.C.M. Smith, *J. Genet. Counsel.* **4**, 115–124 (1995).

3. B. Biesecker, *Kennedy Inst. Ethics J.* **8**, 145–160 (1998).

4. S. Kessler, *Am. J. Med. Genet.* **72**, 164–171 (1997).

5. F. Brunger and A. Lippman, *J. Genet. Counsel.* **4**, 151–167 (1995).

6. S. Suter, *Univ. of Chicago Law School Roundtable* **3**(2), 473–489 (1996).

7. M. White, *J. Genet. Counsel.* **6**(3), 297–313 (1997).

8. S.C. Reed, *Soc. Biol.* **21**, 332–339 (1974).

9. R. Resta, *J. Genet. Counsel.* **6**(2), 255–257 (1997).

10. D. Paul, *Controlling Human Heredity 1865 to the Present*, Humanities Press, Atlantic Highlands, NJ, 1995.

11. B. Fine, in D. Bartels, B. LeRoy, and A. Caplan, eds., *Prescribing Our Future: Ethical Challenges in Genetic Counseling*, Aldine de Gruyer, New York, 1993, pp. 101–117.

12. C. Rogers, *Counseling and Psychotherapy*, Houghton Mifflin, Boston, MA, 1942.

13. M. Yarborough, J. Scott, and L. Dixon, *Theor. Med.* **10**(2), 139–149 (1989).

14. B. LeRoy, in D. Bartels, B. LeRoy, and A. Caplan, eds., *Prescribing our Future: Ethical Challenges in Genetic Counseling*, Aldine de Gruyer, New York, 1993, pp. 39–54.

15. National Society of Genetic Counselors, *J. Genet. Counsel.* **1**(1), 41 (1992).

16. D. Wertz and J. Fletcher, *Am. J. Hum. Genet.* **42**, 592–600 (1988).

17. A. Czeizel, J. Metneki, and M. Osztovics, *J. Med. Genet.* **18**, 91–98 (1981).

18. P. Rowley, S. Loader, and R. Kaplan, *Am. J. Hum. Genet.* **63**(4), 1160–1174 (1998).

19. D. Asch et al., *Med. Decis. Making* **18**(2), 202–212 (1998).

20. D. Asch et al., *Am. J. Public Health* **86**(5), 684–690 (1996).

21. G. Loeben, T. Marteau, and B. Wilfond, *Am. J. Hum. Genet.* **63**, 1181–1189 (1998).

22. N. Press and C. Browner, *Soc. Sci. Med.* **45**(7), 979–989 (1997).

23. D. Wertz and J. Fletcher, *Soc. Sci. Med.* **46**(2), 255–273 (1998).

24. National Society of Genetic Counselors, Resolution Adopted 1987.

25. National Society of Genetic Counselors, Resolution Adopted 1995.

26. W. McKinnon et al., *JAMA* **278**(15), 1217–1220 (1997).

27. K. Nolan, *Hastings Center Rep.* **22**(4), S2–S4 (1992).

28. B. Knoppers et al., *Am. J. Hum. Genet.* **57**(4, suppl.), A296 (1995).

29. J. Botkin, *Hastings Center Rep.* **25**(5), 32–39 (1995).

30. E. Parens and A. Asch, *Hastings Center Rep.* (September–October) spec. suppl. 1–22 (1999).

31. S. Michie et al., *Am. J. Hum. Genet.* **60**, 40–47 (1997).

32. B. Bernhardt, *Am. J. Hum. Genet.* **60**, 17–20 (1997).

33. A. Caplan, in B. LeRoy, D. Bartels, and A. Caplan, eds., *Prescribing our Future: Ethical Challenges in Genetic Counseling*, Walter de Gruyter, New York, 1993, pp. 149–168.

34. M. Rose, J. Benkendorf, and M. Prince, *J. Genet. Counsel.* **7**(6), 497–498 (1998).

35. D. Wertz, *J. Genet. Counsel.* **7**(6), 499–500 (1998).

36. G. Wolff and C. Jung, *J. Genet. Counsel.* **4**(1), 2–25 (1995).

37. B. Bernhardt, B. Biesecker, and C. Mastromarino, *Am. J. Med. Genet.*, in press.

38. S. Kessler, *J. Genet. Counsel.* **8**(6), 333–343 (2000).

See other entries GENETIC COUNSELING; see also
REPRODUCTION, ETHICS entries; REPRODUCTION, LAW, REGULATION OF REPRODUCTIVE TECHNOLOGIES.

REPRODUCTION, LAW, IS INFERTILITY A DISABILITY?

LAURA F. ROTHSTEIN
Louis D. Brandeis School of Law
University of Louisville
Louisville, Kentucky

OUTLINE

Introduction
The Americans with Disabilities Act
    Major Statutory Provisions
    Definition of Who Is Protected
Other Applicable Statutes
    Pregnancy Discrimination Act of 1978
    Family and Medical Leave Act of 1993
    Health Insurance Portability and Accountability Act of 1996
    State Laws
Judicial Interpretation
Probable Future Directions
Bibliography

INTRODUCTION

Infertility is defined as the inability to conceive a child within a period of one year. There are many causes of infertility, and it may be attributable to either partner or a combination of factors related to both partners. Infertility can be costly and time-consuming to treat, and success is not guaranteed or even probable in many cases. The costs to infertile individuals and couples can involve money, time, and physical and mental health.

Working women and men who are infertile want to keep their jobs, even if they require leave time or scheduling changes for fertility treatment. The spouses of infertile partners also may require workplace accommodations to participate in fertility treatment. Both individuals want to have health insurance coverage that provides reimbursement for costly fertility treatments. For these reasons it is important whether infertility is considered a disability under federal or state disability discrimination law.

The Americans with Disabilities Act (ADA) of 1990 (1), the Rehabilitation Act of 1973 (2), and many state laws prohibit discrimination against individuals with disabilities by employers and providers of services, which may include health insurance. It is essential to determine whether a particular condition is a disability before applying the nondiscrimination and reasonable accommodation mandates of various statutes. It has yet to be determined whether infertility is considered a disability.

THE AMERICANS WITH DISABILITIES ACT

Major Statutory Provisions

Employment and Health Insurance Providers. Most employers are subject to one or more federal laws protecting individuals with disabilities from discrimination on the basis of their disabilities. The Americans with Disabilities Act of 1990 applies to private employers with 15 or more employees and to employees of state and local governmental agencies. The Rehabilitation Act of 1973 protects employees of federal agencies, most federal contractors, and recipients of federal financial assistance. Virtually all states have statutes covering public and private employers, although they vary in the number of employees necessary for an employer to be covered.

The application of disability discrimination law to health insurance providers is less clear. Title III of the Americans with Disabilities Act prohibits discrimination by 12 categories of private providers of programs and services to the public. While the weight of opinion is that Title III applies to health insurance providers, this has not been definitively decided. In addition, where health insurance is a benefit of employment, discriminatory treatment in an employer-provided health insurance program would be subject to the employment discrimination prohibitions.

Even if health insurance is covered by disability discrimination laws, insurance companies may be permitted to limit or exclude coverage for certain treatments in appropriate circumstances. The legality of such limitations and exclusions has yet to be clearly defined in the context of infertility treatment (3–6).

Nondiscrimination and Reasonable Accommodation. The major mandate of disability discrimination law is to prohibit discrimination against individuals with disabilities who are otherwise qualified. Lawmakers have recognized that most discrimination is not intentional, particularly in the context of disabilities. For that reason, facially neutral policies and practices that have a disparate impact on individuals with disabilities are subject to challenge as well, although not all will be found to be impermissible. For example, requiring that an employee have a drivers' license could have a disparate impact, and this might be challenged if driving is not an essential function of the position.

In addition to prohibiting discrimination, these laws also require employers to provide reasonable accommodations to known disabilities. Employers are not required to provide accommodation if it would constitute an undue hardship to do so. Undue hardship means significant difficulty or expense. Neither are employers required to lower performance standards or to make fundamental alterations to the program. Employees must be able to perform essential functions of the job if reasonable accommodations are provided, although the employer generally has the burden of showing that a particular function is essential.

Reasonable accommodations in the employment context might be the removal of architectural barriers, acquisition or modification of equipment, and other accommodations such as interpreters or readers. For an employee with fertility problems, the accommodations that might be sought would include job restructuring, part-time or modified work schedules, and reassignment to a vacant position.

Definition of Who Is Protected

Substantial Impairment, Regarded as, Record of. When most people think of disability discrimination laws, they think of individuals who are wheelchair users or those with visual or hearing impairments. Nevertheless, these laws cover a broad range of conditions and require not only nondiscrimination but reasonable accommodation. Whether infertility is to be included in statutory coverage is not clear on the face of the statutory language or the regulations.

Federal discrimination laws and many state discrimination laws define those to be protected similarly (7). Although some have urged that the definition be categorical and that specific impairments be listed to determine coverage, Congress specifically declined to do so. Instead individuals with disabilities are those who have a physical or mental impairment that substantially limits or more major life activities, those who have a record of such an impairment, or those who are regarded as having such an impairment.

The ADA regulations (1630.2 h) define a physical or mental impairment as "any physiological disorder, or condition, cosmetic disfigurement, or anatomical loss affecting one or more of [listed body systems]." These listed body systems include the reproductive system.

Major life activities are defined in the regulations as functions such as caring for oneself, performing manual tasks, walking, seeing, hearing, speaking, breathing, learning, and working. These listed activities are not intended to be all inclusive.

Substantially limited refers to being "unable to perform a major life activity that the average person in the general population can perform" or being "significantly restricted as to the condition, manner or duration under which an individual can perform a particular major life activity as compared to the condition, manner, or duration under which the average person in the general population can perform that same major life activity."

The requirement that the impairment be one that is substantially limiting is an important issue when considering infertility. This is because of the natural physiological changes that occur during the aging process which affect fertility, without an impact from disease, injury, or other condition that would affect fertility during the normal life cycle. This raises the question whether a woman who is in the average menopausal or postmenopausal age range would be considered disabled because she is no longer fertile or whether the definition only covers women in their twenties or thirties who have substantial difficulty conceiving.

Statutory language and the regulations adopted pursuant to statutes are essential starting points for determining what is prohibited and what definitions apply in a particular policy context. Consideration is also generally given to interpretations provided by federal and state agencies charged with enforcement or implementation of statutes. In the case of employment discrimination, the primary agency is the Equal Employment Opportunity Commission (EEOC).

The EEOC was the agency that promulgated the regulations for the employment portion of the ADA, Title I. These regulations include an interpretive appendix. In addition the EEOC has issued a number of separate interpretive guidelines on various aspects of Title I. Not all of the EEOC interpretations have received complete acceptance by the courts and commentators.

The EEOC has published a memorandum providing guidance as to the definition of disability (8). In that memorandum the EEOC did not clarify specifically whether fertility is a disability. It has been argued, however, that the EEOC memorandum indicates an intent that procreation is a major life activity. In this regard, since infertility substantially limits the ability to procreate, infertility thus should be considered a disability under the ADA (9). EEOC does indicate that the ADA should be read broadly. The EEOC's discussion of human immunodeficiency virus (HIV) has been argued to support a determination that infertility is a disability. The EEOC has indicated that even someone with HIV who is asymptomatic would be covered because of the impact of the virus on procreation. Although not all courts initially accepted this interpretation (10), this interpretation was applied by the Supreme Court in 1998 in *Bragdon v. Abbott*, which is discussed later in the section on judicial interpretations (10).

The Department of Justice (DOJ), an agency with major implementation responsibility for the Rehabilitation Act, similarly supports protection for individuals with HIV, even those who are asymptomatic (11). The DOJ position is that a person with HIV cannot procreate without significant fears about the impact of the virus on the child. Again, this reasoning was adopted by the Supreme Court in *Bragdon*.

It has been argued that the DOJ logic on application to individuals with HIV should extend to infertility. The argument is:

> If an asymptomatic HIV-infected individual is protected under the Act because the potential to pass the virus onto a biological child constitutes a substantial limitation of the major life activity of procreation, then an infertile person, whose physical impairment substantially limits his or her ability to procreate in the first instance, likewise should be afforded the protection of the Act (12).

Courts have reached a wide range of conclusions about the coverage of various conditions. In cases involving sensory or mobility impairments, the decisions generally turn on the severity of the condition and the nature of the employment. Cases that are more problematic involve medical conditions, such as cancer, diabetes, obesity, and heart disease. Infertility is one of these problematic conditions.

Associational Disability. In addition to protecting individuals who are themselves impaired, the ADA (and arguably the Rehabilitation Act) also protects individuals from discrimination based on their association with someone with a disability. For example, it would be impermissible for an employer to refuse to hire someone because he or she had a child who is mentally retarded.

While the individual associated with someone with a disability is protected from discrimination, federal law does not require that reasonable accommodations be provided based on "associational disability." For example, while it would be impermissible to fire an employee because it was learned that the employee's daughter had suffered severe brain damage in an automobile accident, the employer is not required under the ADA or the Rehabilitation Act to provide an accommodation of allowing the employee time off to take the daughter for medical treatment or rehabilitation (14). The Family and Medical Leave Act (15), however, may provide relief to the employee in such a case, but nondiscrimination statutes will not.

This is significant with respect to infertility. Even if it were decided that infertility is a disability, accommodations would only be required for the partner with the medical condition, not for the other partner, whose presence may be necessary for certain infertility treatments.

OTHER APPLICABLE STATUTES

The importance of finding protection under disability discrimination statutes is highlighted when viewed in the context of other laws that might provide some protection for individuals and couples with fertility problems. As noted below, while these statutes are of some help, they do not provide the same level of substantive protection that would be available under the ADA or the Rehabilitation Act.

Pregnancy Discrimination Act of 1978

The Pregnancy Discrimination Act (PDA) of 1978 is a amendment to Title VII of the Civil Rights Act of 1964 (16). This statute prohibits employers from discrimination on the basis of "pregnancy, childbirth, or related medical conditions." The PDA does not require reasonable accommodation, so even if it were applied to an individual who is infertile, it is unlikely to be an avenue for the type of remedy being sought, namely accommodations in the work schedule and coverage of fertility treatment by an health insurance provider. It would only provide protection against an employer terminating employment or otherwise adversely treating an employee because of such a condition.

Several courts have found that infertility is a pregnancy-related condition under the PDA. One case involved an employee whose employment was allegedly terminated because of her use of sick leave and vacation days to undergo fertility treatment. The court held that such action was subject to review under both the PDA (because infertility is a pregnancy-related condition) and the ADA (because infertility was determined to be a disability) (17).

Family and Medical Leave Act of 1993

While the PDA is unlikely to be a statutory basis for a leave of absence, the Family and Medical Leave Act (FMLA) of 1993 (18) does provide for such a leave in appropriate circumstances. The FMLA applies to employers with 50 or more employees, and it requires employers to provide up to 12 weeks of unpaid leave in a 12-month period of time. The leave is required only for the birth, adoption, or placement for foster care of a child; for care of a child, spouse, or parent with a *serious* health condition; or for the employee's own *serious* health condition that results in the employee's inability to perform the job. The term *serious health condition* is defined as one that involves inpatient care or continuing treatment by a health care provider. Neither the statute nor interpretations of the statute have discussed the potential applicability of the FMLA to infertility.

Health Insurance Portability and Accountability Act of 1996

The Health Insurance Portability and Accountability Act (HIPAA) of 1996, also known as Kennedy-Kassebaum (19), applies to employer-provided group health plans and group health insurance issued by private providers. HIPAA was intended to allow individuals to change group health insurance coverage without being unduly penalized. After initial eligibility with the first group health insurance plan, a transfer to subsequent plans should not adversely affect the individual. The covered employers and insurers may not deny coverage, or discriminate in eligibility, enrollment, or premium rates based on preexisting conditions. For the individual with infertility, the only benefit would be that if the individual is covered for fertility treatments by a health insurance plan subject to HIPAA, and the individual changes jobs, there would be no preexisting condition exclusion and no waiting period for coverage if fertility treatment is covered by the new employer's health insurance plan.

State Laws

Many states have statutes that are similar to PDA, FMLA, and the ADA. In general, state law interpretation often mirrors federal statutory applicability and interpretation, although there are a few notable exceptions. No state law clearly protects individuals with infertility problems.

JUDICIAL INTERPRETATION

The courts have addressed whether infertility is a disability in several cases. Two early cases have been subject to substantial commentary. Unfortunately, they have reached opposite conclusions, and there is not yet a definitive resolution of this issue as a result.

Both cases involved individuals who had been employed for some time. In *Pacourek v. Inland Steel Co., Inc.* (20) an employee with 10 years of service was dismissed because of her absences related to fertility treatments. The court determined that unexplained fertility is a physical impairment under the ADA. It further decided that reproduction is a major life activity, based on inference from EEOC interpretation and other judicial decisions, and that infertility is substantially limiting to this major life activity. Therefore infertility is a disability under the ADA.

The court relied on an earlier federal appellate court decision, *McWright v. Alexander* (21) in which the court had indicated that the Rehabilitation Act protected individuals with physiological disorders affecting the reproductive system. In *McWright*, the individual was seeking leave time to care for an adopted baby. Ms. McWright was unable to bear children as a result of childhood polio.

Zatarain v. WDSU-Television, Inc. (22) also involved a long-time employee. Ms. Zatarain was a television news anchor whose fertility treatments were initially accommodated. Eventually, however, her contract was not renewed after negotiations involving accommodations to her treatment. The court rejected her ADA claim, deciding that reproduction is not a major life activity. The court's reasoning was that other examples of major life activities in ADA regulations (e.g., walking, seeing, speaking, breathing, learning, and working) are done throughout the day, every day. Because one does not reproduce throughout the day, every day, this is not a major life activity. Neither is she substantially limited in working because she is not "significantly restricted in the ability to perform either a class of jobs or a broad range of jobs in various classes" because of her condition (23). The reasoning in *Zatarain* has been criticized as not being an appropriate interpretation of EEOC guidance (24).

It is noteworthy that these two cases addressing infertility as a disability, cases that result in conflicting definitions, had been decided at the time EEOC issued its 1995 interpretive guidance. The failure of the EEOC to specifically clarify its position on infertility is therefore troubling.

While *Pacourek* and *Zatarain* are the first major cases addressing infertility as a disability, a case decided after these cases is the first to reach a federal appellate court level. In *Krauel v. Iowa Methodist Medical Center* (25), the court considered the denial of health insurance coverage for a surgical procedure for a woman with endometriosis, a condition affecting her fertility. The court held that infertility is not a disability because it does not interfere with performing her job duties as a respiratory therapist, and that it would be a "considerable stretch of federal law" to treat it as a disability (25).

In one of the few cases involving health insurance coverage, rather than termination of employment, a court in the same jurisdiction as the *Pacourek* decision decided that a police officer with an ovarian dysfunction and infertility was protected as disabled under the ADA (26). The claim was that the employer had violated the ADA in denying health insurance coverage for fertility treatments.

The same court also determined that an employee with an incompetent cervix, which compromised her ability to carry a fetus to term, was protected as disabled under the ADA (27).

In 1998 the Supreme Court, in *Bragdon v. Abbott*, answered some of the questions about whether infertility should be treated as a disability under federal discrimination laws (28). The case involved a plaintiff who was a woman of child-bearing age who was HIV positive but asymptomatic. When she sought treatment from a dentist, he examined her in his office but indicated that he would only fill her cavity in a hospital because of her HIV status, and that she would have to bear the additional costs of hospital treatment. She brought suit under Title III of the ADA, claiming discrimination on the basis of disability by a private provider of a public accommodation. The Supreme Court addressed the issue of whether being HIV positive, but asymptomatic, is a disability under the ADA. The Court held that for this plaintiff it is.

The Court first determined that reproduction is a major life activity, by relying on the plain meaning of the statute and congressional intent to construe the statute to be consistent with regulations under the Rehabilitation Act. The second part of the test is whether one is "substantially limited" in that major life activity. Again, the Court held that Ms. Abbott is substantially limited because it affected her decision to conceive because of the significant risk to the partner as well as to the child, and this plaintiff had provided unchallenged testimony that her HIV infection controlled her decision not to have a child. So for Sidney Abbott, her HIV status was a substantial limitation to a major life activity.

Applying this analysis to infertility, it would seem clear that just as in the case of HIV, reproduction would be a major life activity for purposes of determining whether someone who is infertile is protected. The resolution of the question becomes more difficult, however, in applying the second part of the test, that is, whether for a particular individual, infertility is a substantial limitation to a major life activity. At first the answer might seem to be that it clearly is. It is not entirely clear, however, that all individuals who are infertile will be protected, just as the *Bragdon* Court did not hold that all individuals with HIV are automatically protected. Similarly it is not entirely clear that all individuals who are infertile will be automatically covered under the definition.

PROBABLE FUTURE DIRECTIONS

The initial split of opinion by the courts has been reinforced by other subsequent decisions. At the time of this writing, there have been several cases in which the court have determined that infertility is a disability. Most of these decisions, however, have been made by the same federal court in Illinois, so the number of decisions does not necessarily indicate the weight of authority.

The Supreme Court has, however, seemingly resolved the split to some degree in the *Bragdon* decision. The Court has at least decided that reproduction is a major life activity. This decision could certainly be extended to determine that infertility is a substantial impairment to

a major life activity. What is unresolved is whether all individuals who are infertile will be covered.

As was previously noted, substantially limited means being "unable to perform a major life activity that the average person in the general population can perform" or being "significantly restricted as to the condition, manner or duration under which an individual can perform a major life activity as compared to the condition, manner, or duration under which the average person in the general population can perform that same major life activity." This may be interpreted to apply only to individuals who are in the normal age range for child bearing or fertility. This will thus be different for males and females.

Courts applying this analysis to cases involving infertility may examine whether the individual who is infertile is within the normal age range for fertility. Thus the 63-year-old may not be protected as disabled, while a 25-year-old woman would be. The fact that men are generally considered to be fertile under normal conditions throughout their adult lives raises some interesting disparate treatment issues. Would the 63-year-old man be considered disabled if he became impotent as a result of prostate cancer treatment?

Even if it is definitively decided that infertility is a disability under discrimination statutes, that is only the first step in receiving statutory protection. The individual must also be able to carry out the essential functions of the position with or without reasonable accommodations. The employer will generally have the burden of demonstrating what the essential functions of the job are and proving that accommodations such as schedule changes are unduly burdensome or fundamentally alter the program or lower standards. This will necessarily involve an individualized determination. And even if it is reasonable to accommodate an employee who is infertile, employers will not be required to make accommodations for employees where it is the partner with the infertility problem requiring the employee's presence for fertility treatment.

Finally, assuming that infertility is considered a disability and assuming that the accommodations sought are reasonable, there remain unresolved policy questions as to whether employers should be required to accommodate, and insurance companies to provide, health insurance coverage for an individual on an indefinite or undefined basis. Should the protections be extended to the 63-year-old woman who is seeking to conceive a child or to the woman who has already given birth to septuplets and seeks additional pregnancies? How will the application of disability discrimination law apply to individuals seeking access to health insurance coverage for drugs such as Viagra? These questions remain for the policy makers regardless of the direction taken by the courts and regulatory agencies in interpreting existing law.

BIBLIOGRAPHY

1. Americans with Disabilities Act of 1990, Pub. L. No. 101-336, 104 Stat. 327; codified at 42 U.S.C. §§12101 *et seq.* (1994).

2. Rehabilitation Act of 1973, 29 U.S.C. §§706, 791, 793–794 (1994).

3. D. Dallman, *William & Mary Law Rev.* **38**, 371–415 (1996).

4. B. Gilbert, *Defense Counsel J.* 42–57 (January 1996).

5. D. Millsap, *Houston Law Rev.* **32**, 1411–1450 (1996).

6. D. Millsap, *Am. J. Law Med.* **22**, 50–84 (1996).

7. Americans with Disabilities Act Regulations, 29 CFR Part 1630 (1997).

8. EEOC Compliance Manual (BNA) §902 (March 14, 1995).

9. S.M. Tomkowicz, *Syracuse Law Rev.* **46**, 1051–1092 (1996).

10. *Runnebaum v. NationsBank of Maryland*, 123 F.3d 156 (4th Cir. 1997).

11. Department of Justice, Rehabilitation Act Memorandum on the Application of Section 504 of the Rehabilitation Act to HIV-Infected Persons, from Douglas W. Kmiec (September 27, 1988).

12. S.M. Tomkowicz, *Syracuse Law Rev.* **46**, 1051–1092, 1071 (1996).

13. Interpretive Guidance to Title I of the Americans with Disabilities Act, 50 Fed. Reg. 35742 (July 26, 1991).

14. *Tyndall v. National Education Centers*, 31 F.3d 209 (4th Cir. 1994).

15. Family and Medical Leave Act, 29 U.S.C. §§2601–2654 (1994).

16. Pregnancy Discrimination Act of 1978, 42 U.S.C., §2000e(k).

17. *Landfair v. Sheahan*, 911 F. Supp. 323 (N.D. Ill. 1995).

18. Family and Medical Leave Act, 29 U.S.C. §§2601–2654 (1994).

19. Health Insurance Portability and Accountability Act of 1996, 42 U.S.C. §300gg (1997).

20. *Pacourek v. Inland Steel, Co., Inc.*, 916 F. Supp. 797 (N.D. Ill. 1996),

21. *McWright v. Alexander*, 982 F.2d 222 (7th Cir. 1992).

22. *Zatarain v. WDSU-Television, Inc.*, 881 F. Supp. 240 (E.D. La. 1995).

23. 29 CFR 1630.2(j)(3)(i).

24. S.M. Tomkowicz, *Syracuse Law Rev.* **46**, 1051–1092 (1996).

25. *Krauel v. Iowa Methodist Medical Center*, 95 F.3d 674 (8th Cir. 1996).

26. *Bielicki v. City of Chicago*, 10 NDLR ¶28 (N.D. Ill. 1997).

27. *Soodman v. Wildman, Harrold, Allen & Dixon*, 9 NDLR ¶200 (N.D. Ill. 1997).

28. *Bragdon v. Abbott*, 1998 U.S. LEXIS 4212 (1998).

See other entries DISABILITY AND BIOTECHNOLOGY; REPRODUCTION, ETHICS, PRENATAL TESTING, AND THE DISABILITY RIGHTS CRITIQUE; REPRODUCTION, LAW, REGULATION OF REPRODUCTIVE TECHNOLOGIES; REPRODUCTION, LAW, WRONGFUL BIRTH, AND WRONGFUL LIFE ACTIONS.

REPRODUCTION, LAW, REGULATION OF REPRODUCTIVE TECHNOLOGIES

KAYHAN P. PARSI
Institute for Ethics, American Medical Association
Chicago, Illinois

OUTLINE

INTRODUCTION

Almost as fascinating as the reproductive technologies are the cultural responses to the new methods of controlling, enhancing, or limiting individuals' abilities to procreate. Writers have often invoked Aldous Huxley's *Brave New World* as a predictor of things to come, as if a single novel can faithfully capture the complexity and richness of reproductive technology. In reality, reproductive technologies vary a great deal, from the low tech, such as surrogacy and artificial insemination, to more highly sophisticated methods, such as intracytoplasmic sperm injection (ICSI). Regulation of these technologies by professional associations, states, and countries also varies considerably. While these reproductive technologies have created interesting and complicated abstract ethical issues, this article will only examine the regulatory responses to such technologies. Although regulation is often associated with the prohibition of specific activities, it is also enabling by allowing key parties to execute particular agreements in the area of reproductive technologies. This enabling aspect of regulation is in accordance with a liberal notion of procreative rights.

Commentators such as John Robertson generally support an individual's ability to choose a procreation method. Robertson believes that Americans have a procreative liberty interest. The very title of his book, *Children of Choice*, reflects a very American attitude: Choice is an inherently good thing, and this applies to reproductive technologies as well (1). Robertson's perspective epitomizes the viewpoint that reproductive technologies enhance and expand an individual's ability to make choices regarding procreation. Robertson acknowledges that although the technologies may produce ambivalence about their use, limiting persons' reproductive freedoms would curtail one of the most fundamental aspects of our lives. Other commentators, such as Dorothy Roberts, observe a darker side to certain kinds of reproductive technologies. Instead of focusing on the standard set of reproductive technologies, such as in vitro fertilization (IVF) or cryopreservation, Roberts concentrates on technologies that limit the ability to procreate (2). Roberts's concerns focus on utilization of drugs, such as Depo-Provera and Norplant, and sterilization of poor people. She is critical of the general acceptance of reproductive technologies to enhance wealthy people's ability to procreate, while governments employ the aforementioned methods to curtail poor—especially black—women's ability to procreate. As exemplified by Robertson's and Roberts's viewpoints, goals of reproductive technology are quite diverse.

U.S. CASE LAW

In the United States, a number of cases provide a basis for Robertson's notion of procreative liberty. One of the earliest cases to assert a constitutional right to procreate was *Skinner v. Oklahoma* (3). This case overturned the now infamous sterilization case *Buck v. Bell* (4). In *Skinner*, the Supreme Court declared unconstitutional the Oklahoma Habitual Criminal Sterilization Act, which allowed for the sterilization of habitual criminals (2). The Court's language accorded reproductive freedom a high level of deference: "Oklahoma deprives certain individuals of a right which is basic to the perpetuation of a race-the right to have offspring ... we are dealing here with legislation which involves one of the basic civil rights of man." *Griswold v. Connecticut* (6) was another landmark case where the Supreme Court further expanded the notion of procreative liberty. The Court struck down a Connecticut statute that prohibited the distribution of contraceptives to married couples. This statute, the Court held, violated married couples' constitutional privacy rights. Justice Douglas wrote that "the First Amendment has a penumbra where privacy is protected from governmental intrusion. ...We deal with a right of privacy older than the [Bill of Rights]. Marriage ... is an association for as noble a purpose as any involved in our prior decisions" (5). In *Eisenstadt v. Baird* (6), the Court expanded the privacy right beyond *Griswold's* realm of marriage. In his opinion, Justice Brennan wrote:

> If the right of privacy means anything, it is the right of the individual, married or single, to be free from unwarranted governmental intrusion into matters so fundamentally affecting a person as the decision whether to bear or beget a child (7).

The Court again addressed the issue of contraception in *Carey v. Population Services International* (7), by striking down a New York statute that criminalized distribution of contraceptives to minors under 16, prohibited nonpharmacists from distributing contraceptives to people over 16, and banned any advertising or displaying of contraceptives. The *Carey* Court held that:

> [I]t is clear that among the decisions that an individual may make without unjustified government interference are personal decisions relating to marriage...; procreation...; family relationships...; and child rearing and education. The decision whether or not to beget or bear a child is at the very heart of the cluster of constitutionally protected choices (7).

After Carey, procreative liberty rights continued to grow. The now-famous Supreme Court case of *Roe v. Wade* (8) expanded privacy rights to permit women in collaboration with their physician to have an abortion. In *Planned Parenthood v. Casey* (9), the Supreme Court, based on its "liberty" rights reasoning, argued that abortion is an essential liberty. The liberty argument in *Planned Parenthood* focused more on control of one's own body, as opposed to limiting the government's ability to control reproduction, which was the language used in the aforementioned privacy cases (10).

The foregoing cases strongly suggest a basic negative right to procreative decision making and influence a variety of reproductive technologies—such as surrogacy, IVF, artificial insemination, and contraception. The remainder of this article will explain how different entities regulate various reproductive technologies; however, the discussion does not evaluate an exhaustive list of all possible reproductive technologies. First, a description of the guidelines promulgated by certain professional groups, such as the American Medical Association's (AMA) Council on Ethical and Judicial Affairs (CEJA) and the American Society of Reproductive Medicine (ASRM), will be given. The article will then examine state responses in the form of case law and legislation. It will then look at different approaches by countries and then finally different approaches by international bodies (e.g., the Council of Europe) to these technologies. A variety of approaches are represented, from the U.S. free-market approach to the more heavily regulated and centralized UK approach.

PROFESSIONAL GROUP GUIDELINES

Professional guidelines are a useful place to start when examining regulatory aspects of reproductive technologies. Although they do not have the force of law, they often do inform legal cases. Moreover, in the absence of any national legal or ethical consensus regarding these technologies in the United States, professional guidelines provide physicians, researchers, and ethicists with some guidance. The AMA's Code of Medical Ethics addresses the following reproductive technology issues: artificial insemination, IVF, freezing pre-embryos, pre-embryo splitting, and surrogacy.

Artificial Insemination

The AMA Code (11) makes the following requirements for recipients of artificial insemination: counseling, informed consent (e.g., risks, benefits, and alternative treatments), and information regarding the conceived child's legal status. The Code stipulates that sex selection is only allowed to avoid an inheritable sex-linked disease. Posthumous use of frozen sperm, according to the instructions of the decedent, is allowed (see the later California case *Hecht*). The Code also requires rigorous screening of potential donors for infectious or inheritable diseases, recommends the use of frozen semen (to ensure freedom of HIV infection), and advises physicians to use the professional guidelines set out by the ASRM, the Centers for Disease Control and Prevention (CDC), and the Food and Drug Administration (FDA). Physicians are also required to maintain permanent records of the sperm donors, reflecting health and genetic information that is both identifying and nonidentifying. The Code recommends obtaining the consent of the husband if he will become the father through artificial insemination by anonymous donor. Unlike certain European guidelines, the Code does not prohibit single or lesbian women from obtaining artificial insemination by anonymous donor. Last, the Code admonishes compensating donors beyond incidental expenses such as time.

Regarding IVF, the Code prohibits using fertilized ova that will later be implanted to be used for laboratory research. Those fertilized ova that will not be implanted may be used for research purposes, but only in accordance with the Code's fetal research guidelines. The following guidelines are offered as aids to physicians when they are engaged in fetal research:

1. Physicians may participate in fetal research when their activities are part of a competently designed program, under accepted standards of scientific research, to produce data that are scientifically valid and significant.

2. If appropriate, properly performed clinical studies on animals and nongravid humans should precede any particular fetal research project.

3. In fetal research projects, the investigator should demonstrate the same care and concern for the fetus as a physician providing fetal care or treatment in a nonresearch setting.

4. All valid federal or state legal requirements should be followed.

5. There should be no monetary payment to obtain any fetal material for fetal research projects.

6. Competent peer review committees, review boards, or advisory boards should be available, when appropriate, to protect against the possible abuses that could arise in such research.

7. Research on "dead fetus," macerated fetal material, fetal cells, fetal tissue, or fetal organs should be in accord with state laws on autopsy and state laws on organ transplantation or anatomical gifts.

8. In fetal research primarily for treatment of the fetus:

   a. Voluntary and informed consent, in writing, should be given by the gravid woman, acting in the best interest of the fetus.

   b. Alternative treatment or methods of care, if any, should be carefully evaluated and fully explained. If simpler and safer treatment is available, it should be pursued.

9. In research primarily for treatment of the gravid female:

   a. Voluntary and informed consent, in writing, should be given by the patient.

   b. Alternative treatment or methods of care should be carefully evaluated and fully explained to the patient. If simpler and safer treatment is available, it should be pursued.

   c. If possible, the risk to the fetus should be the least possible, consistent with the gravid female's need for treatment.

10. In fetal research involving a fetus in utero, primarily for the accumulation of scientific knowledge:

    a. Voluntary and informed consent, in writing, should be given by the gravid woman under circumstances in which a prudent and informed adult would reasonably be expected to give such consent.

    b. The risk to the fetus imposed by the research should be the least possible.

    c. The purpose of research is the production of data and knowledge that are scientifically significant and that cannot otherwise be obtained.

    d. In this area of research, it is especially important to emphasize that care and concern for the fetus should be demonstrated (12).

The Code makes the gamete providers the primary decision makers when exerting control over a frozen pre-embryo. The providers are prohibited from selling their gametes, but they are allowed to donate them to others. Research is also prohibited on a pre-embryo if it will later be implanted in a woman. Interestingly, however, the Code allows pre-embryos to thaw and deteriorate. With regard to use of pre-embryos, the Code requires consent of both providers and encourages the use of agreements between providers in case the couple divorces. The Code permits pre-embryo splitting with the agreement of both gamete providers. Pre-embryo splitting allows a greater chance for conception while diminishing the number of potentially painful procedures to procure eggs (13).

Surrogacy

Surrogacy is one of the oldest forms of reproductive technologies. It is referred to in the Bible when Abraham's servant Hagar bears Abraham's child to be raised by him and his wife Sarah (14). It is the focus of such recent dystopian novels as *The Handmaid's Tale* by Margaret Atwood. And it has raised a number of ethical and legal concerns in recent years. Accordingly the AMA Code cites certain concerns regarding surrogacy agreements: commodification of children, exploitation of poor women, and interference with the natural maternal–child bond. It also raises potential psychological problems and the possibility that the mother may want to have an abortion or even refuse giving up the child. The intended parents may not even want the child if it happens to be born with disabilities. Despite major criticisms of surrogacy contracts, the Code permits them with certain safeguards. For instance, the birth mother should have:

> the right to void the contract within a reasonable period of time after the birth of the child. If the contract is voided, custody of the child should be determined according to the child's best interests. In gestational surrogacy, in which the surrogate mother has no genetic tie to the fetus, the justification for allowing the surrogate mother to void the contract becomes less clear. Gestational surrogacy contracts should be strictly enforceable (i.e., not voidable by either party) (15).

Cryopreservation

ASRM outlines a number of concerns with regard to cryopreservation. For instance, it encourage couples who are considering storage to put into writing what they wish to happen to their stored embryos in the following instances: death, divorce, separation, failure to pay storage charges, inability to agree on disposition in the future, or lack of contact with the program. ASRM requires that the consent form allow disposal of embryos if the couple lose contact with the program after some period of time and if they do not provide the program with key contact information, such as addresses and phone numbers. The ASRM guidelines allow a couple to revise their initial directions for embryo disposition by drafting a new set of written directions. Moreover, ASRM does not make rigid requirements with regards to embryos that lack written directions for disposal. In the absence of clear legal guidelines, programs may want to store indefinitely or accept the risk of liability by disposing of embryos after attempts to contact the couple have failed and a lengthy period of time has elapsed. ASRM approaches the issue of embryo preservation and storage pragmatically, in that a couple cannot claim an injury if they have failed to provide written directions, have lost contact with the program, and have not provided current address and phone information. Simply put, ASRM does not believe that programs have the ethical obligation to store embryos indefinitely. ASRM does state, however, that a period of five years should elapse and that "diligent effort" by phone and registered mail to contact the couple should be made by the program. Last, ASRM does not allow abandoned embryos to be used for research or donation to another couple without appropriate consent. The embryos must be thawed and allowed to deteriorate (16).

Embryo Splitting

Neither AMA nor ASRM prohibits embryo splitting (a technique whereby in vitro fertilized pre-embryos are split to create genetically identical siblings). Both organizations argue that splitting provides certain benefits: splitting provides a greater number of embryos, increasing the chances of a successful pregnancy and may prevent additional invasive procedures to retrieve more embryos, which are painful and costly. Neither AMA nor ASRM finds persuasive ethical concerns about using split embryos as a source of organs or tissues for an existing child, or the sale of stored embryos with desirable genomes based on the appearance of characteristics of existing children. For instance, the AMA Code recommends a complete prohibition of the sale of pre-embryos. Both the Code and ASRM acknowledge that offspring with identical genomes may be born at different times. ASRM raises the issue of "personal identity and the meaning of being a twin that require further investigation before it can be determined that such transfers are ethical." It recommends that such scenarios can be avoided "by transferring all genetically identical embryos in the same cycle." The Code permits couples to transplant genetically identical siblings in order to harvest their tissue for a needy sibling. The Code discounts any accrual of psychological harm and argues

that the sibling may indeed gain psychological benefits by saving his or her sibling. In sum, both the ASRM and the Code take a consequentialist approach in that they believe that the benefits outweigh any costs in allowing embryo splitting (13,18).

Use of Fetal Oocytes in Assisted Reproduction

ASRM sees a number of problems with fetal oocyte donation: emotional harms to children, informed consent dilemmas, and an impersonalizing influence in assisted reproduction. Because of these concerns ASRM does not believe this technology should be pursued (18).

Posthumous Reproduction

Similar to surrogacy, posthumous births have an ancient lineage. These kinds of births routinely occurred when a woman conceived and her husband or partner died before she delivered the child. This child was commonly considered the legal heir of the father. ASRM, however, recognizes that posthumous reproduction became an issue when semen could be frozen and later implanted after the death of the donor. Although the AMA Code only makes a fleeting commentary regarding posthumous reproduction, ASRM has extensive commentary. ASRM cites a variety of scenarios where such reproduction may take place: A widow may retrieve a dead husband's sperm to procreate, sperm may be retrieved from terminally ill spouses or partners (employing techniques such as stimulated ejaculation, microsurgical epididymal sperm aspiration, MESA, or testicular sperm extraction, TSE), or sperm from a dead anonymous donor may be used. Moreover, a man facing radiation therapy or chemotherapy may want to store his semen for later (possibly posthumous) use.

Cryopreserving ova for posthumous procreation also poses certain concerns, but the inability to successfully freeze ova (as compared to semen and pre-embryos) has limited this particular technology. ASRM permits the designation of frozen gametes or embryos to be used in posthumous procreation as long as the key parties involved are fully informed. However, the absence of clear and written instructions would preclude posthumous reproduction. Moreover, ASRM believes that the requests of a living spouse should not override the express wishes of the deceased spouse.

ASRM also permits a husband to use his deceased wife's ova for implantation in a surrogate. Although the surrogate would not be considered a traditional surrogate, the ASRM requires that the surrogate be made aware of the circumstances and informed that she would be involved in a posthumous pregnancy. The rearing parents may lack genetic ties to the dead donor and may not be involved in the gestating pregnancy. They should, however, be made aware of the deceased status of the donors of gametes and embryos. Although the ASRM guidelines are supportive of posthumous reproduction, they do cite some reservations:

> ...when reproduction takes place as a consequence of a loving relationship in which both partners were desirous of children, but a pregnancy is frustrated by the death of one partner, posthumous reproduction would ordinarily be well accepted both socially and culturally.... There is less certainty of the

impact on the child and more caution should be exercised for posthumous reproduction that occurs with the use of donated gametes from unrelated individuals who are not living and may have been deceased for several years, as may occur with the use of commercial banks as a source for sperm, frozen embryos, or ova (19).

STATE LAW AND REGULATIONS

Reflecting its organization, American state law that regulates reproductive technologies is a hodgepodge of case law, statutes, and administrative regulations. Certain issues, however, have been addressed extensively. One of them is surrogacy. One of the early surrogacy cases was *In The Matter of Baby M* (20). In this case, William Stern entered into a surrogacy contract with Mary Beth Whitehead. Elizabeth, Stern's wife, was infertile, and the Sterns were hoping to be able to have a baby with the assistance of Whitehead. The contract was for the sum of $10,000. In early 1985, Mr. Stern and the Whiteheads executed a surrogacy agreement. To avoid revealing the nature of the agreement, the child's birth certificate listed the Whiteheads as parents. It became clear very quickly that Mrs. Whitehead did not wish to part with her baby. Despite initial misgivings, Mrs. Whitehead gave the baby up to the Sterns. Whitehead underwent a depression and threatened suicide. The Sterns, therefore, returned the baby to Whitehead for a short visit. Thereupon, Whitehead left with the baby to Florida for four months. The Sterns's complaint, in addition to seeking possession and ultimately custody of the child, sought enforcement of the surrogacy contract. Pursuant to the contract, the Stern's asked that the child be permanently placed in their custody, that Mrs. Whitehead's parental rights be terminated, and that Mrs. Stern be allowed to adopt the child.

The Supreme Court of New Jersey invalidated the agreement and named Whitehead the mother of the child. The judge in this case noted certain ethical problems, such as commodification, with regard to surrogacy.

Because of commodification concerns, a number of states do prohibit enforcement of surrogacy agreements. For instance, Arizona, the District of Columbia, Michigan, and Utah prohibit all surrogacy agreements, whereas Kentucky and Louisiana prohibit commercial surrogacy agreements. Some states permit unpaid surrogacy agreements: Florida, Nevada, New Hampshire and Virginia. Florida, New Hampshire and Virginia require that the intended mother be infertile (21).

Another important surrogacy case was *Johnson v. Calvert* (22). In this case Mark and Crispina Calvert entered into a contract with Anna Johnson for Johnson to carry the Calverts's fertilized embryo (Crispina Calvert had underwent a hysterectomy but was able to produce ova). The Calverts agreed to pay Johnson $10,000; in return, Johnson would relinquish her parental rights to the child in favor of the Calverts. Near the end of the pregnancy, Johnson demanded immediate payment or she would not deliver the child after birth to the Calverts. The Johnsons sued, seeking a declaration that they were the legal parents of the child. Calvert countersued. The

case wound its way through the California judicial system, until the Supreme Court of California heard it in 1993. The California court took a very different approach than the New Jersey court in *Baby M*. Here, the court was not concerned about potential exploitation or commodification. Moreover it found such concerns to be paternalistic and condescending toward women. The court held that the woman who intended the birth of a child that she intended to rear was the natural mother under California law.

Another issue that has been addressed by state case law is embryo storage and disposal. Perhaps the most famous case involving frozen embryos is the Tennessee case *Davis v. Davis* (23). In this case, a couple undergoing a divorce were arguing over the disposition of their frozen pre-embryos. The Davises never executed a disposition agreement regarding their embryos. Mary Sue Davis wished to achieve a pregnancy after their divorce, whereas Junior Davis wanted to avoid becoming a father entirely. In the absence of any statutory authority, the Tennessee Supreme Court stated that the pre-embryos occupy an intermediate status between property and persons. The court ruled that the fate of the frozen pre-embryos should be decided by "the party wishing to avoid procreation" if the other party has a reasonable possibility of achieving parenthood by other means and the parties have not made an agreement regarding their disposition. A later New York case, *Kass v. Kass* recognized the legal enforceability of a disposition agreement regarding frozen embryos. The court ruled that the couples' agreement controlled the fate of their embryos (24). Last, a couple of state statutes provide some guidance: Florida law requires that couples execute disposition agreements when undergoing a reproductive technology procedure (25) and Louisiana law defines the embryo as a "juridical person," limiting the ability of progenitors to dispose of their embryos (26).

Finally a case that raised the issue of posthumous use of sperm is *Hecht v. Superior Court of LA County* (27). In this case William Kane "willed" a vial of his own sperm to his girlfriend Deborah Hecht. He thereupon took his own life. Despite the fact that Kane's intent was clear, his ex-wife and children challenged his bequeathal. The probate judge initially ordered that the sperm be divided according to the original property settlement. The Court of Appeals later ruled that the remaining vials of sperm, which were in the custody of the administrator, be delivered to Hecht. The court argued that Kane had a legitimate property interest in his own sperm. Analogous to the Davis court, the Hecht court argued that the sperm occupied an interim category of property because of its potential for life. The court's reasoning in this regard was not dissimilar from the policy guidelines of ASRM.

FEDERAL GUIDELINES

Although federal regulation of assisted reproductive technologies is very weak in the United States, Congress has made some attempts to impose some regulation in this area. For instance, in the early 1990s, Congress passed the Fertility Clinic Success Rate and Certification Act of 1992: A Model Program for the Certification of Embryo Laboratories (28). This act required the Secretary of the Department of Health and Human Services, through CDC, to develop a model program for the certification of embryo laboratories. This program was to be carried out voluntarily by interested states (28).

INTERNATIONAL APPROACHES

In addition to the United States, other countries, such as Canada, the UK and Australia, have all addressed the legal and regulatory aspects of reproductive technologies. Despite Canada's tradition of strong social solidarity reflected in its national health insurance system, no federal guidelines have been created to govern the practice of IVF, artificial insemination, as well as egg and sperm donation. Similarly, although no specific law prohibits surrogacy contracts, they would not stand up in court because they violate Canadian contract and family law principles. Recommendations do exist, however, for storage of gametes and embryos — 10 years and 5 years, respectively (29). The Law Reform Commission's 1992 Working Paper recommends that "[t]he commercialization of donated gametes and embryos must be prohibited outright. Allowing gametes and embryos in the consumer market would constitute a direct assault on human dignity" (30). After many years and millions of dollars, Canada's Royal Commission on New Reproductive Technologies issued its voluminous report *Proceed with Care*. The report's recommendations were essentially prohibitive in nature. In the aftermath of the Commission's issuance of *Proceed with Care*, a liberal government bill, C-47 (The Reproductive and Genetic Technologies Act), was on its way to passage in 1996 but died because a federal election was called (31). This bill would have prohibited the following practices and procedures: commercially exchanging sperm and eggs, cloning, fusing animal and human zygotes, implanting a human embryo into an animal (and vice versa), altering the genetic structure of the germ line, retrieving an ovum or sperm from a cadaver with the intention of using it in a live recipient, maintaining a human embryo outside of a human body, fertilizing an ovum for purposes of research, and commercial surrogacy (32).

In England, the focus shifted from the status of the fetus to the status of the embryo in the aftermath of the birth of Louise Brown, the first "test tube" baby. Public alarm about the untrammeled growth of reproductive technologies motivated Parliament to explore legislative measures (33). The Committee of Inquiry into Human Fertilization and Embryology (the Warnock Committee) took its charge into examining these issues seriously. The Warnock Committee's work in the reproductive health arena left an indelible stamp. The Committee recognized a special status for the embryo but permitted embryonic research up to the fourteenth day after fertilization. The committee permitted research on excess embryos, whether or not the embryos were intentionally developed for research. Britain enacted legislation concerning the reproductive technologies in 1990 with the Human Fertilization and Embryology Act 1990. The Act defines an embryo as "a live human embryo where fertilisation is complete." Moreover, the Act states that "fertilisation is

not complete until the appearance of a two cell zygote." The Act also outlines a number of prohibited practices, as outlined below:

(1) No person shall—
  (a) bring about the creation of an embryo, or
  (b) keep or use an embryo, except in pursuance of a licence.
(2) No person shall place in a woman—
  (a) a live embryo other than a human embryo, or
  (b) any live gametes other than human gametes.
(3) A licence cannot authorise—
  (a) keeping or using an embryo after appearance of the primitive streak,
  (b) placing an embryo in any animal,
  (c) keeping or using an embryo in any circumstances in which regulations prohibit its keeping or use, or
  (d) replacing a nucleus of a cell of an embryo with a nucleus taken from a cell of any person, embryo or subsequent development of an embryo.
(4) For the purposes of subsection (3)(a) above, the primitive streak is to be taken to have appeared in an embryo not later than the end of the period of 14 days beginning with the day when the gametes are mixed, not counting any time during which the embryo is stored (34).

The Act addressed four of the treatments available: artificial insemination using donated gametes, egg donation, embryo donation, and IVF. The act also contained explicit statutory regulations of embryo research, which is permitted until the appearance of the primitive streak ("taken to have appeared in the embryo not later than the end of the period of 14 days beginning with the day when the gametes are mixed") (35). The Act prohibits the creation of hybrids using human gametes, the cloning of embryos by nucleus substitution to produce genetically identical individuals, and genetic engineering to change the structure of an embryo. Despite these regulations commentators have called the Act a "'radical laissez faire' approach; a system of regulated private ordering" (36). Hence sex selection is not explicitly prohibited, as evidenced by an Essex woman who selected her child's sex in 1994 at the London Gender Clinic (36). Moreover, the National Health Service announced in 1994 that it would offer fertility treatment to lesbian couples, a departure from the heterosexual requirements seen in other countries (37).

The Human Fertilisation and Embryology Act of 1990 also amended the Abortion Act of 1967 by reducing the limit for legal abortion to 24 weeks. One commentator noted that the 1988 Alton bill sought to reduce the legal limit for an abortion even further, to 18 weeks (38). Despite the fact that English abortion law was codified in the Abortion Act of 1967, judges were still left with some amount of discretion in their rulings. Moreover certain judges would interpret English legal tradition as affording the fetus a fairly high level of protection. Judge Denning in *Royal College of Nursing* compared the fetus in utero to a child and argued that English law recognized a criminal cause of action if a fetus was intentionally killed. These mixed views toward the status of the fetus, however, did not undermine the United Kingdom's effort in passing the

Human Fertilisation and Embryology Authority. Among the countries surveyed here, the Authority is a rare instance where the government has enough consensus to regulate reproductive technologies on a national level.

In Australia, the Rios case in the early 1980s triggered an interest in the status of embryos ex utero. In this case the Rioses had participated in IVF procedures in Melbourne. They produced three embryos from Ms. Rios's eggs and the sperm of an anonymous donor. One was implanted and the other two were frozen for storage. The Rioses both died in a plane crash, prompting a son from Ms. Rios's previous marriage to declare his share of his stepfather's estate. The embryos were left frozen, following the Australian Waller Committee's declaration that the embryos had no status of their own to be independently unfrozen and implanted in another surrogate mother.

The first official Australian pronouncements on the legality of IVF were issued by the National Health and Medical Research Council in 1982. The Council's guidelines permitted IVF, but with certain constraints: The recipient had to be in an "accepted family relationship," embryos could not be kept beyond the normal implantation time schedule, and cryopreservation could be maintained no longer than usual reproductive need. Cloning was strictly prohibited (39). Each state has modified the Council's rules. For instance, in South Australia and Western Australia, artificial insemination procedures are available to married couples and to couples in de facto relationships of a certain length. In Victoria, access to artificial fertilization procedures is generally limited to married couples. In Western Australia, only married couples or heterosexual de facto couples who have lived together for a total of five years out of the previous six are eligible to be treated with IVF (40).

As far as storage issues are concerned, South Australia prohibits the storage of embryos for more than 10 years (40). In Victoria, there are no statutory time limits on storage, but it is an offense to freeze an embryo unless it is done with the intention of subsequently implanting it in a woman's womb. In Western Australia, storage of reproductive material is prohibited unless undertaken in accordance with a license or exemption. Ova that are being fertilized or embryos must not be stored unless the primary intention of the storage is their "probable future implantation." In any event, they must not be stored for more than three years (40).

In South Australia, Victoria, and Western Australia the rights of control and disposal of gametes and embryos are set down in legislation. In South Australia, the Reproductive Technology Act 1988 (SA) provides that the code of ethical practice will make provision for decisions to be made for the use or disposal of stored embryos (41). Such decisions must be able to be reviewed at least every 12 months. The maximum period of storage for an embryo is 10 years.

In Victoria, under the Infertility (Medical Procedures) Act 1984 (Vic), if an embryo cannot be implanted in a woman's womb due to the woman's death or injury, the embryo may be given to another woman for use in a procedure permitted under the Act in accordance with the consent of the gamete donors. If the donors have died or

cannot be located, the Minister will direct the hospital to make the embryo available for use in a procedure permitted under the Act (42). In Western Australia, under the Human Reproductive Technology Act 1991 (WA), the providers of gametes have all rights of use or disposition of the gametes as if they were personal property until the gametes have been used, although the gametes may not be sold. If gametes are donated to a licensee, rights of control and disposition vest in the licensee who, subject to the consent of the donor, may only use the gametes for in vitro fertilization for a person named or chosen by means specified in the consent, for artificial insemination, for approved research, or for diagnostic procedures. If the gametes are not used for one of these procedures, they must be permitted to succumb subject to the rights of control that may be vested in the recipient couple. Where a donor gives conditional consent for use of gametes, provided that the gametes have not been used, the rights in relation to the gametes revert to the donor if the condition is violated. In Western Australia, consent must be given by a person before his or her gametes or any fertilizing ovum or embryo are used or stored and such use or storage must be in accordance with the consent (43). In Victoria, consent must be given by persons who donate sperm or ova or embryos. Victoria permits embryo experimentation up to 14 days after fertilization.

In South Australia, research using human reproductive material will be governed by the code of ethical practice to be formulated by the South Australian Council on Reproductive Technology. The Code will prohibit development of a human embryo outside the body "beyond the stage of development at which implantation would normally occur." Research using human reproductive material may be undertaken in accordance with a license (40).

In anticipation of such cases as the recent *Kass* case in New York, Australian lawyers at a meeting of the executive of the Law Council's family law section in 1992 proposed a novel rule. Their proposal would require all couples who participate in assisted reproductive technology (ART) to execute a disposition agreement for the future fate of their frozen embryos (44). As Michael Watt, a Melbourne barrister who represented the mother in Australia's first dispute over frozen embryos said: "Trying to squeeze it into custody raises the definition of when does life begin. If it's custody, you have to decide whether an embryo is a child. If it is a human life, it isn't property.... At the moment, every country simply says parties have joint property in the embryos, or an equal say in the future disposal and general responsibility for their future disposal, and there is no deadlock-breaking provision in any legislation. ...It avoids having to have definitions on when life begins" (44).

In early 1995 the Victorian state cabinet approved laws permitting noncommercial surrogacy. The revised laws would make commercial surrogacy a crime (45). In the Australian Capital Territory, the Substitute Parent Agreements Act 1994 prohibits commercial surrogacy. Queensland prohibits commercial surrogacy and the publication of advertisements for surrogacy services. In South Australia, surrogacy contracts and procuration contracts are illegal and void. In Tasmania, the Surrogacy Contracts

Act 1993 (Tas) makes it an offense to introduce potential parties to a surrogacy contract, arrange or negotiate a surrogacy contract, or to give or receive valuable consideration in connection with a surrogacy contract (40). In Victoria, commercial surrogacy is prohibited, and it is an offence to publish or advertise surrogacy services. Even in states without specific legislative provisions governing surrogacy, it appears likely that surrogacy contracts would be unenforceable on the grounds of public policy.

In 1996 the Supreme Court of Tasmania heard the case of *In the Matter of Estate of the Late K* (46). Here the issue was whether the product of the ova of a widow and the semen of her deceased husband are children of the deceased upon being born alive. The deceased died intestate, leaving behind three children from a previous marriage. Justice Slicer in his opinion looked at cases from other common law countries (*Paton v. Trustees of the British Pregnancy Advisory Service, Roe v. Wade, R v. Morgentaler*), as well as in Australia. He determined the following:

- A foetus is not recognized, by the law, as a person in the full legal sense.
- The law has long recognised foetal rights contingent upon a legal personality being acquired upon its subsequent birth alive.
- A child, *en ventre sa mere*, is not a human being. To be human a child must have quitted its mother in a living state.
- A child so born is by a legal fiction treated as having been living at an earlier point of time and as if by being so treated the child would receive a benefit to which it would have been entitled if actually born at that earlier time (46).

Slicer concluded that at the time of the decedent's death there were no human offspring in existence. He asked whether the law should distinguish between a child *en ventre sa mere* and a sibling who was a frozen embryo? He stated that the New South Wales Law Reform Commission 1988 recommended that "children conceived posthumously as a result of IVF procedures and children born from stored embryos should be able to make a claim against the estates of their genetic parents under the Family Provisions Act 1982." Slicer ultimately concluded "[t]hat a child, being the product of his father's semen and mother's ovum, implanted in the mother's womb subsequent to the death of his father is, upon birth, entitled to a right of inheritance afforded by law."

Australia reflects the Anglo-American tradition by respecting certain individual rights (e.g., abortion). Yet Australian law does take an activist approach toward issues such as IVF, surrogacy, and embryo experimentation. Although the rules are primarily procedural, certain substantive values are being promoted. For instance, family stability as a societal goal is embedded in the rules concerning artificial reproduction. A definite preference is given to heterosexual unions or de facto marriages that have existed for a number of years. The law, then, gives legal sanction to a cultural norm. Moreover, the 1989 report concerning the status of the embryo clearly compares its status to that of a living person. Such a statement

in the United States could easily be in the position paper of a "pro-life" organization.

Compared to the other common law countries surveyed, Australian courts have not had to grapple with issues of fetal status nearly as frequently. Yet Australia has been at the forefront in creating innovative legislation with regard to assisted reproductive technologies. Moreover the myriad rules that have been created suggest that embryonic and fetal life is accorded some respect in Australia.

The Council of Europe, a multinational body dedicated to human rights and the rule of law among its member states, has issued its own regulations regarding reproductive technologies. Two articles in the Convention on Human Rights and Biomedicine prohibits two kinds of activities. Article 14 prohibits the use of reproductive technologies to choose a child's sex. The Article makes an exception, however, to avoid a serious hereditary-linked disease. Article 18 of the Convention prohibits the creation of embryos for research purposes. Moreover, although it allows research on embryos in vitro, the Article states that the embryo will be afforded "adequate protection" (47).

Last, a variety of non-English-speaking countries have attempted to address the regulation of reproductive technologies. They have ranged from legislative bills in Argentina that are prohibitive in nature, to bans on embryo research in Norway, to limits on oocyte freezing in Denmark, to French legislation that limits artificial insemination (AI) to heterosexual couples and prohibits embryo experimentation, to German bans on surrogacy contracts. Although this entry has focused on English-speaking countries, reproductive technologies are being used in a number of settings throughout the world. Regulatory responses are as varied as the countries themselves (48).

CONCLUSION

Among English-speaking countries, a great variety of approaches exist in regulating reproductive technologies. The United States has the most laissez-faire approach, with very little federal regulation. Most regulation is left to the private sector or the states; professional organizations such as the AMA and the ASRM have issued the most comprehensive guidelines regarding reproductive technologies. The United States seems to reflect Robertson's view of a strong sense of procreative liberty. Other English-speaking countries are more hesitant to grant such broad negative rights and they have all adopted to varying degrees a greater amount of formal regulation, whether in the form of the Authority in the UK or among the laws of the different states in Australia. In the late 1990s, however, certain international organizations, such as the Council of Europe, have attempted to formalize certain kinds of prohibitions with regards to some of the reproductive technologies available.

ACKNOWLEDGMENT

I would like to thank Patty Sokol and Amy Darby for their editorial assistance in preparing this article.

Unless explicity stated, the views expressed herein do not necessarily reflect official AMA policy.

BIBLIOGRAPHY

1. J.A. Robertson, *Children of Choice*, Princeton University Press, Princeton, NJ, 1994.

2. D. Roberts, *Killing the Black Body*, Vintage Books, New York, 1997.

3. *Skinner v. Oklahoma*, 62 U.S. 1110 (1942).

4. *Buck v. Bell*, 274 U.S. 200 (1927).

5. *Griswold v. Connecticut*, 381 U.S. 479 (1965).

6. *Eisenstadt v. Baird*, 92 U.S. 1029 (1972).

7. *Carey v. Population Services International*, 426 U.S. 918 (1976).

8. *Roe v. Wade*, 410 U.S. 113 (1973).

9. *Planned Parenthood v. Casey*, 502 U.S. 1056 (1992).

10. A. Charo, in *Encyclopedia of Bioethics*, 2241–2247, Simon & Schuster Macmillan, New York, 1995.

11. *Code of Medical Ethics*, Council on Ethical and Judicial Affairs, American Medical Association, Chicago, 1998–1999, Available at: *http://www.ama-assn.org/ethic/ceja.htm*

12. *Code*, Opinion 2.10.

13. *Code*, Opinion 2.145.

14. B. Furrow et al., *Bioethics: Health Care Law and Ethics*, West, St. Paul, MN, 1997.

15. *Code*, Opinion 2.18.

16. American Society for Reproductive Medicine (ASRM), ASRM Ethics Committee, date accessed: 1/26/2000. Available at: *http://www.asrm.org/current/press/abandon.html*

17. American Society for Reproductive Medicine (ASRM), ASRM Ethics Committee, Available at: *http://www.org/current/press/embsplit.html*

18. American Society for Reproductive Medicine (ASRM), ASRM Ethics Committee, Available at: *http://www.asrm.org/current/press/fetalegg.html*

19. American Society for Reproductive Medicine (ASRM), ASRM Ethics Committee, Available at: *http://www.asrm.org/current/press/posthum.html*

20. *In the Matter of Baby M*, 537 A.2d 1227 (1988).

21. H.D. Gabriel and E.B. Davis, *Loyola Law Rev.* **45**, 221 (1999).

22. *Johnson v. Calvert*, 851 P2d 776 (1993).

23. *Davis v. Davis*, 842 S.W.2d 588 (1992).

24. *Kass v. Kass*, 235 A.D.2d 150 (1997).

25. Fl. Stat. Ann. 742.17, 1997.

26. La. Rev. Stat. Ann. 9:123, 9:126, 1991.

27. *Hecht v. Superior Court of LA County*, 20 Cal. Rptr. 2d 275 (Cal. Ct. App. 1993).

28. Pub. L. 102–493, 42 U.S.C. 263a-1 et seq.

29. B.M. Knoppers and S. LeBris, *Current Opin. Obstetrics Gynecol.* **5**, 630–635 (1993).

30. *Commonwealth Law Bull.*, 981 (1992).

31. M. McDonald, *Toronto Sun*, February 25 (1998).

32. Bill C-47, 2nd Session, 35th Parliament, The House of Commons of Canada, 45 Elizabeth II, 1996–97.

33. K. McNorrie, in S. McLaren, ed., *Law Reform and Human Reproduction*, Dartmouth Publishing, Hanover, NH, 1991.

34. *Blackstone's Guide*, 190.

35. D. Morgan and L. Nielson, *J. Law, Med. Ethics* **21**(1), 31 (1993).

36. C. Hall, *The Independent*, March 16 (1994).

37. D. Fletcher, *Daily Telegraph*, July 2 (1994).

38. C. Morris, *UCLA Women's Law J.* **47**(8), 76 (1997).

39. P. Kasimba and P. Singer, *J. Med. Philos.* **14**, 403 (1989).

40. B. Bennett, *Reproductive Technology, Laws of Australia, Title* **20**, 91 (1995).

41. *Commonwealth Law Bull.*, January 1989.

42. The Reproductive Technology Act 1988 (SA).

43. The Human Reproductive Technology Act 1991 (WA).

44. P. Innes, *Melbourne Age*, 10 March 16 (1922).

45. M. Forbes, *Melbourne Age*, 1 April 9 (1995).

46. *In the Matter of Estate of the Late K and In the Matter of the Administration and Probate Act 1935*, No. M25/1996; Judgment No. A16/1996 (Supreme Court of Tasmania, 1996).

47. Council of Europe, Convention on Human Rights and Biomedicine, Oviedo, 04, IV, 1997, Available at: *http://www.coe.fr/eng/legaltxt/164e.htm*

48. E.A. Pitrolo, *Hous. J. Int. Law* **19**, 147 (1996).

See other REPRODUCTION entries.

REPRODUCTION, LAW, WRONGFUL BIRTH, AND WRONGFUL LIFE ACTIONS

JEFFREY R. BOTKIN
University of Utah School of Medicine
Salt Lake City, Utah

OUTLINE

Introduction

Clinical Circumstances Giving Rise to Wrongful Life and Wrongful Birth Claims

Legal History of Wrongful Life

Legal History of Wrongful Birth

Philosophical and Public Policy Issues

Bibliography

INTRODUCTION

Wrongful life and wrongful birth are two closely related medical malpractice actions that have arisen since the 1973 *Roe v. Wade* decision. Both actions typically are brought against health care providers after the birth of a child with congenital malformations or a genetic disease. Wrongful *birth* actions refer to suits by the parents who claim harm from the birth of an impaired child. The claim is that had the parents been adequately informed of their reproductive risk, they would have taken measures to prevent the pregnancy or birth of the affected child. Wrongful *life* claims are brought in similar clinical circumstances; however, these claims arise from the child who claims harm from birth in an impaired condition. The child claims that but for the negligence of the health care provider, she would not have been born to suffer with her condition. Clearly the wrongful life claim poses a complex philosophical challenge. It is important to emphasize that the claims do not allege that the defendant *caused* the impairment through negligent actions. Rather the claims are based on allegations of inadequate or incorrect information that would have permitted the parents to avoid pregnancy or to detect the abnormality prenatally and terminate the pregnancy. (Physicians who are alleged to have caused a congenital malformation through, say, the prescription of a teratogenic drug, are liable under more traditional tort actions.)

The wrongful life and wrongful birth suits have become increasingly common since *Roe v. Wade* for two reasons. First, *Roe v. Wade* established constitutional protection for abortion decisions through the first two trimesters of pregnancy, and, second, technology has offered an expanding array of tests and procedures to evaluate the health of the fetus. In light of these rights and choices, health care providers are seen to have parallel obligations to offer testing in a variety of clinical circumstances, and to adequately warn couples who have an increased risk of bearing a child with a heritable condition or congenital malformation. Failure to provide timely, accurate information according to the standard of care may leave providers liable under wrongful life and/or wrongful birth suits.

The rapid pace of research in human genetics and fetal imaging means that an ever larger number of conditions will be amenable to prenatal diagnosis in the future. Rare conditions, late-onset diseases and relatively mild health conditions may be identifiable early in a pregnancy. In addition future behavioral traits and normal physical characteristics may be predictable to some degree in an embryo or fetus. A clear challenge for the health professions, and for society more broadly, is to articulate the standards for prenatal diagnosis. How much information should prospective parents have access to about the biologic nature of their future children? The wrongful life and wrongful birth suits raise fundamental legal and philosophic issues about reproductive choice in an emerging era of powerful genetic technologies.

As will be discussed below, the wrongful birth suits have been widely successful in the U.S. court system, while the wrongful life claim has met with limited support. The wrongful life and wrongful birth claims should be distinguished from "wrongful pregnancy" suits in which parents claim damages for the birth of a health child following an alleged negligently performed sterilization or abortion procedure. Wrongful pregnancy suits will not be discussed in this article. This discussion will focus on the medical background of the suits, their legal history and the philosophic issues inherent in these claims.

CLINICAL CIRCUMSTANCES GIVING RISE TO WRONGFUL LIFE AND WRONGFUL BIRTH CLAIMS

Wrongful life and wrongful birth claims can arise from a variety of clinical circumstances. In the majority of cases to date, the claims have resulted from allegedly inadequate information provided to pregnant women about risks to their child. The list of conditions prompting wrongful birth or wrongful life suits is included in Table 1. The usual condition is a pregnant woman of an "advanced maternal age," that is, 35 years or older at the anticipated time of delivery, who is not warned of the increased risk

Table 1. Conditions Prompting Wrongful Birth and Wrongful Life Suits

Down syndrome	Fetal hydantoin syndrome
Congenital rubella syndrome	Leber's congenital amaurosis
Spina bifida	Infantile polycystic kidney disease
Tay Sachs disease	Duchenne muscular dystrophy
Cystic fibrosis	Anhidrotic Ectodermal dysplasia
Sickle cell anemia	Pelizaeus-Merzbacher syndrome
Larsen syndrome	Albinism
Retinoblastoma	"Hydrocephalus and multiple congenital defects"
Neurofibromatosis	No arms and other anomalies
Hemophilia B	"Severely deformed and retarded child"

of bearing a child with Trisomy 21 (Down syndrome) and other aneuploidy syndromes (e.g., Trisomy 18 and Trisomy 13). It currently is the standard of care to warn older pregnant women of their increased risk and to offer prenatal diagnosis. Without such a warning, and on the birth of a child with Trisomy 21, the obstetrician would be at risk for a wrongful birth suit by the parents and a wrongful life suit by the child. The suits are based on the assertion that had the warning been provided, prenatal diagnosis would have been pursued, the child's condition would have been detected and the pregnancy would have been terminated.

The second single most common condition giving rise to these suits after Down syndrome is congenital rubella syndrome. Congenital rubella syndrome is due to a prenatal maternal infection with the rubella virus that can cause serious impairments and congenital malformations in the child. If a physician fails to assess the risk of the pregnant woman to rubella or fails to diagnose an active rubella infection, the physician is at risk for suit upon the birth of the impaired child. Again, the allegation here is not that the physician should have prevented the rubella infection, but that she should have provided information sufficient to allow prevention of the birth of the impaired child.

Obstetricians have been subject to the majority of wrongful life and birth suits to date. To the extent that these suits have been successful, obstetricians clearly have an obligation to assess pregnant women for the risk of congenital and hereditary abnormalities and to provide them information accordingly. It is important to note that these suits do not claim that the physician should have provided prenatal diagnosis or pregnancy termination. Providers are free to follow their own ethical standards in the provision of services and there are a substantial number of obstetricians, for example, who are "pro-life" and do not provide these services. In such circumstances physicians still are required to provide risk information according to the standard of care and information about testing options, such that patients can pursue services with other providers, if they wish. Suits may arise either from the failure to provide sufficient information, or the provision of inaccurate information. Clearly, the risk to the child need not be genetic in origin since infections like rubella and other teratogenic infections and agents can cause impairments as well that may be amenable to prenatal diagnosis.

Prenatal care providers other than obstetricians have been subject to wrongful birth and life cases in the context of prenatal diagnosis. Ultrasound imaging of the fetus has become virtually routine in pregnancy. Failure of the radiologist to accurately diagnose a congenital malformation may give rise to these claims. Similarly failure to accurately perform prenatal tests may give rise to suits against laboratories that process the clinical specimens.

While prenatal care has given rise to most of these suits, care providers in other fields of medicine are not immune, and their liability will increase as genetic information increases. The case of *Schroeder v. Perkel* (1) provides an important example. In this case a child with cystic fibrosis was not diagnosed until seven years of age, despite suggestive symptoms for a number of years. Prior to the diagnosis of the child, a sibling was born who also was affected with cystic fibrosis. The parents claimed, successfully, that had an accurate and timely diagnosis of the first child been made, the parents would have been warned of their reproductive risk and they would have taken measures to prevent the birth of a second affected child. Cases also have arisen due to the provision of inadequate genetic information once an accurate diagnosis of a genetic condition was made. In the case of *Ellis v. Sherman* (2), a surgeon correctly diagnosed neurofibromatosis in an adult patient but failed to indicate that this is a hereditary condition. The patient subsequently fathered an affected child and sued the surgeon for failure to provide sufficient information to allow him to make an informed reproductive decision.

The success of this wrongful birth claim has implications for virtually all clinicians caring for patients who may have a genetic component to their illness. Typically physicians focus almost exclusively on the welfare of the individual patient in making decisions about testing and other evaluations. Genetic conditions may be part of a "differential diagnosis" for a patient (the list of possible conditions that might explain the symptoms), but unless there is a specific reason to pursue the diagnosis, other conditions may be considered first. Since genetic conditions are often difficult or impossible to treat, the diagnosis of genetic conditions may be delayed by the physician's desire to initially pursue treatable conditions. There often is little direct benefit to the patient in making a prompt diagnosis of a genetic condition. However, wrongful life and wrongful birth suits illustrate that there may be benefits to the *family* by making a prompt genetic diagnosis. Such a diagnosis will alert family members to their reproductive risks and potentially permit them to avoid the birth of an affected child. Therefore these legal claims herald a significant change in the responsibilities of health care providers from a narrow focus on the health of individual patient to a broader focus on the reproductive interests of the patient and the patient's family members.

LEGAL HISTORY OF WRONGFUL LIFE

Wrongful life and wrongful birth are tort actions or, more specifically, malpractice actions. As such, a successful claim against a health care provider requires that the plaintiff show (1) a duty existed on behalf of the provider to the plaintiff, (2) a breach of duty occurred (i.e., negligent conduct occurred), and (3) the plaintiff was harmed as a result of the negligence. In addition to these three elements, courts may consider broader public policy issues as they attempt to reach a just conclusion. The first case under the wrongful life term is notable in this regard. In *Zapeda v. Zapeda* (3), an illegitimate child brought suit in an Illinois appellate court against his father who had seduced his mother into intimate relations with a promise of marriage. The child sued his father for the harms associated with illegitimacy. The court was willing to recognize the duty, breach of duty, and harm but was unwilling to invite the flood of suits that might arise from children in similar circumstances.

The wrongful life claim has met with limited success in the U.S. judicial system. To date, five state courts have recognized the wrongful life claim (4), while 19 have rejected this tort. The primary stumbling block for the wrongful life claim has been the notion inherent in the suits that a child would prefer nonexistence to existence in an impaired condition. Recall that existence without the condition was never a possibility for these children, so the choice on behalf of the child was existence with impairments or nonexistence through contraception or pregnancy termination. The children in whose name these suits are brought must assert that, but for the negligence of the defendant, they would not exist. In response to this dilemma, most courts have adopted the reasoning first articulated in the New York case of *Becker v. Schwartz* (5), in which two basic problems with wrongful life suits were identified:

> The first, in a sense the more fundamental, is that it does not appear that the infants suffered any legally cognizable injury.... Whether it is better never to have been born at all than to have been born with even gross deficiencies is a mystery more properly to be left to the philosophers and the theologians. Surely the law can assert no competency to resolve the issue. ...Not only is there to be found no predicate at common law or in statutory enactment for judicial recognition of the birth of a defective child as an injury to the child; the implications of any such proposition are staggering.

The second problem identified by the Becker court was the inability to calculate damages on any reasonable basis.

In contrast, the courts that have recognized the wrongful life claims have been willing largely to overlook the philosophical problems inherent in the claim and to support the suits based on the medical needs of the child and/or the public policy advantages of deterring negligent medical care. A California court (6) in 1980 concluded:

> The reality of the "wrongful life" concept is that such a plaintiff both exists and suffers, due to the negligence of others. It is neither necessary nor just to retreat into meditation on the

mysteries of life. We need not be concerned with the fact that had the defendant not been negligent, the plaintiff might not have come into existence at all.

Similarly the New Jersey Supreme Court in 1984 (7), was unwilling to allow the problematic logic of the wrongful life suits to stand in the way of what it judged to be a just outcome for the case. In the case at hand, the parents had been barred from bringing a wrongful birth suit on their own behalf due to the statute of limitations. Since the statute of limitations for suits by children is much longer than for adults, the only available route for the family to receive compensation for the alleged negligence was the wrongful life suit. The court stated that the child should not be denied adequate medical care simply because the parents were unable to sue on their own behalf.

For more than a decade, other states consistently declined to recognize the wrongful life claim. In recent years, however, courts in both Massachusetts and Connecticut have supported the tort. A 1997 decision by the Connecticut Superior Court (8), quotes a 1983 decision: "There is nothing illogical in a plaintiff saying 'I'd rather not be suffering as I am, but since your wrongful conduct preserved my life, I am going to take advantage of my regrettable existence to sue you.'" A Massachusetts court (9) was faced with a case in which a suit was brought against a physician who failed to report abnormalities on a fetal ultrasound and to repeat the examination. The parents of the child, born with heart and bowel abnormalities, relinquished the child for adoption. The court concluded:

> ...Corey's parents are not entitled to recover against the defendant for the ongoing extraordinary costs that Corey will incur because of his defect (due to the fact that they are no longer his legal guardians or official parents). Nor will Corey's adoptive parents be entitled to recover, since they defendant owed them no duty. Therefore, this Court must consider whether Corey should have this cause of action since no one else can recover the extraordinary costs....In this situation, it appears fair ... to require the negligent Doctor to pick up these costs if negligence is proven.

Therefore in order to assure adequate care to a child with disabilities, some courts have been willing to recognize wrongful life claims without explicitly declaring that life with disability can be worse than nonexistence. Some commentators have noted that the incentive to recognize the wrongful life claim in selected courts would decrease if the U.S. health care system assured better services for all children with significant health care needs.

The New York court in *Becker v. Schwartz* deferred to the philosophers and theologians on the basic question of whether existence confers a harm on some children. A range of opinions have been offered on this question from bioethicists, theologians, and physicians. John Lorber (10), a British surgeon, wrote in 1975 of the deliberate nontreatment of some severely affected children with spina bifida:

> There are ethicists and moralists, as well as doctors, who consider that life must be maintained at any cost, because any life is better than no life. It may be legitimate to adhere to such principles within their own family, but is it not right to enforce such a philosophy on others who do not hold with it. To my knowledge none of the world's great religions or religious leaders believe that a severely defective innocent newborn infant would be worse off in heaven or wherever they believe their souls will go after death. Is it therefore humane to inflict an immense amount of suffering on such infants and on their families to ensure that they reach this heaven or haven in the end? ...[Quoting De Lange, a neurosurgeon] "Large numbers of spina bifida children kept alive by early closure of the defect ... are now adolescents, most of them painfully aware of the deficiencies. Some of us feel their presence not as a tribute to a medical achievement, but as an accusation against misuse of medical power."

Margery Shaw, a geneticist and attorney, argues that fetal abuse, through knowingly bringing a child to birth with a genetic condition, should be made analogous to child abuse in the law. She would sanction not only wrongful life suits against negligent physicians but similar suits against parents.

> ...[P]arents should be held accountable to their children if they knowingly and willfully choose to transmit deleterious genes or if the mother waives her right to an abortion if, after prenatal testing, a fetus is discovered to be seriously deformed or mentally defective. They have added to the burdens of the other family members, they have incurred a cost to society, and, most importantly, they have caused needless suffering in their child.

Indeed, the wrongful life claim raises this curious question of the parent's responsibility for the birth of an affected child. When prenatal diagnosis detects a fetus with a genetic condition or congenital malformation, some parents choose to continue the pregnancy. Also parents at risk for bearing a child with a genetic condition may choose to forgo prenatal diagnosis and accept the risk of an affected child. As argued by Shaw, might the affected child have a wrongful life claim against the parents? The State of California was concerned enough about this possibility after the success of a wrongful life claim in the case of *Curlender v. Bioscience* that it passed legislation barring suits by children against parents for the harm of their existence (11).

In contrast to these authors, Bopp et al. (12) argue from a "right to life" perspective that one of the very foundations of modern law and civilized society is that all human life has enormous intrinsic value.

> ...[W]rongful birth/life claims ... require a new legal theory, in that life itself is considered a wrong, and death is preferred over life with disabilities. By deviating from the general principle, historically found in civilized law, that life, even with disabilities, is valuable and that only wrongful death is compensable, wrongful birth/life actions are a radical departure from fundamental legal philosophy.

Similarly authors writing from a disabilities rights perspective assert that it is simply wrong that those with disabilities lead lives of hopeless despair, devoid of the values that all others experience in their lives (13,14).

The greatest difficulties for those with impairments, it is claimed, are often not due to the condition per se, but to the discriminatory attitudes and barriers in society. Wrongful life suits (and wrongful birth) are seen by many of these authors as reflective of an inaccurate and inappropriate attitude in society toward life with a disability.

Finally, some bioethicists claim that the assertion that the life with impairments is worse than nonexistence is only justifiable for a few extremely severe conditions (15,16). From the perspective of the child, even the most rudimentary awareness and existence might be sufficient to experience a life of value. According to these authors, the kinds of conditions for which wrongful life suits have been brought, such as Down syndrome or congenital rubella syndrome, would not be justified from the perspective of the child.

LEGAL HISTORY OF WRONGFUL BIRTH

To date 26 states and the District of Columbia (17) and 3 federal courts (18) have recognized a cause of action for wrongful birth. One state has enacted legislation recognizing the validity of wrongful birth suits (19). In contrast, five appellate courts have rejected the claim (20) and six states have enacted legislation barring wrongful birth suits (21). Legislative bans have been prompted primarily by the philosophy that the birth of a child, even a child with significant impairments, should not be considered a harm to either the child or the parents. Two state laws banning wrongful birth have been upheld as constitutional (22). Although the national trend is clearly toward the recognition of the claim, wrongful birth remains controversial.

Several courts and scholars argue that the wrongful birth concept is an extension of the constitutionally protected right to privacy in abortion decisions (23). The claim is that abortion decisions are dependent on information about the welfare of the fetus. Therefore reproductive choice is limited if inadequate prenatal diagnostic information is provided. It is argued that the harm in these cases is not the birth of the impaired child, but the infringement on free choice in reproductive decisions.

In contrast, other commentators and courts argue that there is no basis for wrongful birth suits under the umbrella of privacy as articulated in *Roe v. Wade* (12,24). The constitutional right of privacy in reproduction and abortion only prevents state interference with abortion decisions, it is argued, and imposes no positive duties on health care providers to provide information about the fetus. Two state courts (Minnesota and Pennsylvania) have examined these arguments and held that the state laws barring wrongful birth suits are constitutional (22). Therefore, to date, the provision of prenatal diagnostic information has not been held to be a protected right under the Constitution.

Other commentators and courts argue that wrongful birth suits fall more appropriately under the patient's right of informed consent (25–27). Informed consent relates specifically to the amount and type of information that health care providers must provide to patients about medical options. It is argued that in the context of the medical condition of pregnancy, couples should be told the risk of a problem for the child in order to decide whether to obtain prenatal diagnosis. Under the current foundation for wrongful birth as recognized by the majority of the courts, physicians are held to the prevailing standard of care for the provision of timely and accurate information about the welfare of the child.

The requirement that the plaintiff (in this case, the parents) demonstrate harm secondary to the negligence of the defendant has not been carefully evaluated by the courts. In many cases, courts presume that the birth of a child with a impairment constitutes a harm. As noted above, many individuals with disabilities strongly contest this notion. In addition pediatricians and hospital ethics committees dealing with pediatric issues are often faced with the dilemma of parents of a severely disabled child who demand full medical support for the child, even when physicians believe such efforts are futile. Therefore whether a disabled child is a harm to the parents is a subjective issue, and this may change for parents between the birth of the child and later life with the child as a unique individual. Bopp et al. have captured the complexity of this issue by inviting us to imagine a woman with an impaired newborn who is driving to court to enter a claim of wrongful birth. An auto accident occurs and the child is killed. The mother now decides to enter a claim of wrongful death against the other driver. This case vignette illustrates the complex mix of benefits and burdens that children bring to families.

The complexity of the issue of harm is reflected in debate over the appropriate calculation of damages in courts recognizing the tort. There are several options that courts have considered that have tried to balance in various ways the benefits and costs of bearing and raising an impaired child. One method of calculation is to award the parents a monetary sum equal to the costs of the continued pregnancy, the delivery, and the medical and other costs incurred by child's impairment. These are seen as the additional costs directly incurred by the claimed negligence of the physician. Courts also may consider an additional award to compensate for the parent's emotional pain and suffering of bearing and raising a child with a disability. A third element that courts have variously considered is an offset to either of these damages for the benefits that a child brings to a family. Therefore the damages for emotional pain might be reduced by the jury's estimate of the child's positive contribution to the family.

Clearly, the concepts of emotional pain from bearing and raising an impaired child and emotional benefits of raising any child are highly value laden. As a result many courts have been unwilling to allow these kinds of calculations (or, in some circumstances, state law does not permit these kinds of awards or offsets). The majority of the courts have permitted damages to be awarded more simply for the medical and other extraordinary costs incurred by the child's unwanted condition (28).

PHILOSOPHICAL AND PUBLIC POLICY ISSUES

The prevalent acceptance of the wrongful birth concept suggests that this kind of malpractice liability will encourage health care providers to conform to the contemporary standard of care. However, the standard of care in this arena is ambiguous and may be a moving target as new tests become available. At the present time it is the standard of care in the United States to inform women of advanced maternal age of their increased risk of trisomy syndromes, to offer Alpha Feto-Protein (AFP) screening to all pregnant women, and to offer couples from specific racial or ethnic backgrounds tests relevant to those backgrounds — Tay Sachs screening for Ashkenasic Jews, for example, and sickle cell screening for African-Americans. Further it can be concluded that any prenatal tests or procedures that are performed must be performed and communicated in a timely and accurate fashion. Currently many pregnant women of all ages are offered "triple screen" blood testing to detect a fetus with Down syndrome. This testing has a relatively poor predictive value, and it is uncertain at the present time whether courts would find a physician liable for failing to offer such testing.

Another specific test that remains the subject of controversy is screening for cystic fibrosis carriers in the general population. Cystic fibrosis (CF) is an autosomal recessive condition that primarily affects Caucasians of Northern European heritage. The carrier frequency in the Caucasian population is approximately 1 in 25 individuals. Genetic testing for CF is possible, however the sensitivity of the test remains less than 100 per cent. With the identification of the CF gene in 1989, there was prevalent speculation that carrier screening of the general population would be forthcoming. To forestall this development, professional societies promptly and clearly articulated an opinion that it was not the standard of care to offer CF carrier screening in the general population (29,30). To date, CF carrier screening has not be widely offered to pregnant women in the absence of a family history, although it is conceivable that a wrongful birth suit could force the issue and accelerate the adoption of this technology into general screening. With the development of each new major test, society will have to struggle with the question of whether, or when, it becomes the standard of care to offer the technology.

The broader philosophic issue raised by the wrongful birth concept is the limits, if any, that should be placed on prenatal diagnosis. If parents have a right to be informed of reproductive risks, what is the extent of such a right? Imagine a woman who carries a mutation in the BRCA1 gene that confers a lifetime risk of up to 85 per cent for cancer of the breast or ovary. Such a risk begins when a woman is in her thirties and forties. Should BRCA1 prenatal testing be offered and made available to this woman? Are parents harmed if their child has a genetic susceptibility to an adult onset disease like cancer or Huntington disease? At the extreme, can parents claim harm if a child of the "wrong gender" is born after an inaccurate prediction by ultrasound? It is unlikely that a flood of such suits will be brought to court, but the responsibilities articulated by successful suits encourages society to consider the appropriate boundaries for prenatal diagnosis.

A number of scholars and authoritative committees have raised concerns over the use of prenatal diagnosis for "mild" conditions or "trivial" indications. The President's Commission for the Study of Ethical Problems in Medicine and Biomedical and Behavioral Research (31) focused primarily on prenatal diagnosis for gender selection of the child, stating:

> The idea that it is morally permissible to terminate pregnancy simply on the grounds that a fetus of that sex is unwanted may also rest on the very dubious notion that virtually any characteristic of an expected child is an appropriate object of appraisal and selection. Taken to an extreme, this attitude treats a child as an artifact and the reproductive process as a chance to design and produce human beings according to parental standards of excellence, which over time are transformed into collective standards. . . .[T]he Commission concludes that although individual physicians are free to follow the dictates of conscience, public policy should discourage the use of amniocentesis for sex selection.

The Committee on Assessing Genetic Risks of the Institute of Medicine (32) took a similar stand, recommending that:

> . . .prenatal diagnosis not be used for minor conditions or characteristics. In particular, the committee felt strongly that the use of fetal diagnosis for determination of fetal sex or use of abortion for the purpose of preferential selection of the sex of the fetus is a misuse of genetic services that is inappropriate and should be discouraged by health professionals. . . .The committee believes this issue warrants careful scrutiny over the next three to five years as the availability of genetic testing becomes more widespread, and especially as simpler, safer technologies for prenatal diagnosis are developed.

A statement by the American Medical Association's Council on Ethical and Judicial Affairs supports limitation of prenatal diagnostic services to more serious conditions. The council suggests: "Selection to avoid genetic disorders would not always be appropriate. . . .[S]election becomes more problematic as the effects of the disease become milder and as they become manifest later in life" (33). Several scholars have taken similar positions. Thomas Murray concludes: "In short, we should not offer to provide prenatally information about traits or afflictions that are not substantial burdens on parent and child. We certainly should not assist couples in a misguided quest for the child that embodies their ideal collection of traits, including gender" (34). Like the President's Commission, other authors have framed the issue of limits of technology use around prenatal sex selection. Wertz and Fletcher state: "[W]e believe that it is important that the medical profession take a stand now against sex selection. A posture of ethical neutrality on this issue could lead to unfortunate precedents in moral thinking about future uses of genetic knowledge. . ." (35).

The courts have, on occasion, addressed the issue of the extent of the physician's obligation in the context of wrongful birth. The Supreme Court of Kansas wrote in a 1990 case: "In recognizing a cause of action for wrongful birth in this state, we assume that the child is severely and permanently handicapped. By handicapped, we mean, in this context, that the child has such gross deformities, not medically correctable, that the child will never be able to function as a normal human being" (36).

Certainly there are scholars and others in the general public who reject the notion of prenatal diagnosis and selective abortion entirely. After considering the rationales for prenatal diagnosis, Leon Kass concludes that there is no convincing "moral justification for the practice of genetic abortion" (37). In contrast, Philip Kitcher (38) writes:

Couples who test their fetuses for the presence of blue eyes or curly hair and who decided to abort otherwise healthy fetuses when an alternative genotype was present would, to say the very least, have a distorted conception of value. Should states therefore limit the liberty of couples to make reproductive decisions? Not necessarily. Freedom of reproductive choice can be defended provided that the fetus is not taken to be a person with rights and interests that the state has a duty to protect. For the moment, we can accept the idea of the testing supermarket, open to free choice with few restrictions.

Similarly John Robertson argues that the legitimate concerns over unlimited access to prenatal diagnostic information do not warrant infringement on the parent's fundamental procreative liberties (39). As noted above, those from a disabilities rights perspective argue that the whole prenatal diagnostic enterprise largely reflects and reinforces negative stereotypes about living with a disability, or parenting a child with a disability. Drawing lines between types of disabling conditions to declare that some are appropriate for targeting with prenatal testing and that some are not is unacceptable for some disability rights advocates. They contend that such a policy decision sends a hurtful message to those living with disabilities who fall on the wrong side of the line.

The broad acceptance of wrongful birth suits and the diversity of opinion on the appropriate uses of prenatal diagnosis leaves medical care providers without clear guidance at the present time. What should the ethical practitioner offer to prospective parents from the expanding menu of tests? What tests should they provide upon request? Should tests for "mild" and late-onset conditions be made available? In the absence of a policy establishing professional standards on this issue, it may be up to the courts to decide if physicians have failed in their professional obligations when wrongful life or birth suits are brought. Wrongful birth and wrongful life suits will provide guidance to the profession on this issue, but malpractice litigation is a painful and inefficient approach to the development of a standard of care.

Prenatal diagnosis promises to be one of the most complex and controversial topics in medicine and in society generally over the next generation. John Fletcher predicts: "[T]he future of prenatal diagnosis and medical genetics will be ethically more complex than its past and present. Practitioners can expect a 'gathering storm' of issues that require societal involvement and establishment of public policy" (40). Society will benefit from mechanisms to reach some measures of consensus on these complex issues that lie at the interface of technology, philosophy, law, and human reproduction.

BIBLIOGRAPHY

1. *Schroeder v. Perkel* 432 A. 2d 834 (NJ 1981).

2. *Ellis v. Sherman* 515 A 2d1327 (Pa 1986).

3. *Zapteda v. Zapeda* 41 Ill App 2d 240, 190 NE 2d 849 (1963).

4. *Harbeson v. Parke Davis, Inc.*, 98 WA 2d 460, 656 P.2d 483 (1983); *Turpin v. Sortini*, 31 CA 3d 220, 643 P2d 954 (1982); *Procanic v. Cillo* 97 NJ 339, 478 A2d 755 (1984); *Rosen v. Katz* 1996 MA Super. Lexus 618; *Quinn v. Blau*, 1997, CT Super Lexis 3319.

5. *Becker v. Schwartz* 46 NY 2d 401, 386 N.E.2d 807 (1978).

6. *Curlender v. Bioscience* 106 CA App 3d 811, 165 CA Rptr 477 (1980).

7. *Procanic v. Cillo* 97 NJ 339, 478 A2d 755 (1984).

8. *Quinn v. Blau*, 1997, CT Super Lexis 3319. A.M. Capron, in A. Milunski and G.D. Annas, eds., *Genetics and the Law II*, Plenum, New York, pp. 81–93.

9. *Rosen v. Katz* 1996, Mass Super. Lexus 618.

10. J. Lorber, *J. R. Coll. Physicians* **10**, 47–60 (1975).

11. CA Civil Code §43.6 (West 1982).

12. J. Bopp, B.A. Bostrom, and D.A. McKinney, *Duquesne Law Rev.* **27**, 461–515 (1989).

13. A. Asch, in S. Cohen and N. Taub, eds., *Reproductive Laws for the 1990s*, Humana Press, Clifton, NJ, 1989, pp. 69–107.

14. D. Kaplan, in K. Rothenberg and E.J. Thomson, eds., *Women and Prenatal Testing: Facing the Challenges of Genetic Technology*, Ohio State University Press, Columbus, 1994, pp. 49–61.

15. J.R. Botkin, *J. Am. Med. Assoc.* **259**, 1541–1545 (1988).

16. B. Steinbock and R. McClamrock, *Hastings Cent. Rep.* **24**(6), 15–21 (1994).

17. *Bader v. Johnson*, 1997, WL 10352 (Ind. 1997); *Greco v. United States*, 1995, 111 Nev. 405; 893 P 2d 345 (NV 1995); *Flanagan v. Williams* 87 Ohio App. 3d 768; 623 N.E. 2d 185 (OH 1993); *Liddington v. Burns* 916 F. Supp 1127 (OK 1995); *Andalon v. Superior Court*, 162 Cal. App. 3d 600 (1984); *Haymon v. Wilkerson*, 535 A.2d 880 (DC 1987); *Fassoula v. Ramey*, 450 So.2d 822 (FL 1984); *Blake v. Cruz*, 108 253, 698 P.2d 315 (ID 1984); *Goldberg v. Ruskin*, 128 Ill. App. 3d 1029, 471 N.E.2d 530 (IL 1984); *Eisbrenner v. Stanley*, 106 Mich. App. 357, 308 N.W.2d 209 (MI 1981); *Smith v. Cote* 513 A.2d 341 (NH 1986); *Berman v. Allen*, 80 N.J. 421, 404 A.2d 8 (NJ 1979); *Becker v. Schwartz*, 46 N.Y.2d 401, 386 N.E.2d 807 (NY 1978); *Speck v. Finegold* 497 PA 77, 439 A.2d 110 (PA 1981); *Jacobs v. Theimer*, 519 S.W.2d 846 (TX 1975); *Naccash v. Burger*, 223 VA 406, 290 S.E.2d 825 (VA 1982); *Harbeson v. Parke Davis, Inc.*, 98 WA 2d 460, 656 P.2d 483 (WA 1983); *James G. v. Caserta*, 332 S.E.2d 872 (WV 1985); *Gallagher v. Duke Univ. Hosp.*, 638 F.Supp. 979 (NC 1986); *Pitre v. Opelousas General Hosp.*, 519 So.2d 105 (LA 1987); *Linenger v. Eisenbaum*, 764 P.2d 1202 (CO 1988); *Garrison by Garrison v. Medical Center of Delaware, Inc.*, 571 A.2d 786 (DE 1989); *Viccara v. Milunsky*, 406 Mass. 777, 551 N.E. 2d 8 (MA 1990); *Arche v. U.S. Dept. of Army* 798 P.2d 477 (Kan 1990); *Walker by Pizano v. Mart* 790 P.2d 735 (AZ 1990); *Reed v. Campagnolo* 630 A.2d 1145 (MD 1993); *Keel v. Banach* 624 So.2d 1022 (A 1993).

18. *Roback v. United States*, 658 F.2d 471 (7th Cir. 1981); *Gildiner v. Thomas Jefferson Univ. Hosp.*, 451 F.Supp. 692 (E.D. PA 1978); *Phillips v. United States*, 575 F. Supp. 1309 (D.S.C. 1989).

19. Me. Rev. Stat Ann. tit. 24, 2931 (Supp. 1988).

20. *Schork v. Huber*, 648 S.W.2d 861 (KY 1983); *Wilson v. Kuenzi*, 751 S.W.2d 741 (MO 1988); *Azzolino v. Dingfelder*, 315 N.C. 103, 337 S.E.2d 528 (1985), cert. denied, 479 U.S. 835 (1986); *Spencer v. Seikel*, 742 P.2d 1126 (OK 1987); *Atlanta Obstetrics & Gynecology Group v. Abelson*, 398 S.E.2d 557 (GA 1990).

21. Idaho Code 5-334 (Supp. 1986); MO Stat. 145.424 (1987 Supp.); MO Ann. Stat 188.130 (Vernon Supp. 1987); S.D. Codefied Laws Annual 21-55-2 (Supp. 1986); Utah Code Ann. 78-11-24 (1986 Supp.); 42 PA Const.Stat.Ann. 8305 [Purdon, 1990, Supp.].

22. *Hickman v. Group Health Plan, Inc.*, 396 N.W.2d 10 (MO 1986); *Dansby v. Thomas Jefferson University Hospital* 623 A.2d 816 (PA Super. 1993).

23. Note. *Harv. Law Rev.* **100**, 2017–2034 (1987). R.K. Johnston, *University of Missouri at Kansas City Law Rev.* **57**, 337–353 (1989). K.J. Jankowski, *Fordham Univ. Law J.* **17**, 27–62 (1989); *Smith v. Cote*, 513 A.2d 341 (N.H. 1986); *Haymon v. Wilkerson*, 535 A.2d 880 (DC App 1987); *Hickman v. Group Health Plan, Inc.*, 396 N.W.2d 10 (Amdahl, dissenting opinion); *Roback v. U.S.*, 658 F.2d 471.

24. J. Lyons, *Villanova Law Rev.* **34**, 681–696 (1989).

25. J.R. Botkin and M.J. Mehlman, *J. Law Med. Ethics* **22**(1), 21–28 (1994).

26. See *Harbison v. Parke Davis, Inc.* 656 P.2d 483 (Wash 1983).

27. P.S. Fox, *Washington and Lee Law Rev.* **44**, 1331–1356 (1987).

28. See *Bader v. Johnson* 1997 WL 10352 (Ind App) for a recent summary.

29. American Society of Human Genetics, *Am. J. Hum. Genet.* **46**, 393 (1990).

30. Workshop on Population Screening for the Cystic Fibrosis Gene, *N. Engl. J. Med.* **323**, 70–71 (1990).

31. The President's Commission for the Study of Ethical Problems in Medicine and Biomedical and Behavioral Research, *Screening and Counseling for Genetic Conditions*, U.S. Government Printing Office, Washington, DC, 1983, pp. 57–58.

32. Committee on Assessing Genetic Risks, Division of Health Sciences Policy, Institute of Medicine, in L.B. Andrew, J.E. Fullarton, N.A. Holtzman, and A. Motulsky, eds., *Assessing Genetic Risks: Implications for Health and Social Policy*, National Academy Press, Washington, DC, 1994, p. 105.

33. American Medical Association Council on Ethical and Judicial Affairs, *Arch. Fam. Med.* **3**, 633–642 (1994).

34. T.H. Murray, *The Worth of a Child*, University of California Press, Berkeley, 1996, p. 139.

35. D.C. Wertz and J.C. Fletcher, *Hastings Cent. Rep.* **19**(3), 21–27 (1989).

36. *Arche v. U.S. Dept. of Army* 798 P.2d 477 (Kan. 1990).

37. L. Kass, *Toward a More Natural Science: Biology and Human Affairs*, Free Press, New York, 1985, pp. 80–98.

38. P. Kitcher, *The Lives to Come: The Genetic Revolution and Human Possibilities*, Touchstone Books, New York, 1997, p. 85.

39. J.A. Robertson, *Children of Choice: Freedom and the New Reproductive Technologies*, Princeton University Press, Princeton, NJ, 1994.

40. J.C. Fletcher, in A. Milunski, ed., *Genetic Disorders and the Fetus: Diagnosis, Prevention and Treatment*, Plenum, New York, 1986, pp. 819–859.

See other entries GENE THERAPY, ETHICS, GENE THERAPY FOR FETUSES AND EMBRYOS; REPRODUCTION, ETHICS, PRENATAL TESTING, AND THE DISABILITY RIGHTS CRITIQUE; REPRODUCTION, LAW, IS INFERTILITY A DISABILITY?; REPRODUCTION, LAW, REGULATION OF REPRODUCTIVE TECHNOLOGIES.

RESEARCH ON ANIMALS, ETHICS, AND THE MORAL STATUS OF ANIMALS

LILLY-MARLENE RUSSOW
Purdue University
West Lafayette, Indiana

OUTLINE

Introduction
The Moral Status of Animals
History and Theory
    Animal Liberation
    The Argument from Marginal Cases
    Rights-Based Arguments
    Other Philosophical Foundations
Animal Welfare and Animal Rights
Animals in Research
    Current Regulatory Structure
    The Three R's
    Teaching and Testing
    Continuing Issues
    On the Horizon
Bibliography

INTRODUCTION

Animal research involves not only scientific but ethical issues. Indeed, these are not two separate, or separable, issues; they are inextricably interwoven. The ethical side of this issue concerns researchers, regulators, concerned citizens, and society at large. Even though questions about animal experimentation are not the exclusive domain of philosophers and ethicists, a careful look at relevant philosophical arguments and moral theories can give us a better understanding of the nature of the debate. The following discussion takes up the issues listed in the title in reverse order.

THE MORAL STATUS OF ANIMALS

The term "moral status" refers to the place of something with respect our ethical reasoning. An alternative way of phrasing the question of moral status is to ask whether something has moral standing. To say that something has no moral standing is simply to say that it does not enter directly into proper ethical deliberations at all. Such things may figure into deliberations indirectly, as a source of concern for those who do have moral standing. My property, for example, does not have moral standing, but *I* can be benefited or harmed, my rights can be respected or violated, depending on how others treat my property. Thus, we might speak of other people's indirect duties to treat my property in certain ways. Traditionally this has been the status accorded to animals by theologians, philosophers, and legal and social practice (1–3). We have duties not to be cruel to animals not because they are worthy of our direct moral concern, but because how we treat animals affects humans — because the animals are their property, because cruelty to animals might lead to mistreatment of humans, and so on.

There are two other possible positions on the moral status of animals, one on either side of the 'indirect duties' view just sketched. One might hold that if we set aside questions of property, human concerns for animal welfare are simply misguided, and should not be part of a well-grounded ethical theory at all. This view is admittedly uncommon, but not unheard of (4–6). As one noted neuroscientist states: "I believe that the inclusion of lower animals in our ethical system is philosophically meaningless and operationally impossible and that, consequently, antivivisectionist theory and practice have no moral or ethical basis" (4, p. 169) The other is to accord animals moral standing, to say that we have direct duties toward them independently of how our actions affect other humans, and that they ought to figure directly in our ethical deliberations. It should be noted that moral standing need not be egalitarian: There is a large gap between saying that animals have moral standing, and saying that their moral standing is in any way equal to that of a human being. This position on the moral status of animals encompasses a wide range of positions, from radical "liberation" theories such as those defended by Peter Singer and Tom Regan (7,8) to those that defend the use of animals for research and agriculture but still argue that we have a direct moral obligation to minimize animal pain and suffering (9,10). In all these cases we must distinguish the question of whether X is a moral agent — the sort of thing which has duties, moral obligations, can be praised or blamed — from the question of whether X has moral standing — whether it is the sort of thing toward which moral agents have direct duties. The term "moral patient" is often used to designate this latter category. While some ethical theories hold that only moral agents can be moral patients, most theories entail two different sets of criteria for the two groups. Thus it will not do to reject the idea that animals have moral standing (are moral patients) simply by noting that "they don't respect *our* rights," or "they don't have any compunction about killing *us*."

Most people who are not deeply involved in the animal rights debate tend to endorse a "middle-of-the-road" position regarding our moral obligations to animals. Thus, it is widely accepted that it is morally wrong to cause an animal pain or suffering without a compelling overriding good reason. Although there is much less agreement about what sorts of things constitute a good reason for overriding this injunction, most people would also agree that the suffering of the animal itself, not just the side effects on humans, is morally relevant. On the other hand, most of these same people would object quite strongly to the view that animals are in any way entitled to the same sorts of moral protection owed to other human beings; they would hold for example, that much less justification is needed for causing an animal to suffer than is required for the same amount of human suffering, that we are justified in using animals in ways

that we ought not use severely retarded orphaned infants, and that injunctions against killing animals are much less stringent and restrictive than similar injunctions against killing humans.

The position (or constellation of positions, for there are many variations within the boundaries sketched) described in the previous paragraph would apparently be endorsed by the vast majority of Americans and Europeans. It is decidedly not, however, the dominant tendency in philosophical writings on ethics and animals. Most of the contemporary philosophical writings argue some variant of the claim that the position described above sanctions much that is morally reprehensible, and that when correctly understood, our moral obligations regarding animals would disallow currently accepted practices such as meat-eating, much of the research in which animals are used, and more (7,8,11–13).

At this point, let us introduce some labels for the sake of convenience. Let us call the sort of position just described "the liberationist view." It is more popularly called "the animal rights position," but this is inaccurate, since at least some of the defenders of the position — notably Peter Singer — do not appeal to rights-talk in their arguments; although the word "rights" crops up occasionally in (7), in his more careful philosophical arguments, he explicitly rejects the notion of animal rights (14,15). However, for better or worse, the label "animal rights" is the one that has been most commonly associated with this strong position; problems with the label will be discussed more fully in a subsequent section. The first sort of position described above admits that there is a basis for direct moral obligations to animals, but holds that it is different, at least in part, from the basis for our obligations to humans; for that reason, it shall be labelled the "differential view." Finally, there is the view that we have no direct obligations or duties to animals, that even the duty to avoid cruel treatment of animals is based on our moral obligations to humans. This shall be refered to this as the "humanist position."

Clearly, these are less-than-ideal terms for views that have been coarsely defined. The arguments within each category differ in important respects that must be considered more carefully. Nonetheless, the labels do identify some clearly distinct trends, and it will be convenient to be able to refer to those trends in a shorthand way. Details and specifics will find their proper place later. It should also be noted that these views cut across the theoretical commitments we will examine presently: One can, for example find utilitarians in all three categories (7,16,17), and while some deontologists (8) subscribe to a liberationist view, others are definitely in the humanist camp (18).

HISTORY AND THEORY

If the "differential view" is the one most people hold, how did the animal liberationist position gain such prominence? At least in the United States, the answer can be traced back to Peter Singer, and the publication of *Animal Liberation* (19).

Animal Liberation

Singer follows Jeremy Bentham and other proponents of "utilitarianism" to support his opposition to animal research or raising animals for meat. Utilitarianism is the view that an action is right if and only if it produces a better balance of benefits and harms than available alternative actions. That is, our ethical evaluation looks only at the consequences of our actions. Utilitarians may disagree about what constitutes a benefit but two widely accepted candidates are pleasure and satisfaction of preferences and interests. Singer argued that sentience — the ability to feel pleasure and pain, and hence to have interests — should be the basis of our ethical assessment of any action, including animal research. The examples he used to support his claim that animal research does not produce the best possible balance of benefits vs. harms have come under serious attack, both for the sketchiness of the descriptions, and for questions of scholarship (20). Nonetheless, they served to make many more people aware of the sorts of research that had been conducted. Even if the specific examples cited are problematic, the theory does give a way of evaluating animal research, both as a general practice, and specific research projects. This approach of trying to weigh the (expected or possible) benefits against the harms, usually to the animals used, is reflected in the current regulations, albeit in an attenuated form.

The Argument from Marginal Cases

More important from the standpoint of theory, Singer launched the first sustained attack on the differential view, arguing that any attempt to justify the use of animals in research would also justify the use of some humans. This argument, which has come to be known as the argument from marginal cases, has become one of the chief weapons in arguments against animal research. It is also intended to support the claim that "speciesism" — a term coined by Richard Ryder (21) but popularized by Singer — is wrong in the same way that racism or sexism is wrong.

The argument is deceptively simple. If we want to justify differential treatment of animals and humans, we must cite a morally relevant difference. However, the differences that have been proposed — such as self-consciousness, autonomy, or the ability to act as moral agents — do not apply to all human beings. Some severely retarded humans, those in a permanent vegetative state, or, to cite the most dramatic example, anencephalic infants, do not possess those qualities now, may never have them, or have possessed them in the past. Thus any argument that attempts to justify animal research on the basis of one of these properties would also justify using these "marginal case" humans for the same procedures. The term "marginal case" may have originally intended to refer to both animals and humans who are at the margins of whatever line we attempt to draw, but it has come to be used to refer to those humans who are severely retarded or otherwise so impaired that they are not capable of very basic perceptions, emotions, and understanding. Since this term is offensive to many, the following discussion will break with tradition and use the term "misfortuned humans" rather than "marginal cases."

There are three possible responses to this argument. The first, best associated with a "humanist" view of the moral standing of animals, is to attempt to identify a morally relevant property that does separate all humans, even "misfortuned humans," from animals. The only candidate for such a property is membership in the human species; anything else, such as the capacity for reason or moral agency, is not going to apply to severely misfortuned humans. This is often disparaged as "speciesism," but has been defended occasionally (22). The second response, defended by R.G. Frey (23), is to argue that there are important differences between normal adults and animals, but also agree that some humans will lack the morally relevant qualities, and accept the other horn of the dilemma: Some research on misfortuned humans would be justified, perhaps even better than animal research. This would be consistent with (although not entailed by) the "differential" view. The third, dictated by the "liberationist" view, is to accept Singer's conclusion, and still insist that research on misfortuned humans would be immoral, but that entails that we are not morally justified in doing research on sentient creatures that meet or exceed the conditions that protect misfortuned humans. Let us look at each of these in a bit more detail.

If one wants to pursue the first strategy and cast about for a difference that might be morally relevant, it is clear, as already noted, that the standard properties — such as autonomy, rationality, the ability to engage in moral reasoning and mutual respect — will not do. Some people have tried to generate a longer list, and even suggest that we ought to adopt a "cluster" approach: none of the qualities individually are necessary, but any small subset of them is sufficient to justify preferential treatment for humans (24). These are exactly the properties that misfortuned humans lack. There are three other possibilities that have been suggested: (1) arguments from potentiality: marginal cases have (or did at one point have) the potential for the morally relevant property, even though they do not actually have it, (2) defenses of "speciesism," or arguments that attempt to justify giving preferences to individuals just because they are members of our own species, and (3) appeals to side effects, or how misfortuned humans must be afforded special protection for the good of others, not necessarily for the misfortuned human. The first of these will give us either too little or too much. Most advocates of this view restrict their attention to the potential that a misfortuned human has now (25), but it is simply false that most of them have the potential in question. An anencephalic baby or someone who is brain dead does not now have such a potential, and appeals to the fact that medicine might someday be able to help such cases is simply otiose; such resources are not available now. Others extend the notion of potentiality even further. They say that the individual had the potential at some point but lost it due to misfortune (26), or a slight variation, that misfortuned humans actually have some moral status in virtue of the fact that they could have possessed some property (27). This may be true, except for those defects that are genetically fixed at conception and manifest during fetal development. However, this casts the net of potentiality too wide; every germ cell (indeed, since the advent of Dolly the sheep, perhaps every human cell) has, under the right conditions, the potential to develop into a normal human being. The idea that every human sperm should be accorded a higher moral status than a chimpanzee, or even a laboratory rat, appears to be a reductio ad absurdum of this approach. Finally, those who seek a morally relevant difference between misfortuned humans and animals might point to the so-called side effects of treating such humans as potential research subjects (28). The side effects include the emotions of the parents, other interested observers, possible changes in attitudes toward the medical profession, and weakening of the familial bond. The problem with these side effects is that cultural factors are the main determinant of how strong or deleterious these effects will be. If our obligations towards misfortuned humans rest solely on these factors, they will embody a "cultural relativist" approach to ethics, an approach for which ethical theorists have long had persuasive refutations.

Consider then the second response to the argument from misfortuned humans, championed by R.G. Frey (16,23). Frey argues that the value of life depends on the quality of life, and the quality of life is determined by the opportunities for rich experiences; he like Singer, is a utilitarian. He also agrees that on this criterion, some animals will fare better—rank higher in terms of quality, and hence value of life—than some misfortuned humans. At the same time he rejects the liberationist claim that animal research is universally unjustified. While he agrees that some work is frivolous, trivial, or simply bad science, he cites numerous cases in which he argues that the benefits outweigh the harms. His solution is to raise the stakes: We can justify such beneficial research, but only if we would be willing to do the same research on a misfortuned human who falls lower on the quality/value of life scale than the animal proposed as the research subject. Although his view is perfectly consistent, most of his critics continue to cast about for another way to avoid grasping this particular horn of the dilemma.

The third option, as already noted, is the one preferred by Singer, Regan, and other critics of animal research. The conclusion here is that since we would not sanction most research on misfortuned humans (although there are some exceptions), we should not allow the same sort of research on animals who match or exceed those humans with respect to morally relevant criteria. This poses a very basic challenge to defenders of animal research, in that it seems to rest solely on a demand for logical consistency, rather than allegiance to one or another arcane-sounding ethical theory. Although Singer situated the argument from misfortuned humans in a utilitarian context, the argument is by no means limited to that context. Tom Regan (8) and James Rachels (29) utilize the same sort of argument to argue that any argument that establishes that all humans have the right to life, liberty, or respect will also apply to animals. Indeed, at least one philosopher, Evelyn Pluhar, makes the argument from misfortuned humans so central to her argument that no specific commitment to a broader moral theory is required (12).

Rights-Based Arguments

Although Singer's arguments are the most generally straightforward, accessible, and therefore best known by the general public, other philosophers have given more complex and sophisticated arguments to examine the contention that animal research (as well as large scale animal agriculture) should be abolished. In doing so, surprisingly, they employ a theory most usually associated with "human rights." Some of these are still utilitarian in flavor, but others adopt a more "deontological" approach.

The history of deonotology, as well as most current versions of deontological theories, emphasize the unique status of rational, autonomous, human agents. This is in marked contrast with the utilitarian tradition, which from its first formal articulation by Jeremy Bentham, has frequently acknowledged the inclusion of animals (30). In contrast to utilitarianism, deontology insists that some actions may be right even if the consequences are not good, or as good as they could be, while other actions are ethically wrong even though they would produce good consequences. In short, consequences are not the only factor in moral evaluation. Examples typically cited are lying (which would be wrong even if you and I were both happier if I lied to you), keeping promises, and justice. The term "deontology" is often used interchangeably with "right-based theory," since the concept of rights has often been used as a guard against sacrificing what is right (e.g., respect for individual autonomy in matters of religion) for what might be good for the group (forcing compliance to the majority choice). It is also sometimes called a "Kantian theory," since the German philosopher Immanuel Kant was its first major proponent. Kant famously intended his theory to apply only to rational beings capable of understanding moral imperatives, and explicitly excluded animals from its scope, and until recently that focus was unquestioned. Indeed, the sorts of rights typically associated with deontology are often called "human rights." As noted above, even contemporary philosophers who develop and defend deontological or rights-based theories rarely grant animals rights. However, that exclusion has been challenged.

Tom Regan, for example, follows Kant in rejecting utilitarianism as an adequate moral theory in favor of a theory that emphasizes rights that cannot be overridden merely because such an override would yield good consequences for everyone affected. Bernard Rollin also defends a rights-based approach, although he is neither strictly Kantian nor an "abolitionist" about either animal research or animal agriculture (31,32). In both arguments, however, the concept of rights is consistent with almost universally accepted social and political philosophy: for example, that a person cannot be killed (perhaps to harvest his heart, lungs, and other tissues and organs) just because more people would benefit from his death than would benefit from his continuing to live. The novelty is extending this concept of rights to animals and that, at least in Regan, is justified by a form of the argument from misfortuned humans.

For Regan, the attribution of rights is closely tied to an obligation to respect another's life and inherent value. He rejects a Schweitzer-style reverence for all life and argues instead that we ought to attribute equal inherent value to what he calls "subjects of a life," individuals who have—

> beliefs and desires, perception, memory, and a sense of the future, including their own future, an emotional life together with feelings of pleasure and pain; preference and welfare interests; an ability to initiate actions in pursuit of their desires and goals; a psychophysical identity over time; and an individual welfare in the sense that their experiential life fares well or ill for them, logically independently of their utility for others, and logically independently of their being the object of anyone else's interest (8, p. 243).

The argument that all subjects of a life have inherent value, and hence deserve respect, depends in part on a version of the argument from misfortuned humans. Regan does not attempt to draw a sharp line between animals that are subjects of a life and those that are not, preferring to leave that issue open pending a better understanding of various animal's psychological abilities. He does claim that *at least* all normal mammals over one year of age are subjects of a life. From the notion of inherent value and what he terms "the respect principle," he concludes that animals have certain rights that are equal in moral weight to those rights possessed by humans. These include, most basically, the right to respectful treatment and the right not be harmed.

The fact that Regan's argument to show that animals have rights represents a minority position raises an interesting question: Do his arguments show that "traditional" deontological theories are merely myopic or biased, or do those traditional restrictions of rights to human beings have a legitimate basis? Both Regan and Rachels rely heavily on the argument from misfortuned humans, but there are at least two other questions to be raised. They can most easily be seen by setting out a condensed form of Regan's argument:

1. All "subjects of a life" have rights.
2. Many animals—at least those noted above—are subjects of a life.
3. Therefore many animals have rights.

Thus the two obvious questions to raise are (*1*) whether being a subject of a life is in fact the appropriate criterion for attribution of rights, and (*2*) if so, whether most animals really are subjects of a life in the sense defined by Regan. Although it might seem as if the first is the domain of philosophers and the second a question to be answered by psychologists, ethologists, and other philosophers, in reality the two are inextricably intertwined. The debate is too complex to pursue here, but it is ongoing (33,34). Quite a lot is at stake here. No matter what Peter Singer says, utilitarianism will necessarily sanction at least some research on animals. Only if Regan can successfully establish his rights-based view can he even approximate the "total abolitionist" stance toward animal research that he advocates.

Other Philosophical Foundations

Although utilitarianism and deontological theories tend to dominate the theoretical landscape, they are by no means the only options. At least two other approaches to ethical theory, contractualism, and what shall be referred to as Humean ethics, have been used to address animal issues explicitly. A third, virtue theory, is enjoying renewed interest among philosophers, but does not seem to have been used by contemporary philosophers to address issues about the use of animals, even though Aristotle, widely regarded as the founder of virtue theory, had quite a bit to say about animals (35). Finally, several philosophers writing on animals do not fit neatly into any of these categories, either because they have theoretical commitments that combine elements of one or more approach, or because their arguments are not as "theory driven" as, for example, Regan's is. The discussion that follows will not attempt a critical analysis of these approaches but will describe them briefly in order to provide a more complete overview.

Contractualism, as the name implies, views moral obligations as the outcome of an implicit or hypothetical contract among members of a society. Such contracts are assumed to have the form: I agree to refrain from doing $x, y,$ and z to you, and to do $a, b,$ and $c,$ if and only if you agree to refrain from doing $x', y',$ and z' to me, and to do $a', b',$ and $c'.$ The "prime" indicators are meant to indicate that while contracts are essentially reciprocal arrangements, they need not be symmetrical. One can have a contract between employer and employee, or sovereign and subject, where the rights and duties of the participants might differ. The essential feature of contractualism is that all parties agree to abide by the rules; hence only creatures who are able to understand an abstract concept of rules and the intention to follow them can enter into the moral sphere and have direct moral standing. Contractualism is sometimes confused with deontological theories, and some philosophers (36) have combined elements of both. But there are important differences between them, some of which will come out in the following paragraphs, and some of which are too complex to address here. What follows will often speak of "contractors" and "contracts" for the sake of convenience, but it is important to remember that these typically refer to hypothetical, not actual, contracts.

It is probably already obvious that animals will not fare well on a contractualist approach to ethics: they, like children, are judged incapable of understanding the abstract rules and reasoning implied by the whole notion of a contract. Being unable to enter into a contractual agreement, they are also ineligible for the protection such contracts provide.

Clearly, contractualism sees moral duties as holding directly only among creatures capable of understanding and abiding by such an abstract contract. As the contrast between Regan and Kant indicates, the basis for obligation or duties in a deontological theory is not necessarily so restricted: Some versions of deontological theories have room for duties to animals, but no contractualist theory can possibly do that: The best it can muster is a "contract" in which participants choose to include certain restrictions on the way we treat animals because the contractors would prefer not to live in a society which, for example, tolerates blatant cruelty. As we have seen, this sort of protection at best generates an indirect duty toward animals. Moreover such protection is to a large extent voluntary, an "option" for the contractors, unlike the strict imperatives generated by a deontological theory. These differences entail that contractualists may differ in their descriptions of what indirect duties, if any, might be included in the hypothetical contract, and we do find the expected variations in the views of contractualists who have written about animals. The two most prominent philosophers in this category are Jan Narveson (37,38) and Peter Carruthers (6). Narveson thinks that contractors might well choose to extend some protection to animals, while Carruthers argues quite vehemently that it would be irrational and misguided to agree to such an extension of contractual protection.

One of several problems with contractualism is the fact that the moral status of children and other "disadvantaged" individuals — those lacking the capacity to understand and hence legitimately agree to the sort of contract required for admission into the moral sphere — are covered only at the whim of the contractors. Some humans in this category might be covered by dint of the contractors' self-interest; if, but only if, they think they will care about the welfare of their children, they might choose to include some protection for children in the contract. However, if they believe that they will only care about their healthy, or "normal," or male offspring, or children born in wedlock, only those children will be protected; no moral agent will have any duties to those "beyond the pale." More generally, since morality is *defined by* the hypothetical contract, there is no way of saying that the contract itself is unethical, unfair or unjust, or fails to capture some real moral obligation; one can only raise a legitimate objection by pointing out that no rational agent would agree to some putative contracts. These features of contractualism have led most philosophers to reject it as an adequate moral theory, and if it is inadequate as a general analysis of morality, it surely cannot be appealed to as a basis of deciding questions about moral obligations towards animals.

Implicit in at least some popular invocations of contractualism are two ineffective justifications: (1) Since animals do not respect our rights, we do not have to respect theirs, and (2) morality is a human invention, and thus applies only to humans. The first is problematic because it assumes that only moral agents can be moral patients, something that needs to be argued for rather than presupposed. The second is even weaker: After all, humans "invented" mathematics and science, but that does not mean we can make them do anything we want nor that they only apply to humans.

The collection of views that may be called *Humean*, in recognition of the British philosopher David Hume, represents another approach to ethical theory in general, and the issue of animals and animal research in particular. The views have this in common: They hold that ethics is not something that requires only abstract, impersonal rationality. Rather, it must also involve emotions, including sympathy or empathy, and avoid the complete detached impartiality and abstractness of overly rationalistic approaches. Hume is famous for claiming "Reason is, and ought to be, the slave of the passions."

The term "passion," in Hume's day covered a wide range of emotional and personal attitudes, but the basic unifying theme is that, in contrast to the impersonal nature of deontological theories and utilitarianism (both of which hold that, all other things being equal one has no ethical justification for giving preferential treatment to one's best friend, spouse, or child), personal, concrete relations do matter as a foundational concern in ethics. Mary Midgeley, for example, justifies her conclusions about our duties toward animals by arguing against excessive emphasis on rationalism, and for the importance of emotion in ethical reasoning (39). Annette Baier also emphasizes Humean themes in her discussions of animals (40). This de-emphasis of rationality has also been a prominent theme in feminist discussions of animal experimentation (41). A rejection of rationality and impartiality seems ripe for ridicule, but both Midgeley and Baier argue extensively that traditional ethical theories have overstated the role of very abstract, detached reasoning.

ANIMAL WELFARE AND ANIMAL RIGHTS

Defenders of animal research often try to draw a deep theoretical distinction between animal welfare and animal rights. They often assert that researchers who use animals, or farmers who raise animals for food, are deeply concerned about animal welfare but ought to reject the notion of animal rights. A more careful analysis of these concepts, however, reveals, that the distinction invoked here is far from clear and often inaccurate (31,34,42), and is usually divisive rather than clarificatory. It is divisive because it frames the discussion in terms of just two sides rather than recognizing the whole spectrum of subtle differences, and because the two sides are presented in an "us versus them" tone rather than looking for points of agreement. It is inaccurate because, as we have seen many "liberationists" such as Peter Singer do not advocate animal *rights*. Moreover many philosophers who argue that animals do have rights reject the idea that these rights afford them the sort of total protection that the label is usually taken to represent (43), and they also explain why talk of animal rights does not entail anything like equal treatment (44). Finally, as reported by Rollin, 80 percent of respondents to a recent poll affirmed that animals have rights (31, p. 149). Thus advocating animal rights is neither necessary nor sufficient for holding the extremist position that the label "animal rights" is often taken to represent. The label "animal welfare" is similarly unhelpful. It usually involves a rejection of the humanist view, as defined at the beginning of this article (although some who invoke the "animal welfare, but not animal rights," slogan seem to attribute only indirect moral status to animals); even so, it covers such a broad spectrum of views that it sometimes tends to become mere window dressing. One end of the spectrum, it could imply a position as strong as Singer's: Animals do not have rights, but their welfare deserves equal consideration, and harms to animals must be weighed against potential human benefits (the converse is true, in principle, but that rarely arises as a moral issue). As we have seen, this would require a major re-evaluation of animal research. On the other end of the spectrum, "animal welfare" is merely an injunction not to harm an animal unnecessarily, unless such harm is dictated by some human interest (including any desire to know, economic considerations, personal taste, and entertainment). But if these labels are unhelpful, what vocabulary should be used? First, we need a more accurate understanding of the term "animal rights," something which is best obtained by going back to the philosophical roots of theories of rights.

There are, of course, philosophical disagreements about how best to understand rights-claims and theories about rights. Accordingly, a brief survey of the available options will be helpful. The most important choice for our understanding here has to do with the force of rights-claims: What do such claims entail or suggest, and how are they different from other sorts of claims about moral obligations? We soon find an important distinction between what may be called "broad" and "narrow" views on the force of rights-claims. Roughly speaking, a broad interpretation of rights-claims sees them as alternative ways of expressing a wide variety of moral obligations. To say that someone has a right, such as to liberty, is merely another way of saying that we (moral agents) have a direct duty not to interfere with her exercise of free movement, free choice, and so on. A narrow interpretation demands something more stringent: Rights provide a foundation for only the more basic obligations, obligations that are much harder legitimately to override.

The basic theoretical difference (admittedly a difference in degree rather than kind) is that "ordinary" duties can and must be balanced and assessed against all sorts of competing demands, interests, and inclinations. We as a community have duties to respect residents' use of their private property, but those duties can and often are assessed against, and sometimes overridden by, other wants and desires: hence zoning restrictions and some environmental legislation. Rights, on the other hand, are not so easily overridden: A person's right to free speech cannot legitimately be overridden no matter how many people dislike what is being said. Rights in this narrow sense are often said to be "basic," "inviolable," "natural," or "inalienable." The justification for this narrower, but stronger, sense of rights is analogous to that in the political arena, in which rights are seen as a way of protecting the minority from the potential abuses of majority rule.

It might be interesting to note in passing that the only legitimate way of characterizing Peter Singer — the author of *Animal Liberation* and the so-called father of the animal rights movement — as "a defender of animal rights" would be to rely on the broadest possible interpretation of rights-claims. All interests, costs, and benefits are to be weighed equally, and none are set aside for special protection. Joel Feinberg (44) and James Rachels (45) have also defended broad versions of an "animal rights thesis." Tom Regan, as we have seen (8), defends a more stringent and radical rights-based position on which animals have rights in the narrow sense.

In using a broad concept of rights, when people say "animals have a right to be treated humanely," they usually mean only that it is morally wrong to treat them inhumanely, and it is wrong because of the harm to the animal,

not just the indirect harm caused to other humans. Understood this way, the position is eminently reasonable, and not necessarily based on any confusion, misunderstanding, or radical propaganda. Accordingly it does not represent a position to be opposed but rather a welcome opportunity for dialogue and better understanding.

When considering the claim that attributing rights to animals is "eminently reasonable," it is important to keep in mind that neither the broad nor the narrow concept of rights *entails* any assumption of equality between animals and humans. Consider, first, the narrow account of rights: Rights bestow a special sort of protection that cannot be overridden by appeals to a greater general good. There is nothing inconsistent in claiming that an obligation to allow an individual to express her political views cannot be overridden (i.e., she has a right to free speech), but her license to vote or drive a car may be revoked or denied for the greater good. We routinely deny both privileges to children on the ground that their immaturity would render their driving or voting harmful to society. Similarly one can, without any inconsistency, argue that animals have the right not to be tortured without thereby being committed to the claim that they have the right not to be killed. Even when the same right is ascribed to two different individuals, narrow views must and do recognize that one such special claim may be stronger than another, or that if only one can be respected, one has an objectively stronger status. Thus, two people may both have the right to inherit someone's estate — a right that cannot be overridden by the fact that more good might be done by distributing the wealth to agencies that would further the public good — and we can still, in many cases, decide that one right is stronger than another. Similarly, even if one argues that both a dog and a human have the right to life, one might legitimately conclude that the human's right is the stronger of the two, if one is in a position where only one can be respected.

The broad view of rights is even clearer on the issue of equality. Even a moment's thought will uncover a wealth of examples in which it is wrong to treat one individual in a certain way and perfectly legitimate to treat another in exactly that way. Since the broad understanding of rights-talk would automatically translate such differential judgments into different rights-claims, ascribing some rights to animals cannot possibly entail that they must have all the same rights as humans, or that the rights they do share with humans have equal weight.

What follows from all of this? Since there is often confusion about the meaning of rights-claims, and since concerns for "animal welfare" overlap significantly with at least some interpretations, perhaps these divisive labels ought to be retired in favor of a more precise statement of what is actually being claimed. If this suggestion is too extreme, at least one ought to be careful to interpret them in a more flexible and open-minded way, rather than to polarize the debate.

ANIMALS IN RESEARCH

There are, of course, some sustained defenses of practically all animal research, without restrictions or qualifications

(4,17,26,46,47). Some are quite thoughtful, but others convey the impression of defensiveness. One litmus test is whether such defenses agree that there is room for improvement, and that not all animal research meets the highest ethical standards. Contrariwise, attacks on animal research that claim that no significant gains have been achieved through animal research, or that all such research could readily be replaced by alternatives such as computer models and in vitro testing, also undermine their own credibility. There are, fortunately, well-reasoned and detailed discussions of all aspects of the debate (48–51), carefully stated and well-documented arguments against animal research (52), other works that concentrate on specific controversies such as research involving primates (53,54), useful literature surveys (55,56), and anthologies that try to present a varied selection of views on animal research in particular (57,58) or broader philosophical debates that have direct implications for research (59).

Current Regulatory Structure

The subject of current regulations is complex, and constantly changing. In addition to official documents such as the National Institutes of Health (NIH) Guide (60) and U.S. Department of Agriculture (USDA) regulations, there are more helpful and detailed studies of these regulations (50,61) as well as the reference library maintained by Animal Welfare Information Center (AWIC) and numerous on-line sites, so what follows is only meant as a quick overview.

Since 1985, any institution which receives federal funding and uses vertebrate animals is subject to NIH regulations, as set forth in the *Guide to the Use and Care of Animals*. Institutions that use mammals other than mice, rats, or common agricultural species are subject to USDA regulations. The strictest level of control is the voluntary Association for Assessment and Accreditation of Laboratory Animal Care (AAALAC) International accreditation. To a large extent, these regulations and guidelines overlap one another, and in some cases even reference one another.

A key feature of current regulations is the demand that each institution covered by the regulations must establish an Institutional Care and Use Committee (IACUC). This committee must have at least five members, including a veterinarian, a "nonscientist" (defined more precisely as someone who does not engage in animal research), and someone who is not affiliated with the institution. This committee is charged with reviewing protocols for all research involving covered species, conducting semiannual reviews of all animal facilities, ensuring that the institution is in compliance with applicable regulations and guidelines, and generally providing mechanisms for the monitoring and control of animal research. Institutions vary widely with regard to the way in which these duties are carried out: how large the committee is, how members are selected, whether reviews are done by the entire committee or a subgroup, the type of information and documentation required from investigators, how much of the meeting is open to the public, and so on (62). USDA regulations also provide for the licensing of dealers in research animals, both those who breed animals

specifically for research — "Class A dealers" — and those who buy and sell animals, typically dogs — "Class B dealers." One of the goals of this licensing process is to answer the public's concerns, fanned by an article in *Life* magazine in 1966, about pets being stolen to be sold to research labs.

In addition to IACUC oversight, covered institutions are subject to regular, unannounced inspections by the USDA. Although these inspections are supposed to take place twice a year, staffing and funding shortages make this more of an ideal than a reality, especially at smaller institutions. Facilities with AAALAC accreditation are also inspected regularly by an independent group of reviewers, and NIH has the right to conduct its own independent inspections, if it chooses to do so. One shortcoming of all these formal reviews is that while they can inspect facilities, animals being held for research, records, and IACUC minutes, they are rarely in a position to monitor ongoing research directly. Thus, it is difficult to guarantee that approved protocols, and only such protocols, are being adhered to. While it is reasonable to believe that the majority of responsible scientists do follow the protocols for which they have received approval, abuses do occur.

The regulatory system just described applies to animal research in the United States; the Canadian system is quite similar. Elsewhere in the world, regulation of animal research ranges from stricter controls than exist in the United States to nonexistent (63–65).

The Three R's

It has become commonplace in justifying animal research to allude to the "three R's," originally formulated by Russell and Burch (66): reduce, replace, and refine.

"Reduction" refers to reducing the numbers of animals used, consistent with obtaining significant results. The latter is an important qualification, since reducing the number of animals too far might render the results statistically questionable and hence a total waste of animals; indeed, IACUCs sometimes find themselves recommending an increase in numbers for just this reason. On the other hand, the goal of reduction could be significantly furthered if the scientific community and research journals were to rethink their definitions of significance. Particularly for preliminary research, a significance level of 90 percent rather than the typical 95 percent would dramatically reduce the number of animals needed for any given test (67).

"Replacement" refers to using nonanimal models, dead animals, or "lower" species whenever possible. In vitro studies are commonly cited as an alternative, but currently tend to be most practical for initial screening of variations on known compounds. Similarly, computer simulation is valuable only when we have enough information to construct an accurate computer model, and thus may be more useful for education than for exploratory research (68,69). The criteria for ranking animals in terms of higher/lower also needs closer scrutiny; it can often mask cultural preferences rather than any objective standard. Thus dogs are often deemed "higher" than pigs, and NIH singles them out, along with cats, primates, and endangered species as worthy of special status, but there is no clear physiological, evolutionary, or psychological evidence for this distinction. In short, while the injunction to "replace" might be a useful maxim, its application is quite problematic.

Thus, while reference to the three R's has become almost obligatory, it is not at all clear that they still provide useful guidance. As just noted, there is much dispute about the practical applicability of the injunctions in specific protocols, and a general lack of theoretical clarity (70). More work on this topic by both scientists and ethicists is clearly needed.

Teaching and Testing

The strongest arguments in favor of research on animals point to the furthering of basic scientific knowledge or biomedical advances. These arguments are much less convincing when applied to the use of animals in teaching and testing (71). The dividing line between any two of these areas may be fuzzy. Is the lab work of a first-year graduate student teaching or research? When in the process of developing a new drug do we switch from research to testing? Nonetheless, looking at paradigm cases of teaching and testing is necessary in order to have a complete picture of the use of research animals.

When animals are used in the classroom, they are typically described as serving one or more of three purposes: (1) as an illustration of a process, event, or state; (2) as part of a project designed to help students learn proper research design and practice; (3) to allow students (e.g., veterinary students) to learn proper surgical or other techniques.

"Illustration" can range from observation without intervention—keeping a gerbil in an elementary school classroom, or watching a tadpole develop into a frog—to demonstrations of acute medical or pharmacological emergencies—infecting a dog with distemper so that veterinary students can observe the progression of the disease, or dosing a rat with cyanide so that pharmacy students can see the symptoms, and the efficacy of various antidotes. Dissection, particularly at the secondary school level, often falls into this category, as does the use of animals in most science fair projects. Those who use such illustrations often defend them by claiming that a live demonstration or actual dissection is a more vivid and effective teaching tool than a textbook illustration or other alternative; it engages the students more. However, the results of those procedures which are invasive and often painful are almost always known ahead of time, unlike the case of basic research. Videotapes, computer programs, and textbook illustrations can all contain the same information. In the face of the suffering and death that more invasive illustrations cause, appeals to students' interest or attention spans seem trivial. Moreover, such demonstrations can have, either deliberately or inadvertently, a desensitizing effect, conveying to students that animals are mere research tools whose suffering should not overly concern us. This desensitization can develop even in the apparently benign case of having an animal in an elementary school classroom. If children are allowed to handle, observe, or otherwise interact

with animals without due understanding of the stress this might cause, or if the issue of animal care over weekends and vacations is trivialized, they may easily pick up a casual, noncaring or nonrespectful attitude toward animals.

As noted above, the dividing line between basic research and testing can easily become blurred. Testing is also easy to trivialize: It is easier to dismiss the claimed need to test the safety of a new mascara than the development of a new treatment for stroke victims. However, many of the basic factors determining toxicity, carcinogenicity, and inflammation are well enough understood that in vitro studies (and, in a few cases, computer modeling) can be substituted. The Johns Hopkins Institute for Alternatives maintains current data on such substitutes. Even when these alternatives are not conclusive—for example, because they fail to fail to detect effects at an organic rather than a cellular level—they can be useful for initial screening. The most infamous tests, including the Draize test and the LD-50, are gradually being replaced by alternatives that use no living animals, or fewer animals in a less invasive way. The Draize test involves inserting the substance to be tested for irritancy into the eye of a rabbit who has been immobilized in a "stock," and then observing changes in the eye over a period of days. Various in vitro tests, or the use of chicken eggs, have often served as reliable alternatives. The use of computers for more sophisticated statistical analysis has allowed researchers to replace the crude LD-50 test, in which increasing doses of a material were given to colonies of mice or other animals until one found the level at which 50 percent of the animals died, with other tests that required fewer animals and did not always use death as an endpoint. Despite the availability of these advances and alternatives, animal testing remains an area in which many advances in animal welfare are still possible.

Continuing Issues

Of course the fundamental continuing issue is when, if at all, animals should be used in research. While many of the relevant arguments have been discussed in the previous section on theory, it will be useful to see how they apply specifically to animal research. Other continuing issues focus more specifically on current regulations and the general contemporary research environment: how well do they provide appropriate protection for the animals (72)?

Another continuing issue reflects society's (including many scientists') demand for a further increase in our moral sensitivity in animal research. The most obvious trend in social pressures is reflected in the growing popularity of animal protection groups, ranging from the radical People for the Ethical Treatment of Animals (PETA) through the more moderate organizations such as the American Society for the Prevention of Cruelty to Animals (ASPCA) or Working for Animals Used in Research, Drugs, and Surgeon (WARDS) whose aim is not total abolition of animal research. More precisely, the moderate groups hold that the use of animals ought to be abandoned when and where it is possible to do so (would any researcher disagree?), but they are more likely to agree with the scientific community about the fact that the range of productive alternatives is today quite limited, thus accepting that animal research must continue for the foreseeable future. In terms of the three R's, moderates tend to see "refine" and "reduce" as more effective immediate options than "replace" — although they are likely to encourage further research on the development of alternatives. In terms of our categories of views about the moral status of animals, radical groups are generally liberationists while moderate groups at least implicitly adopt a differential perspective. The existence of this range of views within what is sometimes called the protectionist movement again illustrates the danger of the "animal rights/animal welfare" dichotomy discussed earlier. If the scientific community insists on viewing all protectionist groups as radical "animal rights people," opportunities for fruitful dialogue and identification of common ground can be missed.

When fruitful dialogue is possible, and common ground is identified, another trend that is just beginning may blossom. This refers to the increased efforts among scientists and regulators to address ethical issues explicitly and directly, with attention to the general moral principles which underlie our decisions about research on animals can and should be conducted (73,74). The American Association for Laboratory Animals (AALAS), the American Veterinary Medical Association (AVMA), the Animal Behavior Society, and the Scientists Center for Animal Welfare are but a few of the professional societies that have included sessions on ethics at their national conferences. This is in marked contrast to earlier, more polarized efforts in which scientists — sometimes with little background in ethical theory — took it upon themselves to demonstrate exactly what was wrong with animal rights, objections to research on animals, or Regan or Singer. By contrast, the efforts mentioned above represent a collaborative effort to formulate and evaluate the various ethical theories and principles that shape (well or badly) specific choices and regulations about animal research. Such collaborative efforts can give us a more solid grounding from which to address some of the more vexing specific questions about the current state of animal research and its evaluation: Should dogs, cats, or primates be singled out for special protection (as they are today), and if so, why? Should rats, mice, birds, or agricultural animals be excluded from USDA regulations, and why or why not? To what extent must considerations of scientific merit be blended with ethical issues, including but not limited to, IACUC reviews and journal publication criteria? After all, one cannot do a cost–benefit analysis of the sort required by NIH and USDA without some consideration of potential benefits (75).

On the Horizon

Crystal balls are notoriously unreliable, but some future trends are already apparent. Two major social influences on the future use of animals in research are apparent, and they pull in opposite directions. As indicated in the previous section, increased public interest in the use of animals will almost certainly demand that the scientific community continue searching for alternatives to whole animal models. At the same time an increased

interest in maintaining good health, especially in an aging population, drives demands for more research on diseases and aging. This apparent inconsistency provides scientists with the opportunity to drive home an essential message: Given the current state of biomedical research, studies will require the use of animals, but such studies can and will be done with ever-increasing sensitivity to the welfare of the animals used.

The largest unknown quantity is the rapidly increasing power and impact of biotechnology and genetic engineering on animal research. Currently the most obvious effect is the ability to develop animal models to aid in the study of human diseases such as cystic fibrosis, Lesch-Nyham's disease, and some forms of cancer (OncoMouse) (76). This raises two sets of issues: The first is simply a new variant of an old problem, and the other poses new problems for researchers, IACUCs, and regulators. The old issue is that of deliberately producing animals with health problems that can sometimes be chronic, debilitating, and painful. While genetic engineering has made it possible for us to produce animals with some new diseases, old-fashioned selective breeding has long been used to produce animals with other equally serious problems. The new issues are that genetic engineering can result in animals that (1) have unpredicted, perhaps unpredictable, health and welfare challenges, (2) are often considered much more valuable than "standard" animals of the same species, and (3) heighten public fears about scientists "playing God" or tampering too much with the "natural" order of things (77).

The variation of the old problem — caring properly for animals that have, or are expected to develop, severe health problems — is related to genetic engineering only insofar as biotechnology affords the possibility of producing animals with diseases hitherto unknown in a given species. This poses important ethical problems for both researchers and IACUCs. The first is when it is justifiable to produce such animals at all, rather than using alternative forms of research such as epidemiological studies of naturally occurring incidents of the disease. Second, one must consider whether the effect on the animals' welfare will require special care — such as different housing or diet, more frequent monitoring by caretakers or veterinarians, or a prescribed regimen of analgesics. While standard procedures are designed to protect an adequate level of welfare for normal animals, they can be inappropriate for animals specifically bred to develop a chronic health problem. This poses a special burden on IACUCs and responsible veterinarians, who must be informed enough about special needs to ensure that those needs are met. Finally, one must consider the problem of identifying an ethically acceptable endpoint. Using death as an endpoint for a procedure, or even waiting until an animal becomes moribund before terminating the procedure and euthanizing the animal, always requires careful scrutiny and detailed justification, but when the research involves chronic, severe, or terminal health problems the issue of determining the appropriate endpoint, and who is responsible for making the necessary assessments, must be fully addressed at the outset.

As noted earlier, the moral questions just described are not unique to genetically engineered animals, even though biotechnology increases the ability to create animals who are likely to develop a targeted health problem. Other difficulties are magnified even more by advances in biotechnology. What follows are a few examples of problems that have been anecdotally identified; there is a definite need for more carefully controlled studies in this area.

One often-noted difference between genetically engineered animals and those produced through traditional methods of selective breeding is that genetically engineered animals often tend to develop unanticipated phenotypic effects, not obviously related to the desired effect. Some of these effects, ranging from severe joint problems in the "Beltsville pigs" to lack of maternal instinct in mice bred as models for Lesch-Nyham's disease, have a direct impact on animal welfare and ethical concerns about whether such animals should be produced. Second, genetically engineered animals are viewed as more valuable than standard laboratory animals, if only from the standpoint of production costs; this can have both a positive and a negative impact on ethical deliberations. On the positive side, more valuable animals are likely to get more intensive care. On the negative side, researchers with a considerable investment in genetically engineered animals may be more reluctant to terminate a study or euthanize an animal on the basis of welfare concerns that do not coincide with research goals. Once again, this poses ethical challenges for both the researcher and the IACUC that must evaluate and monitor the protocol.

The final concern about genetic engineering, usually expressed as "playing God," is unfortunately too often not very well articulated, which makes the ethical concerns hard to assess. However, it is often likely to focus on research that involves introducing human genetic material or patterns into animals. When such research is specifically targeted — such as aimed at developing pigs whose organs can be used for transplantation into humans, or goats whose milk contains hormones useful for treating human diseases — such concerns seem less apparent. Whether or not "playing God" represents a legitimate ethical objection to genetic engineering, it is surely an area in which the research community must improve communication with the public.

BIBLIOGRAPHY

1. R.J. Hoage, ed., *Perception of Animals in American Culture*, Smithsonian Institute Press, Washington, DC, 1989.
2. G. Carson, *Men, Beasts, and Gods*, Scribner's, New York, 1972.
3. T. Regan and P. Singer, eds., *Animal Rights and Human Obligations*, 2nd ed., Prentice Hall, Englewood Cliffs, NJ, 1989.
4. R. White, in T. Regan and P. Singer, eds., *Animal Rights and Human Obligations*, Prentice-Hall, Englewood Cliffs, NJ, 1976, pp. 163–169.
5. R. White, *Hastings Center Rep.* **20**, 43 (1990).
6. P. Carruthers, *The Animals Issue*, Cambridge University Press, New York, 1992.
7. P. Singer, *Animal Liberation*, 2nd ed., Avon, New York, 1990.
8. T. Regan, *The Case for Animal Rights*, University of California Press, Berkeley, 1983.
9. B. Rollin, *Animal Rights and Human Morality*, Prometheus Books, Buffalo, NY, 1981.
10. S. Curtis, in T. Regan and P. Singer, eds., *Animal Rights and Human Obligation*, 2nd ed., Prentice Hall, Englewood Cliffs, NJ, 1989, pp. 169–175.
11. S. Sapontzis, *Morals, Reason, and Animals*, Temple University Press, Philadelphia, PA, 1987.
12. E. Pluhar, *Beyond Prejudice: The Moral Significance of Human and Nonhuman Animals*, Duke University Press, Durham, NC, 1995.
13. D. DeGrazia, *Taking Animals Seriously: Mental Life and Moral Status*, Cambridge University Press, New York, 1996.
14. P. Singer, *Practical Ethics*, Cambridge University Press, New York, 1979.
15. P. Singer, *Monist* **70**(1), 3–14 (1987).
16. R.G. Frey and W. Paton, in T. Regan and P. Singer, eds., *Animal Rights and Human Obligation*, 2nd ed., Prentice Hall, Englewood Cliffs, NJ, 1989, pp. 223–236.
17. M.A. Fox, *The Case for Animal Experimentation*, University of California Press, Berkeley, 1986.
18. I. Kant, reprinted in T. Regan and P. Singer, eds., *Animal Rights and Human Obligations*, 2nd ed., Prentice Hall, Englewood Cliffs, NJ, 1989, pp. 23–24.
19. J.M. Jasper and D. Nelkin, *The Animal Rights Crusade: The Growth of a Moral Protest*, Free Press, New York, 1992.
20. S.M. Russell and C.S. Nicoll, *Proc. Society for Experimental Biology and Medicine*, 1996, pp. 109–138.
21. R. Ryder, *Victims of Science*, David-Poynter, London, 1975.
22. M. Wreen, *Ethics and Animals* **5**, 47–60 (1984).
23. R.G. Frey, *Between the Species* **3**, 191–201 (1987).
24. J.U. Dennis, *J. Am. Vet. Med. Assoc.* **210**(5), 612–618 (1997).
25. J. Stone, *Canadian J. Philos.* **17**, 815–830 (1987).
26. W. Paton, *Man and Mouse*, 2nd ed., Oxford University Press, Oxford, UK, 1993.
27. C. Cohen, *N. Engl. J. Med.* **315**, 865–870 (1986).
28. J. Nelson, *Between the Species* **2**, 116–134 (1986).
29. J. Rachels, *Created from Animals*, Oxford University Press, Oxford, UK, 1990.
30. J. Bentham, *The Principles of Morals and Legislation*, ch. 17, sec. 1.
31. B. Rollin, *The Frankenstein Syndrome*, Cambridge University Press, New York, 1995.
32. B. Rollin, *Farm Animal Welfare*, Iowa State University Press, Ames, IA, 1995.
33. L.-M. Russow, *Between the Species* **8**, 224–229 (1993).
34. L.-M. Russow, *The Question of Animal Rights*, forthcoming.
35. R. Sorabji, *Animal Minds and Human Morals: The Origins of the Western Debate*, Cornell University Press, Ithaca, NY, 1993.
36. J. Rawls, *A Theory of Justice*, Harvard University Press, Cambridge, MA, 1971.
37. J. Narveson, *Canadian J. Philos.* **7**, 161–178 (1977).
38. J. Narveson, *Monist* **70**, 31–49 (1987).
39. M. Midgeley, *Animals and Why They Matter*, University of Georgia Press, Athens, GA, 1983.
40. A. Baier, in H. Miller and W. Williams, eds., *Ethics and Animals*, Humana, Clifton, NJ, 1983, pp. 61–77.
41. L. Birke, *Feminism, Animals, and Science*, Open University Press, Bristol, PA, 1994.

42. J. Tannenbaum, *Veterinary Ethics*, 2nd ed., Mosby, St. Louis, MO, 1995.

43. G. Varner, *Hastings Center Rep.* **24**, 24–28 (1994).

44. J. Feinberg, in J. Feinberg, ed., *Rights, Justice, and the Grounds of Liberty*, Princeton University Press, Princeton, 1980, pp. 185–206.

45. J. Rachels, in T. Regan and P. Singer, eds., *Animal Rights and Human Obligation*, 2nd ed., Prentice Hall, Englewood Cliffs, NJ, 1989, pp. 122–131.

46. F.K. Goodwin, *Contemporary Topics in Laboratory Animal Science* **31**, 6–11 (1992).

47. F.K. Goodwin and A.R. Morrison, *Scientist* **17**, 12 (1993).

48. A. Rowan, *Of Mice, Models, and Men*, SUNY Press, Albany, NY, 1984.

49. J. Tannenbaum and A. Rowan, *Hastings Center Rep.* **25**, 32–43 (1985).

50. B. Orlans, *In the Name of Science*, Oxford University Press, New York, 1993.

51. J. Smith and K. Boyd, eds., *Lives in the Balance: The Ethics of Using Animals in Biomedical Research*, Oxford University Press, New York, 1991.

52. H. LaFollette and N. Shanks, *Brute Science*, Routledge, New York, 1996.

53. D. Blum, *The Monkey Wars*, Oxford University Press, New York, 1994.

54. P. Cavalieri and P. Singer, *The Great Ape Project*, St. Martin's, New York, 1993.

55. C. Magel, *A Bibliography of Animal Rights and Related Matters*, University Presses of America, Latham, MD, 1981.

56. J.L. Nelson, *Am. Philos. Q.* **22**, 13–24 (1985).

57. R. Baird and S. Rosenblum, eds., *Animal Experimentation: The Moral Issues*, Prometheus Books, Buffalo, NY, 1991.

58. G. Langley, ed., *Animal Experimentation*, Chapman and Hall, New York, 1989.

59. H. Miller and W. Williams, eds., *Ethics and Animals*, Humana Press, Clifton, NJ, 1983.

60. National Research Council, *Guide for the Care and Use of Laboratory Animals*, National Academy Press, Washington, DC, 1996.

61. M. Phillips and J. Sechzer, *Animal Research and Ethical Conflict*, Springer Verlag, New York, 1989.

62. K. Bayne et al., *Current Issues and New Frontiers in Animal Research*, Scientists Center for Animal Welfare, Greenbelt, MD, 1995.

63. Available at: *http://www.uel.ac.uk/research/ebra/info.html*

64. Available at: *http://www.labanimal.com/col/ebra3.html*

65. D.E. Blackman, P.N. Humphreys, and P. Todd, *Animal Welfare and the Law*, Cambridge University Press, Cambridge, UK, 1994.

66. W.S. Russell and R.L. Burch, *The Principles of Humane Experimental Technique*, Methane & Co, London, 1959.

67. M.D. Mann, D.A. Crouse, and E.D. Prentice, *Lab. Animal Med.* **41**, 6–14 (1991).

68. D. Pratt, *Alternatives to Pain in Experiments on Animals*, Argus Archives, Washington, DC, 1980.

69. D.M. Foster and R.C. Boston, *ILAR J.* **38**, 58–89 (1997).

70. M. Mukerjee, *Sci. Am.*, 86–93, February (1997).

71. A. Smith et al., *ILAR J.* **38**, 82–88 (1997).

72. L.-M. Russow, *ILAR J.* **40**, 15–21 (1999).

73. *Chimpanzees in Research: Strategies for Their Ethical Care, Management, and Use*, National Academy Press, Washington, DC, 1997.

74. NASA Principles on the Use of Animals; *NASA Policy Directive for Animal Care and USE*, forthcoming, 3/98. Also available at: *http://www.nih.gov/grants/oprr/dearcolleage.htm*

75. B. Orlans et al., *The Human Use of Animals*, Oxford University Press, New York, 1998.

76. S. Donnelley, C. McCarthy, and R. Singleton, *Hastings Center Rep.* **24**(special supp.), S1–S32 (1994).

77. L.R. Batra and W. Klassen, *Public Perceptions of Biotechnology*, Agricultural Research Institute, Bethesda, MD, 1987.

See other entries ANIMAL, MEDICAL BIOTECHNOLOGY, LEGAL, LAWS AND REGULATIONS GOVERNING ANIMALS AS SOURCES OF HUMAN ORGANS; ANIMAL, MEDICAL BIOTECHNOLOGY, POLICY, WOULD TRANSGENIC ANIMALS SOLVE THE ORGAN SHORTAGE PROBLEM?; see also RESEARCH ON ANIMALS entries; TRANSGENIC ANIMALS: AN OVERVIEW.

RESEARCH ON ANIMALS, ETHICS, PRINCIPLES GOVERNING RESEARCH ON ANIMALS

JOHN KLEINIG
City University of New York
New York, New York

OUTLINE

INTRODUCTION

Research using live animals goes back at least as far as Galen, though it was not until the rise of medicoscientific experimentation in the seventeenth and eighteenth centuries that the use of live animals in research became systematic (1). In the past half century the increasing professionalization of research and practice and the growth of product-safety regulation have led to their prevalent use in education and product testing.

From its very beginning the practice of live animal research (or, now inaccurately, vivisection) has attracted controversy, even among those who have defended it. At one end of a wide spectrum are those who have wished to apply to animals the whole range of moral entitlements and protections that govern experimentation with humans. At the other end are those who would ascribe to an animal's cries no more moral significance than one would to the noise made by a squeaky door. Informing this vast range are metaphysical, epistemological, and moral positions that are too complex and controversial to be resolved in the space of a single article.

Most who give the matter consideration would now acknowledge that nonhuman animals make some moral claims on us, even though there are serious disagreements over the nature and extent of those claims (2–4). Should we eat animal flesh? Does the domestication, transgenic breeding, and cloning of animals interfere with their natural integrity? Is hunting morally acceptable? But nowhere are the moral questions more troubling than in the domain of scientific research, experimentation, and testing. For here important values often seem to be in tension, and no simple formula for their resolution appears to be available (5–8). Whereas, in the case of humans, we can at least constrain our endeavors by appealing to the informed consent of research subjects, this is not a serious option in the case of nonhuman animals.

Furthermore, producing ethical guidelines for animal experimentation is not like producing guidelines for the withholding or withdrawal of lifesaving treatment from human beings. Complex though the latter is, there is sufficient congruence of both ends and means to allow the formulation of a reasonably explicit and manageable structure of benchmarks and stipulations for the guidance of those with whom such decisions will lie (9). In the case of animal experimentation, however, such is the plurality of ends and means, not to mention the diversity of subjects, that any general guidelines are likely to be unhelpfully vague or very limited in their application. Participant observation of animals in the wild, genetic experimentation with *Drosophila melanogaster*, and toxicity testing using laboratory-bred rats differ so much in their character and in the specific ethical questions they raise, that no single set of guidelines is likely to be fully responsive to the ethical issues that should be addressed. This is not to gainsay the value of the guidelines that have been produced in recent years (10–12), but it identifies one source of the dissatisfaction they have engendered. At the very least it mandates a plurality of guidelines.

MEANS AND ENDS

The alternative to comprehensive and exhaustive guidelines need not be an absence of moral moorings or of a structure of moral questions that would allow the ordered assessment of animal experimentation and its associated practices. The inadequacies of guidelines need not bespeak the absence of guidance. Indeed, animal experimentation of whatever kind has the form of a means–end relation, and there are established procedures for assessing such relationships. True, there are some very basic questions concerning the moral status of animals on which we are culturally confused and for which there exists no theoretical moral consensus. And these, along with more general theoretical controversies in ethical theory, will continue to bedevil our deliberations. Nevertheless, an ordered structure for identifying and responding to these issues is available and may be articulated as a framework for both primary ethical deliberation and the secondary development of specialized codified guidelines.

To be ethically acceptable, practices that involve a means-end relation must address and perform satisfactorily in relation to each of the following five questions: (*1*) Is the end morally acceptable? (*2*) Are the means appropriate to the end? (*3*) Are the means likely to realize the end? (*4*) Are the means proportionate to the end? and (*5*) Will the means undermine other good ends? Question 1 recognizes that ends, no less than means, need to be scrutinized. Question 2 acknowledges that what may be employed as means may be inappropriate to the end. Question 3 assumes appropriateness, but focuses on the probability that the end will be established. Question 4 allows that the means may establish the end, but it addresses the costs of achieving it. Question 5 concerns itself with external costs—those wider social costs that an institution or practice may have. Although—as will be clear from the discussion that follows—this menu of questions does not represent a simple checklist, in which the questions can be separately considered and okayed, it nevertheless provides a broadly comprehensive and ordered framework for the organizing of moral deliberation. The commentary to follow, though sketchy, will indicate how these questions can function like a main software menu to systematize ethical reflection on animal experimentation. Each menu item then needs to be accompanied by a further set of pull-down menus. We can order even if not simplify the complexity.

Is the End Morally Acceptable?

Research and experimentation are teleologically oriented activities. A subject is investigated or manipulated in the light of some end. Frequently that end is appealed to as a justifying consideration. Although it is commonly—and properly—asserted that "the end does not justify the means," it does not follow that the end is morally irrelevant to the means. The aphoristic prejudice against *any* justificatory appeal to ends tends to reflect the great significance we attach to autonomy or consent in dealings between human beings, and the general presumption against treating them instrumentally or paternalistically. Even so, that should not obscure the fact that an assessment of ends, and human ends in particular, constitutes a relevant and often important determining factor in the appraisal of human practices. Indeed, as the flip side of our concern that humans not be used merely as means, a focus on ends reflects the importance that we ascribe to them as expressions of human purposive activity.

Animal welfare committees, unlike human investigational review boards, will be guided for the most part by considerations of beneficence, largely unmediated by considerations of justice and autonomy, and in their case the evaluation of ends will assume greater significance. To assert this is not to deny that research animals may have ends of their own, or that their ends may not be genuinely competitive with human ends (see question 4), but it acknowledges the considerable justificatory weight that attaches to human ends as human (though not necessarily anthropocentric) ends. If we undervalue such ends, we erode the significance of the very enterprise that allows our questioning of those ends to carry weight.

Even so, we must make important distinctions. Animal experimentation may be directed to a variety of potentially legitimizing ends—the welfare of particular animals or animal populations or species, the welfare of individual human beings or some wider social good, the communication and advancement of human knowledge, organizational profit, some personal benefit to the researcher, and so on. These ends are not exclusive, and they may be given a different value and priority by different researchers, even by researchers engaged in the same series of procedures. And, of course, *we* (the politically potent community) may value some more highly than others. Even if human ends, as human ends, possess an intrinsic value, not all human ends are equally valuable, and some we may think unworthy of human advancement.

Although disagreements about the relative value of ends can make consensus difficult to achieve, we are not left wholly at the mercy or to the vagaries of individual preference. Relevant differentia can be articulated and brought to bear on judgments of priority. For example—though these do not constitute decisive (i.e., lexically ordered) considerations—experimental procedures directed to welfare will generally have better standing than those carried out solely to satisfy our curiosity or to expand or communicate our knowledge; public goods will generally take precedence over private benefits; procedures designed to benefit the subject of those procedures will generally have a stronger claim to our recognition than those designed to benefit others; and human welfare will generally take priority over animal welfare. This is because there is value to benevolence, a communal dimension to value, a special dignity to human life, and an integrity to animal lives that should weigh significantly in our decision making. But these considerations may exist in tension, and it is not possible to read off priorities in a mechanical fashion. Judgments that seek to accommodate them are singular without being arbitrary, and the problem of formalizing them is not peculiar to animal experimentation but reflective of more general problems in the appraisal of human conduct.

Some of the difficulties in making judgments about the relative importance of ends are linked to complexities in the ends themselves. The advancement of human knowledge, for example, may comprehend the satisfaction of curiosity, the exercise and development of our human powers of understanding, an increase in our grasp of the universe and of ourselves within it, and what is sometimes termed "basic research," which, though not directed to some specific application, usually anticipates some later—albeit unspecifiable—usefulness. And where some instrumental benefiting of ourselves is sought, it may be the alleviation or cure of some disease–mild or serious, rare or common, self-induced or unwittingly contracted—some positively enhancing, recreational, labor-saving, or aesthetic end, or some preventive or protective social goal. The possibilities are legion and jointly pursuable, and animal researchers will need to give some detailed account of them. Guidelines that differentiate and categorize ends may assist in this task, even though judgments about their relative importance will require the more sensitive deliberative scrutiny that a review committee may be able to provide. Even then, there is a serious practical problem posed by the fact that the interests of animals can be represented only by proxies who may not be sufficiently sensitive to them.

The fact that different researchers in a single project may have different priorities and may even be in pursuit of different ends highlights one of the difficulties involved in the evaluation of practices independently of their practitioners, and in the development of social policy. Otherwise, justifiable projects may sometimes be compromised by the unworthy, questionable, or only moderately worthy ends of those who engage in them. Dissertation research, the testing of commercially redundant substances whose marketing is designed only to give a company increased profitability, and the development of biological weaponry can represent the compromise or perversion of otherwise justifiable research agendae. Such ends may not justify the moral costs they involve. Unfortunately, we are not usually able to peer into the hidden motives of researchers to determine whether their private motives match their publicly asserted intentions. And social policies can achieve little more than a monitoring of formally stated ends.

As noted earlier, ends, though relevant to the justification of means, are not usually sufficient to justify them. The appropriateness, efficacy, proportionality, and character of the means also need to be taken into account. Means have a "life" of their own which needs to be considered, not just as they are associated with particular ends.

Are the Means Appropriate to the End?

However worthy ends may be, unless the means used to further them are appropriate to their realization, they will fail to provide the justificatory support expected of them. This is particularly important for the use of animals in scientific research, which imposes exacting demands on experimenters if their results are to be valid and reliable.

Experimental design is often viewed only in relation to scientific validity, and the determination of scientific validity is frequently thought to be independent of moral considerations. That, however, is an oversimplification. For one thing, unless an experimental procedure is suited to the realization of scientifically valid results, it will be inappropriate to the ends to which it is supposedly directed, and any justificatory value those ends might have possessed will be forgone. Moral costs will remain uncompensated. Moreover, and more fundamentally, decisions about the level of significance that will be required of experiments involving animals need to take into account the importance of the ends to which the experiments are directed, and the moral costs in suffering or other deprivations that will be caused. The decision to require a p value of .01 rather than, say, .05, has implications for sample size, the level of control that is exercised over variables, and so on, decisions that inevitably confront the researcher with the costs that will be involved in his or her inquiry. In other words, what appears to be only a matter of scientific validity will also involve issues of moral acceptability.

One of the common complaints about experimental procedures involving animals has been that researchers treat their animal subjects in ways that compromise the validity of their results. Rough handling, poor housing, and generally inadequate control over extraneous variables may jeopardize experimental integrity and give rise to misleading or worthless results. Animal life is wasted, needless suffering may be caused, and scarce resources are squandered. The training of those who are to handle experimental animals must encompass not only issues of technique but also sensitivity to the ecology of animal lives. Much of the current concern over experimental conditions was generated as a result of a 1984 raid by the Animal Liberation Front on the Regional Head Injury Center of the University of Pennsylvania, and the subsequent circulation of stolen videotapes by People for the Ethical Treatment of Animals as *Unnecessary Fuss*. Here, worthwhile ends were needlessly compromised by the careless and callous attitudes of researchers and their assistants (13).

But validity is not affected only by careless treatment of the animal subjects. Sample size, species selection, and other elements of research design may also have an important bearing on the credence to be given and conclusions to be drawn from experimental results. The debate over using pound as against purpose-bred animals turned in part on the extent to which the use of one rather than the other would introduce uncontroled variables. And where animal research is intended to have implications for other species or for human beings, there needs to be some assurance that the experimental subjects are similar in relevant respects.

This latter concern is particularly intractable. The evolutionary theory that may seem to allow for some continuity between humans and animal species, such that research results gained from one can be applied to the other, can also cut the other way and be used to suggest that there now exist fundamental discontinuities and that the validity of using animal research for human welfare ends is problematic. Although I think these difficulties have been exaggerated, they are not without force (7). Moreover, how high the probability of transferability needs to be may depend in part on other considerations, such as the importance of the end. The use of simian immunodeficiency virus (SIV) to research AIDS was fairly speculative, but in view of the seriousness of the AIDS problem, there was more to be said for it than would have been the case had the problem been less pressing.

One of the continuing moral dilemmas of human-oriented animal research is the need to affirm two propositions: (1) research animals are sufficiently like humans to allow reasonable inferences to be drawn from data involving one for the other; and (2) research animals are sufficiently unlike humans to justify (morally) our using them for experimental purposes. The two propositions need not be in tension, though the possibility that they are in any particular case must always be considered.

But even the most rigorously designed and monitored experiment is likely to have little to be said for it if there is a very low probability that the data it yields will advance the end it is ultimately intended to serve.

Are the Means Likely to Realize the End?

Though it is true that a number of significant scientific breakthroughs have been the outcome of happenstance and guesswork more than careful planning, such serendipitous occurrences cannot be appealed to as a substitute for the requirement that researchers make some case for the *likelihood* that their investigation will advance the ends to which their work appeals.

Likelihoods of course are always somewhat speculative, and will vary, and so will increments in knowledge and the advancement of particular ends. There is no measurable probability or simple likelihood of success to which all experiments should be required to conform. The degree of probability that might be expected of a particular project will depend on a variety of factors, such as the importance and/or urgency of the ends, the costs in animal life and suffering, the scarcity of resources, and the availability of alternatives.

Trade-offs between these factors are not easy to craft and require a well-rounded sensitivity. Although moral decision making is not a matter of numbers, one of the arguments for requiring committee approval of protocols involving animals is that the diverse interests involved may not otherwise be adequately represented. In theory at least, representative animal care committees (Institutional Animal Care and Use Committees) may provide an environment for the articulation and rational balancing of interests.

The reasonably expected likelihood of an end's furtherance will also depend on the level of research already undertaken. Some experimental ground has already been so thoroughly explored that the likelihood of new discoveries is very slight; other territory may offer only theoretical possibilities, with (as yet) relatively little empirical data. In some cases the more ambitious protocol may be preferable to the conservative one.

Are the Means Proportionate to the End?

Where good ends are sought, and enter into the justification of means, the costs incurred by those means have to be entered against the goodness of the ends. An experiment directed to and likely to advance a good end may nevertheless be unacceptable because its costs are disproportionate to the goodness of the end. Ends are not privileged with respect to means.

In this context, "costs" include not only straightforwardly economic or utilitarian costs, but what we can more broadly term "moral costs"—including the loss or abridgement of certain values, in particular, the value attaching to a life that is allowed to flourish free of burdensome constraints. The tasks of determining and arbitrating between these costs can raise the most intractable problems for judging animal experimentation.

Experiments involving animals may intrude on them in various ways: They may be killed, but even if not, their "lives" may be disrupted or constrained, and pain/suffering may be caused them. Although animals will obviously differ in the sophistication of their lives, we should not disregard the fact that such lives as they have possess an integrity and intrinsic value not entirely disanalogous to the organic integrity and value that is possessed by human bodily life. A functioning organism cannot be reduced to a normatively neutral complex of chemical interactions, but is distinguished by an organic *telos* whose invasion should generate justificatory questions. In assessing such intrusions, not only should the immediate experimental protocol be taken into account, but also the costs incurred by housing and care, aftercare, and the destruction of natural environments or social bonds.

Finding an adequate moral language in which to cast these discussions is not easy. That many—albeit not all—(experimental) animals can experience *pain* is not generally disputed, though the extent to which those animals can *suffer* is somewhat more problematic. We might differentiate pain from distress, discomfort, anxiety, and fear, recognizing that each makes different assumptions about the capacities of its subject, and that some animals will be capable of experiencing some but not other kinds of such suffering. Furthermore, whether such pain and/or suffering should count morally for the same as human pain and suffering is even more problematic. There is little doubt that part of what is problematic about causing pain and suffering to humans is the way in which they tend to subvert autonomy and its fruits, and this is not likely to be an issue in (most) animal experimentation. Whether or not it is an issue is reflected in (thought not simply resolved by) debates about whether animals possess "interests" and/or "rights," and whether we have duties *to* them or simply *regarding* them.

The point of these debates is not usually to establish whether we owe animals any moral consideration, but to decide what kind of consideration is appropriate, and—remembering that we are not involved in a simple consequentialist calculus—to determine how weighty it is with respect to the various human interests that might be realized in experimentation. Even if it is argued that human interests, by virtue of their origination in deliberative activity, have morally relevant features that animal interests lack, it does not follow that every human experimental interest will take priority over any animal interest. One of the common complaints about the Draize test (in which potential irritants are tested on the eye tissues of animals) has been that the cost in animal pain/suffering often far exceeds the worth of adding a new product to an already well-supplied market. Human interests, no less than animal welfare interests, must be scrutinized and ranked.

Problematic though these judgments are, we should not assume too quickly that they are impossible to make. Insofar as the evaluation of alternatives is an enterprise undertaken by practical decision makers, it is ultimately up to *us* as decision makers, as rational beings and normative agents, to determine how much weight we will accord to environmental and/or social destruction as against physical pain, and how important these are with respect to the advancement of knowledge or human welfare. Contrary to those who see something fatally anthropocentric in such judgments, the perspective of those who have to take responsibility for what they do is the only appropriate perspective to take.

Although judgments of proportionality are in some ways too complex to allow of simple codification and commensurability, some assistance to researchers can probably be provided by setting out in a roughly ascending order different degrees of intrusiveness (taking into account not only the kind of intrusion caused, but also its intensity, duration, relievability, and so forth), cross-referencing it with different levels of animal complexity. Human ends might be similarly ranked, taking into account such factors as urgency, whether long- or short-term, those likely to be benefited, whether the ends to be served can be served in other ways, and so on. Some have complained that many of the human medical problems to whose alleviation much animal experimentation is directed are the result of voluntarily adopted lifestyles, and that lifestyle changes would not be an unreasonable expectation (albeit not, perhaps, a sufficient reason for refusing to engage in animal experimentation): The costs to animals, along with others, might be incorporated in a motivational package addressed to the problem of "unhealthy" lifestyles.

In one of its associated expressions, the proportionality requirement mandates the use of the least costly alternative consistent with the ends being realized. Using 200 experimental animals when 100 would do, using experimental animals when computer modeling or tissue cultures would do, causing pain and stress when the use of anesthesia would not compromise the outcome, using untrained animal handlers when experienced handlers would cause less animal distress, and so on, all represent abuses of the proportionality requirement, since there is a less costly way of achieving the same end. The so-called "three Rs" of animal experimentation—replacement, reduction, and refinement (14)—express this dimension of the proportionality requirement.

Some judgments here may be very difficult to make. It may, for example, be hard to determine whether or not the multiple use of single animals is to be preferred to the use of fewer procedures on more animals, or how the "costs" of using pound animals are to be assessed against those of breeding animals specially for experimental purposes. Do numbers count, as well as the amount of suffering? And how do we factor in the production of transgenic animals for experimental purposes? If an animal is bred to be disease prone, does this constitute a violation of species integrity, or does the new animal now have a natural end that makes experimental procedures (e.g., the testing of anticarcinogenic agents) more acceptable? Does the patenting of such animals provide some control over their use, or does it take us too far in the direction of an unacceptable commodification of animals, in which we come to see them as *no more than* commercially exchangeable tools?

Will the Means Undermine Other Good Ends?

Even if the ends to which an experimental procedure is directed are eminently worthwhile, and the procedure is appropriate to those ends and likely to advance them, and even if the direct costs of the procedure are proportionate to the ends sought, there may be other dimensions of its implementation that need to be taken into account and ranged against it. Sometimes, in pursuing one end, we may undermine or jeopardize other ends to which we are independently committed.

Although it would need to be supported by data rather than merely conjectured, the claim that some kinds of animal experimentation tend to brutalize or barbarize researchers suggests how the pursuit of some worthwhile ends — say, human health — might undercut others — say, civilized sensibilities (15,16). One of the deeper anxieties that fueled nineteenth-century opposition to animal experimentation was concern at the dominance of science and its arrogant oligarchy of expertise, the depersonalization of social decision making, a growing detachment from the world of nature of which we are a part (17). It has been followed in the twentieth century by criticisms of human selfishness and profiteering, our concern with self-indulgence and self-advancement without regard to the costs for other living things (18).

A better documented example of the undercutting of other ends might be the effect of primate research on the persistence of an endangered species. The integrity of a species and the value of species diversity may be compromised if primates are used — or used without regard for their survival — for research into cancer or AIDS. Of course, as we noted earlier, ends themselves may be amenable to ranking, and it would not follow merely from the fact that a means of pursuing a good end would subvert another that it should be eschewed. Nevertheless, our consideration of costs should not be limited to those directly associated with the experiment at hand.

Serious though the foregoing claims are, like all claims they have not gone unchallenged. Researchers may just as easily see their activity as one of responsible stewardship rather than one of arrogant domination: the knowledge gained in animal research is seen as serving the good of human and animal well being, a task that falls to humans because of their unique endowments (18).

The resolution of such conflicts is unlikely to be a simple one. Like most human activities, from sport to road building, animal research is likely to be attended with larger social costs, whatever its benefits, and it will behove us to address them as they arise and seek to ensure that they do not fall victim to the political sloganizing that has characterized much of the current debate.

INSTITUTIONALIZATION OF JUDGMENT

In view of the complex nature of the ethical questions confronting the use of live animals in research, a two-part process for evaluating such research can be proposed. The first will consist in the development of formal guidelines for experimentation and research, guidelines that are sufficiently specific to the ends and subjects of the research to avoid the charge of vagueness. Guidelines might, for example, be developed that will be specific to field research (19–22), to biomedical experimentation (10–12), to product testing (23), and to education (24,25), as ends that will tend to generate different requirements and different questions. A further subdivision might have regard to the animals involved, taking into account levels of consciousness, social ecology, replaceability, and so on. Such guidelines will then address questions of intrusiveness with regard to these various "structural" factors.

It is most likely that such guidelines will be seen as the responsibility of national, international, or professional bodies — bodies with a wide enough representation and jurisdiction to ensure not only that a broad spectrum of opinion has been canvased but also that the resultant guidelines will possess the public standing that will allow them to be used as a meaningful standard in holding researchers accountable, whether by assessing eligibility for funding or by informing legal standards of proper use and handling.

Beyond that, however, there will need to be an informed and sensitive application of these guidelines to specific research protocols, and this might be best achieved through the activity of an Institutional Animal Care and Use Committee (IACUC), in which the various interests at stake in animal research may be represented. The task of an IACUC will not be to apply the guidelines in a formulaic or algorithmic manner but to make a judgment, in the light of the guidelines, about the ethical acceptability of a proposed research study or whether existing research facilities meet acceptable standards for animal care. This is important: Bureaucratic guidelines are almost always too crude for the purposes for which they are drawn up (18).

The role of IACUCs is disputed. Some believe that they should provide no more than scientific assessments of research protocols, and even those who believe that their mandate should extend to ethical questions often wish to limit that questioning to means — ends being seen as given — or to questions of general institutional policy. It is argued that the use of IACUCs to monitor specific protocols and even day-today institutional practice allows judgments to made by the inexperienced, diverts valuable resources from research, overburdens the monitoring system, and restricts academic freedom (26). Although these objections are not decisive, they warn that where the power of an IACUC is considerable, there is a corresponding responsibility on the part of institutions to ensure that their memberships are wisely constituted and adequately resourced (13).

Here too, some national guidelines might be appropriate to ensure that the representation of IACUCs does not too easily fall prey to the political winds that often affect even scientific research. Some effort should be made to ensure that the concerns of animal advocates are represented as well as researchers, and that wider public concerns about both scientific research and animal welfare are allowed voice.

Neither the provision of guidelines nor the approval of an IACUC will guarantee that good decisions are made, but they probably represent the best formal steps that fallible and contending humans can take to reach acceptable solutions.

CONCLUSION

The structure of moral deliberation proposed in this article has the merit of providing a framework of questions that forces to raise the basic issues that need to be confronted by researchers wishing to engage in animal experimentation. In its broad outlines, the structure is both comprehensive and rich. It is comprehensive in that provides for all the morally significant questions to be asked. It is rich in that it allows these questions to be pursued at different levels of generality.

It does not, however, provide a simple procedure for cranking out answers to questions that demand sensitivity and judgment more than formulae. There is great diversity in the ends and subjects of animal experimentation, and researchers in each kind of experimental situation will need to determine that situation's moral ecology before they will be able to grapple with the difficult judgments that will often have to be made. Nor does the general structure provided come with ready made "pull-down" submenus. There is little doubt that those sub-menus provide the sites for some of the most difficult and intractable problems.

Nevertheless, this article proposes that within each domain of research—basic research, field research, medical research, product testing, and so on—researchers construct a series of fairly specific questions they will need to ask themselves, based on the general questions canvassed in this article. These questions could then take into account the more particular ends being pursued, the kinds of animals likely to be involved, the costs likely to be encountered by that research, and so forth. If the responses to these questions are then considered by an animal welfare committee or IACUC whose membership is collectively capable of appraising the social value and scientific merit of the proposed experiments, as well as representing the various animal and human interests involved, this might come as close as what can expect to come to a balanced judgment on an issue that will continue to challenge the quality of our moral perception.

BIBLIOGRAPHY

1. N.A. Rupke, ed., *Vivisection in Historical Perspective*, Croom Helm, London, 1987.

2. P. Singer, *Animal Liberation, N.Y. Review*, New York, 1975.

3. T. Regan, *The Case for Animal Rights*, University of California Press, Berkeley, 1983.

4. P. Carruthers, *The Animals Issue*, Cambridge University Press, Cambridge, UK, 1992.

5. B. Rollin, *The Unheeded Cry: Animal Consciousness, Animal Pain, and Science*, Prometheus Books, Buffalo, NY, 1989.

6. A. Rowan, F. Loew, and J. Weir, *The Animal Research Controversy*, Tufts University School of Veterinary Medicine, Medford, MA, 1995.

7. H. LaFollette and N. Shanks, *Brute Science: Dilemmas of Animal Experimentation*, Routledge, London, 1996.

8. F.B. Orlans et al., *The Human Use of Animals: Case Studies in Ethical Choice*, Oxford University Press, Oxford, UK, 1998.

9. Hastings Center, *Guidelines for the Termination of Life-Sustaining Treatment and the Care of the Dying*, Hastings Center, Briarcliff Manor, Westchester, NY, 1987.

10. Council for International Organizations of Medical Sciences (CIOMS), *International Guiding Principles for Biomedical Research Involving Animals*, CIOMS, Geneva, 1985.

11. Australian Government Printing Service, *Australian Code of Practice for the Care and Use of Animals for Scientific Purposes*, AGPS, Canberra, 1990.

12. Institute of Laboratory Animal Resources (ILAR), National Research Council, *Guide for the Care and Use of Laboratory Animals*, National Academy Press, Washington, DC, 1996.

13. F.B. Orlans, *In the Name of Science: Issues in Responsible Animal Experimentation*, Oxford University Press, New York, 1993.

14. W.M.S. Russell and R.L. Burch, *The Principles of Humane Experimental Techniques*, Methuen, London, 1959.

15. T. Aquinas, *Summa Contra Gentiles*, Bk III, Ch. cxiii, University of Notre Dame Press, Notre Dame, IN, 1975.

16. I. Kant, *Lectures on Ethics* (L. Infield, trans.), Century, NY, 1930.

17. R.D. French, *Vivisection and Anti-vivisection in Nineteenth Century Victorian England*, Princeton University Press, Princeton, NJ, 1975.

18. J.McA. Groves, *Hearts and Minds: The Controversy over Laboratory Animals*, Temple University Press, Philadelphia, PA, 1997.

19. American Society of Ichthyologists and Herpetologists, American Fisheries Society, and the American Institute of Fisheries Research Biologists, Guidelines for Use of Fishes in Field Research, *Copeia, Suppl.*, pp. 1–27 (1987).

20. American Society of Ichthyologists and Herpetologists, The Herpetologists' League, and the Society for the Study of Amphibians and Reptiles, Guidelines for Use of Live Amphibians and Reptiles in Field Research, *J. Herpetol.* **4**(Suppl.), 1–14 (1987).

21. Acceptable Field Methods in Mammalogy: Preliminary Guidelines Approved by the American Society of Mammalogists, *J. Mammal* **68**(4, Suppl.), 1–18 (1987).

22. American Ornithologists' Union, Report of Committee on Use of Wild Birds in Research, *Auk* **105**(1, Suppl.), 1A–41A (1988).

23. Foundation for Biomedical Research, *The Use of Animals in Product Safety Testing*, Found. Biomed. Res., Washington, DC, 1988.

24. Schools Animals Care and Ethics Committee, *Animals in Schools: Animals Welfare Guidelines for Teachers*, NSW Department of School Education, New South Wales, 1991.

25. National Science Teachers Association, *Guidelines for Responsible Use of Animals in the Classroom*, NSTA Rep., p. 6, NSTA, Arlington, VA, 1991–1992.

26. N.H. Steneck, *Ethics and Behav.* **7**, 173–84 (1997).

See other entries ANIMAL, MEDICAL BIOTECHNOLOGY, LEGAL, LAWS AND REGULATIONS GOVERNING ANIMALS AS SOURCES OF HUMAN ORGANS; ANIMAL, MEDICAL BIOTECHNOLOGY, POLICY, WOULD TRANSGENIC ANIMALS SOLVE THE ORGAN SHORTAGE PROBLEM?; see also RESEARCH ON ANIMALS entries; TRANSGENIC ANIMALS: AN OVERVIEW.

RESEARCH ON ANIMALS, LAW, LEGISLATIVE, AND WELFARE ISSUES IN THE USE OF ANIMALS FOR GENETIC ENGINEERING AND XENOTRANSPLANTATION

F. BARBARA ORLANS
Kennedy Institute of Ethics, Georgetown University
Washington, District of Columbia

OUTLINE

INTRODUCTION

A new industry has emerged in recent years that uses laboratory animals in ways that have never before been tried. Animals are used as experimental subjects of studies involving genetic manipulation in which there has been a deliberate modification of the genome — the material responsible for inherited characteristics — to produce genetically modified animals or sources of organs and tissues for transplantation into humans (xenotransplantation). Mice are the most-used species in transgenic studies; pigs are the preferred species for xenotransplantation. As a result of these novel technologies, new animal welfare issues have arisen. There is public concern about the sheer increase in the numbers of animals used, the potential increase in animal suffering, and new forms of exploitation of animals. Existing legislation is not able to give these subject animals the quality of consideration and degree of protection to which they are entitled. U.S. federal laws, last amended in 1985, fall behind the current need for protecting the welfare of these animals. For instance, areas of concern are that mice are not included under the Animal Welfare Act, adequate limits are not placed on the invasion of the integrity of an animal, pigs are confined in limited space and barren environments, and more training of laboratory personnel is needed in clinical monitoring of animal pain and distress.

ANIMAL PROTECTIVE LEGISLATION

Two U.S. federal laws govern the use of laboratory animals in biomedical research. The first, now called the Animal Welfare Act (AWA), was passed in 1966. It required registration of animal research facilities with the United States Department of Agriculture (USDA), federal inspections, and the humane treatment and care of certain species of animals (dogs, cats, nonhuman primates, rabbits, hamsters, and guinea pigs). In 1970 Congress changed this wording so that additional warm-blooded species could be included as determined by the Secretary of Agriculture. Any future changes in what

species were included would therefore be up to USDA to announce in its rule making. But no action was taken and numerous other animals were left unprotected. Strengthening amendments to the AWA made in 1970, 1976, and 1985 required the use of pain-relieving drugs, the establishment of Institutional Animal Care and Use Committees (IACUCs) to oversee compliance with the regulations, and promotion of psychological well-being of primates (1).

A second mechanism of control emerged in the 1960s covering the practices of grantees of the National Institutes of Health (NIH), which is part of the Public Health Service (PHS). NIH had provided federal grants for animal experiments since 1946. In 1963, in an effort to forestall the increasing efforts to establish federal legislation, NIH published for the first time voluntary guidelines called the *Guide to the Care and Use of Laboratory Animals* (commonly called the NIH Guide). Under the Health Research Extension Act of 1985 (P.L. 99-158), the NIH Guide is no longer a "guide" but law, and it now covers all federal agencies (e.g., the Department of Defense) and not only PHS. It is now called the Public Health Service Guide for the Care and Use of Laboratory Animals (2). In the 1960s these publications dealt only with husbandry standards — minimal caging size, sanitation, nutrition, and the like. Over time the scope of these publications has broadened to include provisions on experimental procedures as well.

Thus became established the two primary mechanisms for maintaining standards that continue to this day in the United States — AWA and its amendments administered by the Animal and Plant Health Inspection Service (APHIS) of USDA, and the policy of NIH/PHS administered by the Office for Laboratory Animal Welfare, (previously the Office for the Protection from Research Risks) of NIH. These laws and subsequent rule making govern the current conduct of animal research. The two oversight mechanisms cover different constituencies, although there is overlap in their purview. Recent efforts have been undertaken to make the provisions of each compatible with the other.

Funding for enforcement of AWA by USDA has always been a problem. USDA is required to inspect research facilities, dog and cat dealers, and zoos, since these are all covered under AWA. When AWA was first passed in 1966, an appropriation of $300,000 was barely achieved. Annual appropriations rose slowly for several years, but from 1992 to 1999 the appropriation has remained static at $9.2 million. (This compares with a congressional appropriation of over $17 billion to NIH for biomedical research in fiscal year 2000. NIH grants comprise an important national source of funding for animal research.) A shortage of USDA personnel has also been a problem. In Fiscal year 1997, for instance, a staff of about 73 animal care inspectors conducted almost 16,000 inspections to ensure compliance with AWA regulations (3). Currently a consortium of professional scientific and animal advocacy organizations is pressing for an annual increase in funding of at least 3 to 4 million dollars.

In 1998 there were 1227 animal research facilities registered with USDA under the AWA (4). This number compares with some 970 institutions (in February 2000) that must comply with the PHS policy. Some overlap between the two groups exists. Still outside the provisions of any national policies are the academic and commercial institutions that either do not receive federal funding or use species of animals which are exempted. How many such exempt institutions there are is unknown, but the number probably runs to several thousands and includes privately funded facilities that conduct genetic manipulations on mice and rats.

Animal research facilities now have greater responsibilities than previously. The current legislation mandates that each research facility using animals establish an oversight IACUC with members appointed by the chief executive officer of the facility. Each committee is composed of no fewer than three members: one a veterinarian, and another not affiliated with the institution. In practice, animal researchers both chair the committee and dominate its membership. The most recent inquiry by the NIH office that administers the PHS policy found that only two of the approximately 1000 IACUCs have chairpersons who are not animal researchers.

Since 1985, protocol review by IACUCs has been mandatory. For instance, the PHS policy requires IACUCs to review relevant sections of PHS grant applications to ensure that (1) procedures with animals will avoid or minimize discomfort, distress, and pain to the animals, consistent with sound research design; (2) appropriate sedation, analgesia, or anesthesia is used; (3) animals that would otherwise experience severe or chronic pain or distress that cannot be relieved will be painlessly killed; (4) laboratory personnel are appropriately qualified and trained in the procedure(s) they are using; (5) methods of euthanasia are consistent with those prescribed by the American Veterinary Medical Association; (6) procedures involving animals are designed and performed with due consideration of their relevance to human or animal health, the advancement of knowledge, or the good of society; (7) the animals selected are of appropriate species and quality and the minimum number required to obtain valid results; and (8) methods such as mathematical modeling, computer simulation, and in vitro biological systems are "considered." In accomplishing these tasks, each IACUC approves, disapproves, or modifies the proposed animal experiment. The effectiveness of these committees is variable.

The AWA 1985 amendments also added other provisions: Training must be provided to laboratory animal personnel in the humane care and use of animals, the environment in which nonhuman primates are housed must promote the animals' psychological well-being, and dogs must be given exercise. The Secretary of Agriculture was required to issue standards governing these provisions.

The first of these new provisions fared reasonably well in the 1991 USDA rule making implementing the AWA amendments. As a result institutions for the first time began to provide training for their personnel in animal handling, anesthesia, and euthanasia, and what the concept of "Three R" alternative means—to replace animal experiments with nonanimal methods where feasible, to reduce the numbers of animals used, and to refine procedures to minimize or eliminate animal pain and distress, concepts that are all included in the laws. In response, the climate changed appreciably. IACUCs began to be more alert to the qualifications of research investigators to conduct traumatic procedures and to insist on training of the laboratory personnel who were not familiar with the techniques involved. The role of the veterinarian in providing this on-site training became progressively important. The legal requirement that the experiments not be duplicative led to greater use of the computerized library resources at the increasingly influential Animal Welfare Information Center at USDA and at the National Library of Medicine.

Psychological Well-Being

The other two new AWA provisions fared less well. The congressional requirements to promote the psychological well-being of primates and to provide exercise for dogs proved highly controversial. Congress had left unclear exactly how it wanted USDA to write the rules. Researchers protested the inclusion of these requirements in the law, arguing that "well-being" was unmeasurable, exercise for dogs was unnecessary, and any changes would be too costly.

There was considerable delay in USDA's promulgating rules governing primate psychological well-being. What appeared finally was permissive vis-à-vis the biomedical community. Instead of setting specific standards (as wanted by the humane community), the regulations allow each laboratory to determine how it will improve treatment of research animals, and a great deal of discretion is allowed (as wanted by the research community).

Former Congressman John Melcher, a veterinarian and the person responsible for adding the amendments regarding primate well-being and exercise of dogs, wrote of his regret about the 1991 rule making in the *Washington Post* (5). He said: "Imagine a small cube of a cage three feet on a side and three feet high. Within this cube lives a primate—often a baboon or a rhesus monkey—that could weigh as much as 55 pounds. Baboons usually stand on all four feet, but in such a space they cannot walk anywhere. They cannot stand upright or stretch their arms in such a cage. Yet this is a common caging for the animals used for scientific research.... The USDA has failed [in their new regulations]." The USDA rules went into effect in 1994, nine years after the passage of the law.

In July 1999, many years later, USDA reported that research facilities do "not necessarily understand how to develop an environmental enhancement plan that would adequately promote psychological enrichment," and therefore additional policies have been proposed (6). As of February 2000, public comments on these proposals were being assessed before issuance of additional policies.

Despite all these problems, without the 1985 law, funding for research projects to explore environmental enrichment would not likely have been forthcoming. Importantly, NIH started funding projects designed to test the beneficial effects of primate housing that allowed the expression of normal behaviors so that the animals would not be bored and come to express stereotypical obsessive behavior, such as constant rocking or bar chewing—signs of psychological trauma. In fact the expression of stereotypic behaviors may indicate a disordered nervous system, bringing into question the validity of data derived from them (7).

Research has established that enrichment schemes are beneficial to the animals (8–10). Examples include group housing, more space, addition of climbing apparatus and manipulative devices such as chew toys and mirrors, and feeding enrichments. As a result of these efforts there has been a reduction (although not elimination) in the proportion of laboratory chimpanzees, baboons, and other primates that are psychologically damaged and a concomitant improvement in the quality of science.

Expanding Coverage to all Species Used

More than 30 years after the enactment of AWA, several widely used species of laboratory animal are still not covered, notably mice, rats and birds, despite the fact that according to commonly accepted estimates, they comprise 80 to 90 percent of all animals used. Farm animals were not included until 1990.

The animal welfare community has long fought to have *all* species of animals used in experiments included in AWA. No act of Congress is needed, only that USDA amend its rule making. (The Public Health Service policy *does* include all species inasmuch as all vertebrate animals are covered, so at PHS grantee institutions this is not an issue.) With regard to AWA, initially USDA had enough on its hands just to get the law into operation. But as time went on, the exclusions of certain species became a glaring problem.

Over the years individuals and groups concerned about the welfare of animals have exerted pressure to drop exclusion of agricultural farm animals, mice, rats, and birds. USDA has resisted on the ground of financial cost. However, in 1990 USDA finally ruled that horses, sheep, goats, cows, and pigs when used for *biomedical or other nonagricultural research* are covered by AWA. (Excluded still are farm animals used in genetic engineering research to increase productivity for food and fiber purposes and also to produce various biologics and pharmaceuticals.) As of 1990, pigs used for xenotransplantation and other biomedical research must be maintained and cared for in compliance with AWA standards and the facilities are subject to USDA inspection. (Pigs used in such research in PHS-funded facilities have always had to comply with PHS standards.) Between 1990 and 1998, the use of farm animals for biomedical research has more than doubled (66,702 to 157,620 per year) (4). In particular, the use of pigs has grown significantly.

Mice, Rats, and Birds. USDA is the target of complaints about failure to expand coverage of AWA to other species. This is because Congress had given USDA discretion about which species to include in addition to those initially mandated in 1966. After years of trying persuasion, in 1990 the Animal Legal Defense Fund brought suit against USDA to amend the regulations to include mice, rats, and birds. USDA objected to these inclusions because of lack of money and resources. The animal rights group won its case. In a judgment issued January 8, 1992, in the U.S.

Distric Court, Judge Charles R. Richey ruled that USDA's exclusions were "arbitrary and capricious" and that USDA must issue new rules to include these species (11). USDA appealed the court's decision.

Pressures continued to mount. In 1998 the Alternatives Research and Development Foundation (a branch of the American Anti-Vivisection Society) filed a petition with USDA requesting that the agency amend its definition of "animal" to include mice, rats, and birds. The *Federal Register* announced receipt of the petition as well as the agency's response (12). Again USDA's arguments reflected its previous opposition to inclusion on the basis of lack of resources for implementation. Also USDA stated its belief that the majority of these animals were already being afforded certain protections, asserting that 90 percent of these animals are provided oversight by PHS assurance, voluntary accreditation, or both. The agency stated that most biomedical research in the United States is performed in laboratories funded at least in part by PHS. In addition 600 facilities in the United States are accredited by the Association for Assessment and Accreditation of Laboratory Animal Care International (AAALAC) and therefore voluntarily comply with their standards (which, like those of the PHS, cover mice, rats, and birds).

If mice, rats, and birds were included under AWA, the impact would fall primarily on two groups of animal users: (1) two- and four-year liberal arts and community colleges that, in general, use birds, mice, and rats for student education in preference to any other species; and (2) commercial genetic engineering companies.

There are an unknown number of facilities that conduct research on transgenic animals. Such facilities neither fall under the PHS policies nor need to be accredited, since the accreditation system is voluntary.

Mice are the species most frequently used to decode human ills. They are genetically modified to have human diseases such as diabetes, cancer, multiple sclerosis, arthritis, and a host of other ailments. In some experiments, multiple pathologies and serious animal welfare problems such as chronic pain and weak legs resulting in inability to stand up have been reported. Since commercial genetic engineering facilities are outside the law, they are not inspected by federal officials, nor do they have to have IACUCs for protocol review. They are also not required to use approved euthanasia methods. Although some companies are doubtless maintaining acceptable standards of animal care and use, there is no public assurance that they are, and there is no overt public accountability. With the added factor of secrecy in this highly competitive enterprise, public concerns arise. Indeed, major ethical concerns arise from a dangerous combination of factors, including lack of federal oversight in an industry based on the use of procedures which can cause considerable suffering in order to model severe human disorders, and the pursuit of profit.

Arguments for and against Inclusion of Additional Species. The vast majority of animal advocacy organizations have voiced support for including mice, rats, and birds. Among the arguments presented is that expanded coverage is a matter of justice in that animals of similar moral worth should be treated equally (13). On this view, there should be a difference in some morally relevant characteristic if animals are to be treated differently. Mice and rats are so similar to three species currently regulated under AWA (hamsters, guinea pigs, and gerbils) that it is arbitrary, against common sense, and unjust to exclude them from legal protection. Indeed, all five species, mice, rats, hamsters, guinea pigs, and gerbils are similar in many ways. All are commercially purpose-bred for research and widely used; physiologically and anatomically they are, to a large extent, commensurable. Furthermore, and importantly from an ethical stand-point, the burdens they bear as subjects of biomedical experiments are of the same order: They have similar sensibilities in their perception of pain, and all are likely to be killed before the end of their normal life span. It is reasonable to assume that these species have an interest in not being subjected to pain or suffering and not having their lives prematurely foreshortened. Inasmuch as humans cannot be treated differently, unless there is some morally relevant basis on which to do so, dissimilar legal protection cannot be justified for animals that have similar relevant characteristics.

Several organizations have voiced opposition to expansion of coverage to all species used. These include the National Association for Biomedical Research, which states that expansion of AWA to include mice, rats, and birds is "a luxury we can do without" (14). The Federation of American Societies for Experimental Biology argues that expansion "would represent redundant regulation" (15). Other comments underline the fact that this would represent an unnecessary burden and hamper enforcement of existing regulation. However, some scientifically based organizations such as the Association for Accreditation and Assessment of Laboratory Animal Care (the accrediting agency) and the Scientists Center for Animal Welfare have voiced approval of including rats, mice, and birds. As of February 2000, USDA was analyzing the public comments they have received in response to their announcement in the Federal Register to determine what future action to take.

NATURE OF PUBLIC CONCERNS

Public response to new biotechnologies has been both enthusiastic and cautious. Genetically modified animals provide an extremely powerful tool for the development of disease models, since the mechanisms of gene regulation will become better understood. In addition the use of genetically modified mice as models of human diseases closely mimics the human disease and may even, in time, replace the need to use more acutely sentient animals (nonhuman primates) as models. But several reservations have also been expressed. Five issues of public concern are discussed below: the overall increase in (1) numbers of animals used, (2) the sum of animal suffering, (3) the invasions into the integrity of the animal, (4) the standards of housing and care of pigs, and (5) the monitoring of animal pain and distress. These concerns arise despite

acknowledgment that the end result of these novel experiments is likely to be of significant benefit to humans.

Increase in Animal Use

When the U.S. Congress passed AWA in 1966, it recognized that keeping proper records is essential to ensure public accountability. It ordered that the numbers of laboratory animals used be counted and publicly reported each year, and this tabulation has been performed since 1973. But since mice, rats, and birds are not included under the definition of "animal," they are not counted. Data on the most-used species are missing from the statistics. This lack of information is detrimental to animal advocates who wish to track trends in animal use as part of their endeavor to reduce use and to target areas for reduction in animal pain and distress. It hampers commercial estimates of the future need for laboratory animals. It also prevents the public from participating in an informed debate.

Other sources of information show that use of mice is increasing in the United States. A 1999 article reported that Harvard Medical School will probably double its use of mice over the next five years—to about one million mice annually (16). Harvard is no exception. In 1991 NIH reported the use of 294,000 mice in intramural research; this number had increased to 648,000 in 1997. At both institutions, the increased use of mice is attributed to an increased number of experiments involving genetic manipulation.

Even more telling are statistics from the United Kingdom, since these are national data. In UK data both the numbers of animal procedures are reported and their purpose. Of the 1998 total of 2,659,662 animal procedures performed, genetically manipulated animals comprised 447,612 or approximately 20 percent of the total, and their use had doubled since the preceding year (17). In 1998 mice were used in 96.6 percent of the procedures comprising genetic modification. Other species used, in descending order were rat, pig, sheep, and other species.

The data in Table 1 provide some notion of the extent of animal experimentation worldwide. There are over 28 million animals counted in official statistics. This is an underestimate because many countries that use animals for experimentation do not count the numbers used. Not available, for example, are data from South America, Eastern Europe, the Middle East, Russia, Africa, and Asia. Regulations governing the humane use of laboratory animals exist in all geographic areas represented in Table 1—some regulations being more and others less rigorous than those of the United States. In several countries where animal experimentation is conducted, no legislation exists (18).

In all probability the United States is the largest user of laboratory animals worldwide. In fiscal year 1998, a total of 1,213,814 animals were officially reported to have been used in research in the United States (19). When this figure is adjusted for uncounted species, the total comes to over 12 million animals per year in the United States alone (see Table 1). Over the period from 1973 to 1998, the total numbers of animals counted has fluctuated between 1.7 to 1.2 million animals per year, indicating a decline over the last six years.

Table 1. Number of Laboratory Animals Used in Research, by Country

United States (1998)	12,138,000[a]
United Kingdom (1998)	2,660,000
France (1997)	2,609,000
Canada (1997)	1,472,000
Belgium (1996)	1,516,000
Germany (1998)	1,532,000
Australia (New South Wales, South Australia, Victoria, Tasmania, and Western Australia) excludes fish (1996/7)	1,141,000
Italy (1996)	1,094,000
Netherlands (1997)	713,000
Norway (1997)	630,000
Spain (1996)	507,000
Switzerland (1998)	492,000
Denmark (1997)	380,000
New Zealand (1998)	309,000
Sweden (1997)	267,000
Austria (1996)	205,000
Finland (1998)	195,000
Ireland (1998)	69,000
Portugal (1996)	50,000
Hong Kong (1998)	27,000
Greece (1996)	19,000
Total	28,025,000

Note: Numbers represent official statistics of all countries and regions for which information could be found. Numbers are given to the nearest thousand and figures in parenthesis indicate year of count. The data presented are not necessarily comparable from country to country because of differences in animal species included. For instance, fish comprise 91 percent of the animals used in Norway but are not counted in the United States. Also some countries count experimental procedures and others individual animals.

[a]The United States counts only about 10 percent of all animals used in experimentation, since the most used species — rats, mice, and birds — are not protected under the relevant legislation and are exempt from counting. The official count for 1998 was 1,213,814, and this figure has been multiplied by 10 to allow for uncounted species and to achieve approximate comparability with data from other countries.

Increased Animal Suffering

In 1985 scientists at the USDA Beltsville Research Center called in the media to see the first ever genetically modified animals — creatures who became known as "the Beltsville pigs." In an attempt to produce faster-growing animals, these pigs had been genetically modified to express very large quantities of human or bovine growth hormone. The experimental purpose was to bring potentially greater profits to the food industry The public reaction to the pictures was of shock and criticism because of the obvious animal suffering. Some animals had damaged vision or deformed skulls, and others were unable to walk properly. The long-term deleterious effects for these animals were demonstrated two generations later and included gastric ulcers, arthritis, cardiomegaly, and nephritis (20).

Animals are now increasingly used to model human diseases; these studies can involve severe animal suffering. Among the painful diseases that have already been produced by genetic manipulation of mice are cancer, cystic fibrosis, Huntington's disease, and a rare but severe neurological condition called Lesch-Nyhan's syndrome that causes the sufferer to self-mutilate (21). Because the technology is in its infancy, the outcome for the animals is still somewhat hit or miss. Multiple pathologies are frequent: legless animals have been born, and abnormalities in genital organs, liver, kidney, joints, and vision also appear. Until techniques are worked out, survival levels are poor and considerable wastage of animals can occur. Unexpected, uncontrolled, and even undetected animal suffering often result. Particular attention needs to be paid to the ethical justification in terms of likely benefit to human health compared with the likely suffering of the animals.

In some countries statistical data may become available to measure this increased sum of animal suffering. A refinement in national statistics not found in the United States is that data are presented according to the "severity band" or "invasiveness" of the procedure, either minor, moderate, or severe. As of February 2000, six countries mandate that investigators rank their proposed procedures according to the degree of pain and distress (22). Two countries, Canada (23) and the United Kingdom (24), have also begun categorizing the level of adverse state resulting from specific genetic procedures. The Canadian guidelines on genetically modified animals stipulate that proposals to create novel transgenics initially should be assigned CCAC category of invasiveness level D (moderate or severe distress or discomfort) at least until the phenotype has been evaluated (25). In theory, statistical data on genetic modification procedures could be developed that report both on the numbers of animals used and the severity of the procedures used, ranked as minor, moderate, or severe. The rankings of severity would be made by the investigators in concurrence with IACUC. Over years, this would provide information on, for instance, the reduction in severity of effects from genetic modification procedures as the techniques become refined.

In the United States, major reforms would be needed to equal the data already available from other countries. U.S. official statistical data are deficient in not including mice, rats, or birds at all, in not specifying the experiments' purpose, and in not ranking the procedures' severity. Public concerns are fanned by the lack of animal data and lack of disclosure.

Integrity of the Animal

A particularly troublesome issue is interference with the integrity of an animal. The 1997 Experiments on Animals Act of the Netherlands requires that biomedical experiments on animals must be conducted "in recognition of the intrinsic value of animal life" (26). This is the first law in the world with such a statement. But what limits should be placed on preserving the integrity of a life form?

"Naturalness," "integrity," and "intrinsic/inherent value" are concepts that are open to differences in interpretation. For example, are laboratory animals simply tools to exploit, or do substantial alterations to the genome violate species-specific life — their "telos"? What constitutes the pigness of a pig, that is, the telos of an animal? Some critics believe that a pig should not be altered to the point that it ceases to be recognizably a pig. They question whether the biotechnology industry is attempting to

Figure 1. This is an actual photo of a genetically engineered mouse with a human ear on its back. It appeared in the *New York Times*, A11, October 11, 1999, in a full page advertisement protesting this invasion of the boundary between lifeforms. Credit Associated Press.

"capture [control of] the evolutionary process" and distort life forms in ethically unacceptable ways. "Who appointed the biotech industry as the Gods of the 21st Century?" was the heading of a 1998 protest against biotechnological manipulations of living beings. This advertisement, published in the *New York Times*, showed a photograph of a mouse that had been genetically manipulated to cause a human ear to grow on its back, see Figure 1. Many people worldwide were shocked at the sight of this photograph. The sponsors of the protest were a coalition of 19 organizations including the International Center for Technology Assessment, the Council for Responsible Genetics, and the Humane Society of the United States. Amazing mixtures of genes have been tried. For instance, jellyfish genes have been installed in monkey embryos, as well as injected into mouse sperm, and eggs and pregnancies created (27). (The goal of these studies was to test a technique that might eventually be used to create monkeys with added human genes.) The question is: Is there a boundary between life forms that should not be crossed?.

It is true that humans already share many genes with other animals — for instance, about 98 percent of the genes of chimpanzees and humans are identical, so only about 2 percent represent uniquely human characteristics. Because of the similarities, placing a single gene in another species may not, in itself, be so objectionable. But if future developments allow the transfer of uniquely human characteristics to animals, then ethical concerns would be increased. Mench provides a useful discussion of the issues of gene transfer (28).

Standards of Pig Housing and Care

There is considerable public support for the belief that laboratory animals should be provided with humane housing appropriate to their needs. Freedom of movement and the opportunity to express natural behaviors are viewed as basic animal needs. Pigs are intelligent animals with a range of bodily movements and natural

behaviors which, if thwarted, result in manifestations of psychological damage such as obsessive, stereotypic behaviors. In this respect, pigs are like baboons and other primates. As discussed above, the living quarters for primates have been significantly improved over recent years. Will lessons learned about the importance of primate environmental enrichment flow over to other species such as pigs?

Among researchers and ethicists, a consensus is emerging that pigs are the preferred species for the routine supply of organs for xenotransplantation. According to the UK Nuffield Council on Bioethics, such use is "ethically acceptable" (29). The rationale for using pigs rather than nonhuman primates is that pigs have less highly developed mental capacities because they are less closely related to humans and that there is less chance of transmitting diseases from pigs to humans than from primates to humans. The use of pigs as a source of tissue for human use introduces the necessity of genetic modification — the creation of transgenic animals bearing human genes — in order to reduce the risk of hyperacute rejection. Further use of immunosuppressive treatments of the patients, to mitigate the danger of rejections, means that all tissues for transplantation must therefore be produced under sterile conditions in order to reduce the transmission of infectious diseases from animals to humans.

Initial production of transgenic pigs begins with impregnation by artificial insemination and removal of fertilized eggs to be microinjected with the required human gene. The pigs produced through this breeding procedures are used to stock the expansion herd through early weaning procedures. The source animal herd is established as a qualified pathogen-free herd. The pigs are born via hysterectomy or hysterotomy, taken from the sow, and reared in groups in isolated environment. The piglets are kept in isolators for 14 days, having no contact with the sow or the sow's milk (30).

Pathogen-free pig housing conditions are restrictive and closely controlled; they can therefore be stressful. Such housing conditions can mean, in some laboratory facilities, completely barren enclosures. The walls may be stainless steel or sometimes they are like tiled bathrooms. Flooring may be slatted fiberglass, or some other sterilizable matting. The flooring may not be ideal for pigs to walk on but is chosen for its hygienic qualities. Social deprivation is another problem because these pigs are individually housed and may be out of visual contact with other animals. There may be plenty of physical room, but if there is nothing *in* the room, the environment deprives the animal of normal social and play behaviors. The intelligent and social nature of pigs makes such deprived housing stressful.

Are such barren environments essential, or could enrichment alternatives be introduced? An alternative is to maintain the pigs in groups that are treated as a microbiological unit. Pregnant sows show a clear preference for a bedded surface rather than an unbedded surface, and sterilized straw is available (31) or possibly irradiated straw, as the pigs particularly enjoy having straw to root in. On the market are toys suitable for pigs housed in pathogen-free environments, such as teflon balls and other products (32). In cases where pathogen-free environments are not essential, an even greater range of environmental complexity is possible, such as inclusion of wood shavings to provide soft surfaces, scratching posts, chains, footballs, concrete blocks (33), and showers. Often these enrichments are omitted because they are more trouble and take extra time to maintain. Enrichments are being used in some laboratories, but encouragement is needed to make such enrichments standard practice.

A high quality animal welfare system for maintaining pigs, the Nuertinger System (34) developed in Germany, is used by some laboratories conducting xenotranplantation research, such as Imutran Ltd. in Cambridge, UK (35). Neither gestation nor farrowing crates (discussed below) are used. The Nuertinger System has a number of animal-welfare-friendly features: It comprises a warm insulated bed and a cooler area for loafing, feeding and drinking, which gives the pigs a choice of environment and temperature, allowing them to choose where they can rest comfortably; see Figure 2.

Gestation and Farrowing Crates. Another issue concerning the breeding of pigs, not specific to xenotransplantation, is the use of gestation crates (also called gestation stalls) and farrowing crates — controversial practices developed in intensive food production units but now becoming increasingly used in laboratory settings. Both systems involve significant restraint of the sows, so that the animals can only stand up and lie down; they cannot turn around or walk. The animals cannot express their normal behaviors and become severely stressed, especially in gestation crates because the period of confinement is longer. Use of both gestation and farrowing crates is standard practice in the U.S. pork industry. Public protest against the use of gestation crates in particular, and to

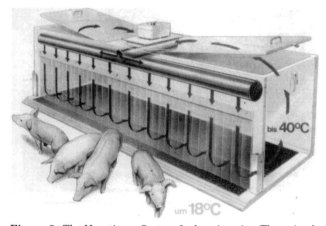

Figure 2. The Neurtinger System for housing pigs. The animals are group housed and can move freely through a plastic curtain between two areas that are maintained at different temperatures, up to 40 °C (104 °F) in one area, and about 18 °C (64 °F) in the other. Among the various environmental variables that can affect the welfare of pigs, temperature is the most important. Often piglets prefer to lie in a warm bed with their heads in the cool fresh air. For suckling, the air can be maintained at a cooler temperature. A controlled ventilation system regulates the temperatures. Credit: HAKA.

a lesser extent of farrowing crates, has been raised both in the United States and Europe. In the United Kingdom, confinement of sows during gestation and pregnancy in gestation crates has been prohibited since January 1, 1999 (36).

Gestation crates are used in food production units because they are economical of space and therefore inexpensive, and because they involve minimal human labor. In the agricultural industry, where profit is a dominating factor, these crates have come into almost universal use. The dimensions and use of gestation crates, as given below, are taken from the *Ag Guide* (37). They have overall dimensions of 5 by 7 feet, with a confining area that is 22 to 24 inches wide that constrains the sow and prevents her from turning around. Swine are large animals and can weight up to 600 pounds. Pregnant females are held in this confinement for four months — the whole period of gestation — and the animals are severely stressed. The animals lack freedom of movement and social interaction with conspecifics: They are unable to root, and they are deprived of expressing their strong maternal instincts of nest building. These deprived housing conditions are beyond the pregnant animals' abilities to cope, and as a result, they frequently exhibit stereotypic behaviors such as bar chewing, vacuum chewing, or head waving. Despite these welfare problems, gestation crates are used to breed animals in some American laboratories. Furthermore pigs are obtained for laboratory use from farms where confinement gestation crates are standard practice.

In laboratories there have been few pressures to avoid using gestation crates despite the fact that humane standards traditionally have been more rigorous for laboratory animals. The one-time clear distinction between farm and laboratory practices has become blurred. Traditionally there have been higher animal welfare standards in laboratories than in farm situations, but this tradition is being broken with the use of gestation crates. The rationale for using gestation crates in laboratory settings is weakened by the fact that the numbers of pigs maintained is relatively small, and there is no pressure to use minimal space to bolster profits as in food production units.

Farrowing crates represent another intensive farming practice that is now in use in laboratories. They too involve restraint of sows and restriction of normal movement but are far less objectionable than gestation crates because they are used for shorter periods, from several days to a few weeks, depending on the period of suckling. The conventional farrowing system, as described and sanctioned in the *Ag Guide*, is rectangular and measures 5 × 7 feet (1.5 × 2.1 m). But the sow resides in a crate within that area that is typically 2 × 7 feet (0.6 × 2.1 m). This width of 24 inches (0.6 m) restrains the sow so that she can, again, do no more than stand up and lie down. The main objective of farrowing crates is to slow the sow as she lies down, so that the piglets can escape to the sides and avoid being crushed. Piglet death is a serious issue, and a balance should be found between allowing the sow postural adjustments and freedom of movement against the crushing deaths of the babies.

John Webster, professor of animal husbandry at the University of Bristol, England, objects to farrowing crates on the ground that they restrict the opportunity for social contact between sow and piglets, and that the sow that is so confined during the period prior to farrowing is unable to satisfy her powerful motivation to build a nest (38). "The farrowing crate is certainly not designed to assist farrowing. The sow is uncomfortable, frustrated and compelled to drop her piglets on the same spot she drops her faeces," he states. The 1996 UK Advisory Group on the Ethics of Xenotranplantation considers the use of farrowing crates "undesirable" and states that "in theory it may be possible to avoid their use in the future" (29, p. 80).

Some refinements on traditional farrowing crates are already in use. For instance, the ellipsoid crate (39) and the Ottawa crate (40) (both developed in Canada) involve lesser degrees of constraint than conventional farrowing crates. Both systems allow the sow to turn around in addition to other movements and also permit easier visual and tactile contact between the dams with their piglets. They do not increase pig crushing rates. The ellipsoid and Ottawa crates are in use in Canadian farms and are suitable for laboratory use. Additional research is currently being conducted, particularly in the United Kingdom and Europe, to find alternative farrowing systems that are more welfare-friendly.

In the laboratory the choice of minipigs (rather than farm pigs) avoids the use of both gestation and farrowing crates. Miniature swine are preferred for this and other reasons. They are bred specifically for research from small wild species who have a mature body weight ranging from 132 to 198 pounds (60–90 kg) compared with domestic farm pigs who typically weigh in excess of 550 pounds (250 kg) at maturity. Recently minipigs have gained popularity for research studies including genetic modification and xenotranplantation.

Monitoring Animal Pain and Distress

Our understanding and recognition of animal pain and distress has advanced significantly in recent decades, and a considerable literature exists (41–45). Currently what is new is the use of scoring systems to keep track of the health status of laboratory animals during intensive periods when adverse effects are developing. With novel experimental techniques to model human diseases, it is especially important to maintain constant clinical monitoring of animal pain and distress. Researchers need to know when to intervene to relieve suffering by the use of drugs, by intense supportive care equivalent to that given to a human in the same state, or other actions.

In the last few years, several useful programs for intensive clinical monitoring of laboratory animals have been proposed. They require that the health status of the animal be evaluated at regular intervals and graded either on a numerical scale or a simple + or − (presence or absence) of physiological conditions or behavior, which together provide an assessment of severity of adverse conditions (46,47).

One such monitoring system in use at NIH includes assessment of the severity of a neurological impairment of mice on a grade from 0 to 5. Grade 0 is normal and grade 5 is moribund. Grades 1 to 4 show increasing signs of incapacity include clumsiness, incontinence, flaccid tail, abnormal plantar response, mild parapareses, trouble initiating movement, inability to move one or both hind legs, noticeable gait disturbance, moderate quadriparesis, and quadriparalysis (48). Qualified staff make frequent clinical assessments of the animals. At each grade, specific interventions have been established such as administration of fluids, dietary changes, expression of bladder, and provision of supplemental heat. The point at which early euthanizing of the animals should occur is specified.

It has long been recognized that a limit should be placed on the suffering of all animals used in experimentation. One way to achieve this is to set early, humane endpoints of experiments. The experimental design should establish the earliest point at which adequate scientific data have been collected and the experiment can be stopped, thus minimizing animal pain, distress, or lasting harm (49–51). In cancer studies, for instance, the endpoint should not be the death of an animal but the earliest point at which adequate scientific data are obtained. New initiatives are needed to foster use of these humane experimental designs (52).

Currently there are a few veterinary surgeons, animal ethologists, and others who are able to make such intensive clinical assessments and determinations of criteria for early endpoint on genetically engineered animals. But there are not enough persons so qualified; additional training programs are needed. However, evaluation of pain and distress for genetically modified mice may be difficult in some laboratory facilities, particularly those using micro-isolator cages and where there are large numbers of knockouts that are being subjected to a variety of breeding strategies to determine the effect of genetic deletions. Knockouts refer to animals who have had one or more genes removed or "knocked out." Defects in transgenic animals can be subtle but still affect welfare. Examples are mismothering, aggression, and spatial disorientation. These require extra special monitoring.

CONCLUSION

It was in 1985 that two important events took place — public awareness of the Beltsville pigs, and the latest amendment to AWA The Beltsville pigs demonstrated not only the scientific potential of genetically modifying animals but also the new welfare problems involved. With no amendment to the law in 15 years, U.S. laws lag behind existing needs because they were implemented before the full welfare implications of genetic modification were recognized. Scientific advances in genetic research challenge ethical norms, and the implications should be carefully considered before such work is approved. There is an expectation in the science community and society generally that genetic modifications should occur within a framework of legislative controls that minimizes the impact on the animals involved. Some national guidance and mandates are needed to help deal with these issues. Animal welfare concerns raised in this article are also being grappled with in other countries: policies are being prepared to address societal issues arising from genetically modified animal experimentation. One such proposal directed toward the European Union has just been published (53). It includes a specific cluster of questions around the issues of justification, scientific relevance, animal suffering, and wider social, economic, and environmental impacts of animal studies involving genetic modification. This proposal may help point the way toward a reconsideration of national policies and a development of fresh initiatives.

ACKNOWLEDGMENT

I wish to sincerely thank the following colleagues for their helpful comments during the preparation of this manuscript: Clément Gauthier, Gilly Griffin, John Gluck, Joy Mench, David Morton, and LeRoy Walters. I also acknowledge the able library assistance of D'Anna Jensen.

BIBLIOGRAPHY

1. Food Security Act of 1985 P.L. 99-198, amendments to AWA Improved Standards for Laboratory Animals Act.
2. National Institutes of Health, *Public Health Service Guide for the Care and Use of Laboratory Animals*, No. 86-23 Rev., NIH, Bethesda, MD, 1985.
3. *Fed. Regist.* **64**(18), 4357 (January 29, 1999).
4. United States Department of Agriculture, *Animal Welfare Report Fiscal Year 1998*, APHIS, Washington, DC, 1998.
5. J. Melcher, *Washington Post*, September 8, C5 (1991).
6. R. DeHaven, *Lab Animal* **29**(1), 44–46 (2000).
7. M.H. Lewis et al., *Brain Res.* **513**, 67–73 (1990).
8. J. Bielitzki et al., *Lab. Animal Sci.* **40**(4), 428–431 (1990).
9. K. Bayne et al., *Am. J. Primatol.* **23**, 23–35 (1991).
10. K.J. Field, J. Denny, and G. Kubicz, *Lab. Primate Newslett.* **36**(2), 5–7 (1992).
11. *Madigan*, 781 F. Suppl. at 805–806.
12. *Fed. Regist.* **64**(18), 4356–4367 (1999).
13. F.B. Orlans, *Lab Animal* **28**(8), 13 (1999).
14. *Foundation Biomedi. Res. News* **16**(1), 1–2 (1999).
15. *FASEB Newslett.* **11** (June 1999).
16. J.F. Lauerman, *Harvard Mag.* **52** (January/February 1999).
17. *Statistics of Scientific Procedures on Living Animals: Great Britain 1998*, Home Office, London, July 1999.
18. F.B. Orlans, in H. Kuhse and P. Singer, eds., *Companion to Bioethics*, Blackwell, London, 1998, pp. 399–410.
19. USDA, APHIS, *Animal Welfare Rep.*, No. 36 (1998).
20. V.G. Pursel et al., *Science* **244**, 1281–1288 (1989).
21. R. Weiss, *Washington Post*, June 7, A1, A12 (1999).
22. F.B. Orlans, *ATLA* (in press 2000).
23. CCAC Guidelines, *Transgenic Animals*, Canadian Council on Animal Care, Ottawa, Canada, 1997.
24. *Supplementary Guidance to Applicants for Project Licences: Projects to Generate and/or Maintain Genetically Modified Animals*, Home Office, London, 1999.
25. *Categories of Invasiveness in Animal Experiments*, Canadian Council on Animal Care, Ottawa, Canada, 1991.
26. Experiments on Animals Act, 1997, Section 1a, Veterinary Public Health Inspectorate, Rijswijk, The Netherlands.
27. G. Kolata, *N.Y. Times*, December 23, pp. A1, A20 (1999).
28. J.A. Mench, in J.D. Murray et al., eds., *Transgenic Animals in Agriculture*. CAB International, New York, 1999, pp. 251–268.
29. *Animal-to-Human Transplants*, Nuffield Council on Bioethics, London, 1996, pp. 49, 51–52.
30. Advisory Group on the Ethics of Xenotransplantation, *Animal Tissue into Humans*, Report, Her Majesty's Stationary Office, Norwich, UK, 1996, pp. 78–80.
31. Available from: David Nunn Ltd., Shrewsbury, Shropshire, UK.
32. Available from: Bioserve, Frenchtown, NJ; Primate Products, Redwood City, CA; and KLASS San Jose, CA; *www.nal.usda.gov/awic/pubs/enrich/supplier.htm*
33. *Animal Welfare Information Newsletter* **7**(3–4), 1–8 (1996–1997).
34. Available from: HAKA, Josef Haeufele GmbH & Co. KG, Robert-Bosch-Str. 6-9, D-89155 Dellmensingen, Germany, Tel: +49/7305/96100; Fax: +49/7305/9610-40.
35. *Animal Welfare: Xenotransplantation*, Imutran Ltd., Cambridge, UK, 1999.
36. Welfare of Livestock Regulations, 1994 (S.I. 1994/2126)-Schedule 3, Home Office, London, UK.
37. *Guide for the Care and Use of Agricultural Animals in Agricultural Research and Teaching*, 1988, rev. ed., Federation of Animal Science Societies, Champaign, IL, 1999.
38. J. Webster, *Animal Welfare: A Cool Eye towards Eden*, Blackwell Science, Oxford, UK, 1994.
39. Z. Lou and J.F. Hurnik, *J. Animal Sci.* **72**(10), 2610–2616 (1994).
40. D. Fraser, P.A. Phillips, and B.K. Thompson, *Livestock Prod. Sci.* **20**, 249–256 (1988).
41. D.B. Morton and P.H.M. Griffiths, *Veterinary Rec.* **116**, 431–436 (1985).
42. D.B. Morton, *ATLA* **18**, 29–39 (1990).
43. FELASA Working Group on Pain and Distress, *Lab. Animals* **28**, 97–112 (1994).
44. P. Flecknell, *AWIC Newslett.* **8**(3–4), 8–14 (1998).
45. M.H. Ullman-Culleré and C.J. Folz, *Lab. Animal Sci.* **49**(3), 319–323 (1999).
46. D.B. Morton, in F.F.M. van Zutphen and M. Balls, eds., *Animal Alternatives, Welfare and Ethics*, Elsevier Science, Amsterdam, The Netherlands, 1997, pp. 235–241.
47. D.B. Morton, in *Proc. Joint ANZCAAT/NAEAC Conference on Ethical Approaches to Animal-Based Science*, Auckland, New Zealand, September 19–20, 1997, pp. 131–148.
48. V.A. Hampshire, J.A. Davis, and C.A. McNickle, *Lab Animal* **29**(5), 40–45 (2000).

49. D. Mellor and D.B. Morton, *ATLA* (in press 2000).

50. G. Griffin and H.B.W.M. Koeter, in *Proc. Third World Congress on Alternatives and Animal Use in the Life Sciences*, ATLA (in press 2000).

51. *Choosing an Appropriate Endpoint in Experiments Using Animals for Research, Teaching and Testing*. Canadian Council on Animal Care, Ottawa, Canada, 1998.

52. C.F.M. Hendriksen and D.B. Morton, in *Proc. International Conference*, November 22–25, 1998, Zeist, The Netherlands Royal Society of Medicine, London, 1998.

53. V.C. Delpire, T.B. Mepham, and M. Balls, *ATLA* **27**, 869–881 (1999).

See other entries ANIMAL, MEDICAL BIOTECHNOLOGY, LEGAL, LAWS AND REGULATIONS GOVERNING ANIMALS AS SOURCES OF HUMAN ORGANS; ANIMAL, MEDICAL BIOTECHNOLOGY, POLICY, WOULD TRANSGENIC ANIMALS SOLVE THE ORGAN SHORTAGE PROBLEM?; FDA REGULATION OF BIOTECHNOLOGY PRODUCTS FOR HUMAN USE; see also RESEARCH ON ANIMALS entries; TRANSGENIC ANIMALS: AN OVERVIEW.

S

SCIENTIFIC RESEARCH, ETHICS, SCIENTIFIC MISCONDUCT

Kenneth J. Ryan
Harvard Medical School
Boston, Massachusetts

OUTLINE

INTRODUCTION

Scientists have traditionally tried to shield their work from ethical scrutiny under the guise that they are involved in a value-free pursuit of the truth, but in the second half of the twentieth century this position became untenable. Ethical concerns were expressed by the public as well as some scientists about such issues as the dangers of atomic warfare, nuclear radiation, human subjects biomedical and behavioral research, recombinant DNA, assisted reproductive technologies, and global warming (1–3). In view of the rapid and massive intrusion of scientific discoveries and their application into human affairs for both good and ill, it was becoming difficult to maintain that science was essentially value free. As Stephen Toulmin suggested "it is meanwhile becoming clear that the professional organization and priorities of scientific work can no longer be concerned solely with considerations of intellectual content and merit, as contrasted with the ethical acceptability and social value, either of the research process itself, or its practical consequences" (1). In addition, by the latter half of the twentieth century a good deal of scientific research was being supported by public funds especially in the United States. This introduced political as well as social concerns with how, and for what purposes, scientists pursued their research. With such ethical concerns already in place, it is not difficult to understand the intense dismay that greeted at first, sporadic and then a flood of reports beginning in the 1970s of scientific research papers that were fraudulent,

particularly when they involved clinical research and matters related to human well-being. The inquiries into this scientific misconduct raised a series of questions about (1) the extent of the problem, (2) the definitions that should be applied to misconduct in contrast to scientific error or chance, (3) the plight of an accused scientist or the whistleblower, (4) the nature of scientific discovery and the role of ethics in the education and practice of scientists, (5) the role of the government and public in the oversight process, and (6) the roles of the academic community, universities, industry, research laboratories, and the individual scientist in dealing with the problem of scientific misconduct (4).

SCIENTIFIC MISCONDUCT

Incidence and Types of Scientific Misconduct

The incidence of fraud or misconduct in science has in the past generally been considered by scientists, science historians, and sociologists to be rare and inconsequential. The scientist has traditionally been regarded as the seeker of true knowledge. Advances in science and technology have resulted in space exploration, modern forms of transportation and communications, and benefits to humankind in agriculture and medicine. The traditional view is that if there is fraud in an important area of science, it will be uncovered quickly by failed attempts to repeat it by other scientists or by inside information from the laboratory concerned. On the other hand, if the published research work is in an unimportant area of science, or is of a trivial nature, little damage is supposedly done to science or the public except for the cluttering of the literature. This line of reasoning has become suspect. The "romantic ideal" that scientists seek the truth and that errors or frauds are uncovered by a continual process of repeated studies has been challenged. While the ideal model may prevail over the long term for primary or critical data, there are many exceptions. A Noble laureate and director of the National Institutes of Health, Harold Varmus, challenged the idea that science is self-correcting in 1994 at a National Academy of Science convocation on scientific misconduct (5). Kiang suggested that errors (or fraud) in the archival record often are never corrected, simply ignored and eventually forgotten (6). The question of the extent of science fraud is, however, of long-standing. Robert Merton, a sociologist of science wrote in 1957 that he believed fraud was rare in science. For an earlier period, he quoted Darwin who "knew of only three falsified statements" in all of science. From the same period, Merton described Charles Babbage's 1830 "inventory of fraud in science" which included cooking and trimming of research data (7). In the 1990s several reports suggested that scientific misconduct was increasing and more widespread than previously believed and that such fraud in science was potentially damaging to both science and the public interest (8). The American Association for the Advancement of Science (AAAS) surveyed its membership and reported in 1992 that 27 percent of those replying had personal knowledge of an average of 2.5 cases of *suspected* fabrication, falsification or theft of research in the prior ten years (9). Thirty-seven percent felt the incidence of misconduct was rising. In 1993 Judith Swazey and colleagues reported on a survey that involved 2000 graduate students and 2000 faculty in chemistry, civil engineering, microbiology, and sociology. While the study could not report on the frequency of misconduct, it did provide insight into the rates of exposure of students and faculty to various forms of misconduct (10). The authors reached the conclusion that although science misconduct is not rampant, it certainly is not rare. Between 6 and 9 percent of students and faculty reported knowledge of faculty who falsified or plagiarized data while one-third of faculty reported incidents of student plagiarism. Review of audits by the Food and Drug Administration of drug research conducted between 1977 and 1988 revealed problems in 12 percent of studies before 1985 and 7 percent thereafter. The problems uncovered included failure to perform studies for which results were given or changing of data (11).

That fraud is not always uncovered quickly and is not always innocuous was demonstrated in the criminal conviction of the psychologist Stephen Breuning in 1988 some years after he reported on the clinical effects of drugs on hyperactive retarded children. Much of his research was simply never performed, although his "work" inappropriately influenced the care of the mentally retarded (12). In the industrial arena there was the case involving the officers of Industrial Bio-Test Laboratories, Inc. who were convicted of fraud in the reporting of toxicity data on which drug companies and pesticide manufacturers relied for the effects of drugs and pesticides on laboratory animals. This ultimately affected the review and approval process of pesticides and drugs by the Environmental Protection Agency and the Food and Drug Administration (12).

When oversight of research misconduct was formalized by the National Science Foundation (NSF) and the National Institutes of Health (NIH) in the 1990s, they reported that some 20 to 40 alleged cases a year of various forms of misconduct reached the investigational stage for each agency (13). Whether one is alarmed by the numbers may depend on whether one takes solace in the small percentage of federal grantees actually involved in misconduct or whether one is concerned with a potential corrosive effect of even a small percentage of cases on the education of research scientists and the trust of the public. The actual numbers of misconduct cases also ultimately depend on what is included as research or scientific misconduct.

The types of research misconduct that have been chronicled in books and articles have varied widely and given rise to the confusion about the frequency and seriousness of the problem. One distinction that has been made is between science in general and research specifically, as when the Public Health Service changed the titles of its oversight committee from the Office of Scientific Integrity to the Office of Research Integrity in

1992 (14). The government makes the obvious distinction in its oversight role between research supported by federal funding and that supported from privates sources, as well as research performed to comply with regulatory agencies. The private sector is obviously responsible for research not funded or regulated by the government and the definition of misconduct could be more rigorous (or less) than federal standards. Other distinctions that have been suggested are between basic and applied research and research involving human subjects. Human Subject Research has generally been under separate government oversight over matters of informed consent, risks to subjects, and equity as distinct from concerns about fraud (15). The main emphasis in government definitions with respect to fraud has been on the scientific record with concern for the information published, formally presented at scientific meetings, or offered in progress reports to grants. In this case examples of misconduct have included reporting on work never performed or manipulating research data to obtain desired results as in adjusting points on a curve, omitting points that fall off the curve, or omitting data that make the conclusions less appealing or untenable. Misconduct might also include the use of inappropriate statistics to achieve significance that is not otherwise demonstrable. While sloppiness or gross errors in research might not warrant government sanctions, they might limit academic advancement or job security (13).

The other major source of misconduct involving the scientific record is in the acts of plagiary, which involve misappropriating the words or ideas of another scientist without giving credit for the source and representing them as one's own. The other forms of misconduct fall into such categories as sexual harassment of colleagues or students in the research environment, secreting data or notebooks that are not exclusively ones own, sabotaging or destroying another scientist's equipment or experimental results. Specific misconducts that might or might not be lumped with protecting the research record include failing to follow federal regulations on the protection of human subjects, treatment of animals, environmental protections especially with radioactivity, or appropriate use of grant funds. The Office of Inspector General of the NSF indicated in 1991 that of the cases of alleged misconduct: 20 were about plagiarism, 9 were for fabrication and falsification, and 8 were in the other "serious deviations" category (13).

The choice of what to include in one's definitions or categories obviously depends on anticipated use. If the list is for educational purposes in academic institutions, inclusiveness might be the goal. If the objective is to check compliance with federal funding or regulatory agencies, then definition of the misconduct is expected to be more restrictive and precise.

Defining Misconduct

Formal definitions for research misconduct have originated largely through government oversight of funded or regulated research. The first definition was published in 1986 in the NIH Guide to Grants and Contracts, as part of an interim policy before final regulations were formulated. Prior U.S. congressional hearings on scientific misconduct resulted in a provision of the Health Research

Extension Act of 1985 which required that research institutions establish procedures to investigate scientific fraud, and that the director of the National Institutes of Health develop an administrative method to respond and deal with it. The first definition of misconduct in science was thus established as "serious deviation such as fabrication, falsification, or plagiarism, from accepted practices in carrying out research or reporting the results of research; or material failure to comply with Federal requirements affecting specific aspects of the conduct of research, for example, the protection of human subjects and the welfare of laboratory animals" (16). The scientific community had an opportunity for critical commentary in response to the publication of proposed policies and procedures of the Public Health Service (PHS) in the Federal Register in 1991 and prior to that to advance the final Notice of Proposed Rulemaking in 1988. A similar procedure was followed by the National Science Foundation (NSF) in 1987 to receive public commentary and establish a definition covering grants under its jurisdiction. The final NSF definition developed in 1991 closely followed that of the PHS: "Misconduct in science and engineering is fabrication, falsification, plagiarism, or other serious deviation from accepted practices in proposing, carrying out, or reporting results from activities funded by NSF: and retaliation of any kind against a person who reported or provided information about suspected or alleged misconduct and who has not acted in bad faith." The second portion of this NSF definition covering retaliation was criticized by outside commentators as not being part of scientific misconduct, but the view of the NSF was that retaliation was a serious deviation from accepted practices and would work against scientific integrity. This clause has remained. The PHS has not had retaliation in its definitions but was later called upon by Congress to develop procedures for protecting whistleblowers. It was also noted that the Whistleblower Protection Act of 1989 protects only federal employees and not whistleblowers who are employees of the grantee institutions. The PHS definition of misconduct was later changed in response to some of the commentary it received. The PHS definition established in 1989 is: "Misconduct or misconduct in science means fabrication, falsification, plagiarism, or other practices that seriously deviate from those that are commonly accepted within the scientific community for proposing, conducting, or reporting research. It does not include honest error or honest differences in interpretations or judgments of data." The final sentence was included in response to commentary from the scientific community. On the other hand there has been no dropping of the clause "practices that seriously deviate from those that are commonly accepted" which has been repeatedly criticized vigorously by scientists for both the PHS and NSF definitions as too vague and a potential basis for unfair treatment of scientists (16).

The U.S. Congress as part of the 1993 NIH Revitalization Act created a Commission on Research Integrity (CRI) with a mandate to make recommendations on a definition for research misconduct. Research misconduct had been a continuing subject for congressional hearings in the 1980s and 1990s, and there was apparent dissatisfaction by the Congress over the way the scientific community, academic institutions and federal agencies were dealing with the problem. The Commission was a 12-member public advisory body composed of scientists, lawyers, sociologists, and ethicists that was asked to make recommendations to the Secretary of Health and Human Services on the definition of research misconduct, an assurance process for compliance of institutions with federal regulations, administrative processes for dealing with misconduct, and, finally, recommendations to protect whistleblowers.

The CRI report did make recommendations to change the definition of research misconduct in an attempt to "provide vital guidance for personal and ethical judgments and decisions concerning the professional behavior of scientists and to provide a legal framework for formal proceedings" (14). From the public record and testimony, the CRI developed a sense of the types of research misconduct which were most common and needed addressing in a definition. The Commission felt that a fundamental principle for scientists was to be truthful and fair in the conduct of research and dissemination of their findings. The recommended definition of the CRI was that "research misconduct is significant misbehavior that improperly appropriates the intellectual property or contributions of others, that intentionally risks corrupting the scientific record or compromising the integrity of scientific practices. Such behaviors are unethical and unacceptable in proposing, conducting, or reporting research or in reviewing the proposals or research reports of others." The commission gave as examples of research misconduct to substitute for fabrication, falsification and plagiarism; the categories of misappropriation, interference, and misrepresentation were explained as follows:

> *Misappropriation* is intentional or reckless plagiary. This means presentation of the documented words or ideas of another without attribution appropriate to the medium of presentation, or to the use of any information in breach of any duty of confidentiality associated with the review of a manuscript or grant application.
>
> *Interference* is intentional and unauthorized taking, sequestering, or material damage to research related property of another. This includes, without limitation, the apparatus, reagents, biological materials, writings, data, hardware, software, or any other substance or device used or produced in the conduct of research.
>
> *Misrepresentation* is an intention to deceive or reckless disregard for the truth. This means stating or presenting a material or significant falsehood or omitting a fact so that what was stated or presented as a whole stated or presented a material or significant falsehood (14).

This somewhat complex and legalistic definition was not well received by the scientific community, but it engendered a broad debate on research integrity. The Commission's definition was in fact a more explicit rendering of fabrication, falsification, and plagiary with a substitution of interference for the "serious deviation from accepted practices" clause that many found objectionable (14,16). The CRI definition did introduce the concept that omission of critical data should be identified as misconduct and made the concept of intent explicit, something that has been discussed by Dresser (17). The Commission further defined two other forms of professional misconduct that included obstruction of research misconduct investigations, which still is in the NSF definition, and noncompliance with federal regulations, which had been in an early PHS definition.

In line with a recommendation of the Commission, a federal interagency task force was appointed to develop a common definition of research misconduct for all government departments. By the end of 1999 they could not agree on a common definition; the secretary had not recommended the CRI definition, and the NSF and PHS definitions from 1991 remained in force.

Plight of the Whistleblower, Plight of the Scientist

The plight of the whistleblower has been championed by the distinguished scientist and journal editor John Edsall who believes that whistleblowers are necessary for the maintenance of honest science. He described cases occurring over a 30-year period in which major harassment and difficulties were faced by the whistleblower, often more severe than the person accused of the misconduct (18). A book entitled *The Whistleblowers* reported on a 6-year study of 64 individuals (19,20). A detailed report on the consequences of whistleblowing for the whistleblower and the exonerated accused was also prepared by the Research Triangle Institute for the Office of Research Integrity in 1995 (21). Sixty-nine percent of whistleblowers and 60 percent of those exonerated of misconduct suffered negative consequences, usually more severe early in the process of investigation and lessening with time. The converse of this is that significant numbers did not suffer adverse outcomes. The public sympathy engendered for the whistleblower from congressional hearings in the 1990s contributed to the inclusion of a mandate to address their protection by the CRI (14). At that time there was much publicity about Margot O'Toole, a junior scientist and whistleblower who had problems with Thereza Imanishi-Kari and David Baltimore at the Massachusetts Institute of Technology. Tom Devine likened whistleblowing to professional suicide. He chronicled the tales of 20 witnesses who appeared before the CRI. They complained of censorship, loss of job, academic expulsion, retaliatory investigations, denial of access to their data and laboratories, as well as threat of deportation and physical harm (22). To quote Tom Devine: "Everyone pays lip service to the ideal that science is the search for the truth, and scientific integrity is a concern for all. But whistleblowers actually live those values." In response to the CRI mandate, a Whistleblower's Bill of Rights was proposed as part of the CRI report (14,22). Even though the National Academy of Science advises beginning researchers that they have an "unmistakable obligation to act" if they suspect someone is violating the ethical standards of science, there is much cynicism. Often the whistleblower must wait years to be vindicated or must resort to the courts to get attention and redress. The scientific community felt that the Whistleblower's Bill of Rights provided too little concern for the accused scientist (4). There is always the possibility

that the whistleblower is wrong or is not acting in good faith.

The plight of the scientist has been typified by two cases settled in 1997 in which after many years of turmoil, and adverse effects on their careers and reputations, the accused scientists were vindicated. The case of Thereza Imanishi-Kari, who was a collaborator of David Baltimore, was settled by an appeal board of the PHS some 10 years after the first charges of misconduct were made. No misconduct was ultimately found although there were criticisms of the research record keeping by Imanishi-Kari (4). All this was chronicled in a *New Yorker* article entitled: "The Assault on David Baltimore" (23) and a book, *The Baltimore Case*, both by Daniel Kevles (24). Baltimore, who initially suffered for the defense of a scientific colleague, became a hero in the Kevles book, and Margot O'Toole, the whistleblower, was recast by Kevles in a less sympathetic light. It is not certain that Kevles gave an evenhanded recounting of the controversy, and at the very least the maintenance of research records was inadequate and contributed to the problem. It is likely that Baltimore, Imanishi Kari, and O'Toole were all victims of a bad system for defining and dealing with misconduct by the scientific community, academic institutions, and the government (4). The other case illustrating the plight of the accused scientist was that of Bernard Fisher who ultimately gained vindication by going to court to receive an apology from the government for its inept oversight and an apology and financial settlement from his institution, the University of Pittsburgh. Fisher was the director of an interinstitutional NIH-funded clinical cancer trial. It was reported that a physician from one of the participating hospitals entered ineligible subjects into the study, and Fisher was caught up in the question of how and when to reveal this information, although it had little effect on the conclusions drawn from the study (25). The CRI also heard testimony from many aggrieved scientists, even professors who claimed to have been unjustly accused of misconduct and had their careers destroyed. This is reviewed in the report noted above on those accused but exonerated of scientific misconduct (21).

ETHICS AND SCIENCE

History, Sociology, and Philosophy of Science: Looking for Causes of Misconduct

The history, sociology, and philosophy of science are so interconnected as to make distinction between these disciplines difficult, but each has contributed to the literature about the values held by the scientist in conducting research, in elaborating scientific theories, in relating to other scientists, and in relating to society at large. Ultimately these values held by scientists contribute to the character of science, and they are a promising place to look for the factors that motivate scientists and the means by which progress in science occurs or is hindered. The so-called normative structure of science with its virtual absence of fraud due to rigorous policing by science itself enunciated in the 1940s by Merton may be an unreal perception of how science works today or for that matter how it operated previously (26).

As noted earlier, the ideal of an incorruptible science based upon timely self-correction is questioned in the real world. Harriet Zuckerman has considered the causes of misconduct under three headings: individual psychopathology, anomie, and alienation (26). The cause of misconduct had been attributed by many scientists simply to individual psychopathology, but this has been criticized as "too convenient and self-serving" to explain what is going on. It is not only the individual guilty of misconduct but also the laboratory environment in which he or she works that affects behavior. Anomie is a theoretical sociological construct for deviance based on a high value being placed on a goal for which the means are not readily available, inducing people to choose dubious routes to try to achieve the highly desired end. Another theoretical cause of misconduct is alienation, which occurs when there is a disconnect between the daily laboratory work and the ultimate goals of the research. Also, when there are gross differences in reward and recognition among the laboratory team, alienation, personal animosity, and misconduct are more possible (26). It reached a point in one prominent laboratory in Rockefeller University where attempts at poisoning were reported (27). The rise in misconduct is also believed due to the change in the way scientific research is organized and funded. Research in the last half of the twentieth century has become more costly, competitive, complex, specialized, and collaborative at the same time that being successful in research creates the opportunity for academic advancement, rewards, and recognition (13). Research may be attracting scientists largely because of external and secondary rewards of fame and fortune, which can corrupt, rather than the incorruptible primary pursuits of asking questions and seeking knowledge for its own sake.

Another interesting sociological perspective on scientists was offered by Bernard Barber in his 1961 *Science* article, "Resistance by Scientists to Scientific Discovery" (28). Although it is generally accepted that there may have been outside religious, political, and ideological impediments to progress in science over the years, the idea that science has cultural and social forces within itself that resist progress is a novel concept. Barber noted that new ideas may be resisted because they clash with existing substantive concepts like the notion of the irreducibility of the atom which clashed with the discoveries of electrolytic dissociation, the discovery of X rays, and the theory of the electronic composition of the atom, all of which were resisted when first proposed. Methodological concepts based on the senses or models of mathematics clashed with the ideas of analyzing colors with prisms, radioactive measuring, electromagnetic theory, or experimentation and provided resistance to their introduction. Religious beliefs of scientists were the basis for resistance to Darwin's theories. Low professional standing can impede acceptance of work, as occurred with resistance to Mendel's theory of inheritance. Such social forces working within science may give rise to suspicions of misconduct when none exist merely because it is conceptually difficult to accept the results of a groundbreaking study. Rivalries based on specialization, societies, schools, or seniority have also impeded acceptance of scientific work, but Barber ends on an optimistic note that scientists even with their human faults are in his view more objective and open-minded than society in general.

A major innovation in the philosophy of science was introduced in 1962 by Thomas Kuhn in his *The Structure of Scientific Revolutions*, which introduced the view that discovery and progress in science are not completely rational and that science is not a steady progression toward an "objective" truth (29). Taking into account the work of Bernard Barber and his own studies in the history of science, Kuhn created a new vocabulary and definitions to cover "normal science," paradigms, incommensurability between paradigms, paradigm shifts, and scientific revolutions. Normal science works within a paradigm with shared rules and standards and is closest to the traditional view of how science works. Normal science works on a historical record, but it does not bring about striking new discovery. When anomalies occur and questions arise that cannot be answered within normal science, there may be a paradigm shift or revolution with different rules and standards and a radical break with the past. The switch from the vision of Ptolemy to Copernicus, from Newton to Einstein, and from creationism to Darwin have all been given as examples of scientific revolutions. Again, the workings of science may raise suspicions of misconduct within the rubric of practices that seriously deviate from those commonly accepted within the scientific community unless there is a broad understanding of the history and sociology of science, of how research is actually conducted in specific fields, and of how theories are developed and tested.

History of Scientific Misconduct

There have over the years been many reports of suspected or proven fraud or misconduct in science, but it is best to divide the cases temporally based on whether they occurred in remote or recent times. This is a selected list of cases and is not meant to be exhaustive (30–34).

Science Misconduct in Ancient Times. Science misconduct as ancient history is probably more speculative than factual, and there are usually accusers and defenders for almost every case involving the legendary scientists of the past. The contributions of most of these scientists are little diminished by these accounts, and there is little evidence of any intent to deceive in cases before the twentieth century. For details, reference should be made to listed sources (30–34).

Claudius Ptolemy of the second century A.D. is accused by both French and American astronomers of not making measurements claimed but extrapolating them from an earlier Greek astronomer or of deriving the data from theoretical projections rather than from personal observation. Historians dispute this suggesting instead that his observations were adjusted in keeping with the standards and methods then in use.

Galileo Galilei of the seventeenth century has been accused of not having performed experiments as described but of creating data to conform to his theories.

Isaac Newton, the great physicist of the late seventeenth and early eighteenth centuries, is believed to have used "fudge factors" after the fact to adjust data to meet theoretical predictions. While some have termed this fraud, others have characterized this as making approximations to test if a theory is in fact feasible.

Gregory Mendel published his work in 1865 that is the basis of modern genetics. It is claimed that the reported observations of the frequency of inherited traits and the expected values are too good to be true. This has been attributed to the manipulation of data by an obliging assistant, experimenter bias, or a more innocent difficulty in sorting the categories.

Robert Millikan published papers in 1910 and 1913 which won him the Nobel Prize in 1923 for determining the electric charge of the electron. Although Millikan claimed to have published all his data points, review of his notebooks reveals that he was selective and left out one-third of his observations. A scientific rival Felix Ehrenhaft of Vienna who lost out in the recognition of the Nobel Prize found fractional charges rather than the exact multiples described by Millikan. With the newer knowledge of subelectronic particles, it is possible that Millikan was wrong not only in his conclusion but also in his selection of data points to publish (31).

Sir Cyril Burt was a famous British psychologist who published studies in the 1950s on identical twins raised together and apart and concluded that IQ is largely inherited. It is claimed that many of the studies were never done and that the conclusions were designed to support his belief that intelligence is determined more by genes than environment (32).

Science Misconduct in Recent Times. Science and research misconduct described in the last 40 years of the twentieth century is based on more substantial evidence than cases dug up from ancient history. In more recent times investigations of misconduct are usually based on laboratory records and often on confessions.

Harold Bates worked in the laboratories of Mel Simpson, Professor at Yale and then with Professor and Noble laureate Fritz Lipmann at Rockefeller University, and published papers with them in the *Journal of Biological Chemistry* in 1960. A paper on the biosynthesis of cytochrome C coauthored with Simpson and a paper on the biosynthesis of glutathione coauthored with Lipmann had to be retracted because the work could not be verified (18).

George Webster, an established investigator of the American Heart Association, joined the Enzyme Institute of the University of Wisconsin and in 1965 published a paper in the *Journal of Biological Chemistry*. His work was challenged by Efraim Racker whose own work was at odds with Webster's. Webster announced that he did not have the original data to back up his work, although the paper has never been formally retracted (18).

William Summerlin was a dermatologist who joined the famous immunologist Robert Good at the University of Minnesota and moved with him to the Sloan Kettering Institute in 1973. Summerlin claimed from his research studies that he could treat mouse skin and human corneal tissue by culture outside the body to abolish immunological rejection when the treated tissue was subsequently transplanted to other unrelated animals.

He claimed to be able to treat skin tissue from black mice, which would keep it from being rejected when transplanted to a different strain of white mice. In 1974 it was discovered by a technician that Summerlin had used a black pen to darken the spots on white animals and the credibility of all his work collapsed. The one mouse with a successful transplant was shown to be a hybrid, which would be expected to have a graft survive from the donor strain. The claim of successful transplantation of human cornea to rabbit eyes was also shown to be a deception. After this case there was a good deal of soul-searching about the pressures put on young scientists in large, highly publicized laboratories and the fact that their work is too readily "accepted" by their supervisors who have expectations of the kind of result they are seeking.

John Long was research pathologist at the Massachusetts General Hospital in the laboratory of Paul Zamecnik and rose to the rank of associate professor. He claimed to have a human tumor cell line from patients with Hodgkin's disease. In 1979, when collaborators asked for primary data on some joint experiments, Long was found to have altered data. He later admitting falsifying the results and it was also discovered that his so-called human Hodgkin's tumor cells were derived from a lymphoid cell line of the owl monkey, possibly by contamination. Long resigned his laboratory position and returned to clinical practice.

Vijay Soman was a scientist from India who started as a postdoctoral fellow in the laboratory of Philip Felig, professor of endocrinology at Yale and worked his way up to the rank of associate professor. In 1980 he and Felig published a paper in the *New England Journal of Medicine* that was shown to be partially plagiarized from a manuscript Felig was asked to review and rejected. Felig had given the paper to Soman to read because it was in their area of research. Soman not only plagiarized exact wording from the rejected manuscript; he made up most of the data he submitted. Review of 14 prior publications by Soman with Felig revealed that only 2 had supporting data and 12 of the questionable papers had to be retracted. Soman returned to India, and Felig who had gone on to be chairman and professor of medicine at Columbia, lost his position and returned to Yale.

Mark Spector was a promising graduate student in the laboratory of Efraim Racker at Cornell University when in 1981 he reported an exciting sequence of events, a cascade hypothesis for the process of transformation of normal animal cells into tumor cells. A collaborator became suspicious when only Spector could repeat certain key gel experiments crucial to the hypothesis. It was found that Spector's work was largely a fabrication. After this, Spector's credentials were checked, and it was found that he had a prior record of forgery and previous research work could not be repeated. The irony of this case is that it was Efraim Racker who exposed the fabrication of George Webster some 20 years earlier as noted above, only to be deceived himself in his own laboratory (18,30).

John Darsee was a promising young cardiologist who worked in the laboratory of Eugene Braunwald of Harvard and the Brigham & Women's Hospital. In 15 months of work at Harvard, Darsee contributed 5 papers coauthored with Braunwald and many abstracts to national meetings. When asked by the laboratory head to submit raw data on some dog experiments, Darsee was observed by other lab workers in the act of fabricating the data in a contrived experiment. Rather than being an isolated event as first believed, it turned out that Darsee had a long trail of fraudulent research stretching back to his medial school and college days. Eight of 10 publications that Darsee had released before coming to Harvard had to be withdrawn or corrected. Darsee left Harvard and obtained a clinical post.

A junior colleague of Dr. Francis S. Collins, head of the Human Genome Project, was accused in 1996 of fabricating data in five research papers on leukemia, which had to be withdrawn. The colleague was identified by the *New York Times* as Amitov Hajra, a graduate student at the University of Michigan who worked in Dr. Collins's laboratory at the NIH. The fraud was uncovered when the reviewer of a subsequent paper submitted for publication by Collins and his student in the journal *Oncogene* questioned the data and suggested intentional deception. Ironically, what was obviously suspicious to an outside reviewer was missed by Collins and others in the laboratory (4,35).

A scientist from Immunex, a biotechnology company, was accused of plagiarizing the gene structure for an interleukin molecule that was obtained by reviewing an article for the journal *Nature* authored by scientists from a competing company Cistron Biotechnology. The paper was rejected, but it is claimed that the reviewer then patented the gene structure presented in the article. The controversy resulted in a suit alleging misappropriation of data during peer review. The case was settled out of court, but at least illustrates the occurrence of alleged misconduct in the biotechnology industry as well as in academic laboratories (36,37).

Plagiarism is a frequently reported form of misconduct. A case often cited is that of Elias Alsabti, an Iraqi who worked in both England and the United States and simply copied previously published works and sent them to obscure journals under his name or took progress reports or grant applications of others as bases for articles. In all he had some 60 fraudulent publications between 1977 and 1980. When found out, he simply went to new positions to ply his deceptions. The checking of references and his past as he went from position to position was obviously seriously deficient.

In 1987 Shervert Frazier, a professor of psychiatry at Harvard, admitted to plagiarizing works of others for review articles between 1960 and 1975. He subsequently resigned his academic post but continued to practice psychiatry at McLean Hospital (33).

The foregoing list of misconduct cases from distinguished laboratories is merely a selection of the many that could be described involving clinical and basic research. While the emphasis is on biological research, misconduct has been a problem in all fields of science, engineering, and technology. It is interesting that in many cases the misconduct was carried out by a junior or mid-level scientist who was inadequately supervised in a high-powered high-profile laboratory. Many of the transgressors were considered brilliant, hard-working, technically skilled, and almost too good to be true. They worked harder and longer and published more than is usually possible or reasonable. These stories are reminiscent of the adage for consumers about deceptive advertising—that attractive offers that seem too good to be true probably are too good to be true.

GOVERNMENTAL ROLES IN SCIENCE AND MISCONDUCT

During World War II scientists were involved in many research projects sponsored by governments in pursuit of advantages in the war effort. Out of such work came the manufacture of Penicillin, the invention of radar, advances in transportation computers, and communications and the use of nuclear energy for the atom bomb. After the war, government science advisor and MIT professor Vannevar Bush in his report to President Truman, "Science: The Endless Frontier" outlined the benefits that would flow from continued governmental support for basic research (38). As a consequence a new system was established that sited the conduct of research at universities, created project grants that could be awarded to scientists for research in their own laboratories at universities, and developed advisory and review committees of private scientists to serve the government on a part-time basis, which gave rise to the peer review system for approving and giving priority to grant applications. David Guston characterized this general arrangement as the "social contract for science" (39). Initially scientists were hesitant to accept government funds lest onerous restrictions and oversight be imposed on their academic and scientific freedoms. The "contract" provided for federal support, but with the responsibility for oversight left to the traditional mechanisms for academic governance at the private universities and institutions. There was considerable faith in the ability of scientists to regulate themselves and to ensure the integrity of the scientific process. Although cases of misconduct in science as described earlier were being reported during the 1960s and 1970s, there seems to have been little interest in the Congress to get involved, trusting instead the mechanisms ordinarily used by scientists and universities to discipline their members. Ironically, the reports of incompetent university and federal investigations of misconduct triggered the interest of congress, rather than the occurrence of the misconduct itself (4,40). The first congressional hearings for oversight on scientific misconduct occurred in 1981, chaired by Albert Gore in the House of Representatives and by Orrin Hatch in the Senate. Gore's committee looked into the Darsee affair and heard testimony from Drs. Philip Felig and John Long about their reactions to their experiences, described in the cases stated above. Philip Handler, president of the National Academy of Sciences, testified before the Gore committee that the problem of science misconduct had been grossly exaggerated by the press and that scientists should be allowed to take care of it by themselves. Senator Hatch was concerned with the institutional responses to the cases and with waste and fraud. He was particularly incensed that a cancer scientist Marc Strauss from Boston University received new funding from the National Cancer Institute while he was under investigation for prior fraudulent practices (34,40). The Gore hearings did lead to legislation on misconduct which was included in the 1985 NIH reauthorization bill. The provision in the bill dealt with the responsibility of universities to develop mechanisms for dealing with misconduct. Congressional hearings were held again in 1988 when the late Ted Weiss from the House Committee on Government Operations looked into abuse of whistleblowers and delays in dealing with allegations of scientific misconduct at the institutional level. Meanwhile John Dingell, chairman of the House Committee on Energy and Commerce Subcommittee on Oversight and Investigation, started exploring the Thereza Imanishi-Kari, David Baltimore case described earlier. These latter hearings would go on for several more years until 1993 and involve Representative Dingell and David Baltimore in acrimonious exchanges on the congressional committee's aggressive investigation (4). In response to the congressional hearings and wide spread coverage in the press about the plight of whistleblowers and the cases of scientific misconduct, the PHS in 1989 set up an Office of Scientific Integrity (OSI) within the NIH to receive reports from institutions and conduct investigations on alleged misconduct. An Office of Scientific Integrity Review (OSIR) was set up within the Office of the Assistant Secretary for Health as an oversight function for the OSI (16). Dissatisfaction with the way this arrangement dealt with the Iminishi-Kari, David Baltimore case and with one involving Robert Gallo of the NIH brought criticisms of these offices. Congressional, scientist, and press criticisms of the OSI and OSIR were that they had inconsistent policies and vague rules, that they were biased against defendants, and used illegal procedures. The most important issue was they were not bringing the high-profile cases to a satisfactory conclusion as far as Dingell's committee was concerned (4,40). In 1992 the Department of Health and Human Services in response to the concerns merged the OSI and OSIR into one Office of Research Integrity (ORI) and placed it in the Office of the Assistant Secretary of Health (13,14). In addition an appeals process for those found guilty was made available with a trial-like hearing before a Research Integrity Adjudication Panel. These panels were appointed by the DHHS Appeals Board and included a scientist. In 1995 the ORI was moved to the Office of the Secretary DHHS (14). Concerns about the process of dealing with alleged misconduct cases continued to be expressed by the scientific community and the congress. In 1993, as part of the NIH Revitalization Act, the Congress as noted earlier (14) authorized a Commission on Research Integrity (CRI). The mandate was to consider a new definition of research misconduct (described earlier) to recommend an assurance process for institutions to comply with PHS regulations on misconduct, to recommend government and institutional policies to deal administratively with investigations, and to make recommendations on the protection of whistleblowers (14). From 1987 the National Science Foundation (NSF) handled cases of misconduct out of their Office of Inspector General, which conducted any investigation. An adjudication process separate from the investigation was available with due process rights. The NSF drew much less criticism than the PHS for the way misconduct cases were investigated. It is not clear whether the more favorable reaction by scientists to the way misconduct cases were handled by NSF was due to the types of cases chosen or the process used. For both the PHS and NSF, investigations were carried out by the scientist's own institution whenever possible as long as they were handled competently (14,40). An interesting phenomenon was the activities of two NIH scientists, Ned Feder and Walter Stewart (41,42). For about 10 years the pair devoted their time, without NIH authorization, to checking out accusations of fraud and plagiarism in science research at the NIH and elsewhere, and urging action by anyone who would listen to them. They championed the side of the whistleblower, Margot O'Toole in the Thereza Imanishi-Kari case and probably increased congressional interest in the problem. They also looked at a neglected aspect of the Darsee case, the responsibility of his many coauthors. They reviewed 109 papers published by Darsee with 47 coauthors, even papers in which fabrication was not alleged. Stewart and Feder reported what they thought were many cases of errors, republication of information, use of common controls for several studies, and lapses from standards that they felt reflected poorly on the coauthors, journal editors, and reviewers. Their paper took four years to be published because of criticisms and even threats of libel by Darsee's coauthors. The article was finally published in the journal *Nature* with an editorial and with a critique of the paper by Dr. Braunwald, Dr. Darsee's mentor at Harvard (43,44). In the four years required to have their paper published, Stewart and Feder made much of the idea that they were being "censored," and had a sympathetic ear in the U.S. Congress. The pair also developed a software package that proved useful to substantiate or refute allegations of plagiarism, the so-called plagiarism machine. Since they were working without authorization while on the government payroll and their zealousness offended many scientists, they were finally moved to new positions at NIH in 1993 and told to restrict work on misconduct investigations to their own time (42).

Office of Research Integrity (ORI)

The role of ORI, which was formed in 1992, is to manage PHS research integrity activities. One function is to investigate misconduct in the NIH intramural programs, but the major activity is to oversee extramural misconduct investigations conducted by grantee institutions. ORI also develops model policies and procedures for handling allegations of misconduct which institutions can adapt for their use. Other responsibilities include evaluation of institutional policies and procedures, and investigation of whistleblower retaliation complaints. The ORI, despite much criticism from scientists and whistleblowers, remains as of 1999 the major federal watchdog for PHS grants (14).

ACADEMIC AND INSTITUTIONAL ROLES IN SCIENTIFIC MISCONDUCT

As the reports of misconduct cases increased and persisted over several years, universities and professional organizations began to study the problem and develop policies and procedures for dealing with alleged cases of misconduct (45,46). Since 1992 the NIH had required that all institutions receiving NIH training grant awards provide educational programs in research integrity for the trainees. As educational materials were developed, common themes emerged for normative rules of behavior. Guidelines for laboratory research practices were developed and shared by many academic institutions. The common features covered the areas where problems had arisen. These included laboratory procedures for recording, storing, and safeguarding research data used for the preparation of scientific papers and progress reports. Also covered were authorship practices such as whose name goes on a paper and in what order; who can be legitimate authors, the responsibility of coauthorship, the problems with honorary authorship, and, finally, mentoring responsibilities for junior research investigators when research is carried out in a training environment (45,46). Unacceptable behaviors were identified in the form of lists with definitions, with some derivations from the PHS and NSF definitions of research misconduct outlined previously. These included falsification and fabrication of data, plagiarism, dishonesty in publications, deliberate violation of regulations, failure to report misconduct, and failure to respect property of others. As factors that might influence behavior were considered, attempts were made to relieve pressures on faculty in the area of promotion and consideration for tenure. For example, some universities began to reduce the number of publications that may be considered in a promotion file for each rank considered (46). With the transfer of basic science findings into biotechnology and the opportunity for scientists to receive stock from joint ventures with industry, academic institutions developed rules about conflicts of interest and disclosure. It became apparent that if scientific integrity were to be considered an important value within an institution, the leaders of the institution would have to pay more than lip service to the concept (47). In addition the PHS and NSF requirement that institutions deal effectively with allegations of misconduct and protection of whistleblowers led to administrative changes within individual universities. In general, a specific official was identified within institutions to receive all allegations and to start the process of response in motion (46).

The Scientific Research Society Sigma Xi prepared a booklet for science students in 1984, with a third printing in 1991, entitled *Honor in Science* (48). This covers misconduct in science and whistleblowing. It starts with a section on "why honesty matters." The Committee on Science, Engineering, and Policy of the National Academy complex updated an educational booklet in 1995, *On Being a Scientist*, as a guide for teaching about scientific integrity (49).

A Panel on Scientific Responsibility and the Conduct of Research was formed under the sponsorship of the Committee on Science, Engineering, and Public Policy of the National Academy complex and published in 1993, in two volumes, a comprehensive report entitled: *Responsible Science, Ensuring the Integrity of the Research Process* (13,46). One critical review of the work noted three shortcomings: (*1*) "It insists on an unworkably narrow definition of misconduct." (The definition was fabrication, falsification, and plagiarism; see the discussion above.) (*2*) "It acknowledges studies of science as a social endeavor without taking to heart their arguments and implications." (*3*) As a consequence of the preceding weaknesses, the report's analysis of putative causes of misconduct and its proposed remedies are inadequate to the challenges now confronting science" (50). The argument has been also made by John Bailar, who works in epidemiology and biostatistics and for many years was statistical consultant to the *New England Journal of Medicine*. He feels that falsification, fabrication, and plagiarism are less a threat to the integrity of science than the day-to-day handling and reporting of data and the use of statistical methods (51). Some professional societies have developed codes of ethics or guidelines for responsible research. The American Society of Biochemistry and Molecular Biology developed a code of ethics in 1998 and seemed to learn a lesson in the process that trust can be expected only by acting responsibly (52). The Society for Neuroscience developed guidelines for scientific communications (53).

Another group that is responsible for publication and authorship practices consists of the editors of scientific publications. They have changed policies to discourage honorary and irresponsible authorship (54). A suggestion has been made by Drummond Rennie that experimental auditing of published manuscripts be undertaken after advance warning to scientists, with an eye to establishing what the real frequency of science fraud is. The information would not be used to investigate or prosecute cases. Although this process was never implemented, there have been calls for much more aggressive audits such as those conducted by the FDA (11) or suggested by the CRI (14). Since 1989 the *Journal of the American Medical Association* requires authors to sign a statement that they will produce data upon which the manuscript is based for examination by the editors or their assignees (55), and many journals now require disclosure of any conflicts of interest.

When all else has failed in bringing cases of misconduct to some satisfactory resolution, whistleblowers have occasionally gone to court. There is a tendency now to use the False Claims Act, which results in triple damages if successfully prosecuted. In one such case Dr. Condie in 1983 felt that a colleague had falsified or fabricated data in published papers and grant applications and brought this to the attention of officials at the University of Utah and University of California, San Diego. The ORI was also involved, and neither they nor the universities found misconduct. In 1989 Condie brought the action to court and asked the Justice Department to take over the case which they did. The universities were ultimately found negligent in 1994 and ordered to pay a total of $1.6 million dollars of which Condie would share 15 percent. After the court findings, the ORI then reached a settlement with the investigator to be excluded from federal grants for 3 years and to publish retractions or corrections of the disputed scientific papers (56). The 11-year time frame certainly indicates that the wheels of justice turn slowly when it comes to dealing with misconduct cases.

The tension between the scientific community and the government that supports it remains, but learning how to deal with scientific misconduct has been recognized by scientists as a serious challenge. In the meantime U.S. government policies are still in the process of evolution (4,57,58), and a common definition for misconduct is still being developed.

No one believes that misconduct can be completely prevented, but there is hope that it can be discouraged, and that if it occurs, it can be detected early and its impact mitigated. The FDA experience suggests that an audit system will reduce misconduct and have the greatest impact on the most serious and flagrant violations (11). Quality control has been used in industry to maintain adequate standards in general, and some forms of this process could be adapted to the research setting to deal with both inadvertent error and misconduct.

BIBLIOGRAPHY

1. S. Toulmin, *Hastings Cent. Rep.* **9**(3), 27–34 (1979).
2. L.R. Graham, *Hastings Cent. Rep.* **9**(3), 35–40 (1979).
3. F. Dyson, *N.Y. Rev. Books* **44**(6), 46–49 (1997).
4. S. Glazer, *CQ Res.* **7**(1), 1–24 (1997).
5. B. Alberts and K. Shine, *Science* **266**, 1660–1661 (1994).
6. N.Y. Kiang, *Sci. Eng. Ethics* **1**(4), 347–356 (1995).
7. R.K. Merton, *The Sociology of Science*, University of Chicago Press, Chicago, IL, 1973.
8. P. Woolf, *Hastings Cent. Rep.* **11**(5), 9–14 (1991).
9. W.J. Broad, *N.Y. Times*, March 27, p. A16 (1992).
10. J.P. Swazey, M.S. Anderson, and K.S. Louis, *Am. Sci.* **81**, 542–553 (1993).
11. M.F. Shapiro, in S. Lock and F. Wells, eds., *Fraud and Misconduct in Medical Research, BMJ* Publishing Group, London, 1993, pp. 128–141.
12. S.M. Kuzma, *Univ. Mich. J. Law Reform* **25**(2), 357–421 (1992).
13. Committee on Science, Engineering and Public Policy (US), *Panel on Scientific Responsibility and the Conduct of Research, Responsible Science*, vol. 1, National Academy Press, Washington, DC, 1992.
14. Report of the Commission on Research Integrity, *Integrity and Misconduct in Research*, U.S. Government Printing Office, Washington, DC, 1996.
15. R.J. Levine, *Ethics and Regulation of Clinical Research*, 2nd ed., Urban & Schwarzenberg, Baltimore, MD, 1986.
16. A.R. Price, *J. Higher Educ.* **65**(3), 286–297 (1994).
17. R. Dresser, *J. Am. Med. Assoc.* **269**(7), 895–897 (1993).
18. J.T. Edsall, *Sci. Eng. Ethics* **1**(4), 329–340 (1995).
19. M.P. Glazer and P.M. Glazer, *The Whistleblowers*, Basic Books, New York, 1990.
20. F. Hoke, *Scientist* **9**(10), 1, 15 (1995).
21. J.S. Lubalin and J.L. Matheson, *Sci. Eng. Ethics* **5**, 229–250 (1999).
22. T. Devine, *Scientist* **9**(10), 11–12 (1995).
23. D.J. Kevles, *The New Yorker*, May 27, pp. 94–109 (1996).
24. D.J. Kevles, *The Baltimore Case*, Norton, New York, 1998.
25. P. Peck, University of Pittsburgh and NCI Settle with Dr. Fisher, *OB.GYN. News*, October 1, P11 (1997).
26. E.J. Hackett, *J. Higher Educ.* **65**(3), 242–260 (1994).
27. W.M. Carley, *Wall Street J.*, July 26, pp. A1, A6 (1994).
28. B. Barber, *Science* **134**, 596–602 (1961).
29. T.S. Kuhn, *The Structure of Scientific Revolutions*, 2nd ed., University of Chicago Press, Chicago, IL, 1970.
30. W. Broad and N. Wade, *Betrayers of the Truth*, Simon & Schuster, New York, 1982.
31. A. Kohn, *False Prophets*, Blackwell, Oxford, UK, 1989.
32. D.J. Miller and M. Hersen, eds., *Research Fraud in the Behavioral and Biomedical Sciences*, Wiley, New York, 1992.
33. M.C. LaFolette, *Stealing into Print*, University of California Press, Berkeley, 1992.
34. S. Lock, in S. Lock and F. Wells, eds., *Fraud and Misconduct in Medical Research*, BMJ Publishing, London, 1993, pp. 5–41.
35. L.K. Altman, *N.Y. Times*, October 30, p. A12 (1996).
36. E. Marshall, *Science* **270**, 1912–1914 (1995).
37. E. Marshall, *Science* **273**, 1162–1164 (1996).
38. D.K. Price, *Daedalus* **107**(2), 75–92 (1978).
39. D.H. Guston, *Centen. Rev.* **38**(2), 215–248 (1994).
40. M.C. LaFollette, *J. Higher Educ.* **65**(3), 261–285 (1994).
41. P. Gwynne, *Scientist* **2**(13), 1, 8 (1988).
42. F. Hoke, *Scientist* **9**(3), 3–15 (1995).
43. W.W. Stewart and N. Feder, *Nature (London)* **325**, 207–214 (1987).
44. E. Braunwald, *Nature (London)* **325**, 215–216 (1987).
45. Institute of Medicine Report, *The Responsible Conduct of Research in the Health Science*, National Academic Press, Washington, DC, 1989.
46. Committee on Science, Engineering and Public Policy, *Panel on Scientific Responsibility and the Conduct of Research, Responsible Science*, vol. 2, National Academic Press, Washington, DC, 1992.
47. C.K. Gunsalus, *Acad. Med.* **68**(9), Suppl. S-33–S-38 (1993).
48. *Honor in Science*, Sigma Xi Press, Research Triangle Park, NC, 1991.
49. Committee on Science, Engineering and Public Policy, *On Being a Scientist, Responsible Conduct in Research*, National Academy Press, Washington, DC, 1995.
50. E.J. Hackett and S.M. Solomon, *BioScience* **43**(10), 717–719 (1993).
51. J.C. Bailar, III, *Chron. Higher Educ.*, April 21, pp. B1–B3 (1995).
52. F. Grinnell, *Sci. Eng. Ethics* **5**, 205–217 (1999).
53. M.J. Zigmond, *Sci. Eng. Ethics* **5**, 219–228 (1999).
54. D. Rennie, *J. Am. Med. Assoc.* **271**(6), 469–471 (1994).
55. D. Rennie, *J. Am. Med. Assoc.* **270**(4), 495–496 (1993).
56. P. Hilts, *N.Y. Times*, July 23, p. 9 (1994).
57. K. Ryan, *Prof. Ethics Rep.* **9**(2), 1, 6–7 (1996).
58. K. Ryan, *Chron. Higher Educ.*, July 19, pp. B1–B2 (1996).

See other entries SCIENTIFIC RESEARCH, ETHICS, VALUES IN SCIENCE; SCIENTIFIC RESEARCH, LAW, AND PENALTIES FOR SCIENTIFIC MISCONDUCT.

SCIENTIFIC RESEARCH, ETHICS, VALUES IN SCIENCE

RACHELLE D. HOLLANDER
National Science Foundation
Arlington, Virginia

OUTLINE

INTRODUCTION

This article reviews recent scholarship to demonstrate ways in which science is value laden and how these value dimensions have ethical importance. It identifies the values scientists generally accept as those that should govern their work, as well as other values that influence support for science, its conduct and outcomes. It discusses some implications of these values for ethical issues in biotechnology. The approach is to review relevant philosophical and social science literature that examines relationships among science, ethics, and values, and to identify and analyze some examples of ethical issues in biotechnology.

The article contains two basic themes. One focuses on values and science. It reviews literature in the philosophy and sociology of science to describe a controversy that occupies the attention of scholars in those fields, as well as that of some scientists. The philosophers and scientists want to answer a question like: Does science tell us something important and true about reality? The sociologists want to answer a question like: Does the scientific community have a distinctive set of norms or values?

This first theme considers such topics as: Is science value laden or value free? What is the relationship of intrinsic and extrinsic values to science? What are constitutive and contextual values in or in relationship to science? From philosophical and sociological perspectives, these are the issues that comprise an attempt to answer the questions: What is science? Is science a set of independent and abstract criteria? Or is it what scientists do, and where they do it, and what they do it with? Or all of the above (1)?

The second theme focuses on the ethical implications of values in science. It takes the position that the embedding of science in society means the conduct of science, and its outcomes always have ethical implications. Furthermore societal commitments to innovation mean that these ethical implications are momentous and complex. Thus people can, and should, make moral judgments about science in its undertaking and with respect to its influences and outcomes. These are judgments about the moral responsibilities of individuals and the social responsibilities of institutions, about the moral and social implications of scientific practices, and about the effects from these endeavors for humans and their communities and environments. Making well-considered judgments can be assisted by moral theories and conceptual analysis, by systematic research to understand phenomena, and by public deliberation that takes care to include a wide variety of views. The proportionality principle in ethics assigns greater responsibility to persons more likely to influence outcomes. It would follow that people in positions of influence in biotechnology have greater responsibility to promote activities that encourage these kinds of considerations and deliberations.

The contemporary world and the political states that comprise it have made a substantial commitment to the production of science and its integration into society. Research institutes and educational, industrial, and governmental organizations house scientific laboratories and research facilities. Thereby the conduct of science has societal impacts in and of itself. To do science requires social institutions and social commitments. The production of this encyclopedia and the attention of its readers testifies also to the societal influences of science. Many entries demonstrate and discuss the changes in individual lives, social institutions, and the environment that result from doing science. It is clear, then, that science in its conduct and impacts cannot be value free. But this is only the beginning of the story.

SCIENCE AND VALUES

Numerous terms are used in the debate over whether or not, and how, science incorporates values. Perhaps one of the earliest and simplest contrasts is between value-free and value-laden science. This contrast goes back to the philosophers of the Vienna Circle in 1924 to 1936 (2). From their perspective, meaning itself was to be limited to value-free statements that lent themselves to sensory proof, or some kind of derivation from sensory proof. Later philosophers could not reconcile this position with quarks and genes, or the logical difficulty of moving from data to evidence to theory, and the philosophers of the Vienna Circle gave up that project. However, the questions of what are the distinctive features of science and whether they lead to results with a special claim to value-free truth remain important to scientists and those studying science. They are of social importance too, since scientific claims and claims about science often underlie large social expenditures and influence political and social outcomes.

In examining this issue further, it is useful to consider the contrast between objectivity and subjectivity, and between constitutive and contextual values. Scientists are content to view science as having constitutive values. These values are also called epistemic, as well as internal or intrinsic to science. They are contrasted with external or extrinsic values. Constitutive values would include the value placed on observation and experiment, on prediction and explanation, on the building and testing of theory and models, and on the development and testing of methods that can reduce the probability of error or self-deception in scientific work. Science remains self-contained in satisfying these values, or most scientists view it to be so. Scientists and others have thought their satisfaction provides objective methods leading to objective truth. For good reviews of recent debates in philosophy of science about the status of scientific claims to privileged standing in understanding the natural world, see Couvalis (3) or Klee (4). Both argue for the preservation of scientific claims to objective understanding of the natural world.

The strength of this perspective on scientific values can be seen in the wide acceptance of the work of sociologist Robert Merton. In 1942 Merton characterized science as satisfying four norms: communism (open sharing of findings), universalism (use of general criteria for judging knowledge claims), disinterestedness (not acting for personal advantage), and organized skepticism (withholding belief). Only after several decades, in the 1970s, was this view challenged. Ian Mitroff proposed that the opposite views operated as counternorms in science, while Aaron Cicourel and Michael Mulkay challenge the sociological foundations of norms as categories, by arguing that norms are comprised in particular acts of individual actors (1, pp. 398–400).

Philosopher Joseph Rouse provides a current critique of both realist views that science tells how things really are and relativist or constructivist views that all scientific beliefs can be explained sociologically. He believes that both of these accounts suffer from their presupposition that a global assessment of science as one kind of knowledge is possible. Instead of a global view, the study of the actual variety of scientific practices might provide better grounds for an assessment of their significance. Understanding science as historically situated responses to past scientific effort in anticipation of and contest over its future development allows people to evaluate particular scientific claims without the presumption of a particular view about the coherence of scientific knowledge or about the relationship of that knowledge to reality (5).

From time to time some scientists and policy makers indicate that science is value free, or objective. Often in doing so, they want to use such statements to provide support for a favored position or to undermine one they do not support. If constitutive, or epistemic, values and the Mertonian norms were the only values in science, perhaps it would be possible for them to do this. Scientists would (sooner or later) find the truth. The hard work involved in making social decisions using that truth would be "handed off" from scientists to policy makers or other social decision makers. Scientists discover what they can, and then societal values and priorities need to take over. This model implies a clean division of labors and categories, value-free science, and value-laden policy.

Before identifying some flaws in this position about the separability of science from values, this entry needs to point out one immediate limitation. No one believes that technologies are value free. Technologies, from abacuses to zylophones, are devised and selected specifically to serve human purposes or values. If science is used in the development of technologies, that development is not value free. Furthermore technologies often make science

possible, so at least in this sense, science is not value free. This encyclopedia is about biotechnology, indicating a kind of science that is not separable in entirety from technology. Thus, even if there is a kind of science that can fulfill the conditions for a value-free science, it does not seem to be the kind under consideration here. Discussing the ways in which science, thought to be value-free, can and does incorporate values is important both for an adequate understanding of what science is, and of its ethical implications. If a strong case can be made that science does incorporate nonconstitutive values, the case for their presence in the science (or sciences) of biotechnology is strengthened.

This entry uses notions of constitutive and contextual values found in the work of Helen Longino (6). Contextual values are values independent of the scientific goals or constitutive values identified above. Contextual values are social, cultural, economic, political, moral, and personal. These values can influence science, while it remains what all might agree is good science according to those constitutive values. Consider, for instance, a social or political commitment to fund materials science and engineering to understand properties of corrosion and fracture; consider the commitment to protection of human or animal subjects. These social values influence whether and what science gets done, but they do not necessarily result in bad science or science whose epistemic value is affected negatively. There are many examples that show that contextual values affect constitutive values, even when such science remains good science according to the constitutive values identified above. Reports of data falsification leading to retractions of published papers and findings provide obvious examples where contextual values produce bad science. A notorious example in the 1970s involved coloring mice to indicate successful skin grafts; the miscreant blamed personal stress and exhaustion (7).

Longino identifies five ways in which contextual values affect science. They affect scientific practices, the questions that get attention, and the description of data. Contextual values also affect the background assumptions and the global assumptions with which science operates (6, p. 86).

A good example of the deep interpenetration of science and contextual values comes in Nelly Oudshoorn's socio-historical analysis of the development of understanding about female sex hormones, which led to the invention of birth control pills (8). Her analysis shows how society influences questions that get attention, scientific practices, and the data that get collected.

This story about the making of sex hormones arose with new developments in the chemical life sciences at the turn of the twentieth century. The actors in the drama were physiologists working in laboratories, gynecologists treating women, and the pharmaceutical industry. The materials these actors needed influenced what would count as knowledge. The episode produced and transformed gender bias in science.

Before the 1920s the three groups began to interact over hormonal products. In the 1920s they ran up against problems in getting the amounts of gonadal material required to do the work they wanted to do. The material was expensive and difficult to find, especially for the laboratory scientists. Because of this need, cooperation began to intensify between the scientists and pharmaceutical companies that supplied hormones from animal gonads. Gynecologists relied on these groups for quality hormonal products — whose benefits for their patients were, by the way, quite unclear.

In 1926 two German scientists happened on the long-sought source — human urine. Urine from pregnant women was particularly good. To gain access, pharmaceutical companies and laboratory scientists relied on the gynecologists' new, inexpensive source. But this was a prolific source of female sex hormone, not male. Urine from men would certainly be a suitable source for male sex hormone, but there was no institutional context for its collection. There were no clinics specializing in men's reproductive systems in the 1920s.

Though laboratory scientists had been interested in the role of both male and female sex hormones in growth and development of the body and in sexual differentiation, the ease of access to certain kinds of research materials lent support to, or privileged, the development of certain knowledge claims rather than others. This material source, along with the institutional context that focused on women's reproductive disorders, saw men disappear as a focus for research. Not until the late 1960s was the study of male reproductive disorders institutionalized as a medical specialty (8, p. 80).

This example demonstrates how contextual factors — ranging from the values different groups placed on social desiderata, such as potency or birth control, to the possibilities created by access to particular material such as a kind of urine — influence the very *constituents* of science. What counts as evidence, and for what — what can be used for the purposes of constitutive values — is intimately wrapped up in the social circumstances and priorities of times and places. This in turn influences what diseases or disorders get so labeled and treated, and when they do. This result has unavoidable ethical implications.

Background and global assumptions in science have implications for the interpretation of data in questionable ways that carry both policy and ethical implications. Longino provides the interesting example of research on human origins. This research uses one of two organizing principles: man-the-hunter and woman-the-gatherer. Each uses a story of the gendered development of tool use to promote a view of the favored sex as the initiator of activities from which defining human traits of intelligence and cooperation evolve. Each story promotes the view that men and women make tools for different purposes. However, while the remnants used to construct these stories provide evidence sufficient to conclude humans shaped them, they do not provide evidence sufficient to conclude if men or women used them, or how (6, pp. 106–108).

These examples demonstrate that contextual values affect science at very fundamental levels, but that these effects might not overthrow claims of scientific objectivity — in the sense of trying to work toward scientific understanding of natural phenomena, no matter how partial or influenced by nonepistemic motives and understandings. Recognizing this kind of partiality places a severe limitation on the adequacy of science or scientific answers at any particular time, for non-scientific purposes. The social responsibility for recognizing this limitation, and figuring out how to respond, falls both to scientists and non-scientists, particularly those in influential policy positions.

Another kind of partiality is less esoteric. Scientists and their academic sponsors, with commercial interests in development of products that follow from their discoveries, can be found using press conferences to promote their latest results. The tragicomic episode announcing the discovery of cold fusion provides a recent example demonstrating the limits of Mertonian norms in influencing scientific behavior (7, pp. 11–12). If the public is somewhat skeptical in its reception of such activities, the skepticism can be regarded as healthy prudence. Scientists who wish to honor the constitutive ideals of replication or peer review would also be likely to withhold approval. Here too, the social responsibility for recognizing this limitation, and figuring out how to respond, falls both to scientists and nonscientists. The proportionality principle would require more from scientists and others in positions to be influential than would be expected or required from those not so placed.

SCIENCE, ETHICS, AND THE STATE

The previous examples demonstrate that and how scientific endeavors require societal support. Science occurs in organizations and institutions and plays an important role in promoting particular social goals and interests. Public and private sector organizations sponsor scientific research with the expectation that their support will lead to public and private gain. These results bring with them questions about the nature and extent of gain; these questions have both utilitarian and distributional components.

Jurgen Schmandt and James Everett Katz identify three ways in which contemporary societies value science — as a product, as evidence, and as method. As a product, science is promoted and controlled in the interests of innovation. As evidence, science is used and interpreted for policy purposes. And scientific methods — analysis, experiment, empirical techniques — are valued in themselves (9). They are valued at least in part because they can serve for purposes of innovation and evidence.

The use of science in these ways means that, inevitably, people in democratic societies that encourage public involvement will call it to account. They will ask whether the use is justified, and consider the ethical implications of the choices that are made. As human interventions become pervasive (ozone holes), mammoth (Three Gorges Dam), and more sophisticated (recombinant DNA technology), so do the scale and requirements of human accountability, including scientific accountability. Two kinds of ethical issues for biotechnology are worth examining in this regard. One concerns ethics and risks; the other, the relationship of science to political consent (10).

Ethics and Risks

As a force for innovation, science brings inevitable risks. When human beings engaged in scientific activities or the use of science-based technologies create risks, those affected will ask whether the activities have been undertaken responsibly. Below are some examples.

Differential Impacts. Equity issues that arise in the management of radioactive or hazardous wastes provide a useful example of the kinds of problems resulting from commitment to science for innovation. Placement of these wastes raises questions of differential impact for locus, labor, and legacy (11). Locus raises the ethical isues of locating facilities that may harm those nearby while benefiting those far away. Legacy raises a similar question for those removed in time. These are issues of distributive justice; they ask about the fairness of the distribution of benefits and costs. Similarly for labor, although here the questions are complicated by issues of consent. Good justification of any state decision requires attention to these ethical issues.

Science and Differential Impacts. Looking more closely at the issue of evidence in this waste disposal example illustrates a particularly important quality in the value dimensions in science and how they raise ethical questions. To determine whether or not people are harmed, a scientific test is used. The decision to use a test has moral dimensions. It is a choice made by human beings, and it may help or harm them or their surroundings. Once the choice is made, decisions about which test to use also involve moral dimensions. Whichever tests are used, they will result in false negatives or false positives, at least to some degree. The choice in either direction will be a moral choice; thus a science-based choice involving constitutive values contains contextual value implications. If the test selected will result in false positives — that is, one that will tell us that some persons are affected who are not — these persons may be subjected to a variety of harms ranging from risky treatments to stigmatization or job loss. If the test selected will result in false negatives — that is, one indicating that affected persons are not affected — persons may suffer unattended illness and premature death. Independent of purposeful wrongdoing, application of science to questions of individual and societal well-being will involve value choices of ethical importance (12).

The previous examples show how science contains values dimensions with ethical implications. They show that these dimensions are not avoidable. Governments as well as other social institutions need to take these implications into account in order to behave responsibly. The sociopolitical context in which biotechnology is developing requires persons and institutions promoting it to pay attention to the contextual and constitutive values that affect biotechnology, as well as the ethical implications that are part of and follow from its promotion.

Value Conflicts. The case of deliberate release of genetically engineered organisms gives examples of direct relevance to this encyclopedia. Using Longino's categories,

Soemini Kasanmoentalib provides a list of examples in which contextual values influence the scientific development and assessment of biotechnologies (13). Under practices, he points out that the multinational operation of companies doing biotechnology R&D allows them to select countries with less rigorous regulations for testing. In this case the commercial values of the company may coincide with the values some scientists place on being free to do their research, but they may conflict with those of the public and other scientists who are more concerned with the potential for harm. Here the conflict arises between the positive constitutive value scientists place on doing research and the negative contextual value placed on potential harm to vulnerable human populations. In another case in which constitutive and contextual values conflict, scientists at a university working with commercial support may find the value they place on open discussions and publication challenged by the need to patent or keep secret their findings. Here the value conflict can involve the constitutive value of open discussion and what a firm might argue is the utilitarian value that its product will provide. The firm might also point to the contextual value of promise keeping, if the scientists have signed an agreement. All of these kinds of cases, where values conflict, need careful ethical consideration.

Science and Short-Term/Long-Term Interests. Commercial interests are not the only interests in economic growth. Governmental desires to foster innovation may result in limiting the kinds of questions that are asked and data that are required before approvals to plant or market products of biotechnology are gained. The Kasanmoentalib article notes that criteria of what should count as risk or damage, or appropriate ecological end points by which environmental stress can be measured, are difficult to establish, and can be limited to the gross and near-term. Under specific assumptions the selection of a model on which to base the risk assessment can be less conservative, ignoring synergistic effects for which it is difficult to devise tests. Thus global assumptions favoring the reductive approaches to assessing risk found in molecular biology and genetics can be favored over those from ecology which incorporate more concern for synergistic, inclusive, and long-term effects.

Here, keep in mind that the decision to delay introducing a new genetically engineered organism may pose ethical risks also. For instance, a plant engineered to resist a pest may require less pesticide and provide more food to an impoverished area. The decision to hold off on its introduction may trade off short-term need for more crop against long-term concern for ecosystem health. How to identify and balance short-term and long-term interests is a difficult ethical question. The stories in June 1999 in the science journal *Nature* and many of the world's newspapers about the lethal effect of bioengineered corn pollen on Monarch butterflies show that these concerns are of more than theoretical importance.

Particular choices in these circumstances are no more scientific than their opposites. Given that this is so, the answers to the questions: What is risk? What is acceptable risk? What is acceptable evidence of risk?, are themselves not just scientific. On the one hand, parties outside the sciences must help to establish standards for acceptable risk. On the other hand, careful scientific or what can better be called meta-scientific considerations of epistemic issues for science, of what scientists are entitled to say they know, are essential (12). Careful analysis of the justifications for the assumptions and findings of various sciences is essential, both to identify and show the limits of science, and to help make scientific progress possible, by challenging the conventional wisdom in particular fields. It is better not to call these choices scientific, since that can connote value-free or objective choices, in a way that they cannot be.

Science and Political Consent

Science and Political Values. Recent controversies over labeling of food products that contain elements that are biotechnologically engineered show another dimension of the intricacies of interactions between science, technology, values, and ethics. The U.S. government and some commercial biotechnology interests are arguing that there are no grounds for labeling bioengineered products as such, since they pose no additional safety risks because of the bioengineering. They are using a narrow definition of a safety risk here. Such risks would be those, for instance, created by incorporating an allergen or toxin into a plant. Engineering an herbicide or herbicide resistance into a plant would not constitute such a risk. Long-term risks to ecosystems are not being considered (13).

Another category of ethical concern needs to be mentioned here. As Paul Thompson points out, people want to know a great many things about food products besides and beyond the kinds of safety concerns identified above (10, p. 75). They want to know where the foods come from. They want to know the processes by which they have been made. It is one thing to say that government should ensure that health-related claims on a product are not false and misleading, and that science should be used to assist in making that determination. It is quite another thing to say that no information except for scientifically validated claims about safety should be allowed on food products. Saying that is to substitute science for the consent of the governed. For persons who wish to buy hot sauce from Louisiana, or free range chickens, or organic produce that is not bioengineered and is raised on farms that use no manufactured pesticides or herbicides, such rulings would abolish choices they currently have and value having. It would substitute scientific rule-making for democratic choice, indirectly silencing a political voice.

Science and Moral Theory. Moral theories, unlike scientific theories, provide judgments about what human beings should do, not what the world is. However, the previous discussion serves to illustrate how science involves values and moral implications. It illustrates how human judgments about the worth of science exist in a context in which questions central to moral discourse—about what promotes human well-being and what is fair—arise. It illustrates how human judgments called scientific may incorporate contextual as well as constitutive values that presuppose particular answers to

the questions: What kinds of scientific research should be done? Who should own the results? How should the results be described in the open market? Who is benefited or harmed? How? And how much?

The previous discussion can also illustrate limits of moral theories in resolving issues. Some scientists and regulators in government agencies and industry spokespeople might insist on utilitarian grounds that no products be labeled not to contain bioengineered components. They might believe that the greatest good for the greatest number will be served by having people buy less expensive safe products than they might otherwise choose, because of unjustified fears, if they see an equivalent, more expensive product with the "no biotech" label. Insisting on the right to have such a label gives pride of place to, or privileges, the rights view. That view insists that persons have rights to make choices that others think they are making for less than satisfactory reasons. Is one of these views the morally right one? On what grounds would that decision be made?

One way to try to answer this question is to try to resolve the facts of the case, where the facts are the empirical claims that are being made. Is the utilitarian claim true? Is it true in all circumstances? There may be circumstances in which ungrounded fears will influence consumer purchases, but these may be relatively few. It is easy to imagine ways to overcome such fears. Further, consumers may begin to feel manipulated and distrustful of a system in which they believe information is being kept from them because of commercial interests. With this scenario, utilitarian theory itself may be better served in a marketplace which allows labeling that includes information in addition to scientific claims about safety. A world in which more rights and freedoms can be honored may be a better world, by utilitarian standards, even when it allows choices that are ostensibly less well grounded by some current scientific standards. While science can be enlisted to serve a particular moral point of view and, in this case, given a utilitarian cast, the claims underlying such an outcome needn't be accepted. If this response is accepted, a moral conflict can be settled by finding a creative middle way in which moral theories can be reconciled (14).

This creative middle way may allow the preservation, or perhaps even the transcendence of utilitarian and rights-based moral theories. Another moral point of view is worth mentioning, one that William Aiken called holistic (15). This view, also referred to as the interconnectedness approach, points out that utilitarian approaches that proceed by examining trade-offs, or costs and benefits, risks and benefits, can lose sight of the connections that mean that the natural world does not operate like a balance sheet. These connections mean that negatives cannot simply be traded off against positives; negatives may be necessary to the maintenance of a desirable whole. Life requires evolution, predation, and death. From this perspective, neither rights views nor trade-off views give due recognition to larger values that need consideration in the relationships between human beings and the larger environment or natural world. This approach demands attention to the values we wish to maintain in our social practices as well as the kinds of long-term consequences that may be overlooked in standard moral theories. The satisfaction of these values for such things as communities and ways of life requires consideration of long-term consequences and the connections that structure and sustain wholes. The interconnections approach may not reduce to either utilitarian or rights-based approaches. However, it may provide a connection between ethical and ontological theories, between theories about what ought to be the case and what is the case. It may require and provide a context which is hospitable to raising new ethical issues and challenges.

Currently scientific approaches to risk in the formulation of environmental policy take what Thompson labels a purist, rather than a hybridization, view. The purist view separates risk into its components: risk to human health, animals, environment. While social consequences could be a component also, the current policy process in the United States does not recognize social consequences as a legitimate topic for discussion. For the public, however, risk is an amalgam or hybridization of at least all of these components (10, pp. 232–236). Additionally, the public view of risk contains a concern for the responsiveness of those with authority and power — be it scientific, political, financial, corporate, organizational — to this amalgam. Unless science and scientists recognize that the approaches they take to risk incorporate these constraints, which have moral dimensions, they are operating under false assumptions with respect to both the ethical and value implications of their work. If scientists and others with the kinds of power identified above can recognize and be responsive to these implications, it may be possible to reconcile scientific and social progress.

SUMMARY

The task of this entry is to show some of the ways in which science is value laden and how its doing and results have ethical implications. The first section shows that both intrinsic or epistemic and extrinsic or contextual values inhere in science. The discussion and examples show how contextual values affect the scientific search for explanations of natural phenomena and the outcomes of that search, and how the search and outcomes incorporate social priorities and biases. The discussion and examples point out that these outcomes have ethical implications. Thus, the material and institutional circumstances surrounding hormonal research in the 1920s led to an emphasis on the study of female rather than male reproduction. Scientific discoveries in this area have been of enormous social benefit. However, the partial understanding that science provided placed limits on its appropriate use as an underpinning for societal decisions. Less esoteric, current examples of scientific partiality — for instance, that arising from scientific promotion of research results for commercial purposes — also give rise to healthy caution. These limits on scientific understanding require careful attention from scientists and others in positions to influence social policy and programs.

The second section continues the discussion of how values enter science and discusses ethical implications in the pursuit of science as a social or national priority. This pursuit affects individuals, organizations, communities, and the environment, and brings with it inevitable ethical questions about the nature and distribution of benefits and harms. The section examines issues of ethics and risks from science-based innovations, and issues of science-based innovations and conflicts over consent. Science-based technologies include and create differential risks. Figuring out what these risks are, and what their ethical implications are, is a complex task. The findings deserve careful attention from those in positions to influence social policy and programs. The same is true with issues raised by the relationships between science and regulation, as the example of food labeling makes clear. The call for science-based labeling gives priority to particular moral and political views as well as a particular view of scientific truth. Once again, these views require careful identification and consideration from those in positions of authority, if they wish to be accountable and responsible for their actions.

BIBLIOGRAPHY

1. T. Gieryn, in S. Jasanoff, G. Markle, J. Peterson, and G. Markle, eds., *Handbook of Science and Technology Studies*, Sage Publ., Thousand Oaks, CA, 1995, pp. 393–443.

2. D.D. Runes, *Dictionary of Philosophy*, Philosophical Library, New York, 1960.

3. G. Couvalis, *The Philosophy of Science, Science and Objectivity*, Sage Publ., London, 1997.

4. R. Klee, *Introduction to the Philosophy of Science, Cutting Nature at its Seams*, Oxford University Press, New York, 1997.

5. J. Rouse, *Engaging Science: How to Understand Its Practices Philosophically*, Cornell University Press, Ithaca, NY, 1996, see "Introduction," and Chapter 6.

6. H.E. Longino, *Science as Social Knowledge*, Princeton University Press, Princeton, NJ, 1990, p. 4.

7. D.B. Resnik, *The Ethics of Science: An Introduction*, Routledge, London and New York, 1998, pp. 75–76.

8. N. Oudshoorn, *Beyond the Natural Body: An Archeology of Sex Hormones*, Routledge, London, 1994, Chapter 4, pp. 65–81.

9. J. Schmandt and J.E. Katz, *Sci. Technol. Hum. Values* **11**(1), 40–52 (1986).

10. P.B. Thompson, *Food Biotechnology in Ethical Perspective*, Blackie Academic & Professional, Chapman & Hall, London, 1997, pp. 57–79.

11. R.E. Kasperson, ed., *Equity Issues in Radioactive Waste Management*, Oelgeschlager, Gunn and Hain, Cambridge, MA, 1983.

12. R.D. Hollander, in D.G. Mayo and R.D. Hollander, eds., *Acceptable Evidence, Science and Values in Risk Management*, Oxford University Press, New York, 1991; see also C.F. Cranor, *Regulating Toxic Substances: A Philosophy of Science and the Law*, Oxford University Press, New York, 1993, pp. 40–48.

13. S. Kasanmoentalib, *J. Agric. Environ. Ethics* **9**(1), 42–60 (1996).

14. C.E. Harris, M. Pritchard, and M. Rabin, *Engineering Ethics: Concepts and Cases*, Wadsworth Press, CA, 1995, pp. 135–138.

15. W. Aiken, in K.A. Dahlberg, ed., *New Directions for Agriculture and Agricultural Research: Neglected Dimensions and Emerging Alternatives*, Rowman & Allanheld, Totowa, NJ, 1986, pp. 31–41.

See other entries SCIENTIFIC RESEARCH, ETHICS, SCIENTIFIC MISCONDUCT; SCIENTIFIC RESEARCH, LAW, AND PENALTIES FOR SCIENTIFIC MISCONDUCT.

SCIENTIFIC RESEARCH, LAW, AND PENALTIES FOR SCIENTIFIC MISCONDUCT

BARBARA MISHKIN
Hogan Hartson L.L.P.
Washington, D.C.

OUTLINE

Introduction

Administrative Sanctions

The Public Health Service and National Science Foundation

The U.S. Food and Drug Administration

Government-wide Debarment and Similar Exclusions

Publication of Misconduct Findings

Retraction or Correction of Publications

Recoupment of Government Funds

Civil Money Penalties

Criminal Prosecution

Statutory Penalties

Examples of Cases

Damage to Personal Health and Professional Reputation

Bibliography

Additional Reading

Federal Agency Actions, Publications

Cases

Miscellaneous

INTRODUCTION

Individual scientists and their research institutions (academic, nonprofit, or industrial) risk a variety of penalties if they engage in scientific misconduct. Penalties may be imposed administratively by the federal agency (if any) that supported the research or by the Food and Drug Administration (FDA) which must approve each new drug, biologic, or device before it may be marketed. Additional penalties may follow criminal prosecution or civil claims pursued through the courts.

Potential penalties range from increased supervision of research to sizable fines and imprisonment. The scientists and their institution may also be excluded from further participation in government funded or federally regulated research for a period of years (or sometimes, permanently). In addition, if the discredited research was supported by a federal grant or contract, the funding agency may demand that the full amount be returned. The agency invariably insists that articles or reports found to be the result of misconduct be formally withdrawn, and the publicity surrounding findings of scientific misconduct can tarnish the reputation of the research institution and destroy the career of the scientists involved. When the research institution is a commercial entity, the misconduct findings and penalties also may affect sales of its products and the value of its stock or its ability to make an initial public offering.

ADMINISTRATIVE SANCTIONS

The two government agencies that provide most of the funding for biomedical research in the United States are the National Institutes of Health (NIH) and the National Science Foundation (NSF). Each has regulations requiring investigation of alleged research misconduct and appropriate action if the allegations are confirmed (1). (Table 2) The FDA has similar authority to impose sanctions for violations of rules governing the development and evaluation of new drugs and medical devices. The actions taken by the agencies supplement any disciplinary action imposed by the research institution.

The NIH and NSF developed their scientific misconduct procedures following a series of congressional hearings in the 1980s which criticized the responses of government agencies and the recipients of federal grants to allegations of fraud in science. The hearings and related press accounts publicized several incidents in which data were fabricated or falsified, and others in which papers submitted to a journal were found to have been plagiarized.

The federal definition of research misconduct has been controversial. Whatever the definition, confirmation that one or more scientists engaged in such misconduct will lead NIH or NSF to impose administrative actions and civil or criminal penalties appropriate to the nature and seriousness of the misconduct that occurred.

The Public Health Service and National Science Foundation

The Public Health Service, of which NIH is a part, and the NSF may take one or more of a range of administrative actions at the conclusion of a scientific misconduct investigation, unless an appeal is filed. In addition the research institution or the funding agency, or both, may impose "interim administrative actions" even before an investigation has been concluded, if necessary to protect human or animal subjects, prevent improper use of federal funds, or safeguard the public interest (2). Although described as administrative actions rather than penalties, the distinction may make little difference to the scientist or institution subjected to the action.

Interim Administrative Actions. The stated government purpose of interim administrative action is to ensure the proper use of public funds, the protection of research subjects, and the fitness of the principal investigator to continue to direct the research project. Although research institutions are expected to take interim administrative actions as appropriate, the funding agency may take one or more actions in addition, to protect the federal interests. Administrative actions once taken are reviewed periodically and may be modified as necessary in light of new information. The possible interim actions that may be taken by a federal agency are as follows:

- Total or partial suspension of ongoing research
- Total or partial suspension of the accused researchers from eligibility to receive additional federal grants or contracts
- Prohibition or restriction of certain research activities (e.g., research involving human subjects or animals)
- Requirements for supervision and prior approvals to ensure/ compliance with federal law and to protect public health and safety
- Delaying the award of pending grants or contracts
- Revoking agency approval of key research personnel to direct or perform research activities

Administrative Sanctions for Scientific Misconduct (Table 1). If allegations of scientific misconduct are confirmed by an investigation, the funding agency may impose one or more of the following sanctions:

- Send letter of reprimand to the scientist's institution
- Require increased supervision of the scientist's research and publications

Table 1. Penalties for Scientific Misconduct: Federal Regulations

FDA regulations	Citation
Administrative actions for noncompliance	21 CFR, pt. 56, Subpart E
Civil money penalties	21 CFR, pt. 17
Disqualification of clinical investigators	21 CFR, §312.70
Disqualification of testing facilities	21 CFR, §§58.200-58.219
HHS regulations	
Government-wide debarment and suspension (nonprocurement)	45 CFR, pt. 76
Responsibility of PHS awardee and applicant institutions for dealing with and reporting possible misconduct in science	42 CFR, pt. 50, Subpart A
NSF regulations	
Government-wide debarment and suspension (nonprocurement)	45 CFR, pt. 620
Misconduct in science and engineering	45 CFR, pt. 689

- Require that a supervisor certify the accuracy and integrity of information submitted to the agency in grant applications and progress reports
- Restrict the use of agency funds to certain activities
- Conduct special reviews of all grant applications from the guilty scientists or their institutions
- Prohibit the scientists from serving on agency advisory committees
- Suspend or terminate ongoing research support
- Debar the scientists or their institution from eligibility for federal research support for a given period of time

The U.S. Food and Drug Administration

Under the Food, Drug, and Cosmetic Act, new drugs, biologics, and devices must be approved by FDA before they may be marketed. Approval is based on data collected in clinical trials demonstrating that the product is safe and effective for its intended use. Sponsors of new products must first apply to FDA for permission to conduct the clinical trials. Sponsors (usually drug or device manufacturers) or the clinical investigators who repeatedly or deliberately fail to follow FDA rules for the conduct of the research, or who fabricate or falsify their data, may be disqualified from further participation in clinical or laboratory research involving investigational products (3). Such disqualification by FDA is similar to debarment, although FDA also may debar research entities, drug and device manufacturers, and individual scientists for serious research misconduct such as submitting false statements to the agency, conviction of a crime related to the development or approval of a drug or device, or involvement in a conspiracy to commit any such crime (4). Debarment is also authorized following conviction of a crime such as bribery, fraud, perjury, falsification or destruction of records, and similar acts related to product development. The period of debarment may be as short as one year or permanent, depending on such factors as the nature and seriousness of the offense, the extent to which management was involved (either in encouraging or participating in the criminal activity or in failing to report it), and the extent to which management tried to correct the causes of or mitigate the offense.

FDA also has authority to take administrative actions similar to those described above for PHS-funded research (e.g., immediate suspension of research in order to protect research subjects or public health and safety). In addition FDA may refuse to accept data from a clinical trial to support an application to market the product being evaluated and may even withdraw approvals previously granted (5).

Finally, FDA may seek criminal convictions or civil money penalties. When a person (individual or corporate) submits a false statement of a material fact to FDA, or knowingly fails to disclose information required to be submitted (e.g., the number and severity of adverse events observed in a clinical trial), that person may be liable for a civil penalty, under the Food, Drug, and Cosmetic Act (6). Individuals may be fined up to $250,000 for each violation, while fines against manufacturers may reach $1 million per violation. Civil fines may be imposed in addition to other authorized civil, criminal, and administrative remedies. Criminal prosecutions are typically based on claims of wire fraud, mail fraud, or submission of false statements to a federal agency. These actions are described below.

Government-wide Debarment and Similar Exclusions

Debarment is an extended exclusion from government grants and contracts, while suspension is a temporary exclusion. Both are viewed by federal agencies as serious actions to be used only to protect the federal government's interests and are not considered to be punishment (7). The actions nevertheless have decidedly punitive effects.

In order to avoid providing government support to anyone found guilty of serious misconduct, President Reagan in 1986 ordered that the debarment of an individual or institution by one agency should have government-wide effect. The order applies as well to suspensions, disqualifications, and "voluntary exclusion agreements," which the agency negotiates with individuals or entities who are willing to settle misconduct charges to avoid the cost and disruption of hearings and appeals. Accordingly debarment of a scientist or technician for research misconduct prevents that individual from receiving research support from any federal agency for the period of debarment (8). The same restrictions apply to any research institution or corporate entity that has been suspended or debarred. As of September 1997, however, no institution had been debarred as a result of a finding by the Department of Health and Human Services (HHS) Office of Research Integrity (ORI) confirming scientific misconduct (9). A suspended, debarred, disqualified, or excluded scientist may not even participate in another scientist's federally funded research without special permission from the funding agency. Periods of debarment or exclusion for scientific misconduct typically range from 3 to 10 years but sometimes are permanent (10). A consolidated list of all agency suspensions, debarments, disqualifications, and voluntary exclusions is maintained by the General Services Administration (GSA) for enforcement purposes and is available to the public (11).

Publication of Misconduct Findings

The PHS, FDA, and NSF policies differ in their approach to publicizing misconduct findings. The PHS publishes the names of the scientists and their research institution, together with a brief summary of their misconduct and the sanctions imposed. These notices appear in both hard copy and Internet versions of the *Federal Register*, the *NIH Guide to Grants and Contracts*, and the *ORI Newsletter*. The PHS also may notify state licensing boards (if the scientist involved is a licensed health practitioner), professional associations, and journals in which the scientist has published reports of past research. FDA publishes notices of disqualifications and debarment in the *Federal Register* and also may notify sponsors of products being tested and collaborating institutions. The list of investigators who are ineligible to receive investigational drugs, or whose use of investigational products is limited, is available to the public. By contrast, NSF publishes summaries of misconduct findings and sanctions, but it does not identify either the scientists or the institutions involved (12). The summaries of NSF cases are included in semiannual reports to Congress from the NSF Inspector General.

Retraction or Correction of Publications

Both PHS and NSF require a formal correction or retraction of journal articles found to contain fabricated, falsified, or plagiarized material. Although previously these notices were submitted as letters to the editor, most biomedical journal editors now agree that retractions should be labeled as such and appear independently on a numbered page of the journal, in order to include references to the retraction or correction in standard bibliographies (13). The National Library of Medicine, for example, annotates in its computerized databases (e.g., MEDLINE) articles that have been corrected or retracted, and provides a citation to the withdrawal or correction notice (14). This practice was challenged in 1994 by a scientist who was then under investigation for scientific misconduct as a result of patients improperly enrolled, by a collaborating researcher, in a multicenter breast cancer clinical trial. The principal investigator challenged the Library of Medicine's annotations of numerous articles from the collaborative trials but was rebuffed by a federal district judge, who ruled that the entries in the Library's databases pertained to publications, not to their authors (15). The district court's ruling was affirmed by a federal appellate court and motions for reconsideration were denied. In a letter to the editor in *Science*, the acting ORI director emphasized that the annotations had not been added to the Library's databases until after there had been a formal finding that the collaborating researcher had committed scientific misconduct. He added: "Scientists should not be concerned that annotations have been in the past or will be in the future placed in databases before a misconduct investigation is completed. They have not and will not be" (16).

Recoupment of Government Funds

Federal agencies have authority to require that institutions return any public funds that have been misused. In the context of research grants, this is typically accomplished by asserting that the funds in question were used improperly, and therefore the institution was not entitled to them and must refund the money (17). When ORI closes a case with a finding of scientific misconduct, it reports its findings to the institute at NIH that awarded the grant or contract. NIH in turn may seek recoupment of the research funds involved. In 1995, for example, NIH recovered $296,478 from an institution after ORI found that a principal investigator had submitted progress reports for three years describing research he had not performed. In 1994 NIH recovered over $1 million from three institutions involved in two scientific misconduct cases. NIH recoveries of research funds are actions independent of ORI's and are not routinely reported in the *ORI Newsletter*.

In late 1996 the Department of Justice sued for restitution of over $100 million from the University of Minnesota which allegedly obtained research grants fraudulently from NIH and illegally sold an antirejection drug that FDA had not approved for marketing (18). In July 1997 a U.S. district judge reduced the amount at stake from $109 million to $60 million by dismissing the False Claims Act portions of the suit (19). That ruling was reversed on appeal (20), and the university ultimately paid $32 million to settle the case (21).

At NSF, recoupments of research support are regularly reported in the Inspector General's Semiannual Reports to Congress and commonly result from findings of scientific misconduct. Restitution often results from criminal or civil litigation but may also result from internal agency determinations.

CIVIL MONEY PENALTIES

In addition to the administrative actions described above, scientists and their employers may be subject to civil money penalties and recoupment of publicly funded research support for offenses related to the development and testing of new drugs and devices. Other agencies have similar authority.

Program Fraud Civil Remedies. The Program Fraud Civil Remedies Act of 1986 was designed to deal with the submission of false statements to the government involving claims of less than $150,000. Under that Act, anyone submitting a claim or statement to a government agency, with knowledge that the claim or statement is false, fictitious, or fraudulent (or acting in deliberate ignorance or reckless disregard of the truth or falsity of the claim), is subject to a civil penalty of up to $5,000 for each such claim and twice the amount of each claim (22).

False Claims Act. If more than $150,000 is involved, the government may proceed under the False Claims Act, 31 U.S.C. §§3729–3730, which authorizes civil fines up to $10,000 plus recoupment of three times the amount of damages suffered by the agency as a result of the false claims. Alternatively, if the person who submitted the false claim cooperates with the government's investigation, the amount assessed may be reduced to two times the amount of damages. Under this provision universities and other research entities have been induced to cooperate and plea bargain, to avoid the treble damages.

In 1994, for example, the University of Utah and the University of California agreed to repay NIH more that $1.5 million in grants allegedly obtained through false data submitted in the grant applications. The universities' alternative to settlement was to risk treble damages (totaling $3.6 million) under the False Claims Act for knowingly presenting a false or fraudulent claim for payment to a government agency (23). Federal research grants are within the Act's definition of a "claim."

Qui Tam Actions. The "qui tam" provisions of the False Claims Act permit private citizens to bring suit in the name of the United States to recover funds paid out by the government of the basis of fraudulent claims. "Qui Tam" means "who as well ..." and denotes actions initiated by private citizens or informers who sue on behalf of the government as well as for themselves. In return for prosecuting the case, or at least alerting the government to the false claims, the informer (called a "relator") is entitled to a significant portion of the amount recovered. If the government successfully prosecutes the case, the relator may receive between 15 and 25 percent of the damages recovered. If the government declines to participate in the litigation, the relator who litigates in the name of the United States — and wins — is entitled to receive up to 30 percent of the damages awarded, in addition to costs and reasonable attorney fees. With damages trebled and potentially reaching millions of dollars, the informer's share can be sizable. Critics of the qui tam provisions say that the process offers an opportunity to settle personal scores while, at the same time, collecting a windfall and posing as a public citizen.

The qui tam provisions were enacted during the Civil War in response to sales of defective supplies to the Union Army. Amendments to the law in 1986 strengthened the role of relators and resulted in a surge of qui tam cases over the next decade. In fact, recoveries under the False Claims Act increased from $2 million in 1988 to over $200 million in fiscal 1995 (24). At the same time the portion of qui tam cases involving fraud related to HHS has surpassed those at the Department of Defense, which dominated the field in the past. HHS fraud cases involve primarily Medicare, Medicaid, and similar third parties who pay for health care services and supplies (24). A growing segment, however, relates to allegations of scientific misconduct in NIH-supported research activities.

A recent case illustrates how the qui tam law operates. In 1994 a former graduate student from Cornell, Pamela Berge, filed a qui tam suit against the University of Alabama, Birmingham, and four of its faculty members for allegedly submitting false statements in grant applications to NIH. A jury returned a verdict favorable to the informer/relator, which resulted in a judgment of just under $2 million (plus costs and attorney fees). Berge's claims were based on allegations of plagiarism, or misappropriation of intellectual property, which had been investigated and found to be meritless by a series of academic, scientific, and administrative reviewers. She therefore transformed her plagiarism charges into a qui tam action on behalf of the United States, asserting that a series of annual reports and grant applications submitted to NIH by the university incorporated plagiarized material and therefore constituted multiple false claims. Following the jury verdict, a federal district court judge awarded Berge 30 percent of the $1.6 million judgment, plus costs and attorney fees. The judgment subsequently was overturned by the Fourth Circuit Court of Appeals (25), which found no evidence on which a reasonable jury could have concluded that the challenged statements were even false, much less the basis for any NIH-funding decisions. The appellate court also ruled that there was no plagiarism and that Berge's claim for misappropriation of intellectual property was preempted by the U.S. Copyright Act. The United States Supreme Court declined Berge's petition for review; thus the Fourth Circuit Court opinion stands as the last word in this case.

The likelihood that personal grievances will generate False Claims Act litigation is increased by the bounty-hunter (qui tam) provisions and the publicity attending successful cases. In addition a Washington-based group called Taxpayers Against Fraud actively solicits and supports potential qui tam plaintiffs, with a Web site on the Internet offering referrals to counsel, loans for litigation, help with legal research, and production of amicus (friend of the court) briefs. Supporters of qui tam view this as a public service, while opponents see disgruntled employees and students being encouraged to file multimillion dollar claims in the name of good citizenship (24). A similar support group has been established in Michigan to encourage and assist "whistleblowers" alleging scientific misconduct more generally. The likelihood is that agency actions and qui tam litigation both will increase in the foreseeable future.

CRIMINAL PROSECUTION

Statutory Penalties

Submitting false statements or information to the federal government, and similar offenses involving deceit, false statements, or fabrication, may result in criminal prosecution. It is a felony knowingly to submit a false statement or false claim to a government agency (26). A statement may be false either through omission (i.e., failure to disclose a material fact) or through submission of a false statement or representation. Conviction may result in incarceration for up to five years, debarment, imposition of fines, and recoupment of grant or contract monies.

When a scientist submits false statements in an application for a research grant, or in an annual report to the granting agency, that constitutes a false claim (request for money) and is punishable under the criminal False Claims Act by imprisonment for up to five years, a fine, or both (27). When the false claim is submitted by telephone or through the mail, the scientist and research institution may also be charged with wire fraud or mail

Table 2. Penalties for Scientific Misconduct: Federal Statutes

False claims and statements	Penalties
Administrative remedies for false claims and statements, 31 U.S.C. §§3801–3802 (applies to claims less than $150,000)	$5000 for each false claim, plus up to twice the amount of the claim
False Claims Act (civil actions), 31 U.S.C. §3729 (false claims exceeding $150,00)	$10,000 for each false claim, plus treble amount of each claim
False claims actions by private persons (qui tam), 31 U.S.C. §3730(d); Informer ("relator") may receive up to 30% of damages, plus attorneys' fees and costs	$10,000 for each false claim, plus treble the amount of each, plus costs and attorneys' fees
False, fictitious, or fraudulent claims (criminal), 18 U.S.C. §287	Fines and/or imprisonment up to 5 years
False statements or entries (criminal), 18 U.S.C. §1001	Fines up to $10,000 and/or imprisonment up to 5 years

Related crimes	Penalties
Conspiracy, 18 U.S.C. §371	Depends on underlying violation(s)
Mail fraud, 18 U.S.C. §1341 and Wire fraud, 18 U.S.C. §1343	Fines (up to $250,000, depending on amount of fraud), restitution, and/or imprisonment up to five years

Presidential order (Ronald Reagan)	Penalties
Debarment and suspension (government-wide effect), Executive Order No. 12549; *Fed. Regist.* **51**, 6370 (1986)	Individual or entity debarred or excluded by one federal agency may not receive grants or contracts from any federal agency

Food, Drug, and Cosmetic Act	Penalties
Withdraw approval of abbreviated drug applications, 21 U.S.C. §335c	Approval withdrawn; drug may not be marketed
Civil penalties (fines), 21 U.S.C. §335b; (informants may receive $250,000 or one-half of penalty, whichever is less)	Fine up to $250,000 for individuals; fine up to $1 million for corporate entities, partnerships, etc.
Debarment, and suspension; also, temporary denial of approval, 21 U.S.C. §335a	For corporation, partnership, etc., exclusion from research involving investigational products for 1–10 years; for individual convicted of felony related to product development, permanent exclusion

Public Health Service Act	Penalties
Office of Research Integrity, 42 U.S.C. §289b	Limitations on use of grant funds, supervision, suspension/termination of grant, debarment (exclusion) from future grants/contracts

fraud, and if more than one person is involved, a conspiracy charge may be added as well (28). Nothing in the law precludes prosecution under multiple federal statutes, and each grant application, status report, or request for payment constitutes a separate claim and, thus, an additional offense.

Examples of Cases

The first scientist to be indicted for research fraud was Stephen Breuning, a psychologist who fabricated data in federally supported studies on the use of stimulant drugs to treat hyperactivity in retarded children. Following indictment for fraud and making false statements to the National Institute of Mental Health, Breuning entered into a plea agreement which resulted in a conviction. Because of the potential impact of his fraud on treatment decisions for vulnerable individuals, Breuning was sentenced to incarceration for two years (suspended except for 60 days on work-release in a halfway house), five years' probation (during which he was required to perform 200 hours of community service), exclusion from federal research support for 10 years, and reimbursement of $11,352 to his university (29,30). He was also forbidden to work as a clinical psychologist for 10 years.

In a case that involved research submitted to FDA but did not involve federal research support, Dr. Robert Fogari, a physician who fabricated results over a period of eight years in a study of investigational anti-inflammatory drugs, was convicted of criminal fraud and obstruction of justice, sentenced to four years in jail, fined over $3.8 million, and ordered to make restitution (29).

Another case, that of Barry Garfinkle, demonstrates the multiple penalties that may be imposed for misconduct in clinical trials of new drugs (31). Dr. Garfinkle was convicted in 1993 of mail fraud and making false statements to the FDA while serving as principal investigator in studies of Anafranil. The drug was being tested as a treatment of obsessive-compulsive disorder in children and adolescents. The prosecution was triggered by complaints from the study coordinator that Garfinkle had ordered her to enter false data about weekly clinical evaluations that either never took place or were conducted by the coordinator rather than a physician. In addition the study coordinator alleged serious breaches of the research protocol. The indictment included charges under the Food, Drug, and Cosmetic Act, False Claims Act, and statutes prohibiting mail fraud.

Garfinkle was sentenced to six months in a halfway house, followed by six months of home detention, 400 hours of community service, and over $200,000 in fines. Based on his conviction of multiple felonies related to drug development, Garfinkle also was permanently debarred by FDA from serving in any capacity in connection with a new drug application. In addition the FDA served notice that it would not accept or review any abbreviated drug applications prepared by, or with the assistance of, Dr. Garfinkle, and that any person with a pending or approved drug application who knowingly used Dr. Garfinkle's services would be subject to a civil money penalty.

Such prosecutions are not limited to research involving NIH or FDA. In 1993, for example, a federal appellate court upheld the conviction of a scientist for conspiracy and fraud in connection with a grant, funded by the Agency for International Development, to create a diagnostic field test for malaria (32). Following a jury trial, the researcher was convicted and sentenced to eight months in prison (five of which were suspended), three years probation, and was ordered to make restitution in the amount of $75,000. The NSF also, together with the U.S. Attorney's Office, has successfully prosecuted individual scientists and biotech companies for fraud and false statements relating to Small Business Innovative Research (SBIR) grants. Within a six month period in 1996, these prosecutions resulted in criminal fines, civil penalties, restitutions, and other savings amounting to over $6 million (33).

DAMAGE TO PERSONAL HEALTH AND PROFESSIONAL REPUTATION

Perhaps the most devastating penalties are the effects of scientific misconduct allegations on the personal and professional life of the accused. Researchers who have been accused and later exonerated report that the investigative process alone has had prolonged and significant adverse effects on their lives (34). Approximately three-fifths of the exonerated scientists who responded to a 1996 survey believed they were stigmatized by the accusations, and nearly 40 percent reported adverse effects on their professional careers, such as damage to their reputation, reduced job mobility, and diminished opportunities for presenting papers. Over three-quarters of the respondents reported negative effects on their mental health, and nearly half reported adverse effects on their physical health. Disruptions of family relationships are not unusual. These outcomes, reported by scientists who were ultimately exonerated, suggest that even more serious personal and professional consequences must follow confirmation of scientific misconduct. Scientists who are found guilty of scientific misconduct, however, have not been surveyed. In many instances, they seem simply to have left the scene.

A widely publicized example of a scientist accused and later cleared is the case of Thereza Imanishi-Kari, who coauthored a paper in 1986 with Nobel prize-winning scientist David Baltimore (35). Imanishi-Kari was accused by a coworker of faking her data. Baltimore was never accused of scientific misconduct, but his name was linked invariably with that of Imanishi-Kari in the scientific and lay press, as well as in congressional hearings. Both scientists suffered personally and professionally throughout a 10 year investigation, although neither ultimately was found to have committed scientific misconduct. Imanishi-Kari's faculty status was suspended and, with it, her eligibility for NIH grants. Baltimore was ultimately forced from his position as President of Rockefeller University and was ostracized for years by many members of the scientific community. In 1997, with his reputation in recovery, he was appointed President of California Institute of Technology. Other scientists, perhaps less conspicuously, have suffered a similar damage to reputation, personal anguish, curtailed career, and diversion of significant emotional, intellectual, and financial resources to defend against the accusations.

Occasionally the accusation of research fraud has led to even worse tragedy. Paul Kammerer, for example, was an Austrian scientist who believed he had developed proof of the heritability of acquired characteristics. As described by Arthur Koestler in *The Case of the Midwife Toad*, Kammerer's controversial support of the Lamarckian theory of inheritance was challenged repeatedly by the followers of Darwin. One vigorous critic ultimately accused Kammerer of having faked his results. Although essential facts about the research remain murky, the effect of the accusation is clearly documented. Despairing of ever proving his innocence to the satisfaction of his critics, Kammerer went into the Vienna woods and shot himself. A note found in his pocket stated:

> Dr. Paul Kammerer requests not to be transported to his home, in order to spare his family the sight. Simplest and cheapest would perhaps be utilization in the dissecting room of one of the university institutes. I would actually prefer to render science at least this small service. Perhaps my worthy academic colleagues will discover in my brain a trace of the quality they found absent from the manifestations of my mental activities while I was alive (36, p. 13).

Kammerer's suicide occurred in 1926. Six decades later, a professor of neurology and neurosurgery at the Montreal Neurological Institute (affiliated with McGill University), together with her husband (a faculty member at another university), committed suicide following publication of anonymous allegations that she had committed research fraud (37). As her lawyer explained: "Given that her work was her life, and she felt that her ability to continue was being seriously undermined, it was obviously more than she could live with" (38).

BIBLIOGRAPHY

1. 42 U.S.C. §289b, as amended June 10, 1993, Pub. L. 103-43, Tit. I, §§161, 163, 107 Stat. 140, 142. NIH regulations are found at 42 CFR, pt. 50; NSF regulations are at 45 CFR, pt. 689. ("CFR" refers to the Code of Federal Regulations.)

2. Public Health Service, *Fed. Regist.* **56**, 27384–27391 (1991).

3. 21 CFR §§58.200, 312.70 (drugs), 812.30(b) (devices).

4. 21 U.S.C. §335a.

5. 21 U.S.C. §§335a(c), 355(e)(5), 360 e(e) (1994).

6. 21 U.S.C. §335b.

7. HHS regulations, 45 CFR §76.115(b).

8. 48 CFR Subparts 9.4 (grants) and 309.4 (contracts); 45 CFR, pt. 76 (PHS); 45 CFR, pt. 620 (NSF).

9. Office of Research Integrity, *ORI Newsl.* **5**(4), 4–5 (1997).

10. 21 U.S.C. §335a(2).

11. 45 CFR §§76.500–76.510. Available at: *www.ARNET.gov/epls/*

12. D.P. Hamilton, *Science* **255**, 1346 (1992).

13. International Committee of Medical Journal Editors, *J. Am. Med. Assoc.* **277**, 927–934 (1997).

14. National Library of Medicine, *Fact Sheet: Errata, Retraction, and Comment Policy*, NLM, Washington, DC, 1989. Available at: *www.nlm.nih.gov/pubs/factsheets/errata.html*

15. *Fisher v. NIH et al.*, 934 F.Supp. 464; *aff'd. per curium* 107 F.3d 922 (1996); *reh'g denied* (D.C. Cir. March 28, 1997). The rulings are described in *ORI Newsl.* **2**, (Sept. 1996) **7** (June 1997).

16. C. Pascal, *Science* **274** (1996).

17. National Institutes of Health (NIH), *NIH Guide to Grants and Contracts* p. 24, (March 10, 1995), Note 2, citing 45 CFR §74.27 *et seq.*

18. *United States ex rel. Zissler v. Regents of the University of Minnesota*, 992 F.Supp. 1097 (D. Minn. 1998).

19. J. Basinger, *Chron. Higher Educ.* August 8, p. A36 (1997).

20. *United States ex rel. Zissler v. Regents of the University of Minnesota*, 1998 WL 560138 (8th Cir. Sept. 4, 1998).

21. U.S. Department of Justice, *University of Minnesota Pays $32 Million to Settle Allegations of Selling an Unlicensed Drug and Mishandling NIH Grant Funds*, press release, November 17, 1998.

22. 31 U.S.C. §3802.

23. 31 U.S.C. §3729.

24. P.R. Budeiri, *The Washington Lawyer*, September/October, 24–29 (1996).

25. *United States ex rel. Berge v. Board of Trustees of the University of Alabama, Birmingham, et al.*, 104 F.3d 1453 (4th Cir.); *cert. denied*, 522 U.S. 916, 118 S. Ct. 301 (1997).

26. 18 U.S.C. §1001 (1994).

27. 18 U.S.C. §287.

28. 18 U.S.C. §§1341, 1343, 371.

29. L.S. Little, *Food Drug Cosmetic Law J.* **46**, 65–69 (1991).

30. F.A. Kaufman, *United States v. Stephen E. Breuning*, Criminal No. K-88-0135 — An example of the applicability of the federal criminal legal process to fraudulent scientific research, in AAAS-ABA National Conference of Lawyers and Scientists, *Project on Scientific Fraud and Misconduct: Report on Workshop Number Three*, AAAS-ABA, Washington, DC, 1989, pp. 165–169.

31. *United States v. Barry Garfinkle*, 29 F.3d 1253 (8th Cir. 1994); Food and Drug Administration, *Fed. Regist.* **62**(63), 15713 (1997).

32. *United States v. Kellerman*, 992 F.2d 177 (8th Cir. 1993).

33. NSF, Office of Inspector General, *Semiannual Report to Congress*, No. 14, for the period Oct. 1, 1995–March 31, 1996, pp. 35–42, 52–54.

34. Research Triangle Institute, *Survey of Accused but Exonerated Individuals in Research Misconduct Cases*, a report submitted to the HHS Office of Research Integrity, Research Triangle Institute, Research Triangle Park, NC, 1996, pp. 45, 53.

35. D.J. Kevles, *The Baltimore Case*, Norton, New York, 1998.

36. A. Koestler, *The Case of the Midwife Toad*, Random House, New York, 1971.

37. L. Ratelle, *Chron. Higher Educ.* April 27, p. A36 (1994).

40. *Id.*, quoting Eric Maldoff, a Montreal attorney.

ADDITIONAL READINGS

Federal Agency Actions, Publications

Food and Drug Administration, Barry D. Garfinkel; Denial of hearing; Final debarment order. *Fed. Regist.* **62**(63), 15713 (1997).

National Institutes of Health Responsibilities of NIH and awardee institutions for the responsible conduct of research. *NIH Guide Grants Contracts* (March 10, 1995), p. 24, Note 2.

National Library of Medicine, *Fact Sheet: Errata, Retraction, and Comment Policy*, NLM, Washington, DC, 1989. Available at: *www.nlm.nih.gov/pubs/factsheets/errata.html*

National Science Foundation, Office of Inspector General, *Semiannual Reports to Congress* (published twice yearly), NSF, Washington, DC.

Office of Research Integrity (HHS), University of Minnesota pays $32 million in qui tam suit; Drops second lawsuit. *ORI Newsl.* December 1998, p. 5.

Ronald Reagan, Executive Order 12549: Debarment and Suspension, *Fed. Regist.* **51**, 6370 (1986), codified at 3 CFR, pts. 100 and 101, Executive Orders (1986 compilation).

U.S. Department of Justice, *University of Minnesota Pays $32 Million to Settle Allegations of Selling an Unlicensed Drug and Mishandling NIH Grant Funds*, Press Release, November 17, 1998.

U.S. Public Health Service, *Policies and procedures for dealing with possible scientific misconduct in extramural research. Fed. Regist.* **56**, 27384–27391 (1991).

Cases

Fisher v. NIH et al., 934 F.Supp. 464, *aff'd. per curium* 107 F.3d 922 (1996), *rehearing denied* (D.C. Cir. March 28, 1997).

United States ex rel. Berge v. University of Alabama et al., 104 F.3d 1453 (4th Cir.), *cert. denied*, 118 S. Ct. 301 (1997).

United States ex rel. Zissler v. University of Minnesota, 154 F.3d 870 (8th Cir. 1998).

United States v. Garfinkle, 822 F.Supp. 1457 (D. Minn.), *remanded*, 29 F.3d 451 (8th Cir. 1994), *aff'd.* 29 F.3d. 1253 (8th Cir. 1994).

United States v. Kellermann, 992 F.2d 177 (8th Cir. 1993).

Miscellaneous

AAAS/ABA National Conference of Lawyers and Scientists, *Project on Scientific Fraud and Misconduct: Report on Workshop Number Three*, AAAS, Washington, DC, 1989.

Agnew, B., False claims: Alabama wins, but stakes soar in Minnesota case. *J. NIH Res.* **9**, 30–31 (1997).

Basinger, J., Judge reduces amount University of Minnesota could have to pay in federal lawsuit. *Chron. Higher Educ.* August 8, 1997, p. A36.

Budeiri, P.R., The return of qui tam. *Washington Lawyer*, September/October, 1996, pp. 24–29.

Herman, K.G. et al., Investigating misconduct in science: The National Science Foundation model. *J. Higher Educ.* **65**(3), 384–400 (Appendix) (1994).

Kevles, D.J., *The Baltimore Case*, Norton, New York, 1998.

Koestler, A., *The Case of the Midwife Toad*, Random House, New York, 1971.

Little, L.S., The role of the HHS Office of Inspector General in the investigation of scientific misconduct. *Food, Drug, Cosmetic Law J.* **46**(1), 65–69 (1991).

Pascal, C.B., Misconduct annotations. *Science* **274**, (November 15 (1996).

Ratelle, L., Montreal researcher commits suicide following allegations of fraud. *Chron. Higher Educ.* April 27 (1994), p. A36.

Research Triangle Institute, *Survey of Accused But Exonerated Individuals in Research Misconduct Cases*, Report submitted to the HHS Office of Research Integrity, June 30, 1996.

Singer, M., Assault on science. *Washington Post*, June 26, 1996, p. A21.

See other entries SCIENTIFIC RESEARCH, ETHICS, SCIENTIFIC MISCONDUCT; SCIENTIFIC RESEARCH, ETHICS, VALUES IN SCIENCE.

SCIENTIFIC RESEARCH, POLICY, TAX TREATMENT OF RESEARCH AND DEVELOPMENT

DAVID L. CAMERON
Willamette University College of Law
Salem, Oregon

OUTLINE

Introduction
The Deduction of Research and Experimental Expenditures Under Section 174
  Overview
  The "In Connection With" Standard
  Definition of Research and Experimental Expenditures
  Section 174 Election
Credit for Increasing Research Activities Under Section 41
  Overview
  Incremental Research Tax Credit
  Alternative Incremental Research Tax Credit
  Basic Research Tax Credit
Limitations on Section 174 Deductions
Orphan Drug Credit Under Section 45C
Bibliography
Additional Reading
  Generally
  Research and Development Expenditures Under Section 174
  The Research Tax Credit Under Section 41

INTRODUCTION

Research and development (R&D) in the field of biotechnology typically requires a significant investment before any marketable products may be produced. In an effort to encourage greater private sector investment in research and development activities, Congress has enacted several provisions under the Internal Revenue Code, including Sections 174, 41, and 45C, that authorize tax benefits for certain types of expenditures incurred in connection with R&D activities. Specifically, Section 174 provides a current tax deduction for research or experimental expenditures while Section 41 permits a tax credit for increases in research expenditures from one tax year to another.

Importantly, a taxpayer can take advantage of both the Section 174 deduction and the Section 41 tax credit with respect to many of the same research expenditures, subject only to the limitations of Section 280C. In addition Section 45C permits a tax credit for qualified clinical testing expenses incurred in the development of so-called orphan drugs. This article explores the requirements of Sections 174, 41, and 45C, as well as the limitations under Section 280C, and the implications of these statutory provisions in the context of R&D activities in biotechnology.

THE DEDUCTION OF RESEARCH AND EXPERIMENTAL EXPENDITURES UNDER SECTION 174

Overview

Under Section 174 a taxpayer may elect to deduct currently all research and experimental expenditures made in connection with the taxpayer's trade or business (1) or to amortize the expenditures over a period of not less than 60 months (2). If the taxpayer fails to treat the expenditures under one of these methods, then the research and experimental expenses are to be capitalized (3). The decision either to deduct Section 174 expenses currently or to defer and amortize them constitutes the adoption of an accounting method, which cannot be changed without the consent of the Commissioner of the Internal Revenue Service (IRS) (4). Although the current deduction of research and experimental expenditures is permitted for regular income tax purposes, for purposes of the alternative minimum tax, research and experimental expenditures must be capitalized and amortized ratably over a 10-year period beginning with the taxable year in which the expenditures were made unless the taxpayer materially participates in the activity within the meaning of Section 469(h) (5).

The "In Connection With" Standard

Importantly Section 174 applies to research or experimental expenditures paid or incurred "in connection with" a trade or business and thus is available to taxpayers who are not yet engaged in a trade or business. In this way the "in connection with" standard of Section 174 is distinguishable from, and less stringent than, the "carrying on" standard of Section 162, which allows the deduction of trade or business expenses more generally. In *Snow v. Commissioner* (6), the Supreme Court held that Section 174 does not require that the taxpayer actually be carrying on a trade or business in order to deduct otherwise qualified research and experimental expenditures. The Supreme Court based its decision on statements contained in the legislative history that Section 174 was intended to equalize the tax treatment of "small and growing businesses" with their "large and well-established competitors" (7). As a result a start-up business may deduct research and experimental expenditures under Section 174. Ordinarily, a start-up business must capitalize its start-up costs and amortize them over not less than a five-year period (8).

The fact that Section 174 applies to allow the deduction of research or experimental expenditures by a start-up business does not completely obviate the requirement that a trade or business exist, however. The courts have concluded that research and experimental expenditures are not deductible under Section 174 if the taxpayer does not have some realistic prospect of entering into a trade or business involving the technology under development or, in fact, never eventually enters into the active conduct of a trade or business. For example, in *Harold J. Green* (9), the Tax Court stated:

> Although the Supreme Court established in *Snow* that the taxpayer need not currently be producing or selling any product in order to obtain a deduction for research expenses, it did not eliminate the "trade or business" requirement of section 174 altogether. For section 174 to apply, the taxpayer must still be engaged in a trade or business *at some time*, and we must still determine, through an examination of the facts of each case, whether the taxpayer's activities in connection with a product are sufficiently substantial and regular to constitute a trade or business for purposes of such section (10).

The requirement that deductions under Section 174 be incurred "in connection with" a trade or business also means that Section 174 is unavailable to taxpayers who merely fund the research of a third party for the development of a product when the taxpayer does not possess the intent or ability to exploit the fruits of that research on its own. These arrangements frequently involve partnerships that enter into agreements to fund research and simultaneously lease or license the rights of that research to the developing party. In such cases the courts closely scrutinize the business arrangements involved and have frequently held that the partnership is not sufficiently engaged in a trade or business to satisfy the requirements of Section 174 (11). This determination, however, must be based on an examination of all the facts of the particular situation (12). As the Tax Court has stated:

> [W]hen a partnership contracts out the performance of the research and development in which it intends to engage, all of the surrounding facts and circumstances are relevant to the inquiry into whether it has any realistic prospect of entering into a trade or business with respect to the technology under development. The inquiry includes consideration of the intentions of the parties to the contract for the performance of the research and development, the amount of capitalization retained by the partnership during the research and development contract period, the exercise of control by the partnership over the person or organization doing the research, the existence of an option to acquire the technology developed by the organization conducting the research and the likelihood of its exercise, the business activities of the partnership during the years in question, and the experience of the partners. Absent a realistic prospect that the partnership will enter a trade or business with respect to the technology, the partnership will be treated as a passive investor, not eligible for deductions under section 174 (13).

For example, in *Kantor v. Commissioner* (14), the Ninth Circuit Court of Appeals affirmed the decision of the Tax Court denying the taxpayer's claimed deductions under Section 174. In *Kantor*, PCS, Ltd. was formed to fund the adaptation of a particular computer software program for use on various types of computer systems. PCS, Ltd. entered into an R&D agreement with PCS, Inc. under which PCS, Inc. agreed to perform the research involving the software program with PCS, Ltd. retaining ownership of the resulting software. The parties also entered into a technology transfer agreement under which PCS, Ltd. granted PCS, Inc. an option to acquire an exclusive worldwide license to market the software on payment of $5,000 and future royalties based on prospective sales. PCS, Ltd. also granted PCS, Inc. a 13-month review period in which to exercise its option.

The Ninth Circuit agreed with the Tax Court that PCS, Ltd. did not possess the objective intent nor the capability of marketing any software that might be developed. Importantly the court noted that by granting PCS, Inc. the right to market the software on payment of a nominal fee, PCS, Ltd. "made it more probable than not that the research firm [PCS, Inc.] would exercise these rights if the software that resulted from the research was at all valuable" (15). The court reached its conclusion that PCS, Ltd. did not posses the capability to market the software despite the fact that the general partner of PCS, Ltd. had significant sales and technical experience in the computer industry, was actively engaged in the research effort, negotiated an arrangement to secure financing, and negotiated licensing agreements for the marketing of the software. Rather, the court likened the general partner's activities to "those of any investor who applies his knowledge and expertise to insure that an investment is successful..." (16). The court concluded that permitting a deduction in such circumstances would simply allow the taxpayer to use Section 174 to deduct its capital investment.

Nevertheless, in *Scoggins v. Commissioner* (17), the Ninth Circuit Court of Appeals concluded that a research partnership was entitled to deductions under Section 174 despite the existence of facts similar to those in *Kantor*. In *Scoggins*, B&B Research and Development Partnership (B&B) contracted with Epitaxy Systems, Inc. (Epitaxy) to perform research to develop an epitaxial reactor. B&B agreed to provide Epitaxy with up to $500,000, consisting of $43,000 to be paid at the time of executing the agreement and additional amounts to be paid at the partnership's discretion. B&B granted Epitaxy a 15-month nonexclusive license to market the technology in return for royalties of 20 percent on net revenue from sales during that period. B&B also granted Epitaxy an option exercisable for one year thereafter to purchase the developed technology for $5 million.

The Ninth Circuit reversed the decision of the Tax Court and permitted the taxpayer's deductions under Section 174. Significantly the two partners in B&B, who also owned the controlling interests in Epitaxy, had previously developed and successfully marketed epitaxial reactors. The court concluded that B&B therefore possessed the technical expertise and financial ability to conduct a trade or business using the developed technology. The court also noted that the license granted to Epitaxy was nonexclusive and that Epitaxy was under no obligation to market the technology. Significant, too, was the amount of the payment to acquire the technology, $5 million, compared to the research cost. The court specifically noted that this fee was a significant impediment to Epitaxy's ability to engage in the marketing of the technology and contrasted this payment to the nominal amount to be paid in *Kantor*.

Definition of Research and Experimental Expenditures

For purposes of Section 174, "research or experimental expenditures" are limited to those expenditures incurred in connection with the taxpayer's trade or business that "represent research and development costs in the experimental or laboratory sense" (18). The regulations expand on this definition through the adoption of an "uncertainty test," conditioned on the quality of the information available to the taxpayer at the time that the expenditures are undertaken. The regulations provide that

> [e]xpenditures represent research and development costs in the experimental or laboratory sense if they are for activities intended to discover information that would eliminate uncertainty concerning the development or improvement of a product. Uncertainty exists if the information available to the taxpayer does not establish the capability or method for developing or improving the product or the appropriate design of the product. Whether expenditures qualify as research or experimental expenditures depends on the nature of the activity to which the expenditures relate, not the nature of the product or improvement being developed or the level of technological advancement the product or improvement represents (19).

Under the regulations, the word "product" is defined to include "any pilot model, process, formula, invention, technique, patent, or similar property, and includes products to be used by the taxpayer in its trade or business as well as products to be held for sale, lease, or license" (20).

While considerable uncertainty exists with respect to the precise contours of the definition of "research or experimental expenditures," the regulations are clear that the costs of obtaining a patent, including such expenditures as attorneys' fees expended to make and perfect a patent application, are included within the definition (21). Because the costs of perfecting a patent are "inextricably a part of the research and development work," such costs reasonably fall within the definition of expenditures for research or experimentation (22).

The application of Section 174 to the development of property protected by copyright law, as opposed to patent law, requires additional consideration. As previously described, research or experimental expenditures must "represent research and development costs in the experimental or laboratory sense" (23). Because such expenditures are limited to the reasonable costs incident to the development or improvement of a product including any pilot model, process, formula, invention, technique, or similar property (24), this definition by itself might exclude from Section 174 any research and development costs typically incurred in creating a work subject to copyright protection. In addition the regulations explicitly state that expenditures incurred for "[r]esearch in connection with literary, historical, or similar projects" do not fall within the definition of research or experimental expenditures (25). In Revenue Ruling 72-395 (26), the IRS expanded on this regulatory provision and concluded that Section 174 did not apply to permit the current deduction of the costs of writing and editing, as well as design and art work for, textbooks and visual teaching aids (27). Importantly, however, the exclusion for literary, historical, and similar projects does not apply to certain expenditures incurred in connection with the development of computer software despite the fact that computer software is frequently protected under the copyright laws (28).

Although the precise contours of the definition of research or experimental expenditures are not fully described under the regulations or in other IRS pronouncements, the regulations are clear that certain costs incurred in the development of a product do not qualify as research or experimental expenditures under Section 174. For example, costs incurred for ordinary testing or inspection of materials for purposes of quality control — in addition to costs incurred for efficiency surveys, management studies, consumer surveys, advertising, or promotions — do not qualify as research or experimental expenditures (29). The regulations also specify that research or experimental expenditures do not include the acquisition costs of another's "patent, model, production or process" (30). The costs of materials or labor used in the construction, installation, acquisition, or improvement of property are also excluded from the definition of research or experimental expenditures (31). In addition Section 174 does not apply to expenditures for the acquisition or improvement of land or depreciable property (32). However, allowances for depreciation may be considered research or experimentation expenditures under Section 174 to the extent that the property is used in connection with activities involving research or experimentation (33).

A significant issue arising under Section 174 concerns the proper treatment of expenditures to a third party to develop a product or process, the expenses of which would otherwise qualify as research or experimental expenditures under Section 174 if incurred directly by the taxpayer. The regulations are clear that the definition of research or experimental expenditures includes expenditures paid or incurred for research or experimentation carried on on the taxpayer's behalf by another person or organization except to the extent that the expenditures are for the acquisition or improvement of land or depreciable property used in connection with the research or experimentation and to which the taxpayer acquires rights of ownership (34). In addition, the regulations under Section 174 provide that research or experimental expenditures made to a third party for the construction or manufacture of depreciable or amortizable property are deductible under Section 174(a) only if "made upon the taxpayer's order and at his risk" (35). Consequently no deduction is allowed if the property acquired is regularly produced or is constructed or manufactured under a performance guarantee. A performance guarantee includes any guarantee, whether express, implied, or imposed by local law, that concerns quality of production or quantity of production in relation to the consumption of raw materials and fuel, unless the guarantee is limited to engineering specifications such that economic utility is not taken into account (36).

Section 174 Election

A taxpayer's election currently to deduct all research or experimental expenditures applies to all qualifying expenditures (37). With the permission of the Commissioner, however, a taxpayer may change to the deferred method with respect to those expenditures attributable to a particular project or projects. In no event will a taxpayer be permitted currently to deduct some of the expenditures and defer and amortize other expenditures relating to the same project (38). Finally, even with the consent of the Commissioner to defer expenses attributable to a particular project, a taxpayer who originally elected currently to deduct research or experimental expenditures may not defer those expenses attributable to a new project without again obtaining the consent of the Commissioner (39).

Instead of currently deducting research or experimental expenditures under Section 174(a), a taxpayer may elect to capitalize and amortize such expenditures over not less than a five-year period under Section 174(b) (40). Under the regulations the amortization period begins with the month in which the taxpayer first realizes benefits from the expenditures. According to the regulations, the taxpayer will typically be deemed to realize benefits from any deferred expenditures in the month in which "the taxpayer first puts the process, formula, invention, or similar property to which the expenditures relate to an income-producing use" (41). Importantly, the taxpayer may select amortization periods of differing lengths for deferred expenditures attributable to different projects.

As a basic rule, deferral under Section 174 is permitted only for those expenditures that are otherwise chargeable to a capital account and that are not chargeable to depreciable or depletable property. However, such expenditures may be deferred only in part under Section 174 if they later become chargeable to depreciable or amortizable property, such as patents (42). In that case any unrecovered expenditures are to be either depreciated or amortized under Section 167 at the time that "the asset becomes depreciable in character" (43). To illustrate this situation, the regulations provide the following example:

> [F]or the taxable year 1954, A, who reports his income on the basis of a calendar year, elects to defer and deduct ratably over a period of 60 months research and experimental expenditures made in connection with a particular project. In 1956, the total of the deferred expenditures amounts to $60,000. At that time, A has developed a process which he seeks to patent. On July 1, 1956, A first realized benefits from the marketing of products resulting from this process. Therefore, the expenditures deferred are deductible ratably over the 60-month period beginning with July 1, 1956 (When A first realized benefits from the project). In his return for the year 1956, A deducted $6,000; in 1957, A deducted $12,000 ($1,000 per month). On July 1, 1958, a patent protecting his process is obtained by A. In his return for 1958, A is entitled to a deduction of $6,000, representing the amortizable portion of the deferred expenses attributable to the period prior to July 1, 1958. The balance of the unrecovered expenditures ($60,000 minus $24,000, or $36,000) is to be recovered as a depreciation deduction over the life of the patent commencing with July 1, 1958. Thus, one-half of the annual depreciation deduction based upon the useful life of the patent is also deductible for 1958 (from July 1 to December 31) (44).

Consequently the opportunity to defer research or experimental expenditures under Section 174 in connection with assets that are otherwise amortizable under Section 167 may be limited.

CREDIT FOR INCREASING RESEARCH ACTIVITIES UNDER SECTION 41

Overview

Section 41 establishes a research tax credit in connection with "qualified research expenses" and "basic research payments," frequently referred to as the incremental research credit and the basic research credit, respectively. The incremental research credit is equal to 20 percent of the excess of the qualified research expenses for the tax year in excess of a base amount, and the basic research credit is equal to 20 percent of any basic research payments in excess of a base amount (45). The purpose of the credit is to provide an incentive for increased research activities in the private sector. According to the legislative history to the research credit as originally enacted in 1981, "Congress concluded that a substantial tax credit for incremental research and experimental expenditures was needed to overcome the reluctance of many ongoing companies to bear the significant costs of staffing and supplies, and certain equipment expenses such as computer charges, which must be incurred to initiate or expand research programs in a trade or business" (46).

The incremental research credit is available in connection with the costs of any qualified research activities incurred in carrying on the trade or business of the taxpayer, including both in-house expenses and contract research expenses (47). Originally the "carrying on" standard of Section 41 generally corresponded to the same requirement found in Section 162 and thus was more restrictive than the "in connection with" standard of Section 174 (48). According to the legislative history,

> it is intended that to be eligible for the credit, research expenditures must be paid or incurred in a particular trade or business being carried on (within the meaning of sec. 162) by the taxpayer; no credit is available for expenditures for research relating to a potential trade or business which the taxpayer is not carrying on at the time the research expenditures are made. Thus, the credit is not available (either for current or carryover use) to a new entity which undertakes research with a view to using the resulting technology through future production and sales, and is not available to an ongoing business which undertakes research with a view to entering a new trade or business (49).

This requirement has been relaxed somewhat in connection with in-house research expenses. A taxpayer will be treated as meeting the trade-or-business requirement of Section 41 with respect to in-house research expenses if, at the time the expenses are paid or incurred, the principal purpose of the taxpayer in making the expenditure is to use the results of the research in the active conduct of a future trade or business of the taxpayer or of another member of the same controlled group (50).

In determining the amount of the research credit, members of a controlled group of corporations are to be treated as a single taxpayer, and the credit, if any, is to be allocated among the members in proportion to their respective increase in qualified research expenses (51). Similar aggregation and allocation rules apply in connection with the research credit available to S corporations, partnerships, estates, and trusts that are under common control (52).

Incremental Research Tax Credit

Definition of Qualified Research Expenses. The incremental research tax credit is equal to 20 percent of the qualified research expenses for the tax year in excess of a base amount (53). The definition of qualified research expenses under Section 41 is more narrow than the definition of research or experimental expenditures under Section 174. This is reflected in the three-part definition of qualified research under Section 41(d):

1. Research that satisfies the requirements of Section 174.
2. Research undertaken for the purpose of discovering information that is technological in nature, whose application is intended to be useful in the development of a new or improved business component of the taxpayer.
3. Research where substantially all of the activities constitute elements of a process of experimentation related to the development of a new or improved function, performance, reliability or quality of a business component (54).

Consequently the second and third parts of the three-part definition limit the range of activities that constitute qualified research as compared to those activities that constitute research or experimental activities for purposes of Section 174.

The second part of the three-part definition requires that the research be designed to discover information that is technological in nature. The legislative history states that qualified research must be within the basic sciences:

> The determination of whether new or improved characteristics of a business item are technological in nature depends on whether the process of experimentation to develop or improve such characteristics fundamentally relies on principles of the physical or biological sciences, engineering, or computer science—in which case the characteristics are deemed technological—or on other principles, such as those of economics—in which case the characteristics are not to be treated as technological. For example, new or improved characteristics of financial services or similar products (such as new types of variable annuities or legal forms) or advertising do not qualify as technological in nature (55).

Importantly research does not rely on principles of computer science merely because a computer is employed (56).

The proposed regulations under Section 41 provide definitions for the terms "discovering information" and "technological in nature." According to the proposed regulations, "discovering information" means "obtaining information that exceeds, expands, or refines the common knowledge of skilled professionals in a particular field of technology or science" (57). The examples contained in the proposed regulations show clearly that research may constitute qualified research when the information is known to others but remains a closely guarded secret and is beyond the common knowledge of skilled professionals in the relevant fields (58). Research may also constitute qualified research even when the taxpayer abandons the project and attempts to develop the technology prove unsuccessful (59). Finally, research may be qualified only in part where research activities advance to the point where further analysis is within the common knowledge of skilled professionals in the relevant field (60). Relying on the language cited from the legislative history, the proposed regulations also provide that information is "technological in nature" if "the process of experimentation used to discover such information fundamentally relies on principles of physical or biological sciences, engineering, or computer science" (61).

The third part of the three-part definition requires that the research constitute a process of experimentation related to the development of a new or improved function, performance, reliability or quality of a business component. In defining the term "process of experimentation," the legislative history refers to

> a process involving the evaluation of more than one alternative designed to achieve a result where the means of achieving that result is uncertain at the start. This may involve developing one or more hypotheses, testing and analyzing those hypotheses (through, for example, modeling or simulation), and refining or discarding the hypotheses as part of a sequential design process to develop the overall component (62).

Although this definition of the term "process of experimentation" suggests that the result itself must be uncertain, subsequent legislative history has emphasized that only the means of reaching the result need be uncertain. "Thus, even though a researcher may know of a particular method of achieving an outcome, the use of the process of experimentation to effect a new or better method of achieving that outcome may be eligible for the credit..." (63). As examples of processes of experimentation, the legislative history specifically notes experiments undertaken by chemists or physicians in developing and testing a new drug and the work of engineers in designing a new computer system or an improved or new integrated circuit.

The proposed regulations under Section 41 follow the legislative history in describing a process of experimentation (64). Interestingly the focus of both the legislative history and the proposed regulations on uncertainty in defining a process of experimentation suggests a certain similarity to the uncertainty test under Section 174 (65). Nevertheless, the proposed regulations provide, as an illustration, that expenditures incurred to resolve uncertainty in manufacturing an improved widget may be treated as research or experimental expenditures under Section 174 but not be undertaken to obtain knowledge that exceeds, expands, or refines the common knowledge of skilled professionals in the relevant

technological fields necessary to satisfy the requirements under Section 41 (66). In addition the proposed regulations provide that, in testing and analyzing one or more hypotheses, the taxpayer must design a "scientific experiment" that, "where appropriate to the particular field of research, is intended to be replicable with an established experimental control" (67). Finally, the statutory definition of qualified research requires that substantially all of the activities that constitute the process of experimentation must relate to the function, performance, reliability, or quality of a business component (68). The proposed regulations provide that the "substantially all" requirement is satisfied only if 80 percent or more of the research activities, measured on a cost or other consistently applied basis, constitute elements of a process of experimentation (69).

Importantly the test of whether particular research is to be treated as qualified research is determined with respect to each business component. A "business component" is defined to include any "product, process, computer software, technique, formula, or invention" held by the taxpayer for sale, lease, or license or used by the taxpayer in its trade or business (70). The proposed regulations under Section 41 provide that the research credit is not available for research activities relating to the development of a manufacturing or other commercial production process unless the activities satisfy the requirements of Section 41 without taking into account the research activities related to the development of the product (71). Similarly the research credit is not available for research activities relating to the development of a product unless the activities satisfy the requirements of Section 41 without taking into account the research activities related to the development of the manufacturing or other commercial production process (72).

The tests to establish eligibility for the credit are applied to each business component or sub-component under the so-called "shrinking-back" concept:

> [T]he requirements for credit eligibility are applied first at the level of the entire product, etc. to be offered for sale, etc. by the taxpayer. If all aspects of such requirements are not met at that level, the test applies at the most significant subset of elements of the product, etc. This shrinking back of the product is to continue until either a subset of elements of the product that satisfies the requirements is reached, or the most basic element of the product is reached and such element fails to satisfy the test (73).

This "shrinking back" concept allows the taxpayer to obtain the benefits of the credit with respect to any portion of its research activities undertaken in the development of a business component that satisfies the requirements of Section 41 (74).

In *Norwest Corporation* (75), the Tax Court considered the definition of qualified research in the context of the development of internal use computer software. The court interpreted Section 41(d) as imposing the following four separate tests:

1. The Section 174 Test (Test 1), which requires that the research expenditures qualify as expenses under Section 174.

2. The Discovery Test (Test 2), which limits the type of information discovered to that which is technological in nature.

3. The Business Component Test (Test 3), which requires that the taxpayer's activities provide some level of functional improvement to a business component of the taxpayer.

4. The Process of Experimentation Test (Test 4), which requires that initial uncertainty concerning the technical ability of the taxpayer to develop the product be eliminated through the development, testing, and analyzing of one or more hypotheses as part of a sequential design process to develop the overall component.

The Tax Court provided content to each of these tests by relying on the legislative history of Section 41 and the policy objectives of the research tax credit.

With respect to Test 2, the Discovery Test, the court concluded that in order to be eligible for the credit, an objective of the taxpayer's activities must be the creation of new knowledge in the field in which the taxpayer is working:

> The legislative history of Section 41 dictates that the knowledge gained from the research and experimentation must be that which exceeds what is known in the field in which the taxpayer is performing the research and experimentation—in this case the computer science field. The fact that the information is new to the taxpayer, but not new to others, is not sufficient for such information to come within the meaning of discovery for purposes of this test. The purpose of the R&E credit was to stimulate capital formation and improve the U.S. economy—not merely the taxpayer's business (76).

The court referred to the legislative history of Section 41 to conclude that the discovery of information must concern the principles of the hard sciences which could result by either "expanding" or "refining" those principles (77). The court also concluded that the Discovery Test of Section 41 differs from the uncertainty test of Section 174. The court pointed out that the uncertainty test under the Section 174 regulations was not adopted until 1994, eight years after the introduction of the Discovery Test under Section 41 in 1986. In addition the court cited the legislative history to Section 41 to suggest that, in amending Section 41 in 1986, "Congress sought to tighten the requirements for obtaining the R&E credit" (78). The court reasoned that because Congress did not change the requirements of Section 174 at that time, the congressional purpose could only be achieved by viewing the two tests as different. Finally, the court viewed the uncertainty test of Section 174 and the Discovery Test of Section 41 as relating to the discovery of different types of information. According to the court the regulations under Section 174 refer to "uncertainty concerning the development or improvement of a product" while Section 41 relates to information that is "technological in nature" and which "fundamentally relies on principles of the hard sciences."

In reaching its decision, the Tax Court also considered Tests 3 and 4. Unfortunately, the court provided only a limited description of Test 3, the Business Component Test. The court only noted that the taxpayer's activities must provide some level of functional improvement, at a minimum, to a business component of the taxpayer. The court provided a greater explanation of its view of Test 4, the Process of Experimentation Test. After quoting from the legislative history, the court stated that this test requires a more structured method of discovery than that required under Section 174, a process in which one or more hypotheses must be developed, tested, and analyzed. According to the court this requirement is to be applied in concert with the shrinking back test until the 80 percent standard is satisfied or the most basic element of the product is reached and that element fails to satisfy the standard. The court concluded by noting that the shrinking back test must be examined on a case-by-case basis to determine which activities are part of the same process or product and which are sufficiently discrete as to warrant separate evaluation.

Exclusion of Certain Research Expenses. Costs incurred in connection with several types of research activities are statutorily excluded from the definition of qualified research expenses under Section 41. Research expenditures are excluded if the research relates only to style, taste, cosmetic, or seasonal design changes (79). More importantly perhaps, qualified research expenses do not include expenses incurred in connection with research conducted after commercial production of a component has started (80). According to the legislative history, "commercial production" is achieved when "the component has been developed to the point where it either meets the basic functional and economic requirements of the taxpayer for such component or is ready for commercial sale or use" (81). The proposed regulations provide that the following activities are deemed to occur after the beginning of commercial production of a business component: pre-production planning for a finished business component, tooling-up for production, trial production runs, troubleshooting involving detecting faults in production equipment or processes, and debugging or correcting flaws in a business component (82). In addition the legislative history states that the credit is not available for the costs of additional clinical testing of a pharmaceutical product after the product is made commercially available, except when the testing is necessary to establish new functional uses for the existing product. For example, "testing a drug currently used to treat hypertension for a new anti-cancer application, and testing an antibiotic in combination with a steroid to determine its therapeutic value as a potential new anti-inflammatory drug, are eligible for the credit" (83).

Qualified research expenses also do not include the costs of research designed to reproduce an existing business component (84). This provision is intended to exclude the costs of "reverse engineering" activities from eligibility for the credit and applies to the reproduction of an existing component by another person based on a physical examination of the business component or on plans, blueprints, detailed specifications, or publicly available information (85). In Private Letter Ruling 9346006 (a Technical Advice Memorandum), the IRS invoked this exclusion and denied the taxpayer's claimed research credit in connection with the development of the generic form of certain drugs that had previously received FDA approval and for which information concerning active and inactive ingredients was publicly available (86). However, this exclusion does not apply if a taxpayer examines a competitor's product in developing its own component through a process of otherwise qualified experimentation.

Section 41 also excludes from the definition of qualified research expenses the costs of research that is fully funded by another entity (87). Under the regulations, research does not constitute qualified research to the extent that it is funded by a grant, contract, or otherwise by another person, including any governmental entity. However, amounts payable under any agreement that are contingent on the success of the research are considered as paid for the results of the research and are not treated as funding (88). In addition research is considered fully funded if the taxpayer retains no substantial rights in the products of the research under the terms of the agreement providing for the performance of the research (89).

The exclusion for fully-funded research was the subject of Private Letter Ruling 9410007 (a Technical Advice Memorandum). In this ruling, the taxpayer was engaged in fixed price contracts with the U.S. government to conduct research and develop certain types of equipment. Under the contracts the taxpayer retained title and rights to any inventions developed as a result of the research, including patents and copyrights, subject to a "non-exclusive, nontransferable, irrevocable" license in favor of the government. The IRS rejected the taxpayer's claimed tax credit under Section 41 for expenditures incurred in conducting the research required under the contracts. Because the contracts provided for progress payments that were subject to little risk of termination or withholding by the government, the IRS viewed the payments for the contracted research as "expected and likely in the normal course of events." As a result the IRS concluded that, because the payments were not contingent on the success of the taxpayer's research, the research was fully funded. In addition the IRS concluded that the research was fully funded because the taxpayer retained no substantial rights in the research. The IRS reasoned that the taxpayer's rights to use and transfer the technology, copyrights, and technical data resulting from the research were subject to "significant restriction" by the U.S. government (90).

The exclusion from the definition of qualified research for research that is considered fully funded was also the subject of review in *Lockheed Martin Corp. v. United States* (91). As previously noted, the regulations provide that research will be treated as fully funded where the taxpayer retains no substantial rights in the research under the agreement to provide the research. In *Lockheed*, the taxpayer performed research under a number of defense contracts with the federal government and claimed a tax refund of over $63 million under Section 41. The court rejected the taxpayer's claim that the "substantial rights" requirement under the regulations was invalid. The court also rejected the claim that a taxpayer fails to satisfy the substantial rights requirement only when the taxpayer retains *no* rights to the research.

Instead, the court looked to Section 1235 for guidance and concluded that the government's unlimited right to use and disclose the research results, as well as the considerable restrictions on the taxpayer's ability to use the research results in the form of security classifications and export restrictions, prevented the taxpayer from claiming that it retained substantial rights in the research. Consequently none of the taxpayer's research expenses qualified for the research credit.

The definition of qualified research expenses also excludes the costs of preparing various types of surveys or studies (92). Under the proposed regulations, this exclusion applies to efficiency surveys; activities (e.g., studies) related to management functions or techniques, market research, market testing, or market development (e.g., advertising or promotions); routine data collections; or routine or ordinary testing or inspection of materials or business components for quality control. Management functions and techniques include the preparation of financial data and analysis, development of employee training programs and management organization plans, and management-based changes in production processes (e.g., rearranging work stations on an assembly line) (93).

Finally, qualified research expenses do not include the costs of research designed to adapt an existing component to a particular customer's needs (94); research involved in the preparation of certain types of computer software (95); research conducted outside the United States, Puerto Rico, or any possession of the United States (96); and research in the social sciences or humanities (97).

Determination of the Incremental Research Tax Credit. Provided that the research at issue satisfies the definition of qualified research, Section 41 allows a tax credit in an amount equal to 20 percent of the qualified research expenses for the taxable year in excess of a base amount (98). Qualified research expenses for the taxable year include both in-house and contract research expenses (99). In-house research expenses include wages paid to employees engaged in qualified research or directly supervising or supporting activities that constitute qualified research (100), the cost of supplies used in the conduct of qualified research (101), and the cost of computers and computer time used in qualified research efforts (102). Contract research expenses, on the other hand, are generally limited to 65 percent of any amount paid or incurred by the taxpayer to a person other than an employee for qualified research (103). However, if the contracting party that will conduct the research is a qualified research consortium, 75 percent of any amounts paid or incurred by the taxpayer to the consortium for qualified research on behalf of the taxpayer and one or more unrelated taxpayers will be treated as qualified research expenses eligible for the credit (104). Contract research expenses that are prepaid are considered as paid or incurred during the period in which the qualified research is actually conducted (105).

With respect to contract research expenses, the regulations specifically require that the qualified research be performed "on behalf of the taxpayer" and that the payments not be contingent on the success of the research (106). The former requirement is satisfied even if the taxpayer retains only a nonexclusive right to the research results (107). In *Norwest Corp.* (108), the IRS argued that the taxpayer's right under a software development agreement to a "perpetual, nontransferable, nonexclusive and ... royalty-free license" to use the developed software did not satisfy the regulatory requirements. The IRS suggested that a difference existed between rights to research results and rights to any final product. The court rejected this argument stating that "the right to use the results of the research without paying for that right is at least a right to the research results as that term is applied [under the regulations] — although it may or may not constitute 'substantial rights in the research' within the purview of the regulations" (109). The IRS also maintained that the taxpayer's ability to terminate the development agreement at selected times violated the regulatory requirement that payments not be contingent on the success of the research. The court rejected this argument as well, noting that the taxpayer had no ability under the agreement to recover any payments that it might have previously made.

Importantly a contract research expense will not be a qualified research expense if the product or result of the research is intended to be transferred to another in return for a license or royalty and the taxpayer does not use the product of the research in the taxpayer's trade or business (110). In such a situation, the taxpayer will not be deemed to be engaged in a trade or business to satisfy the "carrying on" requirement of Section 41(b). The legislative history was emphatic about this point:

> [U]nder the trade or business test of new section [41], the credit generally is not available with regard to a taxpayer's expenditures for "outside" or contract research intended to be transferred by the taxpayer to another in return for license or royalty payments. (Receipt or royalties does not constitute a trade or business under present law, even though expenses attributable to those activities are deductible from gross income in arriving at adjusted gross income.) In such a case, the nexus, if any, between research expenditures of the taxpayer and activities of the transferee to which research results are transferred (e.g., any use by an operating company, that is a general partner in a limited partnership which make the research expenditures, of the research results in the operating company's trade or business) generally will not characterize the taxpayer's expenditures as paid or incurred in carrying on a trade or business *of the taxpayer.* (Under appropriate circumstances, nevertheless, the nexus might be deemed adequate for purposes of the section 174 deduction elections.) If, however, the taxpayer used the product of the research in a trade or business of the taxpayer, as well as licensing use of the product by others, the relationship between the research expenditures of the taxpayer (i.e., those research expenditures paid or incurred after such time as the taxpayer is considered to be carrying on the trade or business in which such expenditures are paid or incurred) and the taxpayer's trade or business in which the research expenditures are paid or incurred generally would be sufficient for credit purposes (111).

As previously noted, Section 41 allows a tax credit in an amount equal to 20 percent of the qualified research expenses for the taxable year in excess of a base amount (112). The base amount is determined by multiplying the average annual gross receipts of the taxpayer for the four taxable years preceding the taxable year for which the credit is being determined by the fixed-base percentage (113). With certain limited exceptions, the term "gross receipts" means "the total amount, as determined under the taxpayer's method of accounting, derived by the taxpayer from all its activities and from all sources (e.g., revenues derived from the sale of inventory before reduction for cost of goods sold)" (114). The "fixed-base percentage" is the ratio of the taxpayer's aggregate qualified research expenses for the taxable years between December 31, 1983, and January 1, 1989, to the taxpayer's aggregate gross receipts for that same period (115). Special rules for the determination of the fixed base percentage apply with respect to start-up companies, under which the fixed-base percentage is typically three percent (116). Two statutory limitations also apply, however, in the determination of the taxpayer's fixed-base percentage and base amount. In no event can the taxpayer's fixed-base percentage exceed a maximum of 16 percent or the base amount be less than 50 percent of the qualified research expenses for the current taxable year (117).

The purpose behind the determination of the tax credit as an amount in excess of a firm specific percentage of gross receipts is described in the following excerpt from the legislative history to the Omnibus Budget Reconciliation Act of 1989:

> Although the committee believes it is important to readjust the base amount annually in a way that does not undercut the incentive effect of the credit (which occurs when a firm's base is adjusted solely by reference to its own prior levels of research spending), the committee also determined it was appropriate that the base adjustment reflect firm-specific factors. By adjusting each taxpayer's base to its own experience, the committee wanted to make the credit widely available at the lowest possible revenue cost.

> Because businesses often determine their research budgets as a fixed percentage of gross receipts, it is appropriate to index each taxpayer's base amount to average growth in its gross receipts. By so adjusting each taxpayer's base amount, the committee believes the credit will be better able to achieve its intended purpose of rewarding taxpayers for research expenses in excess of amounts which would have been expended in any case. Using gross receipts as an index, firms in fast-growing sectors will not be unduly rewarded if their research intensity, as measured by their ratio of qualified research to gross receipts, does not correspondingly increase. Likewise, firms in sectors with slower growth will still be able to earn credits as long as they maintain research expenditures commensurate with their own sales growth.

> Adjusting a taxpayer's base by reference to its gross receipts also has the advantage of effectively indexing the credit for inflation and preventing taxpayers from being rewarded for increases in research spending that are attributable solely to inflation (118).

Alternative Incremental Research Tax Credit

Section 41 also provides for an alternative incremental research tax credit that a taxpayer may elect (119). As described above, the incremental research tax credit is equal to 20 percent of the qualified research expenses in excess of a base amount. The base amount is determined by multiplying the fixed-base percentage by the taxpayer's average annual gross receipts for the four taxable years preceding the taxable year for which the credit is being determined. Because the fixed-base percentage is the ratio of the taxpayer's aggregate qualified research expenses for the taxable years from December 31, 1983, through January 1, 1989, to the taxpayer's aggregate gross receipts for that same period, a taxpayer may not be entitled to the incremental research credit if the growth in the taxpayer's gross receipts has been significantly greater than the growth in its qualified research expenses. For example, assume that a taxpayer had a fixed-base percentage of 10 percent because gross receipts for the taxable years between December 31, 1983, and January 1, 1989, were $10 million and qualified research expenses for that same period were $1 million. If the taxpayer's average annual gross receipts for the four years prior to the taxable year for which the credit is being claimed increased by 20 percent over the average of the taxpayer's annual gross receipts for the 1984 through 1988 taxable years ($2.4 million = 120 percent × [$10 million ÷ 5 years]), but the taxpayer's qualified research expenses for the taxable year increased by only 10 percent as compared to the average annual qualified research expenses over the 1984 through 1988 period ($220,000 = 110 percent × [$1 million ÷ 5 years]), no credit would be available because the taxpayer's qualified research expenses of $220,000 would not exceed the base amount of $240,000 (10 percent of $2.4 million).

To alleviate this problem, the alternative incremental research tax credit dispenses with the fixed-base percentage of the incremental research tax credit and determines the amount of the credit based on the extent to which the qualified research expenses for the taxable year exceed fixed percentages of the taxpayer's average annual gross receipts for the four taxable years preceding the taxable year for which the credit is being determined (the "Section 41(c)(1)(B) amount") (120). The alternative incremental research tax credit is equal to the sum of the following three amount (121):

1. 2.65 percent of the qualified research expenses for the taxable year to the extent that the expenses exceed 1 percent of the Section 41(c)(1)(B) amount but do not exceed 1.5 percent of such amount.
2. 3.2 percent of the qualified research expenses for the taxable year to the extent that the expenses exceed 1.5 percent of the section 41(c)(1)(B) amount but do not exceed 2 percent of such amount.
3. 3.75 percent of the qualified research expenses for the taxable year to the extent that the expenses exceed 2 percent of the section 41(c)(1)(B) amount.

Thus a taxpayer who is not entitled to a credit under the standard incremental research credit may be entitled to relief under the alternative incremental research credit.

Basic Research Tax Credit

Section 41 also permits a basic research tax credit in the amount of 20 percent of any "basic research payment" made during the taxable year in excess of a base amount (122). The basic research credit was enacted in its current form to provide incentives for corporate support of basic scientific research:

> By contrast to other types of research or product development, where expected commercial returns attract private investment, basic research typically does not produce sufficiently immediate commercial applications to make investment in such research self-supporting. Because basic research typically involves greater risks of not achieving a commercially viable result, larger-term projects, and larger capital costs than ordinary product development, the Federal Government traditionally has played a lead role in funding basic research, principally through grants to universities and other nonprofit scientific research organizations. In addition, the research credit as modified by the [Tax Reform Act of 1986] provides increased tax incentives for corporate funding of university basic research to the extent that such expenditures reflect a significant commitment by the taxpayer to basic research (123).

Basic research for purposes of the credit is defined as "any original investigation for the advancement of scientific knowledge not having a specific commercial objective..." (124). A basic research payment includes any amount paid in cash by a corporation to a qualified organization for basic research provided (1) the basic research is performed by the qualified organization and (2) the payment is made pursuant to a written agreement (125). Qualified organizations include colleges and universities, tax-exempt scientific research organizations, and certain tax-exempt organizations operated primarily to promote scientific research by colleges and universities (126).

As previously noted, the credit is equal to 20 percent of the basic research payments in excess of a base amount. The purpose of calculating the tax credit as a percentage of basic research payments in excess of a the base amount is to ensure that the credit is used to encourage increased taxpayer support of basic research and not to encourage taxpayers simply to switch donations from general university giving to forms of support for which the credit is available (127). This base amount is referred to as the "qualified organization base period amount" and is equal to the minimum basic research amount plus the maintenance-of-effort amount (128). In determining the qualified organization base period amount, the minimum basic research amount is an amount equal to the greater of (1) 1 percent of the average amount of any in-house and contract research expenses paid or incurred over the base period or (2) the amount of basic research payments treated as contract research expenses under Section 41(e)(1)(B) during the base period (129). For calendar-year taxpayers, the base period is the three-year period from 1981 to 1983 (130). For taxpayers not in existence during the base period, the minimum basic research amount is not to be less than 50 percent of the basic research payments for the taxable year (131).

The maintenance-of-effort amount is equal to the average of the nondesignated university contributions paid by the taxpayer during the base period in excess of the nondesignated university contributions paid by the taxpayer during the taxable year (132). Nondesignated university contributions are equal to any amount paid by the taxpayer to a qualified organization as defined under Section 41 for which a charitable contribution deduction was allowable under Section 170 and which was not taken into account in determining the basic research credit or as a basic research payment (133). Consequently any reduction in the amount of charitable contributions to qualified organizations from the average amount of contributions made during the base period will offset basic research payments eligible for the basic research credit.

LIMITATIONS ON SECTION 174 DEDUCTIONS

Because qualified research expenses and basic research payments under Section 41 may also be deductible as research or experimental expenditures under Section 174, the Code requires that Section 174 deductions be reduced by the amount of any credit taken under Section 41 (134). The legislative history provides the following example of this requirement:

> For example, assume that a taxpayer makes credit-eligible research expenditures of $1 million during the year, and that the base period amount is $600,000. The taxpayer is allowed a tax credit equal to 20 percent of the $400,000 increase in research expenditures, or $80,000. . . .Under the provision, the taxpayer's deduction is reduced by the $80,000 credit, leaving a deduction of $920,000 (135).

In addition, if research and experimental expenditures are capitalized rather than currently deducted, the capitalized amount must be similarly reduced by the amount of any research credit available under Section 41 (136).

ORPHAN DRUG CREDIT UNDER SECTION 45C

Section 45C of the Code creates the so-called orphan drug credit, which permits a credit equal to 50 percent of the qualified clinical testing expenses paid or incurred for the taxable year (137). The orphan drug credit is so named because it permits a tax credit for certain expenses incurred in the development of drugs used to treat those rare diseases or conditions that affect fewer than 200,000 persons in the United States or that affect more than 200,000 persons but for which the developer of such a drug would have no reasonable expectation of recovering the cost of developing or marketing the drug from its sales in the United States (138). Such diseases and conditions include Huntington's disease, myoclonus, amyotrophic lateral sclerosis (ALS or "Lou Gehrig's disease"), Tourette's syndrome, and Duchenne's dystrophy, a form of muscular dystrophy (139).

Qualified clinical testing expenses are those amounts that would satisfy the definition of qualified research expenses under Section 41(b), with certain modifications (140). One such modification permits 100 percent of any contract research expenses to fall within the definition of qualified clinical testing expenses rather than only 65 percent of such expenses that fall within the definition of qualified research expenses (141). Nevertheless, qualified clinical research expenses do not include any amount otherwise funded under any grant or contract by another person or governmental entity (142). Under the regulations, if the taxpayer conducting the clinical testing for another person retains no substantial rights in the testing, the taxpayer's testing expenses are treated as fully funded (143). Incidental benefits such as increased experience in the field of human clinical testing do not constitute substantial rights in the clinical testing. When the taxpayer conducting the clinical testing retains substantial rights in the testing, the testing expenses are reduced to the extent of any payments and the fair market value of any property to which the taxpayer becomes entitled by conducting the clinical testing (144).

Importantly qualified clinical testing expenses are limited to human clinical testing (145). Human clinical testing requires the use of human subjects to determine the effect of the designated drug on humans necessary to receive approval under Section 505(b) of the Federal Food, Drug, and Cosmetic Act or be licensed under Section 351 of the Public Health Services Act (146). A human subject is an individual who is a participant in research, either as a recipient of the drug or as a control, and may be either a healthy individual or a patient (147). The clinical testing must also occur within the United States unless an insufficient testing population exists within the United States and the testing is performed by a U.S. person or any other person unrelated to the taxpayer (148).

Because qualified clinical testing expenses will also constitute qualified research expenses under Section 41, such expenses are not taken into account for purposes of the research credit under Section 41 if the taxpayer elects the orphan drug credit for the taxable year (149). Nevertheless, such qualified clinical testing expenses are taken into account in determining base period research expenses for purposes of applying Section 41 in any subsequent taxable year (150).

BIBLIOGRAPHY

1. IRC §174(a); Reg. §§174-1 and 1.174-3.

2. IRC §174(b); Reg. §§1.174-1 and 1.174-4.

3. Reg. §1.174-1.

4. Rev. Rul. 58-356, 1958-2 CB 104; Rev. Rul. 83-138, 1983-2 CB 50.

5. IRC §§56(b)(2)(A)(ii) and 56(b)(2)(D); Priv. Ltr. Rul. 9746002 (a Technical Advice Memorandum).

6. 416 U.S. 500 (1974).

7. *Snow v. Commissioner*, 416 U.S. 500, 503–504 (1974).

8. IRC §195.

9. 83 TC 667 (1984).

10. Id. at 686–687 (emphasis in original). Compare *Scoggins v. Commissioner*, 46 F3d 950 (9th Cir. 1995) (taxpayer's objective intent and capability to enter into a business in connection with its research activities demonstrated a "realistic prospect" of subsequently entering into business such that research expenditures were deductible); *Best Universal Lock Co.*, 45 TC 1 (1965), acq. (allowing deductions under §174 for research and experimentation expenses incurred in developing an isothermal air compressor because the expenditures were incurred in connection with taxpayer's business, even though the new product was unrelated to taxpayer's past line of products);

M. Bush, TC Memo 1994-523, 68 TCM 974 (deductions in connection with research and development expenditures permitted despite the taxpayer's failure to sell products during the taxable year at issue); *O.B. Kilroy*, TC Memo 1980-489, 41 TCM 292 (taxpayer engaged in trade or business of exploiting inventions could deduct research or experimentation expenditures under §174 despite fact that gross receipts from the activity were negligible); and Rev. Rul. 71-162, 1971-1 CB 97 (research or experimental expenditures incurred in developing products unrelated to taxpayer's current product line or manufacturing processes may be deductible under §174); with *Mach-Tech, Ltd. Partnership & Serv-Tech, Inc.*, 95-2 USTC ¶50,375 (5th Cir. 1995) (disallowing deductions under §174 because the taxpayer was neither engaged in, nor had any realistic prospect of, entering into a trade or business); *Mayrath v. Commissioner*, 357 F2d 209 (5th Cir. 1966) (disallowing deductions under §174 for costs incurred in developing new techniques for housing construction because no indication of profit motive, as taxpayer was a professional inventor of farm machine products, not construction products); *W.J. Piszczek*, TC Memo 1998-307, 76 TCM 338 (applying the regulatory factors under Regulation §1.183-2(b) to conclude that the taxpayer was not engaged in activities to produce a wind-powered ethanol distillery with the profit motive necessary to satisfy the trade or business requirement of §174); *P.E. Sheehy*, TC Memo 1998-183, 75 TCM 2309 (taxpayer provided no evidence that partnership was actively involved in a trade or business involving the development or manufacture of recyclable plastic containers); *Utah Jojoba I Research*, TC Memo 1998-6, 75 TCM 1524 ("For an investing partnership successfully to claim research and experimental deductions, there must be a realistic prospect that the technology to be developed will be exploited in a trade or business of the partnership claiming deductions under section 174 ... Mere legal entitlement to enter into a trade or business does not satisfy this test."); *H.I. Shaller*, TC Memo 1984-584, 49 TCM 10 (research relating to ocean surf energy did not rise to the level of a trade or business); *Gyro Eng'g Corp.*, TC Memo 1974-288, 33 TCM 1343 (taxpayer not engaged in trade or business of research or inventing). For decisions prior to *Snow* considering the existence of a trade or business, see *O.B. Kilroy*, TC Memo 1973-7, 32 TCM 27; *J.H. Cunningham*, TC Memo 1968-242, 27 TCM 1219; *Stanton v. Commissioner*, TC Memo 1967-137, 26 TCM 618, aff'd, 399 F2d 326 (5th Cir. 1968); *C.H. Schafer*, TC Memo 1964-156, 23 TCM 927; *E.G. Bailey*, TC Memo 1963-251, 22 TCM 1255.

11. *Harris v. Commissioner*, 16 F3d 75 (5th Cir. 1994); *P.D. Martyr*, TC Memo 1990-558, 60 TCM 1115, aff'd sub nom. *Gatto v. Commissioner*, 1 F3d 826 (9th Cir. 1993); *Kantor v. Commissioner*, 998 F2d 1514 (9th Cir. 1993); *United Fibertech, Ltd. v. Commissioner*, 976 F2d 445 (8th Cir. 1992); *Nickeson v. Commissioner*, 962 F2d 973 (10th Cir. 1992); *Diamond v. Commissioner*, 930 F2d 372 (4th Cir. 1991); *Zink v. United States*, 929 F2d 1015 (5th Cir. 1991); *R.C. Jay*, TC Memo 1988-232, 55 TCM 933, aff'd sub nom. *Ben-Porat v. Commissioner*, 908 F2d 976 (9th Cir. 1990); *Property Growth Co. v. Commissioner*, 89-2 USTC ¶9479 (8th Cir. 1989); *Spellman v. Commissioner*, 845 F2d 148 (7th Cir. 1988); *Levin v. Commissioner*, 832 F2d 403 (7th Cir. 1987); *Independent Elec. Supply, Inc. v. Commissioner*, 781 F2d 724 (9th Cir. 1986); *S. Drobny*, 86 TC 1326 (1986); *H.J. Green*, 83 TC 667 (1984); *Utah Jojoba I Research*, TC Memo 1998-6, 75 TCM 1524; *Cactus Wren Jojoba, Ltd.*, TC Memo 1997-504, 74 TCM 1133; *3-Koam Co.*, TC Memo 1997-148, 73 TCM 2415; *S.H. Glassley*, TC Memo 1996-206, 71

TCM 2898; *Digital Accounting Technology, Ltd.*, TC Memo 1995-339, 70 TCM 178; *LDL Research & Dev. II, Ltd.*, TC Memo 1995-172, 69 TCM 2411, aff'd, 124 F3d 1338 (10th Cir. 1997); *Estate of G.B. Cook*, TC Memo 1993-581, 66 TCM 1523; *P.A. Stankevich, Jr.*, TC Memo 1992-458, 64 TCM 460; *Software 16*, TC Memo 1992-247, 63 TCM 2876; *E. Stauber*, TC Memo 1992-128, 63 TCM 2258; *Double Bar Chain Co., Ltd.*, TC Memo 1991-572, 62 TCM 1276; *Scientific Measurement Sys. I, Ltd.*, TC Memo 1991-69, 61 TCM 1951; *J.P. Coleman*, TC Memo 1990-357, 60 TCM 123, upheld on reh'g, TC Memo 1990-511, 60 TCM 889; *C.F. Alexander*, TC Memo 1990-141, 59 TCM 121, aff'd without pub. op. sub nom. *Stell v. Commissioner*, 999 F2d 544 (9th Cir. 1993); *Medical Mobility Ltd. Partnership I*, TC Memo 1993-428, 66 TCM 741; *Active Lipid Dev. Partners, Ltd.*, TC Memo 1991-522, 62 TCM 1046; *N.F. Ben-Avi*, TC Memo 1988-74, 55 TCM 199; *R. Rosenberg*, TC Memo 1987-441, 54 TCM 392; *T.S. Reinke*, TC Memo 1981-120, 41 TCM 1100; Priv. Ltr. Rul. 9604004 (a Technical Advice Memorandum); Field Service Advice 1999-839 (undated), available in LEXIS, 1999 TNT 70-12. But see *Scoggins v. Commissioner*, 46 F3d 950 (9th Cir. 1995); *Smith v. Commissioner*, 937 F2d 1089 (6th Cir. 1991).

12. Several courts have held that this issue involves the application of law to fact and, thus, call for de novo review on appeal. *Scoggins v. Commissioner*, 46 F3d 950 (9th Cir. 1995); *Nickeson v. Commissioner*, 962 F2d 973 (10th Cir. 1992); *Zink v. United States*, 929 F2d 1015 (5th Cir. 1991).

13. *E. Stauber*, TC Memo 1992-128, 63 TCM 2258. In addition, taxpayers who merely invest in an entity engaged in research and development activities are precluded from deducting their investments as research or experimental expenditures. *Safstrom v. Commissioner*, 95-1 USTC ¶50,030 (9th Cir. 1994); *Cleveland v. Commissioner*, 297 F2d 169 (4th Cir. 1961).

14. 998 F2d 1514 (9th Cir. 1993).

15. *Kantor v. Commissioner*, 998 F2d 1514, 1519 (9th Cir. 1993). See also *Diamond v. Commissioner*, 930 F2d 372 (4th Cir. 1991) (based on a review of the financial arrangements between the parties, the court concluded that "if a money-making business should materialize, there exists no reasonable expectation that [the research entity] will permit it to be exploited by one of the partnerships"); *Spellman v. Commissioner*, 845 F2d 148 (7th Cir. 1988) (concluding that an option price of only $20,000 gave the profit potential of any trade or business that might result to the research entity).

16. *Kantor v. Commissioner*, 998 F2d 1514, 1520 (9th Cir. 1993). See also *Levin v. Commissioner*, 832 F2d 403 (7th Cir. 1987) (concluding that the partnership was not engaged in a trade or business where the general partner visited a food machinery plant for the first time when escorted to one after the partnerships were formed).

17. 46 F3d 950 (9th Cir. 1995).

18. Reg. §1.174-2(a)(1).

19. Reg. §1.174-2(a)(1).

20. Reg. §1.174-2(a)(2).

21. Reg. §1.174-2(a)(1).

22. GCM 39527.

23. Reg. §1.174-2(a)(1).

24. See Reg. §§1.174-2(a)(1) and 1.174-2(a)(2).

25. Reg. §1.174-2(a)(3)(vii). See also Rev. Rul. 71-363, 1971-2 CB 156; *Hakim v. Commissioner*, TC Memo 1974-46, 33 TCM 223, aff'd, 512 F2d 1379 (6th Cir. 1975).

26. 1973-2 CB 87.

27. See also *H.F. Crouch*, TC Memo 1990-309, 59 TCM 938; *A. Quinn*, TC Memo 1974-64, 33 TCM 310; Priv. Ltr. Rul. 7004169560A (a Technical Advice Memorandum); GCM 39527.

28. *Yellow Freight Sys., Inc. of Del. v. United States*, 92-1 USTC ¶50,029 (Cl. Ct. 1991).

29. Reg. §§1.174-2(a)(3) and 1.174-2(a)(4); *Utah Jojoba I Research*, TC Memo 1998-6, 75 TCM 1524; *Cactus Wren Jojoba, Ltd.*, TC Memo 1997-504, 74 TCM 1133.

30. Reg. §1.174-2(a)(3)(vi).

31. Reg. §1.174-2(b)(4).

32. Reg. §1.174-2(b)(1); *Ekman v. Commissioner*, TC Memo 1997-318, 74 TCM 72, aff'd, 99-1 USTC ¶50,580 (6th Cir. 1999); *P.F. Sheehy*, TC Memo 1996-334, 72 TCM 178.

33. Reg. §1.174-2(b)(1).

34. Reg. §§1.174-2(a)(8) and 1.174-2(a)(9) examples (1) and (2). See also Rev. Rul. 73-324, 1973-2 CB 72; Rev. Rul. 73-20, 1973-1 CB 133; Rev. Rul. 69-484, 1969-2 CB 38.

35. Reg. §1.174-2(b)(3). See Priv. Ltr. Rul. 8614004. Despite satisfaction of the requirements of Regulation §1.174-2(b)(3), deductible research or experimental expenditures do not include "the costs of the component materials of the depreciable property, the costs of labor or other elements involved in its construction and installation, or costs attributable to the acquisition or improvement of the property." Reg. §1.174-2(b)(4).

36. Reg. §1.174-2(b)(3).

37. IRC §174(a)(3); Reg. §1.174-3(a).

38. Reg. §1.174-3(a).

39. Rev. Rul. 68-144, 1968-1 CB 85.

40. Reg. §1.174-4(a)(2).

41. Reg. §1.174-4(a)(3).

42. IRC §174(b)(1)(C); Reg. §§1.174-4(a)(3) and 1.174-4(a)(4).

43. Reg. §1.174-4(a)(4).

44. Reg. §1.174-4(a)(4).

45. IRC §41(a).

46. Staff of the Joint Comm. on Tax'n, General Explanation of the Economic Recovery Tax Act of 1981, 120 (1981).

47. IRC §41(b). The credit is not available to taxpayers who merely fund research activities, typically of their wholly owned corporations, and do not acquire rights to the product of that research. See *Safstrom v. Commissioner*, TC Memo 1992-587, 64 TCM 971, aff'd, 95-1 USTC ¶50,030 (9th Cir. 1994).

48. See Reg. §1.41-2(a)(1); *N.F. Ben-Avi*, TC Memo 1988-74, 55 TCM 199; *E.H. Allen*, TC Memo 1988-166, 55 TCM 641.

49. Staff of the Joint Comm. on Tax'n, General Explanation of the Economic Recovery Tax Act of 1981, 122 (1981); H.R. Rep. No. 201, 97th Cong., 1st Sess. 112 (1981).

50. IRC §41(b)(4). See also Reg. §1.41-2(a)(4) (extending the availability of the research tax credit to certain partnerships and joint ventures not carrying on a trade or business to which the research relates).

51. IRC §41(f)(1)(A); Reg. §1.41-8.

52. IRC §§41(f)(1)(B) and 41(g); Reg. §§1.41-8 and 1.41-9.

53. IRC §41(a)(1).

54. IRC §§41(d)(1) and 41(d)(3); Prop. Reg. §1.41-4(a)(2).

55. H.R. Rep. No. 426, 99th Cong., 1st Sess. 180 (1985). See also S. Rep. No. 313, 99th Cong., 2d Sess. 696 (1986); Staff of the Joint Comm. on Tax'n, General Explanation of the Tax Reform Act of 1986, 133 (1987).

56. Prop. Reg. §1.41-4(a)(7); H.R. Rep. No. 841, 99th Cong., 2d Sess. II-71, n.3 (1986); Staff of the Joint Comm. on Tax'n, General Explanation of the Tax Reform Act of 1986, 133, n.23 (1987).

57. Prop. Reg. §1.41-4(a)(3). For illustrations of situations in which this requirement is satisfied, see Prop. Reg. §1.41-4(a)(8) examples (1), (5), (6), and (7); for illustrations of situations in which this requirement is not satisfied, see Prop. Reg. §1.41-4(a)(8) examples (2), (3), and (4).

58. Prop. Reg. §1.41-4(a)(8) example (6).

59. Prop. Reg. §1.41-4(a)(8) example (5).

60. Prop. Reg. §1.41-4(a)(8) example (8).

61. Prop. Reg. §1.41-4(a)(4).

62. Staff of the Joint Committee on Taxation, General Explanation of the Tax Reform Act of 1986, 133 (1987). See also H.R. Rep. No. 426, 99th Cong., 1st Sess. 180-181 (1985); S. Rep. No. 313, 99th Cong., 2d Sess. 696 (1986).

63. H.R. Conf. Rep. No. 825, 105th Cong., 2d Sess. 1548-1549 (1998) (describing the provisions of the Tax and Trade Relief Extension Act of 1998, Pub. L. No. 105-277, §1001, 112 Stat. 2681, 2681-888 (1998)); Staff of the Joint Committee on Taxation, General Explanation of Tax Legislation Enacted in 1998, 236 (1998).

64. Prop. Reg. §1.41-4(a)(5).

65. See Reg. §1.174-2(a)(1).

66. Prop. Reg. §1.41-4(a)(8) example (4).

67. Prop. Reg. §1.41-4(a)(5)(ii). See also Prop. Reg. §§1.41-4(a)(5)(iii) and 1.41-4(d).

68. IRC §41(d)(3)(A).

69. Prop. Reg. §1.41-4(a)(6).

70. IRC §41(d)(2); Prop. Reg. §1.41-4(b)(1).

71. Prop. Reg. §1.41-4(b)(1).

72. Prop. Reg. §§1.41-4(c)(2)(iii) and 1.41-4(c)(10) example (3).

73. Staff of the Joint Comm. on Tax'n, General Explanation of the Tax Reform Act of 1986, 134 (1987).

74. Prop. Reg. §1.41-4(b)(2); Prop. Reg. §1.41-4(b)(3).

75. 110 TC 454 (1998).

76. Norwest Corp., 110 TC 454, 493 (1998).

77. Norwest Corp., 110 TC 454, 494 (1998) (referencing H.R. Rep. No. 841, 99th Cong., 2d Sess. II-71, n.3 (1986)).

78. Norwest Corp., 110 TC 454, 493 (1998). See also *United Stationers, Inc. v. United States*, 163 F3d 440 (7th Cir. 1998).

79. IRC §41(d)(3)(B). See also Priv. Ltr. Rul. 9522001 (a Technical Advice Memorandum).

80. IRC §41(d)(4)(A); Prop. Reg. §1.41-4(c)(2).

81. Staff of the Joint Comm. on Tax'n, General Explanation of the Tax Reform Act of 1986, 136 (1987).

82. Prop. Reg. §§1.41-4(c)(2)(ii) and 1.41-4(c)(10) example (3). See Norwest Corp., 110 TC 454 (1998).

83. Staff of the Joint Comm. on Tax'n, General Explanation of the Tax Reform Act of 1986, 136 (1987). See Prop. Reg. §§1.41-4(c)(10) examples (1) and (2).

84. IRC §41(d)(4)(C); Prop. Reg. §1.41-4(c)(4). See Prop. Reg. §1.41-4(c)(10) example (5).

85. Prop. Reg. §1.41-4(c)(4); Staff of the Joint Comm. on Tax'n, General Explanation of the Tax Reform Act of 1986, 137 (1987); Market Segment Specialization Program, "Manufacturing Industry" (May 1, 1998), available in LEXIS, 1999 TNT 81-23.

86. See also Field Service Advice 1999-1023 (dated Oct. 22, 1993), available in LEXIS, 1999 TNT 81-49; Market Segment

Specialization Program, "Manufacturing Industry" (May 1, 1998), available in LEXIS, 1999 TNT 81-23.

87. IRC §41(d)(4)(H); Reg. §1.41-5(d); Prop. Reg. §1.41-4(c)(9).

88. Reg. §1.41-5(d)(1). See also Reg. §1.41-2(e)(2).

89. Reg. §1.41-5(d)(2).

90. But see *Fairchild Indus., Inc. v. United States*, 94-1 USTC ¶50,164 (Fed. Cl. 1994), rev'd, 71 F3d 868 (Fed. Cir. 1995) (permitting a research credit in connection with expenses incurred under a government defense contract where the taxpayer bore the economic risk of loss; "[t]he inquiry turns on who bears the research costs upon failure, not on whether the researcher is likely to succeed in performing the project").

91. 98-2 USTC ¶50,887 (Fed. Cl. 1998).

92. IRC §41(d)(4)(D); Prop. Reg. §1.41-4(c)(5). See also Prop. Reg. §1.41-4(c)(10) example (6).

93. Prop. Reg. §1.41-4(c)(5); Staff of the Joint Comm. on Tax'n, General Explanation of the Tax Reform Act of 1986, 136-137 (1987). See also Norwest Corp., 110 TC 454 (1998).

94. IRC §41(d)(4)(B); Prop. Reg. §1.41-4(c)(3). See also Prop. Reg. §1.41-4(c)(10) example (4); Staff of the Joint Comm. on Tax'n, General Explanation of the Tax Reform Act of 1986, 136 (1987).

95. IRC §41(d)(4)(E); Prop. Reg. §§1.41-4(c)(6) and 1.41-4(e).

96. IRC §41(d)(4)(F); Prop. Reg. §1.41-4(c)(7).

97. IRC §41(d)(4)(G); Prop. Reg. §1.41-4(c)(8).

98. IRC §41(a).

99. IRC §41(b)(1).

100. IRC §41(b)(2)(A)(i); Reg. §1.41-2(d). See also Market Segment Specialization Program, "Manufacturing Industry" (May 1, 1998), available in LEXIS, 1999 TNT 81-23; Market Segment Specialization Program, "Computers, Electronics, and High Tech Industry" (March 15, 1997), available in LEXIS, 98 TNT 199-12; *Apple Computer, Inc.*, 98 TC 232 (1992), acq. in part; *Sun Microsystems, Inc.*, TC Memo 1995-69, 69 TCM 1884, acq.; *E.V. Fudim*, TC Memo 1994-235, 67 TCM 3011; Priv. Ltr. Rul. 8835002 (a Technical Advice Memorandum). Priv. Ltr. Rul. 9018003 (a Technical Advice Memorandum); Internal Revenue Service, Industry Specialization Program Coordinated Issue Paper (Data Processing Industry) Qualifying Wages under Section 41 in Determining Tax Credit for Increasing Research Activities, available in LEXIS, 96 TNT 199-12; Internal Revenue Service, Industry Specialization Program Coordinated Issue Paper (All Industries) Qualifying Wages Under Section 41 in Determining Tax Credit for Increasing Research Activities (February 16, 1999), available in LEXIS, 1999 TNT 74-86.

101. IRC §41(b)(2)(A)(ii); Reg. §1.41-2(b). See Market Segment Specialization Program, "Manufacturing Industry" (May 1, 1998), available in LEXIS, 1999 TNT 81-23.

102. IRC §41(b)(2)(A)(iii).

103. IRC §41(b)(3); Reg. §1.41-2(e).

104. IRC §41(b)(3)(C)(i). A qualified research consortium is a tax-exempt organization organized and operated primarily to conduct scientific research. IRC §41(b)(3)(C)(ii).

105. IRC §41(b)(3)(B); Reg. §1.41-2(e)(4).

106. Reg. §1.41-2(e)(2).

107. Reg. §1.41-2(e)(3).

108. 110 TC 454 (1998).

109. Norwest Corp., 110 TC 454, 519 (1998).

110. Reg. §1.41-2(a)(1).

111. Staff of the Joint Comm. on Tax'n, General Explanation of the Economic Recovery Tax Act of 1981, 122–123 (1981); H.R. Rep. No. 201, 97th Cong., 1st Sess. 113 (1981); S. Rep. No. 144, 97th Cong., 1st Sess. 78 (1981).

112. IRC §41(a). For examples of the calculation of the research tax credit, see Market Segment Specialization Program, "Manufacturing Industry" (May 1, 1998), available in LEXIS, 1999 TNT 81-23, and Market Segment Specialization Program, "Computers, Electronics, and High Tech Industry" (March 15, 1997), available in LEXIS, 98 TNT 199-12.

113. IRC §41(c)(1).

114. Prop. Reg. §§1.41-3(c)(1) and 1.41-3(c)(2).

115. IRC §41(c)(3)(A).

116. IRC §41(c)(3)(B).

117. IRC §§41(c)(2) and 41(c)(3)(C).

118. H.R. Rep. No. 247, 101st Cong., 1st Sess. 1199–1200 (1989).

119. IRC §41(c)(4).

120. IRC §41(c)(4)(A).

121. IRC §41(c)(4).

122. IRC §§41(a)(2) and 41(e)(1)(A).

123. Staff of the Joint Comm. on Tax'n, General Explanation of the Tax Reform Act of 1986, 131 (1986).

124. IRC §41(e)(7)(A).

125. IRC §41(e).

126. IRC §41(e)(6).

127. H.R. Rep. No. 426, 99th Cong., 1st Sess. 178 (1985); S. Rep. No. 313, 99th Cong., 2d Sess. 695 (1986); Staff of the Joint Comm. on Tax'n, General Explanation of the Tax Reform Act of 1986, 131 (1987).

128. IRC §41(e)(3).

129. IRC §41(e)(4)(A).

130. IRC §41(e)(7)(B).

131. IRC §41(e)(4)(B).

132. IRC §41(e)(5)(A).

133. IRC §41(e)(5)(B).

134. IRC §280C(c)(1). See H.R. Rep. No. 795, 100 Cong., 2d Sess. 452 (1988).

135. H.R. Rep. No. 795, 100 Cong., 2d Sess. 453 (1988).

136. IRC §280(c)(2).

137. IRC §45C(a); Reg. §1.28-1(a).

138. IRC §45C(d)(1) (referring to §526 of the Federal Food, Drug, and Cosmetic Act, Pub. L. No. 97-414, §2(a), 96 Stat. 2050 (1983), as amended, codified at 21 USC §360(bb)); Reg. §§1.28-1(c) and 1.28-1(d)(1)(i).

139. Reg. §1.28-1(d)(1)(i). See H.R. Rep. No. 426, 99th Cong., 1st Sess. 230 (1985); Staff of the Joint Comm. on Tax'n, General Explanation of the Tax Reform Act of 1986, 141 (1987).

140. IRC §45C(b)(1); Reg. §1.28-1(b)(1).

141. IRC §45C(b)(1)(B)(ii); Reg. §1.28-1(b)(2).

142. IRC §45C(b)(1)(C); Reg. §1.28-1(b)(3)(i).

143. Reg. §1.28-1(b)(3)(ii).

144. Reg. §1.28-1(b)(3)(iii)(A).

145. IRC §45C(b)(2)(A); Reg. §1.28-1(c)(2).

146. Reg. §1.28-1(c)(2).

147. Reg. §1.28-1(c)(2).

148. IRC §45C(d)(2); Reg. §1.28-1(d)(3).

149. IRC §45C(c)(1).

150. IRC §45C(c)(2).

ADDITIONAL READINGS

Generally

C.E. Falk, Tax Planning for the Development and Licensing of Patents and Know-How, 557 Tax Mgmt. (BNA) (1996).

J.E. Maule, Tax Credits: Concepts and Calculation, 506 Tax Mgmt. (BNA) (1994).

J.E. Maule and L.M. Starczewski, Deductions: Overview and Conceptual Aspects, 503 Tax Mgmt. (BNA) (1995).

J.E. Maule and L.M. Starczewski, Deduction Limitations: General, 504 Tax Mgmt. (BNA) (1995).

P.W. Oosterhuis and J.S. Stanton, Research and Development Expenditures, 42-3rd Tax Mgmt. (BNA) (1987).

P.F. Postlewaite, D.L. Cameron and T. Kittle-Kamp, Federal Income Taxation of Intellectual Properties and Intangible Assets, Warren, Gorham, and Lamont, 1997.

Research and Development Expenditures Under Section 174

J. Bankman, The Structure of Silicon Valley Start-Ups, 41 UCLA L. Rev. 1737 (1994).

J.E. Bischel, Deduction and Allocation of Research and Development Expenditures; The Final Chapter or Just Another Installment? 16 Int 1 Tax J. 225 (1990).

G. Foreman, Research and Experimental Expenditures Sec. 174 and Credit for Increasing Research Activities Sec. 41, 60 CPA J 69 (1990).

D.S. Goldberg, Recent Approaches to the Trade or Business Requirement of Section 174: Unauthorized Snow Removal, 8 Va. Tax Rev. 861 (1989).

D.S. Hudson, The Tax Concept of Research or Experimentation, 45 Tax Law. 85 (1991).

F. Murias, New R&E Regs. Are More Rational Than Prior Rules, but Key Terms Need Clarification, 80 J. Taxation 44 (1994).

B.M. Seltzer, D.A. Golden, and M.E. Monahan, Maximizing Opportunities Under the New Research and Experimentation Regulations, 47 Tax Exec. 102 (1995);.

S.P. Starr and D.S. Shapiro, Current Tax Issues in Research and Experimentation, 52 Inst. on Fed. Taxation 5 (1994).

H.W. Wolosky, IRS Eases Position on R&D, 26 Prac. Acct. 43 (June 1993).

The Research Tax Credit Under Section 41

B.A. Billings and G.A. McGill, The Effect of Base Changes on the Incremental Research and Experimentation Tax Credit, 54 Tax Notes 1155 (1992).

B.A. Billings and J.D. Schroeder, Research and Experimentation as an Investment Alternative: The Effect of the Incremental Tax Credit, 61 CPA J. 60 (1991).

D.L. Champi, Tax Policy and National Security Strategy After Operation Desert Storm: The Role of the Research and Experimentation Credit in Enhancing Expenditures on Basic Research and Technology Development Made by the United States Defense Technology Base, 45 Tax Law. 195 (1991).

Coopers and Lybrand Tax Policy Economics Group, Economic Benefits of the R&D Tax Credit, 78 Tax Notes 1019 (1998).

N.J. Crimm, A Tax Proposal to Promote Pharmacologic Research, to Encourage Conventional Prescription Drug Innovation and Improvement, and To Reduce Product Liability Claims, 29 Wake Forest L. Rev. 1007 (1994).

D.S. Hudson, "The Tax Concept of Research or Experimentation," 45 Tax Law. 85 (1991).

K.J. Lester, Note, Availability of the Research Tax Credit for Government Funded Research: *Fairchild Industries v. United States*, 50 Tax Law. 873 (1997).

National Association of Manufacturers, The R&D Tax Credit: Lasting Gains in Research and GDP, available in LEXIS, 98 TNT 167–22, (April 13, 1998).

P.W. Oosterhuis, International R&D and Technology Transfer Arrangements, 73 Taxes 905 (1995).

B.J.W. Raby and W.L. Raby, Seven Tests for Software Development Research Projects, 80 Tax Notes 583 (1998).

Staff of the Joint Committee on Taxation, General Explanation of the Tax Reform Act of 1986 (1987).

Staff of the Joint Committee on Taxation, General Explanation of Tax Legislation Enacted in 1998 (1998).

S. Stratton and B. Massey, Major Changes to Research Credit Rules Sought at IRS Reg Hearing, 83 Tax Notes 623 (1999).

P.A. Stoffregen, Giving Credit Where Credit Is Due: A Brief History of the Administration of the R&D Credit, 66 Tax Notes 403 (1995).

J.S. Wong, *Fairchild Industries, Inc.:* Was the Court Fair in Denying Research Credit? 64 Tax Notes 1477 (1994).

J.S. Wong, Research Credit Allowed for Some Internal-Use Computer Software, 80 Tax Notes 851 (1998).

H. Watson, The 1990 R&D Tax Credit: A Uniform Tax on Inputs and a Subsidy for R&D, 49 Natl Tax J. 93 (1993).

See other entries FEDERAL REGULATION OF BIOTECHNOLOGY PRODUCTS FOR HUMAN USE, FDA, ORPHAN DRUG ACT; STRATEGIC ALLIANCES AND TECHNOLOGY LICENSING IN BIOTECHNOLOGY.

STRATEGIC ALLIANCES AND TECHNOLOGY LICENSING IN BIOTECHNOLOGY

JOSH LERNER
Harvard University and National Bureau of Economic Research
Boston, Massachusetts

OUTLINE

Introduction

Why are Alliances and Licensing So Important in Biotechnology?

How are Alliances Structured?

What is the Evidence from Field Research?

Where is Further Research Needed?

Acknowledgments

Bibliography

INTRODUCTION

This article explores the roles of strategic alliances and licensing in the biotechnology industry. Two areas that have attracted considerable attention from academics and practitioners are highlighted. First, the article considers the reasons for the prevalence of alliances and licensing agreements in the biotechnology industry. The hypotheses and evidence about the structuring of these agreements are then considered. Corroboratory evidence from recent field research is then summarized. The final section highlights additional issues for future research.

At the outset, it is worth mentioning the complexity of these agreements, which reflects the costly and uncertain nature of biotechnology projects. The complexity and unpredictability of the research presents challenges in drafting enforceable agreements that specify the contributions of each party in the face of all contingencies. A great deal of innovation has consequently been devoted to the design of these contracts, which makes it difficult to generalize about this phenomenon.

Despite these difficulties, the understanding of strategic alliances is critical to those who wish to understand the biotechnology industry, or high technology industries more generally. The availability of equity from public investors for new high technology firms has been variable, with biotechnology a particularly extreme case. The financing activities of biotechnology firms between 1978 and 1995 are summarized in Figure 1 and Table 1 (1,2). During periods with little financing activity, young high technology firms suffer tremendous financial stresses and have few alternatives to raise capital other than strategic alliances.

Furthermore, the economic importance of technology alliances has been increasing. Panel A of Table 2 shows the number of such alliances has been growing in a variety of industries. While obtaining a comprehensive view of alliance financing is exceedingly difficult, tabulations suggest that alliances are the dominant source of external financing for R&D by young firms in many industries, including advanced materials, information technology, and telecommunications (3). Surveys of corporate research managers suggest that alliances will be an increasingly important mechanism through which R&D is financed in the years to come (4).

Nowhere is this trend clearer than in biotechnology, where alliances with pharmaceutical firms have become in recent years the single largest source of financing for biotechnology firms, accounting for several billion dollars of funds annually. Panel B of Table 2 illustrates the growth in the number of alliances involving U.S. biotechnology firms and other private-sector entities. The economic importance of these transactions is also shown by the willingness of firms to spend substantial amounts litigating them and the size of the damage awards: for example, Genentech and Eli Lilly's dispute over their alliance to develop human growth hormone, which led to the filing of at least six suits between 1987 and 1993. One indication of the importance of these agreements is the

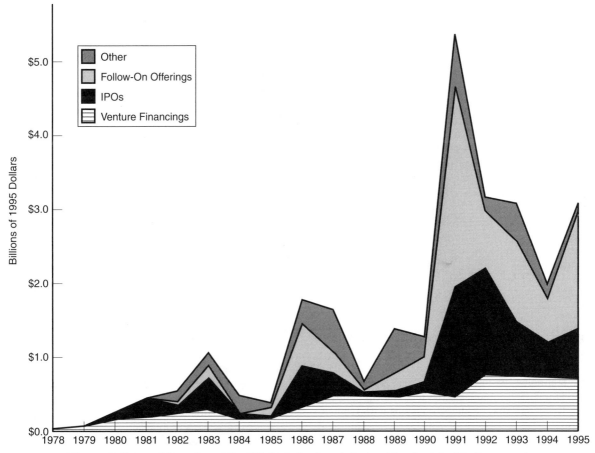

Figure 1. External financing of the U.S. biotechnology industry. The chart depicts the amount raised by U.S. new biotechnology firms through private venture financings, initial public offerings, follow-on public equity offerings, and other sources. (Alliance-related financings are excluded.)

Table 1. External Financing of the U.S. Biotechnology Industry

	Amount (millions of 1995 dollars) raised through:							Biotech equity index
Year	Venture capital	IPOs	Follow-on offerings	Private placements	Debt and convertibles	RDFOs	Total	
1978	23	0	0	0	0	0	23	1.16
1979	71	0	0	0	0	0	71	1.39
1980	161	87	0	0	0	0	248	2.26
1981	185	263	0	13	0	0	461	1.99
1982	247	98	57	0	0	147	549	2.29
1983	292	423	182	0	0	172	1070	1.87
1984	179	55	0	0	0	251	485	1.20
1985	190	10	118	10	0	57	384	1.75
1986	342	538	581	0	148	174	1782	1.60
1987	481	306	309	0	442	113	1651	0.89
1988	467	52	44	35	0	74	671	1.06
1989	469	73	259	24	350	210	1386	1.09
1990	514	152	340	29	130	118	1282	1.14
1991	467	1482	2734	220	301	182	5385	2.48
1992	768	1432	788	12	55	118	3172	2.27
1993	763	716	1092	313	197	11	3091	1.48
1994	737	451	608	184	0	0	1980	1.02
1995	716	670	1605	100	0	0	3091	1.62

Source: The methodology for the construction of the venture financing and biotechnology index series is described in Lerner (1). The IPO and follow-on offering series are from unpublished databases of Recombinant Capital. The private placement, debt and convertible, and RDFO compilations are based on the compilations of Shane (2), extended in time and comprehensiveness through searches of a wide variety of sources.

Note: The table summarizes the total raised (in millions of 1995 dollars) by U.S. new biotechnology firms through several major sources: venture capital investments in private firms, initial public offerings (IPOs), follow-on public equity offerings, private placements by financial investors in public firms, debt and convertible security issues, the issuance of shares in R&D financing organizations (RDFOs), and the sum of these offerings. Alliance-related financings are excluded. The table also indicates the year-end level of an inflation-adjusted index based on the valuation of publicly traded biotechnology firms in this period, normalized to be equal to one on January 1, 1978.

fact that in 1995, the dollar volume of commitments to new alliances in the biotechnology industry was almost equal to venture capital disbursements *in all industries* ($3.4 billion vs. $3.7 billion).

Two limitations of this article should be acknowledged at the outset. First, I do not attempt to duplicate the guides that explain the intricacies of the alliance process to practitioners. Numerous excellent volumes exist (5–7), that document the legal and institutional considerations associated with undertaking biotechnology alliances and technology licenses at much greater depth than could be done here. Second, my focus is primarily on the empirical evidence about these alliances. While a number of works about the theory of technology alliances and licensing are mentioned in passing, the primary focus is on the empirical research.

WHY ARE ALLIANCES AND LICENSING SO IMPORTANT IN BIOTECHNOLOGY?

Academic technology transfer officers and executives at small biotechnology companies often face the challenge of commercializing early-stage biotechnologies with tremendous promise. But a variety of considerations make it difficult to raise financing from traditional sources — such as banks and public investors — for some of the most potentially profitable and exciting technologies. As a result in many cases they are required to turn to strategic alliances

for financing. These difficulties can be sorted into four critical factors: uncertainty, asymmetric information, the nature of firm assets, and the conditions in the relevant financial and product markets.

The first of these four problems, uncertainty, is a measure of the array of potential outcomes for a company or project. The wider the dispersion of potential outcomes, the greater the uncertainty. By their very nature, young biotechnology companies are associated with significant levels of uncertainty: Only a relative handful of drugs actually become commercial products, and a small subset of these proves to be profitable. The extent of intellectual property protection that a new biotechnology product will receive is also often very uncertain. High uncertainty means that entrepreneurs and investors cannot confidently predict what the company will look like in the future. Uncertainty affects the willingness of investors to contribute capital, the desire of larger firms to license unproved technologies, and the decisions of firms' managers.

The second factor, asymmetric information, is distinct from uncertainty. Because of his day-to-day involvement with the technology, a scientist knows vastly more about his discovery's prospects than prospective investors. Various problems develop in settings where asymmetric information is prevalent. For instance, the entrepreneur may take detrimental actions that outsiders cannot observe: perhaps undertaking a riskier strategy than

Table 2. Interfirm Alliances in Three Industries, and by Biotechnology Firms

Panel A: Interfirm alliances by U.S. firms in three industries, 1980–1994

	Number of New Alliances Publicized, by Nationality of Firms		
Year	U.S.–U.S.	U.S.–Europe	U.S.–Japan
1980	42	40	15
1981	48	30	26
1982	57	54	39
1983	51	37	51
1984	88	60	55
1985	86	82	52
1986	118	78	47
1987	133	95	53
1988	141	98	39
1989	122	86	44
1990	121	66	34
1991	106	53	51
1992	155	89	43
1993	192	104	45
1994	235	145	40

Panel B: Interfirm alliances by U.S. biotechnology firms, 1981–1997

		Payments through alliances (millions of 1995 dollars)	
Year	Number of new filed alliances	Precommercial payments promised in new alliances	Actual payments during year to 49 leading firms
1981	30		9
1982	35		111
1983	31		152
1984	42		210
1985	57		149
1986	63		184
1987	62		415
1988	64		298
1989	71		205
1990	81		851
1991	115	741	647
1992	75	931	392
1993	113	1373	806
1994	66	1772	
1995	171	3421	
1996		2334	
1997		4352	

Source: The first panel is from National Science Board (3). The number of new alliances and precommercialization payments series in the second panel are from Recombinant Capital and its unpublished databases. The actual payments series is from Shane (2). It has not been extended beyond 1993 or to include additional firms.

Note: The first panel presents the number of publicized alliances by U.S. firms in three industries — information technology, biotechnology, and advanced materials — between 1980 and 1994. The second panel examines only alliances involving U.S. biotechnology companies between 1981 and 1995 filed with the U.S. Securities and Exchange Commission or state regulatory bodies who make such information public. Presented are the number of new filed alliances each year, the sum of all promised precommercialization payments in the filed alliances that year (the sum of the nominal payments is expressed in millions of 1995 dollars), and the actual payments to a sample of 49 of the largest biotechnology firms in each year (in millions of 1995 dollars).

initially suggested or not working as hard as the investor expects. The entrepreneur might also invest in projects that build up his reputation at the investors' expense.

Asymmetric information can also lead to selection problems. The scientist who makes a potentially important discovery may exploit the fact that he knows more about the project or his abilities than his investors do.

Licensees may find it difficult to distinguish between truly revolutionary technologies and impractical ones. Without the ability to screen out unacceptable projects, outsiders are unable to make efficient and appropriate decisions choices regarding where to invest. These problems have been particularly severe in biotechnology, due to the scientific complexity of the development of new products and processes.

The third factor is the nature of the assets. Firms that have tangible assets — such as machines, buildings, land, or physical inventory — may find financing easier to obtain or may be able to obtain more favorable terms. The ability to abscond with the firm's source of value is more difficult when it relies on physical assets. When the most important assets are intangible, raising outside financing or entering into strategic alliances may be more challenging. In the biotechnology industry, firms have tended to rely on patent protection to protect assets. Those firms that have relied on trade secrets or informal "know-how" have found attracting investors or entering into licensing agreements to be very difficult. For instance, trade secrets offer exceedingly narrow intellectual property protection, only protecting against misappropriation: "the acquisition of a trade secret by a person who knows or has reason to know that the trade secret was acquired by improper means" (8). Thus a firm cannot sue a rival who discovers its trade secret independently or through "reverse engineering" (the disassembly of a device to discover how it works). This is unlike patent protection, which allows the awardee to prosecute others who infringe, regardless of the source of the infringers' ideas. Pooley (9) notes that very few "naked" trade secret licenses are observed, suggesting that the information covered only through this very narrow property right is difficult to transfer in an arm's-length exchange. Further evidence of the importance of broad patent protection in biotechnology is found in Lerner (10).

Market conditions also play a key role in determining the difficulty of financing firms. Both the capital and product markets may be subject to substantial variations. The supply of capital from public investors and the price at which this capital is available may vary dramatically. These changes may be a response to regulatory edicts or shifts in investors' perceptions of future profitability. The availability of equity from public investors for new biotechnology firms has been particularly variable. As Figure 1 indicates, the amount raised by publicly traded biotechnology firms in follow-on offerings (measured in 1995 dollars) went from $340 million in 1990 to $2.7 billion in 1991, then fell again to $788 million in 1992. Since young biotechnology firms face enormous costs while developing new products, they are typically very aggressive in raising capital. Practitioner accounts suggest that during the periods when there are few public equity issues, the markets are essentially "closed" to biotechnology firms.

As a result of all these problems, biotechnology firms often have little choice but to turn to corporations for financing. A pharmaceutical firm or other corporation with a related line of business can overcome many of these information problems, by undertaking extensive due diligence prior to the transaction and monitoring the firm afterwards. As discussed in Lerner (11), this type of intensive oversight is also provided by venture capitalists, though the resources that venture capitalists invest in the typical company are much smaller than those that a major corporation will devote to a substantial technology alliance. The corporation may have assets, such as sales forces and manufacturing know-how, which young biotechnology firms lack and yet are essential

to the successful introduction of a new product. Small, research-intensive firms frequently rely on alliances with larger corporations to avoid having to construct these capabilities, which may take years to develop. Furthermore, the ongoing operations of the corporate partner may enable it to overcome some of the problems associated with the intangible nature of the biotechnology firm's assets.

Described in this manner, these problems may appear to be quite abstract. But they have very real implications for academic technology managers or corporate executives seeking to commercialize early-stage biotechnologies. They may find investors unwilling to invest the time and resources to examine early-stage technologies, or offering only modest payments in exchange for large stakes in innovations that the scientists, technology transfer officers, and company executives believe to be quite valuable. In the remainder of this section, I will summarize the evidence regarding these claims.

This set of suggestions has been most directly examined in two recent Ph.D. dissertations. Both Shane (2) and Majewski (12) examine the decision of biotechnology firms to raise capital through the public markets and through alliances. Through their different methodologies, these works show that firms turn to alliance financing when asymmetric information about the biotechnology industry is particularly high. During these periods — which are measured through such proxies as the variance of the returns of biotechnology securities — firms are likely to delay the time until their next equity issuance, and to rely on alliances rather public offerings as a source of external financing. The authors argue that the greater insight on the part of the pharmaceutical company into the nature of the biotechnology firm's activities allows it to make successful investments at times when uninformed public investors are deterred by information problems.

A challenge to this view is Pisano's (13) examination of the outcome of biotechnology firms' research projects that are and are not developed with the help of alliances with pharmaceutical firms. He examines the probability that a biotechnology company successfully develops a drug being pursued through an alliance, as opposed to one developed by the firm itself. His conclusion that a "lemons problem" leads biotechnology firms to only undertake alliances with pharmaceutical companies that involve inferior technologies, however, seems difficult to reconcile with the large number and dollar volume of these transactions. (A lemons problem can arise when, because of his day-to-day involvement with the firm, an entrepreneur knows more about his company's prospects than investors, suppliers, or strategic partners. Because the counterparty in the exchange is at an informational disadvantage, the potential partner or investor may refuse to enter into a transaction at all.) One way to reconcile these observations is the possibility that biotechnology firms pursue projects through alliances that have lower expected probabilities of success, but whose ultimate payouts are greater.

A related set of work has looked at the decision of the larger company to undertake an alliance. Pisano (14) examines pharmaceutical firms' choices between developing new drugs in-house versus through alliances in the case of 92 drug development projects. He demonstrates that the insights of transaction cost economics (15) and incomplete contracting theory (16) are highly relevant here. In brief, both sets of work highlight the difficulties that two parties may have in undertaking contracts when there is considerable uncertainty. (Among the barriers to writing optimal contracts are the difficulty of negotiating detailed contracts and the problem of being unable to foresee all contingencies.) In particular, when there are few small biotechnology firms working in the field, the pharmaceutical company is more likely to undertake the project in-house. This is consistent, Pisano argues, with the theoretical suggestions that "hold up" problems — that is, efforts by one of the parties to renegotiate the contract on terms more favorable to itself after the agreement is signed — will be greater in this setting. Pisano also finds evidence that firm-specific factors are critical in the decision to undertake alliances.

The determinants of the firm-specific differences in the rate of alliance formation by various pharmaceutical and biotechnology firms are the other major focus of the writings on why firms form alliances. Arora and Gambardella (17) are two of the few economists to examine these firm-specific factors. (Most researchers of these questions have come from an organizational or sociological perspective.) They (17) examine four strategies — alliances with biotechnology firms, research collaborations with universities, purchases of minority interests in biotechnology firms, and acquisitions of these firms — by large pharmaceutical and biotechnology firms. They present a model that suggests that these activities will be complements: If a firm pursues one of these activities, it should be likely to undertake all of them. Furthermore they suggest that these activities should be disproportionately pursued by firms with a greater internal knowledge of biotechnology. On the basis of their analysis of 81 firms, the authors conclude that the four activities are indeed complements, and are disproportionately undertaken by firms with large existing stocks of biotechnology patents. Similar conclusions emerge from a later study by Arora and Gambardella (18). [These studies should be viewed in light of the critique by Athey and Stern (19) of studies of complementarities.] Other studies that have highlighted the importance of firm-specific factors in biotechnology alliance include Argyres and Liebskind (20) and Roberts and Mizouchi (21).

Another approach to firm-specific factors has characterized the organization literature. In particular, these works have highlighted the importance of networks of firms. In industries with rapidly changing and evolving technologies, this literature argues, knowledge is diffused across a variety of firms. As a result firms seek to learn through the formation of alliances. This literature highlights the importance of alliances early in an industry's evolution. These can form the foundation for repeated relationships and further learning.

Powell, Koput, and Smith-Doerr (22) highlight the path-dependent nature of alliance formation in biotechnology. The decision by both biotechnology and pharmaceutical firms to establish alliances seems to be driven less by characteristics such a biotechnology firm's age or growth rate, but rather by earlier experience with such alliances. The authors argue that these firms learn about how to manage and absorb knowledge from their initial strategic alliances. The authors further suggest that firms that are more "centrally" located in alliance networks are more likely to expand their alliance activities.

Some of the consequences of alliances are highlighted in Stuart, Hoang, and Hybels (23) and Koput, Powell, and Smith-Doerr (24). These studies show that biotechnology firms with prestigious sponsors (e.g., financing from well-established venture capital groups and collaborations with prominent pharmaceutical companies) benefit from these relationships. In the former paper, it is shown that biotechnology firms with prestigious partners are likely to complete an IPO sooner and to command a higher market capitalization at the time of the offering. The latter paper documents that companies with such partners are more likely to enter into other collaborations, whether with other elite or non-elite firms.

HOW ARE ALLIANCES STRUCTURED?

The second broad question that this article considers is the manner in which alliances are structured. Unlike the formation of alliances, which as highlighted above has been explored by both economists and sociologists, economists have played a leading role in the study of this issue. Given the dynamic and complex nature of R&D alliances, it is not surprising that there are multiple potential explanations for their structure. This section will begin by highlighting three classes of relevant theoretical research.

The first of these is incomplete contracting theory. A wide range of models, beginning with Grossman and Hart (16) and Hart and Moore (25) and summarized in Hart (26), consider incomplete contracting between a principal and an agent. A typical assumption is that it is impossible for the two parties to write a verifiable contract that could be enforced in a court of law that specifies the effort and final output of the two parties. This is because there are many possible contingencies, all of which cannot be anticipated at the time the contract is drafted. Because of this nonverifiability problem, these models argue that it is optimal for ownership of the project to be assigned to the party with the greatest marginal ability to affect the outcome. This party, who will retain the right to make the decisions that cannot be specified in the contract, should also receive any surplus that results from the project. Because of this incentive, the party will make the decisions that maximize — or come close to maximizing — the returns from the project. An alternative, though complementary, view suggests that firms may write "excessively" incomplete contracts. In particular, Bernheim and Whinston (27) show that in settings where one set of behaviors cannot be contracted upon, it may be optimal to leave other aspects of ownership unsettled. Such "strategic ambiguity" is more likely in settings with greater uncertainty, consistent with the predictions of the Grossman-Hart-Moore class of models.

Aghion and Tirole (28) adapt this general model to a R&D alliance between two firms. As long as the R&D-performing firm has the initial bargaining power or does not face capital constraints, the results are as discussed above: The control rights are assigned to the party whose marginal contribution to the project's success is greatest. When the financing firm has the initial bargaining power and the R&D firm is capital constrained, however, a different pattern may emerge. In particular, if it is optimal for the property rights to be transferred to the R&D firm, the best outcome will not be achieved: The financing firm will be willing to transfer ownership, but the cash-constrained R&D firm will not have enough resources to compensate the financing firm. As a result, an inefficient allocation of the property rights occurs, with the financing firm retaining the rights to the invention.

Biotechnology research has numerous features that resemble the setting depicted in the theoretical literature on incomplete contracts. Biotechnology projects — particularly early-stage efforts — are highly complex and uncertain, making it very difficult to specify the features of the product to be developed. As one biotechnology executive relates:

> Redefining the work when the unexpected happens, as it invariably will, [is essential]. Research is by its very nature an iterative process, requiring constant reassessment depending on its findings. If there is a low risk of unexpected findings requiring program reassessment, then it is probably not much of a research program. (29, pp. 220–221)

Similarly the complexity and unpredictability of the research presents challenges in drafting an enforceable agreement that specifies the contributions of the R&D firm. In particular, firms that contract to perform R&D in alliances frequently have ongoing research projects of their own, in addition to the contracted efforts. In case of a dispute, it may be very difficult for the financing firm to prove that the R&D firm has employed alliance resources to advance projects that are not part of the alliance.

At the same time biotechnology alliances present a more complex picture than many incomplete contract models. Typically these models assume a one-time contracting process between the two parties. Actual alliances reveal more complex contracting patterns. For instance, pairs of firms undertake repeated sets of alliances on different topics. These prior interactions may allow firms to develop reputational capital, and at least partially address some of the contracting problems. Second, these models often assume a vertical relationship: The agent contributes all the effort. In actuality, the relations between some alliance parties are likely to have horizontal elements. For instance, some of the alliances are between pairs of biotechnology concerns, each of which may contribute knowledge. Third, typically parties in these models bargain over a very reduced set of parameters, with a single ownership right being divided between the parties. Real-life agreements, as noted above, are much more complex. Finally, in many models, such as Aghion and Tirole (28), there is no consideration of the impact of asymmetric information on the negotiation of the alliance: Both parties are assumed to be fully informed

at the time the agreement is signed. Rather, the major concern of the negotiating parties is reducing the potential for suboptimal effort after the agreement is signed. In actuality, the situation may be more complex. In particular, the financing party may not be fully informed about the prospects for the project. This may lead to contractual terms that seek to force the informed party to reveal if he has additional information.

This class of models suggests a variety of empirical implications. In general, among alliances where informational asymmetries are greater, the period during which the R&D firm maintains active control over the project should be longer. For instance, alliances that focus on drug development, which is a much more expensive and complex process than the development of diagnostics and other biotechnology applications, are likely to have much greater informational asymmetries. The incomplete contracting hypothesis consequently suggests that these alliances should be associated with longer contract lives.

A second and quite contrasting explanation is the need to provide monitoring. An essential assumption of the incomplete contracting literature is that addressing the agent's (R&D firm's) behavior after the contract is signed is very difficult for the principal (financing firm). In particular, because the contractual agreements cannot foresee every contingency, even if the financier observes problematic behavior on the part of the agent, he cannot address the behavior in a court of law.

The costly monitoring hypothesis, on the other hand, assumes that such behavior can be addressed, but to do so is expensive. Provisions that allow greater oversight and control on the part of the principal are expensive to negotiate and oversee. As a result these terms should be only included in contracts when the danger of opportunistic behavior is high or the costs of oversight modest. The costly monitoring hypothesis [as articulated, for instance, in Smith and Warner (30) and Williamson (15)] predicts that because the ease of monitoring and incentives to pursue opportunistic behavior vary, the optimal degree of restrictiveness will differ across contracts.

One manifestation of these trade-offs should be in the period that the corporation commits to finance the R&D of the smaller firm. If corporations could costlessly monitor the firm, they would monitor and infuse cash continuously. If the project's expected value fell below some value, the corporation would halt funding of the project. (The R&D firm might still wish to continue because the funds are enabling the firm to continue its existing projects, or because the management team itself is enjoying private benefits from the funding.)

The renegotiation or extension of a strategic alliance, however, is a costly process. Particularly within the financing firm, there is likely to be the need for extensive analysis and review prior to the modification of an existing agreement. The renegotiation process is likely to involve a variety of internal and external legal advisors. As a result alliances are negotiated for distinct periods, not continually reviewed and funded.

This view suggests that the duration of these alliances should be a function of the degree of potential agency problems and the cost of providing monitoring. For example, alliances that focus on the complex drug development process are likely to face much greater informational asymmetries. In these settings the costly monitoring hypothesis suggests a greater need for monitoring and shorter alliance lives.

A third explanation is the need for avoiding the costs of financial distress. An extensive corporate finance literature has documented these costs, which can be divided into three classes:

- The liquidation of a biotechnology firm can be highly destructive of value. In particular, it is often difficult to sell "naked" patent awards without the associated know-how and trade secrets. Without the researchers who developed the concept, most patent awards are likely to be worth little. More general evidence of the substantial indirect costs of financial distress is found in Lang and Stultz (31) and Opler and Titman (32).

- Even if the biotechnology firm is not liquidated, concerns about financial distress may cause the firm to be unable to pursue value-creating investment opportunities. Models such as Stultz (33) and Froot, Scharfstein and Stein (34) suggest that if information asymmetries at times preclude external financing or else make it very costly, firms may be unable to pursue value-creating projects.

- Even if financial distress imposes few costs on the firm's shareholders or society as a whole, managers may still seek to avoid distress. In particular, Smith and Stultz (35) formally show that a risk-averse manager who owns shares in a company is likely to engage in socially undesirable levels of risk management. Managers may fear that the bankruptcy of the firm for which they work will be very costly, both in terms of personal wealth and future earning potential (due to the reputational consequences).

This danger of financial distress is very real in the biotechnology industry: Numerous firms have been forced to liquidate or radically trim back promising research programs because of an inability to access external financing.

Motivated by any one of these reasons, the managers of R&D firm may seek to limit the firm's potential exposure to financial distress. The managers may see undertaking one or more corporate alliances as an attractive mechanism to this end. In exchange for giving up much of the eventual profits from an innovation, the R&D firm receives a guaranteed stream of payments from the financing firm. The R&D firm's desire to engage in such "risk management" is likely to be an increasing function of the potential costs of financial distress and the probability that such an event will occur. (Presumably a risk-neutral financing firm would acquiesce to such an agreement in exchange for a lower royalty other concessions.)

While the variation in the extent of the social or managerial costs of financial distress is difficult to observe, it is possible to identify characteristics of the R&D firm that are likely to be associated with a higher probability of distress. For instance, firms whose research focuses on costly drug development are likely to face a greater probability of financial distress, and may be more willing to enter into alliances that guarantee protracted financial payments.

Pisano (36) first examined these questions in a pioneering work. He studied 195 collaborative agreements between biotechnology and pharmaceutical firms, and asked where purchases of equity by the pharmaceutical company were used alongside contractually specified governance rights (e.g., the pharmaceutical company's right to obtain periodic briefings on the progress of the biotechnology firm). Consistent with the costly monitoring view, he found that in settings with greater information problems and information asymmetries (e.g., when the project was R&D intensive or the alliance entailed multiple projects), the pharmaceutical company was more likely to purchase equity as a part of the agreement. The added rights associated with equity ownership can be seen as strengthening the pharmaceutical company's ability to control the biotechnology firm.

Lerner and Merges (37) examine the determinants of control rights within a sample of 200 alliances. They analyze the share of 25 key control rights allocated to the financing firm by regressing the assigned number of rights on independent variables denoting the project stage and financial conditions, as well as controls for a variety of alternative explanations. Consistent with the framework developed by Aghion and Tirole (28), the greater the financial resources of the R&D firm, the fewer control rights are allocated to the financing firm. For instance, a one standard deviation increase in shareholders' equity at the mean of the independent variables leads to an 11 percent drop in the predicted number of control rights assigned to the financing firm. Evidence regarding the relationship between control rights and the stage of the project at the time the alliance is signed is less consistent with existing theory. Projects in their early stages at the time of alliance formation actually assign significantly less control to the R&D firm.

Lerner and Tsai (38) explore the impact of the financing environment at the time was signed on the success of agreements. They show that in periods where financing availability was strong, the agreements were more successful, whether measured by the probability that the drug advanced to the next stage in the clinical trials or was approved. They show that the effect was more pronounced in those agreements where the biotechnology company received little of the control, as Aghion and Tirole predict. This helps address concerns that the result is driven by shifts in an unobserved third factor. Lerner and Tsai also examine the likelihood of renegotiation. If it would maximize innovative output to assign control to the small biotechnology company, though this allocation of control is precluded by financial market conditions, then there should be evident a distinct pattern in renegotiations. In particular, when financing conditions improve for biotechnology firms, it is those agreements assigning the bulk of the control to the major pharmaceutical firm that should be disproportionately renegotiated. The empirical results are consistent with this pattern.

There is also a small but growing empirical sociology literature on the governance of inter-firm transactions. (My thanks to the referee for highlighting this literature. Also related is the much larger literature in the area of social networks, which addresses the communication properties of networks in situations characterized by embedded exchange.) These issues were framed by a theoretical essay by Granovetter (39), who argued that the structure of the alliance network in any given industry plays an important role in diffusing information about the reliability of various industry participants. Access to this information, in turn, will play an important role in influencing how a new alliance will be governed. For instance, a major pharmaceutical company already engaged in a number of alliances will find that its existing relationships provide access to reliable and cheap information about other actors. Information flowing through connections such as shared third parties (i.e., when two firms have an alliance with the same third organization) are quite effective, which allow well-connected contracting parties greater flexibility in structuring collaborative arrangements. The empirical findings in this literature, most importantly Podolny (40) and Gulati (41), are consistent with Granovetter's reputation model. Two of the more robust findings in this literature are that alliances are less likely to have an equity component when two firms have previously formed an alliance or when they are proximately located in the network of prior deals.

Research into contract structure is at an earlier stage than that about the determinants of alliance formation. Thus, the extent of uncertainty about the key drivers of contract structure is not surprising. The proliferation of information on alliances available through database companies such as Recombinant Capital (much of which is publicly accessible on the Internet at *http://www.recap.com*) and the Securities and Exchange Commission's EDGAR database (available at *http://www.sec.gov*) should encourage future researchers.

WHAT IS THE EVIDENCE FROM FIELD RESEARCH?

Another important source of information on biotechnology alliances is case study research. The conclusions of these cases are often not as "neat" as statistical analyses that are crafted to examine a particular question, but field-based analyses can generate a variety of insights. This section highlights the experiences of three companies that have been examined in case studies. While not exhaustive of the case study literature on biotechnology alliances, they suggest the richness of insights that field-based research can provide.

These three young companies all were developing advanced human therapeutics and grappling with the challenges posed by alliances. The biotechnologies pursued by the three firms are quite different: antigen-based allergy drugs (ImmuLogic Pharmaceutical Corporation), advanced drug delivery mechanisms (ALZA Corporation), and monoclonal antibody-based treatments of inflammation (Repligen Corporation). There were considerable differences in the location and sophistication of strategic partners and the stage of development of the technologies.

One point that these cases raise—consistent with the literature discussed above—is how the allocation of control rights is determined both by concerns about behavior after the alliance is signed and by relative bargaining power. One alliance that may be considered successful in many respects was Repligen's May 1992 alliance with Eli Lilly regarding a very early-stage effort to develop a monoclonal antibody-based treatment of inflammation after heart attacks (43). The net-of-market return for Repligen in the three-day window around the announcement of the transaction in May 1992 was +9 percent, and that of Lilly, +2 percent. [These increases can be compared to the +2.1 percent reaction to 55 announcements of R&D initiatives by high-technology firms found by Chan et al. (42).] The early-stage project succeeded in getting its lead product candidate into Phase I trials in just 13 months. (After extending the project in June 1995, however, Lilly canceled its involvement three months later, citing shifting internal priorities.)

In the Repligen-Lilly alliance, three control rights were the subjects of protracted negotiations. The first was the management of clinical trials: the right to decide which drugs would be pursued and when. A second was the control over the marketing strategy, an arena in which Lilly had extensive experience and Repligen only a slight acquaintance. Finally, both parties wished to control the process development and ultimate manufacturing of the drug. Repligen compared favorably on various financial measures to other biotechnology firms at the time that the alliance was signed. Similarly the firm had outperformed an index of biotechnology securities over its history by over 40 percent. (Stock price performance is measured from the close of the day of Repligen's initial public offering to avoid including the "underpricing" of the offering—i.e., the discount at which the underwriters sold the shares to the original investors. Repligen's beta did not differ materially from that of other biotechnology firms.) At the same time, investment banking analysts had expressed concern about the financial pressures that might result if Repligen's earlier alliance with Merck was terminated.

The terms of the alliance that emerged from the negotiations appeared to assign the control rights to the parties whose behavior would have the greatest impact on the product development effort. Repligen was allowed a great deal of control over developing the lead product candidate, an area where it had considerable experience, but tangential product development activities were subject to extensive review by Lilly. Lilly was assigned control over all aspects of marketing, while Repligen was assigned all manufacturing control rights, unless it encountered severe difficulties with regulators.

Other alliances illustrate the importance of the relative bargaining power of the two parties. An example was the January 1978 alliance between ALZA and Ciba-Geigy (44,45). At the time of the alliance, ALZA faced a major financial crisis. The firm had little more than $1 million in the bank, was spending $2 million more per month than it was receiving in revenues, had nearly exhausted its bank credit line, was in violation of several loan covenants, and was precluded from a sale of equity to the public by unfavorable market conditions and the perception that ALZA had been excessively optimistic in its earlier communications with investors and analysts.

The alliance assigned almost total control to the Swiss pharmaceutical giant. Ciba-Geigy was given a super-majority on the joint board that reviewed and approved potential research projects, the right to license and manufacture any of ALZA's current or future products, the ability to block any other alliances that ALZA proposed to enter into, and 8 of the 11 seats on ALZA's board of directors. In addition the Swiss pharmaceutical giant received a new class of preferred shares. If converted into common stock, the new preferred shares would represent 53 percent of the equity in ALZA. Until conversion, however, Ciba-Geigy had 80 percent of the voting rights, an allocation that allowed it to employ ALZA's tax losses.

At the same time it is reasonable to believe that concerns about the postalliance behavior of ALZA also motivated Ciba-Geigy to demand strong control rights. ALZA's leaders had displayed little ability to direct the firm's research effort over the course of the 1970s. This may have led Ciba-Geigy to conclude that the benefits of allocating control rights to ALZA's management were limited. Despite the strict control rights contractually assigned to Ciba-Geigy, there were frequent disputes between the two firms as ALZA researchers sought to either circumvent the pharmaceutical firm's middle management or ignored their instructions outright. Frustrated by these problems, Ciba-Geigy agreed to terminate the alliance and sell back its equity to ALZA in November 1981.

A contrasting illustration is presented by Immu-Logic (46). In March 1991 the firm was considering either entering into an alliance or raising equity in an initial public offering. One concern that led the firm to decide to go public was that a potential strategic partner might exploit its relatively weak financial condition. In other words, ImmuLogic feared that a pharmaceutical company might obtain numerous concessions on key governance and financial issues by protracting the negotiations until ImmuLogic was close to running out of capital. It consequently deferred negotiating an alliance to develop and market its allergy drugs until the firm went public in May 1991. The firm announced an alliance with Marion Merrell Dow in December 1991, which allowed ImmuLogic to retain numerous control rights, such as an equal role in planning marketing strategy in the United States: *In Vivo* magazine hailed the transaction as "push(ing) the limit of the biotech deal . . . a partnership in fact as well as name" (quoted in Ref. 46, Teaching Note 5-293-118, p. 7). Just as ALZA's relinquishment of almost total control to Ciba-Geigy was in large part a consequence of its weak financial position, ImmuLogic's ability to obtain these control rights reflected its financial strength.

These cases also emphasize two issues that are not highlighted in the theoretical literature. One is the interaction between the allocation of control rights and the financial terms of the transactions. For instance, in the negotiations that led to Repligen's retention of control over manufacturing, the firm agreed to an alteration in its compensation. Repligen accepted a lower royalty than originally envisioned, 5 percent of the sales price, but agreed to supply the drug to Lilly at a price (about 15 percent of the sales price) above what it believed its true manufacturing cost would be. Repligen agreed to reduce the price that it charged Lilly if it was able to manufacture the drug for less, but only if its cost was below 8 percent of the sales price.

A second interesting and unexplored aspect is the apparent signal that the allocation of control rights provided to potential investors and other outsiders. Both ImmuLogic and Repligen highlighted their retention of key control rights in the press releases announcing the transactions described here. Their ability to obtain these rights attracted favorable comments in the trade press and analyst reports alike. These patterns suggest a richer set of interactions than theoretical treatments of these issues imply.

WHERE IS FURTHER RESEARCH NEEDED?

This article has sought to suggest the importance, richness, and complexity of alliances and licensing in the biotechnology industry. While much has been learned from the economic and sociological research into biotechnology alliances over the past decade, much more remains to be discovered. This final section will highlight two issues that deserve to be a particular focus of attention.

The first of these relates to the structure of the payments between the financing and the R&D firm. The design and implementation of incentive schemes in general is a major focus of the finance and economics literature, but payments in alliances have been little examined except in theoretical works (47,48). This lack of attention is a reflection of the difficulty in analyzing them. The payments typically are of several types: an initial up-front payment, a purchase of equity (which the financing firm may be able to force the R&D firm to repurchase if the alliance is unfruitful) or warrants, commitments to contract for R&D on specific topics, milestone payments contingent on the achievement of technological and marketing objectives or the renewal of the agreement, and a royalty on the eventual sales generated by the product. Assessing the expected net present value of these payments is very difficult. The magnitude and timing of eventual sales that the project will generate are difficult to anticipate. The amount of the R&D to be contracted for is often ambiguous. Alliances may also include contingent payments for remote outcomes. (The rationale for their inclusion is that firms frequently report — and analysts tabulate when assessing firms — the sum of all precommercialization payments from new alliances, whether the funds are likely to be received or not. These contractually specified contingent payments may thus convey important strategic benefits, even if the probability of payment is very low.) Clearly, this is a difficult but important area for research.

A second question relates to the impact of this contracting regime on the rate and direction of technological innovation in the biotechnology industry. Zucker, Darby, and Brewer (49), in their analysis of the impact of academic research on the development of the biotechnology industry, suggest that academic licensing practices are one of the reasons for geographic clustering of innovative biotechnology firms around top-tier universities. Theoretical work by Gans and Stern (50) provides reasons to believe that the impact of contracting on the pace of innovation may be complex and multidimensional. The impact of intercorporate licensing on innovation is an important issue for future research, given the number and financial significance of these transactions. While a few initial steps along these lines have been taken (51,52), much more remains to be done.

ACKNOWLEDGMENTS

Harvard University and National Bureau of Economic Research. I thank an anonymous referee for helpful comments. Parts of this essay are based on Lerner and Merges (37) and Lerner and Tsai (38). Financial support was provided by the Division of Research at Harvard Business School.

BIBLIOGRAPHY

1. J. Lerner, *J. Financial Econ.* **35**, 293–316 (1994).

2. H.L. Shane, Asymmetric Information and Alliance Financing in the Biotechnology Industry, in *Three Essays in Empirical Finance in High-Technology Firms*, Unpublished Ph.D. Dissertation, Wharton School, University of Pennsylvania, Philadelphia, 1995.

3. National Science Board, *Science and Technology Indicators — 1998*, U.S. Government Printing Office, Washington, DC, 1998.

4. Industrial Research Institute, *Industrial Research Institute's Trends Forecast for 1998*, Industrial Research Institute, Washington, DC, 1997.

5. *Corporate Alliances: Strategies in Biotechnology*, KMPG Peat Marwick, Biotechnology Industry Organization, and Pittiglio, Rabin, Todd & McGrath, San Francisco, CA, 1993.

6. B. Cunningham, *Issues and Trends in Biotech Corporate Partnering*, Cooley, Godward, Castro, Huddleston & Tatum, Palo Alto, CA, 1994.

7. J.W. Schlicher, *Licensing Intellectual Property: Legal, Business, and Market Dynamics*, Wiley, New York, 1996.

8. R.M. Milgrim, *Milgrim on Trade Secrets*, Matthew Bender, New York, 1993.

9. J. Pooley, *Trade Secrets: A Guide to Protecting Proprietary Business Information*, American Management Association, New York, 1989.

10. J. Lerner, *Rand J. Econ.* **25**, 319–333 (1994).

11. J. Lerner, *J. Finance* **50**, 301–318 (1995).

12. S.E. Majewski, Causes and Consequences of Strategic Alliance Formation: The Case of Biotechnology, Unpublished Ph.D. Dissertation, Department of Economics, University of California at Berkeley, 1998.

13. G.P. Pisano, *R&D Performance, Collaborative Arrangements and the Market-for-Know-How: A Test of the 'Lemons' Hypothesis in Biotechnology*, Working Paper No. 97–105, Harvard Business School, Boston, MA, 1997.

14. G.P. Pisano, *Admin. Sci. Q.* **35**, 153–176 (1990).

15. O.E. Williamson, *The Economic Institutions of Capitalism: Firms, Markets, Relational Contracting*, Free Press, New York, 1985.

16. S.J. Grossman and O.D. Hart, *J. Political Econ.* **94**, 691–719 (1986).

17. A. Arora and A. Gambardella, *J. Ind. Econ.* **38**, 361–379 (1990).

18. A. Arora and A. Gambardella, *J. Econ. Behav. Organ.* **24**, 91–114 (1994).

19. S. Athey and S. Stern, *An Empirical Framework for Testing Theories About Complementarity in Organizational Design*, Working Paper No. 6600, National Bureau of Economic Research, Cambridge, MA, 1998.

20. N.S. Argyres and J.P. Liebeskind, *Acad. Manage. Rev.*, (in press).

21. E.B. Roberts and R. Mizouchi, *Int. J. Technol. Manage.* **4**, 43–61 (1989).

22. W.W. Powell, K.W. Koput, and L. Smith-Doerr, *Admin. Sci. Q.* **41**, 116–145 (1996).

23. T.E. Stuart, Ha Hoang, and R.C. Hybels, *Admin. Sci. Q.* **44**, 315–349 (1999).

24. K.W. Koput, W.W. Powell, and L. Smith-Doerr, Interorganizational Relations and Elite Sponsorship: Mobilizing Resources in Biotechnology, Unpublished working paper, University of Arizona, Tucson, 1997.

25. O.D. Hart and J. Moore, *Econometrica* **56**, 755–785 (1988).

26. O.D. Hart, *Firms, Contracts, and Financial Structure*, Oxford University Press, New York, 1995.

27. B.D. Bernheim and M.D. Whinston, *Am. Econ. Rev.* **88**, 902–932 (1998).

28. P. Aghion and J. Tirole, *Q. J. Econ.* **109**, 1185–1207 (1994).

29. J.P. Sherbloom, in R. Dana Ono, ed., *The Business of Biotechnology: From the Bench to the Street*, Butterworth-Heinemann, Stoneham, MA, 1991, pp. 213–224.

30. C.W. Smith, Jr. and J.B. Warner, *J. Financial Econ.* **7**, 117–161 (1979).

31. L.H.P. Lang and R.M. Stultz, *J. Financial Econ.* **32**, 45–60 (1997).

32. T. Opler and S. Titman, *J. Finance.* **49**, 1015–1040 (1994).

33. R.M. Stultz, *J. Financial Econ.* **26**, 3–28 (1990).

34. K.A. Froot, D.S. Scharfstein, and J.C. Stein, *J. Finance* **48**, 1629–1658 (1993).

35. C.W. Smith, Jr. and R.M. Stultz, *J. Financial Quant. Anal.* **20**, 391–405 (1985).

36. G.P. Pisano, *J. Law, Econ., Organ.* **5**, 109–126 (1989).

37. J. Lerner and R.P. Merges, *J. Ind. Econ.* **46**, 125–156 (1998).

38. J. Lerner and A. Tsai, Do Equity Financing Cycles Matter? Evidence from Biotechnology Alliances, Unpublished working paper, Harvard Business School, Boston, MA, 1999.

39. M. Granovetter, *Am. J. Sociol.* **91**, 481–510 (1985).

40. J. Podolny, *Admin. Sci. Q.* **39**, 458–483 (1994).

41. R. Gulati, *Admin. Sci. Q.* **40**, 619–652 (1995).

42. S.H. Chan et al., *J. Financial Econ.* **46**, 199–221 (1997).

43. D. Kane and J. Lerner, *Repligen Corporation: January 1992*, Harvard Business School Case No. 9-294-082 (and Teaching Note 5-295-137), Harvard Business School, Boston, MA, 1994.

44. R. Angelmar and Y. Doz, *Ciba-Geigy/Alza Case Series (including Advanced Drug Delivery Systems: Alza and Ciba-Geigy, A through F; Alza Corporation, A and B; Ciba-Geigy Limited: Pharmaceutical Division, A through C)*, Unnumbered INSEAD case studies, Fontainbleau Cedex, France, 1988.

45. J. Lerner and P. Tufano, *ALZA and Bio-Electro Systems*, Harvard Business School Cases No. 9-293-124 through 9-293-127 (and Teaching Note 5-296-060), Harvard Business School, Boston, MA, 1993.

46. J. Lerner, *ImmuLogic Pharmaceutical Corporation*, Harvard Business School Cases No. 9-293-066 through 9-293-071 (and Teaching Note 5-293-118), Harvard Business School, Boston, MA, 1994.

47. N. Gallini and B.D. Wright, *Rand J. Econ.* **21**, 237–252 (1990).

48. M. Kamien and Y. Tauman, *Q. J. Econ.* **101**, 471–493 (1986).

49. L.G. Zucker, M.R. Darby, and M.B. Brewer, *Am. Econ. Rev.* **88**, 290–306 (1998).

50. J. Gans and S. Stern, Incumbency and R&D Incentives: Licensing the Gale of Creative Destruction, Unpublished working paper, Sloan School of Management, Massachusetts Institute of Technology, Cambridge, MA, 1998.

51. W. Shan, G. Walker, and B. Kogut, *Strategic Manage. J.* **15**, 387–394 (1994).

52. D.C. Mowery, J.E. Oxley, and B.S. Silverman, *Strategic Manage. J.* **17**, 77–91 (1996).

See other entries MEDICAL BIOTECHNOLOGY, UNITED STATES POLICIES INFLUENCING ITS DEVELOPMENT; SCIENTIFIC RESEARCH, POLICY, TAX TREATMENT OF RESEARCH AND DEVELOPMENT.

TRANSFERRING INNOVATIONS FROM ACADEMIC RESEARCH INSTITUTIONS TO INDUSTRY: OVERVIEW

Michael D. Witt
Technology Enterprise Development Company
Troy, Michigan

Susan K. Lehnhardt
Morrison & Foerster LLP
New York, New York

OUTLINE

ACADEMIC RESEARCH INSTITUTIONS

Congress has made a national priority of bringing academe and industry together because of the beneficial effect of biomedical technology transfers on U.S. competitiveness and public health (1).

The importance of this interaction has evolved over time. Traditionally academe viewed its mission as above the fray of industry. Industry, likewise, relied on itself for research and development of new products and viewed academe as the "ivory tower" and not producing anything of commercial value. Since the 1960s academic research centers have attracted the best and brightest scientists and researchers to their laboratories. Large numbers of available and qualified applicants enabled research facilities to be rigorous and selective in their review of medical school and graduate school candidates. The students who survived this process were skilled, exceptionally talented, and dedicated to achieving success in the academic environment. With substantial federal support for research, mostly at the basic research level, a

productive, unique, yet eccentric atmosphere developed where research was not inhibited by commercial or financial restraints.

Although this research capability is still largely in place, tremendous fiscal constraints for the last decade or more have eroded the ability of scientists to pursue basic research without at least some consideration of practical, commercial applications of their research. In addition federal reimbursement levels for medical care at academic medical centers (which also funds teaching, capital improvements and medical research) is increasingly restrictive.

The fiscal drain on academic science is exacerbated by heightened bureaucratic demands on medical research center personnel. Independent yet federally mandated peer review organizations, quality assurance programs, boards of medical examiners, state, federal and private insurance cost containment programs, and malpractice suits, each detract from the institutional mission, and cumulatively extract a profound professional toll on academic scientists. As a result frustration and anger is endemic, even among the most idealistic and dedicated.

In light of these developments, scientists still willing to work in the non-profit sector are turning to industry for research support (2). Research scientists who long eschewed corporate contacts are now more willing to seek out and perform corporate sponsored research.

Many corporations, similarly squeezed by considerable international competition, restrictive federal tax policies, product liability costs, and reduced access to the public financing markets, are looking anew at research centers (3). Corporations sense that internal research and development (R&D) might be productively supplemented with outside research efforts and technology, initially developed at government expense (3).

The resulting relationships between academics and industry require considerable patience. The parties also need to understand each other. Their motivation, stress, pressures and conflicting obligations are quite different.

Academics is Different than Industry

Scientists in the nonprofit sector are accustomed to responding to academic pressures. They are relatively unaccustomed to conducting exploratory research at the applied level as do their industry counterparts. The rate and method of academic research, the selection of research objectives, limited resources, and accountability are fundamentally different.

When an academic scientist and his peer group are convinced of a given result, the result often raises other questions and additional effort, but the given result, even if appropriate, is generally not developed into a product—in industry, this often is only a starting point. The result must be developed into a product, and tested further to the full satisfaction of management and the regulatory and licensing agencies. The scope of the commercial-grade scientist's work may be less purely inventive at times, but the cost and effort required to create a product, then testing it, may be several hundred to thousand-fold the cost of the initial discovery.

Qualitative and Quantitative Differences in Research

Job security in the academic world requires quality teaching, writing, research, and a financial base provided by grants and government research funds. Reduced grant funding threatens this job security. It gives rise to tremendous pressure to generate research funds, and causes scientists to seek funding from any available resource. One source of research funding is the commercial sector. The commercial sector is willing to pay for research, and to pay for intellectual property that may lead to a product. Historically academicians were not financially rewarded for turning their research results into products. Thus they tended to carry the research effort only to the "proof of concept" stage, then moved on to their next activity. Policy changes at most academic centers now allow scientist to receive financial rewards from commercial successes (4).

As nonprofit organizations, academic research institutions are primarily funded by the federal government. Since the federal government has a major interest in promoting the development of innovations, it implemented legislation to facilitate the transfer of new technology from academic research institutions to industry, ensuring that scientists have an added incentive to turn their research results into products. On December 12, 1980, nearly 20 years ago, Congress enacted the Bayh-Dole Act to "reform U.S. patent policy related to government-sponsored research" (4, p. 3). The Bayh-Dole Act has two primary purposes:

- To enable universities, small business, and not-for-profit corporations to "patent and commercialize their federally funded inventions"
- To enable "federal agencies to grant exclusive licenses for their technology to provide more incentive to businesses" (4, p. 3)

The Bayh-Dole Act also has eight key regulations (4):

1. At the time of funding, an agency must make the funding university aware of its intention to hold title to an invention due to extraordinary circumstances or conditions.

2. Within two months of the date the inventor makes the university aware of the invention in writing, the university has to inform the proper federal agency of any invention created with the use of federal funds. This is the disclosure date.

3. To maintain ownership of an invention, the university basically has to inform the agency of its decision to keep the title within two years of the disclosure date. When the one year statutory period is initiated by public use, sale, or publication from which "valid" patent protection can be secured in the United States, the agency may reduce the "period of election to not more than 60 days prior to the end of the statutory period."

4. In order to use its invention, the university must give the U.S. government a paid-up, nontransferable, nonexclusive license (confirmatory license).

5. The government has "march-in rights" over the invention. This means that the government may relieve the university of its title to an invention if the university does not try to develop an invention or if there is a need to "alleviate health or safety concerns" (4,5).

6. The university has to give preference to small businesses when granting licenses for the use of the invention.

7. The university has to make sure that the invention will be developed primarily in the U.S. when it "grants an exclusive license."

8. The university has to share a portion of the royalties with the inventors (4–6). Hopefully this Act and its implementation will exert subtle pressure on the academic researcher to work with companies to develop products.

With the implementation of legislation such as the Bayh-Dole Act, it is not difficult to understand that the historic pressure on the academician to publish, teach, and receive grant funding has increased. Nevertheless, the fundamental differences between academe and industry create challenges for any commercial industry wishing to build on discoveries found in the research center.

Governance and Decision Making

Traditionally research institutions have benefited from corporate and personal charitable donations. These funds are now less available. The trustees of the research centers have responded by requiring management to pursue other forms of support.

To their credit, research institutions have created a variety of constructive programs aimed at addressing the concerns of industry regarding the scope and commercially useful nature of academic center research. In fact some universities have expended considerable funds to facilitate transferring technologies and programs to meet the specifications of the Bayh-Dole Act. Under these programs certain units and personnel are instructed to manage activities relating to inventions" (4).

If the approaches adopted by research centers are well managed, and carefully selected, then the research resulting from the collective endeavors of its scientists will more often lead to products, which in turn, should lead to fees and royalties to the researcher center and its scientists.

Most of these programs are designed to support the educational and basic research mission of the academic center. This is required by their nonprofit charter. The programs identify potential commercial products within the facilities, protect the inventions, and then identify and/or build companies for commercializing the discovery. Researchers are not forced into taking a role in the commercialization process. Through this mechanism academic freedoms traditionally enjoyed by the academician are maintained. It is not clear, however, that in the long term the fundamental mission of the research center will not be subverted, a major risk for and concern of academic governing bodies.

In order to benefit from academic inventions, both academe and industry must develop sensible and affordable cooperative relationships that meet the goals of both parties. While an academician may be aware of commercial tasks required to create a product, commercially oriented research projects are not a typical part of academic research. Through education and by understanding the cultural, administrative and professional constraints on industry, the academic can contribute to the commercial success of the project. One approach which works for some corporations is to recognize the high level of skill and quality research conducted at the academic research center, and effectively integrate them into the corporate R&D process.

PROGRAMS AT UNIVERSITIES TO COMMERCIALIZE INNOVATIONS

Goals of Research Institution Technology Transfer Programs

Licensing executives at most technology transfer offices (TTOs) will state that their primary mission is to enhance the flow of innovations from the institution's laboratories to the commercial sector. This mission statement usually derives from their nonprofit charter, as well as a long history of teaching and research, not for private gain but for the greater good of the community that the research center serves. The logic is that by transferring technologies to the private sector, people will benefit from the innovations, and thus the general welfare of the state is enhanced.

This mission statement also reflects pressure from community leaders to refrain from competing with taxpayers. By conferring nonprofit status on the universities, the federal, state, and local tax codes restrict research facilities from competing with commercial entities. While there is continual and lengthy debate at the trustee level about amounts of unrelated business income derived from for-profit activities conducted under the nonprofit charter, rarely are the amounts of income and commercial activity a serious problem for the research center.

The second most often stated goal of the technology transfer program (TTP) is to make money for the university provided that the activity is consonant with the nonprofit and educational goals of the institution. So long as such is the mission of the institution, the TTP and its staff have complete freedom to conduct business as they wish.

Academic medical centers have been most often criticized for the results of their efforts to maximize health care reimbursement. In pursuit of this effort, hospital corporations restructured, developing series of for-profit and nonprofit corporations. Some of the nonprofit businesses were travel agencies, laundries, janitorial services, or power stations, often far afield from the basic mission and directly in competition with local area businesses. Community reaction and abhorrence held much of this in check. Some state and local tax authorities occasionally successfully levied property taxes on the research centers, pursuant to the logic that the research center was not benefiting the public welfare by offering such services.

The general threat of loss of nonprofit status, which rarely is a serious issue for well-planned research centers, continues to keep the research center focussed on its obligations to its community.

Other stated missions of the TTO include improving patient care by developing innovations that benefit the public, and to reward (and thus retain) talented faculty by allowing them to share in the financial benefits of fees and royalties from their discoveries. Scientists rightly observed that they could make more money by working for private companies, so universities responded by allowing them to consult, for fees and equity, and to share in royalty and fee income. While this raises questions of conflict of interest, most institutions allow and encourage such activity so long as scientists comply with institutional conflict of interest policies.

Probably the most important benefit of transferring a new technology to a local company is to increase community good will. This transfer often translates into donations and additional corporate sponsorship. Many of the research centers are also large area employers, thus further strengthening the local economy.

Effectiveness

Financial performance of the TTOs are difficult to assess. Some of the larger institutions have well-established programs and successfully generate fee income. These programs tend to have substantial revenues — much of those fees derive, however, from only several innovations which are, in turn, used to support an aggressive licensing and technology transfer operation. The surveys conducted by AUTM (Association of University Technology Managers, a nonprofit organization formed to assist university intellectual property administrators in the effective transfer of technology to the public) (4), indicate that some universities have had success with activities involving inventions and the report released by the U.S. General Accounting office on Technology Transfer: The Administration of the Bayh-Dole Act by Research Universities, further indicates that many universities believe that the Bayh-Dole Act is accomplishing its objectives (4).

Licensing Operations

Modern TTPs take a variety of forms. The most common program is a licensing operation (LO). The operational role of this operation is to arrange for intellectual property (IP) protection of discoveries made by university faculty, and to negotiate, prepare, and monitor license agreements with outside companies.

Initially such activities were part of the contracts and grants office. The general role of this office was (and is) to negotiate contracts and maintain relationships with private and government granting agencies.

As federal laws changed to allow title to discoveries to vest in research institutions, the contracts and grants officers were additionally required to arrange for the protection of rights and to create royalty and fee income for the institution by transferring such rights to companies via license agreements. These influences and pressures required different skills and approaches, and the TTO resulted.

TTOs were initiated with high hopes and aspirations. The revenues that were expected to roll in did not. To compound the problem, decisions to patent inventions were made without sound business rational. Patent costs soared.

Additionally the skill set necessary to manage the complex licensing process typically resides at companies accustomed to licensing—there were not many seasoned business executives employed by universities in this role at that time.

It became apparent that the institution undertaking such a program required substantial institutional commitment, a sound business approach, and a long-term view to the process, at least 10 years. Some of the more fortunate institutions capitalized on early discoveries that were highly profitable, encouraging them to pursue other licenses. If these early successes are removed from their revenue streams, it is still clear that they now are investing tremendous resources on the assumption that their current licenses will be profitable. The result of this will not be known for some time, when many hundreds of licensed innovations mature. Although, some of the early licensing programs are beginning to yield substantial returns, it is not a widespread phenomenon (4).

Program Resource Requirements

Underfunding. University TTOs tend to suffer from a variety of dilemmas. Most institutions underestimate the length of time required for their patents to mature and yield returns. If they have a 10 year time horizon, they generally will be able to withstand the criticisms and pressures to perform by university administration.

Most institutions underfund and understaff their TTOs. Usually the cost of maintaining an effective program is greater than the university can afford. The result is that the licensing officers are responsible for handling hundreds of patent disclosures in a variety of technical areas, and the overall effort is diminished.

Patent Process. Because of this lack of personnel and financial resources, licensing officers tend to conduct minimal market, business, and/or technical research to validate their initial decision to patent an invention. This results in waste and inefficiency in the patenting process.

It also results in portfolio of patents that may be too limited for full commercial utility for some valuable discoveries, and an overly optimistic assessment for most of the inventions. The most sensible approach is to invest the requisite time and effort to conduct a business and technical assessment of the invention prior to embarking on the patent process.

Technology Audits. Institutions should take stock of invention inventory through a thorough technology audit.

Each invention is assessed for its market potential, feasibility, time lag to product, and regulatory and financial requirements. The inventions are then ranked according to institutional priorities, and resource allocation requirements. They may also be ranked according to their commercial potential.

This invention assessment allows the institution to make sound business judgments about funding, its level of interest in seeing the products reach the marketplace, and its appetite for pursuing licensing or new enterprise development.

Any subsequent financial decisions with respect to the inventions are weighed against other institutional commitments, and its priority ranking. This is especially valuable in assessing whether to proceed with the patent process, and how seriously the effort should be pursued. Subsequent marketing efforts are also measured against their priority on the list. Those below a certain level, weighed in light of other resource demands, can be returned to the inventor. In this way, the most technologies considered most valuable to the institution are properly and rationally protected.

Marketing Effort. The scope of some discoveries justifies new company formation, with the requisite involvement of capable management and proper funding. For a variety of reasons, however, most successful technologies, perhaps more than 95 percent, are suitable only to be out-licensed. Selecting the proper licensees, unfortunately, requires a considerable amount of time and effort. This effort, due to funding, time, and personnel constraints, is generally beyond the scope of most TTOs. The result then is predictable: The technology licensee is usually the company, any company, that first makes an offer, any offer, to the university.

The more sensible approach is to develop a marketing plan for the technology, beginning with the data that resulted from the technology audit. This information needs to be supplemented with fresh technical, market, business, and regulatory analyses and summarized in both nonconfidential and confidential disclosures. This effort aids the TTO staff in its search for appropriate licensees, which are targeted in a defined marketing program. This defined marketing effort is calculated, if successful, to result in serious and appropriate partners, and new corporate relationships for the institution. Even if the technology is not purchased by the potential licensees, a well-reasoned and sensible approach to a serious company will create a favorable impression. This may well lead to subsequent opportunities—it definitely improves the likelihood that a prospective licensee will take the TTO seriously.

Drafting and Maintaining Licenses. Once a licensee has evaluated the technology, and wishes to enter into a license agreement to acquire the rights to the technology, the license negotiation process is initiated. The drafting and negotiation process is rarely routine. Many treatises have been written, and numerous license agreement forms have been generated, all of which serve as useful tools for the experienced and inexperienced TTO personnel.

Licensing Check List. A typical license agreement should address at least the following:

Exclusive versus nonexclusive

Restrictions in freedom to License and Sublicense

Contamination — e.g., commingled funding by federal and/or other industrial money

Collateral agreements may restrict rights

Confidentiality agreements — any confidential information used from any/other restricted sources?

Any nondisclosure agreements that encumber the invention?

Collaborations with other scientists on site —

    any visitors that were employees of other institutions that signed institutional

    patent agreements

    off-site collaborations

    any scientists who have left.

Financing/Royalty Clause Consideration

License issue fee/varies —

    fully paid up/lump sum

    running royalty, royalty cap, or minimum/maximum

    equity in lieu of or addition to royalties

    credits given against royalties for earlier expenditures

    reimburse for research expenses

Milestones/upfront payments/termination fees

Background rights —

    define patent/other intellectual property with care

    follow on patents, improvement patents

    improvements

    subservient to basic patents

    retain option to these background rights

    negotiate a separate royalty rate, short time window for improvements (6–18 months)

    subject to the rights of other parties

    define field of use

Know-How —

    carefully define and transfer it

    its delivery is very amorphous — if well defined, then know-how transfer is easier

    treat via field of use

    typically nonexclusive

    if exclusive, only if it can be protected, and only if it is no longer needed

    retain the rights to use know-how

    consider using separate agreement

Infringement and patent protection

Indemnity

Disclaim warranty provisions

Territory Infringement

Field of use

March in rights and due diligence

Use periodic payments, with reversion of rights

Automatically for nonperformance

If patents are assigned, pre-execute grant back of patent rights subject to conditions

Once a licensing relationship is in place, it must be regularly and objectively reviewed. Maintaining and monitoring licenses, communicating with licensees, and enforcing the terms of the agreement are time-consuming, and must be planned for in advance, and in light of their ranking in the priority list. Those agreements that have gone awry must be analyzed as to the reasons for their failure, utilizing in-house and independent peer review personnel. Nonperforming or minimally productive licenses should be terminated.

If the licensee does not proceed in a timely and businesslike manner, the technology should revert to the institution, pursuant to defined and clear terms that do not hamper the licensee's ability to raise funding or pursue the technology. Most often, if the company is not willing to commit to financial benchmarks, it is the wrong licensee.

Licensing Agreement Controls. Proper controls to ensure quality license agreements must be in place. Standardized forms are usually a useful starting point, and should be used where possible. Associates preparing the licenses must be subject to performance reviews. In order for the process to result in an appropriate business result, the expectations and demands on both the licensee and licensor must be adhered to, requiring the licensing office to ensure that the institutional commitments are met and that the office is operating in a businesslike fashion.

Patent Decisions. With respect to the decision whether to apply for a patent, claiming certain inventions, the typical licensing officers do not conduct technical analyses of the technology, relying instead on the inventor to do so. With some unusual inventors, this might be adequate provided that they know the industrial side of their research. Most, however, do not have the appropriate skill set. A careful search for and review of relevant technical literature by a scientist other than the inventor may reveal very sound reasons why the patent expenditure does not make any sense. For example, if a use patent is filed on a known compound, but the literature reveals that the use applied for may give rise to an adverse reaction that conclusively eliminates its commercial value, then the licensing office should not expend its resources to patent the invention. Unfortunately, most of these important pieces of information appear when the USPTO examines the patent application, or, more often, when a licensee conducts its due diligence, and it becomes obvious that the money spent on the patent was wasted. It also dilutes the effectiveness of the office staff, who could better use their time on other licensing projects.

If the TTO's decision is not to file the patent application, the university should relinquish its rights to the invention and return such rights to the inventor. It should do so as soon as practicable. It should also describe why it has declined the opportunity to file a patent application and give the details of its business rational.

A corollary to the technical literature search is a careful analysis of the prior art in the patent literature. While this usually is done well by competent patent counsel, the licensing officer can glean valuable information about competing patent estates that impinge on the value of the technology. Learning to operate the computer search software programs is easy; learning to use them well is very, very difficult, and is best done by someone who is in a dedicated service role to the licensing associate.

Another problem is created when the licensing operation fails to plan for the cost of enforcing patents. Usually the decision to prosecute an infringer is made in a hurried manner, without fully appreciating the potential costs and benefits. Prosecuting infringers is a very expensive proposition.

Other Agreements. A typical hazard for the licensee and licensor are "hidden documents" that appear at the last minute, or are discovered very late in the patenting, licensing, or due diligence process. These hidden documents may contain restrictions that limit the university's ability to grant free title to the licensee. Consulting agreements and material transfer agreements may have been executed without appropriate terms and conditions. Federal funding, commingled funding, university collaborations, whether formal or informal, nonconfidential and confidential disclosure agreements, and joint inventorship of patents, each raise case-specific problems that affect the value of the technology. Analyzing these issues and optimizing all aspects of ownership is most effective when done before the TTO approaches or is approached by a potential licensee.

Faculty Communication. The licensing officers must have early and frequent communication with their inventors. By involving them in the decision-making process, the faculty are recruited into the process, learn about the rationale for decisions made, and become a valuable resource. If decisions are made without their involvement, the faculty will be alienated, and resentful of business and administrative decisions, which at many points will detract from the office's effectiveness.

CONFLICT OF INTEREST DILEMMAS IN BIOMEDICAL RESEARCH

Industry–academe collaborations have costs. These collaborations create conflicts among the researchers, the institutions, industry, and the researcher's academic and financial interests. These conflicts of interest threaten the objectivity of science, the integrity of scientists and institutions, and the safety of medical products.

One such example of a conflict of interest between the researcher, the institution, and the industry can be observed from the Synthroid Marketing Litigation case (7). In this case researcher Dr. Betty Dong "discovered" that there were less expensive alternatives to the Synthroid medication (8). Dr. Dong's research was funded by the University of California, San Francisco, and Knoll

Pharmaceutical Company, which manufacturers a thyroid medication called Synthroid. When Dr. Dong attempted to publish the results of her study, the representatives of Knoll informed her that she was barred from doing so because she had signed a contract agreeing to publish her results subject to the approval of the company when she initially began the project (9).

Claiming the right to academic freedom, Dr. Dong took the Knoll Pharmaceutical Company to court (10). This case illustrates a wide range of conflicts. Was it permissible for the company to suppress Dr. Dong's study for its own reasons? Was it permissible for Dr. Dong as a university employee and beneficiary of funding from Knoll to comply with the suppression of her article for seven years? If Dr. Dong were to be penalized for breaching her contract with the company, would the university be liable as well, or would she be treated as an independent contractor? Hopefully these questions will be answered once the case has been resolved (11).

To avoid conflicting interests that undermine scientist's integrity such as was the case with the Dong study, adherence to uniform federal standards should be mandatory. Federal rules are necessary to require disclosure of conflicts, limit the most troublesome forms of conflict, and create uniformity in ethical standards across the country.

Federal Legislation Supporting Industry–University Collaborations

Federal legislation, since 1980, has facilitated industry–university collaborations, and speeded promising new products from the laboratory to the market. This legislation, however, has brought academe and industry together without adequately regulating the consequences of the interactions.

The Stevenson-Wydler Technology Innovation Act of 1980 (12) established a policy of "stimulating improved utilization of federally funded technology developments by state and local governments and the private sector" (13). The Act created an Office of Research and Technology Applications, whose primary purpose was to investigate projects that could be utilized by government or private industry (14). Each Agency implemented this requirement by having it own version of such Office (i.e., the Office of Technology Transfer at the National Institutes of Health, NIH, handles this function for the Department of Health and Human Services, DHHS). The Act also created the Center for the Utilization of Federal Technology, established within the National Technical Information Services, to provide industry with a central source of information on federally owned or developed technologies with potential commercial application (15).

Later in 1980, Congress accelerated technology transfers by amending the patent and trademark laws, and for the purpose of supporting small business. As mentioned previously, the Bayh-Dole Act (16) gave inventors in small business firms and nonprofit organizations the power to retain ownership rights to patents protecting inventions developed with federal funding (17) (Licenses are intended to be granted to small businesses in the United States. If this is not possible, the licensing institution is to use its best efforts to ensure that the licensee, whether foreign or U.S. based, manufactures the product in the United States, and for consumption in the United States). Prior to the Bayh-Dole Act, the federal government owned the rights to most federally-supported inventions and for-profit firms wishing to develop federally supported inventions had to wade through a bureaucratic maze to obtain a license (18). Congress recognized that the federal government had been unsuccessful in nurturing the development of new products to the market, and that it was in the public interest to bring innovative ideas into clinical practice without unnecessary delays (19). A policy statement by the Reagan Administration extended the Bayh-Dole coverage beyond small business firms and nonprofit organizations (20). The Bayh-Dole Act enabled institutions and their investigators to license patents from federally-supported work to companies interested in developing the products for market. This has allowed the institutes and researchers easier access to money, both for research and for personal profits, and opened up to industry a large market for potential commercial advantage (21).

Liability Risks of Research Institutions and Investigators

Despite the progress of the current federal legislation supporting industry–university collaborations, the collaborators still have to contend with the potential for a conflict of interest leading to legal liability. This potential is derived from several sources. First, state tort law imposes liability on institutions for the misconduct of their employees. A research institution could be held liable for the negligence, misrepresentation, or fraud of its investigator or employee. For example, if a researcher misrepresents the quality of an invention that is commercialized, and the company relies on false claims, the university could be held responsible. Considerable damages may be assessed resulting from the delay in marketing the product and/or in wasted or misdirected investments.

Second, research institutions have various obligations under state nonprofit corporation laws and federal tax laws, any breach of which could jeopardize the institution's nonprofit status, or subject the institution to enforcement actions by the state attorney general. Typically these laws require the directors to operate the research institution in a manner consistent with its charitable purpose; forbid certain director conflicts, interlocking or interested director transactions; and require the directors to preserve and prudently invest corporate assets. Although research institutions need to closely adhere to the current regulations, the court has recently given them some flexibility with respect to managing an invention the has been assigned to the institution by an employee. In *Kucharczyk v. Regents of the University of California* (22) addresses the ability of inventors to influence negotiations for the sale of their inventions.

The Regents had negotiated a licensing agreement with Nycomed for the use of a patented medical technique developed by Dr. John Kucharczyk and Dr. Michael Moseley (22). The doctors assigned their rights to the patent to the Regents which then sold these rights to Nycomed for $25,000. Fifty percent of the sale price went to inventors, Drs. Kucharczyk and Moseley in compliance with the Bayh-Dole Act (4,22). The doctors then filed suit against the Regents of the University and Nycomed alleging that the defendants "acted improperly to deprive plaintiffs of their rightful share of the financial rewards of the patented medical technique they developed" (4). In their lawsuit, the plaintiffs claimed that the medical technique invented by them was "worth substantially more than $25,000."

The court ruled that the actions of the University of California and Nycomed did not constitute a breach of contract since the University doctors had contracted out their right to sue when they assigned their inventors rights to the University. The court further ruled that the plaintiffs might have a claim against Nycomed for fraud and interference with contractual relations, because there was sufficient evidence of Nycomed's suggestion to the University doctors that their medical technique had substantial profit-making capacity, and that this information was omitted from the negotiations with the University.

CONCLUSION

With increasing sophistication and skill, technology transfer offices are successfully commercializing useful inventions and generating fee income. The recent court decisions of *Kucharczyk v. Regents* and *Synthroid Marketing Litigation* threaten the delicate balance within and bring unpredictability to the contracting process. Many challenges lie ahead, including conflicting university, faculty, and societal interests; competition; change in structure and focus of the university; congressional intervention; and availability of personnel, among other factors. An enlightened technology transfer office will move past these "speedbumps" and continue down this creative and energetic road.

BIBLIOGRAPHY

1. 35 U.S.C. §§200; 15 U.S.C. §§3701.

2. *Kucharczyk v. Regents of the University of California*, No. C 94-3886, 1999 U.S. Dist. LEXIS 6905 (N.D. Cal. May 6, 1999).

3. J. Tesk, *J. NIH Res.* 3-13–3-14 (1991).

4. *Technology Transfer: Administration of the Bayh-Dole Act by Research Universities* (GAO/RCED-98-126), May 7, 1998.

5. 37 C.F.R. §401.6 (1999).

6. 35 U.S.C.S. §§200–212 (1999).

7. *In re Synthroid Marketing Litigation*, No. 97 C 6017, 1999 U.S. Dist. LEXIS 11195 (N.D.Il. July 19, 1999).

8. J. Kopito, *Synthroid Litigation* (November 30, 1997), Available at: *http://www.medicinegarden.com/archive/00004309.htm*

9. K. Davidson, *USCF Angers Faculty by Quashing Research; Drug Maker Funded Study, Then Exercised Right to Suppress It* (April 26, 1996).

10. L. Marsa, *The Rising Health Costs of Capitalism's Invasion of the Science Lab* (December 20, 1998).

11. Mark Patterson, *Conflicts of Interest in Scientific Expert Testimony*, 40 Wm. and Mary L. Rev 1313, 1346–1350 (1999). *Synthroid Marketing Litigation*, Available at: *www.synthroidsclaims.com/*

12. Stevenson-Wydler Technology Innovation Act, Pub. L. No. 96-480, 94 Stat. 2311 (1980) (codified at 15 U.S.C. §§3701).

13. Pub. L. No. 96-480, 3 (1980) (codified at 15 U.S.C. §3702(3)).

14. Pub. L. No. 96-480, 11 (1980) (codified at 15 U.S.C. §3710(b) & (c)).

15. Pub. L. No. 96-480, 11 (1980) (codified at 15 U.S.C. §3710(d)). (Now operated by the National Technical Information Service, 15 U.S.C.S. §3710(d) [(Deering Supp. 1991)].

16. Pub. L. No. 96-517 (codified at 35 U.S.C. §§200).

17. 35 U.S.C.S. §202(a) (Deering Supp. 1991).

18. E.C. Walterscheid, *Harvard J. Law Tech.* **103**(3), 131–134 (1990).

19. S. Rep. No. 480, 96th Cong., 1st Sess. 19 (1979).

20. Presidential Memorandum to the Heads of Executive Departments and Agencies, Subject: Government Patent Policy. 1983 Pub. Papers 248; Feb. 18, 1983 and Exec. Order No. 12,591, 35 C.F.R. §200 (1987), reprinted in 15 U.S.C.A. §3710, app. at 256-58 (West Supp. 1992).

21. Brown, *Business Goes to College, Forbes* 196, October 11 (1982).

22. 1999 U.S. Dist. Lexis 6905.

See other entries MEDICAL BIOTECHNOLOGY, UNITED STATES POLICIES INFLUENCING ITS DEVELOPMENT; UNIVERSITY-INDUSTRY RESEARCH RELATIONSHIPS, ETHICS, CONFLICT OF INTEREST.

TRANSGENIC ANIMALS: AN OVERVIEW

RIVERS SINGLETON, JR.
University of Delaware
Newark, Delaware

OUTLINE

INTRODUCTION

To present an overview of animal use in modern biotechnology is a difficult task because of both the multitude of procedures and the types of animals involved. Procedures vary from small-scale laboratory research programs, involving a range of different animals from across the evolutionary spectrum to biotechnology product production facilities that may involve herds or colonies of a single animal species. Early biotechnology was restricted to procaryotic organisms, such as insulin production by strains of *Escherichia coli* constructed by recombinant DNA techniques. Since those early 1970s protocols, however, the diversity and complexity of organisms used in biotechnology now varies greatly across taxonomic lines from relatively simple procaryotes to nematodes to fish to mammals. Furthermore, since a major goal of biotechnology is to create organisms possessing genetic properties of other organisms, a pivotal result of these procedures is that taxonomic lines can be blurred as genetic elements of one organism are introduced into another.

The recent development of facile techniques to clone mammals further complicates the view of animals in nature and their role in biotechnology. Previous technologies created chimeric (An organism consisting of two or more tissues of different genetic composition, produced as a result of mutation, grafting, genetic engineering, or the mixture of cell populations from different zygotes.) organisms by combining genetic attributes from differing taxonomic lines. Because cloning bypasses genetic recombination, the technology leads to monophyletic (Of or concerning a single taxon of animals, relating to, descended from, or derived from one stock or source.) organisms, which may have highly unusual genetic traits, that are essentially unique in the biological world.

The point of the previous paragraphs is that both animals and procedures gathered under the rubric of "biotechnology" vary greatly. Because of this complexity, the focus of this article, in large measure, is on transgenic organisms as a model for some of the ways biotechnology uses animals. Thus a goal is not to provide a comprehensive overview of all animal use but rather to provide a paradigm by which the reader may gain insight into other arenas of biotechnology animal use. Some related issues have been discussed elsewhere (1,2).

HUMAN AND NON-HUMAN ANIMALS: SOME HISTORICAL REFLECTIONS

It is a truism to state that humans and animals have always interacted with each other. All living plants and animals are linked together through a shared evolutionary origin. While we are all part of nature's evolutionary web, however, our interactions with the nonhuman world are unique in nature. As human beings have domesticated the world around us, we have interacted with a diverse array of organisms across broad taxonomic lines in pervasive and complex ways. Indeed, our domestication of nature for utilitarian purposes appears a uniquely human activity. Consider, for example, that wheat domestication is viewed as the hallmark of human civilization, yet humans have arguably domesticated dogs for a longer time (3–7). [For a nontechnical but very readable account of canine evolution, see Budiansky (8).] Since those initial domestication forays at our emergence as modern human species, we have radically altered the permanent form and function, through selective breeding, of an immense variety of plants and animals.

Domestication has been driven primarily for human utilitarian benefit, although many organisms have undoubtably benefited from this close human association. While dogs frequently serve as human companions and pets, historically they have also been invaluable for herding sheep, pulling sleds, and other utilitarian purposes. For millennia an immense variety of animals served as both food supply and labor savers (it is not coincidental that a unit of work expenditure is "horse power") for human beings.

Animals have played another vitally important role in human culture. By studying animals, humankind gained an immense knowledge about the natural world. In antiquity Galen speculated about human physiology (often erroneously) and based his observations on animal dissection and vivisection. In the sixteenth century, William Harvey's brilliant description of pulmonary, cardiac, and circulatory physiology was deeply rooted in a variety of animal observations and experiments. As scientific knowledge exploded during the next four centuries, animal study played an important role in that expansion of human knowledge. Since the Renaissance vivisection and animal experimentation have increased human understanding of both basic physiology and pathological disease processes. It is reasonable to conclude that many advances of modern medicine would have been impossible without animal use. Indeed, we can conclude that much of our knowledge of fundamental biology would not have been achieved without recourse to animal experimentation and vivisection (9,10).

The historical streams of both animal domestication and experimentation are important to understand the place of animals in biotechnology. A definition of technology is "the application of scientific discoveries to the production of goods and services that improve the human environment" (11). Biotechnology thus is simply using biological systems, biological processes, or exploiting living organisms as part of the process of producing "goods and services that improve the human environment." The National Agricultural Library defines biotechnology as:

> . . . a set of powerful tools that employ *living organisms* (or parts of organisms) to make or modify products, improve plants or "animals," or develop microorganisms for specific uses. Examples of the "new biotechnology" include the industrial use of recombinant DNA, cell fusion, novel bioprocessing techniques, and bioremediation (12; p. 1 emphasis added).

Thus animals—*living organisms*—are central to biotechnology in several vital aspects. The basic science upon which biotechnology is structured would not exist in the absence of animal experimentation and vivisection. Selective breeding continues to produce a variety of animals, ranging from shrimp to cattle, of great

commercial importance. Modern techniques of molecular biology, such as marker-assisted selection, have enhanced domestic breeding programs to make them more effective and efficient.

Equally important, however, is the notion that—for many people—animal use in biotechnology is a simple extrapolation of human domestication of nature. Domestic breeding can enhance only genetic traits that naturally occur in an organism. Creation and use of transgenic organisms can "leap-frog" these genetic limitations, however, and introduce traits not normally found in particular species. Consequently transgenic animals represent and illustrate notions of domestication and experimentation as well as the general utility of animals in biotechnology.

BIOTECHNOLOGY AND DOMESTIC BREEDING

Traditional Breeding Perspectives

As noted previously, domestication of plants and animals for human utility and companionship has been a characteristic of human beings since our early evolution as a species. The power of selective breeding to bring about radical and relatively stable alterations in the form and function of animals was so well recognized by the nineteenth century that Charles Darwin devoted the first chapter of *Origin of Species* to the subject as a model for natural selection. Modern concepts of molecular biology, such as marker-assisted selection, combined with traditional selective breeding practices have greatly enhanced the power of domestic selection.

In traditional breeding practices, as Darwin noted, a breeder identifies a desirable physical trait in an individual organism within a population, such as increased milk production in a dairy cow. The exemplary animal is then used for breeding purposes. Those progeny exhibiting the desired trait are in turn used as further breeding stock, yielding—after a period of several generations—a population of animals that expresses the desired trait, namely a herd of cows with increased milk production.

Marker-Assisted Selection

A variety of new molecular methods—such as restriction fragment length polymorphism (RFLP) and polymerase chain reaction (PCR)—now allow breeders to identify DNA sequences associated with specific animals. These sequences may or may not be responsible for the particular trait the breeder desires to enhance. For breeding purposes, importance rests in expression of the genotypic "marker sequence" in progeny carrying the desired phenotypic trait; for example, in the case of increased milk capacity, a "marker sequence" of DNA should always be found in progeny expressing increased milk production.

Although the notion rests on numerous molecular methods for cutting, isolating, and analyzing specific segments of DNA, which have evolved over the past several decades, the concept of marker-assisted selection appears to have achieved practical application during the 1990s. For example, a Medline© database shows initial papers, specifically mentioning marker-assisted selection, beginning to appear around the early to mid-1990s. Since then, the number of papers has increased dramatically. Citation analysis also suggests a curious bias in the ways marker-assisted selection is used. A Medline© database search of 80 papers published since 1994 showed approximately 60 percent of the publications dealing with animals. However, another search of 250 publications in several agricultural databases showed only 20 percent dealing with animals. While both searches are anecdotal in nature, they suggest that the concept of marker-assisted selection is receiving wide application in enhancing agricultural plants.

Currently there is an international effort underway to generate a detailed description of the molecular genomic structures in a diverse array of organisms; the Human Genome Project is one part of this effort. As we increase our understanding of these genomic sequence details, and the biological function of particular genetic sequences is clarified, the power of marker-assisted selection will be greatly accelerated and enhanced. Nevertheless, despite advantages introduced by molecular techniques, selection-based breeding programs will always be limited to genetic attributes inherent within a species. There is, for example, a statistical distribution of milk production within cows; all that domestic selection can do is to skew that distribution toward a desired goal. To breach that genetic restriction, the new techniques of transgenic organisms must be employed.

BIOTECHNOLOGY AND TRANSGENIC ORGANISMS

What is a Transgenic Animal?

A simple definition of a transgenic animal is one "to which copies of a gene sequence have been artificially added" (13). However, there is difficulty in such definitions. The *Hastings Center Special Supplement* on animal biotechnology concluded that a transgenic organism "carries and expresses genetic information not normally found in that species of organism" but also noted that such a definition was literal and restrictive. Thus the definition was broadened "to include the purposeful amplification, spread, or dissemination of a *gene* within a *species* at a rate much faster than would have occurred in the absence of artificial interventions" (1; emphasis added). A view of animals possessing *desired* genetic properties, which in turn express novel phenotypes, emerges from this definition. These animals have been intentionally *designed* using modern biotechnological tools.

Some of these terms need further clarification. Our notion of a *gene* emerges from the central dogma of molecular biology:

$$DNA \Rightarrow RNA \Rightarrow protein$$

In traditional biological terms, DNA represents an organism's "genotype," and protein expresses its "phenotype." We can consider a gene as a section of a DNA molecule that provides biological information and is ultimately transcribed and translated into a protein molecule; proteins, in

turn, perform various cellular activities. Changes in DNA molecular structure will alter cellular function because of the resulting changed protein. Thus a transgenic animal carries a novel sequence of DNA [referred to as the *transgene* (13)]. If the transgene is stably incorporated into the animal's chromosomal DNA and its products functionally expressed, the animal — *and its progeny* — will possess an altered phenotype.

Species is more difficult to define, and Ernst Mayr (14) noted that biologists have understood the term in at least three different ways. Historically biologists viewed species from an *essentialist* perspective, which held that "each species is characterized by its unchanging essence and separated from all others by a sharp discontinuity (14, p. 256)." Charles Darwin helped change that view to a more *nominalistic* concept that rejected notions of "essential character" and conceived of species as groups of organisms that shared common attributes with a common descriptive name. Finally, while some biologists might argue with it, the modern notion of species, namely "a reproductive community of populations (reproductively isolated from each other) that occupies a specific niche in nature" (14, p. 273), is acceptable by most of the biological science community.

How are Transgenic Animals Created?

How do we go about this process of "purposefully amplifying, spreading, and disseminating" a gene within a species? Before describing transgenic technology, two brief reflections are important. First, transgenic technology with eucaryotic (also eucaryote, A single-celled or multicellular organism whose cells contain a distinct membrane-bound nucleus.) organisms is a logical and conceptual extrapolation of the recombinant DNA technologies with procaryotes in the 1960s. Intentionally creating a transgenic organism in a laboratory or factory is deeply rooted in the recombinant DNA work of the 1960s (1,2,13). Paul Berg's colleagues created an early transgenic organism when they used restriction enzymes and plasmid vectors to insert genetic elements from Simian Virus 40 into *Escherichia coli*. A significant difference between modern technology and these early techniques is that the latter were unidirectional. One could only introduce genetic material from a foreign source into bacterial (procaryotic) systems. Modern technology allows the manipulation of genetic information between virtually any plant or animal.

Second, clarification of two experimental distinctions is important. A *knockout experiment* is one that creates a mutation in an organism's own genome. Some genetic element native to an organism is inactivated so that the resulting progeny lack the functional capability associated with that gene or genes. In true *transgenic experiments* novel genetic elements, not normally found in that organism, are inserted into an organism's native genome. Thus this type of experiment creates an organism that "carries and expresses genetic information not normally found in that species of organism." Despite the different outcomes, both experiments use similar technological approaches.

A somewhat typical knockout experiment, which involved creation of a mouse with a defective *fosB* mutation (15), illustrates the technology. The knockout experiment discussed here illustrates some of the general technology used to create transgenic organisms. The specific experiment is also important because of the unexpected outcome (see below), which illustrates the serendipitous nature of science. Some background on the *fosB* gene and its protein product is important to understand the experiment and its somewhat unusual results. The FosB protein is one of many transcription factors found in cells; they facilitate the phrase in the central dogma of molecular biology:

$$DNA \Rightarrow RNA$$

The *fos* genes, which produce these proteins, are activated during a variety of adaptive neuronal responses in several brain regions. Despite extensive work that correlated *fos* gene products with mRNA production, their role in nervous system function and development remains unclear (16). Thus Brown et al. decided to create a knockout mouse, which lacked the *fosB* genes, in order to gain insight into its regulatory function; their experimental protocol for creating this *fosB* gene "knockout" mutation is summarized in Figure 1.

Initially murine DNA containing an incomplete fragment (and therefore nonfunctional) of the *fosB* gene was isolated and incorporated into a vector, which permitted three important experimental tasks. First, unique information present on the vector allowed investigators to screen transformed embryonic cells for the presence of the mutant gene. Second, specific sites on the vector, which were easily recognized at a molecular level, facilitated sequencing the incorporated genetic material. Finally, when genetic elements on the vector were phenotypically expressed, they served as a type of "marker-assisted selection" (although not referred to as such). Individual offspring expressing these traits were easily recognized as carrying the vector, and selecting them for further breeding purposes was eased. Once constructed, the vector containing the mutant *fosB* gene was electroporated into embryonic mouse cells.

Transformed embryonic mouse cells, namely those containing the mutant *fosB* gene, were implanted into mouse blastocysts. The blastocysts were then implanted into pseudopregnant female mice, which gave birth to pups expressing various levels of *fosB*. Ultimately the authors derived three strains of mice that exhibited normal Mendelian inheritance of the *fosB* mutation: one group was homozygous for the mutation [*fosB* (−/−)], one group was

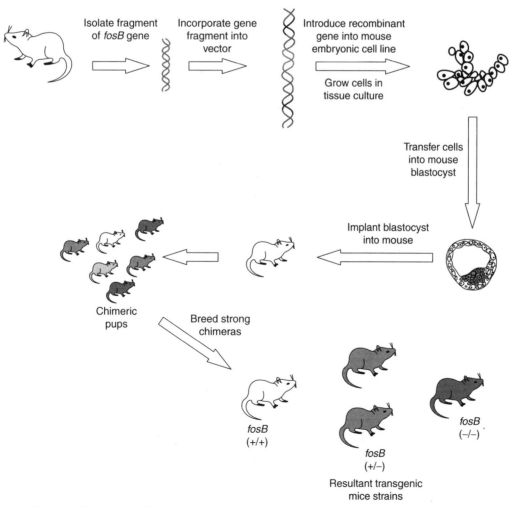

Figure 1. Schematic outline of procedure used to generate the *fosB* knockout mouse strain (15).

heterozygous for the mutation [fosB (+/−)], and a third group was homozygous for the wild-type state [fosB (+/+)].

These mice had interesting phenotypes: (1) the homozygous knockout mutants [fosB (−/−)] were health and viable; (2) there were no apparent histologic abnormalities, suggesting that fosB products are not required for normal mouse development; (3) the mutant strain was about 10 percent smaller than wild-type mice; and (4) pregnancies are normal and carried to full term. However, one phenotypic difference in the fosB mutant was unexpected and very significant. Pup lethality in early postnatal period was eight to ninefold higher in fosB knockout mutants than in wild-type mice. Serendipitously the authors discovered that lethality did not arise from a pup defect but rather arose from a failure of mothers to nurture pups. A graduate student involved in the project was concerned about put lethality. She returned to the lab one evening and discovered that the mother was in one corner of the cage and the pups isolated in another corner. When she moved the mother to the pups, they nursed normally (16). Mutant strain mothers simply abandoned the pups and failed to nurse them properly. Furthermore this was a loss of maternal "nurturing behavior" and was not due to a lactation defect in mothers.

The major significant difference in the knockout experiment discussed here and creation of a transgenic animal is the source of donor DNA. In the above knockout experiment, mouse DNA was incorporated into another mouse. In a transgenic experiment, DNA from *any* source (plant or animal) is incorporated into an organism unrelated to the source. The goal in discussing this experiment is to illustrate the relative ease of transgenic technology. Thus discussion of the ethical implications of the experiment will be deferred. However, two additional reflections are important to remember about this brief sketch. First, the technique and many potential technical complications have been greatly simplified or ignored. Furthermore there are a variety of other technical approaches to create transgenic organisms; this experiment simply serves to illustrate the basic technological concepts. Fundamentally, however, the techniques are relatively easy to perform, and new technological innovation are constantly appearing.

UTILITY OF TRANSGENIC ORGANISMS

Given this relative ease of the technology, what can it be used for? Like the recombinant DNA technology that preceded it, potential applications of transgenic technology is probably limited only by the human imagination (17). The various technological innovations that allowed development of transgenic animals have great potential utility for humankind (18,19). While there is tremendous overlap in various applications, these applications generally fall into three broad areas: commercial utility, improving human health, and furthering basic biological scientific knowledge.

Commercial Applications

We should not forget that the first type of transgenic organisms, that is, organisms intentionally engineered for specific utilitarian applications, were recombinant bacteria. Since the 1960s procaryotic organisms have been designed to carry out numerous industrial and agricultural processes, including (but not limited to) frost control on plants, increased efficiency of nitrogen fixation in soil bacteria, biodegradation and waste treatment, mineral processing, and other processes (20).

Like their procaryotic predecessors, transgenic animals can be used to produce, or are themselves, new commercial products. Utilitarian applications of transgenic organisms promise to provide humans with a variety of new capabilities not readily available by other means. The actual or potential list of commercial applications is immense, a potential briefly suggested by the following examples.

Agricultural Utility. Many early eucaryotic transgenic experiments sought agricultural benefit, and transgenic farm animals, with a variety of potential commercial applications, were created containing various growth hormones. Pigs that expressed enhanced levels of growth hormone exhibited significant daily weight gain, increased efficiency of food utilization, and a decrease in subcutaneous fat (21). While these animals had enhanced growth and improved feed efficiency, their immediate commercial utility was not feasible because of serious health complications including abnormal bone growth, enlarged internal organs, and diabetes.

Despite early enthusiasm over application of transgenic technologies to agricultural animals, that expectation has been tempered by experimental reality. In comparison with laboratory models, production of transgenic livestock is an inefficient process, and research has been hampered by lack of embryonic stem cell lines in farm animal species (22). Ward and Nancarrow (23) noted that, while it was feasible to produce transgenic animals with enhanced agricultural utility by modifying growth hormone levels, this could be done only if ways were found to tightly regulate hormone production. Nevertheless, optimism continues in areas ranging from mariculture (Cultivation of marine organisms in their natural habitats, usually for commercial purposes.) (24) to more conventional livestock (25,26). In the later area, confidence in transgenic technology is such that Murray stated: "Recent advances suggest that within the first decade of the 21st century the first transgenic animals will become available to the livestock industry, with acceptance depending upon their cost versus their potential economic benefit to the producers" (26, p. 149).

Pharmaceutical Utility. Transgenic animals have been effectively and efficiently used to produce many protein pharmaceutical products that are difficult if not impossible to produce by other means (e.g., traditional recombinant DNA techniques). As Wall et al. noted: "The objective of the emerging gene 'pharming' industry is to produce pharmaceuticals for treating human diseases. It is argued that mammary glands are an ideal site for producing complex bioactive proteins that can be cost effectively harvested and purified" (27, p. 2213).

There are two general obstacles to synthesizing these products via many traditional approaches. First, procaryotes (e.g., bacteria *E. coli*) lack the post-translational processing machinery necessary to synthesize many biologically active proteins from eucaryotic organisms. A second major difficulty arises from the difference in genetic organization between procaryotes, where genes are arranged on a single chromosome without interruption, and eucaryotes. Eucaryotic genes are often arranged in fragments so that mRNAs must be joined together before a biologically active protein can be expressed.

Large transgenic animals have been developed and successfully exploited to bypass these difficulties. Such animals are often referred to as "bioreactors" and the process called molecular or gene farming (or "pharming"). Approximately a dozen biotechnology companies now produce a variety of pharmaceuticals in larger amounts this way than can be achieved via other approaches. In addition to obviating the technical difficulties discussed previously, there are distinct advantages to using transgenic large animals to produce valuable human pharmaceutical proteins (27,28).

One advantage of using transgenic animals to produce pharmaceutical proteins is low operating cost, once the transgenic animal strain has been constructed. Furthermore, once the strain has been established, a virtually unlimited bioreactor supply becomes available by way of embryo cloning techniques (29).

However, the major advantage of using transgenic mammals in pharmaceutical protein production is that these genes can be inserted into mammary gland gene control elements so that the transgene product is expressed in milk. Complex pharmaceutical proteins, with correct post-translational modifications and full bioactivity, are correctly expressed and secreted in large amounts (in the order of grams per liter) in the milk. Since a large mammal, such as a cow, can produce 10,000 liters of milk a year, kilogram quantities of pharmaceutical protein can be synthesized per animal annually. As Smith commented, "No other production system can compete with bioreactors in production levels" (28, p. 681).

The technology now uses transgenic sheep, goats, pigs, and cattle to provide a ready supply of previously rare pharmaceutical proteins, such as Alpha 1 Antitrypsin (α-1-AT), Factor IX, and tissue plasminogen activator (t-PA). The later proteins are important therapeutic agents for a variety of human clotting disorders. Estimates of the U.S. market for pharmaceuticals transgenically produced approach $3 billion annually, and several products produced in this way are now in human clinical trials or actual therapeutic use (27).

This technological approach to pharmaceutical production is not without potential health hazard for the animal. Most of the hazards are recognizable, however, and appropriate design considerations can be developed to avoid problems (30).

Applications for Improved Human Health

Second, and of equal importance to their commercial utility, transgenic animals represent a variety of actual or potential improvements for generalized human health and welfare. Of major importance, these techniques allow researchers to develop laboratory organisms that mimic or duplicate many human diseases. Nomura noted that transgenic animals can be valuable models to follow the sequelae and treatment possibilities for these diseases and proposed criteria to evaluate objectives to develop valid animal models (31).

These animals promise exciting models to understand and conceptually intervene in human disease processes and thereby lead to alleviation of human suffering. [For an summary of the diversity of pathologies amenable to study by transgenic technology, see the text edited by Monastersky and Robl (18).] More recently, many individuals have speculated about the possible use of transgenic mammals as organ sources for human transplant. Ultimately, of course, the technology has potential for direct application to human beings. On the one hand, it holds out the promise of correcting debilitating genetic diseases. As our understanding of the interaction of genetics and human personality increases, however, the technology also has the potential to radically alter human nature.

Human Disease Models. Transgenic animals can provide insight into the development and progression of many human diseases, and cystic fibrosis (CF) is a classical example of a disease amenable to such study (1). Cystic fibrosis is a recessive, autosomal disorder and is inherited in classical Mendelian fashion. Disease symptoms originate from abnormal function of epithelial cells in the respiratory, digestive, and reproductive tracts. The abnormal function of these cells is due to a defective chloride ion channel protein called the cystic fibrosis transmembrane conductance regulator (CFTR). Defective chloride transport causes secretory tissues (e.g., lung, pancreas, intestine) to accumulate a thick mucus, characteristic of the disease. The pathology of CF arises from mucus accumulation in affected organ systems, such as lung infections and inadequate intestinal function.

The human CFTR gene has been identified and characterized. Although numerous mutations in this gene have been shown to lead to CF symptoms, defects in a very small region of the gene are responsible for 70 percent of reported cases.

While the biological basis of CF is clear, this understanding did not clarify its pathological basis or progression mechanism, nor did it lead to effective therapeutic approaches. The specific causal relationship between a defective chloride channel in a cell membrane and accumulation of mucus in the surrounding tissue remained unresolved. Consequently development of rational and effective therapies for the underlying disease has been difficult (32).

Previous work demonstrated that mice contain a gene equivalent to the human CFTR gene. Thus creation of a strain of mice with defective CFTR genes seemed to present a reasonable model to study CF in humans (32,33). Since the CF transgenic mouse apparently truly mimics human CF, it can serve as an important means to both clarify disease etiology and facilitate rational therapies. In an animal model we can follow disease sequelae in ways that are impossible or ethically repugnant in humans. Furthermore animal models allow us to develop and explore potential treatment modalities that would also be impossible or would pose ethical questions in humans. The list of genetic diseases that can be studied by such technology is appreciable and accounts for a significant aspect of human suffering (34,35).

Transgenic animals also provide insight into other, nongenetic, diseases. For example, transgenic mice serve two roles in AIDS research (36). First, a transgenic mouse strain was developed with a complete HIV proviral transcript. Progeny mice from this strain, which develop diseaselike symptoms, can potentially serve as a model system to study the etiology of AIDS. Second, transgenic mice have been created that carry only parts of HIV. These strains can play an important role in drug development and allow testing of certain antiretroviral drugs. Both the strengths and weaknesses of using transgenic animals in AIDS research has recently been reviewed by McCune (37).

Organ Transplantation and Xenobiotic Sources. During the past quarter century organ transplantation as a relatively common clinical therapeutic approach for many disease conditions has rapidly increased. However, the procedure has been restricted by two major difficulties: (*1*) organs available for transplant are in extremely limited supply relative to the potential clinical need (38,39); and (*2*) despite use of immunosuppressive drugs, even ideally tissue-matched organs often undergo host rejection (39–41). The latter problem arises from immunological reactions elicited by the recipient's immune system to antigens present on the donor organ. Xenotransplantation initially seemed to provide a means to address the first of these difficulties, namely animals were an unlimited source of organs (39). The procedure was sufficiently promising that *Science* mentioned it as a hot research area in 1996 (42).

While xenobiotic sources seemingly promised an unending organ supply for transplantation, the procedure did nothing to address the second, and equally problematic, phenomenon of host rejection. Transgenic animals, however, have recently been considered as potential sources of organ donors to address both problems, and initial research has focused on the initial, and potentially most overwhelming, transplant immunological barrier, namely hyperacute rejection (39–41).

Hyperacute tissue rejection is triggered by complement system activation in the recipient against proteins recognized as "non-self" on the foreign tissue. Complement activation can be inhibited by a variety of drugs; however, this technique leaves the recipient with a compromised immunological defense against infectious organisms. To obviate complement system mediated host rejection, transgenic animals, such as pigs, that express human complement proteins on their cell surfaces have been created (39–41,43). Animal organs with such cell surfaces do not trigger the complement cascade reaction in an organ recipient. Because complement activation does not occur, hyperacute rejection is prevented; the host's complement defense system is also left intact (39–41). Enthusiasm for

the technology has become so intense that commercial involvement has grown (29,44).

Despite initial encouragement about transgenic xenobiotic organ sources, however, serious concerns have arisen regarding potential transmission of infectious agents, especially retroviruses, from transplanted animal organs (39–41,44–49). These concerns are sufficiently pressing that while the British government gave qualified approval for continued research into the transplantation of pig organs into humans, it ruled out clinical trials until further research demonstrated safety and efficacy of the procedure (50,51). The United States will allow limited clinical trials to go forward under stringent guidelines established by the Food and Drug Administration (FDA), the Centers for Disease Control and Prevention (CDC), and the National Institutes of Health (NIH) (52,53).

Human Gene Therapy. Because the difference between human and animal is a matter of evolutionary degree, ultimately transgenic technology promises direct intervention into human genetic diseases by replacing dysfunctional genes. Although somewhat dated now, Friedmann (34) provided an excellent and very readable summary of this area of research. He noted that diseases arising from bone marrow defects would be most amenable to genetic intervention. Because bone marrow cells are readily susceptible to infection by retroviruses, a common way to introduce foreign genetic material into mammalian cells, these cells would be relatively easy to alter genetically. Furthermore, because these cells can be removed from a patient, altered and manipulated in vitro, and then reintroduced back into the patient, the potential rejection of altered gene products by a recipient's immune system is reduced.

For similar reasons, diseases resulting from hormonal or other diffusible protein deficiencies (e.g., insulin or blood-clotting factor deficiencies) may also be amenable to genetic treatment. Skin cells can be readily transformed by a variety of methods and then reintroduced back into the patient. Since the process again involves a tissue autograft, host immune rejection is again reduced. The technique has theoretical therapeutic implications for many diseases such as hemophilia and insulin-deficient forms of diabetes.

Transgenic Organisms and Basic Science

While it is difficult to draw sharp distinctions between utilitarian goals and the basic science necessary to achieve them, transgenic technology now plays a profound role in fundamental biological science. Like the recombinant DNA technologies that preceded them, undoubtedly the most significant role of the transgenic technologies will be as tools to gain new insights into nature. Reflecting on the current state of biological science, Verbeek noted, "The fast growing knowledge about the complex biology of higher eukaryotic systems demanded new experimental models. Transgenesis of mammals is one of the most fruitful techniques to create these models (54, p. 1)."

Transgenic organisms are radical new tools for fundamental research in all of the biological and medical sciences; they are powerful new instruments that are opening new windows to understand the natural world. Since transgenic technology first began in 1980, it has grown exponentially. Transgenic animals have contributed greatly to elucidating complex biological processes at the molecular level (31). The magnitude of this new technology's impact on research inquiry is illustrated by citation analysis of Medline© references to transgenic organisms. In 1987 Medline© cited 163 references to *transgenic organisms* (including both plants and animals). Since that time the number of references has increased exponentially; by 1992 there were over 1000 Medline© citations in which the word *transgenic* was either in the title or was listed as a keyword (1). A cursory Medline© search indicates that in 1997 alone, there were almost 25,000 references to *transgenic animals*.

Organisms Used and Research Goals. Given the ubiquity of mice or rats in laboratory research, a majority of the studies noted above involved these animals. Nevertheless, these transgenic research programs also used a diversity of other animals. They span a wide evolutionary range and include relatively common laboratory organisms, such as fruit flies, zebrafish, and *Caenorhabditis elegans* (a nematode used extensively in developmental biology studies), to more exotic organisms such as the medaka fish and silkworms. Many animals reflect an ultimate commercial or agricultural interests such as goats, pigs, sheep, and cows.

Research involving transgenic organisms is as complex and diverse as the organisms themselves. Projects range from those with immediate, pragmatic goals — such as increases in agricultural products or solutions to human health concerns — to basic inquiries about the behavior of biological systems, such as antisense RNA, ribozyme action, gene expression in *Drosophila*, zebrafish developmental processes, or the role of molecules like FosB proteins described earlier.

Why Are Transgenic Organisms Important? As previously noted, development of transgenic organisms parallels the use of recombinant DNA in the early 1970s when many researchers were excited by potential commercial applications of recombinant DNA. Despite the utilitarian role, however, these same scientists saw the technology as a powerful tool to explore nature. The technique provided a means to isolate individual genes (or gene clusters) and their products away from a complex organism into a much simpler procaryotic organism. Researchers believed that such isolation would lead to a clearer understanding of gene function, regulation, and interaction (17). Indeed, the history of recent molecular biology has confirmed the validity of that belief, and our understanding of such diverse phenomena as gene action, immunology, ecological processes, or neurobiology has grown immensely during the past 20 or more years as a direct result of recombinant DNA techniques.

Furthermore it would have been difficult to create a list of new things scientists expected to discover with the early recombinant DNA research. One could point to potential societal benefits, such as new understandings of gene function and similar phenomena, expected from the work. For most scientists, however, this was a technique with tremendous investigatory power that could be used to open new vistas of biological knowledge. It was this epistemological dimension of the research that animated many scientists' interest in recombinant DNA technology (17).

In a similar fashion, it is problematic to create a tidy list of projects that transgenic organisms will solve. The dilemma arises from the nature of scientific inquiry that does not permit clear predictions about its own nature and direction, a point marvelously illustrated by the outcome of the *fosB* knockout mouse experiment described previously. Scientific inquiry fundamentally is more than a simple *accumulation of facts* about nature; rather it is a *method to understand* the natural world.

How is this digression into epistemology connected with the technology of transgenic organisms? Simply it is this: Despite their artificial creation, transgenic organisms are now a part of the natural, empirical scientific world. As such, they are important tools to gain a greater understanding of nature. Like the transgenic procaryotes that preceded them in the repertoire of scientific investigative tools, transgenic eucaryotes will ultimately provide humankind with fundamental and valuable knowledge of the natural world.

ANIMALS, BIOTECHNOLOGY, AND ETHICS

Like many technological innovations, development of transgenic organisms presents us with ethical quandaries. In addition to their wondrous utility, their creation also raises potentially troubling ethical questions. While these questions are complex, and often appear refractory to solution, they have been addressed elsewhere (1,2,55,56) and will be extensively dealt with in this volume. Only brief reflection on the issues raised in this article is appropriate here. The ethical questions seemingly fall into four broad areas: breaching species boundaries, potential for animal harm, environmental concerns, and potential human application.

Species Barriers

Transgenic organisms raise obvious concerns about "species barriers." Are *species* physical entities so inherent in the fabric of nature that we are morally culpable in breaking the barrier between them? Is there anything morally significant about being a member of a *species*? These questions might be put another way. Should sheep be allowed to be sheep without carrying burdens of non-ovine genes, some of which are intended only for human benefit?

A scientific perspective suggests negative answers for these questions. The notion of *species* as fixed natural entities is relatively new in human thought and is contrary to modern scientific views. From antiquity all species were seen as eternal and immutable, and this *essentialistic* notion of species dominated Western thinking well into the nineteenth century. From this view, the organisms we encounter in daily existence reflect an essential form that exists within created nature itself.

As noted previously, Charles Darwin began to shift the scientific notion of *species* away from this *essentialist concept*, and post-Darwinian biologists reject all *essentialist* notions of species. Thus there is nothing unique about being a member of a *species* that would seemingly command moral recognition. *Species* do have biological reality, but it is not an *essentialist* reality (14).

Species are not immutable "type-forms" woven into the fabric of nature but are defined in populational terms, a reality that is contextual and is spatiotemporally bounded (14,57). Moreover, as a scientific heuristic device, the *species* concept provides a way to organize and simplify the complex diversity of living organisms; *species* provide what Mayr refers to as a *taxon*, namely an entity in nature with taxonomic significance. Again, it is not readily apparent that taxonomic significance can mandate moral significance.

Transgenic Animal Welfare

Sentient transgenic animals raise serious concerns about potential pain or suffering that we might cause to an animal capable of such experiences. Are we morally permitted to intentionally "create" an organism that we know will ultimately suffer severe debilitation or experience great pain? CF mice exhibit many physical symptoms common to the human disease, including premature death. Creation of the CF mouse, or any animal as a model of human disease, involves potential harm or suffering to individual transgenic animals. So the moral question arises, Do we have a right to intentionally create such an animal capable of experiencing pain that we know will develop such a debilitating and painful disease?

The argument has been made elsewhere (1,2) that transgenic animals are not, in principle, significantly different from other animals. If one accepts this claim, then resolution of questions about pain and suffering are similar to questions regarding animal use in general. Commercial and laboratory use of sentient animals is controlled by regulations and principles established by the Department of Agriculture and by National Institutes of Health Guidelines (9,10). As McCarthy noted, most use of transgenic organisms will be governed by these entities. He also noted, however, that "there are gaps in oversight due in part to whether the particular methods are publicly or privately funded" (58, p. 526).

Like many questions involving research use of animals, the moral parameters for using transgenic organisms must be contextually defined (59). Potential animal suffering must be weighed against potential human suffering alleviated through knowledge gained by animal use. After due consideration we might conclude that creating a strain of mice as models of CF is morally justified because it may ultimately alleviate the acute suffering of a significant number of human children. On the other hand, using similar techniques to create a strain of dogs with some serious physical abnormality (e.g., severely shortened legs) to become novel pets would arguably be morally reprehensible.

Ecological/Environmental Concerns

A third ethical concern entails possible ecological damage arising from intentional or unintentional release of transgenic organisms. How do we ensure that these novel organisms do not unduly disrupt natural habitats and cause serious environmental damage? The potential environmental impact of transgenic organisms has been poorly studied; these questions are probably quite significant, however.

In many respects the environmental impact of organisms used in biotechnology should be minimal, as there is no intent for the animal to be released. Indeed, because of the great expense involved in their creation, extreme precautions are taken to prevent the animal from escaping the laboratory or farm environment. Serious environmental concerns arise, however, in two areas. First, care must be exercised in projects where there is a high potential for a genetically modified organism encountering naturally related organisms, for example, in mariculture. Second, one need only drive through any kudzu-covered forest in the United States to envision the potential ecological havoc that a genetically altered plant might create. (Kudzu, of course, is not a genetically altered plant.) Both of these issues have been fully addressed elsewhere (1,38,55).

Human Application

A fourth ethical issue arising from transgenic technology is perhaps the most serious and the most difficult to resolve, namely application of transgenic technology to humans. The articles by Friedmann and Jaenisch demonstrate that the clear direction of this research is toward human application. As is noted elsewhere in this volume, the ethical issues in this area are complex and often troublesome. On the one hand, for many people we are morally culpable if we have the ability to alleviate the suffering of an individual with a profound genetic defect, such as CF, Tay-Sachs, or sickle cell anemia, and we fail to use that ability. This conclusion, however, clearly places us onto the moral philosopher's slippery slope. Moral distinctions between significant genetic defects (e.g., CF or sickle cell anemia), with their associated suffering, and merely attractive traits that individuals might like for their children to possess (e.g., large body mass so that a son could become a highly paid NFL linebacker) are reasonably clear. Moral distinctions with such extremes, however, are rare. More often we face subtle and less clear choices that are thus more ethically problematic.

These moral considerations become even more problematic when we ask about our obligations to future offspring of individuals suffering from a treatable genetic disease (38). Many people readily find moral obligations to treat individuals with somatic cell deficiencies. However, obligations to treat such deficiencies at the germ-cell level are more complex. While we may be obligated to alleviate the suffering of a person with a genetic disease, are we obligated to ensure that those individuals can produce children who lack the genetic defect? Alleviation of immediate suffering, if we can do so, seems a reasonable obligation. However, does that obligation extend to some future individual not yet conceived? These moral questions intuitively appear problematic, and answers do not appear readily obvious.

CONCLUSIONS

For many people, animal biotechnology promises a powerful new vision of general welfare and health for both humans and other animals. Domestic breeding, enhanced by modern techniques such as marker-assisted selection, and transgenic organisms (and the various technologies associated with their creation) may present a cornucopia of new wealth (both financial and abundance of valuable material possessions or resources).

Creation of transgenic animals, especially, is a new tool for scientific inquiry and has the potential to alter science itself. New questions about nature, which were impossible in the technology's absence, could be asked and radical new answers proposed. Like many new scientific tools, studies with transgenic organisms often have serendipitous turns. The apparent maternal behavior pattern linked to the *fosB* gene discussed in this article is a good example of the unexpected paths this technology can reveal.

Concomitantly the technology's power also creates profound possibilities for moral abuse and environmental chaos. Our most deeply felt sense of human values can be seriously distorted and corrupted by even moderate abuse and potentially could lead to distortions of fundamental and essential aspects of both human and animate nature. From some perspectives, transgenic technologies represent a Frankenstein-like bargain with nature and are the realization of Chargaff's prediction of the "Devil's Doctrine," that what can be done (technologically) will be done [regardless of broader social concerns (60)].

Nevertheless, transgenic organisms are not fictional or creatures of ancient mythology; they are a reality of nature that we humans must deal with. Like most scientific inventions, transgenic animals pose complex moral issues. And, as with any truly moral issue, the fashioning of transgenic organisms present us with moral ambiguities and treacherous slippery slopes. They pose questions about our moral obligations to both our fellow humans as well as the other living beings with whom we share this planet. The moral problems are made more complex as we attempt to discover our obligations to others members of the biological web within which all living beings are intertwined (1).

Science may help enlighten and focus the moral landscape of these questions; however, it does not provide us with adequate tools to derive answers. This difficulty is inherent in the limitations of scientific inquiry, for these answers "are as many as there are different cultural, religious, and philosophic perspectives" (59, p. 518). Because these pluralistic perspectives are not subject to empirical boundaries or testable propositions, science, as a mode of inquiry, is ill prepared to deal with them. Thus answers to questions on our moral obligations must lie in other aspects of the broader human condition. Wrestling to find anchors on the slippery slopes of our moral landscape nevertheless is a natural aspect of our humanity.

ACKNOWLEDGMENT

Part of the work on this article was completed during a sabbatical year at Case Western Reserve University, and the author thanks his many colleagues there for their helpful comments and criticisms. Drs. Melinda Duncan and Milton Stetson, at the University of Delaware, also provided help defining technical terms. Finally, the author thanks the National Science Foundation for financial support (Grant number SBR 9602023).

BIBLIOGRAPHY

1. R. Singleton, Jr., *Hastings Center Rep. (special suppl.)* **24**, S4–S14 (1994).
2. R. Singleton, Jr., in J.C. Gonder and E.D. Prentice, eds., *Genetic Engineering and Animal Welfare: Preparing for the 21st Century*, Scientists Center for Animal Welfare, Washington, DC, 1999.
3. V. Morell, *Science* **276**, 1647–1648 (1997).
4. C. Vilà et al., *Science* **276**, 1687–1689 (1997).
5. C. Vilà et al., *Science* **278**, 208–209 (1997).
6. N.E. Federoff and R.M. Nowak, *Science* **278**, 206 (1997).
7. J.P. Scott, O.S. Elliott, and B.E. Ginsburg, *Science* **278**, 205 (1997).
8. S. Budiansky, *Atlantic Monthly* **284**, 39–49, 52–53 (1999).
9. R. Singleton, Jr., *Persp. Biol. Med.* **37**, 576–594 (1994).
10. R. Singleton, Jr., *Persp. Biol. Med.* **38**, 41–57 (1994).
11. *The Concise Columbia Encyclopedia*, Columbia University Press, 1995.
12. Committee on Biotechnology Research Subcommittee, Fundamental Science, *Biotechnology for the 21st Century: New Horizons*, U.S. Government Printing Office, Washington, DC, 1995.
13. N. Maclean, in N. Maclean, ed., *Animals with Novel Genes*, Cambridge University Press, Cambridge, UK, 1994, pp. 1–20.
14. E. Mayr, *The Growth of Biological Thought*, Harvard University Press, Cambridge, MA, 1982.
15. J.R. Brown et al., *Cell* **86**, 297–309 (1996).
16. J. Cohen, *Science* **273**, 577–578 (1996).
17. R. Singleton, Jr., in D.A. Robins and A.R. Dyer, eds, *Ethical Dimensions of Biomedicine*, Charles C. Thomas, Springfield, IL, 1981.
18. G.M. Monastersky and J.M. Robl, eds., *Strategies in Transgenic Animal Science*, ASM Press, Washington, DC, 1995.
19. N. Maclean, ed., *Animals with Novel Genes*, Cambridge University Press, Cambridge, UK, 1994.
20. S.E. Lindow, N.J. Panopoulos, and B.L. McFarland, *Science* **244**, 1300–1307 (1989).
21. V.G. Pursel et al., *Science* **244**(4910), 1281–1288 (1989).
22. E.R. Cameron, *Mole. Biotechnol.* **7**, 253–265 (1997).
23. K.A. Ward and C.D. Nancarrow, *Mole. Biotechnol.* **4**, 167–178 (1995).
24. M. Gomez-Chiarri, V.L. Kirby, and D.A. Powers, *Mar. Biotechnol.* **1**, 269–278 (1999).
25. E.P. Cunningham, *Livest. Prod. Sci.* **58**, 1–24 (1999).
26. J.D. Murray, *Theriogenol.* **51**, 149–159 (1999).
27. R.J. Wall, D.E. Kerr, and K.R. Bondioli, *J. Dairy Sci.* **80**, 2213–2224 (1997).
28. T.J. Smith, *Biotechnol. Adv.* **12**, 679–686 (1994).

29. E. Pennisi, *Science* **279**, 646–648 (1998).

30. G. Brem and M. Müller, in N. Maclean, ed., *Animals with Novel Genes*, Cambridge University Press, Cambridge, UK, 1994.

31. T. Nomura, *Lab. Animal Sci.* **47**, 113–117 (1997).

32. M. Barinaga, *Science* **257**, 1046–1047 (1992).

33. J.N. Snouwaert et al., *Science* **257**, 1083–1088 (1992).

34. T. Friedmann, *Science* **244**, 1275–1281 (1989).

35. S.M. Weissman, *Proc. Nat. Acad. Sci. (USA)* **89**, 111–112 (1993).

36. A.D. Lewis and P.R. Johnson, *Trends in Biotechnology (Ref. ed.)* **13**, 142–150 (1995).

37. J.M. McCune, *Science* **278**, 2141–2142 (1997).

38. A.L. Caplan, *Am I My Brother's Keeper: The Ethical Frontiers of Biomedicine*, Indiana University Press, Bloomington, IN, 1997.

39. Committee on Xenograft Transplantation: Ethical Issues and Public Policy, Institute of Medicine, *Xenotransplantation: Science, Ethics, and Public Policy*, National Academy Press, Washington, DC, 1996.

40. D.K.C. Cooper, *Frontiers in Bioscience* **1**, 248–265 (1996).

41. J.L. Platt, in J.C. Gonder and E.D. Prentice, eds., *Genetic Engineering and Animal Welfare: Preparing for the 21st Century*, Scientists Center for Animal Welfare, Washington, DC, 1999, pp. 53–75.

42. Anonymous, *Science* **270**, 1902 (1995).

43. M. Schmoeckel et al., *Transplantation* **62**, 729–734 (1996).

44. F. Hoke, *The Scientist* **9**, 1 (1995).

45. X.-J. Meng et al., *Proc. Nat. Acad. Sci. (USA)* **94**, 9860–9865 (1997).

46. C. Patience, Y. Takeuchi, and R.A. Weiss, *Nature Medicine* **3**(3), 282–286 (1997).

47. R. Sikorski and R. Peters, *Science* **276**, 1893 (1997).

48. R.E. Michler, *Emerging Infect. Dis.* **2**, 64–70 (1996).

49. M.J. Hanson, L.-M. Russow, and C.R. McCarthy, *Hastings Center Rep.* **29**, 22–25 (1999).

50. A.S. Daar, *World J. Surg.* **21**, 975–982 (1997).

51. N. Williams, *Science* **275**, 473 (1997).

52. C. O'Brien, *Science* **271**, 1357 (1996).

53. G. Vogel, *Science* **279**, 30 (1998).

54. J.S. Verbeek, in L.F.M. van Zutphen and M. van der Meer, eds., *Welfare Aspects of Transgenic Animals*, Springer Verlag, New York, 1997, pp. 1–17.

55. B.E. Rollin, *The Frankenstein Syndrome: Ethical and Social Issues in the Genetic Engineering of Animals*, Cambridge University Press, Cambridge, England, 1995.

56. L.F.M. van Zutphen and M. van der Meer, eds., *Welfare Aspects of Transgenic Animals*, Springer Verlag, New York, 1997.

57. D.L. Hull, *Science as a Process: An Evolutionary Account of the Social and Conceptual Development of Science*, University of Chicago Press, Chicago, IL, 1988.

58. C.R. McCarthy, *Hastings Center Rep. (special suppl.)* **24**, S24–S29 (1994).

59. S. Donnelley, *Hastings Center Rep. (special suppl.)* **24**, S14–S24 (1994).

60. E. Chargaff, *Persp. Biol. Med.* **16**, 486–502 (1973).

See other entries ANIMAL, MEDICAL BIOTECHNOLOGY, LEGAL, LAWS AND REGULATIONS GOVERNING ANIMALS AS SOURCES OF HUMAN ORGANS; ANIMAL, MEDICAL BIOTECHNOLOGY, POLICY, WOULD TRANSGENIC ANIMALS SOLVE THE ORGAN SHORTAGE PROBLEM?; FDA REGULATION OF BIOTECHNOLOGY PRODUCTS FOR HUMAN USE; see also RESEARCH ON ANIMALS entries.

UNIVERSITY–INDUSTRY RESEARCH RELATIONSHIPS, ETHICS, CONFLICT OF INTEREST

Robert P. Lawry
Thomas W. Anderson
Case Western Reserve University
Cleveland, Ohio

OUTLINE

INTRODUCTION

Since World War II, and increasingly since the early 1980s, there has been a widespread public policy initiative to increase and deepen research and development relationships between industry and academic institutions. The hope is that new technologies will be more rapidly and effectively produced both to promote the general welfare as well as to increase U.S. dominance in the global marketplace. Ethical concerns, however, have arisen simultaneously. Scientific and educational integrity are threatened in a variety of ways by these new academic-industry partnerships. As questions of conflicts of interests increase and deepen, good public policy answers lag behind.

HISTORICAL CONTEXT

Until the late nineteenth century, colleges and universities in the United States were primarily teaching institutions. Their role was largely confined to the transmission of knowledge. The discovery or enlargement of knowledge was secondary to the application of practical subjects that would have utility for students and for the larger society. In the public sector, the Morrill Act of 1862 allocated federal lands to states for the founding of "at least one college where the leading object shall be, without excluding other scientific and classical studies ... to teach such branches of learning as are related to agriculture and the mechanic arts" (1). The Act led to the founding of new state institutions and to the support of recently established state colleges, embodying a vocational, practical educational focus.

Nothing changed the educational landscape in the United States more than the founding of Johns Hopkins University in 1876. It has been described extravagantly but not inaccurately as, "perhaps the single, most decisive event in the history of learning in the Western Hemisphere" (2). Prior to the founding of Johns Hopkins, graduate education and the research associated with it were left to the leading European universities, the German universities in particular. The Hopkins model gave primacy to graduate education over undergraduate instruction and brought together the German concepts of advanced training and the generation of new knowledge, particularly in the natural sciences, into the teaching environment of American universities. During the next 50 years, the uniquely American research university took root and began to flourish.

From 1900 to 1920, private and public universities were participants in the general economic well-being of the era. For private institutions, endowments were established and began to grow. The philanthropies of John D. Rockefeller and of Andrew Carnegie set examples for decades to come. Public institutions secured their places as important and useful state resources requiring more than tuition income to accomplish their increasingly diverse objectives.

It was not yet clear, however, how much of the nation's scientific research would be done in universities. Government agencies and independent research institutions competed with universities for the resources to hire researchers, build facilities, and support research activities. Public universities looked with modest success to their state legislatures for research support and both public and private universities began to appreciate the power of private philanthropy. Early in the century, Harvard President Charles Eliot noted, "it is clear that men of means, who reflect on the uses and results of educational endowments, are more and more inclined to endow research" (3). While the American research university hallmark of combined teaching and research was being firmly established in the first quarter of the twentieth century, the funding structures to adequately support these no longer discretionary activities were not yet in place.

World War I effectively nationalized the research universities, focusing all faculty and student efforts on winning the war. The role of science was enhanced, and the bonding of applied and basic research was seen as important to the war effort. Following the war, the major philanthropic foundations, established in the late nineteenth and early twentieth centuries, began to take notice of scientific research as a means for the "amelioration of the human condition through the advancement of knowledge" (3). Foundation grants enhanced and expanded university research in the years between the two world wars. Fears of foundation control over university educational and research efforts did not materialize as it became clear that the needs of research universities would far surpass the resources of the foundation community.

Following World War I, the advancement of scientific knowledge through university-based research began to require a partnership of universities, private philanthropists, foundations, and now corporations. Major corporate laboratories had been established early in the century for applied research purposes. Interactions between applied industrial researchers and university basic scientists became common. University graduates were recruited to industrial laboratories and faculty members consulted with corporations. Corporate financial support for university research followed naturally from these relationships. Even early on, the differences between universities (dedicated to the advancement of knowledge) and corporations (dedicated to financial gain) raised the potential for misunderstandings. Corporations were not convinced of the importance of basic research and faculty members found that corporate interests were often too narrow to be of educational or scientific interest. But, interactions continued in a variety of forms (graduate fellowships, research contracts, consulting relationships, etc.) in generally ad hoc institutional arrangements.

World War II again found universities deeply involved and effected by the war effort. The Manhattan Project and the Radiation Laboratory at the Massachusetts Institute of Technology were major scientific collaborations between university scientists and the federal government. While it may be overstating the case to say that science won the war, it nevertheless played such a decisive role that neither academic research nor federal scientific interests would ever be the same again. The war not only highlighted the need for a federal science policy, it brought the federal government into a permanent funding relationship with university research. World War II marked the shift from primarily private to primarily public funding for major research projects. Postwar foundation support became focused on three broad objectives: medical and health fields, strengthening the system of university research, and social and behavioral sciences (4). Foundation funding increasingly nudged research universities in the direction of academic excellence rather than the targeted defense related research necessary for a war effort.

Federal Government Involvement

In 1950 the National Science Foundation (NSF) was created to fund basic academic science with public funds. Until this time the postwar research agenda had been driven largely by programmatic funding from the armed services. University administrators and researchers were increasingly concerned about the source and direction of military research. California Institute of Technology President Lee A. DuBridge called the prevailing military authority over the nation's research program "an anomalous and precarious situation to have the future of basic research hang by the thread of continued appropriations to the military agencies, or of their continued interests" (4). In response to this concern, the National Science Foundation was created to fund basic research, by "greasing the wheels of science," funding

scientific research, and developing "a national science policy" (4). Although initially inchoate in its mission, both the role of the NSF and the nation's interest in basic research changed on October 4, 1957, with the Soviet Union's launch of Sputnik, the first space ship to orbit the earth. Because of the threat of an attack from space during the height of the cold war, the days immediately following *Sputnik* were consumed with much national soul-searching. Initial responses included the establishment of the National Aeronautics and Space Administration (NASA) and increased federal appropriations to existing agencies for basic scientific research.

While funding levels would vary over the next 40 years, commitments to basic versus applied research would wax and wane, and the interests of social justice and economic development would often compete, the foundational commitment of public funds for academic research would not be seriously threatened. Basic academic research became a growth industry, largely funded by federal money. This federal commitment also included the beginnings of a federal scientific establishment, initially presided over by the Presidential Science Advisor but eventually permeating all branches and levels of government. As the NSF gained its footing, it was joined (in federal priority) by its sister institution the National Institutes of Health (NIH), the research arm of the Public Health Service. The NIH was to biomedical science and scientists what the NSF was to natural science and scientists.

The 1960s placed significant strains on research universities — student unrest, sluggish economic conditions, governmental oversight, and changing values — leading to a loss of confidence in these institutions and their primary missions of teaching and research. Egalitarian federal programs and the proliferation of institutions seeking research funding led to changes in both the recipients of public funds and the role of private funding sources. In the 1970s one response to these conditions found the leading research university faculty and administrations beginning to seek closer relationships with corporations. While driven partially by funding considerations, these initial efforts by research universities also "implied a break with the cloistered mentality that had flourished in the 1960s" (4).

In the early years of the twentieth century, corporations purchased research capability from universities either through direct support of specific research activities or through joint institutes for applied research. By the beginning of World War II, however, many corporations had established their own laboratories, and thus looked to universities for the theoretical work that would underlie their corporate research interests. But corporate and university interests were inevitably different since universities seek to advance knowledge while industry must apply it. Universities were organized horizontally and corporations were organized vertically. Individual faculty members determine whether they will work with corporations no matter how friendly their university policies may be to corporate interactions. But despite these differences in culture, by the 1980s corporate support of university scientific research was growing at unprecedented rates.

The mid-1970s saw fledgling efforts by the NSF to encourage university–industry interaction for the specific purpose of technology transfer. At about the same time the first large ($23 million) university–corporate contract (Harvard Medical School and the Monsanto Corporation) based on mutual research interests was announced. Unprecedented on both sides, "Monsanto provided funds for endowment, research support, and facilities for a pair of Harvard scientists. In return the company was promised the patent rights to any discoveries that resulted from their research" (4). Viewed with both interest and suspicion, the agreement was only the beginning of similar arrangements at other institutions and corporations, particularly in microelectronics and biotechnology. This highly structured form of technology transfer was supplemented by small start-up companies seeking to commercialize scientific discoveries. Such companies often were founded by scientists, engineers, and graduate students from research universities with funding from venture capital firms.

The blueprint for biotechnology start-up companies was drawn by Genentech, a firm founded in 1976 by a venture capitalist who struck an agreement with a molecular biologist who was interested in commercializing his new technology that led to synthesizing the human gene for insulin. Genentech licensed the discovery to the Eli Lilly Corporation and "thus validated the idea that genetic engineering could produce valuable commercial products..." (4). Venture capitalists, scientists, and universities immediately understood the potential for significant wealth in similar relationships. Scientists could continue their research — either within the academy or at the new firms — with private funding that could lead to both important scientific breakthroughs and the prospect of enormous wealth. Over 200 biotechnology firms were founded between 1980 and 1984, and half of all biotechnology venture capital raised by 1988 was raised during the two years following the Genentech announcement (5). The potential for conflicts of interest and commitment were recognized as enormous, although not always immediately.

The explosion of biotechnology research was coupled with the highly charged business climate of the 1980s. Government deregulation, junk bond financing, and declining federal support for research universities all enhanced the environment for business–university collaborations and partnerships of almost infinite variety. Both state and federal governments enacted legislation to encourage these partnerships in the hope of transferring technology, increasing corporate competitiveness, and retaining business and industry within state boundaries. Universities scrambled to compensate for reduced funding levels and to retain key faculty members who were increasingly being lured outside the academy by entrepreneurial opportunities.

At the federal level, the NSF had two programs (the Industry–University Cooperative Research Projects Program and the Industry–University Cooperative Research Centers Program) which were established in the 1970s

to develop and sustain corporate–university research partnerships. By 1989, 41 Research Centers around the country were operational and 22 were self-sustaining. A second part of the Research Centers program was founded in 1985 to support Engineering Research Centers with 18 established by 1989 (6). All potential economic benefits from the collaborative arrangements encouraged by these centers remain with the centers despite significant funding by the NSF.

Significant Legislation

Perhaps more important were changes by federal agencies allowing universities to retain patent rights from inventions and technologies discovered in federally funded research projects. The Bayh–Dole Act of 1980 specifically granted such rights. The Federal Technology Transfer Act of 1986 permitted federal scientists and university scientists to collaborate with industry to develop commercially patentable ideas. The Act specifically authorized "private companies to gain the exclusive rights to patents, while universities and scientists could receive royalties" (6). These federal efforts, along with complementary state legislation, nudged, if not pushed, universities into the technology transfer business and into increasing collaborations with industry with the objective of commercializing the results of research.

As the Harvard Medical School–Monsanto relationship was followed by equally large and potentially controversial business-university partnerships in the 1980s, issues of academic freedom, freedom to publish, conflicts of interest and commitment, and secrecy began to surface both inside and outside of the academy. Conferences were held with academics, business, and government participants to consider and address such matters. The Government–University–Industry Research Roundtable was founded by the National Academy of Science, the National Academy of Engineering, and the Institute of Medicine in 1984 to "provide a forum where scientists, engineers, administrators, and policy makers from government, university and industry can come together on an ongoing basis to explore ways to improve the productivity of the nation's research enterprise" (6). Although decidedly "pro" business–university relationships, the Roundtable developed a model agreement for business–university research partnerships which set standards for publication, intellectual property ownership, and licensing and patenting procedures.

Despite a climate generally favorable to university–industry collaborations, there was also concern about abuse. Particularly troublesome were ethical issues arising from conflict of interest questions. In 1989 the NIH proposed guidelines that required individual decision makers involved in funded research to disclose "all financial interests and outside professional activities." Any perceived conflict of interest uncovered by these reporting requirements was to be especially noted and resolved prior to funding. Moreover, researchers were prohibited from holding equity or options in any company affected by the outcome of their research. Record keeping was extensive on the part of universities, which bore much of the burden of administering the guidelines. Harsh and wide-ranging criticism of the proposed guidelines resulted in their withdrawal within months of publication. NIH then proceeded to develop formal regulations on the subject, which, together with similar regulations proposed by the NSF, became binding in 1995. Less restrictive, these rules required researchers funded by NSF or NIH to notify their home institution if "they, their spouses, or their dependent children have financial interests — exceeding $10,000 or 5% ownership — in companies that might be affected by their research" (7). But once the researcher has complied with this threshold requirement, it is up to the institution to decide whether it is a conflict of interest and what to do about it. The rules do not cover the situation when an institution has a financial interest in the outcome of federally funded research. Thus it became important to understand both what the concept of conflict of interest entails and to examine proposed remedies for conflict situations.

CONFLICT OF INTEREST

One widely approved definition of conflict of interest is that it is "a set of conditions in which professional judgment concerning a primary interest (such as a patient's welfare or the validity of research) tends to be unduly influenced by a secondary interest (e.g., financial gain)" (8). Although there is some scholarly debate about what is the key concept in conflict of interest analysis (9), a growing consensus seems to find in the risk of impaired professional judgment the locus of the problem. Thus, as one modern legal scholar has put it, "[T]he common feature which brings each of ... (a variety of) questions within the doctrinal niche labeled 'conflict of interest' is concern with the existence of some particular incentive which threatens the effective and ethical functioning of a 'person acting in the role of a fiduciary for the benefit of another or others'" (10). The widespread concern is that an individual, employed by the university to perform one or more of a multiplicity of tasks in its behalf, may be compromised in his or her judgment by an incentive that should be subordinate to what he or she is employed by the university to do. Under the NSF and NIH rules, only a "significant financial interest" counts as a problematic secondary incentive; however, since the university is responsible for determining initially what counts as a conflict of interest and how to deal with it, the university may also enlarge the parameters of their internal policies to deal with incentives beyond the financial. A further complication exists because of institutional conflicts of interest, that is, when the university has a financial stake in the research conducted by those employed by the university. Much less attention has been paid to this topic, and discussion of it will be postponed until later.

Conflict of interest describes a situation of risk, not actual impairment of function. If there is actual impairment, then there is true blameworthiness. Simply being in a conflict situation, however, is usually benign in and of itself (11). Although scholars realistically warn that the phrase as often used is "accusatory" (12), this should not be the case. Since federal public policy strongly supports university–industry partnerships, conflicts of

interest are inevitable, at least at some times and in some ways, for a great many researchers in university settings. It is necessary first to recognize a conflict situation, and then to determine what to do about it. The identification question is logically and realistically prior to the remedy question. For some conflicts, nothing needs to be done. Others, of course, need, in the words of the federal regulation, to be "managed, reduced or eliminated."

Judgment

Although conflict of interest problems are not new, serious academic consideration of them is relatively recent. Using the legal professional literature extensively, Davis was the first professional ethics philosopher to isolate the importance of the risk to judgment impairment as of central importance (13). Previously Margolis had suggested that conflict of interest entailed "an avoidable exploiting of conflicting roles" (14). Davis argued that the issue was really the threat to judgment within a role, rather than a conflict between roles, which suggested a typical but different ethical dilemma. For Thompson, this was the key difference:

> "In ethical dilemmas, both of the competing interests have a presumptive claim to priority, and the problem is in deciding which to choose. In the case of ... conflicts of interest, only one of the interests has a claim to priority, and the problem is to ensure that the other interest does not dominate. This asymmetry between interests is a distinctive characteristic of conflicts of interests (8).

This asymmetry may not be characteristic, however, of all conflicts of interest. In the legal literature, for example, conflicts do occur when lawyers sometimes try to represent two or more clients with "conflicting interests." Nevertheless, in university–industry conflicts, Thompson's point regarding an asymmetry in interests seems well-taken because it is generally understood that the obligation to the university is primary for the researcher.

Luebke challenged Davis's analysis regarding the centrality of judgment, claiming that the issue was not the "correctness of the decision," but "the potential damage to the trust relationship existing between the bearer of the conflict and the person or entity for which the primary interest was to be maintained." (15). Trust is important to the maintenance of the relationship, but as Pritchard noted in defense of Davis, "... The maintenance of trust is what is under threat in conflict-of-interest situations, but precisely, as Davis says, because the reliability of professional judgment is thrown into doubt" (11). So "judgment" is the more precise term. Nevertheless, since the issue of trust is essentially one of trust of the professional judgment, there may be no substantive divergence between the two positions. The appearance issue will be discussed later.

Judgment for Davis "implies discretion" (13). It is not something mechanical or routine, something a clerk could do. Still it is important to stress that the idea must be generalized; it is judgment within a role, not a particular judgment in a role. To suggest otherwise is to fall into the trap Davis himself fell into. In addressing the following hypothetical, Davis declared the answer to depend on whether discretion was called for in the case at hand. The issue was: "Is it a conflict of interest to recommend to one's own company a contract with another firm in which one holds substantial stock?" The question of whether there is a risk of impaired judgment cannot depend on whether the judgment involved discretion in the particular case but only on whether, in general, this role demands judgment-as-discretion at least some of the time. Surely we are as concerned in the above hypothetical with the avaricious person who deliberately seeks to line his pocket at the expense of the person or entity that has a fiduciary claim upon him just as we are concerned with the person who subconsciously or confusedly makes a bad judgment to the detriment of his institution. We want to recognize both kinds of problems as conflict of interest problems because both evidence a risk that there will be impaired performance. In establishing conflict of interest rules or guidelines, there is a need to have everyone treated alike before the fact. This cannot be done if we have to know the subjective answer concerning how discretionary the judgment was before we can determine whether the matter demanded some review. The risk must first be identified category by category before a proper remedy can be determined. Thus, the word judgment might better be replaced by the word "decision" to capture all that we want, especially because the former may imply discretion while the latter may not. Thus McMunigal says it better when he suggests the real concern is with an incentive which threatens "the effective and ethical functioning" of the conflict holder in his fiduciary role (10).

Interests

Thompson's formulation stresses the need for one or more "primary" interests being at risk of subordination to a secondary interest. For an academic working as a researcher in a university setting, there seem to be three primary interests: (1) research integrity, (2) the well-being and education of students, and (3) if in a clinical setting, the welfare of patients (8). Although the NIH and NSF rules focus on financial interests alone, there are surely other secondary interests that university watchdogs will want to be on the lookout for. Pritchard suggests some of these secondary interests: tenure, promotion, satisfaction from supporting one's graduate assistants, or colleagues or institutional connections, or even one's reputation (11). The problem here is that the list can be extended indefinitely. Davis would include "all those influences, loyalties, concerns, emotions, or the like that can make (competent) judgment less reliable than it might otherwise be" (13). Davis even includes "moral constraints." Surely this goes too far. Pritchard finds in Feinberg a more modest and more objective definition: "something one might attempt to advance, protect, or even modify" (11). Although this excludes things like moods and emotions, it may still be too broad to be manageable because it may include rather personal predilections concerning, say, working hours and conditions, irrelevant to real conflicts of interest concerns. But surely financial interests are too narrow a concentration, though Thompson is right

in asserting that financial gain is more "pernicious and more objective and more fungible, and easier to regulate by impartial rules" (8). Since the federal government's regulations focus only on money, universities may lose sight of other secondary interests—some hardly trivial for the individual—that require enumeration and care in determining whether they threaten to impede a good, independent professional judgment/decision which damages a primary interest.

REMEDIES

There are a number of different ways to assess how problematic a conflict of interest might be. Thompson proposes two standards for assessing the severity of a conflict. First, there is the "likelihood" that the judgment will be affected. Rules of thumb under this standard include (1) the greater the value of the secondary gain; the greater the likelihood the judgment will be affected, (2) the longer and closer the association with those connected to the secondary interest, the greater the likelihood the judgment will be affected, and (3) the greater the degree of discretion in judgment, the greater the chance for judgment to be improperly affected. Thompson's second standard is cast in terms of the "seriousness" of the conflict. Crucial concerns here are (1) the value of the primary interest, meaning the potential effects on patient care or on the integrity of the research, (2) the scope of the effects on the project itself, but also on others, including the indirect harm that comes from loss of confidence in the researcher or in his or her institution, and (3) the relative accountability of the researcher. There is presumably less concern if there is reliable review of the work (8).

Another approach to standards can be drawn from the legal literature, which distinguishes at least three kinds of conflict of interest: actual, latent and potential. These categories are distinguished by the closeness of the conflict to the actual impairment in professional judgment that is the underlying concern. An actual conflict therefore is one that is certain to affect the judgment. A latent conflict is one for which there is a reasonable probability that the judgment will be impaired. A potential conflict is one that is reasonably foreseeable (13). Although there is confusion in the legal literature concerning the proper use of these categories (10), if they are simply standards to make the remedies chosen for a given situation more amenable to sorting out, perhaps they can be helpful. Use of these terms in any substantive way is simply confusing.

In any event, it is to standards like the ones articulated by Thompson and in the legal literature that those responsible for determining what to do about conflicts of interest instinctively turn. For example, Blumenthal suggests that the seriousness of the potential harm to patients and to the integrity of research in a clinical research setting require the most restrictive rules. Generally he would ban any conflicts of interest in these settings, except if the financial gain through the secondary interest is de minimus. Blumenthal is equally concerned with situations where students and trainees may be affected by the conflict of interest. Because restrictions on scientific communications are frequent in university–industry partnerships, and the students' careers may be hampered by their lack of publishing results of the work, conflicts affecting educational decisions should be allowed but rarely, and then only under the closest of supervision. More empirical research needs to be done, Blumenthal argues, before it is clear what ought to be the approach to other conflict of interest categories.

In nonclinical settings not involving students or in clinical settings where the secondary interests are "nonexistent or attenuated," it is not clear how restrictive the rules ought to be. Here Blumenthal is thinking of things like "straight-forward academic–industry research relationships" and "patenting and licensing arrangements, in which clinical research may affect whether and how much royalties are received on patented products of research." Preliminary empirical data suggests, on the negative side, that industry sponsorship tends to affect the choice of research topics and also results in scientific information being held back longer, even beyond the time necessary to file a patent. On the positive side, technology transfer activities have increased in a variety of ways, and so far, there have not been reports of actual research misconduct attributable to academic-industry relationships.

Finally, there is a mixed result regarding scientific publications. Generally the relationships seem to spawn more publications, except among researchers who "receive more than two-thirds of their total research budgets from companies or add more than 20 percent to their total salaries from consulting to industries" (16). This last finding seems to confirm Thompson's notion that more intense relationships may have negative effects on researchers. This is based on the supposition that more publications are part of the primary interests on the part of the researcher in that role within the university. While this may be true as a general proposition, it may not be true in any individual case. Quality is, of course, often more important than quantity. This points up the need for careful scrutiny of actual situations, rather than blanket rules, except in the most serious kinds of cases, namely the ones Blumenthal identified as involving risks to patient care, to research integrity, and to students' careers.

Categories of remedies include disclosure, oversight, or some form of escape. Disclosure is usually warranted in all cases where a conflict of interest exists. Obviously no one can investigate and determine how serious is a given risk if they have no knowledge of the conflict to begin with. Even in the more controversial area of disclosure of research funding accompanying publications, one study showed that an overwhelming percentage of researchers who came out positively in favor of a certain type of drug received funding from companies that make the drug, while a much lower percentage of those critical of the class of drugs received such support. The authors of the study did not suggest that the researchers favoring the drugs were dishonest; however, they did recommend that disclosure occur "to avoid suspicion" (17). Critics of disclosure of funding for research that results in a publication claim that the reading public does not know what to do with the information (12). Whatever the merits of the two sides to

that debate, there is no application to the question whether disclosure to university officials ought to be made. The better analogy here is to the lawyer in a conflict situation who must disclose the conflict to the client potentially affected. The client, after being informed, usually has the option to determine whether or not to continue to be represented by the lawyer or to ask for some change to be made. Analogously, the university has the right to know about a conflict situation, to determine whether further action is warranted or not.

Oversight may be by a standing committee within the university or by a person or group outside the university. Here the relationship between the researcher and the company may be welcomed, but the size and nature of the financial arrangement may cause sufficient concern that some additional regulation seems necessary. Since the 1995 federal guidelines require researchers to notify their own institutions if they or their close relatives have a financial interest in companies affected by the research exceeding $10,000 or 5 percent ownership, it is likely that the institutions will want to have some oversight of research that meet the federal criteria. Presumably whether that oversight is to be conducted by a committee within the university or outside it may depend on such variables as the expertise required to perform the oversight function or the manner in which the oversight must be conducted.

The last category of remedies, "escape," is a catchall to gather all those situations where it is deemed necessary or wise to prohibit the arrangement either through divestiture, abstention from decision making or some other mechanism to insulate the researcher from the work or the potential financial gain. This is obviously the most costly and serious remedy with which to handle a conflict situation; but, at times, it will be the only reasonable one.

INSTITUTIONAL CONFLICTS

Most of the literature concerning academic–industry relations has focused on the individual researcher and his or her own conflicts of interest, with the primary interest being to the academic institution and only secondary interests obliging the researcher to industry. Little attention has been paid to the problem of conflicts of interest that may affect the academic institution itself. Universities may have equity interests in companies affiliated with their institutions. Universities may also own patent rights that they license to companies and investigators employed by the institution. Institutional practices may thwart individual investigators or conspire with them to enhance the value of their equity interest or stock holdings to the detriment of their true primary interests. As the university is being asked to develop internal rules and procedures to guard against abuse in conflict situations, who will be the watchdog, guiding the university officials on the firing line from succumbing to the temptation to seek a secondary interest of the entity in preference to one or more of the institution's primary interests (18)?

Clearly, there are different conflict problems when the focus shifts from the individual researcher to the institution. First, the individual may not be at all compensated by a company, but the institution may be benefiting directly from the fruits of the research by increased value to its equity holdings or its licensing agreements. Again, no matter the situation with respect to the researcher, the institution may put subtle or not-so-subtle pressures to prefer a secondary interest in some way, which the individual researcher may be hard-pressed to avoid. The secondary interests of the institution may not be limited to economic gain either. Increased reputation may also be a secondary interest, which complicates the pursuit of the primary missions of teaching and research, and in the case of university clinical matters, the patients' welfare. There have already been suggestions made that any institutional conflict should prima facie be grounds for avoiding the conflict altogether. Of course similar remedies to those put forth for individuals have also been offered, with disclosure — even to individual patients — mandated, while internal or external monitoring providing supplementary remedies (18).

APPEARANCE OF A CONFLICT

A theme that constantly appears in all discussions of conflicts of interests is the problem of "appearance." Since conflicts are simply questions of risk, and what to do to prevent risks from ripening into actual breaches of duty, the question of appearances may arise more often in the remedy area than in determining whether or not a conflict exits. Once a conflict exits, there is, de facto, a risk that some impairment will follow. Even reasonable people may be skeptical that the remedy chosen *will truly prevent* the impairment from taking place. Since the university depends so much on its reputation for integrity in the pursuit of knowledge, these concerns are real and potentially problematic to deal with. However, it is not at all clear what should be done with appearance questions. In the absence of solid empirical evidence, it is hard to know whether any particular conflict category is seriously "risky." Nevertheless, there are those who believe the value of keeping the university's reputation clean requires curtailment of much of the activities that are now underway to foster academic–industry relationships (18).

OBSERVATIONS

Foundational to one's view of the risks and the necessary remedies for conflict of interest in corporate–university relationships is the public policy issue of the value of commercializing scientific research. During the past three decades the federal government has encouraged public–private collaborations through legislation and agency rule making. Scientists and universities, for whom the stakes may be very high economically, have encouraged and facilitated technology transfer. However, even for the strongest advocates of technology transfer and university/corporate relationships, 25 years of experience suggests that government, universities, and corporations

need to remain diligent if they are going to preserve the benefits of commercialization of scientific ideas without sacrificing either scientific rigor or sound educational policy. Corporate–university agreements and federal rules need "to require disclosure of conflicts, limit the most troublesome forms of conflict, and create uniformity in ethical standards across the country" (19). The alternative seems to be an unacceptable slippery slope that threatens both the essence of the scientific process as well as the integrity of educational institutions.

On its face, technology transfer is relatively uncomplicated. Scientific researchers in universities and research institutes pursue new scientific knowledge with financial resources provided largely by the federal government. Commercially viable results are licensed to corporations who support the scientist and the scientist's institution in exchange for the opportunity to develop the idea. Scientists are often given personal consulting contracts or an equity interest or a board seat or a financial interest in future sales or all of these financial incentives by the company developing the scientific idea. The scientist's institution may also receive significant long-term financial incentives for its role in the process.

Inherently the problem is not the financial incentives for scientists or institutions to transfer technology. Within the free-enterprise system, scientists deserve the same opportunity for financial rewards from their work as do other professionals. Nor is the problem the personal gain for individual scientists and their institutions from work financed by public funds. As a matter of public policy, the federal government has determined that the benefits to society of the technology transferred outweigh the costs to taxpayers of allowing financial incentives to scientific researchers and their institutions. It is not an irrational or immoral trade-off.

The problem is the risks that standards of scientific inquiry will be compromised, diminished, or sacrificed for financial gain, or that the primary interests of educational institutions are subordinated to secondary interests. If the scientist's financial enrichment becomes tied to the success of the scientific outcome, then "society runs the risks that researchers will knowingly influence the outcome of 'neutral' scientific inquiries" (19). If the pursuit of economic gain harms students or patients, then the price is too high to pay.

Conflicts of interest threaten the integrity not only of individual scientists and their educational institutions but of the scientific process itself. Scientists caught by the potential for enormous financial gain between research for the public good and research for corporate interests, between their duties as teachers and mentors to graduate students and their duties as corporate consultants or officers, between their employer university or research institute and their corporate sponsor, or between the health interests of their research subjects and patients and the marketing interests of their corporate funders are on precariously thin ice when they record and report scientific results from an alleged position of scientific objectivity.

The classic remedies for conflict of interest of disclosure, oversight, and escape take on new meaning, and the path through the thicket of conflicting claims on the scientist's objectivity and rigor is obscured as the potential for financial gain increases. Many would argue that scientific researchers, "particularly those conducting clinical trials, carry a fiduciary duty to the public. Like fiduciaries, there is a presumption against conflicts of interest" (19).

Current NIH and NSF rules rely on universities to decide whether there is a conflict of interest and what to do about it. Recent experiences, particularly in biomedical research, suggest that the university record to date leaves much to be desired. As pressure mounts on institutions and thus on scientists to seek corporate support and relationships, many argue that "the conflict of interest policies and standards of disclosure that universities rely upon don't do enough to protect academic freedom or the integrity of research in an environment where corporate interests are playing a growing role" (20).

Almost all current university and research institute policies are based on a principle of disclosure. But such disclosure principles are interpreted broadly, and often disclosure documents are not made public. Only a limited number of states require that disclosure documents be available under public record laws.

Very few scientific or lay publications inquire about conflicts or require disclosure by scientists writing on scientific and public policy topics. In 1992 a study of 14 journals showed that one out of three authors of nearly 800 scientific articles had a financial interest in the results of their research. Few if any of these conflicts were disclosed in the articles (20). In a 1997 study of scientists funded by drug companies it was found that 96 percent of the authors of favorable articles on a particular class of drugs had financial ties to the makers of the drugs. These conflicts were reported in only 2 of the 70 articles surveyed. Of those authors who published articles critical of the class of drugs, only 37 percent had financial conflicts of interest (17). While the mechanics of disclosure are neither easy nor obvious, it seems certain that institutions must find ways to make their disclosure requirements more visible and apparent. Moreover disclosure is only the first line of defense in conflict of interest situations when independence of judgment may be compromised.

It is not possible, or even desirable, to remove all conflicts of interest in scientific research. The benefits of collaboration between universities and corporations and the interactions of nonprofit and for profit scientists and investigators are already documented, particularly in biomedical research fields. But, as corporate relationships with universities, research institutes, and research scientists increase in number, size, and complexity, public confidence in the objectivity and rigor of the scientific process will erode rapidly in the face of undisclosed and unresolved conflicts of interest, real or perceived.

Educational values and, in clinical settings, patient care may also be compromised by unattended conflicts of interests. The antidote is vigorous and persistent pursuit of institutional, governmental, corporate, and agency policies and practices that provide disclosure, oversight, and escape from conflicts of interest.

BIBLIOGRAPHY

1. R. Hofstader and W. Smith, eds., *American Higher Education: A Documentary History*, vol. 2, University of Chicago Press, Chicago, IL, 1961, pp. 568–569.

2. E. Shils, in A. Oleson and J. Voss, eds., *The Organization of Knowledge in Modern America*, 1860–1920, Johns Hopkins University Press, Baltimore, MD, 1979, p. 28.

3. R. Geiger, *To Advance Knowledge: The Growth of American Research Universities, 1900–1940*, Oxford University Press, Oxford, UK, 1986.

4. R. Geiger, *Research and Relevant Knowledge: American Research Universities Since World War II*, Oxford University Press, Oxford, UK, 1993.

5. Office of Technology Assessment (OTA), *New Developments in Biotechnology: U.S. Investment in Biotechnology*, OTA, Washington, DC, 1988.

6. N. Bowie, *University-Business Partnerships: An Assessment*, Rowman & Littlefield, Lankan, NJ, 1994.

7. J. Mervis, *Science* **269**, 294 (1995).

8. D. Thompson, *N. Engl. J. Med.* **329**(8), 573–576 (1993).

9. M. Davis, *Bus. Prof. Ethics J.* **12**(4), 21–41 (1993).

10. K. McMunigal, *Geo. J. Legal Ethics* **5**(4), 823–877 (1992).

11. M. Pritchard, *Acad. Med.* **71**(12), 1305–1313 (1996).

12. K. Rothman, *JAMA, J. Am. Med. Assoc.* **269**(21), 2782–2784 (1993).

13. M. Davis, *Bus. Prof. Ethics Jr.* **1**(1), 17–27 (1982).

14. J. Margolis, in T. Beauchamp and N. Bovice, eds., *Ethical Theory and Business*, Prentice-Hall, Englewood Cliffs, NJ, 1996, pp. 361–373.

15. N. Luebke, *Bus. Prof. Ethics J.* **6**(1), 67–81.

16. D. Blumenthal, *Acad. Med.* **71**(12), 1291–1296 (1996).

17. H. Stelfox, G. Choa, K. O'Rourke, and A. Detsky, *N. Engl. J. Med.* **338**(2), 101–106 (1998).

18. E. Emanuel and D. Steiner, *N. Engl. J. Med.* **332**(4), 262–267 (1995).

19. M. Witt and L. Gostin, *JAMA, J. Am. Med. Assoc.* **271**(7), 547–551 (1994).

20. G. Blumenstyk, *Chron. Higher Educ.* **44**(37), May 22, A41 (1998).

See other entries MEDICAL BIOTECHNOLOGY, UNITED STATES POLICIES INFLUENCING ITS DEVELOPMENT; TRANSFERRING INNOVATIONS FROM ACADEMIC RESEARCH INSTITUTIONS TO INDUSTRY: OVERVIEW.

SUPPLEMENT

Every effort was made to obtain articles for this Encyclopedia on all key organizations, government offices, industry groups, interest groups, and so on. However, it was impossible to obtain some of these entries in time for inclusion in this work. The following brief summaries are provided to call the readers' attention to important groups and organizations that treat aspects of the subjects covered by this Encyclopedia, as well as to provide suitable contact information. It is hoped that this will be helpful for the reader seeking additional information.

BIOTECHNOLOGY INDUSTRY ORGANIZATION (BIO)

BIO is a trade association for the biotechnology industry.

> 1625 K Street NW
> Suite 1100
> Washington, DC 20006
> Tel: 202-857-0244
> Fax: 202-857-0357
> *www.bio.org*

The BIO Website includes information on the following topics: the biotechnology industry, the biotechnology record on ethics, legislative issues, biological warfare, biotechnology in agriculture, agricultural biotech products on the market, biotechnology in health care, approved biotechnology drugs, industrial uses of biotechnology, applications of industrial biotechnology, biotechnology for the environment, applications of environmental biotechnology, biotechnology in animal health, and marine biotechnology.

THE FOUNDATION ON ECONOMIC TRENDS (FET), headed by Jeremy Rifkin, is a nonprofit organization whose mission is to examine emerging trends in science and technology and their impacts on the environment, the economy, culture, and society.

> The Foundation on Economic Trends (FET)
> 1660 L Street, NW, Suite 216
> Washington, DC 20036
> Tel: 202-466-2823
> Fax: 202-429-9602
> E-mail: office@biotechcentury.org
> *www.biotechcentury.org*

The FET Website *www.biotechcentury.org* addresses environmental, social, economic, and ethical issues related to biotechnology and provides links to other organizations engaged in biotechnology issues.

HUMAN GENOME ORGANIZATION (HUGO)

HUGO is an international scientific organization.
Contact information (U.S. and Canada):

> HUGO Americas
> Laboratory of Genetics
> National Institute on Aging
> NIH/NIA-IRP. GRC, Box 31
> 5600 Nathan Shock Drive
> Baltimore, MD 21224-6825, USA
> Tel: 410-558-8337
> Fax: 410-558-8331
> E-mail: schlessingerd@grc.nia.nih.gov
> *www.gene.ucl.ac.uk / hugo*

Contact information (international):

> HUGO
> 142-144 Harley Street
> London W1N 1AH
> United Kingdom
> Tel: (44) 171 935 8085
> Fax: (44) 171 935 8341
> E-mail: hugo@hugo-international.org

The Human Genome Organization (HUGO) is the international organization of scientists involved in the Human Genome Project (HGP), the global initiative to map and sequence the human genome. HUGO was established in 1989 by a group of the world's leading genome scientists to promote international collaboration within the project.

HUGO carries out a complex coordinating role within the Human Genome Project.

HUGO activities range from support of data collation for constructing genetic and physical maps of the human genome to the organization of workshops to promote the consideration of a wide range of ethical, legal, social, and intellectual property issues. HUGO fosters the exchange of data and biomaterials, encourages the spreading and sharing of technologies, provides information and advice on aspects of human genome programs, and serves as a coordinating agency for building relationships between various governmental funding agencies and the genome community. HUGO provides an interface between the Human Genome Project and the many groups and organizations interested or involved in the human genome initiative.

PHARMACEUTICAL RESEARCH AND MANUFACTURERS OF AMERICA (PhRMA)

> 1100 Fifteenth St. NW
> Washington, DC 20005
> *www.phrma.org*
> President: Alan F. Holmer

PhRMA membership consists of approximately 100 U.S. companies that have a primary commitment to pharmaceutical research. The mission of the pharmaceutical Research and Manufacturers of America is to help the research-based pharmaceutical industry successfully meet its goal of discovering, developing, and bringing to market medicines to improve human health, patient satisfaction, and the quality of life around the world, as well as to reduce the overall cost of health care.

To achieve its goal, the industry aspires to foster a favorable environment that encourages innovative drug research; swift development and approval of safe and effective drugs; consumer and patient access to medicines in an open and competitive marketplace; support and understanding from the public and other key constituents regarding the critical role and value of the pharmaceutical industry in improving human health and quality of life and in reducing overall health care costs; public policies that allow sufficient returns to foster continued innovation.

U.S. DEPARTMENT OF COMMERCE, TECHNOLOGY ADMINISTRATION

The Technology Administration (TA) (*www.ta.doc.gov*) is a bureau of the U.S. Department of Commerce (*www.doc.gov*). The Technology Administration leads civilian technology for the Department of Commerce and works with U.S. industries to promote U.S. economic competitiveness and growth.

The Undersecretary for Technology supported by the Deputy Undersecretary for Technology, manages the Technology Administration's (TA) three agencies:

(1) The Office of Technology Policy (OTP) is an office of the federal government with the explicit mission of developing and advocating national policies that use technology to build America's economic strength.

(2) The National Institute of Standards and Technology (NIST) promotes economic growth and an improved quality of life by working with industry to develop and apply technology, measurements, and standards.

(3) The National Technical Information Services (NTIS) collects and disseminates scientific, technical, engineering, and related business information produced by the U.S. government and foreign sources.

WHITE HOUSE OFFICE OF SCIENCE AND TECHNOLOGY POLICY (OSTP)

1600 Pennsylvania Ave N.W.
Washington, DC 20502
Tel: 202-395-7347
E-mail: information@ostp.eop.gov
www.whitehouse.gov, then link White House Offices and Agencies

OSTP was established in 1976 to provide the President with policy advice and to coordinate the science and technology investment.

OSTP Divisions: Environmental Division; National Security and International Affairs (NSIA) Division; Science Division; and Technology Division.

INDEX

Page references in **bold** type indicate a main article. Page references followed by italic *t* or italic *f* indicate material in tables or figures respectively.